Singularly Perturbed Boundary Value Problems

Matteo Dalla Riva • Massimo Lanza de Cristoforis
Paolo Musolino

Singularly Perturbed Boundary Value Problems

A Functional Analytic Approach

Springer

Matteo Dalla Riva
College of Engineering and Natural Science
The University of Tulsa
Tulsa, OK, USA

Massimo Lanza de Cristoforis
Dipartimento di Matematica
Università degli Studi di Padova
Padova, Italy

Paolo Musolino
Dipartimento di Scienze Molecolari
e Nanosistemi
Università Ca' Foscari Venezia
Venezia, Italy

ISBN 978-3-030-76258-2 ISBN 978-3-030-76259-9 (eBook)
https://doi.org/10.1007/978-3-030-76259-9

Mathematics Subject Classification: 31B10, 35B10, 35B25, 35B30, 35C15, 35C20, 35J25, 35J66, 35P15, 45P05, 46N20, 47H30, 47G40, 42B20

This Springer imprint is published by the registered company Springer Nature Switzerland AG
The registered company address is: Gewerbestrasse 11, 6330 Cham, Switzerland

To our families

Preface

This book is the first introductory and self-contained presentation of a method for the analysis of singularly perturbed boundary value problems that we have called the Functional Analytic Approach. The method was proposed some 20 years ago by the second author and, since then, it has developed in many directions and been implemented in linear and nonlinear problems, including problems that arise from Continuum Mechanics and Material Sciences.

Before the publication of this book, the presentation of the theoretical aspect and of the applications of the Functional Analytic Approach was disseminated in a number of different journal articles. This book is the first comprehensive introduction to the topic, and it covers both the theoretical material that stands at the basis of the Functional Analytic Approach and its applications to a series of problems that ranges from simple illustrative examples to more involved research results.

The Functional Analytic Approach makes constant use of Potential Theory, of the integral representation method for the solutions of boundary value problems, of the Theory of Analytic Functions both in finite and infinite dimension, and of Nonlinear Functional Analysis. All theoretical results used in the Functional Analytic Approach are collected in the first seven chapters of the book and in an appendix at the end of the book. Despite the fact that these initial chapters and the appendix are mostly self-contained, we did not include proofs that can be found in widely available textbooks and for which we cannot claim any paternity. Some classical results have, however, been revisited in a way that fits our purposes, and the corresponding proofs have been presented in full detail. We begin describing the Functional Analytic Approach in Chapter 8, where we show its application to a Dirichlet problem for the Laplace equation in a domain with a hole that shrinks to a point. In the chapters that follow Chapter 8, we study different kinds of boundary conditions, including nonlinear boundary conditions, and different geometric settings, including unbounded domains with periodic sets of holes or inclusions.

To keep the length of the book within reasonable limits, we excluded some of the directions along which the Functional Analytic Approach has developed. In particular, we do not treat differential equations other than the Laplace and Poisson equations and we stick to domains with shrinking holes. For the analysis involving different equations and, in particular, elliptic systems, and for the study of different domain perturbations, we refer the reader to the works of the authors and their collaborators in the bibliography.

Throughout the book we have tried to keep the presentation at a level which we think adequate for the graduate student. As for prerequisites, we expect the reader to be familiar with basic concepts of Lebesgue integration for functions of several variables and of L^p spaces, with some elementary Functional Analysis, and with Differential Calculus in normed spaces. We have tried to keep the use of Distribution Theory to a minimum. In addition, the book provides references to all accessory results.

The book is designed to serve various purposes.

• The large introductory part can be used as a reference textbook for graduate courses on classical Potential Theory and its applications to boundary value problems for the Laplace and Poisson equations. Students and instructors can find, here, material on Lipschitz and Schauder functions and sets (Chapter 2), Green formulas and layer potentials (Chapter 4), Potential Theory for the Laplace equation in multi-connected domains (Chapter 6), and volume potentials and their applications to the Poisson equation (Chapter 7). The periodic counterpart of these results of Potential Theory is presented later in the book, in Chapter 12.

• The initial chapters also contain results that are rarely presented in the literature and may also, therefore, attract the interest of more expert readers. For example, in Section 4.3 we present a theorem on the Hölder continuity of singular integrals of convolution type. This result was originally proven by Carlo Miranda [212], but the corresponding paper was published in Italian and, to the best of our knowledge, has never appeared in English and in the "modern" form included in the book.

• In Chapter 8 we start introducing the Functional Analytic Approach. A reader looking for a quick introduction to the method can find simple illustrative examples specifically designed for this purpose in Section 8.3, where we study the Dirichlet problem in dimension three or bigger; Section 9.2, where we consider a mixed problem for the Laplace equation; Section 9.3, about a mixed problem for the Poisson equation; Section 10.3, on a problem with two "moderately close" holes; and Section 11.2, for the analysis of a nonlinear problem.

• More expert readers will find a comprehensive presentation of the Functional Analytic Approach, which allows a comparison between our approach and the more classical expansion methods of Asymptotic Analysis and offers insights on the specific features of the approach and its applications to linear boundary value problems (Chapters 8, 9, and 10), nonlinear boundary value problems (Chapter 11), and problems in unbounded periodic domains (Chapter 13).

• Finally, applications of the method to problems that arise in the study of composite or porous materials, and more specifically in the study of the average properties of such materials, can be found in Section 13.5, on the longitudinal flow along a periodic array of thin cylinders, and in Sections 13.6 and 13.7, on the effective conductivity of a composite with a thermal resistance at the materials interface.

We would like to thank the students and colleagues that helped us in writing this book.

In particular, we acknowledge the help received from Prof. Alberto Cialdea, Prof. David Natroshvili, Prof. Sergei V. Rogosin, and Prof. Promarz M. Tamrazov on technical issues related to the use of integral equations and Potential Theory for the solution of boundary value problems. We are indebted to Prof. Joan Verdera for several recent references on issues related to the theory of singular integrals in Hölder spaces. We also mention the help received from Prof. Mark L. Agranovsky on technical issues related to the theory of Analytic Functions, and the help of Prof. Gérard Bourdaud, Prof. Victor I. Burenkov, and Prof. Winfried Sickel on issues related to Function Space Theory.

Finally, we wish to thank Dr. Tuğba Akyel, Dr. Riccardo Falconi, Dr. Paolo Luzzini, Prof. Pier Domenico Lamberti, Dr. Riccardo Molinarolo, Mr. Jonathan Pinkey, and Dr. Roman Pukhtaievych for the generous gift of their time in proofreading preliminary versions of the book and making precious suggestions and remarks. Dr. Paolo Luzzini also prepared the three pictures in the book, for which we owe him additional thanks.

Tulsa, OK, USA — Matteo Dalla Riva

Padova, Italy — Massimo Lanza de Cristoforis

Venezia, Italy — Paolo Musolino

March 2021

Contents

Chapter 1
Introduction

This book is devoted to the analysis of the basic boundary value problems for the Laplace equation in singularly perturbed domains. More precisely, we will focus on domains with small holes and develop an alternative approach to those of Asymptotic Analysis.

While one could develop most of the book's material for Lipschitz domains and a large class of elliptic operators, we have decided to write a primer of sorts, one easily accessible to beginners. Therefore, we have tried to explain our point of view under the most elementary conditions possible: we treat domains of class $C^{1,\alpha}$ for some $\alpha \in]0, 1[$ (the classical Lyapunov sets) and the model of elliptic operators, i.e., the Laplace operator.

In a typical problem found in this book, we have a bounded open subset Ω^o of class $C^{1,\alpha}$ of $\mathbb{R}^n$ and we assume that Ω^o contains the origin. Then we consider a certain boundary value problem in Ω^o, which we assume has a unique solution u^o. Here the superscript '*o*' stands for 'outer.' (Fittingly, the superscript 'i' will stand for 'inner'.)

Next we make in Ω^o a hole of size $\epsilon > 0$ around the origin. For example, we can remove from Ω^o a ball centered at the origin and with radius ϵ. We denote by $\Omega(\epsilon)$ the perforated domain that we obtain and we consider in $\Omega(\epsilon)$ a boundary value problem which is, in a sense, of the same nature of the problem on Ω^o . For example, we can have a Dirichlet problem for the same equation both in $\Omega(\epsilon)$ and in Ω^o. We assume that the problem in $\Omega(\epsilon)$ also has a unique solution u_ϵ.

At this point, we can say that the problem in Ω^o is 'unperturbed', and that the problem in $\Omega(\epsilon)$ has been perturbed by the presence of the hole.

It is then natural to ask how the solution u_ϵ, or its energy integral, or other functionals associated with u_ϵ behave when ϵ approaches 0. The difficulty is that the limiting set $\Omega^o \setminus \{0\}$ has a nature which is different from that of $\Omega(\epsilon)$ and, in particular, $\Omega^o \setminus \{0\}$ is no longer a set of class $C^{1,\alpha}$.

As can be seen in a large body of mathematical literature, problems of this type are not only interesting in themselves, but also have importance in Continuum Mechanics and in applied sciences and engineering. For example, they find several applications in shape and topological optimization problems and in the inverse prob-

M. Dalla Riva et al., *Singularly Perturbed Boundary Value Problems*,
https://doi.org/10.1007/978-3-030-76259-9_1

lems related to nondestructive testing. A description of such applications can be found in the monographs of Novotny and Sokołowski [240] and of Ammari and Kang [8], which contain many further references on the topic.

In an effort to approximate u_ϵ or related functionals, a commonly accepted way to tackle the problem is to figure out an asymptotic expansion of u_ϵ for $\epsilon > 0$ small. So for example, we fix a point p of the unperturbed set, with that point being away from the hole, and we write an equality of the type

$$u_\epsilon(p) = a(\gamma_1(\epsilon), \ldots, \gamma_s(\epsilon)) + R(\epsilon) \qquad \text{as } \epsilon \text{ tends to } 0^+ , \tag{1.1}$$

where s is a natural number, $a(\cdot, \ldots, \cdot)$ is an 'approximating' function of s real variables defined in a neighborhood of the origin in $\mathbb{R}^s$, $\gamma_1, \ldots, \gamma_s$ are s explicitly known functions which tend to zero as ϵ tends to 0^+, such as

$$\epsilon\,, \epsilon \log \epsilon\,, \log^{-1} \epsilon\,, \text{ and so on,}$$

and $R(\epsilon)$ is a reminder term. Here p is kept fixed and thus we did not include it as a variable in the right-hand side of (1.1).

Now the task is to show that the expansion we have written is actually of some use. For example, to show that (1.1) gives an approximation of the unknown $u_\epsilon(p)$ by means of $a(\gamma_1(\epsilon), \ldots, \gamma_s(\epsilon))$, we have to prove that the reminder term $R(\epsilon)$ tends to zero more rapidly than certain functions of $\gamma_1(\epsilon),\ldots, \gamma_s(\epsilon)$ as ϵ tends to zero. The faster $R(\epsilon)$ tends to zero, the better is our approximation.

As is well known, guessing the explicit form of the functions $\gamma_1,\ldots, \gamma_s$ and of the function a is a part of the problem that may be utterly nontrivial, especially if the problem we are looking at is nonlinear. In some cases, one can expect that a is a polynomial and that (1.1) takes the form

$$u_\epsilon(p) = \sum_{|\beta| \leq r} a_\beta \gamma_1(\epsilon)^{\beta_1} \ldots \gamma_s(\epsilon)^{\beta_s} + R_r(\epsilon) \qquad \text{as } \epsilon \text{ tends to } 0^+. \tag{1.2}$$

If that is the case, then the coefficients a_β can usually be computed with means at our disposal, after which we wish to show that the reminder $R_r(\epsilon)$ converges to zero faster than order r with respect to $\gamma_1, \ldots, \gamma_s$.

There has been an enormous effort in the mathematical literature to accomplish the plan described above, not only for problems with a shrinking hole, but also for singular perturbations of Ω^o generated by the formation of peaks, wedges, slits, thin bridges, and so on.

The most common approach adopted in the literature is that of the expansion methods of the so-called Asymptotic Analysis. The large number of authors contributing to the development of such methods prevents us from attempting any complete body of references. Nevertheless, we want to mention the early results in the monographs of Cherepanov [48] and [49] on the formation of cracks and in the books of Nayfeh [234], Van Dyke [270], and Cole [57], which present an extensive review of the expansion methods known at the time. For the rigorous description of the method of matching outer and inner asymptotic expansions we refer to the book of Il'in [127], and for the Compound Expansion Method (also known as Multi-Scale Expansion Method) we mention the two volumes of Mazya et al. [203, 204] where, among other results, the authors introduce a systematic approach for analyzing general Douglis and Nirenberg elliptic boundary value problems in domains with perforations and corners.

Different domain perturbations can be found in Kozlov et al. [146], where the authors consider boundary value problems in domains depending on a small parameter ϵ in such a way that the limit regions as ϵ tends to zero consist of subsets of different space dimensions. For the analysis of problems in domains with two holes that collide into one another while shrinking in size we mention the works of Bonnaillie-Noël et al. [32] and Bonnaillie-Noël et al. [31] and for the analysis of problems in a domain containing 'clouds' of small holes we refer to the book of Maz'ya et al. [200], where the authors exploit Meso-Scale Asymptotic Approximations to compute the asymptotic expansion of Green kernels. We also mention Dauge et al. [85] where the authors consider self-similar perturbations of a corner domain and make a comparison between the Multi-Scale and the Matched Asymptotic Expansion Methods.

Further references on expansion methods and their applications can be found in the books and papers mentioned above. Expansion methods are not, however, the only means available for dealing with domain perturbations. For example, in the two dimensional case one can resort to complex analytic techniques, as in the works of Mityushev et al. [221] and [224] and Rogosin and Vaitekhovich [251].

The Functional Analytic Approach of this book can be considered as a complement to expansion methods, though we take, in a sense, a different standpoint. We no longer look for asymptotic expansions such as (1.1) but, in the example above, we look for representation formulas of the form

$$u_\epsilon(p) = \mathcal{A}(\gamma_1(\epsilon), \ldots, \gamma_s(\epsilon)) \qquad \text{for } \epsilon > 0 \text{ small enough}, \tag{1.3}$$

where $\mathcal{A}(\cdot, \ldots, \cdot)$ is a real analytic function of s real variables defined in a neighborhood of the origin in $\mathbb{R}^s$ and $\gamma_1, \ldots, \gamma_s$ are explicitly known functions which tend to zero as ϵ tends to 0^+ (just as in (1.1)). In particular, formula (1.3) implies that

$$u_\epsilon(p) = \sum_{\beta \in \mathbb{N}^s} \frac{D^\beta \mathcal{A}(0, \ldots, 0)}{\beta!} \gamma_1(\epsilon)^{\beta_1} \ldots \gamma_s(\epsilon)^{\beta_s}$$

for $\epsilon > 0$ sufficiently small, and that accordingly $u_\epsilon(p)$ has an asymptotic expansion in terms of the powers of $\gamma_1(\epsilon), \ldots, \gamma_s(\epsilon)$. Namely, for each natural r, we have

$$u_\epsilon(p) = \sum_{|\beta| \leq r} c_\beta \gamma_1(\epsilon)^{\beta_1} \ldots \gamma_s(\epsilon)^{\beta_s} + R_r(\epsilon) \tag{1.4}$$

for $\epsilon > 0$ sufficiently small, where $R_r(\epsilon)$ converges to zero faster than order r with respect to $\gamma_1, \ldots, \gamma_s$, and where the coefficients c_β are delivered by the equality $c_\beta = D^\beta \mathcal{A}(0, \ldots, 0)/\beta!$.

We emphasize that the main point of this book is demonstrating the existence of the function $\mathcal{A}$ for a suitable choice of $\gamma_1, \ldots, \gamma_s$. The matter of computing $\mathcal{A}$ or the coefficients c_β is secondary from our point of view, although we regard such a problem as being of great importance in various applications, and we do also consider it in some cases.

We also note that a formula such as (1.3) conveys the message that if we look for an asymptotic expansion of the form (1.2) and if we wish to make the approximation actually converge to $u_\epsilon(p)$ when r tends to $+\infty$ and ϵ is fixed, then an effective choice of the functions $\gamma_1, \ldots, \gamma_s$ is that given by formula (1.3).

Especially in the case of nonlinear problems, such information may be of interest for people who look for asymptotic expansions by means of Asymptotic Analysis techniques.

1.1 An Example

To better illustrate our point of view, we now introduce a concrete example where our 'unperturbed' problem is a Dirichlet boundary value problem on Ω^o with a boundary datum $g^o \in C^{1,\alpha}(\partial\Omega^o)$ (cf. [163], [166]). Namely,

$$\begin{cases} \Delta u = 0 \text{ in } \Omega^o, \\ u = g^o \quad \text{on } \partial\Omega^o. \end{cases} \tag{1.5}$$

As is well known, problem (1.5) has one and only one classical solution $u^o \in C^{1,\alpha}(\overline{\Omega^o})$.

In addition to the assumptions introduced above, we assume that the domain Ω^o has a connected exterior. To put it simply, Ω^o has no holes. Then we perturb problem (1.5) by introducing a hole in Ω^o. To do so, we fix another domain Ω^i which satisfies the same assumptions as Ω^o. The set Ω^i will serve as a reference domain for the hole, the letter 'i' standing for 'inner.' Indeed, for $\epsilon_0 > 0$ small enough we have

$$\epsilon\overline{\Omega^i} \subseteq \Omega^o \qquad \forall \epsilon \in [-\epsilon_0, \epsilon_0]$$

(we recall that Ω^i and Ω^o contain the origin). Accordingly, for $\epsilon \in [-\epsilon_0, \epsilon_0]$ we can define the perforated domain

$$\Omega(\epsilon) \equiv \Omega^o \setminus \epsilon\overline{\Omega^i}.$$

We observe that the boundary $\partial\Omega(\epsilon)$ consists of two connected components,

$$\epsilon\partial\Omega^i \quad \text{and} \quad \partial\Omega^o.$$

Also, if ϵ shrinks to 0, then $\Omega(\epsilon)$ degenerates to the punctured domain

$$\Omega(0) = \Omega^o \setminus \{0\}.$$

Our next step is to introduce a Dirichlet problem in $\Omega(\epsilon)$. To assign the Dirichlet datum on the component $\epsilon\partial\Omega^i$ of the boundary, we fix a function $g^i \in C^{1,\alpha}(\partial\Omega^i)$ and we rescale it. Then, for each $\epsilon \in]0, \epsilon_0[$ the 'perturbed' problem is given by

$$\begin{cases} \Delta u = 0 & \text{in } \Omega(\epsilon)\,, \\ u(x) = g^i(x/\epsilon) & \forall x \in \epsilon\partial\Omega^i\,, \\ u = g^o & \text{on } \partial\Omega^o\,. \end{cases} \tag{1.6}$$

Like problem (1.5), problem (1.6) also has a unique classical solution in $C^{1,\alpha}(\overline{\Omega(\epsilon)})$. We denote it by u_ϵ and we wish to study its dependence upon ϵ.

To do so, we fix a point p in $\Omega^o \setminus \{0\}$. Then we observe that, possibly shrinking ϵ_0, we can assume that

$$p \in \Omega(\epsilon) \qquad \forall \epsilon \in]0, \epsilon_0[\,.$$

Accordingly, $u_\epsilon(p)$ is defined for all $\epsilon \in]0, \epsilon_0[$ and we can ask the question:

What can we say about $u_\epsilon(p)$ as $\epsilon > 0$ approaches 0?

The answer given in the book accords with our introductory explanation and turns out to be dependent on the dimension n. In Chapter 8, we will prove that for $n \geq 3$ the map which takes ϵ to $u_\epsilon(p)$ has a real analytic continuation in a neighborhood of zero. Namely, there exist $\epsilon_p > 0$ and a real analytic function $\mathcal{U}_p$ from $] - \epsilon_p, \epsilon_p[$ to $\mathbb{R}$ such that

$$u_\epsilon(p) = \mathcal{U}_p(\epsilon) \qquad \forall \epsilon \in]0, \epsilon_p[. \tag{1.7}$$

Then, with reference to formula (1.3), we have the case where $s = 1$ and $\gamma_1(\epsilon) = \epsilon$. If instead the dimension n is two, then we will show that there are three real analytic functions $\mathcal{U}_p$, $\mathcal{W}_p$, and $\mathcal{R}$ from $] - \epsilon_p, \epsilon_p[$ to $\mathbb{R}$ such that

$$u_\epsilon(p) = \mathcal{U}_p(\epsilon) + \frac{\mathcal{W}_p(\epsilon)}{\mathcal{R}(\epsilon) + \frac{\log\epsilon}{2\pi}} \qquad \forall \epsilon \in]0, \epsilon_p[\,. \tag{1.8}$$

The presence of the logarithmic term implies that the right-hand side cannot be continued analytically in a neighborhood of zero, unless we consider some special cases where $\mathcal{W}_p(\epsilon) = 0$. However, if we divide both the numerator and the denominator of the quotient by $\log\epsilon$, we obtain

$$u_\epsilon(p) = \mathcal{U}_p(\epsilon) + \frac{\mathcal{W}_p(\epsilon)\frac{1}{\log\epsilon}}{\mathcal{R}(\epsilon)\frac{1}{\log\epsilon} + \frac{1}{2\pi}} \qquad \forall \epsilon \in]0, \epsilon_p[\,.$$

Then we observe that for $t_p > 0$ small enough we have

$$|\mathcal{R}(\epsilon)t| < \frac{1}{2\pi} \qquad \forall (\epsilon, t) \in] - \epsilon_p, \epsilon_p[\times] - t_p, t_p[$$

and accordingly

$$\mathcal{R}(\epsilon)t + 1/(2\pi) \neq 0 \qquad \forall (\epsilon, t) \in]-\epsilon_p, \epsilon_p[\times]-t_p, t_p[\,.$$

It follows that the function $\mathcal{A}$ from $]-\epsilon_p, \epsilon_p[\times]-t_p, t_p[$ to $\mathbb{R}$ which takes a pair (ϵ, t) to

$$\mathcal{A}(\epsilon, t) \equiv \mathcal{U}_p(\epsilon) + \frac{\mathcal{W}_p(\epsilon)t}{\mathcal{R}(\epsilon)t + \frac{1}{2\pi}}$$

is real analytic. Also, possibly shrinking ϵ_p, we can assume that

$$\frac{1}{\log \epsilon} \in]-t_p, t_p[\qquad \forall \epsilon \in]0, \epsilon_p[\,.$$

Then we conclude that

$$u_\epsilon(p) = \mathcal{A}\left(\epsilon, \frac{1}{\log \epsilon}\right) \qquad \forall \epsilon \in]0, \epsilon_p[\,.$$

With reference to formula (1.3), we now have the case where $s = 2$, $\gamma_1(\epsilon) = \epsilon$, and $\gamma_2(\epsilon) = 1/\log \epsilon$. As a consequence, the function $u_\epsilon(p)$ can be written as an analytic function of ϵ and $1/\log \epsilon$. So, the message conveyed to us here is that, if we want to write an asymptotic expansion associated with a convergent series, we should try using an expansion of powers of ϵ and $1/\log \epsilon$.

In Chapter 8, we explain how to compute the Taylor coefficients of the analytic functions which appear in the above formulas for $u_\epsilon(p)$ by following the ideas of the paper [82] with Rogosin. In particular, we shall see that rather than using the expansion in ϵ and $1/\log \epsilon$, it is convenient to consider a rearrangement of it. Indeed, it turns out that an expansion in powers of ϵ and

$$\frac{1}{r_0 + \frac{\log \epsilon}{2\pi}}$$

is more efficient. Here r_0 is a real constant that depends only on Ω^o and Ω^i. Then the results that we obtain can be compared with those of Il'in [127] and Maz'ya et al. [203, 204].

1.2 A Selection of Problems

We now indicate some of the problems that have been investigated by means of the Functional Analytic Approach.

1.2.1 Perturbation Problems for the Riemann Map

The first problem to which the Functional Analytic Approach was applied concerns the Riemann map in planar perforated domains. Indeed, when $n = 2$, the set Ω^o and the inner hole $\epsilon\Omega^i$ that we introduced above are Jordan domains. Then the Riemann Mapping Theorem implies that there exist a unique $r_\epsilon \in]0,1[$ and a unique (suitably normalized) holomorphic diffeomorphism from an annulus with inner radius equal to r_ϵ and outer radius equal to 1 to $\Omega(\epsilon)$ (the so-called Riemann map). By the Functional Analytic Approach one can analyze the behavior of both r_ϵ and of the Riemann map as ϵ tends to zero, even in the case where the Jordan curves which bind the domains Ω^o and Ω^i are considered as variables and we have a problem of nonlocal nature. In particular, one can prove that r_ϵ admits a real analytic continuation for small and negative values of ϵ (see [160], [161]).

1.2.2 Linear Elliptic Boundary Value Problems

The analysis of the Dirichlet problem for the Poisson equation, which develops what we have illustrated above for the Laplace operator, has been presented in [162] and [167]. In these papers we have also analyzed the dependence of the solution upon suitable diffeomorphisms that parametrize the boundaries of Ω^o and Ω^i. Then in [74] we have investigated the meaning of equality (1.7) when ϵ is negative. Indeed, one can observe that the solution u_ϵ is defined only for positive values of ϵ, whereas the map $\mathcal{U}_p$ is defined in an open neighborhood of $\epsilon = 0$. Thus, it makes sense to ask what happens for $\epsilon < 0$. In [79] we have considered an analogous question for the case of dimension two and we have investigated the logarithmic behavior that appears in (1.8). The Functional Analytic Approach has also been applied to linear elliptic systems of differential equations. In particular, it has been used to deal with problems for the Lamé equations of linearized elasticity [66], [67] and for the Stokes system, which describes the steady-state flow of an incompressible viscous fluid [62].

1.2.3 Eigenvalues Problems

By the Functional Analytic Approach one can also study the asymptotic behavior of eigenvalues and eigenfunctions in domains with a small hole. For the analysis of a Neumann eigenvalue problem for the Laplace operator, we refer to [169], and for the analysis of the Steklov eigenvalue problem, we refer to the papers [113] and [114] with Gryshchuk and to paper [171].

1.2.4 Nonlinear Boundary Value Problems

Further applications of the Functional Analytic Approach concern nonlinear boundary value problems in domains with small holes and inclusions. Typically, in the applications of the Functional Analytic Approach to such problems, one first shows the existence of an ϵ dependent family of solutions u_ϵ of the perturbed problem in $\Omega(\epsilon)$ (which may not be the only one). Then one looks for representation formulas such as (1.3), again for u_ϵ or related functionals. This being accomplished, one can investigate (local) uniqueness properties of the family of solutions. This approach has been used to study nonlinear boundary value problems for linear differential operators with nonlinear boundary conditions, and certain nonlinear boundary value problems for quasi-linear equations that can be reduced to linear differential equations with nonlinear boundary conditions. For example, a problem for the Laplace equation with a nonlinear Robin boundary condition has been analyzed in [164] and nonlinear transmission conditions are studied in [168], in the paper [72] with Molinarolo, and in the paper [225] of Molinarolo. Nonlinear traction problems for the equations of the linearized elastostatics have been considered in [66] and nonlinear traction problems in which the boundary data are allowed to depend singularly on the parameter ϵ have been studied in [63], [64], and [67].

1.2.5 Problems in Periodic Domains

Along with problems in bounded perforated domains, we have considered problems in periodic domains with an infinite number of holes or inclusions. As a first step, we have considered problems where the periodicity cell is fixed and each of the holes is shrinking to a point. We mention, for example, [228], which concerns a Dirichlet problem for the Laplace equation, and [179], where we consider a Neumann problem for the Poisson equation. Nonlinear Robin boundary conditions have been studied in [176] and [68] and quasilinear heat transmission problems have been investigated in [177] and [70]. Then we have turned to problems in which the periodicity cell is also shrinking: for the analysis of a linear problem, we refer to [178] and [181], and for a nonlinear problem to [182]. Then we have turned to problems in which the periodicity cell is also shrinking: for the analysis of a linear problem, we refer to [178] and [181], and for a nonlinear problem to [182, 183].

1.2.6 Different Boundary Perturbations

Some papers concern applications of the Functional Analytic Approach to boundary perturbations other than holes or inclusions that shrink to interior points. For example, the case of two shrinking holes colliding into one another has been considered in [78], [79], and [80]. The case of a small hole that approaches the outer boundary of a domain has been studied in the paper [29] with Bonnaillie-Noël and Dambrine, and self-similar perturbations of a plane sector domain have been analyzed in [60] with Costabel and Dauge.

1.2.7 Perturbation Results for Integral Operators

To carry out the analysis of singularly perturbed elliptic boundary value problems using the approach introduced in this book, we need to understand the dependence of certain integral operators upon perturbations both of the density and of the support. In this sense, we mention the paper [184] with Preciso on the Cauchy integral; the papers [185] and [186] with Rossi on the layer potentials associated with the fundamental solution of the Laplace and of the Helmholtz operator; the papers [62], [65], and the paper [89] with Dondi on layer potentials associated with a fundamental solution of general elliptic operators with constant coefficients; the paper [174] on periodic layer potentials associated with a fundamental solution of general elliptic operators with constant coefficients; the paper [69] on volume potentials corresponding to parametric families of fundamental solutions; and the papers [172], [173] with Luzzini on the layer potentials associated with the fundamental solution of the heat equation. In addition, special families of parameter-dependent fundamental solutions are introduced and studied in [61] and in paper [73] with Morais, which deal with the case of general elliptic operators with constant real and quaternion coefficients, respectively, and in [180], where we consider the periodic case and we investigate the dependence upon the periodic structure. We also mention the work on nonlinear integral operators with analytic kernels in [175].

1.3 Structure of the Book

In Chapter 2 we present some preliminary material, mainly on spaces of Hölder continuous functions and Schauder spaces, as well as some basics about regular subsets of $\mathbb{R}^n$. Although this part is mostly self-contained, we sometimes refer to other textbooks for basic results of Calculus and of Real and Functional Analysis.

In Chapter 3 we summarize some basic properties of harmonic functions and for most of the proofs we refer to major monographs, in particular to Evans [95], Folland [102], and Gilbarg and Trudinger [107].

In Chapter 4 we present the Green Identities both in a bounded domain and in the corresponding unbounded exterior domain. Then we introduce the notion of single and double layer potentials and show certain mapping properties of these potentials. Although these properties can be found in the classical monographs of Günter [115] and of Kupradze et al. [151], here we prove the corresponding statements with optimal Hölder exponents. To do so, we exploit a result of Miranda [212] on the Hölder continuity of singular integrals of convolution type. We also provide a complete proof of Miranda's result (that was originally published in Italian).

In Chapter 5 we introduce some basic properties of Fredholm operators and the Fredholm Alternative Theorem both in its classical setting and in the version of the duality pairs developed by Wendland [273], [274].

In Chapter 6, we obtain existence and uniqueness results for the classical boundary value problems for the Laplace equation in Schauder spaces using a Potential Theoretic Method and the Fredholm Alternative Theorem. For recent contributions on integral equation methods for the solution of boundary value problems we refer to the monographs of Constanda [58], Constanda et al. [59], Gewinner and Stephan [116], Hsiao and Wendland [126], Maz'ya and Soloviev [205], McLean [206], Medková [208], Mitrea and Mitrea [215], Kohr and Pop [143], Sauter and Schwab [254].

In Chapter 7, we introduce some basic properties of the volume potentials in Schauder spaces and prove existence results for the classical boundary value problems of the Poisson equation. Then we also consider the case of Roumieu spaces. Although the properties of volume potentials in Schauder spaces can be found in the classical monographs of Günter [115] and of Kupradze et al. [151], here we prove the corresponding statements with optimal Hölder exponents by following a proof of Miranda [212] and by exloiting a known lemma (cf. Majda and Bertozzi [196, Prop. 8.12, pp. 348–350]) for which we provide a proof of Mateu et al. [198].

In Chapter 8, we begin to explain the Functional Analytic method of the present book and we do so in detail for the Dirichlet problem for the Laplace operator in a domain with a single hole which shrinks to a point, as we have outlined above.

Then in Chapter 9 we turn to a mixed boundary value problem for the Laplace and Poisson equations and the Steklov eigenvalue problem.

In Chapter 10 we illustrate in detail the case of two holes which shrink to a common point. We focus on the Dirichlet problem for the Laplace equation.

In Chapter 11 we consider the case of nonlinear boundary conditions in domains with a small hole. We do so for nonlinear Robin and nonlinear transmission conditions for the Laplace equation.

In Chapter 12 we consider boundary value problems for the Laplace equation in periodic domains obtained by removing from $\mathbb{R}^n$ a periodic set of holes. To do so, we develop a periodic version of Potential Theory based on layer potentials which are defined by replacing the fundamental solution of the Laplace equation with a periodic analogue.

In Chapter 13, we consider both linear and nonlinear boundary value problems in the entire space with a periodic set of perforations which shrink to points.

In the appendix, we have included a number of known results that are necessary for our purposes throughout the book.

Chapter 2
Preliminaries

Abstract In this chapter we review some preliminary material from Calculus and Functional Analysis and also recall the basic properties of real analytic functions. We introduce the notion of Roumieu classes, then summarize the classical properties of Hölder continuous functions. Next, we introduce the notion of coordinate cylinders and of sets that are hypographs of Cartesian functions locally around the boundary points, including the case of Lipschitz sets. Finally, we turn to Schauder spaces and the corresponding embedding theorems and to the subsets of $\mathbb{R}^n$ of class $C^{m,\alpha}$. Although this part of the book is mostly self-contained, we sometimes refer to other textbooks for basic results of Calculus and of Real and Functional Analysis.

2.1 Basic Notation

In this section we introduce some symbols that we use all throughout the book. We begin with the symbol $=$ that we use for equalities and equations and the symbol $\equiv$ that we use for definitions (instead of $:=$ that other authors use).

We denote by $\mathbb{N}$, $\mathbb{Z}$, $\mathbb{R}$, $\mathbb{C}$, the set of natural numbers including zero, the set of integer numbers, the set of real numbers, and the set of complex numbers, respectively. We denote by $\mathbb{K}$ either the field $\mathbb{R}$ or $\mathbb{C}$.

Let X be a set. Then we denote by D_X the diagonal of $X \times X$, i.e., we set

$$D_X \equiv \{(x_1, x_2) \in X \times X : x_1 = x_2\} .$$

If Y is also a set then we denote by Y^X the set of maps from X to Y.

Let $(X, \|\cdot\|_X)$ and $(Y, \|\cdot\|_Y)$ be normed spaces on $\mathbb{K}$. Let U and V be subsets of X and Y, respectively. Then $\overline{U}$ denotes the closure of U, ∂U denotes the boundary of U, and $\mathrm{diam}\,(U)$ denotes the diameter of U, i.e.,

$$\mathrm{diam}\,(U) \equiv \sup\left\{\|x - y\|_X : x, y \in U\right\} .$$

M. Dalla Riva et al., *Singularly Perturbed Boundary Value Problems*,
https://doi.org/10.1007/978-3-030-76259-9_2

We denote by either U^+ or $\mathring{U}$ the interior of U and by U^- the exterior of U, i.e., we set

$$U^+ \equiv \mathring{U}\,, \qquad U^- \equiv X \setminus \overline{U}\,. \tag{2.1}$$

In particular, $U^+ = U$ if U is open.

We say that a function f from U to Y is locally constant provided that for each $u_0 \in U$ there exists a neighborhood W_0 of u_0 in U such that f is constant in W_0.

We denote by $C^0(U, V)$ the set of continuous functions from U to V. We denote by $B(U, V)$ the set of bounded functions from U to V. As it is well known, if Y is complete then the space $B(U, Y)$ with the sup-norm

$$\|f\|_{B(U,Y)} \equiv \sup_{x\in U} \|f(x)\|_Y$$

is complete and the subspace $C^0_b(U, Y) \equiv C^0(U, Y) \cap B(U, Y)$ of $B(U, Y)$ is also complete. When no ambiguity can arise, we simply write $B(U)$ instead of $B(U, \mathbb{K})$ and similarly for other function spaces defined on U.

If U is a subset of $\mathbb{R}$ and if f is a function from U to $\mathbb{R}$, we say that f is increasing provided that $f(\rho_1) \leq f(\rho_2)$ whenever $\rho_1, \rho_2 \in U$ and $\rho_1 < \rho_2$. Then we say that f is strictly increasing provided that $f(\rho_1) < f(\rho_2)$ whenever $\rho_1, \rho_2 \in U$ and $\rho_1 < \rho_2$. Similarly, we define the decreasing functions and the strictly decreasing functions.

The inverse function of an invertible function f is denoted $f^{(-1)}$, as opposed to the reciprocal of a real-valued function g, or the inverse of a matrix A, which are denoted g^{-1} and A^{-1}, respectively. The preimage of a set E under a function f is denoted by $f^{\leftarrow}(E)$.

Let A be a matrix. Then A^t denotes the transpose matrix of A and A_{ij} denotes the (i, j)-entry of A. If A is invertible, we set

$$A^{-t} \equiv (A^{-1})^t\,.$$

Let $n \in \mathbb{N} \setminus \{0\}$. We denote by $\mathbb{M}_n(\mathbb{K})$ the set of $n \times n$ matrices with entries in the field $\mathbb{K}$. We denote by $\delta_{l,j}$ the Kronecker symbol. Namely,

$$\delta_{l,j} \equiv \begin{cases} 1 & \text{if } l = j\,, \\ 0 & \text{if } l \neq j\,. \end{cases}$$

We denote by I_n or simply by I the identity matrix $I \equiv (\delta_{lj})_{l,j=1,\ldots,n}$. We denote by

$$\mathbb{O}_n(\mathbb{K}) \equiv \left\{A \in \mathbb{M}_n(\mathbb{K}) :\, A^t A = I = A A^t\right\}$$

the set of the orthogonal matrices of $\mathbb{M}_n(\mathbb{K})$.

If S is a set, I_S denotes the identity map from S to itself. If S is clear from the context, we simply write I instead of I_S. We denote by $\mathrm{supp} f$ the support of a complex valued function f. If $x \in \mathbb{R}$, we set

$$\operatorname{sgn}(x) \equiv \begin{cases} 1 & \text{if } x > 0\,, \\ 0 & \text{if } x = 0\,, \\ -1 & \text{if } x < 0\,. \end{cases}$$

For all $\rho > 0$, $x \in \mathbb{R}^n$, x_j denotes the j-th coordinate of x, $|x|$ denotes the Euclidean modulus of x in $\mathbb{R}^n$, i.e.,

$$|x| \equiv \sqrt{\sum_{j=1}^{n} x_j^2} \qquad \forall x \equiv (x_1, \ldots, x_n) \in \mathbb{R}^n\,.$$

Unless otherwise specified, we always equip $\mathbb{R}^n$ with the Euclidean norm $|\cdot|$. Then we set

$$\mathbb{B}_n(x, \rho) \equiv \{y \in \mathbb{R}^n : |x - y| < \rho\}\,.$$

We often abbreviate the unit ball $\mathbb{B}_n(0,1)$ as $\mathbb{B}_n$. We denote by $\{e_1, \ldots, e_n\}$ the canonical basis of $\mathbb{R}^n$. We denote by a dot the scalar product of vectors.

If $(Z, \mathcal{M}, \mu)$ is a measured space, we retain standard abbreviations as

$$\mu - \text{a.a.} = \mu - \text{almost all}\,, \qquad \mu - \text{a.e.} = \mu - \text{almost everywhere}\,.$$

We retain the standard notation for the Lebesgue space $L^p_\mu(Z)$ of p-summable (equivalence classes of) measurable functions for $p \in [1, +\infty]$. In particular, we set

$$\|f\|_{L^p_\mu(Z)} \equiv \left(\int_Z |f|^p \, d\mu\right)^{1/p} \qquad \forall f \in L^p_\mu(Z)$$

if $p \in [1, +\infty[$ and

$$\|f\|_{L^p_\mu(Z)} \equiv \operatorname{ess\,sup}_Z |f| \qquad \forall f \in L^\infty_\mu(Z)$$

if $p = +\infty$ and we abbreviate $\|\cdot\|_{L^p_\mu(Z)}$ as $\|\cdot\|_p$ whenever there is no ambiguity on the space $L^p_\mu(Z)$. For basic inequalities such as the Hölder inequality for Lebesgue spaces, we refer to textbooks such as Folland [103, Chap. 6].

Also, if X is a vector subspace of $L^1_\mu(Z)$, we find convenient to set

$$X_0 \equiv \left\{f \in X : \int_Z f \, d\mu = 0\right\}. \tag{2.2}$$

If $A \in \mathcal{M}$ and $\mu(A) \in]0, +\infty[$, we set

$$\fint_A \equiv \frac{1}{\mu(A)} \int_A\,.$$

By the known inclusions of the $l^p(\mathbb{N}) \equiv L^p_\mu(\mathbb{N})$ spaces with μ equal to the counting measure, we have the known inequality

$$\left(\sum_{j=1}^{\infty}|a_j|^q\right)^{1/q} \leq \left(\sum_{j=1}^{\infty}|a_j|^p\right)^{1/p} \tag{2.3}$$

for all $p, q \in]0, +\infty[$ such that $p \leq q$ and for all sequences $\{a_j\}_{j\in\mathbb{N}}$ of real numbers such that the right-hand side is finite (cf., e.g., Folland [103, Prop. 6.11]). Instead the convexity of the p-power function for $p \in [1, +\infty[$ implies that

$$\left(\sum_{j=1}^{m}|a_j|\right) \leq m^{1-(1/p)} \left(\sum_{j=1}^{m}|a_j|^p\right)^{1/p} \tag{2.4}$$

for all $m \in \mathbb{N} \setminus \{0\}$ and $(a_1, \ldots, a_m) \in \mathbb{R}^m$, and the concavity of the p power function for $p \in]0, 1]$ implies that

$$\left(\sum_{j=1}^{m}|a_j|\right) \geq m^{1-(1/p)} \left(\sum_{j=1}^{m}|a_j|^p\right)^{1/p} \tag{2.5}$$

for all $m \in \mathbb{N} \setminus \{0\}$ and $(a_1, \ldots, a_m) \in \mathbb{R}^m$.

We denote by $\mathcal{L}_n$ the σ-algebra of Lebesgue measurable sets of $\mathbb{R}^n$ and by m_n the Lebesgue measure in $\mathbb{R}^n$. If μ is the Lebesgue measure in $\mathbb{R}^n$, we normally omit the letter μ in the symbol L^p_μ for the Lebesgue spaces and in abbreviations such as $\mu -$ a.a. and $\mu -$ a.e..

A subset M of $\mathbb{R}^n$ is a differential manifold of dimension $s \in \{1, \ldots, n\}$ and of class C^1 embedded in $\mathbb{R}^n$ if, for every $p \in M$, there exists an open neighborhood W of p in $\mathbb{R}^n$ and a map $\psi \in C^1\left(\overline{\mathbb{B}_s(0,1)}, \mathbb{R}^n\right)$ such that ψ is a homeomorphism of $\overline{\mathbb{B}_s}$ onto $W \cap M$, $\psi(0) = p$, and the Jacobian matrix $D\psi$ has rank s at all points of $\mathbb{B}_s(0,1)$, i.e., a ψ is parametrization for M around p. By homeomorphism we understand a continuous bijection which has a continuous inverse. Instead, for the definition of a C^1 function on the closed set $\overline{\mathbb{B}_s(0,1)}$, we refer to Section 2.3.

We denote by $\mathcal{L}_M$ and m_M (or simply m_s) the σ-algebra of Lebesgue surface measurable sets and the Lebesgue surface measure on a manifold M of dimension s embedded in $\mathbb{R}^n$, respectively. We denote by $d\sigma$ the area element of M (cf., e.g., Naumann and Simader [232]). If μ is the Lebesgue surface measure on M, we normally omit the subscript μ in the notation for the Lebesgue spaces and in abbreviations such as a.a. and a.e..

As is well known, if $\alpha \in]0, +\infty[$ then the function $e^{-t}t^{\alpha-1}$ is integrable in $]0, +\infty[$. Then the Euler Gamma function is defined by

$$\Gamma(\alpha) \equiv \int_0^{+\infty} e^{-t}t^{\alpha-1}\,dt \qquad \forall \alpha \in]0, +\infty[\,.$$

By integrating by parts, we have

$$\Gamma(\alpha+1) = \alpha\Gamma(\alpha) \qquad \forall\alpha > 0\,.$$

By a simple inductive argument, we then have

$$\Gamma(n+1) = n! \qquad \forall n \in \mathbb{N}\,.$$

By elementary Calculus, it is known that

$$\Gamma(1/2) = \sqrt{\pi}\,.$$

We denote by s_n the surface measure of $\partial\mathbb{B}_n(0,1)$ and by ω_n the n-dimensional measure of the ball $\mathbb{B}_n(0,1)$. As is well known,

$$\omega_n = \frac{\pi^{n/2}}{\Gamma((n/2)+1)}\,, \qquad s_n = \frac{2\pi^{n/2}}{\Gamma(n/2)}\,, \qquad s_n = n\omega_n\,,$$

for each $n \in \mathbb{N}\setminus\{0\}$ and

$$\int_0^1 t^{x-1}(1-t)^{y-1}\,dt = \frac{\Gamma(x)\Gamma(y)}{\Gamma(x+y)} \qquad \forall x,y \in]0,+\infty[\tag{2.6}$$

(cf., e.g., Lebedev [187, §1.5]). We also mention that if $x \in \mathbb{R}^n$ and $\rho \in]0,+\infty[$, then

$$\int_{\partial\mathbb{B}_n(x,\rho)} f(y)\,d\sigma_y = \int_{\partial\mathbb{B}_n(0,1)} f(x+\rho\xi)\rho^{n-1}\,d\sigma_\xi$$

for each complex-valued integrable function f on $\partial\mathbb{B}_n(x,\rho)$.

2.2 Preliminaries of Linear Functional Analysis

If X and Y a vector spaces on the field $\mathbb{K}$, $L(X,Y)$ denotes the set of linear functions from X to Y.

If X is a vector space on the field $\mathbb{K}$, then a linear topology on X is a topology $\mathcal{T}$ on X such that both the sum from $X \times X$ to X and the product from $\mathbb{K} \times X$ to X are continuous.

We will mainly consider the case in which X is endowed by a norm $\|\cdot\|_X$. As it is well known, a norm on a space X generates a linear topology on X.

A series $\sum_{j=0}^{\infty} x_j$ with terms in a normed space X is said to be normally convergent provided that

$$\sum_{j=0}^{\infty} \|x_j\|_X < +\infty\,.$$

If X is complete, then it is known that a normally convergent series is also convergent in X.

Theorem 2.1. *Let $(X, \|\cdot\|_X)$ be a normed space. Let V be a closed subspace of X. Then the norm on the quotient X/V defined by*

$$\| [x] \|_{X/V} \equiv \inf_{v \in V} \|x + v\|_X \qquad \forall [x] \in X/V$$

generates the quotient topology on X/V, i.e. the strongest topology on X/V such that the canonical projection π of X onto X/V is continuous.

If X is complete, then $(X/V, \|\cdot\|_{X/V})$ is complete. If Y is a normed space and if T is a linear map from X/V to Y, then T is continuous if and only if $T \circ \pi$ is continuous.

For a proof, we refer for example to Schaefer [255, p. 42].

Let X, Y be normed spaces. Then we endow the product space $X \times Y$ with the norm defined by

$$\|(x, y)\|_{X \times Y} \equiv \|x\|_X + \|y\|_Y \qquad \forall (x, y) \in X \times Y\,,$$

while we use the Euclidean norm for $\mathbb{R}^n$. If T is a linear operator from X to Y, then we set

$$\mathrm{Ker}\, T \equiv \{x \in X : T[x] = 0\}$$

for the kernel (or null space) of T and

$$\mathrm{Im}\, T \equiv \{T[x] : x \in X\}$$

for the image (or range) of T. A linear operator T from X to Y is well known to be continuous if and only if it maps bounded subsets of X to bounded subsets of Y. It is customary to say that a linear operator T from X to Y is bounded provided that T maps bounded subsets of X to bounded subsets of Y. With such a terminology, a linear operator T from X to Y is continuous if and only if it is bounded. Then

$$\mathcal{L}(X, Y)$$

denotes the space of linear and continuous operators from X to Y, and we set

$$\|T\|_{\mathcal{L}(X,Y)} \equiv \sup_{x \in B_X(0,1)} \|T[x]\|_Y \qquad \forall T \in \mathcal{L}(X, Y)\,,$$

where

$$B_X(p, r) \equiv \{x \in X : \|x - p\|_X < r\}$$

for all $p \in X$ and $r \in]0, +\infty[$. If Y is complete, then $\left(\mathcal{L}(X, Y), \|\cdot\|_{\mathcal{L}(X,Y)}\right)$ is well known to be a Banach space. We often refer to the following classical result. For a proof, we refer for example to Brezis [35, Theorem 2.6 and Corollary 2.7 p. 35].

Theorem 2.2 (of the Open Mapping). *Let X, Y be Banach spaces. If $T \in \mathcal{L}(X, Y)$ is surjective, then T is open, i.e., T maps open subsets of X to open subsets of Y.*

In particular, if $T \in \mathcal{L}(X,Y)$ is a bijection, then inverse operator $T^{(-1)}$ is continuous and accordingly T is a homeomorphism.

We also note that if $T \in \mathcal{L}(X,Y)$, then the Fundamental Theorem of Homomorphism implies that there exists a unique linear injective map $\tilde{T}$ from $X/\mathrm{Ker}\, T$ to Y such that

$$T = \tilde{T} \circ \pi ,$$

where π denotes the canonical projection of X onto $X/\mathrm{Ker}\, T$ (cf., e.g., Kostrikin and Manin [145]). By Theorem 2.1, $\tilde{T}$ is continuous. If we further assume that T is surjective, then $\tilde{T}$ is a continuous bijection of $X/\mathrm{Ker}\, T$ onto Y and the Open Mapping Theorem implies that $\tilde{T}$ is a homeomorphism.

As is well known a normed space of dimension $n \in \mathbb{N}$ is linearly homeomorphic to $\mathbb{K}^n$ (cf. e.g. Taylor and Lay [265, Chap. 2, Theorem 3.1]) and is therefore complete. Then we also have the following classical result (cf. e.g. Taylor and Lay [265, Chap. 2, Theorem 3.3]).

Theorem 2.3. *Let X be a normed space. If X_1 is a finite dimensional subspace of X, then X_1 is closed.*

Definition 2.4. Let X be a normed space. Let X_1 be a subspace of X. We say that a linear map P_1 from X to X is a projection of X onto X_1 provided that

$$P_1[X] \subseteq X_1 \qquad \text{and} \qquad P_1[\xi] = \xi \qquad \forall \xi \in X_1 .$$

We note that if P is a projection onto X_1, then we necessarily have $X_1 = \mathrm{Im}\, P_1$.

If X_1 and X_2 are subspaces of X and if X equals the algebraic direct sum of X_1 and X_2, i.e., if

$$X = X_1 + X_2 \qquad \text{and} \qquad X_1 \cap X_2 = \{0\} ,$$

then we write $X = X_1 \oplus X_2$, and we say that X_2 is an algebraic direct complement (or algebraic supplement) of X_1 in X.

If $X = X_1 \oplus X_2$, then the map Φ from $X_1 \times X_2$ to X that takes (x_1, x_2) to $x_1 + x_2$ is a linear bijection, and if π_j denotes the canonical projection of $X_1 \times X_2$ onto X_j, then the map

$$P_j \equiv \pi_j \circ \Phi^{(-1)}$$

is a projection of X onto X_j for all $j \in \{1,2\}$ in the sense of Definition 2.4. Moreover, if $x \in X$, then $(P_1[x], P_2[x])$ is the only pair of $X_1 \times X_2$ such that

$$x = P_1[x] + P_2[x] .$$

In particular, the pair of projections (P_1, P_2) is uniquely determined by the condition that P_j is a projection of X onto X_j for all $j \in \{1,2\}$ and that $P_1 + P_2 = I_X$. We say that P_1 is a projection of X onto X_1 along X_2.

Definition 2.5. Let X be a normed space. Let X_1 and X_2 be subspaces of X such that $X = X_1 \oplus X_2$ algebraically. Then we say that the direct sum $X_1 \oplus X_2$ is topological provided that the corresponding projections P_1 and P_2 are continuous. If so, then we say that X_2 is a topological direct complement (or topological supplement) of X_1 in X.

Then we have the following, which we exploit in the sequel.

Theorem 2.6. *Let X be a Banach space. Let X_1 and X_2 be linear subspaces of X such that $X = X_1 \oplus X_2$ algebraically. Then the direct sum $X_1 \oplus X_2$ is topological if and only if both X_1 and X_2 are closed in X.*

Proof. If the direct sum is topological, we note that $X_1 = \{x \in X : P_2[x] = 0\}$. Hence, the continuity of P_2 implies that X_1 is closed in X. Similarly, X_2 is also closed in X.

Conversely, if both X_1 and X_2 are closed in X, then X_1 and X_2 are Banach spaces and $X_1 \times X_2$ is also a Banach space. Then the Open Mapping Theorem implies that the linear and continuous bijection Φ from $X_1 \times X_2$ to X that takes (x_1, x_2) to $x_1 + x_2$ has a continuous inverse, and accordingly $P_j = \pi_j \circ \Phi^{(-1)}$ is continuous for each $j \in \{1, 2\}$ and the direct sum $X = X_1 \oplus X_2$ is topological. □

Then we have the following immediate corollary.

Corollary 2.7. *Let X be a Banach space. Let X_1 be a closed subspace of X of finite codimension, i.e., the dimension r of the quotient space X/X_1 is finite. Then X_1 has a topological direct complement (or topological supplement) X_2 in X.*

Proof. Let π denote the canonical projection of X onto X/X_1. By assumption, there exist $\xi_1, \ldots, \xi_r$ in X such that $\{\pi(\xi_1), \ldots, \pi(\xi_r)\}$ is a basis for X/X_1. Then the subspace X_2 of X generated by $\xi_1, \ldots, \xi_r$ is finite dimensional and X equals the direct algebraic sum $X_1 \oplus X_2$. Since finite dimensional subspaces of a normed space are closed (cf. Theorem 2.3), then both X_1 and X_2 are closed in X and Theorem 2.6 implies that the direct sum $X_1 \oplus X_2$ is topological. Accordingly X_2 is a topological direct complement of X_1 in X. □

Definition 2.8. Let X, Y be normed spaces. A linear map T form X to Y is said to be compact provided that

$$\overline{T(A)}$$

is compact whenever A is a bounded subset of X.

By exploiting the sequential compactness of the compact subsets of a normed space, one can prove that a map $T \in L(X, Y)$ is compact if and only if, whenever $\{x_j\}_{j\in\mathbb{N}}$ is a bounded sequence of X, the sequence $\{T[x_j]\}_{j\in\mathbb{N}}$ has a subsequence which converges in Y.

We often employ the following characterization of those bilinear maps which are continuous. For a proof, we refer to Brezis [35, p. 49].

Proposition 2.9. *Let X_1, X_2, Y be normed spaces. A bilinear map B from $X_1 \times X_2$ to Y is continuous if and only if there exists $b \in]0, +\infty[$ such that*

$$\|B[x_1, x_2]\|_Y \leq b\|x_1\|_{X_1}\|x_2\|_{X_2} \qquad \forall (x_1, x_2) \in X_1 \times X_2 \, .$$

2.3 Spaces of Classically Differentiable Functions

Let $m, n \in \mathbb{N} \setminus \{0\}$. Let Ω be an open subset of $\mathbb{R}^n$. The space of m-times continuously differentiable $\mathbb{R}$-valued functions on Ω is denoted by $C^m(\Omega, \mathbb{R})$, or more simply by $C^m(\Omega)$. We also set $C^\infty(\Omega) \equiv \bigcap_{j \in \mathbb{N}} C^j(\Omega)$.

Let $r \in \mathbb{N} \setminus \{0\}$. Let $f \in (C^m(\Omega))^r$. The s-th component of f is denoted f_s, and Df denotes the Jacobian matrix $\left(\frac{\partial f_s}{\partial x_l}\right)_{\substack{s=1,\dots,r, \\ l=1,\dots,n}}$. Instead, we denote the gradient of a function by ∇. For us the gradient is always a column vector. Let $\beta \equiv (\beta_1, \dots, \beta_n)$ be in $\mathbb{N}^n$, $|\beta| \equiv \beta_1 + \cdots + \beta_n$. Then $D^\beta f$ denotes $\frac{\partial^{|\beta|} f}{\partial x_1^{\beta_1} \dots \partial x_n^{\beta_n}}$, and we also use standard abbreviations as $\partial_{x_l} \equiv \frac{\partial}{\partial x_l}$.

The subspace of $C^m(\Omega)$ of those functions f whose derivatives $D^\beta f$ of order $|\beta| \leq m$ can be extended with continuity to $\overline{\Omega}$ is denoted by $C^m(\overline{\Omega})$. If $f \in C^m(\overline{\Omega})$ and $|\beta| \leq m$, then we set

$$D^\beta f(x) \equiv \lim_{y \to x} D^\beta f(y) \qquad \forall x \in \partial\Omega \, .$$

The reader should be aware that this is not the only definition of $C^m(\overline{\Omega})$ available in the literature. For other authors, $C^m(\overline{\Omega})$ is defined as the space of restrictions to $\overline{\Omega}$ of functions of class C^m in an open neighorhood of $\overline{\Omega}$. Such a definition is easily seen to be more restrictive than the one adopted here. Then we consider the space

$$C^m_b(\overline{\Omega}) \equiv \left\{ f \in C^m(\overline{\Omega}) : \, D^\beta f \in B(\overline{\Omega}), \, \forall \beta \in \mathbb{N}^n, \, |\beta| \leq m \right\}$$

endowed with the norm

$$\|f\|_{C^m_b(\overline{\Omega})} \equiv \sum_{|\beta| \leq m} \sup_{x \in \overline{\Omega}} |D^\beta f(x)| \, .$$

As is well known, the space $\left(C^m_b(\overline{\Omega}), \|\cdot\|_{C^m_b(\overline{\Omega})}\right)$ is complete. If Ω is bounded, then we obviously have $C^m_b(\overline{\Omega}) = C^m(\overline{\Omega})$ (and in this case, we generally drop the subscript b). We note that the inclusion of $C^m_b(\overline{\Omega})$ into $C^0_b(\overline{\Omega})$ has to be understood in the sense that each element $f \in C^m_b(\overline{\Omega})$ has a uniquely determined continuous extension to $\overline{\Omega}$, which we still denote by f. Then by the definition of norm in $C^m_b(\overline{\Omega})$ and in $C^0_b(\overline{\Omega})$, $C^m_b(\overline{\Omega})$ is continuously imbedded into $C^0_b(\overline{\Omega})$.

2.4 Distributions and Weak Derivatives

Let $n \in \mathbb{N} \setminus \{0\}$. Let K be a compact subset of $\mathbb{R}^n$. Then we set

$$\mathcal{D}(K) \equiv C_c^\infty(K) \equiv \{\varphi \in C^\infty(\mathbb{R}^n) : \operatorname{supp} \varphi \subseteq K\}$$

and

$$p_{K,m}(\varphi) \equiv \sup_{|\beta| \le m} \sup_K |D^\beta \varphi| \qquad \forall \varphi \in C_c^\infty(K)$$

for all $m \in \mathbb{N}$. Then $C_c^\infty(K)$ is a vector space and $\{p_{K,m} : m \in \mathbb{N}\}$ is a family of norms on $C_c^\infty(K)$ which generate a topology on $C_c^\infty(K)$. The topology on $C_c^\infty(K)$ generated by the family of norms $\{p_{K,m} : m \in \mathbb{N}\}$ is by definition that generated by the family of open balls

$$\{\psi \in C_c^\infty(K) : p_{K,m}(\varphi - \psi) < r\}$$

with center at $\varphi \in C_c^\infty(K)$, radius $r \in]0, +\infty[$, and $m \in \mathbb{N}$. The topology generated by the family of norms $\{p_{K,m} : m \in \mathbb{N}\}$ is linear. Then one can characterize the linear functions from $C_c^\infty(K)$ to $\mathbb{R}$ which are continuous by means of the following (cf., e.g., Trèves [266]).

Proposition 2.10. *Let K be a compact subset of $\mathbb{R}^n$. A linear function T from $C_c^\infty(K)$ to $\mathbb{R}$ is continuous if and only if there exist $c > 0$ and $m \in \mathbb{N}$ such that*

$$|T[\varphi]| \le c p_{K,m}(\varphi) \qquad \forall \varphi \in C_c^\infty(K).$$

Now let Ω be an open subset of $\mathbb{R}^n$. Then we set

$$\mathcal{D}(\Omega) \equiv \{\varphi \in C_c^\infty(\mathbb{R}^n) : \operatorname{supp} \varphi \text{ is compact}, \operatorname{supp} \varphi \subseteq \Omega\}.$$

The space $\mathcal{D}(\Omega)$ is said to be the space of the test functions in Ω and is a vector space. We now wish to introduce a linear topology on $\mathcal{D}(\Omega)$ that has the property of being locally convex, i.e., such that each element of $\mathcal{D}(\Omega)$ has a basis of convex neighborhoods. To do so, we note that

$$\mathcal{D}(\Omega) = \bigcup_{K \text{ is compact},\ K \subseteq \Omega} \mathcal{D}(K),$$

we denote by i_K the inclusion map from $\mathcal{D}(K)$ into $\mathcal{D}(\Omega)$, and we state the following theorem. For a proof, we refer to Trèves [266].

Theorem 2.11. *There exists a strongest locally convex separated topology $\mathcal{T}$ on $\mathcal{D}(\Omega)$ such that the inclusion i_K is continuous from $\mathcal{D}(K)$ to $(\mathcal{D}(\Omega), \mathcal{T})$ for each compact subset K of Ω.*

If T is a linear map from $(\mathcal{D}(\Omega), \mathcal{T})$ to $\mathbb{R}$, then T is continuous if and only if the map $T \circ i_K$ from $\mathcal{D}(K)$ to $\mathbb{R}$ is continuous for each compact subset K of Ω.

We note that the maps $T \circ i_K$ of the previous statement are precisely the restrictions of T to $\mathcal{D}(K)$.

Unless otherwise specified, we always retain $\mathcal{D}(\Omega)$ as endowed with the topology $\mathcal{T}$ of Theorem 2.11.

Definition 2.12. Let Ω be an open subset of $\mathbb{R}^n$. We define the space of distributions in Ω as the topological dual

$$\mathcal{D}'(\Omega) \equiv \{T \in L(\mathcal{D}(\Omega), \mathbb{R}) : T \text{ is continuous}\}$$

of $\mathcal{D}(\Omega)$.

If $T \in \mathcal{D}'(\Omega)$, it is customary to write

$$\langle T, \varphi \rangle \equiv T[\varphi] \qquad \forall \varphi \in \mathcal{D}(\Omega).$$

We shall consider on $\mathcal{D}'(\Omega)$ the weak* topology. In particular, if $\{u_j\}_{j\in\mathbb{N}}$ is a sequence in $\mathcal{D}'(\Omega)$ and if $\tilde{u} \in \mathcal{D}'(\Omega)$, then we have

$$\lim_{j\to\infty} u_j = \tilde{u} \qquad \text{in } \mathcal{D}'(\Omega)$$

if and only if

$$\lim_{j\to\infty} \langle u_j, \varphi \rangle = \langle \tilde{u}, \varphi \rangle \qquad \forall \varphi \in \mathcal{D}(\Omega).$$

Let $x_0 \in \Omega$. We denote by δ_{x_0} the linear map from $\mathcal{D}(\Omega)$ to $\mathbb{R}$ defined by

$$\langle \delta_{x_0}, \varphi \rangle \equiv \varphi(x_0) \qquad \forall \varphi \in \mathcal{D}(\Omega).$$

The operator δ_{x_0} is said to be the 'Dirac measure with mass concentrated at x_0'. Then δ_{x_0} is a distribution in Ω. The linearity of δ_{x_0} is obvious. In order to prove the continuity of δ_{x_0} it suffices to show that δ_{x_0} is continuous on $\mathcal{D}(K)$ for each compact subset K of Ω. Such a continuity is an immediate consequence of Proposition 2.10 and of the following elementary inequality

$$|\langle \delta_{x_0}, \varphi \rangle| \leq \sup_K |\varphi| = p_{K,0}(\varphi) \qquad \forall \varphi \in \mathcal{D}(K).$$

Let $L^1_{\text{loc}}(\Omega)$ denote the space of locally integrable (equivalence classes of) functions from Ω to $\mathbb{R}$, i.e., the space of measurable (equivalence classes of) functions from Ω to $\mathbb{R}$ such that f is integrable on each compact subset of Ω. If $f \in L^1_{\text{loc}}(\Omega)$, then one can associate to f a distribution $\mathcal{J}f$ by setting

$$\langle \mathcal{J}f, \varphi \rangle \equiv \int_\Omega f\varphi\, dx \qquad \forall \varphi \in \mathcal{D}(\Omega).$$

By the linearity of the integration in Ω, $\mathcal{J}f$ is a linear map from $\mathcal{D}(\Omega)$ to $\mathbb{R}$. Then one can exploit Proposition 2.10 and prove that $\mathcal{J}f \circ i_K$ is actually continuous on $\mathcal{D}(K)$ for all compact subsets K of Ω. Then Theorem 2.11 implies that $\mathcal{J}f$ is continuous on $\mathcal{D}(\Omega)$ and that accordingly $\mathcal{J}f$ is a distribution in Ω.

Then one can assign to each $f \in L^1_{\text{loc}}(\Omega)$ the distribution $\mathcal{J}f$ and the linear map $\mathcal{J}$ from $L^1_{\text{loc}}(\Omega)$ to $\mathcal{D}'(\Omega)$ that takes f to $\mathcal{J}f$ is injective. Indeed, if $f \in L^1_{\text{loc}}(\Omega)$ and $\mathcal{J}f = 0$, then we have

$$\int_{\Omega} f\varphi \, dx = 0 \qquad \forall \varphi \in \mathcal{D}(\Omega)$$

and thus $f = 0$ almost everywhere. The linear injection $\mathcal{J}$ identifies $L^1_{\mathrm{loc}}(\Omega)$ with the subspace $\mathcal{J}[L^1_{\mathrm{loc}}(\Omega)]$ of $\mathcal{D}'(\Omega)$ and it is customary to write simply f instead of $\mathcal{J}f$.

Definition 2.13. Let Ω be an open subset of $\mathbb{R}^n$. Let $\beta \in \mathbb{N}^n$, $u \in \mathcal{D}'(\Omega)$. Then the D^β-derivative of u in the sense of distributions is the element $D^\beta u$ of $\mathcal{D}'(\Omega)$ defined by

$$\langle D^\beta u, \varphi \rangle \equiv (-1)^{|\beta|} \langle u, D^\beta \varphi \rangle \qquad \forall \varphi \in \mathcal{D}(\Omega)\,.$$

If $m \in \mathbb{N} \setminus \{0\}$, $f \in C^m(\Omega)$, then an argument based on the Divergence Theorem shows that the D^β derivative of the distribution $\mathcal{J}f$ associated to f coincides with the distribution $\mathcal{J}D^\beta f$ associated to $D^\beta f$ for each $\beta \in \mathbb{N}^n$ with $|\beta| \leq m$.

If $f \in L^1_{\mathrm{loc}}(\Omega)$, then it is customary to say that $D^\beta \mathcal{J}f$, which we abbreviate as $D^\beta f$, is the D^β-weak derivative of f for each $\beta \in \mathbb{N}^n$. If $P(D)$ is a differential operator with constant coefficients, $u \in \mathcal{D}'(\Omega)$, and

$$P(D)u = 0\,,$$

then it is customary to say that u is a solution in the sense of distributions (or a weak solution) of equation $P(D)u = 0$.

We denote by $\mathcal{S}(\mathbb{R}^n)$ the Schwartz space of rapidly decreasing functions, i.e., the space

$$\mathcal{S}(\mathbb{R}^n) \equiv \left\{ f \in C^\infty(\mathbb{R}^n) : \sup_{x \in \mathbb{R}^n} (1 + |x|^2)^{m/2} |D^\beta f(x)| < +\infty \ \forall m \in \mathbb{N}, \forall \beta \in \mathbb{N}^n \right\}.$$

We set

$$p_{m,\beta}(f) \equiv \sup_{x \in \mathbb{R}^n} (1 + |x|^2)^{m/2} |D^\beta f(x)| \qquad \forall f \in \mathcal{S}(\mathbb{R}^n)$$

for all $m \in \mathbb{N}$ and $\beta \in \mathbb{N}^n$ and we set $\mathcal{P}_{rd} \equiv \{p_{m,\beta} : m \in \mathbb{N}, \beta \in \mathbb{N}^n\}$. Then $\mathcal{S}(\mathbb{R}^n)$ is a linear space and $\mathcal{P}_{rd}$ is a family of seminorms on $\mathcal{S}(\mathbb{R}^n)$ which generate a topology on $\mathcal{S}(\mathbb{R}^n)$. The topology on $\mathcal{S}(\mathbb{R}^n)$ generated by the family of seminorms $\mathcal{P}_{rd}$ is by definition that generated by the family of open balls

$$\{\psi \in \mathcal{S}(\mathbb{R}^n) : p_{m,\beta}(f - \psi) < r\}$$

with center at $f \in \mathcal{S}(\mathbb{R}^n)$, radius $r \in]0, +\infty[$, $m \in \mathbb{N}$, and $\beta \in \mathbb{N}^n$. The topology generated by the family of seminorms $\mathcal{P}_{rd}$ is linear, locally convex, metrizable, and complete, i.e., $\mathcal{S}(\mathbb{R}^n)$ is a Fréchet space. We define the space of tempered distributions in $\mathbb{R}^n$ as the topological dual

$$\mathcal{S}'(\mathbb{R}^n) \equiv \{T \in L(\mathcal{S}(\mathbb{R}^n), \mathbb{R}) : T \text{ is continuous}\}$$

of $\mathcal{S}(\mathbb{R}^n)$. Then one can characterize the continuous linear functions from $\mathcal{S}(\mathbb{R}^n)$ to $\mathbb{R}$ by means of the following proposition (cf., e.g., Trèves [266]).

Proposition 2.14. *A linear function T from $\mathcal{S}(\mathbb{R}^n)$ to $\mathbb{R}$ is continuous if and only if there exist $c > 0$, $m_1, \ldots, m_s \in \mathbb{N}$, and $\beta_1, \ldots, \beta_s \in \mathbb{N}^n$ such that*

$$|\langle T, \psi \rangle| \leq c \sup_{l=1,\ldots,s} p_{m_l,\beta_l}(\psi) \qquad \forall \psi \in \mathcal{S}(\mathbb{R}^n) .$$

We shall consider on $\mathcal{S}'(\mathbb{R}^n)$ the weak* topology. In particular, if $\{u_j\}_{j\in\mathbb{N}}$ is a sequence in $\mathcal{S}'(\mathbb{R}^n)$ and if $\tilde{u} \in \mathcal{S}'(\mathbb{R}^n)$, then we have

$$\lim_{j\to\infty} u_j = \tilde{u} \qquad \text{in } \mathcal{S}'(\mathbb{R}^n)$$

if and only if

$$\lim_{j\to\infty} \langle u_j, \varphi \rangle = \langle \tilde{u}, \varphi \rangle \qquad \forall \varphi \in \mathcal{S}(\mathbb{R}^n) .$$

Finally, we observe that one could embed the space $\mathcal{S}'(\mathbb{R}^n)$ of tempered distributions into the space $\mathcal{D}'(\mathbb{R}^n)$ of distributions in $\mathbb{R}^n$ and that accordingly one can define the derivatives in the sense of distributions also for the tempered distributions.

2.5 Real Analytic Functions and Spaces of Real Analytic Functions

In Section A.6 of the Appendix, we define the notion of analytic function acting between Banach spaces. In the specific case of scalar functions defined on open subsets of $\mathbb{R}^n$ with $n \in \mathbb{N}$ such definition is equivalent to the following.

Definition 2.15. Let Ω be an open subset of $\mathbb{R}^n$. Let $f \in C^\infty(\Omega)$. Let x^o in Ω. Then f is real analytic at x^o provided that there exists a neighborhood V_{x^o} of x^o in Ω such that

(i) $\lim_{j\to\infty} \sum_{|\beta|\leq j} \frac{|D^\beta f(x^o)|}{\beta!} |(x - x^o)^\beta|$ exists in $\mathbb{R}$ for all $x \in V_{x^o}$,

(ii) $\lim_{j\to\infty} \sum_{|\beta|\leq j} \frac{D^\beta f(x^o)}{\beta!} (x - x^o)^\beta = f(x)$ for all $x \in V_{x^o}$.

We say that f is real analytic in Ω, provided that f is real analytic at all points of Ω.

In what follows, we understand that 'analytic' always means 'real analytic'. We denote by $C^\omega(\Omega)$ the set of analytic functions from Ω to $\mathbb{R}$. Then we have

$$C^\omega(\Omega) \subseteq C^\infty(\Omega)$$

and the inclusion is proper whenever Ω is not empty. For example, the function f from $\mathbb{R}$ to itself defined by

$$f(x) \equiv 0 \quad \text{if } x \in]-\infty, 0] , \qquad f(x) \equiv e^{-1/x} \quad \text{if } x \in]0, +\infty[$$

is well known to belong to $C^\infty(\mathbb{R}) \setminus C^\omega(\mathbb{R})$. One could exploit this example to construct functions of $C^\infty(\Omega) \setminus C^\omega(\Omega)$, whenever Ω is a nonempty subset of $\mathbb{R}^n$.

By the classical theorems on the sum and product with scalars of limits, the set $C^\omega(\Omega)$ is a vector space. Next we introduce the known Identity Principle for analytic functions (cf., e.g., John [130, p. 65]).

Proposition 2.16. *Let Ω be a connected open subset of $\mathbb{R}^n$. Let $f \in C^\omega(\Omega)$. Then the following statements hold.*

(i) If there exists $x^o \in \Omega$ such that $D^\alpha f(x^o) = 0$ for all $\alpha \in \mathbb{N}^n$, then f is equal to zero at all points of Ω.
(ii) If the set of zeros $Z(f)$ of f has at least an interior point, then f is equal to zero at all points of Ω.

Next we introduce following known characterization of an analytic function in an open set (cf., e.g., John [130, Chap. 3]).

Theorem 2.17. *Let Ω be an open subset of $\mathbb{R}^n$. Let $f \in C^\infty(\Omega)$. Then the following conditions are equivalent.*

(i) f is analytic in Ω.
(ii) For each compact subset K of Ω, there exist $M_K, r_K \in]0, +\infty[$ such that

$$\sup_{x \in K} |D^\beta f(x)| \le M_K \frac{|\beta|!}{r_K^{|\beta|}} \qquad \forall x \in K, \beta \in \mathbb{N}^n .$$

Next, we turn to introduce the Roumieu classes. For all bounded open subsets Ω of $\mathbb{R}^n$ and $\rho \in]0, +\infty[$, we set

$$C^0_{\omega,\rho}(\overline{\Omega}) \equiv \left\{ u \in C^\infty(\overline{\Omega}) : \sup_{\beta \in \mathbb{N}^n} \frac{\rho^{|\beta|}}{|\beta|!} \|D^\beta u\|_{C^0(\overline{\Omega})} < +\infty \right\} \tag{2.7}$$

and

$$\|u\|_{C^0_{\omega,\rho}(\overline{\Omega})} \equiv \sup_{\beta \in \mathbb{N}^n} \frac{\rho^{|\beta|}}{|\beta|!} \|D^\beta u\|_{C^0(\overline{\Omega})} \qquad \forall u \in C^0_{\omega,\rho}(\overline{\Omega}) .$$

By the characterization of Theorem 2.17, all functions of $C^0_{\omega,\rho}(\overline{\Omega})$ are real analytc in Ω. As is well known, the Roumieu class $\left(C^0_{\omega,\rho}(\overline{\Omega}), \|\cdot\|_{C^0_{\omega,\rho}(\overline{\Omega})}\right)$ is a Banach space.

By the definition of Roumieu norm, if $\tau \in \mathbb{N}^n$ and $\rho \in]0, +\infty[$, then the operator D^τ that maps u in $D^\tau u$ is linear and continuous from $C^0_{\omega,\rho}(\overline{\Omega})$ to $C^0_{\omega,\rho'}(\overline{\Omega})$ for all $\rho' \in]0, \rho[$.

Remark 2.18. Let Ω, Ω_1 be open subsets of $\mathbb{R}^n$. Let $\overline{\Omega}$ be a bounded subset of Ω_1. If $f \in C^\omega(\Omega_1)$, then by applying the condition of the characterization of the analytic functions of Theorem 2.17 (ii) to the compact set $K \equiv \overline{\Omega}$, we deduce that there exists $\rho \in]0, +\infty[$ such that $f_{|\overline{\Omega}} \in C^0_{\omega,\rho}(\overline{\Omega})$.

Then we note that if we integrate a function in a Roumieu class in one of its variables, the resulting function is still in a Roumieu class. Indeed, the following proposition holds.

Proposition 2.19. *Let n, $n_1 \in \mathbb{N} \setminus \{0\}$. Let Ω, Ω_1 be bounded open subsets of $\mathbb{R}^n$, $\mathbb{R}^{n_1}$, respectively. Let $\rho \in]0, +\infty[$. If $F \in C^0_{\omega,\rho}(\overline{\Omega \times \Omega_1})$, then the function $\int_{\Omega_1} F(\cdot, y)\, dy$ belongs to $C^0_{\omega,\rho}(\overline{\Omega})$ and the operator from $C^0_{\omega,\rho}(\overline{\Omega \times \Omega_1})$ to $C^0_{\omega,\rho}(\overline{\Omega})$ that takes F to the function $\int_{\Omega_1} F(\cdot, y)\, dy$ is linear and continuous.*

Proof. By the differentiability theorem for integrals depending on a parameter, we have

$$\begin{aligned} \left| D^\beta_x \int_{\Omega_1} F(x,y)\, dy \right| &= \left| \int_{\Omega_1} D^\beta_x F(x,y)\, dy \right| \\ &\leq \frac{|\beta|!}{\rho^{|\beta|}} m_{n_1}(\Omega_1) \|F\|_{C^0_{\omega,\rho}(\overline{\Omega \times \Omega_1})} \qquad \forall x \in \overline{\Omega} \end{aligned}$$

for all $\beta \in \mathbb{N}^n$. Thus the statement follows by the definition of norm in $C^0_{\omega,\rho}(\overline{\Omega})$. □

2.6 Spaces of Hölder and Lipschitz Continuous Functions

Let ω be a function from $[0, +\infty[$ to itself such that

$$\begin{aligned} &\omega(0) = 0, \qquad \omega(r) > 0 \qquad \forall r \in]0, +\infty[\,, \\ &\omega \text{ is increasing}, \ \lim_{r \to 0^+} \omega(r) = 0\,. \end{aligned} \tag{2.8}$$

Let $n, r \in \mathbb{N}$. If f is a function from a subset $\mathbb{D}$ of $\mathbb{R}^n$ to $\mathbb{R}^r$, then we denote by $|f : \mathbb{D}|_{\omega(\cdot)}$ the $\omega(\cdot)$-Hölder constant of f, which is delivered by the formula

$$|f : \mathbb{D}|_{\omega(\cdot)} \equiv \sup \left\{ \frac{|f(x) - f(y)|}{\omega(|x - y|)} : x, y \in \mathbb{D}, x \neq y \right\}.$$

If $|f : \mathbb{D}|_{\omega(\cdot)} < \infty$, we say that f is $\omega(\cdot)$-Hölder continuous. Sometimes, we simply write $|f|_{\omega(\cdot)}$ instead of $|f : \mathbb{D}|_{\omega(\cdot)}$.

The subset of $C^0(\mathbb{D})$ whose functions are $\omega(\cdot)$-Hölder continuous is denoted by $C^{0,\omega(\cdot)}(\mathbb{D})$. By the triangular inequality, $C^{0,\omega(\cdot)}(\mathbb{D})$ is a vector subspace of $C^0(\mathbb{D})$.

$C^{0,\omega(\cdot)}_{\text{loc}}(\mathbb{D})$ denotes the space of those functions $f \in C^0(\mathbb{D})$ such that $f_{|K}$ belongs to $C^{0,\omega(\cdot)}(K)$ for each compact subset K of $\mathbb{D}$.

Then we consider the space $C^{0,\omega(\cdot)}_b(\mathbb{D}) \equiv C^{0,\omega(\cdot)}(\mathbb{D}) \cap B(\mathbb{D})$ with the norm

$$\|f\|_{C^{0,\omega(\cdot)}_b(\mathbb{D})} \equiv \sup_{x \in \mathbb{D}} |f(x)| + |f|_{\omega(\cdot)} \qquad \forall f \in C^{0,\omega(\cdot)}_b(\mathbb{D})\,.$$

If $\mathbb{D}$ is compact, then we obviously have $C^{0,\omega(\cdot)}_b(\mathbb{D}) = C^{0,\omega(\cdot)}(\mathbb{D})$ (and in this case, we generally drop the subscript b).

Remark 2.20. Let ω be as in (2.8). Let $\mathbb{D}$ be a subset of $\mathbb{R}^n$. Let f be a bounded function from $\mathbb{D}$ to $\mathbb{R}$, $a \in]0, +\infty[$. Then,

$$\sup_{x,y\in\mathbb{D},\ |x-y|\geq a} \frac{|f(x)-f(y)|}{\omega(|x-y|)} \leq \frac{2}{\omega(a)} \sup_{\mathbb{D}} |f| \,.$$

Thus the difficulty of estimating the Hölder quotient $\frac{|f(x)-f(y)|}{\omega(|x-y|)}$ of a bounded function f lies entirely in case $0 < |x-y| < a$.

Then we introduce the following elementary lemma, which enables to infer that the pointwiswe limit of a sequence of $\omega(\cdot)$-Hölder continuous functions is $\omega(\cdot)$-Hölder continuous provided that the $\omega(\cdot)$-Hölder constants of the functions of the sequence are uniformly bounded. We exploit such an elementary lemma in the proof of Theorem 2.22 below, where we prove that $C_b^{0,\omega(\cdot)}(\mathbb{D})$ is complete.

Lemma 2.21. *Let ω be as in (2.8). Let $n \in \mathbb{N}$. Let $\mathbb{D}$ be a subset of $\mathbb{R}^n$. If a sequence $\{f_j\}_{j\in\mathbb{N}}$ of $C^{0,\omega(\cdot)}(\mathbb{D})$ satisfies the inequality $\sup_{l\in\mathbb{N}} |f_l|_{\omega(\cdot)} < +\infty$ and converges pointwise to a function f from $\mathbb{D}$ to $\mathbb{R}$, then $f \in C^{0,\omega(\cdot)}(\mathbb{D})$ and $|f|_{\omega(\cdot)} \leq \sup_{l\in\mathbb{N}} |f_l|_{\omega(\cdot)}$.*

Proof. It suffices to take the limit as j tends to infinity in the inequality

$$|f_j(x) - f_j(y)| \leq (\sup_{l\in\mathbb{N}} |f_l|_{\omega(\cdot)})\omega(|x-y|) \qquad \forall x, y \in \mathbb{D}$$

which holds for all $j \in \mathbb{N}$. □

Next we prove the following.

Theorem 2.22. *Let ω be as in (2.8). Let $n \in \mathbb{N}$. Let $\mathbb{D}$ be a subset of $\mathbb{R}^n$. Then the space $C_b^{0,\omega(\cdot)}(\mathbb{D})$ is complete.*

Proof. Let $\{f_j\}_{j\in\mathbb{N}}$ be a Cauchy sequence in $C_b^{0,\omega(\cdot)}(\mathbb{D})$. Since

$$|f_j(x) - f_l(x)| \leq \|f_j - f_l\|_{C_b^{0,\omega(\cdot)}(\mathbb{D})} \qquad \forall x \in \mathbb{D}$$

for all $j, l \in \mathbb{N}$, $\{f_j(x)\}_{j\in\mathbb{N}}$ is a Cauchy sequence in $\mathbb{R}$ for all $x \in \mathbb{D}$. Then we can define a function f from $\mathbb{D}$ to $\mathbb{R}$ by setting

$$f(x) \equiv \lim_{j\to\infty} f_j(x) \qquad \forall x \in \mathbb{D} \,.$$

Since $\{f_j\}_{j\in\mathbb{N}}$ is a Cauchy sequence in $C_b^{0,\omega(\cdot)}(\mathbb{D})$, the sequence $\{f_j\}_{j\in\mathbb{N}}$ is bounded in $C_b^{0,\omega(\cdot)}(\mathbb{D})$. Hence, we have

$$|f(x)| \leq \limsup_{j\to\infty} |f_j(x)| \leq \sup_{j\in\mathbb{N}} \|f_j\|_{C_b^{0,\omega(\cdot)}(\mathbb{D})} \qquad \forall x \in \mathbb{D}$$

and f is bounded. Moreover, Lemma 2.21 implies that $f \in C^{0,\omega(\cdot)}(\mathbb{D})$. Hence, $f \in C_b^{0,\omega(\cdot)}(\mathbb{D})$. We now prove that $\{f_j\}_{j\in\mathbb{N}}$ converges to f in $C_b^{0,\omega(\cdot)}(\mathbb{D})$. Let

$\epsilon \in]0, +\infty[$. Since $\{f_j\}_{j\in\mathbb{N}}$ is a Cauchy sequence, there exists $k \in \mathbb{N}$ such that

$$j, l \geq k \Rightarrow \|f_j - f_l\|_{C_b^{0,\omega(\cdot)}(\mathbb{D})} \leq \epsilon/2 .$$

We now fix an arbitrary $j \in \mathbb{N}$ such that $j \geq k$, and we prove that

$$\|f_j - f\|_{C_b^{0,\omega(\cdot)}(\mathbb{D})} \leq \epsilon .$$

Since

$$|f_j(x) - f_l(x)| \leq \|f_j - f_l\|_{C_b^{0,\omega(\cdot)}(\mathbb{D})} \leq \epsilon/2 \qquad \forall x \in \mathbb{D}$$

for all $l \in \mathbb{N}$ such that $l \geq k$, we can take the limit as l tends to infinity in the left and right-hand side and deduce that

$$|f_j(x) - f(x)| \leq \epsilon/2 \qquad \forall x \in \mathbb{D} .$$

Since $\{f_j - f_l\}_{l\in\mathbb{N}}$ converges pointwise to $f_j - f$, Lemma 2.21 implies that

$$|f_j - f|_{\omega(\cdot)} \leq \sup_{k\leq l\in\mathbb{N}} |f_j - f_l|_{\omega(\cdot)} \leq \sup_{k\leq l\in\mathbb{N}} \|f_j - f_l\|_{C_b^{0,\omega(\cdot)}(\mathbb{D})} \leq \epsilon/2 .$$

Hence,

$$\|f_j - f\|_{C_b^{0,\omega(\cdot)}(\mathbb{D})} = \sup_{\mathbb{D}} |f_j - f| + |f_j - f|_{\omega(\cdot)} \leq (\epsilon/2) + (\epsilon/2) = \epsilon$$

and thus $C_b^{0,\omega(\cdot)}(\mathbb{D})$ is complete. □

Particularly important is the case in which $\omega(\cdot)$ is the function r^α for some fixed $\alpha \in]0, 1]$ that is called Hölder exponent. In this case, we simply write $|\cdot : \mathbb{D}|_\alpha$ instead of $|\cdot : \mathbb{D}|_{r^\alpha}$, $C^{0,\alpha}(\mathbb{D})$ instead of $C^{0,r^\alpha}(\mathbb{D})$, $C_b^{0,\alpha}(\mathbb{D})$ instead of $C_b^{0,r^\alpha}(\mathbb{D})$, and we say that $|f : \mathbb{D}|_\alpha$ is the α-Hölder constant of f. If $|f : \mathbb{D}|_\alpha < +\infty$, we say that f is α-Hölder continuous.

If $\alpha = 1$, we say that f is Lipschitz continuous provided that f is 1-Hölder continuous and we set $\mathrm{Lip}(f) \equiv |f : \mathbb{D}|_1$. Sometimes, we write Lip $(\mathbb{D})$ instead of $C^{0,1}(\mathbb{D})$.

We now prove the following embedding statement, which shows that the Hölder exponent is actually a smoothness parameter.

Proposition 2.23. *Let $n \in \mathbb{N}$. Let $\mathbb{D}$ be a subset of $\mathbb{R}^n$. Let α, $\beta \in]0, 1]$, $\beta < \alpha$. Then $C_b^{0,\alpha}(\mathbb{D})$ is continuously embedded into $C_b^{0,\beta}(\mathbb{D})$.*

Proof. The statement is an immediate consequence of the following elementary inequality

$$\begin{aligned} |f(x) - f(y)| &= |f(x) - f(y)|^{\beta/\alpha} |f(x) - f(y)|^{1-(\beta/\alpha)} \\ &\leq |f|_\alpha^{\beta/\alpha} |x - y|^\beta (2 \sup_{\mathbb{D}} |f|)^{1-(\beta/\alpha)} \leq 2^{1-(\beta/\alpha)} \|f\|_{C_b^{0,\alpha}(\mathbb{D})} |x - y|^\beta \ \forall x, y \in \mathbb{D} \end{aligned}$$

which holds for all $f \in C_b^{0,\alpha}(\mathbb{D})$. □

Next we prove the continuity of the pointwise product.

Proposition 2.24. *Let $n \in \mathbb{N}$. Let $\mathbb{D}$ be a subset of $\mathbb{R}^n$. Let $\alpha_1, \alpha_2 \in]0,1]$. Then the pointwise multiplication from $C_b^{0,\alpha_1}(\mathbb{D}) \times C_b^{0,\alpha_2}(\mathbb{D})$ to $C_b^{0,\min\{\alpha_1,\alpha_2\}}(\mathbb{D})$ that takes a pair of functions (f,g) to the pointwise product fg is continuous.*

Proof. By the embedding Proposition 2.23, it suffices to consider case $\alpha \equiv \alpha_1 = \alpha_2$, a case which is an immediate consequence of the following elementary inequality

$$\begin{aligned} |f(x)g(x) - f(y)g(y)| &\leq |f(x) - f(y)|\,|g(x)| + |f(y)|\,|g(x) - g(y)| \\ &\leq |f|_\alpha |x-y|^\alpha (\sup_{\mathbb{D}} |g|) + (\sup_{\mathbb{D}} |f|)\,|g|_\alpha |x-y|^\alpha \qquad \forall x, y \in \mathbb{D}\,. \end{aligned}$$

which holds for all $(f,g) \in C_b^{0,\alpha_1}(\mathbb{D}) \times C_b^{0,\alpha_2}(\mathbb{D})$. □

In particular, for all $\alpha \in]0,1]$ the space $C_b^{0,\alpha}(\mathbb{D})$ is a Banach algebra with unity. Next we prove the following.

Proposition 2.25. *Let $n \in \mathbb{N}$. Let $\mathbb{D}$ be a compact subset of $\mathbb{R}^n$. Let $\alpha, \alpha' \in]0,1]$, $\alpha' < \alpha$. Then $C^{0,\alpha}(\mathbb{D})$ is compactly embedded into $C^{0,\alpha'}(\mathbb{D})$.*

Proof. We must show that if $\{f_j\}_{j\in\mathbb{N}}$ is a bounded sequence of $C^{0,\alpha}(\mathbb{D})$, then there exists a subsequence of $\{f_j\}_{j\in\mathbb{N}}$ which converges in $C^{0,\alpha'}(\mathbb{D})$. By the Ascoli-Arzelà Theorem (cf., e.g., Folland [103, Theorem 4.43, p. 137]) and by the inequalities

$$\begin{aligned} &|f_j(x)| \leq \sup_{l\in\mathbb{N}} \|f_l\|_{C^{0,\alpha}(\mathbb{D})}\,, \\ &|f_j(x) - f_j(y)| \leq (\sup_{l\in\mathbb{N}} \|f_l\|_{C^{0,\alpha}(\mathbb{D})})|x-y|^\alpha \qquad \forall x, y \in \mathbb{D}\,, \end{aligned}$$

for all $j \in \mathbb{N}$, there exists a subsequence $\{f_{j_k}\}_{k\in\mathbb{N}}$ of $\{f_j\}_{j\in\mathbb{N}}$ which converges to $f \in C^0(\mathbb{D})$ uniformly in $\mathbb{D}$. Then, by Lemma 2.21, we have $f \in C^{0,\alpha}(\mathbb{D})$.

We now show that $\{f_{j_k}\}_{k\in\mathbb{N}}$ converges to f in the norm of $C^{0,\alpha'}(\mathbb{D})$. Since we already know that $\{f_{j_k}\}_{k\in\mathbb{N}}$ converges to f in $C^0(\mathbb{D})$, it suffices to show that $\limsup_{k\to+\infty} |f_{j_k} - f : \mathbb{D}|_{\alpha'} = 0$. To do so, we fix an arbitrary $\epsilon > 0$ and we show that $\limsup_{k\to+\infty} |f_{j_k} - f : \mathbb{D}|_{\alpha'} \leq \epsilon$. Now, $|f_{j_k} - f : \mathbb{D}|_{\alpha'}$ has been defined as a supremum of a quotient for $x, y \in \mathbb{D}$ and the idea is to fix an appropriate $\delta > 0$ and to analyze separately case $|x-y| \leq \delta$ and case $|x-y| > \delta$. We first try to determine an appropriate δ. By the definition of norm in $C^{0,\alpha}(\mathbb{D})$, we have

$$|(f_{j_k}(x) - f(x)) - (f_{j_k}(y) - f(y))| \leq (\sup_{j\in\mathbb{N}} \|f_j - f\|_{C^{0,\alpha}(\mathbb{D})})|x-y|^\alpha \qquad \forall x, y \in \mathbb{D}\,,$$

for all $k \in \mathbb{N}$. Thus if we set $\delta \equiv \frac{\epsilon^{1/(\alpha-\alpha')}}{(1+\sup_{j\in\mathbb{N}} \|f_j - f\|_{C^{0,\alpha}(\mathbb{D})})^{1/(\alpha-\alpha')}}$, then we have

$$\sup_{k\in\mathbb{N}} \sup \left\{ \frac{|(f_{j_k}(x) - f(x)) - (f_{j_k}(y) - f(y))|}{|x-y|^{\alpha'}} : x, y \in \mathbb{D}, 0 < |x-y| \leq \delta \right\} \leq \epsilon \tag{2.9}$$

and thus we have taken care of the points $x, y \in \mathbb{D}$ such that $|x-y| \leq \delta$. We now turn to consider the points $x, y \in \mathbb{D}$ such that $|x-y| > \delta$. Since $\{f_{j_k}\}_{k\in\mathbb{N}}$ converges to f uniformly in $\mathbb{D}$, we have

$$\lim_{k\to+\infty} \sup\left\{\frac{|(f_{j_k}(x)-f(x))-(f_{j_k}(y)-f(y))|}{|x-y|^{\alpha'}} : x,y\in\mathbb{D}, \delta<|x-y|\right\}=0 \tag{2.10}$$

and thus we have taken care of the points x, $y\in\mathbb{D}$ such that $|x-y|>\delta$. Then by combining the above inequality (2.9) and the above limiting relation (2.10), we conclude that

$$\limsup_{k\to+\infty} \sup\left\{\frac{|(f_{j_k}(x)-f(x))-(f_{j_k}(y)-f(y))|}{|x-y|^{\alpha'}} : x,y\in\mathbb{D}, 0<|x-y|\right\}\leq\epsilon\,.$$

Since $\limsup_{k\to+\infty}\|f_{j_k}-f\|_{C^0(\mathbb{D})}=0$, we have

$$\limsup_{k\to+\infty}\|f_{j_k}-f\|_{C^{0,\alpha'}(\mathbb{D})}\leq\epsilon\,.$$

Since ϵ is arbitrary, we conclude that $\lim_{k\to+\infty}\|f_{j_k}-f\|_{C^{0,\alpha'}(\mathbb{D})}=0$. □

Easily constructed examples show that if $\mathbb{D}$ is unbounded, the embedding of $C^{0,\alpha}_b(\mathbb{D})$ into $C^{0,\alpha'}_b(\mathbb{D})$ for $0<\alpha'<\alpha\leq 1$ is not necessarily compact.

Example 2.26. Let $\mathbb{D}=\mathbb{R}$. Let f be a nonconstant bounded Lipschitz continuous function from $\mathbb{R}$ to itself with compact support. Let $f_j(x)\equiv f(x-j)$ for all $x\in\mathbb{R}$ and $j\in\mathbb{N}$. Then $\sup_{j\in\mathbb{N}}\|f_j\|_{C^{0,1}_b(\mathbb{R})}=\|f\|_{C^{0,1}_b(\mathbb{R})}<+\infty$, but the sequence $\{f_j\}_{j\in\mathbb{N}}$ cannot have any convergent subsequence in $C^{0,\alpha'}_b(\mathbb{R})$ with $\alpha'\in]0,1[$. Indeed, the convergence in $C^{0,\alpha'}_b(\mathbb{R})$ would imply the uniform convergence of the subsequence to its pointwise limit, which equals the 0 constant function. However, the 0 constant function cannot be a uniform limit for any subsequence of $\{f_j\}_{j\in\mathbb{N}}$. Indeed, $\sup_{\mathbb{R}}|f_j|=\sup_{\mathbb{R}}|f|>0$ for all $j\in\mathbb{N}$.

We now turn to the question of whether a function of class C^1 on the closure of an open set is actually Lipschitz continuous. To do so, we need some preliminaries. We assume that Ω is an open connected subset of $\mathbb{R}^n$. For every $x,y\in\Omega$, there exists a path $\gamma_{x,y}\in\mathrm{Lip}\,([0,1],\mathbb{R}^n)$ such that

$$\gamma_{x,y}([0,1])\subseteq\Omega,\quad \gamma_{x,y}(0)=x,\quad \gamma_{x,y}(1)=y. \tag{2.11}$$

The geodesic distance $\lambda_\Omega(x,y)$ of x and y in Ω is defined as

$$\lambda_\Omega(x,y)\equiv\inf\{\mathrm{length}(\gamma_{x,y}):\gamma_{x,y}\in\mathrm{Lip}\,([0,1],\mathbb{R}^n)\text{ and (2.11) holds}\}\,. \tag{2.12}$$

When no ambiguity can arise, we simply write λ instead of λ_Ω. By exploiting the definition of λ_Ω one can easily prove that λ_Ω is actually a distance in Ω. We observe that by exploiting the density of $C^1([0,1])$ in the Sobolev space

$$W^{1,1}(]0,1[)\equiv\{f\in L^1(]0,1[): f'\in L^1(]0,1[)\}\,,$$
$$\|f\|_{W^{1,1}(]0,1[)}\equiv\|f\|_{L^1(]0,1[)}+\|f'\|_{L^1(]0,1[)}\qquad\forall f\in W^{1,1}(]0,1[)$$

and the continuous embedding of $W^{1,1}(]0,1[)$ into $C^0([0,1])$ (cf., e.g., Brezis [35, Theorems 8.2, 8.7]), one could prove that the infimum in (2.12) does not change if we replace $\mathrm{Lip}\,([0,1],\mathbb{R}^n)$ with $C^1([0,1],\mathbb{R}^n)$. We say that

$$c[\Omega] \equiv \sup\left\{\frac{\lambda_\Omega(x,y)}{|x-y|} : x,y \in \Omega,\ x \neq y\right\} \tag{2.13}$$

is the Whitney constant of Ω. Then we introduce the following definition.

Definition 2.27. Let $n \in \mathbb{N}$. Let Ω be an open subset of $\mathbb{R}^n$. We say that Ω is regular in the sense of Whitney provided that Ω is connected and $c[\Omega] < +\infty$.

Note that $c[\Omega] = 1$ if Ω is convex. The following lemma shows the importance of the condition of regularity in the sense of Whitney.

Lemma 2.28. *Let $n \in \mathbb{N}$. Let Ω be an open connected subset of $\mathbb{R}^n$ such that $c[\Omega]$ is finite. If $f \in C^1(\overline{\Omega})$, then*

$$|f(x) - f(y)| \leq c[\Omega]\left(\sup_{\overline{\Omega}} |\nabla f|\right)|x-y| \qquad \forall x,y \in \overline{\Omega}\,.$$

In particular, $C^1_b(\overline{\Omega})$ is continuously embedded into $C^{0,1}_b(\overline{\Omega})$.

Proof. Since f and ∇f are uniquely determined by their values in Ω, it suffices to prove the inequality of the statement for x, $y \in \Omega$. Since Ω is open and connected, there exists $\gamma_{x,y} \in \mathrm{Lip}\,([0,1],\mathbb{R}^n)$ as in (2.11). Then we have

$$\begin{aligned}|f(x) - f(y)| &= |f(\gamma_{x,y}(1)) - f(\gamma_{x,y}(0))| \\ &= \left|\int_0^1 \gamma'_{x,y}(s)\cdot\nabla f(\gamma_{x,y}(s))\,ds\right| \leq \left(\sup_{\overline{\Omega}}|\nabla f|\right)\int_0^1 |\gamma'_{x,y}(s)|\,ds\,.\end{aligned}$$

Then, by taking the infimum on all $\gamma_{x,y} \in \mathrm{Lip}\,([0,1],\mathbb{R}^n)$ as in (2.11), we obtain

$$|f(x) - f(y)| \leq \left(\sup_{\overline{\Omega}}|\nabla f|\right)\lambda_\Omega(x,y)\,.$$

Then the definition of $c[\Omega]$ implies the validity of the inequality of the statement. □

Classical examples show that if $c[\Omega] = +\infty$, then $C^1_b(\overline{\Omega})$ is not necessarily embedded into $C^{0,1}_b(\overline{\Omega})$, not even if Ω is bounded (cf., e.g., Gilbarg and Trudinger [107, p. 53]).

We now introduce the following consequence of Lemma 2.28.

Lemma 2.29. *Let $n \in \mathbb{N}$. Let Ω be a bounded open subset of $\mathbb{R}^n$ with a finite number of connected components, each of which is regular in the sense of Whitney. Then $C^1(\overline{\Omega})$ is compactly embedded into $C^0(\overline{\Omega})$.*

Proof. We first prove the lemma in the case where Ω is connected and regular in the sense of Whitney. By Lemma 2.28, $C^1(\overline{\Omega})$ is continuously embedded into $C^{0,1}(\overline{\Omega})$ and, by Proposition 2.25, $C^{0,1}(\overline{\Omega})$ is compactly embedded into $C^{0,1/2}(\overline{\Omega})$. Since $C^{0,1/2}(\overline{\Omega})$ is continuously embedded into $C^0(\overline{\Omega})$ by definition of $C^{0,1/2}(\overline{\Omega})$, the statement follows. Indeed, the composition of a linear and compact map with a linear and continuous map is linear and compact.

Next we consider the case where Ω has more than one connected component. Let $\Omega_1, \ldots, \Omega_{\varkappa^+}$ be the connected components of Ω and let Ψ be the map from $C^0(\overline{\Omega})$ to $\prod_{l=1}^{\varkappa^+} C^0(\overline{\Omega_l})$ that takes f to the $\varkappa^+$-tuple of restrictions $(f_{|\overline{\Omega_1}}, \ldots, f_{|\overline{\Omega_{\varkappa^+}}})$. We can see that in general Ψ is not surjective. However, by the definition of the involved norms, Ψ is a linear homeomorphism from $C^0(\overline{\Omega})$ onto its image $\Psi(C^0(\overline{\Omega})) \subseteq \prod_{l=1}^{\varkappa^+} C^0(\overline{\Omega_l})$ and restricts to a linear homeomorphism from $C^1(\overline{\Omega})$ onto $\Psi(C^1(\overline{\Omega})) \subseteq \prod_{l=1}^{\varkappa^+} C^1(\overline{\Omega_l})$. Since $C^0(\overline{\Omega})$ is complete, it follows that $\Psi(C^0(\overline{\Omega}))$ is complete and therefore closed in $\prod_{l=1}^{\varkappa^+} C^0(\overline{\Omega_l})$. Now let $\{f_j\}_{j\in\mathbb{N}}$ be a bounded sequence in $C^1(\overline{\Omega})$, then its Ψ-image $\{(f_{j|\overline{\Omega_1}}, \ldots, f_{j|\overline{\Omega_{\varkappa^+}}})\}_{j\in\mathbb{N}}$ is bounded in $\prod_{l=1}^{\varkappa^+} C^1(\overline{\Omega_l})$. By the case in which Ω is connected, there exists a subsequence $\{(f_{j_k|\overline{\Omega_1}}, \ldots, f_{j_k|\overline{\Omega_{\varkappa^+}}})\}_{k\in\mathbb{N}}$ of $\{(f_{j|\overline{\Omega_1}}, \ldots, f_{j|\overline{\Omega_{\varkappa^+}}})\}_{j\in\mathbb{N}}$ that converges to an element $g = (g_1, \ldots, g_{\varkappa^+})$ in $\prod_{l=1}^{\varkappa^+} C^0(\overline{\Omega_l})$. Since $\Psi(C^0(\overline{\Omega}))$ is closed in $\prod_{l=1}^{\varkappa^+} C^0(\overline{\Omega_l})$, there exists $f \in C^0(\overline{\Omega})$ such that $g = \Psi(f)$. Since Ψ is a linear homeomorphism onto its image, we have

$$\lim_{k\to\infty} f_{j_k} = \lim_{k\to\infty} \Psi^{(-1)}(f_{j_k|\overline{\Omega_1}}, \ldots, f_{j_k||\overline{\Omega_{\varkappa^+}}}) = \Psi^{(-1)}(g) = f \qquad \text{in } C^0(\overline{\Omega})$$

and thus the proof is complete. □

As the following example shows, the above Lemma 2.29 does not hold if Ω has an infinite number of connected components which are regular in the sense of Whitney.

Example 2.30. Let $\Omega_j \equiv \mathbb{B}_n((j+1)^{-1}e_1, (j+2)^{-3})$ for all $j \in \mathbb{N}$. Since Ω_j is convex, we have $c[\Omega_j] = 1$ for all $j \in \mathbb{N}$. Since the open balls Ω_j are pairwise disjoint, the open set $\Omega \equiv \bigcup_{j\in\mathbb{N}} \Omega_j$ has a countable family of connected components each of which is regular in the sense of Whitney. Moreover, $\overline{\Omega} = \{0\} \cup \left(\bigcup_{j\in\mathbb{N}} \overline{\Omega_j}\right)$. Now let f_l be the function from $\overline{\Omega}$ to $\mathbb{R}$ defined by

$$f_l(x) = \begin{cases} 1 \text{ if } x \in \bigcup_{0\le j\le l} \overline{\Omega_j}\,, \\ 0 \text{ if } x \in \{0\} \cup \left(\bigcup_{l<j\in\mathbb{N}} \overline{\Omega_j}\right)\,, \end{cases}$$

for all $l \in \mathbb{N}$. Since the closed balls $\overline{\Omega_j}$ are pairwise disjoint, we have $f_l \in C^\infty(\overline{\Omega})$ for all $l \in \mathbb{N}$. Moreover, the sequence $\{f_l\}_{l\in\mathbb{N}}$ is bounded in $C^1(\overline{\Omega})$, but no subsequence of $\{f_l\}_{l\in\mathbb{N}}$ can converge in $C^0(\overline{\Omega})$. Indeed, each subsequence of $\{f_l\}_{l\in\mathbb{N}}$ converges pointwise to the function f which equals 0 at $x = 0$ and 1 on $\bigcup_{j\in\mathbb{N}} \overline{\Omega_j}$ and f is not continuous at 0.

We now show that locally Lipschitz homeomorphisms map relatively compact open Whitney regular sets to Whitney regular sets. More precisely, we prove the following.

Proposition 2.31. *Let $n \in \mathbb{N}$. Let U and W be open subsets of $\mathbb{R}^n$. Let ϕ be a local Lipschitz homeomorphism from U onto W, i.e., ϕ is a bijection and both ϕ and $\phi^{(-1)}$ are Lipschitz continuous on the compact subsets of U and W, respectively.*

If Ω is a bounded open Whitney regular subset of U such that $\overline{\Omega} \subseteq U$, then $\phi(\Omega)$ is also regular in the sense of Whitney.

Proof. Since Ω is bounded, its closure $\overline{\Omega}$ is compact and thus $\phi(\overline{\Omega})$ is compact, ϕ is Lipschitz continuous on $\overline{\Omega}$, and $\phi^{(-1)}$ is Lipschitz continuous on $\phi(\overline{\Omega})$. Since Ω is connected and ϕ is continuous, then $\phi(\Omega)$ is connected. We now prove that $c[\phi(\Omega)] < +\infty$. Let $\xi, \eta \in \phi(\Omega)$. Then there exist $x, y \in \Omega$ such that $\xi = \phi(x)$, $\eta = \phi(y)$. Since Ω is regular in the sense of Whitney, there exists $\gamma_{x,y}$ in $\mathrm{Lip}\,([0,1], \mathbb{R}^n)$ as in (2.11) such that

$$\mathrm{length}(\gamma_{x,y}) \le (c[\Omega] + 1)|x - y| .$$

Then $\phi \circ \gamma_{x,y} \in \mathrm{Lip}\,([0,1], \mathbb{R}^n)$, $\phi \circ \gamma_{x,y}(0) = \xi$, $\phi \circ \gamma_{x,y}(1) = \eta$,

$$\begin{aligned}\mathrm{length}(\phi \circ \gamma_{x,y}) &\le \int_0^1 |(\phi \circ \gamma_{x,y})'|\, ds \le \mathrm{Lip}\,(\phi_{|\overline{\Omega}}) \int_0^1 |\gamma'_{x,y}|\, ds \\ &\le \mathrm{Lip}\,(\phi_{|\overline{\Omega}})(c[\Omega] + 1)|x - y| \le \mathrm{Lip}\,(\phi_{|\overline{\Omega}})(c[\Omega] + 1)\mathrm{Lip}\,(\phi^{(-1)}_{|\phi(\overline{\Omega})})|\xi - \eta| ,\end{aligned}$$

and accordingly

$$c[\phi(\Omega)] \le \mathrm{Lip}\,(\phi_{|\overline{\Omega}})(c[\Omega] + 1)\mathrm{Lip}\,(\phi^{(-1)}_{|\phi(\overline{\Omega})}) < +\infty .$$

□

We now show that a Roumieu function on the closure of a Whitney regular set Ω has a real analytic continuation on an open neighborhood of $\overline{\Omega}$, a result that is, in a sense, a converse of Remark 2.18. The proof that we present follows an argument of Preciso [246, Prop. 4.2.12 (ii)].

Proposition 2.32. *Let $n \in \mathbb{N}$. Let Ω be a bounded open subset of $\mathbb{R}^n$ that is regular in the sense of Whitney. Let $\rho \in]0, +\infty[$. If $f \in C^0_{\omega,\rho}(\overline{\Omega})$ and $\rho' \in]0, \rho/(c[\Omega]\sqrt{n})[$, then there exists a real analytic function $\tilde{f} \in C^\omega(W)$ such that $\tilde{f}_{|\overline{\Omega}} = f$, where*

$$W \equiv \Omega + \mathbb{B}_n(0, \rho') = \{x + y : x \in \Omega ,\ y \in \mathbb{B}_n(0, \rho')\} .$$

Proof. We first prove that if $x^o \in \Omega$, then the series

$$\sum_{\beta \in \mathbb{N}^n} \frac{D^\beta f(x^o)}{\beta!}(x - x^o)^\beta \tag{2.14}$$

converges absolutely for all $x \in \mathbb{B}_n(x^o, \rho/\sqrt{n})$. Indeed, the Multinomial Formula and inequality (2.4) with $p = 2$ imply that

$$\sum_{j=0}^{\infty} \sum_{|\beta|=j} \frac{|D^\beta f(x^o)|}{\beta!} |(x-x^o)^\beta|$$

$$\leq \|f\|_{C^0_{\omega,\rho}(\overline{\Omega})} \sum_{j=0}^{\infty} \sum_{|\beta|=j} \frac{j!}{\rho^j \beta!} \Pi_{l=1}^n |x_l - x_l^o|^{\beta_l}$$

$$= \|f\|_{C^0_{\omega,\rho}(\overline{\Omega})} \sum_{j=0}^{\infty} \left(\sum_{l=1}^n \frac{|x_l - x_l^o|}{\rho} \right)^j$$

$$\leq \|f\|_{C^0_{\omega,\rho}(\overline{\Omega})} \sum_{j=0}^{\infty} \left(\frac{\sqrt{n}|x-x^o|}{\rho} \right)^j .$$

We denote by f_{x^o} the function from $\mathbb{B}_n(x^o, \rho/\sqrt{n})$ to $\mathbb{R}$ that takes x to the sum of the series in (2.14). Since f_{x^o} is a sum of a power series, f_{x^o} is analytic (cf., e.g., John [130, pp. 64, 65]). We now show that

$$f_{x^o}(x) = f(x) \qquad \forall x \in \Omega \cap \mathbb{B}_n(x^o, \rho') . \tag{2.15}$$

To do so, we first show that $\Omega \cap \mathbb{B}_n(x^o, \rho')$ is contained in the same connected component of x^o in $\Omega \cap \mathbb{B}_n(x^o, \rho/\sqrt{n})$. If $x \in \Omega \cap \mathbb{B}_n(x^o, \rho')$, then there exists a path $\gamma_{x^o,x} \in \mathrm{Lip}([0,1], \Omega)$ such that $\gamma_{x^o,x}(0) = x^o$, $\gamma_{x^o,x}(1) = x$, and

$$\mathrm{length}(\gamma_{x^o,x}) < c[\Omega]\rho' < \rho/\sqrt{n} .$$

Then we have $\gamma_{x^o,x}([0,1]) \subseteq \mathbb{B}_n(x^o, \rho/\sqrt{n})$ and thus all points of $\Omega \cap \mathbb{B}_n(x^o, \rho')$ belong to the same connected component of x^o in $\Omega \cap \mathbb{B}_n(x^o, \rho/\sqrt{n})$. Since both f and f_{x^o} are analytic in $\Omega \cap \mathbb{B}_n(x^o, \rho/\sqrt{n})$ and have the same partial derivatives at the point x^o, the Identity Principle of Proposition 2.16 implies that f and f_{x^o} coincide on the connected component of $\Omega \cap \mathbb{B}_n(x^o, \rho/\sqrt{n})$ that contains x^o and thus on its subset $\Omega \cap \mathbb{B}_n(x^o, \rho')$.

Next we show that if x^o and y^o belong to Ω, then

$$f_{x^o}(x) = f_{y^o}(x) \qquad \forall x \in \mathbb{B}_n(x^o, \rho') \cap \mathbb{B}_n(y^o, \rho') .$$

We can clearly assume that $x^o \neq y^o$. Let $x \in \mathbb{B}_n(x^o, \rho') \cap \mathbb{B}_n(y^o, \rho')$. Since both x^o and y^o belong to Ω, $|x^o - y^o| < 2\rho'$ and $c[\Omega] < +\infty$, there exists a path $\gamma_{x^o,y^o} \in \mathrm{Lip}([0,1], \Omega)$ such that $\gamma_{x^o,y^o}(0) = x^o$, $\gamma_{x^o,y^o}(1) = y^o$ and

$$\mathrm{length}(\gamma_{x^o,y^o}) < c[\Omega]2\rho' < 2\rho/\sqrt{n} .$$

Let $\xi \in [0,1]$ be such that

$$\mathrm{length}(\gamma_{x^o,y^o|[0,\xi]}) = \frac{1}{2}\mathrm{length}(\gamma_{x^o,y^o}) .$$

Then for all $s \in [0, \xi]$, we have

$$|x^o - \gamma_{x^o,y^o}(s)| \leq \text{length}(\gamma_{x^o,y^o|[0,s]}) \\ \leq \frac{1}{2}\text{length}(\gamma_{x^o,y^o}) < \frac{1}{2}c[\Omega]2\rho' < \rho/\sqrt{n}$$

i.e., $\gamma_{x^o,y^o}(s) \in \Omega \cap \mathbb{B}_n(x^o, \rho/\sqrt{n})$. Similarly, $\gamma_{x^o,y^o}(s) \in \Omega \cap \mathbb{B}_n(y^o, \rho/\sqrt{n})$ for all $s \in [\xi, 1]$. Now, f and f_{x^o} coincide on the (open) connected component C_{x^o} of $\Omega \cap \mathbb{B}_n(x^o, \rho/\sqrt{n})$ that contains x^o. Since $\gamma_{x^o,y^o}([0, \xi])$ is a connected subset of $\Omega \cap \mathbb{B}_n(x^o, \rho/\sqrt{n})$, we have

$$\gamma_{x^o,y^o}(\xi) \in C_{x^o} .$$

In particular, f and f_{x^o} coincide on an open neighborhood of the point $\gamma_{x^o,y^o}(\xi)$.

Similarly, f and f_{y^o} coincide on the (open) connected component C_{y^o} of y^o in $\Omega \cap \mathbb{B}_n(y^o, \rho/\sqrt{n})$, that contains $\gamma_{x^o,y^o}([\xi, 1])$ and thus $\gamma_{x^o,y^o}(\xi)$. In particular, f and f_{y^o} coincide on an open neighborhood of the point $\gamma_{x^o,y^o}(\xi)$.

Since f_{x^o}, f, and f_{y^o} coincide on an open neighborhood of $\gamma_{x^o,y^o}(\xi)$, the Identity Principle of Proposition 2.16 (ii) implies that f_{x^o} and f_{y^o} coincide on the connected set $\mathbb{B}_n(x^o, \rho/\sqrt{n}) \cap \mathbb{B}_n(y^o, \rho/\sqrt{n})$, which contains $\gamma_{x^o,y^o}(\xi)$. Since

$$\mathbb{B}_n(x^o, \rho') \cap \mathbb{B}_n(y^o, \rho') \subseteq \mathbb{B}_n(x^o, \rho/\sqrt{n}) \cap \mathbb{B}_n(y^o, \rho/\sqrt{n}) ,$$

f_{x^o} and f_{y^o} coincide on $\mathbb{B}_n(x^o, \rho') \cap \mathbb{B}_n(y^o, \rho')$.

We are now ready to define $\tilde{f}$. If $x \in W$, then there exists $x^o \in \Omega$ such that $x \in \mathbb{B}_n(x^o, \rho')$, and we set

$$\tilde{f}(x) \equiv f_{x^o}(x) .$$

By the above argument, the definition of $\tilde{f}(x)$ does not depend on the point $x^o \in \Omega$ such that $x \in \mathbb{B}_n(x^o, \rho')$ and $\tilde{f}(x) = f_{x^o}(x) = f(x)$ if $x \in \Omega \cap \mathbb{B}_n(x^o, \rho')$. □

We also mention the following Tietze Extension Theorem. For a proof, we refer for example to Kuratowski [152] and Folland [103, Theorem 4.16, exercise 3, p. 118].

Theorem 2.33. *Let $n \in \mathbb{N}$. Let $\mathbb{D}$ be a closed subset of $\mathbb{R}^n$. If $f \in C_b^0(\mathbb{D})$, then there exists $\tilde{f} \in C_b^0(\mathbb{R}^n)$ such that $\tilde{f}_{|\mathbb{D}} = f$ and $\sup_{\mathbb{R}^n} |\tilde{f}| = \sup_{\mathbb{D}} |f|$.*

Finally, we mention the following extension theorem for Hölder continuous functions of McShane [207] and Björk [27]. For a proof, we refer to Kufner et al. [150, Theorem 1.8.3, p. 48] and to Troianiello [268, Theorem 1.2].

Theorem 2.34. *Let $n \in \mathbb{N}$. Let $\mathbb{D}$ be a closed subset of $\mathbb{R}^n$. Let $\alpha \in]0, 1]$. If f belongs to $C_b^{0,\alpha}(\mathbb{D})$, then there exists $\tilde{f} \in C_b^{0,\alpha}(\mathbb{R}^n)$ such that $\tilde{f}_{|\mathbb{D}} = f$ and $\|\tilde{f}\|_{C_b^{0,\alpha}(\mathbb{R}^n)} = \|f\|_{C_b^{0,\alpha}(\mathbb{D})}$.*

2.7 Coordinate Cylinders and Local Strict Hypographs

We wish to introduce the notion of a set which is locally around each of its boundary point a rotation of a translation of a strict hypograph of a continuous function. We do so by requiring that if p is a boundary point of Ω, then the translated set $\Omega - p$ can be rotated around the origin by means of a rotation R (or more generally by $R \in \mathbb{O}_n(\mathbb{R})$) so that the rotated set $R(\Omega - p)$ is, at least around the point 0, the strict hypograph of a continuous function γ.

In order to perform the above task, we need to introduce the following definition, which is a variant of a corresponding terminology used by Burenkov in his book [38] on the analysis of Sobolev spaces on domains.

Definition 2.35. Let $n \in \mathbb{N} \setminus \{0,1\}$. Let Ω be an open subset of $\mathbb{R}^n$. Let $p \in \partial\Omega$, $R \in \mathbb{O}_n(\mathbb{R})$, r, $\delta \in]0,+\infty[$. We say that the set

$$C(p,R,r,\delta) \equiv p + R^t(\mathbb{B}_{n-1}(0,r)\times]-\delta,\delta[)$$

is a coordinate cylinder for Ω around p, provided that the intersection

$$R(\Omega - p) \cap (\mathbb{B}_{n-1}(0,r)\times]-\delta,\delta[)$$

is the strict hypograph of a continuous function γ from $\mathbb{B}_{n-1}(0,r)$ to $]-\delta,\delta[$ which vanishes at 0 and such that $|\gamma(\eta)| < \delta/2$ for all $\eta \in \mathbb{B}_{n-1}(0,r)$, i.e., provided that there exists $\gamma \in C^0(\mathbb{B}_{n-1}(0,r),]-\delta,\delta[)$ such that

$$\begin{aligned} R(\Omega - p) \cap (\mathbb{B}_{n-1}(0,r)\times]-\delta,\delta[) \\ = \{(\eta,y) \in \mathbb{B}_{n-1}(0,r)\times]-\delta,\delta[:\ y < \gamma(\eta)\} \\ \equiv \text{hypograph}_s(\gamma)\,, \\ |\gamma(\eta)| < \delta/2 \qquad \forall \eta \in \mathbb{B}_{n-1}(0,r)\,, \qquad \gamma(0) = 0\,. \end{aligned} \tag{2.16}$$

Here the subscript 's' of 'hypograph$_s$' stands for 'strict'. Given a coordinate cylinder $C(p,R,r,\delta)$ for Ω around p, the corresponding function γ is uniquely determined and

$$\gamma(\eta) = \sup\{y \in]-\delta,\delta[:\ (\eta,y) \in R(\Omega - p) \cap (\mathbb{B}_{n-1}(0,r)\times]-\delta,\delta[)\}$$

for all $\eta \in \mathbb{B}_{n-1}(0,r)$. We also note that the continuity of γ implies that

$$\begin{aligned} R((\partial\Omega) - p) \cap (\mathbb{B}_{n-1}(0,r)\times]-\delta,\delta[) \\ = \{(\eta,y) \in \mathbb{B}_{n-1}(0,r)\times]-\delta,\delta[:\ y = \gamma(\eta)\} \\ \equiv \text{graph}(\gamma)\,. \end{aligned} \tag{2.17}$$

We say that γ is the function which represents $\partial\Omega$ in the coordinate cylinder $C(p,R,r,\delta)$ as a graph and that the function ψ_p from $\mathbb{B}_{n-1}(0,r)$ to $\mathbb{R}^n$ defined by

$$\psi_p(\eta) \equiv p + R^t \begin{pmatrix} \eta \\ \gamma(\eta) \end{pmatrix} \qquad \forall \eta \in \mathbb{B}_{n-1}(0,r) \tag{2.18}$$

is the parametrization of $\partial\Omega$ around p in the coordinate cylinder $C(p,R,r,\delta)$.

Since the continuous function γ induces the homeomorphism $(\cdot,\gamma(\cdot))$ from its domain onto its graph, the map ψ_p is a homeomorphism of $\mathbb{B}_{n-1}(0,r)$ onto $\psi_p(\mathbb{B}_{n-1}(0,r)) = (\partial\Omega) \cap C(p,R,r,\delta)$.

We also note that $\text{hypograph}_s(\gamma)$ is path connected and that accordingly

$$\Omega \cap C(p,R,r,\delta) = p + R^t(\text{hypograph}_s(\gamma))$$

is path connected. Hence, $\Omega \cap C(p,R,r,\delta)$ is contained in at most one connected component of Ω. It is sometimes useful to know that by shrinking r we still obtain a coordinate cylinder around the point p. More precisely, we have the following.

Remark 2.36. If $C(p,R,r,\delta)$ is a coordinate cylinder around the point p of $\partial\Omega$, then also $C(p,R,\rho,\delta)$ is a coordinate cylinder around the point p of $\partial\Omega$ for each $\rho \in]0,r[$, and the restriction $\gamma_{|\mathbb{B}_{n-1}(0,\rho)}$ represents $\partial\Omega$ in $C(p,R,\rho,\delta)$ as a graph.

We are now ready to introduce the following.

Definition 2.37. Let $n \in \mathbb{N} \setminus \{0,1\}$. We say that an open subset Ω of $\mathbb{R}^n$ is a local strict hypograph of class C^0 (of $\mathbb{R}^n$) provided that, for every point $p \in \partial\Omega$, there exist $R \in \mathbb{O}_n(\mathbb{R})$ and $r,\delta \in]0,+\infty[$ such that $C(p,R,r,\delta)$ is a coordinate cylinder for Ω around p.

If Ω is a local strict hypograph of class C^0 and if $p \in \partial\Omega$, then possibly shrinking r, we can always assume that $r < \delta$ and that the corresponding function γ, which represents $\partial\Omega$ in the coordinate cylinder $C(p,R,r,\delta)$ as a graph, has a continuous extension to $\overline{\mathbb{B}_{n-1}(0,r)}$ (cf. Remark 2.36). It is also customary to denote such extension by the same symbol γ.

We note that if an open subset Ω of $\mathbb{R}^n$ is a local strict hypograph of class C^0, then the exterior domain Ω^- of Ω is also a local strict hypograph of class C^0.

We now show that if a bounded open subset Ω of $\mathbb{R}^n$ is a local strict hypograph of class C^0, then the number of connected components of Ω and of its exterior Ω^- are necessarily finite. We do so by means of the following.

Lemma 2.38. *Let $n \in \mathbb{N} \setminus \{0,1\}$. Let Ω be a bounded open local strict hypograph of class C^0 of $\mathbb{R}^n$. Then the following statements hold.*

(i) The set Ω has at most a finite number of pairwise disjoint connected components, which we denote by $\Omega_1,\ldots, \Omega_{\varkappa^+}$ and which are open.
If $j \in \{1,\ldots,\varkappa^+\}$ and $p \in \partial\Omega_j$, then $p \in \partial\Omega$ and a coordinate cylinder for Ω around p is also a coordinate cylinder for Ω_j around p. In particular, Ω_j is a local strict hypograph of class C^0 of $\mathbb{R}^n$.
Moreover, the sets of the family $\{\overline{\Omega_j}\}_{j=1}^{\varkappa^+}$ are pairwise disjoint.

(ii) The set Ω^- has at most a finite number of pairwise disjoint connected components, which we denote by $(\Omega^-)_0$, $(\Omega^-)_1$, $\ldots$, $(\Omega^-)_{\varkappa^-}$ and which are open.

One and only one of such connected components is unbounded. We denote it by $(\Omega^-)_0$.
If $j \in \{0, \dots, \varkappa^-\}$ *and* $p \in \partial(\Omega^-)_j$, *then* $p \in \partial\Omega = \partial\Omega^-$ *and a coordinate cylinder for* Ω^- *around* p *is also a coordinate cylinder for* $(\Omega^-)_j$ *around* p. *In particular,* $(\Omega^-)_j$ *is a local strict hypograph of class* C^0 *of* $\mathbb{R}^n$.
Moreover, the sets of the family $\{\overline{(\Omega^-)_j}\}_{j=0}^{\varkappa^-}$ *are pairwise disjoint.*

Proof. We first consider statement (i). The connected components of an open subset of $\mathbb{R}^n$ are well known to be open.

If Ω_j is a connected component of Ω and if $p \in \partial\Omega_j$, then each ball centered at p intersects Ω_j and thus Ω. Hence, $p \in \overline{\Omega}$. We now show that $p \in \partial\Omega$. If $p \in \Omega$, then there exists $r > 0$ such that $\mathbb{B}_n(p,r) \subseteq \Omega$. Since both $\mathbb{B}_n(p,r)$ and Ω_j are connected and $\mathbb{B}_n(p,r) \cap \Omega_j \neq \emptyset$, the union $\mathbb{B}_n(p,r) \cup \Omega_j$ is connected. Since $\mathbb{B}_n(p,r) \cup \Omega_j \subseteq \Omega$, the maximality of the connected component Ω_j implies that $\mathbb{B}_n(p,r) \cup \Omega_j \subseteq \Omega_j$, and accordingly $\mathbb{B}_n(p,r)$ cannot intersect the complement of Ω_j, a contradiction. Hence, $p \in \partial\Omega$.

We now show that Ω_j is a local strict hypograph of class C^0. If $p \in \partial\Omega_j$, then we have just shown that $p \in \partial\Omega$ and thus there exist $R \in \mathbb{O}_n(\mathbb{R})$, $r, \delta \in]0, +\infty[$ such that $C(p, R, r, \delta)$ is a coordinate cylinder for Ω around p. Since $\Omega \cap C(p, R, r, \delta)$ is path connected and contained in Ω, it can intersect only one connected component of Ω, and it must be contained in such a component. Since $p \in \partial\Omega_j$, $C(p, R, r, \delta)$ must intersect Ω_j. Accordingly $\Omega \cap C(p, R, r, \delta) \subseteq \Omega_j \subseteq \Omega$ and thus

$$\Omega \cap C(p, R, r, \delta) = \Omega_j \cap C(p, R, r, \delta) .$$

Hence,

$$R(\Omega_j - p) \cap (\mathbb{B}_{n-1}(0,r) \times] - \delta, \delta[) = R(\Omega - p) \cap (\mathbb{B}_{n-1}(0,r) \times] - \delta, \delta[)$$

is the strict hypograph of some $\gamma \in C^0(\mathbb{B}_{n-1}(0,r),] - \delta, \delta[)$ such that $\gamma(0) = 0$ and $|\gamma(\eta)| < \delta/2$ for all $\eta \in \mathbb{B}_{n-1}(0,r)$. Then $C(p, R, r, \delta)$ is a coordinate cylinder for Ω_j around p and Ω_j is a local strict hypograph of class C^0.

In order to show that the number of connected components of Ω is finite, we assume by contradiction that Ω has a sequence $\{\Omega_j\}_{j \in \mathbb{N}}$ of (pairwise disjoint) connected components. Then we choose a point $x_j \in \Omega_j$ for each $j \in \mathbb{N}$. By the compactness of $\overline{\Omega}$ the sequence $\{x_j\}_{j \in \mathbb{N}}$ has a subsequence, which we still denote by the same symbol, which converges to a point $\tilde{x} \in \overline{\Omega}$.

If $\tilde{x} \in \Omega$, then $\tilde{x}$ can belong to a single connected component of Ω and such a component is open. Hence, x_j is contained in the very same component of $\tilde{x}$ for all j large enough, a contradiction. Indeed, each connected component can contain x_j only for one index j. Hence, $\tilde{x}$ must belong to $\partial\Omega$.

Then there exist $R \in \mathbb{O}_n(\mathbb{R})$, $r, \delta \in]0, +\infty[$ such that $C(\tilde{x}, R, r, \delta)$ is a coordinate cylinder for Ω around $\tilde{x}$. Since $\Omega \cap C(\tilde{x}, R, r, \delta)$ is path connected, there exists one and only one $l \in \mathbb{N}$ such that

$$\Omega_l \cap C(\tilde{x}, R, r, \delta) = \Omega \cap C(\tilde{x}, R, r, \delta)$$

(see the above argument). Since $C(\tilde{x}, R, r, \delta)$ is a neighborhood of $\tilde{x} = \lim_{j\to\infty} x_j$, the set $\Omega \cap C(\tilde{x}, R, r, \delta)$ and thus the set $\Omega_l \cap C(\tilde{x}, R, r, \delta)$ should contain x_j for all j large enough and not only for the index $j = l$, a contradiction. Hence, the number of connected components of Ω is finite.

Next we prove that the sets of the family $\{\overline{\Omega_j}\}_{j=1}^{\varkappa^+}$ are pairwise disjoint. Assume that there exist indices $j_1, j_2 \in \{1, \ldots, \varkappa^+\}$ and a point $p \in \overline{\Omega}$ such that $j_1 \neq j_2$ and $p \in \overline{\Omega_{j_1}} \cap \overline{\Omega_{j_2}}$. Since Ω_{j_1} and Ω_{j_2} are disjoint and open, we must have $p \in \partial\Omega_{j_1} \cap \partial\Omega_{j_2}$. Since $p \in \partial\Omega_{j_1} \cap \partial\Omega_{j_2} \subseteq \partial\Omega$, there exist $R \in \mathbb{O}_n(\mathbb{R})$, $r, \delta \in]0, +\infty[$ such that $C(p, R, r, \delta)$ is a coordinate cylinder for Ω around p. Next we note that $C(p, R, r, \delta)$ is a neighborhood of p and that p is a boundary point for both Ω_{j_1} and Ω_{j_2}. Then we must have

$$\Omega_{j_1} \cap C(p, R, r, \delta) \neq \emptyset \quad \text{and} \quad \Omega_{j_2} \cap C(p, R, r, \delta) \neq \emptyset\,.$$

Since $\Omega \cap C(p, R, r, \delta)$ is path connected, it must be contained in exactly one connected component of Ω and thus there exists a unique $l \in \mathbb{N}$ such that

$$\Omega \cap C(p, R, r, \delta) = \Omega_l \cap C(p, R, r, \delta)$$

(see the above argument). Then the inclusion $\Omega_{j_1}, \Omega_{j_2} \subseteq \Omega$ implies that

$$\emptyset \neq \Omega_{j_s} \cap C(p, R, r, \delta) \subseteq \Omega \cap C(p, R, r, \delta) = \Omega_l \cap C(p, R, r, \delta) \quad \forall s \in \{1, 2\}$$

and $\Omega_{j_s} \cap \Omega_l \neq \emptyset$ for all $s \in \{1, 2\}$. Since the connected components are pairwise disjoint, it follows that $\Omega_{j_1} = \Omega_l = \Omega_{j_2}$ and $j_1 = j_2$, a contradiction. Hence, $\overline{\Omega_{j_1}} \cap \overline{\Omega_{j_2}} = \emptyset$.

We now prove statement (ii). Let $r \in]0, +\infty[$ be such that $\overline{\Omega} \subseteq \mathbb{B}_n(0, r)$. Then we note that $\mathbb{R}^n \setminus \overline{\mathbb{B}_n(0, r)}$ is contained in Ω^- and intersects all unbounded connected components of Ω^-. Moreover, $\mathbb{R}^n \setminus \overline{\mathbb{B}_n(0, r)}$ is connected. Then any unbounded connected component of Ω^- must contain $\mathbb{R}^n \setminus \overline{\mathbb{B}_n(0, r)}$. It follows that Ω^- has only one unbounded connected component. Then, by the same argument used to prove statement (i), we conclude that statement (ii) holds true. □

2.8 Tangent Space to a Local Strict Hypograph

Definition 2.39. Let $n \in \mathbb{N} \setminus \{0\}$. Let M be a subset of $\mathbb{R}^n$, $p \in M$. We say that the vector $w \in \mathbb{R}^n$ is semi-tangent to M at the point p provided that either $w = 0$ or there exists a sequence $\{x_j\}_{j\in\mathbb{N}}$ in $M \setminus \{p\}$ which converges to p and such that

$$\frac{w}{|w|} = \lim_{j\to\infty} \frac{x_j - p}{|x_j - p|}\,.$$

We say that the vector $w \in \mathbb{R}^n$ is tangent to M at the point p provided that both w and $-w$ are semi-tangent to M at the point p.

We denote by T_pM the set of all semi-tangent vectors to M at p. One can easily check that T_pM is a cone of $\mathbb{R}^n$, i.e., that

$$\lambda w \in T_pM \qquad \text{whenever} \quad (\lambda, w) \in]0, +\infty[\times T_pM .$$

We say that T_pM is the cone of semi-tangent vectors to M at p. If T_pM is also a subspace of $\mathbb{R}^n$, then we say that M has a tangent space at p, that T_pM is the tangent space to M at p and that $p + T_pM$ is the affine tangent space to M at p.

Lemma 2.40. *Let $n \in \mathbb{N} \setminus \{0\}$. Let M be a subset of $\mathbb{R}^n$. Let γ be a function from an interval J of $\mathbb{R}$ to M. Let $\tilde{t} \in J$.*

(i) If $J^+_{\tilde{t}} \equiv J \cap [\tilde{t}, +\infty[$ is nondegenerate and if γ is right differentiable at $\tilde{t}$ then its right derivative $\gamma'_r(\tilde{t})$ is semi-tangent to M at $\gamma(\tilde{t})$.
(ii) If $J^-_{\tilde{t}} \equiv J \cap]-\infty, \tilde{t}]$ is nondegenerate, γ is left differentiable at $\tilde{t}$, and $\gamma'_l(\tilde{t})$ denotes its left derivative, then the vector $-\gamma'_l(\tilde{t})$ is semi-tangent to M at $\gamma(\tilde{t})$.

Proof. If either $\gamma'_r(\tilde{t})$ or $\gamma'_l(\tilde{t})$ are equal to 0, then they are semi-tangent to M at $\gamma(\tilde{t})$ by definition. Thus we can assume that $\gamma'_r(\tilde{t}) \neq 0$ and that $\gamma'_l(\tilde{t}) \neq 0$.

Now let $\{t_j\}_{j \in \mathbb{N}}$ be a sequence in $J \setminus \{\tilde{t}\}$ which converges to $\tilde{t}$. In case (i) we assume that the terms of the sequence belong to $J^+_{\tilde{t}} \setminus \{\tilde{t}\}$ and in case (ii) we assume that the terms of the sequence belong to $J^-_{\tilde{t}} \setminus \{\tilde{t}\}$.

Since the limit $\gamma'_r(\tilde{t})$ or $\gamma'_l(\tilde{t})$ of the incremental ratio of γ cannot be equal to zero, possibly neglecting a finite number of indices, we can assume that $\gamma(t_j) - \gamma(\tilde{t}) \neq 0$ for all $j \in \mathbb{N}$. Then we have

$$\lim_{j \to \infty} \frac{\gamma(t_j) - \gamma(\tilde{t})}{|\gamma(t_j) - \gamma(\tilde{t})|} = \lim_{j \to \infty} \frac{\frac{\gamma(t_j) - \gamma(\tilde{t})}{t_j - \tilde{t}}}{\left|\frac{\gamma(t_j) - \gamma(\tilde{t})}{t_j - \tilde{t}}\right|} \operatorname{sgn}(t_j - \tilde{t}) = \begin{cases} \frac{\gamma'_r(\tilde{t})}{|\gamma'_r(\tilde{t})|} & \text{in case } (i), \\ -\frac{\gamma'_l(\tilde{t})}{|\gamma'_l(\tilde{t})|} & \text{in case } (ii). \end{cases}$$

It follows that $\gamma'_r(\tilde{t})$ is semi-tangent to M at $\gamma(\tilde{t})$ in case (i) and that $-\gamma'_l(\tilde{t})$ is semi-tangent to M at $\gamma(\tilde{t})$ in case (ii). □

Lemma 2.41. *Let $m, n \in \mathbb{N}$, $0 < m \leq n$. Let U be an open subset of $\mathbb{R}^m$, $x \in U$. Let M be a subset of $\mathbb{R}^n$. Let φ be a function from U to M. Let φ be differentiable at x. Then each vector of*

$$d\varphi(x)[\mathbb{R}^m]$$

is semi-tangent to M at $p \equiv \varphi(x)$.

Proof. Indeed, if $v \in \mathbb{R}^m$, then the continuity of the product by a scalar number implies that there exists $\delta \in]0, +\infty[$ such that

$$x + tv \in U \qquad \forall t \in [-\delta, \delta]$$

and we can consider the map γ from $[-\delta, \delta]$ to M defined by

$$\gamma(t) \equiv \varphi(x + tv) \qquad \forall t \in [-\delta, \delta] .$$

By the chain rule

$$\gamma'(0) = d\varphi(x)[v],$$

and the previous Lemma 2.40 implies that $\gamma'(0)$ is semi-tangent to M at the point $\gamma(0) = \varphi(x) = p$. □

Proposition 2.42. *Let $n \in \mathbb{N} \setminus \{0, 1\}$. Let Ω be an open local strict hypograph of class C^0 of $\mathbb{R}^n$. Let $p \in \partial\Omega$. Let $R \in \mathbb{O}_n(\mathbb{R})$, $r, \delta \in]0, +\infty[$ be such that $C(p, R, r, \delta)$ is a coordinate cylinder for Ω around p. Let the function γ in $C^0(\mathbb{B}_{n-1}(0, r))$ represent $\partial\Omega$ as a graph in $C(p, R, r, \delta)$. Let $\tilde{\eta} \in \mathbb{B}_{n-1}(0, r)$. Then the following statements hold.*

(i) If γ is differentiable at $\tilde{\eta}$, then the cone of semi-tangent vectors $T_{\psi_p(\tilde{\eta})}(\partial\Omega)$ to $\partial\Omega$ at the point

$$\psi_p(\tilde{\eta}) \equiv p + R^t \begin{pmatrix} \tilde{\eta} \\ \gamma(\tilde{\eta}) \end{pmatrix}$$

is a vector subspace of $\mathbb{R}^n$ of dimension $n - 1$ and the unit vector

$$\nu_\Omega(\psi_p(\tilde{\eta})) \equiv R^t \begin{pmatrix} -\nabla\gamma(\tilde{\eta}) \\ 1 \end{pmatrix} \frac{1}{\sqrt{1 + |\nabla\gamma(\tilde{\eta})|^2}}$$

is normal to $T_{\psi_p(\tilde{\eta})}(\partial\Omega)$.

(ii) If there exists $r_{\tilde{\eta}} \in]0, +\infty[$ such that $\mathbb{B}_{n-1}(\tilde{\eta}, r_{\tilde{\eta}}) \subseteq \mathbb{B}_{n-1}(0, r)$ and

$$q_{\tilde{\eta},1}(\gamma) \equiv \sup_{\eta \in \mathbb{B}_{n-1}(\tilde{\eta}, r_{\tilde{\eta}}) \setminus \{\tilde{\eta}\}} \frac{|\gamma(\eta) - \gamma(\tilde{\eta})|}{|\eta - \tilde{\eta}|} < +\infty,$$

and if the cone $T_{\psi_p(\tilde{\eta})}(\partial\Omega)$ of semi-tangent vectors to $\partial\Omega$ at the point $\psi_p(\tilde{\eta})$ is a vector subspace of $\mathbb{R}^n$ of dimension $n - 1$, then γ is differentiable at $\tilde{\eta}$ and the unit vector $\nu_\Omega(\psi_p(\tilde{\eta}))$ of statement (i) is normal to $T_{\psi_p(\tilde{\eta})}(\partial\Omega)$.

Proof. (i) Since the map from $\mathbb{R}^n$ to itself that takes w to $p + R^t w$ is an affine isometry, it suffices to show that the cone $T_{(\tilde{\eta},\gamma(\tilde{\eta}))}\mathrm{graph}(\gamma)$ of semi-tangent vectors is a subspace of $\mathbb{R}^n$ of dimension $(n - 1)$ and that

$$\nu_{\mathrm{graph}(\gamma)}(\tilde{\eta}, \gamma(\tilde{\eta})) \equiv \begin{pmatrix} -\nabla\gamma(\tilde{\eta}) \\ 1 \end{pmatrix} \frac{1}{\sqrt{1 + |\nabla\gamma(\tilde{\eta})|^2}}$$

is normal to $T_{(\tilde{\eta},\gamma(\tilde{\eta}))}\mathrm{graph}(\gamma)$. We set $g(\eta, y) \equiv y - \gamma(\eta)$ for all (η, y) in $\mathbb{B}_{n-1}(0, r) \times] - \delta, \delta[$. Then g is continuous, is differentiable at $(\tilde{\eta}, \gamma(\tilde{\eta}))$, and

$$\nabla g(\tilde{\eta}, \gamma(\tilde{\eta})) = \begin{pmatrix} -\nabla\gamma(\tilde{\eta}) \\ 1 \end{pmatrix} \neq 0.$$

Then the kernel $\mathrm{Ker}\, dg(\tilde{\eta}, \gamma(\tilde{\eta}))$ of the differential $dg(\tilde{\eta}, \gamma(\tilde{\eta}))$ of g at $(\tilde{\eta}, \gamma(\tilde{\eta}))$ has dimension $(n - 1)$. We now show that

$$T_{(\tilde{\eta},\gamma(\tilde{\eta}))}\mathrm{graph}(\gamma) \subseteq \mathrm{Ker}\, dg(\tilde{\eta},\gamma(\tilde{\eta}))\,. \tag{2.19}$$

To do so, we take $v \in T_{(\tilde{\eta},\gamma(\tilde{\eta}))}\mathrm{graph}(\gamma)$ and we show that $\nabla g(\tilde{\eta},\gamma(\tilde{\eta})) \cdot v = 0$. We can clearly assume that $v \neq 0$. Since v is semi-tangent to $\mathrm{graph}(\gamma)$ at the point $(\tilde{\eta},\gamma(\tilde{\eta}))$, there exists a sequence $\{\eta_j\}_{j\in\mathbb{N}}$ in $\mathbb{B}_{n-1}(0,r) \setminus \{\tilde{\eta}\}$ such that

$$(\tilde{\eta},\gamma(\tilde{\eta})) = \lim_{j\to\infty}(\eta_j,\gamma(\eta_j))\,,$$
$$\frac{v}{|v|} = \lim_{j\to\infty}\frac{(\eta_j,\gamma(\eta_j)) - (\tilde{\eta},\gamma(\tilde{\eta}))}{|(\eta_j,\gamma(\eta_j)) - (\tilde{\eta},\gamma(\tilde{\eta}))|}\,.$$

Since g is differentiable at $(\tilde{\eta},\gamma(\tilde{\eta}))$, we have

$$\lim_{j\to\infty}\frac{|g(\eta_j,\gamma(\eta_j)) - g(\tilde{\eta},\gamma(\tilde{\eta})) - dg(\tilde{\eta},\gamma(\tilde{\eta}))[\eta_j-\tilde{\eta},\gamma(\eta_j)-\gamma(\tilde{\eta})]|}{|(\eta_j-\tilde{\eta},\gamma(\eta_j)-\gamma(\tilde{\eta}))|} = 0\,.$$

Moreover, $g(\eta_j,\gamma(\eta_j)) = g(\tilde{\eta},\gamma(\tilde{\eta})) = 0$ for all $j \in \mathbb{N}$ and thus we have

$$\lim_{j\to\infty} dg(\tilde{\eta},\gamma(\tilde{\eta}))\left[\frac{(\eta_j-\tilde{\eta},\gamma(\eta_j)-\gamma(\tilde{\eta}))}{|(\eta_j-\tilde{\eta},\gamma(\eta_j)-\gamma(\tilde{\eta}))|}\right] = 0\,.$$

Since the argument of $dg(\tilde{\eta},\gamma(\tilde{\eta}))$ tends to $v/|v|$ as j tends to infinity and the linear map $dg(\tilde{\eta},\gamma(\tilde{\eta}))$ is continuous, we deduce that

$$dg(\tilde{\eta},\gamma(\tilde{\eta}))\left[\frac{v}{|v|}\right] = 0\,.$$

Accordingly, $v \in \mathrm{Ker}\, dg(\tilde{\eta},\gamma(\tilde{\eta}))$. Hence, we have proved that inclusion (2.19) holds true. Next we set

$$\varphi(\eta) \equiv (\eta,\gamma(\eta)) \qquad \forall \eta \in \mathbb{B}_{n-1}(0,r)\,.$$

By the previous Lemma 2.41, we have

$$d\varphi(\tilde{\eta})[\mathbb{R}^{n-1}] \subseteq T_{(\tilde{\eta},\gamma(\tilde{\eta}))}\mathrm{graph}(\gamma)$$

and thus we have

$$d\varphi(\tilde{\eta})[\mathbb{R}^{n-1}] \subseteq T_{(\tilde{\eta},\gamma(\tilde{\eta}))}\mathrm{graph}(\gamma) \subseteq \mathrm{Ker}\, dg(\tilde{\eta},\gamma(\tilde{\eta}))\,.$$

Since $D\varphi(\tilde{\eta}) = \begin{pmatrix} I_{n-1} \\ D\gamma(\tilde{\eta}) \end{pmatrix}$ has rank $n-1$, the map $d\varphi(\tilde{\eta})$ is injective and thus $d\varphi(\tilde{\eta})[\mathbb{R}^{n-1}]$ is a vector space of dimension $(n-1)$. Since $\mathrm{Ker}\, dg(\tilde{\eta},\gamma(\tilde{\eta}))$ is also a vector space of dimension $(n-1)$, we conclude that

$$d\varphi(\tilde{\eta})[\mathbb{R}^{n-1}] = T_{(\tilde{\eta},\gamma(\tilde{\eta}))}\mathrm{graph}(\gamma) = \mathrm{Ker}\, dg(\tilde{\eta},\gamma(\tilde{\eta}))$$

and that $T_{(\tilde{\eta},\gamma(\tilde{\eta}))}\mathrm{graph}(\gamma)$ is a vector space of dimension $(n-1)$. Since the vector

$$\begin{pmatrix} -\nabla\gamma(\tilde{\eta}) \\ 1 \end{pmatrix}$$

is normal to $\mathrm{Ker}\, dg(\tilde{\eta}, \gamma(\tilde{\eta}))$, we conclude that $\nu_{\mathrm{graph}(\gamma)}(\tilde{\eta}, \gamma(\tilde{\eta}))$ is normal to the space $T_{(\tilde{\eta},\gamma(\tilde{\eta}))}\mathrm{graph}(\gamma)$.

We now turn to the proof of statement (ii). Since the map from $\mathbb{R}^n$ to itself that takes w to $p + R^t w$ is an affine isometry, the cone $T_{(\tilde{\eta},\gamma(\tilde{\eta}))}\mathrm{graph}(\gamma)$ of semi-tangent vectors is a subspace of $\mathbb{R}^n$ of dimension $(n-1)$. Let the unit vector $\nu \equiv (\nu_1, \dots, \nu_n)$ be orthogonal to $T_{(\tilde{\eta},\gamma(\tilde{\eta}))}\mathrm{graph}(\gamma)$. We first show that $\nu_n \neq 0$. Assume by contradiction that $\nu_n = 0$. Then we have $\nu_n = \nu \cdot e_n = 0$ and accordingly $e_n \in T_{(\tilde{\eta},\gamma(\tilde{\eta}))}\mathrm{graph}(\gamma)$. Then there exists a sequence $\{\eta_j\}_{j\in\mathbb{N}}$ in $\mathbb{B}_{n-1}(\tilde{\eta}, r_{\tilde{\eta}}) \setminus \{\tilde{\eta}\}$ which converges to $\tilde{\eta}$ and such that

$$\lim_{j\to\infty} (\eta_j, \gamma(\eta_j)) = (\tilde{\eta}, \gamma(\tilde{\eta})),$$

$$\lim_{j\to\infty} \frac{(\eta_j, \gamma(\eta_j)) - (\tilde{\eta}, \gamma(\tilde{\eta}))}{|(\eta_j, \gamma(\eta_j)) - (\tilde{\eta}, \gamma(\tilde{\eta}))|} = e_n .$$

In particular,

$$\lim_{j\to\infty} \frac{\eta_j - \tilde{\eta}}{|(\eta_j, \gamma(\eta_j)) - (\tilde{\eta}, \gamma(\tilde{\eta}))|} = 0 .$$

By assumption, we have

$$\frac{|(\eta_j, \gamma(\eta_j)) - (\tilde{\eta}, \gamma(\tilde{\eta}))|}{|\eta_j - \tilde{\eta}|} = \frac{\sqrt{|\eta_j - \tilde{\eta}|^2 + |\gamma(\eta_j) - \gamma(\tilde{\eta})|^2}}{|\eta_j - \tilde{\eta}|} \leq \sqrt{1 + q_{\tilde{\eta},1}^2} \qquad \forall j \in \mathbb{N},$$

with $q_{\tilde{\eta},1}$ as in statement (ii) of the proposition. Thus the above limiting relation implies that

$$\lim_{j\to\infty} \frac{\eta_j - \tilde{\eta}}{|\eta_j - \tilde{\eta}|} = \lim_{j\to\infty} \frac{\eta_j - \tilde{\eta}}{|(\eta_j, \gamma(\eta_j)) - (\tilde{\eta}, \gamma(\tilde{\eta}))|} \frac{|(\eta_j, \gamma(\eta_j)) - (\tilde{\eta}, \gamma(\tilde{\eta}))|}{|\eta_j - \tilde{\eta}|} = 0 ,$$

a contradiction. Hence, $\nu_n \neq 0$. Next we turn to show that γ is differentiable at $\tilde{\eta}$ and that

$$d\gamma(\tilde{\eta})[v] = -\frac{1}{\nu_n} v \cdot (\nu_1, \dots, \nu_{n-1}) \qquad \forall v \in \mathbb{R}^{n-1} .$$

Clearly, the operator in the right-hand side is linear and continuous in v. Thus, we have to show that

$$\lim_{v\to 0} \frac{\gamma(\tilde{\eta} + v) - \gamma(\tilde{\eta}) + \frac{1}{\nu_n} v \cdot (\nu_1, \dots, \nu_{n-1})}{|v|} = 0 .$$

It suffices to show that if $\{v_j\}_{j\in\mathbb{N}}$ is a sequence in $\mathbb{B}_{n-1}(0,r_{\tilde{\eta}})\setminus\{0\}$ which converges to 0, then each subsequence $\{v_{j_k}\}_{k\in\mathbb{N}}$ of $\{v_j\}_{j\in\mathbb{N}}$ has a further subsequence $\{v_{j_{k_l}}\}_{l\in\mathbb{N}}$ such that

$$\lim_{l\to\infty}\frac{\gamma(\tilde{\eta}+v_{j_{k_l}})-\gamma(\tilde{\eta})+\frac{1}{\nu_n}v_{j_{k_l}}\cdot(\nu_1,\ldots,\nu_{n-1})}{|v_{j_{k_l}}|}=0\,.$$

Since the sequence

$$\left\{\frac{((\tilde{\eta}+v_{j_k})-\tilde{\eta},\gamma(\tilde{\eta}+v_{j_k})-\gamma(\tilde{\eta}))}{|(\tilde{\eta}+v_{j_k})-\tilde{\eta},\gamma(\tilde{\eta}+v_{j_k})-\gamma(\tilde{\eta}))|}\right\}_{k\in\mathbb{N}}$$

is bounded, then there exists a further subsequence $\{v_{j_{k_l}}\}_{l\in\mathbb{N}}$ and a unit vector w in $T_{(\tilde{\eta},\gamma(\tilde{\eta}))}\mathrm{graph}(\gamma)$ such that

$$\lim_{l\to\infty}\frac{((\tilde{\eta}+v_{j_{k_l}})-\tilde{\eta},\gamma(\tilde{\eta}+v_{j_{k_l}})-\gamma(\tilde{\eta}))}{|(\tilde{\eta}+v_{j_{k_l}})-\tilde{\eta},\gamma(\tilde{\eta}+v_{j_{k_l}})-\gamma(\tilde{\eta}))|}=w\,.$$

Then, by taking the scalar product with ν, we have

$$\lim_{l\to\infty}\frac{v_{j_{k_l}}\cdot(\nu_1,\ldots,\nu_{n-1})+(\gamma(\tilde{\eta}+v_{j_{k_l}})-\gamma(\tilde{\eta}))\nu_n}{|(v_{j_{k_l}},\gamma(\tilde{\eta}+v_{j_{k_l}})-\gamma(\tilde{\eta}))|}=w\cdot\nu=0\,.$$

Next we note that

$$\frac{|(v_{j_{k_l}},\gamma(\tilde{\eta}+v_{j_{k_l}})-\gamma(\tilde{\eta}))|}{|v_{j_{k_l}}|}=\frac{\sqrt{|v_{j_{k_l}}|^2+(\gamma(\tilde{\eta}+v_{j_{k_l}})-\gamma(\tilde{\eta}))^2}}{|v_{j_{k_l}}|}\leq\sqrt{1+q^2_{\tilde{\eta},1}}$$

for all $l\in\mathbb{N}$, and thus the above limiting relation implies that

$$\begin{aligned}&\lim_{l\to\infty}\frac{v_{j_{k_l}}\cdot(\nu_1,\ldots,\nu_{n-1})+(\gamma(\tilde{\eta}+v_{j_{k_l}})-\gamma(\tilde{\eta}))\nu_n}{|v_{j_{k_l}}|}\\&\qquad=\lim_{l\to\infty}\frac{v_{j_{k_l}}\cdot(\nu_1,\ldots,\nu_{n-1})+(\gamma(\tilde{\eta}+v_{j_{k_l}})-\gamma(\tilde{\eta}))\nu_n}{|(v_{j_{k_l}},\gamma(\tilde{\eta}+v_{j_{k_l}})-\gamma(\tilde{\eta}))|}\\&\qquad\quad\times\frac{|(v_{j_{k_l}},\gamma(\tilde{\eta}+v_{j_{k_l}})-\gamma(\tilde{\eta}))|}{|v_{j_{k_l}}|}=0\end{aligned}$$

and γ is differentiable at $\tilde{\eta}$. Then statement (i) implies that the unit vector $\nu_\Omega(\psi_p(\tilde{\eta}))$ of statement (i) is normal to $T_{\psi_p(\tilde{\eta})}(\partial\Omega)$ and the proof is complete. □

Definition 2.43. Let Ω be an open subset of $\mathbb{R}^n$. Let $x\in\partial\Omega$. Let $v\in\mathbb{R}^n\setminus\{0\}$.

(i) We say that v points to the interior of Ω at x provided that there exists $\varsigma_{x,v}>0$ such that

$$x+tv\in\Omega\qquad\forall t\in]0,\varsigma_{x,v}[\,.$$

(ii) We say that v points to the exterior of Ω at x provided that there exists $\varsigma_{x,v} > 0$ such that

$$x + tv \in \mathbb{R}^n \setminus \overline{\Omega} \qquad \forall t \in]0, \varsigma_{x,v}[\,.$$

Then we have the following proposition, which gives a sufficient condition in order that a vector points to the interior or to the exterior of a local strict hypograph at a boundary point at which the tangent space does exist.

Proposition 2.44. *Let $n \in \mathbb{N} \setminus \{0,1\}$. Let Ω be an open local strict hypograph of class C^0. Let $p \in \partial\Omega$. Let $R \in \mathbb{O}_n(\mathbb{R})$, $r, \delta \in]0,+\infty[$ be such that $C(p,R,r,\delta)$ is a coordinate cylinder for Ω around p. Let the function $\gamma \in C^0(\mathbb{B}_{n-1}(0,r))$ represent $\partial\Omega$ as a graph in $C(p,R,r,\delta)$. If there exists $r_0 \in]0,r[$ such that*

$$q_1(\gamma) \equiv \sup_{\eta \in \mathbb{B}_{n-1}(0,r_0)\setminus\{0\}} \frac{|\gamma(\eta)|}{|\eta|} < +\infty$$

and if the cone $T_p(\partial\Omega)$ of semi-tangent vectors to $\partial\Omega$ at the point p is a vector subspace of $\mathbb{R}^n$ of dimension $n-1$, then γ is differentiable at 0, the unit vector

$$\nu_\Omega(p) \equiv R^t \begin{pmatrix} -\nabla\gamma(0) \\ 1 \end{pmatrix} \frac{1}{\sqrt{1+|\nabla\gamma(0)|^2}}$$

is normal to $T_p(\partial\Omega)$ and points to the exterior of Ω, and $-\nu_\Omega(p)$ points to the interior of Ω.

If $v \in \mathbb{R}^n$ and if $v \cdot \nu_\Omega(p) > 0$, then v points to the exterior of Ω. If instead $v \cdot \nu_\Omega(p) < 0$, then v points to the interior of Ω.

Proof. By Proposition 2.42 (ii) we already know that γ is differentiable at 0 and that $\nu_\Omega(p)$ is normal to $T_p(\partial\Omega)$. So, it suffices to prove the last statement on v and then take $v = \nu_\Omega(p)$ to deduce the rest. Let

$$\xi \equiv (\xi', \xi_n) \equiv Rv\,, \qquad \nu_\gamma(p) \equiv R\nu_\Omega(p)\,.$$

Since the map that takes z to $p + R^t z$ is the composition of a translation with an orthogonal transformation, it suffices to show that if $\xi \cdot \nu_\gamma(p) > 0$, then ξ points to the exterior of $\mathrm{hypograph}_s(\gamma)$ at $(0,\gamma(0))$ and if $\xi \cdot \nu_\gamma(p) < 0$ then ξ points to the interior of $\mathrm{hypograph}_s(\gamma)$ at $(0,\gamma(0))$. We first consider the case in which $\xi \cdot \nu_\gamma(p) > 0$. Since γ is differentiable at 0, then we have

$$\begin{aligned}
\lim_{t\to 0} \frac{\gamma(t\xi') - \gamma(0)}{t} - \xi_n &= \nabla\gamma(0)\cdot\xi' - \xi_n \\
&= \sqrt{1+|\nabla\gamma(0)|^2}\frac{\nabla\gamma(0)\cdot\xi'}{\sqrt{1+|\nabla\gamma(0)|^2}} - \xi_n \\
&= -\xi\cdot\nu_\gamma(p)\sqrt{1+|\nabla\gamma(0)|^2}\,.
\end{aligned}$$

If $\xi \cdot \nu_\gamma(p) > 0$, then there exists $t_\xi \in]0,+\infty[$ such that

$$\gamma(t\xi') - \gamma(0) < t\xi_n \qquad \forall t \in]0, t_\xi[\,,$$

i.e., such that $(0, \gamma(0)) + t\xi \in \mathbb{R}^n \setminus \overline{\mathrm{hypograph}_s(\gamma)}$ for all $t \in]0, t_\xi[$. Hence, ξ points to the exterior of $\mathrm{hypograph}_s(\gamma)$ at $(0, \gamma(0))$. A proof in case $\xi \cdot \nu_\gamma(x) < 0$ can be effected by the same argument. □

2.9 Lipschitz Subsets of $\mathbb{R}^n$

Definition 2.45. Let $n \in \mathbb{N} \setminus \{0, 1\}$. We say that an open subset Ω of $\mathbb{R}^n$ is a Lipschitz subset of $\mathbb{R}^n$ provided that for every $p \in \partial\Omega$, there exist $R \in \mathbb{O}_n(\mathbb{R})$ and $r, \delta \in]0, +\infty[$ such that $C(p, R, r, \delta)$ is a coordinate cylinder for Ω around p and that the corresponding function γ which represents $\partial\Omega$ as a graph in $C(p, R, r, \delta)$ is Lipschitz continuous.

By definition, a Lipschitz subset of $\mathbb{R}^n$ is also a local strict hypograph of class C^0.

If a Lipschitz subset Ω of $\mathbb{R}^n$ is also bounded, then Lemma 2.38 implies that Ω and of Ω^- have a finite number of connected components which are also Lipschitz subsets of $\mathbb{R}^n$.

We now consider the question of the Whitney regularity of Lipschitz sets and we first introduce the following.

Lemma 2.46. *Let $n \in \mathbb{N} \setminus \{0, 1\}$, $r, \delta, \delta_1 \in]0, +\infty[$, $\delta_1 < \delta$. Let γ belong to $\mathrm{Lip}\,(\mathbb{B}_{n-1}(0, r),]-\delta, \delta[)$ and $|\gamma(\eta)| < \delta_1$ for all $\eta \in \mathbb{B}_{n-1}(0, r)$. Then the strict hypograph*

$$\mathrm{hypograph}_s(\gamma) \equiv \{(\eta, y) \in \mathbb{B}_{n-1}(0, r) \times]-\delta, \delta[:\, y < \gamma(\eta)\}$$

is regular in the sense of Whitney (cf. Definition 2.16).

Proof. Let $(\tilde{\eta}, \tilde{y}), (\eta^\sharp, y^\sharp) \in \mathrm{hypograph}_s(\gamma)$. In order to define a path which joins the two points $(\tilde{\eta}, \tilde{y})$ and $(\eta^\sharp, y^\sharp)$, we fix $\delta' \in]0, \min\{\delta - \delta_1, \gamma(\tilde{\eta}) - \tilde{y}, \gamma(\eta^\sharp) - y^\sharp\}[$ and we define the function φ by setting

$$\varphi(t) \equiv (a(t), b(t)) \equiv \left(\tilde{\eta} + t(\eta^\sharp - \tilde{\eta}), \gamma(\tilde{\eta} + t(\eta^\sharp - \tilde{\eta})) - \delta'\right) \qquad \forall t \in [0, 1]\,.$$

Clearly,

$$\begin{aligned} &\varphi(0) = (\tilde{\eta}, \gamma(\tilde{\eta}) - \delta')\,, \qquad \varphi(1) = (\eta^\sharp, \gamma(\eta^\sharp) - \delta')\,, \\ &-\delta < -\delta_1 - \delta' \leq \gamma(\tilde{\eta} + t(\eta^\sharp - \tilde{\eta})) - \delta' < \gamma(\tilde{\eta} + t(\eta^\sharp - \tilde{\eta})) \qquad \forall t \in [0, 1]\,. \end{aligned}$$

In particular, $\varphi(t) \in \mathrm{hypograph}_s(\gamma)$ for all $t \in [0, 1]$. Since φ is a composition of Lipschitz continuous maps, φ is Lipschitz continuous. The length of φ satisfies the following inequality

$$\text{length}(\varphi) \le \int_0^1 \sqrt{|\tilde\eta-\eta^\sharp|^2+(\text{Lip}(\gamma))^2|\tilde\eta-\eta^\sharp|^2}\,dt = |\tilde\eta-\eta^\sharp|\sqrt{1+(\text{Lip}(\gamma))^2}\,.$$

If the set

$$B \equiv \{t \in [0,1] : \tilde y + t(y^\sharp - \tilde y) \ge b(t)\}$$

is empty, then

$$\tilde y + t(y^\sharp - \tilde y) < b(t) < \gamma(\tilde\eta + t(\eta^\sharp - \tilde\eta)) \qquad \forall t \in [0,1]$$

and the segment with end points $(\tilde\eta, \tilde y)$ and $(\eta^\sharp, y^\sharp)$ lies entirely in $\text{hypograph}_s(\gamma)$. Then we have the following equality for the geodesic distance $\lambda \equiv \lambda_{\text{hypograph}_s(\gamma)}$ in $\text{hypograph}_s(\gamma)$

$$\lambda((\tilde\eta,\tilde y),(\eta^\sharp,y^\sharp)) = |(\tilde\eta,\tilde y)-(\eta^\sharp,y^\sharp)|\,. \tag{2.20}$$

If instead $B \ne \emptyset$, then we set

$$t_1 \equiv \min B\,, \qquad t_2 \equiv \max B\,.$$

Since δ' is smaller than $\gamma(\tilde\eta) - \tilde y$ and $\gamma(\eta^\sharp) - y^\sharp$, we have $0, 1 \notin B$. Then we have $0 < t_1 < t_2 < 1$, and the continuity of γ implies that

$$\tilde y + t_r(y^\sharp - \tilde y) = b(t_r) \qquad r \in \{1,2\}\,.$$

Since the points $(\tilde\eta, \tilde y)$ and $(\eta^\sharp, y^\sharp)$ can be joined by the piecewise Lipschitz path in $\text{hypograph}_s(\gamma)$ obtained by joining $(\tilde\eta, \tilde y)$ and $\varphi(t_1)$ by a segment, $\varphi(t_1)$ and $\varphi(t_2)$ by the restriction of φ to $[t_1, t_2]$, and $\varphi(t_2)$ and $(\eta^\sharp, y^\sharp)$ by a segment, then the following inequality for the geodesic distance $\lambda((\tilde\eta,\tilde y),(\eta^\sharp,y^\sharp))$ holds

$$\begin{aligned}\lambda((\tilde\eta,\tilde y),(\eta^\sharp,y^\sharp)) & \qquad\qquad (2.21)\\ \le |(\tilde\eta,\tilde y)-\varphi(t_1)| &+ |\tilde\eta-\eta^\sharp|\sqrt{1+(\text{Lip}(\gamma))^2} + |\varphi(t_2)-(\eta^\sharp,y^\sharp)|\\ \le 2|(\tilde\eta,\tilde y)-(\eta^\sharp,y^\sharp)| &+ |(\tilde\eta,\tilde y)-(\eta^\sharp,y^\sharp)|\sqrt{1+(\text{Lip}(\gamma))^2}\,.\end{aligned}$$

By (2.20) and (2.21), we conclude that the Whitney constant $c[\text{hypograph}_s(\gamma)]$ of $\text{hypograph}_s(\gamma)$ satisfies the following inequality

$$c[\text{hypograph}_s(\gamma)] \le 2 + \sqrt{1+(\text{Lip}(\gamma))^2}\,.$$

□

Thus we are ready to prove the following.

Proposition 2.47. *Let $n \in \mathbb{N} \setminus \{0,1\}$. A bounded open connected Lipschitz subset Ω of $\mathbb{R}^n$ is regular in the sense of Whitney.*

Proof. As usual, we denote by $\lambda_\Omega(p,q)$ the geodesic distance in Ω between two points p and q of Ω. Assume by contradiction that Ω is not regular in the sense of Whitney. Then there exists a sequence $\{(p_j,q_j)\}_{j\in\mathbb{N}}$ in $\Omega \times \Omega$ such that

$$p_j \neq q_j \qquad \forall j \in \mathbb{N}, \qquad \lim_{j\to\infty} \frac{\lambda_\Omega(p_j,q_j)}{|p_j-q_j|} = +\infty. \tag{2.22}$$

Since $\overline{\Omega} \times \overline{\Omega}$ is compact, there exist a subsequence of $\{(p_j,q_j)\}_{j\in\mathbb{N}}$ which we still denote by the same symbol and $(P,Q) \in \overline{\Omega} \times \overline{\Omega}$ such that

$$\lim_{j\to\infty} p_j = P, \qquad \lim_{j\to\infty} q_j = Q. \tag{2.23}$$

We now define a set $A(x)$ for each point x of $\overline{\Omega}$. If $x \in \Omega$, then we denote by $A(x)$ an open ball with center at x and closure contained in Ω. Since $A(x)$ is convex, $A(x)$ is regular in the sense of Whitney. If instead $x \in \partial\Omega$, then we denote by $A(x)$ the intersection of Ω with a coordinate cylinder for Ω around x. Since $A(x)$ is a translation of a rotation of a strict hypograph, Lemma 2.46 implies that $A(x)$ is regular in the sense of Whitney. Hence, $A(x)$ is regular in the sense of Whitney for all $x \in \overline{\Omega}$. In particular, we have

$$c[A(P)] < +\infty, \qquad c[A(Q)] < +\infty. \tag{2.24}$$

Now let $P_1 \in A(P)\setminus\{P\}$, $Q_1 \in A(Q)\setminus\{Q\}$. We can clearly assume that $P_1 \neq Q_1$. By (2.23) there exists $j_0 \in \mathbb{N}$ such that

$$p_j \in A(P), \; q_j \in A(Q) \qquad \forall j \in \mathbb{N}, j \geq j_0.$$

By (2.24) there exist a path $\gamma_{p_j,P_1} \in \mathrm{Lip}([0,1],A(P))$ and a path $\gamma_{Q_1,q_j} \in \mathrm{Lip}([0,1],A(Q))$ such that

$$\gamma_{p_j,P_1}(0) = p_j, \; \gamma_{p_j,P_1}(1) = P_1, \; \gamma_{Q_1,q_j}(0) = Q_1, \; \gamma_{Q_1,q_j}(1) = q_j,$$

and

$$\mathrm{length}(\gamma_{p_j,P_1}) \leq (c[A(P)]+1)|p_j - P_1|,$$
$$\mathrm{length}(\gamma_{Q_1,q_j}) \leq (c[A(Q)]+1)|Q_1 - q_j|$$

for each $j \geq j_0$. Since Ω is connected, there exists a path $\gamma_{P_1,Q_1} \in \mathrm{Lip}([0,1],\Omega)$ such that

$$\gamma_{P_1,Q_1}(0) = P_1, \qquad \gamma_{P_1,Q_1}(1) = Q_1.$$

Then we have

$$\frac{\lambda_\Omega(p_j,q_j)}{|p_j-q_j|} \leq (c[A(P)]+1)\frac{|p_j-P_1|}{|p_j-q_j|} + \frac{\mathrm{length}(\gamma_{P_1,Q_1})}{|p_j-q_j|} + (c[A(Q)]+1)\frac{|Q_1-q_j|}{|p_j-q_j|} \qquad j \in \mathbb{N}, j \geq j_0.$$

Since

$$\lim_{j\to\infty} |p_j - P_1| = |P - P_1| \in]0, +\infty[\text{ and } \lim_{j\to\infty} |Q_1 - q_j| = |Q_1 - Q| \in]0, +\infty[\,, \tag{2.25}$$

the limiting relation in (2.22) implies that $\lim_{j\to\infty} |p_j - q_j| = 0$ and accordingly that $P = Q$. Since $p_j, q_j \in A(P) = A(Q)$ for $j \geq j_0$ and $A(P)$ is regular in the sense of Whitney and contained in Ω, we must have

$$\frac{\lambda_\Omega(p_j, q_j)}{|p_j - q_j|} \leq \frac{\lambda_{A(P)}(p_j, q_j)}{|p_j - q_j|} \leq c[A(P)] \qquad \forall j \geq j_0\,,$$

in contradiction with (2.22). □

Next we prove the following technical lemma.

Lemma 2.48. *Let* $n \in \mathbb{N} \setminus \{0, 1\}$, $r \in]0, +\infty[$. *Then* $\mathbb{R}^n \setminus \overline{\mathbb{B}_n(0, r)}$ *is regular in the sense of Whitney.*

Proof. Since $\mathbb{R}^n \setminus \overline{\mathbb{B}_n(0, r)}$ is pathwise connected, it suffices to show that there exists $c \in [1, +\infty[$ such that, for all given $x, y \in \mathbb{R}^n \setminus \overline{\mathbb{B}_n(0, r)}$ with $x \neq y$, we can find a path $\gamma_{x,y} \in \text{Lip}([0, 1], \mathbb{R}^n)$ as in (2.11) with $\Omega = \mathbb{R}^n \setminus \overline{\mathbb{B}_n(0, r)}$ such that

$$\text{length}\,(\gamma_{x,y}) \leq c|x - y|\,.$$

We claim that we can take $c = \pi/\sqrt{2}$ (although we are not saying that this is the best constant). Indeed, let Π be the two-dimensional plane that contains x, y, and the origin of $\mathbb{R}^n$. Since Π is linearly isometric to $\mathbb{C}$, which we identify with $\mathbb{R}^2$, and orthogonal transformations in $\mathbb{C}$ are isometries, it suffices to show that if $\xi = \rho_1$ and $\eta = \rho_2 e^{i\theta}$ with ρ_1, ρ_2, and θ real numbers such that $r < \rho_2 \leq \rho_1$ and $\theta \in [0, \pi]$, then we can find a path $\tilde{\gamma}_{\xi,\eta} \in \text{Lip}([0, 1], \mathbb{C})$ as in (2.11) with $\Omega = \mathbb{C} \setminus \overline{\mathbb{B}_2(0, r)}$ such that

$$\text{length}\,(\tilde{\gamma}_{\xi,\eta}) \leq (\pi/\sqrt{2})|\xi - \eta|\,.$$

To do so, we join ξ and η with the path $\tilde{\gamma}_{\xi,\eta}$ that is the composite of

$$t \in [0, 1] \mapsto \rho_1 + t(\rho_2 - \rho_1) \in \mathbb{C}$$

and

$$t \in [0, 1] \mapsto \rho_2 e^{it\theta} \in \mathbb{C}\,.$$

We can verify that $\tilde{\gamma}_{\xi,\eta}$ is a Lipschitz path, that $\tilde{\gamma}_{\xi,\eta}$ is contained in $\mathbb{C} \setminus \overline{\mathbb{B}_n(0, r)}$, and that we have

$$\text{length}(\tilde{\gamma}_{\xi,\eta}) = (\rho_1 - \rho_2) + \rho_2\theta\,.$$

To compare $\text{length}(\tilde{\gamma}_{\xi,\eta})$ with $|\xi - \eta|$, we note that

$$|\xi - \eta|^2 = \rho_1^2 + \rho_2^2 - 2\rho_1\rho_2 \cos\theta = (\rho_1 - \rho_2)^2 + 2\rho_1\rho_2(1 - \cos\theta)\,,$$

that

$$\frac{1 - \cos\theta}{2} = \sin^2(\theta/2) \geq (\theta/\pi)^2 \qquad \forall \theta \in [0, \pi]\,,$$

and we compute

$$
\begin{aligned}
(\mathrm{length}(\tilde{\gamma}_{\xi,\eta}))^2 &\le 2(\rho_1-\rho_2)^2+2(\rho_2\theta)^2 \\
&\le 2(\rho_1-\rho_2)^2+2\rho_1\rho_2\pi^2\left(\frac{1-\cos\theta}{2}\right)\le\frac{\pi^2}{2}|\xi-\eta|^2 .
\end{aligned}
$$

Accordingly,

$$
\mathrm{length}(\tilde{\gamma}_{\xi,\eta})\le\frac{\pi}{\sqrt{2}}|\xi-\eta|
$$

and the proof is complete. □

Then we can prove the following.

Proposition 2.49. *Let $n\in\mathbb{N}\setminus\{0,1\}$. If Ω is a bounded open Lipschitz subset of $\mathbb{R}^n$, then all the connected components of Ω and of Ω^- are Lipschitz sets and are regular in the sense of Whitney.*

Proof. Since every bounded open Lipschitz subset of $\mathbb{R}^n$ has at most a finite number of connected components, each of which is a Lipschitz subset of $\mathbb{R}^n$, we conclude that each one of such components is regular in the sense of Whitney.

Since the exterior Ω^- of Ω has also at most a finite number of connected components, each of which is a Lipschitz subset of $\mathbb{R}^n$, all the bounded connected components of the exterior Ω^- are regular in the sense of Whitney, and thus it suffices to show that the unbounded connected component $(\Omega^-)_0$ of Ω^- is regular in the sense of Whitney.

Now let $r\in]1,+\infty[$ be such that $\overline{\Omega}\subseteq\mathbb{B}_n(0,r-1)$. Then both

$$
(\Omega^-)_0\cap\mathbb{B}_n(0,r)\qquad\text{and}\qquad(\Omega^-)_0\cap\mathbb{B}_n(0,r+1)
$$

are bounded open connected Lipschitz subsets of $\mathbb{R}^n$ and are accordingly regular in the sense of Whitney. We also note that the sets

$$
\begin{aligned}
(\Omega^-)_0\cap(\mathbb{R}^n\setminus\overline{\mathbb{B}_n(0,r-1)}) &= \mathbb{R}^n\setminus\overline{\mathbb{B}_n(0,r-1)}\,, \\
(\Omega^-)_0\cap(\mathbb{R}^n\setminus\overline{\mathbb{B}_n(0,r)}) &= \mathbb{R}^n\setminus\overline{\mathbb{B}_n(0,r)}
\end{aligned}
$$

are regular in the sense of Whitney by Lemma 2.48.

Now let $x,y\in(\Omega^-)_0$. If both x and y belong to $(\Omega^-)_0\cap\mathbb{B}_n(0,r)$, then we have $\lambda_{(\Omega^-)_0}(x,y)\le\lambda_{(\Omega^-)_0\cap\mathbb{B}_n(0,r)}(x,y)\le c[(\Omega^-)_0\cap\mathbb{B}_n(0,r)]|x-y|$.

If both x and y belong to $(\Omega^-)_0\cap(\mathbb{R}^n\setminus\mathbb{B}_n(0,r))$, then both x and y belong to $(\Omega^-)_0\cap(\mathbb{R}^n\setminus\overline{\mathbb{B}_n(0,r-1)})$ and thus we have

$$
\begin{aligned}
\lambda_{(\Omega^-)_0}(x,y) &\le \lambda_{(\Omega^-)_0\cap(\mathbb{R}^n\setminus\overline{\mathbb{B}_n(0,r-1)})}(x,y) \\
&\le c[(\Omega^-)_0\cap(\mathbb{R}^n\setminus\overline{\mathbb{B}_n(0,r-1)})]|x-y|\,.
\end{aligned}
$$

If instead $x\in(\Omega^-)_0\cap\mathbb{B}_n(0,r)$ and $y\in(\Omega^-)_0\cap(\mathbb{R}^n\setminus\mathbb{B}_n(0,r))$, then there exists a point ξ in the segment $[x,y]$ such that $\xi\in\partial\mathbb{B}_n(0,r)$. On the other hand $\partial\mathbb{B}_n(0,r)\subseteq(\Omega^-)_0$ and thus $\partial\mathbb{B}_n(0,r)\subseteq(\Omega^-)_0\cap\mathbb{B}_n(0,r+1)$. Hence,

$$
\lambda_{(\Omega^-)_0}(x,\xi)\le\lambda_{(\Omega^-)_0\cap\mathbb{B}_n(0,r+1)}(x,\xi)\le c[(\Omega^-)_0\cap\mathbb{B}_n(0,r+1)]|x-\xi|\,.
$$

Since $\partial\mathbb{B}_n(0,r) \subseteq (\Omega^-)_0 \cap (\mathbb{R}^n \setminus \overline{\mathbb{B}_n(0,r-1)})$, we also have

$$\lambda_{(\Omega^-)_0}(\xi,y) \\ \leq \lambda_{(\Omega^-)_0 \cap (\mathbb{R}^n \setminus \overline{\mathbb{B}_n(0,r-1)})}(\xi,y) \leq c[(\Omega^-)_0 \cap (\mathbb{R}^n \setminus \overline{\mathbb{B}_n(0,r-1)})]|\xi - y| .$$

Since $\xi \in [x,y]$, we have $|x-y| = |x-\xi| + |\xi - y|$ and thus

$$\begin{aligned}\lambda_{(\Omega^-)_0}(x,y) &\leq \lambda_{(\Omega^-)_0}(x,\xi) + \lambda_{(\Omega^-)_0}(\xi,y) \\ &\leq \max\left\{c[(\Omega^-)_0 \cap \mathbb{B}_n(0,r+1)], c[(\Omega^-)_0 \cap (\mathbb{R}^n \setminus \overline{\mathbb{B}_n(0,r-1)})]\right\} \\ &\quad \times (|x-\xi| + |\xi - y|) \\ &= \max\left\{c[(\Omega^-)_0 \cap \mathbb{B}_n(0,r+1)], c[(\Omega^-)_0 \cap (\mathbb{R}^n \setminus \overline{\mathbb{B}_n(0,r-1)})]\right\} |x-y| .\end{aligned}$$

Then the above inequalities imply that $(\Omega^-)_0$ is regular in the sense of Whitney. □

We note that if γ is the Lipschitz continuous map associated to a coordinate cylinder $C(p,R,r,\delta)$ around a boundary point p of a Lipschitz subset Ω of $\mathbb{R}^n$, then the Rademacher Theorem A.12 implies that γ is differentiable at almost every point of its domain $\mathbb{B}_{n-1}(0,r)$ and the area element corresponding to the parametrization

$$\psi_p(\eta) \equiv p + R^t \begin{pmatrix} \eta \\ \gamma(\eta) \end{pmatrix} \qquad \forall \eta \in \mathbb{B}_{n-1}(0,r)$$

is given by

$$\sqrt{\det\left((D\psi_p(\eta))^t (D\psi_p(\eta))\right)} = \sqrt{1 + |\nabla\gamma(\eta)|^2} \qquad \text{for a.e. } \eta \in \mathbb{B}_{n-1}(0,r) .$$

Then we can define the $(n-1)$-dimensional measure on the image of the map ψ_p by saying that a subset E of the image of the map ψ_p is measurable if and only if $\psi_p^{\leftarrow}(E)$ is measurable in $\mathbb{B}_{n-1}(0,r)$ and has measure

$$m_{\psi_p(\mathbb{B}_{n-1}(0,r))}(E) \equiv \int_{\psi_p^{\leftarrow}(E)} \sqrt{1 + |\nabla\gamma(\eta)|^2}\, d\eta .$$

Then, by exploiting a partition of unity, we can define the $(n-1)$-dimensional measure on $\partial\Omega$ (cf., e.g., Naumann and Simader [232]). We denote by $\mathcal{L}_{\partial\Omega}$ and by $m_{\partial\Omega}$ (or simply m_{n-1}) the σ-algebra of measurable subsets of $\partial\Omega$ and the $(n-1)$-dimensional measure on $\partial\Omega$, respectively.

Theorem 2.50 (of Rademacher for Lipschitz Sets). *Let $n \in \mathbb{N} \setminus \{0,1\}$. Let Ω be an open Lipschitz subset of $\mathbb{R}^n$. Then there exists a subset N of $\partial\Omega$ of $m_{\partial\Omega}$-measure zero such that if $p \in (\partial\Omega) \setminus N$, then the cone $T_p(\partial\Omega)$ of semi-tangent vectors is a vector space of dimension $(n-1)$, i.e., $\partial\Omega$ has a tangent space of dimension $(n-1)$ at p.*

Proof. Since Ω is a Lipschitz set, for every point $x \in \partial\Omega$ there exist r_x, δ_x in $]0,+\infty[$, $R_x \in \mathbb{O}_n(\mathbb{R})$ and a coordinate cylinder $C(x,R_x,r_x,\delta_x)$ for Ω around x.

Then

$$\partial\Omega \subseteq \bigcup_{x\in\partial\Omega} C(x, R_x, r_x, \delta_x)\,.$$

Now each cylinder $C(x, R_x, r_x, \delta_x)$ is a neighborhood of x and accordingly it contains an open cube Q_x which contains x and which equals the cartesian product of intervals with rational endpoints. Since the set of rational numbers is countable, the set of cubes $\{Q_x\}_{x\in\partial\Omega}$ is either finite or countable and there exists a sequence $\{x_j\}_{j\in\mathbb{N}}$ in $\partial\Omega$ such that

$$\{Q_x\}_{x\in\partial\Omega} = \{Q_{x_j}\}_{j\in\mathbb{N}}\,.$$

Since $\{Q_x\}_{x\in\partial\Omega}$ covers $\partial\Omega$, we have

$$\partial\Omega \subseteq \bigcup_{j\in\mathbb{N}} Q_{x_j} \subseteq \bigcup_{j\in\mathbb{N}} C(x_j, R_{x_j}, r_{x_j}, \delta_{x_j})\,.$$

Let the Lipschitz continuous function γ_{x_j} from $\mathbb{B}_{n-1}(0, r_j)$ to $]-\delta_{x_j}, \delta_{x_j}[$ represent $\partial\Omega$ in $C(x_j, R_{x_j}, r_{x_j}, \delta_{x_j})$ as a graph, for each $j\in\mathbb{N}$. Since γ_{x_j} is Lipschitz continuous, the Rademacher Theorem implies that there exists a subset N_j of measure zero of $\mathbb{B}_{n-1}(0, r_j)$ such that γ_{x_j} is differentiable at all points of $\mathbb{B}_{n-1}(0, r_j)\setminus N_j$. Now let ψ_{x_j} be the parametrization corresponding to γ_{x_j}. By definition of $(n-1)$-dimensional measure on $\partial\Omega$, the set $\psi_{x_j}(N_j)$ has $m_{\partial\Omega}$-measure equal to zero. Moreover, Proposition 2.42 (i) implies that the cone of semi-tangent vectors to $\partial\Omega$ at the points of $\psi_{x_j}(\mathbb{B}_{n-1}(0, r_j)\setminus N_j)$ is a vector space of dimension $(n-1)$. Now let

$$N \equiv \bigcup_{j\in\mathbb{N}} \psi_{x_j}(N_j)\,.$$

Since N is a finite or countable union of sets of $m_{\partial\Omega}$-measure zero, then $m_{\partial\Omega}(N)$ equals 0. Moreover, if $p\in\partial\Omega\setminus N$, then the cone of semi-tangent vectors to $\partial\Omega$ at p is a vector space of dimension $(n-1)$. Thus the proof is complete. □

Proposition 2.51. *Let $n\in\mathbb{N}\setminus\{0,1\}$. Let Ω be an open Lipschitz subset of $\mathbb{R}^n$. If $p\in\partial\Omega$ and if $\partial\Omega$ has a tangent space $T_p(\partial\Omega)$ at p of dimension $(n-1)$, then there exists one and only one unit vector $\nu_\Omega(p)$ which is normal to $T_p(\partial\Omega)$ and such that $\nu_\Omega(p)$ points to the exterior of Ω and $-\nu_\Omega(p)$ points to the interior of Ω.*

Proof. Since Ω is a Lipschitz subset of $\mathbb{R}^n$, then there exist r, $\delta\in]0,+\infty[$ such that $C(p, R, r, \delta)$ is a coordinate cylinder for Ω around p and the function γ from $\mathbb{B}_{n-1}(0, r)$ to $]-\delta, \delta[$ which represents $\partial\Omega$ in $C(p, R, r, \delta)$ is Lipschitz continuous. Since $T_p(\partial\Omega)$ is a vector space of dimension $(n-1)$ and γ is Lipschitz continuous, Proposition 2.44 implies that γ is differentiable at 0, that

$$\nu_\Omega(p) \equiv R^t \begin{pmatrix} -\nabla\gamma(0) \\ 1 \end{pmatrix} \frac{1}{\sqrt{1+|\nabla\gamma(0)|^2}}$$

is normal to $T_p(\partial\Omega)$ and points to the exterior of Ω, and that $-\nu_\Omega(p)$ points to the interior of Ω. Since the unit vectors which are normal to a vector subspace of dimension $(n-1)$ are precisely 2 and $\Omega \cap \Omega^- = \emptyset$. We conclude that $\nu_\Omega(p)$ is unique and the proof is complete. □

We say that the unit vector $\nu_\Omega(p)$ of the previous statement is the outward unit normal to $\partial\Omega$ at p. By the Rademacher Theorem 2.50, $\nu_\Omega(p)$ exists at almost all $p \in \partial\Omega$. If no ambiguity can arise, then we simply write $\nu(p)$ instead of $\nu_\Omega(p)$.

2.10 Elementary Inequalities on the Boundary of a Lipschitz Subset of $\mathbb{R}^n$

We first introduce the following elementary lemma, which collects either known inequalities or variants of known inequalities (cf., e.g., Cialdea [53, §8], and the paper [89, Lemma 3.2] with Dondi). We exploit such inequalities in the next chapters of this book.

Lemma 2.52. *Let $n \in \mathbb{N}$, $\lambda \in \mathbb{R}$. Then the following statements hold.*

(i)

$$\frac{1}{2}|x'-y| \le |x''-y| \le 2|x'-y|$$

for all $x', x'' \in \mathbb{R}^n$, $x' \neq x''$, $y \in \mathbb{R}^n \setminus \mathbb{B}_n(x', 2|x'-x''|)$.

(ii)

$$|x'-y|^\lambda \le 2^{|\lambda|}|x''-y|^\lambda, \quad |x''-y|^\lambda \le 2^{|\lambda|}|x'-y|^\lambda$$

for all $x', x'' \in \mathbb{R}^n$, $x' \neq x''$, $y \in \mathbb{R}^n \setminus \mathbb{B}_n(x', 2|x'-x''|)$.

(iii) There exists $q_\lambda \in]0, +\infty[$ such that

$$\left||x'-y|^\lambda - |x''-y|^\lambda\right| \le q_\lambda |x'-x''|\, |x'-y|^{\lambda-1}$$

for all $x', x'' \in \mathbb{R}^n$, $x' \neq x''$, $y \in \mathbb{R}^n \setminus \mathbb{B}_n(x', 2|x'-x''|)$.

Proof. Statement (i) is an immediate consequence of the triangular inequality. Indeed,

$$|x''-y| \ge |x'-y| - |x'-x''| \ge |x'-y| - \frac{1}{2}|x'-y| = \frac{1}{2}|x'-y|,$$
$$|x''-y| \le |x'-x''| + |x'-y| \le \frac{1}{2}|x'-y| + |x'-y| \le 2|x'-y|.$$

The first inequality in (ii) for $\lambda \ge 0$ follows by raising inequality $|x'-y| \le 2|x''-y|$ of statement (i) to the power λ. Instead, for $\lambda < 0$ the same inequality follows by raising inequality $|x''-y| \le 2|x'-y|$ of statement (i) to the power λ. Since statement (i) implies that

$$|x'-y| \le 2|x''-y|, \qquad |x''-y| \le 2|x'-y|,$$

the second inequality of (ii) can be proved by interchanging the roles of x' and x''.

To prove (iii), we follow Cialdea [53, §8]. We first assume that $|x'-y| < |x''-y|$. By the Lagrange's Mean Value Theorem, there exists $\zeta \in [|x'-y|, |x''-y|]$ such that

$$||x'-y|^\lambda - |x''-y|^\lambda| \leq |\lambda|\zeta^{\lambda-1}\,||x'-y| - |x''-y||\,.$$

If $\lambda \geq 1$, inequality $\zeta \leq |x''-y|$ and (i) imply that

$$\zeta^{\lambda-1} \leq |x''-y|^{\lambda-1} \leq 2^{|\lambda-1|}|x'-y|^{\lambda-1}\,.$$

If $\lambda < 1$, inequality $\zeta \geq |x'-y|$ implies that

$$\zeta^{\lambda-1} \leq |x'-y|^{\lambda-1} \leq 2^{|\lambda-1|}|x'-y|^{\lambda-1}\,.$$

Then we have

$$\begin{aligned} &||x'-y|^\lambda - |x''-y|^\lambda| \\ &\qquad \leq |\lambda|2^{|\lambda-1|}\,||x'-y| - |x''-y||\,|x'-y|^{\lambda-1}\,, \end{aligned} \tag{2.26}$$

which implies the validity of (iii). Similarly, in case $|x'-y| > |x''-y|$, we can prove that (2.26) holds with x' and x'' interchanged. Then (ii) implies the validity of (iii). □

Then we have the following classical lemma.

Lemma 2.53. *Let $n \in \mathbb{N} \setminus \{0,1\}$. Let Ω be a bounded open Lipschitz subset of $\mathbb{R}^n$. Then the following statements hold.*

(i) If $\lambda \in]-\infty, n-1[$, then

$$c'_{\Omega,\lambda} \equiv \sup_{x\in\partial\Omega} \int_{\partial\Omega} \frac{d\sigma_y}{|x-y|^\lambda}$$

is finite.

(ii) If $\lambda \in]-\infty, n-1[$, then

$$c''_{\Omega,\lambda} \equiv \sup_{x'\in\partial\Omega, s\in]0,+\infty[} s^{\lambda-(n-1)} \int_{\mathbb{B}_n(x',s)\cap\partial\Omega} \frac{d\sigma_y}{|x'-y|^\lambda}$$

is finite.

(iii) If $\lambda \in]n-1, +\infty[$, then

$$c'''_{\Omega,\lambda} \equiv \sup_{x'\in\partial\Omega, s\in]0,+\infty[} s^{\lambda-(n-1)} \int_{\partial\Omega\setminus\mathbb{B}_n(x',s)} \frac{d\sigma_y}{|x'-y|^\lambda}$$

is finite.

Proof. For each $x \in \partial\Omega$, there exists a coordinate cylinder $C(x, R_x, r_x, \delta_x)$ around x, and we can assume that $r_x + \delta_x < \mathrm{diam}(\partial\Omega)$ so that $(\partial\Omega)\setminus C(x, R_x, r_x, \delta_x) \neq \emptyset$.

Since $\{C(x, R_x, r_x/2, \delta_x) \cap \partial\Omega\}_{x\in\partial\Omega}$ is an open cover of $\partial\Omega$ and $\partial\Omega$ is compact, there exist finite families $\{x^{(j)}\}_{j=1}^q$, $\{R_j\}_{j=1}^q$, $\{r_j\}_{j=1}^q$, $\{\delta_j\}_{j=1}^q$ such that

$$\partial\Omega \subseteq \bigcup_{j=1}^{q} (C(x^{(j)}, R_j, r_j/2, \delta_j) \cap \partial\Omega) .$$

We denote by $\{\gamma_j\}_{j=1}^q$ and $\{\psi_{x^{(j)}}\}_{j=1}^q$, the corresponding families of functions which represent $\partial\Omega$ in the cylinders $\{C(x^{(j)}, R_j, r_j, \delta_j)\}_{j=1}^q$ as a graph and which parametrize $\partial\Omega$ around the points $\{x^{(j)}\}_{j=1}^q$, respectively.

Let $\omega \in]0, +\infty[$ be a Lebesgue number for the above finite cover of $\partial\Omega$ (cf. Theorem A.6). There is no loss of generality in assuming that

$$\omega < \frac{1}{2} \min_{j=1,\dots,q} r_j . \tag{2.27}$$

We now prove statement (i). It suffices to show that if $j \in \{1, \dots, q\}$, then

$$\sup_{x\in C(x^{(j)},R_j,r_j/2,\delta_j)\cap\partial\Omega} \int_{\partial\Omega} \frac{d\sigma_y}{|x-y|^\lambda} < +\infty .$$

Since

$$\left(\overline{C(x^{(j)}, R_j, r_j/2, \delta_j) \cap \partial\Omega}\right) \cap \left((\partial\Omega) \setminus \left(C(x^{(j)}, R_j, r_j, \delta_j) \cap \partial\Omega\right)\right) = \emptyset ,$$

the (finite) distance b_j between the set $\overline{C(x^{(j)}, R_j, r_j/2, \delta_j) \cap \partial\Omega}$ and the nonempty set $(\partial\Omega) \setminus \left(C(x^{(j)}, R_j, r_j, \delta_j) \cap \partial\Omega\right)$ is not zero and thus

$$\begin{aligned}
&\sup_{x\in C(x^{(j)},R_j,r_j/2,\delta_j)\cap\partial\Omega} \int_{\partial\Omega} \frac{d\sigma_y}{|x-y|^\lambda} \\
&\quad \le B_j m_{n-1}(\partial\Omega) + \sup_{x\in C(x^{(j)},R_j,r_j/2,\delta_j)\cap\partial\Omega} \int_{C(x^{(j)},R_j,r_j,\delta_j)\cap\partial\Omega} \frac{d\sigma_y}{|x-y|^\lambda} \\
&\quad \le B_j m_{n-1}(\partial\Omega) + \sup_{\xi\in\mathbb{B}_{n-1}(0,r_j/2)} \int_{\mathbb{B}_{n-1}(0,r_j)} \frac{\sqrt{1+\|\,|\nabla\gamma_j|\,\|_\infty^2}}{|\xi-\eta|^\lambda}\, d\sigma_\eta \\
&\quad = B_j m_{n-1}(\partial\Omega) + \sup_{\xi\in\mathbb{B}_{n-1}(0,r_j/2)} \int_{\mathbb{B}_{n-1}(\xi,r_j)} \frac{\sqrt{1+\|\,|\nabla\gamma_j|\,\|_\infty^2}}{|\zeta|^\lambda}\, d\sigma_\zeta \\
&\quad \le B_j m_{n-1}(\partial\Omega) + \int_{\mathbb{B}_{n-1}(0,3r_j/2)} \frac{\sqrt{1+\|\,|\nabla\gamma_j|\,\|_\infty^2}}{|\zeta|^\lambda}\, d\sigma_\zeta < +\infty ,
\end{aligned}$$

where $B_j \equiv \max\{b_j^{-\lambda}, (\mathrm{diam}(\partial\Omega))^{-\lambda}\}$. Indeed,

$$\mathbb{B}_{n-1}(\xi, r_j) \subseteq \mathbb{B}_{n-1}(0, 3r_j/2) \qquad \forall \xi \in \mathbb{B}_{n-1}(0, r_j/2)$$

and

$$\frac{1}{|x-y|^\lambda} \leq \begin{cases} b_j^{-\lambda} & \text{if } \lambda \geq 0\,, \\ (\mathrm{diam}(\partial\Omega))^{-\lambda} & \text{if } \lambda < 0\,, \end{cases}$$

for all $x \in (\partial\Omega)\setminus C(x^{(j)}, R_j, r_j, \delta_j)$, $y \in C(x^{(j)}, R_j, r_j/2, \delta_j)\cap\partial\Omega$. We now turn to the proof of statements (ii), (iii). By statement (i), we have

$$\sup_{x'\in\partial\Omega, s\in[\omega/6,+\infty[} \left\{ s^{\lambda-(n-1)} \int_{\mathbb{B}_n(x',s)\cap\partial\Omega} \frac{d\sigma_y}{|x'-y|^\lambda} \right\} \leq (\omega/6)^{\lambda-(n-1)} c'_{\Omega,\lambda}$$

in case of statement (ii). Since $(\partial\Omega)\setminus\mathbb{B}_n(x',s) = \emptyset$ for $s \geq \mathrm{diam}(\Omega)$, we also have

$$\sup_{x'\in\partial\Omega, s\in[\omega/6,+\infty[} \left\{ s^{\lambda-(n-1)} \int_{(\partial\Omega)\setminus\mathbb{B}_n(x',s)} \frac{d\sigma_y}{|x'-y|^\lambda} \right\}$$
$$\leq \mathrm{diam}(\Omega)^{\lambda-(n-1)} \frac{m_{n-1}(\partial\Omega)}{(\omega/6)^\lambda}$$

in case of statement (iii). Then statements (ii) and (iii) follow by the validity of the following two statements.

(j) If $\lambda \in]-\infty, n-1[$, then the supremum

$$\sup_{x'\in\partial\Omega, s\in]0,\omega/6[} \left\{ s^{\lambda-(n-1)} \int_{\mathbb{B}_n(x',s)\cap\partial\Omega} \frac{d\sigma_y}{|x'-y|^\lambda} \right\}$$

is finite.

(jj) If $\lambda \in]n-1, +\infty[$, then the supremum

$$\sup_{x'\in\partial\Omega, s\in]0,\omega/6[} \left\{ s^{\lambda-(n-1)} \int_{(\partial\Omega)\setminus\mathbb{B}_n(x',s)} \frac{d\sigma_y}{|x'-y|^\lambda} \right\}$$

is finite.

We note that if $x' \in \partial\Omega$, $s < \omega/6$, then there exists $j \in \{1,\dots,q\}$ such that

$$\mathbb{B}_n(x',s)\cap\partial\Omega \subseteq \psi_{x^{(j)}}(\mathbb{B}_n(0, r_j/2)) \equiv A_j\,.$$

Indeed, the diameter of $\mathbb{B}_n(x',s)\cap\partial\Omega$ is less than the Lebesgue number. Hence, statements (j)–(jj) follow by the validity of the following two statements.

(k) If $\lambda \in]-\infty, n-1[$ and $j \in \{1,\dots,q\}$, then the supremum

$$\sup\left\{ s^{\lambda-(n-1)} \int_{\mathbb{B}_n(x',s)\cap\partial\Omega} \frac{d\sigma_y}{|x'-y|^\lambda} : \right.$$
$$\left. x' \in \partial\Omega,\ 0 < s < \omega/6, \mathbb{B}_n(x',s)\cap\partial\Omega \subseteq A_j \right\}$$

is finite.

(kk) If $\lambda \in]n-1,+\infty[$ and $j \in \{1,\dots,q\}$, then the supremum

$$\sup\Big\{ s^{\lambda-(n-1)} \int_{\partial\Omega\setminus\mathbb{B}_n(x',s)} \frac{d\sigma_y}{|x'-y|^\lambda} : \\ x' \in \partial\Omega,\ 0<s<\omega/6, \mathbb{B}_n(x',s)\cap\partial\Omega \subseteq A_j \Big\}$$

is finite.

We start by proving statement (k). Since $\mathbb{B}_n(x',s)\cap\partial\Omega \subseteq A_j$, then there exists $\eta' \in \mathbb{B}_{n-1}(0,r_j/2)$ such that $x' = \psi_{x^{(j)}}(\eta')$. Then we note that

$$\mathbb{B}_{n-1}(\eta',s) \subseteq \mathbb{B}_{n-1}(0,r_j)\,.$$

Indeed, if $\eta \in \mathbb{R}^{n-1}$, $|\eta-\eta'| \le s$, then we have

$$|\eta| \le |\eta-\eta'| + |\eta'| \le s + \frac{1}{2}r_j \le \frac{\omega}{6} + \frac{1}{2}r_j \le \frac{1}{12}r_j + \frac{1}{2}r_j < r_j\,.$$

In particular, it makes sense to consider $\psi_{x^{(j)}}(\eta)$ when $\eta \in \mathbb{B}_{n-1}(\eta',s)$. By the definition of the function $\psi_{x^{(j)}}$, we have

$$|x'-\psi_{x^{(j)}}(\eta)| = |\psi_{x^{(j)}}(\eta') - \psi_{x^{(j)}}(\eta)| \ge |\eta-\eta'| \qquad \forall \eta \in \mathbb{B}_{n-1}(0,r_j)\,.$$

We also note that

$$\mathbb{B}_n(x',s)\cap\partial\Omega \\ = \{\psi_{x^{(j)}}(\eta) : \eta \in \mathbb{B}_{n-1}(0,r_j),\ |\psi_{x^{(j)}}(\eta') - \psi_{x^{(j)}}(\eta)| < s\}\,.$$

Thus if $\lambda \in]0,(n-1)[$, we have

$$\begin{aligned}
&\int_{\mathbb{B}_n(x',s)\cap\partial\Omega} |x'-y|^{-\lambda}\,d\sigma_y \\
&\quad\le \int_{\mathbb{B}_{n-1}(\eta',s)} |\psi_{x^{(j)}}(\eta') - \psi_{x^{(j)}}(\eta)|^{-\lambda} \sqrt{1+\|\,|\nabla\gamma_j|\,\|_\infty^2}\,d\eta \\
&\quad\le \int_{\mathbb{B}_{n-1}(\eta',s)} |\eta-\eta'|^{-\lambda}\,d\eta \sqrt{1+\|\,|\nabla\gamma_j|\,\|_\infty^2} \\
&\quad= \int_{\mathbb{B}_{n-1}(0,s)} |\eta|^{-\lambda}\,d\eta \sqrt{1+\|\,|\nabla\gamma_j|\,\|_\infty^2} \\
&\quad= \int_0^s \rho^{-\lambda} \int_{\partial\mathbb{B}_{n-1}(0,\rho)} d\sigma_\eta\,d\rho \sqrt{1+\|\,|\nabla\gamma_j|\,\|_\infty^2} \\
&\quad= \int_0^s \rho^{-\lambda}\rho^{n-2} s_{n-1}\,d\rho \sqrt{1+\|\,|\nabla\gamma_j|\,\|_\infty^2}
\end{aligned}$$

$$= \frac{s_{n-1}}{n-1-\lambda} s^{n-1-\lambda} \sqrt{1 + \| \, |\nabla \gamma_j| \, \|_\infty^2}$$

and if $\lambda \in]-\infty, 0]$, we have

$$\int_{\mathbb{B}_n(x',s)\cap\partial\Omega} |x'-y|^{-\lambda} \, d\sigma_y$$
$$\leq \int_{\mathbb{B}_{n-1}(\eta',s)} s^{-\lambda} \sqrt{1 + \| \, |\nabla \gamma_j| \, \|_\infty^2} \, d\eta = s^{-\lambda} \omega_{n-1} s^{n-1} \sqrt{1 + \| \, |\nabla \gamma_j| \, \|_\infty^2} \, .$$

Then the last two inequalities imply the validity of statement (k).

We now turn to the proof of statement (kk). Since $\mathbb{B}_n(x',s)\cap\partial\Omega \subseteq A_j$, we have

$$(\partial\Omega) \setminus \mathbb{B}_n(x',s) = ((\partial\Omega) \setminus \psi_{x^{(j)}}(\mathbb{B}_{n-1}(0,r_j))) \qquad (2.28)$$
$$\cup \{\psi_{x^{(j)}}(\eta) : \eta \in \mathbb{B}_{n-1}(0,r_j), |\psi_{x^{(j)}}(\eta') - \psi_{x^{(j)}}(\eta)| \geq s\} \, .$$

Since $\psi_{x^{(j)}}$ is a parametrization of $\partial\Omega$ defined on the open set $\mathbb{B}_{n-1}(0,r_j)$, $\psi_{x^{(j)}}(\mathbb{B}_{n-1}(0,r_j)) \subsetneq \partial\Omega$, and $\partial\Omega$ is compact, then $(\partial\Omega) \setminus \psi_{x^{(j)}}(\mathbb{B}_{n-1}(0,r_j))$ is not empty and closed in $\partial\Omega$ and has a positive distance from $\psi_{x^{(j)}}(\mathbb{B}_{n-1}(0,r_j/2))$. Hence, there exists

$$\omega' \in \left]0, \inf_{j\in\{1,\dots,q\}} \left\{ \inf_{y\in(\partial\Omega)\setminus\psi_{x^{(j)}}(\mathbb{B}_{n-1}(0,r_j)),y'\in A_j} |y-y'| \right\}\right[\, . \qquad (2.29)$$

Since $\lambda > (n-1) \geq 0$ and $x' \in \mathbb{B}_n(x',s)\cap\partial\Omega \subseteq A_j$, we have

$$\int_{(\partial\Omega)\setminus\psi_{x^{(j)}}(\mathbb{B}_{n-1}(0,r_j))} |x'-y|^{-\lambda} \, d\sigma_y \qquad (2.30)$$
$$\leq \int_{(\partial\Omega)\setminus\psi_{x^{(j)}}(\mathbb{B}_{n-1}(0,r_j))} d\sigma (\omega')^{-\lambda} \leq m_{n-1}(\partial\Omega)(\omega')^{-\lambda} \, ,$$

an inequality which enables to estimate the integral of $\frac{1}{|x'-y|^\lambda}$ for y in the first set in the right-hand side of equality (2.28). We now estimate the integral of $\frac{1}{|x'-y|^\lambda}$ for y in the second set in the right-hand side of equality (2.28). If $y = \psi_{x^{(j)}}(\eta)$ with $\eta \in \mathbb{B}_{n-1}(0,r_j)$ and $|\psi_{x^{(j)}}(\eta') - \psi_{x^{(j)}}(\eta)| \geq s$, then the elementary inequality

$$|\psi_{x^{(j)}}(\eta') - \psi_{x^{(j)}}(\eta)| = \sqrt{|\eta-\eta'|^2 + |\gamma_j(\eta') - \gamma_j(\eta)|^2}$$
$$\leq \sqrt{1 + \| \, |\nabla \gamma_j| \, \|_\infty^2} |\eta - \eta'|$$

implies that

$$|\eta - \eta'| \geq \frac{s}{\sqrt{1 + \| \, |\nabla \gamma_j| \, \|_\infty^2}} \, .$$

Since $s \leq \frac{\omega}{6} \leq \frac{1}{12} \min_{j=1,\dots,s} r_j$, we have

$$2r_j \geq 2s \geq \frac{s}{\sqrt{1+\| |\nabla\gamma_j| \|_\infty^2}} .$$

Moreover, if $\xi \in \mathbb{B}_{n-1}(0, r_j)$, then we have

$$|\eta' - \xi| \leq |\eta'| + |\xi| \leq \frac{1}{2} r_j + r_j < 2r_j .$$

Hence,

$$\begin{aligned}
&\int_{\psi_{x^{(j)}}(\mathbb{B}_{n-1}(0,r_j))\setminus\{\psi_{x^{(j)}}(\eta):\eta\in\mathbb{B}_{n-1}(0,r_j),|\psi_{x^{(j)}}(\eta')-\psi_{x^{(j)}}(\eta)|\geq s\}} |x'-y|^{-\lambda}\, d\sigma_y \\
&\leq \sqrt{1+\| |\nabla\gamma_j| \|_\infty^2} \int_{\mathbb{B}_{n-1}(0,r_j)\setminus\mathbb{B}_{n-1}\left(\eta', \frac{s}{\sqrt{1+\| |\nabla\gamma_j| \|_\infty^2}}\right)} |\eta-\eta'|^{-\lambda}\, d\eta \\
&\leq \sqrt{1+\| |\nabla\gamma_j| \|_\infty^2} \int_{\mathbb{B}_{n-1}(\eta',2r_j)\setminus\mathbb{B}_{n-1}\left(\eta', \frac{s}{\sqrt{1+\| |\nabla\gamma_j| \|_\infty^2}}\right)} |\eta-\eta'|^{-\lambda}\, d\eta \\
&\leq \sqrt{1+\| |\nabla\gamma_j| \|_\infty^2} \int_{\frac{s}{\sqrt{1+\| |\nabla\gamma_j| \|_\infty^2}}}^{2r_j} \rho^{-\lambda}\rho^{n-2}\, d\rho s_{n-1} \qquad (2.31)\\
&= \frac{s_{n-1}\sqrt{1+\| |\nabla\gamma_j| \|_\infty^2}}{(n-1)-\lambda}\left[(2r_j)^{(n-1)-\lambda} - \left(\frac{s}{\sqrt{1+\| |\nabla\gamma_j| \|_\infty^2}}\right)^{(n-1)-\lambda}\right] \\
&\leq \frac{s_{n-1}\sqrt{1+\| |\nabla\gamma_j| \|_\infty^2}}{\lambda-(n-1)} \sqrt{1+\| |\nabla\gamma_j| \|_\infty^2}^{\,\lambda-(n-1)} s^{(n-1)-\lambda} .
\end{aligned}$$

Then inequalities (2.30) and (2.31) imply the validity of statement (kk). □

2.11 Schauder Spaces in Open Subsets of $\mathbb{R}^n$

Let $m \in \mathbb{N}\setminus\{0\}$, $n \in \mathbb{N}$, $\alpha \in]0,1]$. Let Ω be an open subset of $\mathbb{R}^n$. Then the Schauder space of exponents m, α in $\overline{\Omega}$ is defined as

$$C^{m,\alpha}(\overline{\Omega}) \equiv \left\{ f \in C^m(\overline{\Omega}) :\, D^\eta f \in C^{0,\alpha}(\overline{\Omega}),\ \forall \eta \in \mathbb{N}^n,\ |\eta| = m \right\} .$$

As we have already said in Section 2.3, $C^m(\overline{\Omega})$ is the subspace of $C^m(\Omega)$ of those functions f from Ω to $\mathbb{R}$ whose derivatives $D^\eta f$ of order $|\eta| \leq m$ can be extended with continuity to $\overline{\Omega}$, and if $x \in \partial\Omega$, then we set

$$D^\eta f(x) \equiv \lim_{y\to x} D^\eta f(y) .$$

Then $C^{m,\alpha}_{\mathrm{loc}}(\overline{\Omega})$ denotes the space of those functions $f \in C^m(\overline{\Omega})$ such that $f_{|\overline{\Omega\cap\mathbb{B}_n(0,\rho)}}$ belongs to $C^{m,\alpha}(\overline{\Omega\cap\mathbb{B}_n(0,\rho)})$ for all $\rho \in]0,+\infty[$ while $C^{m,\alpha}_{\mathrm{loc}}(\Omega)$ denotes the space of those functions $f \in C^m(\Omega)$ such that $f_{|\overline{\Omega'}}$ belongs to $C^{m,\alpha}(\overline{\Omega'})$ for each bounded open subset Ω' of Ω such that $\overline{\Omega'} \subseteq \Omega$.

We set

$$C^{m,\alpha}_b(\overline{\Omega}) \equiv C^m_b(\overline{\Omega}) \cap C^{m,\alpha}(\overline{\Omega})$$

and we equip $C^{m,\alpha}_b(\overline{\Omega})$ with the norm

$$\|f\|_{C^{m,\alpha}_b(\overline{\Omega})} \equiv \|f\|_{C^m_b(\overline{\Omega})} + \sum_{|\eta|=m} |D^\eta f|_\alpha \qquad \forall f \in C^{m,\alpha}_b(\overline{\Omega}).$$

By exploiting the completeness of $C^m_b(\overline{\Omega})$ and of $C^{0,\alpha}_b(\overline{\Omega})$, we can show that $C^{m,\alpha}_b(\overline{\Omega})$ is complete. If Ω is bounded, then we obviously have $C^{m,\alpha}_b(\overline{\Omega}) = C^{m,\alpha}(\overline{\Omega})$ (and in this case we may drop the subscript b).

We note that the inclusion of $C^{m,\alpha}_b(\overline{\Omega})$ into $C^0_b(\overline{\Omega})$ has to be understood in the sense that each element $f \in C^{m,\alpha}_b(\overline{\Omega})$ has a uniquely determined continuous extension to $\overline{\Omega}$, which we still denote f. Then by the definition of norm in $C^{m,\alpha}_b(\overline{\Omega})$ and in $C^0_b(\overline{\Omega})$, $C^{m,\alpha}_b(\overline{\Omega})$ is continuously imbedded into $C^0_b(\overline{\Omega})$.

In case the exponent m is replaced with 0 and $\alpha \in]0,1]$, the Schauder space of exponents 0, α in $\overline{\Omega}$ is defined as the Hölder space $C^{0,\alpha}(\overline{\Omega})$ that we had already introduced.

If $m \in \mathbb{N}\setminus\{0\}$, then the definition of norm in $C^{m,\alpha}_b(\overline{\Omega})$ implies that the following two elementary inequalities hold

$$\|\partial_{x_r} u\|_{C^{m-1,\alpha}_b(\overline{\Omega})} \leq \|u\|_{C^{m,\alpha}_b(\overline{\Omega})} \qquad \forall r \in \{1,\dots,n\}, \tag{2.32}$$

$$\|u\|_{C^{m,\alpha}_b(\overline{\Omega})} \leq \|u\|_{C^0_b(\overline{\Omega})} + \sum_{j=1}^{n} \|\partial_{x_j} u\|_{C^{m-1,\alpha}_b(\overline{\Omega})} \leq n\|u\|_{C^{m,\alpha}_b(\overline{\Omega})}$$

for all $u \in C^{m,\alpha}_b(\overline{\Omega})$. In particular, ∂_{x_r} is linear and continuous from $C^{m,\alpha}_b(\overline{\Omega})$ to $C^{m-1,\alpha}_b(\overline{\Omega})$ for all $r \in \{1,\dots,n\}$. In order to compactify our notation when we deal with Schauder spaces in the closure of an open set, we find convenient to set

$$C^{0,0} \equiv C^0, \qquad C^{m,0} \equiv C^m \qquad \text{if } m \in \mathbb{N}\setminus\{0\}. \tag{2.33}$$

Then we have the following.

Proposition 2.54. *Let $m, n \in \mathbb{N}$. Let Ω be an open subset of $\mathbb{R}^n$. Then the following statements hold.*

(i) If α, $\alpha' \in]0,1]$, $\alpha > \alpha'$, then $C^{m,\alpha}_b(\overline{\Omega})$ is continuously embedded into $C^{m,\alpha'}_b(\overline{\Omega})$.

(ii) If Ω is regular in the sense of Whitney, then $C^{m+1}_b(\overline{\Omega})$ is continuously embedded into $C^{m,1}_b(\overline{\Omega})$ (cf. Definition 2.27).

(iii) If Ω is regular in the sense of Whitney and if $\alpha_1, \alpha_2 \in]0,1]$, then the pointwise multiplication from $C_b^{m,\alpha_1}(\overline{\Omega}) \times C_b^{m,\alpha_2}(\overline{\Omega})$ to $C_b^{m,\min\{\alpha_1,\alpha_2\}}(\overline{\Omega})$ that takes a pair of functions (f,g) to the pointwise product fg is continuous.

(iv) If Ω is bounded and if Ω has a finite number of connected components, each of which is regular in the sense of Whitney, and if $\alpha, \alpha' \in]0,1]$, $\alpha > \alpha'$, then $C^{m,\alpha}(\overline{\Omega})$ is compactly embedded into $C^{m,\alpha'}(\overline{\Omega})$.

Proof. Statement (i) follows by the definition of norm in $C_b^{m,\alpha}(\overline{\Omega})$ and by the continuity of the embedding of $C_b^{0,\alpha}(\overline{\Omega})$ into $C_b^{0,\alpha'}(\overline{\Omega})$ (cf. Proposition 2.23).

Since Ω is regular in the sense of Whitney, statement (ii) follows by the inequality

$$\begin{aligned} \|u\|_{C_b^{m,1}(\overline{\Omega})} &= \|u\|_{C_b^m(\overline{\Omega})} + \sum_{|\eta|=m} |D^\eta u|_1 \\ &\leq \|u\|_{C_b^m(\overline{\Omega})} + \sum_{|\eta|=m} c[\Omega] \sup_{\overline{\Omega}} |\nabla(D^\eta u)| \\ &\leq \|u\|_{C_b^m(\overline{\Omega})} + nc[\Omega] \sum_{|\eta|=m+1} \sup_{\overline{\Omega}} |D^\eta u| \leq nc[\Omega] \|u\|_{C_b^{m+1}(\overline{\Omega})}, \end{aligned}$$

for all $u \in C_b^{m+1}(\overline{\Omega})$ (cf. Lemma 2.28).

We now consider statement (iii). By statement (i), it suffices to consider case $\alpha \equiv \alpha_1 = \alpha_2$. We proceed by induction on m. Case $m = 0$ follows by Proposition 2.24. Next we assume that the statement holds for m and we prove it for $m+1$. If u, $v \in C_b^{m+1,\alpha}(\overline{\Omega})$, then we have

$$\begin{aligned} &\|uv\|_{C_b^{m+1,\alpha}(\overline{\Omega})} \\ &\qquad \leq \|uv\|_{C_b^0(\overline{\Omega})} + \sum_{j=1}^n \|(\partial_{x_j} u) v\|_{C_b^{m,\alpha}(\overline{\Omega})} + \sum_{j=1}^n \|u(\partial_{x_j} v)\|_{C_b^{m,\alpha}(\overline{\Omega})}, \\ &\|uv\|_{C_b^0(\overline{\Omega})} \leq \|u\|_{C_b^0(\overline{\Omega})} \|v\|_{C_b^0(\overline{\Omega})} \leq \|u\|_{C_b^{m+1,\alpha}(\overline{\Omega})} \|v\|_{C_b^{m+1,\alpha}(\overline{\Omega})} \end{aligned}$$

(cf. (2.32)). Since ∂_{x_j} is continuous from $C_b^{m+1,\alpha}(\overline{\Omega})$ to $C_b^{m,\alpha}(\overline{\Omega})$ and $C_b^{m+1,\alpha}(\overline{\Omega})$ is continuously embedded into $C_b^{m,\alpha}(\overline{\Omega})$, the inductive assumption implies the validity of statement (iii) (cf. Proposition 2.9).

Next we consider statement (iv). Again, we argue by induction on m. If $m = 0$, then the statement holds by Proposition 2.25. Next we assume that the statement holds for m and we prove it for $m+1$. Let $\{f_l\}_{l \in \mathbb{N}}$ be a bounded sequence in $C^{m+1,\alpha}(\overline{\Omega})$. The continuity of first order partial derivatives from $C^{m+1,\alpha}(\overline{\Omega})$ to $C^{m,\alpha}(\overline{\Omega})$ implies that the sequence $\{\partial_{x_j} f_l\}_{l \in \mathbb{N}}$ is bounded in $C^{m,\alpha}(\overline{\Omega})$ for all $j \in \{1, \ldots, n\}$. Then the inductive assumption implies that there exist a subsequence $\{\partial_{x_j} f_{l_k}\}_{k \in \mathbb{N}}$ of $\{\partial_{x_j} f_l\}_{l \in \mathbb{N}}$ and $g_j \in C^{m,\alpha'}(\overline{\Omega})$ such that

$$\lim_{k \to \infty} \partial_{x_j} f_{l_k} = g_j \qquad \text{in } C^{m,\alpha'}(\overline{\Omega}), \tag{2.34}$$

for all $j \in \{1,\ldots,n\}$. Since $\{f_{l_k}\}_{k\in\mathbb{N}}$ is bounded in $C^1(\overline{\Omega})$, which is compactly embedded into $C^0(\overline{\Omega})$, possibly selecting a subsequence, we can assume that $\{f_{l_k}\}_{k\in\mathbb{N}}$ has a uniform limit $f \in C^0(\overline{\Omega})$ (cf. Lemma 2.29). Then the limiting relation (2.34) and classical differentiation theorems for sequences of differentiable functions imply that

$$\partial_{x_j} f = g_j \qquad \text{in } \Omega \qquad \forall j \in \{1,\ldots,n\}.$$

Accordingly $f \in C^{m+1,\alpha'}(\overline{\Omega})$. Moreover, the inequality

$$\|f_{l_k} - f\|_{C^{m+1,\alpha'}(\overline{\Omega})} \leq \|f_{l_k} - f\|_{C^0(\overline{\Omega})} + \sum_{j=1}^{n} \|\partial_{x_j} f_{l_k} - \partial_{x_j} f\|_{C^{m,\alpha'}(\overline{\Omega})} \qquad \forall k \in \mathbb{N}$$

implies that $\{f_{l_k}\}_{k\in\mathbb{N}}$ converges to f in $C^{m+1,\alpha'}(\overline{\Omega})$ (cf. (2.32)). □

Remark 2.55. Statement (ii) can be shown to hold in case Ω has a finite number of Whitney regular connected components with pairwise disjoint closures. Instead if we assume that Ω has a finite number of Whitney regular connected components whose closures are not pairwise disjoint the embedding of statement (ii) is not necessarily true (to construct an example, cf., e.g., Gilbarg and Trudinger [107, p. 53]).

Also statement (iii) can be shown to hold in case Ω has a finite number of Whitney regular connected components with pairwise disjoint closures.

Remark 2.56. As Example 2.30 shows, statement (iv) is no longer true for $m > 0$ if we remove the assumption that Ω has a finite number of connected components.

Corollary 2.57. *Let $n \in \mathbb{N}\setminus\{0,1\}$. Let $m \in \mathbb{N}$. Let Ω be a bounded open Lipschitz subset of $\mathbb{R}^n$. Then the following statements hold.*

(i) $C^{m+1}(\overline{\Omega})$ is continuously embedded into $C^{m,1}(\overline{\Omega})$.
(ii) If $\alpha_1, \alpha_2 \in]0,1]$, then the pointwise multiplication from $C^{m,\alpha_1}(\overline{\Omega}) \times C^{m,\alpha_2}(\overline{\Omega})$ to $C^{m,\min\{\alpha_1,\alpha_2\}}(\overline{\Omega})$ that takes a pair of functions (f,g) to the pointwise product fg is continuous.
(iii) If $\alpha, \alpha' \in]0,1]$, $\alpha > \alpha'$, then $C^{m,\alpha}(\overline{\Omega})$ is compactly embedded into $C^{m,\alpha'}(\overline{\Omega})$.

Proof. Since Ω is a bounded open Lipschitz subset of $\mathbb{R}^n$, it has a finite number of connected components which are bounded open Lipschitz sets and which have pairwise disjoint closures. Hence, Ω has a finite number of Whitney regular bounded open connected components with pairwise disjoint closures. Then the statement is an immediate consequence of Proposition 2.54 and of the above Remark 2.55. □

We conclude the present section with an elementary inequality for functions in Schauder spaces.

Lemma 2.58. *Let $m, n \in \mathbb{N}$, $\alpha \in]0,1]$. Let Ω be a convex open subset of $\mathbb{R}^n$. Then we have*

$$\left| f(y) - \sum_{j=0}^{m} \frac{d^j f(x)}{j!}[y-x]^j \right| \\ \leq |y-x|^{m+\alpha} |d^m f(\cdot) : \Omega|_\alpha \frac{\Gamma(\alpha+1)}{\Gamma(m+\alpha+1)} \qquad \forall x, y \in \Omega$$

for all $f \in C_b^{m,\alpha}(\overline{\Omega})$, where Γ is the Euler Gamma function and

$$|d^m f(\cdot) : \Omega|_\alpha \equiv \left\{ \frac{\|d^m f(\xi') - d^m f(\xi'')\|_{\mathcal{L}^{(m)}(\mathbb{R}^n,\mathbb{R})}}{|\xi' - \xi''|^\alpha} : \xi', \xi'' \in \Omega, \xi' \neq \xi'' \right\},$$

where

$$\|d^m f(\xi') - d^m f(\xi'')\|_{\mathcal{L}^{(m)}(\mathbb{R}^n,\mathbb{R})} \equiv |f(\xi') - f(\xi'')| \qquad \forall \xi', \xi'' \in \Omega \quad \text{if } m = 0,$$
$$\|d^m f(\xi') - d^m f(\xi'')\|_{\mathcal{L}^{(m)}(\mathbb{R}^n;\mathbb{R})} \\ = \sup_{(u_1,\dots,u_m) \in \mathbb{B}_n(0,1)^m} |[d^m f(\xi') - d^m f(\xi'')][u_1, \dots, u_m]| \ \forall \xi', \xi'' \in \Omega \ \text{if } m > 0$$

(see Appendix A.6 for the notation relative to high order differentials and for the corresponding norms).

Proof. Let x, $y \in \Omega$. If $m = 0$, then the inequality follows by the definition of α-Hölder continuous function. If $m \geq 1$, then the Taylor formula with remainder term in integral form implies that

$$f(y) = \sum_{j=0}^{m-1} \frac{d^j f(x)}{j!}[y-x]^j \\ + \frac{1}{(m-1)!} \int_0^1 (1-t)^{m-1} d^m f(x + t(y-x))[y-x]^m \, dt$$

and thus

$$\left| f(y) - \sum_{j=0}^{m} \frac{d^j f(x)}{j!}[y-x]^j \right| \\ \leq \left| \frac{1}{(m-1)!} \int_0^1 (1-t)^{m-1} d^m f(x + t(y-x))[y-x]^m \, dt \right. \\ \left. - \frac{d^m f(x)}{m!}[y-x]^m \right| \\ = \left| \frac{1}{(m-1)!} \int_0^1 (1-t)^{m-1} [d^m f(x + t(y-x)) - d^m f(x)][y-x]^m \, dt \right| \\ \leq \frac{1}{(m-1)!} \int_0^1 (1-t)^{m-1} \\ \times \|d^m f(x + t(y-x)) - d^m f(x)\|_{\mathcal{L}^{(m)}(\mathbb{R}^n,\mathbb{R})} |y-x|^m \, dt$$

$$\leq \frac{1}{(m-1)!}\int_0^1 (1-t)^{m-1}|d^m f(\cdot):\Omega|_\alpha |y-x|^m |t(y-x)|^\alpha \,dt$$
$$= |x-y|^{m+\alpha}|d^m f(\cdot):\Omega|_\alpha \frac{1}{\Gamma(m)}\int_0^1 t^{(\alpha+1)-1}(1-t)^{m-1}\,dt$$
$$= |x-y|^{m+\alpha}|d^m f(\cdot):\Omega|_\alpha \frac{1}{\Gamma(m)}\frac{\Gamma(\alpha+1)\Gamma(m)}{\Gamma(m+\alpha+1)},$$

and thus the proof is complete (cf. (2.6)). □

2.12 Composition of Functions in Schauder Spaces

We now wish to prove that the composition of two functions in a Schauder space is still a function in a Schauder space. Thus we introduce the following.

Theorem 2.59. *Let n, $n_1 \in \mathbb{N}\setminus\{0\}$. Let $m \in \mathbb{N}$. Let α, $\beta \in [0,1]$. Let*

$$\mu_m(\alpha,\beta) \equiv \begin{cases} \alpha\beta & \text{if } m = 0\,, \\ \min\{\alpha,\beta\} & \text{if } m \geq 1\,. \end{cases}$$

Let Ω be a bounded open subset of $\mathbb{R}^n$. We assume that Ω is regular in the sense of Whitney. Let Ω_1 be a bounded open subset of $\mathbb{R}^{n_1}$. Then the following statements hold.

(i) If $(f,g) \in C^{0,\alpha}(\overline{\Omega_1}) \times C^{0,\beta}(\overline{\Omega},\overline{\Omega_1})$, then the composite function

$$T[f,g] \equiv f \circ g$$

belongs to $C^{0,\mu_0(\alpha,\beta)}(\overline{\Omega})$.
If $m \geq 1$ and if $(f,g) \in C^{m,\alpha}(\overline{\Omega_1}) \times C^{m,\beta}(\overline{\Omega},\overline{\Omega_1})$ and $g(\Omega) \subseteq \Omega_1$, then the composite function $T[f,g]$ belongs to $C^{m,\mu_m(\alpha,\beta)}(\overline{\Omega})$.

(ii) The (nonlinear) map T from $C^{0,\alpha}(\overline{\Omega_1}) \times C^{0,\beta}(\overline{\Omega})$ to $C^{0,\mu_0(\alpha,\beta)}(\overline{\Omega})$, that takes (f,g) to $T[f,g]$, maps bounded sets to bounded sets.
If $m \geq 1$, then the (nonlinear) map T from

$$\{(f,g) \in C^{m,\alpha}(\overline{\Omega_1}) \times C^{m,\beta}(\overline{\Omega},\overline{\Omega_1}) : g(\Omega) \subseteq \Omega_1\}$$

to $C^{m,\mu_m(\alpha,\beta)}(\overline{\Omega})$, that takes (f,g) to $T[f,g]$, maps bounded sets to bounded sets.

Proof. First we choose a norm on $C^{m,\beta}(\overline{\Omega},\mathbb{R}^{n_1})$ and we set

$$\|g\|_{C^{m,\beta}(\overline{\Omega},\mathbb{R}^{n_1})} \equiv \sum_{l=1}^{n_1} \|g_l\|_{C^{m,\beta}(\overline{\Omega})} \qquad \forall g \equiv (g_1,\ldots,g_{n_1}) \in C^{m,\beta}(\overline{\Omega},\mathbb{R}^{n_1})\,.$$

We proceed by induction on m. If $m = 0$, then we have

$$\sup_{\overline{\Omega}} |f \circ g| \leq \sup_{\overline{\Omega_1}} |f| \leq \|f\|_{C^{m,\alpha}(\overline{\Omega_1})} \,.$$

If $\mu_0(\alpha, \beta) > 0$, then we have

$$\begin{aligned}|f \circ g(x) - f \circ g(y)| &\leq |f : \overline{\Omega_1}|_\alpha |g(x) - g(y)|^\alpha \\ &\leq |f : \overline{\Omega_1}|_\alpha \sum_{l=1}^{n_1} |g_l(x) - g_l(y)|^\alpha \\ &\leq |f : \overline{\Omega_1}|_\alpha \sum_{l=1}^{n_1} |g_l : \overline{\Omega}|_\beta^\alpha |x - y|^{\alpha\beta} \qquad \forall x, y \in \overline{\Omega}\end{aligned}$$

for all $(f, g) \in C^{m,\alpha}(\overline{\Omega_1}) \times C^{m,\beta}(\overline{\Omega}, \overline{\Omega_1})$ (see inequality (2.3)). Thus (i) and (ii) hold true for $m = 0$. We now assume that (i) and (ii) hold true for $m \geq 0$ and we prove them for $m+1$. If $(f, g) \in C^{m+1,\alpha}(\overline{\Omega_1}) \times C^{m+1,\beta}(\overline{\Omega}, \overline{\Omega_1})$ and $g(\Omega) \subseteq \Omega_1$, then

$$\frac{\partial}{\partial x_j} f \circ g(x) = \sum_{l=1}^{n_1} \left(\frac{\partial f}{\partial y_l} \circ g(x) \right) \frac{\partial g_l}{\partial x_j}(x) \qquad \forall x \in \overline{\Omega}$$

for all $j \in \{1, \ldots, n\}$. Indeed, if $x \in \Omega$, we have $g(x) \in \Omega_1$ and the equality follows by the chain rule. If instead $x \in \partial\Omega$, then there exists a sequence $\{x_s\}_{s \in \mathbb{N}}$ in Ω which converges to x and the continuity of g at x implies that the sequence $\{g(x_s)\}_{s \in \mathbb{N}}$ of Ω_1 converges to $g(x)$. Since the above equality holds at the points x_s, the continuity of $\frac{\partial f}{\partial y_l}$ at $g(x)$ and the continuity of $\frac{\partial g_l}{\partial x_j}$ at x imply that the above equality holds at x. Then we observe that Proposition 2.54 (ii) implies that

$$C^{m+1,\beta}(\overline{\Omega}) \text{ is continuously embedded into } C^{m,1}(\overline{\Omega})\,, \tag{2.35}$$

that the inductive assumption implies that

$$\begin{aligned}&T \text{ maps bounded subsets of} \\ &\{(f, g) \in C^{m,\alpha}(\overline{\Omega_1}) \times C^{m,1}(\overline{\Omega}, \overline{\Omega_1}) : g(\Omega) \subseteq \Omega_1\} \\ &\text{to bounded subsets of } C^{m,\mu_m(\alpha,1)}(\overline{\Omega}) = C^{m,\alpha}(\overline{\Omega})\,,\end{aligned} \tag{2.36}$$

and that Proposition 2.54 (iii) implies that

$$\begin{aligned}&\text{the pointwise product is continuous from } C^{m,\alpha}(\overline{\Omega}) \times C^{m,\beta}(\overline{\Omega}) \\ &\text{to } C^{m,\mu_{m+1}(\alpha,\beta)}(\overline{\Omega})\,.\end{aligned} \tag{2.37}$$

Then $\left(\frac{\partial f}{\partial y_l} \circ g \right) \frac{\partial g_l}{\partial x_j} \in C^{m,\mu_{m+1}(\alpha,\beta)}(\overline{\Omega})$ for all $l \in \{1, \ldots, n_1\}$ and $j \in \{1, \ldots, n\}$ and accordingly $f \circ g \in C^{m+1,\mu_{m+1}(\alpha,\beta)}(\overline{\Omega})$. Next we denote by $c_{m,\alpha,\beta}$ the norm of the (bilinear) pointwise product of (2.37) (for the norm of a multilinear operator, see (A.3) of the Appendix). Then we have

$$\|f \circ g\|_{C^{m+1,\mu_{m+1}(\alpha,\beta)}(\overline{\Omega})}$$

$$\leq \sup_{\overline{\Omega}} |f \circ g| + \sum_{j=1}^{n} \sum_{l=1}^{n_1} \left\| \left(\frac{\partial f}{\partial y_l} \circ g \right) \frac{\partial g_l}{\partial x_j} \right\|_{C^{m,\mu_{m+1}(\alpha,\beta)}(\overline{\Omega})}$$

$$\leq \|f\|_{C^{m+1,\alpha}(\overline{\Omega_1})} + \sum_{j=1}^{n} \sum_{l=1}^{n_1} c_{m,\alpha,\beta} \left\| \frac{\partial f}{\partial y_l} \circ g \right\|_{C^{m,\alpha}(\overline{\Omega})} \left\| \frac{\partial g_l}{\partial x_j} \right\|_{C^{m,\beta}(\overline{\Omega})}$$

$$\leq \|f\|_{C^{m+1,\alpha}(\overline{\Omega_1})} + \sum_{j=1}^{n} \sum_{l=1}^{n_1} c_{m,\alpha,\beta} \left\| \frac{\partial f}{\partial y_l} \circ g \right\|_{C^{m,\alpha}(\overline{\Omega})} \|g\|_{C^{m+1,\beta}(\overline{\Omega})}$$

for all $(f, g) \in C^{m+1,\alpha}(\overline{\Omega_1}) \times C^{m+1,\beta}(\overline{\Omega}, \overline{\Omega_1})$ such that $g(\Omega) \subseteq \Omega_1$. Then statements (2.35), (2.36) imply the validity of statement (ii) and the proof is complete. □

We note that one cannot expect the operator T of Theorem 2.59 to be continuous at all points. For a discussion on the regularity properties of T and for corresponding references, we refer to Appell and Zabreiko [14] and to [157].

2.13 Local Strict Hypographs of a Schauder Class

Definition 2.60. Let $n \in \mathbb{N} \setminus \{0, 1\}$. Let $m \in \mathbb{N}$, $\alpha \in [0, 1]$. We say that an open subset Ω of $\mathbb{R}^n$ is a local strict hypograph of class $C^{m,\alpha}$ provided that, for every point $p \in \partial\Omega$, there exist $R \in \mathbb{O}_n(\mathbb{R})$ and $r, \delta \in]0, +\infty[$ such that $C(p, R, r, \delta)$ is a coordinate cylinder for Ω around p and that the corresponding function γ, which represents $\partial\Omega$ as a graph in $C(p, R, r, \delta)$ is of class $C^{m,\alpha}(\overline{\mathbb{B}_{n-1}(0, r)})$

Since the function γ that represents $\partial\Omega$ as a graph in $C(p, R, r, \delta)$ is defined on $\mathbb{B}_{n-1}(0, r)$, the membership of γ in $C^{m,\alpha}(\overline{\mathbb{B}_{n-1}(0, r)})$ in case $m = 0$ of Definition 2.60 has to be understood in the sense that γ has a continuous extension to $\overline{\mathbb{B}_{n-1}(0, r)}$ and that such extension is of class $C^{m,\alpha}(\overline{\mathbb{B}_{n-1}(0, r)})$.

Since $C^{m,\alpha}(\overline{\mathbb{B}_{n-1}(0, r)})$ is a subspace of $C^{m,\alpha_1}(\overline{\mathbb{B}_{n-1}(0, r)})$, which in turn is a subspace of $C^{h,\beta}(\overline{\mathbb{B}_{n-1}(0, r)})$, for all $\alpha_1 \in [0, \alpha]$, $h \in \{0, \dots, m-1\}$, $\beta \in [0, 1]$, $m \in \mathbb{N} \setminus \{0\}$, a local strict hypograph of class $C^{m,\alpha}$ is also of class C^{m,α_1} and of class $C^{h,\beta}$ for all $\alpha_1 \in [0, \alpha]$, $h \in \{0, \dots, m-1\}$, $\beta \in [0, 1]$, $m \in \mathbb{N} \setminus \{0\}$.

If Ω is a local strict hypograph of class $C^{m,\alpha}$, then Lemma 2.38 implies that the connected components of both Ω and Ω^- are of class $C^{m,\alpha}$. We note that if Ω is a local strict hypograph of class C^1, $p \in \partial\Omega$, and $C(p, R, r, \delta)$ is a coordinate cylinder for Ω around p with a corresponding function γ, then the map

$$\psi_p(\eta) \equiv p + R^t \begin{pmatrix} \eta \\ \gamma(\eta) \end{pmatrix} \qquad \forall \eta \in \mathbb{B}_{n-1}(0, r)$$

is a homeomorphism of $\mathbb{B}_{n-1}(0, r)$ onto $\psi_p(\mathbb{B}_{n-1}(0, r)) = (\partial\Omega) \cap C(p, R, r, \delta)$ and

$$\sqrt{\det\left((D\psi_p(\eta))^t(D\psi_p(\eta))\right)} = \sqrt{1+|\nabla\gamma(\eta)|^2} \qquad \forall \eta \in \mathbb{B}_{n-1}(0,r)\,.$$

In particular, ψ_p is a parametrization for $(\partial\Omega)\cap C(p,R,r,\delta)$. If $\eta \in \mathbb{B}_{n-1}(0,r)$, then the vector $\begin{pmatrix} -\nabla\gamma(\eta) \\ 1 \end{pmatrix} \frac{1}{\sqrt{1+|\nabla\gamma(\eta)|^2}}$ is normal to the graph of γ at the point $(\eta,\gamma(\eta))$ and the vector

$$\nu_\Omega(\psi_p(\eta)) = R^t \begin{pmatrix} -\nabla\gamma(\eta) \\ 1 \end{pmatrix} \frac{1}{\sqrt{1+|\nabla\gamma(\eta)|^2}}$$

is the outward unit normal to $\partial\Omega$ at p (cf. Propositions 2.44 and 2.51). Thus, in case $\nabla\gamma(0)$ equals 0, we have

$$\nu_\Omega(p) = \nu_\Omega(\psi_p(0)) = R^t e_n\,.$$

Namely, the vector $R^t e_n$, which has the direction of the axis of the cylinder $C(p,R,r,\delta)$, coincides with $\nu_\Omega(p)$. We find convenient to introduce the following definition.

Definition 2.61. Let $n \in \mathbb{N}\setminus\{0,1\}$. Let Ω be an open subset of $\mathbb{R}^n$. Let $p \in \partial\Omega$, $R \in \mathbb{O}_n(\mathbb{R})$, r, $\delta \in]0,+\infty[$. We say that a coordinate cylinder $C(p,R,r,\delta)$ for Ω around p is normal provided that the corresponding function γ which represents $\partial\Omega$ in $C(p,R,r,\delta)$ as a graph is differentiable at 0 and

$$\nabla\gamma(0) = 0\,.$$

Next we note that if Ω is a local strict hypograph of class $C^{m,\alpha}$ with $m \geq 1$ and $\alpha \in [0,1]$, then for each $p \in \partial\Omega$, there exists a coordinate cylinder $C(p,R,r,\delta)$ around p and thus

$$\begin{aligned}
&\Omega \cap \left(p + R^t(\mathbb{B}_{n-1}(0,r)\times]-\delta,\delta[)\right) \\
&\quad = p + R^t\{(\eta,y)\in\mathbb{B}_{n-1}(0,r)\times]-\delta,\delta[:\ y<\gamma(\eta)\} \\
&\quad = \Big\{x \in p + R^t\{(\eta,y)\in\mathbb{B}_{n-1}(0,r)\times]-\delta,\delta[\}: \\
&\qquad\qquad\qquad R(x-p)\in \mathrm{hypograph}_s(\gamma)\Big\} \\
&\quad = \Big\{x \in p + R^t\{(\eta,y)\in\mathbb{B}_{n-1}(0,r)\times]-\delta,\delta[\}:\ h(R(x-p))<0\Big\}\,,
\end{aligned}$$

where $h(\eta,y) \equiv y-\gamma(\eta)$ for all $(\eta,y)\in\mathbb{B}_{n-1}(0,r)\times]-\delta,\delta[$. Then, if we set

$$\begin{aligned}
W &\equiv p + R^t(\mathbb{B}_{n-1}(0,r)\times]-\delta,\delta[)\,, \\
g(x) &\equiv h(R(x-p)) \qquad \forall x \in W\,,
\end{aligned}$$

we know that W is an open and convex neighborhood of p, that $g \in C^{m,\alpha}(\overline{W})$, that

$$\Omega \cap W = \{x \in W : g(x) < 0\},$$

and that

$$\nabla g(x) = (\nabla h(R(x-p)))^t R \neq 0 \qquad \forall x \in W.$$

Indeed, $\nabla h = \begin{pmatrix} -\nabla\gamma \\ 1 \end{pmatrix}$ cannot vanish. Hence, a local strict hypograph of class $C^{m,\alpha}$ is an open set of class $C^{m,\alpha}$ in the sense of the following definition in the form with the local constraints.

Definition 2.62. Let $n \in \mathbb{N} \setminus \{0,1\}$. Let $m \in \mathbb{N} \setminus \{0\}$, $\alpha \in [0,1]$. We say that an open subset Ω of $\mathbb{R}^n$ is of class $C^{m,\alpha}$ (in the form with the local constraints), if for every $p \in \partial\Omega$, there exist an open neighborhood W of p in $\mathbb{R}^n$ and $g \in C^{m,\alpha}(\overline{W})$ such that

$$\Omega \cap W = \{x \in W : g(x) < 0\},$$
$$\nabla g(x) \neq 0 \qquad \forall x \in W.$$

The function g is said to be a local constraint for Ω around p.

We can verify that if Ω, p, W, g are as in the above definition, then

$$(\partial\Omega) \cap W = \{x \in W : g(x) = 0\}$$

and

$$\nu_\Omega(x) = \frac{\nabla g(x)}{|\nabla g(x)|} \qquad \forall x \in (\partial\Omega) \cap W.$$

Conversely, we now prove that if Ω is of class $C^{m,\alpha}$ with $m \geq 1$ in the sense of the above definition in the form with the local constraints, then Ω is a local strict hypograph of class $C^{m,\alpha}$. Actually, we prove that we can choose the coordinate cylinder, so that $\nabla\gamma(0) = 0$. Moreover, we show that if Ω is bounded, then we can make a uniform choice for r, δ and we can take $\nabla\gamma$ as small as we please. These facts are presented in the following Lemma 2.63. For a weaker version in case $m = 0$ we refer for example to [170, Lemma 10.1].

Lemma 2.63 (of the Uniform Cylinders). *Let $n \in \mathbb{N} \setminus \{0,1\}$. Let $m \in \mathbb{N} \setminus \{0\}$, $\alpha \in [0,1]$. Let Ω be an open subset of $\mathbb{R}^n$ of class $C^{m,\alpha}$ (in the form with the local constraints of Definition 2.62). Let r_*, $\delta_* \in]0,+\infty[$, $\varsigma \in]0,1[$.*

(i) If $p \in \partial\Omega$, then there exist $r_p \in]0,r_*[$, $\delta_p \in]0,\delta_*[$ with $r_p < \delta_p$, and $R_p \in \mathbb{O}_n(\mathbb{R})$ such that $C(p,R_p,r_p,\delta_p)$ is a coordinate cylinder for Ω around p, and such that the corresponding function γ_p is of class $C^{m,\alpha}(\overline{\mathbb{B}_{n-1}(0,r_p)})$ and satisfies the condition $\nabla\gamma_p(0) = 0$ and the inequality

$$|\nabla\gamma_p(\eta)| \leq \varsigma \qquad \forall \eta \in \overline{\mathbb{B}_{n-1}(0,r_p)}.$$

(ii) If we further assume that Ω is bounded, then there exist $r \in]0,r_*[$, $\delta \in]0,\delta_*[$ with $r < \delta$ such that if $p \in \partial\Omega$, then there exists $R_p \in \mathbb{O}_n(\mathbb{R})$ such that

$C(p, R_p, r, \delta)$ is a coordinate cylinder for Ω around p and the corresponding function γ_p which represents $\partial\Omega$ in $C(p, R_p, r, \delta)$ satisfies the condition $\nabla\gamma_p(0) = 0$ and the inequalities

$$|\nabla\gamma_p(\eta)| \leq \varsigma \qquad \forall \eta \in \overline{\mathbb{B}_{n-1}(0, r)}\,,$$
$$\sup_{p\in\partial\Omega} \|\gamma_p\|_{C^{m,\alpha}(\overline{\mathbb{B}_{n-1}(0,r)})} < +\infty\,.$$

Proof. If $p \in \partial\Omega$, then there exists

$$\rho_p \in]0, \min\{r_*, \delta_*, \varsigma, 1\}[\,,$$

and a local constraint $g_p \in C^{m,\alpha}(\overline{\mathbb{B}_n(p, \rho_p)})$ for Ω around p. In particular,

$$\Omega \cap \mathbb{B}_n(p, \rho_p) = \{x \in \mathbb{B}_n(p, \rho_p) : g_p(x) < 0\}\,, \quad \nabla g_p(x) \neq 0 \quad \forall x \in \mathbb{B}_n(p, \rho_p)\,.$$

Possibly shrinking ρ_p, we can assume that

$$\inf\{|\nabla g_p(x) \cdot \nu_\Omega(y)| : (x, y) \in \mathbb{B}_n(p, \rho_p) \times ((\partial\Omega) \cap \mathbb{B}_n(p, \rho_p))\} > 0\,. \quad (2.38)$$

Indeed,

$$\nabla g_p(p) \cdot \nu_\Omega(p) = |\nabla g_p(p)| > 0\,.$$

We first prove statement (ii) and then we turn to consider the proof of statement (i), which can be deduced by a straightforward modification of that of statement (ii). Since $\partial\Omega$ is compact, there exist a natural number k and, for each $j \in \{1, \ldots, k\}$, a point $p^{(j)} \in \partial\Omega$, a real number $\rho_{p^{(j)}} \in]0, \min\{r_*, \delta_*, \varsigma, 1\}[$, and a function $g_{p^{(j)}}$ in $C^{m,\alpha}(\overline{\mathbb{B}_n(p^{(j)}, \rho_{p^{(j)}})})$ as above such that

$$\partial\Omega \subseteq \bigcup_{j=1}^{k} \mathbb{B}_n(p^{(j)}, \rho_{p^{(j)}}/2)\,. \quad (2.39)$$

Next we choose $\iota \in]0, 1/2[$ and $c \in]0, +\infty[$ such that

$$\begin{aligned} \iota < \inf\{|\nabla g_{p^{(j)}}(x) \cdot \nu_\Omega(y)| :& \\ &(x, y) \in \mathbb{B}_n(p^{(j)}, \rho_{p^{(j)}}) \times ((\partial\Omega) \cap \mathbb{B}_n(p^{(j)}, \rho_{p^{(j)}}))\}\,, \\ \|g_{p^{(j)}}\|_{C^{m,\alpha}(\overline{\mathbb{B}_n(p^{(j)}, \rho_{p^{(j)}})})} \leq c\,,& \end{aligned} \quad (2.40)$$

for all $j \in \{1, \ldots, k\}$. Since the functions $\nabla g_{p^{(j)}}$ are uniformly continuous, there exists an increasing function ω from $]0, +\infty[$ to itself such that $\lim_{t\to 0^+} \omega(t) = 0$ and

$$|\nabla g_{p^{(j)}}(x') - \nabla g_{p^{(j)}}(x'')| \leq \omega(|x' - x''|) \qquad \forall x', x'' \in \overline{\mathbb{B}_n(p^{(j)}, \rho_{p^{(j)}})}\,, \quad (2.41)$$

for all $j \in \{1, \ldots, k\}$. Next we set

$$\rho \equiv \min_{j\in\{1,\dots,k\}} \frac{1}{2}\rho_{p^{(j)}} \tag{2.42}$$

and we choose an arbitrary $q \in \partial\Omega$. By the inclusion in (2.39), there exists $j_q \in \{1,\dots,k\}$ such that $q \in \mathbb{B}_n(p^{(j_q)}, \rho_{p^{(j_q)}}/2)$ and accordingly

$$\mathbb{B}_n(q,\rho) \subseteq \mathbb{B}_n(q, \rho_{p^{(j_q)}}/2) \subseteq \mathbb{B}_n(p^{(j_q)}, \rho_{p^{(j_q)}}) . \tag{2.43}$$

Then the map

$$g_q \equiv g_{p^{(j_q)}|\overline{\mathbb{B}_n(q,\rho)}} \tag{2.44}$$

is a local constraint for Ω around q. Next we turn to define a matrix $R_q \in \mathbb{O}_n(\mathbb{R})$ so that the map

$$\mathbb{R}^n \ni (\eta, y) \mapsto q + R_q^t \begin{pmatrix} \eta \\ y \end{pmatrix} \in \mathbb{R}^n$$

maps $\mathbb{B}_n(0,\rho)$ onto $\mathbb{B}_n(q,\rho)$ and the hyperplane $\mathbb{R}^{n-1}\times\{0\}$ onto the affine tangent space $q + T_q(\partial\Omega)$ to $\partial\Omega$ at q, which coincides with $q + \operatorname{Ker} dg_q(q)$. To do so, we choose an orthonormal basis $v^{(1)}, \dots, v^{(n-1)}$ of $\operatorname{Ker} dg_q(q)$, and we define R_q by the equalities

$$R_q v^{(l)} = e_l \qquad \forall l \in \{1,\dots,n-1\}, \quad R_q \nu_\Omega(q) = e_n . \tag{2.45}$$

Then we set

$$f_q(\eta,y) = g_q\left(q + R_q^t \begin{pmatrix} \eta \\ y \end{pmatrix}\right) \qquad \forall (\eta,y) \in \mathbb{B}_n(0,\rho) . \tag{2.46}$$

By the definition of f_q, we have

$$\begin{aligned} & f_q(0,0) = 0 , \qquad (\nabla f_q(0,0))^t = (\nabla g_q(q))^t R_q^t , \\ & \frac{\partial f_q}{\partial \eta_l}(0,0) = \nabla f_q(0,0)\cdot e_l \\ & \qquad = (R_q \nabla g_q(q))\cdot e_l = \nabla g_q(q)\cdot v^{(l)} = 0 \qquad \forall l \in \{1,\dots,n-1\} , \\ & \frac{\partial f_q}{\partial y}(0,0) = \nabla f_q(0,0)\cdot e_n = (R_q\nabla g_q(q))\cdot e_n \\ & \qquad = \nabla g_q(q)\cdot \nu_\Omega(q) = |\nabla g_q(q)| > 0 . \end{aligned} \tag{2.47}$$

Next we note that

$$\begin{aligned} \iota < \inf\{ & |\nabla g_{p^{(j_q)}}(x)\cdot \nu_\Omega(y)| : \\ & (x,y) \in \mathbb{B}_n(p^{(j_q)}, \rho_{p^{(j_q)}}) \times ((\partial\Omega)\cap \mathbb{B}_n(p^{(j_q)}, \rho_{p^{(j_q)}}))\} \\ \le & \inf_{x\in\mathbb{B}_n(q,\rho)} |\nabla g_{p^{(j_q)}}(x)\cdot \nu_\Omega(q)| \\ = & \inf_{(\eta,y)\in\mathbb{B}_n(0,\rho)} \left|\nabla g_{p^{(j_q)}}\left(q + R_q^t\begin{pmatrix} \eta \\ y \end{pmatrix}\right)\cdot R_q^t e_n\right| \end{aligned}$$

$$= \inf_{(\eta,y)\in\overline{\mathbb{B}_n(0,\rho)}} \left|\frac{\partial f_q}{\partial y}(\eta,y)\right| . \tag{2.48}$$

Since all entries of the orthogonal matrices are bounded by 1, the definition of norm in $C^{m,\alpha}(\overline{\mathbb{B}_n(0,\rho)})$ implies that

$$\sup_{(q,R)\in(\partial\Omega)\times\mathbb{O}_n(\mathbb{R})} \|q + R^t(\eta,y)^t\|_{C^{m,\alpha}(\overline{\mathbb{B}_n(0,\rho)})} < +\infty .$$

Then inequality (2.40), equalities (2.44), (2.46), and Theorem 2.59 on the composition of functions in Schauder spaces with $\Omega = \mathbb{B}_n(0,\rho)$ and with $\Omega_1 = \mathbb{B}(p^{(j)}, \rho_{p^{(j)}})$ for each $j \in \{1,\ldots,k\}$ ensure the existence of $\sigma \in]\iota, +\infty[$ such that

$$\|f_q\|_{C^{m,\alpha}(\overline{\mathbb{B}_n(0,\rho)})} \leq \sigma \qquad \forall q \in \partial\Omega . \tag{2.49}$$

Moreover,

$$\begin{aligned} &|\nabla f_q(\eta',y') - \nabla f_q(\eta'',y'')| \qquad (2.50)\\ &= \left|\nabla g_{p^{(j_q)}}(q + R_q^t(\eta',y')^t) - \nabla g_{p^{(j_q)}}(q + R_q^t(\eta'',y'')^t)\right| \\ &\leq \omega(|R_q^t(\eta',y')^t - R_q^t(\eta'',y'')^t|) \\ &= \omega(|(\eta',y') - (\eta'',y'')|) \qquad \forall (\eta',y'),(\eta'',y'') \in \overline{\mathbb{B}_n(0,\rho)} \end{aligned}$$

(cf. (2.41)). Next we choose δ, $r \in]0,\rho[$ such that the following conditions hold

$$r(n-1) < \delta/2 < \delta , \quad r + \delta < \rho , \quad \omega(r+\delta) < \varsigma\iota < \iota . \tag{2.51}$$

In particular, we note that the inequality $r + \delta < \rho$ of (2.51) ensures that

$$\overline{\mathbb{B}_{n-1}(0,r)} \times [-\delta,\delta] \subseteq \mathbb{B}_n(0,\rho) .$$

Possibly taking a smaller δ and r, we can also assume that δ is as in Corollary A.22 of the Implicit Function Theorem with $\Omega = \mathbb{B}_n(0,\rho)$, $(x^o, y^o) = (0,0)$, $\delta^* = \delta_*$, $\omega_{1,0} = \omega$ and that $r < r_1$, where r_1 is as in the same Corollary A.22.

Then inequalities (2.48), (2.49), and Corollary A.22 of the Implicit Function Theorem with $\Omega = \mathbb{B}_n(0,\rho)$, $(x^o,y^o) = (0,0)$, $\delta^* = \delta_*$, $\omega_{1,0} = \omega$ imply that there exist $M \in]0,+\infty[$ and for each $q \in \partial\Omega$ a function γ_q in $C^{m,\alpha}(\overline{\mathbb{B}_{n-1}(0,r)},]-\delta,\delta[)$ such that

$$\begin{aligned} &\{(\eta,y) \in \mathbb{B}_{n-1}(0,r)\times]-\delta,\delta[:\ f_q(\eta,y) = 0\} \\ &\qquad = \{(\eta,\gamma_q(\eta)) : \eta \in \mathbb{B}_{n-1}(0,r)\} , \qquad \|\gamma_q\|_{C^{m,\alpha}(\overline{\mathbb{B}_{n-1}(0,r)})} \leq M . \end{aligned}$$

In particular, $\gamma_q(0) = 0$. We also note that the known formula for the partial derivatives of an implicitly defined function implies that

$$\frac{\partial \gamma_l}{\partial \eta_l}(0) = -\frac{\frac{\partial f_q}{\partial \eta_l}(0,\gamma_q(0))}{\frac{\partial f_q}{\partial y}(0,\gamma_q(0))} = 0$$

$$\left|\frac{\partial \gamma_l}{\partial \eta_l}(\eta)\right| = \left|-\frac{\frac{\partial f_q}{\partial \eta_l}(\eta,\gamma_q(\eta))}{\frac{\partial f_q}{\partial y}(\eta,\gamma_q(\eta))}\right| \leq \iota^{-1}\left|\frac{\partial f_q}{\partial \eta_l}(\eta,\gamma_q(\eta)) - \frac{\partial f_q}{\partial \eta_l}(0,\gamma_q(0))\right|$$
$$\leq \iota^{-1}\omega(|(\eta,\gamma_q(\eta))|) \leq \iota^{-1}\omega(\delta + r) \leq \iota^{-1}\varsigma\iota = \varsigma \tag{2.52}$$

for all $l \in \{1,\ldots,n-1\}$ and for all $q \in \partial\Omega$ (cf. the third inequality of (2.51)). Moreover, the Mean Value inequality and the following conditions

$$\gamma_q(0) = 0\,, \qquad r(n-1)\varsigma < \varsigma\delta/2 < \delta/2$$

(cf. the first inequality of (2.51)) imply that

$$\gamma_q(\overline{\mathbb{B}_{n-1}(0,r)}) \subseteq]-\delta/2,\delta/2[$$

(cf. inequality (2.52)). A simple topological argument shows that the sets

$$\{(\eta,y) \in \mathbb{B}_{n-1}(0,r)\times]-\delta,\delta[:\ y < \gamma_q(\eta)\}\,,$$
$$\{(\eta,y) \in \mathbb{B}_{n-1}(0,r)\times]-\delta,\delta[:\ y > \gamma_q(\eta)\}\,,$$

are path connected. Since the zeros of f_q are located on the graph of γ_q, then f_q does not vanish on such sets. Accordingly, the sign of f_q is constant on such sets. Since $f_q(0,0) = 0$ and $\frac{\partial f_q}{\partial y_n}(0,0) > 0$, we conclude that f_q is negative on $\{(\eta,y) \in \mathbb{B}_{n-1}(0,r)\times]-\delta,\delta[:\ y < \gamma_q(\eta)\}$ and positive on

$$\{(\eta,y) \in \mathbb{B}_{n-1}(0,r)\times]-\delta,\delta[:\ y > \gamma_q(\eta)\}$$

and that accordingly

$$\{(\eta,y) \in \mathbb{B}_{n-1}(0,r)\times]-\delta,\delta[:\ f_q(y,\eta) < 0\}$$
$$= \{(\eta,y) \in \mathbb{B}_{n-1}(0,r)\times]-\delta,\delta[:\ y < \gamma_q(\eta)\}\,.$$

Since g_q is a local constraint for Ω around q and

$$q + R_q^t(\mathbb{B}_{n-1}(0,r)\times]-\delta,\delta[) \subseteq \mathbb{B}_n(0,\rho)\,,$$

we have

$$\begin{aligned}(q + R_q^t(\mathbb{B}_{n-1}(0,r)\times]-\delta,\delta[)) \cap \Omega \\ &= \left\{x \in q + R_q^t(\mathbb{B}_{n-1}(0,r)\times]-\delta,\delta[)\ :\ g_q(x) < 0\right\} \\ &= q + R_q^t\left(\{(\eta,y) \in \mathbb{B}_{n-1}(0,r)\times]-\delta,\delta[:\ f_q(y,\eta) < 0\}\right) \\ &= q + R_q^t(\mathrm{hypograph}_s(\gamma_q))\,.\end{aligned}$$

Hence, $C(q, R_q, r, \delta)$ is a coordinate cylinder for Ω around q and the proof of statement (ii) is complete.

In order to prove statement (i), we fix $p \in \partial\Omega$, we consider a local constraint g_p as in (i), we define a corresponding R_p as in (2.45), and we deduce the existence of $\gamma_p \in C^{m,\alpha}(\overline{\mathbb{B}_{n-1}(0, r_p)},]-\delta_p, \delta_p[)$ by means of the Implicit Function Theorem. Then the same argument above shows that $C(p, R_p, r_p, \delta_p)$ is a coordinate cylinder for Ω around p. □

By the remarks right before Definition 2.62 of set of class $C^{m,\alpha}$ in the form with the local constraints and by the Lemma 2.63 of the uniform cylinders, we have the following.

Proposition 2.64. *Let $n \in \mathbb{N} \setminus \{0, 1\}$. Let $m \in \mathbb{N} \setminus \{0\}$, $\alpha \in]0, 1]$. Let Ω be an open subset of $\mathbb{R}^n$. Then Ω is of class $C^{m,\alpha}$ (in the form with the local constraints) if and only if Ω is a local strict hypograph of class $C^{m,\alpha}$.*

We also introduce the following form of the definition of open set of class $C^{m,\alpha}$.

Definition 2.65. Let $n \in \mathbb{N} \setminus \{0, 1\}$. Let $m \in \mathbb{N} \setminus \{0\}$, $\alpha \in [0, 1]$. We say that an open subset Ω of $\mathbb{R}^n$ is of class $C^{m,\alpha}$ (in the form with the diffeomorphisms) if, for every $p \in \partial\Omega$, there exist an open neighborhood U of 0 in $\mathbb{R}^n$, an open neighborhood W of p in $\mathbb{R}^n$, and a homeomorphism ϕ_p from U onto W such that $\phi_p \in C^{m,\alpha}\left(\overline{U}, \mathbb{R}^n\right)$, $\phi_p^{(-1)} \in C^{m,\alpha}\left(\overline{W}, \mathbb{R}^n\right)$, and

$$\begin{aligned} &\phi_p(\{x \in U : x_n < 0\}) = W \cap \Omega\,, \\ &\phi_p(\{x \in U : x_n = 0\}) = W \cap \partial\Omega\,, \\ &\phi_p(0) = p\,. \end{aligned} \tag{2.53}$$

The map ϕ_p is said to be a parametrization of $\overline{\Omega}$ around p.

Remark 2.66. Since ϕ_p is continuous, possibly shrinking the open neighborhood U, we can take the open neighborhood W to be as small as we wish. Since a diffeomorphism is a locally Lipschitz homeomorphism, Proposition 2.31 implies that possibly shrinking U in the above definition, we can assume that U is an open ball and that $W = \phi_p(U)$ is regular in the sense of Whitney. Possibly shrinking the radius of the ball we can assume that ϕ_p extends to a homeomorphism from $\overline{U}$ onto $\overline{W}$. Then by taking the composition of ϕ_p with a dilation, we can assume that $U = \mathbb{B}_n(0, 1)$.

Then a classical argument based on the Implicit Function Theorem shows the validity of the following.

Proposition 2.67. *Let $n \in \mathbb{N} \setminus \{0, 1\}$. Let $m \in \mathbb{N} \setminus \{0\}$, $\alpha \in [0, 1]$. Let Ω be an open subset of $\mathbb{R}^n$. Then Ω is of class $C^{m,\alpha}$ (in the form with the diffeomorphisms) if and only if it is a local strict hypograph of class $C^{m,\alpha}$.*

Proof. Let Ω be of class $C^{m,\alpha}$ (in the form with the diffeomorphisms). Let $p \in \partial\Omega$. Let U, W, ϕ_p be as in Definition 2.65. Then we define g_p to be the n-th component of the map $\phi_p^{(-1)}$. We observe that $(\nabla g_p)^t$ coincides with the n-th row of the Jacobian matrix of $\phi_p^{(-1)}$, whose determinant cannot vanish at any point of W. Then ∇g_p cannot be equal to zero at any point of W and the map g_p is a local constraint for Ω around p. Thus Ω is of class $C^{m,\alpha}$ in the form with the local constraints. Then Proposition 2.64 implies that Ω is a local strict hypograph of class $C^{m,\alpha}$.

We now prove the converse. Let Ω be a local strict hypograph of class $C^{m,\alpha}$. Let $p \in \partial\Omega$. Then there exist $R \in \mathbb{O}_n(\mathbb{R})$, r, $\delta \in]0,+\infty[$ such that $C(p,R,r,\delta)$ is a coordinate cylinder for Ω around p, and the corresponding function γ which represents $\partial\Omega$ in $C(p,R,r,\delta)$ as a graph is of class $C^{m,\alpha}(\overline{\mathbb{B}_{n-1}(0,r)})$. Then the function ϕ_p from $U \equiv \mathbb{B}_{n-1}(0,r)\times]-\delta/2,\delta/2[$ onto $W \equiv p+R^t(U)$ defined by

$$\phi_p(\eta,y) \equiv p+R^t(\eta,y+\gamma(\eta))^t \qquad \forall(\eta,y)\in U\,,$$

can be verified to be a parametrization of $\overline{\Omega}$ around p of class $C^{m,\alpha}$. Indeed, ϕ_p is injective, satisfies (2.53), and its inverse function

$$\begin{aligned}&\phi_p^{(-1)}(\xi)\\&\quad\equiv(\pi_{\mathbb{R}^{n-1}}\circ R(\xi-p),\pi_n\circ R(\xi-p)-\gamma\circ\pi_{\mathbb{R}^{n-1}}\circ R(\xi-p))\quad\forall\xi\in W\end{aligned}$$

is of class $C^{m,\alpha}$. Here $\pi_{\mathbb{R}^{n-1}}$ denotes the projection of the first $(n-1)$ components and π_n denotes the projection of the last component. □

By Propositions 2.64 and 2.67 we can see that the so far introduced different definitions of sets of class $C^{m,\alpha}$ are equivalent when $m \geq 1$. Namely, we have the following.

Remark 2.68. Let $n \in \mathbb{N}\setminus\{0,1\}$. If $m \in \mathbb{N}\setminus\{0\}$, $\alpha \in [0,1]$, and Ω is an open subset of $\mathbb{R}^n$, then the following statements are equivalent:

(i) Ω is of class $C^{m,\alpha}$ in the form with the local constraints,
(ii) Ω is of class $C^{m,\alpha}$ in the form with the diffeomorphisms,
(iii) Ω is a local strict hypograph of class $C^{m,\alpha}$.

Thus, from now on we simply say that Ω is of class $C^{m,\alpha}$ for a certain $m \geq 1$ and $\alpha \in [0,1]$, without specifying in which of the above forms. Moreover, we say that Ω is of class C^m if Ω is of class $C^{m,0}$, and that Ω is of class C^∞ provided that Ω is of class C^m for all $m \geq 1$.

If instead $m = 0$, then we understand that a set Ω is of class $C^{0,\alpha}$ when it is a local strict hypograph of class $C^{0,\alpha}$.

By Lemma 2.63 on the uniform coordinate cylinders, the following statement holds.

Corollary 2.69. *Let $n \in \mathbb{N}\setminus\{0,1\}$. Let Ω be a bounded open subset of $\mathbb{R}^n$ of class C^1. Then there exists $t_0 \in]0,+\infty[$ such that*

$$\begin{aligned}x+t\nu_\Omega(x)&\in\Omega&&\forall t\in]-t_0,0[\,,\\x+t\nu_\Omega(x)&\in\Omega^-&&\forall t\in]0,t_0[\,,\end{aligned}$$

for all $x \in \partial\Omega$.

Proof. By Lemma 2.63, there exist $r, \delta \in]0, +\infty[$ such that, for each $p \in \partial\Omega$, there exists $R_p \in \mathbb{O}_n(\mathbb{R})$ such that $C(p, R_p, r, \delta)$ is a coordinate cylinder for Ω around p. Then it suffices to take $t_0 = \delta/2$. Indeed, $p + t\nu_\Omega(p) = p + tR_p^t e_n$ belongs to $C(p, R_p, r, \delta) \cap \Omega$ for all $t \in] - \delta, 0[$ and belongs to $C(p, R_p, r, \delta) \cap \Omega^-$ for all $t \in]0, \delta[$. □

Then we have the following classical lemma, which is often useful.

Lemma 2.70. *Let* $n \in \mathbb{N} \setminus \{0, 1\}$. *Let* Ω *be an open subset of* $\mathbb{R}^n$. *Let* K *be a compact subset of* Ω. *Then there exists an open bounded subset* Ω_1 *of* Ω *of class* C^∞ *such that*

$$K \subseteq \Omega_1 \subseteq \overline{\Omega_1} \subseteq \Omega .$$

If we further assume that K *is connected, then we can take* Ω_1 *to be connected.*

Proof. Here we follow an argument suggested by G. De Marco.[1] By the classical Lemma A.36, there exists a cutoff function $\phi \in \mathcal{D}(\Omega)$ such that $0 \leq \phi \leq 1$ in Ω, and $\phi = 1$ on K. By Sard's Theorem A.31, there exists a regular value $c \in]0, 1[$ for ϕ, i.e., a constant $c \in]0, 1[$ such that

$$\nabla\phi(x) \neq 0 \qquad \forall x \in \{\xi \in \Omega : \phi(\xi) = c\} .$$

Then the open and bounded subset

$$\Omega_1 \equiv \{x \in \Omega : \phi(\xi) > c\}$$

of Ω is of class C^∞ and contains K. Moreover

$$\overline{\Omega_1} \subseteq \operatorname{supp} \phi \subseteq \Omega .$$

If we further assume that K is connected, then K must be contained in a single connected component of Ω_1. Then we can replace Ω_1 with the only connected component of Ω_1 which contains K. Indeed, Proposition 2.64 ensures that Ω_1 is a local strict hypograph of class C^∞, Lemma 2.38 ensures that its connected components are also local strict hypographs of class C^∞. □

2.14 Extendibility of Functions of Schauder Spaces on an Open Subset of Class $C^{m,\alpha}$

In this section, we plan to extend functions of the Schauder class $C^{h,\alpha}(\overline{\Omega})$ for some $h \in \{0, \ldots, m\}$ to the whole of $\mathbb{R}^n$ in case Ω is of class $C^{m,\alpha}$ with $m \geq 1$ and $\alpha \in [0, 1]$. Since locally around the boundary points, sets of class $C^{m,\alpha}$ can be mapped diffeomorphically onto a half-ball, we begin with the following technical

[1] Professor at the University of Padua, Italy.

extension lemma for functions defined on a half-ball. We present here a proof based on an argument of Hestenes [123].

Lemma 2.71. *Let* $n \in \mathbb{N} \setminus \{0,1\}$, $h \in \mathbb{N}$, $\alpha \in [0,1]$ *(cf. (2.33)). Let* $\rho \in]0,+\infty[$. *Let*

$$\mathbb{B}_n^+(0,\rho) \equiv \mathbb{B}_n(0,\rho) \cap \{x \in \mathbb{R}^n : x_n > 0\}, \qquad (2.54)$$
$$\mathbb{B}_n^-(0,\rho) \equiv \mathbb{B}_n(0,\rho) \cap \{x \in \mathbb{R}^n : x_n < 0\}.$$

Then there exists a linear and continuous extension operator E from

$$\{f \in C^{h,\alpha}(\overline{\mathbb{B}_n^-(0,\rho)}) : \operatorname{supp}(f) \subseteq \overline{\mathbb{B}_n^-(0,\rho/2)}\} \qquad (2.55)$$

to

$$\{g \in C^{h,\alpha}(\overline{\mathbb{B}_n(0,\rho)}) : \operatorname{supp}(g) \subseteq \overline{\mathbb{B}_n(0,\rho/2)}\} \qquad (2.56)$$

such that

$$E[f]_{|\overline{\mathbb{B}_n^-(0,\rho)}} = f$$

for all $f \in C^{h,\alpha}(\overline{\mathbb{B}_n^-(0,\rho)})$ *such that* $\operatorname{supp}(f) \subseteq \overline{\mathbb{B}_n^-(0,\rho/2)}$ *(cf. (2.33)).*

Proof. The Vandermonde determinant associated to the real numbers

$$(-1), \ldots, (-h-1),$$

is not equal to zero and accordingly there exists a unique solution $(\lambda_1, \ldots, \lambda_{h+1})$ of the linear system

$$\sum_{j=1}^{h+1} \lambda_j (-j)^s = 1 \qquad \forall s \in \{0, \ldots, h\}. \qquad (2.57)$$

Then we set

$$E[f](x_1, \ldots, x_n)$$
$$\equiv \begin{cases} f(x_1, \ldots, x_n) & \text{if } x \in \overline{\mathbb{B}_n(0,\rho)}, x_n \leq 0, \\ \sum_{j=1}^{h+1} \lambda_j f(x_1, \ldots, x_{n-1}, -jx_n) & \text{if } x \in \overline{\mathbb{B}_n(0,\rho)}, x_n > 0, \end{cases}$$

for all $f \in C^{h,\alpha}(\overline{\mathbb{B}_n^-(0,\rho)})$ such that $\operatorname{supp}(f) \subseteq \overline{\mathbb{B}_n^-(0,\rho/2)}$. In the definition of $E[f]$, we understand that f has value 0 at all points outside of $\overline{\mathbb{B}_n^-(0,\rho)}$. We also note that if $|(x_1, \ldots, x_n)| > \rho/2$ and $x_n > 0$, then $|(x_1, \ldots, x_{n-1}, -jx_n)| > \rho/2$ and thus $f(x_1, \ldots, x_{n-1}, -jx_n) = 0$ for all $j \in \{1, \ldots, h+1\}$.

If $f \in C^{h,\alpha}(\overline{\mathbb{B}_n^-(0,\rho)})$ and $\operatorname{supp}(f) \subseteq \overline{\mathbb{B}_n^-(0,\rho/2)}$, then by the definition of $E[f]$, we can see that the support of $E[f]$ is contained in $\overline{\mathbb{B}_n(0,\rho/2)}$. Moreover, we can verify that the function $E[f]$ and its derivatives up to order h are continuous at all points of $\overline{\mathbb{B}_n(0,\rho)}$ with $x_n \neq 0$. Finally, by the equalities (2.57) and by a straightforward computation, we can prove that $E[f]$ and its derivatives up to order

h are continuous at all points of $\mathbb{B}_n(0,\rho)$ with $x_n = 0$. Hence, $E[f]$ is of class C^h and has support in $\overline{\mathbb{B}_n(0,\rho/2)}$.

We now prove that E is linear and continuous from the space in (2.55) to the space in (2.56). To do so, we estimate the $C^{h,\alpha}$ norm of $E[f]$ in terms of the $C^{h,\alpha}$ norm of f. By the inequality (2.57), we have $1 \leq \sum_{j=1}^{h+1} j^s|\lambda_j|$ for all $s \in \{0,\ldots,h\}$. Then the chain rule implies that

$$|D^\beta(E[f])(x)| \leq \left(\sum_{j=1}^{h+1} j^{\beta_n}|\lambda_j|\right) \sup_{\overline{\mathbb{B}_n^-(0,\rho))}} |D^\beta f| \quad \forall x \in \overline{\mathbb{B}_n(0,\rho)} \tag{2.58}$$

for all $f \in C^{h,\alpha}(\overline{\mathbb{B}_n^-(0,\rho)})$ such that $\mathrm{supp}\,(f) \subseteq \overline{\mathbb{B}_n^-(0,\rho/2)}$ and $\beta \in \mathbb{N}^n$ such that $|\beta| \leq h$. Now let $\alpha > 0$. If $\beta \in \mathbb{N}^n$ and $|\beta| = h$, then

$$|D^\beta(E[f])(x) - D^\beta(E[f])(y)| \leq |D^\beta f : \overline{\mathbb{B}_n^-(0,\rho)}|_\alpha |x-y|^\alpha$$

for all $x, y \in \overline{\mathbb{B}_n^-(0,\rho)}$. By Theorem 2.59 on the composition of functions in Shauder spaces and by the continuity of the pointwise product in Schauder spaces, there exists $c_\beta > 1$ such that

$$|D^\beta(E[f])(x) - D^\beta(E[f])(y)| \leq c_\beta |D^\beta f : \overline{\mathbb{B}_n^-(0,\rho)}|_\alpha |x-y|^\alpha$$

for all $x, y \in \overline{\mathbb{B}_n^+(0,\rho)}$. If $x \in \overline{\mathbb{B}_n^-(0,\rho)}$ and $y \in \overline{\mathbb{B}_n^+(0,\rho)}$, then there exists ξ in the segment $[x,y]$ such that $\xi_n = 0$ and $[x,\xi] \subseteq \overline{\mathbb{B}_n^-(0,\rho)}$, $[\xi,y] \subseteq \overline{\mathbb{B}_n^+(0,\rho)}$. Then we have

$$\begin{aligned}
&|D^\beta(E[f])(x) - D^\beta(E[f])(y)| \\
&\quad \leq |D^\beta f : \overline{\mathbb{B}_n^-(0,\rho)}|_\alpha |x-\xi|^\alpha + c_\beta |D^\beta f : \overline{\mathbb{B}_n^-(0,\rho)}|_\alpha |\xi-y|^\alpha \\
&\quad \leq c_\beta |D^\beta f : \overline{\mathbb{B}_n^-(0,\rho)}|_\alpha (|x-\xi|^\alpha + |\xi-y|^\alpha) \\
&\quad \leq c_\beta |D^\beta f : \overline{\mathbb{B}_n^-(0,\rho)}|_\alpha (|x-y|^\alpha + |x-y|^\alpha)
\end{aligned}$$

and we deduce that

$$|D^\beta(E[f])(x) - D^\beta(E[f])(y)| \leq 2c_\beta |D^\beta f : \overline{\mathbb{B}_n^-(0,\rho)}|_\alpha |x-y|^\alpha \tag{2.59}$$

for all $x, y \in \overline{\mathbb{B}_n(0,\rho)}$. Inequalities (2.58) and (2.59) imply the continuity of the linear map E and thus the proof is complete. □

Then we are ready to prove the following extension theorem, which we formulate as in Troianiello [268, Theorem 1.3].

Theorem 2.72. *Let $n \in \mathbb{N} \setminus \{0,1\}$. Let $m \in \mathbb{N} \setminus \{0\}$, $h \in \{0,\ldots,m\}$, $\alpha \in [0,1]$. Let Ω be a bounded open subset of $\mathbb{R}^n$ of class $C^{m,\alpha}$. Let Ω' be a bounded open subset of $\mathbb{R}^n$ such that $\overline{\Omega} \subseteq \Omega'$. Then there exists a linear and continuous* *extension operator* $F_{\Omega'}$ *from* $C^{h,\alpha}(\overline{\Omega})$ *to* $C^{h,\alpha}(\overline{\Omega'})$ *such that* $F_{\Omega'}[f]_{|\overline{\Omega}} = f$ *and* $\text{supp}\, F_{\Omega'}[f] \subseteq \Omega'$ *for all* $f \in C^{h,\alpha}(\overline{\Omega})$ *(cf. (2.33)).*

Proof. Let $\{\phi_a\}_{a=1,\dots,b}$ with $\phi_a \in C^{m,\alpha}\left(\overline{\mathbb{B}_n(0,1)}, \mathbb{R}^n\right)$ be parametrizations of $\overline{\Omega}$ around points of $\partial\Omega$ such that

$$\partial\Omega \subseteq \bigcup_{a=1}^{b} \phi_a(\mathbb{B}_n(0,1/2))$$

(see Definition 2.65). By Remark 2.66, we can assume that $\overline{\phi_a(\mathbb{B}_n(0,1))} \subseteq \Omega'$ and that $\phi_a(\mathbb{B}_n(0,1))$ is regular in the sense of Whitney for all $a \in \{1,\dots,b\}$. By Lemma A.35 on the existence of the partition of unity, there exists a finite family $\{\theta_a\}_{a\in\{1,\dots,b\}}$ of functions of class $C^\infty_c(\mathbb{R}^n)$ such that

$$\text{supp}\,\theta_a \subseteq \phi_a(\mathbb{B}_n(0,1/2)) \quad \forall a \in \{1,\dots,b\}, \qquad \sum_{a=1}^{b} \theta_a(x) = 1 \quad \forall x \in \partial\Omega .$$

Next we note that $\phi_a \in C^{m,\alpha}\left(\overline{\mathbb{B}_n(0,1)}, \mathbb{R}^n\right) \subseteq C^{h,1}\left(\overline{\mathbb{B}_n(0,1)}, \mathbb{R}^n\right)$ when $h < m$. Then, by Proposition 2.54 (iii) on the continuity of the pointwise product, by the inclusion $\phi_a(\mathbb{B}^-_n(0,1)) \subseteq \Omega$, and by Theorem 2.59 on the composition of functions in Schauder spaces, there exists $c_1 > 0$ such that

$$\|(\theta_a f) \circ \phi_a\|_{C^{h,\alpha}(\overline{\mathbb{B}^-_n(0,1)})} \leq c_1 \|f\|_{C^{h,\alpha}(\overline{\Omega})} \qquad \forall f \in C^{h,\alpha}(\overline{\Omega}),$$

for all $a \in \{1,\dots,b\}$ both in case $h < m$ and $h = m$. We also note that

$$\text{supp}\,(\theta_a f) \circ \phi_a \subseteq \overline{\mathbb{B}^-_n(0,1/2)} \qquad \forall f \in C^{h,\alpha}(\overline{\Omega})$$

for all $a \in \{1,\dots,b\}$. Let E be the extension operator for the unit ball of Lemma 2.71. Then we have

$$\text{supp}\, E[(\theta_a f) \circ \phi_a] \subseteq \overline{\mathbb{B}_n(0,1/2)} \subseteq \mathbb{B}_n(0,1)$$

for all $f \in C^{h,\alpha}(\overline{\Omega})$ and for all $a \in \{1,\dots,b\}$. Since $\phi_a(\mathbb{B}_n(0,1))$ is regular in the sense of Whitney and $\phi_a^{(-1)} \in C^{m,\alpha}(\overline{\phi_a(\mathbb{B}_n(0,1))}) \subseteq C^{h,1}(\overline{\phi_a(\mathbb{B}_n(0,1))})$ when $h < m$, the composition Theorem 2.59 for functions in a Schauder space implies that there exists $c_2 > 0$ such that

$$\begin{aligned} &\|E[(\theta_a f) \circ \phi_a] \circ \phi_a^{(-1)}\|_{C^{h,\alpha}(\overline{\phi_a(\mathbb{B}_n(0,1))})} \\ &\quad \leq c_2 \|E[(\theta_a f) \circ \phi_a]\|_{C^{h,\alpha}(\overline{\mathbb{B}_n(0,1)})} \leq c_1 c_2 \|f\|_{C^{h,\alpha}(\overline{\Omega})} \qquad \forall f \in C^{h,\alpha}(\overline{\Omega}) \end{aligned}$$

for all $a \in \{1,\dots,b\}$ both in case $h < m$ and $h = m$. The operator G_a from

$$\{g \in C^{h,\alpha}(\overline{\phi_a(\mathbb{B}_n(0,1))}) : \text{supp}\,(g) \subseteq \phi_a(\overline{\mathbb{B}_n(0,1/2)})\}$$

to $C^{h,\alpha}(\overline{\Omega'})$, that takes a function g to

$$G_a[g](y) \equiv \begin{cases} g(y) & \text{if } y \in \phi_a(\overline{\mathbb{B}_n(0,1/2)}) , \\ 0 & \text{if } y \in \overline{\Omega'} \setminus \phi_a(\overline{\mathbb{B}_n(0,1/2)}) , \end{cases}$$

is linear and continuous for all $a \in \{1, \ldots, b\}$. Since the support of the function $E[(\theta_a f) \circ \phi_a] \circ \phi_a^{(-1)}$ is contained in $\phi_a(\overline{\mathbb{B}_n(0,1/2)}) \subseteq \Omega'$ for all $f \in C^{h,\alpha}(\overline{\Omega})$ and for all $a \in \{1, \ldots, b\}$, then the operator $F_{\Omega'}$ defined by

$$F_{\Omega'}[f] \equiv \sum_{a=1}^{b} G_a \left[E[(\theta_a f) \circ \phi_a] \circ \phi_a^{(-1)} \right] \qquad \forall f \in C^{h,\alpha}(\overline{\Omega})$$

satisfies the properties of the statement. □

2.15 On the Extendibility of Continuous Functions to the Closure of Open Sets of Class C^1

We first introduce the following technical elementary lemma.

Lemma 2.73. *Let $n \in \mathbb{N} \setminus \{0, 1\}$. Let M be a $(n-1)$-dimensional manifold of class C^1 embedded into $\mathbb{R}^n$. If $x \in \mathbb{R}^n \setminus M$ and if $\tilde{x} \in M$ satisfies the equality*

$$|x - \tilde{x}| = \inf_{y \in M} |x - y| ,$$

then the vector $(x - \tilde{x})$ is orthogonal to the tangent space $T_{\tilde{x}} M$ of M at $\tilde{x}$.

Proof. Let f be the function from M to $\mathbb{R}$ defined by $f(y) \equiv (y - x) \cdot (y - x)$ for all $y \in M$. Clearly, f has a local minimum at $y = \tilde{x}$ and $\nabla f(\tilde{x}) = 2(\tilde{x} - x)$. By the Lagrange Multiplier Theorem, $\nabla f(\tilde{x})$ must equal a multiple of the gradient of a local constraint for M around $\tilde{x}$. Since the gradient of a local constraint for M around $\tilde{x}$ is orthogonal to $T_{\tilde{x}} M$, we deduce that $\nabla f(\tilde{x}) = 2(\tilde{x} - x)$ is orthogonal to $T_{\tilde{x}} M$. □

We are now ready to introduce a sufficient condition for the extendibility to the boundary of a continuous function on an open subset of class C^1.

Lemma 2.74. *Let $n \in \mathbb{N} \setminus \{0, 1\}$. Let Ω be a bounded open subset of $\mathbb{R}^n$ of class C^1. Then the following statements hold.*

(i) If $u \in C^0(\Omega)$ and if

$$\lim_{t \to 0^-} u(x + t\nu_\Omega(x)) \equiv f(x)$$

exists uniformly in $x \in \partial\Omega$, then $f \in C^0(\partial\Omega)$ and the function $\tilde{u}$ from $\overline{\Omega}$ to $\mathbb{R}$ defined by

$$\tilde{u}(x) \equiv \begin{cases} u(x) & \text{if } x \in \Omega , \\ f(x) & \text{if } x \in \partial\Omega , \end{cases}$$

is continuous.

(ii) *If* $u \in C^0(\Omega^-)$ *and if*

$$\lim_{t \to 0^+} u(x + t\nu_\Omega(x)) \equiv f(x)$$

exists uniformly in $x \in \partial\Omega$, *then* $f \in C^0(\partial\Omega)$ *and the function* $\tilde{u}^-$ *from* $\overline{\Omega^-}$ *to* $\mathbb{R}$ *defined by*

$$\tilde{u}^-(x) \equiv \begin{cases} u(x) \text{ if } x \in \Omega^- , \\ f(x) \text{ if } x \in \partial\Omega , \end{cases}$$

is continuous.

Proof. We consider for example statement (i). Since f is a uniform limit of continuous functions, then f is continuous. We now fix $x_0 \in \partial\Omega$ and prove the continuity of $\tilde{u}$ at x_0. Let $\epsilon > 0$. By the uniform continuity of the function f on the compact set $\partial\Omega$, there exists $\delta > 0$ such that

$$|f(x_1) - f(x_2)| \leq \epsilon/2 \qquad \text{if} \qquad x_1, x_2 \in \partial\Omega\,,\ |x_1 - x_2| \leq \delta\,.$$

Possibly shrinking δ, we can assume that $y - t\nu_\Omega(y) \in \Omega$ and that $y + t\nu_\Omega(y) \in \Omega^-$ for all $t \in]0, \delta]$ and for all $y \in \partial\Omega$ (see Corollary 2.69). We now assume that $x \in \overline{\Omega}$ and that $|x - x_0| \leq \delta/2$. If $x \in \partial\Omega$, we have $|\tilde{u}(x) - \tilde{u}(x_0)| = |f(x) - f(x_0)| \leq \epsilon$. We now consider case $x \in \Omega$. Since $\partial\Omega$ is compact, there exists $\tilde{x} \in \partial\Omega$ such that

$$|x - \tilde{x}| = \inf_{y \in \partial\Omega} |x - y|$$

and we have

$$|x - x_0| \geq |x - \tilde{x}| > 0\,, \qquad (x - \tilde{x}) \text{ is orthogonal to } T_{\tilde{x}}(\partial\Omega)$$

(cf. Lemma 2.73), and

$$|\tilde{x} - x_0| \leq |\tilde{x} - x| + |x - x_0| \leq \frac{\delta}{2} + \frac{\delta}{2} = \delta\,.$$

Since $(x - \tilde{x})$ is orthogonal to $T_{\tilde{x}}(\partial\Omega)$, there exists $\lambda \in \{1, -1\}$ such that $\nu_\Omega(\tilde{x}) = \lambda\frac{x - \tilde{x}}{|x - \tilde{x}|}$. If $\lambda = 1$, then $x = \tilde{x} + |x - \tilde{x}|\nu_\Omega(\tilde{x})$. Since $|x - \tilde{x}| \leq |x - x_0| \leq \delta$, our choice of δ implies that $x = \tilde{x} + |x - \tilde{x}|\nu_\Omega(\tilde{x}) \in \Omega^-$ contrary to our assumption $x \in \Omega$. Hence, $\lambda = -1$ and $\nu_\Omega(\tilde{x}) = -\frac{x - \tilde{x}}{|x - \tilde{x}|}$. Then $x = \tilde{x} - |x - \tilde{x}|\nu_\Omega(\tilde{x})$ and

$$\begin{aligned} &|\tilde{u}(x) - \tilde{u}(x_0)| = |u(x) - f(x_0)| \\ &\quad \leq |u(x) - f(\tilde{x})| + |f(\tilde{x}) - f(x_0)| \leq |u(\tilde{x} - |x - \tilde{x}|\nu_\Omega(\tilde{x})) - f(\tilde{x})| + \epsilon/2\,. \end{aligned}$$

By assumption, we have

$$\lim_{t \to 0^-} u(x + t\nu_\Omega(x)) \equiv f(x) \qquad \forall x \in \partial\Omega\,,$$

uniformly in $x \in \partial\Omega$. Accordingly, there exists $\delta' \in]0, \delta/2[$ such that

$$|u(x + t\nu_\Omega(x)) - f(x)| \leq \epsilon/2 \qquad \forall x \in \partial\Omega\,,$$

for all $t \in]-\delta', 0[$. In particular,

$$|u(\tilde{x} - |x - \tilde{x}|\nu_\Omega(\tilde{x})) - f(\tilde{x})| \leq \epsilon/2\,,$$

if $|x - x_0| \leq \delta'$. Then the inequality above implies that

$$|\tilde{u}(x) - \tilde{u}(x_0)| \leq \epsilon/2 + \epsilon/2 = \epsilon \qquad \text{if } |x - x_0| \leq \delta'\,,$$

and $\tilde{u}$ is continuous at x_0. □

2.16 A Consequence of the Rule of Change of Variables for Diffeomorphisms

We now introduce the following consequence of the chain rule and of the rule of change of variables in Lebesgue integrals. For the convenience of the reader, we include a proof.

Lemma 2.75. *Let $n \in \mathbb{N} \setminus \{0,1\}$. Let Ω be a bounded open subset of $\mathbb{R}^n$ of class C^1. Let $p \in \mathbb{R}^n$, $\epsilon \in \mathbb{R} \setminus \{0\}$. Then the set $\Omega_{p,\epsilon} \equiv p + \epsilon\Omega$ is of class C^1 and the following statements hold.*

(i)

$$\nu_{\Omega_{p,\epsilon}}(p + \epsilon\xi) = \operatorname{sgn}(\epsilon)\nu_\Omega(\xi) \qquad \forall \xi \in \partial\Omega\,. \tag{2.60}$$

(ii) If $\omega \in L^1(\partial\Omega_{p,\epsilon})$, then

$$\int_{p+\epsilon\partial\Omega} \omega(y)\, d\sigma_y = |\epsilon|^{n-1} \int_{\partial\Omega} \omega(p + \epsilon\eta)\, d\sigma_\eta \tag{2.61}$$

for all $\omega \in L^1(p + \epsilon\partial\Omega)$.

Proof. Since dilations and translations are homeomorphisms in $\mathbb{R}^n$, we can easily see that $\partial\Omega_{p,\epsilon} = p + \epsilon\partial\Omega$.

Now let $\xi \in \partial\Omega$, $x \equiv p + \epsilon\xi$. Since Ω is of class C^1, there exist $r_\xi > 0$ and a local constraint $g_\xi \in C^1(\overline{\mathbb{B}_n(\xi, r_\xi)})$ for $\overline{\Omega}$ around ξ. Then the function

$$g_x^\sharp(y) \equiv g_\xi\left((y - p)/\epsilon\right) \qquad \forall y \in p + \epsilon\mathbb{B}_n(\xi, r_\xi)$$

is easily seen to be a local constraint for x. Hence, $\Omega_{p,\epsilon}$ is of class C^1. We now prove the formula of (i). Since g_ξ is a local constraint for $\overline{\Omega}$ around ξ and $g_x^\sharp$ is a local constraint for $\overline{\Omega_{p,\epsilon}}$ around x, we have

$$\nu_\Omega(\xi) = \frac{\nabla g_\xi(\xi)}{|\nabla g_\xi(\xi)|}, \qquad \nu_{\Omega_{p,\epsilon}}(p+\epsilon\xi) = \frac{\nabla g^\sharp_x(p+\epsilon\xi)}{|\nabla g^\sharp_x(p+\epsilon\xi)|}.$$

By the chain rule, we have

$$(\nabla g^\sharp_x(x)) = (\nabla g_\xi)((x-p)/\epsilon)\epsilon^{-1} = (\nabla g_\xi)(\xi)\epsilon^{-1}.$$

Hence the formula in (i) holds true.

We now consider statement (ii). If $\psi \in C^1(\mathbb{B}_{n-1}(0,1), \partial\Omega)$ is a local parametrization for the C^1 manifold $\partial\Omega$, then $p + \epsilon\psi$ is a parametrization for the C^1 manifold $p + \epsilon\partial\Omega = \partial\Omega_{p,\epsilon}$ and the area element corresponding to $p + \epsilon\psi$ is given by

$$\begin{aligned}
&\sqrt{\det\left((D(p+\epsilon\psi)(\eta))^t(D(p+\epsilon\psi)(\eta))\right)} \\
&\quad = \left|\frac{\partial(p+\epsilon\psi)}{\partial\eta_1}(\eta)\bar{\wedge}\dots\bar{\wedge}\frac{\partial(p+\epsilon\psi)}{\partial\eta_{n-1}}(\eta)\right| \\
&\quad = |\epsilon|^{n-1}\left|\frac{\partial\psi}{\partial\eta_1}(\eta)\bar{\wedge}\dots\bar{\wedge}\frac{\partial\psi}{\partial\eta_{n-1}}(\eta)\right| \\
&\quad = |\epsilon|^{n-1}\sqrt{\det\left(D\psi(\eta))^t(D\psi(\eta)\right)} \qquad \forall\eta\in\mathbb{B}_{n-1}(0,1),
\end{aligned}$$

where $\cdot\bar{\wedge}\dots\bar{\wedge}\cdot$ denotes the vector product of $(n-1)$ vectors in $\mathbb{R}^n$ (which is a vector orthogonal to the space generated by its factors). Then statement (ii) can be verified by exploiting the definition of integral over $\partial\Omega$ and $\partial\Omega_{p,\epsilon}$, the corresponding formulas for the area element, and the rule of change of variables in integrals. We leave the details to the reader. □

2.17 A Fundamental Inequality of the Unit Normal on the Boundary of a Set of Class $C^{1,\alpha}$

We now introduce the following two lemmas, which can be verified by elementary Calculus and by exploiting the Lebesgue number of a cover (see Theorem A.6). In particular, the following classical lemma is relevant in the analysis of singular integrals.

Lemma 2.76. *Let $n \in \mathbb{N}\setminus\{0,1\}$. Let Ω be a bounded open subset $\mathbb{R}^n$ of class C^1. Then the following statements hold.*

(i) Let $x \in \partial\Omega$. Let $C(x, R, r, \delta)$ be a normal coordinate cylinder for Ω around x. Let the function $\gamma \in C^1(\overline{\mathbb{B}_{n-1}(0,r)},]-\delta,\delta[)$ represent $\partial\Omega$ in the cylinder as a graph. Let ω_γ be a function from $[0,+\infty[$ to itself such that

$$\begin{aligned}
&\omega_\gamma(0) = 0, \qquad \omega_\gamma(t) > 0 \qquad \forall t\in]0,+\infty[, \\
&\omega_\gamma \text{ is increasing, and } \lim_{t\to 0^+}\omega_\gamma(t) = 0,
\end{aligned} \tag{2.62}$$

and such that

$$\left|\frac{\partial\gamma}{\partial\eta_l}(\eta)-\frac{\partial\gamma}{\partial\eta_l}(\xi)\right|\le\omega_\gamma(|\eta-\xi|)\qquad\forall\eta,\xi\in\mathbb{B}_{n-1}(0,r)\tag{2.63}$$

for all $l\in\{1,\dots,n-1\}$. *Then*

$$\left|\sum_{l=1}^{n-1}(\eta_l-\xi_l)\frac{\partial\gamma}{\partial\eta_l}(\eta)-(\gamma(\eta)-\gamma(\xi))\right|\le(n-1)\omega_\gamma(|\xi-\eta|)|\xi-\eta|$$

for all ξ, $\eta\in\overline{\mathbb{B}_{n-1}(0,r)}$.

(ii) The function ω_Ω *from* $[0,+\infty[$ *to itself defined by*

$$\omega_\Omega(t)\equiv\sup\left\{\frac{|(y-x)\cdot\nu_\Omega(y)|}{|x-y|}:x,y\in\partial\Omega\,,\ 0<|x-y|\le t\right\}\qquad\forall t\in]0,+\infty[\,,$$

$$\omega_\Omega(0)\equiv0$$

satisfies the conditions in (2.62) and the inequality

$$|(y-x)\cdot\nu_\Omega(y)|\le\omega_\Omega(|x-y|)|x-y|\qquad\forall x,y\in\partial\Omega\,.$$

(iii) If $\alpha\in]0,1]$ *and if we further assume that* Ω *is of class* $C^{1,\alpha}$, *then*

$$c_{\Omega,\alpha}\equiv\sup\left\{\frac{|(y-x)\cdot\nu_\Omega(y)|}{|x-y|^{1+\alpha}}:x,y\in\partial\Omega\,,\ x\ne y\right\}<+\infty\,.$$

Proof. We first note that ω_γ as in (2.62) and (2.63) certainly exists. Indeed, the function $\frac{\partial\gamma}{\partial\eta_l}$ is uniformly continuous in $\overline{\mathbb{B}_{n-1}(0,r)}$ and it suffices to take the supremum of the moduli of continuity of the functions $\frac{\partial\gamma}{\partial\eta_l}$ in $\overline{\mathbb{B}_{n-1}(0,r)}$ as $l\in\{1,\dots,n-1\}$. We now prove the inequality in statement (i). If ξ, $\eta\in\mathbb{B}_{n-1}(0,r)$, then we can assume that $\xi\ne\eta$ and the segment $[\xi,\eta]$ with endpoints ξ and η is contained in $\mathbb{B}_{n-1}(0,r)$. Thus the Mean Value Theorem implies that there exists $\tau\in]\xi,\eta[$ such that

$$\gamma(\eta)-\gamma(\xi)=(\eta-\xi)\cdot\nabla\gamma(\tau)=\sum_{l=1}^{n-1}\frac{\partial\gamma}{\partial\eta_l}(\tau)(\eta_l-\xi_l)\,.$$

Then we deduce that

$$\begin{aligned}\left|\sum_{l=1}^{n-1}(\eta_l-\xi_l)\frac{\partial\gamma}{\partial\eta_l}(\eta)-(\gamma(\eta)-\gamma(\xi))\right|&=\left|\sum_{l=1}^{n-1}(\eta_l-\xi_l)\left(\frac{\partial\gamma}{\partial\eta_l}(\eta)-\frac{\partial\gamma}{\partial\eta_l}(\tau)\right)\right|\\&\le(n-1)\omega_\gamma(|\eta-\tau|)|\eta-\xi|\le(n-1)\omega_\gamma(|\eta-\xi|)|\eta-\xi|\end{aligned}$$

and thus the proof of statement (i) is complete.

We now consider statement (ii). For each $x \in \partial\Omega$, there exists a coordinate cylinder $C(x, R_x, r_x, \delta_x)$ around x. Then $\{C(x, R_x, r_x, \delta_x) \cap \partial\Omega\}_{x\in\partial\Omega}$ is an open cover of $\partial\Omega$. Possibly shrinking r_x, we can assume that the function which represents $\partial\Omega$ in $C(x, R_x, r_x, \delta_x)$ is of class C^1 on $\overline{\mathbb{B}_{n-1}(0, r_x)}$. Since $\partial\Omega$ is compact, there exist finite families $\{x^{(j)}\}_{j=1}^m$, $\{R_j\}_{j=1}^m$, $\{r_j\}_{j=1}^m$, $\{\delta_j\}_{j=1}^m$, $\{\gamma_j\}_{j=1}^m$, such that every $C(x^{(j)}, R_j, r_j, \delta_j)$ is a coordinate cylinder around $x^{(j)}$, γ_j represents $\partial\Omega$ in $C(x^{(j)}, R_j, r_j, \delta_j)$, and

$$\partial\Omega = \bigcup_{j=1}^{m} (C(x^{(j)}, R_j, r_j, \delta_j) \cap \partial\Omega) .$$

Let Λ be a Lebesgue number for such a finite cover of $\partial\Omega$ (cf. Theorem A.6). Let ω_{γ_j} be as in (i) for the coordinate cylinder $C(x^{(j)}, R_j, r_j, \delta_j)$ for all $j \in \{1, \ldots, m\}$. Then we set

$$\tilde{\omega}(t) \equiv (n-1) \sup_{j\in\{1,\ldots,m\}} \omega_{\gamma_j}(t) \qquad \forall t \in]0, +\infty[.$$

Since each ω_{γ_j} satisfies the conditions in (2.62), then also $\tilde{\omega}$ satisfies the conditions in (2.62). We now estimate ω_Ω. By the Schwarz inequality, we have

$$M_1 \equiv \sup_{(x,y)\in(\partial\Omega)^2, |x-y|\geq\Lambda} \left| \frac{(y-x)\cdot \nu_\Omega(y)}{|x-y|} \right| \leq 1 < +\infty . \tag{2.64}$$

Then we turn to show that

$$|(y-x)\cdot\nu_\Omega(y)| \leq \tilde{\omega}(|x-y|)|x-y| \qquad \forall (x,y) \in (\partial\Omega)^2 , |x-y| \leq \Lambda . \tag{2.65}$$

Let x, $y \in \partial\Omega$ be such that $|x-y| \leq \Lambda$. Since Λ is a Lebesgue number, there exists $j \in \{1, \ldots, m\}$ such that $\{x, y\} \subseteq C(x^{(j)}, R_j, r_j, \delta_j) \cap \partial\Omega$. Now let ξ, η in $\mathbb{B}_{n-1}(0, r_j)$ be such that

$$x = \psi_{x^{(j)}}(\xi) \qquad y = \psi_{x^{(j)}}(\eta) ,$$

where $\psi_{x^{(j)}}$ is the parametrization of $\partial\Omega$ in $C(x^{(j)}, R_j, r_j, \delta_j)$ associated to γ_j. Then we have

$$\begin{aligned}
&|(y-x)\cdot\nu_\Omega(y)| \\
&= \left| \left(x^{(j)} + R_j^t \begin{pmatrix} \eta \\ \gamma_j(\eta) \end{pmatrix} - x^{(j)} - R_j^t \begin{pmatrix} \xi \\ \gamma_j(\xi) \end{pmatrix} \right) \right. \\
&\qquad\qquad \left. \cdot \left(R_j^t \begin{pmatrix} -\nabla\gamma_j(\eta) \\ 1 \end{pmatrix} \frac{1}{\sqrt{1+|\nabla\gamma_j(\eta)|^2}} \right) \right| \\
&= \left| \begin{pmatrix} \eta - \xi \\ \gamma_j(\eta) - \gamma_j(\xi) \end{pmatrix} \cdot \left(\begin{pmatrix} -\nabla\gamma_j(\eta) \\ 1 \end{pmatrix} \frac{1}{\sqrt{1+|\nabla\gamma_j(\eta)|^2}} \right) \right|
\end{aligned}$$

$$= |-(\eta - \xi) \cdot \nabla\gamma_j(\eta) + (\gamma_j(\eta) - \gamma_j(\xi))| \frac{1}{\sqrt{1 + |\nabla\gamma_j(\eta)|^2}} .$$

Now by statement (i), we have

$$\begin{aligned} &|-(\eta - \xi) \cdot \nabla\gamma_j(\eta) + (\gamma_j(\eta) - \gamma_j(\xi))| \frac{1}{\sqrt{1 + |\nabla\gamma_j(\eta)|^2}} \\ &\quad \leq (n-1)\omega_{\gamma_j}(|\xi - \eta|)|\xi - \eta| \\ &\quad \leq (n-1)\omega_{\gamma_j}(|\psi_{x^{(j)}}(\xi) - \psi_{x^{(j)}}(\eta)|)|\psi_{x^{(j)}}(\xi) - \psi_{x^{(j)}}(\eta)| \\ &\quad = (n-1)\omega_{\gamma_j}(|x - y|)|x - y| \leq \tilde{\omega}(|x - y|)|x - y| . \end{aligned}$$

Indeed,

$$\begin{aligned} |\xi - \eta| &\leq |(\xi - \eta, \gamma_j(\xi) - \gamma_j(\eta))| \\ &= |R_j^t(\xi - \eta, \gamma_j(\xi) - \gamma_j(\eta))^t| = |\psi_{x^{(j)}}(\xi) - \psi_{x^{(j)}}(\eta)| . \end{aligned}$$

Hence, inequality (2.65) holds true. By inequalities (2.64) and (2.65), we have

$$\begin{aligned} &\omega_\Omega(t) \leq \tilde{\omega}(t) \qquad \forall t \in]0, \Lambda] , \\ &\omega_\Omega(t) \leq \sup\{M_1, \tilde{\omega}(t)\} \qquad \forall t \in [\Lambda, +\infty[. \end{aligned}$$

In particular, $\omega_\Omega(t) < +\infty$ for all $t \in]0, +\infty[$. By definition of ω_Ω, the function ω_Ω is increasing. Since $\lim_{t\to 0} \tilde{\omega}(t) = 0$, we have $\lim_{t\to 0} \omega_\Omega(t) = 0$ and thus ω_Ω satisfies all conditions in (2.62).

Finally, we prove statement (iii). If Ω is of class $C^{1,\alpha}$, then each function γ_j is of cass $C^{1,\alpha}$ and we can take

$$\omega_{\gamma_j}(t) = \max_{l=1,\dots,n-1} \left| \frac{\partial\gamma_j}{\partial\eta_l} : \overline{\mathbb{B}_{n-1}(0, r_j)} \right|_\alpha t^\alpha \qquad \forall t \in]0, +\infty[$$

for all $j \in \{1, \dots, m\}$. Then we have

$$\tilde{\omega}(t) = (n-1) \max_{\substack{l=1,\dots,n-1, \\ j=1,\dots,m}} \left| \frac{\partial\gamma_j}{\partial\eta_l} : \overline{\mathbb{B}_{n-1}(0, r_j)} \right|_\alpha t^\alpha , \quad \omega_\Omega(t) \leq \tilde{\omega}(t)$$

for all $t \in]0, \Lambda]$, and the above proof of statement (ii) ensures that

$$c_{\Omega,\alpha} \leq \max\left\{ \frac{M_1}{\Lambda^\alpha}, \sup_{(x,y)\in(\partial\Omega)^2, 0<|x-y|\leq\Lambda} \frac{\tilde{\omega}(|x-y|)}{|x-y|^\alpha} \right\} \leq \max\left\{ \frac{M_1}{\Lambda^\alpha}, \tilde{\omega}(1) \right\} .$$

□

Lemma 2.77. *Let $n \in \mathbb{N} \setminus \{0, 1\}$. Let $\alpha \in]0, 1]$. Let Ω be a bounded open subset $\mathbb{R}^n$ of class $C^{1,\alpha}$. Then the outward unit normal field ν_Ω is Hölder continuous on $\partial\Omega$ with exponent α.*

Proof. It suffices to exploit the covering argument of the proof of statement (ii) of Lemma 2.76, and the α-Hölder continuity of

$$\nu_\Omega(\psi_{x_j}(\cdot)) = R_j^t \begin{pmatrix} -\nabla\gamma_j(\cdot) \\ 1 \end{pmatrix} \frac{1}{\sqrt{1+|\nabla\gamma_j(\cdot)|^2}}$$

in the ball $\overline{\mathbb{B}_{n-1}(0,r_j)}$, which follows by the α–Hölder continuity of $\nabla\gamma_j$. Indeed, the pointwise product of α-Hölder continuous functions in $\overline{\mathbb{B}_{n-1}(0,r_j)}$ is α-Hölder continuous, the function $\sqrt{1+t}$ is Lipschitz continuous in $[0,+\infty[$, and the reciprocal of a nonvanishing α-Hölder continuous function in $\overline{\mathbb{B}_{n-1}(0,r_j)}$ is α-Hölder continuous. □

2.18 Existence of Tubular Neighborhoods of the Boundary of Bounded Open Sets

In Section 2.19, we plan to prove a sufficient condition for a function of class C^1 in a bounded open subset Ω of $\mathbb{R}^n$ of class $C^{1,\alpha}$ to be Hölder continuous and we plan to do so by analyzing separately the set of points which belong to a compact subset of Ω and the set of points of Ω which belong to an open neighborhood of $\partial\Omega$ in $\mathbb{R}^n$.

Thus we need to introduce an appropriate neighborhood of $\partial\Omega$ in $\mathbb{R}^n$ and we want such a neighborhood to be 'globally parametrized'. Thus it is natural to think of the set

$$\{x + t\nu_\Omega(x) : t \in]-t_0, t_0[\,,\ x \in \partial\Omega\},$$

for some small $t_0 \in]0,+\infty[$. Unfortunately however, the map ν_Ω is only of class $C^{0,\alpha}$ if Ω of class $C^{1,\alpha}$ (cf. Lemma 2.77). Thus, the corresponding 'global correspondence'

$$(t,x) \mapsto x + t\nu_\Omega(x)$$

between the points of $]-t_0,t_0[\times(\partial\Omega)$ and the points of the above set cannot be expected to be Lipschitz continuous as we need it to be both to prove that such a correspondence is injective and to carry out our proof of the Hölder continuity of a suitable class of C^1 functions on Ω. Then we follow the paper [186] with Rossi and we prove the following two lemmas, which show the existence of a sufficiently small $t_0 \in]0,+\infty[$ and of a unit vector field a of class C^∞ in a neighborhood of $\partial\Omega$ such that the map from $]-t_0,t_0[\times(\partial\Omega)$ to $\mathbb{R}^n$ that takes (t,x) to $x+ta(x)$ is injective and Lipschitz continuous and such that the set

$$\{x + ta(x) : t \in]-t_0, t_0[\,,\ x \in \partial\Omega\}$$

is an open neighborhood of $\partial\Omega$ in $\mathbb{R}^n$.

Lemma 2.78. *Let $n \in \mathbb{N}\setminus\{0,1\}$. Let Ω be a bounded open subset of $\mathbb{R}^n$ of class C^1. Let $\vartheta \in]0,1[$. Then there exist an open neighborhood U of class C^∞ of $\partial\Omega$, $a \in C^\infty(\overline{U},\mathbb{R}^n)$, and $\tau \in]0,+\infty[$ such that the following conditions are fulfilled*

$$
\begin{aligned}
&|a(x)| = 1 \qquad \forall x \in \overline{U}\,, \\
&\sup_{\partial\Omega} |a - \nu_\Omega| < \vartheta\,, \\
&\inf_{\partial\Omega} a \cdot \nu_\Omega > 1 - \frac{\vartheta^2}{2} > 1 - \vartheta\,, \\
&|a(x) \cdot (y - x)| \le \vartheta |x - y| \qquad \forall x, y \in \partial\Omega\,,\ |x - y| < \tau\,.
\end{aligned}
\tag{2.66}
$$

Proof. Since $\nu_\Omega \in C^0(\partial\Omega, \mathbb{R}^n)$, the Tietze Extension Theorem 2.33 (see also Theorem 2.85 below) implies that there exists $\tilde{\nu} \in C^0_c(\mathbb{R}^n)$ such that

$$\tilde{\nu}_{|\partial\Omega} = \nu_\Omega \qquad \text{on } \partial\Omega\,.$$

Then the set $\{x \in \mathbb{R}^n : |\tilde{\nu}(x)| > \vartheta\}$ is an open neighborhood of $\partial\Omega$ and Lemma 2.70 implies that there exists an open bounded subset U of $\mathbb{R}^n$ of class C^∞ such that

$$\partial\Omega \subseteq U \subseteq \overline{U} \subseteq \{x \in \mathbb{R}^n : |\tilde{\nu}(x)| > \vartheta\}\,.$$

Let $\{\varphi_\epsilon\}_{\epsilon>0}$ be a family of standard mollifiers. Then we have

$$\lim_{\epsilon \to 0} \varphi_\epsilon * \tilde{\nu} = \tilde{\nu}$$

uniformly on $\overline{U}$ (cf. Proposition A.33). Then there exists $\epsilon_1 \in]0, 1[$ such that

$$\left| \frac{\varphi_\epsilon * \tilde{\nu}}{|\varphi_\epsilon * \tilde{\nu}|} - \frac{\tilde{\nu}}{|\tilde{\nu}|} \right| < \vartheta/4 \qquad \text{on } \overline{U}$$

for all $\epsilon \in]0, \epsilon_1]$ and we can set

$$a(x) \equiv \frac{\varphi_{\epsilon_1} * \tilde{\nu}(x)}{|\varphi_{\epsilon_1} * \tilde{\nu}(x)|} \qquad \forall x \in \overline{U}\,.$$

Since $|a(x) - \nu_\Omega(x)| < \vartheta/4 < \vartheta$ for all $x \in \partial\Omega$, we have

$$|a(x)|^2 + |\nu_\Omega(x)|^2 - 2a(x) \cdot \nu_\Omega(x) < \vartheta^2 \qquad \forall x \in \partial\Omega$$

and accordingly

$$2 - \vartheta^2 < 2a(x) \cdot \nu_\Omega(x) \qquad \forall x \in \partial\Omega\,.$$

Hence, $\min_{\partial\Omega} a \cdot \nu_\Omega > 1 - \frac{\vartheta^2}{2}$ and the second last inequality in (2.66) holds true.

By Lemma 2.76, the function ω_Ω from $[0, +\infty[$ to itself is increasing and satisfies the equality $\omega_\Omega(0) = 0$ and the inequality

$$|\nu_\Omega(x) \cdot (y - x)| \le \omega_\Omega(|x - y|)|x - y| \qquad \forall x, y \in \partial\Omega\,.$$

Then we have

$$|a(x) \cdot (y - x)| \le |a(x) - \nu_\Omega(x)||y - x| + |\nu_\Omega(x) \cdot (y - x)|$$

$$\leq ((\vartheta/4) + \omega_\Omega(|x-y|))|x-y| \qquad \forall x, y \in \partial\Omega\,.$$

Since $\lim_{t\to 0} \omega_\Omega(t) = 0$, there exists $\tau \in]0, +\infty[$ such that

$$\omega_\Omega(t) < \vartheta/2 \qquad \forall t \in [0, \tau]$$

and thus all the conditions in (2.66) are satisfied. □

We note that Fichera [98, pp. 207-208] already recognized the importance of the existence of a vector field a such that the essential infimum of $a \cdot \nu_\Omega$ is bounded away from 0 for sets with a piecewise smooth boundary in the analysis of boundary value problems for systems of partial differential equations, and in particular for the analysis of boundary value problems with unilateral constraints (see also Fichera [99, p. 413]). For the introduction of a vector field such that the essential infimum of $a \cdot \nu_\Omega$ is bounded away from 0 in the case of Lipschitz sets, we mention Grisvard [112, Lemma 1.5.1.9].

A unit vector field a as in the previous lemma can play the role of a normal in the definition of a tubular neighborhood of $\partial\Omega$, as the following statement shows (see also reference [186] with Rossi for a related result which holds under stronger assumptions on Ω).

Lemma 2.79. *Let $n \in \mathbb{N} \setminus \{0, 1\}$. Let $m \in \mathbb{N} \setminus \{0\}$. Let Ω be a bounded open subset of $\mathbb{R}^n$ of class C^1. If an open neighborhood U of $\partial\Omega$ and $a \in C^1(\overline{U}, \mathbb{R}^n)$ satisfy conditions (2.66) for some $\vartheta \in]0, 1[$ and $\tau \in]0, +\infty[$, then there exists $t_1 \in]0, +\infty[$ such that the following statements hold.*

(i) The map Ψ from $(\partial\Omega)\times] - t_1, t_1[$ to $\mathbb{R}^n$ defined by

$$\Psi(x, t) = x + ta(x) \qquad \forall (x, t) \in (\partial\Omega)\times] - t_1, t_1[$$

is injective, and the set

$$A(t_2) \equiv \Psi((\partial\Omega)\times] - t_2, t_2[)$$

is an open neighborhood of $\partial\Omega$ for all $t_2 \in]0, t_1]$.

(ii) We have

$$x + ta(x) \in \Omega \qquad \forall t \in] - t_1, 0[\quad \text{and} \quad x + ta(x) \in \Omega^- \qquad \forall t \in]0, t_1[\,.$$

Proof. By Lemma 2.63 of the uniform cylinders, there exist $r, \delta \in]0, +\infty[$ such that

$$r < \delta \leq \tau/2\,,$$

and such that, for each $x \in \partial\Omega$, there exists $R_x \in \mathbb{O}_n(\mathbb{R})$ such that $C(x, R_x, r, \delta)$ is a coordinate cylinder for Ω around x. Possibly shrinking r we can assume that the function γ_x which represents $\partial\Omega$ in $C(x, R_x, r, \delta)$ as a graph is of class $C^1(\overline{\mathbb{B}_n(0, r)},] - \delta/2, \delta/2[)$. We first show the existence of t_1 as in (i). Assume by contradiction that Ψ is not injective for any choice of t_1. Then for each $j \in \mathbb{N}$ there exist $(x'_j, t'_j), (x''_j, t''_j) \in \partial\Omega\times] - 2^{-j}, 2^{-j}[$ such that

$$(x'_j, t'_j) \neq (x''_j, t''_j), \qquad \Psi(x'_j, t'_j) = \Psi(x''_j, t''_j).$$

In particular,

$$x'_j \neq x''_j \qquad \forall j \in \mathbb{N}.$$

Indeed, if $x'_j = x''_j$, then $a(x'_j) = a(x''_j)$ and the equality

$$x'_j + t'_j a(x'_j) = x''_j + t''_j a(x''_j)$$

implies that

$$t'_j a(x'_j) = t''_j a(x'_j)$$

and thus $t'_j = t''_j$, contrary to $(x'_j, t'_j) \neq (x''_j, t''_j)$. Possibly selecting a subsequence, we can assume that there exist x', $x'' \in \partial\Omega$ such that

$$x' = \lim_{j\to\infty} x'_j, \qquad x'' = \lim_{j\to\infty} x''_j.$$

Since a is continuous and $\lim_{j\to\infty} t'_j = \lim_{j\to\infty} t''_j = 0$, we have

$$x' = \lim_{j\to\infty} x'_j + t'_j a(x'_j) = \lim_{j\to\infty} x''_j + t''_j a(x''_j) = x''$$

and we set $\tilde{x} \equiv x' = x''$. By our contradiction assumption, we have the equality

$$0 = x''_j + t''_j a(x''_j) - x'_j - t'_j a(x'_j) \qquad \forall j \in \mathbb{N}$$

which we rewrite as

$$0 = (x''_j - x'_j) + (t''_j - t'_j) a(x''_j) + t'_j (a(x''_j) - a(x'_j)) \qquad \forall j \in \mathbb{N} \tag{2.67}$$

and, by dividing by $|x'_j - x''_j|$, we obtain

$$\frac{t''_j - t'_j}{|x''_j - x'_j|} a(x''_j) = -\frac{x''_j - x'_j}{|x''_j - x'_j|} - t'_j \frac{a(x''_j) - a(x'_j)}{|x''_j - x'_j|} \qquad \forall j \in \mathbb{N}.$$

Then the triangular inequality implies that

$$\left| \frac{t''_j - t'_j}{|x''_j - x'_j|} \right| \leq 1 + |t'_j| \mathrm{Lip}(a) \qquad \forall j \in \mathbb{N}. \tag{2.68}$$

Next, we go back to the above equality (2.67), we take the scalar product with $\frac{x''_j - x'_j}{|x''_j - x'_j|^2}$, and obtain

$$0 = 1 + \frac{t''_j - t'_j}{|x''_j - x'_j|} a(x''_j) \cdot \frac{x''_j - x'_j}{|x''_j - x'_j|} + t'_j \frac{a(x''_j) - a(x'_j)}{|x''_j - x'_j|} \cdot \frac{x''_j - x'_j}{|x''_j - x'_j|} \qquad \forall j \in \mathbb{N}. \tag{2.69}$$

In order to exploit the last inequality in (2.66), we choose $j_0 \in \mathbb{N}$ so that

$$|x''_j - x'_j| \le |x''_j - \tilde{x}| + |\tilde{x} - x'_j| < \tau \qquad \forall j_0 \le j \in \mathbb{N}$$

and thus inequality (2.68) and equality (2.69) imply that

$$1 \le (1 + |t'_j|\mathrm{Lip}(a))\vartheta + |t'_j|\mathrm{Lip}(a) \qquad \forall j_0 \le j \in \mathbb{N}.$$

Since $\lim_{j\to\infty} t'_j = 0$, we obtain $1 \le \vartheta$, a contradiction. Hence, there exists t_1 in $]0, +\infty[$ such that Ψ is injective on $(\partial\Omega)\times] - t_1, t_1[$.

We now turn to show that $A(t_2)$ is open for all $t_2 \in]0, t_1]$. Let $t_2 \in]0, t_1]$ and $(x^\sharp, t^\sharp) \in A(t_2)$. Since Ψ is injective, then the composition $\Psi^\sharp$ of Ψ with the continuous and injective map $(x^\sharp + R^t_{x^\sharp}(\eta, \gamma(\eta))^t, s)$ of the variable (η, s), i.e., the map

$$\begin{aligned}\Psi^\sharp(\eta, s) &\equiv \Psi(x^\sharp + R^t_{x^\sharp}(\eta, \gamma(\eta))^t, s)\\ &= x^\sharp + R^t_{x^\sharp}(\eta, \gamma(\eta))^t + sa(x^\sharp + R^t_{x^\sharp}(\eta, \gamma(\eta))^t)\\ &\qquad\qquad \forall(\eta, s) \in \mathbb{B}_{n-1}(0, r)\times] - t_2, t_2[\,,\end{aligned}$$

is continuous and injective. Then the Theorem of Invariance of Domain A.32 implies that $\Psi^\sharp(\mathbb{B}_{n-1}(0, r)\times] - t_2, t_2[)$ is an open subset of $\mathbb{R}^n$. Since $(x^\sharp, t^\sharp) \in \Psi^\sharp(\mathbb{B}_{n-1}(0, r)\times] - t_2, t_2[) \subseteq A(t_2)$, it follows that $(x^\sharp, t^\sharp)$ is interior to $A(t_2)$. Hence, $A(t_2)$ is open.

Next we turn to prove statement (ii). Let $x \in \partial\Omega$. Then we set

$$A^+(x, t_1) \equiv \{x + ta(x) : t \in]-t_1, 0[\}, \qquad A^-(x, t_1) \equiv \{x + ta(x) : t \in]0, t_1[\}.$$

Since Ψ is injective, we have

$$\Psi(\{x\}\times] - t_1, t_1[) \cap \partial\Omega = \{x\}.$$

Indeed, if there exists $t \in] - t_1, t_1[$ such that

$$y = x + ta(x) \in \partial\Omega,$$

then

$$\Psi(y, 0) = \Psi(x, t)$$

and the injectivity of Ψ implies that $x = y$, $t = 0$. Hence,

$$A^\pm(x, t_1) \subseteq \mathbb{R}^n \setminus \partial\Omega.$$

Since the sets $A^\pm(x, t_1)$ are connected, they cannot intersect both Ω and Ω^-. Indeed, the connected components of Ω and Ω^- are disjoint. We now show that

$$A^+(x, t_1) \subseteq \Omega \quad \text{and} \quad A^-(x, t_1) \subseteq \Omega^-.$$

By the third condition of (2.66) and by Proposition 2.44, $-a(x)$ points to the interior of Ω and thus there exists $\varsigma_{x,-a(x)} > 0$ such that

$$x + ta(x) \in \Omega \qquad \forall t \in] - \varsigma_{x,-a(x)}, 0[\,,$$

and thus in particular for all $t \in] - \min\{\varsigma_{x,-a(x)}, t_1\}, 0[$. Hence, the connected set $A^+(x, t_1)$ contains points of Ω and thus it cannot contain points of Ω^-. As a consequence, we must have $A^+(x, t_1) \subseteq \Omega$. The inclusion $A^-(x, t_1) \subseteq \mathbb{R}^n \setminus \overline{\Omega}$ can be proved similarly. □

2.19 A Sufficient Condition for the Hölder Continuity of Continuously Differentiable Functions, in the Wake of the Work of Carlo Miranda

In this section we plan to prove a sufficient condition for a real valued continuously differentiable function f in an open subset Ω of $\mathbb{R}^n$ to be α-Hölder continuous in Ω for some $\alpha \in]0, 1]$.

If $\Omega =]0, 1[$, then an elementary sufficient condition is that $|y|^{1-\alpha}|f'(y)|$ is bounded for $y \in]0, 1[$.

If $\Omega = \mathbb{B}_{n-1}(0, 1) \times]0, 1[$, then a sufficient condition of the α-Hölder continuity of $f \in C^1(\Omega)$ is that

$$|y|^{1-\alpha}|\nabla f(\eta, y)| \qquad \text{is bounded in } (\eta, y) \in \mathbb{B}_{n-1}(0, 1) \times]0, 1[$$

(cf. Agmon, Douglis and Nirenberg [2, p. 717]).

If $\gamma \in C^1(\overline{\mathbb{B}_{n-1}(0, r)},] - \delta, \delta[)$ for some r, $\delta \in]0, +\infty[$ and if

$$\Omega = \text{hypograph}_s(\gamma)\,,$$

then Miranda [212, p. 308] has shown that a sufficient condition for the α-Hölder continuity of $f \in C^1(\Omega)$ is that

$$|\gamma(\eta) - y|^{1-\alpha}|\nabla f(\eta, y)| \qquad \text{is bounded in } (\eta, y) \in \text{hypograph}_s(\gamma)\,.$$

Then in order to obtain a sufficient condition for a function defined in an open subset Ω of $\mathbb{R}^n$ of class C^1, one can exploit the above result of Miranda and a standard localization argument based on the existence of a finite cover of $\partial\Omega$ consisting of coordinate cylinders and of a corresponding partition of unity. However, later in the book it will be convenient to have a condition that is independent from the choice of a specific atlas of $\partial\Omega$. For this reason, we prove a global condition in the following Proposition 2.82, where we exploit the approximate outer unit normal a of Lemma 2.78. For a treatment in the case where Ω is a Lipschitz set, we refer to [170]. To prove Proposition 2.82, we need the following two preliminary technical Lemmas 2.80 and 2.81.

Lemma 2.80. *Let $n \in \mathbb{N} \setminus \{0, 1\}$. Let $\vartheta \in]0, 1[$. If v, $w \in \mathbb{R}^n$ and if*

$$|v \cdot w| \leq \vartheta |v|\, |w|\,,$$

then $|v+w|^2 \geq (1-\vartheta)(|v|^2+|w|^2)+\vartheta(|v|-|w|)^2$.

Proof. It suffices to note that

$$\begin{aligned}|v+w|^2 &= |v|^2+|w|^2+2v\cdot w\\ &\geq |v|^2+|w|^2-2|v\cdot w| \geq |v|^2+|w|^2-2\vartheta|v||w|\\ &= (1-\vartheta)(|v|^2+|w|^2)+\vartheta(|v|^2+|w|^2-2|v||w|)\\ &= (1-\vartheta)(|v|^2+|w|^2)+\vartheta(|v|-|w|)^2\,.\end{aligned}$$

□

Lemma 2.81. *Let* $n \in \mathbb{N}\setminus\{0,1\}$. *Let* Ω *be a bounded open subset of* $\mathbb{R}^n$ *of class* C^1. *Let* $\vartheta \in]0,1[$. *Let* U *be an open neighborhood of* $\partial\Omega$ *of class* C^∞. *Let* a *in* $C^\infty(\overline{U},\mathbb{R}^n)$ *satisfy the conditions in (2.66) for some* $\vartheta \in]0,1[$ *and* $\tau \in]0,+\infty[$. *Let* $t_1 \in]0,+\infty[$ *be as in Lemma 2.79. Let* $p \in \partial\Omega$. *Let* r, $\delta \in]0,+\infty[$,

$$2\delta < \min\{\tau/2, t_1\}\,, \qquad r<\delta\,,$$

and $R_p \in \mathbb{O}_n(\mathbb{R})$ *be such that* $C(p,R_p,r,\delta)$ *is a coordinate cylinder for* Ω *around* p. *Let* $\gamma_p \in C^1(\overline{\mathbb{B}_{n-1}(0,r)})$ *represent* $\partial\Omega$ *in* $C(p,R_p,r,\delta)$ *as a graph. Let*

$$C(p,R_p,r,\delta)\cap\Omega \subseteq A^+(t_1) \equiv \{x+ta(x)\,:\, t\in]-t_1,0[, x\in\partial\Omega\}\,.$$

Let

$$t_2 \in \left]0, \min\left\{\frac{r}{4}, \frac{(1-\vartheta)^{1/2}}{2\sqrt{2}(\mathrm{Lip}(a)+1)}, \frac{t_1}{2}\right\}\right[\,.$$

If $\alpha \in]0,1]$, *then there exists* $B \in]0,+\infty[$ *such that*

$$|f\,:\,[p+R_p^t(\mathbb{B}_{n-1}(0,r/4)\times]-\delta/4,\delta/4[)]\cap A^+(t_2)|_\alpha \leq BM_{t_1,\alpha}(f) \tag{2.70}$$

for all $f\in C^1(\Omega)$ *such that*

$$M_{t_1,\alpha}(f) \equiv \sup\{|t|^{1-\alpha}|\nabla f(x+ta(x))|\,:\,(x,t)\in(\partial\Omega)\times]-t_1,0[\} < +\infty\,. \tag{2.71}$$

Proof. Let p', $p'' \in [p+R_p^t(\mathbb{B}_{n-1}(0,r/4)\times]-\delta/4,\delta/4[)]\cap A^+(t_2)$. We plan to estimate the difference $|f(p')-f(p'')|$ and show that it is smaller than a constant times $M_{t_1,\alpha}(f)|p'-p''|^\alpha$. In order to exploit condition (2.71) on f, we plan to to define a C^1 path in $A^+(t_1)$ that joins p' and p''. Lemma 2.79 implies that there exist $x', x'' \in \partial\Omega$, $t', t'' \in]-t_2,0[$ such that

$$p' = x'+t'a(x')\,, \qquad p''=x''+t''a(x'')\,.$$

By the triangular inequality and by the equality $|a(x')|=1$, and by the inequality $t_2<r/4<\delta/4$, we have

$$x' \in \mathbb{B}_n(p',t_2) \subseteq [p+R_p^t(\mathbb{B}_{n-1}(0,r/4)\times]-\delta/4,\delta/4[)]+\mathbb{B}_n(0,t_2)$$

$$\subseteq [p + R_p^t(\mathbb{B}_{n-1}(0, r/2) \times] - \delta/2, \delta/2[)]$$

and similarly

$$x'' \in [p + R_p^t(\mathbb{B}_{n-1}(0, r/2) \times] - \delta/2, \delta/2[)] .$$

Next we set

$$\psi_p(\eta) \equiv p + R_p^t(\eta, \gamma_p(\eta))^t \qquad \forall \eta \in \mathbb{B}_{n-1}(0, r) .$$

Since x', $x'' \in \partial\Omega$, then there exist ξ', $\xi'' \in \mathbb{B}_{n-1}(0, r/2)$ such that

$$x' = \psi_p(\xi') = p + R_p^t(\xi', \gamma_p(\xi'))^t ,$$
$$x'' = \psi_p(\xi'') = p + R_p^t(\xi'', \gamma_p(\xi''))^t .$$

Then we note that the convexity of $\mathbb{B}_{n-1}(0, r/2)$ implies that

$$y(s) \equiv (1 - s)\xi' + s\xi'' \in \mathbb{B}_{n-1}(0, r/2) \qquad \forall s \in [0, 1]$$

and that the convexity of $] - t_2, 0[$ implies that

$$(1 - s)t' + st'' \in] - t_2, 0[\qquad \forall s \in [0, 1] .$$

Then we have

$$\begin{aligned} G(s) &\equiv \psi_p(y(s)) \\ &\quad +[(1 - s)t' + st'']a(\psi_p(y(s))) \in A^+(t_2) \qquad \forall s \in [0, 1] , \\ G(s) &\in [p + R_p^t(\mathbb{B}_{n-1}(0, r/2) \times] - \delta/2, \delta/2[)] + \mathbb{B}_n(0, t_2) \\ &\quad \subseteq [p + R_p^t(\mathbb{B}_{n-1}(0, 3r/4) \times] - 3\delta/4, 3\delta/4[)] \quad \forall s \in [0, 1] , \\ G(0) &= p' , \qquad G(1) = p'' . \end{aligned}$$

Moreover,

$$\begin{aligned} \text{length}(G) &\leq \int_0^1 \sqrt{|\xi' - \xi''|^2 + \text{Lip}^2(\gamma_p)|\xi' - \xi''|^2}\, ds + |t' - t''| \\ &+|t_2|\text{Lip}\,(a) \int_0^1 \sqrt{|\xi' - \xi''|^2 + \text{Lip}^2(\gamma_p)|\xi' - \xi''|^2}\, ds \\ &\leq |\xi' - \xi''|\sqrt{1 + \text{Lip}^2(\gamma_p)} + |t' - t''| + |t_2|\text{Lip}\,(a)|\xi' - \xi''|\sqrt{1 + \text{Lip}^2(\gamma_p)} \\ &\leq |x' - x''|\sqrt{1 + \text{Lip}^2(\gamma_p)}(1 + |t_2|\text{Lip}\,(a)) + |t' - t''| . \end{aligned}$$

We now wish to estimate $|x' - x''|$ and $|t' - t''|$ in terms of $|p' - p''|$. There is no loss of generality in assuming that

$$t'' < t' .$$

By assumption, we know that

$$\left|a(x) \cdot \frac{y-x}{|y-x|}\right| \leq \vartheta \qquad \forall x, y \in C(p, R_p, r, \delta) \cap (\partial \Omega), \; x \neq y .$$

Indeed, $|x-y| \leq 2r+\delta < \tau$ for all $x, y \in C(p, R_p, r, \delta) \cap (\partial \Omega)$. Then Lemma 2.80 implies that

$$\begin{aligned}
|p' - p''| &= |(x' + t'a(x')) - (x'' + t''a(x''))| \\
&= |(x' - x'') + (t' - t'')a(x') - t''(a(x'') - a(x'))| \\
&\geq |(x' - x'') + (t' - t'')a(x')| - |t''|\,|(a(x'') - a(x'))| \\
&\geq (1-\vartheta)^{1/2}\sqrt{|x' - x''|^2 + |t' - t''|^2 |a(x')|^2} - t_2 \mathrm{Lip}(a)|x' - x''| .
\end{aligned}$$

By the elementary inequality

$$a_1 + a_2 \leq \sqrt{2}(a_1^2 + a_2^2)^{1/2} \qquad \forall a_1, a_2 \in [0, +\infty[,$$

we have

$$\begin{aligned}
&(1-\vartheta)^{1/2}\sqrt{|x' - x''|^2 + |t' - t''|^2 |a(x')|^2} - t_2 \mathrm{Lip}(a)|x' - x''| \\
&\quad \geq \frac{(1-\vartheta)^{1/2}}{\sqrt{2}}(|x' - x''| + |t' - t''|) - t_2 \mathrm{Lip}(a)|x' - x''| \\
&\quad = \left(\frac{(1-\vartheta)^{1/2}}{\sqrt{2}} - t_2 \mathrm{Lip}(a)\right)|x' - x''| + \frac{(1-\vartheta)^{1/2}}{\sqrt{2}}|t' - t''| \\
&\quad \geq \frac{(1-\vartheta)^{1/2}}{2\sqrt{2}}|x' - x''| + \frac{(1-\vartheta)^{1/2}}{\sqrt{2}}|t' - t''| .
\end{aligned}$$

Indeed, $t_2 \leq \frac{(1-\vartheta)^{1/2}}{2\sqrt{2}(1+\mathrm{Lip}(a))}$. Hence, we deduce that

$$|t' - t''| \leq \frac{\sqrt{2}}{(1-\vartheta)^{1/2}}|p' - p''| , \qquad |x' - x''| \leq \frac{2\sqrt{2}}{(1-\vartheta)^{1/2}}|p' - p''| . \quad (2.72)$$

At this point we may try to use the path G to estimate the difference $|f(p') - f(p'')|$ for a function $f \in C^1(\Omega)$ that satisfies condition (2.71). Indeed, we have

$$\begin{aligned}
&|f(p') - f(p'')| \leq \int_0^1 |\nabla f(G(s))|\,|G'(s)|\,ds \\
&\leq \int_0^1 M_{t_1,\alpha}(f)\,|(1-s)t' - st''|^{\alpha-1}|G'(s)|\,ds \\
&\leq M_{t_1,\alpha}(f)\,|t'|^{\alpha-1}\mathrm{length}(G) \\
&\leq M_{t_1,\alpha}(f)\,|t'|^{\alpha-1}
\end{aligned}$$

$$\times \left(\frac{2\sqrt{2}}{(1-\vartheta)^{1/2}} \sqrt{1+\mathrm{Lip}^2(\gamma_p)}(1+|t_2|\mathrm{Lip}\,(a)) + \frac{\sqrt{2}}{(1-\vartheta)^{1/2}} \right) |p'-p''| \,.$$

However, we cannot estimate $|t'|$ from below and, as a consequence, the factor $|t'|^{\alpha-1}$ may be arbitrarily large, making useless the inequality above. To overcome this difficulty we have to modify the path G. Then, we take

$$d_1 \equiv \min\{|p'-p''|, r/4\}$$

and we define a path with endpoints

$$x' + (t'-d_1)a(x') \,, \qquad x'' + (t''-d_1)a(x'') \,.$$

To do so, we set

$$G_{d_1}(s) \equiv G(s) - d_1 a(\psi_p(y(s))) \qquad \forall s \in [0,1] \,.$$

Since

$$\begin{aligned}(1-s)t' + st'' - d_1 &\in]-t_2-d_1, 0[\\ &\subseteq]-(t_1/2)-(r/4), 0[\subseteq]-(t_1/2)-(t_1/8), 0[\subseteq]-t_1, 0[\qquad \forall s \in]0,1[\,,\end{aligned}$$

we have

$$G_{d_1}(s) \in A^+(t_1) \qquad \forall s \in]0,1[\,.$$

Then we have

$$\begin{aligned} G_{d_1}(s) \in [p + R_p^t(\mathbb{B}_{n-1}(0,3r/4) \times]-3\delta/4, 3\delta/4[)] + \mathbb{B}_n(0,d_1) \\ \subseteq C(p, R_p, r, \delta) \end{aligned}$$

for all $s \in [0,1]$ and

$$G_{d_1}(0) = x' + (t'-d_1)a(x') \,, \qquad G_{d_1}(1) = x'' + (t''-d_1)a(x'') \,.$$

Then we have the following inequality for the length of G_{d_1},

$$\begin{aligned} &\mathrm{length}(G_{d_1}) \\ &\quad \leq \mathrm{length}(G) + d_1\mathrm{Lip}(a) \int_0^1 \sqrt{|\xi'-\xi''|^2 + \mathrm{Lip}^2(\gamma_p)|\xi'-\xi''|^2}\, ds \\ &\quad \leq (1+d_1\mathrm{Lip}(a)) \left[|x'-x''| \sqrt{1+\mathrm{Lip}^2(\gamma_p)}(1+|t_2|\mathrm{Lip}\,(a)) + |t'-t''| \right] \\ &\quad \leq (1+d_1\mathrm{Lip}(a)) \\ &\qquad \times \left[\frac{2\sqrt{2}}{(1-\vartheta)^{1/2}} \sqrt{1+\mathrm{Lip}^2(\gamma_p)}(1+|t_2|\mathrm{Lip}\,(a)) + \frac{\sqrt{2}}{(1-\vartheta)^{1/2}} \right] |p'-p''|. \end{aligned}$$

Next we note that

$$x' + (t' - sd_1)a(x') \in [p + R_p^t(\mathbb{B}_{n-1}(0, 3r/4)\times] - 3\delta/4, 3\delta/4[)] + \mathbb{B}_n(0, d_1)$$
$$\subseteq C(p, R_p, r, \delta) \qquad \forall s \in [0, 1]$$

and similarly

$$x'' + (t'' - sd_1)a(x'') \in C(p, R_p, r, \delta) \qquad \forall s \in [0, 1] .$$

Then by the memberships

$$t' - sd_1, t'' - sd_1 \in] - t_2 - d_1, 0[$$
$$\subseteq] - (t_1/2) - (r/4), 0[\subseteq] - (t_1/2) - (t_1/8), 0[\subseteq] - t_1, 0[\qquad \forall s \in [0, 1] ,$$

we have

$$x' + (t' - sd_1)a(x') \in A^+(t_1) , \quad x'' + (t'' - sd_1)a(x'') \in A^+(t_1) \qquad \forall s \in [0, 1] .$$

We now assume that $f \in C^1(\Omega)$ satisfies condition (2.71) and we turn to estimate $|f(p') - f(p'')|$. We have

$$\begin{aligned}
|f(p') - f(p'')| &= |f(x' + t'a(x')) - f(x'' + t''a(x''))| \\
&\leq |f(x' + t'a(x')) - f(x' + (t' - d_1)a(x'))| \\
&\quad + |f(x' + (t' - d_1)a(x')) - f(x'' + (t'' - d_1)a(x''))| \\
&\quad + |f(x'' + (t'' - d_1)a(x'')) - f(x'' + t''a(x''))| \\
&\leq \int_{t'-d_1}^{t'} |a(x') \cdot \nabla f(x' + sa(x'))| \, ds + \int_0^1 |\nabla f(G_{d_1}(s))| \, |G'_{d_1}(s)| \, ds \\
&\quad + \int_{t''-d_1}^{t''} |a(x'') \cdot \nabla f(x'' + sa(x''))| \, ds \\
&\leq \int_{t'-d_1}^{t'} M_{t_1,\alpha}(f)|s|^{\alpha-1} \, ds \\
&\quad + \int_0^1 M_{t_1,\alpha}(f)|(1 - s)t' + st'' - d_1|^{\alpha-1}|G'_{d_1}(s)| \, ds \\
&\quad + \int_{t''-d_1}^{t''} M_{t_1,\alpha}(f)|s|^{\alpha-1} \, ds .
\end{aligned}$$

Since

$$(1 - s)t' + st'' - d_1 \in] - t_2 - d_1, -d_1[\qquad \forall s \in [0, 1] ,$$

we have

$$|f(p') - f(p'')| \leq M_{t_1,\alpha}(f) \int_{|t'|}^{|t'|+d_1} s^{\alpha-1} \, ds \tag{2.73}$$

$$
\begin{aligned}
&+\mathrm{length}(G_{d_1})d_1^{\alpha-1}M_{t_1,\alpha}(f)+M_{t_1,\alpha}(f)\int_{|t''|}^{|t''|+d_1} s^{\alpha-1}\,ds \\
&\le M_{t_1,\alpha}(f)\Bigg\{\frac{d_1^\alpha}{\alpha}+|p'-p''|d_1^{\alpha-1}(1+d_1\mathrm{Lip}(a)) \\
&\quad\times\left[\frac{2\sqrt{2}}{(1-\vartheta)^{1/2}}(1+\mathrm{Lip}(\gamma_p))(1+|t_1|\mathrm{Lip}(a))+\frac{\sqrt{2}}{(1-\vartheta)^{1/2}}\right]+\frac{d_1^\alpha}{\alpha}\Bigg\}
\end{aligned}
$$

(cf. (2.72)). Next we observe that

$$
d_1^\alpha \le |p'-p''|^\alpha
$$

and that

$$
\begin{aligned}
\frac{|p'-p''|}{d_1^{1-\alpha}} &= \frac{|p'-p''|}{\min^{1-\alpha}\{|p'-p''|,r/4\}} \\
&\le \begin{cases} |p'-p''|^\alpha & \text{if } |p'-p''|\le r/4\,, \\ \frac{|p'-p''|}{(r/4)^{1-\alpha}} \le \frac{\mathrm{diam}(\Omega)^{1-\alpha}}{(r/4)^{1-\alpha}}|p'-p''|^\alpha & \text{if } |p'-p''|\ge r/4\,. \end{cases}
\end{aligned}
$$

Hence, inequality (2.73) implies the validity of inequality (2.70) and the proof is complete. □

We are now ready to prove a sufficient condition for the Hölder continuity of a continuously differentiable real valued function defined on an open subset of $\mathbb{R}^n$ of class C^1.

Proposition 2.82. *Let $n\in\mathbb{N}\setminus\{0,1\}$. Let Ω be a bounded open subset of $\mathbb{R}^n$ of class C^1. Let U be an open neighborhood of $\partial\Omega$ of class C^∞. Let $a\in C^\infty(\overline{U},\mathbb{R}^n)$ satisfy the conditions in (2.66) for some $\vartheta\in]0,1[$ and $\tau\in]0,+\infty[$. Let $t_1\in]0,+\infty[$ be as in Lemma 2.79. Let $\alpha\in]0,1]$. Then there exist $B\in]0,+\infty[$ and a compact subset H of Ω such that*

$$
\sup_\Omega |f| + |f:\Omega|_\alpha \le B\max\left\{\sup_H |f|, \sup_H|\nabla f|, M_{t_1,\alpha}(f)\right\}
$$

for all $f\in C^1(\Omega)$ such that

$$
M_{t_1,\alpha}(f)\equiv\sup\{|t|^{1-\alpha}|\nabla f(x+ta(x))| : (x,t)\in(\partial\Omega)\times]-t_1,0[\}<+\infty\,. \tag{2.74}
$$

Proof. Since Ω is of class C^1, for each point $x\in\partial\Omega$ there exists a coordinate cylinder $C(x,R_x,r_x,\delta_x)$ for Ω around x and if $r_x+\delta_x$ is less than the distance between $\partial\Omega$ and $\mathbb{R}^n\setminus A(t_1)$, then we have $C(x,R_x,r_x,\delta_x)\subseteq A(t_1)$. Thus we now choose r_*, $\delta_*\in]0,+\infty[$ such that

$$
r_*+\delta_* < \mathrm{dist}(\partial\Omega,\mathbb{R}^n\setminus A(t_1))\,.
$$

Since we plan to invoke Lemma 2.81, we also assume that

$$2\delta_* < \min\{\tau/2, t_1\},$$

where τ is as in (2.66). Then, by Lemma 2.63 of the uniform cylinders, there exist $r \in]0, r_*[$, $\delta \in]0, \delta_*[$, with $r < \delta$, such that if $x \in \partial\Omega$, then there exists R_x in $\mathbb{O}_n(\mathbb{R})$ such that $C(x, R_p, r, \delta)$ is a coordinate cylinder for Ω around x and the corresponding function γ_x satisfies the condition $\nabla\gamma_x(0) = 0$ and the inequalities

$$\begin{aligned} &|\nabla\gamma_x(\eta)| \leq 1/3 \qquad \forall \eta \in \overline{\mathbb{B}_{n-1}(0, r)}, \\ &\sup_{x\in\partial\Omega} \|\gamma_x\|_{C^1(\overline{\mathbb{B}_{n-1}(0,r)})} < +\infty. \end{aligned} \tag{2.75}$$

Since $r + \delta < \mathrm{dist}(\partial\Omega, \mathbb{R}^n \setminus A(t_1))$, we have

$$C(x, R_x, r, \delta) \cap \Omega \subseteq A^+(t_1) = A(t_1) \cap \Omega \qquad \forall x \in \partial\Omega.$$

Since $\partial\Omega$ is compact, there exists a finite family $\{x^{(j)}\}_{j=1}^m$ of points of $\partial\Omega$ such that

$$\partial\Omega \subseteq \bigcup_{j=1}^m [x^{(j)} + R^t_{x^{(j)}}(\mathbb{B}_{n-1}(0, r/4)\times]-\delta/4, \delta/4[)],$$

and we note that the right-hand side is an open neighborhood of $\partial\Omega$. We now set

$$\mu \equiv \min_{j=1,\dots,m} \Big\{ r/4, \frac{(1-\vartheta)^{1/2}}{2\sqrt{2}(\mathrm{Lip}(a)+1)}, \frac{t_1}{2}, \\ \mathrm{dist}\Big(\partial\Omega, \mathbb{R}^n \setminus \bigcup_{j=1}^m [x^{(j)} + R^t_{x^{(j)}}(\mathbb{B}_{n-1}(0, r/4)\times]-\delta/4, \delta/4[)]\Big)\Big\},$$

and we choose $t_2 \in]0, \mu[$. In particular, we have

$$\begin{aligned} A^+(t_2) \subseteq A(t_2) &\subseteq \{x \in \Omega : \mathrm{dist}(x, \partial\Omega) < \mu\} \\ &\subseteq \bigcup_{j=1}^m [x^{(j)} + R^t_{x^{(j)}}(\mathbb{B}_{n-1}(0, r/4)\times]-\delta/4, \delta/4[)]. \end{aligned}$$

Then Lemma 2.81 implies that there exists $B_j \in]0, +\infty[$ such that

$$|f : [x^{(j)} + R^t_{x^{(j)}}(\mathbb{B}_{n-1}(0, r/4)\times]-\delta/4, \delta/4[)] \cap A^+(t_2)|_\alpha \leq B_j M_{t_1,\alpha}(f) \tag{2.76}$$

for all $j \in \{1, \dots, m\}$ and for all $f \in C^1(\Omega)$ such that $M_{t_1,\alpha}(f) < +\infty$. Now let Ω_1 be an open subset of class C^∞ of Ω such that

$$\overline{\Omega \setminus A^+(t_2)} \subseteq \Omega_1 \subseteq \overline{\Omega_1} \subseteq \Omega$$

(cf. Lemma 2.70). Since Ω_1 is of class C^1, $C^1(\overline{\Omega_1})$ is continuously embedded into $C^{0,\alpha}(\overline{\Omega_1})$ and thus there exists $c_\alpha(\Omega_1) \in]0, +\infty[$ such that

$$|f : \overline{\Omega_1}|_\alpha \leq c_\alpha(\Omega_1) \sup \left\{ \sup_{\overline{\Omega_1}} |f|, \sup_{\overline{\Omega_1}} |\nabla f| \right\} \tag{2.77}$$

for all $f \in C^1(\overline{\Omega_1})$ (cf. Corollary 2.57 (i), (iii)). Then we take $f \in C^1(\Omega)$ such that $M_{t_1,\alpha}(f) < +\infty$ and we turn to estimate $|f : \Omega|_\alpha$. To do so, we observe that

$$\overline{\Omega} \subseteq \Omega_1 \cup \bigcup_{j=1}^{m} \left\{ [x^{(j)} + R^t_{x^{(j)}}(\mathbb{B}_{n-1}(0, r/4) \times] - \delta/4, \delta/4[)] \cap A(t_2) \right\} . \tag{2.78}$$

Let Λ be a Lebesgue number corresponding to the open cover of $\overline{\Omega}$ in the right-hand side of (2.78). We can clearly assume that

$$\Lambda < \delta/12 .$$

If p', $p'' \in \Omega$ and $|p' - p''| \leq \Lambda$, then both p' and p'' belong to at least one of the open sets in the right-hand side of (2.78) and thus inequalities (2.76) and (2.77) imply that

$$|f(p') - f(p'')| \leq \max \left\{ \tilde{B} M_{t_1,\alpha}(f), c_\alpha(\Omega_1) \sup_{\overline{\Omega_1}} |f|, c_\alpha(\Omega_1) \sup_{\overline{\Omega_1}} |\nabla f| \right\} |p' - p''|^\alpha , \tag{2.79}$$

where

$$\tilde{B} \equiv \max_{j \in \{1,\dots,m\}} B_j .$$

In order to estimate $|f(p') - f(p'')|$ in case $|p' - p''| > \Lambda$, we need to estimate $\sup_\Omega |f|$ (cf. Remark 2.20). To do so, we note that

$$x^{(j)} - \frac{t_2}{2} a(x^{(j)}) \in A(t_2)^+ \subseteq \Omega \qquad \forall j \in \{1, \dots, m\} ,$$

and that our assumptions $t_2 < \mu < r/4$, $r < \delta$ imply that

$$x^{(j)} - \frac{t_2}{2} a(x^{(j)}) \in \mathbb{B}_n(x^{(j)}, r/8) \subseteq x^{(j)} + R^t_{x^{(j)}}(\mathbb{B}_{n-1}(0, r/4) \times] - \delta/4, \delta/4[)$$

for all $j \in \{1, \dots, m\}$. By Lemma 2.70 of the Appendix, there exists an open subset Ω_2 of class C^∞ of Ω such that

$$\overline{\Omega_1} \cup \left\{ x^{(j)} - \frac{t_2}{2} a(x^{(j)}) : j \in \{1, \dots, m\} \right\} \subseteq \Omega_2 \subseteq \overline{\Omega_2} \subseteq \Omega .$$

If $p \in \Omega \setminus \overline{\Omega_2}$, then $p \in \Omega \setminus \overline{\Omega_1}$ and

$$p \in A(t_2)^+ \subseteq \bigcup_{j=1}^{m} [x^{(j)} + R^t_{x^{(j)}}(\mathbb{B}_{n-1}(0, r/4) \times] - \delta/4, \delta/4[)]$$

and accordingly, there exists $\tilde{j} \in \{1, \ldots, m\}$ such that

$$p \in A(t_2)^+ \cap [x^{(\tilde{j})} + R^t_{x^{(\tilde{j})}}(\mathbb{B}_{n-1}(0, r/4) \times] - \delta/4, \delta/4[)] .$$

Since both p and $x^{(\tilde{j})} - \frac{t_2}{2} a(x^{(\tilde{j})})$ belong to

$$A(t_2)^+ \cap [x^{(\tilde{j})} + R^t_{x^{(\tilde{j})}}(\mathbb{B}_{n-1}(0, r/4) \times] - \delta/4, \delta/4[)]$$

and $x^{(\tilde{j})} - \frac{t_2}{2} a(x^{(\tilde{j})})$ belongs to Ω_2, we have

$$\begin{aligned} |f(p)| &\leq |f(p) - f(x^{(\tilde{j})} - \frac{t_2}{2} a(x^{(\tilde{j})}))| + |f(x^{(\tilde{j})} - \frac{t_2}{2} a(x^{(\tilde{j})}))| \\ &\leq \tilde{B} M_{t_1,\alpha}(f) \left| p - x^{(\tilde{j})} - \frac{t_2}{2} a(x^{(\tilde{j})}) \right|^\alpha + \sup_{\Omega_2} |f| . \end{aligned}$$

Since both p and $x^{(\tilde{j})} - \frac{t_2}{2} a(x^{(\tilde{j})})$ belong to

$$x^{(\tilde{j})} + R^t_{x^{(\tilde{j})}}(\mathbb{B}_{n-1}(0, r/4) \times] - \delta/4, \delta/4[)$$

that has a diameter less than or equal to $2(r/4) + 2(\delta/4) < \delta$, we have

$$|f(p)| \leq \tilde{B} M_{t_1,\alpha}(f) \delta^\alpha + \sup_{\Omega_2} |f| .$$

If instead $p \in \overline{\Omega_2}$, we certainly have $|f(p)| \leq \sup_{\overline{\Omega_2}} |f|$. Hence,

$$\sup_{\Omega} |f| \leq \tilde{B} M_{t_1,\alpha}(f) \delta^\alpha + \sup_{\overline{\Omega_2}} |f|$$

and

$$\begin{aligned} &|f : \Omega|_\alpha \\ &\leq \max\left\{ \max\left\{ \tilde{B} M_{t_1,\alpha}(f), c_\alpha(\Omega_1) \sup_{\overline{\Omega_1}} |f|, c_\alpha(\Omega_1) \sup_{\overline{\Omega_1}} |\nabla f| \right\}, \frac{2}{\Lambda^\alpha} \sup_{\Omega} |f| \right\} . \end{aligned}$$

Then by taking $H = \overline{\Omega_2}$, we conclude that B as in the statement does exist. □

2.20 Schauder Spaces on a Compact Manifold and on the Boundary of a Bounded Open Subset of $\mathbb{R}^n$

Let $n \in \mathbb{N} \setminus \{0, 1\}$. Let $m \in \mathbb{N} \setminus \{0\}$ and $\alpha \in [0, 1]$. A subset M of $\mathbb{R}^n$ is a differential manifold of dimension $s \in \{1, \ldots, n-1\}$ and of class $C^{m,\alpha}$ embedded in $\mathbb{R}^n$ if, for every $p \in M$, there exist an open neighborhood W of p in $\mathbb{R}^n$ and a map $\psi \in C^{m,\alpha}\left(\overline{\mathbb{B}_s(0,1)}, \mathbb{R}^n\right)$ such that ψ is a homeomorphism of $\mathbb{B}_s(0,1)$ onto $W \cap M$, $\psi(0) = p$, and the Jacobian matrix $D\psi$ has rank s at all points of $\overline{\mathbb{B}_s(0,1)}$, i.e., if there exists a parametrization ψ of class $C^{m,\alpha}$ for M around p. If ψ is such a parametrization, then Lemma 2.41 implies that the operator $d\psi(u)$ maps $\mathbb{R}^s$ onto the s-dimensional tangent space $T_{\psi(u)}M$ at $\psi(u)$, for all $u \in \mathbb{B}_s(0,1)$.

If we further assume that M is compact, then there exist $p^{(1)},\dots, p^{(r)} \in M$, and parametrizations $\{\psi_l\}_{l=1,\dots,r}$, with $\psi_l \in C^{m,\alpha}\left(\overline{\mathbb{B}_s(0,1)}, \mathbb{R}^n\right)$, such that $\cup_{l=1}^r \psi_l(\mathbb{B}_s(0,1)) = M$.

Since $C^{m,\alpha}(\overline{\mathbb{B}_{n-1}(0,1)})$ is a subspace of $C^{m,\alpha_1}(\overline{\mathbb{B}_{n-1}(0,1)})$ for all $\alpha_1 \in [0,\alpha]$, a differential manifold of class $C^{m,\alpha}$ embedded in $\mathbb{R}^n$ is also of class C^{m,α_1} for all $\alpha_1 \in [0,\alpha]$. If in addition $m > 1$, then $C^{m,\alpha}(\overline{\mathbb{B}_{n-1}(0,1)})$ is a subspace of $C^{h,\beta}(\overline{\mathbb{B}_{n-1}(0,1)})$ for all $h \in \{1,\dots,m-1\}$ and $\beta \in [0,1]$, and thus a differential manifold of class $C^{m,\alpha}$ embedded in $\mathbb{R}^n$ is also of class $C^{h,\beta}$ for all $h \in \{1,\dots,m-1\}$ and $\beta \in [0,1]$.

Let $m \in \mathbb{N}\setminus\{0\}$, $\alpha \in [0,1]$, $\beta \in [0,\alpha]$, $h \in \{0,\dots,m\}$. We denote by $C^{h,\beta}(M)$ the linear space of functions f from M to $\mathbb{R}$ such that $f \circ \psi_l \in C^{h,\beta}\left(\overline{\mathbb{B}_s(0,1)}\right)$ for all $l \in \{1,\dots,r\}$, and we introduce the norm

$$\|f\|_{C^{h,\beta}(M)} \equiv \sup_{l=1,\dots,r} \|f \circ \psi_l\|_{C^{h,\beta}\left(\overline{\mathbb{B}_s(0,1)}\right)} \qquad \forall f \in C^{h,\beta}(M).$$

It is well known that choosing a different finite family of parametrizations as $\{\psi_l\}_{l=1,\dots,r}$, we would obtain an equivalent norm. Moreover, if $h = 0$, then $C^{h,\beta}(M)$ coincides with the space of β-Hölder continuous functions on the set M and the above norm is equivalent to the one that we have already introduced in Section 2.6.

If $1 \leq h < m$, we note that the assumption that M is of class $C^{m,\alpha}$ implies that M is of class $C^{m-1,1}$ and that accordingly we can define $C^{h,\beta}(M)$ as above for all $\beta \in [0,1]$, and not only for $\beta \in [0,\alpha]$, while instead for $h = m$ we have defined $C^{h,\beta}(M)$ only for $\beta \in [0,\alpha]$.

By the properties of $C^{h,\alpha}\left(\overline{\mathbb{B}_s(0,1)}\right)$ of Proposition 2.54, we deduce the validity of the following.

Theorem 2.83. *Let $n \in \mathbb{N} \setminus \{0,1\}$, $s \in \{1,\dots,n-1\}$, $m \in \mathbb{N} \setminus \{0\}$, $\alpha \in [0,1]$ (cf. (2.33)). Let M be a compact manifold of dimension s, of class $C^{m,\alpha}$ and embedded in $\mathbb{R}^n$. Let $h \in \{1,\dots,m\}$. Then the following statements hold.*

(i) The space $\left(C^{h,\alpha}(M), \|\cdot\|_{C^{h,\alpha}(M)}\right)$ is complete.
(ii) $C^h(M)$ is continuously embedded into $C^{h-1,1}(M)$.
(iii) If $\alpha' \in [0,\alpha[$, then $C^{h,\alpha}(M)$ is compactly embedded into $C^{h,\alpha'}(M)$.
(iv) Let α_1, $\alpha_2 \in [0,\alpha]$. Then the pointwise multiplication from $C^{h,\alpha_1}(M) \times C^{h,\alpha_2}(M)$ to $C^{h,\min\{\alpha_1,\alpha_2\}}(M)$ that takes a pair of functions (f,g) to the pointwise product fg is continuous.

Under the assumptions of Theorem 2.83, if $1 \leq h < m$, then Ω is of class $C^{h,1}$ and thus by applying Theorem 2.83 with h instead of m we deduce that statements (iii) and (iv) of Theorem 2.83 as it is hold also by replacing α with some $\beta \in]0,1]$.

We retain the same notation for C^1 manifolds that are embedded in $\mathbb{R}^n$ of Section 2.1. Accordingly, we denote by $d\sigma$ the area element of a manifold, by $\mathcal{L}_M$ the σ-algebra of measurable subsets of M, and by m_M (or simply by m_s) the s-dimensional measure of M (cf. e.g., Naumann and Simader [232]) .

We plan to show that if Ω is of class $C^{m,\alpha}$ with $m \geq 1$ and $\alpha \in [0,1]$, then the functions of $C^{h,\alpha}(\partial\Omega)$ with $h \in \{0,\ldots,m\}$ can be extended to functions of class $C^{h,\alpha}$ in the whole of $\mathbb{R}^n$. Since locally around the boundary points, the boundary of sets of class $C^{m,\alpha}$ can be mapped onto $(n-1)$-dimensional balls by means of a parametrization, we first prove the following well-known technical lemma for functions defined on an $(n-1)$-dimensional ball.

Lemma 2.84. *Let $n \in \mathbb{N} \setminus \{0,1\}$, $h \in \mathbb{N}$, $\alpha \in [0,1]$ (cf. (2.33)). Let $\rho \in]0,+\infty[$. Then there exists a linear and continuous operator G from*

$$\{f \in C^{h,\alpha}(\overline{\mathbb{B}_{n-1}(0,\rho)}) : \operatorname{supp}(f) \subseteq \overline{\mathbb{B}_{n-1}(0,\rho/2)}\}$$

to

$$\{g \in C^{h,\alpha}(\overline{\mathbb{B}_{n}(0,\rho)}) : \operatorname{supp}(g) \subseteq \mathbb{B}_{n}(0,3\rho/4)\}$$

such that

$$G[f](x_1,\ldots,x_{n-1},0) = f(x_1,\ldots,x_{n-1}) \qquad \forall (x_1,\ldots,x_{n-1}) \in \overline{\mathbb{B}_{n-1}(0,\rho)}$$

for all $f \in C^{h,\alpha}(\overline{\mathbb{B}_{n-1}(0,\rho)})$ with $\operatorname{supp}(f) \subseteq \overline{\mathbb{B}_{n-1}(0,\rho/2)}$.

Proof. Let

$$f^\sharp(x_1,\ldots,x_{n-1},x_n) \equiv f(x_1,\ldots,x_{n-1}) \qquad \forall x \in \overline{\mathbb{B}_{n}(0,\rho)}$$

for all $f \in C^{h,\alpha}(\overline{\mathbb{B}_{n-1}(0,\rho)})$. By the definition of the $C^{h,\alpha}$-norm, we have

$$\|f^\sharp\|_{C^{h,\alpha}(\overline{\mathbb{B}_{n}(0,\rho)})} \leq \|f\|_{C^{h,\alpha}(\overline{\mathbb{B}_{n-1}(0,\rho)})} .$$

By the cut-off function Lemma A.36, there exists a function $\theta \in C^\infty_c(\mathbb{R}^n)$ such that $\theta = 1$ on $\overline{\mathbb{B}_{n}(0,5\rho/8)}$ and $\operatorname{supp}\theta \subseteq \mathbb{B}_n(0,3\rho/4)$. Then we set

$$G[f] \equiv \theta f^\sharp \qquad \forall f \in C^{h,\alpha}(\overline{\mathbb{B}_{n-1}(0,\rho)}) .$$

By the continuity of the pointwise product in $C^{h,\alpha}(\overline{\mathbb{B}_{n}(0,\rho)})$ and by the above inequality, the operator G is continuous from $C^{h,\alpha}(\overline{\mathbb{B}_{n-1}(0,\rho)})$ to $C^{h,\alpha}(\overline{\mathbb{B}_{n}(0,\rho)})$. Moreover,

$$\begin{aligned}
&G[f](x_1,\ldots,x_{n-1},0) = f(x_1,\ldots,x_{n-1}) \qquad \forall (x_1,\ldots,x_{n-1}) \in \overline{\mathbb{B}_{n-1}(0,\rho)} ,\\
&\operatorname{supp} G[f] \subseteq \mathbb{B}_n(0,3\rho/4)
\end{aligned}$$

for all $f \in C^{h,\alpha}(\overline{\mathbb{B}_{n-1}(0,\rho)})$ such that $\operatorname{supp}(f) \subseteq \overline{\mathbb{B}_{n-1}(0,\rho/2)}$. □

We are now ready to prove the following.

Theorem 2.85. *Let $n \in \mathbb{N} \setminus \{0,1\}$. Let $m \in \mathbb{N} \setminus \{0\}$, $h \in \{0,\dots,m\}$, $\alpha \in [0,1]$ (cf. (2.33)). Let Ω be a bounded open subset of $\mathbb{R}^n$ of class $C^{m,\alpha}$. Then*

(i) $\partial\Omega$ is a manifold of class $C^{m,\alpha}$ and dimension $n-1$ embedded in $\mathbb{R}^n$, and the restriction map is linear and continuous from $C^{h,\alpha}\left(\overline{\Omega}\right)$ to $C^{h,\alpha}(\partial\Omega)$ for all $h = 0,\dots,m$.

(ii) Let Ω' be a bounded open subset of $\mathbb{R}^n$ such that $\overline{\Omega} \subseteq \Omega'$. Then there exists a linear and continuous extension operator $F_{\Omega'}$ from $C^{h,\alpha}(\partial\Omega)$ to $C^{h,\alpha}(\overline{\Omega'})$ such that

$$F_{\Omega'}[f]_{|\partial\Omega} = f\,, \qquad \operatorname{supp} F_{\Omega'}[f] \textit{ is compact and contained in } \Omega'$$

for all $f \in C^{h,\alpha}(\partial\Omega)$.

(iii) There exists a linear and continuous extension operator F of $C^{h,\alpha}(\partial\Omega)$ to $C^{h,\alpha}(\overline{\Omega})$ such that $F[f]_{|\partial\Omega} = f$ for all $f \in C^{h,\alpha}(\partial\Omega)$.

Proof. Let $\{\phi_a\}_{a\in\{1,\dots,b\}}$ with $\phi_a \in C^{m,\alpha}\left(\overline{\mathbb{B}_n(0,1)},\mathbb{R}^n\right)$ be parametrizations of $\overline{\Omega}$ around points of $\partial\Omega$ such that

$$\partial\Omega \subseteq \bigcup_{a=1}^{b} \phi_a(\mathbb{B}_n(0,1/2))$$

(see Definition 2.65). By Remark 2.66, we can assume that $\overline{\phi_a(\mathbb{B}_n(0,1))} \subseteq \Omega'$ and that $\phi_a(\mathbb{B}_n(0,1))$ is regular in the sense of Whitney for all $a \in \{1,\dots,b\}$. Next we set

$$\tilde{\phi}_a(\eta) \equiv \phi_a(\eta,0) \qquad \forall \eta \in \mathbb{B}_{n-1}(0,1)$$

for all $a \in \{1,\dots,b\}$. Then $\{\tilde{\phi}_a\}_{a\in\{1,\dots,b\}}$ is a family of parametrizations for $\partial\Omega$ such that $\partial\Omega \subseteq \bigcup_{a=1}^{b} \tilde{\phi}_a(\mathbb{B}_{n-1}(0,1/2))$.

We now turn to prove statement (i). We first point out that if $h > 0$, a function g of $C^{h,\alpha}\left(\overline{\Omega}\right)$ is defined in Ω and admits a continuous extension to $\overline{\Omega}$ by definition (see Section 2.11). Thus we understand that the restriction to $\partial\Omega$ of g of statement (i) is referred to the restriction to $\partial\Omega$ of the continuous extension of g to $\overline{\Omega}$. By Definition 2.65 of set of class $C^{m,\alpha}$ (with the diffeomorphisms), it follows that $\partial\Omega$ is a manifold of class $C^{m,\alpha}$ and dimension $n-1$ embedded in $\mathbb{R}^n$. By the definition of norm in $C^{h,\alpha}(\partial\Omega)$, we can take

$$\|f\|_{C^{h,\alpha}(\partial\Omega)} = \sup_{a\in\{1,\dots,b\}} \|f \circ \tilde{\phi}_a\|_{C^{h,\alpha}(\overline{\mathbb{B}_{n-1}(0,1)})} \qquad \forall f \in C^{h,\alpha}(\partial\Omega)\,.$$

Next we note that $\phi_a \in C^{m,\alpha}\left(\overline{\mathbb{B}_n(0,1)},\mathbb{R}^n\right) \subseteq C^{h,1}\left(\overline{\mathbb{B}_n(0,1)},\mathbb{R}^n\right)$ when $h < m$. Then Theorem 2.59 on the composition of functions in Schauder spaces implies that there exists $c_1 > 0$ such that

$$\|g_{|\partial\Omega}\|_{C^{h,\alpha}(\partial\Omega)} = \sup_{a\in\{1,\dots,b\}} \|g_{|\partial\Omega} \circ \tilde{\phi}_a\|_{C^{h,\alpha}(\overline{\mathbb{B}_{n-1}(0,1)})}$$

$$\leq \sup_{a\in\{1,\dots,b\}} \|g\circ\phi_a\|_{C^{h,\alpha}(\overline{\mathbb{B}_n^+(0,1)})} \leq c_1\|g\|_{C^{h,\alpha}(\overline{\Omega})}$$

for all $g \in C^{h,\alpha}(\overline{\Omega})$ both in case $h < m$ and $h = m$. Hence, statement (i) holds true.

We now turn to prove statement (ii). By Lemma A.35 on the partition of unity, there exists a finite family $\{\theta_a\}_{a=1}^b$ of functions of class $C_c^\infty(\mathbb{R}^n)$ such that

$$\operatorname{supp}\theta_a \subseteq \phi_a(\mathbb{B}_n(0,1/2)) \quad \forall a \in \{1,\dots,b\} \quad \text{and} \quad \sum_{a=1}^{b}\theta_a(x) = 1 \quad \forall x \in \partial\Omega\,.$$

If $a \in \{1,\dots,b\}$, then by the definition of norm in $C^{h,\alpha}(\partial\Omega)$, we can take

$$\|\theta_a f\|_{C^{h,\alpha}(\partial\Omega)} = \sup_{l\in\{1,\dots,b\}} \|(\theta_a f)\circ\tilde{\phi}_l\|_{C^{h,\alpha}(\overline{\mathbb{B}_{n-1}(0,1)})} \qquad \forall f \in C^{h,\alpha}(\partial\Omega)$$

and we note that

$$\operatorname{supp}(\theta_a f) \subseteq \tilde{\phi}_a(\mathbb{B}_{n-1}(0,1/2)) \qquad \forall f \in C^{h,\alpha}(\partial\Omega)\,.$$

Let G be the operator for the unit ball of Lemma 2.84. Let $\|G\|$ be the norm of G. Since $\overline{\phi_a(\mathbb{B}_n(0,1))}$ is regular in the sense of Whitney and $\phi_a^{(-1)}$ belongs to $C^{m,\alpha}(\overline{\phi_a(\mathbb{B}_n(0,1))}) \subseteq C^{h,1}(\overline{\phi_a(\mathbb{B}_n(0,1))})$ when $h < m$, the composition Theorem 2.59 for functions in a Schauder space implies that there exists $c' > 0$ such that

$$\begin{aligned} &\|G[(\theta_a f)\circ\tilde{\phi}_a]\circ\phi_a^{(-1)}\|_{C^{h,\alpha}(\overline{\phi_a(\mathbb{B}_n(0,1))})} \\ &\quad\leq c'\|G[(\theta_a f)\circ\tilde{\phi}_a]\|_{C^{h,\alpha}(\overline{\mathbb{B}_n(0,1)})} \leq c'\|G\|\|(\theta_a f)\circ\tilde{\phi}_a\|_{C^{h,\alpha}(\overline{\mathbb{B}_{n-1}(0,1)})} \\ &\quad\leq c'\|G\|\|\theta_a f\|_{C^{h,\alpha}(\partial\Omega)} \qquad \forall f \in C^{h,\alpha}(\partial\Omega) \end{aligned} \tag{2.80}$$

for all $a \in \{1,\dots,b\}$ both in case $h < m$ and $h = m$. Since the pointwise product is continuous in $C^{h,\alpha}(\partial\Omega)$, there exists $c'' > 0$ such that

$$\|\theta_a f\|_{C^{h,\alpha}(\partial\Omega)} \leq c''\|f\|_{C^{h,\alpha}(\partial\Omega)} \qquad \forall f \in C^{h,\alpha}(\partial\Omega) \tag{2.81}$$

for all $a \in \{1,\dots,b\}$. We also note that

$$\begin{aligned} &\operatorname{supp} G[(\theta_a f)\circ\tilde{\phi}_a] \subseteq \mathbb{B}_n(0,3/4) \subseteq \overline{\mathbb{B}_n(0,3/4)} \subseteq \mathbb{B}_n(0,1) \\ &\operatorname{supp}[G[(\theta_a f)\circ\tilde{\phi}_a]\circ\phi_a^{(-1)}] \subseteq \phi_a(\mathbb{B}_n(0,3/4)) \end{aligned}$$

for all $f \in C^{h,\alpha}(\partial\Omega)$ and for all $a \in \{1,\dots,b\}$. The operator E_a from

$$\{g \in C^{h,\alpha}(\overline{\phi_a(\mathbb{B}_n(0,1))}) : \operatorname{supp}(g) \subseteq \phi_a(\mathbb{B}_n(0,3/4))\}$$

to $C^{h,\alpha}(\overline{\Omega'})$ defined by

$$E_a[g](y) \equiv \begin{cases} g(y) & \text{if } y \in \phi_a(\mathbb{B}_n(0,3/4))\,, \\ 0 & \text{if } y \in \overline{\Omega'} \setminus \phi_a(\mathbb{B}_n(0,3/4))\,, \end{cases}$$

for all $g \in C^{h,\alpha}(\overline{\phi_a(\mathbb{B}_n(0,1))})$ is linear and continuous for all $a \in \{1,\dots,b\}$. Since the support of $G[(\theta_a f) \circ \tilde{\phi}_a] \circ \phi_a^{(-1)}$ is compact and contained in the set $\phi_a(\mathbb{B}_n(0,3/4)) \subseteq \Omega'$ for all $f \in C^{h,\alpha}(\partial\Omega)$ and for all $a \in \{1,\dots,b\}$, then the operator F defined by

$$F[f] \equiv \sum_{a=1}^{b} E_a\left[G[(\theta_a f) \circ \tilde{\phi}_a] \circ \phi_a^{(-1)}\right] \qquad \forall f \in C^{h,\alpha}(\partial\Omega)$$

satisfies the properties of the statement. To prove statement (iii), it suffices to take an open ball Ω' that contains $\overline{\Omega}$ and to take F equal to the composition of the corresponding operator $F_{\Omega'}$ of statement (ii) with the restriction operator from $C^{h,\alpha}(\overline{\Omega'})$ to $C^{h,\alpha}(\overline{\Omega})$. □

2.21 Tangential Derivatives

Let $n \in \mathbb{N} \setminus \{0,1\}$. Let Ω be a bounded open subset of $\mathbb{R}^n$ of class C^1. Let

$$\nu \equiv (\nu_1, \dots, \nu_n) \equiv \nu_\Omega$$

be the outward unit normal to $\partial\Omega$. The tangential gradient $\nabla_{\partial\Omega} f$ of $f \in C^1(\partial\Omega)$ is defined as

$$\nabla_{\partial\Omega} f \equiv \nabla\tilde{f} - (\nu \cdot \nabla\tilde{f})\nu \qquad \text{on } \partial\Omega\,, \tag{2.82}$$

where $\tilde{f}$ is an extension of f of class C^1 in an open neighborhood of $\partial\Omega$ (see Theorem 2.85). The vector field $\nabla_{\partial\Omega} f$ is independent of the specific choice of the extension $\tilde{f}$ of f.

To prove it, it suffices to show that if $\tilde{f}$ vanishes on $\partial\Omega$ and $p \in \partial\Omega$, then the vector

$$a \equiv \nabla\tilde{f}(p) - (\nu(p) \cdot \nabla\tilde{f}(p))\nu(p)$$

equals zero. To do so, we note that if $p \in M$ and $\psi \in C^1\left(\overline{\mathbb{B}_{n-1}(0,1)}, \mathbb{R}^n\right)$ is a parametrization for the $(n-1)$-dimensional manifold $\partial\Omega$ around p, with $\psi(0) = p$, then Lemma 2.41 implies that $d\psi(0)$ maps $\mathbb{R}^{n-1}$ onto the $(n-1)$-dimensional tangent space $T_{\psi(0)}(\partial\Omega)$ to $\partial\Omega$ at $\psi(0)$ and accordingly $\nu(\psi(0))^t D\psi(0) = 0$. Since $\tilde{f} \circ \psi = 0$ on $\mathbb{B}_{n-1}(0,1)$, we have

$$\begin{aligned} a^t D\psi(0) &= (\nabla\tilde{f}(\psi(0)))^t D\psi(0) - (\nu(\psi(0)) \cdot \nabla\tilde{f}(\psi(0)))\nu(\psi(0))^t D\psi(0) \\ &= \nabla\left(\tilde{f} \circ \psi\right)(0) - 0 = 0\,. \end{aligned}$$

In particular, a is orthogonal to the image of $d\psi(0)$, i.e., to the $(n-1)$-dimensional space $T_p(\partial\Omega)$. Now, by the definition of a we have $a \cdot \nu(p) = 0$, and thus a belongs to $T_p(\partial\Omega)$. Hence, $a = 0$ and $\nabla_{\partial\Omega} f$ is independent on the specific extension of f.

If $l, r \in \{1, \ldots, n\}$, then M_{lr} denotes the tangential derivative operator from $C^1(\partial\Omega)$ to $C^0(\partial\Omega)$ that takes f to

$$M_{lr}[f] \equiv \nu_l \frac{\partial \tilde{f}}{\partial x_r} - \nu_r \frac{\partial \tilde{f}}{\partial x_l} \qquad \text{on } \partial\Omega\,, \tag{2.83}$$

where $\tilde{f}$ is any continuously differentiable extension of f to an open neighborhood of $\partial\Omega$. The tangential derivative M_{lr} is related to the tangential gradient $\nabla_{\partial\Omega}$ by the formulas

$$(\nabla_{\partial\Omega} f)_r = \frac{\partial \tilde{f}}{\partial x_r} - (\nu \cdot \nabla \tilde{f})\nu_r = \sum_{l=1}^{n} M_{lr}[f]\nu_l \tag{2.84}$$

and

$$M_{lr}[f] = \nu_l (\nabla_{\partial\Omega} f)_r - \nu_r (\nabla_{\partial\Omega} f)_l\,.$$

Since $\nabla_{\partial\Omega} f$ does not depend on the specific choice of the extension $\tilde{f}$ of f, the last equality implies that also $M_{lr}[f]$ is independent of the specific choice of $\tilde{f}$. We also need the following consequence of the Divergence Theorem.

Lemma 2.86. *Let $n \in \mathbb{N} \setminus \{0, 1\}$. Let Ω be a bounded open subset of $\mathbb{R}^n$ of class C^1. If φ, $\psi \in C^1(\partial\Omega)$, then*

$$\int_{\partial\Omega} M_{lr}[\varphi]\psi \, d\sigma = -\int_{\partial\Omega} \varphi M_{lr}[\psi] \, d\sigma$$

for all $l, r \in \{1, \ldots, n\}$.

Proof. By Theorem 2.85, there exist Φ, $\Psi \in C^1_c(\mathbb{R}^n)$ such that $\varphi = \Phi_{|\partial\Omega}$ and $\psi = \Psi_{|\partial\Omega}$. By taking the convolution of Φ and Ψ with a family of mollifiers, we can show that there exist sequences $\{\Phi_j\}_{j\in\mathbb{N}}$ and $\{\Psi_j\}_{j\in\mathbb{N}}$ in $C^\infty_c(\mathbb{R}^n)$ such that

$$\lim_{j\to\infty} \Phi_j = \Phi \quad \text{and} \quad \lim_{j\to\infty} \Psi_j = \Psi$$

uniformly on the compact subsets of $\mathbb{R}^n$, together with their first order partial derivatives (see Proposition A.33). By the Divergence Theorem, we have

$$\begin{aligned}
&\int_{\partial\Omega} M_{lr}[\varphi]\psi \, d\sigma + \int_{\partial\Omega} \varphi M_{lr}[\psi] \, d\sigma \\
&= \int_{\partial\Omega} \left(\nu_l \frac{\partial\Phi}{\partial x_r} - \nu_r \frac{\partial\Phi}{\partial x_l} \right) \Psi \, d\sigma + \int_{\partial\Omega} \left(\nu_l \frac{\partial\Psi}{\partial x_r} - \nu_r \frac{\partial\Psi}{\partial x_l} \right) \Phi \, d\sigma \\
&= \int_{\partial\Omega} \nu_l \frac{\partial(\Phi\Psi)}{\partial x_r} - \nu_r \frac{\partial(\Phi\Psi)}{\partial x_l} \, d\sigma = \lim_{j\to\infty} \int_{\partial\Omega} \nu_l \frac{\partial(\Phi_j\Psi_j)}{\partial x_r} - \nu_r \frac{\partial(\Phi_j\Psi_j)}{\partial x_l} \, d\sigma \\
&= \lim_{j\to\infty} \int_{\Omega} \frac{\partial}{\partial x_l} \frac{\partial(\Phi_j\Psi_j)}{\partial x_r} - \frac{\partial}{\partial x_r} \frac{\partial(\Phi_j\Psi_j)}{\partial x_l} \, dx = 0
\end{aligned}$$

and thus the proof is complete. □

Next we introduce the following auxiliary lemma, whose proof is based on the definition of norm in a Schauder space.

Lemma 2.87. *Let $n \in \mathbb{N} \setminus \{0,1\}$. Let $m \in \mathbb{N} \setminus \{0\}$, $\alpha \in [0,1]$. Let Ω be a bounded open connected subset of $\mathbb{R}^n$ of class $C^{m,\alpha}$. Then the following statements hold.*

(i) M_{lr} is a linear and continuous operator from $C^{m,\alpha}(\partial\Omega)$ to $C^{m-1,\alpha}(\partial\Omega)$ for all $l, r \in \{1,\ldots,n\}$.
(ii) A function $f \in C^1(\partial\Omega)$ belongs to $C^{m,\alpha}(\partial\Omega)$ if and only if $M_{lr}[f]$ belongs to $C^{m-1,\alpha}(\partial\Omega)$ for all $l, r \in \{1,\ldots,n\}$.
(iii) The norm $\|\cdot\|_{C^{m,\alpha}(\partial\Omega)}$ is equivalent to the norm on $C^{m,\alpha}(\partial\Omega)$ defined by

$$\|f\|_{C^0(\partial\Omega)} + \sum_{l,r=1}^{n} \|M_{lr}[f]\|_{C^{m-1,\alpha}(\partial\Omega)} \qquad \forall f \in C^{m,\alpha}(\partial\Omega)\,. \tag{2.85}$$

Proof. (i) Since $\nu_\Omega \in C^{m-1,\alpha}(\partial\Omega, \mathbb{R}^n)$ and the extension and restriction operators of Theorem 2.85 (i), (ii) are continuous, the continuity of the pointwise product in $C^{m-1,\alpha}(\partial\Omega)$ of Theorem 2.83 ensures the continuity of M_{lr} (see the definition (2.83) of M_{lr}).

(ii) The necessity is an immediate consequence of statement (i). We now prove the sufficiency. Let $f \in C^1(\partial\Omega)$ be such that $M_{lr}[f] \in C^{m-1,\alpha}(\partial\Omega)$ for all $l, r \in \{1,\ldots,n\}$. By the definition of $C^{m,\alpha}(\partial\Omega)$, it suffices to show that if ψ is a parametrization of class $C^{m,\alpha}\left(\overline{\mathbb{B}_{n-1}(0,1)}, \mathbb{R}^n\right)$ for the $(n-1)$-dimensional manifold $\partial\Omega$, then $f \circ \psi \in C^{m,\alpha}(\overline{\mathbb{B}_{n-1}(0,1)})$. Since $f \circ \psi$ belongs to $C^1(\overline{\mathbb{B}_{n-1}(0,1)})$, it suffices to show that the first order partial derivatives of $f \circ \psi$ belong to the space $C^{m-1,\alpha}(\overline{\mathbb{B}_{n-1}(0,1)})$. Since the vector $\nu \circ \psi(\xi)$ is normal to the tangent space $T_{\psi(\xi)}(\partial\Omega) = d\psi(\xi)[\mathbb{R}^{n-1}]$ to $\partial\Omega$ at the point $\psi(\xi)$, we have

$$\nu^t \circ \psi(\xi)\, D\psi(\xi) = 0\,.$$

By the extension Theorem 2.85 (ii), there exists $\tilde{f} \in C^1_c(\mathbb{R}^n)$ such that $\tilde{f}_{|\partial\Omega} = f$, and thus we have

$$\begin{aligned} \nabla(f \circ \psi)(\xi) &= \nabla(\tilde{f} \circ \psi)(\xi) = (\nabla \tilde{f})^t \circ \psi(\xi) D\psi(\xi) \\ &= \left[(\nabla \tilde{f})^t \circ \psi(\xi) - \left((\nabla \tilde{f}) \circ \psi(\xi) \cdot \nu \circ \psi(\xi)\right) \nu^t \circ \psi(\xi)\right] D\psi(\xi) \\ &= (\nabla_{\partial\Omega} f)^t \circ \psi(\xi) D\psi(\xi) \\ &= \nu^t \circ \psi(\xi)\, (M_{lr}[f] \circ \psi(\xi))_{l,r=1,\ldots,n}\, D\psi(\xi) \qquad \forall \xi \in \mathbb{B}_{n-1}(0,1)\,, \end{aligned} \tag{2.86}$$

(cf. (2.82) and (2.84) for the definition of the tangential gradient $\nabla_{\partial\Omega} f$ and for the formula in terms of the tangential derivatives). Since $M_{lr}[f]$ belongs to $C^{m-1,\alpha}(\partial\Omega)$, ν belongs to $C^{m-1,\alpha}(\partial\Omega, \mathbb{R}^n)$, and ψ is a parametrization of class

$$C^{m,\alpha}\left(\overline{\mathbb{B}_{n-1}(0,1)}, \mathbb{R}^n\right)$$

of $\partial\Omega$, then $M_{lr}[f]\circ\psi$ belongs to the space $C^{m-1,\alpha}\left(\overline{\mathbb{B}_{n-1}(0,1)}\right)$ and $\nu\circ\psi$ belongs to $C^{m-1,\alpha}(\overline{\mathbb{B}_{n-1}(0,1)},\mathbb{R}^n)$ for all l,r in $\{1,\ldots,n\}$. Then the components of the function in the right-hand side of (2.86) are sums of products of functions of class $C^{m-1,\alpha}(\overline{\mathbb{B}_{n-1}(0,1)})$ and thus belong to $C^{m-1,\alpha}(\overline{\mathbb{B}_{n-1}(0,1)})$. Hence, the first order partial derivatives of $f\circ\psi$ belong to $C^{m-1,\alpha}(\overline{\mathbb{B}_{n-1}(0,1)})$ and we conclude that $f\circ\psi$ belongs to $C^{m,\alpha}(\overline{\mathbb{B}_{n-1}(0,1)})$.

(iii) Let $\|\cdot\|^{\dagger}_{C^{m,\alpha}(\partial\Omega)}$ be the norm on $C^{m,\alpha}(\partial\Omega)$ defined by (2.85). By the definition of the norm $\|\cdot\|_{C^{m,\alpha}(\partial\Omega)}$ and by statement (i), the identity map from $(C^{m,\alpha}(\partial\Omega),\|\cdot\|_{C^{m,\alpha}(\partial\Omega)})$ to $(C^{m,\alpha}(\partial\Omega),\|\cdot\|^{\dagger}_{C^{m,\alpha}(\partial\Omega)})$ is continuous. Thus it suffices to show that we can estimate the norm $\|\cdot\|_{C^{m,\alpha}(\partial\Omega)}$ in terms of a constant multiple of the norm $\|\cdot\|^{\dagger}_{C^{m,\alpha}(\partial\Omega)}$. Let $\{\psi_a\}_{a\in\{1,\ldots,b\}}$, with ψ_a in $C^{m,\alpha}\left(\overline{\mathbb{B}_{n-1}(0,1)},\mathbb{R}^n\right)$, be parametrizations of $\partial\Omega$ such that

$$\bigcup_{a=1}^{b}\psi_a(\mathbb{B}_{n-1}(0,1))=\partial\Omega$$

and

$$\|f\|_{C^{m,\alpha}(\partial\Omega)}=\sup_{a\in\{1,\ldots,b\}}\|f\circ\psi_a\|_{C^{m,\alpha}\left(\overline{\mathbb{B}_{n-1}(0,1)}\right)}\qquad\forall f\in C^{m,\alpha}(\partial\Omega).\ (2.87)$$

By the second inequality in (2.32) for the $C^{m,\alpha}\left(\overline{\mathbb{B}_{n-1}(0,1)}\right)$-norm and by the definition of the $\|\cdot\|^{\dagger}_{C^{m,\alpha}(\partial\Omega)}$ norm, we have

$$\begin{aligned}&\|f\circ\psi_a\|_{C^{m,\alpha}\left(\overline{\mathbb{B}_{n-1}(0,1)}\right)}\\&\quad\le\|f\circ\psi_a\|_{C^{0}\left(\overline{\mathbb{B}_{n-1}(0,1)}\right)}+\sum_{j=1}^{n}\|\partial_{\xi_j}(f\circ\psi_a)\|_{C^{m-1,\alpha}\left(\overline{\mathbb{B}_{n-1}(0,1)}\right)}.\end{aligned}\qquad(2.88)$$

Moreover,

$$\|f\circ\psi_a\|_{C^{0}\left(\overline{\mathbb{B}_{n-1}(0,1)}\right)}\le\|f\|_{C^{0}(\partial\Omega)}\le\|f\|^{\dagger}_{C^{m,\alpha}(\partial\Omega)}.\qquad(2.89)$$

Thus it suffices to estimate the norms $\|\partial_{\xi_j}(f\circ\psi_a)\|_{C^{m-1,\alpha}\left(\overline{\mathbb{B}_{n-1}(0,1)}\right)}$ in terms of $\|f\|^{\dagger}_{C^{m,\alpha}(\partial\Omega)}$ for all $f\in C^{m,\alpha}(\partial\Omega)$, $a\in\{1,\ldots,b\}$, and $j\in\{1,\ldots,n-1\}$. By formula (2.86) with $\psi=\psi_a$, we have

$$\begin{aligned}&\|\partial_{\xi_j}(f\circ\psi_a)\|_{C^{m-1,\alpha}\left(\overline{\mathbb{B}_{n-1}(0,1)}\right)}\\&\quad\le\sum_{l,r=1}^{n}\left\|\nu_l\circ\psi_a M_{lr}[f]\circ\psi_a\frac{\partial(\psi_a)_r}{\partial\xi_j}\right\|_{C^{m-1,\alpha}\left(\overline{\mathbb{B}_{n-1}(0,1)}\right)}.\end{aligned}\qquad(2.90)$$

By the continuity of the pointwise product in $C^{m-1,\alpha}\left(\overline{\mathbb{B}_{n-1}(0,1)}\right)$, there exists $c_1 > 0$ such that

$$\sum_{l,r=1}^{n} \left\| \nu_l \circ \psi_a M_{lr}[f] \circ \psi_a \frac{\partial(\psi_a)_r}{\partial \xi_j} \right\|_{C^{m-1,\alpha}\left(\overline{\mathbb{B}_{n-1}(0,1)}\right)} \tag{2.91}$$
$$\leq c_1 \sum_{l,r=1}^{n} \| \nu_l \circ \psi_a M_{lr}[f] \circ \psi_a \|_{C^{m-1,\alpha}\left(\overline{\mathbb{B}_{n-1}(0,1)}\right)} \left\| \frac{\partial(\psi_a)_r}{\partial \xi_j} \right\|_{C^{m-1,\alpha}\left(\overline{\mathbb{B}_{n-1}(0,1)}\right)}$$
$$\leq c_1 \sum_{l,r=1}^{n} \| \nu_l M_{lr}[f] \|_{C^{m-1,\alpha}(\partial\Omega)} \left\| \frac{\partial(\psi_a)_r}{\partial \xi_j} \right\|_{C^{m-1,\alpha}\left(\overline{\mathbb{B}_{n-1}(0,1)}\right)}$$

for all $f \in C^{m,\alpha}(\partial\Omega)$, $a \in \{1, \ldots, b\}$, and $j \in \{1, \ldots, n-1\}$. By the continuity of the pointwise product in $C^{m-1,\alpha}(\partial\Omega)$, there exists $c_2 > 0$ such that

$$c_1 \sum_{l,r=1}^{n} \| \nu_l M_{lr}[f] \|_{C^{m-1,\alpha}(\partial\Omega)} \left\| \frac{\partial(\psi_a)_r}{\partial \xi_j} \right\|_{C^{m-1,\alpha}\left(\overline{\mathbb{B}_{n-1}(0,1)}\right)} \tag{2.92}$$
$$\leq c_1 c_2 \sum_{l,r=1}^{n} \| \nu_l \|_{C^{m-1,\alpha}(\partial\Omega)} \| M_{lr}[f] \|_{C^{m-1,\alpha}(\partial\Omega)} \left\| \frac{\partial(\psi_a)_r}{\partial \xi_j} \right\|_{C^{m-1,\alpha}\left(\overline{\mathbb{B}_{n-1}(0,1)}\right)}$$

for all $f \in C^{m,\alpha}(\partial\Omega)$, $a \in \{1, \ldots, b\}$, and $j \in \{1, \ldots, n-1\}$. Then, by combining inequalities (2.87)–(2.92), we conclude that the identity map from the space $(C^{m,\alpha}(\partial\Omega), \| \cdot \|^{\dagger}_{C^{m,\alpha}(\partial\Omega)})$ to the space $(C^{m,\alpha}(\partial\Omega), \| \cdot \|_{C^{m,\alpha}(\partial\Omega)})$ is continuous. Now the proof is complete. □

2.22 Schauder Spaces in Open Subsets of $\mathbb{R}^n$, a Case of a Negative Exponent

So far, we have defined and considered Schauder spaces only with a positive exponent $m \geq 0$. We now consider a case with negative exponent. Namely, the case with $m = -1$, which we need later in the book. We note that Schauder spaces with negative exponents have been known for a long time and have been used in the analysis of elliptic and parabolic partial differential equations (cf. Triebel [267], Gilbarg and Trudinger [107], Lunardi and Vespri [191]).

Definition 2.88. Let $n \in \mathbb{N} \setminus \{0, 1\}$. Let $\alpha \in]0, 1]$. Let Ω be a bounded open subset of $\mathbb{R}^n$. We denote by $C^{-1,\alpha}(\overline{\Omega})$ the subspace

$$\left\{ f_0 + \sum_{j=1}^{n} \frac{\partial}{\partial x_j} f_j : f_j \in C^{0,\alpha}(\overline{\Omega})\ \forall j \in \{0, \ldots, n\} \right\},$$

of the space of distributions $\mathcal{D}'(\Omega)$ in Ω.

According to the above definition, the space $C^{-1,\alpha}(\overline{\Omega})$ is the image of the linear and continuous map

$$\Xi : (C^{0,\alpha}(\overline{\Omega}))^{n+1} \to \mathcal{D}'(\Omega)$$

that takes an $(n+1)$-tuple $(f_0, \ldots, f_n)$ to $f_0 + \sum_{j=1}^{n} \frac{\partial}{\partial x_j} f_j$. Let π denote the canonical projection

$$\pi : (C^{0,\alpha}(\overline{\Omega}))^{n+1} \to (C^{0,\alpha}(\overline{\Omega}))^{n+1}/\mathrm{Ker}\, \Xi$$

of $(C^{0,\alpha}(\overline{\Omega}))^{n+1}$ onto the quotient space $(C^{0,\alpha}(\overline{\Omega}))^{n+1}/\mathrm{Ker}\, \Xi$. Let $\tilde{\Xi}$ be the unique linear injective map from $(C^{0,\alpha}(\overline{\Omega}))^{n+1}/\mathrm{Ker}\, \Xi$ onto the image $C^{-1,\alpha}(\overline{\Omega})$ of Ξ such that

$$\Xi = \tilde{\Xi} \circ \pi .$$

Then, $\tilde{\Xi}$ is a linear bijection from $(C^{0,\alpha}(\overline{\Omega}))^{n+1}/\mathrm{Ker}\, \Xi$ onto $C^{-1,\alpha}(\overline{\Omega})$.

Since $(C^{0,\alpha}(\overline{\Omega}))^{n+1}$ is a Banach space and $\mathrm{Ker}\, \Xi$ is a closed subspace of $(C^{0,\alpha}(\overline{\Omega}))^{n+1}$, we know that $(C^{0,\alpha}(\overline{\Omega}))^{n+1}/\mathrm{Ker}\, \Xi$ is a Banach space (cf. Theorem 2.1).

We endow $C^{-1,\alpha}(\overline{\Omega})$ with the norm induced by $\tilde{\Xi}$, i.e., we set

$$\|f\|_{C^{-1,\alpha}(\overline{\Omega})} \equiv \inf\Biggl\{ \sum_{j=0}^{n} \|f_j\|_{C^{0,\alpha}(\overline{\Omega})} : \qquad (2.93)$$

$$f = f_0 + \sum_{j=1}^{n} \frac{\partial}{\partial x_j} f_j \,,\ f_j \in C^{0,\alpha}(\overline{\Omega})\ \forall j \in \{0, \ldots, n\} \Biggr\} .$$

By definition of the norm $\|\cdot\|_{C^{-1,\alpha}(\overline{\Omega})}$, the linear bijection $\tilde{\Xi}$ is an isometry of the space $(C^{0,\alpha}(\overline{\Omega}))^{n+1}/\mathrm{Ker}\, \Xi$ onto the space $(C^{-1,\alpha}(\overline{\Omega}), \|\cdot\|_{C^{-1,\alpha}(\overline{\Omega})})$.

Since the quotient $(C^{0,\alpha}(\overline{\Omega}))^{n+1}/\mathrm{Ker}\, \Xi$ is a Banach space, it follows that $(C^{-1,\alpha}(\overline{\Omega}), \|\cdot\|_{C^{-1,\alpha}(\overline{\Omega})})$ is also a Banach space.

Since Ξ is continuous from $(C^{0,\alpha}(\overline{\Omega}))^{n+1}$ to $\mathcal{D}'(\Omega)$, a fundamental property of the quotient topology implies that the map $\tilde{\Xi}$ is continuous from the quotient space $(C^{0,\alpha}(\overline{\Omega}))^{n+1}/\mathrm{Ker}\, \Xi$ to $\mathcal{D}'(\Omega)$ (cf. Proposition A.5).

Hence, $(C^{-1,\alpha}(\overline{\Omega}), \|\cdot\|_{C^{-1,\alpha}(\overline{\Omega})})$ is continuously embedded into $\mathcal{D}'(\Omega)$. Also, the definition of the norm $\|\cdot\|_{C^{-1,\alpha}(\overline{\Omega})}$ implies that $C^{0,\alpha}(\overline{\Omega})$ is continuously embedded into $C^{-1,\alpha}(\overline{\Omega})$ and that the partial derivation $\frac{\partial}{\partial x_j}$ is continuous from $C^{0,\alpha}(\overline{\Omega})$ to $C^{-1,\alpha}(\overline{\Omega})$ for all $j \in \{1, \ldots, n\}$.

The elements of $C^{-1,\alpha}(\overline{\Omega})$ are not integrable functions, but distributions in Ω. We now define a linear functional $\mathcal{I}_\Omega$ on $C^{-1,\alpha}(\overline{\Omega})$ which extends the integration in Ω to all elements of $C^{-1,\alpha}(\overline{\Omega})$. We do so by means of the following proposition.

Proposition 2.89. *Let $n \in \mathbb{N} \setminus \{0, 1\}$. Let $\alpha \in]0, 1]$. Let Ω be a bounded open Lipschitz subset of $\mathbb{R}^n$. Then the following statements hold.*

(i) If $(f_0, \ldots, f_n) \in C^{0,\alpha}(\overline{\Omega})^{n+1}$ *and*

$$f_0 + \sum_{j=1}^{n} \frac{\partial}{\partial x_j} f_j = 0$$

in the sense of distributions in Ω*, then*

$$\int_{\Omega} f_0 \, dx + \int_{\partial\Omega} \sum_{j=1}^{n} (\nu_\Omega)_j f_j \, d\sigma = 0 \,.$$

(ii) There exists one and only one linear and continuous operator $\mathcal{I}_\Omega$ *from* $C^{-1,\alpha}(\overline{\Omega})$ *to* $\mathbb{R}$ *such that*

$$\mathcal{I}_\Omega[f] = \int_{\Omega} f_0 \, dx + \int_{\partial\Omega} \sum_{j=1}^{n} (\nu_\Omega)_j f_j \, d\sigma \tag{2.94}$$

for all $f = f_0 + \sum_{j=1}^{n} \frac{\partial}{\partial x_j} f_j \in C^{-1,\alpha}(\overline{\Omega})$*. Moreover,*

$$\mathcal{I}_\Omega[f] = \int_{\Omega} f \, dx \qquad \forall f \in C^{0,\alpha}(\overline{\Omega}) \,.$$

Proof. (i) Since all components of the vector valued function $(f_1, \ldots, f_n)$ and its divergence $-f_0$ belong to $C^{0,\alpha}(\overline{\Omega})$, which is continuously embedded into $L^2(\Omega)$, there exists a sequence $\{(f_{l1}, \ldots, f_{ln})\}_{l \in \mathbb{N}}$ in $C^\infty(\overline{\Omega}, \mathbb{R}^n)$ such that

$$\lim_{l\to\infty} (f_{l1}, \ldots, f_{ln}) = (f_1, \ldots, f_n) \qquad \text{in } L^2(\Omega, \mathbb{R}^n) \,,$$

$$\lim_{l\to\infty} \sum_{j=1}^{n} \frac{\partial}{\partial x_j} f_{lj} = \sum_{j=1}^{n} \frac{\partial}{\partial x_j} f_j \qquad \text{in } L^2(\Omega) \,,$$

$$\lim_{l\to\infty} \int_{\partial\Omega} \sum_{j=1}^{n} (\nu_\Omega)_j f_{lj} \, d\sigma = \int_{\partial\Omega} \sum_{j=1}^{n} (\nu_\Omega)_j f_j \, d\sigma$$

(cf., e.g., Tartar [264, p. 101]). By the Divergence Theorem on Lipschitz domains, we have

$$\int_{\Omega} \sum_{j=1}^{n} \frac{\partial}{\partial x_j} f_{lj} \, dx = \int_{\partial\Omega} \sum_{j=1}^{n} (\nu_\Omega)_j f_{lj} \, d\sigma \qquad \forall l \in \mathbb{N}$$

(cf., e.g., Ziemer [277, Theorem 5.8.2, Remark 5.8.3]). Then, by taking the limit as l tends to infinity, we obtain

$$-\int_{\Omega} f_0 \, dx = \int_{\Omega} \sum_{j=1}^{n} \frac{\partial}{\partial x_j} f_j \, dx = \int_{\partial\Omega} \sum_{j=1}^{n} (\nu_\Omega)_j f_j \, d\sigma$$

and the proof of statement (i) is complete.

(ii) Let L be the linear operator from $C^{0,\alpha}(\overline{\Omega})^{n+1}$ to $\mathbb{R}$ that takes $(f_0, \ldots, f_n)$ to the function in the right-hand side of (2.94). By (i), we have $\operatorname{Ker} \Xi \subseteq \operatorname{Ker} L$. Since the operator Ξ from $C^{0,\alpha}(\overline{\Omega})^{n+1}$ to $C^{-1,\alpha}(\overline{\Omega})$ is surjective, the Homomorphism Theorem A.1 for linear maps between vector spaces implies the existence of a unique linear map $\mathcal{I}_\Omega$ from $C^{-1,\alpha}(\overline{\Omega})$ to $\mathbb{R}$ such that $L = \mathcal{I}_\Omega \circ \Xi$, i.e., such that (2.94) holds true. Then we note that if $f = f_0 + \sum_{j=1}^n \frac{\partial}{\partial x_j} f_j$, we have

$$|\mathcal{I}_\Omega[f]| \leq m_n(\Omega) \|f_0\|_{C^{0,\alpha}(\overline{\Omega})} + \sum_{j=1}^n \|f_j\|_{C^{0,\alpha}(\overline{\Omega})} m_{n-1}(\partial\Omega)\,.$$

By taking the infimum on all $(f_0, \ldots, f_n) \in C^{0,\alpha}(\overline{\Omega})^{n+1}$ such that

$$f = f_0 + \sum_{j=1}^n \frac{\partial}{\partial x_j} f_j\,,$$

we obtain

$$|\mathcal{I}_\Omega[f]| \leq \max\{m_n(\Omega), m_{n-1}(\partial\Omega)\} \|f\|_{C^{-1,\alpha}(\overline{\Omega})}\,.$$

Hence, the linear map $\mathcal{I}_\Omega$ is continuous. The last equality of the statement follows by taking $f_0 = f$, $f_1 = \cdots = f_n = 0$ in (2.94). □

Chapter 3
Preliminaries on Harmonic Functions

Abstract This chapter is devoted to the properties of harmonic functions that we exploit in this book. For most of the proofs we refer to classic monographs and in particular to Evans (*Partial differential equations*, volume 19 of *Graduate Studies in Mathematics*. American Mathematical Society, Providence, RI, second edition, 2010), Folland (*Introduction to partial differential equations*. Princeton University Press, Princeton, NJ, second edition, 1995), and Gilbarg and Trudinger (*Elliptic partial differential equations of second order*, volume 224 of *Grundlehren der Mathematischen Wissenschaften [Fundamental Principles of Mathematical Sciences]*. Springer-Verlag, Berlin, second edition, 1983). We present the Maximum Principle, the Liouville Theorem, the interior inequalities for the derivatives of a harmonic function, Weyl's Lemma, the Lemma of Hopf-Oleĭnik, some properties of harmonic functions around isolated singularities, the Kelvin transform, and some properties of harmonic functions at infinity.

Unless otherwise specified, we understand that throughout this chapter n is a natural number bigger than or equal to two. Namely,

$$n \in \mathbb{N}, \, n \geq 2 \, .$$

3.1 Basic Properties of Harmonic Functions

Definition 3.1. Let Ω be an open subset of $\mathbb{R}^n$. We say that a function u from Ω to $\mathbb{R}$ is harmonic in Ω provided that $u \in C^2(\Omega)$ and that $\Delta u = 0$ in Ω, where

$$\Delta u \equiv \sum_{j=1}^{n} \frac{\partial^2 u}{\partial x_j^2} \, .$$

M. Dalla Riva et al., *Singularly Perturbed Boundary Value Problems*,
https://doi.org/10.1007/978-3-030-76259-9_3

Example 3.2. (j) If $n \geq 3$ and $x_0 \in \mathbb{R}^n$, then the function $|x - x_0|^{2-n}$ is harmonic in $\mathbb{R}^n \setminus \{x_0\}$.
(jj) If $n = 2$ and $x_0 \in \mathbb{R}^2$, then the function $\log|x - x_0|$ is harmonic in $\mathbb{R}^2 \setminus \{x_0\}$.

As we shall see in Weyl's Lemma 3.14, we can replace the assumption that $u \in C^2(\Omega)$ in the definition of harmonic function with the assumption that u is locally integrable in Ω, provided that we ask that equality $\Delta u = 0$ is satisfied in the sense of distributions.

Definition 3.3. Let Ω be an open subset of $\mathbb{R}^n$. Let u be a function from Ω to $\mathbb{R}$. We say that u

(i) satisfies the strong Maximum Principle in Ω if either

$$u(x) < \sup_{\Omega} u \qquad \forall x \in \Omega\,,$$

or u is constantly equal to $\sup_\Omega u$, i.e.,

$$u(x) = \sup_{\Omega} u \qquad \forall x \in \Omega\,,$$

(ii) satisfies the strong Minimum Principle in Ω if either

$$u(x) > \inf_{\Omega} u \qquad \forall x \in \Omega\,,$$

or

$$u(x) = \inf_{\Omega} u \qquad \forall x \in \Omega\,.$$

As shown by the following statement, harmonic functions satisfy the strong Maximum Principles in an open connected set. For a proof, we refer for example to Folland [102, (2.13), p. 72] or Gilbarg and Trudinger [107, Cor. 3.2, p. 33]).

Theorem 3.4. *Let Ω be an open connected subset of $\mathbb{R}^n$. If u is a real valued harmonic function in Ω, then u satisfies both the strong Maximum Principle and the strong Minimum Principle in Ω.*

Then we have the following corollary (cf., e.g., Folland [102, (2.14), p. 72]).

Corollary 3.5. *Let Ω be a bounded open connected subset of $\mathbb{R}^n$. If $u \in C^0(\overline{\Omega})$ is a real valued harmonic function in Ω, then*

$$\max_{\overline{\Omega}} u = \max_{\partial\Omega} u\,, \qquad \min_{\overline{\Omega}} u = \min_{\partial\Omega} u\,, \qquad \max_{\overline{\Omega}} |u| = \max_{\partial\Omega} |u|\,.$$

In particular, Corollary 3.5 says that a harmonic function u attains its maximum and minimum on the boundary. Then we can immediately derive the following uniqueness theorem (cf., e.g., Folland [102, (2.15), p. 72]).

Theorem 3.6. *Let Ω be an open bounded connected subset of $\mathbb{R}^n$. Let $u, v \in C^0(\overline{\Omega})$. If both u, and v are real valued harmonic functions in Ω and if $u = v$ on $\partial\Omega$, then $u = v$ on $\overline{\Omega}$.*

Example 3.7. As the following examples show, the previous corollary no longer holds if we omit the assumption that Ω is bounded.

(i) Let $n \geq 3$. Then the functions 1 and $|x|^{2-n}$ are both harmonic in $\mathbb{R}^n \setminus \overline{\mathbb{B}_n(0,1)}$ continuous in $\mathbb{R}^n \setminus \mathbb{B}_n(0,1)$ and coincide on $\partial\mathbb{B}_n(0,1)$.
(ii) Let $n = 2$. Then the functions 0 and $\log|x|$ are both harmonic $\mathbb{R}^2 \setminus \overline{\mathbb{B}_2(0,1)}$ continuous in $\mathbb{R}^2 \setminus \mathbb{B}_2(0,1)$ and coincide on $\partial\mathbb{B}_2(0,1)$.

Then we have the following theorem, which says that harmonic functions are infinitely many times differentiable. For a proof we refer for example to Folland [102, (2.11), p. 71].

Theorem 3.8 (of Köbe). *Let Ω be an open subset of $\mathbb{R}^n$. If a real valued function u is harmonic in Ω, then u is of class C^∞ in Ω.*

We also mention that a uniform limit on compact sets of a sequence of harmonic functions is harmonic. Indeed, the following theorem holds (cf., e.g., Folland [102, (2.12), p. 71]).

Theorem 3.9. *Let Ω be an open subset of $\mathbb{R}^n$. If $\{u_j\}_{j\in\mathbb{N}}$ is a sequence of harmonic functions in Ω and if the sequence $\{u_j\}_{j\in\mathbb{N}}$ converges to a function u uniformly on the compact subsets of Ω, then u is harmonic in Ω.*

We also need to know that an entire bounded harmonic function is constant. Indeed, the following holds (cf., e.g., Folland [102, (2.16), p. 72]).

Theorem 3.10 (of Liouville). *If u is a real valued harmonic function in $\mathbb{R}^n$ and if u is bounded, then u is constant.*

By the Köbe Theorem 3.8, a harmonic function is of class C^∞. However, a lot more holds. Indeed, the following statement holds true (cf., e.g., Gilbarg and Trudinger [107, Theorem 2.10, p. 23]).

Theorem 3.11 (of the Interior Inequalities). *Let Ω be an open subset of $\mathbb{R}^n$. Let u be a real valued harmonic function in Ω. If $\overline{\mathbb{B}_n(x,r)} \subseteq \Omega$, then*

$$|D^\alpha u(x)| \leq \left(\frac{n|\alpha|}{r}\right)^{|\alpha|} \sup_{\partial\mathbb{B}_n(x,r)} |u| \qquad \forall \alpha \in \mathbb{N}^n \setminus \{0\}.$$

Then we have the following corollary.

Corollary 3.12. *Let Ω be an open subset of $\mathbb{R}^n$ with $\Omega \neq \mathbb{R}^n$. Let Ω' be an open subset of Ω such that $\overline{\Omega'}$ is contained in Ω and is compact. Let d be the distance between Ω' and $\partial\Omega$. If u is a harmonic function in Ω, then*

$$\sup_{\Omega'} |D^\alpha u| \leq \left(\frac{n|\alpha|}{d}\right)^{|\alpha|} \sup_{\Omega} |u| \qquad \forall \alpha \in \mathbb{N}^n \setminus \{0\}.$$

Here we understand that the right-hand side equals $+\infty$ if $\sup_\Omega |u| = +\infty$.

By Stirling's inequality, we have

$$|\alpha|^{|\alpha|} \leq e^{|\alpha|}|\alpha|! \qquad \forall \alpha \in \mathbb{N}^n \setminus \{0\}.$$

Thus, the characterization Theorem 2.17 of real analytic functions and Corollary 3.12 imply the validity of the following theorem.

Theorem 3.13. *Let Ω be an open subset of $\mathbb{R}^n$. If u is a real valued harmonic function in Ω, then u is real analytic.*

We also need the following well known result of Weyl [275].

Lemma 3.14 (of Weyl). *Let Ω be an open subset of $\mathbb{R}^n$. If u is a locally integrable function from Ω to $\mathbb{R}$ and if*

$$\int_\Omega u \Delta\varphi \, dx = 0 \qquad \forall \varphi \in \mathcal{D}(\Omega),$$

i.e., if $\Delta u = 0$ in the sense of distributions in Ω, then u equals almost everywhere a harmonic function in Ω.

Then we have the following refined form of the Hopf-Oleĭnik Lemma. The Hopf-Oleĭnik Lemma in the form we present here is due to Kamynin and Himčenko [132], [133, p. 84 of the English translation] (see also Lieberman [188, Theorem 4.1] for a simpler proof; for a historical account and for further developments see Alvarado et al. [5, p. 321]).

Lemma 3.15 (of Hopf-Oleĭnik). *Let $\alpha \in]0,1]$. Let Ω be an open subset of $\mathbb{R}^n$ of class $C^{1,\alpha}$. Let $x_0 \in \partial\Omega$. Let $u \in C^0(\overline{\Omega})$ be a real valued harmonic function in Ω. Let $u(x) < u(x_0)$ for all $x \in \Omega$. If the limit*

$$\frac{\partial u}{\partial \nu_\Omega}(x_0) \equiv \lim_{t \to 0^-} \frac{u(x_0 + t\nu_\Omega(x_0)) - u(x_0)}{t}$$

exists in $\mathbb{R}$, then $\frac{\partial u}{\partial \nu_\Omega}(x_0) > 0$.

3.2 A Fundamental Solution for the Laplace Operator

We say that a function $E \in L^1_{\mathrm{loc}}(\mathbb{R}^n)$ is a fundamental solution for Δ provided that

$$\int_{\mathbb{R}^n} E \Delta\varphi \, dx = \varphi(0) \qquad \forall \varphi \in C_c^\infty(\mathbb{R}^n).$$

In terms of the theory of distributions, it means that

$$\Delta E = \delta_0,$$

where δ_0 is the Dirac distribution with mass at 0. We now introduce a specific fundamental solution of Δ, which is commonly addressed to as 'the fundamental solution of the Laplace operator'. Let Υ_n be the function from $]0, +\infty[$ to $\mathbb{R}$ defined by

$$\Upsilon_n(r) \equiv \begin{cases} \frac{1}{s_2} \log r & \forall r \in]0, +\infty[, \quad \text{if } n = 2\,, \\ \frac{1}{(2-n)s_n} r^{2-n} & \forall r \in]0, +\infty[, \quad \text{if } n > 2\,, \end{cases} \tag{3.1}$$

where s_n denotes the $(n-1)$-dimensional measure of $\partial\mathbb{B}_n(0,1)$. We denote by S_n the function from $\mathbb{R}^n \setminus \{0\}$ to $\mathbb{R}$ defined by

$$S_n(x) \equiv \Upsilon_n(|x|) \qquad \forall x \in \mathbb{R}^n \setminus \{0\}\,. \tag{3.2}$$

It is well known that S_n is a fundamental solution of Δ. Actually, one can prove that the set of all radial fundamental solutions of Δ is delivered by the set

$$\{S_n + c \colon c \in \mathbb{R}\}\,.$$

Clearly,

$$\frac{\partial}{\partial x_j} S_n(x) = \frac{1}{s_n} \frac{x_j}{|x|^n} \qquad \forall j \in \{1, \ldots, n\}\,, \tag{3.3}$$

$$\nabla S_n(x) = \frac{1}{s_n} \frac{x}{|x|^n}\,, \qquad |\nabla S_n(x)| = \frac{1}{s_n} \frac{1}{|x|^{n-1}}$$

for all $x \in \mathbb{R}^n \setminus \{0\}$. We note that contrary to S_n, the gradient ∇S_n has the same form for all $n \geq 2$. A simple computation shows that

$$\left| \frac{\partial^2 S_n}{\partial x_i \partial x_j}(x) \right| = \left| \frac{1}{s_n} \left(\frac{\delta_{ij}}{|x|^n} - \frac{n x_i x_j}{|x|^{n+2}} \right) \right| \leq \frac{1}{\omega_n} \frac{1}{|x|^n} \qquad \forall x \in \mathbb{R}^n \setminus \{0\}\,, \tag{3.4}$$

for all $i, j \in \{1, \ldots, n\}$, where ω_n denotes the n-dimensional measure of $\mathbb{B}_n(0,1)$.

3.3 Isolated Singularities of Harmonic Functions

Definition 3.16. Let Ω be an open subset of $\mathbb{R}^n$. Let $x_0 \in \Omega$. Let u be a harmonic function from $\Omega \setminus \{x_0\}$ to $\mathbb{R}$. We say that x_0 is a removable singularity for u provided that there exists a harmonic function v in Ω such that

$$u(x) = v(x) \qquad \forall x \in \Omega \setminus \{x_0\}\,.$$

We now introduce a criterium to establish whether a singularity is removable for a harmonic function. For a proof, we refer for example to Folland [102, (2.69), p. 111].

Theorem 3.17. *Let Ω be an open subset of $\mathbb{R}^n$. Let $x_0 \in \Omega$. Let u be a harmonic function from $\Omega \setminus \{x_0\}$ to $\mathbb{R}$. Then x_0 is removable for u if and only if*

$$u(x) = o(S_n(x - x_0)) \qquad \text{as } x \to x_0$$

i.e., if and only if

$$\lim_{x \to x_0} \frac{u(x)}{S_n(x - x_0)} = 0 .$$

Remark 3.18. By exploiting the previous theorem, one can easily show that u has a removable singularity at x_0 if and only if there exists at least one punctured neighborhood of x_0 on which u is bounded. (A punctured neighborhood of a point is by definition a neighborhood of the point minus the point itself.)

3.4 Behavior at Infinity of Harmonic Functions

In order to analyze the behavior at infinity of a harmonic function, we need to introduce the inversion with respect to a sphere and the Kelvin transform.

Definition 3.19. Let $r \in]0, +\infty[$. We define the inversion map with respect to the sphere $\partial\mathbb{B}_n(0, r)$ to be the map $(\cdot)^*$ from $\mathbb{R}^n \setminus \{0\}$ to itself defined by

$$y^* \equiv y \frac{r^2}{|y|^2} \qquad \forall y \in \mathbb{R}^n \setminus \{0\} .$$

As one can easily verify, the inversion with respect to the sphere $\partial\mathbb{B}_n(0, r)$ is a diffeomorphism of class C^∞ from $\mathbb{R}^n \setminus \{0\}$ onto $\mathbb{R}^n \setminus \{0\}$ of involutory character, i.e., $y^{**} = y$ for all $y \in \mathbb{R}^n \setminus \{0\}$. We are now ready to introduce the Kelvin transform.

Definition 3.20. Let A be an open subset of $\mathbb{R}^n \setminus \{0\}$. Let $(\cdot)^*$ be the inversion map with respect to the unit sphere $\partial\mathbb{B}_n(0, 1)$. Let

$$A^* \equiv \left\{ \frac{y}{|y|^2} : y \in A \right\} .$$

Let u be a function from A to $\mathbb{R}$. Then we define the Kelvin transform of u to be the function $\tilde{u}$ from A^* to $\mathbb{R}$ delivered by

$$\tilde{u}(y) \equiv |y|^{2-n} u\left(\frac{y}{|y|^2}\right) \qquad \forall y \in A^* .$$

By the chain rule, one can verify that a function u from A to $\mathbb{R}$ is harmonic if and only if its Kelvin transform $\tilde{u}$ is harmonic in A^*.

We also note that there exists $\rho \in]0, +\infty[$ such that $\mathbb{R}^n \setminus \overline{\mathbb{B}_n(0, \rho)} \subseteq A$ if and only if $\mathbb{B}_n(0, \rho^{-1}) \setminus \{0\} \subseteq A^*$. In particular, $A \cup \{\infty\}$ is an open neighborhood of ∞ if and only if $A^* \cup \{0\}$ is an open neighborhood of 0. We are now ready to introduce the following.

Definition 3.21. Let Ω be an open subset of $\mathbb{R}^n$ such that there exists a compact subset K of $\mathbb{R}^n$ with $\Omega \supseteq \mathbb{R}^n \setminus K$. We say that a function u from Ω to $\mathbb{R}$ is harmonic at infinity provided that there exists $\rho > 0$ such that $\Omega \supseteq \mathbb{R}^n \setminus \overline{\mathbb{B}_n(0,\rho)}$, u is harmonic in $\mathbb{R}^n \setminus \overline{\mathbb{B}_n(0,\rho)}$, and the Kelvin transform of $u_{|\mathbb{R}^n \setminus \overline{\mathbb{B}_n(0,\rho)}}$ has a removable singularity at 0.

Then we have the following characterization of functions which are harmonic at infinity. For a proof we refer for example to Folland [102, (2.74), p. 114].

Theorem 3.22. *Let Ω be an open subset of $\mathbb{R}^n$ such that there exists a compact subset K of $\mathbb{R}^n$ with $\Omega \supseteq \mathbb{R}^n \setminus K$. Let u be a harmonic function from Ω to $\mathbb{R}$. Then the following statements are equivalent.*

(i) u is harmonic at infinity.
(ii) $u(x) = O(|x|^{2-n})$ as x tends to ∞, i.e., $u(x)|x|^{n-2}$ is bounded in at least a punctured neighborhood of infinity.
(iii) If $n \geq 3$, then $\lim_{x\to\infty} u(x) = 0$. If $n = 2$, then $u(x) = o(\log|x|)$ as x tends to ∞, i.e., $\lim_{x\to\infty} \frac{u(x)}{\log|x|} = 0$.

Remark 3.23. A constant function $c \in \mathbb{R} \setminus \{0\}$ in $\mathbb{R}^n$ is harmonic at infinity if and only if $n = 2$.

Corollary 3.24. *Let Ω be an open subset of $\mathbb{R}^n$ such that there exists $\rho \in]0,+\infty[$ with $\Omega \supseteq \mathbb{R}^n \setminus \overline{\mathbb{B}_n(0,\rho)}$. If u is harmonic in Ω and is harmonic at infinity, then the following statements hold.*

(i) If $n \geq 3$ then $\lim_{x\to\infty} u(x) = 0$.
(ii) If $n = 2$ and $r > \rho$, then $\lim_{x\to\infty} u(x) = \fint_{\partial\mathbb{B}_n(0,r)} u\, d\sigma$.

Proof. Statement (i) is an immediate consequence of the previous statement. We now consider statement (ii). We know that the Kelvin transform $\tilde{u}(y) \equiv u(\frac{y}{|y|^2})$ is harmonic in $y \in \mathbb{B}_2(0,\rho^{-1}) \setminus \{0\}$ and has a removable singularity at 0. Let v denote an extension of $\tilde{u}$ in $\mathbb{B}_2(0,\rho^{-1})$. Then by the Spherical Mean Value Property for harmonic functions, we have

$$v(0) = \fint_{\partial\mathbb{B}_2(0,1/r)} u\left(\frac{y}{|y|^2}\right) d\sigma_y$$

(cf., e.g., Folland [102, Theorem 2.8, p. 69]). Then we can change the variable in the integral by setting $x = \frac{y}{|y|^2} = r^2 y$ and we obtain

$$v(0) = \frac{1}{r^{-1}2\pi} \int_{\partial\mathbb{B}_2(0,r)} u(x) r^{-2}\, d\sigma_x = \frac{1}{2\pi r} \int_{\partial\mathbb{B}_2(0,r)} u\, d\sigma\,.$$

Since $v(0) = \lim_{y\to 0} u(\frac{y}{|y|^2}) = \lim_{x\to\infty} u(x)$, the proof is complete. □

Let Ω be an open subset of $\mathbb{R}^n$. Let $u \in C^1(\Omega)$. Then we set

$$\partial_r u(x) = \nabla u(x) \cdot \frac{x}{|x|} \qquad \forall x \in \Omega \setminus \{0\}.$$

Clearly,

$$\partial_r u(x) = \lim_{h \to 0} \frac{u(x + h\frac{x}{|x|}) - u(x)}{h}$$

and thus $\partial_r u(x)$ coincides with the derivative of u in the direction $\frac{x}{|x|}$ for all $x \in \Omega \setminus \{0\}$ (the radial derivative of u). Then we have the following result on the decay of the radial derivative of a harmonic function at infinity. For a proof, we refer for example to Folland [102, (2.75), p. 114].

Proposition 3.25. *Let Ω be an open subset of $\mathbb{R}^n$ such that there exists $\rho > 0$ with $\Omega \supseteq \mathbb{R}^n \setminus \overline{\mathbb{B}_n(0, \rho)}$. Let $u \in C^2(\Omega)$ be harmonic at infinity.*

(i) If $n \geq 3$, then $\partial_r u(x) = O(|x|^{1-n})$ as x tends to ∞, i.e., $\partial_r u(x)|x|^{n-1}$ is bounded in at least a punctured neighborhood of infinity.
(ii) If $n = 2$, then $\partial_r u(x) = O(|x|^{-2})$ as x tends to ∞, i.e., $\partial_r u(x)|x|^2$ is bounded in at least a punctured neighborhood of infinity.

The above properties at infinity of harmonic functions imply the validity of the following technical proposition that we exploit later in the book.

Proposition 3.26. *Let Ω be an open subset of $\mathbb{R}^n$ such that there exists $\rho > 0$ with $\Omega \supseteq \mathbb{R}^n \setminus \overline{\mathbb{B}_n(0, \rho)}$. Let u, v be real valued, harmonic in Ω and harmonic at infinity. Then we have*

$$\lim_{s \to +\infty} \int_{\partial \mathbb{B}_n(0,s)} u \frac{\partial}{\partial \nu_{\mathbb{B}_n(0,s)}} v \, d\sigma = 0. \tag{3.5}$$

Proof. Since $\frac{\partial v}{\partial \nu_{\mathbb{B}_n(0,s)}}(y) = \frac{y}{|y|} \cdot \nabla v(y) = \partial_r v(y)$ for all $y \in \partial \mathbb{B}_n(0, s)$, we have

$$\int_{\partial \mathbb{B}_n(0,s)} u \frac{\partial v}{\partial \nu_{\mathbb{B}_n(0,s)}} \, d\sigma = \int_{\partial \mathbb{B}_n(0,s)} u \partial_r v \, d\sigma ,$$

for all $s \in]\rho, +\infty[$. Since u and v are harmonic at infinity, Theorem 3.22 and Proposition 3.25 ensure the existence of $C > 0$ and $\rho_1 > \rho$ such that

$$|u(x)| \leq C|x|^{2-n} \qquad \text{if } x \in \mathbb{R}^n \setminus \overline{\mathbb{B}_n(0, \rho_1)} \qquad \text{if } n \geq 2,$$

and such that

$$|\partial_r v(x)| \leq C|x|^{1-n} \text{ if } x \in \mathbb{R}^n \setminus \overline{\mathbb{B}_n(0, \rho_1)} \qquad \text{and } n \geq 3,$$
$$|\partial_r v(x)| \leq C|x|^{-2} \text{ if } x \in \mathbb{R}^n \setminus \overline{\mathbb{B}_n(0, \rho_1)} \qquad \text{and } n = 2.$$

Then we have

$$\left| \int_{\partial \mathbb{B}_n(0,s)} u \partial_r v \, d\sigma \right| \leq s_n s^{n-1} C^2 s^{2-n} s^{1-n} = s_n C^2 s^{2-n} \qquad \forall s \in]\rho_1, +\infty[$$

for $n \geq 3$ and

$$\left| \int_{\partial \mathbb{B}_n(0,s)} u \partial_r v \, d\sigma \right| \leq 2\pi s C^2 s^{-2} = 2\pi C^2 s^{-1} \qquad \forall s \in]\rho_1, +\infty[$$

for $n = 2$. Thus the validity of the limiting relation (3.5) follows. □

Chapter 4
Green Identities and Layer Potentials

Abstract This chapter is devoted to the Green Identities and to the layer potentials corresponding to the fundamental solution of the Laplace operator, cf. (3.2). We first consider the Green Identities for bounded domains and then the case of exterior domains for functions which are harmonic at infinity. Next, we introduce the single and double layer potentials and the corresponding mapping properties, especially in spaces of Hölder continuous functions and in Schauder spaces. Although such properties can be found in the classic monographs of Günter (*Potential theory and its applications to basic problems of mathematical physics*. Translated from the Russian by John R. Schulenberger. Frederick Ungar Publishing, New York, 1967) and Kupradze et al. (*Three-dimensional problems of the mathematical theory of elasticity and thermoelasticity*, volume 25 of *North-Holland Series in Applied Mathematics and Mechanics*. North-Holland Publishing, Amsterdam-New York, Russian edition, 1979. Edited by V. D. Kupradze.), here we prove the corresponding statements with optimal Hölder exponents. To do so, we exploit a result of Miranda (Atti Accad Naz Lincei Mem Cl Sci Fis Mat Natur Sez I (8), 7:303–336, 1965) on the Hölder continuity of singular integrals of convolution type, for which we provide a complete proof.

Unless otherwise specified, we understand that throughout this chapter n is a natural number bigger than or equal to two. Namely,

$$n \in \mathbb{N}, \, n \geq 2 .$$

4.1 Green Identities for Bounded Domains

The Green Identities play an important role in the representation of the solutions of boundary value problems. Before introducing the Green Identities, we show a variant of the Divergence Theorem. The difference between the statement here below and the standard formulation which can be found in most textbooks is that, instead

M. Dalla Riva et al., *Singularly Perturbed Boundary Value Problems*,
https://doi.org/10.1007/978-3-030-76259-9_4

of having a vector field that is C^1 on the closure of a certain open set, we deal with a vector field which is C^1 in the interior and only continuous on the closure. The proof is performed by adapting an argument of Dautray and Lions [86, pp. 226–229] and is here included for the sake of completeness. For a more general statement on open sets with Ahlfors regular boundary, we refer to Hofmann et al. [125, Prop. 2.13].

Theorem 4.1 (of the Divergence). *Let Ω be a bounded open Lipschitz subset of $\mathbb{R}^n$. Let $v \in C^0(\overline{\Omega}, \mathbb{R}^n)$ be a continuous vector field on $\overline{\Omega}$ such that $v_{|\Omega} \in C^1(\Omega, \mathbb{R}^n)$ and $\operatorname{div} v \in L^1(\Omega)$. Then we have*

$$\int_\Omega \operatorname{div} v \, dx = \int_{\partial\Omega} v \cdot \nu_\Omega \, d\sigma \,. \tag{4.1}$$

Proof. Since $\overline{\Omega}$ is compact, we can exploit Lemma A.35 on the partition of unity and write v as a finite sum of functions which have compact support in Ω and of functions which have support in a coordinate cylinder. Now, the terms which have compact support in Ω can be extended to functions of class C^1 in $\mathbb{R}^n$ and thus satisfy equality (4.1) by the classical Divergence Theorem in bounded open Lipschitz sets (cf., e.g., Ziemer [277, Theorem 5.8.2, Remark. 5.8.3]). Thus it suffices to prove equality (4.1) in the case in which the support of v is contained in a coordinate cylinder. Namely, we can assume that there exist $p \in \partial\Omega$, $R \in \mathbb{O}_n(\mathbb{R})$, and $r, \delta \in]0, +\infty[$ such that

$$\operatorname{supp} v \subseteq C(p, R, r, \delta) \cap \overline{\Omega} \,,$$

where $C(p, R, r, \delta)$ is a coordinate cylinder for Ω around p. Then we have

$$\int_\Omega \operatorname{div} v \, dx = \int_{C(p,R,r,\delta)\cap\Omega} \operatorname{div} v \, dx$$

and

$$\int_{\partial\Omega} v \cdot \nu_\Omega \, d\sigma = \int_{C(p,R,r,\delta)\cap\partial\Omega} v \cdot \nu_\Omega \, d\sigma \,,$$

and thus it suffices to prove that

$$\int_{C(p,R,r,\delta)\cap\Omega} \operatorname{div} v \, dx = \int_{C(p,R,r,\delta)\cap\partial\Omega} v \cdot \nu_\Omega \, d\sigma \,. \tag{4.2}$$

Without loss of generality we can assume that p is the point at the origin and that R is the identity matrix. Then we denote by $\gamma \in C^{0,1}(\overline{\mathbb{B}_{n-1}(0,r)},]-\delta, \delta[)$ the function that represents $\partial\Omega$ in $C(p, R, r, \delta)$. We have

$$C(p, R, r, \delta) \cap \Omega = \{(\eta, y) \in \mathbb{B}_{n-1}(0, r) \times]-\delta, \delta[: \ y < \gamma(\eta)\} \,.$$

and, for all $\epsilon \in [0, \delta/2[$, we set

$$A_\epsilon = \{(\eta, y) \in \mathbb{B}_{n-1}(0, r) \times]-\delta, \delta[: \ y < \gamma(\eta) - \epsilon\} \,.$$

We can assume that δ and r have been chosen small enough so that $\overline{A_\epsilon} \subseteq \Omega$ for all $\epsilon \in]0, \delta/2[$. Thus v is of class C^1 on the neighborhood Ω of $\overline{A_\epsilon}$. Moreover, the support of the restriction $v_{|A_\epsilon}$ is contained in

$$A_\epsilon \cup \partial' A_\epsilon \,,$$

where $\partial' A_\epsilon$ is the portion of the boundary of A_ϵ where $y = \gamma(\eta) - \epsilon$ and $\eta \in \mathbb{B}_{n-1}(0, r)$. Namely,

$$\partial' A_\epsilon = \{(\eta, \gamma(\eta) - \epsilon) \,:\, \eta \in \mathbb{B}_{n-1}(0, r)\} \,.$$

By the Divergence Theorem on bounded open Lipschitz sets for C^1 functions on a neighborhood of the Lipschitz domain $\overline{A_\epsilon}$ it follows that

$$\int_{A_\epsilon} \operatorname{div} v \, dx = \int_{\partial' A_\epsilon} v \cdot \nu_{A_\epsilon} \, d\sigma \qquad \forall \epsilon \in]0, \delta/2[\,, \tag{4.3}$$

(cf., e.g., Ziemer [277, Theorem. 5.8.2, Remark. 5.8.3]). We now take the limit as $\epsilon \to 0^+$ of the integrals in the left and right-hand sides of (4.3). Since

$$A_\epsilon \subseteq A_0 = C(p, R, r, \delta) \cap \Omega \qquad \forall \epsilon \in [0, \delta/2[$$

and $\operatorname{div} v \in L^1(\Omega)$, the Dominated Convergence Theorem implies that

$$\lim_{\epsilon \to 0^+} \int_{A_\epsilon} \operatorname{div} v \, dx = \int_{C(p,R,r,\delta) \cap \Omega} \operatorname{div} v \, dx \,. \tag{4.4}$$

Moreover, we have

$$\nu_{A_\epsilon}(\eta, \gamma(\eta) - \epsilon) = \begin{pmatrix} -\nabla\gamma(\eta) \\ 1 \end{pmatrix} \frac{1}{\sqrt{1 + |\nabla\gamma(\eta)|^2}} = \nu_\Omega(\eta, \gamma(\eta))$$

for almost all $\eta \in \mathbb{B}_{n-1}(0, r)$. Then we compute

$$\int_{\partial' A_\epsilon} v \cdot \nu_{A_\epsilon} \, d\sigma = \int_{\mathbb{B}_{n-1}(0,r)} v(\eta, \gamma(\eta) - \epsilon) \cdot \begin{pmatrix} -\nabla\gamma(\eta) \\ 1 \end{pmatrix} d\eta \,.$$

Since v is (uniformly) continuous on $\overline{\Omega}$, it follows that

$$\begin{aligned} \lim_{\epsilon \to 0^+} \int_{\partial' A_\epsilon} v \cdot \nu_{A_\epsilon} \, d\sigma &= \lim_{\epsilon \to 0^+} \int_{\mathbb{B}_{n-1}(0,r)} v(\eta, \gamma(\eta) - \epsilon) \cdot \begin{pmatrix} -\nabla\gamma(\eta) \\ 1 \end{pmatrix} d\eta \\ &= \int_{\mathbb{B}_{n-1}(0,r)} v(\eta, \gamma(\eta)) \cdot \begin{pmatrix} -\nabla\gamma(\eta) \\ 1 \end{pmatrix} d\eta = \int_{C(p,R,r,\delta) \cap \partial\Omega} v \cdot \nu_\Omega \, d\sigma \,. \end{aligned} \tag{4.5}$$

Now the validity of (4.2) follows by (4.3), (4.4), and (4.5) and the theorem is proven. □

As a consequence of Theorem 4.1 we deduce the validity of the following.

Theorem 4.2 (First Green Identity). *Let Ω be a bounded open Lipschitz subset of $\mathbb{R}^n$. If $u, v \in C^1(\overline{\Omega})$, $v \in C^2(\Omega)$, and $u\Delta v \in L^1(\Omega)$, then*

$$\int_{\partial\Omega} u \frac{\partial v}{\partial \nu_\Omega}\, d\sigma = \int_\Omega \sum_{j=1}^n \frac{\partial u}{\partial x_j}\frac{\partial v}{\partial x_j}\, dx + \int_\Omega u\Delta v\, dx\,.$$

The above first Green Identity holds in particular if $u, v \in C^2(\overline{\Omega})$. As an immediate corollary, we have the following.

Theorem 4.3 (Second Green Identity). *Let Ω be a bounded open Lipschitz subset of $\mathbb{R}^n$. If $u, v \in C^1(\overline{\Omega}) \cap C^2(\Omega)$ and $v\Delta u$, $u\Delta v \in L^1(\Omega)$, then the following identity holds*

$$\int_\Omega (\Delta u)\, v - u\,(\Delta v)\, dx = \int_{\partial\Omega} \frac{\partial u}{\partial \nu_\Omega} v - u \frac{\partial v}{\partial \nu_\Omega}\, d\sigma\,. \tag{4.6}$$

The above second Green Identity holds in particular if $u, v \in C^2(\overline{\Omega})$.

Remark 4.4. We note that the first Green Identity of Theorem 4.2 and the second Green Identity of Theorem 4.3 hold also in case u and v are complex-valued.

Next we prove the following third Green Identity, which enables us to write a (sufficiently regular) function on $\overline{\Omega}$ in terms of an integral operator applied to Δu in Ω and of integral operators applied to $u_{|\partial\Omega}$ and $\frac{\partial u}{\partial \nu_\Omega}$ on $\partial\Omega$.

Theorem 4.5 (Third Green Identity). *Let Ω be a bounded open Lipschitz subset of $\mathbb{R}^n$. If $u \in C^1(\overline{\Omega}) \cap C^2(\Omega)$ and Δu is integrable in Ω, then the following formula holds.*

$$\begin{aligned} u(x) = {} & \int_\Omega \Delta u(y) S_n(x-y)\, dy \\ & + \int_{\partial\Omega} u(y) \frac{\partial}{\partial \nu_{\Omega_y}} (S_n(x-y)) - \frac{\partial u}{\partial \nu_\Omega}(y) S_n(x-y)\, d\sigma_y \qquad \forall x \in \Omega\,, \end{aligned} \tag{4.7}$$

where

$$\frac{\partial}{\partial \nu_{\Omega_y}} (S_n(x-y)) \equiv -\nu_\Omega(y) \cdot \nabla S_n(x-y) \tag{4.8}$$

for all $x \in \mathbb{R}^n$ and $y \in \partial\Omega$ such that $y \neq x$ and $\nu_\Omega(y)$ exists.

Proof. Let $\epsilon_0 > 0$ be such that $\overline{\mathbb{B}_n(x, \epsilon_0)} \subseteq \Omega$. We note that if $\epsilon \in]0, \epsilon_0]$, then $\Omega_\epsilon \equiv \Omega \setminus \overline{\mathbb{B}_n(x, \epsilon)}$ is a bounded open Lipschitz subset of $\mathbb{R}^n$.

Since $x \notin \overline{\Omega_\epsilon}$, we have $S_n(x - \cdot) \in C^\infty(\overline{\Omega_\epsilon})$. Moreover, $u \in C^1(\overline{\Omega_\epsilon})$. Since $S_n(x - \cdot)$ is continuous and bounded in $\overline{\Omega_\epsilon}$ and Δu is integrable in Ω, then $\Delta u(y) S_n(x-y)$ is integrable in $y \in \Omega_\epsilon$. Since $S_n(x-y)$ is harmonic in $y \in \Omega_\epsilon$, then $u(y)\Delta_y S_n(x-y) = 0$ for all $y \in \Omega_\epsilon$. Hence, we can apply the second Green

Identity to the functions $u(y)$ and $v(y) = S_n(x-y)$. Next we denote by ν_ϵ the outward unit normal to $\partial\Omega_\epsilon$. Then we have $\partial\Omega_\epsilon = \partial\Omega \cup \partial\mathbb{B}_n(x,\epsilon)$ and

$$\nu_\epsilon(y) \equiv \begin{cases} \nu_\Omega(y) \text{ if } y \in \partial\Omega \text{ and if } \nu_\Omega(y) \text{ exists,} \\ \frac{x-y}{|x-y|} \text{ if } y \in \partial\mathbb{B}_n(x,\epsilon)\,. \end{cases}$$

Since $\Delta v = 0$ in Ω_ϵ, the second Green Identity implies that

$$\int_{\Omega_\epsilon} -S_n(x-y)\Delta u(y)\,dy = \int_{\partial\Omega_\epsilon} u(y)\frac{\partial}{\partial\nu_{\epsilon_y}}\left(S_n(x-y)\right) - \frac{\partial u}{\partial\nu_\epsilon}(y)S_n(x-y)\,d\sigma_y$$

and thus

$$\begin{aligned} &-\int_{\partial\mathbb{B}_n(x,\epsilon)} u(y)\frac{\partial}{\partial\nu_{\epsilon_y}}\left(S_n(x-y)\right) - \frac{\partial u}{\partial\nu_\epsilon}(y)S_n(x-y)\,d\sigma_y \\ &= \int_{\Omega_\epsilon} S_n(x-y)\Delta u(y)\,dy \\ &\quad + \int_{\partial\Omega} u(y)\frac{\partial}{\partial\nu_{\Omega_y}}\left(S_n(x-y)\right) - \frac{\partial u}{\partial\nu_\Omega}(y)S_n(x-y)\,d\sigma_y\,. \end{aligned}$$

Thus it suffices to show that

(i)

$$\lim_{\epsilon\to 0}\int_{\Omega_\epsilon} S_n(x-y)\Delta u(y)\,dy = \int_\Omega S_n(x-y)\Delta u(y)\,dy\,,$$

(ii)

$$\lim_{\epsilon\to 0}\int_{\partial\mathbb{B}_n(x,\epsilon)} u(y)\frac{\partial}{\partial\nu_{\epsilon_y}} S_n(x-y) - \frac{\partial u}{\partial\nu_\epsilon}(y)S_n(x-y)\,d\sigma_y = -u(x)\,.$$

We first consider (i). We note that

$$\int_{\Omega_\epsilon} S_n(x-y)\Delta u(y)\,dy = \int_\Omega \chi_{\Omega_\epsilon}(y)S_n(x-y)\Delta u(y)\,dy\,,$$

and that

$$\begin{aligned} &|\chi_{\Omega_\epsilon}(y)S_n(x-y)\Delta u(y)| \le |\chi_\Omega(y)S_n(x-y)\Delta u(y)|\,, \\ &\lim_{\epsilon\to 0}\chi_{\Omega_\epsilon}(y)S_n(x-y)\Delta u(y) = \chi_\Omega(y)S_n(x-y)\Delta u(y) \end{aligned}$$

for all $y \in \Omega\setminus\{x\}$. Since $\Delta u \in C^0(\overline{\mathbb{B}_n(x,\epsilon_0)})$ and $S_n(x-\cdot) \in L^1(\mathbb{B}_n(x,\epsilon_0))$, the function $S_n(x-y)\Delta u(y)$ is integrable in $y \in \mathbb{B}_n(x,\epsilon_0)$. On the other hand, we have already observed that the same function is integrable in $\Omega\setminus\mathbb{B}_n(x,\epsilon_0)$, and accordingly $\chi_\Omega(y)S_n(x-y)\Delta u(y) \in L^1(\Omega)$. Then the Dominated Convergence Theorem implies that

$$\lim_{\epsilon\to 0}\int_{\Omega}\chi_{\Omega_\epsilon}(y)S_n(x-y)\Delta u(y)\,dy=\int_{\Omega}\chi_{\Omega}(y)S_n(x-y)\Delta u(y)\,dy\,.$$

Hence, (i) holds true. Next we turn to the proof of (ii). It clearly suffices to prove that

(j) $\lim_{\epsilon\to 0}\int_{\partial\mathbb{B}_n(x,\epsilon)}u(y)\frac{\partial}{\partial\nu_{\epsilon_y}}\left(S_n(x-y)\right)\,d\sigma_y=-u(x)$,

(jj) $\lim_{\epsilon\to 0}\int_{\partial\mathbb{B}_n(x,\epsilon)}\frac{\partial u}{\partial\nu_\epsilon}(y)S_n(x-y)\,d\sigma_y=0$.

We first consider (j). Clearly,

$$\begin{aligned}&\int_{\partial\mathbb{B}_n(x,\epsilon)}u(y)\frac{\partial}{\partial\nu_{\epsilon_y}}\left(S_n(x-y)\right)\,d\sigma_y\\&=-\int_{\partial\mathbb{B}_n(x,\epsilon)}u(y)\frac{x-y}{|x-y|}\cdot\frac{x-y}{|x-y|^n}\frac{d\sigma_y}{s_n}\\&=-\frac{1}{\epsilon^{n-1}s_n}\int_{\partial\mathbb{B}_n(x,\epsilon)}u(y)d\sigma_y=-\fint_{\partial\mathbb{B}_n(x,\epsilon)}u(y)d\sigma_y\qquad\forall\epsilon\in]0,\epsilon_0[\,.\end{aligned}$$

Since u is continuous at x, we have

$$\lim_{\epsilon\to 0}\fint_{\partial\mathbb{B}_n(x,\epsilon)}u(y)d\sigma_y=u(x)$$

and thus (j) holds. We now prove (jj). Clearly,

$$\left|\int_{\partial\mathbb{B}_n(x,\epsilon)}\frac{\partial u}{\partial\nu_\epsilon}(y)S_n(x-y)\,d\sigma_y\right|\le\epsilon^{n-1}s_n|\Upsilon_n(\epsilon)|\sup_{\mathbb{B}_n(x,\epsilon_0)}|\nabla u|\quad\forall\epsilon\in]0,\epsilon_0[$$

(see (3.1) for the definition of Υ_n).Then, by taking the limit as ϵ tends to zero, we conclude that (jj) holds true and thus the proof is complete. □

The above third Green Identity holds in particular if $u\in C^2(\overline{\Omega})$. We also have the following corollary.

Corollary 4.6. *Let Ω be a bounded open Lipschitz subset of $\mathbb{R}^n$. If $u\in C^1(\overline{\Omega})$ is harmonic in Ω, then*

$$\int_{\partial\Omega}u(y)\frac{\partial}{\partial\nu_{\Omega_y}}\left(S_n(x-y)\right)-\frac{\partial u}{\partial\nu_\Omega}(y)S_n(x-y)\,d\sigma_y=\begin{cases}u(x)&\text{if }x\in\Omega\,,\\0&\text{if }x\in\mathbb{R}^n\setminus\overline{\Omega}\,.\end{cases}$$

In particular,

$$\int_{\partial\Omega}\frac{\partial}{\partial\nu_{\Omega_y}}\left(S_n(x-y)\right)\,d\sigma_y=\begin{cases}1\text{ if }x\in\Omega\,,\\0\text{ if }x\in\mathbb{R}^n\setminus\overline{\Omega}\end{cases}$$

(see also (4.8)).

Proof. If $x\in\Omega$, the statement is an immediate consequence of the third Green Identity. If $x\in\mathbb{R}^n\setminus\overline{\Omega}$, then both the functions $u(y)$ and $v(y)\equiv S_n(x-y)$ are of class $C^1(\overline{\Omega}) \cap C^2(\Omega)$ and harmonic in Ω. Then the statement follows by the second Green Identity. The second part of the statement follows by setting $u = 1$. □

4.2 Green Identities for Harmonic Functions on Exterior Domains

In this section we introduce some Green Identities for functions in the exterior Ω^- of an open bounded subset Ω of $\mathbb{R}^n$. In particular, we will deal with functions that are harmonic and harmonic at infinity. We begin with the following Corollary 4.7, where we present a counterpart in Ω^- of the first Green Identity.

Corollary 4.7 (First Green Identity in Exterior Domains). *Let Ω be a bounded open Lipschitz subset of $\mathbb{R}^n$. Let $u, v \in C^1(\overline{\Omega^-})$ be harmonic in Ω^- and harmonic at infinity. Then we have*

$$\lim_{R\to+\infty} \int_{\Omega^- \cap \mathbb{B}_n(0,R)} \sum_{j=1}^{n} \frac{\partial u}{\partial x_j}\frac{\partial v}{\partial x_j}\, dx = -\int_{\partial\Omega} u \frac{\partial v}{\partial \nu_\Omega}\, d\sigma \,. \tag{4.9}$$

In particular, $|\nabla u|^2$ is integrable in Ω^- and

$$\int_{\Omega^-} |\nabla u|^2\, dx = -\int_{\partial\Omega} u \frac{\partial u}{\partial \nu_\Omega}\, d\sigma \,. \tag{4.10}$$

Proof. Let $R \in]0, +\infty[$ be such that $\overline{\Omega} \subseteq \mathbb{B}_n(0,R)$. Let $\Omega_R^- \equiv \Omega^- \cap \mathbb{B}_n(0,R)$. Clearly, Ω_R^- is a bounded open Lipschitz subset of $\mathbb{R}^n$ and

$$\partial\Omega_R^- = \partial\Omega \cup \partial\mathbb{B}_n(0,R)\,, \qquad \partial\Omega \cap \partial\mathbb{B}_n(0,R) = \emptyset\,,$$
$$\nu_{\Omega_R^-} = -\nu_\Omega \quad \text{a.e. on } \partial\Omega = \partial\Omega^-\,, \qquad \nu_{\Omega_R^-}(y) = \frac{y}{|y|} \quad \forall y \in \partial\mathbb{B}_n(0,R)\,.$$

Moreover, $u, v \in C^1(\overline{\Omega_R^-}) \cap C^2(\Omega_R^-)$. Then the first Green Identity implies that

$$\begin{aligned}\int_{\Omega_R^-} \sum_{j=1}^{n} \frac{\partial u}{\partial x_j}\frac{\partial v}{\partial x_j}\, dx &= \int_{\partial\Omega_R^-} u \frac{\partial v}{\partial \nu_{\Omega_R^-}}\, d\sigma \\ &= -\int_{\partial\Omega} u \frac{\partial v}{\partial \nu_\Omega}\, d\sigma + \int_{\partial\mathbb{B}_n(0,R)} u \frac{\partial v}{\partial \nu_{\mathbb{B}_n(0,R)}}\, d\sigma \,.\end{aligned}$$

Since both u and v are harmonic at infinity, the limit as R tends to infinity of the last integral in the right-hand side exists and equals zero (cf. Proposition 3.26). Then the limit in the statement exists finite, and formula (4.9) holds.

If $v = u$, then the Monotone Convergence Theorem implies that

$$\int_{\Omega^-} |\nabla u|^2 \, dx = \lim_{R\to\infty} \int_{\Omega^-} \chi_{\Omega_R^-} |\nabla u|^2 \, dx = \lim_{R\to\infty} \int_{\Omega_R^-} |\nabla u|^2 \, dx$$

and formula (4.9) for $v = u$ implies the validity of formula (4.10). □

By the previous corollary, we deduce immediately the following form of the second Green Identity for harmonic functions in exterior domains.

Corollary 4.8 (Second Green Identity in Exterior Domains). *Let Ω be a bounded open Lipschitz subset of $\mathbb{R}^n$. Let $u, v \in C^1(\overline{\Omega^-})$ be harmonic in Ω^- and harmonic at infinity. Then we have*

$$\int_{\partial\Omega} u \frac{\partial v}{\partial \nu_\Omega} - v \frac{\partial u}{\partial \nu_\Omega} \, d\sigma = 0 \, . \tag{4.11}$$

Remark 4.9. We note that the first Green Identity of (4.9) and the second Green Identity of (4.11) hold also in case u and v are complex-valued. Then by taking $v = \overline{u}$ in (4.9) one can deduce that the left-hand side of (4.10) equals $-\int_{\partial\Omega} u \frac{\partial \overline{u}}{\partial \nu_\Omega} \, d\sigma$ when u is complex-valued.

We now consider the third Green Identity for harmonic functions in Ω^-.

Theorem 4.10 (Third Green Identity in Exterior Domains). *Let Ω be a bounded open Lipschitz subset of $\mathbb{R}^n$. Let $u \in C^1(\overline{\Omega^-})$ be harmonic in Ω^- and harmonic at infinity. Let u_∞ be the limiting value of u at infinity. Then we have*

$$u(x) = -\int_{\partial\Omega} u(y) \frac{\partial}{\partial \nu_{\Omega_y}} (S_n(x-y)) - \frac{\partial u}{\partial \nu_\Omega}(y) S_n(x-y) \, d\sigma_y + u_\infty \quad \forall x \in \Omega^- .$$

If $x \in (\Omega^-)_j$ for some $j \in \{1, \ldots, \varkappa^-\}$, then we also have

$$u(x) = -\int_{\partial(\Omega^-)_j} u(y) \frac{\partial}{\partial \nu_{\Omega_y}} (S_n(x-y)) - \frac{\partial u}{\partial \nu_\Omega}(y) S_n(x-y) \, d\sigma_y$$

(see (4.8)). In addition,

$$0 = -\int_{\partial\Omega} u(y) \frac{\partial}{\partial \nu_{\Omega_y}} (S_n(x-y)) - \frac{\partial u}{\partial \nu_\Omega}(y) S_n(x-y) \, d\sigma_y + u_\infty \qquad \forall x \in \Omega .$$

Proof. Let $x \in \Omega^-$. Let $R_0 \in]0, +\infty[$ be such that $\overline{\Omega} \cup \{x\} \subseteq \mathbb{B}_n(0, R_0)$. Let $\Omega_R^- \equiv \Omega^- \cap \mathbb{B}_n(0, R)$ for all $R \geq R_0$. Clearly, Ω_R^- is a bounded open subset of class C^1 and

$$\begin{aligned} \partial\Omega_R^- &= \partial\Omega \cup \partial\mathbb{B}_n(0, R) \, , \\ \nu_{\Omega_R^-} &= -\nu_\Omega \quad \text{a.e. on } \partial\Omega = \partial\Omega^- \, , \qquad \nu_{\mathbb{B}_n(0,R)}(y) = \frac{y}{|y|} \quad \forall y \in \partial\mathbb{B}_n(0, R) \, , \end{aligned} \tag{4.12}$$

for all $R \geq R_0$. By assumption, $u \in C^1(\overline{\Omega^- \cap \mathbb{B}_n(0, R)}) \cap C^2(\Omega^- \cap \mathbb{B}_n(0, R))$. Then we can apply the third Green Identity in the bounded open Lipschitz set Ω_R^-

and obtain

$$u(x) = -\int_{\partial\Omega} u(y)\frac{\partial}{\partial\nu_{\Omega_y}}(S_n(x-y)) - \frac{\partial u}{\partial\nu_\Omega}(y)S_n(x-y)\,d\sigma_y$$
$$+\int_{\partial\mathbb{B}_n(0,R)} u(y)\frac{\partial}{\partial\nu_{\mathbb{B}_n(0,R)_y}}(S_n(x-y)) - \frac{\partial u}{\partial\nu_{\mathbb{B}_n(0,R)}}(y)S_n(x-y)\,d\sigma_y$$

for all $R \geq R_0$. Thus it suffices to prove that

(i) $\lim_{R\to+\infty}\int_{\partial\mathbb{B}_n(0,R)} u(y)\frac{\partial}{\partial\nu_{\mathbb{B}_n(0,R)_y}}(S_n(x-y))\,d\sigma_y = u_\infty$,
(ii) $\lim_{R\to+\infty}\int_{\partial\mathbb{B}_n(0,R)} \frac{\partial u}{\partial\nu_{\mathbb{B}_n(0,R)}}(y)S_n(x-y)\,d\sigma_y = 0$.

By the decay properties of harmonic functions at infinity of Proposition 3.25, we know that possibly choosing a larger R_0, there exists $C > 0$ such that

$$\begin{aligned} |\partial_r u(y)| &\leq C|y|^{1-n} && \text{if } |y| \geq R_0\,,\ n \geq 3\,, \\ |\partial_r u(y)| &\leq C|y|^{-2} && \text{if } |y| \geq R_0\,,\ n = 2\,. \end{aligned} \tag{4.13}$$

Then we note that

$$\left|\frac{\partial}{\partial\nu_{\mathbb{B}_n(0,R)_y}}(S_n(x-y))\right| = \left|\frac{y}{|y|}\nabla S_n(x-y)\right| \leq \frac{1}{s_n}\frac{1}{|x-y|^{n-1}} \leq \frac{1}{s_n}\frac{1}{(R-R_0)^{n-1}}\,, \tag{4.14}$$

$$|S_n(x-y)| \leq \frac{1}{(n-2)s_n}\frac{1}{|x-y|^{n-2}} \leq \frac{1}{(n-2)s_n}\frac{1}{(R-R_0)^{n-2}} \quad \text{if } n \geq 3\,, \tag{4.15}$$

$$|S_2(x-y)| \leq \frac{1}{2\pi}\left|\log|x-y|\right| \leq \frac{1}{2\pi}\log(R+R_0) \tag{4.16}$$

for all $y \in \partial\mathbb{B}_n(0,R)$ and $R > R_0 + 1$.

Then we are ready to prove (i). By the third Green Identity applied to the constant function 1, we know that

$$1 = \int_{\partial\mathbb{B}_n(0,R)} \frac{\partial}{\partial\nu_{\mathbb{B}_n(0,R)_y}}(S_n(x-y))\,d\sigma_y\,.$$

Then we have

$$\int_{\partial\mathbb{B}_n(0,R)} u(y)\frac{\partial}{\partial\nu_{\mathbb{B}_n(0,R)_y}}(S_n(x-y))\,d\sigma_y - u_\infty$$
$$= \int_{\partial\mathbb{B}_n(0,R)} (u(y)-u_\infty)\frac{\partial}{\partial\nu_{\mathbb{B}_n(0,R)_y}}(S_n(x-y))\,d\sigma_y\,,$$

and it suffices to prove that

$$\lim_{R\to+\infty}\int_{\partial\mathbb{B}_n(0,R)}(u(y)-u_\infty)\frac{\partial}{\partial\nu_{\mathbb{B}_n(0,R)_y}}(S_n(x-y))\,d\sigma_y=0\,.$$

To do so, we note that the above inequality (4.14) implies that

$$\begin{aligned}&\left|\int_{\partial\mathbb{B}_n(0,R)}(u(y)-u_\infty)\frac{\partial}{\partial\nu_{\mathbb{B}_n(0,R)_y}}(S_n(x-y))\,d\sigma_y\right|\\&\qquad\le\frac{s_nR^{n-1}}{s_n(R-R_0)^{n-1}}\sup_{y\in\partial\mathbb{B}_n(0,R)}|u(y)-u_\infty|\qquad\forall R\ge R_0+1\,.\end{aligned}$$

On the other hand the limiting relation $\lim_{x\to\infty}u(x)=u_\infty$ implies that

$$\lim_{R\to+\infty}\sup_{y\in\partial\mathbb{B}_n(0,R)}|u(y)-u_\infty|=0\,.$$

Hence, statement (i) follows.

We now consider statement (ii). If $n\ge 3$, the above inequalities (4.13) and (4.15) imply that

$$\begin{aligned}&\left|\int_{\partial\mathbb{B}_n(0,R)}\frac{\partial u}{\partial\nu_{\mathbb{B}_n(0,R)}}(y)S_n(x-y)\,d\sigma_y\right|\\&\qquad\le\frac{s_nR^{n-1}}{(n-2)s_n(R-R_0)^{n-2}}\sup_{y\in\partial\mathbb{B}_n(0,R)}\left|\frac{\partial u}{\partial\nu_{\mathbb{B}_n(0,R)}}(y)\right|\\&\qquad\le\frac{s_nR^{n-1}C}{(n-2)s_n(R-R_0)^{n-2}R^{n-1}}\qquad\forall R\ge R_0+1\,.\end{aligned}$$

If instead $n=2$, , the above inequalities (4.13) and (4.16) imply that

$$\begin{aligned}&\left|\int_{\partial\mathbb{B}_n(0,R)}\frac{\partial u}{\partial\nu_{\mathbb{B}_n(0,R)}}(y)S_n(x-y)\,d\sigma_y\right|\\&\qquad\le\frac{2\pi R}{2\pi}\log(R+R_0)\sup_{y\in\partial\mathbb{B}_n(0,R)}\left|\frac{\partial u}{\partial\nu_{\mathbb{B}_n(0,R)}}(y)\right|\\&\qquad\le CR^{-1}\log(R+R_0)\qquad\forall R>R_0+1\,.\end{aligned}$$

Accordingly, (ii) holds both if $n\ge 3$ and if $n=2$. In case $n\ge 3$, one could prove statement (ii) also by observing that that both $u(\cdot)$ and $S_n(x-\cdot)$ are harmonic at infinity and by invoking Proposition 3.26.

To prove the equality in case $x\in(\Omega^-)_j$ with $j\ge 1$, it suffices to apply the third Green Identity to $(\Omega^-)_j$ and to note that $\nu_{(\Omega^-)_j}(x)=-\nu_\Omega(x)$ for all $x\in\partial(\Omega^-)_j$.

We now prove the last part of the statement. Let $x\in\Omega$. Let $R_0\in]0,+\infty[$ be such that $\overline{\Omega}\subseteq\mathbb{B}_n(0,R_0)$. Let $\Omega_R^-\equiv\Omega^-\cap\mathbb{B}_n(0,R)$ for all $R\ge R_0$. Since $u(\cdot)$ and $S_n(x-\cdot)$ are harmonic in Ω_R^- and of class $C^1(\overline{\Omega_R^-})$, the second Green Identity in Ω_R^- and formulas (4.12) for $\nu_{\Omega_R^-}$ imply that

$$
\begin{aligned}
0 &= \int_{\partial\Omega_R^-} u(y)\frac{\partial}{\partial\nu_{\Omega^-_{R_y}}}(S_n(x-y)) - \frac{\partial u}{\partial\nu_{\Omega_R^-}}(y)S_n(x-y)\,d\sigma_y \\
&= -\int_{\partial\Omega} u(y)\frac{\partial}{\partial\nu_{\Omega_y}}(S_n(x-y)) - \frac{\partial u}{\partial\nu_{\Omega}}(y)S_n(x-y)\,d\sigma_y \\
&\quad + \int_{\partial\mathbb{B}_n(0,R)} u(y)\frac{\partial}{\partial\nu_{\mathbb{B}_n(0,R)_y}}(S_n(x-y)) - \frac{\partial u}{\partial\nu_{\mathbb{B}_n(0,R)}}(y)S_n(x-y)\,d\sigma_y
\end{aligned}
$$

for all $R \geq R_0$. Arguing as above, we can prove the limiting relations (i) and (ii) for $x \in \Omega$ and deduce the validity of the last formula of the statement. □

In conclusion of this section, we note that the third Green Identities in Ω and in Ω^- show that we can write a large class of functions u in terms of integrals of the form

$$
\int_{\partial\Omega} \phi(y)S_n(x-y)\,d\sigma_y\,, \qquad \int_{\partial\Omega} \psi(y)\frac{\partial}{\partial\nu_{\Omega_y}}(S_n(x-y))\,d\sigma_y\,,
$$

which are known as single (or simple) layer potential of moment (or density) ϕ and double layer potential of moment (or density) ψ, respectively, and in terms of integrals of the form

$$
\int_{\Omega} S_n(x-y)f(y)\,dy\,,
$$

which are known as volume potentials with density f. In the next sections, we will analyze some properties of the single and double layer potentials. In Chapter 6, we will exploit these results to prove existence theorems for the solutions of boundary value problems for the Laplace equation. Later on, in Chapter 7, we will consider volume potentials and prove existence results for the solutions of boundary value problems for the Poisson equation.

4.3 Preliminaries on Singular Integrals and Layer Potentials

We first note that both the single and the double layer potentials can be written in terms of integrals of the form

$$
\int_{\partial\Omega} k(x-y)\mu(y)\,d\sigma_y\,,
$$

where k is a continuous function in $\mathbb{R}^n \setminus \{0\}$. Indeed, for the single layer potential it suffices to take $k = S_n$ while we can write the double layer potential as

$$
\int_{\partial\Omega} \omega(y)\frac{\partial}{\partial\nu_{\Omega_y}}S_n(x-y)\,d\sigma_y = -\sum_{l=1}^{n}\int_{\partial\Omega}\frac{\partial}{\partial x_l}S_n(x-y)(\nu_\Omega)_l(y)\omega(y)\,d\sigma_y
$$

and each integral of the sum in the right-hand side can be written in the above form with $k = \frac{\partial}{\partial x_l} S_n$ and $\mu = (\nu_\Omega)_l \omega$. Thus our goal is to analyze the properties of integrals of the form

$$K[k,\mu](x) \equiv \int_{\partial\Omega} k(x-y)\mu(y)\, d\sigma_y \qquad \forall x \in \mathbb{R}^n\,,$$

which are called single layer potentials with support in $\partial\Omega$, (convolution) kernel k, and density, or moment, μ, and we will focus on the case where k equals the fundamental solution S_n or a first order partial derivative of S_n.

If k equals S_n and if $x \in \partial\Omega$, then $k(x-y)$ is integrable in $y \in \partial\Omega$. Instead, if k equals one of the first order partial derivatives of S_n, then $k(x-y)$ is not integrable in $y \in \partial\Omega$ and the above integral may exist only in the sense of the principal value, i.e., as the limit

$$\text{p.v.} \int_{\partial\Omega} k(x-y)\mu(y)\, d\sigma_y \equiv \lim_{\epsilon\to 0} \int_{\partial\Omega\setminus\mathbb{B}_n(x,\epsilon)} k(x-y)\mu(y)\, d\sigma_y\,.$$

Since the first order partial derivatives of S_n are positively homogeneous of degree $-(n-1)$, we are interested in the properties of the above principal value in case where k is positively homogeneous of degree $-(n-1)$. We now briefly summarize some properties of the positively homogeneous functions and we introduce some notation.

Definition 4.11. Let $n \in \mathbb{N} \setminus \{0\}$, $h \in \mathbb{R}$. We say that a function f from $\mathbb{R}^n \setminus \{0\}$ to $\mathbb{R}$ is positively homogeneous of degree h provided that

$$f(tx) = t^h f(x) \qquad \forall (t,x) \in]0,+\infty[\times(\mathbb{R}^n \setminus \{0\})\,.$$

By setting $t = |x|$, we have

$$f(x) = f(|x|\frac{x}{|x|}) = |x|^h f(\frac{x}{|x|}) \qquad \forall x \in \mathbb{R}^n \setminus \{0\}\,,$$

an equality which shows that a positively homogeneous function of degree h is entirely determined by its values on the unit sphere $\partial\mathbb{B}_n(0,1)$.

Proposition 4.12. *Let $n \in \mathbb{N} \setminus \{0\}$, $h \in \mathbb{R}$. Let $f \in C^0(\mathbb{R}^n \setminus \{0\})$ be positively homogeneous of degree h. Then the following statements hold.*

(i) $|f(x)| \leq \left(\sup_{\partial\mathbb{B}_n(0,1)} |f|\right) |x|^h$ *for all* $x \in \mathbb{R}^n \setminus \{0\}$.

(ii) If $m \in \mathbb{N} \setminus \{0\}$ and if we further assume that $f \in C^m(\mathbb{R}^n \setminus \{0\})$, then $D^\beta f$ is positively homogeneous of degree $h - |\beta|$ for all $\beta \in \mathbb{N}^n$ such that $|\beta| \leq m$.

The proof is an advanced Calculus exercise and it is left to the reader. If $m \in \mathbb{N}$ and $\alpha \in [0,1]$, then we set

$$\mathcal{K}_{m,\alpha} \tag{4.17}$$

$$\equiv \{k \in C^{m,\alpha}_{\mathrm{loc}}(\mathbb{R}^n \setminus \{0\}) : k \text{ is positively homogeneous of degree } -(n-1)\}$$

and

$$\|k\|_{\mathcal{K}_{m,\alpha}} \equiv \left\|k_{|\partial\mathbb{B}_n(0,1)}\right\|_{C^{m,\alpha}(\partial\mathbb{B}_n(0,1))} \qquad \forall k \in \mathcal{K}_{m,\alpha} \tag{4.18}$$

(see Section 2.11 for the definition of $C^{m,\alpha}_{\mathrm{loc}}(\mathbb{R}^n \setminus \{0\})$). Then one can easily verify that $\left(\mathcal{K}_{m,\alpha}, \|\cdot\|_{\mathcal{K}_{m,\alpha}}\right)$ is a Banach space. We will also consider the subspace

$$\mathcal{K}_{m,\alpha;o} \equiv \{k \in \mathcal{K}_{m,\alpha} : k \text{ is odd}\} \tag{4.19}$$

of $\mathcal{K}_{m,\alpha}$. Since positively homogeneous functions are uniquely determined by their values on the unit sphere, we can prove the following lemma.

Lemma 4.13. *Let $\alpha \in [0,1]$, $j \in \{1,\ldots,n\}$. Then the linear operator from $\mathcal{K}_{1,\alpha}$ to $C^{0,\alpha}(\partial\mathbb{B}_n(0,1))$ that takes k to $\frac{\partial k}{\partial x_j}_{|\partial\mathbb{B}_n(0,1)}$ is linear and continuous.*

Proof. Let $\mathbb{A}_n(1/2,1) \equiv \mathbb{B}_n(0,1) \setminus \overline{\mathbb{B}_n(0,1/2)}$. By Theorem 2.85 (i), the restriction operator $\mathcal{R}$ from the subspace

$$C^{1,\alpha}_{-(n-1)}(\overline{\mathbb{A}_n(1/2,1)}) \equiv \Big\{ f_{|\overline{\mathbb{A}_n(1/2,1)}} : f \in C^{1,\alpha}_{\mathrm{loc}}(\mathbb{R}^n \setminus \{0\}),$$
$$f \text{ is positively homogeneous of degree } -(n-1) \Big\}$$

of $C^{1,\alpha}(\overline{\mathbb{A}_n(1/2,1)})$ to $C^{1,\alpha}(\partial\mathbb{B}_n(0,1))$ is linear and continuous.

Since positively homogeneous functions are uniquely determined by their values on the unit sphere, the restriction operator $\mathcal{R}$ is a bijection. We now prove that $C^{1,\alpha}_{-(n-1)}(\overline{\mathbb{A}_n(1/2,1)})$ is closed in the space $C^{1,\alpha}(\overline{\mathbb{A}_n(1/2,1)})$. Let $\{g_j\}_{j\in\mathbb{N}}$ be a sequence of $C^{1,\alpha}_{-(n-1)}(\overline{\mathbb{A}_n(1/2,1)})$ that converges to g in $C^{1,\alpha}(\overline{\mathbb{A}_n(1/2,1)})$. We need to show that g belongs to $C^{1,\alpha}_{-(n-1)}(\overline{\mathbb{A}_n(1/2,1)})$. To do so, it suffices note that g is the restriction to $\overline{\mathbb{A}_n(1/2,1)}$ of the positively homogeneous function $|x|^{-(n-1)}g(x/|x|)$ and to prove that $|x|^{-(n-1)}g(x/|x|)$ is of class $C^{1,\alpha}_{\mathrm{loc}}(\mathbb{R}^n \setminus \{0\})$. Since the composition of a function of $C^{1,\alpha}(\overline{\mathbb{A}_n(1/2,1)})$ and of the function $x/|x|$ is of class $C^{1,\alpha}_{\mathrm{loc}}(\mathbb{R}^n \setminus \{0\})$ and the pointwise product of functions in Schauder spaces is continuous, we deduce that the function $|x|^{-(n-1)}g(x/|x|)$ belongs to $C^{1,\alpha}_{\mathrm{loc}}(\mathbb{R}^n \setminus \{0\})$ (cf. Proposition 2.54, Theorem 2.59). Hence, $C^{1,\alpha}_{-(n-1)}(\overline{\mathbb{A}_n(1/2,1)})$ is a Banach space. Then the Open Mapping Theorem 2.2 implies that $\mathcal{R}$ is a homeomorphism.

Then the definition of norm in $\mathcal{K}_{1,\alpha}$ implies that the map from $\mathcal{K}_{1,\alpha}$ to the space $C^{1,\alpha}_{-(n-1)}(\overline{\mathbb{A}_n(1/2,1)})$ that takes k to $k_{|\overline{\mathbb{A}_n(1/2,1)}}$ is linear and continuous. Since $\frac{\partial}{\partial x_j}$ is linear and continuous from $C^{1,\alpha}_{-(n-1)}(\overline{\mathbb{A}_n(1/2,1)})$ to $C^{0,\alpha}(\overline{\mathbb{A}_n(1/2,1)})$ and the restriction operator is linear and continuous from $C^{0,\alpha}(\overline{\mathbb{A}_n(1/2,1)})$ to $C^{0,\alpha}(\partial\mathbb{B}_n(0,1))$ (cf. Theorem 2.85 (i)), we conclude that $\frac{\partial}{\partial x_j}_{|\partial\mathbb{B}_n(0,1)}$ is linear and continuous from $\mathcal{K}_{1,\alpha}$ to $C^{0,\alpha}(\partial\mathbb{B}_n(0,1))$. □

Then we introduce the following elementary Hölder continuity lemma for positively homogeneous functions, that we prove by an argument of Cialdea [52, VIII, p. 47] who has considered case $\alpha = 1$.

Lemma 4.14. *Let $n \in \mathbb{N} \setminus \{0\}$, $h \in \mathbb{N}$. Let $\alpha \in]0,1]$. Let $k \in C^{0,\alpha}_{\mathrm{loc}}(\mathbb{R}^n \setminus \{0\})$ be a positively homogeneous function of degree $-h$. Then*

$$\begin{aligned}
&|k(x) - k(y)| \\
&\qquad \le (2^\alpha + 2h) \max\{ \sup_{\partial\mathbb{B}_n(0,1)} |k|, |k : \partial\mathbb{B}_n(0,1)|_\alpha \} |x-y|^\alpha (\min\{|x|,|y|\})^{-h-\alpha}
\end{aligned}$$

for all $x, y \in \mathbb{R}^n \setminus \{0\}$.

Proof. Let $c' \equiv \max\{\sup_{\partial\mathbb{B}_n(0,1)} |k|, |k : \partial\mathbb{B}_n(0,1)|_\alpha\}$. We first consider case $h > 0$. By the triangular inequality, we have

$$\begin{aligned}
|k(x) - k(y)| &\le \left| k\left(\frac{x}{|x|}\right) |x|^{-h} - k\left(\frac{y}{|y|}\right) |y|^{-h} \right| \\
&\le \left| k\left(\frac{x}{|x|}\right) - k\left(\frac{y}{|y|}\right) \right| |x|^{-h} + \left| k\left(\frac{y}{|y|}\right) \right| \left| |x|^{-h} - |y|^{-h} \right| \\
&\le c' \left| \frac{x}{|x|} - \frac{y}{|y|} \right|^\alpha |x|^{-h} + c' \left| |x|^{-h} - |y|^{-h} \right| \\
&\le c' \left| \frac{x|y| - y|x|}{|x||y|} \right|^\alpha |x|^{-h} + c' \left| |x|^{-1} - |y|^{-1} \right| \sum_{l=0}^{h-1} |x|^{-l} |y|^{l-(h-1)} \\
&\le c' \left| \frac{(x-y)|y| + y(|y| - |x|)}{|x||y|} \right|^\alpha |x|^{-h} \\
&\quad + c'|x-y|^\alpha (|x| + |y|)^{1-\alpha} \sum_{l=0}^{h-1} |x|^{-l-1} |y|^{l-h}
\end{aligned}$$

for all $x, y \in \mathbb{R}^n \setminus \{0\}$. By inequality (2.3) with $p = 1 - \alpha$, $q = 1$, we have

$$(|x| + |y|)^{1-\alpha} \le |x|^{1-\alpha} + |y|^{1-\alpha} \qquad \forall x, y \in \mathbb{R}^n \setminus \{0\}$$

and thus

$$\begin{aligned}
|k(x) - k(y)| &\le 2^\alpha c' |x-y|^\alpha |x|^{-\alpha-h} \\
&\quad + c'|x-y|^\alpha \left(\sum_{l=0}^{h-1} |x|^{-l-\alpha} |y|^{l-h} + \sum_{l=0}^{h-1} |x|^{-l-1} |y|^{l-h+1-\alpha} \right) \\
&\le 2^\alpha c' |x-y|^\alpha (\min\{|x|,|y|\})^{-(h+\alpha)} + c'|x-y|^\alpha 2h (\min\{|x|,|y|\})^{-(h+\alpha)}
\end{aligned}$$

for all $x, y \in \mathbb{R}^n \setminus \{0\}$. Indeed, $-\alpha - h$ and all the exponents which appear in the two above summation signs are negative and accordingly

$$|x|^{-\alpha-h} \le (\min\{|x|,|y|\})^{-(h+\alpha)}, \qquad |x|^{-l-\alpha} |y|^{l-h} \le (\min\{|x|,|y|\})^{-(h+\alpha)},$$

$$|x|^{-l-1}|y|^{l-h+1-\alpha} \leq (\min\{|x|,|y|\})^{-(h+\alpha)}$$

for all $x, y \in \mathbb{R}^n \setminus \{0\}$ and $l \in \{1, \ldots, h-1\}$. Hence, the proof of case $h > 0$ is complete. If $h = 0$, then $c' \left||x|^{-h} - |y|^{-h}\right| = 0$ and the above inequalities hold with $h = 0$ in the right-hand side and without the terms which contain the summation sign. □

Then we introduce the following two preliminary lemmas, which we prove by an argument of Agmon et al. [2, Lemma 3.1, p. 644]. We use the following notation,

$$(\eta, t) \text{ denotes a vector of } \mathbb{R}^n \text{ with } \eta \in \mathbb{R}^{n-1} \text{ and } t \in \mathbb{R}. \tag{4.20}$$

Lemma 4.15. *Let $k \in C^1(\mathbb{R}^n \setminus \{0\})$ be a positively homogeneous function of degree $-(n-1)$. If*

$$\int_{\partial\mathbb{B}_{n-1}(0,1)} k(\eta, 0)\, d\sigma_\eta = 0\,, \tag{4.21}$$

then $\frac{\partial}{\partial t}k(\eta, t)$ is integrable in the variable $\eta \in \mathbb{R}^{n-1}$ for all $t \in \mathbb{R} \setminus \{0\}$ and

$$\int_{\mathbb{R}^{n-1}} \frac{\partial}{\partial t} k(\eta, t)\, d\eta = 0 \qquad \forall t \in \mathbb{R} \setminus \{0\}\,. \tag{4.22}$$

Proof. Since $\frac{\partial}{\partial t}k(\eta, t)$ is positively homogeneous of degree

$$-(n-1) - 1 < -(n-1)\,,$$

the continuous function $\frac{\partial}{\partial t}k(\cdot, t)$ is integrable in $\mathbb{R}^{n-1}$ for all $t \in \mathbb{R}\setminus\{0\}$ (cf. Proposition 4.12). Next we note that

$$k(\eta, t) = k_1(\eta, t) + k_2(\eta, t) \qquad \forall (\eta, t) \in \mathbb{R}^n \setminus \{0\}\,,$$

where

$$k_1(\eta, t) \equiv \frac{k(\frac{\eta}{|\eta|}, 0)}{(|\eta|^2 + t^2)^{\frac{(n-1)}{2}}}\,, \qquad k_2(\eta, t) \equiv \frac{k(\frac{\eta}{|(\eta,t)|}, \frac{t}{|(\eta,t)|}) - k(\frac{\eta}{|\eta|}, 0)}{(|\eta|^2 + t^2)^{\frac{(n-1)}{2}}}\,,$$

for all $(\eta, t) \in \mathbb{R}^n \setminus \{0\}$. Since $k \in C^1(\mathbb{R}^n \setminus \{0\})$ and k is Lipschitz continuous on $\partial\mathbb{B}_n(0, 1)$, then we have

$$\begin{aligned} &|k_2(\eta, t)| \\ &\quad \leq |k : \partial\mathbb{B}_n(0,1)|_1 \left(\left| \frac{\eta}{|(\eta,t)|} - \frac{\eta}{|\eta|} \right| + \frac{|t|}{|(\eta,t)|} \right) \frac{1}{|(\eta,t)|^{n-1}} \\ &\quad \leq |k : \partial\mathbb{B}_n(0,1)|_1 \left(\frac{|\eta|}{|(\eta,t)|^n} \left| \frac{|(\eta,t)| - |\eta|}{|\eta|} \right| + \frac{|t|}{|(\eta,t)|^n} \right) \\ &\quad \leq |k : \partial\mathbb{B}_n(0,1)|_1 \left(\frac{|t|}{|(\eta,t)|^n} + \frac{|t|}{|(\eta,t)|^n} \right) = |k : \partial\mathbb{B}_n(0,1)|_1 \frac{2|t|}{|(\eta,t)|^n} \end{aligned} \tag{4.23}$$

and the right-hand side is integrable in $\eta \in \mathbb{R}^{n-1}$ for all $t \in \mathbb{R} \setminus \{0\}$. Moreover, by setting $\eta = \xi t$, we have

$$\int_{\mathbb{R}^{n-1}} k_2(\eta, t)\, d\eta$$
$$= \int_{\mathbb{R}^{n-1}} \frac{k(\frac{\text{sgn}\,(t)\xi}{|(\xi,1)|}, \frac{\text{sgn}\,(t)}{|(\xi,1)|}) - k(\frac{\text{sgn}\,(t)\xi}{|\xi|}, 0)}{(|\xi|^2+1)^{\frac{(n-1)}{2}}}\, d\xi \qquad \forall t \in \mathbb{R} \setminus \{0\}$$

and accordingly

$$\frac{\partial}{\partial t} \int_{\mathbb{R}^{n-1}} k_2(\eta, t)\, d\eta = 0 \qquad \forall t \in \mathbb{R} \setminus \{0\}.$$

Moreover,

$$k_2(\eta, t) = k(\eta, t) - k(\frac{\eta}{|\eta|}|(\eta, t)|, 0)$$

and thus

$$\left|\frac{\partial}{\partial t} k_2(\eta, t)\right| = \left|\frac{\partial}{\partial t}\left(k(\eta, t) - k(\frac{\eta}{|\eta|}|(\eta, t)|, 0)\right)\right| \tag{4.24}$$
$$= \left|\frac{\partial k}{\partial t}(\eta, t) - \sum_{j=1}^{n-1} \frac{\partial k}{\partial \eta_j}(\frac{\eta}{|\eta|}|(\eta, t)|, 0) \frac{\eta_j}{|\eta|} \frac{t}{|(\eta, t)|}\right|$$
$$\leq \left(\sup_{\partial \mathbb{B}_n(0,1)} \left|\frac{\partial k}{\partial t}\right| + \sum_{j=1}^{n-1} \sup_{\partial \mathbb{B}_n(0,1)} \left|\frac{\partial k}{\partial \eta_j}\right|\right) |(\eta, t)|^{-n} \qquad \forall \eta \in \mathbb{R}^{n-1},$$

for all $t \in \mathbb{R} \setminus \{0\}$. Then (4.23) and (4.24) and the classical differentiability theorem for integrals depending on a parameter imply that

$$0 = \frac{\partial}{\partial t} \int_{\mathbb{R}^{n-1}} k_2(\eta, t)\, d\eta = \int_{\mathbb{R}^{n-1}} \frac{\partial}{\partial t} k_2(\eta, t)\, d\eta \qquad \forall t \in \mathbb{R} \setminus \{0\}. \tag{4.25}$$

We now turn to consider k_1. By integrating on the spheres and by assumption (4.21), we have

$$\int_{\mathbb{R}^{n-1}} \frac{\partial}{\partial t} k_1(\eta, t)\, d\eta$$
$$= -(n-1)t \int_{\mathbb{R}^{n-1}} \frac{k(\frac{\eta}{|\eta|}, 0)}{|(\eta, t)|^{n+1}}\, d\eta$$
$$= -(n-1)t \int_0^{+\infty} \int_{\partial \mathbb{B}_{n-1}(0,\rho)} \frac{k(\frac{\eta}{|\eta|}, 0)}{|(\eta, t)|^{n+1}}\, d\sigma_\eta\, d\rho$$
$$= -(n-1)t \int_0^{+\infty} \rho^{n-2} \int_{\partial \mathbb{B}_{n-1}(0,1)} \frac{k(\frac{\xi}{|\xi|}, 0)}{(\rho^2 + t^2)^{\frac{n+1}{2}}}\, d\sigma_\xi\, d\rho$$

$$= -(n-1)t \int_0^{+\infty} \frac{\rho^{n-2}}{(\rho^2+t^2)^{\frac{n+1}{2}}}\, d\rho \int_{\partial\mathbb{B}_{n-1}(0,1)} k(\xi,0)\, d\sigma_\xi = 0$$

and thus $\int_{\mathbb{R}^{n-1}} \frac{\partial}{\partial t} k(\eta,t)\, d\eta = \int_{\mathbb{R}^{n-1}} \frac{\partial}{\partial t} k_1(\eta,t)\, d\eta + \int_{\mathbb{R}^{n-1}} \frac{\partial}{\partial t} k_2(\eta,t)\, d\eta = 0$ for all $t \in \mathbb{R}\setminus\{0\}$. Thus the proof is complete. □

Lemma 4.16. *Let $k \in C^1(\mathbb{R}^n\setminus\{0\})$ be positively homogeneous of degree $-(n-1)$. If $j \in \{1,\dots,n-1\}$ and $t \in \mathbb{R}\setminus\{0\}$, then $\frac{\partial}{\partial\eta_j}k(\cdot,t)$ is integrable in $\mathbb{R}^{n-1}$ and*

$$\int_{\mathbb{R}^{n-1}} \frac{\partial}{\partial\eta_j} k(\eta,t)\, d\eta = 0 \tag{4.26}$$

(cf. (4.20)).

Proof. Since k is positively homogeneous of degree $-(n-1)$, then $\frac{\partial k}{\partial\eta_j}$ is positively homogeneous of degree $-n$ and thus $\frac{\partial}{\partial\eta_j}k(\eta,t)$ is integrable in $\eta \in \mathbb{R}^{n-1}$ for all t in $\mathbb{R}\setminus\{0\}$ (cf. Proposition 4.12). Then the Dominated Convergence Theorem together with the Divergence Theorem 4.1 imply that

$$\begin{aligned}
\int_{\mathbb{R}^{n-1}} \frac{\partial}{\partial\eta_j} k(\eta,t)\, d\eta &= \lim_{\rho\to+\infty} \int_{\mathbb{B}_{n-1}(0,\rho)} \frac{\partial}{\partial\eta_j} k(\eta,t)\, d\eta \\
&= \lim_{\rho\to+\infty} \int_{\partial\mathbb{B}_{n-1}(0,\rho)} k(\eta,t)\frac{\eta_j}{|\eta|}\, d\sigma_\eta = \lim_{\rho\to+\infty} \int_{\partial\mathbb{B}_{n-1}(0,1)} k(\rho\xi,t)\frac{\xi_j}{|\xi|}\rho^{n-2}\, d\sigma_\xi \\
&= \lim_{\rho\to+\infty} \int_{\partial\mathbb{B}_{n-1}(0,1)} k(\xi,t/\rho)\xi_j \rho^{n-2-(n-1)}\, d\sigma_\xi = 0\,,
\end{aligned}$$

for all $t \in \mathbb{R}\setminus\{0\}$. Indeed, $\lim_{\rho\to+\infty}\rho^{-1} = 0$ and the continuity of k in $\mathbb{R}^n\setminus\{0\}$ implies that the limit

$$\lim_{\rho\to+\infty} \int_{\partial\mathbb{B}_{n-1}(0,1)} k(\xi,t/\rho)\xi_j\, d\sigma_\xi = \int_{\partial\mathbb{B}_{n-1}(0,1)} k(\xi,0)\xi_j\, d\sigma_\xi\,,$$

exists finite for all $t \in \mathbb{R}\setminus\{0\}$. Hence, the proof is complete. □

We are now ready to prove the main Theorem 4.17 of this section. The result of the theorem is due to Miranda and has been presented in [212]. Our proof is similar to that of Miranda, but we are able to shorten some parts of it by exploiting the sufficient condition for Hölder continuity of our Proposition 2.82 and we can improve the regularity assumptions on the kernel k which is here of cass $C^{1,1}$ and was of class C^2 in the paper of Miranda. A reader interested in understanding all details of the proof may also read Sections 2.13, 2.18, and 2.19 of the Preliminaries (see also Section A.6 in the Appendix).

Theorem 4.17. *Let $\alpha \in]0,1[$. Let Ω be a bounded open subset of $\mathbb{R}^n$ of class $C^{1,\alpha}$. Then the following statements hold.*

(i) For each $(k,\mu) \in \mathcal{K}_{1,1;o} \times C^{0,\alpha}(\partial\Omega)$*, the map*

$$K[k,\mu]_{|\Omega}(x) = \int_{\partial\Omega} k(x-y)\mu(y)\,d\sigma_y \qquad \forall x \in \Omega$$

can be extended to a unique α*-Hölder continuous function* $K[k,\mu]^+$ *on* $\overline{\Omega}$*. Moreover, the map from* $\mathcal{K}_{1,1;o} \times C^{0,\alpha}(\partial\Omega)$ *to* $C^{0,\alpha}(\overline{\Omega})$ *that takes* (k,μ) *to* $K[k,\mu]^+$ *is bilinear and continuous.*

(ii) Let $r \in]0,+\infty[$ *be such that* $\overline{\Omega} \subseteq \mathbb{B}_n(0,r)$*. Then for each* $(k,\mu) \in \mathcal{K}_{1,1;o} \times C^{0,\alpha}(\partial\Omega)$ *the map*

$$K[k,\mu]_{|\mathbb{R}^n\setminus\overline{\Omega}}(x) = \int_{\partial\Omega} k(x-y)\mu(y)\,d\sigma_y \qquad \forall x \in \mathbb{R}^n \setminus \overline{\Omega},$$

can be extended to a unique continuous function $K[k,\mu]^-$ *on* $\mathbb{R}^n \setminus \Omega$ *such that the restriction* $K[k,\mu]^-_{|\overline{\mathbb{B}_n(0,r)}\setminus\Omega}$ *is* α*-Hölder continuous. Moreover, the map from* $\mathcal{K}_{1,1;o} \times C^{0,\alpha}(\partial\Omega)$ *to* $C^{0,\alpha}(\overline{\mathbb{B}_n(0,r)} \setminus \Omega)$ *that takes* (k,μ) *to* $K[k,\mu]^-_{|\overline{\mathbb{B}_n(0,r)}\setminus\Omega}$ *is bilinear and continuous.*

Proof. We first prove statement (i). The classical theorem of continuity for integrals depending on a parameter implies that the integral $\int_{\partial\Omega} k(x-y)\mu(y)\,d\sigma_y$ is continuous in $x \in \mathbb{R}^n \setminus \partial\Omega$. Since Hölder continuous functions are uniformly continuous and uniformly continuous functions in Ω can be uniquely extended to the closure $\overline{\Omega}$, it suffices to show that there exists $C > 0$ such that

$$\sup_{x\in\Omega} |K[k,\mu](x)| + |K[k,\mu] : \Omega|_\alpha \leq C\|k\|_{\mathcal{K}_{1,1}}\|\mu\|_{C^{0,\alpha}(\partial\Omega)}$$

for all $(k,\mu) \in \mathcal{K}_{1,1;o} \times C^{0,\alpha}(\partial\Omega)$. Let $\vartheta \in]0,1[$. By Lemma 2.78 there exist an open neighborhood U of class C^∞ of $\partial\Omega$ and $a \in C^\infty(\overline{U},\mathbb{R}^n)$ such that conditions (2.66) hold true for some $\tau \in]0,+\infty[$. Then, by the sufficient condition for Hölder continuity of Proposition 2.82, it suffices to show that

(j) If H is a compact subset of Ω, then there exists $C_H > 0$ such that

$$\sup_{x\in H} |K[k,\mu](x)| + \sup_{x\in H} |\nabla K[k,\mu]| \leq C_H\|k\|_{\mathcal{K}_{1,1}}\|\mu\|_{C^{0,\alpha}(\partial\Omega)}$$

for all $(k,\mu) \in \mathcal{K}_{1,1;o} \times C^{0,\alpha}(\partial\Omega)$.

(jj) If $t_1 \in]0,+\infty[$ is as in Lemma 2.79, then there exist $t' \in]0,t_1[$ and $C_2 > 0$ such that

$$M_{t',\alpha}(K[k,\mu]_{|\Omega}) \leq C_2\|k\|_{\mathcal{K}_{1,1}}\|\mu\|_{C^{0,\alpha}(\partial\Omega)} \tag{4.27}$$

for all $(k,\mu) \in \mathcal{K}_{1,1;o} \times C^{0,\alpha}(\partial\Omega)$ (see (2.74) for the definition of $M_{t',\alpha}(\cdot)$).

Now, the validity of (j) follows by the classical theorem of continuity for integrals depending on a parameter. Indeed, the distance from H to $\partial\Omega$ is positive. (See also Lemma 4.13). Thus it remains to prove statement (jj). In order to estimate $\nabla K[k,\mu](x + ta(x))$ for $x \in \partial\Omega$ and $t \in]-t_1, 0[$ we fix $j \in \{1,\ldots,n\}$ and we estimate

$$\partial_{x_j} K[k,\mu](x+ta(x)) = \int_{\partial\Omega} \partial_{x_j} k(x+ta(x)-y)\mu(y)\, d\sigma_y$$

when $(t,x) \in]-t_1, 0[\times\partial\Omega$ for a perhaps smaller t_1 and when $(k,\mu) \in \mathcal{K}_{1,1;o} \times C^{0,\alpha}(\partial\Omega)$. By Lemma 2.80 and by the properties of a in (2.66), we have

$$|x-y+ta(x)| \geq (1-\vartheta)^{1/2}(|x-y|^2+|t|^2)^{1/2} \tag{4.28}$$

for all x, $y \in \partial\Omega$ such that $|x-y| < \tau$ and for all $t \in]-t_1, 0[$. Next we note that if $\mu \in C^{0,\alpha}(\partial\Omega)$, then

$$\begin{aligned}
&\int_{\partial\Omega} \partial_{x_j} k(x+ta(x)-y)\mu(y)\, d\sigma_y \\
&\quad = \int_{\partial\Omega} \partial_{x_j} k(x+ta(x)-y)(\mu(y)-\mu(x))\, d\sigma_y \\
&\quad\quad + \mu(x)\int_{\partial\Omega} \partial_{x_j} k(x+ta(x)-y)\, d\sigma_y \qquad \forall x \in \partial\Omega
\end{aligned} \tag{4.29}$$

for all $t \in]-t_1, 0[$. In order to make the proof more readable, we now find convenient to split it into two parts, in which we consider separately the first and the second integral in the right-hand side of (4.29) and into a series of intermediate steps.

Part 1. Here we estimate the first integral in the right-hand side of (4.29). To do so, we split the domain of integration $\partial\Omega$ into the two parts

$$(\partial\Omega) \setminus \mathbb{B}_n(x,\tau), \qquad (\partial\Omega) \cap \mathbb{B}_n(x,\tau).$$

Since $\partial_{x_j} k$ is positively homogeneous of degree $-n$, we have

$$\begin{aligned}
&\left| \int_{(\partial\Omega)\setminus\mathbb{B}_n(x,\tau)} \partial_{x_j} k(x+ta(x)-y)(\mu(y)-\mu(x))\, d\sigma_y \right| \\
&\quad \leq \|\partial_{x_j} k\|_{C^0(\partial\mathbb{B}_n(0,1))} |\mu : \partial\Omega|_\alpha \int_{(\partial\Omega)\setminus\mathbb{B}_n(x,\tau)} \frac{|x-y|^\alpha\, d\sigma_y}{|x+ta(x)-y|^n} \\
&\quad \leq \|\partial_{x_j} k\|_{C^0(\partial\mathbb{B}_n(0,1))} |\mu : \partial\Omega|_\alpha \int_{\partial\Omega} \frac{\mathrm{diam}(\partial\Omega)^\alpha\, d\sigma_y}{(\tau-|t|)^n} \\
&\quad \leq \|\partial_{x_j} k\|_{C^0(\partial\mathbb{B}_n(0,1))} |\mu : \partial\Omega|_\alpha m_{n-1}(\partial\Omega)\mathrm{diam}(\partial\Omega)^\alpha (\tau/2)^{-n}
\end{aligned} \tag{4.30}$$

for all $t \in]-\min\{t_1, \tau/2\}, 0[$.

• We now turn to consider the integral on $(\partial\Omega) \cap \mathbb{B}_n(x,\tau)$. By inequality (4.28), we have

$$\left| \int_{(\partial\Omega)\cap\mathbb{B}_n(x,\tau)} \partial_{x_j} k(x+ta(x)-y)(\mu(y)-\mu(x))\, d\sigma_y \right| \tag{4.31}$$

$$\leq \|\partial_{x_j} k\|_{C^0(\partial\mathbb{B}_n(0,1))} |\mu : \partial\Omega|_\alpha \int_{(\partial\Omega)\cap\mathbb{B}_n(x,\tau)} \frac{|x-y|^\alpha \, d\sigma_y}{|x+ta(x)-y|^n}$$
$$\leq \frac{\|\partial_{x_j} k\|_{C^0(\partial\mathbb{B}_n(0,1))} |\mu : \partial\Omega|_\alpha}{(1-\vartheta)^{n/2}} \int_{(\partial\Omega)\cap\mathbb{B}_n(x,\tau)} \frac{|x-y|^\alpha \, d\sigma_y}{(|x-y|^2+t^2)^{n/2}} .$$

• We now estimate the last integral in the right-hand side of (4.31) by considering separately the case where

$$|x-y| \geq |t|$$

and the case where

$$|x-y| < |t|$$

for $t \in]-\min\{t_1, \tau/2\}, 0[$. Then we have

$$\int_{(\partial\Omega)\cap\mathbb{B}_n(x,\tau)} \frac{|x-y|^\alpha \, d\sigma_y}{(|x-y|^2+t^2)^{n/2}} \tag{4.32}$$
$$\leq \int_{\{y\in(\partial\Omega)\cap\mathbb{B}_n(x,\tau):\, |x-y|\geq|t|\}} \frac{|x-y|^\alpha \, d\sigma_y}{(|x-y|^2+t^2)^{n/2}}$$
$$+ \int_{\{y\in(\partial\Omega)\cap\mathbb{B}_n(x,\tau):\, |x-y|<|t|\}} \frac{|x-y|^\alpha \, d\sigma_y}{(|x-y|^2+t^2)^{n/2}} .$$

We note that Lemma 2.53 (iii) implies that

$$\int_{\{y\in(\partial\Omega)\cap\mathbb{B}_n(x,\tau):\, |x-y|\geq|t|\}} \frac{|x-y|^\alpha \, d\sigma_y}{(|x-y|^2+t^2)^{n/2}} \tag{4.33}$$
$$\leq \int_{\{y\in(\partial\Omega)\cap\mathbb{B}_n(x,\tau):\, |x-y|\geq|t|\}} |x-y|^{\alpha-n} \, d\sigma_y$$
$$\leq c'''_{\Omega,n-\alpha} |t|^{(n-1)-(n-\alpha)} = c'''_{\Omega,n-\alpha} |t|^{\alpha-1}$$

for all $t \in]-\min\{t_1, \tau/2\}, 0[$ and that Lemma 2.53 (ii) implies that

$$\int_{\{y\in(\partial\Omega)\cap\mathbb{B}_n(x,\tau):\, |x-y|<|t|\}} \frac{|x-y|^\alpha \, d\sigma_y}{(|x-y|^2+t^2)^{n/2}} \tag{4.34}$$
$$\leq |t|^{-n} \int_{\{y\in(\partial\Omega)\cap\mathbb{B}_n(x,\tau):\, |x-y|<|t|\}} \frac{d\sigma_y}{|x-y|^{-\alpha}}$$
$$\leq |t|^{-n} c''_{\Omega,-\alpha} |t|^{(n-1)-(-\alpha)} = c''_{\Omega,-\alpha} |t|^{\alpha-1}$$

for all $t \in]-\min\{t_1, \tau/2\}, 0[$.

• Finally, combining inequalities (4.30)–(4.34), we obtain

$$\left| \int_{\partial\Omega} \partial_{x_j} k(x+ta(x)-y)(\mu(y)-\mu(x)) \, d\sigma_y \right| \tag{4.35}$$
$$\leq \|\partial_{x_j} k\|_{C^0(\partial\mathbb{B}_n(0,1))} |\mu : \partial\Omega|_\alpha \Big\{ m_{n-1}(\partial\Omega) \mathrm{diam}(\partial\Omega)^\alpha (\tau/2)^{-n}$$
$$\times (\min\{t_1, \tau/2\})^{1-\alpha} + \frac{c'''_{\Omega,n-\alpha} + c''_{\Omega,-\alpha}}{(1-\vartheta)^{n/2}} \Big\} |t|^{\alpha-1}$$

for all $t \in]-\min\{t_1, \tau/2\}, 0[$ and $x \in \partial\Omega$.

Part 2. We now turn to consider the second integral in the right-hand side of (4.29). By Lemma 2.63 of the uniform cylinders, there exist

$$r \in]0, \tau/4[\,, \qquad \delta \in]0, \tau/4[\,, \qquad r < \delta$$

such that if $p \in \partial\Omega$, then there exists $R_p \in \mathbb{O}_n(\mathbb{R})$ such that $C(p, R_p, r, \delta)$ is a coordinate cylinder for Ω around p and the corresponding function γ_p satisfies the condition $\nabla\gamma_p(0) = 0$ and the inequality

$$\sup_{p \in \partial\Omega} \|\gamma_p\|_{C^{1,\alpha}(\overline{\mathbb{B}_{n-1}(0,r)})} < +\infty\,. \tag{4.36}$$

Moreover, we know that

$$R_p^t e_n = \nu_\Omega(p)\,, \qquad e_n = R_p \nu_\Omega(p)\,.$$

Then we introduce the vector field

$$b^t \equiv (b_*, b_n) : \partial\Omega \to \mathbb{R}^{n-1} \times \mathbb{R}$$

by setting

$$(b_*(p), b_n(p))^t \equiv R_p a(p) \qquad \forall p \in \partial\Omega \tag{4.37}$$

and we note that

$$b_n(p) = b(p) \cdot e_n = R_p a(p) \cdot e_n = a(p) \cdot R_p^t e_n = a(p) \cdot \nu_\Omega(p) > 1 - \vartheta > 0 \tag{4.38}$$

(cf. (2.66)). We denote by $M_x(\partial\Omega)$ the affine hyperplane through $x \in \partial\Omega$ and orthogonal to $\nu_\Omega(x)$. Namely, we set

$$M_x(\partial\Omega) \equiv \{y \in \mathbb{R}^n \,:\, \nu_\Omega(x) \cdot (y - x) = 0\}\,.$$

The key point of this second part of the proof will be the comparison between the integrals

$$\int_{\partial\Omega} \partial_{x_j} k(x + ta(x) - y)\, d\sigma_y$$

and

$$\int_{M_x(\partial\Omega)} \partial_{x_j} k(x + ta(x) - y)\, d\sigma_y\,.$$

Then we write the inequality

$$\begin{aligned} &\left| \int_{\partial\Omega} \partial_{x_j} k(x + ta(x) - y)\, d\sigma_y \right| \\ &\quad \leq \left| \int_{\partial\Omega} \partial_{x_j} k(x + ta(x) - y)\, d\sigma_y - \int_{M_x(\partial\Omega)} \partial_{x_j} k(x + ta(x) - y)\, d\sigma_y \right| \end{aligned} \tag{4.39}$$

$$+\left|\int_{M_x(\partial\Omega)} \partial_{x_j} k(x+ta(x)-y)\,d\sigma_y\right| \qquad \forall x \in \partial\Omega$$

that holds for all $t \in]-\min\{t_1, \tau/2\}, 0[$.

• We first consider the second term in the right-hand side of (4.39) for some fixed $x \in \partial\Omega$. We observe that condition $\nu_\Omega(x) \cdot (y-x) = 0$ implies that there exists $\eta \in \mathbb{R}^{n-1}$ such that

$$y - x = R_x^t(\eta, 0)^t .$$

Then we have

$$\begin{aligned}\int_{M_x(\partial\Omega)} \partial_{x_j} k(x+ta(x)-y)\,d\sigma_y &= \int_{\mathbb{R}^{n-1}} \partial_{x_j} k(ta(x) - R_x^t(\eta,0)^t)\,d\eta \\ &= \int_{\mathbb{R}^{n-1}} \partial_{x_j} k(R_x^t(t(b_*(x), b_n(x))^t - (\eta,0)^t))\,d\eta \\ &= \int_{\mathbb{R}^{n-1}} (\partial_{x_j} k) \circ R_x^t\left((tb_*(x)-\eta, tb_n(x))^t\right)\,d\eta \quad (4.40)\end{aligned}$$

(cf. the identification (4.20)). Next we set

$$\tilde{k}_x \equiv k \circ R_x^t .$$

Since $k \in \mathcal{K}_{1,1;o}$, then also $\tilde{k}_x \in \mathcal{K}_{1,1;o}$. By (4.38) we know that $b_n > 0$ and thus Lemma 4.15 implies that

$$\begin{aligned}&\int_{\mathbb{R}^{n-1}} \partial_t \tilde{k}_x(tb_*(x)-\eta, tb_n(x))\,d\eta \\ &\qquad = \int_{\mathbb{R}^{n-1}} \partial_t \tilde{k}_x(\eta, tb_n(x))\,d\eta = 0 \qquad \forall t \in \mathbb{R}\setminus\{0\},\end{aligned}$$

and Lemma 4.16 implies that

$$\begin{aligned}&\int_{\mathbb{R}^{n-1}} \partial_{\eta_l} \tilde{k}_x(tb_*(x)-\eta, tb_n(x))\,d\eta \\ &\qquad = \int_{\mathbb{R}^{n-1}} \partial_{\eta_l} \tilde{k}_x(\eta, tb_n(x))\,d\eta = 0 \qquad \forall t \in \mathbb{R}\setminus\{0\}\end{aligned}$$

for all $l \in \{1, \ldots, n-1\}$. Hence,

$$\begin{aligned}&\int_{\mathbb{R}^{n-1}} (\nabla k) \circ R_x^t(tb_*(x)-\eta, tb_n(x))\,d\eta \\ &\qquad = R_x^t \int_{\mathbb{R}^{n-1}} \nabla \tilde{k}_x(tb_*(x)-\eta, tb_n(x))\,d\eta = 0 \qquad \forall t \in \mathbb{R}\setminus\{0\}\end{aligned}$$

and accordingly equality (4.40) implies that

$$\int_{M_x(\partial\Omega)} \partial_{x_j} k(x + ta(x) - y)\, d\sigma_y = 0 \qquad \forall t \in \mathbb{R} \setminus \{0\}. \tag{4.41}$$

• We now turn to estimate the first term in the right-hand side of (4.39). We have

$$\begin{aligned} &\left| \int_{\partial\Omega} \partial_{x_j} k(x + ta(x) - y)\, d\sigma_y - \int_{M_x(\partial\Omega)} \partial_{x_j} k(x + ta(x) - y)\, d\sigma_y \right| \\ &\quad \leq \left| \int_{C(x,R_x,r,\delta)\cap\partial\Omega} \partial_{x_j} k(x + ta(x) - y)\, d\sigma_y \right. \\ &\qquad \left. - \int_{C(x,R_x,r,\delta)\cap M_x(\partial\Omega)} \partial_{x_j} k(x + ta(x) - y)\, d\sigma_y \right| \\ &\qquad + \left| \int_{(\partial\Omega)\setminus C(x,R_x,r,\delta)} \partial_{x_j} k(x + ta(x) - y)\, d\sigma_y \right| \\ &\qquad + \left| \int_{M_x(\partial\Omega)\setminus C(x,R_x,r,\delta)} \partial_{x_j} k(x + ta(x) - y)\, d\sigma_y \right|. \end{aligned} \tag{4.42}$$

• Then we estimate the first term in the right-hand side of (4.42). By setting

$$y = x + R_x^t(\eta, \gamma_x(\eta))^t \qquad \forall \eta \in \mathbb{B}_{n-1}(0, r)$$

in the first integral and

$$y = x + R_x^t(\eta, 0)^t \qquad \forall \eta \in \mathbb{B}_{n-1}(0, r)$$

in the second integral and by exploiting definition (4.37) of $b^t \equiv (b_*, b_n)$, we obtain

$$\begin{aligned} &\left| \int_{C(x,R_x,r,\delta)\cap\partial\Omega} \partial_{x_j} k(x + ta(x) - y)\, d\sigma_y \right. \\ &\qquad \left. - \int_{C(x,R_x,r,\delta)\cap M_x(\partial\Omega)} \partial_{x_j} k(x + ta(x) - y)\, d\sigma_y \right| \\ &\leq \left| \int_{\mathbb{B}_{n-1}(0,r)} \partial_{x_j} k(R_x^t(b_*(x)t - \eta, b_n(x)t - \gamma_x(\eta))^t) \sqrt{1 + |\nabla\gamma_x(\eta)|^2}\, d\eta \right. \\ &\quad \left. - \int_{\mathbb{B}_{n-1}(0,r)} \partial_{x_j} k(R_x^t(b_*(x)t - \eta, b_n(x)t)^t)\, d\eta \right| \\ &\leq \left| \int_{\{\eta\in\mathbb{B}_{n-1}(0,r):\, |\eta|\geq|t|\}} \partial_{x_j} k(R_x^t(b_*(x)t - \eta, b_n(x)t - \gamma_x(\eta))^t) \right. \\ &\quad \times \sqrt{1 + |\nabla\gamma_x(\eta)|^2}\, d\eta \\ &\quad \left. - \int_{\{\eta\in\mathbb{B}_{n-1}(0,r):\, |\eta|\geq|t|\}} \partial_{x_j} k(R_x^t(b_*(x)t - \eta, b_n(x)t)^t)\, d\eta \right| \\ &\quad + \left| \int_{\{\eta\in\mathbb{B}_{n-1}(0,r):\, |\eta|<|t|\}} \partial_{x_j} k(R_x^t(b_*(x)t - \eta, b_n(x)t - \gamma_x(\eta))^t) \right. \end{aligned} \tag{4.43}$$

$$\times\sqrt{1+|\nabla\gamma_x(\eta)|^2}\,d\eta$$
$$\left.-\int_{\{\eta\in\mathbb{B}_{n-1}(0,r):\,|\eta|<|t|\}}\partial_{x_j}k(R_x^t(b_*(x)t-\eta,b_n(x)t)^t)\,d\eta\right| .$$

- We now consider the first term in the right-hand side of (4.43). We have

$$\begin{aligned}
&\Bigg|\int_{\{\eta\in\mathbb{B}_{n-1}(0,r):\,|\eta|\geq|t|\}}\partial_{x_j}k(R_x^t(b_*(x)t-\eta,b_n(x)t-\gamma_x(\eta))^t)\\
&\qquad\times\sqrt{1+|\nabla\gamma_x(\eta)|^2}\,d\eta\\
&\qquad-\int_{\{\eta\in\mathbb{B}_{n-1}(0,r):\,|\eta|\geq|t|\}}\partial_{x_j}k(R_x^t(b_*(x)t-\eta,b_n(x)t)^t)\,d\eta\Bigg| \qquad (4.44)\\
&\leq\int_{\{\eta\in\mathbb{B}_{n-1}(0,r):\,|\eta|\geq|t|\}}\Big|\partial_{x_j}k(R_x^t(b_*(x)t-\eta,b_n(x)t-\gamma_x(\eta))^t)\\
&\qquad-\partial_{x_j}k(R_x^t(b_*(x)t-\eta,b_n(x)t)^t)\Big|\sqrt{1+|\nabla\gamma_x(\eta)|^2}\,d\eta\\
&\qquad+\int_{\{\eta\in\mathbb{B}_{n-1}(0,r):\,|\eta|\geq|t|\}}\Big|\partial_{x_j}k(R_x^t(b_*(x)t-\eta,b_n(x)t)^t)\Big|\\
&\qquad\times\Big|\sqrt{1+|\nabla\gamma_x(\eta)|^2}-1\Big|\,d\eta\,.
\end{aligned}$$

Since $\partial_{x_j}k\in C^{0,1}_{\mathrm{loc}}(\mathbb{R}^n\setminus\{0\})$ is positively homogeneous of degree $-n$, Lemma 4.14 implies that

$$\begin{aligned}
&|\partial_{x_j}k(x')-\partial_{x_j}k(x'')| \qquad (4.45)\\
&\qquad\leq 2(n+1)\|\partial_{x_j}k\|_{C^{0,1}(\partial\mathbb{B}_n(0,1))}|x'-x''|(\min\{|x'|,|x''|\})^{-(n+1)}
\end{aligned}$$

for all $x',x''\in\mathbb{R}^n\setminus\{0\}$. Since the Lipschitz constant of $\sqrt{1+t^2}$ in $t\in[0,+\infty[$ is one, the Lipschitz constant of the modulus in $\mathbb{R}^{n-1}$ is one and $\nabla\gamma_x$ is α-Hölder continuous in $\mathbb{B}_{n-1}(0,r)$, we have

$$|\sqrt{1+|\nabla\gamma_x|^2}:\mathbb{B}_{n-1}(0,r)|_\alpha\leq|\nabla\gamma_x|_\alpha\,.$$

Hence,

$$\begin{aligned}
&\int_{\{\eta\in\mathbb{B}_{n-1}(0,r):\,|\eta|\geq|t|\}}\Big|\partial_{x_j}k(R_x^t(b_*(x)t-\eta,b_n(x)t-\gamma_x(\eta))^t) \qquad (4.46)\\
&\qquad-\partial_{x_j}k(R_x^t(b_*(x)t-\eta,b_n(x)t)^t)\Big|\sqrt{1+|\nabla\gamma_x(\eta)|^2}\,d\eta\\
&\qquad+\int_{\{\eta\in\mathbb{B}_{n-1}(0,r):\,|\eta|\geq|t|\}}\Big|\partial_{x_j}k(R_x^t(b_*(x)t-\eta,b_n(x)t)^t)\Big|\\
&\qquad\times\Big|\sqrt{1+|\nabla\gamma_x(\eta)|^2}-1\Big|\,d\eta\\
&\leq 2(n+1)\|\partial_{x_j}k\|_{C^{0,1}(\partial\mathbb{B}_n(0,1))}\sup\sqrt{1+|\nabla\gamma_x|^2}
\end{aligned}$$

$$\times \int_{\{\eta\in\mathbb{B}_{n-1}(0,r):\,|\eta|\geq|t|\}} \Big| R_x^t(b_*(x)t-\eta, b_n(x)t-\gamma_x(\eta))^t$$
$$-R_x^t(b_*(x)t-\eta, b_n(x)t)^t\Big| \Big(\min\{|R_x^t(b_*(x)t-\eta, b_n(x)t-\gamma_x(\eta))^t|,$$
$$|R_x^t(b_*(x)t-\eta, b_n(x)t)^t|\}\Big)^{-(n+1)} \, d\eta$$
$$+\|\partial_{x_j} k\|_{C^0(\partial\mathbb{B}_n(0,1))} \int_{\{\eta\in\mathbb{B}_{n-1}(0,r):\,|\eta|\geq|t|\}} \frac{|\nabla\gamma_x|_\alpha |\eta|^\alpha}{|R_x^t(b_*(x)t-\eta, b_n(x)t)^t|^n}\, d\eta$$

for all $x \in \partial\Omega$ and $t \in]-\min\{t_1, \tau/2\}, 0[$.

- By the inequality (4.28), we have

$$\begin{aligned}
|x - [x + R_x^t(\eta, \gamma_x(\eta))^t] + ta(x)| \\
&\geq (1-\vartheta)^{1/2}\sqrt{|x-[x+R_x^t(\eta,\gamma_x(\eta))^t]|^2+|t|^2} \\
&= (1-\vartheta)^{1/2}\sqrt{|\eta|^2+|\gamma_x(\eta)|^2+t^2}
\end{aligned}$$

provided that $|x-[x+R_x^t(\eta,\gamma_x(\eta))^t]| = \sqrt{|\eta|^2+|\gamma_x(\eta)|^2} < \tau$. Since

$$\begin{aligned}
&|x - [x + R_x^t(\eta, \gamma_x(\eta))^t] + ta(x)| \\
&\quad = |R_x^t(b_*(x)t-\eta, b_n(x)t-\gamma_x(\eta))^t| = |(b_*(x)t-\eta, b_n(x)t-\gamma_x(\eta))|\,,
\end{aligned}$$

we have

$$|(b_*(x)t-\eta, b_n(x)t-\gamma_x(\eta))| \geq (1-\vartheta)^{1/2}\sqrt{|\eta|^2+|\gamma_x(\eta)|^2+t^2}\,, \tag{4.47}$$

whenever $\eta \in \mathbb{B}_{n-1}(0,r)$ and $t \in]-\min\{t_1,\tau/2\},0[$. Indeed,

$$\sqrt{|\eta|^2+|\gamma_x(\eta)|^2} \leq r + (\delta/2) \leq (\tau/4) + (\tau/8) < \tau$$

for all $\eta \in \mathbb{B}_{n-1}(0,r)$ and $t \in]-\min\{t_1,\tau/2\},0[$. Next we turn to prove that

$$|(b_*(x)t-\eta, b_n(x)t)| \geq (1-\vartheta)^{1/2}\sqrt{|\eta|^2+t^2} \tag{4.48}$$

for all $\eta \in \mathbb{B}_{n-1}(0,r)$ and $t \in \mathbb{R}$. By (2.66), we have $\sup_{\partial\Omega}|a-\nu_\Omega| < \vartheta$ and thus

$$|a(x)\cdot(y-x)| \leq |(a(x)-\nu_\Omega(x))\cdot(y-x)| + |\nu_\Omega(x)\cdot(y-x)| = \vartheta|y-x| + 0\,,$$

for all $y \in M_x(\partial\Omega)$. Then Lemma 2.80 implies that

$$|(x-y)+ta(x)| \geq (1-\vartheta)^{1/2}\sqrt{|x-y|^2+t^2} \qquad \forall y \in M_x(\partial\Omega), t \in \mathbb{R}\setminus\{0\}\,. \tag{4.49}$$

In particular, by taking $y = x + R_x^t(\eta, 0)^t$ with $\eta \in \mathbb{B}_{n-1}(0,r)$, we have

$$|R_x^t(b_*(x)t-\eta, b_n(x)t)^t| \geq (1-\vartheta)^{1/2}\sqrt{|R_x(\eta,0)^t|^2+t^2} \qquad \forall t \in \mathbb{R}\setminus\{0\}$$

and thus inequality (4.48) holds true.

• We now turn to estimate the two integrals in the right-hand side of (4.46). By the elementary inequality of Lemma 2.58 for the function γ_x and by equalities $\gamma_x(0) = 0$ and $\nabla\gamma_x(0) = 0$, we have

$$|\gamma_x(\eta)| \leq |\nabla\gamma_x|_\alpha |\eta|^{\alpha+1} \qquad \forall \eta \in \mathbb{B}_{n-1}(0,r)\,.$$

Hence,

$$\begin{aligned}
&\int_{\{\eta\in\mathbb{B}_{n-1}(0,r):\,|\eta|\geq|t|\}} \Big| R_x^t(b_*(x)t-\eta, b_n(x)t-\gamma_x(\eta))^t \\
&\quad -R_x^t(b_*(x)t-\eta, b_n(x)t)^t \Big| \Big(\min\{|R_x^t(b_*(x)t-\eta, b_n(x)t-\gamma_x(\eta))^t|, \\
&\quad |R_x^t(b_*(x)t-\eta, b_n(x)t)^t|\}\Big)^{-(n+1)} d\eta \\
&\quad \leq \int_{\{\eta\in\mathbb{B}_{n-1}(0,r):\,|\eta|\geq|t|\}} \frac{(1-\vartheta)^{-(n+1)/2}|\gamma_x(\eta)|\,d\eta}{\Big(\min\{\sqrt{|\eta|^2+t^2}, \sqrt{|\eta|^2+|\gamma_x(\eta)|^2+t^2}\}\Big)^{(n+1)}} \\
&\quad \leq |\nabla\gamma_x|_\alpha (1-\vartheta)^{-(n+1)/2} \int_{\{\eta\in\mathbb{B}_{n-1}(0,r):\,|\eta|\geq|t|\}} |\eta|^{1+\alpha-(n+1)}\,d\eta \\
&\quad \leq |\nabla\gamma_x|_\alpha (1-\vartheta)^{-(n+1)/2} s_{n-1} \int_{\{\rho\in]0,+\infty[:\,|t|\leq\rho<r\}} \rho^{\alpha-n}\rho^{n-2}\,d\rho \\
&\quad \leq |\nabla\gamma_x|_\alpha (1-\vartheta)^{-(n+1)/2} s_{n-1} \frac{|t|^{\alpha-1}}{1-\alpha}
\end{aligned} \tag{4.50}$$

for all $t \in]-\min\{t_1, \tau/2\}, 0[$ and $x \in \partial\Omega$ and

$$\begin{aligned}
&\int_{\{\eta\in\mathbb{B}_{n-1}(0,r):\,|\eta|\geq|t|\}} \frac{|\nabla\gamma_x|_\alpha|\eta|^\alpha}{|R_x^t(b_*(x)t-\eta, b_n(x)t)^t|^n}\,d\eta \\
&\quad \leq \int_{\{\eta\in\mathbb{B}_{n-1}(0,r):\,|\eta|\geq|t|\}} \frac{|\nabla\gamma_x|_\alpha|\eta|^\alpha}{(1-\vartheta)^{n/2}(|\eta|^2+t^2)^{n/2}}\,d\eta \\
&\quad \leq \frac{|\nabla\gamma_x|_\alpha}{(1-\vartheta)^{n/2}} \int_{\{\eta\in\mathbb{B}_{n-1}(0,r):\,|\eta|\geq|t|\}} |\eta|^{\alpha-n}\,d\eta \\
&\quad \leq \frac{|\nabla\gamma_x|_\alpha}{(1-\vartheta)^{n/2}} s_{n-1} \frac{|t|^{\alpha-1}}{1-\alpha}
\end{aligned} \tag{4.51}$$

for all $t \in]-\min\{t_1, \tau/2\}, 0[$ and $x \in \partial\Omega$.

• Next we estimate the second term in the right-hand side of (4.43). By the very same steps of the above case $|\eta| \geq t$ in (4.46), (4.50), and (4.51), we obtain

$$\Big| \int_{\{\eta\in\mathbb{B}_{n-1}(0,r):\,|\eta|<|t|\}} \partial_{x_j} k(R_x^t(b_*(x)t-\eta, b_n(x)t-\gamma_x(\eta))^t)$$

$$
\begin{aligned}
&\times\sqrt{1+|\nabla\gamma_x(\eta)|^2}\,d\eta \\
&-\int_{\{\eta\in\mathbb{B}_{n-1}(0,r):\,|\eta|<|t|\}}\partial_{x_j}k(R_x^t(b_*(x)t-\eta,b_n(x)t)^t)\,d\eta\Big| \qquad (4.52)\\
\le\ & 2(n+1)\|\partial_{x_j}k\|_{C^{0,1}(\partial\mathbb{B}_n(0,1))}\sup\sqrt{1+|\nabla\gamma_x|^2}\,|\nabla\gamma_x|_\alpha(1-\vartheta)^{-(n+1)/2}\\
&\int_{\{\eta\in\mathbb{B}_{n-1}(0,r):\,|\eta|<|t|\}}\frac{|\eta|^{1+\alpha}}{(|\eta|^2+t^2)^{(n+1)/2}}\,d\eta\\
&+\|\partial_{x_j}k\|_{C^0(\partial\mathbb{B}_n(0,1))}|\nabla\gamma_x|_\alpha(1-\vartheta)^{-n/2}\\
&\times\int_{\{\eta\in\mathbb{B}_{n-1}(0,r):\,|\eta|<|t|\}}\frac{|\eta|^{\alpha}}{(|\eta|^2+t^2)^{n/2}}\,d\eta
\end{aligned}
$$

for all $t\in]-\min\{t_1,\tau/2\},0[$ and $x\in\partial\Omega$. Then we consider the two integrals in the right-hand side of (4.52). We have

$$
\begin{aligned}
&\int_{\{\eta\in\mathbb{B}_{n-1}(0,r):\,|\eta|<|t|\}}\frac{|\eta|^{1+\alpha}}{(|\eta|^2+t^2)^{(n+1)/2}}\,d\eta \qquad (4.53)\\
&\le|t|^{-(n+1)}\int_{\{\eta\in\mathbb{B}_{n-1}(0,r):\,|\eta|<|t|\}}|\eta|^{1+\alpha}\,d\eta\\
&\le|t|^{-(n+1)}\int_0^{|t|}\rho^{1+\alpha}s_{n-1}\rho^{n-2}\,d\rho=|t|^{-(n+1)}s_{n-1}\frac{|t|^{n+\alpha}}{n+\alpha}=\frac{s_{n-1}}{n+\alpha}|t|^{\alpha-1}
\end{aligned}
$$

and

$$
\begin{aligned}
&\int_{\{\eta\in\mathbb{B}_{n-1}(0,r):\,|\eta|<|t|\}}\frac{|\eta|^{\alpha}}{(|\eta|^2+t^2)^{n/2}}\,d\eta \qquad (4.54)\\
&\le|t|^{-n}\int_{\{\eta\in\mathbb{B}_{n-1}(0,r):\,|\eta|<|t|\}}|\eta|^{\alpha}\,d\eta\\
&\le|t|^{-n}\int_0^{|t|}\rho^{\alpha}s_{n-1}\rho^{n-2}\,d\rho=|t|^{-n}s_{n-1}\frac{|t|^{n+\alpha-1}}{n+\alpha-1}=s_{n-1}\frac{|t|^{\alpha-1}}{n+\alpha-1}
\end{aligned}
$$

for all $t\in]-\min\{t_1,\tau/2\},0[$.

• We now turn to estimate the second term which appears in the right-hand side of inequality (4.42). We have

$$
\begin{aligned}
&\left|\int_{(\partial\Omega)\setminus C(x,R_x,r,\delta)}\partial_{x_j}k(x+ta(x)-y)\,d\sigma_y\right|\\
&\quad\le\|\partial_{x_j}k\|_{C^0(\partial\mathbb{B}_n(0,1))}m_{n-1}(\partial\Omega)\sup_{y\in(\partial\Omega)\setminus C(x,R_x,r,\delta)}|x+ta(x)-y|^{-n}.
\end{aligned}
$$

Next we note that

$$
|y-x|\ge\min\{r,\delta\}\qquad\forall y\in(\partial\Omega)\setminus C(x,R_x,r,\delta)
$$

and that accordingly

$$|y-(x+ta(x))| > |y-x|-|ta(x)| \geq r-|t| \geq r/2 \qquad \forall y \in (\partial\Omega)\setminus C(x,R_x,r,\delta) \tag{4.55}$$

for all $t \in]-\min\{t_1,\tau/2,r/2\},0[$. Then we have

$$\left|\int_{(\partial\Omega)\setminus C(x,R_x,r,\delta)} \partial_{x_j}k(x+ta(x)-y)\,d\sigma_y\right| \tag{4.56}$$
$$\leq \|\partial_{x_j}k\|_{C^0(\partial\mathbb{B}_n(0,1))} m_{n-1}(\partial\Omega)(r/2)^{-n}$$

for all $t \in]-\min\{t_1,\tau/2,r/2\},0[$.

• Finally, we estimate the third term which appears in the right-hand side of inequality (4.42). By taking $y = x + R_x(\eta,0)^t$ with $\eta \in \mathbb{R}^{n-1}\setminus\mathbb{B}_{n-1}(0,r)$ in inequality (4.49), we have

$$|R_x^t(b_*(x)-\eta,b_n(x))^t| \geq (1-\vartheta)^{1/2}\sqrt{|R_x(\eta,0)^t|^2+t^2}$$
$$= (1-\vartheta)^{1/2}\sqrt{|\eta|^2+t^2} \qquad \forall t \in \mathbb{R}\setminus\{0\}.$$

Then we obtain

$$\left|\int_{M_x(\partial\Omega)\setminus C(x,R_x,r,\delta)} \partial_{x_j}k(x+ta(x)-y)\,d\sigma_y\right| \tag{4.57}$$
$$\leq \|\partial_{x_j}k\|_{C^0(\partial\mathbb{B}_n(0,1))}\int_{M_x(\partial\Omega)\setminus C(x,R_x,r,\delta)} |x+ta(x)-y|^{-n}\,d\sigma_y$$
$$\leq \|\partial_{x_j}k\|_{C^0(\partial\mathbb{B}_n(0,1))}\int_{\mathbb{R}^{n-1}\setminus\mathbb{B}_{n-1}(0,r)} |R_x^t(tb_*(x)-\eta,tb_n(x))^t|^{-n}\,d\eta$$
$$\leq \|\partial_{x_j}k\|_{C^0(\partial\mathbb{B}_n(0,1))}\int_{\mathbb{R}^{n-1}\setminus\mathbb{B}_{n-1}(0,r)} \frac{d\eta}{(1-\vartheta)^{n/2}\sqrt{|\eta|^2+t^2}^{\,n}}$$
$$\leq \|\partial_{x_j}k\|_{C^0(\partial\mathbb{B}_n(0,1))}\int_{\mathbb{R}^{n-1}\setminus\mathbb{B}_{n-1}(0,r)} \frac{d\eta}{(1-\vartheta)^{n/2}|\eta|^n}.$$

Thus by equality (4.29), inequalities (4.35), (4.39), equality (4.41), inequalities (4.42), (4.43), (4.44), (4.46), (4.50), (4.51), (4.52), (4.53), (4.54), (4.56), (4.57), and Lemma 4.13, we conclude that there exists a constant $C_2 > 0$ as in (4.27) for all $t \in]-\min\{t_1,\tau/2,r/2\},0[=]-\min\{t_1,r/2\},0[$ and that accordingly statement (jj) holds true. Thus the proof of (i) is complete.

Then statement (ii) can be deduced applying statement (i) to the open bounded set $\mathbb{B}_n(0,r)\setminus\overline{\Omega}$. Indeed the map from $C^{0,\alpha}(\partial\Omega)$ to $C^{0,\alpha}(\partial(\mathbb{B}_n(0,r)\setminus\overline{\Omega}))$ that takes μ to

$$\tilde{\mu}(x) \equiv \begin{cases} \mu(x) & \text{if } x \in \partial\Omega\,, \\ 0 & \text{if } x \in \partial\mathbb{B}_n(0,r) \end{cases}$$

is linear and continuous and we have

$$K[k,\mu]^-_{|\overline{\mathbb{B}_n(0,r)\setminus\Omega}} = K[k,\tilde{\mu}]^+$$

for all $(k,\mu) \in \mathcal{K}_{1,1;o} \times C^{0,\alpha}(\partial\Omega)$. □

4.4 The Single Layer Potential

Definition 4.18. Let Ω be a bounded open Lipschitz subset of $\mathbb{R}^n$. If $\phi \in C^0(\partial\Omega)$, then we denote by $v_\Omega[\phi]$ the single (or simple) layer potential with moment (or density) ϕ, i.e., the function from $\mathbb{R}^n$ to $\mathbb{R}$ defined by

$$v_\Omega[\phi](x) \equiv \int_{\partial\Omega} S_n(x-y)\phi(y)\, d\sigma_y \qquad \forall x \in \mathbb{R}^n .$$

We note that inequality

$$|S_n(x-y)\phi(y)| \\ \leq \sup_{(a,b)\in(\partial\Omega)^2, a\neq b} \left(|S_n(a-b)|\,|a-b|^{\max\{n-2,1/2\}}\right) \frac{\sup_{\partial\Omega}|\phi|}{|x-y|^{\max\{n-2,1/2\}}},$$

which holds for all $x, y \in \partial\Omega$ with $x \neq y$ and inequality $\max\{n-2, 1/2\} < n-1$ ensure the integrability of $S_n(x-y)\phi(y)$ in the variable $y \in \partial\Omega$ for all $x \in \partial\Omega$.

We wish to prove that the single layer potential is continuous in $\mathbb{R}^n$. To do so, we need some preliminary lemmas which enable the application of the Vitali Convergence Theorem for integrals (see Theorem A.11).

Lemma 4.19. *Let $f \in L^1_{\mathrm{loc}}(\mathbb{R}^n)$. If there exists $r \in]0,+\infty[$ such that $f_{|\mathbb{R}^n\setminus\mathbb{B}_n(0,r)}$ is essentially bounded, then for each $\epsilon > 0$, there exists $\delta > 0$ such that*

$$\int_E |f|\, dx \leq \epsilon \qquad \forall E \in \mathcal{L}_n, m_n(E) \leq \delta .$$

Proof. Clearly,

$$f = f\chi_{\mathbb{B}_n(0,r)} + f\chi_{\mathbb{R}^n\setminus\mathbb{B}_n(0,r)} .$$

Since $f\chi_{\mathbb{B}_n(0,r)} \in L^1(\mathbb{R}^n)$, there exists $\delta_1 \in]0,+\infty[$ such that

$$\int_E \left|f\chi_{\mathbb{B}_n(0,r)}\right| dx \leq \epsilon/2 \qquad \forall E \in \mathcal{L}_n, m_n(E) \leq \delta_1$$

(cf., e.g., Folland [103, Corollary 3.6, p. 89]). Next we note that if $E \in \mathcal{L}_n$, $m_n(E) < +\infty$, then

$$\int_E \left|f\chi_{\mathbb{R}^n\setminus\mathbb{B}_n(0,r)}\right| dx \\ \leq \|f\|_{L^\infty(\mathbb{R}^n\setminus\mathbb{B}_n(0,r))} m_n(E \cap (\mathbb{R}^n \setminus \mathbb{B}_n(0,r))) \leq \|f\|_{L^\infty(\mathbb{R}^n\setminus\mathbb{B}_n(0,r))} m_n(E) .$$

Hence, if we take

$$\delta \in \left]0, \min\left\{\delta_1, \frac{\epsilon}{2(\|f\|_{L^\infty(\mathbb{R}^n \setminus \mathbb{B}_n(0,r))}+1)}\right\}\right[,$$

then we conclude that

$$\begin{aligned}\int_E |f|\,dx &\le \int_{E\cap \mathbb{B}_n(0,r)} |f|\,dx + \int_{E\cap(\mathbb{R}^n\setminus \mathbb{B}_n(0,r))} |f|\,dx \\ &\le \epsilon/2 + \|f\|_{L^\infty(\mathbb{R}^n\setminus\mathbb{B}_n(0,r))}\frac{\epsilon}{2(\|f\|_{L^\infty(\mathbb{R}^n\setminus\mathbb{B}_n(0,r))}+1)} \le \epsilon\end{aligned}$$

whenever $m_n(E) \le \delta$. □

Then we have the following.

Corollary 4.20. *Let $n \in \mathbb{N}\setminus\{0\}$. Let $\lambda \in]0,n[$. For each $\epsilon > 0$, there exists $\delta > 0$ such that*

$$\int_E \frac{1}{|\xi-\eta|^\lambda}\,d\eta \le \epsilon$$

for all $E \in \mathcal{L}_n$ with $m_n(E) \le \delta$ and for all $\xi \in \mathbb{R}^n$.

Proof. The function $|\cdot|^{-\lambda}$ is locally integrable in $\mathbb{R}^n$ and is bounded outside $\mathbb{B}_n(0,1)$. Then Lemma 4.19 implies that there exists $\delta > 0$ such that

$$\int_E \frac{1}{|y|^\lambda}\,dy \le \epsilon \qquad \text{if } E \in \mathcal{L}_n, m_n(E) \le \delta\,.$$

Since $m_n(E) = m_n(\xi - E)$ and $\int_{\xi-E}\frac{1}{|y|^\lambda}\,dy = \int_E \frac{1}{|\xi-\eta|^\lambda}\,d\eta$ for all $\xi \in \mathbb{R}^n$, we conclude that the statement holds true. □

We now prove a result corresponding to the previous lemma on the boundary of a Lipschitz set. As elsewhere, we denote by $\mathcal{L}_{\partial\Omega}$ and by $m_{\partial\Omega}$ the σ-algebra of measurable subsets of $\partial\Omega$ and the $(n-1)$-dimensional measure of $\partial\Omega$, respectively.

Lemma 4.21. *Let Ω be a bounded open Lipschitz subset of $\mathbb{R}^n$. Let $\lambda \in]0, n-1[$. Then for each $\epsilon \in]0,+\infty[$ there exists $\delta \in]0,+\infty[$ such that*

$$\int_E \frac{1}{|x-y|^\lambda}\,d\sigma_y \le \epsilon \qquad \text{if } E \in \mathcal{L}_{\partial\Omega},\ m_{\partial\Omega}(E) \le \delta\,,\ x \in \mathbb{R}^n\,. \tag{4.58}$$

Moreover, $\sup_{x\in\mathbb{R}^n}\int_{\partial\Omega}\frac{1}{|x-y|^\lambda}\,d\sigma_y < +\infty$.

Proof. For each $x \in \partial\Omega$, there exists a coordinate cylinder $C(x, R_x, r_x, \tau_x)$ around x. Then $\{[x + R_x^t(\mathbb{B}_{n-1}(0,r_x/8)\times]-3\tau_x/4, 3\tau_x/4[)]\cap\partial\Omega\}_{x\in\partial\Omega}$ is an open cover of $\partial\Omega$. Since $\partial\Omega$ is compact, there exist finite families $\{x^{(l)}\}_{l=1}^m$, $\{R_l\}_{l=1}^m$, $\{r_l\}_{l=1}^m$, $\{\tau_l\}_{l=1}^m$ such that

$$\partial\Omega \subseteq \bigcup_{l=1}^m Q_l$$

where

$$Q_l \equiv x^{(l)} + R_l^t(\mathbb{B}_{n-1}(0, r_l/8)\times] - 3\tau_l/4, 3\tau_l/4[) \quad \forall l \in \{1, \dots, m\}.$$

We denote by $\{\gamma_l\}_{l=1}^m$ and $\{\psi_{x^{(l)}}\}_{l=1}^m$ the corresponding families of functions which represent $\partial\Omega$ in the cylinders $\{C(x^{(l)}, R_l, r_l, \tau_l)\}_{l=1}^m$ as a graph and which parametrize $\partial\Omega$ around the points $\{x^{(l)}\}_{l=1}^m$, respectively.

Now let $d > 0$ denote the distance between the closed set

$$W \equiv \mathbb{R}^n \setminus \bigcup_{l=1}^m Q_l$$

and the compact set $\partial\Omega$. We can show that there exists $\delta > 0$ as in (4.58) for all points x of W. Indeed, if $x \in W$, then we have

$$\int_E \frac{1}{|x-y|^\lambda}\, d\sigma_y \le d^{-\lambda} \int_E d\sigma_y = d^{-\lambda} m_{\partial\Omega}(E)\,, \tag{4.59}$$

and it suffices to take $\delta \in]0, d^\lambda \epsilon[$ in (4.58).

Next we show that there exists $\delta > 0$ as in (4.58) for all points x of the open set $\bigcup_{l=1}^m Q_l$. It clearly suffices to show that if $l \in \{1, \dots, m\}$ and $\epsilon \in]0, +\infty[$ then there exists $\delta_l > 0$ such that

$$\int_E \frac{1}{|x-y|^\lambda}\, d\sigma_y \le \epsilon \qquad \text{if } E \in \mathcal{L}_{\partial\Omega},\ m_{\partial\Omega}(E) \le \delta_l\,,\ x \in Q_l\,. \tag{4.60}$$

So let $x \in Q_l$. Let $(\xi, \xi') \in \mathbb{B}_{n-1}(0, r_l/8)\times] - 3\tau_l/4, 3\tau_l/4[$ be such that

$$x = x^{(l)} + R_l^t \begin{pmatrix} \xi \\ \xi' \end{pmatrix}.$$

Then we note that $\mathbb{B}_{n-1}(\xi, r_l/8) \subseteq \mathbb{B}_{n-1}(0, r_l/4)$ and that

$$\begin{aligned} &\int_E \frac{d\sigma_y}{|x-y|^\lambda} \\ &= \int_{E\cap((\partial\Omega)\setminus\psi_{x^{(l)}}(\mathbb{B}_{n-1}(0,r_l/2)))} \frac{d\sigma_y}{|x-y|^\lambda} + \int_{E\cap(\psi_{x^{(l)}}(\mathbb{B}_{n-1}(0,r_l/2)))} \frac{d\sigma_y}{|x-y|^\lambda}\,. \end{aligned} \tag{4.61}$$

If $y \in (\partial\Omega) \setminus \psi_{x^{(l)}}(\mathbb{B}_{n-1}(0, r_l/2))$, then $y \notin C(x^{(l)}, R_l, r_l/2, \tau_l)$. Since x belongs to Q_l and the distance between Q_l and the complement of the cylinder $C(x^{(l)}, R_l, r_l/2, \tau_l)$ is at least $\min\{r_l/4, \tau_l/4\}$, we have $|x-y| \ge \min\{r_l/4, \tau_l/4\}$, and accordingly

$$\begin{aligned} &\int_{E\cap((\partial\Omega)\setminus\psi_{x^{(l)}}(\mathbb{B}_{n-1}(0,r_l/2)))} \frac{d\sigma_y}{|x-y|^\lambda} \\ &\qquad \le [\min\{r_l/4, \tau_l/4\}]^{-\lambda} m_{\partial\Omega}\left(E \cap ((\partial\Omega) \setminus \psi_{x^{(l)}}(\mathbb{B}_{n-1}(0, r_l/2)))\right) \end{aligned} \tag{4.62}$$

$$\leq [\min\{r_l/4, \tau_l/4\}]^{-\lambda} m_{\partial\Omega}(E).$$

Next we note that

$$\begin{aligned}
&\int_{E\cap(\psi_{x^{(l)}}(\mathbb{B}_{n-1}(0,r_l/2)))} \frac{d\sigma_y}{|x-y|^\lambda} \\
&\qquad \leq \sqrt{1+\| |\nabla\gamma_l| \|_\infty^2} \int_{\psi^{\leftarrow}_{x^{(l)}}(E)} \frac{d\eta}{\sqrt{|\xi-\eta|^2+|\xi'-\gamma_l(\eta)|^2}^\lambda} \\
&\qquad \leq \sqrt{1+\| |\nabla\gamma_l| \|_\infty^2} \int_{\psi^{\leftarrow}_{x^{(l)}}(E)} \frac{d\eta}{|\xi-\eta|^\lambda}
\end{aligned} \tag{4.63}$$

and that

$$\begin{aligned}
m_{n-1}(\psi^{\leftarrow}_{x^{(l)}}(E)) = \int_{\psi^{\leftarrow}_{x^{(l)}}(E)} d\eta \leq \int_{\psi^{\leftarrow}_{x^{(l)}}(E)} \sqrt{1+|\nabla\gamma_l(\eta)|^2}\, d\eta \\
= m_{\partial\Omega}(E\cap\psi_{x^{(l)}}(\mathbb{B}_{n-1}(0,r_l))) \leq m_{\partial\Omega}(E).
\end{aligned} \tag{4.64}$$

Hence, Corollary 4.20, implies the existence of $\delta_{l1} > 0$ such that

$$\int_{\psi^{\leftarrow}_{x^{(l)}}(E)} \frac{d\eta}{|\xi-\eta|^\lambda} \leq \frac{\epsilon}{2\sqrt{1+\| |\nabla\gamma_l| \|_\infty^2}} \qquad \text{whenever } m_{\partial\Omega}(E) \leq \delta_{l1}. \tag{4.65}$$

Then if we choose

$$0 < \delta_l \leq \min\left\{\delta_{l1}, [\min\{r_l/4, \tau_l/4\}]^\lambda \epsilon/2\right\},$$

inequalities (4.61)–(4.63), (4.65) imply that (4.60) holds for all x in Q_l.

We now turn to prove the last inequality of the statement. By inequality (4.59) with $E = \partial\Omega$, we have $\sup_{x\in W} \int_{\partial\Omega} \frac{1}{|x-y|^\lambda}\, d\sigma_y < +\infty$. By the inclusion

$$\mathbb{B}_{n-1}(\xi, r_l) \subseteq \mathbb{B}_{n-1}(0, 2r_l),$$

which holds for all $\xi \in \mathbb{B}_{n-1}(0, r_l/8)$, we have

$$\begin{aligned}
\int_{\psi^{\leftarrow}_{x^{(l)}}(\partial\Omega)} \frac{d\eta}{|\xi-\eta|^\lambda} &\leq \int_{\mathbb{B}_{n-1}(0,r_l)} \frac{d\eta}{|\xi-\eta|^\lambda} = \int_{\mathbb{B}_{n-1}(\xi,r_l)} \frac{d\eta}{|\eta|^\lambda} \\
&\leq \int_{\mathbb{B}_{n-1}(0,2r_l)} \frac{d\eta}{|\eta|^\lambda} \leq \int_{\mathbb{B}_{n-1}(0,2\max_{l\in\{1,\ldots,m\}} r_l)} \frac{d\eta}{|\eta|^\lambda} < +\infty
\end{aligned}$$

for all $\xi \in \mathbb{B}_{n-1}(0, r_l/8)$ and $l \in \{1,\ldots,m\}$. Then inequalities (4.61)–(4.63) with $E = \partial\Omega$ imply that $\sup_{x\in\bigcup_{l=1}^m Q_l} \int_{\partial\Omega} \frac{1}{|x-y|^\lambda}\, d\sigma_y < +\infty$ and accordingly the last inequality of the statement holds true. □

We are now ready to prove the following.

Theorem 4.22. *Let Ω be a bounded open Lipschitz subset of $\mathbb{R}^n$. If $\phi \in C^0(\partial\Omega)$, then the following statement holds.*

(i) $v_\Omega[\phi]$ is harmonic in $\mathbb{R}^n \setminus \partial\Omega$.
(ii) $v_\Omega[\phi]$ is continuous in $\mathbb{R}^n$.

Proof. Since $D_x^\gamma S_n(x-y)\phi(y)$ is continuous in $(\mathbb{R}^n \setminus \partial\Omega) \times \partial\Omega$ for all $\gamma \in \mathbb{N}^n$, then the classical differentiation theorem for integrals depending on a parameter implies that $v_\Omega[\phi] \in C^\infty(\mathbb{R}^n \setminus \partial\Omega)$ and that

$$D^\gamma v_\Omega[\phi](x) = \int_{\partial\Omega} D_x^\gamma S_n(x-y)\phi(y)\,d\sigma_y \qquad \forall x \in \mathbb{R}^n \setminus \partial\Omega$$

for all $\gamma \in \mathbb{N}^n$. Since $\Delta_x S_n(x-y) = 0$ for all $x \in \mathbb{R}^n \setminus \partial\Omega$ and $y \in \partial\Omega$, we conclude that $v_\Omega[\phi]$ is harmonic in $\mathbb{R}^n \setminus \partial\Omega$. In particular, $v_\Omega[\phi]$ is continuous at all points of $\mathbb{R}^n \setminus \partial\Omega$.

We now prove that $v_\Omega[\phi]$ is continuous at all points of $\partial\Omega$. Let $\tilde{x} \in \partial\Omega$. We want to verify that

$$\lim_{x \to \tilde{x}} \int_{\partial\Omega} \phi(y) S_n(x-y)\,d\sigma_y = \int_{\partial\Omega} \phi(y) S_n(\tilde{x}-y)\,d\sigma_y\,. \qquad (4.66)$$

By the continuity of S_n in $\mathbb{R}^n \setminus \{0\}$, we have

$$\lim_{x \to \tilde{x}} \phi(y) S_n(x-y) = \phi(y) S_n(\tilde{x}-y) \qquad \forall y \in (\partial\Omega) \setminus \{\tilde{x}\}\,.$$

Now let $r > 0$ be such that $\overline{\Omega} \subseteq \mathbb{B}_n(0,r)$. In order to invoke the Vitali Convergence Theorem A.11, we show that, for each $\epsilon > 0$, there exists $\delta > 0$ such that

$$\int_E |\phi(y) S_n(x-y)|\,d\sigma_y \le \epsilon \qquad \text{for all } E \in \mathcal{L}_{\partial\Omega},\ m_{\partial\Omega}(E) \le \delta\,,\ x \in \mathbb{B}_n(0,r)\,.$$

If $n \ge 3$, then we have

$$\int_E |\phi(y) S_n(x-y)|\,d\sigma_y \le \frac{\|\phi\|_\infty}{(n-2)s_n} \int_E \frac{1}{|x-y|^{n-2}}\,d\sigma_y \qquad \forall x \in \mathbb{R}^n$$

and Lemma 4.21 implies that there exists $\delta > 0$ such that

$$\int_E \frac{1}{|x-y|^{n-2}}\,d\sigma_y \le \frac{\epsilon}{1 + \frac{\|\phi\|_\infty}{(n-2)s_n}} \qquad \text{for all } E \in \mathcal{L}_{\partial\Omega},\ m_{\partial\Omega}(E) \le \delta\,,\ x \in \mathbb{R}^n\,.$$

If $n = 2$, then we have

$$\int_E |\phi(y) S_n(x-y)|\,d\sigma_y \le \frac{\|\phi\|_\infty}{2\pi} \int_E |\log|x-y||\,d\sigma_y$$
$$\le \frac{\|\phi\|_\infty}{2\pi} \Big(\sup_{\rho \in]0,2r]} \rho^{1/2} |\log \rho| \Big) \int_E |x-y|^{-1/2}\,d\sigma_y \qquad \forall x \in \mathbb{B}_n(0,r)$$

and Lemma 4.21 implies that there exists $\delta > 0$ such that

$$\int_E |x-y|^{-1/2}\, d\sigma_y \le \frac{\epsilon}{1+\frac{\|\phi\|_\infty}{2\pi}(\sup_{\rho\in]0,2r]}\rho^{1/2}|\log\rho|)}$$
$$\text{for all } E \in \mathcal{L}_{\partial\Omega},\ m_{\partial\Omega}(E) \le \delta\,,\ x \in \mathbb{B}_n(0,r)\,.$$

Then the Vitali Convergence Theorem A.11 implies that, for each sequence $\{x_j\}_{j\in\mathbb{N}}$ in $\mathbb{B}_n(0,r)\setminus\{\tilde{x}\}$ which converges to $\tilde{x}$, we have

$$\lim_{j\to\infty}\int_{\partial\Omega}\phi(y)S_n(x_j-y)\,d\sigma_y = \int_{\partial\Omega}\phi(y)S_n(\tilde{x}-y)\,d\sigma_y\,.$$

Therefore, the limiting relation in (4.66) holds true. □

Under the assumptions of the previous theorem, we set

$$v_\Omega^+[\phi] \equiv v_\Omega[\phi]_{|\overline{\Omega}} \qquad v_\Omega^-[\phi] \equiv v_\Omega[\phi]_{|\overline{\Omega^-}}\,.$$

Then we have the following theorem, which clarifies the behavior of $v_\Omega^-[\phi]$ at infinity.

Theorem 4.23. *Let Ω be a bounded open Lipschitz subset of $\mathbb{R}^n$. Let $\phi \in C^0(\partial\Omega)$.*

(i) If $n \ge 3$, then $\lim_{x\to\infty} v_\Omega^-[\phi](x) = 0$ and accordingly $v_\Omega^-[\phi]$ is harmonic at infinity.

(ii) Let $n = 2$. If $\int_{\partial\Omega}\phi\,d\sigma = 0$, then $\lim_{x\to\infty} v_\Omega^-[\phi](x) = 0$ and accordingly $v_\Omega^-[\phi]$ is harmonic at infinity. If, instead, $\int_{\partial\Omega}\phi\,d\sigma \ne 0$, then $v_\Omega^-[\phi]$ is not harmonic at infinity.

Proof. Let $r_0 > 0$ be such that $\overline{\Omega} \subseteq \mathbb{B}_n(0,r_0)$. Clearly,

$$|x-y| \ge ||x|-|y|| \ge |x| - r_0 > 0 \qquad \forall y \in \partial\Omega,\ x \in \mathbb{R}^n\setminus\overline{\mathbb{B}_n(0,r_0)}\,.$$

If $n \ge 3$, we have

$$\begin{aligned} |v_\Omega^-[\phi](x)| &\le \int_{\partial\Omega}|\phi(y)||S_n(x-y)|\,d\sigma_y \qquad (4.67)\\ &\le \frac{1}{(n-2)s_n}\int_{\partial\Omega}\frac{|\phi(y)|}{|x-y|^{n-2}}\,d\sigma_y\\ &\le \frac{1}{(n-2)s_n}\frac{1}{||x|-r_0|^{n-2}}\int_{\partial\Omega}|\phi(y)|\,d\sigma_y \qquad \forall x\in\mathbb{R}^n\setminus\overline{\mathbb{B}_n(0,r_0)}\,. \end{aligned}$$

Thus $\lim_{x\to\infty} v_\Omega^-[\phi](x) = 0$ and we conclude that $v_\Omega^-[\phi]$ is harmonic at infinity. If instead $n = 2$, then we note that

$$\begin{aligned} v_\Omega^-[\phi](x) &= \int_{\partial\Omega}\phi(y)S_2(x-y)\,d\sigma_y \qquad (4.68)\\ &= \frac{1}{2\pi}\int_{\partial\Omega}\phi(y)\log\left(1+\left(\frac{|x-y|}{|x|}-1\right)\right)d\sigma_y + \frac{1}{2\pi}(\log|x|)\int_{\partial\Omega}\phi(y)\,d\sigma_y \end{aligned}$$

for all $x \in \Omega^- \setminus \{0\}$. We now show that

$$\lim_{x\to\infty} \int_{\partial\Omega} \phi(y) \log\left(1 + \left(\frac{|x-y|}{|x|} - 1\right)\right) d\sigma_y = 0 . \tag{4.69}$$

Since $\lim_{\rho\to 0} \rho^{-1} \log(1+\rho) = 1$, there exists $\delta \in]0,1[$ such that

$$|\log(1+\rho)| \leq \frac{3}{2}|\rho| \qquad \forall \rho \in [-\delta, +\delta] .$$

Moreover, we have

$$\left|\frac{|x-y|}{|x|} - 1\right| = \frac{||x-y| - |x||}{|x|} \leq \frac{|y|}{|x|} \leq \frac{r_0}{|x|} \qquad \forall y \in \partial\Omega, x \in \mathbb{R}^2 \setminus \mathbb{B}_2(0, r_0) .$$

Then we choose $r_1 > r_0$ such that $r_0/r_1 < \delta$ and we obtain

$$\begin{aligned} &\left|\frac{|x-y|}{|x|} - 1\right| \leq \frac{r_0}{|x|} \leq \frac{r_0}{r_1} < \delta , \\ &\left|\log\left(1 + \left(\frac{|x-y|}{|x|} - 1\right)\right)\right| \leq \frac{3}{2}\left|\frac{|x-y|}{|x|} - 1\right| \leq \frac{3}{2}\frac{r_0}{|x|} \end{aligned}$$

for all $y \in \partial\Omega$, $x \in \mathbb{R}^2 \setminus \mathbb{B}_2(0, r_1)$. Hence

$$\left|\int_{\partial\Omega} \phi(y) \log\left(1 + \left(\frac{|x-y|}{|x|} - 1\right)\right) d\sigma_y\right| \leq \frac{3}{2}\frac{r_0}{|x|} \int_{\partial\Omega} |\phi(y)|\, d\sigma_y$$

for all $x \in \mathbb{R}^2 \setminus \mathbb{B}_2(0, r_1)$. By taking the limit as x tends to infinity, we deduce that the left-hand side has limit 0 as x tends to infinity. Hencc, (4.69) holds.

Now if $\int_{\partial\Omega} \phi\, d\sigma = 0$, then (4.68) and (4.69) imply that

$$\lim_{x\to\infty} v_\Omega^-[\phi](x) = 0$$

and accordingly $v_\Omega^-[\phi]$ is harmonic at infinity. Conversely, if $v_\Omega^-[\phi]$ is harmonic at infinity, then $v_\Omega^-[\phi]$ is bounded in a punctured neighborhood of infinity. Hence, (4.68) and (4.69) imply that $\int_{\partial\Omega} \phi\, d\sigma = 0$. □

In the next theorem, we introduce a known formula for the first order partial derivatives of a single layer potential.

Theorem 4.24. *Let $\alpha \in]0,1[$. Let Ω be a bounded open subset of $\mathbb{R}^n$ of class $C^{1,\alpha}$. Let $\phi \in C^{0,\alpha}(\partial\Omega)$. Let $\tilde{x} \in \partial\Omega$. Then the following two statements hold.*

(i) Let $j \in \{1, \ldots, n\}$. Then the principal value

$$\text{p.v.} \int_{\partial\Omega} \frac{\partial S_n}{\partial x_j}(\tilde{x} - y)\phi(y)\, d\sigma_y \equiv \lim_{\epsilon\to 0} \int_{\partial\Omega\setminus\mathbb{B}_n(\tilde{x},\epsilon)} \frac{\partial S_n}{\partial x_j}(\tilde{x} - y)\phi(y)\, d\sigma_y$$

exists finite and we have the following jump formula for the first order partial derivatives of a single layer potential

$$\lim_{t\to 0^\mp} \frac{\partial}{\partial x_j} v_\Omega[\phi](\tilde{x}+t\nu_\Omega(\tilde{x})) = \mp\frac{1}{2}\phi(\tilde{x})(\nu_\Omega)_j(\tilde{x})+\text{p.v.}\int_{\partial\Omega}\frac{\partial S_n}{\partial x_j}(\tilde{x}-y)\phi(y)\,d\sigma_y\,.$$

(ii) The function $\nu_\Omega(\tilde{x})\cdot\nabla S_n(\tilde{x}-y)\phi(y)$ is integrable in $y\in\partial\Omega$ and we have the following jump formula for the normal derivative of a single layer potential

$$\frac{\partial}{\partial\nu_\Omega} v_\Omega^\pm[\phi](\tilde{x}) = \mp\frac{1}{2}\phi(\tilde{x}) + \int_{\partial\Omega}\nu_\Omega(\tilde{x})\cdot\nabla S_n(\tilde{x}-y)\phi(y)\,d\sigma_y\,.$$

Proof. For a proof of (i), we refer to Miranda [212, 2.II p. 314] or to Cialdea [52, Theorem XI, p. 52], Hackbusch [117, Corollary 8.1.14, p. 280]. By Lemma 2.76 (iii), the function $\nu_\Omega(\tilde{x})\cdot\nabla S_n(\tilde{x}-y)\phi(y)$ is integrable in $y\in\partial\Omega$. By the L'Hôpital's Rule, we have

$$\begin{aligned}\frac{\partial}{\partial\nu_\Omega} v_\Omega^\pm[\phi](\tilde{x}) &= \lim_{t\to 0^\mp}\frac{v_\Omega^\pm[\phi](\tilde{x}+t\nu_\Omega(\tilde{x})) - v_\Omega^\pm[\phi](\tilde{x})}{t}\\ &= \lim_{t\to 0^\mp}\nabla v_\Omega^\pm[\phi](\tilde{x}+t\nu_\Omega(\tilde{x}))\cdot\nu_\Omega(\tilde{x})\,.\end{aligned}$$

Then statement (ii) follows by statement (i). □

Then we have the following classical Schauder regularity result for the restriction of the single layer potential to $\Omega^+\equiv\Omega$ and to Ω^-.

Theorem 4.25. *Let $\alpha\in]0,1[$. Let Ω be a bounded open subset of $\mathbb{R}^n$ of class $C^{1,\alpha}$.*

(i) If $\phi\in C^{0,\alpha}(\partial\Omega)$, then $v_\Omega^+[\phi]$ belongs to $C^{1,\alpha}(\overline{\Omega})$. Moreover, the map $v_\Omega^+[\cdot]$ is linear and continuous from $C^{0,\alpha}(\partial\Omega)$ to $C^{1,\alpha}(\overline{\Omega})$.
(ii) If $\phi\in C^{0,\alpha}(\partial\Omega)$, then $v_\Omega^-[\phi]$ belongs $C^{1,\alpha}_{\text{loc}}(\overline{\Omega^-})$. Moreover, if $r\in]0,+\infty[$ and $\overline{\Omega}\subseteq\mathbb{B}_n(0,r)$, then the map $v_\Omega^-[\cdot]_{|\overline{\mathbb{B}_n(0,r)}\setminus\Omega}$ is linear and continuous from $C^{0,\alpha}(\partial\Omega)$ to $C^{1,\alpha}(\overline{\mathbb{B}_n(0,r)}\setminus\Omega)$.

Proof. (i) We already know that $v_\Omega^+[\phi]$ belongs to $C^0(\overline{\Omega})\cap C^\infty(\Omega)$. Moreover, if we set $\mu_n\equiv\max\{n-2,1/2\}$, then we have $\mu_n\in]0,n-1[$ and

$$\begin{aligned}|v_\Omega^+[\phi](x)| &\le \int_{\partial\Omega}|x-y|^{\mu_n}|S_n(x-y)|\frac{d\sigma_y}{|x-y|^{\mu_n}}\sup_{\partial\Omega}|\phi|\\ &\le \sup_{(a,b)\in(\partial\Omega)^2,\,a\neq b}\left(|S_n(a-b)|\,|a-b|^{\mu_n}\right)\sup_{x\in\mathbb{R}^n}\int_{\partial\Omega}\frac{d\sigma_y}{|x-y|^{\mu_n}}\sup_{\partial\Omega}|\phi|\,.\end{aligned}$$

By the definition of S_n in (3.2) and by Lemma 4.21, we can verify that the first and the second supremum in the right-hand side are finite. Accordingly, $v_\Omega^+[\cdot]$ is linear and continuous from $C^0(\partial\Omega)$ to $C^0(\overline{\Omega})$.

To complete the proof of (i), we observe that the kernels $k_j\equiv\frac{\partial S_n}{\partial x_j}$ are positively homogeneous of degree $-(n-1)$, odd, and of class C^∞. Then Theorem 4.17 (i) of

Miranda implies that $\frac{\partial}{\partial x_j} v^+_\Omega[\phi]$ belongs to $C^{0,\alpha}(\overline{\Omega})$ for all $j \in \{1, \ldots, n\}$ and that the map from $C^{0,\alpha}(\partial\Omega)$ to $C^{0,\alpha}(\overline{\Omega})$ that takes ϕ to $\frac{\partial}{\partial x_j} v^+_\Omega[\phi] \in C^{0,\alpha}(\overline{\Omega})$ is linear and continuous for all $j \in \{1, \ldots, n\}$. Hence, statement (i) holds true.

Then statement (ii) can be deduced applying statement (i) to the open bounded set $\mathbb{B}_n(0,r)\setminus\overline{\Omega}$ of class $C^{1,\alpha}$. Indeed the map from $C^{0,\alpha}(\partial\Omega)$ to $C^{0,\alpha}(\partial(\mathbb{B}_n(0,r)\setminus\overline{\Omega}))$ that takes ϕ to

$$\tilde{\phi}(x) \equiv \begin{cases} \phi(x) & \text{if } x \in \partial\Omega\,, \\ 0 & \text{if } x \in \partial\mathbb{B}_n(0,r) \end{cases}$$

is linear and continuous and we have

$$v^-_\Omega[\phi]_{|\overline{\mathbb{B}_n(0,r)}\setminus\Omega} = v^+_{\mathbb{B}_n(0,r)\setminus\overline{\Omega}}[\tilde{\phi}]$$

for all $\phi \in C^{0,\alpha}(\partial\Omega)$. □

When $n \geq 3$, we have the following consequence of the previous statement.

Corollary 4.26. *Let $n \in \mathbb{N}$, $n \geq 3$, $\alpha \in]0,1[$. Let Ω be a bounded open subset of $\mathbb{R}^n$ of class $C^{1,\alpha}$. Then the map that takes ϕ to $v^-_\Omega[\phi]$ is linear and continuous from $C^{0,\alpha}(\partial\Omega)$ to $C^{1,\alpha}_b(\overline{\Omega^-})$.*

Proof. Let $r \in]0, +\infty[$ be such that $\overline{\Omega} \subseteq \mathbb{B}_n(0, r-1)$. By Theorem 4.25 (ii), the map $v^-_\Omega[\cdot]_{|\overline{\mathbb{B}_n(0,r)}\setminus\Omega}$ from $C^{0,\alpha}(\partial\Omega)$ to $C^{1,\alpha}(\overline{\mathbb{B}_n(0,r)}\setminus\Omega)$ is linear and continuous.

We now show that the map $v^-_\Omega[\cdot]_{|\mathbb{R}^n\setminus\mathbb{B}_n(0,r)}$ is linear and continuous from $C^{0,\alpha}(\partial\Omega)$ to $C^{1,\alpha}_b(\mathbb{R}^n \setminus \mathbb{B}_n(0,r))$. By Lemma 2.48, $\mathbb{R}^n \setminus \overline{\mathbb{B}_n(0,r)}$ is regular in the sense of Whitney and thus Proposition 2.54 (i), (ii) implies that $C^2_b(\mathbb{R}^n \setminus \mathbb{B}_n(0,r))$ is continuously imbedded into $C^{1,\alpha}_b(\mathbb{R}^n \setminus \mathbb{B}_n(0,r))$. Thus it suffices to show that $v^-_\Omega[\cdot]_{|\mathbb{R}^n\setminus\mathbb{B}_n(0,r)}$ is linear and continuous from $C^{0,\alpha}(\partial\Omega)$ to $C^2_b(\mathbb{R}^n \setminus \mathbb{B}_n(0,r))$.

By the classical differentiability theorem for integrals depending on a parameter, by the equalities (3.2), (3.3) for S_n and ∇S_n, by inequality (3.4) for the partial derivatives of order two of S_n, and by the elementary inequality

$$|x-y| \geq |x| - (r-1) \geq r - (r-1) = 1 \qquad \forall (x,y) \in (\mathbb{R}^n \setminus \mathbb{B}_n(0,r)) \times (\partial\Omega)\,,$$

we have

$$\begin{aligned} |D^\eta v^-_\Omega[\phi](x)| &= \left| \int_{\partial\Omega} D^\eta S_n(x-y)\phi(y)\, d\sigma_y \right| \\ &\leq \max\{(n-2)^{-1}s_n^{-1}, s_n^{-1}, \omega_n^{-1}\} m_{n-1}(\partial\Omega)\|\phi\|_{C^{0,\alpha}(\partial\Omega)} \ \forall x \in \mathbb{R}^n \setminus \mathbb{B}_n(0,r) \end{aligned}$$

for all $\beta \in \mathbb{N}^n$ such that $|\beta| \leq 2$ and $\phi \in C^{0,\alpha}(\partial\Omega)$. Hence, $v^-_\Omega[\cdot]_{|\mathbb{R}^n\setminus\mathbb{B}_n(0,r)}$ is linear and continuous from $C^{0,\alpha}(\partial\Omega)$ to $C^2_b(\mathbb{R}^n \setminus \mathbb{B}_n(0,r))$.

Next we note that if $x \in \overline{\mathbb{B}_n(0,r)} \setminus \Omega$ and $y \in \mathbb{R}^n \setminus \mathbb{B}_n(0,r)$, then there exists ξ in the segment $[x,y]$ such that $\xi \in \partial\mathbb{B}_n(0,r)$. Thus, if g belongs to $C^0_b(\overline{\Omega^-})$,

$$g_{|\overline{\mathbb{B}_n(0,r)}\setminus\Omega} \in C^{0,\alpha}(\overline{\mathbb{B}_n(0,r)} \setminus \Omega)\,, \quad \text{and} \quad g_{|\mathbb{R}^n\setminus\mathbb{B}_n(0,r)} \in C^{0,\alpha}_b(\mathbb{R}^n \setminus \mathbb{B}_n(0,r))\,,$$

then we have

$$
\begin{aligned}
|g(x)-g(y)| &\leq |g(x)-g(\xi)|+|g(\xi)-g(y)| \\
&\leq |g:\overline{\mathbb{B}_n(0,r)}\setminus\Omega|_\alpha |x-\xi|^\alpha + |g:\mathbb{R}^n\setminus\mathbb{B}_n(0,r)|_\alpha|\xi-y|^\alpha \\
&\leq \max\{|g:\overline{\mathbb{B}_n(0,r)}\setminus\Omega|_\alpha, |g:\mathbb{R}^n\setminus\mathbb{B}_n(0,r)|_\alpha\}(|x-\xi|^\alpha+|\xi-y|^\alpha) \\
&\leq 2\max\{|g:\overline{\mathbb{B}_n(0,r)}\setminus\Omega|_\alpha, |g:\mathbb{R}^n\setminus\mathbb{B}_n(0,r)|_\alpha\}|x-y|^\alpha .
\end{aligned}
$$

By applying the last inequality to the first order partial derivatives of $v_\Omega^-[\phi]$, we obtain

$$
\begin{aligned}
&\|v_\Omega^-[\phi]\|_{C_b^{1,\alpha}(\overline{\Omega^-})} \\
&\quad\leq 2\max\{\|v_\Omega^-[\phi]_{|\overline{\mathbb{B}_n(0,r)}\setminus\Omega}\|_{C^{1,\alpha}(\overline{\mathbb{B}_n(0,r)}\setminus\Omega)}, \|v_\Omega^-[\phi]_{|\mathbb{R}^n\setminus\mathbb{B}_n(0,r)}\|_{C_b^{1,\alpha}(\mathbb{R}^n\setminus\mathbb{B}_n(0,r))}\}
\end{aligned}
$$

for all $\phi\in C^{0,\alpha}(\partial\Omega)$. Hence, the continuity of $v_\Omega^-[\cdot]_{|\mathbb{R}^n\setminus\mathbb{B}_n(0,r)}$ from $C^{0,\alpha}(\partial\Omega)$ to $C_b^{1,\alpha}(\mathbb{R}^n\setminus\mathbb{B}_n(0,r))$ and Theorem 4.25 (ii) imply that the linear map $v_\Omega^-[\cdot]$ is continuous from $C^{0,\alpha}(\partial\Omega)$ to $C_b^{1,\alpha}(\overline{\Omega^-})$. Thus the proof is complete. □

4.5 The Double Layer Potential

Definition 4.27. Let $\alpha\in]0,1]$. Let Ω be a bounded open subset of $\mathbb{R}^n$ of class $C^{1,\alpha}$. If $\psi\in C^0(\partial\Omega)$, then we denote by $w_\Omega[\psi]$ the double layer potential with moment (or density) ψ, i.e., the function from $\mathbb{R}^n$ to $\mathbb{R}$ defined by

$$
w_\Omega[\psi](x)\equiv\int_{\partial\Omega}\psi(y)\frac{\partial}{\partial\nu_{\Omega_y}}(S_n(x-y))\,d\sigma_y \qquad \forall x\in\mathbb{R}^n ,
$$

where

$$
\frac{\partial}{\partial\nu_{\Omega_y}}(S_n(x-y))\equiv-\nu_\Omega(y)\cdot\nabla S_n(x-y) \qquad \forall(x,y)\in\mathbb{R}^n\times\partial\Omega\,, x\neq y .
$$

We note that if $x\in\mathbb{R}^n\setminus\partial\Omega$, then $\nu_\Omega(y)\cdot\nabla S_n(x-y)$ is continuous in $y\in\partial\Omega$ and that accordingly $\psi(y)\nu_\Omega(y)\cdot\nabla S_n(x-y)$ is integrable in $y\in\partial\Omega$. If instead $x\in\partial\Omega$, then Lemma 2.76 (iii) implies the existence of a constant $c_{\Omega,\alpha}>0$ such that

$$
|\nu_\Omega(y)\cdot\nabla S_n(x-y)|\leq\frac{c_{\Omega,\alpha}}{s_n}\frac{1}{|x-y|^{(n-1)-\alpha}} \qquad \forall y\in\partial\Omega\setminus\{x\} . \tag{4.70}
$$

Since ψ is measurable and essentially bounded, inequality $(n-1)-\alpha<n-1$ ensures the integrability of the function $\psi(y)\nu_\Omega(y)\cdot\nabla S_n(x-y)$ in $y\in\partial\Omega$. We now introduce the following preliminary result.

Proposition 4.28. *Let $\alpha \in]0,1]$. Let Ω be a bounded open subset of $\mathbb{R}^n$ of class $C^{1,\alpha}$. If $\psi \in C^0(\partial\Omega)$, then the following statements hold.*

(i) $w_\Omega[\psi]$ is harmonic in $\mathbb{R}^n \setminus \partial\Omega$.
(ii) $\lim_{x\to\infty} w_\Omega[\psi](x) = 0$ and accordingly $w_\Omega[\psi]$ is harmonic at infinity.

Proof. (i) Since $D_x^\eta\left(\nu_\Omega(y) \cdot \nabla S_n(x-y)\psi(y)\right)$ is continuous in $(\mathbb{R}^n \setminus \partial\Omega) \times \partial\Omega$ for all $\eta \in \mathbb{N}^n$, then the classical differentiation theorem for integrals depending on a parameter implies that $w_\Omega[\psi] \in C^\infty(\mathbb{R}^n \setminus \partial\Omega)$ and that

$$D^\eta w_\Omega[\psi](x) = -\int_{\partial\Omega} D_x^\eta\left(\nu_\Omega(y) \cdot \nabla S_n(x-y)\psi(y)\right) d\sigma_y \quad \forall x \in \mathbb{R}^n \setminus \partial\Omega$$

for all $\eta \in \mathbb{N}^n$. Since $\Delta_x\left(\nu_\Omega(y) \cdot \nabla S_n(x-y)\psi(y)\right) = 0$ for all $x \in \mathbb{R}^n \setminus \partial\Omega$ and $y \in \partial\Omega$, we conclude that $w_\Omega[\psi]$ is harmonic in $\mathbb{R}^n \setminus \partial\Omega$. Now let $r > 0$ be such that $\overline{\Omega} \subseteq \mathbb{B}_n(0,r)$. Then we have

$$\begin{aligned} |w_{\overline{\Omega}}[\psi](x)| &\leq \int_{\partial\Omega} |\psi(y)| \left| \frac{\partial}{\partial \nu_{\Omega_y}} (S_n(x-y)) \right| d\sigma_y \\ &\leq \frac{1}{s_n} \int_{\partial\Omega} \frac{|\psi(y)|}{|x-y|^{n-1}}\, d\sigma_y \leq \frac{1}{s_n} \frac{1}{||x|-r|^{n-1}} \int_{\partial\Omega} |\psi(y)|\, d\sigma_y \end{aligned}$$

for all $x \in \mathbb{R}^n \setminus \overline{\mathbb{B}_n(0,r)}$, and thus $\lim_{x\to\infty} w_{\overline{\Omega}}[\psi](x) = 0$ and we conclude that $w_{\overline{\Omega}}[\psi]$ is harmonic at infinity. □

We also note that the following known technical lemma holds (see Cialdea [51, p. 187], where (ii) is proved and interpreted in the theory of conjugate differential forms).

Lemma 4.29. *Let $\alpha \in]0,1]$. Let Ω be a bounded open subset of $\mathbb{R}^n$ of class $C^{1,\alpha}$. Then the following statements hold.*

(i) If $\psi \in C^{0,\alpha}(\partial\Omega)$, then we have

$$w_\Omega[\psi](x) = -\sum_{j=1}^n \frac{\partial}{\partial x_j}\left\{ \int_{\partial\Omega} \psi(y)(\nu_\Omega(y))_j S_n(x-y)\, d\sigma_y \right\}$$

for all $x \in \mathbb{R}^n \setminus \partial\Omega$.
(ii) If $\psi \in C^{1,\alpha}(\partial\Omega)$, then we have

$$\frac{\partial}{\partial x_l} w_\Omega[\psi](x) = \sum_{j=1}^n \frac{\partial}{\partial x_j}\left\{ \int_{\partial\Omega} M_{lj}[\psi](y) S_n(x-y)\, d\sigma_y \right\}$$

for all $x \in \mathbb{R}^n \setminus \partial\Omega$ and $l \in \{1,\dots,n\}$.

Proof. (i) Since $D_x^\eta\left(\nu_\Omega(y) \cdot \nabla S_n(x-y)\psi(y)\right)$ is continuous in $(\mathbb{R}^n \setminus \partial\Omega) \times \partial\Omega$ for all $\eta \in \mathbb{N}^n$, then the classical differentiation theorem for integrals depending on a parameter and the obvious equality

$$\frac{\partial}{\partial y_j}(S_n(x-y)) = -\frac{\partial}{\partial x_j}(S_n(x-y)) \qquad \forall (x,y) \in \mathbb{R}^n \times \mathbb{R}^n,\ x \neq y\,,$$

imply the validity of statement (i).

(ii) Since $S_n(x - \cdot)$ is harmonic in $\mathbb{R}^n \setminus \{x\}$, the above equality implies that

$$\begin{aligned}
&\frac{\partial}{\partial x_l}\left[\sum_{j=1}^{n}(\nu_\Omega(y))_j \frac{\partial}{\partial y_j}(S_n(x-y))\right] \\
&\quad = \sum_{j=1}^{n}\left[(\nu_\Omega(y))_j \frac{\partial}{\partial y_l} - (\nu_\Omega(y))_l \frac{\partial}{\partial y_j}\right]\left(\frac{\partial}{\partial x_j}(S_n(x-y))\right)
\end{aligned}$$

for all $x \in \mathbb{R}^n \setminus \partial\Omega$ and $y \in \partial\Omega$. Now we fix $x \in \mathbb{R}^n \setminus \partial\Omega$. Since both ψ and $\frac{\partial}{\partial x_j} S_n(x - \cdot)$ are of class $C^1(\partial\Omega)$, the above equality, and Lemma 2.86, and the classical differentiation theorem for integrals depending on a parameter imply that

$$\begin{aligned}
\frac{\partial}{\partial x_l} w_\Omega[\psi](x) &= \int_{\partial\Omega} \psi(y) \sum_{j=1}^{n} M_{jl,y}\left[\frac{\partial}{\partial x_j}(S_n(x-y))\right](y)\, d\sigma_y \\
&= \sum_{j=1}^{n} \int_{\partial\Omega} M_{lj}[\psi](y) \frac{\partial}{\partial x_j} S_n(x-y)\, d\sigma_y \\
&= \sum_{j=1}^{n} \frac{\partial}{\partial x_j} \int_{\partial\Omega} M_{lj}[\psi](y) S_n(x-y)\, d\sigma_y
\end{aligned}$$

and thus the proof is complete. □

Next we introduce the following known jump formula for the limiting value of the double layer potential on the boundary.

Theorem 4.30. *Let $\alpha \in]0,1]$. Let Ω be a bounded open subset of $\mathbb{R}^n$ of class $C^{1,\alpha}$. Let $\psi \in C^{0,\alpha}(\partial\Omega)$. Let $\tilde{x} \in \partial\Omega$. Then we have the following jump formula for the double layer potential*

$$\lim_{t \to 0^{\mp}} w_\Omega[\psi](\tilde{x} + t\nu_\Omega(\tilde{x})) = \pm\frac{1}{2}\psi(\tilde{x}) + w_\Omega[\psi](\tilde{x})$$

(cf. (2.1)).

Proof. We only prove the formula for the limit in case $t < 0$, i.e., in the case where $\tilde{x} + t\nu_\Omega(\tilde{x}) \in \Omega^+ = \Omega$. The proof for $t > 0$ follows same the lines and is accordingly omitted. By Lemma 4.29 (i), we have

$$w_\Omega[\psi](x) = -\sum_{j=1}^{n} \frac{\partial}{\partial x_j} v_\Omega[\psi \cdot (\nu_\Omega)_j](x) \qquad \forall x \in \Omega\,.$$

Hence, the jump formula for the jth-partial derivative of a single layer potential of Theorem 4.24 (i) implies that

$$\begin{aligned}
\lim_{t\to 0^-} w_\Omega[\psi](\tilde{x}+t\nu_\Omega(\tilde{x})) &= -\sum_{j=1}^n \lim_{t\to 0^-} \frac{\partial}{\partial x_j} v_\Omega[\psi\cdot(\nu_\Omega)_j](\tilde{x}+t\nu_\Omega(\tilde{x})) \\
&= -\sum_{j=1}^n \left[-\frac{1}{2}\psi(\tilde{x})(\nu_\Omega(\tilde{x}))_j^2 + \int_{\partial\Omega} \frac{\partial}{\partial x_j} S_n(\tilde{x}-y)\psi(y)(\nu_\Omega(y))_j\, d\sigma_y\right] \\
&= \frac{1}{2}\psi(\tilde{x}) - \int_{\partial\Omega}\sum_{j=1}^n \frac{\partial}{\partial x_j} S_n(\tilde{x}-y)\psi(y)(\nu_\Omega(y))_j\, d\sigma_y \\
&= \frac{1}{2}\psi(\tilde{x}) + w_\Omega[\psi](\tilde{x})\,.
\end{aligned}$$

□

Then we have the following classical Schauder regularity result for the restriction of the double layer potential to $\Omega^+ \equiv \Omega$ and to the exterior Ω^- of Ω.

Theorem 4.31. *Let $\alpha \in]0,1[$. Let Ω be a bounded open subset of $\mathbb{R}^n$ of class $C^{1,\alpha}$.*

(i) If $\psi \in C^{1,\alpha}(\partial\Omega)$, then the restriction $w_\Omega[\psi]_{|\Omega}$ extends to a function $w_\Omega^+[\psi]$ of class $C^{1,\alpha}(\overline{\Omega})$. Moreover, the map from $C^{1,\alpha}(\partial\Omega)$ to $C^{1,\alpha}(\overline{\Omega})$ that takes ψ to $w_\Omega^+[\psi]$ is linear and continuous.

(ii) If $\psi \in C^{1,\alpha}(\partial\Omega)$, then the restriction $w_\Omega[\psi]_{|\Omega^-}$ extends to a function $w_\Omega^-[\psi]$ of $C^{1,\alpha}_{\mathrm{loc}}(\overline{\Omega^-})$. Moreover, if $r \in]0,+\infty[$ and $\overline{\Omega} \subseteq \mathbb{B}_n(0,r)$, then the map from $C^{1,\alpha}(\partial\Omega)$ to $C^{1,\alpha}(\overline{\mathbb{B}_n(0,r)}\setminus\Omega)$ that takes ψ to $w_\Omega^-[\psi]_{|\overline{\mathbb{B}_n(0,r)}\setminus\Omega}$ is linear and continuous.

(iii) If $\psi \in C^{1,\alpha}(\partial\Omega)$ and $\tilde{x} \in \partial\Omega$, then

$$\begin{aligned}
\nu_\Omega(\tilde{x})\cdot\nabla w_\Omega^+[\psi](\tilde{x}) &= \nu_\Omega(\tilde{x})\cdot\nabla w_\Omega^-[\psi](\tilde{x}) \\
&= \sum_{l,j=1}^n \text{p.v.}\int_{\partial\Omega} (\nu_\Omega(\tilde{x}))_l M_{lj}[\psi](y)\frac{\partial}{\partial x_j} S_n(\tilde{x}-y)\, d\sigma_y\,.
\end{aligned}$$

Proof. (i) Since $\psi \in C^{1,\alpha}(\partial\Omega) \subseteq C^{0,\alpha}(\partial\Omega)$ and $(\nu_\Omega)_j \in C^{0,\alpha}(\partial\Omega)$, Theorem 4.25 (i) on the regularity of the single layer potential implies that the single layer potential $v_\Omega^+[\psi\cdot(\nu_\Omega)_j]$ belongs to $C^{1,\alpha}(\overline{\Omega})$. Then, by the formula of Lemma 4.29 (i), the function $w_\Omega[\psi]_{|\Omega}$ equals the restriction to Ω of a function of class $C^{0,\alpha}(\overline{\Omega})$. In particular, $w_\Omega[\psi]_{|\Omega}$ admits a (unique) continuous extension to $\overline{\Omega}$ and such an extension is of class $C^{0,\alpha}(\overline{\Omega})$. Since the pointwise product by $(\nu_\Omega)_j$ is linear and continuous in $C^{0,\alpha}(\partial\Omega)$ and the Schauder regularity Theorem 4.25 (i) implies that v_Ω^+ is linear and continuous from $C^{0,\alpha}(\partial\Omega)$ to $C^{1,\alpha}(\overline{\Omega})$, the formula of Lemma 4.29 (i) implies that the map from $C^{1,\alpha}(\partial\Omega)$ to $C^{0,\alpha}(\overline{\Omega})$ that takes ψ to $w_\Omega^+[\psi]$ is linear and continuous.

If $\psi \in C^{1,\alpha}(\partial\Omega)$, then $M_{lj}[\psi] \in C^{0,\alpha}(\partial\Omega)$ and the Schauder regularity Theorem 4.25 (i) implies that $v_\Omega^+[M_{lj}[\psi]] \in C^{1,\alpha}(\overline{\Omega})$. Then the formula of Lemma 4.29

(ii) implies that $\frac{\partial}{\partial x_l} w_\Omega[\psi]_{|\Omega}$ equals the restriction to Ω of a function of class $C^{0,\alpha}(\overline{\Omega})$ for all $l \in \{1, \dots, n\}$. Hence, $w^+_\Omega[\psi] \in C^{1,\alpha}(\overline{\Omega})$. Since $M_{lj}[\cdot]$ is linear and continuous from $C^{1,\alpha}(\partial\Omega)$ to $C^{0,\alpha}(\partial\Omega)$ and v^+_Ω is linear and continuous from $C^{0,\alpha}(\partial\Omega)$ to $C^{1,\alpha}(\overline{\Omega})$, the formula of Lemma 4.29 (ii) implies that $\frac{\partial}{\partial x_l} w^+_\Omega[\cdot]$ is continuous from $C^{1,\alpha}(\partial\Omega)$ to $C^{0,\alpha}(\overline{\Omega})$ for all $l \in \{1, \dots, n\}$. Then the above proved continuity of $w^+_\Omega[\cdot]$ implies that $w^+_\Omega[\cdot]$ is linear and continuous from $C^{1,\alpha}(\partial\Omega)$ to $C^{1,\alpha}(\overline{\Omega})$.

Then statement (ii) can be deduced applying statement (i) to the open bounded set $\mathbb{B}_n(0,r)\setminus\overline{\Omega}$ of class $C^{1,\alpha}$. Indeed the map from $C^{1,\alpha}(\partial\Omega)$ to $C^{1,\alpha}(\partial(\mathbb{B}_n(0,r)\setminus\overline{\Omega}))$ that takes ψ to

$$\tilde{\psi}(x) \equiv \begin{cases} \psi(x) & \text{if } x \in \partial\Omega\,, \\ 0 & \text{if } x \in \partial\mathbb{B}_n(0,r) \end{cases}$$

is linear and continuous and we have

$$w^-_\Omega[\psi]_{|\overline{\mathbb{B}_n(0,r)}\setminus\Omega} = -w^+_{\mathbb{B}_n(0,r)\setminus\overline{\Omega}}[\tilde{\psi}]$$

for all $\psi \in C^{1,\alpha}(\partial\Omega)$.

(iii) By statement (i), by Lemma 4.29 (ii), by the jump formula for the derivatives of the single layer potential of Theorem 4.24, and by the skew-symmetry of the matrix $(M_{lj}[\psi])_{l,j=1,\dots,n}$ we have

$$\begin{aligned}
\nu_\Omega(\tilde{x}) \cdot \nabla w^+_\Omega[\psi](\tilde{x}) &= \sum_{l=1}^n (\nu_\Omega(\tilde{x}))_l \lim_{t\to 0^-} \frac{\partial}{\partial x_l} w^+_\Omega[\psi](\tilde{x} + t\nu_\Omega(\tilde{x})) \\
&= \sum_{l,j=1}^n \lim_{t\to 0^-} (\nu_\Omega(\tilde{x}))_l \frac{\partial}{\partial x_j} v^+_\Omega[M_{lj}[\psi]](\tilde{x} + t\nu_\Omega(\tilde{x})) \\
&= -\sum_{l,j=1}^n \frac{1}{2} (\nu_\Omega(\tilde{x}))_l (\nu_\Omega(\tilde{x}))_j M_{lj}[\psi](\tilde{x}) \\
&\quad + \sum_{l,j=1}^n (\nu_\Omega(\tilde{x}))_l \, \text{p.v.} \int_{\partial\Omega} \frac{\partial}{\partial x_j} S_n(\tilde{x} - y) M_{lj}[\psi](y)\, d\sigma_y \\
&= \text{p.v.} \int_{\partial\Omega} \sum_{lj=1}^n (\nu_\Omega(\tilde{x}))_l M_{lj}[\psi](y) \frac{\partial}{\partial x_j} S_n(\tilde{x} - y)\, d\sigma_y\,.
\end{aligned}$$

The proof of the corresponding equality for $w^-_\Omega[\psi]$ is similar and is accordingly omitted. □

By arguing exactly as in the proof of Corollary 4.26, we can verify the following consequence of the previous statement.

Corollary 4.32. *Let $\alpha \in]0,1[$. Let Ω be a bounded open subset of $\mathbb{R}^n$ of class $C^{1,\alpha}$. Then the map from $C^{1,\alpha}(\partial\Omega)$ to $C^{1,\alpha}_b(\overline{\Omega^-})$ that takes ψ to $w^-_\Omega[\psi]$ is linear and continuous.*

4.6 A Regularizing Property of the Double Layer Potential on the Boundary

In this section, we present some results of Schauder [256, 257] on the restriction of the double layer potential on the boundary. For extensions to potentials corresponding to arbitrary second order elliptic operators with constant coefficients and for an account of other contributions on the topic, we refer to the paper [89] with Dondi. In particular, we wish to mention the contribution of von Wahl [272] in the frame of Sobolev spaces and the corresponding developments of Heinemann [120] for Schauder spaces. We begin with the following result of Schauder [256, Hilfsatz VII, p. 112]. For the sake of completeness, we include a proof.

Theorem 4.33. *Let $\alpha \in]0,1[$. Let Ω be a bounded open subset of $\mathbb{R}^n$ of class $C^{1,\alpha}$. Then the following statements hold.*

(i) Let $\psi \in L^\infty(\partial\Omega)$. Then the function $W_\Omega[\psi]$ from $\partial\Omega$ to $\mathbb{R}$ defined by

$$W_\Omega[\psi](x) \equiv -\int_{\partial\Omega} \psi(y)\, \nu_\Omega(y) \cdot \nabla S_n(x-y)\, d\sigma_y \qquad \forall x \in \partial\Omega \tag{4.71}$$

belongs to $C^{0,\alpha}(\partial\Omega)$.

(ii) The operator W_Ω from $L^\infty(\partial\Omega)$ to $C^{0,\alpha}(\partial\Omega)$ that takes ψ to $W_\Omega[\psi]$ is linear and continuous.

Proof. We first prove statement (i). Since $\psi \in L^\infty(\partial\Omega)$, inequality (4.70) implies that the function $\frac{\partial}{\partial\nu_{\Omega_y}}(S_n(x-y))\psi(y)$ is Lebesgue integrable in $y \in \partial\Omega$ for all $x \in \partial\Omega$ and the Vitali Convergence Theorem A.11 and Lemma 4.21 imply that $W_\Omega[\psi]$ is continuous on $\partial\Omega$. By inequality (4.70) and by Lemma 2.53 (i), we have

$$\sup_{\partial\Omega} |W_\Omega[\psi]| \leq \frac{c_{\Omega,\alpha} c'_{\Omega,(n-1)-\alpha}}{s_n} \|\psi\|_{L^\infty(\partial\Omega)} \, . \tag{4.72}$$

We now turn to estimate the Hölder quotient of $W_\Omega[\psi]$. Let $x', x'' \in \partial\Omega$. By Remark 2.20, there is no loss of generality in assuming that $0 < |x'-x''| \leq 1$. Then we have

$$\begin{aligned}
&|W_\Omega[\psi](x') - W_\Omega[\psi](x'')| \\
&\leq \left| \int_{\mathbb{B}_n(x',2|x'-x''|)\cap\partial\Omega} \frac{\partial}{\partial\nu_{\Omega_y}}(S_n(x'-y))\psi(y)\, d\sigma_y \right| \\
&+ \left| \int_{\mathbb{B}_n(x',2|x'-x''|)\cap\partial\Omega} \frac{\partial}{\partial\nu_{\Omega_y}}(S_n(x''-y))\psi(y)\, d\sigma_y \right| \\
&+ \left| \int_{\partial\Omega\setminus\mathbb{B}_n(x',2|x'-x''|)} \left[\frac{\partial}{\partial\nu_{\Omega_y}}(S_n(x'-y)) - \frac{\partial}{\partial\nu_{\Omega_y}}(S_n(x''-y)) \right] \psi(y)\, d\sigma_y \right| .
\end{aligned} \tag{4.73}$$

By inequality (4.70) and Lemma 2.53 (ii) with $\lambda \equiv (n-1)-\alpha$, we have

$$\left|\int_{\mathbb{B}_n(x',2|x'-x''|)\cap\partial\Omega}\frac{\partial}{\partial\nu_{\Omega_y}}(S_n(x'-y))\psi(y)\,d\sigma_y\right|$$
$$\leq \|\psi\|_{L^\infty(\partial\Omega)}\frac{c_{\Omega,\alpha}c''_{\Omega,(n-1)-\alpha}}{s_n}2^\alpha|x'-x''|^\alpha \qquad \text{and}$$
$$\left|\int_{\mathbb{B}_n(x',2|x'-x''|)\cap\partial\Omega}\frac{\partial}{\partial\nu_{\Omega_y}}(S_n(x''-y))\psi(y)\,d\sigma_y\right|$$
$$\leq \int_{\mathbb{B}_n(x'',3|x'-x''|)\cap\partial\Omega}\left|\frac{\partial}{\partial\nu_{\Omega_y}}(S_n(x''-y))\right||\psi(y)|\,d\sigma_y$$
$$\leq \|\psi\|_{L^\infty(\partial\Omega)}\frac{c_{\Omega,\alpha}c''_{\Omega,(n-1)-\alpha}}{s_n}3^\alpha|x'-x''|^\alpha\,. \tag{4.74}$$

Now we consider the third term in the right-hand side of inequality (4.73), and we note that

$$\left|\int_{\partial\Omega\setminus\mathbb{B}_n(x',2|x'-x''|)}\left[\frac{\partial}{\partial\nu_{\Omega_y}}(S_n(x'-y))-\frac{\partial}{\partial\nu_{\Omega_y}}(S_n(x''-y))\right]\psi(y)\,d\sigma_y\right|$$
$$\leq\frac{1}{s_n}\left|\int_{\partial\Omega\setminus\mathbb{B}_n(x',2|x'-x''|)}\frac{\nu_\Omega(y)\cdot(x'-x'')}{|x'-y|^n}\psi(y)\,d\sigma_y\right| \tag{4.75}$$
$$+\frac{1}{s_n}\left|\int_{\partial\Omega\setminus\mathbb{B}_n(x',2|x'-x''|)}\left[\frac{1}{|x'-y|^n}-\frac{1}{|x''-y|^n}\right]\nu_\Omega(y)\cdot(x''-y)\psi(y)\,d\sigma_y\right|.$$

By Lemmas 2.76 (iii), 2.77, and 2.53 (iii), we have

$$\frac{1}{s_n}\left|\int_{\partial\Omega\setminus\mathbb{B}_n(x',2|x'-x''|)}\frac{\nu_\Omega(y)\cdot(x'-x'')}{|x'-y|^n}\psi(y)\,d\sigma_y\right| \tag{4.76}$$
$$\leq\frac{1}{s_n}\int_{\partial\Omega\setminus\mathbb{B}_n(x',2|x'-x''|)}\left(\frac{|\nu_\Omega(y)-\nu_\Omega(x')||x'-x''|}{|x'-y|^n}\right.$$
$$\left.+\frac{|\nu_\Omega(x')\cdot(x'-x'')|}{|x'-y|^n}\right)|\psi(y)|\,d\sigma_y$$
$$\leq\frac{\|\psi\|_{L^\infty(\partial\Omega)}}{s_n}\int_{\partial\Omega\setminus\mathbb{B}_n(x',2|x'-x''|)}\frac{|\nu_\Omega|_\alpha|x'-y|^\alpha}{|x'-y|^n}$$
$$+\frac{c_{\Omega,\alpha}|x'-x''|^\alpha}{|x'-y|^n}\,d\sigma_y|x'-x''|$$
$$\leq\frac{\|\psi\|_{L^\infty(\partial\Omega)}}{s_n}|x'-x''|(|\nu_\Omega|_\alpha+c_{\Omega,\alpha})\int_{\partial\Omega\setminus\mathbb{B}_n(x',2|x'-x''|)}\frac{d\sigma_y}{|x'-y|^{n-\alpha}}$$
$$\leq\frac{\|\psi\|_{L^\infty(\partial\Omega)}}{s_n}(|\nu_\Omega|_\alpha+c_{\Omega,\alpha})c'''_{\Omega,n-\alpha}2^{\alpha-1}|x'-x''|^\alpha\,.$$

By Lemma 2.52 (i), we have

$$|x''-y|\leq 2|x'-y|$$

for all $y \in \partial\Omega \setminus \mathbb{B}_n(x', 2|x' - x''|)$ and thus Lemma 2.52 (iii) with $\lambda = -n$, Lemma 2.53 (iii) with $\lambda = n - \alpha$, and Lemma 2.76 imply that there exists $q_{-n} \in]0, +\infty[$ such that

$$\begin{aligned}
&\frac{1}{s_n}\left|\int_{\partial\Omega\setminus\mathbb{B}_n(x',2|x'-x''|)}\left[\frac{1}{|x'-y|^n}-\frac{1}{|x''-y|^n}\right]\nu_\Omega(y)\cdot(x''-y)\psi(y)\,d\sigma_y\right| \\
&\quad\le \frac{\|\psi\|_{L^\infty(\partial\Omega)}}{s_n}2^{1+\alpha}q_{-n}|x'-x''|c_{\Omega,\alpha}\int_{\partial\Omega\setminus\mathbb{B}_n(x',2|x'-x''|)}\frac{1}{|x'-y|^{n-\alpha}}\,d\sigma_y \\
&\quad\le \frac{c_{\Omega,\alpha}\|\psi\|_{L^\infty(\partial\Omega)}}{s_n}2^{1+\alpha}q_{-n}c'''_{\Omega,n-\alpha}2^{\alpha-1}|x'-x''|^\alpha\,.
\end{aligned} \tag{4.77}$$

Then, by combining inequalities (4.73)–(4.77), we deduce that $W_\Omega[\psi]$ is α-Hölder continuous and that statement (i) holds. Statement (ii) follows by the same inequalities and by inequality (4.72) (see also Remark 2.20). □

In particular, Theorem 4.33 implies that W_Ω maps continuous functions to Hölder continuous functions. We note that if $\psi \in C^0(\partial\Omega)$, then

$$W_\Omega[\psi](x) = w_\Omega[\psi](x) \qquad \forall x \in \partial\Omega\,.$$

We now investigate what happens when we apply W_Ω to continuously differentiable functions and to functions of class $C^{1,\beta}$, with $\beta \in]0, 1]$.

To do so, we first introduce an auxiliary operator in the following Lemma 4.34. In what follows, we denote by $D_{\partial\Omega}$ the subset of $\partial\Omega \times \partial\Omega$ consisting of the pairs (x, y) with $x = y$. Namely,

$$D_{\partial\Omega} \equiv \{(x, x) \in \partial\Omega \times \partial\Omega\}\,.$$

Lemma 4.34. *Let $r \in \{1, \ldots, n\}$. Then the following statements hold.*

(i) *Let $\theta \in]0, 1]$. Let Ω be a bounded open Lipschitz subset of $\mathbb{R}^n$. Let (f, μ) belong to $C^{0,\theta}(\overline{\Omega}) \times C^0(\partial\Omega)$. Let $H_r^\sharp[f]$ be the function from $(\overline{\Omega} \times \partial\Omega) \setminus D_{\partial\Omega}$ to $\mathbb{R}$ defined by*

$$H_r^\sharp[f](x, y) \equiv (f(x) - f(y))\frac{\partial}{\partial x_r}(S_n(x - y)) \;\; \forall (x, y) \in (\overline{\Omega} \times \partial\Omega) \setminus D_{\partial\Omega}\,.$$

If $x \in \overline{\Omega}$, then the function $H_r^\sharp[f](x, \cdot)$ is Lebesgue integrable in $\partial\Omega$ and the function $Q_r^\sharp[f, \mu]$ from $\overline{\Omega}$ to $\mathbb{R}$ defined by

$$Q_r^\sharp[f, \mu](x) \equiv \int_{\partial\Omega} H_r^\sharp[f](x, y)\mu(y)\,d\sigma_y \qquad \forall x \in \overline{\Omega}$$

is continuous.

(ii) *Let $\alpha \in]0, 1[$, and $\beta, \theta \in]0, 1]$. Let Ω be of class $C^{1,\alpha}$. If (f, μ) belongs to $C^{0,\theta}(\overline{\Omega}) \times C^{0,\beta}(\partial\Omega)$, then $Q_r^\sharp[f, \mu] \in C^{0,\min\{\alpha,\beta,\theta\}}(\overline{\Omega})$.*

Proof. Since $\frac{\partial S_n}{\partial x_r}(x-y) = \frac{1}{s_n}\frac{x_r-y_r}{|x-y|^n}$ and f is Hölder continuous with exponent θ, we have

$$|H_r^\sharp[f](x,y)| \leq \frac{|f|_\theta}{s_n}\frac{1}{|x-y|^{(n-1)-\theta}} \qquad \forall (x,y) \in (\overline{\Omega} \times \partial\Omega) \setminus D_{\partial\Omega}\,.$$

Then $H_r^\sharp[f](x,\cdot)$ is Lebesgue integrable in $\partial\Omega$ for all $x \in \overline{\Omega}$ and the Vitali Convergence Theorem A.11 and Lemma 4.21 imply that $Q_r^\sharp[f,\mu]$ is continuous in $\overline{\Omega}$.

We now consider statement (ii). By treating separately the case where $x \in \partial\Omega$ and the case where $x \in \Omega$ and by exploiting the jump formulas for the partial derivatives of the single layer potential of Theorem 4.24 and the Schauder regularity Theorem 4.25 for the single layer potential, we have

$$Q_r^\sharp[f,\mu](x) = f(x)\frac{\partial}{\partial x_r}v_\Omega^+[\mu](x) - \frac{\partial}{\partial x_r}v_\Omega^+[f\mu](x) \qquad \forall x \in \overline{\Omega}\,.$$

Then, by the continuity of the pointwise product in Hölder spaces of Proposition 2.24 and by Theorem 4.25 (i), we have $Q_r^\sharp[f,\mu] \in C^{0,\min\{\alpha,\beta,\theta\}}(\overline{\Omega})$. □

Then we can prove the following theorem on the action of $Q_r^\sharp$ on the boundary.

Theorem 4.35. *Let $r \in \{1,\ldots,n\}$. Let $\alpha \in]0,1[$ and $\beta \in]0,1]$. Let Ω be a bounded open subset of $\mathbb{R}^n$ of class $C^{1,\alpha}$. Then the following statements hold.*

(i) Let (g,μ) be a pair of functions in $C^{0,\alpha}(\partial\Omega) \times C^{0,\beta}(\partial\Omega)$. Let $H_r[g]$ be the function from $(\partial\Omega \times \partial\Omega) \setminus D_{\partial\Omega}$ to $\mathbb{R}$ defined by

$$H_r[g](x,y) \equiv (g(x)-g(y))\frac{\partial}{\partial x_r}(S_n(x-y)) \; \forall (x,y) \in (\partial\Omega \times \partial\Omega) \setminus D_{\partial\Omega}\,.$$

If $x \in \partial\Omega$, then the function $H_r[g](x,\cdot)$ is Lebesgue integrable in $\partial\Omega$, and the function $Q_r[g,\mu]$ from $\partial\Omega$ to $\mathbb{R}$ defined by

$$Q_r[g,\mu](x) \equiv \int_{\partial\Omega} H_r[g](x,y)\mu(y)\,d\sigma_y \qquad \forall x \in \partial\Omega$$

belongs to $C^{0,\alpha}(\partial\Omega)$.

(ii) The bilinear map Q_r from $C^{0,\alpha}(\partial\Omega) \times C^{0,\beta}(\partial\Omega)$ to $C^{0,\alpha}(\partial\Omega)$ that takes a pair (g,μ) to $Q_r[g,\mu]$ is continuous.

Proof. Let $\rho \in]0,+\infty[$ be such that $\overline{\Omega} \subseteq \mathbb{B}_n(0,\rho)$. Let '$\tilde{\ }$' be the extension operator defined on $C^{1,\alpha}(\partial\Omega)$ as in Theorem 2.85 (ii) with $\Omega' = \mathbb{B}_n(0,\rho)$, $h = 0$. Since $Q_r[g,\mu] = Q_r^\sharp[\tilde{g},\mu]$ on $\partial\Omega$, the integrability of $H_r[g](x,\cdot)$ and the continuity of $Q_r[g,\mu]$ follow by Lemma 4.34 (i). Since

$$\frac{\partial S_n}{\partial x_r}(x-y) = \frac{1}{s_n}\frac{x_r-y_r}{|x-y|^n}$$

and g is Hölder continuous with exponent α, Lemma 2.53 (i) implies that

$$\sup_{\partial\Omega} |Q_r[g,\mu]| \leq \frac{c'_{\Omega,(n-1)-\alpha}}{s_n} |g|_\alpha \|\mu\|_{C^0(\partial\Omega)} \,. \tag{4.78}$$

We now turn to estimate the Hölder quotient of $Q_r[g,\mu]$. By the Hölder continuity of g, we have

$$\begin{aligned} &|(g(x') - g(y))(x'_r - y_r) - (g(x'') - g(y))(x''_r - y_r)| \\ &\qquad \leq |g|_\alpha \left\{ |x' - y|^\alpha |x' - y| + |x'' - y|^\alpha |x'' - y| \right\} \end{aligned} \tag{4.79}$$

for all $x', x'', y \in \partial\Omega$. Now let $x', x'' \in \partial\Omega$. Then we have

$$\begin{aligned} &|Q_r[g,\mu](x') - Q_r[g,\mu](x'')| \\ &\quad \leq \left| \int_{\mathbb{B}_n(x',2|x'-x''|)\cap\partial\Omega} H_r[g](x',y)\mu(y)\,d\sigma_y \right| \\ &\qquad + \left| \int_{\mathbb{B}_n(x',2|x'-x''|)\cap\partial\Omega} H_r[g](x'',y)\mu(y)\,d\sigma_y \right| \\ &\qquad + \left| \int_{\partial\Omega\setminus\mathbb{B}_n(x',2|x'-x''|)} [H_r[g](x',y) - H_r[g](x'',y)]\mu(y)\,d\sigma_y \right| . \end{aligned} \tag{4.80}$$

By Lemma 2.53 (ii) with $\lambda \equiv (n-1) - \alpha$, we have

$$\begin{aligned} &\left| \int_{\mathbb{B}_n(x',2|x'-x''|)\cap\partial\Omega} H_r[g](x',y)\mu(y)\,d\sigma_y \right| \\ &\qquad \leq \|\mu\|_{L^\infty(\partial\Omega)} \frac{|g|_\alpha}{s_n} c''_{\Omega,(n-1)-\alpha} 2^\alpha |x' - x''|^\alpha \end{aligned} \tag{4.81}$$

and

$$\begin{aligned} &\left| \int_{\mathbb{B}_n(x',2|x'-x''|)\cap\partial\Omega} H_r[g](x'',y)\mu(y)\,d\sigma_y \right| \\ &\qquad \leq \int_{\mathbb{B}_n(x'',3|x'-x''|)\cap\partial\Omega} \left| H_r[g](x'',y) \right| |\mu(y)|\,d\sigma_y \\ &\qquad \leq \|\mu\|_{L^\infty(\partial\Omega)} \frac{|g|_\alpha}{s_n} c''_{\Omega,(n-1)-\alpha} 3^\alpha |x' - x''|^\alpha \,. \end{aligned} \tag{4.82}$$

Next we turn to estimate the third integral in the right-hand side of inequality (4.80). Since $C^{0,\beta_1}(\partial\Omega)$ is continuously embedded into $C^{0,\beta_2}(\partial\Omega)$ whenever $0 < \beta_2 \leq \beta_1$, then we can assume that $\alpha + \beta < 1$. We observe that the third integral in the right-hand side of inequality (4.80) equals

$$J_1(x',x'') + \mu(x')J_2(x',x'') \tag{4.83}$$

where

$J_1(x',x'')$

$$\equiv \int_{\partial\Omega\setminus\mathbb{B}_n(x',2|x'-x''|)} [H_r[g](x',y) - H_r[g](x'',y)](\mu(y) - \mu(x'))\, d\sigma_y \,,$$

$$J_2(x',x'') \equiv \int_{\partial\Omega\setminus\mathbb{B}_n(x',2|x'-x''|)} [H_r[g](x',y) - H_r[g](x'',y)]\, d\sigma_y \,. \tag{4.84}$$

In order to estimate J_1, we observe that

$$\begin{aligned} &\Big| H_r[g](x',y) - H_r[g](x'',y) \Big| \\ &\qquad \leq \frac{1}{s_n} |(g(x') - g(y))(x'_r - y_r) - (g(x'') - g(y))(x''_r - y_r)| \frac{1}{|x'-y|^n} \\ &\qquad + \frac{1}{s_n} |(g(x'') - g(y))(x''_r - y_r)| \left| \frac{1}{|x'-y|^n} - \frac{1}{|x''-y|^n} \right| . \end{aligned} \tag{4.85}$$

Now, we have $|x'-x''| \leq |x'-y|$ and Lemma 2.52 (i) implies that $|x''-y| \leq 2|x'-y|$. Then, combining Lemma 2.52 (iii) with $\lambda = -n$ and inequality (4.79), we conclude that there exists $q_{-n} \in]0,+\infty[$ such that the right-hand side of (4.85) is less or equal to

$$\begin{aligned} &\frac{|g|_\alpha}{s_n} \frac{|x'-y|^\alpha |x'-y| + |x''-y|^\alpha |x''-y|}{|x'-y|^n} \\ &\qquad + \frac{|g|_\alpha}{s_n} |x''-y|^{\alpha+1} q_{-n} |x'-x''| |x'-y|^{-n-1} \\ &\qquad \leq \frac{2^{1+\alpha}|g|_\alpha}{s_n} (2 + q_{-n}) \frac{1}{|x'-y|^{(n-1)-\alpha}} . \end{aligned} \tag{4.86}$$

Then Lemma 2.53 (iii) with $\lambda = n - (\alpha + \beta)$ and inequalities (4.85), (4.86) imply that

$$\begin{aligned} &\Big| J_1(x',x'') \Big| \\ &\leq |\mu|_\beta \int_{\partial\Omega\setminus\mathbb{B}_n(x',2|x'-x''|)} \frac{2^{1+\alpha}|g|_\alpha}{s_n} (2 + q_{-n}) \frac{1}{|x'-y|^{(n-1)-\alpha-\beta}}\, d\sigma_y \\ &\leq \frac{2^{1+\alpha}(2+q_{-n})}{s_n} |g|_\alpha |\mu|_\beta c'''_{\Omega,n-(\alpha+\beta)} 2^{\alpha+\beta} |x'-x''|^{\alpha+\beta} . \end{aligned} \tag{4.87}$$

We now consider $J_2(x',x'')$ and we note that

$$\begin{aligned} Q_r[g,1](x) &= \int_{\partial\Omega} (g(x) - g(y)) \frac{\partial}{\partial x_r}(S_n(x-y))\, d\sigma_y \\ &= g(x)\text{p.v.} \int_{\partial\Omega} \frac{\partial}{\partial x_r}(S_n(x-y))\, d\sigma_y - \text{p.v.} \int_{\partial\Omega} \frac{\partial}{\partial x_r}(S_n(x-y)) g(y)\, d\sigma_y \end{aligned} \tag{4.88}$$

for all $x \in \partial\Omega$, where the last two integrals exist in the sense of the principal value. Since Ω is of class $C^{1,\alpha}$ and $g \in C^{0,\alpha}(\partial\Omega)$, the following equality holds

$$\frac{\partial}{\partial x_r} v_\Omega^+[g](x) = -\frac{1}{2}(\nu_\Omega)_r(x)g(x) + \text{p.v.} \int_{\partial\Omega} \frac{\partial}{\partial x_r}(S_n(x-y))g(y)\, d\sigma_y \quad (4.89)$$

for all $x \in \partial\Omega$ (see Theorems 4.24, 4.25). By the Schauder regularity Theorem 4.25 (i) for the single layer potential, the function $\frac{\partial}{\partial x_r} v_\Omega^+[g](x)$ of the variable $x \in \partial\Omega$ is of class $C^{0,\alpha}$. Since $(\nu_\Omega)_r g \in C^{0,\alpha}(\partial\Omega)$, then equality (4.89) implies that the principal value

$$\text{p.v.} \int_{\partial\Omega} \frac{\partial}{\partial x_r}(S_n(x-y))g(y)\, d\sigma_y$$

defines a function of class $C^{0,\alpha}$ in the variable $x \in \partial\Omega$. The same clearly holds if we replace g with the constant 1. Hence, equalities (4.88) and (4.89) imply that $Q_r[g,1] \in C^{0,\alpha}(\partial\Omega)$. Next we note that

$$\begin{aligned} |J_2(x',x'')| \leq {} & |Q_r[g,1](x') - Q_r[g,1](x'')| \\ & + \left| \int_{\mathbb{B}_n(x',2|x'-x''|)\cap\partial\Omega} H_r[g](x',y)\, d\sigma_y \right| \\ & + \left| \int_{\mathbb{B}_n(x',2|x'-x''|)\cap\partial\Omega} H_r[g](x'',y)\, d\sigma_y \right| . \end{aligned} \quad (4.90)$$

Since $\mathbb{B}_n(x',2|x'-x''|) \subseteq \mathbb{B}_n(x'',3|x'-x''|)$, inequalities (4.81) and (4.82) with $\mu = 1$ imply that

$$|J_2(x',x'')| \leq |Q_r[g,1] : \partial\Omega|_\alpha |x'-x''|^\alpha + 2\frac{|g|_\alpha}{s_n} c''_{\Omega,(n-1)-\alpha} 3^\alpha |x'-x''|^\alpha . \quad (4.91)$$

Combining inequalities (4.87) and (4.91), we conclude that

$$\begin{aligned} & \left| \int_{\partial\Omega\setminus\mathbb{B}_n(x',2|x'-x''|)} [H_r[g](x',y) - H_r[g](x'',y)]\mu(y)\, d\sigma_y \right| \\ & \leq \left\{ \frac{2^{1+\alpha}(2+q_{-n})}{s_n} |g|_\alpha |\mu|_\beta c'''_{\Omega,n-(\alpha+\beta)} 2^{\alpha+\beta} (\operatorname{diam}\Omega)^\beta \right. \\ & \left. + \|\mu\|_{L^\infty(\partial\Omega)} \left[|Q_r[g,1] : \partial\Omega|_\alpha + 3^{1+\alpha}\frac{|g|_\alpha}{s_n} c''_{\Omega,(n-1)-\alpha} \right] \right\} |x'-x''|^\alpha . \end{aligned} \quad (4.92)$$

Combining inequalities (4.78), (4.80), (4.81), (4.82), and (4.92), we conclude that $Q_r[g,\mu]$ is α-Hölder continuous and that statement (ii) holds true. □

We now exploit Theorem 4.33 to prove a classical formula for the tangential derivatives of W_Ω.

Proposition 4.36. *Let $\alpha \in]0,1[$, $\beta \in]0,\alpha]$. Let Ω be an open bounded subset of $\mathbb{R}^n$ of class $C^{1,\alpha}$. If $\psi \in C^{1,\beta}(\partial\Omega)$, then $W_\Omega[\psi] \in C^{1,\beta}(\partial\Omega)$ and*

$$\begin{aligned} M_{ij}[W_\Omega[\psi]] = {} & \sum_{r=1}^n \Big\{ Q_r[(\nu_\Omega)_i, M_{jr}[\psi]] - Q_r[(\nu_\Omega)_j, M_{ir}[\psi]] \Big\} \\ & + W_\Omega[M_{ij}[\psi]] \qquad \text{on } \partial\Omega , \end{aligned} \quad (4.93)$$

for all $i, j \in \{1, \ldots, n\}$.

Proof. In this proof we find convenient to simplify our notation and set

$$\nu_i \equiv (\nu_\Omega)_i \qquad \forall i \in \{1, \ldots, n\} .$$

By the jump formula of Theorem 4.30, by the Schauder regularity Theorem 4.31 for the double layer potential, and by the formula of Lemma 4.29 (ii) for the partial derivatives of the double layer potential in Ω, we have $w_\Omega^+[\psi] \in C^{1,\beta}(\overline{\Omega})$ and

$$W_\Omega[\psi] = -\frac{1}{2}\psi + w_\Omega^+[\psi]_{|\partial\Omega} \qquad \text{on } \partial\Omega ,$$
$$\frac{\partial}{\partial x_i} w_\Omega^+[\psi] = \sum_{j=1}^{n} \frac{\partial}{\partial x_j} v_\Omega^+[M_{ij}[\psi]] \qquad \text{in } \overline{\Omega}$$

for all $i \in \{1, \ldots, n\}$. In particular, $W_\Omega[\psi] \in C^{1,\beta}(\partial\Omega)$ (cf. Theorem 4.31 (i)). Hence,

$$\begin{aligned} M_{ij}[W_\Omega[\psi]] & \qquad (4.94) \\ &= -\frac{1}{2} M_{ij}[\psi] + M_{ij}[w_\Omega^+[\psi]_{|\partial\Omega}] \\ &= -\frac{1}{2} M_{ij}[\psi] + \nu_i \frac{\partial}{\partial x_j} w_\Omega^+[\psi] - \nu_j \frac{\partial}{\partial x_i} w_\Omega^+[\psi] \\ &= -\frac{1}{2} M_{ij}[\psi] + \sum_{r=1}^{n} \left\{ \nu_i \frac{\partial}{\partial x_r} v_\Omega^+[M_{jr}[\psi]] - \nu_j \frac{\partial}{\partial x_r} v_\Omega^+[M_{ir}[\psi]] \right\} , \end{aligned}$$

on $\partial\Omega$. By the jump formulas for the partial derivatives of the single layer potential of Theorem 4.24, we have

$$\frac{\partial}{\partial x_r} v_\Omega^+[\phi](x) = -\frac{1}{2}\nu_r(x)\phi(x) + \text{p.v.} \int_{\partial\Omega} \frac{\partial}{\partial x_r}(S_n(x-y))\phi(y)\, d\sigma_y \qquad \forall x \in \partial\Omega ,$$

for all $\phi \in C^{0,\beta}(\partial\Omega)$. Hence,

$$\begin{aligned} & M_{ij}[W_\Omega[\psi]](x) \qquad (4.95) \\ &= -\frac{1}{2} M_{ij}[\psi](x) - \frac{1}{2} \left\{ \sum_{r=1}^{n} M_{jr}[\psi](x)\nu_r(x)\nu_i(x) - \sum_{r=1}^{n} M_{ir}[\psi](x)\nu_r(x)\nu_j(x) \right\} \\ &\quad + \sum_{r=1}^{n} \Big\{ \nu_i(x)\text{p.v.} \int_{\partial\Omega} \frac{\partial}{\partial x_r}(S_n(x-y)) M_{jr}[\psi](y)\, d\sigma_y \\ &\quad - \nu_j(x)\text{p.v.} \int_{\partial\Omega} \frac{\partial}{\partial x_r}(S_n(x-y)) M_{ir}[\psi](y)\, d\sigma_y \Big\} \qquad \forall x \in \partial\Omega . \end{aligned}$$

We now consider the first term in braces in the right-hand side of (4.95) and we note that

$$
\begin{aligned}
&\left\{\sum_{r=1}^{n} M_{jr}[\psi]\nu_r\nu_i - \sum_{r=1}^{n} M_{ir}[\psi]\nu_r\nu_j\right\} \qquad (4.96)\\
&\quad = \sum_{r=1}^{n}\left[\nu_j\frac{\partial\psi}{\partial x_r}\nu_r\nu_i - \nu_r^2\frac{\partial\psi}{\partial x_j}\nu_i - \left(\nu_i\frac{\partial\psi}{\partial x_r}\nu_r\nu_j - \nu_r^2\frac{\partial\psi}{\partial x_i}\nu_j\right)\right]\\
&\quad = \sum_{r=1}^{n}\left[-\nu_r^2\frac{\partial\psi}{\partial x_j}\nu_i + \nu_r^2\frac{\partial\psi}{\partial x_i}\nu_j\right]\\
&\quad = -M_{ij}[\psi] \qquad \text{on } \partial\Omega\,.
\end{aligned}
$$

Next we consider the second term in braces in the right-hand side of (4.95) and we note that

$$
\begin{aligned}
&\nu_i(x)M_{jr}[\psi](y) - \nu_j(x)M_{ir}[\psi](y) \qquad (4.97)\\
&\quad = [\nu_i(x)M_{jr}[\psi](y) - \nu_i(y)M_{jr}[\psi](y)] + [\nu_i(y)M_{jr}[\psi](y) - \nu_j(y)M_{ir}[\psi](y)]\\
&\qquad +[\nu_j(y)M_{ir}[\psi](y) - \nu_j(x)M_{ir}[\psi](y)] \qquad \forall x, y \in \partial\Omega
\end{aligned}
$$

and that

$$
\begin{aligned}
&[\nu_i M_{jr}[\psi] - \nu_j M_{ir}[\psi]] \qquad (4.98)\\
&\quad = \nu_i\nu_j\frac{\partial\psi}{\partial x_r} - \nu_i\nu_r\frac{\partial\psi}{\partial x_j} - \nu_j\nu_i\frac{\partial\psi}{\partial x_r} + \nu_j\nu_r\frac{\partial\psi}{\partial x_i} = -\nu_r M_{ij}[\psi] \qquad \text{on } \partial\Omega\,.
\end{aligned}
$$

Then, combining (4.95)–(4.98), we deduce that

$$
\begin{aligned}
&M_{ij}[W_\Omega[\psi]](x)\\
&\quad = \sum_{r=1}^{n}\int_{\partial\Omega}\frac{\partial}{\partial x_r}(S_n(x-y))M_{jr}[\psi](y)(\nu_i(x) - \nu_i(y))\,d\sigma_y\\
&\qquad - \sum_{r=1}^{n}\int_{\partial\Omega}\frac{\partial}{\partial x_r}(S_n(x-y))M_{ij}[\psi](y)\nu_r(y)\,d\sigma_y\\
&\qquad + \sum_{r=1}^{n}\int_{\partial\Omega}\frac{\partial}{\partial x_r}(S_n(x-y))M_{ir}[\psi](y)(\nu_j(y) - \nu_j(x))\,d\sigma_y\\
&\quad = \sum_{r=1}^{n} Q_r[\nu_i, M_{jr}[\psi]](x) + W_\Omega[M_{ij}[\psi]](x)\\
&\qquad - \sum_{r=1}^{n} Q_r[\nu_j, M_{ir}[\psi]](x) \qquad \forall x \in \partial\Omega
\end{aligned}
$$

and thus formula (4.93) holds true. □

Formulas as (4.93) are known and could be deduced by others in the literature. In particular, we mention Cialdea [51, Proof of Theorem I, p. 186], Cialdea et al. [54,

(11) p. 70]. See also Cialdea et al. [55, (66), p. 7] for the double layer hydrodynamic potential.

For a generalization of formula (4.93) for the tangential derivatives of the double layer potential corresponding to second order elliptic homogeneous operators, we refer to Hofmann et al. [125, (6.2.6)]. For inhomogeneous second order elliptic operators, we refer to the paper [89, (9.2)] with Dondi.

For the double layer potential corresponding to the fundamental solution of the heat equation, we refer to the papers [172, 173] with Luzzini.

We now exploit formula (4.93) for the tangential derivatives of W_Ω to prove that the operator W_Ω has a regularizing property on functions of class $C^{1,\beta}(\partial\Omega)$. We do so by means of the following known result (see also Giraud [108, 109], Miranda [213, Theorem. 15.VI p. 42], and the paper [89, Theorem. 9.2] with Dondi).

Theorem 4.37. *Let $\alpha \in]0,1[$, $\beta \in]0,\alpha]$. Let Ω be an open bounded subset of $\mathbb{R}^n$ of class $C^{1,\alpha}$. If $\psi \in C^{1,\beta}(\partial\Omega)$, then $W_\Omega[\psi] \in C^{1,\alpha}(\partial\Omega)$. Moreover, the operator W_Ω from $C^{1,\beta}(\partial\Omega)$ to $C^{1,\alpha}(\partial\Omega)$ that takes ψ to $W_\Omega[\psi]$ is linear and continuous.*

Proof. We denote by $T_{ij}[\psi]$ the right-hand side of formula (4.93) for the tangential derivatives of $W_\Omega[\psi]$. By Lemma 2.87 (iii), it suffices to prove that the following two statements hold.

(j) W_Ω is continuous from $C^{1,\beta}(\partial\Omega)$ to $C^0(\partial\Omega)$.
(jj) $T_{ij}[\cdot]$ is continuous from $C^{1,\beta}(\partial\Omega)$ to $C^{0,\alpha}(\partial\Omega)$ for all $i, j \in \{1,\dots,n\}$.

Theorem 4.33 (ii) implies the validity of statement (j). Since M_{ij} is linear and continuous from $C^{1,\beta}(\partial\Omega)$ to $C^{0,\beta}(\partial\Omega)$ and $C^{0,\beta}(\partial\Omega)$ is continuously embedded into $L^\infty(\partial\Omega)$, Theorem 4.33 (ii) implies that $W_\Omega[M_{ij}[\cdot]]$ is linear and continuous from $C^{1,\beta}(\partial\Omega)$ to $C^{0,\alpha}(\partial\Omega)$. By the membership of $(\nu_\Omega)_i$, $(\nu_\Omega)_j$ in $C^{0,\alpha}(\partial\Omega)$ and by Theorem 4.35 (ii), the operator $Q_r[(\nu_\Omega)_i,\cdot] - Q_r[(\nu_\Omega)_j,\cdot]$ is linear and continuous from $C^{0,\beta}(\partial\Omega)$ to $C^{0,\alpha}(\partial\Omega)$. Since M_{ij} is linear and continuous from $C^{1,\beta}(\partial\Omega)$ to $C^{0,\beta}(\partial\Omega)$, we conclude that the operator $Q_r[(\nu_\Omega)_i, M_{jr}[\cdot]] - Q_r[(\nu_\Omega)_j, M_{ir}[\cdot]]$ is linear and continuous from $C^{1,\beta}(\partial\Omega)$ to $C^{0,\alpha}(\partial\Omega)$, and accordingly statement (jj) holds true. □

Chapter 5
Preliminaries on the Fredholm Alternative Principle

Abstract In the chapters that follow this one, we will exploit certain results of Wendland (Math. Z., 101:61–64, 1967; Methoden Verfahren Math. Phys. 3:141–176,1970), on Fredholm operators in dual systems. The present chapter's purpose is to introduce these results. We also recall some classical theorems of Functional Analysis in Banach spaces. We do not include proofs, however, as these can be found in monographs such as those of Kress (*Linear integral equations*, volume 82 of *Applied Mathematical Sciences*. Springer, New York, third edition, 2014.), Wloka et al. (*Boundary value problems for elliptic systems*. Cambridge University Press, Cambridge, 1995), as well as in the above-mentioned papers of Wendland.

5.1 Fredholm Alternative

Let X be a normed space on $\mathbb{R}$. We denote by X' the topological dual space $\mathcal{L}(X, \mathbb{R})$ of X, so that X' consists of the linear continuos operators from X to $\mathbb{R}$. As usual, the elements of X' are also called functionals.

Definition 5.1. Let X and Y be normed spaces. If L is a linear map from X to Y, then the transpose L^t of L is the linear map from Y' to X' that takes a functional $\psi \in Y'$ to the functional $L^t[\psi] \in X'$ defined by

$$\left(L^t[\psi]\right)[\phi] = \psi\left[L[\phi]\right] \qquad \forall \phi \in X.$$

As is well known, for every bounded operator L there exists a unique transpose operator L^t and it is bounded. In addition, if L is compact (i.e. if it maps bounded subsets of X to subsets of Y with compact closure), then also L^t is compact. We now introduce the definition of Fredholm operator.

Definition 5.2 (Fredholm Operator). We say that a bounded linear operator L from a Banach space X to a Banach space Y is Fredholm if and only if both the following conditions hold.

M. Dalla Riva et al., *Singularly Perturbed Boundary Value Problems*,
https://doi.org/10.1007/978-3-030-76259-9_5

(i) The null space $\operatorname{Ker} L$ has finite dimension.
(ii) The co-kernel $Y/\operatorname{Im} L$ has finite dimension.

If L is a Fredholm operator, then the index of L is the integer defined by

$$\operatorname{index} L \equiv \dim \operatorname{Ker} L - \dim(Y/\operatorname{Im} L)\,.$$

So for example, if a bounded operator L is an isomorphism from X onto Y, then L is a Fredholm operator and has index equal to zero. We observe that the condition that $Y/\operatorname{Im} L$ has finite dimension implies that the image $\operatorname{Im} L$ of L is closed in Y (see, e.g., Wloka et al. [276, Prop. 9.2]).

We shall exploit the following theorems (for a proof we refer, e.g., to Wloka et al. [276, Chap. 9]).

Theorem 5.3 (Compact Perturbation of a Fredholm Operator). *Let X and Y be Banach spaces. If L is a Fredholm operator from X to Y and K is a compact operator from X to Y, then $L+K$ is a Fredholm operator from X to Y and* $\operatorname{index}(L+K) = \operatorname{index} L$.

So for example, if a bounded operator L is an isomorphism from X onto Y and K is a compact operator from X to Y, then $L+K$ is a Fredholm operator of index zero.

Theorem 5.4 (Fredholm Alternative). *Let X and Y be Banach spaces. Let L be a Fredholm operator of index 0 from X to Y. Then exactly one of the following statements hold true.*

(i) The operator L is an isomorphism (i.e., a linear homeomorphism) from X onto Y and the transpose operator L^t is an isomorphism from Y' onto X'.
(ii) The null spaces $\operatorname{Ker} L$ and $\operatorname{Ker} L^t$ have the same nonzero finite dimension and

$$\operatorname{Im} L = \{\phi \in Y \,:\, \psi[\phi] = 0 \text{ for all } \psi \in \operatorname{Ker} L^t\}\,,$$
$$\operatorname{Im} L^t = \{\psi \in X' \,:\, \psi[\phi] = 0 \text{ for all } \phi \in \operatorname{Ker} L\}\,.$$

5.2 Fredholm Alternative in a Dual System

In the following Definitions 5.5, 5.6, and 5.7 we introduce the notion of bilinear form, dual system, and transpose with respect to a dual system, respectively.

Definition 5.5. Let X_1 and X_2 be real Banach spaces. A map $\langle\cdot,\cdot\rangle$ from $X_1 \times X_2$ to $\mathbb{R}$ is a bilinear form if

$$\langle c_1\psi_1 + c_2\psi_2, \phi_1\rangle = c_1\langle\psi_1,\phi_1\rangle + c_2\langle\psi_2,\phi_1\rangle$$

and

$$\langle\psi_1, c_1\phi_1 + c_2\phi_2\rangle = c_1\langle\psi_1,\phi_1\rangle + c_2\langle\psi_1,\phi_2\rangle$$

for all $c_1, c_2 \in \mathbb{R}$, $\psi_1, \psi_2 \in X_1$, and $\phi_1, \phi_2 \in X_2$. A bilinear form is non-degenerate if for all $\psi \in X_1 \setminus \{0\}$ there exists $\phi \in X_2$ such that $\langle \psi, \phi \rangle \neq 0$ and for all $\phi \in X_2 \setminus \{0\}$ there exists $\psi \in X_1$ such that $\langle \psi, \phi \rangle \neq 0$.

Definition 5.6. A dual system consists of a pair of Banach spaces X_1 and X_2 and of a non-degenerate bilinear form $\langle \cdot, \cdot \rangle$ from $X_1 \times X_2$ to $\mathbb{R}$ and it is denoted by $\langle X_1, X_2 \rangle$.

So for example, $\langle X_2', X_2 \rangle$ together with the evaluation map that takes a pair (ψ, ϕ) of $X_2' \times X_2$ to $\psi[\phi]$ is a dual system, the so-called canonical dual system between X_2 and its dual X_2'.

Definition 5.7. Let $\langle X_1, X_2 \rangle$ be a dual system. Let A be a linear operator from X_1 to itself. Let B be a linear operator from X_2 to itself. Then the operator A is said to be transpose to the operator B with respect to the dual system $\langle X_1, X_2 \rangle$ if

$$\langle A[\psi], \phi \rangle = \langle \psi, B[\phi] \rangle$$

for all $\psi \in X_1$ and all $\phi \in X_2$.

So for example, if $X_1 = X_2'$, then B^t is transpose to B with respect to the canonical dual system $\langle X_2', X_2 \rangle$.

Then, in a dual system we have the following Fredholm Alternative Theorem. For a proof, we refer the reader to Wendland [273], [274] and to Kress [149, Chapt. 4].

Theorem 5.8 (Fredholm Alternative in a Dual System). *Let $\langle X_1, X_2 \rangle$ be a dual system. Let K_1 be a linear compact operator from X_1 to itself. Let K_2 be a linear compact operator from X_2 to itself. Assume that K_1 is transpose to K_2 with respect to the dual system $\langle X_1, X_2 \rangle$. Then exactly one of the following statements holds true.*

(i) The operator $I_{X_1} + K_1$ is an isomorphism from X_1 to itself and the operator $I_{X_2} + K_2$ is an isomorphism from X_2 to itself.
(ii) The null spaces $\mathrm{Ker}(I_{X_1} + K_1)$ *and* $\mathrm{Ker}(I_{X_2} + K_2)$ *have the same nonzero finite dimension and*

$$\mathrm{Im}(I_{X_1} + K_1) = \{\psi \in X_1 \,:\, \langle \psi, \phi \rangle = 0 \textit{ for all } \phi \in \mathrm{Ker}(I_{X_2} + K_2)\},$$
$$\mathrm{Im}(I_{X_2} + K_2) = \{\phi \in X_2 \,:\, \langle \psi, \phi \rangle = 0 \textit{ for all } \psi \in \mathrm{Ker}(I_{X_1} + K_1)\}.$$

Chapter 6
Boundary Value Problems and Boundary Integral Operators

Abstract In this chapter we show the existence of solutions to the basic boundary value problems for the Laplace equation using a classical approach based on Potential Theory. Specifically, we consider the Dirichlet problem, the Neumann problem, the Robin problem, the transmission problem, and a mixed problem. To do so, we carry out an analysis of the boundary integral operators associated with the single and double layer potentials in a Schauder space setting. Our presentation of the topic stems from that of Folland (*Introduction to partial differential equations*. Princeton University Press, Princeton, NJ, second edition, 1995, Chap. 3).

6.1 The Geometric Setting

We introduce here some notation that will be used all throughout the chapter. We fix a natural number

$$n \in \mathbb{N},\ n \geq 2\,,$$

a real number

$$\alpha \in]0,1[\,,$$

and an open bounded subset Ω of $\mathbb{R}^n$ of class $C^{1,\alpha}$ (cf. Section 2.13). As usual, we denote by

$$\Omega^- \equiv \mathbb{R}^n \setminus \overline{\Omega}$$

the exterior of Ω. Moreover, we assume that Ω has $\varkappa^+$ connected components and that Ω^- has $\varkappa^- + 1$ connected components. Then, the (bounded) connected components of Ω are denoted by $\Omega_1, \ldots, \Omega_{\varkappa^+}$, the unbounded connected component of Ω^- is denoted by $(\Omega^-)_0$, and the bounded connected components of Ω^- are denoted by $(\Omega^-)_1, \ldots, (\Omega^-)_{\varkappa^-}$ (see Lemma 2.38).

We also remark that, when no misunderstanding is possible, we will use the symbol I instead of $I_{C^{0,\alpha}(\partial\Omega)}$ or $I_{C^{1,\alpha}(\partial\Omega)}$ for the identity operator from $C^{0,\alpha}(\partial\Omega)$ to itself and from $C^{1,\alpha}(\partial\Omega)$ to itself, respectively.

M. Dalla Riva et al., *Singularly Perturbed Boundary Value Problems*,
https://doi.org/10.1007/978-3-030-76259-9_6

6.2 The Dirichlet and Neumann Boundary Value Problems

We plan to study the following boundary value problems for the Laplace operator in the bounded open set Ω and on its exterior Ω^-.
The interior Dirichlet boundary value problem. Given $g \in C^{1,\alpha}(\partial\Omega)$, find u in $C^{1,\alpha}(\overline{\Omega}) \cap C^2(\Omega)$ such that

$$\begin{cases} \Delta u = 0 & \text{in } \Omega\,, \\ u = g & \text{on } \partial\Omega\,. \end{cases}$$

The exterior Dirichlet boundary value problem. Given $g \in C^{1,\alpha}(\partial\Omega)$, find u in $C^{1,\alpha}_{\text{loc}}(\overline{\Omega^-}) \cap C^2(\Omega^-)$ such that

$$\begin{cases} \Delta u = 0 & \text{in } \Omega^-\,, \\ u = g & \text{on } \partial\Omega^- = \partial\Omega\,. \end{cases}$$

The interior Neumann boundary value problem. Given $g \in C^{0,\alpha}(\partial\Omega)$, find u in $C^{1,\alpha}(\overline{\Omega}) \cap C^2(\Omega)$ such that

$$\begin{cases} \Delta u = 0 & \text{in } \Omega\,, \\ \frac{\partial u}{\partial \nu_\Omega} = g & \text{on } \partial\Omega\,. \end{cases}$$

The exterior Neumann boundary value problem. Given $g \in C^{0,\alpha}(\partial\Omega)$, find u in $C^{1,\alpha}_{\text{loc}}(\overline{\Omega^-}) \cap C^2(\Omega^-)$ such that

$$\begin{cases} \Delta u = 0 & \text{in } \Omega^-\,, \\ \frac{\partial u}{\partial \nu_\Omega} = g & \text{on } \partial\Omega^- = \partial\Omega\,. \end{cases}$$

Later, we will consider other boundary value problems.

We observe that the requirement that the solutions of the problems above are of class C^2 in the interior of their domain could be omitted, provided that we understand the equation $\Delta u = 0$ in the sense of distributions. Indeed, by Weyl's Lemma 3.14, continuous functions that satisfy the (weak) Laplace equation $\Delta u = 0$ are harmonic, and therefore real analytic, in the interior of their domain (see also Theorem 3.13). We will henceforth omit to write explicitly the membership in C^2 of the interior and we will just write that the solutions are of class $C^{1,\alpha}$ (or $C^{1,\alpha}_{\text{loc}}$) on the closure of their domain. In particular, we look for solutions that satisfy the boundary conditions in the classical sense.

In the next section we investigate the uniqueness of the solutions of the above boundary value problems.

6.3 Uniqueness for the Interior and Exterior Dirichlet and Neumann Boundary Value Problems

We first make some remarks on the uniqueness for the Dirichlet problem. As we have already seen, we have a uniqueness theorem based on the Maximum Principle for the interior Dirichlet problem (cf. Theorem 3.6). We have also seen that for the exterior Dirichlet problem, we cannot expect uniqueness (cf. Example 3.7). We can prove, however, a uniqueness result for the exterior Dirichlet problem provided that we restrict the class of solutions by imposing a condition at infinity as in the following statement.

Proposition 6.1. *Let* $u_1, u_2 \in C^0(\overline{\Omega^-})$. *If* u_1, u_2 *are harmonic in* Ω^- *and harmonic at infinity and if* $u_1 = u_2$ *on* $\partial\Omega^-$, *then* $u_1 = u_2$ *in* $\overline{\Omega^-}$.

Proof. Possibly scaling and translating, we can assume that $\overline{\mathbb{B}_n(0,1)} \subseteq \Omega$. Then we apply an inversion with respect to the sphere $\partial\mathbb{B}_n(0,1)$ to the set Ω^- and obtain a set $(\Omega^-)^*$ such that

$$\overline{(\Omega^-)^*} \subseteq \mathbb{B}_n(0,1)$$

(cf. Definition 3.19 with $r = 1$). Moreover, $U \equiv (\Omega^-)^* \cup \{0\}$ is an open subset of $\mathbb{R}^n$ containing 0, and

$$\partial U = (\partial\Omega)^* .$$

Since u_1, u_2 are harmonic at infinity, we know that the Kelvin transforms

$$\tilde{u}_j(x) \equiv |x|^{2-n} u_j\left(\frac{x}{|x|^2}\right) \qquad \forall x \in \overline{U} \setminus \{0\}, \qquad \forall j = 1, 2,$$

have a removable singularity at zero, and we denote the corresponding harmonic extensions to $\overline{U}$ by the same symbols $\tilde{u}_1, \tilde{u}_2$ (cf. Definition 3.21). Since $u_1 = u_2$ on $\partial\Omega^-$, we have

$$\tilde{u}_1 = \tilde{u}_2 \text{ on } \partial U ,$$

and thus $\tilde{u}_1 = \tilde{u}_2$ in $\overline{U}$ by the uniqueness of the interior Dirichlet problem of Theorem 3.6. Accordingly, $u_1 = u_2$ in Ω^-. □

Incidentally, we observe that in the proof of Proposition 6.1 we did not use the $C^{1,\alpha}$ regularity of Ω and the statement could be proved without any regularity assumption on Ω.

We now consider the Neumann problem. Since by adding a constant to a solution of a Neumann problem we obtain another solution of the same problem, we cannot expect uniqueness, but we can prove the following.

Theorem 6.2. *Let* $g \in C^{0,\alpha}(\partial\Omega)$. *Then the following statements hold.*

(i) *If* $u_1, u_2 \in C^{1,\alpha}(\overline{\Omega})$ *solve the interior Neumann problem, then* $u_1 - u_2$ *is constant in each connected component of* Ω. *In particular, all solutions of the interior Neumann problem in* $C^{1,\alpha}(\overline{\Omega})$ *can be obtained by adding to* u_1 *an arbitrary function which is constant on the closure of each connected component of* Ω.

(ii) If $u_1, u_2 \in C^{1,\alpha}_{\mathrm{loc}}(\overline{\Omega^-})$ solve the exterior Neumann problem, and if u_1, u_2 are harmonic at infinity, then $u_1 - u_2$ is constant in each connected component of Ω^-. Moreover, if $n \geq 3$, then $u_1 - u_2 = 0$ in the unbounded connected component $(\Omega^-)_0$. In particular, all the solutions of the exterior Neumann problem in $C^{1,\alpha}_{\mathrm{loc}}(\overline{\Omega^-})$ which are harmonic at infinity can be obtained by adding to u_1 an arbitrary function which is constant on the closure of each connected component of Ω^- and equals zero in $\overline{(\Omega^-)_0}$ when $n \geq 3$.

Proof. We present two different proofs of statement (i), the first one based on the Hopf-Oleĭnik Lemma 3.15 and the second one based on an energy estimate.

So, let $u \equiv u_1 - u_2$ and let Ω_1 be a connected component of Ω. We claim that u is constant on Ω_1. Indeed, if u is not constant then, by Corollary 3.5, it attains its maximum on a point x_0 of the boundary $\partial\Omega$ and, by the strong Maximum Principle 3.4, we have $u(x) < u(x_0)$ for all $x \in \Omega_1$. Accordingly, the Hopf-Oleĭnik Lemma 3.15 implies that

$$\frac{\partial u}{\partial \nu_\Omega}(x_0) < 0\,.$$

This inequality yields a contradiction, because

$$\frac{\partial u}{\partial \nu_\Omega}(x_0) = \frac{\partial u_1}{\partial \nu_\Omega}(x_0) - \frac{\partial u_2}{\partial \nu_\Omega}(x_0) = 0\,.$$

We now prove statement (i) with an energy estimate. By applying the first Green Identity of Theorem 4.2 to the function $u \equiv u_1 - u_2$, we obtain

$$0 = \int_{\partial\Omega} u \frac{\partial u}{\partial \nu_\Omega}\, d\sigma = \int_\Omega |\nabla u|^2\, dx\,.$$

Hence, $|\nabla u| = 0$ in Ω and thus u is constant in each connected component of Ω.

Finally, we show statement (ii) with an energy argument. As for (i), we set $u \equiv u_1 - u_2$. Since u is harmonic at infinity, the first Green Identity for exterior domains of Corollary 4.7 implies that

$$\int_{\Omega^-} |\nabla u|^2\, dx = -\int_{\partial\Omega} u \frac{\partial u}{\partial \nu_\Omega}\, d\sigma = 0\,.$$

Since $|\nabla u|$ is continuous in Ω^-, we have $\nabla u = 0$ in Ω^- and thus u is constant in each connected component of Ω^-. Moreover, if $n \geq 3$ the assumption that u is harmonic at infinity implies that u must vanish in $(\Omega^-)_0$ (cf. Theorem 3.22). □

On account of the previous Proposition 6.1 and of statement (ii) of Theorem 6.2, we will consider the exterior Dirichlet and Neumann problems with the condition of harmonicity at infinity on the solutions.

For a mixed problem with a Dirichlet condition on a non-empty portion of the boundary and a Neumann condition on the rest of it, one can prove uniqueness results that resemble those for the Dirichlet problem. We need, however, to introduce some assumptions on the connectedness of Ω and Ω^-.

Proposition 6.3. *Let $\Sigma \subseteq \partial\Omega$, $\Sigma \neq 0$. Let $f \in C^{1,\alpha}(\partial\Omega)$ and $g \in C^{0,\alpha}(\partial\Omega)$. Then the following statements hold.*

(i) Assume that u_1, $u_2 \in C^{1,\alpha}(\overline{\Omega})$ solve the mixed boundary value problem

$$\begin{cases} \Delta u = 0 & \text{in } \Omega\,, \\ u = f & \text{on } \Sigma\,, \\ \frac{\partial u}{\partial \nu_\Omega} = g & \text{on } \partial\Omega \setminus \Sigma\,. \end{cases}$$

If Ω is connected, then $u_1 = u_2$.

(ii) Assume that u_1, $u_2 \in C^{1,\alpha}_{\mathrm{loc}}(\overline{\Omega^-})$ solve the exterior mixed boundary value problem

$$\begin{cases} \Delta u = 0 & \text{in } \Omega^-\,, \\ u = f & \text{on } \Sigma\,, \\ \frac{\partial u}{\partial \nu_\Omega} = g & \text{on } \partial\Omega \setminus \Sigma\,, \\ u \text{ is harmonic at infinity}\,. \end{cases}$$

If Ω^- is connected, then $u_1 = u_2$.

Proof. Let u_1, u_2 be as in (i). Since $(u_1 - u_2)\frac{\partial}{\partial \nu_\Omega}(u_1 - u_2)$ vanishes on the whole of $\partial\Omega$, the energy argument used in the proof of Theorem 6.2 implies that the difference $u_1 - u_2$ is locally constant on $\overline{\Omega}$. Then the condition that Ω is connected implies that $u_1 - u_2$ is constant on $\overline{\Omega}$. Since there exists at least a point $x \in \Sigma \subseteq \partial\Omega$ where $u_1(x) = u_2(x) = f(x)$, it follows that $u_1 - u_2 = 0$ in $\overline{\Omega}$ and statement (i) holds true. Statement (ii) can be verified by a similar argument. □

6.4 The Boundary Integral Operators Associated to the Single and Double Layer Potentials

In this section we introduce the boundary integral operators that we use to study the boundary value problems of Section 6.2. We begin with

$$V_\Omega[\phi] \equiv v_\Omega[\phi]_{|\partial\Omega} \qquad \forall \phi \in C^0(\partial\Omega)\,.$$

Then Theorem 4.25 implies the validity of the following.

Theorem 6.4. *The linear operator V_Ω from $C^{0,\alpha}(\partial\Omega)$ to $C^{1,\alpha}(\partial\Omega)$ that takes ϕ to $V_\Omega[\phi]$ is bounded.*

Then we set

$$W_\Omega[\psi] \equiv w_\Omega[\psi]_{|\partial\Omega} \qquad \forall \psi \in C^{1,\alpha}(\partial\Omega)\,,$$

i.e.,

$$W_\Omega[\psi](x) \equiv -\int_{\partial\Omega} \psi(y)\, \nu_\Omega(y) \cdot \nabla S_n(x - y)\, d\sigma_y \qquad \forall x \in \partial\Omega$$

for all $\psi \in C^{1,\alpha}(\partial\Omega)$ (see also (4.71)). By Theorems 4.30 and 4.31, the operator W_Ω maps $C^{1,\alpha}(\partial\Omega)$ to itself and is bounded. We observe that in Definition 4.27 the operator W_Ω was defined in a larger domain. In this chapter and in the rest of the book, however, we will always understand that, if not otherwise specified, the domain of W_Ω is $C^{1,\alpha}(\partial\Omega)$.

In the following theorem, we introduce a transpose of W_Ω.

Theorem 6.5. *The transpose operator of W_Ω with respect to the dual system*

$$\langle C^{0,\alpha}(\partial\Omega), C^{1,\alpha}(\partial\Omega)\rangle$$

induced by the non-degenerate bilinear form that takes a pair $(\phi,\psi) \in C^{0,\alpha}(\partial\Omega) \times C^{1,\alpha}(\partial\Omega)$ to

$$\langle \phi, \psi\rangle = \int_{\partial\Omega} \phi\psi \, d\sigma \tag{6.1}$$

is the operator W^t_Ω from $C^{0,\alpha}(\partial\Omega)$ to itself defined by the formula

$$W^t_\Omega[\phi](x) \equiv \int_{\partial\Omega} \phi(y)\, \nu_\Omega(x) \cdot \nabla S_n(x-y)\, d\sigma_y \qquad \forall x \in \partial\Omega \tag{6.2}$$

for all $\phi \in C^{0,\alpha}(\partial\Omega)$.

Proof. By Theorem 4.24, the integral which defines $W^t_\Omega[\phi](x)$ is convergent for all $x \in \partial\Omega$. By the jump formula for the normal derivative of a single layer potential of Theorem 4.24, we have

$$W^t_\Omega[\phi] = \frac{\partial}{\partial\nu_\Omega} v^+_\Omega[\phi] + \frac{1}{2}\phi \qquad \forall \phi \in C^{0,\alpha}(\partial\Omega)\,. \tag{6.3}$$

By Theorem 4.25 (i), the operator $v^+_\Omega[\cdot]$ is bounded from $C^{0,\alpha}(\partial\Omega)$ to $C^{1,\alpha}(\overline{\Omega})$ and accordingly the right-hand side of the above equation (6.3) defines a bounded operator from $C^{0,\alpha}(\partial\Omega)$ to $C^{0,\alpha}(\partial\Omega)$. By Lemma 2.76 (iii), there exists $c_{\Omega,\alpha} > 0$ such that

$$\begin{aligned} &|\phi(x)\psi(y)\nu_\Omega(x) \cdot \nabla S_n(x-y)| \\ &\qquad \le \frac{c_{\Omega,\alpha} \sup_{\partial\Omega}|\phi| \sup_{\partial\Omega}|\psi|}{|x-y|^{(n-1)-\alpha}} \qquad \forall (x,y) \in (\partial\Omega)^2,\ x \neq y\,. \end{aligned}$$

Then the inequality $(n-1)-\alpha < (n-1)$ and the Fubini-Tonelli Theorem imply that

$$\begin{aligned} \langle \phi, W_\Omega[\psi]\rangle &= -\int_{\partial\Omega} \phi(x) \int_{\partial\Omega} \nu_\Omega(y) \cdot \nabla S_n(x-y)\psi(y)\, d\sigma_y\, d\sigma_x \\ &= \int_{\partial\Omega} \psi(y) \int_{\partial\Omega} \nu_\Omega(y) \cdot \nabla S_n(y-x)\phi(x)\, d\sigma_x\, d\sigma_y = \langle W^t_\Omega[\phi], \psi\rangle \end{aligned}$$

for all $(\phi, \psi) \in C^{0,\alpha}(\partial\Omega) \times C^{1,\alpha}(\partial\Omega)$. Hence, W^t_Ω is transpose to W_Ω as in the statement. □

Then we can rewrite the jump formulas of Theorems 4.24 (ii), 4.30, and 4.31 (iii) in terms of the boundary operators W_Ω and W^t_Ω.

Theorem 6.6. *Let $\phi \in C^{0,\alpha}(\partial\Omega)$ and $\psi \in C^{1,\alpha}(\partial\Omega)$, then the following equalities hold on $\partial\Omega$.*

$$\begin{aligned}
\nu_\Omega \cdot \nabla v^+_\Omega[\phi]_{|\partial\Omega} &= -\frac{1}{2}\phi + W^t_\Omega[\phi]\,,\\
\nu_\Omega \cdot \nabla v^-_\Omega[\phi]_{|\partial\Omega} &= \frac{1}{2}\phi + W^t_\Omega[\phi]\,,\\
w^+_\Omega[\psi]_{|\partial\Omega} &= \frac{1}{2}\psi + W_\Omega[\psi]\,,\\
w^-_\Omega[\psi]_{|\partial\Omega} &= -\frac{1}{2}\psi + W_\Omega[\psi]\,,\\
\nu_\Omega \cdot \nabla w^+_\Omega[\psi]_{|\partial\Omega} &= \nu_\Omega \cdot \nabla w^-_\Omega[\psi]_{|\partial\Omega}\,.
\end{aligned}$$

Theorem 6.6 relates the operators $\pm\frac{1}{2}I + W_\Omega$ and their transpose operators $\pm\frac{1}{2}I + W^t_\Omega$ to the boundary conditions of the interior and exterior Dirichlet and Neumann problems of Section 6.2. Our plan now is to use these operators and the Fredholm Theory of Chapter 5 to deduce existence results for the boundary value problems. As a first step, we show that W_Ω and W^t_Ω are compact in $C^{1,\alpha}(\partial\Omega)$ and $C^{0,\alpha}(\partial\Omega)$, respectively, and that accordingly we can apply the Fredholm Alternative of Theorem 5.8 in the dual systems $\langle C^{0,\alpha}(\partial\Omega)\,,\ C^{1,\alpha}(\partial\Omega)\rangle$ with $K_1 = 2W_\Omega$ and $K_2 = 2W^t_\Omega$. To prove that W_Ω and W^t_Ω are compact we exploit the compact embedding of $C^{0,\alpha}(\partial\Omega)$ in $C^{0,\beta}(\partial\Omega)$ and of $C^{1,\alpha}(\partial\Omega)$ in $C^{1,\beta}(\partial\Omega)$ which holds for all $\beta \in]0,\alpha[$ and the regularizing properties of W_Ω and W^t_Ω. Indeed, we have already proved that the operator W_Ω has a regularizing effect on the functions of $L^\infty(\partial\Omega)$ and $C^{1,\beta}(\partial\Omega)$ (cf. Theorems 4.33 and 4.37). We now show that also the transpose operator W^t_Ω has a regularizing effect on $C^{0,\beta}(\partial\Omega)$. Namely, we prove the following.

Theorem 6.7. *Let $\beta \in]0,\alpha]$. Then the operator W^t_Ω is linear and continuous from $C^{0,\beta}(\partial\Omega)$ to $C^{0,\alpha}(\partial\Omega)$.*

Proof. If $\mu \in C^{0,\beta}(\partial\Omega)$, then we have

$$\begin{aligned}
W^t_\Omega[\mu](x) &= \int_{\partial\Omega} \nu_\Omega(x) \cdot \nabla S_n(x-y)\mu(y)\, d\sigma_y\\
&= \sum_{l=1}^{n} \int_{\partial\Omega} ((\nu_\Omega)_l(x) - (\nu_\Omega)_l(y)) \frac{\partial S_n}{\partial x_l}(x-y)\mu(y)\, d\sigma_y\\
&\quad + \sum_{l=1}^{n} \int_{\partial\Omega} (\nu_\Omega)_l(y) \frac{\partial S_n}{\partial x_l}(x-y)\mu(y)\, d\sigma_y\\
&= \sum_{l=1}^{n} Q_l[(\nu_\Omega)_l, \mu](x) - W_\Omega[\mu](x) \qquad \forall x \in \partial\Omega
\end{aligned}$$

(for the definition of Q_l, see Theorem 4.35). Since the components of ν_Ω are of class $C^{0,\alpha}$, the validity of the statement follows by Theorem 4.35, by the continuity of the embedding of $C^{0,\beta}(\partial\Omega)$ into $L^\infty(\partial\Omega)$, and by Theorem 4.33. □

For an extension of the previous statement to potentials corresponding to the fundamental solution of an arbitrary elliptic operator with constant coefficients, we refer to the paper [89, Theorem 10.1] with Dondi.

By Theorems 4.37 and 6.7 on the regularizing properties of W_Ω and W_Ω^t and by the compactness of the embeddings of $C^{1,\alpha}(\partial\Omega)$ into $C^{1,\beta}(\partial\Omega)$ and of $C^{0,\alpha}(\partial\Omega)$ into $C^{0,\beta}(\partial\Omega)$ for all $\beta \in]0,\alpha[$ (cf. Proposition 2.25 and Theorem 2.83), we can deduce the validity of the following classical result of Schauder [256, 257].

Theorem 6.8. *The following statements hold.*

(i) The operator W_Ω is compact from $C^{1,\alpha}(\partial\Omega)$ to itself.
(ii) The operator W_Ω^t is compact from $C^{0,\alpha}(\partial\Omega)$ to itself.

We conclude this section with the following Theorems 6.9 and 6.10. The statements are an immediate consequence of the third Green Identities in Ω and Ω^- of Corollary 4.6 and Theorem 4.10, respectively, and of the jump formulas of Theorem 6.6.

Theorem 6.9. *If $u \in C^{1,\alpha}(\overline{\Omega})$ is harmonic in Ω then*

$$w_\Omega[u_{|\partial\Omega}](x) - v_\Omega[\nu_\Omega \cdot \nabla u_{|\partial\Omega}](x) = \begin{cases} u(x) & \text{if } x \in \Omega\,, \\ \frac{1}{2}u(x) & \text{if } x \in \partial\Omega\,, \\ 0 & \text{if } x \in \Omega^-\,. \end{cases}$$

Theorem 6.10. *Let $u \in C^{1,\alpha}_{\mathrm{loc}}(\overline{\Omega^-})$ be harmonic in Ω^- and be harmonic at infinity. Let $u_\infty \equiv \lim_{x\to\infty} u(x)$. (Note that $u_\infty = 0$ if $n \geq 3$.) Then*

$$-w_\Omega[u_{|\partial\Omega}](x) + v_\Omega[\nu_\Omega \cdot \nabla u_{|\partial\Omega}](x) + u_\infty = \begin{cases} u(x) & \text{if } x \in \Omega^-\,, \\ \frac{1}{2}u(x) & \text{if } x \in \partial\Omega\,, \\ 0 & \text{if } x \in \Omega\,. \end{cases}$$

6.5 The Null Spaces of $\frac{1}{2}I + W_\Omega$ and $\frac{1}{2}I + W_\Omega^t$

As mentioned in the previous section, we plan to use the operators $\pm\frac{1}{2}I + W_\Omega$ and $\pm\frac{1}{2}I + W_\Omega^t$ and the Fredholm Alternative of Theorem 5.8 to obtain existence results for the interior and exterior Dirichlet and Neumann boundary value problems and accordingly we want to better understand the properties of these operators. If, for example, we can prove that $\frac{1}{2}I + W_\Omega$ is an isomorphism (that is, a linear homeomorphim) from $C^{1,\alpha}(\partial\Omega)$ to itself, then the jump formula for the double layer potential of Theorem 6.6 guarantees that a solution of the interior Dirichlet problem can be found in the form

$$u = w_\Omega^+[\psi]$$

for some $\psi \in C^{1,\alpha}(\partial\Omega)$. We shall, however, see that $\frac{1}{2}I + W_\Omega$ is an isomorphism only when Ω^- is connected and, in general, the operators $\pm\frac{1}{2}I + W_\Omega$ and $\pm\frac{1}{2}I + W_\Omega^t$ have nontrivial null spaces. Therefore, we will exploit statement (ii) of the Fredholm Alternative Theorem 5.8 –instead of statement (i)– and, to handle the corresponding orthogonality conditions, we will have to characterize the null spaces of these operators.

In this section we consider the operators with the plus sign, that is $+\frac{1}{2}I + W_\Omega$ and $+\frac{1}{2}I + W_\Omega^t$, which are transpose to one another. Then, in the next Section 6.6, we turn to $-\frac{1}{2}I + W_\Omega$ and $-\frac{1}{2}I + W_\Omega^t$, which are also transpose to one another.

Our plan is the following. First we show that V_Ω maps $\mathrm{Ker}(\frac{1}{2}I + W_\Omega^t)$ into $\mathrm{Ker}(\frac{1}{2}I + W_\Omega)$, then we prove that V_Ω is actually an isomorphism (i.e., a linear homeomorphism) from $\mathrm{Ker}(\frac{1}{2}I + W_\Omega^t)$ to $\mathrm{Ker}(\frac{1}{2}I + W_\Omega)$, and finally we will be ready to introduce an explicit characterization of $\mathrm{Ker}(\frac{1}{2}I + W_\Omega)$.

We begin with the following technical lemma.

Lemma 6.11. *Let $\phi \in C^{0,\alpha}(\partial\Omega)$. Then*

$$\int_{\partial\Omega} \left(\frac{1}{2}I + W_\Omega^t\right)[\phi]\, d\sigma = \int_{\partial\Omega} \phi\, d\sigma\,.$$

Proof. We have

$$\int_{\partial\Omega} \left(\frac{1}{2}I + W_\Omega^t\right)[\phi]\, d\sigma = \int_{\partial\Omega} \phi\, d\sigma + \int_{\partial\Omega} \left(-\frac{1}{2}I + W_\Omega^t\right)[\phi]\, d\sigma$$

and

$$\int_{\partial\Omega} \left(-\frac{1}{2}I + W_\Omega^t\right)[\phi]\, d\sigma = \int_{\partial\Omega} \nu_\Omega \cdot \nabla v_\Omega^+[\phi]\, d\sigma = 0$$

by the first Green Identity 4.2 (see also Theorem 6.6). □

We now show that V_Ω maps $\mathrm{Ker}(\frac{1}{2}I + W_\Omega^t)$ into $\mathrm{Ker}(\frac{1}{2}I + W_\Omega)$.

Proposition 6.12. *If $\mu \in \mathrm{Ker}(\frac{1}{2}I + W_\Omega^t)$ then $V_\Omega[\mu] \in \mathrm{Ker}(\frac{1}{2}I + W_\Omega)$.*

Proof. If $\mu \in \mathrm{Ker}(\frac{1}{2}I + W_\Omega^t)$ then Lemma 6.11 implies that $\int_{\partial\Omega} \mu\, d\sigma = 0$ and accordingly $v_\Omega^-[\mu]$ is harmonic at infinity and has limit equal to 0 at infinity (cf. Theorem 4.23). Then the third Green Identity of Theorem 6.10 implies that

$$-w_\Omega[V_\Omega[\mu]](x) + v_\Omega[\nu_\Omega \cdot \nabla v_\Omega^-[\mu]_{|\partial\Omega}](x) = 0$$

for all $x \in \Omega$. Then, by Theorem 4.31 on the continuous extendibility of the double layer to $\overline{\Omega}$, we have

$$-w_\Omega^+[V_\Omega[\mu]](x) + v_\Omega^+[\nu_\Omega \cdot \nabla v_\Omega^-[\mu]_{|\partial\Omega}](x) = 0$$

for all $x \in \overline{\Omega}$ and, by the membership of μ in $\mathrm{Ker}(\frac{1}{2}I+W_{\Omega}^{t})$ and the jump formulas in Theorem 6.6, it follows that

$$\begin{aligned}&\left(\frac{1}{2}I + W_{\Omega}\right)[V_{\Omega}[\mu]]\\&\qquad = w_{\Omega}^{+}[V_{\Omega}[\mu]]_{|\partial\Omega} = v_{\Omega}^{+}[\nu_{\Omega} \cdot \nabla v_{\Omega}^{-}[\mu]_{|\partial\Omega}]_{|\partial\Omega} = v_{\Omega}^{+}[\frac{1}{2}\mu + W_{\Omega}^{t}[\mu]] = 0\,.\end{aligned}$$

□

Then we prove that V_{Ω} is an isomorphism from $\mathrm{Ker}(\frac{1}{2}I+W_{\Omega}^{t})$ to $\mathrm{Ker}(\frac{1}{2}I+W_{\Omega})$.

Proposition 6.13. *The map from* $\mathrm{Ker}(\frac{1}{2}I + W_{\Omega}^{t})$ *to* $\mathrm{Ker}(\frac{1}{2}I + W_{\Omega})$ *that takes* μ *to* $V_{\Omega}[\mu]$ *is an isomorphism.*

Proof. By Proposition 6.12, V_{Ω} maps $\mathrm{Ker}(\frac{1}{2}I+W_{\Omega}^{t})$ into $\mathrm{Ker}(\frac{1}{2}I+W_{\Omega})$. By Theorem 6.8, W_{Ω} and W_{Ω}^{t} are compact, and thus the Fredholm Alternative Theorem 5.8 implies that the null spaces $\mathrm{Ker}(\frac{1}{2}I + W_{\Omega}^{t})$ and $\mathrm{Ker}(\frac{1}{2}I + W_{\Omega})$ have the same finite dimension. As a consequence, it suffices to show that the map that takes μ to $V_{\Omega}[\mu]$ is injective to prove that it is an isomorphism. If $\mu \in \mathrm{Ker}(\frac{1}{2}I + W_{\Omega}^{t})$ and $V_{\Omega}[\mu] = 0$, then by the uniqueness Theorem 3.6 we have $v_{\Omega}^{+}[\mu] = 0$ on the whole of $\overline{\Omega}$ and accordingly $\nu_{\Omega} \cdot \nabla v_{\Omega}^{+}[\mu] = 0$ on $\partial\Omega$. Moreover, by the jump formula for the normal derivative of a single layer of Theorem 6.6 and by the membership of μ in $\mathrm{Ker}(\frac{1}{2}I + W_{\Omega}^{t})$, we have $\nu_{\Omega} \cdot \nabla v_{\Omega}^{-}[\mu] = 0$. Then, again by the jump formula for the normal derivatives of a single layer of Theorem 6.6, we deduce that

$$\mu = \nu_{\Omega} \cdot \nabla v_{\Omega}^{-}[\mu] - \nu_{\Omega} \cdot \nabla v_{\Omega}^{+}[\mu] = 0\,.$$

□

We are now ready to describe the null space of $\frac{1}{2}I + W_{\Omega}$.

Theorem 6.14. *If* $\varkappa^{-} > 0$*, then the null space* $\mathrm{Ker}(\frac{1}{2}I + W_{\Omega})$ *consists of the functions from* $\partial\Omega$ *to* $\mathbb{R}$ *that are constant on* $\partial(\Omega^{-})_j$ *for all* $j \in \{1, \dots, \varkappa^{-}\}$ *and that are identically equal to* 0 *on* $\partial(\Omega^{-})_0$*. If* $\varkappa^{-} = 0$*, then* $\mathrm{Ker}(\frac{1}{2}I + W_{\Omega}) = \{0\}$.

Proof. If $\psi \in \mathrm{Ker}(\frac{1}{2}I + W_{\Omega})$, then Proposition 6.13 implies that there exists $\mu \in \mathrm{Ker}(\frac{1}{2}I + W_{\Omega}^{t})$ such that $\psi = V_{\Omega}[\mu]$. Then the jump formula for the normal derivative of a single layer potential of Theorem 6.6 implies that $\nu_{\Omega} \cdot \nabla v_{\Omega}^{-}[\mu] = 0$ on $\partial\Omega$. Moreover, Lemma 6.11 implies that $\int_{\partial\Omega} \mu \, d\sigma = \int_{\partial\Omega} (\frac{1}{2}I + W_{\Omega}^{t})\mu \, d\sigma = 0$. Then, by Theorem 4.23, $v_{\Omega}^{-}[\mu]$ vanishes at infinity. As a consequence, the uniqueness Theorem 6.2 (ii) for the exterior Neumann problem implies that $v_{\Omega}^{-}[\mu]$ is locally constant on Ω^{-} and equal to 0 on $(\Omega^{-})_0$. Accordingly, when $\varkappa^{-} = 0$ we have that $\psi = V_{\Omega}[\mu]$ is 0 and we conclude that $\mathrm{Ker}(\frac{1}{2}I + W_{\Omega})$ is trivial. If instead $\varkappa^{-} > 0$, then $\psi = V_{\Omega}[\mu]$ is constant on $\partial(\Omega^{-})_j$ for all $j \in \{1, \dots, \varkappa^{-}\}$ and equal to 0 on $\partial(\Omega^{-})_0$.

It remains to show that, for $\varkappa^{-} > 0$, all functions that are constant on $\partial(\Omega^{-})_j$ for all $j \in \{1, \dots, \varkappa^{-}\}$ and equal to 0 on $\partial(\Omega^{-})_0$ belong to $\mathrm{Ker}(\frac{1}{2}I + W_{\Omega})$. So, let ψ be one such function. Then there exists a function u constant on $\overline{(\Omega^-)_j}$ for all $j \in \{1, \ldots, \varkappa^-\}$ and equal to 0 on $\overline{(\Omega^-)_0}$ such that $\psi = u_{|\partial\Omega}$. Hence $\nabla u = 0$ and, by the Green Identity in Ω^- of Theorem 6.10, we have

$$-w_\Omega[\psi](x) = -w_\Omega[u_{|\partial\Omega}](x) + v_\Omega[\nu_\Omega \cdot \nabla u_{|\partial\Omega}](x) = 0 \qquad \forall x \in \Omega\,.$$

Then $(\frac{1}{2}I + W_\Omega)[\psi] = w^+_\Omega[\psi]_{|\partial\Omega} = 0$ by the jump formulas of Theorem 6.6. □

We observe that Proposition 6.13 and Theorem 6.14 imply that $\mathrm{Ker}(\frac{1}{2}I + W_\Omega)$ and $\mathrm{Ker}(\frac{1}{2}I + W^t_\Omega)$ have dimension $\varkappa^-$. Since W_Ω and W^t_Ω are compact (cf. Theorem 6.8), the Fredholm Alternative of Theorem 5.8 implies the validity of the following.

Corollary 6.15. *If $\varkappa^- = 0$, then $\frac{1}{2}I + W_\Omega$ is an isomorphism from $C^{1,\alpha}(\partial\Omega)$ onto itself and $\frac{1}{2}I + W^t_\Omega$ is an isomorphism from $C^{0,\alpha}(\partial\Omega)$ onto itself.*

We conclude this section with the following result, which shows that there are no nonzero elements of $\mathrm{Ker}(\frac{1}{2}I + W^t_\Omega)$ that are orthogonal to $\mathrm{Ker}(\frac{1}{2}I + W_\Omega)$.

Theorem 6.16. *Assume that $\varkappa^- > 0$. If $\phi \in \mathrm{Ker}(\frac{1}{2}I + W^t_\Omega)$ and $\int_{\partial(\Omega^-)_j} \phi \, d\sigma = 0$ for all j in $\{1, \ldots, \varkappa^-\}$ (i.e., ϕ is orthogonal to $\mathrm{Ker}(\frac{1}{2}I + W_\Omega)$), then $\phi = 0$.*

Proof. Let $\phi \in \mathrm{Ker}(\frac{1}{2}I + W^t_\Omega)$ and $\int_{\partial(\Omega^-)_j} \phi \, d\sigma = 0$ for all $j \in \{1, \ldots, \varkappa^-\}$. By the membership of ϕ in $\mathrm{Ker}(\frac{1}{2}I + W^t_\Omega)$ and by the jump properties of the single layer potential of Theorem 6.6, we have

$$\nu_\Omega \cdot \nabla v^-_\Omega[\phi] = 0 \quad \text{and} \quad \nu_\Omega \cdot \nabla v^+_\Omega[\phi] = -\phi\,. \tag{6.4}$$

Moreover, by the characterization of $\mathrm{Ker}(\frac{1}{2}I + W_\Omega)$ of Theorem 6.14 we have $\int_{\partial\Omega} \phi\psi \, d\sigma = 0$ for all $\psi \in \mathrm{Ker}(\frac{1}{2}I + W_\Omega)$. Accordingly, the Fredholm Alternative Theorem 5.8 implies that there exists $\mu \in C^{0,\alpha}(\partial\Omega)$ such that

$$\phi = \left(\frac{1}{2}I + W^t_\Omega\right)[\mu]\,.$$

By the jump properties of the single layer potential of Theorem 6.6 it follows that

$$\nu_\Omega \cdot \nabla v^-_\Omega[\mu] = \phi\,. \tag{6.5}$$

In addition, Lemma 6.11 together with the membership of ϕ in $\mathrm{Ker}(\frac{1}{2}I + W^t_\Omega)$ implies that $\int_{\partial\Omega} \phi \, d\sigma = 0$ and that

$$\int_{\partial\Omega} \mu \, d\sigma = \int_{\partial\Omega} \left(\frac{1}{2}I + W^t_\Omega\right)[\mu] \, d\sigma = \int_{\partial\Omega} \phi \, d\sigma = 0\,.$$

Hence, both the single layers $v^-_\Omega[\phi]$ and $v^-_\Omega[\mu]$ are harmonic at infinity (cf. Theorem 4.23), and thus the second Green Identity for exterior domains of Corollary 4.8 implies that

$$\int_{\partial\Omega} (\nu_\Omega \cdot \nabla v^-_\Omega[\phi])\, v^-_\Omega[\mu] - v^-_\Omega[\phi]\, (\nu_\Omega \cdot \nabla v^-_\Omega[\mu])\, d\sigma = 0\,. \tag{6.6}$$

Now, by (6.4)–(6.6), by equality $v^-_\Omega[\phi]_{|\partial\Omega} = v^+_\Omega[\phi]_{|\partial\Omega}$, and by the first Green Identity in Ω, we deduce that

$$\begin{aligned} 0 &= \int_{\partial\Omega} (\nu_\Omega \cdot \nabla v^-_\Omega[\phi])\, v^-_\Omega[\mu] - v^-_\Omega[\phi]\, (\nu_\Omega \cdot \nabla v^-_\Omega[\mu])\, d\sigma \\ &= -\int_{\partial\Omega} v^-_\Omega[\phi]\, (\nu_\Omega \cdot \nabla v^-_\Omega[\mu])\, d\sigma = -\int_{\partial\Omega} v^-_\Omega[\phi]\phi\, d\sigma \\ &= \int_{\partial\Omega} v^+_\Omega[\phi]\, (\nu_\Omega \cdot \nabla v^+_\Omega[\phi])\, d\sigma = \int_\Omega \left|\nabla v^+_\Omega[\phi]\right|^2 dx\,. \end{aligned}$$

Hence, $v^+_\Omega[\phi]$ is locally constant on Ω. Thus $\nu_\Omega \cdot \nabla v^+_\Omega[\phi] = 0$ and by the second equality of (6.4) we deduce that $\phi = 0$. □

As a consequence of Theorem 6.16, we have the following Corollary 6.17 (for a similar statement in $L^2(\partial\Omega)$ we refer to Folland [102, Corollary 3.39]).

Corollary 6.17. *We have*

$$C^{0,\alpha}(\partial\Omega) = \mathrm{Im}\left(\frac{1}{2}I + W^t_\Omega\right) \oplus \mathrm{Ker}\left(\frac{1}{2}I + W^t_\Omega\right) \tag{6.7}$$

and

$$C^{1,\alpha}(\partial\Omega) = \mathrm{Im}\left(\frac{1}{2}I + W_\Omega\right) \oplus \mathrm{Ker}\left(\frac{1}{2}I + W_\Omega\right)\,, \tag{6.8}$$

where the sums are direct but not necessarily orthogonal.

Proof. By Theorems 5.3 and 6.8 (ii) we can see that $\frac{1}{2}I + W^t_\Omega$ is a Fredholm operator of index zero and, by Theorem 6.14, the null space of $\frac{1}{2}I + W^t_\Omega$ is a closed subspace $C^{0,\alpha}(\partial\Omega)$ of dimension $\varkappa^-$. We deduce that the image of $\frac{1}{2}I + W^t_\Omega$ has co-dimension $\varkappa^-$ and is closed in $C^{0,\alpha}(\partial\Omega)$ (see Section 5.1). Then, to prove (6.7), it suffices to show that $\mathrm{Im}\left(\frac{1}{2}I + W^t_\Omega\right)$ and $\mathrm{Ker}\left(\frac{1}{2}I + W^t_\Omega\right)$ have a trivial intersection, a fact that follows from Theorem 6.16 and by noting that $\mathrm{Im}\left(\frac{1}{2}I + W^t_\Omega\right)$ coincides with the subspace of $C^{0,\alpha}(\partial\Omega)$ of the functions that are orthogonal to $\mathrm{Ker}\left(\frac{1}{2}I + W^t_\Omega\right)$ (see also the Fredholm Alternative Theorem 5.8).

To prove (6.8) we observe that, by Theorems 5.3 and 6.8 (i), $\frac{1}{2}I + W_\Omega$ is a Fredholm operator of index zero. Moreover, $\mathrm{Ker}\left(\frac{1}{2}I + W_\Omega\right)$ is closed and has dimension $\varkappa^-$ by Theorem 6.14. Then the image of $\frac{1}{2}I + W_\Omega$ is a closed subspace of $C^{1,\alpha}(\partial\Omega)$ and has co-dimension $\varkappa^-$ (see Section 5.1). So, it remains to show that $\mathrm{Im}\left(\frac{1}{2}I + W_\Omega\right)$ and $\mathrm{Ker}\left(\frac{1}{2}I + W_\Omega\right)$ have a trivial intersection. Let

$$\phi \in \mathrm{Im}\left(\frac{1}{2}I + W_\Omega\right) \cap \mathrm{Ker}\left(\frac{1}{2}I + W_\Omega\right)\,.$$

Since $\phi \in C^{1,\alpha}(\partial\Omega) \subseteq C^{0,\alpha}(\partial\Omega)$, equality (6.7) implies that there exist (and are unique) $\phi_1 \in \mathrm{Im}\left(\frac{1}{2}I + W_\Omega^t\right)$ and $\phi_2 \in \mathrm{Ker}\left(\frac{1}{2}I + W_\Omega^t\right)$ such that $\phi = \phi_1 + \phi_2$. But then $\langle\phi_1, \phi\rangle = \int_{\partial\Omega} \phi_1 \phi \, d\sigma = 0$ by the membership of ϕ in $\mathrm{Ker}\left(\frac{1}{2}I + W_\Omega\right)$ and $\langle\phi_2, \phi\rangle = \int_{\partial\Omega} \phi_2 \phi \, d\sigma = 0$ by the membership of ϕ in $\mathrm{Im}\left(\frac{1}{2}I + W_\Omega\right)$ (see also the Fredholm Alternative of Theorem 5.8). Hence, $\langle\phi, \phi\rangle = 0$ and $\phi = 0$. □

6.6 The Null Spaces of $-\frac{1}{2}I + W_\Omega$ and $-\frac{1}{2}I + W_\Omega^t$

We turn to the operator $-\frac{1}{2}I + W_\Omega$ and its transpose $-\frac{1}{2}I + W_\Omega^t$. Our plan to describe their null spaces is similar to the one used in Section 6.5 and accordingly we begin by showing that the image of $\mathrm{Ker}(-\frac{1}{2}I + W_\Omega^t)$ under V_Ω is contained in $\mathrm{Ker}(-\frac{1}{2}I + W_\Omega)$. In this case, however, we can prove that V_Ω restricts to an isomorphism between $\mathrm{Ker}(-\frac{1}{2}I + W_\Omega^t)$ and $\mathrm{Ker}(-\frac{1}{2}I + W_\Omega)$ only when the dimension n is ≥ 3. For $n = 2$ we can just show that V_Ω is injective on the subspace of $\mathrm{Ker}(-\frac{1}{2}I + W_\Omega^t)$ of the functions with zero integral mean. Then, to obtain an explicit description of $\mathrm{Ker}(-\frac{1}{2}I + W_\Omega)$ we will pass through an intermediate proposition where we show that $\mathrm{Ker}(-\frac{1}{2}I + W_\Omega)$ is the direct sum of the space of constant functions on $\partial\Omega$ and of the image under V_Ω of the subspace of $\mathrm{Ker}(-\frac{1}{2}I + W_\Omega^t)$ consisting of functions with zero integral mean. The reason why this extra difficulty appears in dimension $n = 2$ is that for a function $\mu \in \mathrm{Ker}(-\frac{1}{2}I + W_\Omega^t)$ we cannot claim that $v_\Omega^-[\mu]$ is harmonic at infinity unless $\int_{\partial\Omega} \mu \, d\sigma = 0$ (cf. Theorem 4.23). Instead, if $\mu \in \mathrm{Ker}(\frac{1}{2}I + W_\Omega^t)$, then Lemma 6.11 ensures that $v_\Omega^-[\mu]$ is harmonic at infinity, a fact that we could exploit in Section 6.5 but not in this section.

Then we begin with the following.

Proposition 6.18. *If* $\mu \in \mathrm{Ker}(-\frac{1}{2}I + W_\Omega^t)$ *then* $V_\Omega[\mu] \in \mathrm{Ker}(-\frac{1}{2}I + W_\Omega)$.

Proof. By applying the third Green Identity of Theorem 6.9 to the function $v_\Omega^+[\mu]$ in Ω, we have

$$w_\Omega[V_\Omega[\mu]](x) - v_\Omega[\nu_\Omega \cdot \nabla v_\Omega^+[\mu]_{|\partial\Omega}](x) = 0$$

for all $x \in \Omega^-$. Then, by Theorem 4.31 we have

$$w_\Omega^-[V_\Omega[\mu]](x) - v_\Omega^-[\nu_\Omega \cdot \nabla v_\Omega^+[\mu]_{|\partial\Omega}](x) = 0$$

for all $x \in \overline{\Omega^-}$ and, by the membership of μ in $\mathrm{Ker}(-\frac{1}{2}I + W_\Omega^t)$ and the jump formulas in Theorem 6.6, it follows that

$$\begin{aligned}\left(-\frac{1}{2}I + W_\Omega\right)[V_\Omega[\mu]] &= w_\Omega^-[V_\Omega[\mu]]_{|\partial\Omega} \\ &= V_\Omega[\nu_\Omega \cdot \nabla v_\Omega^+[\mu]_{|\partial\Omega}] = V_\Omega\left[\left(-\frac{1}{2}I + W_\Omega^t\right)[\mu]\right] = 0\,.\end{aligned}$$

□

In addition, if $n \geq 3$, then V_Ω is an isomorphism between $\mathrm{Ker}(-\frac{1}{2}I + W^t_\Omega)$ and $\mathrm{Ker}(-\frac{1}{2}I + W_\Omega)$.

Proposition 6.19. *If $n \geq 3$, then the map from $\mathrm{Ker}(-\frac{1}{2}I+W^t_\Omega)$ to $\mathrm{Ker}(-\frac{1}{2}I+W_\Omega)$ that takes μ to $V_\Omega[\mu]$ is an isomorphism.*

Proof. By Proposition 6.18, V_Ω maps $\mathrm{Ker}(-\frac{1}{2}I + W^t_\Omega)$ into $\mathrm{Ker}(-\frac{1}{2}I + W_\Omega)$. By Theorem 6.8, W_Ω and W^t_Ω are compact, and thus the Fredholm Alternative of Theorem 5.8 implies that the null spaces $\mathrm{Ker}(-\frac{1}{2}I + W^t_\Omega)$ and $\mathrm{Ker}(-\frac{1}{2}I + W_\Omega)$ have the same finite dimension. As a consequence, it suffices to show that the map that takes μ to $V_\Omega[\mu]$ is injective to show that it is an isomorphism. If $\mu \in \mathrm{Ker}(-\frac{1}{2}I+W^t_\Omega)$ and $V_\Omega[\mu] = 0$, then Theorem 4.23 implies that $v^-_\Omega[\mu]$ is harmonic at infinity and the uniqueness of the solution of the exterior Dirichlet problem of Proposition 6.1 implies that $v^-_\Omega[\mu] = 0$ on the whole of $\overline{\Omega^-}$. Accordingly, we have $\nu_\Omega \cdot \nabla v^-_\Omega[\mu] = 0$. Then the jump formulas of Theorem 6.6 and the membership of μ in $\mathrm{Ker}(-\frac{1}{2}I + W^t_\Omega)$ imply that

$$\mu = \nu_\Omega \cdot \nabla v^-_\Omega[\mu] - \nu_\Omega \cdot \nabla v^+_\Omega[\mu] = 0\,.$$

□

Let us recall the definition of formula (2.2): if X is a subspace of $L^1(\partial\Omega)$, then X_0 is the subspace of X consisting of the functions which have 0 integral mean. In general, for $n \geq 2$, we have to restrict V_Ω to $\mathrm{Ker}(-\frac{1}{2}I + W^t_\Omega)_0$ to obtain an injective map.

Proposition 6.20. *The map from $\mathrm{Ker}(-\frac{1}{2}I + W^t_\Omega)_0$ to $\mathrm{Ker}(-\frac{1}{2}I + W_\Omega)$ that takes μ to $V_\Omega[\mu]$ is injective.*

Proof. By the membership of μ in $\mathrm{Ker}(-\frac{1}{2}I + W^t_\Omega)_0$ we have $\int_{\partial\Omega} \mu\, d\sigma = 0$, and thus $\lim_{x\to\infty} v^-_\Omega[\mu](x) = 0$ by Theorem 4.23. Then the proof follows by the same argument of the proof of Proposition 6.19. □

In Proposition 6.23 we will describe $\mathrm{Ker}(-\frac{1}{2}I + W_\Omega)$ as the direct sum of the space of constant functions on $\partial\Omega$ and of $V_\Omega[\mathrm{Ker}(-\frac{1}{2}I + W^t_\Omega)_0]$. Meanwhile, we introduce two preliminary lemmas. In the first one we show that locally constant functions on $\overline{\Omega}$ have a restriction to $\partial\Omega$ that belongs to $\mathrm{Ker}(-\frac{1}{2}I + W_\Omega)$.

Lemma 6.21. *The null space $\mathrm{Ker}(-\frac{1}{2}I + W_\Omega)$ contains the functions from $\partial\Omega$ to $\mathbb{R}$ which are constant on $\partial\Omega_j$ for all $j \in \{1, \dots, \varkappa^+\}$.*

Proof. If ψ is constant on $\partial\Omega_j$ for all $j \in \{1, \dots, \varkappa^+\}$, then there exists a function u that is constant on $\overline{\Omega_j}$ for all $j \in \{1, \dots, \varkappa^+\}$ and such that $\psi = u_{|\partial\Omega}$. Hence, by the third Green Identity of Theorem 6.9 we have

$$w_\Omega[\psi](x) = w_\Omega[\psi](x) - v_\Omega[\nu_\Omega \cdot \nabla u_{|\partial\Omega}](x) = 0$$

for all $x \in \Omega^-$ and accordingly $w^-_\Omega[\psi](x) = 0$ for all $x \in \overline{\Omega^-}$ (cf. Theorem 4.31). Then we have $(-\frac{1}{2}I + W_\Omega)\psi = w^-_\Omega[\psi]_{|\partial\Omega} = 0$ by the jump formulas in Theorem 6.6. □

Moreover, we have the following.

Lemma 6.22. *If $\phi \in \mathrm{Ker}(-\frac{1}{2}I + W_\Omega^t)_0$ and $V_\Omega[\phi]$ is constant on $\partial\Omega$, then $\phi = 0$.*

Proof. By the jump formula for the normal derivative of a single layer potential of Theorem 6.6, we have

$$\nu_\Omega \cdot \nabla v^-[\phi]_{|\partial\Omega} = \left(\frac{1}{2}I + W_\Omega^t\right)[\phi] = \phi\,. \tag{6.9}$$

Since $\int_{\partial\Omega} \phi \, d\sigma = 0$, Theorem 4.23 implies that $v_\Omega^-[\phi]$ is harmonic at infinity, and thus the first Green Identity for exterior domains of Corollary 4.7 implies that

$$\int_{\Omega^-} |\nabla v_\Omega^-[\phi]|^2 \, dx = -\int_{\partial\Omega} V_\Omega[\phi]\, \nu_\Omega \cdot \nabla v_\Omega^-[\phi] \, d\sigma = -V_\Omega[\phi] \int_{\partial\Omega} \phi \, d\sigma = 0\,.$$

It follows that $v^-[\phi]$ is locally constant on $\overline{\Omega^-}$. Then equality (6.9) implies that $\phi = 0$. □

Then we are ready to prove the following.

Proposition 6.23. *The null space $\mathrm{Ker}(-\frac{1}{2}I + W_\Omega)$ is the topological direct sum of*

$$V_\Omega\left[\mathrm{Ker}\left(-\frac{1}{2}I + W_\Omega^t\right)_0\right]$$

and of the space of constant functions from $\partial\Omega$ to $\mathbb{R}$.

Proof. The closed subspace $\mathrm{Ker}(-\frac{1}{2}I + W_\Omega^t)_0$ of $\mathrm{Ker}(-\frac{1}{2}I + W_\Omega^t)$ is defined by the vanishing of the linear functional that takes ϕ to $\int_{\partial\Omega} \phi \, d\sigma$, therefore it has co-dimension smaller than or equal to one. By Theorem 6.8, W_Ω and W_Ω^t are compact, and thus the Fredholm Alternative of Theorem 5.8 implies that the null spaces $\mathrm{Ker}(-\frac{1}{2}I + W_\Omega^t)$ and $\mathrm{Ker}(-\frac{1}{2}I + W_\Omega)$ have the same finite dimension. Then by Proposition 6.20, the subspace $\mathrm{Ker}(-\frac{1}{2}I + W_\Omega^t)_0$ of $\mathrm{Ker}(-\frac{1}{2}I + W_\Omega^t)$ and the subspace $V_\Omega[\mathrm{Ker}(-\frac{1}{2}I + W_\Omega^t)_0]$ of $\mathrm{Ker}(-\frac{1}{2}I + W_\Omega)$ have the same (finite) dimension and co-dimension. In particular, $V_\Omega[\mathrm{Ker}(-\frac{1}{2}I + W_\Omega^t)_0]$ has co-dimension less than or equal to one in $\mathrm{Ker}(-\frac{1}{2}I + W_\Omega)$. By Lemma 6.21, the constant functions on $\partial\Omega$ are contained in $\mathrm{Ker}(-\frac{1}{2}I + W_\Omega)$. Instead, Lemma 6.22 implies that $V_\Omega[\mathrm{Ker}(-\frac{1}{2}I + W_\Omega^t)_0]$ does not contain non-trivial constant functions. Since the space of constant functions on $\partial\Omega$ has dimension one, we conclude that $V_\Omega[\mathrm{Ker}(-\frac{1}{2}I + W_\Omega^t)_0]$ has co-dimension exactly equal to one and that $\mathrm{Ker}(-\frac{1}{2}I + W_\Omega)$ is the algebraic direct sum of $V_\Omega[\mathrm{Ker}(-\frac{1}{2}I + W_\Omega^t)_0]$ and of the space of constant functions. Finally, since all the spaces involved have finite dimension, they are closed and the algebraic direct sum is topological by Theorem 2.6 (see also Theorem 2.3). □

In the following Theorem 6.24 we obtain an explicit description of the null space of $-\frac{1}{2}I + W_\Omega$.

Theorem 6.24. *The null space* $\mathrm{Ker}(-\frac{1}{2}I + W_\Omega)$ *consists of the functions from* $\partial\Omega$ *to* $\mathbb{R}$ *which are constant on* $\partial\Omega_j$ *for all* $j \in \{1, \dots, \varkappa^+\}$.

Proof. If $\psi \in \mathrm{Ker}(-\frac{1}{2}I + W_\Omega)$, then Proposition 6.23 implies that there exist $\mu \in \mathrm{Ker}(-\frac{1}{2}I + W_\Omega^t)_0$ and a constant function ρ on $\partial\Omega$ such that $\psi = V_\Omega[\mu] + \rho$. Then, by the jump properties of the single layer potential (cf. Theorem 6.6) we have $\nu_\Omega \cdot \nabla v_\Omega^+[\mu] = 0$ on $\partial\Omega$ and, by the first Green Identity of Theorem 4.2, it follows that $v_\Omega^+[\mu]$ is locally constant on Ω. Accordingly, $\psi = V_\Omega[\mu] + \rho$ is constant on $\partial\Omega_j$ for all $j \in \{1, \dots, \varkappa^+\}$. Conversely, Lemma 6.21 implies that every function on $\partial\Omega$ which is constant on the $\partial\Omega_j$'s belongs to $\mathrm{Ker}(-\frac{1}{2}I + W_\Omega)$. Thus, the theorem is proved. □

We observe that Theorem 6.24 implies that the null spaces $\mathrm{Ker}(-\frac{1}{2}I + W_\Omega)$ and $\mathrm{Ker}(-\frac{1}{2}I + W_\Omega^t)$ have dimension $\varkappa^+$ (see also Theorems 5.8 and 6.8).

We conclude this subsection with the following theorem, which shows that there are no nonzero elements of $\mathrm{Ker}(-\frac{1}{2}I + W_\Omega^t)$ orthogonal to $\mathrm{Ker}(-\frac{1}{2}I + W_\Omega)$.

Theorem 6.25. *If* $\phi \in \mathrm{Ker}(-\frac{1}{2}I + W_\Omega^t)$ *and* $\int_{\partial\Omega_j} \phi \, d\sigma = 0$ *for all* $j \in \{1, \dots, \varkappa^+\}$ *(i.e.,* ϕ *is orthogonal to* $\mathrm{Ker}(-\frac{1}{2}I + W_\Omega)$*), then* $\phi = 0$.

Proof. Let $\phi \in \mathrm{Ker}(-\frac{1}{2}I + W_\Omega^t)$ and $\int_{\partial\Omega_j} \phi \, d\sigma = 0$ for all $j \in \{1, \dots, \varkappa^+\}$. By the membership of $\phi \in \mathrm{Ker}(-\frac{1}{2}I + W_\Omega^t)$ and by the jump properties of the single layer potential of Theorem 6.6, we have

$$\nu_\Omega \cdot \nabla v_\Omega^+[\phi] = 0 \quad \text{and} \quad \nu_\Omega \cdot \nabla v_\Omega^-[\phi] = \phi \,. \tag{6.10}$$

Moreover, by the characterization of $\mathrm{Ker}(-\frac{1}{2}I + W_\Omega)$ of Theorem 6.24 we have $\int_{\partial\Omega} \phi\psi \, d\sigma = 0$ for all $\psi \in \mathrm{Ker}(-\frac{1}{2}I + W_\Omega)$. Accordingly, by the Fredholm Alternative of Theorem 5.8 there exists $\mu \in C^{0,\alpha}(\partial\Omega)$ such that

$$\phi = \left(-\frac{1}{2}I + W_\Omega^t\right)[\mu] \,.$$

By the jump properties of the single layer potential of Theorem 6.6 it follows that

$$\nu_\Omega \cdot \nabla v_\Omega^+[\mu] = \phi \,. \tag{6.11}$$

Then, by the second Green Identity of Theorem 4.5, we have

$$\begin{aligned} \int_{\partial\Omega} (\nu_\Omega \cdot \nabla v_\Omega^+[\phi]) \, v_\Omega^+[\mu] - v_\Omega^+[\phi] \, (\nu_\Omega \cdot \nabla v_\Omega^+[\mu]) \, d\sigma & \\ = \int_\Omega (\Delta v_\Omega^+[\phi]) v_\Omega^+[\mu] - v_\Omega^+[\phi] \Delta v_\Omega^+[\mu] \, dx = 0 \,. & \end{aligned} \tag{6.12}$$

Moreover, we have

$$\int_{\partial\Omega} \phi \, d\sigma = \sum_{j=1}^{\varkappa^+} \int_{\partial\Omega_j} \phi \, d\sigma = 0 \,,$$

and by Theorem 4.23 we deduce that $v_\Omega^-[\phi]$ is harmonic at infinity. Hence, by (6.10), (6.11), and (6.12) and by the first Green Identity for exterior domains of Corollary 4.7 it follows that

$$0 = \int_{\partial\Omega} (\nu_\Omega \cdot \nabla v_\Omega^+[\phi])\, v_\Omega^+[\mu] - v_\Omega^+[\phi]\,(\nu_\Omega \cdot \nabla v_\Omega^+[\mu])\, d\sigma$$
$$= -\int_{\partial\Omega} v_\Omega^+[\phi]\,\phi\, d\sigma = -\int_{\partial\Omega} v_\Omega^-[\phi]\,(\nu_\Omega \cdot \nabla v_\Omega^-[\phi])\, d\sigma = \int_{\Omega^-} \left|\nabla v_\Omega^-[\phi]\right|^2 dx\,.$$

As a consequence, $v_\Omega^-[\phi]$ is locally constant on Ω^-. Thus, $\nu_\Omega \cdot \nabla v_\Omega^-[\phi] = 0$ and the second equality of (6.10) implies that $\phi = 0$. □

By Theorem 6.25 we can deduce the following Corollary 6.26. The proof can be obtained by a straightforward modification of that of Corollary 6.17 and using Theorems 6.24 and 6.25 instead of Theorems 6.14 and 6.16 (for a similar statement in $L^2(\partial\Omega)$ we refer to Folland [102, Corollary 3.39]).

Corollary 6.26. *We have*

$$C^{0,\alpha}(\partial\Omega) = \mathrm{Im}\left(-\frac{1}{2}I + W_\Omega^t\right) \oplus \mathrm{Ker}\left(-\frac{1}{2}I + W_\Omega^t\right)$$

and

$$C^{1,\alpha}(\partial\Omega) = \mathrm{Im}\left(-\frac{1}{2}I + W_\Omega\right) \oplus \mathrm{Ker}\left(-\frac{1}{2}I + W_\Omega\right), \qquad (6.13)$$

where the sums are direct but not necessarily orthogonal.

6.7 The Dirichlet Problem in Ω

We are now ready to prove the existence of a solution for the Dirichlet problem in the bounded domain Ω. As observed in the introduction of Section 6.5, when the exterior of Ω is connected and $\varkappa^- = 0$, we can obtain the solution in the form of a double layer potential. Indeed, if that is the case, then the operator $\frac{1}{2}I + W_\Omega$ is an isomorphism from $C^{1,\alpha}(\partial\Omega)$ to itself by Corollary 6.15. Instead, if Ω^- has more than one connected component (that is, if $\varkappa^- > 0$), then the null space of $\frac{1}{2}I + W_\Omega$ is nontrivial (cf. Theorem 6.14). Then, the Fredholm Alternative of Theorem 5.8 tells us that also the null space of the transpose operator $\frac{1}{2}I + W_\Omega^t$ is nontrivial and, as a consequence, we cannot expect every function of $C^{1,\alpha}(\partial\Omega)$ to be in the image of $\frac{1}{2}I + W_\Omega$. In such a case, the double layer potential alone will not suffice to write the solution of the Dirichlet problem for any possible boundary datum $g \in C^{1,\alpha}(\partial\Omega)$. To overcome this difficulty we will exploit the direct sum in (6.8). Accordingly, we will split a function $g \in C^{1,\alpha}(\partial\Omega)$ into the sum of a function g_1 in the image of $\frac{1}{2}I + W_\Omega$ and a function g_2 in the null of space of $\frac{1}{2}I + W_\Omega$. Then the Dirichlet problem with boundary datum g_1 has a solution that can be written as a double layer potential, and, by Proposition 6.13, the Dirichlet problem with boundary datum g_2 has a solution that can be written as a single layer potential with a density that belongs to the null space of $\frac{1}{2}I + W^t_\Omega$. By linearity, the sum of these two solutions is the solution of the problem with boundary datum g. We use this argument in the proof the following Theorem 6.27, where we show an existence result for the interior Dirichlet problem.

Theorem 6.27. *If $g \in C^{1,\alpha}(\partial\Omega)$, then the boundary value problem*

$$\begin{cases} \Delta u = 0 & \text{in } \Omega\,, \\ u = g & \text{on } \partial\Omega \end{cases} \tag{6.14}$$

has one and only one solution $u \in C^{1,\alpha}(\overline{\Omega})$.

Proof. By the direct sum in (6.8) there exist a function $g_1 \in \mathrm{Im}(\frac{1}{2}I + W_\Omega)$ and a function $g_2 \in \mathrm{Ker}(\frac{1}{2}I + W_\Omega)$ such that

$$g = g_1 + g_2$$

(here $g_2 = 0$ when $\varkappa^- = 0$). Then there exists $\psi \in C^{1,\alpha}(\partial\Omega)$ such that

$$\frac{1}{2}\psi + W_\Omega[\psi] = g_1$$

and by the jump formula for the double layer potential of Theorem 6.6 it follows that

$$w^+_\Omega[\psi]_{|\partial\Omega} = g_1\,. \tag{6.15}$$

Moreover, by the membership of g_2 in $\mathrm{Ker}(\frac{1}{2}I + W_\Omega)$ and by Proposition 6.13, there exists a function $\mu \in \mathrm{Ker}(\frac{1}{2}I + W^t_\Omega) \subseteq C^{0,\alpha}(\partial\Omega)$ such that

$$V_\Omega[\mu] = v_\Omega[\mu]_{|\partial\Omega} = g_2\,. \tag{6.16}$$

Then, by equalities (6.15) and (6.16), by Theorem 4.22 (i) and Proposition 4.28 (i) on the harmonicity of the single and double layer potentials, and by the regularity Theorems 4.25 (i) and 4.31 (i) for the single and double layer potentials, we can verify that

$$u \equiv w^+_\Omega[\psi] + v^+_\Omega[\mu]$$

is a solution of (6.14) and belongs to $C^{1,\alpha}(\overline{\Omega})$. The uniqueness of u is a consequence of Theorem 3.6. □

To work out some of the perturbation problems that are the objective of this book, we need to show in a constructive way how to obtain the solution of a Dirichlet problem solving a specific set of boundary integral equations. For this purpose, we can exploit the argument in the proof of Theorem 6.27 and write the solution as the sum of a single and a double layer potential. We still have, however, to identify the corresponding densities as the solution of a specific system of integral equations, and to do so we have to write a suitable expression for the functions g_1 and g_2. Then, with this goal in mind, we introduce some convenient bases for $\mathrm{Ker}(\frac{1}{2}I + W_\Omega)$ and $\mathrm{Ker}(\frac{1}{2}I + W_\Omega^t)$.

We begin with some notation. If Σ is a subset of $\partial\Omega$, then we denote by ζ_Σ the map from $\partial\Omega$ to $\mathbb{R}$ defined by

$$\zeta_\Sigma(x) \equiv \begin{cases} 1 & \text{if } x \in \Sigma, \\ 0 & \text{if } x \in \partial\Omega \setminus \Sigma. \end{cases} \tag{6.17}$$

So that

$$\zeta_{\partial(\Omega^-)_l}(x) = \delta_{l,j} \qquad \forall x \in \partial(\Omega^-)_j\,,\ l,j \in \{1,\dots,\varkappa^-\},$$

where $\delta_{l,j}$ denotes the Kronecker delta function defined by $\delta_{l,j} = 1$ if $l = j$ and $\delta_{l,j} = 0$ if $l \neq j$, for all $l,j \in \mathbb{N}$. Then Theorem 6.14 implies that for $\varkappa^- > 0$ the set $\{\zeta_{\partial(\Omega^-)_1},\dots,\zeta_{\partial(\Omega^-)_{\varkappa^-}}\}$ is a basis for $\mathrm{Ker}(\frac{1}{2}I + W_\Omega)$. Moreover, by Theorem 6.16, the linear map from the $\varkappa^-$-dimensional space $\mathrm{Ker}(\frac{1}{2}I + W_\Omega^t)$ to $\mathbb{R}^{\varkappa^-}$ that takes τ to $\left(\int_{\partial(\Omega^-)_j} \tau\, d\sigma\right)_{j=1,\dots,\varkappa^-}$ is an isomorphism. Then we deduce the validity of the following.

Lemma 6.28. *If $\varkappa^- > 0$, then for each $l \in \{1,\dots,\varkappa^-\}$ there exists a unique function $\tau_l \in C^{0,\alpha}(\partial\Omega)$ such that*

$$(\frac{1}{2}I + W_\Omega^t)[\tau_l] = 0 \quad \text{and} \quad \int_{\partial(\Omega^-)_j} \tau_l\, d\sigma = \delta_{l,j} \quad \forall j \in \{1,\dots,\varkappa^-\}$$

and the set $\{\tau_1,\dots,\tau_{\varkappa^-}\}$ is a basis for $\mathrm{Ker}(\frac{1}{2}I + W_\Omega^t)$.

Moreover, by Proposition 6.13, we have the following.

Lemma 6.29. *If $\varkappa^- > 0$, then $\{V_\Omega[\tau_1],\dots,V_\Omega[\tau_{\varkappa^-}]\}$ is a basis for* $\mathrm{Ker}(\frac{1}{2}I+W_\Omega)$.

Now we have two bases for $\mathrm{Ker}(\frac{1}{2}I + W_\Omega)$. Namely,

$$\{\zeta_{\partial(\Omega^-)_1},\dots,\zeta_{\partial(\Omega^-)_{\varkappa^-}}\} \quad \text{and} \quad \{V_\Omega[\tau_1],\dots,V_\Omega[\tau_{\varkappa^-}]\}\,.$$

Thus, we can consider the corresponding transition matrix from one basis to the other. Since the basis $\{\zeta_{\partial(\Omega^-)_1},\dots,\zeta_{\partial(\Omega^-)_{\varkappa^-}}\}$ is orthogonal in $L^2(\partial\Omega)$, an elementary argument of linear algebra shows the validity of the following Lemma 6.30.

Lemma 6.30. *Let $\varkappa^- > 0$. Let $\Lambda_{\Omega^-} \equiv (\lambda_{\Omega^-}^{l,j})_{(l,j)\in\{1,\dots,\varkappa^-\}^2}$ be the real $\varkappa^- \times \varkappa^-$ matrix with entries $\lambda_{\Omega^-}^{l,j}$ defined by*

$$\lambda_{\Omega^-}^{l,j} \equiv \frac{1}{m_{n-1}(\partial(\Omega^-)_j)} \int_{\partial\Omega} V_\Omega[\tau_l]\, \zeta_{\partial(\Omega^-)_j} d\sigma = \fint_{\partial(\Omega^-)_j} V_\Omega[\tau_l]\, d\sigma$$

for all $(l,j) \in \{1,\dots,\varkappa^-\}^2$. Then Λ_{Ω^-} is invertible and we have

$$V_\Omega[\tau_l] = \sum_{j=1}^{\varkappa^-} \lambda^{l,j}_{\Omega^-} \zeta_{\partial(\Omega^-)_j} \qquad on\ \partial\Omega$$

for all $l \in \{1, \dots, \varkappa^-\}$.

(We recall that $m_{n-1}(M) = \int_M d\sigma$ for all $(n-1)$-dimensional manifolds M of class C^1 embedded in $\mathbb{R}^n$.)

We are now ready to deduce the validity of the following Theorem 6.31, where we write the solution of the interior Dirichlet problem as a linear combination of single and double layer potentials whose densities solve a certain system of boundary integral equations.

Theorem 6.31. *Let* $g \in C^{1,\alpha}(\partial\Omega)$. *Then the following statements hold.*

(i) If $\varkappa^- = 0$, *then there exists one and only one function* $\psi_g \in C^{1,\alpha}(\partial\Omega)$ *such that*

$$\left(\frac{1}{2}I + W_\Omega\right)[\psi_g] = g\,.$$

If $\varkappa^- > 0$, *then there exists one and only one function* $\psi_g \in C^{1,\alpha}(\partial\Omega)$ *such that*

$$\begin{cases} (\frac{1}{2}I + W_\Omega)[\psi_g] = g - \sum_{l=1}^{\varkappa^-}\left(\int_{\partial\Omega} g\tau_l\, d\sigma\right)\zeta_{\partial(\Omega^-)_l}\,, & \\ \int_{\partial(\Omega^-)_j} \psi_g\, d\sigma = 0 & \forall j \in \{1, \dots, \varkappa^-\}. \end{cases} \tag{6.18}$$

(ii) If $u \in C^{1,\alpha}(\overline{\Omega})$ *is the unique solution of problem* (6.14), *then we have*

$$u = w^+_\Omega[\psi_g] + \sum_{l,j=1}^{\varkappa^-}\left(\int_{\partial\Omega} g\tau_l\, d\sigma\right)(\Lambda^{-1}_{\Omega^-})_{l,j} v^+_\Omega[\tau_j]\,, \tag{6.19}$$

with the τ_l*'s as in Lemma 6.28 and* Λ_{Ω^-} *as in Lemma 6.30 and where we understand that the sum is zero for* $\varkappa^- = 0$.

Proof. (i) The first part of the statement on the case where $\varkappa^- = 0$ is a consequence of Corollary 6.15. To prove the second part for $\varkappa^- > 0$ we observe that, by Lemma 6.28, the function

$$g_1 \equiv g - \sum_{l=1}^{\varkappa^-}\left(\int_{\partial\Omega} g\tau_l\, d\sigma\right)\zeta_{\partial(\Omega^-)_l}$$

in the right-hand side of the first equation of (6.18) satisfies the equalities

$$\int_{\partial\Omega} \tau_j g_1\, d\sigma = 0 \qquad \forall j \in \{1, \dots, \varkappa^-\}$$

and it is therefore orthogonal to $\mathrm{Ker}(\frac{1}{2}I + W^t_\Omega)$ with respect to the dual system $\langle C^{0,\alpha}(\partial\Omega), C^{1,\alpha}(\partial\Omega)\rangle$ induced by the non-degenerate bilinear form (6.1). Accordingly, the Fredholm Alternative of Theorem 5.8 implies that g_1 is in the image of $\frac{1}{2}I + W_\Omega$.

Since $\{\zeta_{\partial(\Omega^-)_1}, \ldots, \zeta_{\partial(\Omega^-)_{\varkappa^-}}\}$ is a basis of $\mathrm{Ker}(\frac{1}{2}I + W_\Omega)$ and is orthogonal in $L^2(\partial\Omega)$ (cf. Theorem 6.14), we can also verify that a function ψ_g that satisfies both the equations of (6.18) exists and is unique. (We observe that here we have introduced an expression for the functions g_1 and $g_2 = g - g_1$ of the proof of Theorem 6.27.)

(ii) Let $\tilde{u}$ be the function in the right-hand side of (6.19). By the regularity Theorems 4.25 and 4.31 for the single and the double layer potentials, $\tilde{u}$ is of class $C^{1,\alpha}(\overline{\Omega})$. By the definition of ψ_g in statement (i) and by the jump formula for the double layer potential of Theorem 6.6 we have

$$w^+_\Omega[\psi_g]_{|\partial\Omega} = g - \sum_{l=1}^{\varkappa^-} \Big(\int_{\partial\Omega} g\tau_l \, d\sigma\Big) \zeta_{\partial(\Omega^-)_l} ,$$

where we omit the sum for $\varkappa^- = 0$. If $\varkappa^- > 0$, then Lemma 6.30 implies that

$$\sum_{l,j=1}^{\varkappa^-} \Big(\int_{\partial\Omega} g\tau_l \, d\sigma\Big) (\Lambda^{-1}_{\Omega^-})_{l,j} v^+_\Omega[\tau_j]_{|\partial\Omega} = \sum_{l=1}^{\varkappa^-} \Big(\int_{\partial\Omega} g\tau_l \, d\sigma\Big) \zeta_{\partial(\Omega^-)_l} .$$

Then $\tilde{u}_{|\partial\Omega} = g$. We conclude that $\tilde{u}$ is a solution of the Dirichlet problem (6.14) and thus $u = \tilde{u}$ by the uniqueness Theorem 3.6. □

We shall see in what follows that there are many ways to represent the solution of a boundary value problem for the Laplace equation $\Delta u = 0$ by means of layer potentials. Depending on the problem under consideration and the specific properties that one wishes to emphasize, one might choose one representation or another. For example, the expression (6.19) obtained in Theorem 6.31 will be very useful to study a Dirichlet problem in a planar domain with one or more small holes which shrink to points as we do in Sections 8.4 and 10.4.

As a consequence of Theorem 6.31, when $\varkappa^- = 0$ the image of the map

$$C^{1,\alpha}(\partial\Omega) \ni \psi \mapsto w^+_\Omega[\psi] \in C^{1,\alpha}(\overline{\Omega}) \tag{6.20}$$

coincides with the set of the harmonic functions of $C^{1,\alpha}(\overline{\Omega})$. By Theorem 6.31 we can also prove the following result, which characterizes the image of the map in (6.20) for $\varkappa^- > 0$.

Theorem 6.32. *Let $\varkappa^- > 0$. Let $u \in C^{1,\alpha}(\overline{\Omega})$ be harmonic in Ω. Then $u = w^+_\Omega[\psi]$ for some $\psi \in C^{1,\alpha}(\partial\Omega)$ if and only if*

$$\int_{\partial(\Omega^-)_j} \nu_{(\Omega^-)_j} \cdot \nabla u \, d\sigma = 0 \qquad \forall j \in \{1, \ldots, \varkappa^-\} . \tag{6.21}$$

Proof. We first prove that (6.21) holds for all $u = w^+_\Omega[\psi]$ with $\psi \in C^{1,\alpha}(\partial\Omega)$. To do so, we verify that

$$\int_{\partial(\Omega^-)_j} \nu_{(\Omega^-)_j} \cdot \nabla w^+_\Omega[\psi]\, d\sigma = 0 \qquad \forall j \in \{1,\ldots,\varkappa^-\},\ \psi \in C^{1,\alpha}(\partial\Omega). \tag{6.22}$$

Indeed, if $j \in \{1,\ldots,\varkappa^-\}$, then by the jump formulas of Theorem 6.6 we have

$$\nu_{(\Omega^-)_j} \cdot \nabla w^+_\Omega[\psi] = \nu_{(\Omega^-)_j} \cdot \nabla w^-_\Omega[\psi] \qquad \text{on } \partial(\Omega^-)_j\,. \tag{6.23}$$

By Proposition 4.28 (i) the function $w^-_\Omega[\psi]$ is harmonic in Ω^- and, by Theorem 4.31 (ii), it belongs to $C^{1,\alpha}_{\mathrm{loc}}(\overline{\Omega^-})$. Then the restriction of $w^-_\Omega[\psi]$ to $\overline{(\Omega^-)_j}$ is harmonic in $(\Omega^-)_j$ and belongs to $C^{1,\alpha}(\overline{(\Omega^-)_j})$ for all $j \in \{1,\ldots,\varkappa^-\}$. Here we note that we can omit the subscript 'loc' of $C^{1,\alpha}$ because $\overline{(\Omega^-)_j}$ is bounded for $j \neq 0$. Then the first Green Identity of Theorem 4.2 implies that

$$\int_{\partial(\Omega^-)_j} \nu_{(\Omega^-)_j} \cdot \nabla w^-_\Omega[\psi]\, d\sigma = 0 \tag{6.24}$$

for all $j \in \{1,\ldots,\varkappa^-\}$. The validity of (6.22) follows by (6.23) and (6.24).

We now prove the other implication: we assume that (6.21) holds true and we show that $u = w^+_\Omega[\psi]$ for some $\psi \in C^{1,\alpha}(\partial\Omega)$. We shall exploit formula (6.19) for the solution of the Dirichlet problem in Ω in the case where $g = u_{|\partial\Omega}$. If $i \in \{1,\ldots,\varkappa^-\}$, then we verify that

$$\begin{aligned} 0 = \int_{\partial(\Omega^-)_i} \nu_{(\Omega^-)_i} \cdot \nabla u\, d\sigma &= \int_{\partial(\Omega^-)_i} \nu_{(\Omega^-)_i} \cdot \nabla w^+_\Omega[\psi_g]\, d\sigma \\ &+ \sum_{l,j=1}^{\varkappa^-} \Big(\int_{\partial\Omega} g\tau_l\, d\sigma\Big) (\Lambda^{-1}_{\Omega^-})_{l,j} \int_{\partial(\Omega^-)_i} \nu_{(\Omega^-)_i} \cdot \nabla v^+_\Omega[\tau_j]\, d\sigma\,. \end{aligned} \tag{6.25}$$

By (6.22) we have

$$\int_{\partial(\Omega^-)_i} \nu_{(\Omega^-)_i} \cdot \nabla w^+_\Omega[\psi_g]\, d\sigma = 0\,. \tag{6.26}$$

Moreover, by the jump relations of Theorem 6.6 and by the definition of τ_j in Lemma 6.28 we have

$$\begin{aligned} \int_{\partial(\Omega^-)_i} \nu_{(\Omega^-)_i} \cdot \nabla v^+_\Omega[\tau_j]\, d\sigma &= -\int_{\partial(\Omega^-)_i} \nu_\Omega \cdot \nabla v^+_\Omega[\tau_j]\, d\sigma \\ &= -\int_{\partial(\Omega^-)_i} \nu_\Omega \cdot \nabla v^-_\Omega[\tau_j]\, d\sigma + \int_{\partial(\Omega^-)_i} \tau_j\, d\sigma \\ &= \int_{\partial(\Omega^-)_i} \nu_{(\Omega^-)_i} \cdot \nabla v^-_\Omega[\tau_j]\, d\sigma + \delta_{ij}\,. \end{aligned} \tag{6.27}$$

Since the function $v_\Omega^-[\tau_j]$ is harmonic in Ω^- and belongs to $C^{1,\alpha}_{\mathrm{loc}}(\overline{\Omega^-})$ (cf. Theorems 4.22 (i) and 4.25 (ii)), the restriction of $v_\Omega^-[\tau_j]$ to the bounded set $\overline{(\Omega^-)_i}$ is harmonic in $(\Omega^-)_i$ and belongs to $C^{1,\alpha}(\overline{(\Omega^-)_i})$. Then the first Green Identity of Theorem 4.2 implies that

$$\int_{\partial(\Omega^-)_i} \nu_{(\Omega^-)_i} \cdot \nabla v_\Omega^-[\tau_j]\, d\sigma = 0 \tag{6.28}$$

for all $i \in \{1, \dots, \varkappa-\}$. Then, by (6.27) and (6.28), we can write

$$\int_{\partial(\Omega^-)_i} \nu_{(\Omega^-)_i} \cdot \nabla v_\Omega^+[\tau_j]\, d\sigma = \delta_{ji} \tag{6.29}$$

and, by (6.25), (6.26), and (6.29), we conclude that

$$\sum_{l=1}^{\varkappa^-} \Big(\int_{\partial\Omega} g\tau_l\, d\sigma \Big) (\Lambda_{\Omega^-}^{-1})_{l,j} = 0 \qquad \forall j \in \{1, \dots, \varkappa^-\}.$$

Then

$$\sum_{l,j=1}^{\varkappa^-} \Big(\int_{\partial\Omega} g\tau_l\, d\sigma \Big) (\Lambda_{\Omega^-}^{-1})_{l,j} v_\Omega^+[\tau_j] = 0$$

and formula (6.19) of Theorem 6.31 (ii) implies that $u = w_\Omega^+[\psi_g]$ with $\psi_g \in C^{1,\alpha}(\partial\Omega)$. □

A variant of Theorem 6.32 can be obtained from Theorem 8.20 of McLean [206], where the author characterizes the null space of the (bounded self-adjoint) boundary operator corresponding to the normal derivative of the double layer potential (see, e.g., [60] with Costabel and Dauge).

6.8 The Dirichlet Problem in Ω^-

We now turn to the Dirichlet problem in the exterior domain Ω^-. We have the same kind of difficulty we met with the interior problem: If the boundary datum g is in the image of $-\frac{1}{2}I + W_\Omega$, then we can obtain the solution as a double layer potential. The image of $-\frac{1}{2}I + W_\Omega$, however, never fills the whole of $C^{1,\alpha}(\partial\Omega)$ and so the double layer potential is not in general sufficient (this is different for the interior problem, where $\frac{1}{2}I + W_\Omega$ is surjective for $\varkappa^- = 0$). Indeed, any non-empty domain Ω has at least one connected component, thus we always have $\varkappa^+ > 0$ and, by Theorem 6.24 and by the Fredholm Alternative of Theorem 5.8, we can see that the null spaces of $-\frac{1}{2}I + W_\Omega$ and $-\frac{1}{2}I + W_\Omega^t$ are non-trivial and that $\mathrm{Im}(-\frac{1}{2}I + W_\Omega)$ is properly contained in $C^{1,\alpha}(\partial\Omega)$. Then we resort to the same strategy employed in the interior case and we exploit the direct sum of (6.13) to split g into a sum $g_1 + g_2$

with g_1 in the image of $-\frac{1}{2}I + W_\Omega$ and g_2 in the null space of the same operator. The solution of the Dirichlet problem with boundary datum g_1 can be obtained as a double layer potential and the solution of the problem with datum g_2 as a single layer if $n \geq 3$, and as the sum of a single layer and a constant function if $n = 2$ (cf. Propositions 6.19 and 6.23). Then the solution with boundary datum g is given by the sum of the solutions for g_1 and g_2. With this argument we now prove the following existence theorem.

Theorem 6.33. *Let $g \in C^{1,\alpha}(\partial\Omega)$. The boundary value problem*

$$\begin{cases} \Delta u = 0 & \text{in } \Omega^-, \\ u = g & \text{on } \partial\Omega, \\ u \text{ is harmonic at infinity} \end{cases} \tag{6.30}$$

has one and only one solution $u \in C^{1,\alpha}_{\mathrm{loc}}(\overline{\Omega^-})$.

Proof. By the direct sum in (6.13) there exist a function $g_1 \in \mathrm{Im}(-\frac{1}{2}I + W_\Omega)$ and a function $g_2 \in \mathrm{Ker}(-\frac{1}{2}I + W_\Omega)$ such that

$$g = g_1 + g_2 .$$

We first consider the Dirichlet problem with boundary datum g_1. Since g_1 belongs to $\mathrm{Im}(-\frac{1}{2}I + W_\Omega)$, there exists $\psi \in C^{1,\alpha}(\partial\Omega)$ such that

$$-\frac{1}{2}\psi + W_\Omega[\psi] = g_1$$

and the jump formula for the double layer potential of Theorem 6.6 implies that

$$w^-_\Omega[\psi]_{|\partial\Omega} = g_1 .$$

Moreover, the function $w^-_\Omega[\psi]$ is harmonic and harmonic at infinity by Proposition 4.28 and belongs to $C^{1,\alpha}_{\mathrm{loc}}(\overline{\Omega^-})$ by the regularity Theorem 4.31 (ii).

We now turn to the problem with datum g_2 and we consider separately the cases of dimension $n \geq 3$ and $n = 2$. For $n \geq 3$, the membership of g_2 in the null space of $-\frac{1}{2}I + W_\Omega$ and Proposition 6.19 imply that there exists a function μ in $\mathrm{Ker}(-\frac{1}{2}I + W^t_\Omega) \subseteq C^{0,\alpha}(\partial\Omega)$ such that

$$V_\Omega[\mu] = v_\Omega[\mu]_{|\partial\Omega} = g_2 .$$

Then, by Theorems 4.22 and 4.23 (i) the function $v^-_\Omega[\mu]$ is harmonic, harmonic at infinity, and solves the Dirichlet problem with boundary datum g_2. Moreover, $v^-_\Omega[\mu] \in C^{1,\alpha}_{\mathrm{loc}}(\overline{\Omega^-})$ by the regularity Theorems 4.25 (ii). Hence,

$$u \equiv w^-_\Omega[\psi] + v^-_\Omega[\mu]$$

is a solution of (6.30) and belongs to $C^{1,\alpha}_{\mathrm{loc}}(\overline{\Omega^-})$.

Finally, we consider the case when $n = 2$. By Proposition 6.23, there exist a function $\mu \in \mathrm{Ker}(-\frac{1}{2}I + W_\Omega^t)_0 \subseteq C^{0,\alpha}(\partial\Omega)_0$ and a constant function ρ on $\partial\Omega$ such that

$$V_\Omega[\mu] + \rho = v_\Omega[\mu]_{|\partial\Omega} + \rho = g_2 .$$

By Theorems 4.22 and 4.23 (ii), the function $v_\Omega^-[\mu]$ is harmonic in Ω^-, harmonic at infinity, and equals $g_2 - \rho$ on $\partial\Omega$ (we note that $\int_{\partial\Omega} \mu \, d\sigma = 0$). Moreover, $v_\Omega^-[\mu]$ belongs to $C^{1,\alpha}_{\mathrm{loc}}(\overline{\Omega^-})$ by the regularity Theorem 4.25 (ii). Then, if we still denote by ρ the constant extension of ρ to $\overline{\Omega^-}$, we have that

$$u \equiv w_\Omega^-[\psi] + v_\Omega^-[\mu] + \rho$$

is a solution of (6.30) and belongs to $C^{1,\alpha}_{\mathrm{loc}}(\overline{\Omega^-})$.

Both for $n \geq 3$ and for $n = 2$, the uniqueness of the solution u is a consequence of Theorem 6.1 □

We can exploit the argument in the proof of Theorem 6.33 to write the solution of an exterior Dirichlet problem as the sum of a single and a double layer potential and, for $n = 2$, a constant function. But, to identify the corresponding densities as the solutions of a specific set of boundary integral equations, we need to write a suitable expression for the functions g_1 and g_2.

To do so, we now introduce some convenient bases for the null spaces of the operators $-\frac{1}{2}I + W_\Omega$ and $-\frac{1}{2}I + W_\Omega^t$. With the notation introduced in (6.17) we have

$$\zeta_{\partial\Omega_l}(x) = \delta_{l,j} \qquad \forall x \in \partial\Omega_j \,,\, l, j \in \{1, \ldots, \varkappa^+\}$$

and Theorem 6.24 implies that $\{\zeta_{\partial\Omega_1}, \ldots, \zeta_{\partial\Omega_{\varkappa^+}}\}$ is a basis for the null space of $-\frac{1}{2}I + W_\Omega$. Moreover, by Theorem 6.25 the linear map from the $\varkappa^+$-dimensional space $\mathrm{Ker}(-\frac{1}{2}I + W_\Omega^t)$ to $\mathbb{R}^{\varkappa^+}$ that takes τ to $\left(\int_{\partial\Omega_j} \tau \, d\sigma\right)_{j=1,\ldots,\varkappa^+}$ is an isomorphism and thus we deduce the validity of the following.

Lemma 6.34. *For each $l \in \{1, \ldots, \varkappa^+\}$ there exists a unique function ς_l in $C^{0,\alpha}(\partial\Omega)$ such that*

$$(-\frac{1}{2}I + W_\Omega^t)[\varsigma_l] = 0 \quad \textit{and} \quad \int_{\partial\Omega_j} \varsigma_l \, d\sigma = \delta_{l,j} \quad \forall j \in \{1, \ldots, \varkappa^+\}$$

and the set $\{\varsigma_1, \ldots, \varsigma_{\varkappa^+}\}$ is a basis for $\mathrm{Ker}(-\frac{1}{2}I + W_\Omega^t)$.

Then, for $n \geq 3$ we have the following Lemma 6.35, which follows from Proposition 6.19

Lemma 6.35. *If $n \geq 3$, then for each $l \in \{1, \ldots, \varkappa^+\}$ there exists a unique function $\psi_l \in \mathrm{Ker}(-\frac{1}{2}I + W_\Omega^t)$ such that $V_\Omega[\psi_l] = \zeta_{\partial\Omega_l}$.*

In addition, by Proposition 6.23 we have the following.

Lemma 6.36. *For each $l \in \{1, \ldots, \varkappa^+\}$ there exists a unique function ϕ_l in $\mathrm{Ker}(-\frac{1}{2}I + W^t_\Omega)_0$ and a unique constant $\rho_l \in \mathbb{R}$ such that $V_\Omega[\phi_l] + \rho_l = \zeta_{\partial\Omega_l}$.*

We are now ready to deduce the validity of the following.

Theorem 6.37. *Let $g \in C^{1,\alpha}(\partial\Omega)$. Then the following statements hold.*

(i) There exists one and only one function $\psi_g \in C^{1,\alpha}(\partial\Omega)$ such that

$$\begin{cases} (-\frac{1}{2}I + W_\Omega)[\psi_g] = g - \sum_{l=1}^{\varkappa^+}\left(\int_{\partial\Omega} g\varsigma_l \, d\sigma\right)\zeta_{\partial\Omega_l} \,, \\ \int_{\partial\Omega_j} \psi_g \, d\sigma = 0 & \forall j \in \{1, \ldots, \varkappa^+\}. \end{cases} \tag{6.31}$$

(ii) If $n \geq 3$ and $u \in C^{1,\alpha}_{\mathrm{loc}}(\overline{\Omega^-})$ is the unique solution of (6.30), *then we have*

$$u = w^-_\Omega[\psi_g] + \sum_{l=1}^{\varkappa^+}\left(\int_{\partial\Omega} g\varsigma_l \, d\sigma\right) v^-_\Omega[\psi_l] \,, \tag{6.32}$$

where the ψ_l are as in Lemma 6.35.

(iii) If $n = 2$ and $u \in C^{1,\alpha}_{\mathrm{loc}}(\overline{\Omega^-})$ is the unique solution of (6.30), *then we have*

$$u = w^-_\Omega[\psi_g] + \sum_{l=1}^{\varkappa^+}\left(\int_{\partial\Omega} g\varsigma_l \, d\sigma\right)(v^-_\Omega[\phi_l] + \rho_l) \,, \tag{6.33}$$

where ϕ_l and ρ_l are as in Lemma 6.36.

Proof. (i) By Lemma 6.34 we can verify that the function

$$g_1 \equiv g - \sum_{l=1}^{\varkappa^+}\left(\int_{\partial\Omega} g\varsigma_l \, d\sigma\right)\zeta_{\partial\Omega_l}$$

in the right-hand side of the first equation in (6.31) is orthogonal to the null space $\mathrm{Ker}(-\frac{1}{2}I + W^t_\Omega)$ with respect to the dual system $\langle C^{0,\alpha}(\partial\Omega), C^{1,\alpha}(\partial\Omega)\rangle$ induced by the non-degenerate bilinear form (6.1). Then by Theorem 6.5 and by the Fredholm Alternative of Theorem 5.8 we deduce the existence of ψ_g. The uniqueness of ψ_g is a consequence of Theorem 6.24. (We observe that g_1 and $g_2 \equiv g - g_1$ are the functions that appear in the proof of Theorem 6.33.)

(ii) Let $\tilde{u}$ be the function in the right hand side of (6.32). By the regularity Theorems 4.25 (ii) and 4.31 (ii), $\tilde{u}$ is of class $C^{1,\alpha}_{\mathrm{loc}}(\overline{\Omega^-})$. By Theorems 4.22 (i), 4.23 (i), and Proposition 4.28, $\tilde{u}$ is harmonic in Ω^- and harmonic at infinity. By statement (i) and by the jump formula for the double layer potential of Theorem 6.6 we have

$$w^-_\Omega[\psi_g]_{|\partial\Omega} = g - \sum_{l=1}^{\varkappa^+}\left(\int_{\partial\Omega} g\varsigma_l \, d\sigma\right)\zeta_{\partial\Omega_l} \,.$$

By Theorem 4.22 (ii) and by Lemma 6.35, we have

$$\sum_{l=1}^{\varkappa^+}\Big(\int_{\partial\Omega} g\varsigma_l \, d\sigma\Big) v_\Omega^-[\psi_l]_{|\partial\Omega} = \sum_{l=1}^{\varkappa^+}\Big(\int_{\partial\Omega} g\varsigma_l \, d\sigma\Big)\zeta_{\partial\Omega_l} \, .$$

Hence, $\tilde{u}_{|\partial\Omega} = g$. Then $\tilde{u}$ is a solution of (6.30) and by the uniqueness Proposition 6.1 it follows that $\tilde{u}$ coincide with the solution u of Theorem 6.33.

(iii) The proof of statement (iii) is very similar to that of statement (ii), we just need to use Lemma 6.36 instead of Lemma 6.35 and Theorem 4.23 (ii) instead of Theorem 4.23 (i). □

We observe that, to prove Theorem 6.37, we did not follow the same path as for the analogous Theorem 6.31 for the interior problem. In particular, we did not introduce a transition matrix similar to Λ_{Ω^-}. Theorem 6.37 is, however, presented only for the sake of completeness, and so we opted for the fastest proof. On the contrary, Theorem 6.31 is largely used in the book and we exploit the specific form of formula (6.19) to analyze certain perturbation problems (cf. Chapters 8 and 10).

If Ω is connected (i.e., $\varkappa^+ = 1$), then we have the following variant of Theorem 6.37 (i).

Lemma 6.38. *Let Ω be connected. Let $g \in C^{1,\alpha}(\partial\Omega)$. Then there exists one and only pair $(\psi_g, c_g) \in C^{1,\alpha}(\partial\Omega)_0 \times \mathbb{R}$ such that*

$$(-\frac{1}{2}I + W_\Omega)[\psi_g] + c_g = g \, . \tag{6.34}$$

Proof. By Theorem 6.24 the null space of $-\frac{1}{2}I + W_\Omega$ consists of the constant functions on $\partial\Omega$. Then, by the direct sum in (6.13), there exist a pair $(\phi_g, c_g) \in C^{1,\alpha}(\partial\Omega) \times \mathbb{R}$ that satisfies (6.34). Taking $\psi_g \equiv \phi_g - \fint_{\partial\Omega} \phi_g \, d\sigma$, we obtain a pair (ψ_g, c_g) as in the statement. The uniqueness of (ψ_g, c_g) follows by the direct sum (6.13), by Theorem 6.24, and by condition $\int_{\partial\Omega} \psi_g d\sigma = 0$. □

Then, when $n = 2$ and Ω is connected we do not need the single layer potential to write the solution of the exterior Dirichlet problem. Indeed, we have the following.

Proposition 6.39. *Assume that $n = 2$ and Ω is connected. Let $g \in C^{1,\alpha}(\partial\Omega)$ and let $u \in C^{1,\alpha}_{\mathrm{loc}}(\overline{\Omega^-})$ be the unique solution of the boundary value problem* (6.30). *Then we have*

$$u = w_\Omega^-[\psi_g] + c_g = w_\Omega^-[\psi_g] + \lim_{y\to\infty} u(y) \, ,$$

where $(\psi_g, c_g) \in C^{1,\alpha}(\partial\Omega)_0 \times \mathbb{R}$ is the unique solution of (6.34).

Proof. The first equality follows from Lemma 6.38, from the jump formula for $w_\Omega^-[\psi_g]$ in Theorem 6.6, and from the uniqueness of the solution of the Dirichlet problem in Ω^- of Proposition 6.1. For the second equality we note that $w_\Omega^-[\psi_g](y)$ vanishes as $y \to \infty$ by Proposition 4.28 (ii). □

6.9 The Neumann Problem in Ω and Ω^-

For the interior and exterior Neumann problems to have solutions, the boundary datum g has to satisfy certain compatibility conditions, as we show in the following Lemmas 6.40 and 6.41.

Lemma 6.40. *Let $g \in C^{0,\alpha}(\partial\Omega)$. If the interior Neumann problem*

$$\begin{cases} \Delta u = 0 & \text{in } \Omega\,, \\ \nu_\Omega \cdot \nabla u = g & \text{on } \partial\Omega\,, \end{cases} \tag{6.35}$$

has a solution $u \in C^{1,\alpha}(\overline{\Omega})$, then

$$\int_{\partial\Omega_j} g\, d\sigma = 0 \qquad \forall j \in \{1, \ldots, \kappa^+\}\,. \tag{6.36}$$

Proof. If $u \in C^{1,\alpha}(\overline{\Omega})$ is a solution of (6.35) and $j \in \{1, \ldots, \kappa^+\}$, then by the first Green Identity of Theorem 4.2 we have

$$\int_{\partial\Omega_j} g\, d\sigma = \int_{\partial\Omega_j} 1\, \nu_\Omega \cdot \nabla u\, d\sigma = \int_{\Omega_j} \nabla 1 \cdot \nabla u\, dx + \int_{\Omega_j} \Delta u\, dx = 0$$

and the lemma is proved. □

Lemma 6.41. *Let $g \in C^{0,\alpha}(\partial\Omega)$. If the exterior Neumann problem*

$$\begin{cases} \Delta u = 0 & \text{in } \Omega^-\,, \\ \nu_\Omega \cdot \nabla u = g & \text{on } \partial\Omega\,, \\ u \text{ is harmonic at infinity,} \end{cases} \tag{6.37}$$

has a solution $u \in C^{1,\alpha}_{\mathrm{loc}}(\overline{\Omega^-})$, then

$$\int_{\partial(\Omega^-)_j} g\, d\sigma = 0 \qquad \forall j \in \{1, \ldots, \kappa^-\}\,. \tag{6.38}$$

In addition, if $n = 2$, then we also have

$$\int_{\partial(\Omega^-)_0} g\, d\sigma = 0\,. \tag{6.39}$$

Proof. The proof of (6.38) is very similar to that of (6.36). To show the validity of (6.39) we observe that, when $n = 2$, the constant function identically equal to 1 on $(\Omega^-)_0$ is harmonic and harmonic at infinity (cf. Theorem 3.22). Then, by the first Green Identity in exterior domains of Corollary 4.7, we deduce that

$$\int_{\partial(\Omega^-)_0} g\,d\sigma = \int_{\partial(\Omega^-)_0} 1\,\nu_\Omega\cdot\nabla u\,d\sigma = -\lim_{R\to+\infty}\int_{(\Omega^-)_0\cap\mathbb{B}_2(0,R)} \nabla 1\cdot\nabla u\,dx = 0\,.$$

□

We now prove the existence of solutions to the interior and exterior Neumann problems. We begin with Theorem 6.42, where we consider the interior case.

Theorem 6.42. *If $g \in C^{0,\alpha}(\partial\Omega)$ satisfies the compatibility condition* (6.36), *then the integral equation*

$$\left(-\frac{1}{2}I + W^t_\Omega\right)[\phi] = g \tag{6.40}$$

has a solution $\phi \in C^{0,\alpha}(\partial\Omega)$ and the single layer

$$v^+_\Omega[\phi] \in C^{1,\alpha}(\overline{\Omega})$$

is a solution of the interior Neumann boundary value problem (6.35). *The set of all solutions of* (6.35) *consists of the functions $u \in C^{1,\alpha}(\overline{\Omega})$ such that the difference $u - v^+_\Omega[\phi]$ is constant on each connected component $\Omega_1, \ldots, \Omega_{\kappa^+}$ of Ω.*

Proof. If $g \in C^{0,\alpha}(\partial\Omega)$ satisfies the compatibility condition (6.36), then, by Theorem 6.24, it is orthogonal to the null space of the integral operator $-\frac{1}{2}I + W_\Omega$. Accordingly, the Fredholm Alternative of Theorem 5.8 implies that equation (6.40) has a solution $\phi \in C^{0,\alpha}(\partial\Omega)$. Then, Theorem 4.22 implies that $v^+_\Omega[\phi]$ is harmonic in Ω and Theorem 4.25 implies that it is a function of $C^{1,\alpha}(\overline{\Omega})$. By the jump formula for the normal derivative of single layer potential of Theorem 6.6, it follows that $v^+_\Omega[\phi]$ is a solution of (6.35). The last statement of the theorem is a consequence of the uniqueness result of Theorem 6.2 (i). □

For the exterior Neumann problem we find convenient to describe separately the case of dimension $n \geq 3$ and the case of dimension $n = 2$.

Theorem 6.43. *If $n \geq 3$ and $g \in C^{0,\alpha}(\partial\Omega)$ satisfies the compatibility condition* (6.38), *then the integral equation*

$$\left(\frac{1}{2}I + W^t_\Omega\right)[\phi] = g$$

has a solution $\phi \in C^{0,\alpha}(\partial\Omega)$ and the single layer

$$v^-_\Omega[\phi] \in C^{1,\alpha}_{\mathrm{loc}}(\overline{\Omega^-})$$

is a solution of the exterior Neumann boundary value problem (6.37). *The set of all solutions of* (6.37) *consists of the functions $u \in C^{1,\alpha}_{\mathrm{loc}}(\overline{\Omega^-})$ such that the difference $u - v^-_\Omega[\phi]$ is constant on each bounded connected component $(\Omega^-)_1, \ldots, (\Omega^-)_{\kappa^-}$ of Ω^- and is identically equal to 0 on the unbounded connected component $(\Omega^-)_0$.*

Proof. The proof is very similar to that of Theorem 6.42. We just have to use Theorem 6.14 instead of Theorem 6.24 and notice that $v^-_\Omega[\phi]$ is harmonic at infinity by

Theorem 4.23 (i). The last statement of the theorem follows by the uniqueness result of Theorem 6.2 (ii). □

Finally, for the two-dimensional case we have the following.

Theorem 6.44. *If* $n = 2$ *and* $g \in C^{0,\alpha}(\partial\Omega)$ *satisfies the compatibility conditions* (6.38) *and* (6.39), *then the integral equation*

$$\left(\frac{1}{2}I + W_\Omega^t\right)[\phi] = g \tag{6.41}$$

has a solution $\phi \in C^{0,\alpha}(\partial\Omega)_0$ *and the single layer*

$$v_\Omega^-[\phi] \in C^{1,\alpha}_{\mathrm{loc}}(\overline{\Omega})$$

is a solution of the exterior Neumann boundary value problem (6.37). *Moreover, the set of all solutions of* (6.37) *consists of the functions* $u \in C^{1,\alpha}_{\mathrm{loc}}(\overline{\Omega^-})$ *such that the difference* $u - v_\Omega^-[\phi]$ *is constant on each (bounded and unbounded) connected component* $(\Omega^-)_0, \ldots, (\Omega^-)_{\kappa^-}$ *of* Ω^-.

Proof. The proof is very similar to that of Theorems 6.42 and 6.43. We only observe that, if g satisfies (6.38) and (6.39) and ϕ is a solution of (6.41), then Lemma 6.11 implies that

$$0 = \int_{\partial\Omega} g\, d\sigma = \int_{\partial\Omega} \left(\frac{1}{2}I + W_\Omega^t\right)[\phi]\, d\sigma = \int_{\partial\Omega} \phi\, d\sigma\,.$$

Accordingly, $v_\Omega^-[\phi]$ is harmonic at infinity by Theorem 4.23 (ii). □

6.10 Further Mapping Properties of V_Ω

We have already seen in Theorem 6.4 that V_Ω is bounded from $C^{0,\alpha}(\partial\Omega)$ to $C^{1,\alpha}(\partial\Omega)$. In this section we show that if $n \geq 3$, then V_Ω is actually an isomorphism and that, both for $n = 2$ and $n \geq 3$, the map that takes a function $\phi \in C^{0,\alpha}(\partial\Omega)$ with zero integral mean and a real number ρ to the sum $V_\Omega[\phi] + \rho$ is an isomorphism between $C^{0,\alpha}(\partial\Omega)_0 \times \mathbb{R}$ and $C^{1,\alpha}(\partial\Omega)$. The results of this section will be useful to analyze the Dirichlet problem in a domain with a small hole, especially in the case where the dimension n is three or bigger, problems with mixed boundary conditions, transmission problems, and problems with Robin boundary conditions.

We begin with the following lemma, where we show that any function of $C^{1,\alpha}(\partial\Omega)$ can be written as the sum of a single layer potential and a real number, which can be 0 for $n \geq 3$.

Lemma 6.45. *If* $f \in C^{1,\alpha}(\partial\Omega)$, *then there exist* $\phi \in C^{0,\alpha}(\partial\Omega)$ *and* $\rho \in \mathbb{R}$ *such that* $V_\Omega[\phi] + \rho = f$. *If* $n = 2$, *then we can choose* ϕ *and* ρ *so that* $\int_{\partial\Omega} \phi\, d\sigma = 0$. *If* $n \geq 3$, *then we can take* ϕ *and* ρ *so that* $\rho = 0$.

Proof. By the solvability Theorem 6.27 for the interior Dirichlet problem, there exists a function $u^+ \in C^{1,\alpha}(\overline{\Omega})$ that is harmonic in Ω and such that $u^+ = f$ on $\partial\Omega$. Hence, the third Green Identity in the form of Theorem 6.9 implies that

$$\frac{1}{2}f = W_\Omega[f] - V_\Omega[\nu_\Omega \cdot \nabla u^+_{|\partial\Omega}] \,. \tag{6.42}$$

By the solvability Theorem 6.33 for the exterior Dirichlet problem, there exists a function $u^- \in C^{1,\alpha}_{\mathrm{loc}}(\overline{\Omega^-})$ that is harmonic in Ω^-, harmonic at infinity, and such that $u^- = f$ on $\partial\Omega$. Then the third Green Identity for exterior domains in the form of Theorem 6.10 implies that

$$\frac{1}{2}f = -W_\Omega[f] + V_\Omega[\nu_\Omega \cdot \nabla u^-_{|\partial\Omega}] + \rho \tag{6.43}$$

with $\rho = \lim_{x\to\infty} u^-(x)$. Now, taking the sum of (6.42) and (6.43) we verify that $V_\Omega[\phi] + \rho = f$ with $\phi \equiv \nu_\Omega \cdot \nabla u^-_{|\partial\Omega} - \nu_\Omega \cdot \nabla u^+_{|\partial\Omega}$. In addition, $\rho = 0$ for $n \geq 3$. Thus, to complete the proof we have to show that $\int_{\partial\Omega} \phi \, d\sigma = 0$ when $n = 2$. By the first Green Identity of Theorem 4.2 we have $\int_{\partial\Omega} \nu_\Omega \cdot \nabla u^+_{|\partial\Omega} \, d\sigma = 0$. Then we observe that for $n = 2$ the function identically equal to 1 is harmonic in Ω^- and harmonic at infinity. Thus $\int_{\partial\Omega} \nu_\Omega \cdot \nabla u^-_{|\partial\Omega} \, d\sigma = 0$ by the Green Identity for exterior domains of Corollary 4.7. Hence,

$$\int_{\partial\Omega} \phi \, d\sigma = \int_{\partial\Omega} \nu_\Omega \cdot \nabla u^-_{|\partial\Omega} \, d\sigma - \int_{\partial\Omega} \nu_\Omega \cdot \nabla u^+_{|\partial\Omega} \, d\sigma = 0 \,.$$

□

For $n \geq 3$ we can exploit Lemma 6.45 to show that V_Ω is an isomorphism.

Theorem 6.46. *If $n \geq 3$, then V_Ω is an isomorphism from $C^{0,\alpha}(\partial\Omega)$ to $C^{1,\alpha}(\partial\Omega)$.*

Proof. By the Open Mapping Theorem 2.2 and by the boundedness of V_Ω (cf. Theorem 4.25), it suffices to show that V_Ω is a bijection. By Lemma 6.45 we already know that V_Ω is surjective. Thus it suffices to prove that V_Ω is injective. If $V_\Omega[\phi] = 0$, then $v^+_\Omega[\phi] = 0$ on Ω by the uniqueness of the solution of the Dirichlet problem of Theorem 3.6. By the jump formula for the normal derivative of $v^+_\Omega[\phi]$ of Theorem 6.6, it follows that $(-\frac{1}{2}I + W^t_\Omega)[\phi] = \nu_\Omega \cdot \nabla v^+_\Omega[\phi]_{|\partial\Omega} = 0$. Then Proposition 6.19 implies that $\phi = 0$. □

In general, for all dimensions $n \geq 2$ we have the following.

Theorem 6.47. *The map from $C^{0,\alpha}(\partial\Omega)_0 \times \mathbb{R}$ to $C^{1,\alpha}(\partial\Omega)$ that takes (ϕ, ρ) to $V_\Omega[\phi] + \rho$ is an isomorphism.*

Proof. Since V_Ω is bounded from $C^{0,\alpha}(\partial\Omega)$ to $C^{1,\alpha}(\partial\Omega)$ (cf. Theorem 4.25), the map in the statement is bounded from $C^{0,\alpha}(\partial\Omega)_0 \times \mathbb{R}$ to $C^{1,\alpha}(\partial\Omega)$. Then by the Open Mapping Theorem 2.2 it suffices to show that it is a bijection to conclude that it is an isomorphism. We first prove that it is injective. We observe that if $V_\Omega[\phi]+\rho=0$, then $v_\Omega^+[\phi]=-\rho$ on $\overline{\Omega}$ by the uniqueness of the solution of the Dirichlet problem of Theorem 3.6. By the jump formula of Theorem 6.6 for the normal derivative of the single layer $v_\Omega^+[\phi]$, it follows that $(-\frac{1}{2}I+W_\Omega^t)[\phi]=\nu_\Omega\cdot\nabla v_\Omega^+[\phi]_{|\partial\Omega}=0$. Then $\phi\in\mathrm{Ker}(-\frac{1}{2}I+W_\Omega^t)_0$ and we have $\phi=0$ by Lemma 6.22. Hence, $0=v_\Omega^+[\phi]=-\rho$ and $(\phi,\rho)=(0,0)$. To show that the map in the statement is surjective, we consider separately the case when $n=2$ and when $n\geq 3$. Indeed, for $n=2$ the surjectivity follows directly from Lemma 6.45. Instead, for $n\geq 3$ we argue as follows. If $f\in C^{1,\alpha}(\partial\Omega)$, then Lemma 6.45 implies that there exists $\phi\in C^{0,\alpha}(\partial\Omega)$ such that $V_\Omega[\phi]=f$. By Theorem 6.24 (or Lemma 6.21) the constant function that is identically equal to 1 belongs to $\mathrm{Ker}(-\frac{1}{2}I+W_\Omega)$. By Proposition 6.19 it follows that there exists $\tilde{\phi}\in\mathrm{Ker}(-\frac{1}{2}I+W_\Omega^t)$ such that $V_\Omega[\tilde{\phi}]$ is equal to the constant 1 on $\partial\Omega$. Then Lemma 6.22 implies that $\int_{\partial\Omega}\tilde{\phi}\,d\sigma\neq 0$, otherwise it would be $\tilde{\phi}=0$, contrary to $V_\Omega[\tilde{\phi}]=1$. Then we can take

$$\rho\equiv\frac{\int_{\partial\Omega}\phi\,d\sigma}{\int_{\partial\Omega}\tilde{\phi}\,d\sigma}$$

and we have

$$V_\Omega[\phi-\rho\tilde{\phi}]+V_\Omega[\rho\tilde{\phi}]=V_\Omega[\phi-\rho\tilde{\phi}]+\rho=f\quad\text{and}\quad\int_{\partial\Omega}(\phi-\rho\tilde{\phi})\,d\sigma=0.$$

□

We conclude this section with the following Theorem 6.48, where we show that harmonic functions in a bounded open set and in the exterior of a bounded open set can be written as a sum of a single layer and a constant function. The proof can be deduced by Theorems 6.46 and 6.47 and by the uniqueness of the solution of the interior and exterior Dirichlet problems of Theorems 3.6 and 6.1, respectively.

Theorem 6.48. *The following statements hold.*

(i) The map that takes a pair (μ,c) to $v_\Omega^+[\mu]+c$ is a linear bijection from $C^{0,\alpha}(\partial\Omega)_0\times\mathbb{R}$ to the space of the functions of $C^{1,\alpha}(\overline{\Omega})$ that are harmonic in Ω.

(ii) Let $n\geq 3$. The map that takes μ to $v_\Omega^-[\mu]$ is a linear bijection from $C^{0,\alpha}(\partial\Omega)$ to the space of those functions of $C^{1,\alpha}_{\mathrm{loc}}(\overline{\Omega^-})$ that are harmonic in Ω^- and harmonic at infinity.

(iii) Let $n=2$. The map that takes a pair (μ,c) to $v_\Omega^-[\mu]+c$ is a linear bijection from $C^{0,\alpha}(\partial\Omega)_0\times\mathbb{R}$ to the space of those functions of $C^{1,\alpha}_{\mathrm{loc}}(\overline{\Omega^-})$ that are harmonic in Ω^- and harmonic at infinity.

6.11 A Mixed Boundary Value Problem

We see here an application of the results obtained in Section 6.10. We consider two (nonempty) open subsets Ω^i and Ω^o of $\mathbb{R}^n$, both of class $C^{1,\alpha}$ and with connected exteriors Ω^{i-} and Ω^{o-}. We also assume that $\overline{\Omega^i} \subseteq \Omega^o$. So that the superscript 'i' stands for 'inner' and 'o' stands for 'outer.' Then we take a function $g \in C^{0,\alpha}(\partial\Omega^i)$ and a function $f \in C^{1,\alpha}(\partial\Omega^o)$ and we consider the following mixed boundary value problem for a function $u \in C^{1,\alpha}(\overline{\Omega^o} \setminus \Omega^i)$,

$$\begin{cases} \Delta u = 0 & \text{in } \Omega^o \setminus \overline{\Omega^i}\,, \\ \nu_{\Omega^i} \cdot \nabla u = g & \text{on } \partial\Omega^i\,, \\ u = f & \text{on } \partial\Omega^o\,. \end{cases} \tag{6.44}$$

By Proposition 6.3 we know that if a solution u of (6.44) exists, then it is unique. In the following Proposition 6.49 we show that such a solution actually exists. We observe that the specific integral representation of u that we use in the proof of Proposition 6.49 (cf. formula (6.45)) will present some advantage when we will study a perturbed version of problem (6.44) in Section 9.2.

Proposition 6.49. *There exists a solution $u \in C^{1,\alpha}(\overline{\Omega^o} \setminus \Omega^i)$ of problem* (6.44).

Proof. We look for a solution u in the form

$$u(x) = v^-_{\Omega^i}[\phi^i](x) + v^+_{\Omega^o}[\phi^o](x) + \rho \qquad \forall x \in \overline{\Omega^o} \setminus \Omega^i \tag{6.45}$$

with $\phi^i \in C^{0,\alpha}(\partial\Omega^i)$, $\phi^o \in C^{0,\alpha}(\partial\Omega^o)_0$, and $\rho \in \mathbb{R}$. Then, by the jump formulas for the single layer potential of Theorem 6.6, u is a solution of (6.44) if and only if both the following equations hold:

$$\left(\frac{1}{2}I + W^t_{\Omega^i}\right)[\phi^i](x) \tag{6.46}$$

$$+ \nu_{\Omega^i}(x) \cdot \int_{\partial\Omega^o} \phi^o(y)\, \nabla S_n(x-y)\, d\sigma_y = g(x) \qquad \forall x \in \partial\Omega^i\,,$$

$$V_{\Omega^o}[\phi^o](x) + \rho + \int_{\partial\Omega^i} S_n(x-y)\, \phi^i(y)\, d\sigma_y = f(x) \qquad \forall x \in \partial\Omega^o\,. \tag{6.47}$$

We denote by L_1 the linear operator from $C^{0,\alpha}(\partial\Omega^i) \times C^{0,\alpha}(\partial\Omega^o)_0 \times \mathbb{R}$ to $C^{0,\alpha}(\partial\Omega^i)$ that takes (ϕ^i, ϕ^o, ρ) to the left-hand side of (6.46). We denote by L_2 the linear operator from $C^{0,\alpha}(\partial\Omega^i) \times C^{0,\alpha}(\partial\Omega^o)_0 \times \mathbb{R}$ to $C^{1,\alpha}(\partial\Omega^o)$ that takes (ϕ^i, ϕ^o, ρ) to the left-hand side of (6.47). Then we let L be the linear operator from $C^{0,\alpha}(\partial\Omega^i) \times C^{0,\alpha}(\partial\Omega^o)_0 \times \mathbb{R}$ to $C^{0,\alpha}(\partial\Omega^i) \times C^{1,\alpha}(\partial\Omega^o)$ defined by

$$L \equiv (L_1, L_2)\,.$$

We observe that $\frac{1}{2}I + W^t_{\Omega^i}$ is an isomorphism from $C^{0,\alpha}(\partial\Omega^i)$ to itself (cf. Corollary 6.15) and $(\phi^o, \rho) \mapsto V_{\Omega^o}[\phi^o] + \rho$ is an isomorphism from $C^{0,\alpha}(\partial\Omega^o)_0 \times \mathbb{R}$ to

$C^{1,\alpha}(\partial\Omega^o)$ (cf. Theorem 6.47). Hence, the map from $C^{0,\alpha}(\partial\Omega^i)\times C^{0,\alpha}(\partial\Omega^o)_0\times\mathbb{R}$ to $C^{0,\alpha}(\partial\Omega^i)\times C^{1,\alpha}(\partial\Omega^o)$ that takes (ϕ^i,ϕ^o,ρ) to $\left(\frac{1}{2}\phi^i+W^t_{\Omega^i}[\phi^i], V_{\Omega^o}[\phi^o]+\rho\right)$ is an isomorphism. Moreover, by the compact embedding of $C^{0,\alpha}(\partial\Omega^o)$ into $L^1(\partial\Omega^o)$ (cf. Proposition 2.25) and by Theorem A.28 (ii) on the properties of integral operators with real analytic kernel, the map from $C^{0,\alpha}(\partial\Omega^o)_0$ to $C^{1,\alpha}(\partial\Omega^i)$ that takes ϕ^i to the function

$$\nu_{\Omega^i}(x)\cdot\int_{\partial\Omega^o}\phi^o(y)\,\nabla S_n(x-y)\,d\sigma_y \qquad \forall x\in\partial\Omega^i$$

is compact. Since $C^{1,\alpha}(\partial\Omega^i)$ is continuously imbedded into $C^{0,\alpha}(\partial\Omega^i)$, the same map is compact from $C^{0,\alpha}(\partial\Omega^o)_0$ to $C^{0,\alpha}(\partial\Omega^i)$. Similarly, we can show that the map from $C^{0,\alpha}(\partial\Omega^i)$ to $C^{1,\alpha}(\partial\Omega^o)$ that takes ϕ^i to

$$\int_{\partial\Omega^i} S_n(x-y)\,\phi^i(y)\,d\sigma_y \qquad \forall x\in\partial\Omega^o$$

is compact. Accordingly, L is a compact perturbation of an isomorphism and therefore a Fredholm operator of index zero (cf. Theorem 5.3). Then we can show that it is a linear homeomorphism by proving that it is injective (cf. Theorem 5.4). So we assume that $L[\phi^i,\phi^o,\rho]=(0,0)$. Then, following backward the argument above, we can see that the corresponding function u, defined by (6.45), is the unique solution of the homogeneous mixed problem

$$\begin{cases}\Delta u=0 & \text{in } \Omega^o\setminus\overline{\Omega^i}\,,\\ \nu_{\Omega^i}\cdot\nabla u=0 & \text{on } \partial\Omega^i\,,\\ u=0 & \text{on } \partial\Omega^o\,.\end{cases}$$

Proposition 6.3 implies that $u=0$ and, as a consequence, we have that

$$v^-_{\Omega^i}[\phi^i](x)+v^+_{\Omega^o}[\phi^o](x)+\rho=0 \qquad \forall x\in\overline{\Omega^o}\setminus\Omega^i\,.$$

Since $v_{\Omega^i}[\phi^i]$ and $v_{\Omega^o}[\phi^o]$ are continuous in $\mathbb{R}^n$ and harmonic on Ω^i, we deduce by the uniqueness of the solution of the interior Dirichlet problem of Theorem 6.1 that

$$v^+_{\Omega^i}[\phi^i](x)+v^+_{\Omega^o}[\phi^o](x)+\rho=0 \qquad \forall x\in\overline{\Omega^i}\,.$$

Then, by the jump formulas for the single layer potential of Theorem 6.6, we have

$$\begin{aligned}\phi^i &= \nu^i_\Omega\cdot\nabla(v^-_{\Omega^i}[\phi^i](x)+v^+_{\Omega^o}[\phi^o](x)+\rho)\\ &\quad-\nu^i_\Omega\cdot\nabla(v^+_{\Omega^i}[\phi^i](x)+v^+_{\Omega^o}[\phi^o](x)+\rho)=0\,.\end{aligned}$$

It follows that $v^+_{\Omega^o}[\phi^o]+\rho=0$ on $\overline{\Omega^o}\setminus\Omega^i$ and thus that $V_{\Omega^o}[\phi^o]+\rho=0$, which in turn implies that $(\phi^o,\rho)=(0,0)$ by Theorem 6.47. We conclude that L is an isomorphism and we deduce the existence of (ϕ^i,ϕ^o,ρ) as in (6.46) and (6.47). □

6.12 The Operators $I + \lambda W_\Omega$ and $I + \lambda W_\Omega^t$

In the following Lemma 6.50 we show that $I + \lambda W_\Omega^t$ is injective for all $\lambda \in]-2, 2[$. The result is due to Kellogg and the proof that we present is adapted from that in [137, Chap. XI, Sect. 11].

Lemma 6.50. *Let* $\lambda \in]-2, 2[$. *If* $\phi \in C^{0,\alpha}(\partial\Omega)$ *and* $(I + \lambda W_\Omega^t)[\phi] = 0$, *then* $\phi = 0$.

Proof. Since $(I + \lambda W_\Omega^t)[\phi] = 0$, Lemma 6.11 implies that

$$\begin{aligned} 0 &= \int_{\partial\Omega} (I + \lambda W_\Omega^t)[\phi]\, d\sigma \\ &= \left(1 - \frac{\lambda}{2}\right) \int_{\partial\Omega} \phi\, d\sigma + \lambda \int_{\partial\Omega} \left(\frac{1}{2} I + W_\Omega^t\right)[\phi]\, d\sigma \\ &= \left(1 - \frac{\lambda}{2}\right) \int_{\partial\Omega} \phi\, d\sigma + \lambda \int_{\partial\Omega} \phi\, d\sigma = \left(1 + \frac{\lambda}{2}\right) \int_{\partial\Omega} \phi\, d\sigma\,, \end{aligned}$$

and accordingly, $\int_{\partial\Omega} \phi\, d\sigma = 0$. Then $v_\Omega^-[\phi]$ is harmonic at infinity (cf. Theorem 4.23) and there exist λ^+ and λ^- in $[0, +\infty[$ such that

$$\lambda^+ = \int_\Omega |\nabla v_\Omega^+[\phi]|^2\, dx \qquad \text{and} \qquad \lambda^- = \int_{\Omega^-} |\nabla v_\Omega^-[\phi]|^2\, dx\,. \tag{6.48}$$

(We note that the first Green Identity for exterior domains of Corollary 4.7 implies that λ^- is actually finite.) In addition, by the first Green Identity (cf. Theorem 4.2 and Corollary 4.7) and by the jump formula for the normal derivative of the single layer potential of Theorem 6.6 we have

$$\begin{aligned} \lambda^+ &= \int_{\partial\Omega} V_\Omega[\phi] \left(-\frac{1}{2} I + W_\Omega^t\right)[\phi]\, d\sigma\,, \\ \lambda^- &= \int_{\partial\Omega} V_\Omega[\phi] \left(-\frac{1}{2} I - W_\Omega^t\right)[\phi]\, d\sigma\,. \end{aligned}$$

Then, a straightforward computation shows that

$$\begin{aligned} \lambda^+ + \lambda^- &= -\int_{\partial\Omega} V_\Omega[\phi]\, \phi\, d\sigma\,, \\ \lambda^+ - \lambda^- &= 2 \int_{\partial\Omega} V_\Omega[\phi]\, W_\Omega^t[\phi]\, d\sigma \end{aligned}$$

and thus equality $(I + \lambda W_\Omega^t)[\phi] = 0$ implies that

$$2(\lambda^+ + \lambda^-) = \lambda(\lambda^+ - \lambda^-)\,.$$

Hence, $(2 - \lambda)\lambda^+ + (2 + \lambda)\lambda^- = 0$. Since λ belongs to $]-2, 2[$, both $(2 - \lambda)$ and $(2 + \lambda)$ are strictly positive and accordingly both λ^+ and λ^- must be equal to zero.

Then by (6.48) we deduce that $v^+_\Omega[\phi]$ and $v^-_\Omega[\phi]$ are locally constant in Ω and Ω^-, respectively. Finally, the jump formula for the normal derivative of the single layer potential of Theorem 6.6 implies that $\phi = \nu_\Omega \cdot \nabla v^-[\phi]_{|\partial\Omega} - \nu_\Omega \cdot \nabla v^-[\phi]_{|\partial\Omega} = 0$.
□

Then, by the compactness of W_Ω and W^t_Ω of Theorem 6.8, by the Fredholm Alternative of Theorem 5.8, and by Lemma 6.50 that we have just proved, we have the following.

Theorem 6.51. *If $\lambda \in]-2,2[$, then $I + \lambda W_\Omega$ is an isomorphism from $C^{1,\alpha}(\partial\Omega)$ onto itself and $I + \lambda W^t_\Omega$ is an isomorphism from $C^{0,\alpha}(\partial\Omega)$ onto itself.*

Theorem 6.51 could be reformulated by saying that every real number ρ with $|\rho| > 1/2$ is in the resolvent set of W_Ω and W^t_Ω. With an argument similar to the one in the proof of Lemma 6.50, Kellogg also proved that all complex numbers with nonzero imaginary part are in the resolvent set of W_Ω and W^t_Ω (cf. [137, Chap. XI, Sect. 11]). It follows that the spectrum of W_Ω and W^t_Ω is contained in the real interval $[-1/2, 1/2]$ and, thanks to Theorems 6.14 and 6.24, we can also say that $1/2$ is an eigenvalue whereas $-1/2$ is an eigenvalue only when Ω^- has at least two connected components, that is, if $\varkappa^- \geq 1$. (Then, for $\varkappa^- = 0$, the spectrum is contained in $]-1/2, 1/2]$.)

At this point, it would be natural to ask ourselves what are the other eigenvalues of W_Ω and W^t_Ω and how they are related to the shape of the domain Ω. This is a classical question and there is a large literature about it, also in view of the applications to conformal mappings and inverse scattering theory. The early works on the topic date back to the seminal paper of Poincaré [244] and later contributions include those of Plemelj [243] and Carleman [39]. We also mention the works of Ahlfors [3] and the papers of Schiffer, as for example [258]. We will not, however, attempt a complete list. The interested reader can find a 'modern' treatment of the problem in the paper of Khavinson, Putinar, and Shapiro [138] and for applications to scattering theory we refer, for example, to Ammari and Kang [8].

6.13 A Linear Transmission Problem

To illustrate some applications of the results in Sections 6.10 and 6.12, we study in this section a transmission problem in Ω. We fix a strictly positive real number a and we take two functions $f \in C^{1,\alpha}(\partial\Omega)$ and $g \in C^{0,\alpha}(\partial\Omega)$. Then we consider the following transmission problem for a pair $(u^+, u^-) \in C^{1,\alpha}(\overline{\Omega}) \times C^{1,\alpha}_{\mathrm{loc}}(\overline{\Omega^-})$,

$$\begin{cases} \Delta u^+ = 0 & \text{in } \Omega\,, \\ \Delta u^- = 0 & \text{in } \Omega^-\,, \\ a\,u^+ - u^- = f & \text{on } \partial\Omega\,, \\ \nu_\Omega \cdot \nabla u^+ - \nu_\Omega \cdot \nabla u^- = g & \text{on } \partial\Omega\,, \\ u^- \text{ is harmonic at infinity.} \end{cases} \tag{6.49}$$

Problems like (6.49) play a role in the mathematical description of composite materials with a physical field governed by the Laplace equation. Typical examples are those of the electric field, the temperature field, and the longitudinal flow. The boundary conditions model the contact between different media. The constant a expresses the contrast parameter of the physical properties (e.g., the conductivity or the permeability) of the materials that occupy the domain Ω and the domain Ω^-.

We observe that, in dimension two, problem (6.49) can have a solution only if the function g satisfies a compatibility condition. Indeed we have the following.

Proposition 6.52. *If* $n = 2$ *and the transmission problem* (6.49) *has a solution, then*

$$\int_{\partial\Omega} g \, d\sigma = 0 \, .$$

Proof. If problem (6.49) has a solution, then we have

$$\begin{aligned} \int_{\partial\Omega} g \, d\sigma &= \int_{\partial\Omega} \left(\nu_\Omega \cdot \nabla u^+ - \nu_\Omega \cdot \nabla u^-\right) d\sigma \\ &= \int_{\partial\Omega} \nu_\Omega \cdot \nabla u^+ \, d\sigma - \sum_{j=1}^{\varkappa^-} \int_{\partial\Omega_j^-} \nu_{\Omega_j^-} \cdot \nabla u^- \, d\sigma - \int_{\partial\Omega_0^-} \nu_{\Omega_0^-} \cdot \nabla u^- \, d\sigma \, . \end{aligned}$$

By the first Green Identity of Theorem 4.2 we have $\int_{\partial\Omega} \nu_\Omega \cdot \nabla u^+ \, d\sigma = 0$ and $\int_{\partial\Omega_j^-} \nu_{\Omega_j^-} \cdot \nabla u^- \, d\sigma = 0$ for all $j \in \{1, \ldots, \varkappa^-\}$. Since u^- and the function identically equal to 1 are harmonic at infinity (in dimension two), the second Green Identity in exterior domains of Corollary 4.8 implies that $\int_{\partial\Omega_0^-} \nu_{\Omega_0^-} \cdot \nabla u^- \, d\sigma = 0$. Hence, the validity of the statement follows. □

Remark 6.53. If $n = 2$, then the solution of the transmission problem (6.49), cannot be unique. Indeed, if $(u^+, u^-) \in C^{1,\alpha}(\overline{\Omega}) \times C^{1,\alpha}_{\mathrm{loc}}(\overline{\Omega^-})$ is a solutions of the transmission problem (6.49), then $(u^+ + c, u^- + ac)$ solves the same problem for all $c \in \mathbb{R}$.

Then we have the following uniqueness result.

Proposition 6.54. *If* (u_1^+, u_1^-) *and* (u_2^+, u_2^-) *are solutions of the transmission problem* (6.49) *in* $C^{1,\alpha}(\overline{\Omega}) \times C^{1,\alpha}_{\mathrm{loc}}(\overline{\Omega^-})$*, then there exists* $c \in \mathbb{R}$ *such that* $u_1^+ = u_2^+ + c$ *and* $u_1^- = u_2^- + ac$*. If* $n \geq 3$*, then* $c = 0$.

Proof. Let $u^+ \equiv u_1^+ - u_2^+$ and $u^- \equiv u_1^- - u_2^-$. Then (u^+, u^-) is a solution of (6.49) with $f = g = 0$. Since u^- is harmonic at infinity, the first Green Identities of Theorem 4.2 and Corollary 4.7 imply that

$$\begin{aligned} a \int_\Omega |\nabla u^+|^2 \, dx &+ \int_{\Omega^-} |\nabla u^-|^2 \, dx \\ &= \int_{\partial\Omega} \left(a \, u^+ \, \nu_\Omega \cdot \nabla u^+ - u^- \, \nu_\Omega \cdot \nabla u^-\right) dx = 0 \, . \end{aligned}$$

Since a is strictly positive, we deduce that both $\int_{\Omega} |\nabla u^+|^2\,dx$ and $\int_{\Omega^-} |\nabla u^-|^2\,dx$ are equal to zero. It follows that u^+ and u^- are locally constant on Ω and on Ω^-, respectively.

We now find convenient to denote by u the function from $\mathbb{R}^n$ to $\mathbb{R}$ that is equal to au^+ on $\overline{\Omega}$ and to u^- on Ω^-. We show that u is constant on $\mathbb{R}^n$. Since $\mathbb{R}^n$ is connected it suffices to prove that u is locally constant. So let $x \in \mathbb{R}^n$. We may either have $x \in \mathbb{R}^n \setminus \partial\Omega$ or $x \in \partial\Omega$. If $x \in \mathbb{R}^n \setminus \partial\Omega$ then there exists $i \in \{1, \dots, \varkappa^+\}$ or $j \in \{0, \dots, \varkappa^-\}$ such that $x \in \Omega_i$ or $x \in (\Omega^-)_j$. Since u is constant on each Ω_i and $(\Omega^-)_j$, we deduce that there exists an open neighborhood of x where u is constant. If instead $x \in \partial\Omega$, then there exist a coordinate cylinder C around x (see Lemma 2.38) and indexes $i \in \{1, \dots, \varkappa^+\}$ and $j \in \{0, \dots, \varkappa^-\}$ such that $C \cap \Omega \subseteq \Omega_i$ and $C \cap \Omega^- \subseteq (\Omega^-)_j$. Since u is constant on $\overline{\Omega_i}$ it follows that u is identically equal to $au^+(x)$ on $C \cap \overline{\Omega} = C \cap \overline{\Omega_i}$. Since $u = u^-$ on $(\Omega^-)_j$, u^- is constant on $\overline{(\Omega^-)_j}$, and the transmission condition $au^+ - u^- = 0$ holds on $\partial\Omega$, it follows that u is identically equal to $au^+(x)$ on $C \cap \Omega^- = C \cap (\Omega^-)_j$. Hence u is equal to the constant value $au^+(x)$ on the whole of C. Doing so, we have shown the existence of an open neighborhood of x where u is constant and we can conclude that u is constant on $\mathbb{R}^n$.

Now, if we take $c \in \mathbb{R}$ such that $u(x) = ac$ for all $x \in \mathbb{R}^n$ we deduce that $u_1^+ = u_2^+ + c$ and $u_1^- = u_2^- + ac$. To complete the proof we observe that if the dimension n is bigger than or equal to three, then the condition that u^- is harmonic at infinity implies that $\lim_{x\to\infty} u^-(x) = 0$ (cf. Theorem 3.22). Hence $ac = \lim_{x\to\infty} u^-(x) = 0$ and, since $a > 0$, we conclude that $c = 0$. □

Finally, exploiting Theorem 6.47 on the representation of a function on the boundary as a sum of a single layer and a constant and the isomorphism Theorem 6.51 for $I + \lambda W_{\Omega}^t$, we can prove the existence of solutions for the transmission problem (6.49) and that such solutions can be written in terms of single layer potentials and constant functions.

Theorem 6.55. *If the dimension n is two, assume that $\int_{\partial\Omega} g\,d\sigma = 0$. Then there exists one and only one triple (ϕ, μ, ρ) in $C^{0,\alpha}(\partial\Omega) \times C^{0,\alpha}(\partial\Omega)_0 \times \mathbb{R}$ such that the pair*

$$(u^+, u^-) \equiv (v_{\Omega}^+[\phi + \mu] + \rho, av_{\Omega}^-[\phi])$$

is a solution of the transmission problem (6.49).

Proof. By the continuity of the single layer potential (cf. Theorem 4.22) the pair $(u^+, u^-) \equiv (v_{\Omega}^+[\phi+\mu]+\rho, av_{\Omega}^-[\phi])$ satisfies the boundary condition $au^+ - u^- = f$ if and only if

$$aV_{\Omega}[\mu] + a\rho = f\,. \tag{6.50}$$

Since $a > 0$, Theorem 6.47 implies that (6.50) has a unique solution $(\mu, \rho) \in C^{0,\alpha}(\partial\Omega)_0 \times \mathbb{R}$. By the jump properties of Theorem 6.6, the pair of functions

$$(v_{\Omega}^+[\phi + \mu] + \rho, av_{\Omega}^-[\phi])$$

satisfies the boundary condition $\nu_{\Omega} \cdot \nabla u^+ - \nu_{\Omega} \cdot \nabla u^- = g$ if and only if

$$\left(-\frac{1}{2}(1+a)I + (1-a)W^t_\Omega\right)[\phi] = g - \left(-\frac{1}{2}I + W^t_\Omega\right)[\mu]\,. \tag{6.51}$$

Since $2(1-a)/(1+a) \in]-2, 2[$, Theorem 6.51 implies that $-\frac{1}{2}(1+a)I+(1-a)W^t_\Omega$ is an isomorphism from $C^{0,\alpha}(\partial\Omega)$ to itself. As a consequence, we deduce that there exists a unique $\phi \in C^{0,\alpha}(\partial\Omega)$ that satisfies (6.51). Moreover, if $\int_{\partial\Omega} g\,d\sigma = 0$, then a computation based on Lemma 6.11 and equality (6.51) shows that $\int_{\partial\Omega} \phi\,d\sigma = 0$. Accordingly $\lim_{x\to\infty} v^-_\Omega[\phi] = 0$ both in dimension $n = 2$ and in dimension $n \geq 3$ (cf. Theorem 4.23). It follows that $v^-_\Omega[\phi]$ is harmonic at infinity and the theorem is proved. □

Incidentally, we observe that in Theorem 6.55 condition $\int_{\partial\Omega} g\,d\sigma = 0$ implies that $\int_{\partial\Omega} \phi\,d\sigma = 0$ both in dimension $n = 2$ and in dimension $n \geq 3$. In addition, in dimension $n \geq 3$ the constant ρ can be omitted if we allow $\int_{\partial\Omega} \mu\,d\sigma$ to be different from zero (cf. Theorem 6.46). In such a case, however, we could no longer infer that $\int_{\partial\Omega} \phi\,d\sigma$ is necessarily equal to 0.

6.14 A Robin Problem

We fix two functions $g, b \in C^{0,\alpha}(\partial\Omega)$ and then we consider the following interior Robin problem

$$\begin{cases} \Delta u(x) = 0 & \forall x \in \Omega\,, \\ \frac{\partial}{\partial\nu_\Omega}u(x) + b(x)u(x) = g(x) & \forall x \in \partial\Omega \end{cases} \tag{6.52}$$

in the unknown $u \in C^{1,\alpha}(\overline{\Omega})$ and the corresponding exterior Robin problem

$$\begin{cases} \Delta u(x) = 0 & \forall x \in \Omega^-\,, \\ \frac{\partial}{\partial\nu_\Omega}u(x) + b(x)u(x) = g(x) & \forall x \in \partial\Omega\,, \\ u \text{ is harmonic at infinity,} \end{cases} \tag{6.53}$$

in the unknown $u \in C^{1,\alpha}_{\mathrm{loc}}(\overline{\Omega^-})$. As a first step, we prove the following theorem, which ensures the uniqueness of a solution of the Robin problems (6.52) and (6.53) under suitable assumptions.

Theorem 6.56. *The following statements hold.*

(i) If $b \geq 0$ on $\partial\Omega$, $\int_{\partial\Omega_j} b\,d\sigma \neq 0$ for all $j \in \{1,\ldots,\varkappa^+\}$, and u_1, $u_2 \in C^{1,\alpha}(\overline{\Omega})$ solve the interior Robin problem (6.52), then $u_1 = u_2$.

(ii) If $n \geq 3$, $b \leq 0$ on $\partial\Omega$, $\int_{\partial(\Omega^-)_j} b\,d\sigma \neq 0$ for all $j \in \{1,\ldots,\varkappa^-\}$, and u_1, $u_2 \in C^{1,\alpha}_{\mathrm{loc}}(\overline{\Omega^-})$ solve the exterior Robin problem (6.53), then $u_1 = u_2$.

(iii) If $n = 2$, $b \leq 0$ on $\partial\Omega$, $\int_{\partial(\Omega^-)_j} b\,d\sigma \neq 0$ for all $j \in \{0,\ldots,\varkappa^-\}$ (including for $j = 0$), and u_1, $u_2 \in C^{1,\alpha}_{\mathrm{loc}}(\overline{\Omega^-})$ solve the exterior Robin problem (6.53), then $u_1 = u_2$.

Proof. We first prove (i). We set $u \equiv u_1 - u_2$, and we show that $\nabla u = 0$ in Ω. By the first Green Identity of Theorem 4.2, we have

$$0 \le \int_{\Omega_j} |\nabla u|^2 \, dx = \int_{\partial\Omega_j} u \, \nu_\Omega \cdot \nabla u d\sigma = - \int_{\partial\Omega_j} b \, u^2 \, d\sigma \le 0 \tag{6.54}$$

for all $j \in \{1, \ldots, \varkappa^+\}$ (here we have used the assumption that $b \ge 0$). Then we have $\nabla u = 0$ in Ω_j and there exists $c_j \in \mathbb{R}$ such that $u(x) = c_j$ for all $x \in \Omega_j$ and all $j \in \{1, \ldots, \varkappa^+\}$. By (6.54) and, again, by inequality $b \ge 0$, we have

$$0 = \int_{\partial\Omega_j} b \, u^2 \, d\sigma = c_j^2 \int_{\partial\Omega_j} b \, d\sigma \, .$$

Since we assume that $\int_{\partial\Omega_j} b \, d\sigma \neq 0$, we deduce that $c_j = 0$ for all $j \in \{1, \ldots, \varkappa^+\}$.

We now consider statements (ii) and (iii). As in (i), we set $u \equiv u_1 - u_2$. Since u is harmonic at infinity, the first Green Identity for exterior domains of Corollary 4.7 in Ω^- implies that

$$\begin{aligned} 0 \le \int_{\Omega^-} |\nabla u|^2 \, dx &= - \int_{\partial\Omega} u \, \nu_\Omega \cdot \nabla u \, d\sigma \\ &= \int_{\partial\Omega} b \, u^2 \, d\sigma = \sum_{j=0}^{\varkappa^-} \int_{\partial(\Omega^-)_j} b \, u^2 \, d\sigma \le 0 \end{aligned}$$

(we note that $b \le 0$ by assumption). Accordingly, we have $\nabla u = 0$ in Ω^- and thus u is constant in all connected components of Ω^-. Then, arguing as in the proof of statement (i), we conclude that $u = 0$ in all the bounded connected components of Ω^-. If $n \ge 3$ the condition that u is harmonic at infinity implies that u vanishes on the unbounded connected component $(\Omega^-)_0$. □

Inspired by the representation Theorem 6.48 for harmonic functions, we look for a solution of the interior Robin problem (6.52) in the form of a sum $v_\Omega^+[\mu] + c$ for some $\mu \in C^{0,\alpha}(\partial\Omega)_0$ and $c \in \mathbb{R}$. We note that if $v_\Omega^+[\mu] + c$ solves problem (6.52), then the jump formula for the normal derivative of a single layer potential and the boundary condition of the interior Robin problem (6.52) imply that μ must solve the following integral equation

$$\left(-\frac{1}{2} I + W_\Omega^t \right) [\mu] + b \, V_\Omega[\mu] + b \, c = g \qquad \text{on } \partial\Omega \, . \tag{6.55}$$

Conversely, if $(\mu, c) \in C^{0,\alpha}(\partial\Omega)_0 \times \mathbb{R}$ is a solution of equation (6.55), then $v_\Omega^+[\mu] + c$ solves the Robin problem in (6.52). Then, we find convenient to introduce the operator $L_{\Omega,b,+}$ from $C^{0,\alpha}(\partial\Omega) \times \mathbb{R}$ to $C^{0,\alpha}(\partial\Omega)$, defined by

$$L_{\Omega,b,+}[\mu, c] \equiv \left(-\frac{1}{2} I + W_\Omega^t \right) [\mu] + b \, V_\Omega[\mu] + b \, c \tag{6.56}$$

for all $(\mu, c) \in C^{0,\alpha}(\partial\Omega) \times \mathbb{R}$.

Similarly, if $n \ge 3$, we introduce the operator $L_{\Omega,b,-}$ from $C^{0,\alpha}(\partial\Omega)$ to $C^{0,\alpha}(\partial\Omega)$ defined by

$$L_{\Omega,b,-}[\mu] \equiv \left(\frac{1}{2}I + W^t_\Omega\right)[\mu] + b\,V_\Omega[\mu] \tag{6.57}$$

for all $\mu \in C^{0,\alpha}(\partial\Omega)$, and if $n = 2$, we introduce the operator $L_{\Omega,b,-}$ from $C^{0,\alpha}(\partial\Omega) \times \mathbb{R}$ to $C^{0,\alpha}(\partial\Omega)$ defined by

$$L_{\Omega,b,-}[\mu, c] \equiv \left(\frac{1}{2}I + W^t_\Omega\right)[\mu] + b\,V_\Omega[\mu] + b\,c \tag{6.58}$$

for all $(\mu, c) \in C^{0,\alpha}(\partial\Omega) \times \mathbb{R}$. Then, if $n \geq 3$, a function $\mu \in C^{0,\alpha}(\partial\Omega)$ satisfies equation

$$L_{\Omega,b,-}[\mu] = g\,, \tag{6.59}$$

if and only if $v^-_\Omega[\mu]$ is a solution of the exterior Robin problem (6.53). If $n = 2$, a pair $(\mu, c) \in C^{0,\alpha}(\partial\Omega)_0 \times \mathbb{R}$ satisfies equation

$$L_{\Omega,b,-}[\mu, c] = g \tag{6.60}$$

if and only if $v^-_\Omega[\mu] + c$ is a solution of the exterior Robin problem (6.53).

To analyze the integral equations (6.55), (6.59) and (6.60), we introduce the following statement.

Proposition 6.57. *The following statements hold.*

(i) If $n \geq 2$, $b \geq 0$ on $\partial\Omega$, and $\int_{\partial\Omega_j} b\,d\sigma \neq 0$ for all $j \in \{1, \dots, \varkappa^+\}$, then the operator from $C^{0,\alpha}(\partial\Omega) \times \mathbb{R}$ to itself that takes a pair (μ, c) to

$$\left(L_{\Omega,b,+}[\mu, c], \int_{\partial\Omega} \mu\,d\sigma\right)$$

is an isomorphism. In particular, $L_{\Omega,b,+}$ is an isomorphism from $C^{0,\alpha}(\partial\Omega)_0 \times \mathbb{R}$ onto $C^{0,\alpha}(\partial\Omega)$.

(ii) If $n \geq 3$, $b \leq 0$ on $\partial\Omega$, and $\int_{\partial(\Omega^-)_j} b\,d\sigma \neq 0$ for all $j \in \{1, \dots, \varkappa^-\}$, then $L_{\Omega,b,-}$ is an isomorphism from $C^{0,\alpha}(\partial\Omega)$ to itself.

(iii) If $n = 2$, $b \leq 0$ on $\partial\Omega$, and $\int_{\partial(\Omega^-)_j} b\,d\sigma \neq 0$ for all $j \in \{0, \dots, \varkappa^-\}$ (including $j = 0$), then the operator from $C^{0,\alpha}(\partial\Omega) \times \mathbb{R}$ to itself that takes a pair (μ, c) to

$$\left(L_{\Omega,b,-}[\mu, c], \int_{\partial\Omega} \mu\,d\sigma\right)$$

is an isomorphism. In particular, $L_{\Omega,b,-}$ is an isomorphism from $C^{0,\alpha}(\partial\Omega)_0 \times \mathbb{R}$ onto $C^{0,\alpha}(\partial\Omega)$.

Proof. Let us denote by $J_{\Omega,b,+}$ the operator of statement (i). We can write

$$J_{\Omega,b,+} = \hat{J}_{\Omega,b,+} + \tilde{J}_{\Omega,b,+}$$

with

$$\hat{J}_{\Omega,b,+}[\mu, c] \equiv \left(-\frac{1}{2}\mu\,,\; c\right)$$

and

$$\tilde{J}_{\Omega,b,+}[\mu, c] \equiv \left(W^t_\Omega[\mu] + bV_\Omega[\mu] + bc\,,\; -c + \int_{\partial\Omega} \mu\, d\sigma\right)$$

for all $(\mu, c) \in C^{0,\alpha}(\partial\Omega) \times \mathbb{R}$. Then we can see that $\hat{J}_{\Omega,b,+}$ is an isomorphism from $C^{0,\alpha}(\partial\Omega) \times \mathbb{R}$ to itself. Moreover, $\tilde{J}_{\Omega,b,+}$ is compact from $C^{0,\alpha}(\partial\Omega) \times \mathbb{R}$ to itself. Indeed, by Theorem 6.8 (ii), W^t_Ω is compact in $C^{0,\alpha}(\partial\Omega)$ and, by Theorem 4.25, V_Ω is bounded from $C^{0,\alpha}(\partial\Omega)$ to $C^{1,\alpha}(\partial\Omega)$, which is compactly embedded into $C^{0,\alpha}(\partial\Omega)$ (cf. Proposition 2.25 and Theorem 2.83). Being the sum of an isomorphism and a compact operator, it follows that $J_{\Omega,b,+}$ is a Fredholm operator of index zero (cf. Theorem 5.3). Accordingly, to show that it is a isomorphism it suffices to prove that it is injective. So, let $(\mu, c) \in C^{0,\alpha}(\partial\Omega) \times \mathbb{R}$ satisfy equations

$$L_{\Omega,b,+}[\mu, c] = 0\,, \qquad \int_{\partial\Omega} \mu\, d\sigma = 0\,.$$

We claim that $(\mu, c) = (0, 0)$. Indeed, the jump formula for the normal derivative of the single layer potential of Theorem 6.6 implies that the function $v = v^+_\Omega[\mu] + c$ satisfies the interior Robin problem (6.52) with $g = 0$. Then the uniqueness Theorem 6.56 (i) implies that $v = 0$. Since $\int_{\partial\Omega} \mu\, d\sigma = 0$, the representation Theorem 6.48 (i) for harmonic functions implies that $(\mu, c) = (0, 0)$. It follows that $J_{\Omega,b,+}$ is an isomorphism and we can verify the validity of statement (i).

The proof of statements (ii) and (iii) follows the lines of that of statement (i), we only need to use Theorem 6.56 (ii), (iii) instead of Theorem 6.56 (i) and Theorem 6.48 (ii), (iii) instead of Theorem 6.48 (i) (see also Theorem 4.23). □

We are now ready to state the following.

Theorem 6.58 (Existence and Uniqueness for the Robin Problem). *The following statements hold.*

(i) If $n \geq 2$, $b \geq 0$ on $\partial\Omega$, and $\int_{\partial\Omega_j} b\, d\sigma \neq 0$ for all $j \in \{1, \ldots, \varkappa^+\}$, then the interior Robin problem (6.52) has a unique solution $u \in C^{1,\alpha}(\overline{\Omega})$. Such solution is delivered by the formula

$$u = v^+_\Omega[\mu] + c\,, \tag{6.61}$$

where (μ, c) is the only solution in $C^{0,\alpha}(\partial\Omega)_0 \times \mathbb{R}$ of the equation $L_{\Omega,b,+}[\mu, c] = g$.

(ii) If $n \geq 3$, $b \leq 0$ on $\partial\Omega$, and $\int_{\partial(\Omega^-)_j} b\, d\sigma \neq 0$ for all $j \in \{1, \ldots, \varkappa^-\}$, then the exterior Robin problem (6.53) has a unique solution $u \in C^{1,\alpha}_{\mathrm{loc}}(\overline{\Omega^-})$. Such solution is delivered by the formula

$$u = v^-_\Omega[\mu]\,, \tag{6.62}$$

where μ is the only solution in $C^{0,\alpha}(\partial\Omega)$ of the equation $L_{\Omega,b,-}[\mu] = g$.

(iii) If $n = 2$, $b \leq 0$ *on* $\partial\Omega$, *and* $\int_{\partial(\Omega^-)_j} b\, d\sigma \neq 0$ *for all* $j \in \{0, \dots, \varkappa^-\}$ *(including* $j = 0$*), then the exterior Robin problem (6.53) has a unique solution* $u \in C^{1,\alpha}_{\text{loc}}(\overline{\Omega^-})$*. Such solution is delivered by the formula*

$$u = v^-_\Omega[\mu] + c\,, \tag{6.63}$$

where the pair (μ, c) *is the only solution in* $C^{0,\alpha}(\partial\Omega)_0 \times \mathbb{R}$ *of the equation* $L_{\Omega,b,-}[\mu, c] = g$.

Proof. By Proposition 6.57, the integral equations (6.55), (6.59), and (6.60) have a unique solution. Then the jump formulas form the normal derivative of the single layer potential of Theorem 6.6 imply that the functions in (6.61)–(6.63) are solutions of the interior Robin problem (6.52) and of the exterior Robin problem (6.53) both for $n \geq 3$ and $n = 2$, respectively. Finally, the uniqueness of such solutions follows by Theorem 6.56. □

In conclusion of this chapter we observe that, when dealing with the two-dimensional case, one could exploit the fact that harmonic functions are real (or imaginary) parts of holomorphic functions. Then, to obtain integral representations of the solutions of boundary value problems for the Laplace equation, one could resort to the Cauchy formula and other complex analysis techniques. This program has been carried out for a variety of boundary value problems. Classic references on the topic are the books of Gakhov [105] and Muskhelishvili [227]. More recent references can be found in the monograph by Mityushev and Rogosin [224].

We also note that in the present book we focus exclusively on boundary value problems in domains of class $C^{1,\alpha}$, but the regularity of the boundary may be relaxed. Indeed, there exists a large literature on boundary value problems in piecewise smooth domains and in Lipschitz domains. The reader interested in boundary value problems in piecewise smooth domains is referred, for example, to Borsuk and Kondratiev [33], Dauge [84], Grisvard [112], Plamenevskij [242], and to references therein. For the development of potential theoretic methods in Lipschitz domains we refer, for example, to Gewinner and Stephan [116], Maz'ya and Soloviev [205], McLean [206], Medková [208], Mitrea and Mitrea [215], and to references therein.

Chapter 7
Poisson Equation and Volume Potentials

Abstract This chapter analyzes volume potentials and the solvability of the Poisson equation by means of such volume potentials. To go into detail, we first consider the case of weakly singular kernels, and then the case in which the first order partial derivatives of the kernel are also weakly singular. That been done, we turn to singular kernels and, after that, to weakly singular kernels with singular first order partial derivatives. Finally, we prove the basic properties of the corresponding volume potentials in spaces of Hölder continuous functions, in Schauder spaces, and in the Roumieu classes. Although similar properties in Schauder spaces can be found in the classical monographs of Günter (Potential theory and its applications to basic problems of mathematical physics. Translated from the Russian by John R. Schulenberger. Frederick Ungar Publishing Co., New York, 1967) and of Kupradze, Gegelia, Basheleishvili, and Burchuladze (Three-dimensional problems of the mathematical theory of elasticity and thermoelasticity. North-Holland series in applied mathematics and mechanics, vol 25, Russian edition. North-Holland Publishing Co., Amsterdam, New York, 1979. Edited by V. D. Kupradze), here we prove the corresponding statements with optimal Hölder exponents. To do so, we follow a proof of Miranda (Atti Accad Naz Lincei Mem Cl Sci Fis Mat Natur Sez I (8), 7, 303–336, 1965) and we exploit a known lemma (cf. Majda and Bertozzi, Vorticity and incompressible flow. Cambridge texts in applied mathematics, vol 27, Prop. 8.12, pp 348–350. Cambridge University Press, Cambridge, 2002) for which we provide a proof of Mateu, Orobitg, and Verdera (J Math Pures Appl (9), 91(4), 402–431, 2009). To conclude the chapter, we use the properties proven for the volume potentials to study the solvability of the basic boundary value problems for the Poisson equation.

Unless otherwise specified, we understand that throughout this chapter n is a natural number bigger than or equal to two. Namely,

$$n \in \mathbb{N}, \; n \geq 2 .$$

We also fix a real number

$$\alpha \in]0, 1[.$$

M. Dalla Riva et al., *Singularly Perturbed Boundary Value Problems*,
https://doi.org/10.1007/978-3-030-76259-9_7

7.1 Preliminary Remarks on the Poisson Equation

We are interested in the existence of solutions of boundary value problems for the Poisson equation

$$\Delta u = f \tag{7.1}$$

in an open subset Ω, which we assume to be at least of class $C^{1,\alpha}$. The function f is given and the unknown of the problem is u. A good starting point to study the existence of solutions for the Poisson equation is the third Green Identity of Theorem 4.5, which shows that any $u \in C^{1,\alpha}(\overline{\Omega}) \cap C^2(\Omega)$ such that Δu is integrable can be written as a sum of a harmonic function and of the Newtonian potential

$$\int_\Omega S_n(x-y)\Delta u(y)\,dy \qquad \forall x \in \Omega\,.$$

Thus it is natural to think of looking for a solution of the Poisson equation (7.1) in the form of the Newtonian potential

$$P_n[\Omega, f](x) \equiv \int_\Omega S_n(x-y)f(y)\,dy \qquad \forall x \in \Omega\,, \tag{7.2}$$

which is a particular case of a volume potential (as in Günter [115]). By the classical Kellogg Theorem, if $f \in C^{0,\alpha}(\overline{\Omega})$, then the corresponding Newtonian potential belongs to $C^{2,\alpha}(\overline{\Omega})$ and solves the Poisson equation classically (see also Theorem 7.18). Then, if we have to show the existence of a solution for a boundary value problem for the Poisson equation (7.1), we can always introduce a new unknown $v = u - P_n[\Omega, f]$ and obtain a new boundary value problem for the Laplace equation $\Delta v = 0$, as for example one of those considered in the previous chapter. Thus, we now turn to analyze volume potentials.

7.2 Volume Potentials

As we shall see in Section 7.3, we can use the Newtonian potential (7.2) associated to the fundamental solution S_n of the Laplace operator and the single and double layer potentials of the previous Chapter 6 to represent the solutions of the basic boundary value problems for the Poisson equation in Ω. However, later in the book we need to consider volume potentials with integral kernels that are different from S_n. For this reason, we study in this section volume potentials associated to a general class of functions h, and not only to S_n. We begin with two subsections of preliminaries, where we follow the presentation of [69, §3]. In the first one we consider volume potentials for a class of weakly singular integral kernels and in the second one volume potentials for a class of kernels that are weakly singular together with their first order partial derivatives.

7.2.1 Volume Potentials with Weakly Singular Kernels

We first introduce a class of continuous functions in a punctured ball that are singular at the center.

Definition 7.1. Let $\lambda \in]0,+\infty[$. Let $R \in]0,+\infty[$. Then we denote by $A^0_\lambda(R)$ the set of functions $k \in C^0((\overline{\mathbb{B}_n(0,R)}) \setminus \{0\})$ such that

$$\sup_{x\in(\overline{\mathbb{B}_n(0,R)})\setminus\{0\}} |k(x)|\,|x|^\lambda < +\infty\,,$$

and we set

$$\|k\|_{A^0_\lambda(R)} \equiv \sup_{x\in(\overline{\mathbb{B}_n(0,R)})\setminus\{0\}} |k(x)|\,|x|^\lambda \qquad \forall k \in A^0_\lambda(R)\,.$$

We can readily verify that $(A^0_\lambda(R), \|\cdot\|_{A^0_\lambda(R)})$ is a Banach space.

Then we say that the singularity of a function k defined in a punctured neighborhood of 0 is weak or that k is weakly singular (convolution) kernel at 0 provided that k is integrable in a punctured neighborhood of 0. If instead k is not integrable in any punctured neighborhood of zero, we say that k is a singular (convolution) kernel, or that k has a strong singularity at 0.

If $\lambda < n$ then the elements of $A^0_\lambda(R)$ are integrable in a punctured neighborhood of 0 and accordingly have a weak singularity at 0. If instead $\lambda \geq n$, then the elements of $A^0_\lambda(R)$ may be singular.

We now introduce the volume potential associated to a function of class $A^0_\lambda(R)$ in the case where $\lambda < n$. We do so by means of the following.

Proposition 7.2. *Let $\lambda \in]0,n[$. Let Ω be a bounded open subset of $\mathbb{R}^n$. Then the following statements hold.*

(i) If $(k,\varphi) \in A^0_\lambda(\mathrm{diam}\,(\Omega)) \times L^\infty(\Omega)$ and if $x \in \overline{\Omega}$, then the function from Ω to $\mathbb{R}$ that takes $y \in \Omega$ to $k(x-y)\varphi(y)$ is integrable.

(ii) If $(k,\varphi) \in A^0_\lambda(\mathrm{diam}\,(\Omega)) \times L^\infty(\Omega)$, then the function

$$\mathcal{P}[k,\varphi](x) \equiv \int_\Omega k(x-y)\varphi(y)\,dy \qquad \forall x \in \overline{\Omega} \tag{7.3}$$

is continuous on $\overline{\Omega}$.

(iii) $\mathcal{P}[k,\varphi]$ is bounded and

$$\sup_{\overline{\Omega}} |\mathcal{P}[k,\varphi]| \leq s_n \frac{(\mathrm{diam}\,(\Omega))^{n-\lambda}}{n-\lambda} \|k\|_{A^0_\lambda(\mathrm{diam}\,(\Omega))} \|\varphi\|_{L^\infty(\Omega)} \tag{7.4}$$

for all $(k,\varphi) \in A^0_\lambda(\mathrm{diam}\,(\Omega)) \times L^\infty(\Omega)$.

Proof. If $(k,\varphi) \in A^0_\lambda(\mathrm{diam}\,(\Omega)) \times L^\infty(\Omega)$, then we have

$$|k(x-y)\varphi(y)| \leq |k(x-y)|\|\varphi\|_{L^\infty(\Omega)} \qquad \text{for a.a. } y \in \Omega$$

for all $x \in \overline{\Omega}$. Then $k(x-\cdot)\varphi(\cdot)$ is integrable in Ω. Since $\Omega \subseteq \mathbb{B}_n(x, \mathrm{diam}\,(\Omega))$ for all $x \in \overline{\Omega}$, we have

$$\begin{aligned}\left|\int_\Omega k(x-y)\varphi(y)\,dy\right| &\leq \int_{\mathbb{B}_n(x,\mathrm{diam}\,(\Omega))} |k(x-y)|\,dy \|\varphi\|_{L^\infty(\Omega)} \\ &\leq \|k\|_{A^0_\lambda(\mathrm{diam}\,(\Omega))} \int_{\mathbb{B}_n(x,\mathrm{diam}\,(\Omega))} \frac{dy}{|x-y|^\lambda} \|\varphi\|_{L^\infty(\Omega)} \\ &= \|k\|_{A^0_\lambda(\mathrm{diam}\,(\Omega))} s_n \frac{(\mathrm{diam}\,(\Omega))^{n-\lambda}}{n-\lambda} \|\varphi\|_{L^\infty(\Omega)} \qquad \forall x \in \overline{\Omega}\,.\end{aligned}$$

Hence, inequality (7.4) follows.

To show that $\mathcal{P}[k,\varphi]$ is continuous, we proceed by a standard argument. We take a function $g \in C^\infty(\mathbb{R})$ such that

$$g(t) = 0 \qquad \forall t \in]-\infty, 1]\,, \qquad g(t) = 1 \qquad \forall t \in [2, +\infty[\,. \tag{7.5}$$

Then we set

$$g_\delta(t) \equiv g(t/\delta) \qquad \forall t \in \mathbb{R}\,, \tag{7.6}$$

and

$$u_\delta(x) \equiv \int_\Omega g_\delta(|x-y|)k(x-y)\varphi(y)\,dy \qquad \forall x \in \overline{\Omega}\,, \tag{7.7}$$

for all $\delta \in]0, +\infty[$. We observe that the function that takes $(x,y) \in \overline{\Omega} \times \overline{\Omega}$ to $g_\delta(|x-y|)k(x-y)$ is continuous. By the definition of g_δ, we have

$$|g_\delta(|x-y|)k(x-y)\varphi(y)| \leq \|g\|_{L^\infty(\mathbb{R})} \left(\sup_{\overline{\mathbb{B}_n(0,\mathrm{diam}\,(\Omega))\setminus\mathbb{B}_n(0,\delta)}} |k| \right) |\varphi(y)|\,, \tag{7.8}$$

for all $x \in \overline{\Omega}$ and for almost all $y \in \Omega$. Since $\varphi \in L^1(\Omega)$, inequality (7.8) and the continuity theorem for integrals depending on a parameter imply that u_δ is continuous in $\overline{\Omega}$. In order to prove that $\mathcal{P}[k,\varphi]$ is continuous, it suffices to show that

$$\lim_{\delta\to 0} u_\delta = \mathcal{P}[k,\varphi] \qquad \text{uniformly in } \overline{\Omega}\,. \tag{7.9}$$

Since $1 - g_\delta(|x-y|) = 0$ for $|x-y| \geq 2\delta$, we have

$$\begin{aligned}&|\mathcal{P}[k,\varphi](x) - u_\delta(x)| \\ &= \left| \int_{\mathbb{B}_n(x,2\delta)\cap\Omega} (1 - g_\delta(|x-y|))k(x-y)\varphi(y)\,dy \right| \\ &\leq (1 + \|g\|_{L^\infty(\mathbb{R})})\|\varphi\|_{L^\infty(\Omega)} \int_{\mathbb{B}_n(x,2\delta)\cap\mathbb{B}_n(x,\mathrm{diam}\,(\Omega))} |k(x-y)|\,dy \\ &= (1 + \|g\|_{L^\infty(\mathbb{R})})\|\varphi\|_{L^\infty(\Omega)} \int_{\mathbb{B}_n(0,2\delta)\cap\mathbb{B}_n(0,\mathrm{diam}\,(\Omega))} |k(y)|\,dy\end{aligned}$$

$$\leq (1+\|g\|_{L^\infty(\mathbb{R})})\|\varphi\|_{L^\infty(\Omega)}$$
$$\times \left(\sup_{y\in\mathbb{B}_n(0,\mathrm{diam}\,(\Omega))\setminus\{0\}} |k(y)|\,|y|^\lambda\right)\int_{\mathbb{B}_n(0,2\delta)} |y|^{-\lambda}\,dy$$
$$= (1+\|g\|_{L^\infty(\mathbb{R})})\|\varphi\|_{L^\infty(\Omega)}\|k\|_{A^0_\lambda(\mathrm{diam}\,(\Omega))}s_n\frac{(2\delta)^{n-\lambda}}{n-\lambda},$$

for all $x\in\overline{\Omega}$ and for all $\delta\in]0,\mathrm{diam}\,(\Omega)/2]$. Hence, (7.9) holds true and $\mathcal{P}[k,\varphi]$ is continuous. □

The function $\mathcal{P}[k,\varphi]$ of (7.3) is said to be a volume potential of (convolution) kernel k and density (or moment) φ. Whenever necessary, we may write $\mathcal{P}_\Omega[k,\varphi]$ instead of $\mathcal{P}[k,\varphi]$ in order to emphasize that the domain of integration is Ω.

As we can easily verify, the fundamental solution S_n of the Laplace operator belongs to $A^0_{\max\{n-2,1/2\}}(R)$ for all $R\in]0,+\infty[$ and accordingly Proposition 7.2 implies the validity of the following.

Proposition 7.3. *Let Ω be a bounded open subset of $\mathbb{R}^n$. Then the operator from $L^\infty(\Omega)$ to $C^0(\overline{\Omega})$ that takes φ to $\mathcal{P}[S_n,\varphi]$ $(= P_n[\Omega,\varphi])$ is linear and continuous (see (7.2)).*

7.2.2 Volume Potentials with Kernels Which are Weakly Singular Together with Their First Order Partial Derivatives

We now introduce a subclass $A^1_\lambda(R)$ of $A^0_\lambda(R)$ consisting of continuously differentiable functions with a controlled singular behavior at the origin (cf. Definition 7.1).

Definition 7.4. Let $\lambda\in]0,+\infty[$. Let $R\in]0,+\infty[$. Then we denote by $A^1_\lambda(R)$ the set of functions $k\in C^1((\overline{\mathbb{B}_n(0,R)})\setminus\{0\})$ such that

$$k\in A^0_\lambda(R)\,,\qquad \frac{\partial k}{\partial x_j}\in A^0_{\lambda+1}(R)\qquad \forall j\in\{1,\ldots,n\}\,,$$

and we set

$$\|k\|_{A^1_\lambda(R)}\equiv\|k\|_{A^0_\lambda(R)}+\sum_{j=1}^n\left\|\frac{\partial k}{\partial x_j}\right\|_{A^0_{\lambda+1}(R)}\qquad \forall k\in A^1_\lambda(R)\,.$$

One can easily verify that $\left(A^1_\lambda(R),\|\cdot\|_{A^1_\lambda(R)}\right)$ is a Banach space. In the following proposition we consider the function $\mathcal{P}[k,\varphi]$ with (k,φ) in $A^1_\lambda(\mathrm{diam}\,(\Omega))\times L^\infty(\Omega)$ and λ smaller than $(n-1)$.

Proposition 7.5. *Let $\lambda\in]0,n-1[$. Let Ω be a bounded open subset of $\mathbb{R}^n$. Then the following statements hold.*

(i) *If* $(k,\varphi) \in A^1_\lambda(\mathrm{diam}\,(\Omega)) \times L^\infty(\Omega)$ *and if* $x \in \overline{\Omega}$, *then the functions from* Ω *to* $\mathbb{R}$ *which take* $y \in \Omega$ *to* $k(x-y)\varphi(y)$ *and to* $\frac{\partial k}{\partial x_j}(x-y)\varphi(y)$, *with* $j \in \{1,\dots,n\}$, *are integrable.*

(ii) *If* $(k,\varphi) \in A^1_\lambda(\mathrm{diam}\,(\Omega)) \times L^\infty(\Omega)$, *then* $\mathcal{P}[k,\varphi] \in C^1(\overline{\Omega})$ *and*

$$\frac{\partial}{\partial x_j}\mathcal{P}[k,\varphi] = \mathcal{P}[\frac{\partial k}{\partial x_j},\varphi] \qquad \text{in } \overline{\Omega}\,. \tag{7.10}$$

Proof. Statement (i) is an immediate consequence of Proposition 7.2 applied to k and $\frac{\partial k}{\partial x_j}$.

We now consider statement (ii). By Proposition 7.2 (ii), $\mathcal{P}[k,\varphi]$ and $\mathcal{P}[\frac{\partial k}{\partial x_j},\varphi]$ are continuous in $\overline{\Omega}$ for all $j \in \{1,\dots,n\}$. Thus it suffices to show that $\frac{\partial}{\partial x_j}\mathcal{P}[k,\varphi]$ exists in Ω and that (7.10) holds in Ω. We proceed by a standard argument, $g \in C^\infty(\mathbb{R})$, $g_\delta \in C^\infty(\mathbb{R})$, u_δ be as in (7.5), (7.6), (7.7) for all $\delta \in]0,+\infty[$. We already know that $u_\delta \in C^0(\overline{\Omega})$. We also observe that the function that takes $(x,y) \in \overline{\Omega}\times\overline{\Omega}$ to $g_\delta(|x-y|)k(x-y)$ is of class C^1. We now show that $u_\delta \in C^1(\overline{\Omega})$ by applying the classical theorem of differentiation for integrals depending on a parameter. We have

$$\begin{aligned}&\left|\frac{\partial}{\partial x_j}\{g_\delta(|x-y|)k(x-y)\varphi(y)\}\right| \\ &\qquad \le \left|g'_\delta(|x-y|)\frac{x_j-y_j}{|x-y|}k(x-y)\varphi(y)\right| + \left|g_\delta(|x-y|)\frac{\partial k}{\partial x_j}(x-y)\varphi(y)\right|\end{aligned} \tag{7.11}$$

for all $x \in \overline{\Omega}$ and for almost all $y \in \Omega$. The functions $\frac{\partial k}{\partial x_j}$ are continuous in the set $\overline{\mathbb{B}_n(0,\mathrm{diam}\,(\Omega))} \setminus \{0\}$. Hence, the functions $\frac{\partial k}{\partial x_j}$ are bounded in the compact set $\overline{\mathbb{B}_n(0,\mathrm{diam}\,(\Omega))} \setminus \mathbb{B}_n(0,\delta)$. Then the right-hand side of (7.11) is less than or equal to

$$\begin{aligned}&\frac{1}{\delta}\|g'\|_{L^\infty(\mathbb{R})}\left(\sup_{\overline{\mathbb{B}_n(0,\mathrm{diam}\,(\Omega))}\setminus\mathbb{B}_n(0,\delta)}|k|\right)|\varphi(y)| \\ &\qquad +\|g\|_{L^\infty(\mathbb{R})}\left(\sup_{\overline{\mathbb{B}_n(0,\mathrm{diam}\,(\Omega))}\setminus\mathbb{B}_n(0,\delta)}\left|\frac{\partial k}{\partial x_j}\right|\right)|\varphi(y)|\end{aligned} \tag{7.12}$$

for all $x \in \overline{\Omega}$ and for almost all $y \in \Omega$. Since $\varphi \in L^1(\Omega)$, inequalities (7.11), (7.12) and the differentiability theorem for integrals depending on a parameter imply that

$$\frac{\partial u_\delta}{\partial x_j}(x) = \int_\Omega \frac{\partial}{\partial x_j}\left[g_\delta(|x-y|)k(x-y)\right]\varphi(y)\,dy \qquad \forall x \in \Omega$$

and that $\frac{\partial u_\delta}{\partial x_j}$ has a continuous extension to $\overline{\Omega}$. Hence, $u_\delta \in C^1(\overline{\Omega})$. In order to prove that $\mathcal{P}[k,\varphi]$ belongs to $C^1(\overline{\Omega})$, it suffices to remember the validity of the limiting relation (7.9) and to show that

$$\lim_{\delta\to 0}\frac{\partial u_\delta}{\partial x_j} = \mathcal{P}[\frac{\partial k}{\partial x_j},\varphi] \quad \text{uniformly in } \overline{\Omega} \tag{7.13}$$

for all $j \in \{1,\ldots,n\}$. Since the support of g'_δ is contained in $[\delta, 2\delta]$, the same argument we have exploited to prove (7.9) implies that

$$\begin{aligned}
&\left|\mathcal{P}[\frac{\partial k}{\partial x_j},\varphi](x) - \frac{\partial u_\delta}{\partial x_j}(x)\right| \\
&\leq \left|\int_\Omega (1-g_\delta(|x-y|))\frac{\partial k}{\partial x_j}(x-y)\varphi(y)\,dy\right| \\
&\quad + \left|\int_\Omega \frac{1}{\delta}g'\left(\frac{|x-y|}{\delta}\right)\frac{x_j-y_j}{|x-y|}k(x-y)\varphi(y)\,dy\right| \\
&\leq (1+\|g\|_{L^\infty(\mathbb{R})})\|\varphi\|_{L^\infty(\Omega)} \\
&\quad \times \left(\sup_{y\in\mathbb{B}_n(0,\mathrm{diam}\,(\Omega))\setminus\{0\}}\left|\frac{\partial k}{\partial x_j}(y)\right|\,|y|^{\lambda+1}\right) s_n\frac{(2\delta)^{n-\lambda-1}}{n-\lambda-1} \\
&\quad + \frac{1}{\delta}\|g'\|_{L^\infty(\mathbb{R})}\int_{\mathbb{B}_n(x,2\delta)\setminus\mathbb{B}_n(x,\delta)}|k(x-y)|\,dy\|\varphi\|_{L^\infty(\Omega)} \\
&\leq s_n(1+\|g\|_{L^\infty(\mathbb{R})}+\|g'\|_{L^\infty(\mathbb{R})})\|\varphi\|_{L^\infty(\Omega)} \\
&\quad \times\left\{\left(\sup_{x\in\mathbb{B}_n(0,\mathrm{diam}\,(\Omega))\setminus\{0\}}\left|\frac{\partial k}{\partial x_j}(y)\right|\,|y|^{\lambda+1}\right)\frac{(2\delta)^{n-\lambda-1}}{n-\lambda-1}\right. \\
&\quad \left. + \frac{1}{\delta}\left(\sup_{y\in\mathbb{B}_n(0,\mathrm{diam}\,(\Omega))\setminus\{0\}}|k(y)|\,|y|^\lambda\right)\frac{(2\delta)^{n-\lambda}}{n-\lambda}\right\} \qquad \forall x\in\Omega\,,
\end{aligned}$$

for all $\delta \in]0, \mathrm{diam}\,(\Omega)/2[$. Since $n-\lambda-1>0$, the limiting relation (7.13) follows. □

As we can easily verify, the fundamental solution S_n of the Laplace operator belongs to $A^1_{\max\{n-2,1/2\}}(R)$ for all $R \in]0,+\infty[$, and accordingly Propositions 7.2 and 7.5 imply the validity of the following.

Proposition 7.6. *Let Ω be a bounded open subset of $\mathbb{R}^n$. Then the operator from $L^\infty(\Omega)$ to $C^1(\overline{\Omega})$ that takes φ to $\mathcal{P}[S_n,\varphi] = P_n[\Omega,\varphi]$ is linear and continuous (see (7.2)) and*

$$\frac{\partial}{\partial x_j}\mathcal{P}[S_n,\varphi] = \mathcal{P}[\frac{\partial S_n}{\partial x_j},\varphi] \qquad \text{in } \overline{\Omega}\,,$$

for all $j \in \{1,\ldots,n\}$.

Then we introduce the following classical formula for the first order partial derivatives of a volume potential with a differentiable density.

Lemma 7.7. *Let $\lambda \in]0, n-1[$. Let Ω be a bounded open Lipschitz subset of $\mathbb{R}^n$. If $(k,\varphi) \in A^1_\lambda(\mathrm{diam}\,(\Omega)) \times C^1(\overline{\Omega})$ and $j \in \{1,\ldots,n\}$, then*

$$\frac{\partial}{\partial x_j}\mathcal{P}[k,\varphi](x) = \mathcal{P}[k,\frac{\partial \varphi}{\partial x_j}](x) - \int_{\partial\Omega} k(x-y)\varphi(y)(\nu_\Omega)_j(y)\,d\sigma_y \qquad \forall x \in \Omega\,. \tag{7.14}$$

Proof. By Proposition 7.5 and by the Leibniz rule, we have

$$\begin{aligned}\frac{\partial}{\partial x_j}\mathcal{P}[k,\varphi](x) &= \int_\Omega \frac{\partial k}{\partial x_j}(x-y)\varphi(y)\,dy \qquad (7.15)\\ &= -\int_\Omega \frac{\partial}{\partial y_j}(k(x-y))\varphi(y)\,dy\\ &= -\int_\Omega \frac{\partial}{\partial y_j}\left(k(x-y)\varphi(y)\right)\,dy + \int_\Omega k(x-y)\frac{\partial\varphi}{\partial y_j}(y)\,dy \qquad \forall x\in\overline{\Omega}\,.\end{aligned}$$

Next we fix $x\in\Omega$ and we take $\epsilon_x \in]0, \text{dist}\,(x,\partial\Omega)[$. Then $\overline{\mathbb{B}_n(x,\epsilon)} \subseteq \Omega$ and the set

$$\Omega_\epsilon \equiv \Omega \setminus \overline{\mathbb{B}_n(x,\epsilon)}$$

is of Lipschitz class for all $\epsilon\in]0,\epsilon_x[$. By the Divergence Theorem 4.1, we have

$$\begin{aligned}&\int_\Omega \frac{\partial}{\partial y_j}\left(k(x-y)\varphi(y)\right)\,dy \qquad (7.16)\\ &= \int_{\Omega_\epsilon} \frac{\partial}{\partial y_j}\left(k(x-y)\varphi(y)\right)\,dy + \int_{\mathbb{B}_n(x,\epsilon)} \frac{\partial}{\partial y_j}\left(k(x-y)\varphi(y)\right)\,dy\\ &= \int_{\partial\Omega} k(x-y)\varphi(y)(\nu_\Omega)_j(y)\,d\sigma_y + \int_{\partial\mathbb{B}_n(x,\epsilon)} k(x-y)\varphi(y)\frac{x_j-y_j}{|x-y|}\,d\sigma_y\\ &\quad - \int_{\mathbb{B}_n(x,\epsilon)} \frac{\partial k}{\partial x_j}(x-y)\varphi(y)\,dy + \int_{\mathbb{B}_n(x,\epsilon)} k(x-y)\frac{\partial\varphi}{\partial y_j}(y)\,dy\end{aligned}$$

for all $\epsilon\in]0,\epsilon_x[$. We note that

$$\begin{aligned}&\left|\int_{\partial\mathbb{B}_n(x,\epsilon)} k(x-y)\varphi(y)\frac{x_j-y_j}{|x-y|}\,d\sigma_y\right| \qquad (7.17)\\ &\le \left(\sup_{y\in\mathbb{B}_n(0,\text{diam}\,(\Omega))\setminus\{0\}} |k(y)|\,|y|^\lambda\right)\|\varphi\|_{L^\infty(\Omega)}\int_{\partial\mathbb{B}_n(0,\epsilon)} |y|^{-\lambda}\,d\sigma_y\\ &= \|k\|_{A^0_\lambda(\text{diam}\,(\Omega))}\|\varphi\|_{L^\infty(\Omega)} s_n \epsilon^{n-1-\lambda}\end{aligned}$$

for all $\epsilon\in]0,\epsilon_x[$. By Proposition 7.5 (i), both the functions $\frac{\partial k}{\partial x_j}(x-\cdot)\varphi(\cdot)$ and $k(x-\cdot)\frac{\partial\varphi}{\partial x_j}(\cdot)$ are integrable in Ω and accordingly

$$\begin{aligned}&\lim_{\epsilon\to 0}\int_{\mathbb{B}_n(x,\epsilon)} \frac{\partial k}{\partial x_j}(x-y)\varphi(y)\,dy = 0\,, \qquad (7.18)\\ &\lim_{\epsilon\to 0}\int_{\mathbb{B}_n(x,\epsilon)} k(x-y)\frac{\partial\varphi}{\partial y_j}(y)\,dy = 0\,.\end{aligned}$$

By (7.17) and (7.18), we can take the limit as ϵ tends to 0 in (7.16) and deduce that

$$\int_{\Omega} \frac{\partial}{\partial y_j} (k(x-y)\varphi(y))\, dy = \int_{\partial\Omega} k(x-y)\varphi(y)(\nu_\Omega)_j(y)\, d\sigma_y \,.$$

Then equality (7.15) implies the validity of formula (7.14). □

7.2.3 Volume Potentials with Singular Kernels and with a Constant Density

In Section 7.2.1 we have considered volume potentials of the form

$$\int_{\Omega} k(x-y)\varphi(y)\, dy$$

where φ is essentially bounded and k has a weak singularity at the origin, i.e., k is integrable in a neighborhood of the origin, a case in which $k(x-y)\varphi(y)$ is integrable in the sense of Lebesgue in the variable $y \in \Omega$ for all $x \in \mathbb{R}^n$.

We now want to consider the case in which the singularity of k at the origin is strong, i.e., the case in which k is not integrable in a neighborhood of the origin. In such a case, the function $k(x-y)\varphi(y)$ may not have an integral in the sense of Lebesgue in the variable $y \in \Omega$, but the integral $\int_\Omega k(x-y)\varphi(y)\, dy$ may exist in the sense of the principal value. In the present subsection we mainly concentrate on the case where φ is constant.

Moreover, we consider only the specific class of kernels which are positively homogeneous of degree $-n$, a class which includes the important case of the second order partial derivatives of the fundamental solution S_n of the Laplace operator, which is relevant for the present book. We start by introducing some preliminaries on the principal value.

Definition 7.8. Let Ω be an open subset of $\mathbb{R}^n$. Let $x^o \in \Omega$. Let f be a function from $\Omega \setminus \{x^o\}$ to $\mathbb{R}$ such that the restriction $f_{|\Omega\setminus\mathbb{B}_n(x^o,\epsilon)}$ is integrable for all $\epsilon \in]0,+\infty[$.

We say that f admits an integral in the sense of the principal value in Ω provided that the limit

$$\text{p.v.} \int_{\Omega} f\, dx \equiv \lim_{\epsilon\to 0} \int_{\Omega\setminus\mathbb{B}_n(x^o,\epsilon)} f\, dx$$

exists in $\mathbb{R} \cup \{-\infty\} \cup \{+\infty\}$.

We now look closely at the case of positively homogeneous functions of degree $-n$.

Proposition 7.9. *Let $n \in \mathbb{N} \setminus \{0\}$. Let $f \in C^0(\mathbb{R}^n \setminus \{0\})$ be a positively homogeneous function of degree $-n$. Then the following statements hold.*

(i) If $a, b \in]0,+\infty[$, $a < b$, then

$$\int_{\mathbb{B}_n(0,b)\setminus\mathbb{B}_n(0,a)} f(x)\,dx = \int_{\partial\mathbb{B}_n(0,1)} f(u)\,d\sigma_u \log(b/a)\,, \tag{7.19}$$

where we understand that $\int_{\partial\mathbb{B}_n(0,1)} f\,d\sigma \equiv f(1)+f(-1)$ *in case* $n=1$. *In particular, if*

$$\int_{\partial\mathbb{B}_n(0,1)} f\,d\sigma = 0\,, \tag{7.20}$$

then the integral of f *on the circular annulus* $\mathbb{B}_n(0,b)\setminus\mathbb{B}_n(0,a)$ *equals* 0.

(ii) If condition (7.20) holds true, then

$$\text{p.v.}\int_{\mathbb{B}_n(0,b)} f(x)\,dx = \lim_{\epsilon\to 0}\int_{\mathbb{B}_n(0,b)\setminus\mathbb{B}_n(0,\epsilon)} f(x)\,dx = 0$$

for all $b\in]0,+\infty[$. *Moreover,* p.v. $\int_{\mathbb{B}_n(0,b)} f(x)\,dx$ *can exist finite for some* b *in* $]0,+\infty[$ *only if condition (7.20) holds true.*

Proof. Let $n\geq 2$. By integrating on the spheres, we have

$$\begin{aligned}
\int_{\mathbb{B}_n(0,b)\setminus\mathbb{B}_n(0,a)} f(x)\,dx &= \int_a^b\int_{\partial\mathbb{B}_n(0,\rho)} f(\xi)\,d\sigma_\xi\,d\rho \\
&= \int_a^b\int_{\partial\mathbb{B}_n(0,1)} f(\rho\eta)\,d\sigma_\eta\rho^{n-1}\,d\rho = \int_a^b\int_{\partial\mathbb{B}_n(0,1)} f(\eta)\,d\sigma_\eta\rho^{-n+n-1}\,d\rho \\
&= \int_{\partial\mathbb{B}_n(0,1)} f(\eta)\,d\sigma_\eta \log(b/a)\,.
\end{aligned}$$

and thus statement (i) holds true. If instead $n=1$, then $\mathbb{B}_n(0,b)\setminus\mathbb{B}_n(0,a) =]-b,-a]\cup[a,b[$ and we have

$$\begin{aligned}
\int_{\mathbb{B}_n(0,b)\setminus\mathbb{B}_n(0,a)} f(x)\,dx &= \int_{-b}^{-a} f(x)\,dx + \int_a^b f(x)\,dx \\
&= \int_a^b f(-x)\,dx + \int_a^b f(x)\,dx \\
&= \int_a^b f(-1)|x|^{-1}\,dx + \int_a^b f(1)|x|^{-1}\,dx \\
&= (f(-1)+f(1))\log(b/a)\,,
\end{aligned}$$

and thus statement (i) holds true also for $n=1$. Statement (ii) is an immediate consequence of statement (i). □

By the previous statement, one can prove the following lemma for volume potentials with constant density.

Lemma 7.10. *Let* Ω *be an open subset of* $\mathbb{R}^n$ *of finite measure. Let the function* $k\in C^0(\mathbb{R}^n\setminus\{0\})$ *be positively homogeneous of degree* $-n$. *Let*

$$\int_{\partial \mathbb{B}_n(0,1)} k \, d\sigma = 0 \,. \tag{7.21}$$

Then the following statements hold.

(i) (**Cancellation property for singular integrals**) *If $x \in \Omega$, then the function $k(x-y)$ is Lebesgue integrable in the variable $y \in \Omega \setminus \mathbb{B}_n(x,\epsilon)$ for all ϵ in $]0, d(x)]$ and*

$$\int_{\Omega \setminus \mathbb{B}_n(x,\epsilon)} k(x-y) \, dy = \int_{\Omega \setminus \mathbb{B}_n(x,d(x))} k(x-y) \, dy \qquad \forall \epsilon \in]0, d(x)] \,,$$

where

$$d(x) \equiv \mathrm{dist}(x, \partial\Omega) \,.$$

If $x \in \mathbb{R}^n \setminus \overline{\Omega}$, then the function $k(x-y)$ is Lebesgue integrable in the variable $y \in \Omega$.

(ii) *If $x \in \Omega$, then*

$$\mathrm{p.v.} \int_{\Omega} k(x-y) \, dy = \int_{\Omega \setminus \mathbb{B}_n(x,d(x))} k(x-y) \, dy \,.$$

If $x \in \mathbb{R}^n \setminus \overline{\Omega}$, then $\mathrm{p.v.} \int_{\Omega} k(x-y) \, dy = \int_{\Omega} k(x-y) \, dy$.

Proof. (i) If $x \in \Omega$, then

$$\mathbb{B}_n(x, d(x)) \setminus \mathbb{B}_n(x, \epsilon) \subseteq \Omega \qquad \forall \epsilon \in]0, d(x)]$$

and

$$\left| \chi_{\Omega \setminus \mathbb{B}_n(x,\epsilon)}(y) k(x-y) \right| \leq \sup_{\partial \mathbb{B}_n(0,1)} |k| \, |x-y|^{-n} \leq \epsilon^{-n} \sup_{\partial \mathbb{B}_n(0,1)} |k|$$

for all $y \in \Omega \setminus \mathbb{B}_n(x,\epsilon)$ and $\epsilon \in]0, d(x)]$. Since the measure of Ω is finite, the function $k(x-y)$ is Lebesgue integrable in the variable $y \in \Omega \setminus \mathbb{B}_n(x,\epsilon)$ for all $\epsilon \in]0, d(x)]$. Then formula (7.19) and assumption (7.21) imply that

$$\begin{aligned} &\int_{\Omega \setminus \mathbb{B}_n(x,\epsilon)} k(x-y) \, dy \\ &\quad = \int_{\Omega \setminus \mathbb{B}_n(x,d(x))} k(x-y) \, dy + \int_{\mathbb{B}_n(x,d(x)) \setminus \mathbb{B}_n(x,\epsilon)} k(x-y) \, dy \\ &\quad = \int_{\Omega \setminus \mathbb{B}_n(x,d(x))} k(x-y) \, dy + \int_{\partial \mathbb{B}_n(0,1)} k \, d\sigma \log(d(x)/\epsilon) \\ &\quad = \int_{\Omega \setminus \mathbb{B}_n(x,d(x))} k(x-y) \, dy + 0 \qquad \forall \epsilon \in]0, d(x)] \,. \end{aligned}$$

If instead $x \in \mathbb{R}^n \setminus \overline{\Omega}$, then the assumption that Ω has finite measure and the inequality

$$|k(x-y)| \leq \left(\sup_{\partial\mathbb{B}_n(0,1)} |k|\right) |x-y|^{-n} \leq \left(\sup_{\partial\mathbb{B}_n(0,1)} |k|\right) d(x)^{-n} \qquad \forall y \in \Omega\,,$$

ensure that $k(x-y)$ is Lebesgue integrable in the variable $y \in \Omega$. Hence, the proof of statement (i) is complete. Statement (ii) is an immediate consequence of statement (i). □

Next we set

$$\mathcal{K}'_{0,1} \equiv \left\{ k \in C^{0,1}_{\mathrm{loc}}(\mathbb{R}^n \setminus \{0\}) : k \text{ is positively homogeneous of degree } -n \right\},$$

and

$$\|k\|_{\mathcal{K}'_{0,1}} \equiv \|k\|_{C^{0,1}(\partial\mathbb{B}_n(0,1))} \qquad \forall k \in \mathcal{K}'_{0,1}\,.$$

We can easily verify that $\left(\mathcal{K}'_{0,1}, \|k\|_{\mathcal{K}'_{0,1}}\right)$ is a Banach space and we consider the subspace

$$\mathcal{K}'_{0,1;e,0} \equiv \left\{ k \in \mathcal{K}'_{0,1} \,:\, k \text{ is even}, \int_{\partial\mathbb{B}_n(0,1)} k\, d\sigma = 0 \right\} \tag{7.22}$$

of $\mathcal{K}'_{0,1}$. Then we introduce the following known lemma (cf. Majda and Bertozzi [196, Prop. 8.12, pp. 348–350]), which we exploit later in the analysis of volume potentials in Hölder spaces. For the sake of completeness, we include a proof that follows the lines of that of Mateu, Orobitg, Verdera [198, estimate of $(IV)_\delta$, p. 408].

Lemma 7.11. *Let Ω be a bounded open subset of $\mathbb{R}^n$ of class $C^{1,\alpha}$. Then there exists $c^*_\Omega \in]0,+\infty[$ such that*

$$\sup_{(x,\epsilon)\in\mathbb{R}^n\times]0,+\infty[} \left| \int_{\Omega\setminus\mathbb{B}_n(x,\epsilon)} k(x-y)\, dy \right| \leq c^*_\Omega \, \|k\|_{\mathcal{K}'_{0,1}} \qquad \forall k \in \mathcal{K}'_{0,1;e,0}\,. \tag{7.23}$$

Proof. Since Ω is bounded and of class $C^{1,\alpha}$, Lemma 2.63 of the uniform cylinders implies that there exist

$$0 < r < \delta < \delta_* \equiv \min\{1, \mathrm{diam}(\Omega)\}$$

with the property that for all $p \in \partial\Omega$ there exists $R_p \in \mathbb{O}_n(\mathbb{R})$ such that $C(p, R_p, r, \delta)$ is a coordinate cylinder for Ω around p and the corresponding function γ_p that represents $\partial\Omega$ in $C(p, R_p, r, \delta)$ satisfies the condition $\nabla\gamma_p(0) = 0$ and the inequality

$$a \equiv \sup_{p\in\partial\Omega} \|\gamma_p\|_{C^{1,\alpha}(\overline{\mathbb{B}_{n-1}(0,r)})} < +\infty\,. \tag{7.24}$$

Next we note that if $\epsilon \in [r, +\infty[$, then

$$\left|\int_{\Omega\setminus\mathbb{B}_n(x,\epsilon)} k(x-y)\,dy\right| \leq \int_{\Omega\setminus\mathbb{B}_n(x,\epsilon)} |k(x-y)|\,dy$$
$$\leq \sup_{\partial\mathbb{B}_n(0,1)} |k| \int_{\Omega\setminus\mathbb{B}_n(x,\epsilon)} \frac{dy}{|x-y|^n} \leq r^{-n} m_n(\Omega)\|k\|_{\mathcal{K}'_{0,1}}$$

for all $x \in \mathbb{R}^n$ and $k \in \mathcal{K}'_{0,1;e,0}$. On the other hand, we have

$$\Omega \setminus \mathbb{B}_n(x,\epsilon) = (\Omega \setminus \mathbb{B}_n(x,r)) \cup [\Omega \cap (\mathbb{B}_n(x,r) \setminus \mathbb{B}_n(x,\epsilon))] \qquad \forall \epsilon \in]0,r[$$

and accordingly

$$\left|\int_{\Omega\setminus\mathbb{B}_n(x,\epsilon)} k(x-y)\,dy\right|$$
$$\leq \left|\int_{\Omega\setminus\mathbb{B}_n(x,r)} k(x-y)\,dy\right| + \left|\int_{\Omega\cap(\mathbb{B}_n(x,r)\setminus\mathbb{B}_n(x,\epsilon))} k(x-y)\,dy\right|$$
$$\leq r^{-n} m_n(\Omega)\|k\|_{\mathcal{K}'_{0,1}} + \left|\int_{\Omega\cap(\mathbb{B}_n(x,r)\setminus\mathbb{B}_n(x,\epsilon))} k(x-y)\,dy\right| \quad \forall \epsilon \in]0,r[$$

for all $x \in \mathbb{R}^n$ and $k \in \mathcal{K}'_{0,1;e,0}$. Thus it suffices to show that there exists $c^*_{0,\Omega}$ in $]0,+\infty[$ such that

$$\sup_{(x,\epsilon)\in\mathbb{R}^n\times]0,r[} \left|\int_{\Omega\cap(\mathbb{B}_n(x,r)\setminus\mathbb{B}_n(x,\epsilon))} k(x-y)\,dy\right| \tag{7.25}$$
$$\leq c^*_{0,\Omega}\,\|k\|_{\mathcal{K}'_{0,1}} \qquad \forall k \in \mathcal{K}'_{0,1;e,0}\,.$$

To shorten our notation, we set

$$\mathbb{A}_n(x,\epsilon,r) \equiv \mathbb{B}_n(x,r) \setminus \mathbb{B}_n(x,\epsilon) \qquad \forall x \in \mathbb{R}^n, \epsilon \in]0,r[\,.$$

We first estimate the integral in (7.25) in case $x \in \partial\Omega$. Since $r < \delta$, we have

$$\mathbb{B}_n(0,r) \subseteq \mathbb{B}_{n-1}(0,r)\times]-\delta,\delta[$$

and accordingly

$$\mathbb{B}_n(x,r) = x + R_x^t(\mathbb{B}_n(0,r)) \subseteq C(x,R_x,r,\delta)\,.$$

Then, by exploiting the change of variables $y = x + R_x^t(\eta,\zeta)^t$, we obtain

$$\int_{\Omega\cap\mathbb{A}_n(x,\epsilon,r)} k(x-y)\,dy \tag{7.26}$$
$$= \int_{[R_x(\Omega-x)]\cap\mathbb{A}_n(0,\epsilon,r)} k(-R_x^t(\eta,\zeta)^t)\,d\eta d\zeta$$

$$= \int_{[R_x(\Omega - x)] \cap [\mathbb{B}_{n-1}(0,r) \times]-\delta,\delta[] \cap \mathbb{A}_n(0,\epsilon,r)} \tilde{k}_x(\eta,\zeta)\, d\eta d\zeta$$

$$= \int_{\text{hypograph}_s(\gamma_x) \cap \mathbb{A}_n(0,\epsilon,r)} \tilde{k}_x(\eta,\zeta)\, d\eta d\zeta \qquad \forall \epsilon \in]0,r[\,,$$

where

$$\tilde{k}_x(\eta,\zeta) \equiv k(-R_x^t(\eta,\zeta)^t) \qquad \forall (\eta,\zeta) \in (\mathbb{R}^{n-1} \times \mathbb{R}) \setminus \{(0,0)\}\,,$$

and

$$\int_{\text{hypograph}_s(\gamma_x) \cap \mathbb{A}_n(0,\epsilon,r)} \tilde{k}_x(\eta,\zeta)\, d\eta d\zeta \tag{7.27}$$

$$= \int_{\text{hypograph}_s(\gamma_x) \cap (\mathbb{B}_n^+(0,r) \setminus \mathbb{B}_n^+(0,\epsilon))} \tilde{k}_x(\eta,\zeta)\, d\eta d\zeta$$

$$+ \int_{\text{hypograph}_s(\gamma_x) \cap (\mathbb{B}_n^-(0,r) \setminus \mathbb{B}_n^-(0,\epsilon))} \tilde{k}_x(\eta,\zeta)\, d\eta d\zeta$$

$$= \int_{\text{hypograph}_s(\gamma_x) \cap (\mathbb{B}_n^+(0,r) \setminus \mathbb{B}_n^+(0,\epsilon))} \tilde{k}_x(\eta,\zeta)\, d\eta d\zeta$$

$$+ \int_{\mathbb{B}_n^-(0,r) \setminus \mathbb{B}_n^-(0,\epsilon)} \tilde{k}_x(\eta,\zeta)\, d\eta d\zeta$$

$$- \int_{\{(\eta,\zeta) \in \mathbb{B}_n^-(0,r) \setminus \mathbb{B}_n^-(0,\epsilon) :\, \gamma_x(\eta) \leq \zeta < 0\}} \tilde{k}_x(\eta,\zeta)\, d\eta d\zeta$$

for all $\epsilon \in]0,r[$ (cf. (2.54) for the symbol $\mathbb{B}_n^\pm(0,r)$). By our assumption on k, we have

$$\int_{\partial \mathbb{B}_n(0,1)} \tilde{k}_x\, d\sigma = \int_{\partial \mathbb{B}_n(0,1)} k\, d\sigma = 0\,,$$

and thus Proposition 7.9 (i) implies that

$$\int_{\mathbb{B}_n(0,r) \setminus \mathbb{B}_n(0,\epsilon)} \tilde{k}_x(\eta,\zeta)\, d\eta d\zeta = 0 \qquad \forall \epsilon \in]0,r[\,.$$

Since k is even, then $\tilde{k}_x$ is even and thus

$$\int_{\mathbb{B}_n^-(0,r) \setminus \mathbb{B}_n^-(0,\epsilon)} \tilde{k}_x(\eta,\zeta)\, d\eta d\zeta = \frac{1}{2} \int_{\mathbb{B}_n(0,r) \setminus \mathbb{B}_n(0,\epsilon)} \tilde{k}_x(\eta,\zeta)\, d\eta d\zeta = 0 \tag{7.28}$$

for all $\epsilon \in]0,r[$. Moreover,

$$\begin{aligned} &\left[\text{hypograph}_s(\gamma_x) \cap (\mathbb{B}_n^+(0,r) \setminus \mathbb{B}_n^+(0,\epsilon))\right] \\ &\qquad \cup \{(\eta,\zeta) \in \mathbb{B}_n^-(0,r) \setminus \mathbb{B}_n^-(0,\epsilon) :\, \gamma_x(\eta) \leq \zeta < 0\} \\ &\quad \subseteq \{(\eta,\zeta) \in \mathbb{B}_{n-1}(0,r) \times]-\delta,\delta[:\, |\zeta| < |\gamma_x(\eta)|\} \qquad \forall \epsilon \in]0,r[\,. \end{aligned}$$

Then equalities (7.26), (7.27), (7.28) imply that

$$
\begin{aligned}
&\left|\int_{\Omega\cap\mathbb{A}_n(x,\epsilon,r)} k(x-y)\,dy\right| \qquad (7.29)\\
&\quad \le \int_{\{(\eta,\zeta)\in\mathbb{B}_{n-1}(0,r)\times]-\delta,\delta[:\,|\zeta|<|\gamma_x(\eta)|\}} |\tilde{k}_x(\eta,\zeta)|\,d\eta d\zeta\\
&\quad \le \left(\sup_{\partial\mathbb{B}_n(0,1)}|k|\right)\int_{\{(\eta,\zeta)\in\mathbb{B}_{n-1}(0,r)\times]-\delta,\delta[:\,|\zeta|<|\gamma_x(\eta)|\}} \frac{d\eta d\zeta}{(|\eta|^2+|\zeta|^2)^{n/2}}\\
&\quad \le 2\left(\sup_{\partial\mathbb{B}_n(0,1)}|k|\right)\int_{\mathbb{B}_{n-1}(0,r)}\int_0^{|\gamma_x(\eta)|}\frac{d\zeta}{|\eta|^n}\,d\eta\\
&\quad = 2\left(\sup_{\partial\mathbb{B}_n(0,1)}|k|\right)\int_{\mathbb{B}_{n-1}(0,r)} |\eta|^{-n}|\gamma_x(\eta)|\,d\eta \qquad \forall\epsilon\in]0,r[\,.
\end{aligned}
$$

By the elementary inequality of Lemma 2.58 for the function γ_x, by equalities $\gamma_x(0)=0$, $\nabla\gamma_x(0)=0$, and by inequality (7.24), we have

$$
|\gamma_x(\eta)| \le |\nabla\gamma_x|_\alpha|\eta|^{\alpha+1} \le a|\eta|^{\alpha+1} \qquad \forall\eta\in\mathbb{B}_{n-1}(0,r)\,.
$$

Then we have

$$
\int_{\mathbb{B}_{n-1}(0,r)} |\eta|^{-n}|\gamma_x(\eta)|\,d\eta \le a\int_{\mathbb{B}_{n-1}(0,r)}\frac{d\eta}{|\eta|^{(n-1)-\alpha}} = a s_{n-1}\frac{r^\alpha}{\alpha}\,. \qquad (7.30)
$$

By inequalities (7.29) and (7.30), we conclude that there exists $c^*_{1,\Omega}\in]0,+\infty[$ such that

$$
\sup_{(x,\epsilon)\in(\partial\Omega)\times]0,r[}\left|\int_{\Omega\cap\mathbb{A}_n(x,\epsilon,r)} k(x-y)\,dy\right| \le c^*_{1,\Omega}\,\|k\|_{\mathcal{K}'_{0,1}} \qquad \forall k\in\mathcal{K}'_{0,1;e,0}\,. \qquad (7.31)
$$

Next we consider the case of the points $x\in\mathbb{R}^n\setminus\partial\Omega$. To do so, we set

$$
d(x)\equiv\operatorname{dist}(x,\partial\Omega)
$$

and we note that

$$
\begin{aligned}
\int_{\Omega\cap\mathbb{A}_n(x,\epsilon,r)} k(x-y)\,dy &= \int_{\Omega\cap\mathbb{A}_n(x,d(x),r)} k(x-y)\,dy\\
&\quad + \int_{\Omega\cap\mathbb{A}_n(x,\epsilon,\min\{d(x),r\})} k(x-y)\,dy \qquad \forall\epsilon\in]0,d(x)]\,.
\end{aligned}
$$

We note that if $x\in\Omega$, then $\mathbb{A}_n(x,\epsilon,d(x))\subseteq\mathbb{B}_n(x,d(x))\subseteq\Omega$ and Proposition 7.9 (i) implies that

$$\int_{\Omega\cap\mathbb{A}_n(x,\epsilon,\min\{d(x),r\})} k(x-y)\,dy = \int_{\mathbb{A}_n(x,\epsilon,\min\{d(x),r\})} k(x-y)\,dy = 0$$

for all $\epsilon \in]0, d(x)]$. If instead $x \in \mathbb{R}^n \setminus \overline{\Omega}$, then

$$\Omega \cap \mathbb{A}_n(x,\epsilon,\min\{d(x),r\}) \subseteq \Omega \cap \mathbb{B}_n(x,d(x)) = \emptyset$$

and again

$$\int_{\Omega\cap\mathbb{A}_n(x,\epsilon,\min\{d(x),r\})} k(x-y)\,dy = 0\,.$$

Then we have

$$\int_{\Omega\cap\mathbb{A}_n(x,\epsilon,r)} k(x-y)\,dy = \int_{\Omega\cap\mathbb{A}_n(x,d(x),r)} k(x-y)\,dy \qquad \forall \epsilon \in]0,d(x)] \quad (7.32)$$

for all $x \in \mathbb{R}^n \setminus \partial\Omega$. Next we consider separately case $d(x) \geq r/2$ and case $d(x) < r/2$. If $d(x) \geq r/2$ and $x \in \mathbb{R}^n \setminus \partial\Omega$, equality (7.32) implies that

$$\begin{aligned}
\sup_{\epsilon\in]0,r[}\left|\int_{\Omega\cap\mathbb{A}_n(x,\epsilon,r)} k(x-y)\,dy\right| &= \sup_{\epsilon\in[d(x),r[}\left|\int_{\Omega\cap\mathbb{A}_n(x,\epsilon,r)} k(x-y)\,dy\right| \\
&\leq \left(\sup_{\partial\mathbb{B}_n(0,1)} |k|\right) \sup_{\epsilon\in[d(x),r[}\int_{\Omega\cap\mathbb{A}_n(x,\epsilon,r)} \frac{dy}{|x-y|^n} \\
&\leq \sup_{\partial\mathbb{B}_n(0,1)} |k|(r/2)^{-n} m_n(\Omega) \qquad (7.33)
\end{aligned}$$

and we conclude that there exists $c^*_{2,\Omega} \in]0,+\infty[$ such that

$$\begin{aligned}
\sup_{(x,\epsilon)\in\{x\in\mathbb{R}^n\setminus\partial\Omega:\, d(x)\geq r/2\}\times]0,r[}\left|\int_{\Omega\cap\mathbb{A}_n(x,\epsilon,r)} k(x-y)\,dy\right| \qquad (7.34)\\
\leq c^*_{2,\Omega}\,\|k\|_{\mathcal{K}'_{0,1}} \quad \forall k \in \mathcal{K}'_{0,1;e,0}\,.
\end{aligned}$$

Thus it remains to consider case $d(x) < r/2$. By equality (7.32), we have

$$\begin{aligned}
&\sup_{(x,\epsilon)\in\{x\in\mathbb{R}^n\setminus\partial\Omega:\, d(x)<r/2\}\times]0,r[}\left|\int_{\Omega\cap\mathbb{A}_n(x,\epsilon,r)} k(x-y)\,dy\right| \qquad (7.35)\\
&\leq \sup_{(x,\epsilon)\in\{(x,\epsilon)\in(\mathbb{R}^n\setminus\partial\Omega)\times]0,r[:\, d(x)<r/2,\ \epsilon\in[d(x),r[\}}\left|\int_{\Omega\cap\mathbb{A}_n(x,\epsilon,r)} k(x-y)\,dy\right|\,.
\end{aligned}$$

Thus it suffices to consider case $\epsilon \in [d(x), r[$ and we turn to prove that there exists $c^*_{3,\Omega} \in]0,+\infty[$ such that

$$\sup_{(x,\epsilon)\in\{(x,\epsilon)\in(\mathbb{R}^n\setminus\partial\Omega)\times]0,r[:\, d(x)<r/2,\ \epsilon\in[d(x),r[\}} \left| \int_{\Omega\cap\mathbb{A}_n(x,\epsilon,r)} k(x-y)\,dy \right| \tag{7.36}$$
$$\leq c^*_{3,\Omega}\, \|k\|_{\mathcal{K}'_{0,1}} \qquad \forall k\in\mathcal{K}'_{0,1;e,0}\,.$$

Let $x^*\in\partial\Omega$ be such that

$$d(x)\equiv \mathrm{dist}(x,\partial\Omega)=|x-x^*|\,,$$

for all $x\in\mathbb{R}^n\setminus\partial\Omega$. If $\epsilon\in[d(x),r[$, we plan to estimate $\int_{\Omega\cap\mathbb{A}_n(x,\epsilon,r)} k(x-y)\,dy$ by exploiting inequality (7.31) and by estimating the difference

$$\left| \int_{\Omega\cap\mathbb{A}_n(x,\epsilon,r)} k(x-y)\,dy - \int_{\Omega\cap\mathbb{A}_n(x^*,\epsilon,r)} k(x^*-y)\,dy \right| \tag{7.37}$$
$$\leq \int_{\Omega\cap\mathbb{A}_n(x,\epsilon,r)\cap\mathbb{A}_n(x^*,\epsilon,r)} |k(x-y)-k(x^*-y)|\,dy$$
$$+\int_{(\Omega\cap\mathbb{A}_n(x,\epsilon,r))\setminus(\Omega\cap\mathbb{A}_n(x^*,\epsilon,r))} |k(x-y)|\,dy$$
$$+\int_{(\Omega\cap\mathbb{A}_n(x^*,\epsilon,r))\setminus(\Omega\cap\mathbb{A}_n(x,\epsilon,r))} |k(x^*-y)|\,dy\,.$$

So, we now to estimate the three summands in the right-hand side of (7.37). Since $\epsilon\geq d(x)$, we have

$$\Omega\cap\mathbb{A}_n(x,\epsilon,r)\cap\mathbb{A}_n(x^*,\epsilon,r)\subseteq \Omega\cap\mathbb{A}_n(x,d(x),r)\cap\mathbb{A}_n(x^*,d(x),r)$$

and we can estimate the first term in the right-hand side of (7.37) as follows

$$\int_{\Omega\cap\mathbb{A}_n(x,\epsilon,r)\cap\mathbb{A}_n(x^*,\epsilon,r)} |k(x-y)-k(x^*-y)|\,dy$$
$$\leq \int_{\Omega\cap\mathbb{A}_n(x,d(x),r)\cap\mathbb{A}_n(x^*,d(x),r)} |k(x-y)-k(x^*-y)|\,dy$$

for all $\epsilon\in[d(x),r[$. By Lemma 4.14, we have

$$\int_{\Omega\cap\mathbb{A}_n(x,d(x),r)\cap\mathbb{A}_n(x^*,d(x),r)} |k(x-y)-k(x^*-y)|\,dy$$
$$\leq (2+2n)\max\left\{\max_{\partial\mathbb{B}_n(0,1)}|k|, |k:\partial\mathbb{B}_n(0,1)|_1\right\}|x-x^*|$$
$$\times\int_{\Omega\cap\mathbb{A}_n(x,d(x),r)\cap\mathbb{A}_n(x^*,d(x),r)} (\min\{|x-y|,|x^*-y|\})^{-n-1}\,dy\,.$$

Next we note that if $y\in\Omega\cap\mathbb{A}_n(x,d(x),r)\cap\mathbb{A}_n(x^*,d(x),r)$, then

$$|x-y| \leq |x^*-y| + |x-x^*| = |x^*-y| + d(x) \leq |x^*-y| + |x^*-y| = 2|x^*-y|$$

and accordingly

$$|x^*-y| \geq \frac{1}{2}|x-y| .$$

Hence,

$$\begin{aligned}
&\int_{\Omega\cap\mathbb{A}_n(x,\epsilon,r)\cap\mathbb{A}_n(x^*,\epsilon,r)} |k(x-y)-k(x^*-y)|\,dy \qquad (7.38)\\
&\quad\leq \int_{\Omega\cap\mathbb{A}_n(x,d(x),r)\cap\mathbb{A}_n(x^*,d(x),r)} |k(x-y)-k(x^*-y)|\,dy\\
&\quad\leq (2+2n)\,\|k\|_{\mathcal{K}'_{0,1}}\,|x-x^*|\,2^{n+1}\int_{\Omega\cap\mathbb{A}_n(x,d(x),r)\cap\mathbb{A}_n(x^*,d(x),r)} \frac{dy}{|x-y|^{n+1}}\\
&\quad\leq 2^{n+2}(n+1)\,\|k\|_{\mathcal{K}'_{0,1}}\,|x-x^*|\int_{|x-x^*|}^{r} \rho^{n-1}s_n\rho^{-n-1}\,d\rho\\
&\quad\leq 2^{n+2}(n+1)s_n\,\|k\|_{\mathcal{K}'_{0,1}} \qquad \forall\epsilon\in[d(x),r[\,,\ k\in\mathcal{K}'_{0,1;e,0}\,.
\end{aligned}$$

Indeed, $r<1$, $|x-x^*|=d(x)<r/2<1/2<1$. In order to estimate the second term in the right-hand side of (7.37), we note that

$$\begin{aligned}
&(\Omega\cap\mathbb{A}_n(x,\epsilon,r))\setminus(\Omega\cap\mathbb{A}_n(x^*,\epsilon,r)) \qquad (7.39)\\
&\quad=\Omega\cap\mathbb{A}_n(x,\epsilon,r)\cap(\mathbb{R}^n\setminus\mathbb{A}_n(x^*,\epsilon,r))\\
&\quad=\{\Omega\cap\mathbb{A}_n(x,\epsilon,r)\cap\mathbb{B}_n(x^*,\epsilon)\}\\
&\qquad\cup\{\Omega\cap[\mathbb{A}_n(x,\epsilon,r)\cap(\mathbb{R}^n\setminus\mathbb{B}_n(x^*,r))]\}\\
&\quad\subseteq[(\mathbb{R}^n\setminus\mathbb{B}_n(x,\epsilon))\cap\mathbb{B}_n(x^*,\epsilon)]\cup[\Omega\cap(\mathbb{R}^n\setminus\mathbb{B}_n(x^*,r))]
\end{aligned}$$

for all $\epsilon\in[d(x),r[$. Since $d(x)<r/2$, we have

$$|y-x| \geq |y-x^*| - |x-x^*| = |y-x^*| - d(x) \geq r-(r/2) = r/2$$

for all $y\in\mathbb{R}^n\setminus\mathbb{B}_n(x^*,r)$ and thus

$$\begin{aligned}
&\int_{(\Omega\cap\mathbb{A}_n(x,\epsilon,r))\setminus(\Omega\cap\mathbb{A}_n(x^*,\epsilon,r))} |k(x-y)|\,dy \qquad (7.40)\\
&\quad\leq\left(\sup_{\partial\mathbb{B}_n(0,1)}|k|\right)\int_{(\mathbb{R}^n\setminus\mathbb{B}_n(x,\epsilon))\cap\mathbb{B}_n(x^*,\epsilon)}\frac{dy}{|x-y|^n}\\
&\qquad+\left(\sup_{\partial\mathbb{B}_n(0,1)}|k|\right)\int_{\Omega\cap(\mathbb{R}^n\setminus\mathbb{B}_n(x^*,r))}\frac{dy}{|x-y|^n}\\
&\quad\leq\left(\sup_{\partial\mathbb{B}_n(0,1)}|k|\right)\int_{\mathbb{B}_n(x^*,\epsilon)}\frac{dy}{\epsilon^n}+\left(\sup_{\partial\mathbb{B}_n(0,1)}|k|\right)\int_{\Omega}\frac{dy}{(r/2)^n}
\end{aligned}$$

$$\leq \left(\sup_{\partial \mathbb{B}_n(0,1)} |k| \right) \left(\omega_n + \frac{m_n(\Omega)}{(r/2)^n} \right) \qquad \forall \epsilon \in [d(x), r[\, .$$

Similarly, in order to estimate the third term in the right-hand side of (7.37), we note that

$$\begin{aligned} &(\Omega \cap \mathbb{A}_n(x^*, \epsilon, r)) \setminus (\Omega \cap \mathbb{A}_n(x, \epsilon, r)) \\ &\quad = \Omega \cap \mathbb{A}_n(x^*, \epsilon, r) \cap (\mathbb{R}^n \setminus \mathbb{A}_n(x, \epsilon, r)) \\ &\quad = \{\Omega \cap \mathbb{A}_n(x^*, \epsilon, r) \cap \mathbb{B}_n(x, \epsilon)\} \\ &\qquad \cup \{\Omega \cap [\mathbb{A}_n(x^*, \epsilon, r) \cap (\mathbb{R}^n \setminus \mathbb{B}_n(x, r))]\} \\ &\quad \subseteq [(\mathbb{R}^n \setminus \mathbb{B}_n(x^*, \epsilon)) \cap \mathbb{B}_n(x, \epsilon)] \cup [\Omega \cap (\mathbb{R}^n \setminus \mathbb{B}_n(x, r))] \, . \end{aligned} \tag{7.41}$$

Since $d(x) < r/2$, we have

$$|y - x^*| \geq |y - x| - |x - x^*| = |y - x| - d(x) \geq r - (r/2) = r/2 \quad \forall y \in \mathbb{R}^n \setminus \mathbb{B}_n(x, r)$$

and thus

$$\begin{aligned} &\int_{(\Omega \cap \mathbb{A}_n(x^*, \epsilon, r)) \setminus (\Omega \cap \mathbb{A}_n(x, \epsilon, r))} |k(x^* - y)| \, dy \\ &\quad \leq \left(\sup_{\partial \mathbb{B}_n(0,1)} |k| \right) \int_{(\mathbb{R}^n \setminus \mathbb{B}_n(x^*, \epsilon)) \cap \mathbb{B}_n(x, \epsilon)} \frac{dy}{|x^* - y|^n} \\ &\qquad + \left(\sup_{\partial \mathbb{B}_n(0,1)} |k| \right) \int_{\Omega \cap (\mathbb{R}^n \setminus \mathbb{B}_n(x, r))} \frac{dy}{|x^* - y|^n} \\ &\quad \leq \left(\sup_{\partial \mathbb{B}_n(0,1)} |k| \right) \int_{\mathbb{B}_n(x, \epsilon)} \frac{dy}{\epsilon^n} + \left(\sup_{\partial \mathbb{B}_n(0,1)} |k| \right) \int_{\Omega} \frac{dy}{(r/2)^n} \\ &\quad \leq \left(\sup_{\partial \mathbb{B}_n(0,1)} |k| \right) \left(\omega_n + \frac{m_n(\Omega)}{(r/2)^n} \right) . \end{aligned} \tag{7.42}$$

By combining inequalities (7.31), (7.37), (7.38), (7.40), (7.42), we conclude that there exists $c^*_{3,\Omega} \in]0, +\infty[$ such that inequality (7.36) holds true.

Finally, combining inequalities (7.31), (7.34), (7.35), (7.36) we conclude that if

$$c^*_{0,\Omega} \geq \max\{c^*_{1,\Omega}, c^*_{2,\Omega}, c^*_{3,\Omega}\} ,$$

then inequality (7.25) holds true, and thus the proof is complete. □

7.2.4 Volume Potentials with Kernels Which are Weakly Singular and Which Have a Strong Singularity in the First Order Partial Derivatives

In this subsection, we consider volume potentials with kernels which belong to the class $\mathcal{K}_{1,1;o}$ of odd positively homogeneous functions of degree $-(n-1)$ and class $C^{1,1}_{\mathrm{loc}}(\mathbb{R}^n \setminus \{0\})$ introduced in (4.19) and kernels that are equal to first order partial derivatives of kernels of class $\mathcal{K}_{1,1;o}$ and that are generally not integrable in a neighborhood of 0, i.e., they may have a strong singularity. The general scheme is that of Miranda [212, pp. 320–322], but we also exploit the classical Lemma 7.11. We first prove the following technical lemma on positively homogeneous functions.

Lemma 7.12. *Let $k \in C^1(\mathbb{R}^n \setminus \{0\})$ be a positively homogeneous function of degree $-(n-1)$. Let $j \in \{1,\ldots,n\}$. Then the partial derivative $\frac{\partial k}{\partial x_j}$ is positively homogeneous of degree $-n$ and $\int_{\partial\mathbb{B}_n(0,1)} \frac{\partial k}{\partial x_j}\,d\sigma = 0$.*

Proof. By Lemma 4.12, $\frac{\partial k}{\partial x_j}$ is positively homogeneous of degree $-n$. Let r belong to $]1,+\infty[$. By applying the Divergence Theorem 4.1 in the open set

$$\mathbb{A}_n(1,r) \equiv \mathbb{B}_n(0,r) \setminus \overline{\mathbb{B}_n(0,1)}\,,$$

we obtain

$$\begin{aligned}
\int_{\mathbb{A}_n(1,r)} \frac{\partial k}{\partial x_j}\,dx &= \int_{\partial\mathbb{A}_n(1,r)} (\nu_{\mathbb{A}_n(1,r)})_j k\,d\sigma \\
&= -\int_{\partial\mathbb{B}_n(0,1)} (\nu_{\mathbb{B}_n(0,1)})_j k\,d\sigma + \int_{\partial\mathbb{B}_n(0,r)} (\nu_{\mathbb{B}_n(0,r)})_j k\,d\sigma \\
&= -\int_{\partial\mathbb{B}_n(0,1)} x_j k(x)\,d\sigma_x + \int_{\partial\mathbb{B}_n(0,r)} \frac{x_j}{r} k(x)\,d\sigma_x \\
&= -\int_{\partial\mathbb{B}_n(0,1)} x_j k(x)\,d\sigma_x + r^{n-1}\int_{\partial\mathbb{B}_n(0,1)} \frac{r y_j}{r} k(ry)\,d\sigma_y \\
&= -\int_{\partial\mathbb{B}_n(0,1)} x_j k(x)\,d\sigma_x + r^{-(n-1)+(n-1)}\int_{\partial\mathbb{B}_n(0,1)} y_j k(y)\,d\sigma_y = 0\,.
\end{aligned}$$

On the other hand, $\frac{\partial k}{\partial x_j}$ is positively homogeneous of degree $-n$ and thus Proposition 7.9 (i) implies that

$$\int_{\mathbb{A}_n(1,r)} \frac{\partial k}{\partial x_j}\,dx = \int_{\partial\mathbb{B}_n(0,1)} \frac{\partial k}{\partial x_j}\,d\sigma \log r\,.$$

Then we must have $\int_{\partial\mathbb{B}_n(0,1)} \frac{\partial k}{\partial x_j}\,d\sigma = 0$. □

Next we prove the following lemma.

Lemma 7.13. *Let $\Omega \subseteq \mathbb{R}^n$ be a bounded open set of class C^1. Let $k \in C^1(\mathbb{R}^n \setminus \{0\})$ be a positively homogeneous function of degree $-(n-1)$. Let $j \in \{1, \dots, n\}$. If $x \in \Omega$, then*

$$\text{p.v.} \int_\Omega \frac{\partial k}{\partial x_j}(x-y)\, dy \\ = -\int_{\partial \mathbb{B}_n(0,1)} k(\xi)\xi_j \, d\sigma_\xi - \int_{\partial\Omega} k(x-y)(\nu_\Omega)_j(y)\, d\sigma_y$$

(cf. Definition 7.8).

Proof. If $\epsilon \in]0, \text{dist}(x, \partial\Omega)[$, then by applying the Divergence Theorem 4.1 in the set $\Omega \setminus \mathbb{B}_n(x, \epsilon)$, we obtain

$$\begin{aligned}
\int_{\Omega\setminus\mathbb{B}_n(x,\epsilon)} \frac{\partial k}{\partial x_j}(x-y)\, dy &= -\int_{\Omega\setminus\mathbb{B}_n(x,\epsilon)} \frac{\partial}{\partial y_j}(k(x-y))\, dy \\
&= -\int_{\partial\Omega} k(x-y)(\nu_\Omega)_j(y)\, d\sigma_y - \int_{\partial\mathbb{B}_n(x,\epsilon)} k(x-y)\frac{x_j - y_j}{|x-y|}\, d\sigma_y \\
&= -\int_{\partial\Omega} k(x-y)(\nu_\Omega)_j(y)\, d\sigma_y - \int_{\partial\mathbb{B}_n(0,1)} k(\epsilon\xi)\frac{\xi_j}{|\xi|}\epsilon^{n-1}\, d\sigma_\xi \\
&= -\int_{\partial\Omega} k(x-y)(\nu_\Omega)_j(y)\, d\sigma_y - \int_{\partial\mathbb{B}_n(0,1)} k(\xi)\xi_j\, d\sigma_\xi \,.
\end{aligned}$$

Indeed, k is positively homogeneous of degree $-(n-1)$. Then by taking the limit as ϵ tends to 0, the limit exists and equals the principal value of the statement. □

Next we prove the following proposition, which provides the first order partial derivatives of a volume potential with a kernel of class $\mathcal{K}_{1,0;o}$ (cf. Definition 4.19).

Proposition 7.14. *Let Ω be a bounded open subset of $\mathbb{R}^n$ of class $C^{1,\alpha}$. Then the following statements hold.*

(i) If $(k, \varphi) \in \mathcal{K}_{0,0} \times C^{0,\alpha}(\overline{\Omega})$, then the map

$$\mathcal{P}[k, \varphi](x) \equiv \int_\Omega k(x-y)\varphi(y)\, dy \qquad \forall x \in \overline{\Omega} \tag{7.43}$$

is continuous. Moreover, the bilinear operator $\mathcal{P}$ from $\mathcal{K}_{0,0} \times C^{0,\alpha}(\overline{\Omega})$ to $C^0(\overline{\Omega})$ that takes (k, φ) to $\mathcal{P}[k, \varphi]$ is continuous.

(ii) If $(k, \varphi) \in \mathcal{K}_{1,0} \times C^{0,\alpha}(\overline{\Omega})$, $j \in \{1, \dots, n\}$, and $x \in \overline{\Omega}$, then the function

$$\frac{\partial k}{\partial x_j}(x-y)(\varphi(y) - \varphi(x)) \qquad \forall y \in \Omega$$

is Lebesgue integrable and the map $G_j[k, \varphi]$ from $\overline{\Omega}$ to $\mathbb{R}$ defined by

$$G_j[k, \varphi](x) \equiv \int_\Omega \frac{\partial k}{\partial x_j}(x-y)(\varphi(y) - \varphi(x))\, dy \qquad \forall x \in \overline{\Omega} \tag{7.44}$$

is continuous. Moreover, the bilinear operator G_j from $\mathcal{K}_{1,0} \times C^{0,\alpha}(\overline{\Omega})$ to $C^0(\overline{\Omega})$ that takes (k, φ) to $G_j[k, \varphi]$ is continuous.

(iii) If $(k, \varphi) \in \mathcal{K}_{1,0} \times C^{0,\alpha}(\overline{\Omega})$ and $j \in \{1, \ldots, n\}$, then

$$\begin{aligned}
&\text{p.v.} \int_\Omega \frac{\partial k}{\partial x_j}(x-y)\varphi(y)\,dy \\
&\quad = \int_\Omega \frac{\partial k}{\partial x_j}(x-y)(\varphi(y)-\varphi(x))\,dy + \text{p.v.} \int_\Omega \frac{\partial k}{\partial x_j}(x-y)\,dy\,\varphi(x)
\end{aligned}$$

for all $x \in \Omega$.

(iv) If $(k, \varphi) \in \mathcal{K}_{1,0} \times C^{0,\alpha}(\overline{\Omega})$ and $j \in \{1, \ldots, n\}$, then

$$\begin{aligned}
&\frac{\partial}{\partial x_j}\mathcal{P}[k, \varphi](x) \qquad (7.45)\\
&\quad = \int_{\partial\mathbb{B}_n(0,1)} k(\xi)\xi_j\,d\sigma_\xi \varphi(x) + \text{p.v.} \int_\Omega \frac{\partial k}{\partial x_j}(x-y)\varphi(y)\,dy \\
&\quad = \int_\Omega \frac{\partial k}{\partial x_j}(x-y)(\varphi(y)-\varphi(x))\,dy - \int_{\partial\Omega} k(x-y)(\nu_\Omega)_j(y)\,d\sigma_y \varphi(x)
\end{aligned}$$

for all $x \in \Omega$.

Proof. (i) Since φ is bounded and k is continuous and positively homogeneous of degree $-(n-1)$ and $(n-1) < n$, Proposition 7.2 on volume potentials with a weakly singular kernel implies that $\mathcal{P}[k, \varphi]$ is continuous in $\overline{\Omega}$ and that

$$\begin{aligned}
\sup_{\overline{\Omega}} |\mathcal{P}[k, \varphi]| &\leq s_n \text{diam}(\Omega) \|k\|_{A^0_{n-1}(\text{diam}\,(\Omega))} \|\varphi\|_{L^\infty(\Omega)} \\
&\leq s_n \text{diam}(\Omega) \sup_{x \in (\overline{\mathbb{B}_n(0,\text{diam}\,(\Omega))}) \setminus \{0\}} |k(x)|\,|x|^{n-1} \|\varphi\|_{L^\infty(\Omega)} \\
&= s_n \text{diam}(\Omega) \sup_{\partial\mathbb{B}_n(0,1)} |k| \|\varphi\|_{L^\infty(\Omega)} \leq s_n \text{diam}(\Omega) \|k\|_{\mathcal{K}_{0,0}} \|\varphi\|_{L^\infty(\Omega)}
\end{aligned}$$

(see (4.18) for the definition of norm in $\mathcal{K}_{0,0}$). Thus the proof of statement (i) is complete.

(ii) Since $\frac{\partial k}{\partial x_j}$ is positively homogeneous of degree $-n$ and φ is α-Hölder continuous, the following inequality holds,

$$\begin{aligned}
&\left| \frac{\partial k}{\partial x_j}(x-y)(\varphi(y)-\varphi(x)) \right| \qquad (7.46)\\
&\quad \leq \left\| \frac{\partial k}{\partial x_j} \right\|_{C^0(\partial\mathbb{B}_n(0,1))} \|\varphi\|_{C^{0,\alpha}(\overline{\Omega})} |x-y|^{-n+\alpha} \qquad \forall x, y \in \overline{\Omega}, \ x \neq y\,.
\end{aligned}$$

By inequality (7.46), the function $\frac{\partial k}{\partial x_j}(x-y)(\varphi(y)-\varphi(x))$ is integrable in $y \in \Omega$ for all $x \in \overline{\Omega}$ and we have

$$\left|\int_\Omega \frac{\partial k}{\partial x_j}(x-y)(\varphi(y)-\varphi(x))\,dy\right|$$
$$\leq \left\|\frac{\partial k}{\partial x_j}\right\|_{C^0(\partial\mathbb{B}_n(0,1))} \|\varphi\|_{C^{0,\alpha}(\overline{\Omega})} \sup_{x\in\overline{\Omega}}\int_\Omega \frac{dy}{|x-y|^{n-\alpha}} \qquad \forall x\in\overline{\Omega}\,.$$

Moreover, if $r \in]0,+\infty[$ and $\overline{\Omega}-\overline{\Omega}\subseteq \mathbb{B}_n(0,r)$, then

$$\int_\Omega \frac{dy}{|x-y|^{n-\alpha}} = \int_{x-\Omega}\frac{d\xi}{|\xi|^{n-\alpha}} \leq \int_{\mathbb{B}_n(0,r)}\frac{d\xi}{|\xi|^{n-\alpha}} = s_n\frac{r^\alpha}{\alpha} \qquad \forall x\in\overline{\Omega}\,.$$

Hence, the function $G_j[k,\varphi]$ is bounded and

$$\sup_{\overline{\Omega}}|G_j[k,\varphi]| \leq s_n\frac{r^\alpha}{\alpha}\left\|\frac{\partial k}{\partial x_j}\right\|_{C^0(\partial\mathbb{B}_n(0,1))}\|\varphi\|_{C^{0,\alpha}(\overline{\Omega})} \tag{7.47}$$

for all $(k,\varphi)\in\mathcal{K}_{1,0}\times C^{0,\alpha}(\overline{\Omega})$. We now turn to show that $G_j[k,\varphi]$ is continuous at all points of $\overline{\Omega}$. Let $\tilde{x}\in\overline{\Omega}$. We want to prove that

$$\lim_{x\to\tilde{x}}\int_\Omega \frac{\partial k}{\partial x_j}(x-y)(\varphi(y)-\varphi(x))\,dy = \int_\Omega \frac{\partial k}{\partial x_j}(\tilde{x}-y)(\varphi(y)-\varphi(\tilde{x}))\,dy\,. \tag{7.48}$$

By the continuity of $\frac{\partial k}{\partial x_j}$ in $\mathbb{R}^n\setminus\{0\}$ and of φ at $\tilde{x}$, we have

$$\lim_{x\to\tilde{x}}\frac{\partial k}{\partial x_j}(x-y)(\varphi(y)-\varphi(x)) = \frac{\partial k}{\partial x_j}(\tilde{x}-y)(\varphi(y)-\varphi(\tilde{x})) \qquad \forall y\in\overline{\Omega}\setminus\{\tilde{x}\}\,.$$

In order to invoke the Vitali Convergence Theorem A.11, we now prove that for each $\epsilon>0$, there exists $\delta>0$ such that

$$\int_E\left|\frac{\partial k}{\partial x_j}(x-y)(\varphi(y)-\varphi(x))\right|\,dy \leq \epsilon \text{ if } E\subseteq\Omega,\ E\in\mathcal{L}_n,\ m_n(E)\leq\delta,\ x\in\overline{\Omega}\,.$$

By inequality (7.46), we have

$$\int_E\left|\frac{\partial k}{\partial x_j}(x-y)(\varphi(y)-\varphi(x))\right|\,dy$$
$$\leq \left\|\frac{\partial k}{\partial x_j}\right\|_{C^0(\partial\mathbb{B}_n(0,1))}\|\varphi\|_{C^{0,\alpha}(\overline{\Omega})}\int_E\frac{dy}{|x-y|^{n-\alpha}} \qquad \forall x\in\overline{\Omega}\,.$$

Then Corollary 4.20 implies that there exists $\delta>0$ such that

$$\int_E \frac{1}{|x-y|^{n-\alpha}}\,d\sigma_y \leq \frac{\epsilon}{1+\left\|\frac{\partial k}{\partial x_j}\right\|_{C^0(\partial\mathbb{B}_n(0,1))}\|\varphi\|_{C^{0,\alpha}(\overline{\Omega})}}$$

for all $E \in \mathcal{L}_n$, $m_n(E) \leq \delta$, $x \in \overline{\Omega}$. Then the Vitali Convergence Theorem A.11 implies that for each sequence $\{x_l\}_{l\in\mathbb{N}}$ in $\overline{\Omega} \setminus \{\tilde{x}\}$ that converges to $\tilde{x}$, we have

$$\lim_{l\to\infty} \int_\Omega \frac{\partial k}{\partial x_j}(x_l - y)(\varphi(y) - \varphi(x_l))\, dy = \int_\Omega \frac{\partial k}{\partial x_j}(\tilde{x} - y)(\varphi(y) - \varphi(\tilde{x}))\, dy\,,$$

and thus the limiting relation in (7.48) holds true. By Lemma 4.13 and by inequality (7.47), the bilinear operator G_j from $\mathcal{K}_{1,0} \times C^{0,\alpha}(\overline{\Omega})$ to $C^0(\overline{\Omega})$ is continuous (cf. Proposition 2.9).

(iii) We note that

$$\begin{aligned}
\text{p.v.} \int_\Omega \frac{\partial k}{\partial x_j}(x - y)\varphi(y)\, dy &= \lim_{\epsilon\to 0} \int_{\Omega\setminus\mathbb{B}_n(x,\epsilon)} \frac{\partial k}{\partial x_j}(x - y)\varphi(y)\, dy \\
&= \lim_{\epsilon\to 0} \int_{\Omega\setminus\mathbb{B}_n(x,\epsilon)} \frac{\partial k}{\partial x_j}(x - y)(\varphi(y) - \varphi(x))\, dy \\
&\quad +\varphi(x) \lim_{\epsilon\to 0} \int_{\Omega\setminus\mathbb{B}_n(x,\epsilon)} \frac{\partial k}{\partial x_j}(x - y)\, dy\,.
\end{aligned}$$

Then inequality (7.46) and the Lebesgue Dominated Convergence Theorem imply the validity of statement (iii) (see also Lemmas 7.10, 7.12).

(iv) By Lemma 7.13 and by statement (iii), we have

$$\begin{aligned}
&\int_{\partial\mathbb{B}_n(0,1)} k(\xi)\xi_j\, d\sigma_\xi \varphi(x) + \text{p.v.} \int_\Omega \frac{\partial k}{\partial x_j}(x - y)\varphi(y)\, dy \qquad (7.49) \\
&= \int_{\partial\mathbb{B}_n(0,1)} k(\xi)\xi_j\, d\sigma_\xi \varphi(x) + G_j[k, \varphi](x) + \varphi(x)\, \text{p.v.} \int_\Omega \frac{\partial k}{\partial x_j}(x - y)\, dy \\
&= G_j[k, \varphi](x) - \int_{\partial\Omega} k(x - y)(\nu_\Omega)_j(y)\, d\sigma_y \varphi(x) \qquad \forall x \in \Omega\,,
\end{aligned}$$

and thus the second equality in the right-hand side of (7.45) holds true. By a result of differentiation of singular integrals, the distributional first order partial derivatives of $\mathcal{P}[k, \varphi]$ are delivered by the first equality of (7.45). For the result on the distributional partial derivatives of singular integrals, we refer to Mitrea [214, Theorem. 4.100, p. 178]. Since both $\mathcal{P}[k, \varphi]$ and its first order distributional partial derivatives are continuous, the first equality of formula (7.45) holds true for the classical partial derivatives as well (cf., e.g., Mitrea [214, Ex. 2.46, p. 36]).

Alternatively, one could have invoked directly the classical result of differentiation of singular integrals for the classical partial derivatives (see Mikhlin [210, §8], Kupradze, Gegelia, Basheleishvili, and Burchuladze [151, Theorem 2.9, p. 185], Cialdea [53, Theorem 26]). □

In order to understand the mapping properties of $\frac{\partial}{\partial x_j}\mathcal{P}[\cdot, \cdot]$ in Hölder spaces, we now turn to analyze the mapping properties of the operator $G_j[\cdot, \cdot]$ of the previous statement.

Proposition 7.15. *Let Ω be a bounded open subset of $\mathbb{R}^n$ of class $C^{1,\alpha}$. Let $j \in \{1,\ldots,n\}$. Then the map G_j defined in (7.44) is bilinear and continuous from $\mathcal{K}_{1,1;o} \times C^{0,\alpha}(\overline{\Omega})$ to $C^{0,\alpha}(\overline{\Omega})$,*

Proof. By Proposition 7.14 (ii) and by the continuity of the embedding of $\mathcal{K}_{1,1;o}$ into $\mathcal{K}_{1,0}$, the map G_j is bilinear and continuous from $\mathcal{K}_{1,1;o} \times C^{0,\alpha}(\overline{\Omega})$ to $C^0(\overline{\Omega})$. Thus it suffices to estimate the Hölder constant of $G_j[k,\varphi]$ for a given (k,φ) in $\mathcal{K}_{1,1;o} \times C^{0,\alpha}(\overline{\Omega})$. We first note that if $k \in \mathcal{K}_{1,1;o}$, then Lemma 7.12 implies that the function $\frac{\partial k}{\partial x_j}$ is positively homogeneous of degree $-n$ and satisfies condition $\int_{\partial\mathbb{B}_n(0,1)} \frac{\partial k}{\partial x_j}\, d\sigma = 0$. Moreover, k is odd and thus $\frac{\partial k}{\partial x_j}$ is even. Then $\frac{\partial k}{\partial x_j}$ belongs to the space $\mathcal{K}'_{0,1;e,0}$ that was introduced in (7.22) and we can invoke the classical Lemma 7.11 for the kernel $\frac{\partial k}{\partial x_j}$ and conclude that there exists $c^*_\Omega \in]0,+\infty[$ such that

$$\left| \int_{\Omega\setminus\mathbb{B}_n(x,\epsilon)} \frac{\partial k}{\partial x_j}(x-y)\, dy \right| \leq c^*_\Omega \left\| \frac{\partial k}{\partial x_j} \right\|_{\mathcal{K}'_{0,1}}$$

for all $k \in \mathcal{K}_{1,1;o}$, $x \in \mathbb{R}^n$, $\epsilon \in]0,+\infty[$ $j \in \{1,\ldots,n\}$. We are now ready to estimate the Hölder constant of $G_j[k,\varphi]$. To do so, we take two arbitrary points x', $x'' \in \Omega$ and we note that Lemmas 4.14 and 7.11 and the inclusion

$$\mathbb{B}_n(x', 2|x'-x''|) \subseteq \mathbb{B}_n(x'', 3|x'-x''|)$$

imply that

$$\begin{aligned}
&|G_j[k,\varphi](x') - G_j[k,\varphi](x'')| \qquad (7.50)\\
&= \left| \int_\Omega \frac{\partial k}{\partial x_j}(x'-y)(\varphi(y)-\varphi(x'))\, dy - \int_\Omega \frac{\partial k}{\partial x_j}(x''-y)(\varphi(y)-\varphi(x''))\, dy \right|\\
&\leq \int_{\Omega\cap\mathbb{B}_n(x',2|x'-x''|)} \left| \frac{\partial k}{\partial x_j}(x'-y) \right| |\varphi(y)-\varphi(x')|\, dy\\
&\quad + \int_{\Omega\cap\mathbb{B}_n(x',2|x'-x''|)} \left| \frac{\partial k}{\partial x_j}(x''-y) \right| |\varphi(y)-\varphi(x'')|\, dy\\
&\quad + \left| \int_{\Omega\setminus\mathbb{B}_n(x',2|x'-x''|)} \frac{\partial k}{\partial x_j}(x'-y)(\varphi(y)-\varphi(x')) \right.\\
&\qquad \left. - \frac{\partial k}{\partial x_j}(x''-y)(\varphi(y)-\varphi(x''))\, dy \right|\\
&\leq \left\| \frac{\partial k}{\partial x_j} \right\|_{C^0(\partial\mathbb{B}_n(0,1))} \|\varphi\|_{C^{0,\alpha}(\overline{\Omega})} \int_{\mathbb{B}_n(x',2|x'-x''|)} \frac{dy}{|x'-y|^{n-\alpha}}\\
&\quad + \left\| \frac{\partial k}{\partial x_j} \right\|_{C^0(\partial\mathbb{B}_n(0,1))} \|\varphi\|_{C^{0,\alpha}(\overline{\Omega})} \int_{\mathbb{B}_n(x'',3|x'-x''|)} \frac{dy}{|x''-y|^{n-\alpha}}\\
&\quad + \left| \int_{\Omega\setminus\mathbb{B}_n(x',2|x'-x''|)} \frac{\partial k}{\partial x_j}(x'-y)[(\varphi(y)-\varphi(x')) - (\varphi(y)-\varphi(x''))]\, dy \right|
\end{aligned}$$

$$
\begin{aligned}
&+\left|\int_{\Omega\setminus\mathbb{B}_n(x',2|x'-x''|)}\left[\frac{\partial k}{\partial x_j}(x'-y)-\frac{\partial k}{\partial x_j}(x''-y)\right](\varphi(y)-\varphi(x''))\,dy\right| \\
&\leq \left\|\frac{\partial k}{\partial x_j}\right\|_{C^0(\partial\mathbb{B}_n(0,1))}\|\varphi\|_{C^{0,\alpha}(\overline{\Omega})}\left\{(2|x'-x''|)^\alpha\frac{s_n}{\alpha}+(3|x'-x''|)^\alpha\frac{s_n}{\alpha}\right\} \\
&+c^*_\Omega\left\|\frac{\partial k}{\partial x_j}\right\|_{\mathcal{K}'_{0,1}}\|\varphi\|_{C^{0,\alpha}(\overline{\Omega})}|x'-x''|^\alpha \\
&+(2+2n)\max\left\{\sup_{\partial\mathbb{B}_n(0,1)}\left|\frac{\partial k}{\partial x_j}\right|,\left|\frac{\partial k}{\partial x_j}:\partial\mathbb{B}_n(0,1)\right|_1\right\}\|\varphi\|_{C^{0,\alpha}(\overline{\Omega})} \\
&\times\int_{\Omega\setminus\mathbb{B}_n(x',2|x'-x''|)}(\min\{|x'-y|,|x''-y|\})^{-n-1}|x'-x''||y-x''|^\alpha\,dy\,.
\end{aligned}
$$

By Lemma 2.52 (i), we have

$$
|x''-y|\geq\frac{1}{2}|x'-y|\,,\quad |y-x''|\leq 2|x'-y|\quad\forall y\in\Omega\setminus\mathbb{B}_n(x',2|x'-x''|)\,.
$$

Then we conclude that

$$
\begin{aligned}
&\int_{\Omega\setminus\mathbb{B}_n(x',2|x'-x''|)}(\min\{|x'-y|,|x''-y|\})^{-n-1}|x'-x''||y-x''|^\alpha\,dy \qquad (7.51)\\
&\qquad\leq 2^{n+1+\alpha}\int_{\Omega\setminus\mathbb{B}_n(x',2|x'-x''|)}|x'-y|^{-n-1}|x'-x''||y-x'|^\alpha\,dy
\end{aligned}
$$

and that

$$
\begin{aligned}
&\int_{\Omega\setminus\mathbb{B}_n(x',2|x'-x''|)}|x'-y|^{-n-1+\alpha}\,dy|x'-x''| \qquad (7.52)\\
&\qquad\leq\int_{2|x'-x''|}^{\mathrm{diam}(\Omega)}s_n\rho^{n-1}\rho^{-n-1+\alpha}\,d\rho|x'-x''| \\
&\qquad=\frac{s_n}{1-\alpha}((2|x'-x''|)^{-1+\alpha}-(\mathrm{diam}(\Omega))^{-1+\alpha})|x'-x''| \\
&\qquad\leq\frac{s_n}{1-\alpha}2^{-1+\alpha}|x'-x''|^\alpha\,.
\end{aligned}
$$

By Lemma 4.13, the map from $\mathcal{K}_{1,1;o}$ to $\mathcal{K}'_{1,0;e,0}$ that takes k to $\frac{\partial k}{\partial x_j}$ is linear and continuous. Then inequalities (7.50)–(7.52) and Proposition 7.14 (ii) imply that the bilinear map G_j is continuous from $\mathcal{K}_{1,1;o}\times C^{0,\alpha}(\overline{\Omega})$ to $C^{0,\alpha}(\overline{\Omega})$ (cf. Proposition 2.9). □

We are now ready to prove the following continuity result for the volume potential $\mathcal{P}$ in Schauder spaces, which is a refinement of Miranda [212, 3.II, p. 320, case $m=1$] who has considered kernels k of class C^2.

Theorem 7.16. *Let Ω be a bounded open subset of $\mathbb{R}^n$ of class $C^{1,\alpha}$.*

If $(k,\varphi) \in \mathcal{K}_{1,1;o} \times C^{0,\alpha}(\overline{\Omega})$, *then the map* $\mathcal{P}[k,\varphi]$ *which has been defined in (7.43) belongs to* $C^{1,\alpha}(\overline{\Omega})$.

Moreover, the bilinear operator $\mathcal{P}$ *from* $\mathcal{K}_{1,1;o} \times C^{0,\alpha}(\overline{\Omega})$ *to* $C^{1,\alpha}(\overline{\Omega})$ *that takes* (k,φ) *to* $\mathcal{P}[k,\varphi]$ *is continuous.*

Proof. By Proposition 7.14 (i), (iv), it suffices to show that the right-hand side of (7.45) defines a bilinear and continuous map from $\mathcal{K}_{1,1;o} \times C^{0,\alpha}(\overline{\Omega})$ to $C^{0,\alpha}(\overline{\Omega})$. We now rewrite the right-hand side of (7.45) in the form

$$G_j[k,\varphi] - K[k,(\nu_\Omega)_j]\varphi \tag{7.53}$$

for all $(k,\varphi) \in \mathcal{K}_{1,1;o} \times C^{0,\alpha}(\overline{\Omega})$ where

$$K[k,(\nu_\Omega)_j](x) = \int_{\partial\Omega} k(x-y)(\nu_\Omega)_j(y)\,d\sigma_y \qquad \forall x \in \Omega\,.$$

Here we note that Theorem 4.17 (i) ensures that $K[k,(\nu_\Omega)_j]$ defines a continuous function in Ω that can be extended to a α-Hölder continuous function on $\overline{\Omega}$. By Proposition 7.15 the bilinear operator G_j is continuous from $\mathcal{K}_{1,1;o} \times C^{0,\alpha}(\overline{\Omega})$ to $C^{0,\alpha}(\overline{\Omega})$. By Theorem 4.17 (i), $K[\cdot,(\nu_\Omega)_j]$ is continuous from $\mathcal{K}_{1,1;o}$ to $C^{0,\alpha}(\overline{\Omega})$. Then the continuity of the pointwise product in $C^{0,\alpha}(\overline{\Omega})$ ensures the continuity of the second summand in (7.53) from $\mathcal{K}_{1,1;o} \times C^{0,\alpha}(\overline{\Omega})$ to $C^{0,\alpha}(\overline{\Omega})$. Thus the proof is complete. □

7.2.5 The Newtonian Potential in Schauder Spaces

By applying Theorem 7.16 on volume potentials in Schauder spaces to the specific case where the kernel is a first order partial derivative of the fundamental solution S_n of the Laplace operator, we can readily deduce the following classical result.

Theorem 7.17. *Let* Ω *be a bounded open subset of* $\mathbb{R}^n$ *of class* $C^{1,\alpha}$. *Let* $l \in \{1,\ldots,n\}$. *Then the volume potential operator* $\mathcal{P}[\frac{\partial S_n}{\partial x_l},\cdot]$ *is linear and continuous from* $C^{0,\alpha}(\overline{\Omega})$ *to* $C^{1,\alpha}(\overline{\Omega})$. *Moreover, if* $j \in \{1,\ldots,n\}$, *we have*

$$\frac{\partial}{\partial x_j}\mathcal{P}[\frac{\partial S_n}{\partial x_l},\varphi](x) = \frac{\delta_{jl}}{n}\varphi(x) + \mathrm{p.v.}\int_\Omega \frac{\partial^2 S_n}{\partial x_j \partial x_l}(x-y)\varphi(y)\,dy \qquad \forall x \in \Omega\,, \tag{7.54}$$

where

$$\mathrm{p.v.}\int_\Omega \frac{\partial^2 S_n}{\partial x_j \partial x_l}(x-y)\varphi(y)\,dy = \lim_{\epsilon\to 0}\int_{\Omega\setminus\mathbb{B}_n(x,\epsilon)} \frac{\partial^2 S_n}{\partial x_j \partial x_l}(x-y)\varphi(y)\,dy\,.$$

Proof. The kernel $k(x) \equiv \frac{\partial S_n}{\partial x_l}(x) = \frac{x_l}{s_n|x|^n}$ is odd, positively homogeneous of degree $-(n-1)$ and of class $C^\infty(\mathbb{R}^n \setminus \{0\})$. Accordingly $\frac{\partial S_n}{\partial x_l}$ belongs to $\mathcal{K}_{1,1;o}$.

Then Theorem 7.16 implies that $\mathcal{P}[\frac{\partial S_n}{\partial x_l},\cdot]$ is linear and continuous from $C^{0,\alpha}(\overline{\Omega})$ to $C^{1,\alpha}(\overline{\Omega})$. By Proposition 7.14 (iv), the principal value p.v. $\int_\Omega \frac{\partial^2 S_n}{\partial x_j \partial x_l}(x-y)\varphi(y)\,dy$ exists finite for all $x \in \Omega$ and

$$\frac{\partial}{\partial x_j}\mathcal{P}[\frac{\partial S_n}{\partial x_l},\varphi](x)$$
$$= \int_{\partial \mathbb{B}_n(0,1)} \frac{\partial S_n}{\partial x_l}(\xi)\xi_j\,d\sigma_\xi \varphi(x) + \text{p.v.}\int_\Omega \frac{\partial^2 S_n}{\partial x_j \partial x_l}(x-y)\varphi(y)\,dy\,.$$

By the Divergence Theorem 4.1, we have

$$\int_{\partial \mathbb{B}_n(0,1)} \frac{\partial S_n}{\partial x_l}(\xi)\xi_j\,d\sigma_\xi = \int_{\partial \mathbb{B}_n(0,1)} \frac{\xi_l \xi_j}{s_n|\xi|^n}\,d\sigma_\xi$$
$$= \int_{\partial \mathbb{B}_n(0,1)} \frac{\xi_l \xi_j}{s_n}\,d\sigma_\xi = \int_{\mathbb{B}_n(0,1)} \frac{1}{s_n}\frac{\partial}{\partial \xi_l}\xi_j\,d\xi = \frac{\omega_n \delta_{jl}}{s_n} = \frac{\delta_{jl}}{n}\,,$$

and thus the proof is complete. □

Then we prove the following classical result of Kellog, which shows that $\mathcal{P}[S_n,\varphi]$ is a solution of the Poisson equation $\Delta u = \varphi$ for all $\varphi \in C^{0,\alpha}(\overline{\Omega})$.

Theorem 7.18. *Let Ω be a bounded open subset of $\mathbb{R}^n$ of class $C^{1,\alpha}$. Then the Newtonian potential $\mathcal{P}[S_n,\cdot]$ ($= P_n[\Omega,\cdot]$) is linear and continuous from $C^{0,\alpha}(\overline{\Omega})$ to $C^{2,\alpha}(\overline{\Omega})$. Moreover,*

$$\Delta\left(\mathcal{P}[S_n,\varphi]\right) = \varphi \qquad \forall \varphi \in C^{0,\alpha}(\overline{\Omega})\,. \tag{7.55}$$

Proof. Since $C^{0,\alpha}(\overline{\Omega})$ is continuously embedded into $L^\infty(\Omega)$, the continuity Proposition 7.3 implies that $\mathcal{P}[S_n,\cdot]$ is continuous from $C^{0,\alpha}(\overline{\Omega})$ to $C^0(\overline{\Omega})$. By Proposition 7.6, we have

$$\frac{\partial}{\partial x_l}\mathcal{P}[S_n,\varphi] = \mathcal{P}[\frac{\partial S_n}{\partial x_l},\varphi] \qquad \text{in } \overline{\Omega}\,, \tag{7.56}$$

for all $\varphi \in C^{0,\alpha}(\overline{\Omega})$ and for all $l \in \{1,\ldots,n\}$. Then Theorem 7.17 implies that $\frac{\partial}{\partial x_l}\mathcal{P}[S_n,\cdot]$ is linear and continuous from $C^{0,\alpha}(\overline{\Omega})$ to $C^{1,\alpha}(\overline{\Omega})$ for all $l \in \{1,\ldots,n\}$. By the definition of norm in $C^{2,\alpha}(\overline{\Omega})$, we conclude that $\mathcal{P}[S_n,\cdot]$ is linear and continuous from $C^{0,\alpha}(\overline{\Omega})$ to $C^{2,\alpha}(\overline{\Omega})$. Equality (7.55) follows by equalities (7.54) and (7.56). □

Next we turn to consider the case in which the domain of the volume potential operator is $C^{-1,\alpha}(\overline{\Omega})$ (cf. Section 2.22). Since $C^{-1,\alpha}(\overline{\Omega})$ is not a subspace of $L^\infty(\Omega)$, and we have defined the Newtonian potential only for functions of $L^\infty(\Omega)$, we must define $\mathcal{P}[S_n,\varphi]$ for $\varphi \in C^{-1,\alpha}(\overline{\Omega})$. We consider only the case where Ω is of class $C^{1,\alpha}$ for some $\alpha \in]0,1[$. We first observe that in the specific case in which

$$\varphi = \varphi_0 + \sum_{j=1}^{n} \frac{\partial}{\partial x_j}\varphi_j$$

with $\varphi_0 \in C^{0,\alpha}(\overline{\Omega})$ and $\varphi_j \in C^{1,\alpha}(\overline{\Omega})$ for all $j \in \{1,\ldots,n\}$, Lemma 7.7 implies that

$$\mathcal{P}[S_n,\varphi] = \mathcal{P}[S_n,\varphi_0] + \sum_{j=1}^{n} \mathcal{P}[S_n, \frac{\partial}{\partial x_j}\varphi_j] \tag{7.57}$$
$$= \mathcal{P}[S_n,\varphi_0] + \sum_{j=1}^{n} \frac{\partial}{\partial x_j}\mathcal{P}[S_n,\varphi_j] + \sum_{j=1}^{n} v_\Omega^+[(\nu_\Omega)_j\varphi_j] .$$

Now, contrary to the left-hand side, which so far has not been defined for an arbitrary element $\varphi \in C^{-1,\alpha}(\overline{\Omega})$, the right-hand side makes perfectly sense, not only in the specific case above with $\varphi_0 \in C^{0,\alpha}(\overline{\Omega})$ and $\varphi_j \in C^{1,\alpha}(\overline{\Omega})$ for all $j \in \{1,\ldots,n\}$, but also in the general case in which $\varphi_j \in C^{0,\alpha}(\overline{\Omega})$ for all $j \in \{0,\ldots,n\}$. Hence, it makes sense to define $\mathcal{P}[S_n,\varphi]$ with $\varphi = \varphi_0 + \sum_{j=1}^{n} \frac{\partial}{\partial x_j}\varphi_j \in C^{-1,\alpha}(\overline{\Omega})$ as the right-hand side of equality (7.57). To do so however, we must verify that the right-hand side of equality (7.57) does not depend on the specific choice of $(\varphi_0,\ldots,\varphi_n)$ such that $\varphi = \varphi_0 + \sum_{j=1}^{n} \frac{\partial}{\partial x_j}\varphi_j$. We do so in the following.

Theorem 7.19. *Let Ω be a bounded open subset of $\mathbb{R}^n$ of class $C^{1,\alpha}$. Then the following statements hold.*

(i) If $(\varphi_0,\ldots,\varphi_n) \in C^{0,\alpha}(\overline{\Omega})^{n+1}$ and $\varphi_0 + \sum_{j=1}^{n} \frac{\partial}{\partial x_j}\varphi_j = 0$ in the sense of distributions in Ω, then

$$\mathcal{P}[S_n,\varphi_0] + \sum_{j=1}^{n} \frac{\partial}{\partial x_j}\mathcal{P}[S_n,\varphi_j] + \sum_{j=1}^{n} v_\Omega^+[\varphi_j(\nu_\Omega)_j] = 0 \qquad \text{in } \overline{\Omega} .$$

(ii) There exists one and only one linear and continuous operator $\mathcal{P}[S_n,\cdot]$ from $C^{-1,\alpha}(\overline{\Omega})$ to $C^{1,\alpha}(\overline{\Omega})$ that satisfies the equality

$$\mathcal{P}[S_n,\varphi] = \mathcal{P}[S_n,\varphi_0] + \sum_{j=1}^{n} \frac{\partial}{\partial x_j}\mathcal{P}[S_n,\varphi_j] + \sum_{j=1}^{n} v_\Omega^+[\varphi_j(\nu_\Omega)_j] \qquad \text{in } \overline{\Omega} , \tag{7.58}$$

for all $\varphi = \varphi_0 + \sum_{j=1}^{n} \frac{\partial}{\partial x_j}\varphi_j \in C^{-1,\alpha}(\overline{\Omega})$.

(iii) If $\phi \in C^{-1,\alpha}(\overline{\Omega})$, then $\Delta(\mathcal{P}[S_n,\varphi]) = \varphi$ in the sense of distributions in Ω.

Proof. The Schauder regularity Theorem 4.25 (i) for the single layer potential, the membership of ν_Ω in $C^{0,\alpha}(\partial\Omega,\mathbb{R}^n)$, the continuity of the pointwise product in $C^{0,\alpha}(\partial\Omega)$, and the continuity Theorem 7.18 for the Newtonian potential imply that the operator B that takes $(\varphi_0,\ldots,\varphi_n)$ to

$$B[\varphi_0,\ldots,\varphi_n] \equiv \mathcal{P}[S_n,\varphi_0] + \sum_{j=1}^{n} \frac{\partial}{\partial x_j}\mathcal{P}[S_n,\varphi_j] + \sum_{j=1}^{n} v_\Omega^+[\varphi_j(\nu_\Omega)_j] ,$$

is linear and continuous from $(C^{0,\alpha}(\overline{\Omega}))^{n+1}$ to $C^{1,\alpha}(\overline{\Omega})$. By definition, the space $C^{-1,\alpha}(\overline{\Omega})$ is the image of the linear and continuous map Ξ from $(C^{0,\alpha}(\overline{\Omega}))^{n+1}$

to $\mathcal{D}'(\Omega)$ that takes an $(n+1)$-tuple $(\varphi_0,\dots,\varphi_n)$ to $\varphi_0+\sum_{j=1}^n \frac{\partial}{\partial x_j}\varphi_j$ (cf. Section 2.22). We denote by π the canonical projection from $(C^{0,\alpha}(\overline{\Omega}))^{n+1}$ onto the quotient space $(C^{0,\alpha}(\overline{\Omega}))^{n+1}/\mathrm{Ker}\,\Xi$ and we denote by $\tilde{\Xi}$ the unique linear map from $(C^{0,\alpha}(\overline{\Omega}))^{n+1}/\mathrm{Ker}\,\Xi$ onto the image $C^{-1,\alpha}(\overline{\Omega})$ of Ξ such that $\Xi=\tilde{\Xi}\circ\pi$.

We now turn to prove statement (i), i.e., we assume that $(\varphi_0,\dots,\varphi_n)\in\mathrm{Ker}\,\Xi$, and we prove that $(\varphi_0,\dots,\varphi_n)\in\mathrm{Ker}\,B$. Since $B[\varphi_0,\dots,\varphi_n]$ is continuous on $\overline{\Omega}$, it suffices to show that $B[\varphi_0,\dots,\varphi_n]=0$ in Ω. Let $x\in\Omega$. The function $S_n(x-y)(\varphi_j(y))_{j=1,\dots,n}$ is continuous in $y\in\overline{\Omega}\setminus\{x\}$ and has a divergence in the sense of distributions in $\Omega\setminus\{x\}$ equal to

$$-\nabla S_n(x-y)\cdot(\varphi_1(y),\dots,\varphi_n(y))^t+S_n(x-y)\sum_{j=1}^n\frac{\partial}{\partial x_j}\varphi_j$$
$$=-\nabla S_n(x-y)\cdot(\varphi_1(y),\dots,\varphi_n(y))^t-S_n(x-y)\varphi_0(y)\,.$$

Then the variant of the Divergence Theorem of Proposition 2.89 (i) implies that

$$\begin{aligned}
-\int_{\Omega\setminus\overline{\mathbb{B}_n(x,\epsilon)}}&\nabla S_n(x-y)\cdot(\varphi_1(y),\dots,\varphi_n(y))^t\,dy \\
&-\int_{\Omega\setminus\overline{\mathbb{B}_n(x,\epsilon)}}S_n(x-y)\varphi_0(y)\,dy \\
&=\int_{\partial\Omega}S_n(x-y)(\varphi_1(y),\dots,\varphi_n(y))\cdot\nu_\Omega(y)\,d\sigma_y \\
&-\int_{\partial\mathbb{B}_n(x,\epsilon)}S_n(x-y)(\varphi_1(y),\dots,\varphi_n(y))\cdot\frac{y-x}{|y-x|}\,d\sigma_y
\end{aligned}\tag{7.59}$$

for all $\epsilon\in]0,\mathrm{dist}\,(x,\partial\Omega)[$. By the local integrability of ∇S_n and S_n in Ω and by the Dominated Convergence Theorem, we have

$$\lim_{\epsilon\to 0}\int_{\Omega\setminus\overline{\mathbb{B}_n(x,\epsilon)}}\nabla S_n(x-y)\cdot(\varphi_1(y),\dots,\varphi_n(y))^t\,dy$$
$$=\int_\Omega\nabla S_n(x-y)\cdot(\varphi_1(y),\dots,\varphi_n(y))^t\,dy\,,$$
$$\lim_{\epsilon\to 0}\int_{\Omega\setminus\overline{\mathbb{B}_n(x,\epsilon)}}S_n(x-y)\varphi_0(y)\,dy=\int_\Omega S_n(x-y)\varphi_0(y)\,dy\,.$$

By the limiting relation $\lim_{\epsilon\to 0}\epsilon^{n-1}\Upsilon_n(\epsilon)=0$, we have

$$\lim_{\epsilon\to 0}\int_{\partial\mathbb{B}_n(x,\epsilon)}S_n(x-y)(\varphi_1(y),\dots,\varphi_n(y))\cdot\frac{y-x}{|y-x|}\,d\sigma_y=0$$

(cf. (3.1), (3.2)). Then we can let ϵ tend to zero in equality (7.59) and obtain that

$$-\int_\Omega\nabla S_n(x-y)\cdot(\varphi_1(y),\dots,\varphi_n(y))^t\,dy$$

$$-\mathcal{P}[S_n,\varphi_0](x) - \int_{\partial\Omega} S_n(x-y)(\varphi_1(y),\ldots,\varphi_n(y))\cdot\nu_\Omega(y)\,d\sigma_y = 0\,.$$

Hence, statement (i) holds true (cf. Proposition 7.6). We now prove statement (ii). Since the linear map Ξ from $(C^{0,\alpha}(\overline{\Omega}))^{n+1}$ to $C^{-1,\alpha}(\overline{\Omega})$ is surjective and $\mathrm{Ker}\,\Xi$ is a subspace of $\mathrm{Ker}\,B$, the Homomorphism Theorem A.1 implies that there exists a unique linear operator $\mathcal{P}[S_n,\cdot]$ from $C^{-1,\alpha}(\overline{\Omega})$ to $C^{1,\alpha}(\overline{\Omega})$ that satisfies the equality

$$\mathcal{P}[S_n,\Xi[\varphi_0,\ldots,\varphi_n]] = B[\varphi_0,\ldots,\varphi_n] \qquad \forall(\varphi_0,\ldots,\varphi_n)\in(C^{0,\alpha}(\overline{\Omega}))^{n+1}\,,$$

i.e., equality (7.58). To prove that $\mathcal{P}[S_n,\cdot]$ is continuous, we show that $\mathcal{P}[S_n,\cdot]$ coincides with a linear and continuous map that we now define.

By statement (i), we have $\mathrm{Ker}\,\pi = \mathrm{Ker}\,\Xi \subseteq \mathrm{Ker}\,B$ and thus the Homomorphism Theorem A.1 for linear maps between vector spaces implies the existence of a linear map Λ from $(C^{0,\alpha}(\overline{\Omega}))^{n+1}/\mathrm{Ker}\,\Xi$ to $C^{1,\alpha}(\overline{\Omega})$ such that $B = \Lambda\circ\pi$. Since B is continuous, a known property of the topology on the quotient space $(C^{0,\alpha}(\overline{\Omega}))^{n+1}/\mathrm{Ker}\,\Xi$ implies that Λ is continuous (see Theorem 2.1).

We now show that $\mathcal{P}[S_n,\cdot] = \Lambda\circ\tilde{\Xi}^{(-1)}$, an equality which implies that the map $\mathcal{P}[S_n,\cdot]$ is continuous. Indeed, Λ is continuous and $\tilde{\Xi}$ is an isometry by definition of norm in $C^{-1,\alpha}(\overline{\Omega})$.

Let $\varphi\in C^{-1,\alpha}(\overline{\Omega})$. Let $(\varphi_0,\ldots,\varphi_n)\in C^{0,\alpha}(\overline{\Omega})^{n+1}$ be such that $\varphi = \varphi_0 + \sum_{j=1}^n \frac{\partial}{\partial x_j}\varphi_j$. Then we have

$$\begin{aligned}\mathcal{P}[S_n,\varphi] &= \mathcal{P}[S_n,\Xi[\varphi_0,\ldots,\varphi_n]] = B[\varphi_0,\ldots,\varphi_n]\\ &= \Lambda\circ\pi[\varphi_0,\ldots,\varphi_n] = \Lambda\circ\tilde{\Xi}^{(-1)}\circ\tilde{\Xi}\circ\pi[\varphi_0,\ldots,\varphi_n]\\ &= \Lambda\circ\tilde{\Xi}^{(-1)}\circ\Xi[\varphi_0,\ldots,\varphi_n] = \Lambda\circ\tilde{\Xi}^{(-1)}[\varphi]\,.\end{aligned}$$

Finally, to prove statement (iii), we note that if $\varphi\in C^{-1,\alpha}(\overline{\Omega})$, and if $\varphi = \varphi_0 + \sum_{j=1}^n \frac{\partial}{\partial x_j}\varphi_j$, with $(\varphi_0,\ldots,\varphi_n)\in C^{0,\alpha}(\overline{\Omega})^{n+1}$, then Theorem 7.18 implies that

$$\begin{aligned}\Delta(\mathcal{P}[S_n,\varphi]) &= \Delta(\mathcal{P}[S_n,\varphi_0]) + \sum_{j=1}^n \frac{\partial}{\partial x_j}\Delta(\mathcal{P}[S_n,\varphi_j]) + \sum_{j=1}^n \Delta v_\Omega^+[\varphi_j(\nu_\Omega)_j]\\ &= \varphi_0 + \sum_{j=1}^n \frac{\partial}{\partial x_j}\varphi_j + 0 = \varphi \qquad \text{in } \Omega\,.\end{aligned}$$

□

Thus under the assumptions of the previous Theorem, we define $\mathcal{P}[S_n,\varphi]$ to be the Newtonian potential of $\varphi\in C^{-1,\alpha}(\overline{\Omega})$.

7.2.6 Volume Potentials in Roumieu Classes

In order to estimate the Roumieu norm of a volume potential in terms of a norm of the kernel and of a norm of the density, we introduce the following class of functions which are singular at the origin and analytic away from the origin.

Definition 7.20. Let $\delta_1, \delta_2 \in]0,+\infty[$ with $\delta_1 < \delta_2$. Let $\lambda, \rho \in]0,+\infty[$. Then we set

$$H^{\lambda,\rho}(\delta_1,\delta_2) \equiv \left\{k \in A^1_\lambda(\delta_2) : k_{|\overline{\mathbb{B}_n(0,\delta_2)}\setminus\mathbb{B}_n(0,\delta_1)} \in C^0_{\omega,\rho}(\overline{\mathbb{B}_n(0,\delta_2)}\setminus\mathbb{B}_n(0,\delta_1))\right\},$$

and we set

$$\|k\|_{H^{\lambda,\rho}(\delta_1,\delta_2)} \equiv \|k\|_{A^1_\lambda(\delta_2)} + \|k\|_{C^0_{\omega,\rho}(\overline{\mathbb{B}_n(0,\delta_2)}\setminus\mathbb{B}_n(0,\delta_1))} \qquad \forall k \in H^{\lambda,\rho}(\delta_1,\delta_2).$$

One can readily verify that $(H^{\lambda,\rho}(\delta_1,\delta_2), \|\cdot\|_{H^{\lambda,\rho}(\delta_1,\delta_2)})$ is a Banach space. Then we can prove the following (see also [162, Prop. 2.4], [69, Prop, 4.2]).

Proposition 7.21. *Let $\rho \in]0,+\infty[$, $\lambda \in]0,n-1[$. Let Ω be a bounded open Lipschitz subset of $\mathbb{R}^n$. Let Ω_1 be a nonempty open subset of $\mathbb{R}^n$ such that $\overline{\Omega_1} \subseteq \Omega$. Let*

$$\delta^* \equiv \operatorname{diam}(\Omega), \qquad \delta_* \equiv \min\left\{|x-y| : x \in \overline{\Omega_1}, y \in \partial\Omega\right\}. \tag{7.60}$$

Then the restriction of $\mathcal{P}[k,\varphi]$ to $\overline{\Omega_1}$ belongs to the set $C^0_{\omega,\rho}(\overline{\Omega_1})$ for all (k,φ) in $H^{\lambda,\rho}(\delta_*,\delta^*) \times C^0_{\omega,\rho}(\overline{\Omega})$.

Moreover, the map from $H^{\lambda,\rho}(\delta_*,\delta^*) \times C^0_{\omega,\rho}(\overline{\Omega})$ to $C^0_{\omega,\rho}(\overline{\Omega_1})$ that takes (k,φ) to $\mathcal{P}[k,\varphi]_{|\overline{\Omega_1}}$ is bilinear and continuous.

Proof. We first prove that if $m \in \mathbb{N}\setminus\{0\}$ and if $(k,\varphi) \in H^{\lambda,\rho}(\delta_*,\delta^*) \times C^m(\overline{\Omega})$, then $\mathcal{P}[k,\varphi]_{|\overline{\Omega_1}} \in C^m(\overline{\Omega_1})$ and

$$\begin{aligned}\partial^\beta\mathcal{P}[k,\varphi](x) &= \mathcal{P}[k,\partial^\beta\varphi](x) \\ &\quad - \sum_{s=1}^n \sum_{l_s=0}^{\beta_s-1} \partial^{\beta_n}_{x_n}\dots\partial^{\beta_{s+1}}_{x_{s+1}}\partial^{l_s}_{x_s}\left\{\int_{\partial\Omega} k(x-y)\right. \\ &\qquad \left.\times(\partial^{\beta_s-1-l_s}_{y_s}\partial^{\beta_{s-1}}_{y_{s-1}}\dots\partial^{\beta_1}_{y_1}\varphi(y))(\nu_\Omega)_s(y)\,d\sigma_y\right\},\end{aligned} \tag{7.61}$$

for all $x \in \overline{\Omega_1}$ and for all $\beta \in \mathbb{N}^n$ such that $|\beta| \le m$, where we understand that $\sum_{l_s=0}^{\beta_s-1}$ is omitted if $\beta_s = 0$. We proceed by induction on m. If $m = 1$, then the statement follows by Lemma 7.7. Next we assume that the statement holds for m and we prove it for $m+1$. Let $(k,\varphi) \in H^{\lambda,\rho}(\delta_*,\delta^*) \times C^{m+1}(\overline{\Omega})$. By the inductive assumption, we have $\mathcal{P}[k,\frac{\partial\varphi}{\partial x_j}]_{|\overline{\Omega_1}} \in C^m(\overline{\Omega_1})$ for all $j \in \{1,\dots,n\}$. Since $k_{|\overline{\mathbb{B}_n(0,\delta^*)}\setminus\mathbb{B}_n(0,\delta_*)} \in C^m(\overline{\mathbb{B}_n(0,\delta^*)}\setminus\mathbb{B}_n(0,\delta_*))$ and φ, $(\nu_\Omega)_j \in C^0(\partial\Omega)$, the classical differentiability theorem for integrals depending on a parameter implies that the second term in the right-hand side of formula (7.14) defines a function of class $C^m(\overline{\Omega_1})$. Then formula (7.14) implies that $\frac{\partial}{\partial x_j}\mathcal{P}[k,\varphi]_{|\overline{\Omega_1}}$ belongs to $C^m(\overline{\Omega_1})$. Hence, $\mathcal{P}[k,\varphi]_{|\overline{\Omega_1}} \in C^{m+1}(\overline{\Omega_1})$. To prove formula (7.61) for $|\beta| \leq m+1$, we first prove the formula for $\partial^\beta = \partial_{x_j}^{\beta_j}$ by finite induction on the length of β and then we prove the formula for $\partial^\beta = \partial_{x_1}^{\beta_1}\dots\partial_{x_j}^{\beta_j}$ by finite induction on $j \in \{1,\dots,n\}$. For the details, we refer to [162, p. 856].

If $(k,\varphi) \in H^{\lambda,\rho}(\delta_*,\delta^*) \times C^\infty(\overline{\Omega})$, then by applying the above statement for all $m \in \mathbb{N}\setminus\{0\}$ we deduce that $\mathcal{P}[k,\varphi]_{|\overline{\Omega_1}}$ belongs to $C^\infty(\overline{\Omega_1})$ and that formula (7.61) holds for all order partial derivatives.

We now assume that $(k,\varphi) \in H^{\lambda,\rho}(\delta_*,\delta^*) \times C^0_{\omega,\rho}(\overline{\Omega})$ and we turn to estimate the sup-norm in $\overline{\Omega_1}$ of the double summation in the right-hand side of (7.61), which we denote by I. To do so, we abbreviate by $I(s,l_s)$ the (s,l_s)-th term in the sum I and we estimate the supremum of $I(s,l_s)$ in $\overline{\Omega_1}$. We can clearly assume that $\beta_s > 0$. Then we have

$$\begin{aligned}
&\sup_{\overline{\Omega_1}} |I(s,l_s)| \\
&\quad = \sup_{x\in\overline{\Omega_1}} \Bigg| \partial_{x_n}^{\beta_n}\dots\partial_{x_{s+1}}^{\beta_{s+1}}\partial_{x_s}^{l_s}\Bigg\{\int_{\partial\Omega} k(x-y) \\
&\qquad\qquad \times \partial_{y_s}^{\beta_s-1-l_s}\partial_{y_{s-1}}^{\beta_{s-1}}\dots\partial_{y_1}^{\beta_1}\varphi(y)(\nu_\Omega)_s(y)\,d\sigma_y\Bigg\}\Bigg| \\
&\quad \leq \int_{\partial\Omega}\sup_{\xi\in A}\left|\partial_{\xi_n}^{\beta_n}\dots\partial_{\xi_{s+1}}^{\beta_{s+1}}\partial_{\xi_s}^{l_s}k(\xi)\right|\,\left|\partial_{y_s}^{\beta_s-1-l_s}\partial_{y_{s-1}}^{\beta_{s-1}}\dots\partial_{y_1}^{\beta_1}\varphi(y)\right|\,d\sigma_y\,,
\end{aligned}$$

where $A \equiv \{x-y : x\in\overline{\Omega_1},\ y\in\partial\Omega\}$. Since $k \in H^{\lambda,\rho}(\delta_*,\delta^*)$, we have

$$\sup_{\xi\in A}\left|\partial_{\xi_n}^{\beta_n}\dots\partial_{\xi_{s+1}}^{\beta_{s+1}}\partial_{\xi_s}^{l_s}k(\xi)\right| \leq \|k\|_{H^{\lambda,\rho}(\delta_*,\delta^*)}\frac{(\beta_n+\dots+\beta_{s+1}+l_s)!}{\rho^{\beta_n+\dots+\beta_{s+1}+l_s}}\,.$$

Moreover,

$$\left|\partial_{y_s}^{\beta_s-1-l_s}\partial_{y_{s-1}}^{\beta_{s-1}}\dots\partial_{y_1}^{\beta_1}\varphi(y)\right| \leq \|\varphi\|_{C^0_{\omega,\rho}(\overline{\Omega})}\frac{(\beta_1+\dots+\beta_{s-1}+\beta_s-1-l_s)!}{\rho^{\beta_1+\dots+\beta_{s-1}+\beta_s-1-l_s}}\,,$$

for all $y \in \overline{\Omega}$. Then we have

$$\begin{aligned}
\sup_{\overline{\Omega_1}}|I(s,l_s)| &\leq m_{n-1}(\partial\Omega)\|k\|_{H^{\lambda,\rho}(\delta_*,\delta^*)}\|\varphi\|_{C^0_{\omega,\rho}(\overline{\Omega})} \\
&\quad\times\frac{(\beta_n+\dots+\beta_{s+1}+l_s)!(\beta_1+\dots+\beta_{s-1}+\beta_s-1-l_s)!}{\rho^{|\beta|-1}}\,.
\end{aligned} \tag{7.62}$$

Next we note that

$$m_1!m_2! \leq (m_1+m_2)!\,,$$

for all $m_1, m_2 \in \mathbb{N}$. Indeed,

$$1 \leq \binom{m_1+m_2}{m_1} = \frac{(m_1+m_2)!}{m_1!m_2!}\,.$$

Then (7.62) implies that

$$\sup_{\overline{\Omega_1}} |I(s,l_s)| \leq m_{n-1}(\partial\Omega)\|k\|_{H^{\lambda,\rho}(\delta_*,\delta^*)}\|\varphi\|_{C^0_{\omega,\rho}(\overline{\Omega})}\frac{(|\beta|-1)!}{\rho^{|\beta|-1}}\,.$$

Hence,

$$\sup_{\overline{\Omega_1}} |I| \leq n\rho m_{n-1}(\partial\Omega)\|k\|_{H^{\lambda,\rho}(\delta_*,\delta^*)}\|\varphi\|_{C^0_{\omega,\rho}(\overline{\Omega})}\frac{|\beta|!}{\rho^{|\beta|}}\,, \tag{7.63}$$

and we have

$$\|\mathcal{P}[k,\partial^\beta\varphi]\|_{L^\infty(\Omega)} \leq s_n\frac{(\mathrm{diam}\,(\Omega))^{n-\lambda}}{n-\lambda}\|k\|_{A^0_\lambda(\delta^*)}\|\partial^\beta\varphi\|_{L^\infty(\Omega)} \tag{7.64}$$

by Proposition 7.2 (iii). Then equality (7.61) and inequalities (7.63) and (7.64) imply that there exists $C \in]0,+\infty[$ such that

$$\|\partial^\beta\mathcal{P}[k,\varphi]_{|\Omega_1}\|_{L^\infty(\Omega_1)} \leq C\|h\|_{H^{\lambda,\rho}(\delta_*,\delta^*)}\|\varphi\|_{C^0_{\omega,\rho}(\overline{\Omega})}\frac{|\beta|!}{\rho^{|\beta|}} \qquad \forall\beta\in\mathbb{N}^n\,,$$

for all $(k,\varphi) \in H^{\lambda,\rho}(\delta_*,\delta^*) \times C^0_{\omega,\rho}(\overline{\Omega})$. □

Proposition 7.21 can be applied in case k is replaced with a fundamental solution of a second order elliptic operator. As shown in John [129], if S is a fundamental solution of a second order elliptic operator and if $\delta \in]0,+\infty[$, then for $n-2>0$ we have

$$\sup_{x\in\mathbb{B}_n(0,\delta)\setminus\{0\}} |S(x)|\,|x|^{n-2} < +\infty\,, \qquad \sup_{x\in\mathbb{B}_n(0,\delta)\setminus\{0\}} \left|\frac{\partial S}{\partial x_j}(x)\right|\,|x|^{n-1} < +\infty\,,$$

for all $j \in \{1,\dots,n\}$ and for $n-2=0$ we have

$$\sup_{x\in\mathbb{B}_n(0,\delta)\setminus\{0\}} |S(x)|\,|x|^{1/2} < +\infty\,, \qquad \sup_{x\in\mathbb{B}_n(0,\delta)\setminus\{0\}} \left|\frac{\partial S}{\partial x_j}(x)\right|\,|x|^{3/2} < +\infty\,,$$

for all $j \in \{1,\dots,n\}$. Moreover, S is analytic in $\mathbb{R}^n \setminus \{0\}$ and thus, if δ_1, $\delta_2 \in]0,+\infty[$ and $\delta_1 < \delta_2$, then the classical Cauchy inequalities for the partial derivatives of S on a compact set imply that there exists $\rho \in]0,+\infty[$ sufficiently small such that $S \in C^0_{\omega,\rho}(\overline{\mathbb{B}_n(0,\delta_2)} \setminus \mathbb{B}_n(0,\delta_1))$ (see Remark 2.18). Hence,

$$S \in H^{\max\{n-2,\frac{1}{2}\},\rho}(\delta_1,\delta_2)\,. \tag{7.65}$$

Thus, if we plan to apply Proposition 7.21 with h replaced by a fundamental solution of a second order elliptic operator, we can choose $\lambda = \max\{n-2,\frac{1}{2}\}$.

7.3 Boundary Value Problems for the Poisson Equation in Schauder Spaces

We now turn to consider the basic interior boundary value problems for the Poisson equation, and we show that if the data are in Schauder spaces, then also the solutions are in Schauder spaces. To do so, we exploit the results of the previous sections.

7.3.1 The Interior Dirichlet Problem for the Poisson Equation in Schauder Spaces

Theorem 7.22. *Let Ω be a bounded open subset of $\mathbb{R}^n$ of class $C^{1,\alpha}$. If (f,g) belongs to $C^{-1,\alpha}(\overline{\Omega}) \times C^{1,\alpha}(\partial\Omega)$, then there exists a unique solution $u \in C^{1,\alpha}(\overline{\Omega})$ of the interior Dirichlet problem*

$$\begin{cases} \Delta u = f & \text{in } \Omega\,, \\ u = g & \text{on } \partial\Omega\,. \end{cases}$$

Moreover, the operator $(\Delta, \mathrm{Tr}_{|\partial\Omega})$ from $C^{1,\alpha}(\overline{\Omega})$ to $C^{-1,\alpha}(\overline{\Omega}) \times C^{1,\alpha}(\partial\Omega)$ that takes u to $(\Delta u, u_{|\partial\Omega})$ is a linear homeomorphism.

Proof. If $f \in C^{-1,\alpha}(\overline{\Omega})$, then $\mathcal{P}[S_n, f] \in C^{1,\alpha}(\overline{\Omega})$ and $\Delta\mathcal{P}[S_n, f] = f$ in Ω (see Theorem 7.19). Since $g - \mathcal{P}[S_n, f]_{|\partial\Omega} \in C^{1,\alpha}(\partial\Omega)$, the Dirichlet problem

$$\begin{cases} \Delta v = 0 & \text{in } \Omega\,, \\ v_{|\partial\Omega} = g - \mathcal{P}[S_n, f]_{|\partial\Omega} & \text{on } \partial\Omega \end{cases}$$

has a unique solution $v \in C^{1,\alpha}(\overline{\Omega})$ (see Theorem 6.27). Then

$$u \equiv v + \mathcal{P}[S_n, f] \in C^{1,\alpha}(\overline{\Omega})$$

solves the Dirichlet problem of the statement, i.e., $(\Delta u, u_{|\partial\Omega}) = (f, g)$. In particular, the operator $(\Delta, \mathrm{Tr}_{|\partial\Omega})$ is surjective. The uniqueness of the solution u for the problem of the statement follows by the uniqueness of the solutions for the interior Dirichlet problem for the Laplace operator of Theorem 3.6. As a consequence $(\Delta, \mathrm{Tr}_{|\partial\Omega})$ is a linear bijection. Since $(\Delta, \mathrm{Tr}_{|\partial\Omega})$ is continuous, the Open Mapping Theorem 2.2 implies that $(\Delta, \mathrm{Tr}_{|\partial\Omega})$ is a homeomorphism. □

7.3.2 The Interior Neumann Problem for the Poisson Equation in Schauder Spaces

Theorem 7.23. *Let Ω be a bounded open connected subset of $\mathbb{R}^n$ of class $C^{1,\alpha}$. If $(f, g) \in C^{-1,\alpha}(\overline{\Omega}) \times C^{0,\alpha}(\partial\Omega)$ and if*

$$\int_{\partial\Omega} g \, d\sigma = \mathcal{I}_\Omega[f] , \tag{7.66}$$

then the interior Neumann problem

$$\begin{cases} \Delta u = f \ \text{ in } \Omega , \\ \frac{\partial u}{\partial \nu_\Omega} = g \text{ on } \partial\Omega \end{cases}$$

has at least a solution $u \in C^{1,\alpha}(\overline{\Omega})$ (for the definition of $\mathcal{I}_\Omega[\cdot]$, see Proposition 2.89). All other solutions in $C^{1,\alpha}(\overline{\Omega})$ can be obtained by adding to u a constant function. Moreover, the operator $(\Delta, \frac{\partial}{\partial \nu_\Omega})$ from $C^{1,\alpha}(\overline{\Omega})$ to the closed subspace

$$\left\{ (f, g) \in C^{-1,\alpha}(\overline{\Omega}) \times C^{0,\alpha}(\partial\Omega) : \int_{\partial\Omega} g \, d\sigma = \mathcal{I}_\Omega[f] \right\} \tag{7.67}$$

of $C^{-1,\alpha}(\overline{\Omega}) \times C^{0,\alpha}(\partial\Omega)$ that takes u to $(\Delta u, \frac{\partial u}{\partial \nu_\Omega})$ is a linear and continuous surjection and the null space Ker $(\Delta, \frac{\partial}{\partial \nu_\Omega})$ *consists of the functions which are constant in $\overline{\Omega}$.*

Proof. If $f = f_0 + \sum_{j=1}^n \frac{\partial}{\partial x_j} f_j \in C^{-1,\alpha}(\overline{\Omega})$, then $\mathcal{P}[S_n, f] \in C^{1,\alpha}(\overline{\Omega})$ and $\Delta \mathcal{P}[S_n, f] = f$ in Ω (see Theorem 7.19). We now show that the interior Neumann problem

$$\begin{cases} \Delta v = 0 & \text{in } \Omega , \\ \frac{\partial}{\partial \nu_\Omega} v = g - \frac{\partial}{\partial \nu_\Omega} \mathcal{P}[S_n, f] & \text{on } \partial\Omega \end{cases} \tag{7.68}$$

has a solution $v \in C^{1,\alpha}(\overline{\Omega})$. To do so, we must show that g satisfies the compatibility condition $\int_{\partial\Omega} g - \frac{\partial}{\partial \nu_\Omega} \mathcal{P}[S_n, f] \, d\sigma = 0$ (cf. Theorem 6.42).

Since $\mathcal{P}[S_n, f_j] \in C^{2,\alpha}(\overline{\Omega})$, $\Delta \mathcal{P}[S_n, f_j] = f_j$ in Ω (see Theorem 7.18), and $v_\Omega^+[f_j(\nu_\Omega)_j] \in C^{1,\alpha}(\partial\Omega)$ is harmonic in Ω for all $j \in \{1, \ldots, n\}$ (see Theorem 4.25), the definition of $\mathcal{P}[S_n, f]$ of Thorem 7.19 (ii), the Divergence Theorem 4.1, the first Green Identity of Theorem 4.2, and Proposition 2.89 (ii) on the definition of $\mathcal{I}_\Omega$ imply that

$$\int_{\partial\Omega} g - \frac{\partial}{\partial \nu_\Omega} \mathcal{P}[S_n, f] \, d\sigma = \int_{\partial\Omega} g \, d\sigma$$

$$- \int_{\partial\Omega} \frac{\partial}{\partial \nu_\Omega} \mathcal{P}[S_n, f_0] \, d\sigma - \int_{\partial\Omega} \sum_{l=1}^n (\nu_\Omega)_l \frac{\partial}{\partial x_l} \left(\sum_{j=1}^n \frac{\partial}{\partial x_j} \mathcal{P}[S_n, f_j] \right) d\sigma$$

$$
\begin{aligned}
&-\sum_{j=1}^{n}\int_{\partial\Omega}\frac{\partial}{\partial\nu_\Omega}v_\Omega^+[f_j(\nu_\Omega)_j]\,d\sigma \\
&=\int_{\partial\Omega} g\,d\sigma-\int_{\partial\Omega} f_0\,d\sigma-\int_{\partial\Omega}\sum_{l=1}^{n}(\nu_\Omega)_l\frac{\partial}{\partial x_l}\left(\sum_{j=1}^{n}\frac{\partial}{\partial x_j}\mathcal{P}[S_n,f_j]\right)d\sigma \\
&=\int_{\partial\Omega} g\,d\sigma-\mathcal{I}_\Omega\left[f_0+\sum_{l=1}^{n}\frac{\partial}{\partial x_l}\frac{\partial}{\partial x_l}\left(\sum_{j=1}^{n}\frac{\partial}{\partial x_j}\mathcal{P}[S_n,f_j]\right)\right] \\
&=\int_{\partial\Omega} g\,d\sigma-\mathcal{I}_\Omega\left[f_0+\sum_{j=1}^{n}\frac{\partial}{\partial x_j}\Delta\mathcal{P}[S_n,f_j]\right] \\
&=\int_{\partial\Omega} g\,d\sigma-\mathcal{I}_\Omega\left[f_0+\sum_{j=1}^{n}\frac{\partial}{\partial x_j}f_j\right] \\
&=\int_{\partial\Omega} g\,d\sigma-\mathcal{I}_\Omega[f]=0\,.
\end{aligned}
$$

Hence, the interior Neumann problem (7.68) has a solution $v\in C^{1,\alpha}(\overline{\Omega})$ (see Theorem 6.42). Then $u\equiv v+\mathcal{P}[S_n,f]$ belongs to $C^{1,\alpha}(\overline{\Omega})$ and solves the Neumann problem of the statement, i.e., $(\Delta u,\frac{\partial}{\partial\nu_\Omega}u)=(f,g)$.

Since the components of $(\Delta,\frac{\partial}{\partial\nu_\Omega})$ are linear and continuous, $(\Delta,\frac{\partial}{\partial\nu_\Omega})$ is linear and continuous from $C^{1,\alpha}(\overline{\Omega})$ to $C^{-1,\alpha}(\overline{\Omega})\times C^{0,\alpha}(\partial\Omega)$. Next we show that if $u\in C^{1,\alpha}(\overline{\Omega})$, then the pair $(\Delta u,\frac{\partial}{\partial\nu_\Omega}u)$ belongs to the space in (7.67). Since

$$
\Delta u=\sum_{j=1}^{n}\frac{\partial}{\partial x_j}\left(\frac{\partial u}{\partial x_j}\right)
$$

and $\frac{\partial u}{\partial x_j}\in C^{0,\alpha}(\overline{\Omega})$ for all $j\in\{1,\ldots,n\}$, the definition of $\mathcal{I}_\Omega$ of Proposition 2.89 (ii) implies that

$$
\mathcal{I}_\Omega[\Delta u]=\sum_{j=1}^{n}\int_{\partial\Omega}(\nu_\Omega)_j\frac{\partial u}{\partial x_j}\,d\sigma=\int_{\partial\Omega}\frac{\partial u}{\partial\nu_\Omega}\,d\sigma\,,
$$

and accordingly $(\Delta u,\frac{\partial}{\partial\nu_\Omega}u)$ belongs to the space in (7.67). By the above argument, the operator $(\Delta,\frac{\partial}{\partial\nu_\Omega})$ is surjective onto the space in (7.67). By the solvabilty Theorem 6.42 for the interior Neumann problem for the Laplace equation, we know that all other solutions in $C^{1,\alpha}(\overline{\Omega})$ can be obtained by adding to u a constant function. In particular, Ker $(\Delta,\frac{\partial}{\partial\nu_\Omega})$ consists of the functions which are constant in $\overline{\Omega}$.

By Proposition 2.89, the operator $\mathcal{I}_\Omega$ from $C^{-1,\alpha}(\overline{\Omega})$ to $\mathbb{R}$ is linear and continuous. Since the operator from $C^{0,\alpha}(\partial\Omega)$ to $\mathbb{R}$ that takes g to $\int_{\partial\Omega} g\,d\sigma$ is linear and continuous, then the map from $C^{-1,\alpha}(\overline{\Omega})\times C^{0,\alpha}(\partial\Omega)$ to $\mathbb{R}$ that takes (f,g)

to $\int_{\partial\Omega} g\,d\sigma - \mathcal{I}_\Omega[f]$ is linear and continuous. Thus the space in (7.66) is closed in $C^{-1,\alpha}(\overline{\Omega}) \times C^{0,\alpha}(\partial\Omega)$. □

We note that if $f \in C^{0,\alpha}(\overline{\Omega})$, then $\mathcal{I}_\Omega[f] = \int_\Omega f\,dx$ and thus the compatibility condition (7.67) takes the form

$$\int_{\partial\Omega} g\,d\sigma = \int_\Omega f\,dx$$

(cf. Proposition 2.89 (ii)).

7.3.3 The Interior Robin Problem for the Poisson Equation in Schauder Spaces

Theorem 7.24. *Let Ω be a bounded open connected subset of $\mathbb{R}^n$ of class $C^{1,\alpha}$. Let $b \in C^{0,\alpha}(\partial\Omega)$, $b \geq 0$ on $\partial\Omega$. Let $\int_{\partial\Omega} b\,d\sigma \neq 0$. If $(f,g) \in C^{-1,\alpha}(\overline{\Omega}) \times C^{0,\alpha}(\partial\Omega)$, then the interior Robin problem*

$$\begin{cases} \Delta u = f & \text{in } \Omega\,, \\ \frac{\partial}{\partial\nu_\Omega}u + bu = g & \text{on } \partial\Omega\,, \end{cases}$$

has a unique solution $u \in C^{1,\alpha}(\overline{\Omega})$. Moreover, the operator $(\Delta, \frac{\partial}{\partial\nu_\Omega} + b\mathrm{Tr}_{|\partial\Omega})$ from $C^{1,\alpha}(\overline{\Omega})$ to $C^{-1,\alpha}(\overline{\Omega}) \times C^{0,\alpha}(\partial\Omega)$, that takes u to $(\Delta u, \frac{\partial}{\partial\nu_\Omega}u + bu_{|\partial\Omega})$, is a linear homeomorphism.

Proof. If $f \in C^{-1,\alpha}(\overline{\Omega})$, then $\mathcal{P}[S_n, f] \in C^{1,\alpha}(\overline{\Omega})$ and $\Delta\mathcal{P}[S_n, f] = f$ in Ω (see Theorem 7.19). Since $g - \frac{\partial}{\partial\nu_\Omega}\mathcal{P}[S_n, f] - b\mathcal{P}[S_n, f]_{|\partial\Omega} \in C^{0,\alpha}(\partial\Omega)$, the interior Robin problem

$$\begin{cases} \Delta v = 0 & \text{in } \Omega\,, \\ \frac{\partial}{\partial\nu_\Omega}v + bv = g - \frac{\partial}{\partial\nu_\Omega}\mathcal{P}[S_n, f] - b\mathcal{P}[S_n, f] & \text{on } \partial\Omega\,, \end{cases}$$

has a unique solution v in $C^{1,\alpha}(\overline{\Omega})$ (see Theorem 6.58). Then

$$u \equiv v + \mathcal{P}[S_n, f] \in C^{1,\alpha}(\overline{\Omega})$$

solves the Robin problem of the statement, i.e., $(\Delta u, \frac{\partial}{\partial\nu_\Omega}u + bu_{|\partial\Omega}) = (f, g)$. In particular, the operator $(\Delta, \frac{\partial}{\partial\nu_\Omega} + b\mathrm{Tr}_{|\partial\Omega})$ is surjective. The uniqueness of the solution u for the problem of the statement follows by the uniqueness of the solutions for the interior Robin problem for the Laplace operator (see Theorem 6.56). As a consequence $(\Delta, \frac{\partial}{\partial\nu_\Omega} + b\mathrm{Tr}_{|\partial\Omega})$ is a linear bijection. Since $(\Delta, \frac{\partial}{\partial\nu_\Omega} + b\mathrm{Tr}_{|\partial\Omega})$ is continuous, the Open Mapping Theorem 2.2 implies that $(\Delta, \frac{\partial}{\partial\nu_\Omega} + b\mathrm{Tr}_{|\partial\Omega})$ is a homeomorphism. □

Chapter 8
A Dirichlet Problem in a Domain with a Small Hole

Abstract In this chapter we introduce the Functional Analytic Approach for the analysis of elliptic boundary value problems in singularly perturbed domains. To illustrate how the method works, we closely study a boundary value problem for the Laplace equation in a bounded domain with a small hole that shrinks to an interior point. We impose Dirichlet conditions both on the outer boundary of the domain and on the boundary of the hole. One peculiarity of this specific problem is that, in the case of dimension two, the solution displays a logarithmic behavior. We shall see in Chapter 9 that such behavior might not appear with other boundary conditions.

8.1 The Geometric Setting

We fix once for all a natural number

$$n \in \mathbb{N}\,,\; n \geq 2$$

and a real number

$$\alpha \in]0,1[\,.$$

Then we take two sets Ω^i and Ω^o in the n-dimensional Euclidean space $\mathbb{R}^n$. The letter 'i' stands for 'inner domain' and the letter 'o' stands for 'outer domain.' We assume that Ω^i and Ω^o satisfy the following condition

$$\begin{aligned}&\Omega^i \text{ and } \Omega^o \text{ are bounded open connected subsets of } \mathbb{R}^n\\ &\text{of class } C^{1,\alpha}\text{, the exterior domains } \Omega^{i-} \text{ and } \Omega^{o-} \text{ are connected,}\\ &\text{and the origin 0 of } \mathbb{R}^n \text{ belongs both to } \Omega^i \text{ and to } \Omega^o.\end{aligned} \tag{8.1}$$

Condition (8.1) implies that Ω^i and Ω^o have no holes and that there exists a positive real number ϵ_0 such that

M. Dalla Riva et al., *Singularly Perturbed Boundary Value Problems*,
https://doi.org/10.1007/978-3-030-76259-9_8

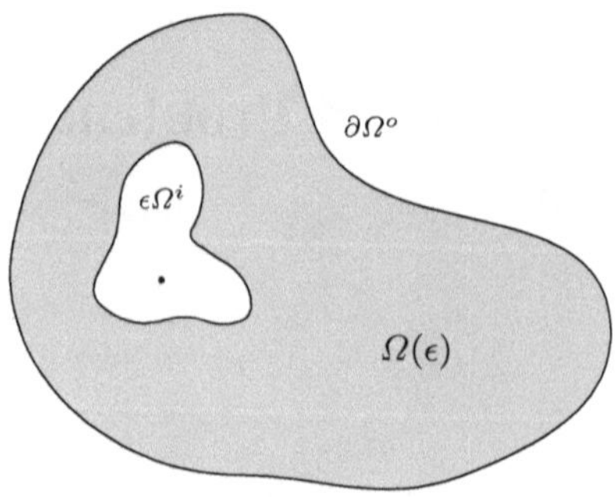

Fig. 8.1 The perforated domain $\Omega(\epsilon)$ ($n = 2$)

$$\epsilon\overline{\Omega^i} \subseteq \Omega^o \text{ for all } \epsilon \in]-\epsilon_0, \epsilon_0[\,. \tag{8.2}$$

Then we denote by $\Omega(\epsilon)$ the perforated domain defined by

$$\Omega(\epsilon) \equiv \Omega^o \setminus (\epsilon\overline{\Omega^i}) \qquad \forall \epsilon \in]-\epsilon_0, \epsilon_0[\tag{8.3}$$

(see Figure 8.1). A topological argument shows that, for all $\epsilon \in]-\epsilon_0, \epsilon_0[$, the domain $\Omega(\epsilon)$ is an open bounded connected subset of $\mathbb{R}^n$ with boundary consisting of the disjoint union of $\partial\Omega^o$ and $\epsilon\partial\Omega^i$ (for a proof that $\Omega(\epsilon)$ is connected see, e.g., Theorem A.10). If $\epsilon \in]-\epsilon_0, \epsilon_0[$ is different from 0, then we can also verify that $\Omega(\epsilon)$ is a domain of class $C^{1,\alpha}$ (cf. Section 2.13). If instead $\epsilon = 0$, then we have $\Omega(0) = \Omega^o \setminus \{0\}$, the boundary of $\Omega(0)$ consists of the disjoint union of $\partial\Omega^o$ and $\{0\}$, and the domain $\Omega(0)$ is no longer of class $C^{1,\alpha}$. Actually, $\Omega(0)$ is not even a local strict hypograph of class C^0 in the sense of Definition 2.37, because it is not a hypograph in any neighborhood of the boundary point 0. Then we can say that when ϵ approaches zero the perforated domain $\Omega(\epsilon)$ loses some of its regularity and, in this sense, we are dealing with a 'singularly perturbed domain.'

8.2 A Dirichlet Problem for the Laplace Equation

We now fix $f^i \in C^{1,\alpha}(\partial\Omega^i)$ and $f^o \in C^{1,\alpha}(\partial\Omega^o)$. Then, for $\epsilon \in]0, \epsilon_0[$ we consider the following Dirichlet boundary value problem

$$\begin{cases} \Delta u = 0 & \text{in } \Omega(\epsilon)\,, \\ u(x) = f^i(x/\epsilon) & \text{for } x \in \epsilon\partial\Omega^i\,, \\ u(x) = f^o(x) & \text{for } x \in \partial\Omega^o\,. \end{cases} \tag{8.4}$$

By Theorem 6.27, problem (8.4) has a unique solution in $C^{1,\alpha}(\overline{\Omega(\epsilon)})$ (see also Theorem 3.6). We denote it by u_ϵ. Then we want to describe how u_ϵ depends upon ϵ, especially when ϵ is close to 0. To formulate a precise question we can, for example, fix a point p in $\Omega^o \setminus \{0\}$ and take $\epsilon_p \in]0, \epsilon_0[$ such that $p \in \Omega(\epsilon)$ for all $\epsilon \in]0, \epsilon_p[$.

Then it makes sense to consider $u_\epsilon(p)$ for all $\epsilon \in]0, \epsilon_p[$ and we can ask the following question.

$$\text{What can be said of the map from }]0, \epsilon_p[\text{ to } \mathbb{R} \text{ that takes } \epsilon \text{ to } u_\epsilon(p)? \tag{8.5}$$

In the following Sections 8.3 and 8.4, we consider separately the case when $n \geq 3$ and when $n = 2$.

8.3 Analysis for $n \geq 3$

In case of dimension $n \geq 3$, we will answer to the question in (8.5) by proving that the map that takes ϵ to $u_\epsilon(p)$ can be continued real analytically in a neighborhood of $\epsilon = 0$. As opposed to the case when $n = 2$, the analysis in dimension $n \geq 3$ is facilitated by the fact that we can represent the solution u_ϵ as a single layer potential and we can use Theorem 6.46. We first observe that, for an $\epsilon \in]0, \epsilon_0[$ fixed, the single layer potential $v^+_{\Omega(\epsilon)}[\theta_\epsilon]$ with $\theta_\epsilon \in C^{0,\alpha}(\partial\Omega(\epsilon))$ is the solution of (8.4) if we have

$$\begin{aligned} v_{\Omega(\epsilon)}[\theta_\epsilon](x) &= f^i(x/\epsilon) \qquad &&\text{for } x \in \epsilon\partial\Omega^i\,, \\ v_{\Omega(\epsilon)}[\theta_\epsilon](x) &= f^o(x) \qquad &&\text{for } x \in \partial\Omega^o\,. \end{aligned} \tag{8.6}$$

Then we transform the equation on the ϵ-depending boundary $\epsilon\partial\Omega^i$ into an equation on the fixed boundary $\partial\Omega^i$. To do so, we introduce the (so-called) 'fast' variable $\xi = x/\epsilon$ and we define

$$\begin{aligned} \theta^i_\epsilon(\xi) &\equiv \epsilon\,\theta_\epsilon(\epsilon\xi) \qquad &&\forall \xi \in \partial\Omega^i\,, \\ \theta^o_\epsilon(x) &\equiv \theta_\epsilon(x) \qquad &&\forall x \in \partial\Omega^o \end{aligned}$$

(note that θ^o_ϵ is just the restriction of θ_ϵ to $\partial\Omega^o$). By changing variables of integration in the integrals supported on $\epsilon\partial\Omega^i$ and by exploiting the equality

$$S_n(\epsilon\xi - \epsilon\eta) = \epsilon^{2-n} S_n(\xi - \eta)\,,$$

which holds for all $\epsilon > 0$ and all $\xi, \eta \in \partial\Omega^i$ with $\xi \neq \eta$, we obtain a new set of conditions, defined on the fixed boundaries $\partial\Omega^i$ and $\partial\Omega^o$ and equivalent to (8.6):

$$\begin{aligned} &\int_{\partial\Omega^i} S_n(\xi - \eta)\theta^i_\epsilon(\eta)\,d\sigma_\eta + \int_{\partial\Omega^o} S_n(\epsilon\xi - y)\theta^o_\epsilon(y)\,d\sigma_y = f^i(\xi) \quad \forall \xi \in \partial\Omega^i\,, \\ &\epsilon^{n-2}\int_{\partial\Omega^i} S_n(x - \epsilon\eta)\theta^i_\epsilon(\eta)\,d\sigma_\eta + \int_{\partial\Omega^o} S_n(x - y)\theta^o_\epsilon(y)\,d\sigma_y = f^o(x) \quad \forall x \in \partial\Omega^o\,. \end{aligned} \tag{8.7}$$

Now for all $\epsilon \in]-\epsilon_0, \epsilon_0[$ we define the auxiliary operator

$$\mathcal{V}(\epsilon) \equiv (\mathcal{V}^i(\epsilon), \mathcal{V}^o(\epsilon))$$

from $C^{0,\alpha}(\partial\Omega^i) \times C^{0,\alpha}(\partial\Omega^o)$ to $C^{1,\alpha}(\partial\Omega^i) \times C^{1,\alpha}(\partial\Omega^o)$ by setting

$$\begin{aligned}
\mathcal{V}^i(\epsilon)[\theta^i, \theta^o](\xi) \equiv & \int_{\partial\Omega^i} S_n(\xi - \eta)\theta^i(\eta)\, d\sigma_\eta \\
& + \int_{\partial\Omega^o} S_n(\epsilon\xi - y)\theta^o(y)\, d\sigma_y \qquad \forall \xi \in \partial\Omega^i, \\
\mathcal{V}^o(\epsilon)[\theta^i, \theta^o](x) \equiv & \, \epsilon^{n-2} \int_{\partial\Omega^i} S_n(x - \epsilon\eta)\theta^i(\eta)\, d\sigma_\eta \\
& + \int_{\partial\Omega^o} S_n(x - y)\theta^o(y)\, d\sigma_y \qquad \forall x \in \partial\Omega^o
\end{aligned} \tag{8.8}$$

for all (θ^i, θ^o) in $C^{0,\alpha}(\partial\Omega^i) \times C^{0,\alpha}(\partial\Omega^o)$. It is apparent that the system of equations in (8.7) is equivalent to the operator equation

$$\mathcal{V}(\epsilon)[\theta^i_\epsilon, \theta^o_\epsilon] = (f^i, f^o) \tag{8.9}$$

and thus, by Theorem 6.46, we deduce the validity of the following.

Proposition 8.1. *Let* $\epsilon \in]0, \epsilon_0[$. *There exists one and only one pair of functions* $(\theta^i_\epsilon, \theta^o_\epsilon) \in C^{0,\alpha}(\partial\Omega^i) \times C^{0,\alpha}(\partial\Omega^o)$ *such that* (8.9) *holds true. Moreover, the function* u *defined by*

$$u(x) \equiv \epsilon^{n-2} \int_{\partial\Omega^i} S_n(x - \epsilon\eta)\theta^i_\epsilon(\eta)\, d\sigma_\eta + \int_{\partial\Omega^o} S_n(x - y)\theta^o_\epsilon(y)\, d\sigma_y \quad \forall x \in \overline{\Omega(\epsilon)} \tag{8.10}$$

is the unique solution in $C^{1,\alpha}(\overline{\Omega(\epsilon)})$ *of the boundary value problem* (8.4).

By Proposition 8.1, the map from $]0, \epsilon_0[$ to $C^{0,\alpha}(\partial\Omega^i) \times C^{0,\alpha}(\partial\Omega^o)$ that takes ϵ to $(\theta^i_\epsilon, \theta^o_\epsilon)$ is a well-defined function. Our aim now is to show that such map has a real analytic continuation in a open neighborhood of 0. To do so, we will first prove that the map that takes ϵ to the operator $\mathcal{V}(\epsilon)$ is real analytic in $]-\epsilon_0, \epsilon_0[$, then we will show that $\mathcal{V}(0)$ is an isomorphism (i.e., a linear homeomorphism), and, finally, we will deduce the existence of the real analytic continuation of $\epsilon \mapsto (\theta^i_\epsilon, \theta^o_\epsilon)$ by exploiting the equality

$$(\theta^i_\epsilon, \theta^o_\epsilon) = \mathcal{V}(\epsilon)^{(-1)}[f^i, f^o],$$

which holds for ϵ positive and small. Once we prove that $\epsilon \mapsto (\theta^i_\epsilon, \theta^o_\epsilon)$ has a real analytic continuation, we can derive real analyticity results for the solution u_ϵ by exploiting the representation (8.10). At this point, we can give an answer to the question in 8.5.

So, we start by showing that $\epsilon \mapsto \mathcal{V}(\epsilon)$ is real analytic.

Lemma 8.2. *The map from* $]-\epsilon_0, \epsilon_0[$ *to*

$$\mathcal{L}(C^{0,\alpha}(\partial\Omega^i) \times C^{0,\alpha}(\partial\Omega^o), C^{1,\alpha}(\partial\Omega^i) \times C^{1,\alpha}(\partial\Omega^o))$$

that takes ϵ *to* $\mathcal{V}(\epsilon)$ *is real analytic.*

We remind that $\mathcal{L}(X,Y)$ denotes the space of linear bounded operators between the Banach spaces X and Y equipped with the usual operator norm (cf. Section 2.2).
Proof. We first consider $\mathcal{V}^i$. By (8.8) we can write

$$\mathcal{V}^i(\epsilon)[\theta^i, \theta^o] = V_{\Omega^i}[\theta^i] + \tilde{\mathcal{V}}^i(\epsilon)[\theta^o]$$

with

$$\tilde{\mathcal{V}}^i(\epsilon)[\theta^o](\xi) \equiv \int_{\partial\Omega^o} S_n(\epsilon\xi - y)\theta^o(y)\, d\sigma_y \qquad \forall \xi \in \partial\Omega^i . \tag{8.11}$$

By Theorem 6.4, V_{Ω^i} is a bounded linear operator from $C^{0,\alpha}(\partial\Omega^i)$ to $C^{1,\alpha}(\partial\Omega^i)$ and it does not depend on ϵ. By Lemma A.26 in the Appendix, $\tilde{\mathcal{V}}^i$ is real analytic from $]-\epsilon_0, \epsilon_0[$ to $\mathcal{L}(C^{0,\alpha}(\partial\Omega^o), C^{1,\alpha}(\partial\Omega^i))$. It follows that $\mathcal{V}^i$ is real analytic from $]-\epsilon_0, \epsilon_0[$ to $\mathcal{L}(C^{0,\alpha}(\partial\Omega^i) \times C^{0,\alpha}(\partial\Omega^o), C^{1,\alpha}(\partial\Omega^i))$.

Similarly, we can write

$$\mathcal{V}^o(\epsilon)[\theta^i, \theta^o] = \epsilon\tilde{\mathcal{V}}^o(\epsilon)[\theta^i] + V_{\Omega^o}[\theta^o]$$

with

$$\tilde{\mathcal{V}}^o(\epsilon)[\theta^i](x) \equiv \int_{\partial\Omega^i} S_n(x - \epsilon\eta)\theta^i(\eta)\, d\sigma_\eta \qquad \forall x \in \partial\Omega^o .$$

We note that V_{Ω^o} is linear and bounded from $C^{0,\alpha}(\partial\Omega^o)$ to $C^{1,\alpha}(\partial\Omega^o)$ by Theorem 6.4 while $\tilde{\mathcal{V}}^o$ is real analytic from $]-\epsilon_0, \epsilon_0[$ to $\mathcal{L}(C^{0,\alpha}(\partial\Omega^i), C^{1,\alpha}(\partial\Omega^o))$ by Corollary A.29 and by the continuous embedding of $C^{0,\alpha}(\partial\Omega^i)$ in $L^1(\partial\Omega^i)$. Then also $\mathcal{V}^o$ is real analytic from $]-\epsilon_0, \epsilon_0[$ to $\mathcal{L}(C^{0,\alpha}(\partial\Omega^i) \times C^{0,\alpha}(\partial\Omega^o), C^{1,\alpha}(\partial\Omega^o))$ and the lemma is proven. □

We now consider equation $\mathcal{V}(0)[\theta^i, \theta^o] = (g^i, g^o)$, write an equivalent system of boundary integral equations, and deduce that both the system and the equation have unique solutions. We will use this result to prove that $\mathcal{V}(0)$ is an isomorphism.

Lemma 8.3. *For all* $(g^i, g^o) \in C^{1,\alpha}(\partial\Omega^i) \times C^{1,\alpha}(\partial\Omega^o)$ *and for all* $(\theta^i, \theta^o) \in C^{0,\alpha}(\partial\Omega^i) \times C^{0,\alpha}(\partial\Omega^o)$ *we have*

$$\mathcal{V}(0)[\theta^i, \theta^o] = (g^i, g^o) \tag{8.12}$$

if and only if

$$\begin{cases} V_{\Omega^i}[\theta^i] + v_{\Omega^o}[\theta^o](0) = g^i , \\ V_{\Omega^o}[\theta^o] = g^o . \end{cases} \tag{8.13}$$

In particular, for all $(g^i, g^o) \in C^{1,\alpha}(\partial\Omega^i) \times C^{1,\alpha}(\partial\Omega^o)$ *there exists one and only one solution* $(\theta^i, \theta^o) \in C^{0,\alpha}(\partial\Omega^i) \times C^{0,\alpha}(\partial\Omega^o)$ *of* (8.12).

Proof. The equivalence of (8.12) and (8.13) follows by the definition of $\mathcal{V}$ in (8.8). To prove the last statement of the lemma we observe that, by Theorem 6.46, the second equation in (8.13) has a unique solution $\theta^o \in C^{0,\alpha}(\partial\Omega^o)$, then, again by Theorem 6.46, also the first equation in (8.13) has a unique solution $\theta^i \in C^{0,\alpha}(\partial\Omega^i)$. □

Then we deduce that $\mathcal{V}(0)$ is an isomorphism.

Lemma 8.4. *The operator $\mathcal{V}(0)$ is an isomorphism from $C^{0,\alpha}(\partial\Omega^i) \times C^{0,\alpha}(\partial\Omega^o)$ to $C^{1,\alpha}(\partial\Omega^i) \times C^{1,\alpha}(\partial\Omega^o)$.*

Proof. As a consequence of the previous Lemma 8.2 we know that $\mathcal{V}(0)$ is a linear bounded operator. Then by the Open Mapping Theorem 2.2 it is an isomorphism if it is one to one and onto. Hence, we can conclude the proof by Lemma 8.3. □

Since the set of invertible operators is open in the space of bounded operators and since the map that takes an invertible operator to its inverse is real analytic (cf. Proposition A.15), we now deduce the following.

Lemma 8.5. *There exists $\epsilon_* \in]0, \epsilon_0[$ such that the operator $\mathcal{V}(\epsilon)$ is invertible for all $\epsilon \in]-\epsilon_*, \epsilon_*[$ and the map from $]-\epsilon_*, \epsilon_*[$ to*

$$\mathcal{L}(C^{1,\alpha}(\partial\Omega^i) \times C^{1,\alpha}(\partial\Omega^o), C^{0,\alpha}(\partial\Omega^i) \times C^{0,\alpha}(\partial\Omega^o))$$

that takes ϵ to $\mathcal{V}(\epsilon)^{(-1)}$ is real analytic.

Proof. It follows by Lemmas 8.2 and 8.4 and Proposition A.15. □

Then by Lemma 8.5 we can show the existence of a real analytic continuation of the map $\epsilon \mapsto (\theta^i_\epsilon, \theta^o_\epsilon)$.

Proposition 8.6. *There exists a real analytic map*

$$\Theta \equiv (\Theta^i, \Theta^o) :]-\epsilon_*, \epsilon_*[\to C^{0,\alpha}(\partial\Omega^i) \times C^{0,\alpha}(\partial\Omega^o)$$

such that

$$\mathcal{V}(\epsilon)[\Theta^i[\epsilon], \Theta^o[\epsilon]] = (f^i, f^o) \qquad \forall \epsilon \in]-\epsilon_*, \epsilon_*[. \tag{8.14}$$

In particular,

$$(\theta^i_\epsilon, \theta^o_\epsilon) = \Theta[\epsilon] \qquad \forall \epsilon \in]0, \epsilon_*[.$$

Proof. It suffices to take

$$\Theta[\epsilon] \equiv \mathcal{V}(\epsilon)^{(-1)}[f^i, f^o] \qquad \forall \epsilon \in]-\epsilon_*, \epsilon_*[$$

(with ϵ_* as in Lemma 8.5). The last equality follows by Proposition 8.1. □

We can now exploit Proposition 8.6 to deduce real analytic continuation results for the solution u_ϵ of problem (8.4). In the next Theorem 8.7 we examine the behavior of u_ϵ away from the hole $\epsilon\Omega^i$. Instead of looking at the evaluation of u_ϵ at a single point $p \in \Omega^o$, as in question (8.5), we consider the restriction of u_ϵ to a subset Ω_M of $\Omega^o \setminus \{0\}$ with a nonzero distance from 0. The capital 'M' stands for 'macroscopic' and it is opposed to the small 'm,' which corresponds to the 'microscopic' behavior of the forthcoming Theorem 8.8.

Theorem 8.7. *Let Ω_M be an open subset of Ω^o such that $0 \notin \overline{\Omega_M}$. Let $\epsilon_M \in]0, \epsilon_*[$ be such that*

$$\overline{\Omega_M} \cap \epsilon\overline{\Omega^i} = \emptyset \qquad \forall \epsilon \in]-\epsilon_M, \epsilon_M[.$$

Then, there exists a real analytic map

$$\mathcal{U}_M :]-\epsilon_M, \epsilon_M[\to C^{1,\alpha}(\overline{\Omega_M})$$

such that

$$u_{\epsilon|\overline{\Omega_M}} = \mathcal{U}_M[\epsilon] \qquad \forall \epsilon \in]0, \epsilon_M[. \tag{8.15}$$

Moreover,

$$\mathcal{U}_M[0] = u^o_{|\overline{\Omega_M}} , \tag{8.16}$$

where $u^o \in C^{1,\alpha}(\overline{\Omega^o})$ is the unique solution of the unperturbed problem

$$\begin{cases} \Delta u^o = 0 & \text{in } \Omega^o , \\ u^o = f^o & \text{on } \partial\Omega^o . \end{cases} \tag{8.17}$$

Proof. By Proposition 8.1, we have a formula for u_ϵ when $\epsilon \in]0, \epsilon_0[$. Namely, formula (8.10). Then we replace the densities θ^i_ϵ, θ^o_ϵ that appear in the right-hand side of (8.10) with $\Theta^i[\epsilon]$, $\Theta^o[\epsilon]$, which are actually equal to θ^i_ϵ, θ^o_ϵ for $\epsilon \in]0, \epsilon_0[$, and we set $\mathcal{U}_M[\epsilon]$ equal to the resulting expression. Namely, we set

$$\mathcal{U}_M[\epsilon](x) \equiv \epsilon^{n-2} \int_{\partial\Omega^i} S_n(x - \epsilon\eta)\Theta^i[\epsilon](\eta)\, d\sigma_\eta + v_{\Omega^o}[\Theta^o[\epsilon]](x)$$

for all $x \in \overline{\Omega_M}$ and for all $\epsilon \in]-\epsilon_M, \epsilon_M[$. Then by definition, equality (8.15) holds true. To prove (8.16), we observe that $\mathcal{U}_M[0](x) = v_{\Omega^o}[\Theta^o[0]](x)$ for all $x \in \overline{\Omega_M}$ and that

$$V_{\Omega^o}[\Theta^o[0]] = f^o$$

by Lemma 8.3. In particular, $v^+_{\Omega^o}[\Theta^o[0]]$ is the unique solution of the unperturbed problem (8.17). To show that $\epsilon \mapsto \mathcal{U}_M[\epsilon]$ is real analytic we first observe that, by Theorem 4.25 and Proposition 8.6, the map from $]-\epsilon_M, \epsilon_M[$ to $C^{1,\alpha}(\overline{\Omega_M})$ that takes ϵ to $v_{\Omega^o}[\Theta^o[\epsilon]]_{|\overline{\Omega_M}}$ is real analytic. Thus, it remains to prove that the map from $]-\epsilon_M, \epsilon_M[$ to $C^{1,\alpha}(\overline{\Omega_M})$ that takes ϵ to the function

$$\epsilon^{n-2} \int_{\partial\Omega^i} S_n(x - \epsilon\eta)\Theta^i[\epsilon](\eta)\, d\sigma_\eta$$

of $x \in \overline{\Omega_M}$ is also real analytic, a fact that follows from the properties of integral operators with real analytic kernels of Theorem A.28 (i). □

We now turn to the behavior of u_ϵ near the boundary of the hole $\epsilon\Omega^i$. To do so, we consider the rescaled function $u_\epsilon(\epsilon\xi)$ of $\xi \in (\epsilon^{-1}\overline{\Omega^o}) \setminus \Omega^i$ and we prove a real analytic continuation result for the restriction of $u_\epsilon(\epsilon\cdot)$ to the closure of an open subset Ω_m of $\mathbb{R}^n \setminus \overline{\Omega^i}$. As mentioned above, here the letter 'm' stands for 'microscopic,' as opposed to the capital 'M' of Theorem 8.7 that stands for 'macroscopic.'

Theorem 8.8. *Let Ω_m be an open bounded subset of $\mathbb{R}^n \setminus \overline{\Omega^i}$. Let $\epsilon_m \in]0, \epsilon_*[$ be such that*

$$\epsilon\overline{\Omega_m} \subseteq \Omega^o \qquad \forall \epsilon \in]-\epsilon_m, \epsilon_m[.$$

Then, there exists a real analytic map

$$\mathcal{U}_m :]-\epsilon_m, \epsilon_m[\to C^{1,\alpha}(\overline{\Omega_m})$$

such that

$$u_\epsilon(\epsilon\xi) = \mathcal{U}_m[\epsilon](\xi) \qquad \forall \xi \in \overline{\Omega_m}\,,\ \epsilon \in]0, \epsilon_m[. \tag{8.18}$$

Moreover,

$$\mathcal{U}_m[0] = u^i_{|\overline{\Omega_m}} + u^o(0) \tag{8.19}$$

where $u^o \in C^{1,\alpha}(\overline{\Omega^o})$ is the unique solution of (8.17) *and $u^i \in C^{1,\alpha}_{\mathrm{loc}}(\mathbb{R}^n \setminus \Omega^i)$ is the unique solution of*

$$\begin{cases} \Delta u^i = 0 & \text{in } \mathbb{R}^n \setminus \overline{\Omega^i}\,, \\ u^i = f^i - u^o(0) & \text{on } \partial\Omega\,, \\ u^i \text{ is harmonic at infinity.} \end{cases} \tag{8.20}$$

Proof. By Proposition 8.1, we have a formula for u_ϵ when $\epsilon \in]0, \epsilon_0[$. Namely, formula (8.10). Then we compute u_ϵ at $\epsilon\xi$, we change the variable in the first integral, we replace the densities θ^i_ϵ, θ^o_ϵ that appear in the right-hand side of (8.10) with $\Theta^i[\epsilon]$, $\Theta^o[\epsilon]$, which are actually equal to θ^i_ϵ, θ^o_ϵ for $\epsilon \in]0, \epsilon_0[$, and we set $\mathcal{U}_m[\epsilon](\xi)$ equal to the resulting expression. Namely, we set

$$\mathcal{U}_m[\epsilon](\xi) \equiv v_{\Omega^i}[\Theta^i[\epsilon]](\xi) + \int_{\partial\Omega^o} S_n(\epsilon\xi - y)\Theta^o[\epsilon](y)\, d\sigma_y$$

for all $\xi \in \overline{\Omega_m}$ and for all $\epsilon \in]-\epsilon_m, \epsilon_m[$. Then, by the definition of $\mathcal{U}_m[\epsilon]$, equality (8.18) holds true. To prove (8.19), we observe that

$$\mathcal{U}_m[0](\xi) = v_{\Omega^i}[\Theta^i[0]](\xi) + v_{\Omega^o}[\Theta^o[0]](0) \qquad \forall \xi \in \overline{\Omega_m} .$$

Then equality (8.14) with $\epsilon = 0$ and Lemma 8.3 imply that

$$V_{\Omega^i}[\Theta^i[0]] + v_{\Omega^o}[\Theta^o[0]](0) = f^i .$$

We note that $v^+_{\Omega^o}[\Theta^o[0]] = u^o$ by the proof of Theorem 8.7 and that $v^-_{\Omega^i}[\Theta^i[0]]$ is harmonic at infinity by Theorem 4.23 (i). Hence, Theorem 6.1 implies that $v^-_{\Omega^i}[\Theta^i[0]]$ is the unique solution of problem (8.20). To show that $\epsilon \mapsto \mathcal{U}_m[\epsilon]$ is real analytic we first observe that Theorem 4.25 on the continuity of the linear operator $v_{\Omega^o}[\cdot]$ and Proposition 8.6 on the analyticity of $\Theta^i[\cdot]$, imply that the map from $]-\epsilon_m, \epsilon_m[$ to $C^{1,\alpha}(\overline{\Omega_m})$ that takes ϵ to $v_{\Omega^i}[\Theta^i[\epsilon]]_{|\overline{\Omega_m}}$ is real analytic. Thus, it remains to prove that the map from $]-\epsilon_m, \epsilon_m[$ to $C^{1,\alpha}(\overline{\Omega_m})$ that takes ϵ to the function

$$\int_{\partial\Omega^o} S_n(\epsilon\xi - y)\Theta^o[\epsilon](y)\, d\sigma_y$$

of $\xi \in \overline{\Omega_m}$ is also real analytic, a fact that follows from the properties of the integral operators with a real analytic kernel of Theorem A.28 (i). □

Theorems 8.7 and 8.8 can be used to study quantities related to the solution u_ϵ. For example, in the following Theorem 8.9 we consider the energy integral.

Theorem 8.9. *There exist $\epsilon_E \in]0, \epsilon_*[$ and a real analytic function*

$$\mathcal{E} :]-\epsilon_E, \epsilon_E[\to \mathbb{R}$$

such that

$$\int_{\Omega(\epsilon)} |\nabla u_\epsilon|^2\, dx = \mathcal{E}(\epsilon) \qquad \forall \epsilon \in]0, \epsilon_E[\,. \tag{8.21}$$

Moreover,

$$\mathcal{E}(0) = \int_{\Omega^o} |\nabla u^o|^2\, dx\,, \tag{8.22}$$

where $u^o \in C^{1,\alpha}(\overline{\Omega^o})$ is the unique solution of problem (8.17).

Proof. By the first Green Identity of Theorem 4.2 and by the theorem of change of variable in integrals we have

$$\begin{aligned}\int_{\Omega(\epsilon)} |\nabla u_\epsilon|^2\, dx &= \int_{\partial\Omega^o} u_\epsilon\, \nu_{\Omega^o} \cdot \nabla u_\epsilon\, d\sigma - \int_{\epsilon\partial\Omega^i} u_\epsilon\, \nu_{\epsilon\Omega^i} \cdot \nabla u_\epsilon\, d\sigma \\ &= \int_{\partial\Omega^o} u_\epsilon\, \nu_{\Omega^o} \cdot \nabla u_\epsilon\, d\sigma - \epsilon^{n-2} \int_{\partial\Omega^i} u_\epsilon(\epsilon\eta)\, \nu_{\Omega^i}(\eta) \cdot \nabla_\eta\left(u_\epsilon(\epsilon\eta)\right) d\sigma_\eta\end{aligned}$$

for all $\epsilon \in]0, \epsilon_0[$. Then, by recalling that u_ϵ is a solution of (8.4), we have

$$\begin{aligned}&\int_{\Omega(\epsilon)} |\nabla u_\epsilon|^2\, dx \\ &\quad = \int_{\partial\Omega^o} f^o\, \nu_{\Omega^o} \cdot \nabla u_\epsilon\, d\sigma - \epsilon^{n-2} \int_{\partial\Omega^i} f^i(\eta)\, \nu_{\Omega^i}(\eta) \cdot \nabla_\eta\left(u_\epsilon(\epsilon\eta)\right) d\sigma_\eta\end{aligned}$$

for all $\epsilon \in]0, \epsilon_0[$. We now take Ω_M as in Theorem 8.7 and Ω_m as in Theorem 8.8 and we also assume that

$$\partial\Omega^o \subseteq \overline{\Omega_M} \qquad \text{and} \qquad \partial\Omega^i \subseteq \overline{\Omega_m}\,.$$

By exploiting the corresponding maps $\mathcal{U}_M$ and $\mathcal{U}_m$, we can write

$$\int_{\Omega(\epsilon)} |\nabla u_\epsilon|^2 \, dx = \int_{\partial\Omega^o} f^o \, \nu_{\Omega^o} \cdot \nabla \mathcal{U}_M[\epsilon] \, d\sigma - \epsilon^{n-2} \int_{\partial\Omega^i} f^i \, \nu_{\Omega^i} \cdot \nabla \mathcal{U}_m[\epsilon] \, d\sigma$$

for all $\epsilon \in]0, \epsilon_E[$, where $\epsilon_E \equiv \min\{\epsilon_M, \epsilon_m\}$. It is apparent that (8.21) holds with $\mathcal{E}$ defined by

$$\mathcal{E}(\epsilon) \equiv \int_{\partial\Omega^o} f^o \, \nu_{\Omega^o} \cdot \nabla \mathcal{U}_M[\epsilon] \, d\sigma - \epsilon^{n-2} \int_{\partial\Omega^i} f^i \, \nu_{\Omega^i} \cdot \nabla \mathcal{U}_m[\epsilon] \, d\sigma$$

for all $\epsilon \in] - \epsilon_E, \epsilon_E[$.

Then by (8.16) we have

$$\mathcal{E}(0) = \int_{\partial\Omega^o} f^o \, \nu_{\Omega^o} \cdot \nabla u^o \, d\sigma$$

and the validity of (8.22) follows by the first Green Identity of Theorem 4.2 (see also (8.17)).

To verify that $\mathcal{E}$ is real analytic we note that Theorem 8.7 implies that the map $\mathcal{U}_M$ is real analytic from $] - \epsilon_E, \epsilon_E[$ to $C^{1,\alpha}(\overline{\Omega_M})$. Since the map from $C^{1,\alpha}(\overline{\Omega_M})$ to $C^{0,\alpha}(\overline{\Omega_M})^n$ that takes a function to its gradient is linear and continuous, we deduce that the map $\epsilon \mapsto \nabla \mathcal{U}_M[\epsilon]$ is real analytic from $] - \epsilon_E, \epsilon_E[$ to $C^{0,\alpha}(\overline{\Omega_M})^n$. Similarly, we can show that Theorem 8.8 implies that $\epsilon \mapsto \nabla \mathcal{U}_m[\epsilon]$ is real analytic from $] - \epsilon_E, \epsilon_E[$ to $C^{0,\alpha}(\overline{\Omega_m})^n$. Then, by the continuity of the pointwise product in a Hölder space and of the linear map that takes a function of $C^{0,\alpha}(\overline{\Omega_M})$ to its integral over $\partial\Omega^o$ and of the linear map that takes a function of $C^{0,\alpha}(\overline{\Omega_m})$ to its integral over $\partial\Omega^i$, we conclude that $\mathcal{E}$ is real analytic. □

In the specific case where $f^o = 0$ we can also prove that the energy integral vanishes as ϵ^{n-2}.

Theorem 8.10. *If $f^o = 0$, then there exist $\epsilon_E \in]0, \epsilon_*[$ and a real analytic function*

$$\mathcal{H} :] - \epsilon_E, \epsilon_E[\to \mathbb{R}$$

such that

$$\int_{\Omega(\epsilon)} |\nabla u_\epsilon|^2 \, dx = \epsilon^{n-2} \mathcal{H}(\epsilon) \qquad \forall \epsilon \in]0, \epsilon_E[\,. \tag{8.23}$$

Moreover,

$$\mathcal{H}(0) = \int_{\mathbb{R}^n \setminus \Omega^i} |\nabla u^i|^2 \, d\xi \,, \tag{8.24}$$

where $u^i \in C^{1,\alpha}_{\mathrm{loc}}(\mathbb{R}^n \setminus \Omega^i)$ is the unique solution of (8.20).

Proof. The proof can be effected following the footsteps of the proof of Theorem 8.9 and taking

$$\mathcal{H}(\epsilon) \equiv - \int_{\partial\Omega^i} f^i \, \nu_{\Omega^i} \cdot \nabla \mathcal{U}_m[\epsilon] \, d\sigma \qquad \forall \epsilon \in] - \epsilon_E, \epsilon_E[\,.$$

Then, by (8.19) and noting that $u^o = 0$, we have

$$\mathcal{H}(0) = \int_{\partial\Omega^i} f^i \, \nu_{\Omega^i} \cdot \nabla u^i \, d\sigma \,.$$

The validity of (8.24) follows by a computation based on the first Green Identity for exterior domains of Corollary 4.7 (indeed u^i is harmonic, harmonic at infinity, and $u^i = f^i$ on $\partial\Omega^i$, cf. (8.20)). To show that $\mathcal{H}$ is real analytic we argue as in the proof of Theorem 8.9 for $\mathcal{E}$. □

We can now use Theorem 8.10 to study the electrostatic capacity of $\epsilon\overline{\Omega^i}$ with respect to Ω^o, which is defined as the energy integral of the solution of (8.4) with $f^i = 1$ and $f^o = 0$. Namely,

$$\mathrm{Cap}[\epsilon\overline{\Omega^i}, \Omega^o] \equiv \int_{\Omega(\epsilon)} |\nabla v_\epsilon|^2 \, dx \tag{8.25}$$

where v_ϵ is the so-called capacity solution, i.e., the unique solution of

$$\begin{cases} \Delta v_\epsilon = 0 & \text{in } \Omega(\epsilon)\,, \\ v_\epsilon = 1 & \text{on } \epsilon\partial\Omega^i\,, \\ v_\epsilon = 0 & \text{on } \partial\Omega^o\,. \end{cases}$$

We denote by $\mathrm{Cap}[\overline{\Omega^i}, \mathbb{R}^n]$ the electrostatic capacity of $\overline{\Omega^i}$ in $\mathbb{R}^n$. Namely, we set

$$\mathrm{Cap}[\overline{\Omega^i}, \mathbb{R}^n] \equiv \int_{\mathbb{R}^n \setminus \Omega^i} |\nabla v^i|^2 \, dx$$

where v^i is the unique solution of

$$\begin{cases} \Delta v^i = 0 & \text{in } \mathbb{R}^n \setminus \Omega^i\,, \\ v^i = 1 & \text{on } \partial\Omega\,, \\ v^i \text{ is harmonic at infinity.} & \end{cases}$$

Then Theorem 8.10 implies the validity of the following.

Corollary 8.11. *There exist* $\epsilon_C \in]0, \epsilon_*[$ *and a real analytic function*

$$\mathcal{C} :]-\epsilon_C, \epsilon_C[\to \mathbb{R}$$

such that

$$\mathrm{Cap}[\epsilon\overline{\Omega^i}, \Omega^o] = \epsilon^{n-2}\mathcal{C}(\epsilon) \qquad \forall \epsilon \in]0, \epsilon_C[\,.$$

Moreover,

$$\mathcal{C}(0) = \mathrm{Cap}[\overline{\Omega^i}, \mathbb{R}^n]$$

Finally, we wish to present a global representation of u_ϵ in a fashion that we can directly compare with the results obtained by the expansion methods of Asymptotic

Analysis. To do so, we need to employ two real analytic maps $\mathcal{U}^o$ and $\mathcal{U}^i$ instead of just one, as we have done in Theorems 8.7 and 8.8. The map $\mathcal{U}^o$ will provide functions in the 'slow' variable x while $\mathcal{U}^i$ takes care of the 'fast' variable $\xi = x/\epsilon$.

Theorem 8.12. *There exist two uniquely determined real analytic maps*

$$\mathcal{U}^o :]-\epsilon_*, \epsilon_*[\to C^{1,\alpha}(\overline{\Omega^o})$$

and

$$\mathcal{U}^i :]-\epsilon_*, \epsilon_*[\to C_b^{1,\alpha}(\mathbb{R}^n \setminus \Omega^i)$$

such that

- $\mathcal{U}^o[\epsilon]$ *is harmonic in* Ω^o *and* $\mathcal{U}^i[\epsilon]$ *is harmonic in* $\mathbb{R}^n \setminus \overline{\Omega^i}$ *and at infinity for all* $\epsilon \in]-\epsilon_*, \epsilon_*[$,
- *we have*

$$u_\epsilon(x) = \mathcal{U}^o[\epsilon](x) + \mathcal{U}^i[\epsilon](x/\epsilon) \qquad \forall x \in \overline{\Omega(\epsilon)} \tag{8.26}$$

  for all $\epsilon \in]0, \epsilon_*[$.

Moreover,

$$\mathcal{U}^o[0] = u^o \qquad \text{and} \qquad \mathcal{U}^i[0] = u^i\,, \tag{8.27}$$

where u^o *and* u^i *are as in* (8.17) *and* (8.20).

Proof. We define

$$\mathcal{U}^o[\epsilon] \equiv v^+_{\Omega^o}[\Theta^o[\epsilon]] \quad \text{and} \quad \mathcal{U}^i[\epsilon] \equiv v^-_{\Omega^i}[\Theta^i[\epsilon]] \qquad \forall \epsilon \in]-\epsilon_*, \epsilon_*[\,.$$

Then, Theorem 4.25 and Corollary 4.26 on the continuity of the linear operators $v^+_{\Omega^o}[\cdot]$, $v^-_{\Omega^i}[\cdot]$ and Proposition 8.6 on the analyticity of Θ^o, Θ^i imply that $\mathcal{U}^o$ and $\mathcal{U}^i$ are real analytic. The validity of (8.26) follows by the representation formula (8.10) and by a computation based on the theorem of change of variables in integrals. Equality (8.27) follows by Lemma 8.3 and by the definition (8.17) of u^o and (8.20) of u^i. To show that $\mathcal{U}^o$ and $\mathcal{U}^i$ are unique we observe that, if $u \in C^{1,\alpha}(\overline{\Omega^o})$ is harmonic in Ω^o and $v \in C_b^{1,\alpha}(\mathbb{R}^n \setminus \Omega^i)$ is harmonic in $\mathbb{R}^n \setminus \overline{\Omega^i}$ and at infinity, and we have

$$u(x) + v(x/\epsilon) = u_\epsilon(x) \qquad \forall x \in \Omega(\epsilon)$$

for some $\epsilon \in]0, \epsilon_0[$, then

$$\mathcal{U}^i[\epsilon](x/\epsilon) - v(x/\epsilon) = u(x) - \mathcal{U}^o[\epsilon](x) \qquad \forall x \in \Omega(\epsilon)\,.$$

We deduce that the function $\mathcal{U}^i[\epsilon](\cdot/\epsilon) - v(\cdot/\epsilon)$ has an harmonic continuation to the whole of $\mathbb{R}^n$. Since such function is harmonic at infinity, it must be identically equal to zero by the Liouville Theorem 3.10. Hence $u = \mathcal{U}^o[\epsilon]$ and $v = \mathcal{U}^i[\epsilon]$. □

As a consequence of Theorem 8.12 there exist $\epsilon_{**} \in]0, \epsilon_*[$, a sequence of functions $\{u^o_j\}_{j\in\mathbb{N}}$ in $C^{1,\alpha}(\overline{\Omega^o})$, and a sequence of functions $\{u^i_j\}_{j\in\mathbb{N}}$ in $C_b^{1,\alpha}(\mathbb{R}^n \setminus \Omega^i)$ such that the series

$$\sum_{j=0}^{\infty} \epsilon^j u_j^o \qquad \text{and} \qquad \sum_{j=0}^{\infty} \epsilon^j\, u_j^i$$

converge in $C^{1,\alpha}(\overline{\Omega^o})$ and in $C^{1,\alpha}_{\mathrm{b}}(\mathbb{R}^n \setminus \Omega^i)$, respectively, for all $\epsilon \in]-\epsilon_{**}, \epsilon_{**}[$ and such that

$$u_\epsilon(x) = \sum_{j=0}^{\infty} \epsilon^j u_j^o(x) + \sum_{j=0}^{\infty} \epsilon^j\, u_j^i(x/\epsilon) \qquad \forall x \in \overline{\Omega(\epsilon)} \tag{8.28}$$

for all $\epsilon \in]0, \epsilon_{**}[$. Moreover, by the uniqueness of $\mathcal{U}^o$ and $\mathcal{U}^i$ we can verify that such families $\{u_j^o\}_{j\in\mathbb{N}}$ and $\{u_j^i\}_{j\in\mathbb{N}}$ are uniquely determined by (8.28). (We could also prove that the u_j^o's are harmonic and the u_j^i's are harmonic and harmonic at infinity.)

At this point, we may want to compute explicitly the coefficients $\{u_j^o\}_{j\in\mathbb{N}}$ and $\{u_j^i\}_{j\in\mathbb{N}}$. To do so, one possibility is to exploit equation (8.9) and the representation formula (8.10) and obtain the coefficients by solving a recursive system of integral equations, as we shall see in Section 8.5 for the two-dimensional case. We can also, however, exploit the approximation methods that are usually employed in Asymptotic Analysis. Indeed, we have the following Proposition 8.13, that can be proved by a standard inductive argument (see, e.g., [30]).

Proposition 8.13. *Let $N \in \mathbb{N}$. Let $\{\tilde{u}_j^o\}_{j\in\{1,\dots,N\}}$ be a set of continuous functions on Ω^o and $\{\tilde{u}_j^i\}_{j\in\{1,\dots,N\}}$ be a set consisting of continuous functions on $\mathbb{R}^n \setminus \overline{\Omega^i}$ that vanish at infinity. If the asymptotic approximation formula*

$$\left\| u_\epsilon - \sum_{j=0}^{N} \epsilon^j \tilde{u}_j^o + \sum_{j=0}^{N} \epsilon^j\, \tilde{u}_j^i(\cdot/\epsilon) \right\|_{L^\infty(\Omega(\epsilon))} = o(\epsilon^N) \qquad \text{as } \epsilon \to 0^+ \tag{8.29}$$

holds, then $\tilde{u}_j^o = u_j^o$ and $\tilde{u}_j^i = u_j^i$ for all $j \in \{1, \dots, N\}$.

In a sense, Proposition 8.13 says that any suitable approximation of u_ϵ given by the sum of continuous function of the slow variable x and continuous functions of the fast variable x/ϵ has to coincide with the one that we can deduce taking partial sums in the expansion (8.28). Asymptotic approximations like the one in (8.29) can be obtained for example by the Compound Expansions Method of Maz'ya, Nazarov and Plamenewskii [203, 204]. We also remark that the validity of (8.29) alone does not imply that the series

$$\sum_{j=0}^{\infty} \epsilon^j \tilde{u}_j^o \qquad \text{and} \qquad \sum_{j=0}^{\infty} \epsilon^j\, \tilde{u}_j^i$$

converge. A result that instead follows by Theorem 8.12 and Proposition 8.13.

8.4 Analysis for $n = 2$

We now turn to the two-dimensional case. The analysis will be more involved than the one seen in the previous Section 8.3 for $n = 3$ or greater. The reason is that we will have to deal with a logarithmic behavior of the solution u_ϵ generated by the particular form of the fundamental solution $S_2(x) = (2\pi)^{-1} \log |x|$. Indeed, for $n = 2$ we have

$$S_2(\epsilon x) = S_2(x) + \frac{\log |\epsilon|}{2\pi} \tag{8.30}$$

for all $x \in \mathbb{R}^2 \setminus \{0\}$ and all $\epsilon \neq 0$, whereas for $n \geq 3$ the fundamental solution S_n is a homogeneous function of degree $2-n$. To handle such logarithmic behavior we find convenient to adopt the integral representation (6.19) for the solution u_ϵ. Then, writing (6.19) for the specific case when $\Omega = \Omega(\epsilon)$ has just one hole, we obtain

$$u_\epsilon = w^+_{\Omega(\epsilon)}[\mu_\epsilon] + \frac{\left(\int_{\partial\Omega(\epsilon)} f_\epsilon \, \tau_\epsilon \, d\sigma\right) v^+_{\Omega(\epsilon)}[\tau_\epsilon]}{\int_{\epsilon\partial\Omega^i} V_{\Omega(\epsilon)}[\tau_\epsilon] \, d\sigma}, \tag{8.31}$$

where $f_\epsilon \in C^{1,\alpha}(\partial\Omega(\epsilon))$ is defined by

$$f_\epsilon(x) \equiv \begin{cases} f^i(x/\epsilon) & \text{if } x \in \epsilon\partial\Omega^i, \\ f^o(x) & \text{if } x \in \partial\Omega^o \end{cases}$$

and where $\tau_\epsilon \in C^{0,\alpha}(\partial\Omega(\epsilon))$ and $\mu_\epsilon \in C^{1,\alpha}(\partial\Omega(\epsilon))$ are the (unique) solutions of the systems of integral equations

$$\begin{cases} \left(\frac{1}{2}I + W^t_{\Omega(\epsilon)}\right)[\tau_\epsilon] = 0, \\ \int_{\epsilon\partial\Omega^i} \tau_\epsilon \, d\sigma = 1 \end{cases} \tag{8.32}$$

and

$$\begin{cases} \left(\frac{1}{2}I + W_{\Omega(\epsilon)}\right)[\mu_\epsilon] = f_\epsilon - \int_{\partial\Omega(\epsilon)} f_\epsilon \tau_\epsilon \, d\sigma \, \zeta_{\epsilon\partial\Omega^i}, \\ \int_{\epsilon\partial\Omega^i} \mu_\epsilon \, d\sigma = 0, \end{cases} \tag{8.33}$$

respectively. Here the function $\zeta_{\epsilon\partial\Omega^i}$ is defined as

$$\zeta_{\epsilon\partial\Omega^i}(x) \equiv \begin{cases} 1 & \text{if } x \in \epsilon\partial\Omega^i, \\ 0 & \text{if } x \in \partial\Omega^o \end{cases}$$

(cf. (6.17)). Incidentally, we observe that by Lemma 6.30 the denominator in (8.31) is different from 0, namely we have

$$\int_{\epsilon\partial\Omega^i} V_{\Omega(\epsilon)}[\tau_\epsilon] \, d\sigma \neq 0 \qquad \forall \epsilon \in]0, \epsilon_0[\,. \tag{8.34}$$

Our analysis will proceed in two steps. First we will analyze system (8.32) and prove a real analytic continuation result for τ_ϵ. Then we will consider system (8.33) and prove a similar result also for μ_ϵ. At this point we will be ready to study the behavior of the solution u_ϵ by exploiting the representation (8.31).

8.4.1 Analysis of System (8.32)

By splitting the first equation of (8.32) into an equation on $\epsilon\partial\Omega^i$ and one on $\partial\Omega^o$, we obtain the following two equations

$$\begin{aligned}
&\frac{1}{2}\tau_\epsilon(x) - \int_{\epsilon\partial\Omega^i} \tau_\epsilon(y)\, \nu_{\epsilon\Omega^i}(x)\cdot\nabla S_2(x-y)\, d\sigma_y \\
&\qquad - \int_{\partial\Omega^o} \tau_\epsilon(y)\, \nu_{\epsilon\Omega^i}(x)\cdot\nabla S_2(x-y)\, d\sigma_y = 0 \qquad \forall x \in \epsilon\partial\Omega^i\,, \\
&\frac{1}{2}\tau_\epsilon(x) + \int_{\partial\Omega^o} \tau_\epsilon(y)\, \nu_{\Omega^o}(x)\cdot\nabla S_2(x-y)\, d\sigma_y \\
&\qquad + \int_{\epsilon\partial\Omega^i} \tau_\epsilon(y)\, \nu_{\Omega^o}(x)\cdot\nabla S_2(x-y)\, d\sigma_y = 0 \qquad \forall x \in \partial\Omega^o
\end{aligned} \tag{8.35}$$

(note that $\nu_{\Omega(\epsilon)}(x) = -\nu_{\epsilon\Omega^i}(x)$ for $x \in \epsilon\partial\Omega^i$). Then, by passing to the variable

$$\xi = x/\epsilon \in \partial\Omega^i$$

in the first equation of (8.35), by changing the variable of integration in order to transform the integrals over $\epsilon\partial\Omega^i$ into integrals over $\partial\Omega^i$, and by exploiting the equality

$$\nabla S_2(\epsilon\xi - \epsilon\eta) = \epsilon^{-1}\nabla S_2(\xi-\eta) \qquad \forall \xi,\eta \in \partial\Omega^i\,,\ \xi \neq \eta\,,\ \text{and } \epsilon > 0\,, \tag{8.36}$$

we verify that the system in (8.32) is equivalent to the three equations

$$\begin{aligned}
&\frac{1}{2}\tau_\epsilon(\epsilon\xi) - \int_{\partial\Omega^i} \tau_\epsilon(\epsilon\eta)\, \nu_{\Omega^i}(\xi)\cdot\nabla S_2(\xi-\eta)\, d\sigma_\eta \\
&\qquad - \int_{\partial\Omega^o} \tau_\epsilon(y)\, \nu_{\Omega^i}(\xi)\cdot\nabla S_2(\epsilon\xi-y)\, d\sigma_y = 0 \qquad \forall \xi \in \partial\Omega^i\,, \\
&\frac{1}{2}\tau_\epsilon(x) + \int_{\partial\Omega^o} \tau_\epsilon(y)\, \nu_{\Omega^o}(x)\cdot\nabla S_2(x-y)\, d\sigma_y \\
&\qquad + \epsilon\int_{\partial\Omega^i} \tau_\epsilon(\epsilon\eta)\, \nu_{\Omega^o}(x)\cdot\nabla S_2(x-\epsilon\eta)\, d\sigma_\eta = 0 \quad \forall x \in \partial\Omega^o\,, \\
&\int_{\partial\Omega^i} \epsilon\tau_\epsilon(\epsilon\eta)\, d\sigma_\eta = 1\,.
\end{aligned}$$

Now, taking

$$\phi^i_\epsilon(\xi) \equiv \epsilon \tau_\epsilon(\epsilon\xi) \quad \forall \xi \in \partial\Omega^i \qquad \text{and} \qquad \phi^o_\epsilon(x) \equiv \tau_\epsilon(x) \quad \forall x \in \partial\Omega^o\,, \tag{8.37}$$

we can rewrite the system as

$$\begin{aligned}
&\frac{1}{2}\phi^i_\epsilon(\xi) - W^t_{\Omega^i}[\phi^i_\epsilon](\xi) - \epsilon \int_{\partial\Omega^o} \phi^o_\epsilon(y)\, \nu_{\Omega^i}(\xi) \cdot \nabla S_2(\epsilon\xi - y)\, d\sigma_y = 0 \\
&\qquad\qquad\qquad\qquad\qquad\qquad\qquad\qquad\qquad \forall \xi \in \partial\Omega^i\,, \\
&\frac{1}{2}\phi^o_\epsilon(x) + W^t_{\Omega^o}[\phi^o_\epsilon](x) + \int_{\partial\Omega^i} \phi^i_\epsilon(\eta)\, \nu_{\Omega^o}(x) \cdot \nabla S_2(x - \epsilon\eta)\, d\sigma_\eta = 0 \\
&\qquad\qquad\qquad\qquad\qquad\qquad\qquad\qquad\qquad \forall x \in \partial\Omega^o\,, \\
&\int_{\partial\Omega^i} \phi^i_\epsilon\, d\sigma = 1\,.
\end{aligned} \tag{8.38}$$

Then, to write system (8.38) in a more convenient form, we introduce the auxiliary operator

$$\mathcal{N}(\epsilon) \equiv (\mathcal{N}^i(\epsilon), \mathcal{N}^o(\epsilon), \mathcal{N}^s(\epsilon))$$

defined by

$$\begin{aligned}
\mathcal{N}^i(\epsilon)[\phi^i, \phi^o](\xi) \equiv& \frac{1}{2}\phi^i(\xi) - W^t_{\Omega^i}[\phi^i](\xi) \\
&- \epsilon \int_{\partial\Omega^o} \phi^o(y)\, \nu_{\Omega^i}(\xi) \cdot \nabla S(\epsilon\xi - y)\, d\sigma_y \qquad \forall \xi \in \partial\Omega^i\,,
\end{aligned} \tag{8.39}$$

$$\begin{aligned}
\mathcal{N}^o(\epsilon)[\phi^i, \phi^o](x) \equiv& \frac{1}{2}\phi^o(x) + W^t_{\Omega^o}[\phi^o](x) \\
&+ \int_{\partial\Omega^i} \phi^i(\eta)\, \nu_{\Omega^o}(x) \cdot \nabla S(x - \epsilon\eta)\, d\sigma_\eta \qquad \forall x \in \partial\Omega^o\,,
\end{aligned} \tag{8.40}$$

$$\mathcal{N}^s(\epsilon)[\phi^i, \phi^o] \equiv \int_{\partial\Omega^i} \phi^i\, d\sigma\,, \tag{8.41}$$

for all $\epsilon \in]-\epsilon_0, \epsilon_0[$ and all $(\phi^i, \phi^o) \in C^{0,\alpha}(\partial\Omega^i) \times C^{0,\alpha}(\partial\Omega^o)$ (the superscript 's' in $\mathcal{N}^s$ stands for 'scalar'). It is apparent that system (8.32) is equivalent to the equation

$$\mathcal{N}(\epsilon)[\phi^i_\epsilon, \phi^o_\epsilon] = (0, 0, 1) \tag{8.42}$$

with $\phi^i_\epsilon, \phi^o_\epsilon$ as in (8.37). Hence, Lemma 6.28 implies the validity of the following.

Proposition 8.14. *For all $\epsilon \in]0, \epsilon_0[$ there is one and only one solution $(\phi^i_\epsilon, \phi^o_\epsilon)$ in $C^{0,\alpha}(\partial\Omega^i) \times C^{0,\alpha}(\partial\Omega^o)$ of equation* (8.42).

It follows that the map from $]0, \epsilon_0[$ to $C^{0,\alpha}(\partial\Omega^i) \times C^{0,\alpha}(\partial\Omega^o)$ that takes ϵ to $(\phi^i_\epsilon, \phi^o_\epsilon)$ is well-defined. Our aim in this subsection is to prove that such map has a real analytic continuation in a open neighborhood of $\epsilon = 0$. To do so, we first show that $\mathcal{N}$ is real analytic, then we prove that $\mathcal{N}(0)$ is an isomorphism, and finally we exploit the equality $(\phi^i_\epsilon, \phi^o_\epsilon) = \mathcal{N}(\epsilon)^{(-1)}(0, 0, 1)$, in a fashion that resembles what we have done in Section 8.3 to prove that $\epsilon \mapsto (\theta^i_\epsilon, \theta^o_\epsilon)$ has a real analytic continuation. We begin with the following technical lemma.

Lemma 8.15. *For all $\epsilon \in]-\epsilon_0, \epsilon_0[$ and all $(\phi^i, \phi^o) \in C^{0,\alpha}(\partial\Omega^i) \times C^{0,\alpha}(\partial\Omega^o)$ the function $\mathcal{N}^i(\epsilon)[\phi^i, \phi^o]$ belongs to $C^{0,\alpha}(\partial\Omega^i)_0$.*

Proof. The membership of $\mathcal{N}^i(\epsilon)[\phi^i, \phi^o]$ in $C^{0,\alpha}(\partial\Omega^i)$ is a consequence of the continuity of the linear operator W^t of Theorem 6.7 and of the properties of the integral operators with real analytic kernel of Theorem A.28 (ii). To prove that

$$\int_{\partial\Omega^i} \mathcal{N}^i(\epsilon)[\phi^i, \phi^o]\, d\sigma = 0 \tag{8.43}$$

we observe that, by the jump formulas for the normal derivative of the single layer potential of Theorem 6.6 and by the first Green Identity of Theorem 4.2, we have

$$\int_{\partial\Omega^i} \left(\frac{1}{2}\phi^i - W^t_{\Omega^i}[\phi^i]\right) d\sigma = -\int_{\partial\Omega^i} \nu_{\Omega^i} \cdot \nabla v^+_{\Omega^i}[\phi^i]\, d\sigma = 0\,. \tag{8.44}$$

Hence, the above equality (8.43) holds true for $\epsilon = 0$. If $\epsilon \in]-\epsilon_0, \epsilon_0[\setminus \{0\}$, then the classical theorem of differentiability for integrals depending on a parameter and the chain rule imply that

$$\begin{aligned}
&\epsilon \int_{\partial\Omega^o} \phi^o(y)\, \nu_{\Omega^i}(x) \cdot \nabla S(\epsilon\xi - y)\, d\sigma_y \\
&\qquad = \nu_{\Omega^i}(\xi) \cdot \nabla_\xi \left(\int_{\partial\Omega^o} \phi^o(y)\, S(\epsilon\xi - y)\, d\sigma_y \right) \qquad \forall \xi \in \partial\Omega^i\,.
\end{aligned}$$

Since the function that takes ξ to

$$\int_{\partial\Omega^o} \phi^o(y)\, S(\epsilon\xi - y)\, d\sigma_y = v^+_{\Omega^o}[\phi^o](\epsilon\xi)$$

is harmonic in a neighborhood of $\overline{\Omega^i}$ for $\epsilon \in]-\epsilon_0, \epsilon_0[$, the first Green Identity of Theorem 4.2 implies that

$$\begin{aligned}
&\int_{\partial\Omega^i} \epsilon \int_{\partial\Omega^o} \phi^o(y)\, \nu_{\Omega^i}(x) \cdot \nabla S(\epsilon\xi - y)\, d\sigma_y\, d\sigma_\xi \\
&\qquad = \int_{\partial\Omega^i} \nu_{\Omega^i}(\xi) \cdot \nabla_\xi \left(\int_{\partial\Omega^o} \phi^o(y)\, S(\epsilon\xi - y)\, d\sigma_y \right) d\sigma_\xi = 0\,.
\end{aligned}$$

Then equality (8.44) and the definition of $\mathcal{N}^i$ in (8.39) imply that (8.43) holds true, and the proof is complete. □

Then we can show that the map that takes ϵ to $\mathcal{N}(\epsilon)$ is real analytic.

Lemma 8.16. *The map from* $]-\epsilon_0,\epsilon_0[$ *to*

$$\mathcal{L}(C^{0,\alpha}(\partial\Omega^i)\times C^{0,\alpha}(\partial\Omega^o)\,,\,C^{0,\alpha}(\partial\Omega^i)_0\times C^{0,\alpha}(\partial\Omega^o)\times\mathbb{R})$$

that takes ϵ *to* $\mathcal{N}(\epsilon)$ *is real analytic.*

Proof. We first consider $\mathcal{N}^i$. We can write

$$\mathcal{N}^i(\epsilon)[\phi^i,\phi^o]=\left(\frac{1}{2}I-W^t_{\Omega^i}\right)[\phi^i]-\epsilon\tilde{\mathcal{N}}^i(\epsilon)[\phi^o]$$

with $\tilde{\mathcal{N}}^i$ defined by

$$\tilde{\mathcal{N}}^i(\epsilon)[\phi^o](\xi)\equiv\int_{\partial\Omega^o}\phi^o(y)\,\nu_{\Omega^i}(\xi)\cdot\nabla S(\epsilon\xi-y)\,d\sigma_y\qquad\forall\xi\in\partial\Omega^i$$

for all $\epsilon\in]-\epsilon_0,\epsilon_0[$ and all $\phi^o\in C^{0,\alpha}(\partial\Omega^o)$. By Theorem 6.7, $\frac{1}{2}I-W^t_{\Omega^i}$ is a bounded linear operator from $C^{0,\alpha}(\partial\Omega^i)$ to itself. Moreover, it does not depend on ϵ. By Corollary A.29 in the Appendix, the map from $]-\epsilon_0,\epsilon_0[$ to $\mathcal{L}(L^1(\partial\Omega^o),C^{1,\alpha}(\partial\Omega^i))$ that takes ϵ to $\tilde{\mathcal{N}}^i(\epsilon)$ is real analytic. Then, by the continuous embedding of $C^{0,\alpha}(\partial\Omega^o)$ into $L^1(\partial\Omega^o)$ and of $C^{1,\alpha}(\partial\Omega^i)$ into $C^{0,\alpha}(\partial\Omega^i)$ we deduce that $\tilde{\mathcal{N}}^i$ is real analytic from $]-\epsilon_0,\epsilon_0[$ to $\mathcal{L}(C^{0,\alpha}(\partial\Omega^o),C^{0,\alpha}(\partial\Omega^i))$. Since $C^{0,\alpha}(\partial\Omega^i)_0$ is a closed subspace of $C^{0,\alpha}(\partial\Omega^i)$, Lemma 8.15 implies that $\mathcal{N}^i$ is real analytic from $]-\epsilon_0,\epsilon_0[$ to $\mathcal{L}(C^{0,\alpha}(\partial\Omega^i)\times C^{0,\alpha}(\partial\Omega^o),C^{0,\alpha}(\partial\Omega^i)_0)$.

Then we verify that $\mathcal{N}^o$ is real analytic from $]-\epsilon_0,\epsilon_0[$ to

$$\mathcal{L}(C^{0,\alpha}(\partial\Omega^i)\times C^{0,\alpha}(\partial\Omega^o),C^{0,\alpha}(\partial\Omega^o))\,.$$

Indeed, $\frac{1}{2}I+W^t_{\Omega^o}$ is linear and bounded from $C^{0,\alpha}(\partial\Omega^o)$ to itself by Theorem 6.7 (and it does not depend on ϵ) and the map from the interval $]-\epsilon_0,\epsilon_0[$ to $\mathcal{L}(C^{0,\alpha}(\partial\Omega^i),C^{0,\alpha}(\partial\Omega^o))$ that takes ϵ to the operator

$$\tilde{\mathcal{N}}^o(\epsilon)[\phi^i](x)\equiv\int_{\partial\Omega^i}\phi^i(\eta)\,\nu_{\Omega^o}(x)\cdot\nabla S(x-\epsilon\eta)\,d\sigma_\eta\qquad\forall x\in\partial\Omega^o$$

is real analytic by Corollary A.29 in the Appendix and by the continuous embedding of $C^{0,\alpha}(\partial\Omega^i)$ into $L^1(\partial\Omega^i)$ and of $C^{1,\alpha}(\partial\Omega^o)$ into $C^{0,\alpha}(\partial\Omega^o)$.

Finally, the map that takes a function of $C^{0,\alpha}(\partial\Omega^i)$ into its integral over $\partial\Omega^i$ is linear and continuous and we deduce that also $\mathcal{N}^s$ is real analytic. The lemma is now proven. □

We now consider the equation $\mathcal{N}(0)[\phi^i,\phi^o]=(g^i,g^o,\rho)$, write an equivalent system of boundary integral equations, and deduce that both the system and the equation have unique solutions. Later, we will use this result to prove that $\mathcal{N}(0)$ is an isomorphism.

Lemma 8.17. *For all* (g^i,g^o,ρ) *in* $C^{0,\alpha}(\partial\Omega^i)_0\times C^{0,\alpha}(\partial\Omega^o)\times\mathbb{R}$ *and* (ϕ^i,ϕ^o) *in* $C^{0,\alpha}(\partial\Omega^i)\times C^{0,\alpha}(\partial\Omega^o)$ *we have*

$$\mathcal{N}(0)[\phi^i, \phi^o] = (g^i, g^o, \rho) \tag{8.45}$$

if and only if

$$\begin{cases} -\frac{1}{2}\phi^i + W^t_{\Omega^i}[\phi^i] = -g^i\,, \\ \frac{1}{2}\phi^o + W^t_{\Omega^o}[\phi^o] = -\rho\,\nu_{\Omega^o} \cdot (\nabla S_2)_{|\partial\Omega^o} + g^o\,, \\ \int_{\partial\Omega^i} \phi^i\, d\sigma = \rho\,. \end{cases} \tag{8.46}$$

In particular, for all $(g^i, g^o, \rho) \in C^{0,\alpha}(\partial\Omega^i)_0 \times C^{0,\alpha}(\partial\Omega^o) \times \mathbb{R}$ *there exists one and only one solution* $(\phi^i, \phi^o) \in C^{0,\alpha}(\partial\Omega^i) \times C^{0,\alpha}(\partial\Omega^o)$ *of* (8.45).

Proof. The equivalence of (8.45) and (8.46) follows by the definition of $\mathcal{N}$ in (8.39)–(8.41). We now prove the last statement of the lemma. Since $\mathbb{R}^2 \setminus \Omega^o$ is connected, Corollary 6.15 implies that the operator $\frac{1}{2}I + W^t_{\Omega^o}$ is an isomorphism from $C^{0,\alpha}(\partial\Omega^o)$ to itself. As a consequence, there exists one and only one solution ϕ^o of the second equation of (8.46) (also note that $\nu^o \in C^{0,\alpha}(\partial\Omega^o, \mathbb{R}^n)$). We now consider the first equation of (8.46). By the membership of g^i in $C^{0,\alpha}(\partial\Omega^i)_0$ we have $\int_{\partial\Omega^i} g^i\, d\sigma = 0$. Since Ω^i is connected, it follows that $-g^i$ is orthogonal to the null space of $-\frac{1}{2}I + W_{\Omega^i}$, which consists of the constant functions on $\partial\Omega^i$ (cf. Lemma 6.21). Thus, by the Fredholm Alternative of Theorem 5.8, there exists $\tilde{\phi}^i \in C^{0,\alpha}(\partial\Omega^i)$ that satisfies the first equation of (8.46). The Fredholm Alternative of Theorem 5.8 also implies that the null space of $-\frac{1}{2}I + W^t_{\Omega^i}$ is one-dimensional and is accordingly spanned by a certain function $\phi \in C^{0,\alpha}(\partial\Omega^i)$. By Theorem 6.25, we have $\int_{\partial\Omega^i} \phi\, d\sigma \neq 0$. Then we can verify that the function

$$\phi^i \equiv \tilde{\phi}^i + \frac{\rho - \int_{\partial\Omega^i} \tilde{\phi}^i\, d\sigma}{\int_{\partial\Omega^i} \phi\, d\sigma}\phi$$

is a solution of the first and third equation of (8.46). Finally, Theorem 6.25 implies that ϕ^i is unique. □

Then we can prove that $\mathcal{N}(0)$ is an isomorphism.

Lemma 8.18. *The operator* $\mathcal{N}(0)$ *is an isomorphism from* $C^{0,\alpha}(\partial\Omega^i) \times C^{0,\alpha}(\partial\Omega^o)$ *to* $C^{0,\alpha}(\partial\Omega^i)_0 \times C^{0,\alpha}(\partial\Omega^o) \times \mathbb{R}$.

Proof. $\mathcal{N}(0)$ is a linear bounded operator by Lemma 8.16. By the Open Mapping Theorem 2.2, it is sufficient to verify that it is one to one and onto in order to prove that it is an isomorphism. Then the statement follows by Lemma 8.17. □

Now, since the set of invertible operators is open in the space of bounded operators and since the map that takes an invertible operator to its inverse is real analytic (cf. Proposition A.15), we have the following.

Lemma 8.19. *There is $\epsilon_{*1} \in]0, \epsilon_0[$ such that $\mathcal{N}(\epsilon)$ is invertible for all $\epsilon \in]-\epsilon_{*1}, \epsilon_{*1}[$ and the map from $]-\epsilon_{*1}, \epsilon_{*1}[$ to*

$$\mathcal{L}(C^{0,\alpha}(\partial\Omega^i)_0 \times C^{0,\alpha}(\partial\Omega^o) \times \mathbb{R}\,,\, C^{0,\alpha}(\partial\Omega^i) \times C^{0,\alpha}(\partial\Omega^o))$$

that takes ϵ to $\mathcal{N}(\epsilon)^{(-1)}$ is real analytic.

Proof. It follows by Lemmas 8.16 and 8.18 and by Proposition A.15. □

Then we can exploit Lemma 8.19 to prove that the map $\epsilon \mapsto (\phi^i_\epsilon, \phi^o_\epsilon)$ has a real analytic continuation.

Proposition 8.20. *There exists a real analytic map*

$$\Phi \equiv (\Phi^i, \Phi^o) \,:\,]-\epsilon_{*1}, \epsilon_{*1}[\to C^{0,\alpha}(\partial\Omega^i) \times C^{0,\alpha}(\partial\Omega^o)$$

such that

$$\mathcal{N}(\epsilon)[\Phi^i[\epsilon], \Phi^o[\epsilon]] = (0,0,1) \qquad \forall \epsilon \in]-\epsilon_{*1}, \epsilon_{*1}[\,.$$

In particular,

$$(\phi^i_\epsilon, \phi^o_\epsilon) = \Phi[\epsilon] \qquad \forall \epsilon \in]0, \epsilon_{*1}[\,. \tag{8.47}$$

Proof. We define

$$\Phi[\epsilon] \equiv \mathcal{N}(\epsilon)^{(-1)}[0,0,1] \qquad \forall \epsilon \in]-\epsilon_{*1}, \epsilon_{*1}[$$

with ϵ_{*1} as in Lemma 8.19. Then the statement follows by Lemma 8.19 and Proposition 8.14. □

8.4.2 Analysis of System (8.33)

We now turn to system (8.33). As we did in Section 8.4.1 to study system (8.32), we begin by splitting the first equation in (8.33) into an equation on $\epsilon\partial\Omega^i$ and one on $\partial\Omega^i$. We obtain

$$\begin{aligned}
&\frac{1}{2}\mu_\epsilon(x) + \int_{\epsilon\partial\Omega^i} \mu_\epsilon(y)\, \nu_{\epsilon\Omega^i}(y) \cdot \nabla S_2(x-y)\, d\sigma_y \\
&\qquad - \int_{\partial\Omega^o} \mu_\epsilon(y)\, \nu_{\Omega^o}(y) \cdot \nabla S_2(x-y)\, d\sigma_y \\
&\qquad\qquad = f^i(x/\epsilon) - \int_{\epsilon\partial\Omega^i} f^i(y/\epsilon)\tau_\epsilon(y)\, d\sigma_y - \int_{\partial\Omega^o} f^o \tau_\epsilon\, d\sigma \quad \forall x \in \epsilon\partial\Omega^i\,, \\
&\frac{1}{2}\mu_\epsilon(x) - \int_{\partial\Omega^o} \mu_\epsilon(y)\, \nu_{\Omega^o}(y) \cdot \nabla S_2(x-y)\, d\sigma_y \\
&\qquad + \int_{\epsilon\partial\Omega^i} \mu_\epsilon(y)\, \nu_{\epsilon\Omega^i}(y) \cdot \nabla S_2(x-y)\, d\sigma_y = f^o(x) \qquad \forall x \in \partial\Omega^o
\end{aligned} \tag{8.48}$$

for all $\epsilon \in]0, \epsilon_0[$. Then, exploiting the substitution $\xi = x/\epsilon \in \partial\Omega^i$ in (8.48), changing variable of integration in order to transform the integrals over $\epsilon\partial\Omega^i$ into integrals over $\partial\Omega^i$, and using equalities (8.36) and (8.37), we get

$$\begin{aligned}
&\frac{1}{2}\mu_\epsilon(\epsilon\xi) + \int_{\partial\Omega^i} \mu_\epsilon(\epsilon\eta)\, \nu_{\Omega^i}(\eta) \cdot \nabla S_2(\xi - \eta)\, d\sigma_\eta \\
&\qquad - \int_{\partial\Omega^o} \mu_\epsilon(y)\, \nu_{\Omega^o}(y) \cdot \nabla S_2(\epsilon\xi - y)\, d\sigma_y \\
&\qquad\quad = f^i(\xi) - \int_{\partial\Omega^i} f^i \phi^i_\epsilon \, d\sigma - \int_{\partial\Omega^o} f^o \phi^o_\epsilon \, d\sigma \qquad \forall \xi \in \partial\Omega^i , \\
&\frac{1}{2}\mu_\epsilon(x) - \int_{\partial\Omega^o} \mu_\epsilon(y)\, \nu_{\Omega^o}(y) \cdot \nabla S_2(x - y)\, d\sigma_y \\
&\qquad + \epsilon \int_{\partial\Omega^i} \mu_\epsilon(\epsilon\eta)\, \nu_{\Omega^i}(\eta) \cdot \nabla S_2(x - \epsilon\eta)\, d\sigma_\eta = f^o(x) \qquad \forall x \in \partial\Omega^o
\end{aligned}$$

for all $\epsilon \in]0, \epsilon_0[$. Hence, taking

$$\psi^i_\epsilon(\xi) \equiv \mu_\epsilon(\epsilon\xi) \quad \forall \xi \in \partial\Omega^i \qquad \text{and} \qquad \psi^o_\epsilon(x) \equiv \mu_\epsilon(x) \quad \forall x \in \partial\Omega^o , \tag{8.49}$$

we deduce that (8.33) is equivalent to the three equations

$$\begin{aligned}
&\frac{1}{2}\psi^i_\epsilon(\xi) - W_{\Omega^i}[\psi^i_\epsilon](\xi) - \int_{\partial\Omega^o} \psi^o_\epsilon(y)\, \nu_{\Omega^o}(y) \cdot \nabla S_2(\epsilon\xi - y)\, d\sigma_y \\
&\qquad = f^i(\xi) - \int_{\partial\Omega^i} f^i \phi^i_\epsilon \, d\sigma - \int_{\partial\Omega^o} f^o \phi^o_\epsilon \, d\sigma \qquad \forall \xi \in \epsilon\partial\Omega^i , \\
&\frac{1}{2}\psi^o_\epsilon(x) + W_{\Omega^o}[\psi^o_\epsilon](x) + \epsilon \int_{\partial\Omega^i} \psi^i_\epsilon(\eta)\, \nu_{\Omega^i}(\eta) \cdot \nabla S_2(x - \epsilon\eta)\, d\sigma_\eta = f^o(x) \\
&\qquad \forall x \in \partial\Omega^o , \\
&\int_{\partial\Omega^i} \psi^i_\epsilon \, d\sigma = 0
\end{aligned} \tag{8.50}$$

for all $\epsilon \in]0, \epsilon_0[$. We also observe that the third equation of (8.50) can be omitted if we require that the solution $(\psi^i_\epsilon, \psi^o_\epsilon)$ belongs to $C^{1,\alpha}(\partial\Omega^i)_0 \times C^{1,\alpha}(\partial\Omega^o)$. (We remaind that $C^{1,\alpha}(\partial\Omega^i)_0$ is the subspace of $C^{1,\alpha}(\partial\Omega^i)$ of the functions with zero integral mean, cf. (2.2).) Then we introduce the auxiliary operator

$$\mathcal{M}(\epsilon) \equiv (\mathcal{M}^i(\epsilon), \mathcal{M}^o(\epsilon))$$

from $C^{1,\alpha}(\partial\Omega^i)_0 \times C^{1,\alpha}(\partial\Omega^o)$ to $C^{1,\alpha}(\partial\Omega^i) \times C^{1,\alpha}(\partial\Omega^o)$ defined by

$$\begin{aligned}
\mathcal{M}^i(\epsilon)[\psi^i, \psi^o](\xi) \equiv &\frac{1}{2}\psi^i(\xi) - W_{\Omega^i}[\psi^i](\xi) \\
&- \int_{\partial\Omega^o} \psi^o(y)\, \nu_{\Omega^o}(y) \cdot \nabla S_2(\epsilon\xi - y)\, d\sigma_y \qquad \forall \xi \in \partial\Omega^i ,
\end{aligned} \tag{8.51}$$

$$\mathcal{M}^o(\epsilon)[\psi^i,\psi^o](x) \equiv \frac{1}{2}\psi^o(x) + W_{\Omega^o}[\psi^o](x) \qquad (8.52)$$
$$+\epsilon \int_{\partial\Omega^i} \psi^i(\eta)\, \nu_{\Omega^i}(\eta)\cdot \nabla S_2(x-\epsilon\eta)\, d\sigma_\eta \quad \forall x \in \partial\Omega^o$$

for all $\epsilon \in]-\epsilon_0,\epsilon_0[$ and all $(\psi^i,\psi^o) \in C^{1,\alpha}(\partial\Omega^i)_0 \times C^{1,\alpha}(\partial\Omega^o)$. If $\epsilon \in]0,\epsilon_0[$, then the system in (8.33) is equivalent to the functional equation

$$\mathcal{M}(\epsilon)[\psi^i_\epsilon,\psi^o_\epsilon] = \left(f^i - \int_{\partial\Omega^i} f^i\phi^i_\epsilon\, d\sigma - \int_{\partial\Omega^o} f^o\phi^o_\epsilon\, d\sigma\,,\, f^o\right) \qquad (8.53)$$

for the pair $(\psi^i_\epsilon,\psi^o_\epsilon) \in C^{1,\alpha}(\partial\Omega^i)_0 \times C^{1,\alpha}(\partial\Omega^o)$. In particular, Theorem 6.31 (i) implies the validity of the following.

Proposition 8.21. *For all $\epsilon \in]0,\epsilon_0[$ there is one and only one solution $(\psi^i_\epsilon,\psi^o_\epsilon)$ in $C^{1,\alpha}(\partial\Omega^i)_0 \times C^{1,\alpha}(\partial\Omega^o)$ of* (8.53).

As a consequence, we can consider the function from $]0,\epsilon_0[$ to $C^{1,\alpha}(\partial\Omega^i)_0 \times C^{1,\alpha}(\partial\Omega^o)$ that takes ϵ to the solution $(\psi^i_\epsilon,\psi^o_\epsilon)$. Our aim in this subsection is to show that such map $\epsilon \mapsto (\psi^i_\epsilon,\psi^o_\epsilon)$ has a real analytic continuation in a neighborhood of $\epsilon = 0$. Similarly to what we have done in Sections 8.3 and 8.4.1 to prove that the maps $\epsilon \mapsto (\theta^i_\epsilon,\theta^o_\epsilon)$ and $\epsilon \mapsto (\phi^i_\epsilon,\phi^o_\epsilon)$ have an analytic continuation, our first step is to prove that $\mathcal{M}$ is real analytic.

Lemma 8.22. *The map from $]-\epsilon_0,\epsilon_0[$ to*

$$\mathcal{L}(C^{1,\alpha}(\partial\Omega^i)_0 \times C^{1,\alpha}(\partial\Omega^o), C^{1,\alpha}(\partial\Omega^i) \times C^{1,\alpha}(\partial\Omega^o))$$

that takes ϵ to $\mathcal{M}(\epsilon)$ is real analytic.

Proof. We confine to prove that $\mathcal{M}^i$ is real analytic from $]-\epsilon_0,\epsilon_0[$ to

$$\mathcal{L}(C^{1,\alpha}(\partial\Omega^i)_0 \times C^{1,\alpha}(\partial\Omega^o), C^{1,\alpha}(\partial\Omega^i) \times C^{1,\alpha}(\partial\Omega^o))\,.$$

The proof for $\mathcal{M}^o$ is very similar and we omit it. We observe that we can write

$$\mathcal{M}^i(\epsilon)[\psi^i,\psi^o] = \left(\frac{1}{2}I - W_{\Omega^i}\right)[\psi^i] - \tilde{\mathcal{M}}^i(\epsilon)[\psi^o]$$

with

$$\tilde{\mathcal{M}}^i(\epsilon)[\psi^o](\xi) = \int_{\partial\Omega^o} \psi^o(y)\, \nu_{\Omega^o}(y)\cdot\nabla S_2(\epsilon\xi - y)\, d\sigma_y \qquad \forall \xi \in \partial\Omega^i$$

for all $\epsilon \in]-\epsilon_0,\epsilon_0[$ and all $\psi^o \in C^{1,\alpha}(\partial\Omega^o)$. By Theorem 4.37, $\frac{1}{2}I - W_{\Omega^i}$ is a bounded linear operator from $C^{1,\alpha}(\partial\Omega^i)$ to itself. Moreover, it does not depend on ϵ. By Corollary A.29 in the Appendix, the map from $]-\epsilon_0,\epsilon_0[$ to $\mathcal{L}(L^1(\partial\Omega^o), C^{1,\alpha}(\partial\Omega^i))$ that takes ϵ to $\tilde{\mathcal{M}}^i(\epsilon)$ is real analytic. Then, by the continuous embedding of $C^{1,\alpha}(\partial\Omega^o)$ into $L^1(\partial\Omega^o)$, we deduce that $\tilde{\mathcal{M}}^i$ is real analytic from $]-\epsilon_0, \epsilon_0[$ to $\mathcal{L}(C^{1,\alpha}(\partial\Omega^o), C^{1,\alpha}(\partial\Omega^i))$. We conclude that $\mathcal{M}^i$ is real analytic from $]-\epsilon_0, \epsilon_0[$ to $\mathcal{L}(C^{1,\alpha}(\partial\Omega^i)_0 \times C^{1,\alpha}(\partial\Omega^o), C^{1,\alpha}(\partial\Omega^i) \times C^{1,\alpha}(\partial\Omega^o))$. □

Then, by the definition of $\mathcal{M}$ in (8.51) and (8.52) we can verify the following.

Lemma 8.23. *For all (g^i, g^o) in $C^{1,\alpha}(\partial\Omega^i) \times C^{1,\alpha}(\partial\Omega^o)$ and all (ψ^i, ψ^o) in $C^{1,\alpha}(\partial\Omega^i)_0 \times C^{1,\alpha}(\partial\Omega^o)$ we have*

$$\mathcal{M}(0)[\psi^i, \psi^o] = (g^i, g^o) \tag{8.54}$$

if and only if

$$\begin{cases} -\dfrac{1}{2}\psi^i + W_{\Omega^i}[\psi^i] = w_{\Omega^o}[\psi^o](0) - g^i \,, \\ \dfrac{1}{2}\psi^o + W_{\Omega^o}[\psi^o] = g^o \,. \end{cases} \tag{8.55}$$

Having in mind the argument used in Sections 8.3 and 8.4.1, we now try to understand whether or not $\mathcal{M}(0)$ is invertible. As we prove in the following Lemma 8.24, $\mathcal{M}(0)$ is injective but not surjective.

Lemma 8.24. *The operator $\mathcal{M}(0)$ is injective and its image* $\mathrm{Im}\,\mathcal{M}(0)$ *is the closed subspace of $C^{1,\alpha}(\partial\Omega^i) \times C^{1,\alpha}(\partial\Omega^o)$ consisting of the pairs (g^i, g^o) such that*

$$\int_{\partial\Omega^i} g^i \, \Phi^i[0] \, d\sigma + \int_{\partial\Omega^o} g^o \, \Phi^o[0] \, d\sigma = 0 \tag{8.56}$$

(cf. Proposition 8.20).

Proof. To prove the lemma we show that equation (8.54) has a solution (ψ^i, ψ^o) in $C^{1,\alpha}(\partial\Omega^i)_0 \times C^{1,\alpha}(\partial\Omega^o)$ if and only if $(g^i, g^o) \in C^{1,\alpha}(\partial\Omega^i) \times C^{1,\alpha}(\partial\Omega^o)$ satisfies condition (8.56). Moreover, such a solution is unique.

We note that (8.54) is a system consisting of two equations. Namely,

$$\mathcal{M}^i(0)[\psi^i, \psi^o] = g^i \quad \text{and} \quad \mathcal{M}^o(0)[\psi^i, \psi^o] = g^o .$$

Moreover, the first of such equations is equivalent to the first equation of (8.55) (cf. (8.51)). By the Fredholm Alternative of Theorem 5.8 such equation has a solution in $C^{1,\alpha}(\partial\Omega^i)$ if and only if $w_{\Omega^o}[\psi^o](0) - g^i$ is orthogonal to the null space of $-\frac{1}{2}I + W^t_{\Omega^i}$. By Lemmas 6.34 and 8.17 and by Proposition 8.20 the null space of $-\frac{1}{2}I + W^t_{\Omega^i}$ is spanned by the function $\Phi^i[0]$. It follows that the first equation in (8.55) has a solution if and only if

$$\int_{\partial\Omega^i} \left(w_{\Omega^o}[\psi^o](0) - g^i\right) \Phi^i[0] \, d\sigma = 0 \,.$$

Since $\int_{\partial\Omega^i} \Phi^i[0] \, d\sigma = 1$ (cf. Lemma 8.17 and Proposition 8.20) the last condition is equivalent to

$$\int_{\partial\Omega^i} g^i \, \Phi^i[0] \, d\sigma = w_{\Omega^o}[\psi^o](0) \,. \tag{8.57}$$

If condition (8.57) is satisfied, then Theorem 6.24, which characterizes the kernel of $-\frac{1}{2}I + W_{\Omega^i}$, implies that the first equation in (8.55) has one and only one solution ψ^i in $C^{1,\alpha}(\partial\Omega^i)_0$.

We now turn to equation $\mathcal{M}^o(0)[\psi^i, \psi^o] = g^o$, which is equivalent to the second equation in (8.55). By Corollary 6.15, such equation has a unique solution ψ^o in $C^{1,\alpha}(\partial\Omega^o)$ for all given $g^o \in C^{1,\alpha}(\partial\Omega^o)$. Moreover, by Theorem 6.5 the operator $\frac{1}{2}I + W^t_{\Omega^o}$ is transpose to $\frac{1}{2}I + W_{\Omega^o}$. As a consequence, if ψ^o is the solution of the second equation in (8.55), then we have

$$\begin{aligned}\int_{\partial\Omega^o} g^o\, \Phi^o[0]\, d\sigma &= \int_{\partial\Omega^o} \left(\frac{1}{2}\psi^o + W_{\Omega^o}[\psi^o]\right) \Phi^o[0]\, d\sigma \\ &= \int_{\partial\Omega^o} \psi^o \left(\frac{1}{2}I + W^t_{\Omega^o}\right) [\Phi^o[0]]\, d\sigma\,.\end{aligned} \tag{8.58}$$

By Lemma 8.17 and Proposition 8.20 the last integral equals

$$-\int_{\partial\Omega^o} \psi^o\, \nu_{\Omega^o} \cdot \nabla S_2\, d\sigma\,.$$

Then by (8.58) and by the Definition 4.27 of the double layer potential we obtain that

$$\int_{\partial\Omega^o} g^o\, \Phi^o[0]\, d\sigma = -w_{\Omega^o}[\psi^o](0)\,.$$

Hence, condition (8.57) is equivalent to (8.56). The lemma is now proved. □

As a consequence of Lemma 8.24, the operator $\mathcal{M}(0)$ is not invertible and therefore we cannot proceed to prove that $\epsilon \mapsto (\psi^i_\epsilon, \psi^o_\epsilon)$ has a real analytic continuation following the footsteps of the argument used in Section 8.3 for the map $\epsilon \mapsto (\theta^i_\epsilon, \theta^o_\epsilon)$ and in Section 8.4.1 for the map $\epsilon \mapsto (\phi^i_\epsilon, \phi^o_\epsilon)$. To overcome this difficulty, we study equation (8.53) by means of the following Lemma 8.25, which is in a sense a specific variant of the Implicit Function Theorem (for a similar result see also [165, Theorem 13]).

Lemma 8.25. *Let E, X, Y, and Z be Banach spaces. Let U be an open neighborhood of 0 in E. Let*

$$\mathcal{A} : U \to \mathcal{L}(X,Y) \qquad \text{and} \qquad f : U \to Y$$

be real analytic. Let

$$\mathcal{B} : U \to \mathcal{L}(Y,Z)$$

be continuous. Assume that

$$\mathcal{A}(0) : X \to Y \qquad \text{is injective,}$$

that

$$\mathcal{B}(0) : Y \to Z \qquad \text{is surjective,}$$

and that

$$\mathrm{Ker}\,\mathcal{B}(0) = \mathrm{Im}\,\mathcal{A}(0)\,. \tag{8.59}$$

Also assume that

$$\mathcal{B}(\varepsilon)[\mathcal{A}(\varepsilon)[a] - f(\varepsilon)] = 0 \qquad \forall \varepsilon \in U\,,\ a \in X \tag{8.60}$$

and that

$$\mathrm{Im}\,\mathcal{A}(0) \textit{ has a closed direct complement } Y_1 \textit{ in } Y. \tag{8.61}$$

Then there exist an open neighborhood $\tilde{U} \subseteq U$ *of* 0 *in* E *and a real analytic map*

$$g : \tilde{U} \to X$$

such that, for an element $a \in X$ *and* $\varepsilon \in \tilde{U}$, *we have*

$$\mathcal{A}(\varepsilon)[a] = f(\varepsilon) \quad \textit{if and only if} \quad a = g(\varepsilon)\,. \tag{8.62}$$

Proof. By assumptions (8.59) and (8.61), the subspace

$$Y_2 \equiv \mathrm{Im}\,\mathcal{A}(0) \subseteq Y$$

is closed and has a closed direct complement Y_1 in Y. As a consequence, the projection operator

$$\pi : Y \to Y_2$$

is continuous (see Theorem 2.6). Then we set

$$\mathcal{P}(\varepsilon) \equiv \pi \circ \mathcal{A}(\varepsilon) \qquad \forall \varepsilon \in U\,.$$

Since $\mathcal{A}$ is real analytic, it follows that

$$\mathcal{P} : U \to \mathcal{L}(X, Y_2)$$

is real analytic. Moreover, since $\mathcal{A}(0)$ is assumed to be injective and

$$\mathcal{P}(0) = \pi \circ \mathcal{A}(0)$$

has image equal to Y_2, the operator

$$\mathcal{P}(0) : X \to Y_2$$

is invertible. Hence, since the set of invertible operators is open in $\mathcal{L}(X, Y_2)$, there exists $\tilde{U} \subseteq U$ such that $\mathcal{P}(\varepsilon)$ is invertible for all $\varepsilon \in \tilde{U}$. Moreover, the map from $\tilde{U}$ to $\mathcal{L}(X, Y_2)$ that takes ε to $\mathcal{P}(\varepsilon)^{(-1)}$ is real analytic. Then we define

$$g(\varepsilon) \equiv \mathcal{P}(\varepsilon)^{(-1)}[\pi[f(\varepsilon)]] \qquad \forall \varepsilon \in \tilde{U}\,.$$

Since the composition of real analytic maps is real analytic, it follows that g is real analytic from $\tilde{U}$ to X. By the definition of $\mathcal{P}(\varepsilon)$ and of $g(\varepsilon)$ and by the invertibility of $\mathcal{P}(\varepsilon)$, we have

$$\pi\left[\mathcal{A}(\varepsilon)[a]\right] = \pi[f(\varepsilon)] \quad \text{if and only if} \quad a = g(\varepsilon)\,. \tag{8.63}$$

To complete the proof we wish to show that (8.63) implies (8.62). To do so, we set

$$\mathcal{J}(\varepsilon) \equiv \mathcal{B}(\varepsilon) \circ j \qquad \forall \epsilon \in U\,, \tag{8.64}$$

where

$$j : Y_1 \to Y$$

is the canonical injection from Y_1 into Y. Since $\mathcal{B}$ is continuous, it follows that

$$\mathcal{J} : U \to \mathcal{L}(Y_1, Z)$$

is continuous. Moreover, since $\mathcal{B}(0)$ is surjective and Ker $\mathcal{B}(0) = Y_2$ is a direct complement of Y_1, we deduce that $\mathcal{J}(0)$ is invertible. Thus, possibly taking a smaller open neighborhood $\tilde{U} \subseteq U$ of 0 in E, we can say that $\mathcal{J}(\varepsilon)$ is invertible for all $\varepsilon \in \tilde{U}$. Then we observe that the equality $\pi\left[\mathcal{A}(\varepsilon)[a]\right] = \pi[f(\varepsilon)]$ is equivalent to $\pi[\mathcal{A}(\varepsilon)[a] - f(\varepsilon)] = 0$ and thus to the inclusion

$$\mathcal{A}(\varepsilon)[a] - f(\varepsilon) \in Y_1\,.$$

Then, by condition (8.60) and by the definition (8.64) of $\mathcal{J}$ we have that

$$\pi\left[\mathcal{A}(\varepsilon)[a]\right] = \pi[f(\varepsilon)] \quad \text{implies that} \quad \mathcal{J}(\varepsilon)[\mathcal{A}(\epsilon)[a] - f(\varepsilon)] = 0\,.$$

For $\varepsilon \in \tilde{U}$ the operator $\mathcal{J}(\varepsilon)$ is invertible and we have

$$\mathcal{J}(\varepsilon)[\mathcal{A}(\varepsilon)[a] - f(\varepsilon)] = 0 \quad \text{if and only if} \quad \mathcal{A}(\varepsilon)[a] - f(\varepsilon) = 0\,.$$

Since $\mathcal{A}(\varepsilon)[a] = f(\varepsilon)$ clearly implies that $\pi\left[\mathcal{A}(\varepsilon)[a]\right] = \pi[f(\varepsilon)]$, we deduce that

$$\pi\left[\mathcal{A}(\varepsilon)[a]\right] = \pi[f(\varepsilon)] \quad \text{if and only if} \quad \mathcal{A}(\varepsilon)[a] = f(\varepsilon)\,.$$

Then, (8.63) implies (8.62) and the lemma is proved. □

We can now prove that the map $\epsilon \mapsto (\psi^i_\epsilon, \psi^o_\epsilon)$ has a real analytic continuation.

Proposition 8.26. *There exist $\epsilon_{*2} \in]0, \epsilon_0]$ and a real analytic map*

$$\Psi \equiv (\Psi^i, \Psi^o) \,:] - \epsilon_{*2}, \epsilon_{*2}[\to C^{1,\alpha}(\partial\Omega^i)_0 \times C^{1,\alpha}(\partial\Omega^o)$$

such that

$$\mathcal{M}(\epsilon)[\Psi^i[\epsilon], \Psi^o[\epsilon]] = \left(f^i - \int_{\partial\Omega^i} f^i\, \Phi^i[\epsilon]\, d\sigma - \int_{\partial\Omega^o} f^o\, \Phi^o[\epsilon]\, d\sigma\,,\, f^o \right) \tag{8.65}$$

for all $\epsilon \in]-\epsilon_{*2}, \epsilon_{*2}[$. *In particular,*

$$(\psi^i_\epsilon, \psi^o_\epsilon) = \Psi[\epsilon] \qquad \forall \epsilon \in]0, \epsilon_{*2}[\,. \tag{8.66}$$

Proof. Let ϵ_{*1} be as in Proposition 8.20. To show the existence of Ψ, we plan to apply Lemma 8.25 with $E = \mathbb{R}$, $U =]-\epsilon_{*1}, \epsilon_{*1}[$, $X = C^{1,\alpha}(\partial\Omega^i)_0 \times C^{1,\alpha}(\partial\Omega^o)$, $Y = C^{1,\alpha}(\partial\Omega^i) \times C^{1,\alpha}(\partial\Omega^o)$, $Z = \mathbb{R}$, $\mathcal{A} = \mathcal{M}$,

$$f(\epsilon) = \left(f^i - \int_{\partial\Omega^i} f^i\, \Phi^i[\epsilon]\, d\sigma - \int_{\partial\Omega^o} f^o\, \Phi^o[\epsilon]\, d\sigma\,,\, f^o\right) \quad \forall \epsilon \in]-\epsilon_{*1}, \epsilon_{*1}[\,,$$

and

$$\mathcal{B}(\epsilon)[g^i, g^o] = \int_{\partial\Omega^i} g^i\, \Phi^i[\epsilon]\, d\sigma + \int_{\partial\Omega^o} g^o\, \Phi^o[\epsilon]\, d\sigma$$

for all $\epsilon \in]-\epsilon_{*1}, \epsilon_{*1}[$ and all $(g^i, g^o) \in C^{1,\alpha}(\partial\Omega^i) \times C^{1,\alpha}(\partial\Omega^o)$. Then we have to verify that the assumptions of Lemma 8.25 are satisfied with this specific choice. As a first step, we observe that our map $\mathcal{A}$ is real analytic by Lemma 8.22. Moreover, Proposition 8.20 implies that Φ is real analytic, and thus the maps f and $\mathcal{B}$ are real analytic. In particular, $\mathcal{B}$ is continuous. The validity of (8.59) is a consequence of Lemma 8.24. To prove that (8.60) holds true we observe that, by equality $\int_{\partial\Omega^i} \Phi^i[\epsilon]\, d\sigma = 1$ (cf. (8.41) and Proposition 8.20) and by a straightforward computation, we have

$$\mathcal{B}(\epsilon)[f(\epsilon)] = 0 \qquad \forall \epsilon \in]-\epsilon_{*1}, \epsilon_{*1}[\,. \tag{8.67}$$

Moreover,

$$\begin{aligned} &\mathcal{B}(\epsilon)[\mathcal{A}(\epsilon)[\psi^i, \psi^o]] \\ &\quad = \int_{\partial\Omega^i} \mathcal{M}^i(\epsilon)[\psi^i, \psi^o]\, \Phi^i[\epsilon]\, d\sigma + \int_{\partial\Omega^o} \mathcal{M}^o(\epsilon)[\psi^i, \psi^o]\, \Phi^o[\epsilon]\, d\sigma\,. \end{aligned}$$

By the computation in (8.48)–(8.50) we have

$$\mathcal{M}^i(\epsilon)[\psi^i, \psi^o](\xi) = \frac{1}{2}\mu_\epsilon(\epsilon\xi) + W_{\Omega(\epsilon)}[\mu_\epsilon](\epsilon\xi) \qquad \forall \xi \in \partial\Omega^i$$

and

$$\mathcal{M}^o(\epsilon)[\psi^i, \psi^o](x) = \frac{1}{2}\mu_\epsilon(x) + W_{\Omega(\epsilon)}[\mu_\epsilon](x) \qquad \forall x \in \partial\Omega^o$$

with $\mu_\epsilon \in C^{1,\alpha}(\partial\Omega(\epsilon))$ which satisfies the conditions

$$\mu_\epsilon(\epsilon\xi) = \psi^i(\xi) \quad \forall \xi \in \partial\Omega^i \qquad \text{and} \qquad \mu_\epsilon(x) = \psi^o(x) \quad \forall x \in \partial\Omega^o$$

(see also (8.49)). By equalities (8.37) and (8.47), we have

$$\Phi^i[\epsilon](\xi) = \epsilon\tau_\epsilon(\epsilon\xi) \quad \forall \xi \in \partial\Omega^i \qquad \text{and} \qquad \Phi^o[\epsilon](x) = \tau_\epsilon(x) \quad \forall x \in \partial\Omega^o$$

for all $\epsilon \in]0, \epsilon_{*1}[$, where $\tau_\epsilon \in C^{0,\alpha}(\partial\Omega(\epsilon))$ is the unique solution of system (8.32). It follows that

$$\mathcal{B}(\epsilon)[\mathcal{A}(\epsilon)[\psi^i,\psi^o]] = \epsilon \int_{\partial\Omega^i} \left(\frac{1}{2}\mu_\epsilon(\epsilon\xi) + W_{\Omega(\epsilon)}[\mu_\epsilon](\epsilon\xi)\right) \tau_\epsilon(\epsilon\xi)\, d\sigma_\xi$$
$$+ \int_{\partial\Omega^o} \left(\frac{1}{2}\mu_\epsilon(x) + W_{\Omega(\epsilon)}[\mu_\epsilon](x)\right) \tau_\epsilon(x)\, d\sigma_x\,.$$

Then, changing the variable of integration in the integral over $\partial\Omega^i$, we verify that

$$\mathcal{B}(\epsilon)[\mathcal{A}(\epsilon)[\psi^i,\psi^o]] = \int_{\partial\Omega(\epsilon)} \left(\frac{1}{2}\mu_\epsilon + W_{\Omega(\epsilon)}[\mu_\epsilon]\right) \tau_\epsilon\, d\sigma \qquad \forall \epsilon \in]0,\epsilon_{*1}[\,.$$

Since the operator $\frac{1}{2}I + W_{\Omega(\epsilon)}$ is transpose to $\frac{1}{2}I + W^t_{\Omega(\epsilon)}$ (cf. Theorem 6.5), we deduce that

$$\mathcal{B}(\epsilon)[\mathcal{A}(\epsilon)[\psi^i,\psi^o]] = \int_{\partial\Omega(\epsilon)} \mu_\epsilon \left(\frac{1}{2}I + W^t_{\Omega(\epsilon)}\right) [\tau_\epsilon]\, d\sigma \qquad \forall \epsilon \in]0,\epsilon_{*1}[\,.$$

Hence, we have

$$\mathcal{B}(\epsilon)[\mathcal{A}(\epsilon)[\psi^i,\psi^o]] = 0 \qquad \forall \epsilon \in]0,\epsilon_{*1}[$$

by the first equation of (8.32). We now observe that the function from $]-\epsilon_{*1},\epsilon_{*1}[$ to $\mathbb{R}$ that takes ϵ to $\mathcal{B}(\epsilon)[\mathcal{A}(\epsilon)[\psi^i,\psi^o]]$ is real analytic. Then, by the Identity Principle for real analytic functions (cf. Proposition 2.16), we have

$$\mathcal{B}(\epsilon)[\mathcal{A}(\epsilon)[\psi^i,\psi^o]] = 0 \qquad \forall \epsilon \in]-\epsilon_{*1},\epsilon_{*1}[\,. \tag{8.68}$$

Now (8.60) follows by (8.67) and (8.68).

Finally, to verify that $\operatorname{Im}\mathcal{A}(0)$ has a closed direct complement in Y, we observe that $\mathcal{B}(0)$ is a non-trivial real-valued bounded functional on Y and that $\operatorname{Im}\mathcal{A}(0) = \operatorname{Ker}\mathcal{B}(0)$ by Lemma 8.24. It follows that $\operatorname{Im}\mathcal{A}(0)$ has co-dimension one in Y. Accordingly, it has a closed direct complement of dimension one in Y (see Corollary 2.7).

Now that the assumptions of Lemma 8.24 have been verified, we can deduce that there exist $\epsilon_{*2} \in]0,\epsilon_1]$ and a real analytic map Ψ from $]-\epsilon_{*2},\epsilon_{*2}[$ to $C^{1,\alpha}(\partial\Omega^i) \times C^{1,\alpha}(\partial\Omega^o)$ which satisfies (8.65). Then (8.66) follows by Proposition 8.21. □

8.4.3 Real Analytic Representation of the Map $\epsilon \mapsto u_\epsilon$.

We are now ready to draw some conclusions from Propositions 8.20 and 8.26 and derive real analytic representation results for the map that takes ϵ to the solution u_ϵ. In what follows, we set

$$\epsilon_* \equiv \min\{\epsilon_{*1},\epsilon_{*2}\}$$

with ϵ_{*1} and ϵ_{*2} as in Lemma 8.19 and Proposition 8.26, respectively. Then we resort to the representation formula (8.31) for u_ϵ. Using formulas (8.49) and (8.66) for μ_ϵ and formulas (8.37) and (8.47) for τ_ϵ, we see that

$$
\begin{aligned}
u_\epsilon(x) = {}& w^+_{\Omega^o}[\Psi^o[\epsilon]](x) + \int_{\epsilon\partial\Omega^i} \Psi^i[\epsilon](y/\epsilon)\, \nu_{\epsilon\Omega^i}(y) \cdot \nabla S_2(x-y)\, d\sigma_y \\
&+ \left(\frac{1}{\epsilon} \int_{\epsilon\partial\Omega^i} f^i(y/\epsilon)\, \Phi^i[\epsilon](y/\epsilon)\, d\sigma_y + \int_{\partial\Omega^o} f^o(y)\, \Phi^o[\epsilon](y)\, d\sigma_y \right) \\
&\times \left(\frac{1}{\epsilon} \int_{\partial\Omega^i} \Phi^i[\epsilon](y/\epsilon)\, S_2(x-y)\, d\sigma_y + v^+_{\Omega^o}[\Phi^o[\epsilon]](x) \right) \\
&\times \left(\fint_{\epsilon\partial\Omega^i} \frac{1}{\epsilon} \int_{\epsilon\partial\Omega^i} S_2(y-z)\, \Phi^i[\epsilon](z/\epsilon)\, d\sigma_z + v_{\partial\Omega^o}[\Phi^o[\epsilon]](y)\, d\sigma_y \right)^{-1}
\end{aligned}
$$

for all $x \in \overline{\Omega^o} \setminus \epsilon\Omega^i$ and for all $\epsilon \in]0, \epsilon_*[$ (see also Definitions 4.18 and 4.27 of the single and double layer potentials, respectively). Then, changing the variable of integration in the integrals over $\epsilon\partial\Omega^i$ and exploiting equality (8.30), we obtain the following integral representation for u_ϵ:

$$
\begin{aligned}
u_\epsilon(x) = {}& w^+_{\Omega^o}[\Psi^o[\epsilon]](x) + \epsilon \int_{\partial\Omega^i} \Psi^i[\epsilon](\eta)\, \nu_{\Omega^i}(\eta) \cdot \nabla S_2(x-\epsilon\eta)\, d\sigma_\eta \\
&+ \left(\int_{\partial\Omega^i} f^i\, \Phi^i[\epsilon]\, d\sigma + \int_{\partial\Omega^o} f^o\, \Phi^o[\epsilon]\, d\sigma \right) \\
&\times \left(\int_{\partial\Omega^i} \Phi^i[\epsilon](\eta)\, S_2(x-\epsilon\eta)\, d\sigma_\eta + v^+_{\Omega^o}[\Phi^o[\epsilon]](x) \right) \\
&\times \left(\fint_{\partial\Omega^i} v_{\Omega^i}[\Phi^i[\epsilon]](\eta) + v_{\partial\Omega^o}[\Phi^o[\epsilon]](\epsilon\eta)\, d\sigma_\eta + \frac{\log\epsilon}{2\pi} \right)^{-1}
\end{aligned}
\tag{8.69}
$$

for all $x \in \overline{\Omega^o} \setminus \epsilon\Omega^i$ and for all $\epsilon \in]0, \epsilon_*[$. We observe that in the representation (8.69) all the integrals are defined on the boundary of fixed domains, which do not depend on ϵ, and all the maps involved are either completely known or depend real analytically on ϵ.

In view of (8.69), we now introduce the auxiliary functions $\mathcal{F}$ and $\mathcal{R}$.

Proposition 8.27. *Let $\mathcal{F}$ and $\mathcal{R}$ be the functions from $] - \epsilon_*, \epsilon_*[$ to $\mathbb{R}$ defined by*

$$
\mathcal{F}(\epsilon) \equiv \int_{\partial\Omega^i} f^i\, \Phi^i[\epsilon]\, d\sigma + \int_{\partial\Omega^o} f^o\, \Phi^o[\epsilon]\, d\sigma
$$

and

$$
\mathcal{R}(\epsilon) \equiv \fint_{\partial\Omega^i} v_{\Omega^i}[\Phi^i[\epsilon]](\eta) + v_{\Omega^o}[\Phi^o[\epsilon]](\epsilon\eta)\, d\sigma_\eta\,,
$$

for all $\epsilon \in] - \epsilon_*, \epsilon_*[$. Then $\mathcal{F}$ and $\mathcal{R}$ are real analytic.

Proof. We first prove the statement for $\mathcal{F}$. By Proposition 8.20, the maps Φ^i and Φ^o are analytic. Moreover, the map from $C^{0,\alpha}(\partial\Omega^i)$ to itself that takes a function to its product with f^i is real analytic and, similarly, the map from $C^{0,\alpha}(\partial\Omega^o)$ to itself that takes a function to its product with f^o is real analytic. Since the maps that take functions of $C^{0,\alpha}(\partial\Omega^i)$ and $C^{0,\alpha}(\partial\Omega^o)$ to their integrals are linear and continuous, it follows that $\mathcal{F}$ is real analytic. We now turn to $\mathcal{R}$. We can write

$$\mathcal{R} = \mathcal{R}_1 + \mathcal{R}_2$$

with

$$\mathcal{R}_1(\epsilon) \equiv \fint_{\partial\Omega^i} v_{\Omega^i}[\Phi^i[\epsilon]](\eta)\, d\sigma_\eta$$

and

$$\mathcal{R}_2(\epsilon) \equiv \fint_{\partial\Omega^i} v_{\Omega^o}[\Phi^o[\epsilon]](\epsilon\eta)\, d\sigma_\eta$$

for all $\epsilon \in]-\epsilon_*, \epsilon_*[$. To prove that $\mathcal{R}_1$ is real analytic, we observe that Φ^i is real analytic from $]-\epsilon_*, \epsilon_*[$ to $C^{0,\alpha}(\partial\Omega^i)$ by Proposition 8.20, that $V_{\Omega^i} = v_{\Omega^i}[\cdot]_{|\partial\Omega^i}$ is linear and continuous from $C^{0,\alpha}(\partial\Omega^i)$ to $C^{1,\alpha}(\partial\Omega^i)$ by Theorem 6.4, and that the map that takes a function of $C^{1,\alpha}(\partial\Omega^i)$ to its integral mean is linear and continuous. To prove that $\mathcal{R}_2$ is real analytic, we observe that

$$v_{\Omega^o}[\Phi^o[\epsilon]](\epsilon\eta) = \int_{\partial\Omega^o} S_2(\epsilon\eta - y)\Phi^o[\epsilon](y)\, d\sigma_y \qquad \forall \eta \in \partial\Omega^i\,. \tag{8.70}$$

Then, Theorem A.28 (ii) on the properties of integral operators with real analytic kernels and Proposition 8.20 on the real analyticity of Φ^o imply that the map from $]-\epsilon_*, \epsilon_*[$ to $C^{1,\alpha}(\partial\Omega^i)$ that takes ϵ to the function in (8.70) is real analytic. Since the map that takes a function of $C^{1,\alpha}(\partial\Omega^i)$ to its integral mean is linear and continuous, we deduce that $\mathcal{R}_2$ is real analytic and we conclude that also $\mathcal{R}$ is real analytic. □

We observe that for $\epsilon \in]0, \epsilon_*[$ we can characterize the case when $\mathcal{F}(\epsilon) = 0$ by the following.

Remark 8.28. If $\epsilon \in]0, \epsilon_*[$, then we have $\mathcal{F}(\epsilon) = 0$ if and only if

$$\int_{\epsilon\partial\Omega^i} \nu_{\epsilon\Omega^i} \cdot \nabla u_\epsilon = 0\,. \tag{8.71}$$

Proof. By (8.4) and (8.37) and by the definition of $\mathcal{F}(\epsilon)$ we can see that for ϵ in $]0, \epsilon_*[$ we have

$$\mathcal{F}(\epsilon) = \int_{\partial\Omega(\epsilon)} \tau_\epsilon u_\epsilon\, d\sigma\,.$$

Then $\mathcal{F}(\epsilon) = 0$ if and only if $u_{\epsilon|\partial\Omega(\epsilon)}$ is orthogonal to the null space

$$\mathrm{Ker}\left(\frac{1}{2}I + W^t_{\Omega(\epsilon)}\right)$$

generated by τ_ϵ (cf. Lemma 6.28 and (8.32)). Hence, by the Fredholm Alternative of Theorem 5.8 we have $\mathcal{F}(\epsilon) = 0$ if and only if

$$u_{\epsilon|\partial\Omega(\epsilon)} \in \mathrm{Im}\left(\frac{1}{2}I + W_{\Omega(\epsilon)}\right)\,.$$

By the jump formulas in Theorem 6.6 and by Theorem 6.32 the last inclusion is equivalent to (8.71). □

We also observe that

$$\mathcal{R}(\epsilon) + \frac{\log \epsilon}{2\pi} = \int_{\epsilon\partial\Omega^i} V_{\Omega(\epsilon)}[\tau_\epsilon]\, d\sigma \qquad \forall \epsilon \in]0, \epsilon_*[$$

(see (8.37) and Proposition 8.20). Thus, by (8.34) we have

$$\mathcal{R}(\epsilon) + \frac{\log \epsilon}{2\pi} \neq 0 \qquad \forall \epsilon \in]0, \epsilon_*[\,. \tag{8.72}$$

In the following Theorem 8.29 we consider the restriction of u_ϵ to the closure of an open subset Ω_M of Ω^o with $0 \notin \overline{\Omega^o}$ and prove a real analytic representation formula for the map $\epsilon \mapsto u_{\epsilon|\overline{\Omega^o}}$. As in the $(n \geq 3)$-dimensional case, the capital letter 'M' stands for 'macroscopic,' as opposed to the small letter 'm' of Theorem 8.30 that stands for 'microscopic.'

Theorem 8.29. *Let Ω_M be an open subset of Ω^o such that $0 \notin \overline{\Omega_M}$. Let $\epsilon_M \in]0, \epsilon_*[$ be such that*

$$\overline{\Omega_M} \cap \epsilon\overline{\Omega^i} = \emptyset \qquad \forall \epsilon \in]-\epsilon_M, \epsilon_M[\,.$$

Then there exist real analytic maps

$$\mathcal{U}_M :]-\epsilon_M, \epsilon_M[\to C^{1,\alpha}(\overline{\Omega_M})$$

and

$$\mathcal{V}_M :]-\epsilon_M, \epsilon_M[\to C^{1,\alpha}(\overline{\Omega_M})$$

such that the following representation formula holds:

$$u_{\epsilon|\overline{\Omega_M}} = \mathcal{U}_M[\epsilon] + \mathcal{F}(\epsilon)\frac{\mathcal{V}_M[\epsilon]}{\mathcal{R}(\epsilon) + \frac{\log \epsilon}{2\pi}} \qquad \forall \epsilon \in]0, \epsilon_M[\,. \tag{8.73}$$

Moreover,

$$\mathcal{U}_M[0] = u^o_{|\overline{\Omega_M}} \tag{8.74}$$

where $u^o \in C^{1,\alpha}(\overline{\Omega^o})$ is the unique solution of the unperturbed problem

$$\begin{cases} \Delta u^o = 0 & \text{in } \Omega^o \,, \\ u^o = f^o & \text{on } \partial\Omega^o \,. \end{cases} \tag{8.75}$$

Proof. We set

$$\mathcal{U}_M[\epsilon](x) \equiv w^+_{\Omega^o}[\Psi^o[\epsilon]](x) + \epsilon \int_{\partial\Omega^i} \Psi^i[\epsilon](y)\, \nu_{\Omega^i}(\eta) \cdot \nabla S(x - \epsilon\eta)\, d\sigma_\eta \tag{8.76}$$

and

$$\mathcal{V}_M[\epsilon](x) \equiv \int_{\partial\Omega^i} \Phi^i[\epsilon](\eta)\, S(x-\epsilon\eta)\, d\sigma_\eta + v^+_{\Omega^o}[\Phi^o[\epsilon]](x) \tag{8.77}$$

for all $x \in \overline{\Omega_M}$ and for all $\epsilon \in]-\epsilon_M, \epsilon_M[$. Then the validity of (8.73) follows by (8.69) and by the definition of $\mathcal{F}$ and $\mathcal{R}$ in Proposition 8.27.

To show that $\epsilon \mapsto \mathcal{U}_M[\epsilon]$ is real analytic we first observe that Theorem 4.31 (i) on the linearity and continuity of $w^+_{\Omega^o}$ and Proposition 8.26 on the analyticity of Ψ^o imply that the map from $]-\epsilon_M, \epsilon_M[$ to $C^{1,\alpha}(\overline{\Omega_M})$ that takes ϵ to $w^+_{\Omega^o}[\Psi^o[\epsilon]]_{|\overline{\Omega_M}}$ is real analytic. Thus, it remains to prove that the map from $]-\epsilon_M, \epsilon_M[$ to $C^{1,\alpha}(\overline{\Omega_M})$ that takes ϵ to the function

$$\epsilon \int_{\partial\Omega^i} \Psi^i[\epsilon](y)\, \nu_{\Omega^i}(\eta) \cdot \nabla S(x-\epsilon\eta)\, d\sigma_\eta$$

of $x \in \overline{\Omega_M}$ is also real analytic, a fact that follows from the real analyticity of Ψ^i of Proposition 8.26 and from the properties of integral operators with real analytic kernels of Theorem A.28 (i).

Similarly, to prove that $\epsilon \mapsto \mathcal{V}_M[\epsilon]$ is real analytic we note that Theorem 4.25 on the linearity and continuity of $v^+_{\Omega^o}$ and Proposition 8.20 on the analyticity of Φ^o imply that the map from $]-\epsilon_M, \epsilon_M[$ to $C^{1,\alpha}(\overline{\Omega_M})$ that takes ϵ to $v^+_{\Omega^o}[\Phi^o[\epsilon]]_{|\overline{\Omega_M}}$ is real analytic. Thus, it remains to show that the map from $]-\epsilon_M, \epsilon_M[$ to $C^{1,\alpha}(\overline{\Omega_M})$ that takes ϵ to the function

$$\int_{\partial\Omega^i} \Phi^i[\epsilon](\eta)\, S(x-\epsilon\eta)\, d\sigma_\eta$$

of $x \in \overline{\Omega_M}$ is real analytic, a fact that follows from the real analyticity of Φ^i of Proposition 8.20 and from the properties of integral operators with real analytic kernels of Theorem A.28 (i).

To verify (8.74) we observe that by Lemma 8.23 and Proposition 8.26 we have

$$\mathcal{U}_M[0](x) = w^+_{\Omega^o}[\Psi^o[0]]_{|\overline{\Omega_M}} = u^o_{|\overline{\Omega_M}}.$$

The theorem is now proved. □

By the representation formula (8.73) for u_ϵ and by the continuity at $\epsilon = 0$ of the maps $\mathcal{U}_M$, $\mathcal{F}$, $\mathcal{V}_M$, and $\mathcal{R}$, we have

$$\lim_{\epsilon\to 0^+} u_\epsilon(x) = u^o(x) \qquad \forall x \in \Omega^o \setminus \{0\}, \tag{8.78}$$

a well-known limiting relation in the literature on Asymptotic Analysis.

We now turn to the 'microscopic' behavior of u_ϵ in the proximity of the boundary of the hole $\epsilon\Omega^i$, namely, we study the rescaled function $u_\epsilon(\epsilon\cdot)$. As a first step we resort to the integral representation formula (8.69) for u_ϵ and, changing the variable of integration in the integrals over $\epsilon\partial\Omega^i$, we obtain the following integral representation formula for $u_\epsilon(\epsilon\cdot)$,

$$
\begin{aligned}
u_\epsilon(\epsilon\xi) = & -w_{\overline{\Omega^i}}[\Psi^i[\epsilon]](\xi) + w_{\Omega^o}[\Psi^o[\epsilon]](\epsilon\xi) \\
& + \left(\int_{\partial\Omega^i} f^i\, \Phi^i[\epsilon]\, d\sigma + \int_{\partial\Omega^o} f^o\, \Phi^o[\epsilon]\, d\sigma\right) \\
& \times \left(v_{\overline{\Omega^i}}[\Phi^i[\epsilon]](\xi) + v_{\Omega^o}[\Phi^o[\epsilon]](\epsilon\xi) + \frac{\log\epsilon}{2\pi}\right) \\
& \times \left(\fint_{\partial\Omega^i} v_{\Omega^i}[\Phi^i[\epsilon]](\eta) + v_{\partial\Omega^o}[\Phi^o[\epsilon]](\epsilon\eta)\, d\sigma_\eta + \frac{\log\epsilon}{2\pi}\right)^{-1}
\end{aligned} \tag{8.79}
$$

for all $\xi \in (\epsilon^{-1}\Omega^o) \setminus \Omega^i$ and for all $\epsilon \in]0, \epsilon_*[$ (to compute (8.79) we have also used equality (8.30)). Then we deduce the validity of the following Theorem 8.30, where we prove a real analytic representation formula for the map that takes ϵ to the restriction of $u_\epsilon(\epsilon\cdot)$ to the closure of a bounded subset Ω_m of $\mathbb{R}^2 \setminus \overline{\Omega^i}$. As usual the subscript 'm' stands for 'microscopic.'

Theorem 8.30. *Let Ω_m be an open bounded subset of $\mathbb{R}^2 \setminus \overline{\Omega^i}$. Let $\epsilon_m \in]0, \epsilon_*[$ be such that*

$$\epsilon\overline{\Omega_m} \subseteq \Omega^o \qquad \forall \epsilon \in]-\epsilon_m, \epsilon_m[\,.$$

Then, there exist real analytic maps

$$\mathcal{U}_m :]-\epsilon_m, \epsilon_m[\to C^{1,\alpha}(\overline{\Omega_m})$$

and

$$\mathcal{V}_m :]-\epsilon_m, \epsilon_m[\to C^{1,\alpha}(\overline{\Omega_m})$$

such that

$$u_\epsilon(\epsilon\cdot)_{|\overline{\Omega_m}} = \mathcal{U}_m[\epsilon] + \mathcal{F}(\epsilon)\frac{\mathcal{V}_m[\epsilon] + \frac{\log\epsilon}{2\pi}}{\mathcal{R}(\epsilon) + \frac{\log\epsilon}{2\pi}} \qquad \forall \epsilon \in]0, \epsilon_m[\,. \tag{8.80}$$

Moreover,

$$\mathcal{U}_m[0] = u^i_{|\overline{\Omega_m}} - \int_{\partial\Omega^i} f^i\, \Phi^i[0]\, d\sigma - \int_{\partial\Omega^o} f^o\, \Phi^o[0]\, d\sigma \tag{8.81}$$

where $u^i \in C^{1,\alpha}_{\mathrm{loc}}(\mathbb{R}^n \setminus \Omega^i)$ is the unique solution of the problem

$$
\begin{cases}
\Delta u^i = 0 & \text{in } \mathbb{R}^2 \setminus \overline{\Omega^i}\,, \\
u^i = f^i & \text{on } \partial\Omega^i\,, \\
u^i \text{ is harmonic at infinity}\,.
\end{cases} \tag{8.82}
$$

Proof. Inspired by formula (8.79), we set

$$\mathcal{U}_m[\epsilon](\xi) \equiv -w_{\overline{\Omega^i}}[\Psi^i[\epsilon]](\xi) + w_{\Omega^o}[\Psi^o[\epsilon]](\epsilon\xi) \tag{8.83}$$

and

$$\mathcal{V}_m[\epsilon](\xi) \equiv v_{\overline{\Omega^i}}[\Phi^i[\epsilon]](\xi) + v_{\Omega^o}[\Phi^o[\epsilon]](\epsilon\xi) \tag{8.84}$$

for all $\xi \in \overline{\Omega_m}$ and for all $\epsilon \in]-\epsilon_m, \epsilon_m[$. Then the validity of (8.80) follows by (8.79) and by the definition of $\mathcal{F}$ and $\mathcal{R}$ in Proposition 8.27. To prove (8.81) we observe that for $\epsilon = 0$ we have

$$\mathcal{U}_m[0](\xi) = -w^-_{\Omega^i}[\Psi^i[0]](\xi) + w_{\Omega^o}[\Psi^o[0]](0)\,.$$

By Lemma 8.23 and Proposition 8.26, the function $-w^-_{\Omega^i}[\Psi^i[0]]$ is the solution of the exterior Dirichlet problem with boundary datum

$$-w_{\Omega^o}[\Psi^o[0]](0) + f^i(\xi) - \int_{\partial\Omega^i} f^i\,\Phi^i[0]\,d\sigma - \int_{\partial\Omega^o} f^o\,\Phi^o[0]\,d\sigma \qquad \forall \xi \in \partial\Omega^i\,.$$

Then the validity of (8.81) follows by a straightforward computation.

To show that $\epsilon \mapsto \mathcal{U}_m[\epsilon]$ is real analytic we first observe that Theorem 4.31 (ii) on the continuity of the linear operator $w^-_{\Omega^i}$ and Proposition 8.26 on the analyticity of Ψ^i imply that the map from $]-\epsilon_m, \epsilon_m[$ to $C^{1,\alpha}(\overline{\Omega_m})$ that takes ϵ to $w^-_{\Omega^i}[\Psi^i[\epsilon]]_{|\overline{\Omega_m}}$ is real analytic. Then we can prove that the map from $]-\epsilon_m, \epsilon_m[$ to $C^{1,\alpha}(\overline{\Omega_m})$ that takes ϵ to the function

$$w_{\Omega^o}[\Psi^o[\epsilon]](\epsilon\xi) = -\int_{\partial\Omega^o} \Psi^o[\epsilon](y)\,\nu_\Omega(y)\cdot\nabla S_2(\epsilon\xi - y)\,d\sigma_y$$

of $\xi \in \overline{\Omega_m}$ is real analytic by the real analyticity of Ψ^o of Proposition 8.26 and by the properties of integral operators with real analytic kernes of Theorem A.28 (i).

Similarly, to show that $\epsilon \mapsto \mathcal{V}_m[\epsilon]$ is real analytic we first observe that Theorem 4.25 on the continuity of the linear operator $v^-_{\Omega^i}$ and Proposition 8.20 on the analyticity of Φ^i imply that map from $]-\epsilon_m, \epsilon_m[$ to $C^{1,\alpha}(\overline{\Omega_M})$ that takes ϵ to $v^-_{\Omega^i}[\Phi^i[\epsilon]]_{|\overline{\Omega_m}}$ is real analytic. Then we are left to show that the map from $]-\epsilon_m, \epsilon_m[$ to $C^{1,\alpha}(\overline{\Omega_m})$ that takes ϵ to the function

$$v_{\Omega^o}[\Phi^o[\epsilon]](\epsilon\xi) = \int_{\partial\Omega^o} \Phi^o[\epsilon](y) S_2(\epsilon\xi - y)\,d\sigma_y$$

of $\xi \in \overline{\Omega_m}$ is real analytic. A fact that follows from the real analyticity of Φ^o of Proposition 8.20 and from the properties of integral operators with real analytic kernel of Theorem A.28 (i). The theorem is now proved. □

By the representation formula (8.80) for $u_\epsilon(\epsilon\cdot)$ and the continuity of the functions $\mathcal{U}_m$, $\mathcal{F}$, $\mathcal{V}_m$, and $\mathcal{R}$ at $\epsilon = 0$, we can see that

$$\lim_{\epsilon\to 0^+} u_\epsilon(\epsilon\xi) = \mathcal{U}_m[0](\xi) + \mathcal{F}(0) \tag{8.85}$$

for all $\xi \in \mathbb{R}^2 \setminus \Omega^i$. Then we observe that by the definition of $\mathcal{F}$ in Proposition 8.27 we have

$$\mathcal{F}(0) = \int_{\partial\Omega^i} f^i\,\Phi^i[0]\,d\sigma + \int_{\partial\Omega^o} f^o\,\Phi^o[0]\,d\sigma\,.$$

The last equality, together with formula (8.81) for $\mathcal{U}_m[0]$ and (8.85), yields the (well-known) limiting relation

$$\lim_{\epsilon \to 0+} u_\epsilon(\epsilon\xi) = u^i(\xi) \tag{8.86}$$

for all $\xi \in \mathbb{R}^2 \setminus \Omega^i$.

We can now exploit Theorems 8.29 and 8.30 to prove a real analytic representation formula for the energy integral of u_ϵ.

Theorem 8.31. *There exist* $\epsilon_E \in]0, \epsilon_*[$ *and real analytic functions*

$$\mathcal{E}_1\,,\ \mathcal{E}_2 :]-\epsilon_E, \epsilon_E[\to \mathbb{R}$$

such that

$$\int_{\Omega(\epsilon)} |\nabla u_\epsilon|^2\, dx = \mathcal{E}_1(\epsilon) + \mathcal{F}(\epsilon)\frac{\mathcal{E}_2(\epsilon)}{\mathcal{R}(\epsilon) + \frac{\log \epsilon}{2\pi}} \qquad \forall \epsilon \in]0, \epsilon_E[\,. \tag{8.87}$$

Moreover,

$$\mathcal{E}_1(0) = \int_{\Omega^o} |\nabla u^o|^2\, dx + \int_{\mathbb{R}^2 \setminus \Omega^i} |\nabla u^i|^2\, d\xi\,. \tag{8.88}$$

(For the definition of u^o, u^i, *see* (8.75) *and* (8.82).*)*

Proof. As we did to prove Theorem 8.9 for $n \geq 3$, we begin by observing that the first Green Identity of Theorem 4.2 and the theorem of change of variable in integrals imply that

$$\begin{aligned}\int_{\Omega(\epsilon)} |\nabla u_\epsilon|^2\, dx &= \int_{\partial\Omega^o} u_\epsilon\, \nu_{\Omega^o} \cdot \nabla u_\epsilon\, d\sigma - \int_{\epsilon\partial\Omega^i} u_\epsilon\, \nu_{\epsilon\Omega^i} \cdot \nabla u_\epsilon\, d\sigma \\ &= \int_{\partial\Omega^o} u_\epsilon\, \nu_{\Omega^o} \cdot \nabla u_\epsilon\, d\sigma - \int_{\partial\Omega^i} u_\epsilon(\epsilon\eta)\, \nu_{\Omega^i}(\eta) \cdot \nabla_\eta\left(u_\epsilon(\epsilon\eta)\right) d\sigma_\eta\end{aligned}$$

for all $\epsilon \in]0, \epsilon_0[$. Since u_ϵ solves the Dirichlet problem (8.4), we can write

$$\int_{\Omega(\epsilon)} |\nabla u_\epsilon|^2\, dx = \int_{\partial\Omega^o} f^o\, \nu_{\Omega^o} \cdot \nabla u_\epsilon\, d\sigma - \int_{\partial\Omega^i} f^i(\eta)\, \nu_{\Omega^i}(\eta) \cdot \nabla_\eta\left(u_\epsilon(\epsilon\eta)\right) d\sigma_\eta \tag{8.89}$$

for all $\epsilon \in]0, \epsilon_0[$. Then we take Ω_M as in Theorem 8.29 and Ω_m as in Theorem 8.30 and we also assume that

$$\partial\Omega^o \subseteq \overline{\Omega_M} \qquad \text{and} \qquad \partial\Omega^i \subseteq \overline{\Omega_m}\,.$$

By exploiting the corresponding maps $\mathcal{U}_M$, $\mathcal{V}_M$, $\mathcal{U}_m$, and $\mathcal{V}_m$ and the representation formulas (8.73) and (8.80) for u_ϵ and $u_\epsilon(\epsilon\cdot)$, we can write

$$\int_{\Omega(\epsilon)} |\nabla u_\epsilon|^2\, dx = \int_{\partial\Omega^o} f^o\, \nu_{\Omega^o} \cdot \nabla\mathcal{U}_M[\epsilon]\, d\sigma$$

$$
\begin{aligned}
&+ \frac{\mathcal{F}(\epsilon)}{\mathcal{R}(\epsilon) + \frac{\log \epsilon}{2\pi}} \int_{\partial\Omega^o} f^o \, \nu_{\Omega^o} \cdot \nabla \mathcal{V}_M[\epsilon] \, d\sigma \\
&- \int_{\partial\Omega^i} f^i \, \nu_{\Omega^i} \cdot \nabla \mathcal{U}_m[\epsilon] \, d\sigma \\
&- \frac{\mathcal{F}(\epsilon)}{\mathcal{R}(\epsilon) + \frac{\log \epsilon}{2\pi}} \int_{\partial\Omega^i} f^i \, \nu_{\Omega^i} \cdot \nabla \left(\mathcal{V}_m[\epsilon] + \frac{\log \epsilon}{2\pi} \right) d\sigma
\end{aligned}
$$

for all $\epsilon \in]0, \epsilon_E[$, with $\epsilon_E \equiv \min\{\epsilon_M, \epsilon_m\}$. Since $\nabla \left(\mathcal{V}_m[\epsilon] + \frac{\log \epsilon}{2\pi} \right) = \nabla \mathcal{V}_m[\epsilon]$, we can deduce that (8.87) holds with $\mathcal{E}_1$ and $\mathcal{E}_2$ defined by

$$
\mathcal{E}_1(\epsilon) \equiv \int_{\partial\Omega^o} f^o \, \nu_{\Omega^o} \cdot \nabla \mathcal{U}_M[\epsilon] \, d\sigma - \int_{\partial\Omega^i} f^i \, \nu_{\Omega^i} \cdot \nabla \mathcal{U}_m[\epsilon] \, d\sigma \, , \tag{8.90}
$$

$$
\mathcal{E}_2(\epsilon) \equiv \int_{\partial\Omega^o} f^o \, \nu_{\Omega^o} \cdot \nabla \mathcal{V}_M[\epsilon] \, d\sigma - \int_{\partial\Omega^i} f^i \, \nu_{\Omega^i} \cdot \nabla \mathcal{V}_m[\epsilon] \, d\sigma \tag{8.91}
$$

for all $\epsilon \in] - \epsilon_E, \epsilon_E[$.

To prove (8.88) we observe that formulas (8.74) and (8.81) for $\mathcal{U}_M[0]$ and $\mathcal{U}_m[0]$ imply that

$$
\begin{aligned}
\mathcal{E}_1(0) = &\int_{\partial\Omega^o} f^o \, \nu_{\Omega^o} \cdot \nabla u^o \, d\sigma \\
&- \int_{\partial\Omega^i} f^i \, \nu_{\Omega^i} \cdot \nabla \left(u^i - \int_{\partial\Omega^i} f^i \, \Phi^i[0] \, d\sigma - \int_{\partial\Omega^o} f^o \, \Phi^o[0] \, d\sigma \right) d\sigma \\
= &\int_{\partial\Omega^o} f^o \, \nu_{\Omega^o} \cdot \nabla u^o \, d\sigma - \int_{\partial\Omega^i} f^i \, \nu_{\Omega^i} \cdot \nabla u^i \, d\sigma \, .
\end{aligned}
$$

Then (8.88) follows by the first Green Identity of Theorem 4.2 and of Corollary 4.7.

To verify that $\mathcal{E}_1$ and $\mathcal{E}_2$ are real analytic we note that Theorem 8.29 implies that the maps $\mathcal{U}_M$ and $\mathcal{V}_M$ are real analytic from $] - \epsilon_E, \epsilon_E[$ to $C^{1,\alpha}(\overline{\Omega_M})$. Hence, the membership of (f^o, ν_{Ω^o}) in $C^{0,\alpha}(\partial\Omega^o) \times C^{0,\alpha}(\partial\Omega^o, \mathbb{R}^n)$ and the continuity of the pointwise product in Hölder spaces imply that the maps $\epsilon \mapsto f^o \, \nu_{\Omega^o} \cdot \nabla \mathcal{U}_M[\epsilon]$ and $\epsilon \mapsto f^o \, \nu_{\Omega^o} \cdot \nabla \mathcal{V}_M[\epsilon]$ are real analytic from $] - \epsilon_E, \epsilon_E[$ to $C^{0,\alpha}(\partial\Omega^o)$. By Theorem 8.30, we can show by a similar argument that the maps $\epsilon \mapsto f^i \, \nu_{\Omega^i} \cdot \nabla \mathcal{U}_m[\epsilon]$ and $\epsilon \mapsto f^i \, \nu_{\Omega^i} \cdot \nabla \mathcal{V}_m[\epsilon]$ are real analytic from $] - \epsilon_E, \epsilon_E[$ to $C^{0,\alpha}(\partial\Omega^i)$. Then, by the continuity of the linear map that takes a function of $C^{0,\alpha}(\partial\Omega^o)$ to its integral over $\partial\Omega^o$ and of the linear map that takes a function of $C^{0,\alpha}(\partial\Omega^i)$ to its integral over $\partial\Omega^i$, we can conclude that $\mathcal{E}_1$ and $\mathcal{E}_2$ are real analytic. □

By the representation formula (8.87) for the energy integral, by the continuity at $\epsilon = 0$ of the corresponding functions $\mathcal{E}_1$, $\mathcal{E}_2$, $\mathcal{F}$, and $\mathcal{R}$, and by formula (8.88) for $\mathcal{E}_1(0)$, we have the (well-known) limiting relation

$$
\lim_{\epsilon \to 0^+} \int_{\Omega(\epsilon)} |\nabla u_\epsilon|^2 \, dx = \int_{\Omega^o} |\nabla u^o|^2 \, dx + \int_{\mathbb{R}^2 \setminus \Omega^i} |\nabla u^i|^2 \, d\xi \, . \tag{8.92}
$$

In particular, if $n = 2$, then the energy integral might not vanish for $\epsilon \to 0^+$ also when $f^o = 0$, as opposed to the case of dimension $n \geq 3$, where for $f^o = 0$ the energy vanishes as ϵ^{n-2} (cf. Theorem 8.10).

To conclude this section, we observe that the factor $(\mathcal{R}(\epsilon) + \frac{\log \epsilon}{2\pi})^{-1}$ which appears in the previous Theorems 8.29, 8.30, and 8.31 is the opposite of the electrostatic capacity of $\epsilon\overline{\Omega^i}$ in Ω^o (cf. definition (8.25)). Namely, we have the following.

Theorem 8.32. *There exists* $\epsilon_C \in]0, \epsilon_*[$ *such that*

$$\mathrm{Cap}[\epsilon\overline{\Omega^i}, \Omega^o] = -\frac{1}{\mathcal{R}(\epsilon) + \frac{\log \epsilon}{2\pi}} \qquad \forall \epsilon \in]0, \epsilon_C[\,.$$

Proof. We consider the definition of $\mathcal{F}$, $\mathcal{E}_1$, and $\mathcal{E}_2$ in the specific case where $f^i = 1$ and $f^o = 0$. By Proposition 8.27 we have

$$\mathcal{F}(\epsilon) = \int_{\partial\Omega^i} \Phi^i[\epsilon]\, d\sigma \qquad \forall \epsilon \in] - \epsilon_*, \epsilon_*[\,.$$

Since $\int_{\partial\Omega^i} \Phi^i[\epsilon]\, d\sigma = 1$ (cf. Proposition 8.20) we deduce that

$$\mathcal{F}(\epsilon) = 1 \qquad \forall \epsilon \in] - \epsilon_*, \epsilon_*[\,. \tag{8.93}$$

Now let $\epsilon_C \equiv \epsilon_E$, with ϵ_E as in Theorem 8.31. By the definition (8.90) of $\mathcal{E}_1$ we have

$$\mathcal{E}_1(\epsilon) = -\int_{\partial\Omega^i} \nu_{\Omega^i} \cdot \nabla\mathcal{U}_m[\epsilon]\, d\sigma \qquad \forall \epsilon \in] - \epsilon_C, \epsilon_C[\,.$$

Hence, the definition (8.83) of $\mathcal{U}_m$ and the formula

$$\nu_{\Omega^i} \cdot \nabla w^-_{\Omega^i}[\Psi^i[\epsilon]] = \nu_{\Omega^i} \cdot \nabla w^+_{\Omega^i}[\Psi^i[\epsilon]]$$

of Theorem 6.6 imply that

$$\begin{aligned}\mathcal{E}_1(\epsilon) &= -\int_{\partial\Omega^i} \nu_{\Omega^i} \cdot \nabla w^-_{\Omega^i}[\Psi^i[\epsilon]]\, d\sigma + \int_{\partial\Omega^i} \nu_{\Omega^i} \cdot \nabla_\xi \left(w_{\Omega^o}[\Psi^o[\epsilon]](\epsilon\xi)\right) d\sigma_\xi \\ &= -\int_{\partial\Omega^i} \nu_{\Omega^i} \cdot \nabla w^+_{\Omega^i}[\Psi^i[\epsilon]]\, d\sigma + \int_{\partial\Omega^i} \nu_{\Omega^i} \cdot \nabla_\xi \left(w_{\Omega^o}[\Psi^o[\epsilon]](\epsilon\xi)\right) d\sigma_\xi\end{aligned}$$

for all $\epsilon \in] - \epsilon_C, \epsilon_C[$. Since both the functions $w^+_{\Omega^i}[\Psi^i[\epsilon]]$ and $w_{\Omega^o}[\Psi^o[\epsilon]](\epsilon\cdot)$ are harmonic in Ω^i, a computation based on the first Green Identity of Theorem 4.2 shows that

$$\mathcal{E}_1(\epsilon) = 0 \qquad \forall \epsilon \in] - \epsilon_C, \epsilon_C[\,. \tag{8.94}$$

We now turn to $\mathcal{E}_2$. By (8.91) we have

$$\mathcal{E}_2(\epsilon) = -\int_{\partial\Omega^i} \nu_{\Omega^i} \cdot \nabla\mathcal{V}_m[\epsilon]\, d\sigma \qquad \forall \epsilon \in] - \epsilon_C, \epsilon_C[\,.$$

Then, the representation formula (8.84) for $\mathcal{V}_m$ implies that

$$\mathcal{E}_2(\epsilon) = -\int_{\partial\Omega^i} \nu_{\Omega^i} \cdot \nabla v^-_{\Omega^i}[\Phi^i[\epsilon]]\, d\sigma - \int_{\partial\Omega^i} \nu_{\Omega^i} \cdot \nabla_\xi \left(v_{\Omega^o}[\Phi^o[\epsilon]](\epsilon\xi)\right) d\sigma_\xi$$

for all $\epsilon \in]-\epsilon_C, \epsilon_C[$. Since $v_{\Omega^o}[\Phi^o[\epsilon]](\epsilon\cdot)$ is harmonic on Ω^i, a computation based on the first Green Identity of Theorem 4.2 shows that

$$\int_{\partial\Omega^i} \nu_{\Omega^i} \cdot \nabla_\xi \left(v_{\Omega^o}[\Phi^o[\epsilon]](\epsilon\xi)\right) d\sigma_\xi = 0\,.$$

Then, by the jump formula for $\nu_{\Omega^i} \cdot \nabla v^-_{\Omega^i}[\Phi^i[\epsilon]]$ of Theorem 6.6 we deduce that

$$\mathcal{E}_2(\epsilon) = -\int_{\partial\Omega^i} \frac{1}{2}\Phi^i[\epsilon] + W^*_\Omega\left[\Phi^i[\epsilon]\right] d\sigma \qquad \forall \epsilon \in]-\epsilon_C, \epsilon_C[\,.$$

Hence

$$\mathcal{E}_2(\epsilon) = -\int_{\partial\Omega^i} \Phi^i[\epsilon]\, d\sigma \qquad \forall \epsilon \in]-\epsilon_C, \epsilon_C[\,,$$

by Lemma 6.11, and thus

$$\mathcal{E}_2(\epsilon) = -1 \qquad \forall \epsilon \in]-\epsilon_C, \epsilon_C[\tag{8.95}$$

by Proposition 8.20. Then the validity of the theorem follows by (8.93), (8.94), (8.95), and by Theorem 8.31. □

By the representation formula of Theorem 8.32 and by the continuity of the function $\mathcal{R}$ at $\epsilon = 0$, $\mathrm{Cap}[\epsilon\overline{\Omega^i}, \Omega^o]$ vanishes as $1/\log\epsilon$ when $\epsilon \to 0^+$. Indeed, we have the (well-known) limiting relation

$$\lim_{\epsilon\to 0^+} (\log\epsilon)\, \mathrm{Cap}[\epsilon\overline{\Omega^i}, \Omega^o] = -2\pi\,.$$

8.4.4 Some Remarks on the Logarithmic Behavior

As a corollary of the representation Theorem 8.29 for u_ϵ, we can show that the map that takes ϵ to $u_\epsilon(x)$ can be described as a real analytic function of two variables defined in a neighborhood of $(0,0)$ in $\mathbb{R}^2$ and evaluated at $(\epsilon, 1/\log\epsilon)$. Namely, we have the following.

Corollary 8.33. *Let Ω_M and ϵ_M be as in Theorem 8.29. Then there exist ϵ'_M in $]0, \max\{1, \epsilon_M\}[$, ϵ''_M in $]-1/\log\epsilon'_M, +\infty[$, and a real analytic map*

$$\mathcal{W}_M :]-\epsilon'_M, \epsilon'_M[\times]-\epsilon''_M, \epsilon''_M[\to C^{1,\alpha}(\overline{\Omega_M})$$

such that

$$u_{\epsilon|\overline{\Omega_M}} = \mathcal{W}_M\left[\epsilon, \frac{1}{\log\epsilon}\right] \qquad \forall \epsilon \in]0, \epsilon'_M[\,. \tag{8.96}$$

Moreover,

$$\mathcal{W}_M[0,0] = u^o_{|\overline{\Omega_M}}\,. \tag{8.97}$$

Proof. We take ϵ'_M and ϵ''_M small enough so that

$$\mathcal{R}(\epsilon_1)\epsilon_2 + 2\pi \neq 0 \qquad \forall(\epsilon_1,\epsilon_2) \in]-\epsilon'_M,\epsilon'_M[\times]-\epsilon''_M,\epsilon''_M[$$

and then we define

$$\mathcal{W}_M[\epsilon_1,\epsilon_2] \equiv \mathcal{U}_M[\epsilon_1] + \mathcal{F}(\epsilon_1)\epsilon_2\frac{\mathcal{V}_M[\epsilon_1]}{\mathcal{R}(\epsilon_1)\epsilon_2 + 2\pi}$$

for all $(\epsilon_1,\epsilon_2) \in]-\epsilon'_M,\epsilon'_M[\times]-\epsilon''_M,\epsilon''_M[$. Since the reciprocal of a nonzero real analytic function is real analytic, Proposition 8.27 on the analyticity of $\mathcal{F}$, $\mathcal{R}$, and the analyticity of $\mathcal{U}_M$, which follows by Theorem 8.29, imply that $\mathcal{W}_M$ is real analytic. Moreover, condition $\epsilon''_M > -1/\log\epsilon'_M$ guarantees that $(\epsilon, 1/\log\epsilon) \in]-\epsilon'_M,\epsilon'_M[\times]-\epsilon''_M,\epsilon''_M[$ for all $\epsilon \in]0,\epsilon'_M[$. Then we can evaluate $\mathcal{W}_M$ at $(\epsilon, 1/\log\epsilon)$ and write equality (8.96). Since $\mathcal{W}_M[0,0] = \mathcal{U}_M[0]$, formula (8.97) for $\mathcal{W}_M[0,0]$ follows from formula (8.74) for $\mathcal{U}_M[0]$. □

We can see that results similar to Corollary 8.33 hold also for the behavior of u_ϵ close to the boundary of the hole $\epsilon\Omega^i$ and for its energy integral. Namely we have the following Corollaries 8.34 and 8.35 which can be deduced by Theorems 8.30 and 8.31, respectively, and by arguing as in the proof of Corollary 8.33.

Corollary 8.34. *Let Ω_m and ϵ_m be as in Theorem 8.30. Then there exist ϵ'_m in $]0,\max\{1,\epsilon_m\}[$, ϵ''_m in $]-1/\log\epsilon'_m,+\infty[$, and a real analytic map*

$$\mathcal{W}_m :]-\epsilon'_m,\epsilon'_m[\times]-\epsilon''_m,\epsilon''_m[\to C^{1,\alpha}(\overline{\Omega_m})$$

such that

$$u_\epsilon(\epsilon\cdot)_{|\overline{\Omega_m}} = \mathcal{W}_m\left[\epsilon, \frac{1}{\log\epsilon}\right] \qquad \forall\epsilon \in]0,\epsilon'_m[\,. \tag{8.98}$$

Moreover,

$$\mathcal{W}_m[0,0] = u^i_{|\overline{\Omega_m}}\,.$$

Corollary 8.35. *Let ϵ_E be as in Theorem 8.31. Then there exist ϵ'_E in $]0,\max\{1,\epsilon_E\}[$, ϵ''_E in $]-1/\log\epsilon'_E,+\infty[$, and a real analytic function*

$$\mathcal{E} :]-\epsilon'_E,\epsilon'_E[\times]-\epsilon''_E,\epsilon''_E[\to \mathbb{R}$$

such that

$$\int_{\Omega(\epsilon)} |\nabla u_\epsilon|^2\,dx = \mathcal{E}\left(\epsilon, \frac{1}{\log\epsilon}\right) \qquad \forall\epsilon \in]0,\epsilon'_E[\,. \tag{8.99}$$

Moreover,

$$\mathcal{E}(0,0) = \int_{\Omega^o} |\nabla u^o|^2\,dx + \int_{\mathbb{R}^2\setminus\Omega^i} |\nabla u^i|^2\,d\xi\,.$$

Then we may ask ourselves if equality (8.96) determines uniquely the real analytic map $\mathcal{W}_M$, and similarly, if equalities (8.98) and (8.99) determine uniquely $\mathcal{W}_m$

and $\mathcal{E}$, respectively. Indeed, if the dimension n is bigger than or equal to 3, then we can invoke the Identity Principle for real analytic maps of Proposition 2.16 to conclude that the map $\mathcal{U}_M$ that satisfies (8.15) is essentially unique, all other analytic maps that satisfy (8.15) being extension or restrictions of $\mathcal{U}_M$ (a similar result holds also for the other real analytic maps that we have introduced in dimension $n \geq 3$ to describe u_ϵ and its energy).

Instead, in dimension $n = 2$, the classical Identity Principle for real analytic maps does not help us, because it requires that two functions coincide on a nonempty open subset of their domain to be identical (cf. Proposition 2.16), and equality (8.96), as well as equalities (8.98) and (8.99), holds only on the graph of $\epsilon \mapsto 1/\log \epsilon$, which is a nowhere dense subset of $\mathbb{R}^2$ (i.e., a set whose closure has empty interior).

It turns out, however, that the set $\{(\epsilon, 1/\log \epsilon) \mid \epsilon \in]0, \epsilon'_M[\}$ has some particular features related to the fact that it is not subanalytic (cf., e.g., Krantz and Parks [147, Chap. 6]). In particular, we have the following folklore Lemma 8.36, which implies that the analytic map $\mathcal{W}_M$ that satisfies (8.96) is essentially unique, all other possible maps being extensions or restrictions of $\mathcal{W}_M$, and similarly $\mathcal{W}_m$ and $\mathcal{E}$ are essentially identified by (8.98) and (8.99), respectively.

Lemma 8.36. *Let $\Omega \subseteq \mathbb{R}^2$ be an open connected neighborhood of $(0,0)$. Let $\mathcal{A}$ be a real analytic function from Ω to a Banach space X. Assume that $\mathcal{A}(\epsilon, 1/\log \epsilon) = 0$ for all $(\epsilon, 1/\log \epsilon) \in \Omega$ with $\epsilon > 0$. Then $\mathcal{A}(\epsilon_1, \epsilon_2) = 0$ for all $(\epsilon_1, \epsilon_2) \in \Omega$.*

Proof. Since $\mathcal{A}$ is real analytic, there exist $\delta_1, \delta_2 > 0$ and a family $\{a_{(i,j)}\}_{(i,j)\in\mathbb{N}^2}$ in X such that $]-\delta_1, \delta_1[\times]-\delta_2, \delta_2[\subseteq \Omega$ and

$$\mathcal{A}(\epsilon_1, \epsilon_2) = \sum_{(i,j)\in\mathbb{N}^2} a_{(i,j)} \epsilon_1^i \epsilon_2^j \qquad \forall (\epsilon_1, \epsilon_2) \in]-\delta_1, \delta_1[\times]-\delta_2, \delta_2[, \tag{8.100}$$

where the series converges normally in X. We will prove that $a_{(i,j)} = 0$ for all $(i,j) \in \mathbb{N}^2$. Then the validity of the lemma follows by the Identity Principle for real analytic functions of several variables (cf. Proposition 2.16).

Possibly shrinking δ_1, we can assume that $1/\log \epsilon \in]-\delta_2, 0[$ for all $\epsilon \in]0, \delta_1[$. Hence,

$$\sum_{(i,j)\in\mathbb{N}^2} a_{(i,j)} \epsilon^i (1/\log \epsilon)^j = \mathcal{A}(\epsilon, 1/\log \epsilon) = 0$$

for all $\epsilon \in]0, \delta_1[$. It follows that $a_{(0,0)} = \lim_{\epsilon\to 0^+} \mathcal{A}(\epsilon, 1/\log \epsilon) = 0$. Then, an inductive argument on j shows that $a_{(0,j)} = 0$ for all $j \in \mathbb{N}$. Indeed, if $k \in \mathbb{N}$ and $a_{(0,j)} = 0$ for all $j \leq k$, then we have

$$\mathcal{A}(\epsilon_1, \epsilon_2) = \epsilon_2^{k+1} \mathcal{A}_{(0,k+1)}(\epsilon_1, \epsilon_2) + \epsilon_1 \mathcal{A}_1(\epsilon_1, \epsilon_2)$$

with

$$\mathcal{A}_{(0,k+1)}(\epsilon_1, \epsilon_2) \equiv \sum_{j\in\mathbb{N}\setminus\{0,\dots,k\}} a_{(0,j)}\, \epsilon_2^{j-k-1}$$

and

$$\mathcal{A}_1(\epsilon_1, \epsilon_2) \equiv \sum_{(i,j)\in(\mathbb{N}\setminus\{0\})\times\mathbb{N}} a_{(i,j)}\epsilon_1^{i-1}\epsilon_2^j$$

for all $(\epsilon_1, \epsilon_2) \in]-\delta_1, \delta_1[\times]-\delta_2, \delta_2[$. Since the series in (8.100) converges normally in X, we can verify that also the series for $\mathcal{A}_{(0,k+1)}$ and $\mathcal{A}_1$ converge normally for (ϵ_1, ϵ_2) in a neighborhood of $(0, 0)$. Thus Lemma A.13 implies that the maps $\mathcal{A}_{(0,k+1)}$ and $\mathcal{A}_1$ are analytic, and therefore continuous, in a neighborhood of $(0, 0)$. In particular, we have

$$\lim_{(\epsilon_1,\epsilon_2)\to(0,0)} \mathcal{A}_{(0,k+1)}(\epsilon_1, \epsilon_2) = a_{(0,k+1)} \, .$$

Since

$$(\log \epsilon)^{k+1}\mathcal{A}(\epsilon, 1/\log \epsilon) = 0 \qquad \forall \epsilon \in]0, \delta_1[\,,$$

it follows that

$$\begin{aligned} a_{(0,k+1)} &= \lim_{\epsilon\to 0^+} \mathcal{A}_{(0,k+1)}(\epsilon, 1/\log \epsilon) + \epsilon(\log \epsilon)^{k+1}\mathcal{A}_1(\epsilon, 1/\log \epsilon) \\ &= \lim_{\epsilon\to 0^+} (\log \epsilon)^{k+1}\mathcal{A}(\epsilon, 1/\log \epsilon) = 0\, . \end{aligned}$$

Now we argue by induction on i. Let $k \in \mathbb{N}$ and assume that $a_{(i,j)} = 0$ for all $(i, j) \in \mathbb{N}^2$ with $i \leq k$. Let $\mathcal{A}_{k+1}$ denote the function from $] - \delta_1, \delta_1[\times] - \delta_2, \delta_2[$ to $\mathbb{R}$ defined by

$$\mathcal{A}_{k+1}(\epsilon_1, \epsilon_2) \equiv \epsilon_1^{-(k+1)}\mathcal{A}(\epsilon_1, \epsilon_2) \qquad \forall(\epsilon_1, \epsilon_2) \in] - \delta_1, \delta_1[\times] - \delta_2, \delta_2[\,.$$

Then

$$\mathcal{A}_{k+1}(\epsilon_1, \epsilon_2) = \sum_{(i,j)\in(\mathbb{N}\setminus\{0,\dots,k\})\times\mathbb{N}} a_{(i,j)}\epsilon_1^{i-(k+1)}\epsilon_2^j \qquad \forall(\epsilon_1, \epsilon_2) \in] - \delta_1, \delta_1[\times] - \delta_2, \delta_2[$$

and, by taking $b_{(i,j)} = a_{(i+k+1,j)}$ for all $(i, j) \in \mathbb{N}^2$, we can write

$$\mathcal{A}_{k+1}(\epsilon_1, \epsilon_2) = \sum_{(i,j)\in\mathbb{N}^2} b_{(i,j)}\epsilon_1^i\epsilon_2^j \qquad \forall(\epsilon_1, \epsilon_2) \in] - \delta_1, \delta_1[\times] - \delta_2, \delta_2[\,.$$

Since the series in (8.100) converges normally in X, we can verify that also the series above converges normally in X for (ϵ_1, ϵ_2) in a neighborhood of $(0, 0)$. Moreover, $\mathcal{A}_{k+1}(\epsilon, 1/\log \epsilon) = 0$ for all $\epsilon \in]0, \delta_1[$. Then, by arguing as above for $\mathcal{A}$, we can verify that $b_{(0,j)} = a_{(k+1,j)} = 0$ for all $j \in \mathbb{N}$. Hence $a_{(i,j)} = 0$ for all $(i, j) \in \mathbb{N}^2$ and the proof is complete. □

Now that we know that the maps $\mathcal{W}_M$, $\mathcal{W}_m$, and $\mathcal{E}$ are essentially unique, we want to understand if also in dimension $n = 2$ there are cases where the asymptotic behavior of u_ϵ can be described by a real analytic map of ϵ, as it happens for $n \geq 3$. In other words, we wish to understand if in certain particular cases we can get rid of the logarithmic terms in the representation formulas for u_ϵ, $u_\epsilon(\epsilon\cdot)$, and the energy integral of u_ϵ.

In the following Theorem 8.38 we present a list of equivalent conditions. In particular, we prove that either $\epsilon \mapsto u_\epsilon(x)$ and $\epsilon \mapsto u_\epsilon(\epsilon\xi)$ have a real analytic continuation on a neighborhood of $\epsilon = 0$ for all $x \in \Omega^o \setminus \{0\}$ and all $\xi \in \mathbb{R}^2 \setminus \overline{\Omega^i}$, or the logarithmic terms appear for each and every point $x \in \Omega^o \setminus \{0\}$ and $\xi \in \mathbb{R}^2 \setminus \overline{\Omega^i}$. Moreover, the representation formula for u_ϵ contains logarithmic terms if and only if that for its energy integral does. The key point in the proof of Theorem 8.38 is to show that the absence of a logarithmic behavior at a point $x \in \Omega^o \setminus \{0\}$, or a point $\xi \in \mathbb{R}^2 \setminus \overline{\Omega^i}$, or for the energy integral, implies that $\mathcal{F}(\epsilon) = 0$ for all $\epsilon \in]-\epsilon_*, \epsilon_*[$.

To prove Theorem 8.38 we shall also need the following Lemma 8.37, which is an elementary consequence of the asymptotic behavior of $\log \epsilon$ as ϵ tends to zero and of standard properties of real analytic functions.

Lemma 8.37. *Let $\epsilon' > 0$. Let f and g be real analytic maps from $]-\epsilon', \epsilon'[$ to $\mathbb{R}$ such that $f(\epsilon) \log \epsilon = g(\epsilon)$ for all $\epsilon \in]0, \epsilon'[$. Then $f = g = 0$.*

Theorem 8.38. *The following statements are equivalent.*

(i) For all open subsets Ω_M of Ω^o such that $0 \notin \overline{\Omega_M}$ and all $\epsilon_M \in]0, \epsilon_*[$ small enough that $\overline{\Omega_M} \cap \epsilon\overline{\Omega^i} = \emptyset$ for all $\epsilon \in]-\epsilon_M, \epsilon_M[$, there exists a real analytic map

$$\tilde{\mathcal{W}}_M :]-\epsilon_M, \epsilon_M[\to C^{1,\alpha}(\overline{\Omega_M})$$

such that

$$u_{\epsilon|\overline{\Omega_M}} = \tilde{\mathcal{W}}_M[\epsilon] \qquad \forall \epsilon \in]0, \epsilon_M[.$$

(ii) For all open and bounded subsets Ω_m of $\mathbb{R}^n \setminus \overline{\Omega^i}$ and all $\epsilon_m \in]0, \epsilon_0[$ small enough that $\epsilon\overline{\Omega_m} \subseteq \Omega^o$ for all $\epsilon \in]-\epsilon_m, \epsilon_m[$, there exists a real analytic map

$$\tilde{\mathcal{W}}_m :]-\epsilon_m, \epsilon_m[\to C^{1,\alpha}(\overline{\Omega_m})$$

such that

$$u_\epsilon(\epsilon\cdot)_{|\overline{\Omega_m}} = \tilde{\mathcal{W}}_m[\epsilon] \qquad \forall \epsilon \in]0, \epsilon_m[.$$

(iii) There exist $x^o \in \Omega^o \setminus \{0\}$, $\epsilon^o \in]0, \epsilon_0[$, and a real analytic function

$$\tilde{\mathcal{W}}^o :]-\epsilon^o, \epsilon^o[\to \mathbb{R}$$

such that $x^o \in \Omega(\epsilon)$ for all $\epsilon \in]-\epsilon^o, \epsilon^o[$ and

$$u_\epsilon(x^o) = \tilde{\mathcal{W}}^o(\epsilon) \qquad \forall \epsilon \in]0, \epsilon^o[.$$

(iv) There exist $\xi^i \in \mathbb{R}^2 \setminus \overline{\Omega^i}$, $\epsilon^i \in]0, \epsilon_0[$, and a real analytic function

$$\tilde{\mathcal{W}}^i :]-\epsilon^i, \epsilon^i[\to \mathbb{R}$$

such that $\epsilon\xi^i \in \Omega^o$ for all $\epsilon \in]-\epsilon^i, \epsilon^i[$ and

$$u_\epsilon(\epsilon\xi^i) = \tilde{\mathcal{W}}^i(\epsilon) \qquad \forall \epsilon \in]0, \epsilon^i[\,.$$

(v) There exist $\epsilon^e \in]0, \epsilon_0[$ *and a real analytic function*

$$\tilde{\mathcal{E}} :]-\epsilon^e, \epsilon^e[\to \mathbb{R}$$

such that

$$\int_{\Omega(\epsilon)} |\nabla u_\epsilon|^2 \, dx = \tilde{\mathcal{E}}(\epsilon) \qquad \forall \epsilon \in]0, \epsilon^e[\,.$$

(vi) We have

$$\mathcal{F}(\epsilon) = 0 \qquad \forall \epsilon \in]-\epsilon_*, \epsilon_*[\,.$$

Proof. Clearly (i) implies (iii) and (ii) implies (iv). By Theorems 8.29, 8.30, and 8.31 it follows that (vi) implies (i), (ii), and (v). Then, to complete the proof, it suffices to show that each of (iii), (iv), and (v) implies (vi).

First we assume that (iii) holds true and we prove (vi). We take an open subset Ω_M of Ω^o such that $x^o \in \Omega_M$ and $0 \notin \overline{\Omega_M}$ and we consider the real analytic maps $\mathcal{U}_M$ and $\mathcal{V}_M$ of Theorem 8.29. Then the functions $\epsilon \mapsto \mathcal{U}_M[\epsilon](x^o)$ and $\epsilon \mapsto \mathcal{V}_M[\epsilon](x^o)$ are real analytic from $]-\epsilon_M, \epsilon_M[$ to $\mathbb{R}$. Moreover, the representation formula (8.73) for $u_\epsilon(x^o)$ implies that

$$\left(\tilde{\mathcal{W}}^o(\epsilon) - \mathcal{U}_M[\epsilon](x^o)\right) \frac{\log \epsilon}{2\pi} = \mathcal{F}(\epsilon)\mathcal{V}_M[\epsilon](x^o) - \left(\tilde{\mathcal{W}}^o(\epsilon) - \mathcal{U}_M[\epsilon](x^o)\right) \mathcal{R}(\epsilon)$$

for all $\epsilon \in]0, \epsilon^o_M[$, with $\epsilon^o_M \equiv \min\{\epsilon_M, \epsilon^o\}$ (see also (8.72)). Thus, by Lemma 8.37 we have

$$\tilde{\mathcal{W}}^o(\epsilon) - \mathcal{U}_M[\epsilon](x^o) = 0$$

and

$$\mathcal{F}(\epsilon)\mathcal{V}_M[\epsilon](x^o) - \left(\tilde{\mathcal{W}}^o(\epsilon) - \mathcal{U}_M[\epsilon](x^o)\right) \mathcal{R}(\epsilon) = 0$$

for all $\epsilon \in]0, \epsilon^o_M[$. Hence,

$$\mathcal{F}(\epsilon)\mathcal{V}_M[\epsilon](x^o) = 0 \qquad \forall \epsilon \in]0, \epsilon^o_M[\,. \tag{8.101}$$

To prove that $\mathcal{F}(\epsilon) = 0$, we show that $\mathcal{V}_M[\epsilon](x^o) \neq 0$ for all $\epsilon \in]0, \epsilon^o_M[$. To do so, we note that the definition of $\mathcal{V}_M[\epsilon]$ in (8.77), equality (8.47) for $\Phi[\epsilon]$, and equality (8.37) for τ_ϵ imply that

$$\mathcal{V}_M[\epsilon](x^o) = v_{\Omega(\epsilon)}[\tau_\epsilon](x^o)$$

for all $\epsilon \in]0, \epsilon^o_M[$. Let now $\epsilon \in]0, \epsilon^o_M[$ be fixed. By system (8.32), which defines τ_ϵ and by Lemmas 6.28 and 6.29, τ_ϵ is a generator of $\mathrm{Ker}(\frac{1}{2}I + W^t_{\Omega(\epsilon)})$ and $v_{\Omega(\epsilon)}[\tau_\epsilon]_{|\partial\Omega(\epsilon)}$ is a generator of $\mathrm{Ker}(\frac{1}{2}I + W_{\Omega(\epsilon)})$. Then by Theorem 6.14 there is a nonzero real number c_ϵ such that

$$v_{\Omega(\epsilon)}[\tau_\epsilon]_{|\partial\Omega^o} = 0 \qquad \text{and} \qquad v_{\Omega(\epsilon)}[\tau_\epsilon]_{|\epsilon\partial\Omega^i} = c_\epsilon\,. \tag{8.102}$$

Since $v_{\Omega(\epsilon)}[\tau_\epsilon]$ is harmonic in $\Omega(\epsilon)$, the Maximum Principle implies that 0 is a maximum or minimum value for $v_{\Omega(\epsilon)}[\tau_\epsilon]$ in $\overline{\Omega(\epsilon)}$, depending whether $c_\epsilon > 0$ or $c_\epsilon < 0$ (cf. Definition 3.3 and Theorem 3.4). Moreover, $v_{\Omega(\epsilon)}[\tau_\epsilon]$ can attain the value 0 only on the boundary of $\Omega(\epsilon)$. In particular, we have

$$v_{\Omega(\epsilon)}[\tau_\epsilon](x^o) \neq 0$$

because x^o belongs to the interior of $\Omega(\epsilon)$. Then (8.101) implies that $\mathcal{F}(\epsilon) = 0$ for all $\epsilon \in]0, \epsilon_M^o[$. To prove (v) it remains to observe that equality $\mathcal{F}(\epsilon) = 0$ extends to the whole of $] - \epsilon_*, \epsilon_*[$ by the Identity Principle for real analytic functions of Proposition 2.16.

We now assume that (iv) holds true, and we prove (vi). Let Ω_m be an open subset of $\mathbb{R}^2 \setminus \overline{\Omega^i}$ that contains ξ^i. Let $\mathcal{U}_m$ and $\mathcal{U}_m$ be the real analytic maps of Theorem 8.30. Then the functions $\epsilon \mapsto \mathcal{U}_m[\epsilon](x^o)$ and $\epsilon \mapsto \mathcal{V}_m[\epsilon](x^o)$ are real analytic from $] - \epsilon_m, \epsilon_m[$ to $\mathbb{R}$. Moreover, the representation formula (8.80) for $u_\epsilon(\epsilon\xi^i)$ implies that

$$\begin{aligned}\left(\tilde{\mathcal{W}}^i(\epsilon) - \mathcal{U}_m[\epsilon](\xi^i) - \mathcal{F}(\epsilon)\right) \frac{\log \epsilon}{2\pi} \\ = \mathcal{F}(\epsilon)\mathcal{V}_m[\epsilon](\xi^i) - \left(\tilde{\mathcal{W}}^i(\epsilon) - \mathcal{U}_m[\epsilon](\xi^i)\right) \mathcal{R}(\epsilon)\end{aligned}$$

for all $\epsilon \in]0, \epsilon_m^i[$, with $\epsilon_m^i \equiv \min\{\epsilon_m, \epsilon^i\}$ (see also (8.72)). Then, by Lemma 8.37, we have

$$\tilde{\mathcal{W}}^i(\epsilon) - \mathcal{U}_m[\epsilon](\xi^i) - \mathcal{F}(\epsilon) = 0$$

and

$$\mathcal{F}(\epsilon)\mathcal{V}_m[\epsilon](\xi^i) - \left(\tilde{\mathcal{W}}^i(\epsilon) - \mathcal{U}_m[\epsilon](\xi^i)\right) \mathcal{R}(\epsilon) = 0$$

for all $\epsilon \in]0, \epsilon_m^i[$. It follows that

$$\mathcal{F}(\epsilon) \left(\mathcal{V}_m[\epsilon](\xi^i) - \mathcal{R}(\epsilon)\right) = 0 \qquad \forall \epsilon \in]0, \epsilon_m^i[. \tag{8.103}$$

Let now $\epsilon \in]0, \epsilon_m^i[$ be fixed. We prove that $\mathcal{V}_m[\epsilon](\xi^i) - \mathcal{R}(\epsilon) \neq 0$. To do so, we assume for the sake of contradiction that $\mathcal{V}_m[\epsilon](\xi^i) - \mathcal{R}(\epsilon) = 0$. Then we have

$$\mathcal{V}_m[\epsilon](\xi^i) + \frac{\log \epsilon}{2\pi} = \mathcal{R}(\epsilon) + \frac{\log \epsilon}{2\pi}$$

and by the definition of $\mathcal{V}_m[\epsilon]$ in (8.84) and of $\mathcal{R}(\epsilon)$ in Proposition 8.27 we can write

$$\begin{aligned}v_{\Omega^i}[\Phi^i[\epsilon]](\xi^i) + v_{\Omega^o}[\Phi^o[\epsilon]](\epsilon\xi^i) + \frac{\log \epsilon}{2\pi} \\ = \fint_{\partial\Omega^i} v_{\Omega^i}[\Phi^i[\epsilon]](\eta) + v_{\partial\Omega^o}[\Phi^o[\epsilon]](\epsilon\eta)\, d\sigma_\eta + \frac{\log \epsilon}{2\pi} .\end{aligned}$$

Then, by equality (8.47) of $\Phi[\epsilon]$ and $(\phi_\epsilon^i, \phi_\epsilon^o)$, by definition (8.37) of ϕ_ϵ^i and ϕ_ϵ^o in terms of τ_ϵ, and by changing the variable of integration on the integrals over $\partial\Omega^i$, we can verify that

$$v_{\Omega(\epsilon)}[\tau_\epsilon](\epsilon\xi^i) = \fint_{\epsilon\partial\Omega^i} v_{\Omega(\epsilon)}[\tau_\epsilon]\, d\sigma\,.$$

Then (8.102) implies that the right-hand side of the last equality is equal to c_ϵ and we have

$$v_{\Omega(\epsilon)}[\tau_\epsilon](\epsilon\xi^i) = c_\epsilon.$$

But this is a contradiction, because c_ϵ is either a maximum or a minimum for $v_{\Omega(\epsilon)}[\tau_\epsilon]$ in $\overline{\Omega(\epsilon)}$, $v_{\Omega(\epsilon)}[\tau_\epsilon]$ attains its maximum and minimum on $\partial\Omega(\epsilon)$, and $\epsilon\xi^i$ belongs to the interior of $\Omega(\epsilon)$ for all $\epsilon \in]0, \epsilon^i_m[$ (cf. Definition 3.3 and Theorem 3.4). Hence, $\mathcal{V}_m[\epsilon](\xi^i) - \mathcal{R}(\epsilon) \neq 0$ and (8.103) implies that $\mathcal{F}(\epsilon) = 0$ for all $\epsilon \in]0, \epsilon^i_m[$. We conclude by the Identity Principle that $\mathcal{F}(\epsilon) = 0$ for all $\epsilon \in]-\epsilon_*, \epsilon_*[$ (cf. Proposition 2.16).

Finally, we show that (v) implies (vi). If (v) holds true then the representation formula (8.87) for the energy integral of u_ϵ implies that

$$\left(\tilde{\mathcal{E}}(\epsilon) - \mathcal{E}_1(\epsilon)\right) \frac{\log\epsilon}{2\pi} = \mathcal{F}(\epsilon)\mathcal{E}_2(\epsilon) - \left(\tilde{\mathcal{E}}(\epsilon) - \mathcal{E}_1(\epsilon)\right) \mathcal{R}(\epsilon)$$

for all $\epsilon \in]0, \epsilon^e_E[$, with $\epsilon^e_E \equiv \min\{\epsilon_E, \epsilon^e\}$ (see also (8.72)). By Lemma 8.37 it follows that

$$\tilde{\mathcal{E}}(\epsilon) - \mathcal{E}_1(\epsilon) = 0 \qquad \text{and} \qquad \mathcal{F}(\epsilon)\mathcal{E}_2(\epsilon) - \left(\tilde{\mathcal{E}}(\epsilon) - \mathcal{E}_1(\epsilon)\right) \mathcal{R}(\epsilon) = 0$$

for all $\epsilon \in]0, \epsilon^e_E[$. Hence,

$$\mathcal{F}(\epsilon)\mathcal{E}_2(\epsilon) = 0 \qquad \forall \epsilon \in]0, \epsilon^e_E[\,. \tag{8.104}$$

We now prove that if $\mathcal{E}_2(\epsilon)$ vanishes, then also $\mathcal{F}(\epsilon)$ must vanish. To do so, we observe that by the definition of $\mathcal{E}_2(\epsilon)$ in (8.91) and by those of $\mathcal{V}_M[\epsilon]$ and $\mathcal{V}_m[\epsilon]$ in (8.77) and (8.84), respectively, we have

$$\begin{aligned}
&\mathcal{E}_2(\epsilon) \\
&= \int_{\partial\Omega^o} f^o(x)\, \nu_{\Omega^o}(x) \cdot \nabla_x \left(\int_{\partial\Omega^i} \Phi^i[\epsilon](\eta)\, S(x - \epsilon\eta)\, d\sigma_\eta + v^+_{\Omega^o}[\Phi^o[\epsilon]](x) \right) d\sigma_x \\
&\quad - \int_{\partial\Omega^i} f^i(\xi)\, \nu_{\Omega^i}(\xi) \cdot \nabla_\xi \left(v^-_{\Omega^i}[\Phi^i[\epsilon]](\xi) + v_{\Omega^o}[\Phi^o[\epsilon]](\epsilon\xi) \right) d\sigma_\xi
\end{aligned}$$

for all $\epsilon \in]0, \epsilon^e_E[$. Then, changing the variable of integration in the integrals over $\partial\Omega^i$, we obtain

$$\begin{aligned}
&\mathcal{E}_2(\epsilon) \\
&= \int_{\partial\Omega^o} f^o(x)\, \nu_{\Omega^o}(x) \cdot \nabla_x \left(\int_{\epsilon\partial\Omega^i} \epsilon^{-1}\Phi^i[\epsilon](y/\epsilon)\, S(x - y)\, d\sigma_y + v^+_{\Omega^o}[\Phi^o[\epsilon]](x) \right) d\sigma_x \\
&\quad - \int_{\epsilon\partial\Omega^i} f^i(x/\epsilon)\, \nu_{\epsilon\Omega^i}(x) \cdot \nabla_x \Big(v^-_{\epsilon\Omega^i}[\epsilon^{-1}\Phi^i[\epsilon](\cdot/\epsilon)](x)
\end{aligned}$$

$$-\frac{\log\epsilon}{2\pi}\int_{\epsilon\partial\Omega^i}\epsilon^{-1}\Phi^i[\epsilon](y/\epsilon)\,d\sigma_y+v_{\Omega^o}[\Phi^o[\epsilon]](x)\Big)\,d\sigma_x$$

for all $\epsilon\in]0,\epsilon_E^e[$. By the definition of ϕ^i_ϵ in terms of τ_ϵ of (8.37), by the equality (8.47) of ϕ^i_ϵ and $\Phi^i[\epsilon]$, and by the second equation of (8.32) we have

$$\int_{\epsilon\partial\Omega^i}\epsilon^{-1}\Phi^i[\epsilon](y/\epsilon)\,d\sigma_y=\int_{\epsilon\partial\Omega^i}\tau_\epsilon\,d\sigma=1\,.$$

Then, by the definition of $(\phi^i_\epsilon,\phi^o_\epsilon)$ in terms of τ_ϵ of (8.37), by the equality (8.47) of $(\phi^i_\epsilon,\phi^o_\epsilon)$ and $(\Phi^i[\epsilon],\Phi^o[\epsilon])$, and by the boundary condition of (8.4) for u_ϵ on $\partial\Omega(\epsilon)$, we obtain

$$\mathcal{E}_2(\epsilon)=\int_{\partial\Omega(\epsilon)}u_\epsilon\,\nu_{\Omega(\epsilon)}\cdot\nabla v^+_{\Omega(\epsilon)}[\tau_\epsilon]\,d\sigma$$

for all $\epsilon\in]0,\epsilon_E^e[$. Then, by the second Green Identity of Theorem 4.3, we have

$$\mathcal{E}_2(\epsilon)=\int_{\partial\Omega(\epsilon)}v^+_{\Omega(\epsilon)}[\tau_\epsilon]\,\nu_{\Omega(\epsilon)}\cdot\nabla u_\epsilon\,d\sigma$$

for all $\epsilon\in]0,\epsilon_E^e[$. By (8.102) we deduce that

$$\mathcal{E}_2(\epsilon)=-c_\epsilon\int_{\epsilon\partial\Omega^i}\nu_{\epsilon\Omega^i}\cdot\nabla u_\epsilon\,d\sigma$$

for all $\epsilon\in]0,\epsilon_E^e[$. Now let $\epsilon\in]0,\epsilon_E^e[$ and assume that $\mathcal{E}_2(\epsilon)=0$. Since $c_\epsilon\neq 0$ it follows that

$$\int_{\epsilon\partial\Omega^i}\nu_{\epsilon\Omega^i}\cdot\nabla u_\epsilon\,d\sigma=0$$

which in turn implies that $\mathcal{F}(\epsilon)=0$ by Remark 8.28. Since (8.104) implies that at least one of $\mathcal{F}(\epsilon)$ and $\mathcal{E}_2(\epsilon)$ is 0, we conclude that $\mathcal{F}(\epsilon)=0$ for all $\epsilon\in]0,\epsilon_E^e[$. Then (vi) follows by the Identity Principle for real analytic maps of Proposition 2.16. □

By the definition of $\mathcal{F}(\epsilon)$ in Proposition 8.27, condition (vi) can be written in the form

$$\int_{\partial\Omega^i}f^i\,\Phi^i[\epsilon]\,d\sigma+\int_{\partial\Omega^o}f^o\,\Phi^o[\epsilon]\,d\sigma=0\qquad\forall\epsilon\in]-\epsilon_*,\epsilon_*[\,. \tag{8.105}$$

Then, the advantage of (vi) with respect to (i)–(v) is that (vi) can be verified on the boundary data f^i and f^o and does not require to study the solution u_ϵ. In some simple cases, we can make such condition much more explicit. This is for example the case when both f^i and f^o are constant functions.

Example 8.39. Assume that there exist real numbers c^i and c^o such that $f^i(\xi)=c^i$ for all $\xi\in\partial\Omega^i$ and $f^o(x)=c^o$ for all $x\in\partial\Omega^o$. Then statements (i)–(vi) of Theorem 8.38 hold if and only if $c^i=c^o$.

Proof. If $c^o=c^i$ then $u_\epsilon(x)=c^o$ for all $\epsilon\in]0,\epsilon_0[$ and all $x\in\overline{\Omega(\epsilon)}$. Then the validity of (i)–(v) is apparent and (vi) follows by Theorem 8.38. In particular, by taking $c^o = c^i = 1$ in (vi) we obtain

$$\int_{\partial\Omega^i} \Phi^i[\epsilon]\, d\sigma + \int_{\partial\Omega^o} \Phi^o[\epsilon]\, d\sigma = 0 \qquad \forall \epsilon \in]-\epsilon_*, \epsilon_*[\tag{8.106}$$

(see also (8.105)). To prove that (i)–(vi) imply $c^i = c^o$, we observe that with our assumptions on f^i and f^o condition (vi) is equivalent to

$$c^i \int_{\partial\Omega^i} \Phi^i[\epsilon]\, d\sigma + c^o \int_{\partial\Omega^o} \Phi^o[\epsilon]\, d\sigma = 0 \qquad \forall \epsilon \in]-\epsilon_*, \epsilon_*[\tag{8.107}$$

(see also (8.105)). Then, combining (8.106) (which does not depend on c^i and c^o) with (8.107) we have

$$(c^i - c^o) \int_{\partial\Omega^i} \Phi^i[\epsilon]\, d\sigma = 0 \qquad \forall \epsilon \in]-\epsilon_*, \epsilon_*[\,.$$

Since $\int_{\partial\Omega^i} \Phi^i[\epsilon]\, d\sigma = 1$ (see (8.42) and Proposition 8.20) we deduce that $c^i = c^o$. □

In particular, by Example 8.39 it follows that the capacity $\mathrm{Cap}[\epsilon\overline{\Omega^i}, \Omega^o]$ defined by (8.25) always displays a logarithmic behavior in dimension two (a fact that can be proved also by Theorem 8.32).

Finally, we consider the case when Ω^i and Ω^o are open balls.

Example 8.40. Assume that $\Omega^o = \Omega^i = \mathbb{B}_2$ and $\Omega(\epsilon)$ is the annular domain $\mathbb{B}_2 \setminus \epsilon\overline{\mathbb{B}_2}$ for all $\epsilon \in]-1, 1[$. Then (i)–(vi) holds true if and only if

$$\int_{\partial\mathbb{B}_2} f^i\, d\sigma = \int_{\partial\mathbb{B}_2} f^o\, d\sigma \,. \tag{8.108}$$

Proof. By (8.42) and Proposition 8.20 and by a standard symmetry argument we verify that $\Phi^i[\epsilon](x) = \Phi^i[\epsilon](Tx)$ and $\Phi^o[\epsilon](x) = \Phi^o[\epsilon](Tx)$ for all $x \in \partial\Omega^o = \partial\mathbb{B}_2$ and for all orthogonal transformation T on $\mathbb{R}^2$. It follows that $\Phi^i[\epsilon](x)$ and $\Phi^o[\epsilon](x)$ are constant functions of $x \in \partial\mathbb{B}_2$. Then, by equality $\int_{\partial\Omega^i} \Phi^i[\epsilon]\, d\sigma = 1$ (cf. (8.42) and Proposition 8.20) and by (8.106) we deduce that

$$\Phi^i[\epsilon](x) = \frac{1}{2\pi} \quad \text{and} \quad \Phi^o[\epsilon](x) = -\frac{1}{2\pi} \qquad \forall x \in \partial\mathbb{B}_2\,,\ \epsilon \in]-1, 1[\,.$$

Hence, (vi) and (8.108) are equivalent (see also (8.105)). □

8.5 How to Compute the Coefficients (in Dimension 2)

As we have seen in Corollary 8.33, the asymptotic behavior of the solution u_ϵ can be described by means of a real analytic map of two variables evaluated at $(\epsilon, 1/\log\epsilon)$. In particular, for $x \in \Omega^o \setminus \{0\}$ and $\epsilon > 0$ small enough we have

$$u_\epsilon(x) = \sum_{i,j=0}^{\infty} u_{i,j}(x)\epsilon^i \left(\frac{1}{\log \epsilon}\right)^j \tag{8.109}$$

for some coefficients $\{u_{i,j}(x)\}_{(i,j)\in\mathbb{N}^2}$ in $\mathbb{R}$, where the power series

$$\sum_{i,j=0}^{\infty} u_{ij}(x)\epsilon_1^i \epsilon_2^j$$

is absolutely convergent for (ϵ_1, ϵ_2) in a neighborhood of $(0,0)$. In this section we show how to reduce the problem of computing the coefficients of a power series expansion of $u_\epsilon(x)$ to that of solving suitable recursive systems of boundary integral equations. However, instead of the series in (8.109) we consider a rearrangement in the form

$$u_\epsilon(x) = \sum_{k=0}^{\infty} \epsilon^k \sum_{l=0}^{k+1} \frac{\lambda_{M,(k,l)}(x)}{\left(r_0 + \frac{\log \epsilon}{2\pi}\right)^l} \tag{8.110}$$

for a certain real number r_0 and suitable real coefficients $\lambda_{M,(k,l)}(x)$. The advantage of (8.110) with respect to (8.109) is that for any $N \in \mathbb{N}$ it yields a finite approximations of $u_\epsilon(x)$ as ϵ approaches 0^+ with reminder in $O(\epsilon^N)$. For example, we can obtain a two terms approximation with a reminder in $O(\epsilon)$:

$$u_\epsilon(x) = \lambda_{M,(0,0)}(x) + \frac{\lambda_{M,(0,1)}(x)}{r_0 + \frac{\log \epsilon}{2\pi}} + O(\epsilon) \qquad \text{as } \epsilon \to 0^+ .$$

Instead, the approximation with reminder in $O(\epsilon)$ that we can deduce from (8.109) involves an infinite sum of logarithmic terms. Namely, we have to write

$$u_\epsilon(x) = \sum_{j=0}^{\infty} u_{0,j}(x) \left(\frac{1}{\log \epsilon}\right)^j + O(\epsilon) \qquad \text{as } \epsilon \to 0^+ .$$

In this sense the expansion in (8.110) is much more computational cost effective than the one in (8.109).

We will obtain expansions similar to (8.110) also for the rescaled function $u_\epsilon(\epsilon\cdot)$ and for the energy integral of u_ϵ. Moreover, we will compute the real number r_0 and the first coefficients of the asymptotic expansions in terms of the auxiliary functions u^o and u^i of (8.75) and (8.82) and of the Green functions of Ω^o and $\mathbb{R}^2 \setminus \overline{\Omega^i}$. When Ω^o and Ω^i are concentric balls in $\mathbb{R}^2$ (so that $\Omega(\epsilon)$ is a circular annulus) the expansions for u_ϵ, $u_\epsilon(\epsilon\cdot)$, and the energy integral can be notably simplified and we can easily carry out explicit computations using the fact that $W_{\mathbb{B}_2}[\phi] = \frac{1}{2}\fint_{\partial\mathbb{B}_2} \psi \, d\sigma$ for all $\psi \in C^{1,\alpha}(\partial\mathbb{B}_2)$.

The expansions that we write in this section can be compared with those obtained by Il'in [127] with the Matched Asymptotic Expansions Method and by Maz'ya et al. [203, 204] with the Compound (Multi-Scale) Asymptotic Expansions Method.

8.5.1 Series Expansions of $(\Phi^i[\epsilon], \Phi^o[\epsilon])$ and $(\Psi^i[\epsilon], \Psi^o[\epsilon])$

As a first step, we consider the power series expansions around 0 of the real analytic maps $\epsilon \mapsto (\Phi^i[\epsilon], \Phi^o[\epsilon])$ and $\epsilon \mapsto (\Psi^i[\epsilon], \Psi^o[\epsilon])$ of Propositions 8.20 and 8.26, respectively. In our calculations we will often use the chain rule formula

$$\partial_\epsilon^j (f(\epsilon x)) = \sum_{h=0}^{j} \binom{j}{h} x_1^h x_2^{j-h} (\partial_1^h \partial_2^{j-h} f)(\epsilon x) \tag{8.111}$$

which holds for all $j \in \mathbb{N}$, $\epsilon \in \mathbb{R}$, $x \equiv (x_1, x_2) \in \mathbb{R}^2$, and for all functions f that are analytic in a neighbourhood of ϵx. Here and in what follows $(\partial_j f)(y)$ with $j \in \{1, 2\}$ denotes the partial derivative with respect to x_j of the function $f(x) \equiv f(x_1, x_2)$ evaluated at $y \equiv (y_1, y_2) \in \mathbb{R}^2$.

In the following Proposition 8.41 we show that the coefficients of the power series expansion of the map $\epsilon \mapsto (\Phi^i[\epsilon], \Phi^o[\epsilon])$ can be computed as the solutions of a certain recursive system of boundary integral equations.

Proposition 8.41. *There exist $\epsilon_\Phi \in]0, \epsilon_{*1}]$ and a sequence of pairs $\{(\Phi^i_k, \Phi^o_k)\}_{k\in\mathbb{N}}$ in $C^{0,\alpha}(\partial\Omega^i) \times C^{0,\alpha}(\partial\Omega^o)$ such that*

$$\Phi^i[\epsilon] = \sum_{k=0}^{+\infty} \frac{\Phi^i_k}{k!}\epsilon^k \quad \textit{and} \quad \Phi^o[\epsilon] = \sum_{k=0}^{+\infty} \frac{\Phi^o_k}{k!}\epsilon^k \qquad \forall \epsilon \in]-\epsilon_\Phi, \epsilon_\Phi[\,, \tag{8.112}$$

where the two series converge normally in $C^{0,\alpha}(\partial\Omega^o)$ and in $C^{0,\alpha}(\partial\Omega^i)$, respectively, for all $\epsilon \in]-\epsilon_\Phi, \epsilon_\Phi[$. Moreover, the pair of functions (Φ^i_0, Φ^o_0) is the unique solution in $C^{0,\alpha}(\partial\Omega^i) \times C^{0,\alpha}(\partial\Omega^o)$ of the following system of integral equations

$$\frac{1}{2}\Phi^i_0(\xi) - W^t_{\Omega^i}[\Phi^i_0](\xi) = 0 \qquad \forall \xi \in \partial\Omega^i\,, \tag{8.113}$$

$$\frac{1}{2}\Phi^o_0(x) + W^t_{\Omega^o}[\Phi^o_0](x) = -\nu_{\Omega^o}(x)\cdot\nabla S_2(x) \qquad \forall x \in \partial\Omega^o\,, \tag{8.114}$$

$$\int_{\partial\Omega^i} \Phi^i_0 \, d\sigma = 1 \tag{8.115}$$

and for each $k \in \mathbb{N}\setminus\{0\}$ the pair (Φ^i_k, Φ^o_k) is the unique solution in $C^{0,\alpha}(\partial\Omega^i) \times C^{0,\alpha}(\partial\Omega^o)$ of the following system of integral equations which involves the preceding terms $\{(\Phi^i_j, \Phi^k_j)\}_{j=0}^{k-1}$,

$$\begin{aligned} &\frac{1}{2}\Phi^i_k(\xi) - W^t_{\Omega^i}[\Phi^i_k](\xi) \\ &= k\sum_{j=0}^{k-1}\binom{k-1}{j}(-1)^{j+1}\sum_{h=0}^{j}\binom{j}{h}\xi_1^h\xi_2^{j-h} \\ &\times \nu_{\Omega^i}(\xi)\cdot\int_{\partial\Omega^o}\Phi^o_{k-1-j}(\nabla\partial_1^h\partial_2^{j-h}S_2)\,d\sigma \qquad \forall \xi \in \partial\Omega^i\,, \end{aligned} \tag{8.116}$$

$$\frac{1}{2}\Phi^o_k(x) + W^t_{\Omega^o}[\Phi^o_k](x) \tag{8.117}$$
$$= \sum_{j=0}^{k} \binom{k}{j} (-1)^{j+1} \sum_{h=0}^{j} \binom{j}{h}$$
$$\times\, \nu_{\Omega^o}(x) \cdot (\nabla \partial_1^h \partial_2^{j-h} S_2)(x) \int_{\partial\Omega^i} \Phi^i_{k-j}(\eta) \eta_1^h \eta_2^{j-h}\, d\sigma_\eta \quad \forall x \in \partial\Omega^o\,,$$

$$\int_{\partial\Omega^i} \Phi^i_k \, d\sigma = 0\,. \tag{8.118}$$

Proof. The existence of ϵ_Φ and of the sequence $\{(\Phi^i_k, \Phi^o_k)\}_{k\in\mathbb{N}}$ such that (8.112) holds true follows by the real analyticity of the map that takes ϵ to $(\Phi^i[\epsilon], \Phi^o[\epsilon])$ (cf. Proposition 8.20). Then we observe that $\mathcal{N}(\epsilon)[\Phi^i[\epsilon], \Phi^o[\epsilon]] = (0,0,1)$ for all $\epsilon \in]-\epsilon_\Phi, \epsilon_\Phi[$ (cf. Proposition 8.20). Accordingly, the map that takes ϵ to $\mathcal{N}(\epsilon)[\Phi^i[\epsilon], \Phi^o[\epsilon]]$ has all derivatives equal to 0. Namely, $\partial^k_\epsilon(\mathcal{N}(\epsilon)[\Phi^i[\epsilon], \Phi^o[\epsilon]]) = 0$ for all $\epsilon \in]-\epsilon_\Phi, \epsilon_\Phi[$ and all $k \in \mathbb{N} \setminus \{0\}$. Then by the chain rule formula (8.111) and by a straightforward calculation we have

$$\partial^k_\epsilon(\mathcal{N}^i(\epsilon)[\Phi^i[\epsilon], \Phi^o[\epsilon]])(\xi) = \frac{1}{2}\partial^k_\epsilon \Phi^i[\epsilon](\xi) - W^t_{\Omega^i}[\partial^k_\epsilon \Phi^i[\epsilon]](\xi) \tag{8.119}$$
$$- \epsilon \sum_{j=0}^{k} \binom{k}{j} \sum_{h=0}^{j} \binom{j}{h} \xi_1^h \xi_2^{j-h}$$
$$\times \int_{\partial\Omega^o} \partial^{k-j}_\epsilon \Phi^o[\epsilon](y)\, \nu_{\Omega^i}(\xi) \cdot (\nabla \partial_1^h \partial_2^{j-h} S_2)(\epsilon\xi - y)\, d\sigma_y$$
$$- k \sum_{j=0}^{k-1} \binom{k-1}{j} \sum_{h=0}^{j} \binom{j}{h} \xi_1^h \xi_2^{j-h}$$
$$\times\, \nu_{\Omega^i}(\xi) \cdot \int_{\partial\Omega^o} \partial^{k-1-j}_\epsilon \Phi^o[\epsilon](y)\, (\nabla \partial_1^h \partial_2^{j-h} S_2)(\epsilon\xi - y)\, d\sigma_y = 0$$
$$\forall \xi \in \partial\Omega^i\,,$$

$$\partial^k_\epsilon(\mathcal{N}^o(\epsilon)[\Phi^i[\epsilon], \Phi^o[\epsilon]])(x) = \frac{1}{2}\partial^k_\epsilon \Phi^o[\epsilon](x) + W^t_{\Omega^o}[\partial^k_\epsilon \Phi^o[\epsilon]](x) \tag{8.120}$$
$$+ \sum_{j=0}^{k} \binom{k}{j} (-1)^j \sum_{h=0}^{j} \binom{j}{h}$$
$$\times\, \nu_{\Omega^o}(x) \cdot \int_{\partial\Omega^i} \partial^{k-j}_\epsilon \Phi^i[\epsilon](y)\, \eta_1^h \eta_2^{j-h} (\nabla \partial_1^h \partial_2^{j-h} S_2)(x - \epsilon\eta)\, d\sigma_\eta = 0$$
$$\forall x \in \partial\Omega^o\,,$$

$$\partial^k_\epsilon(\mathcal{N}^s(\epsilon)[\Phi^i[\epsilon], \Phi^o[\epsilon]]) = \int_{\partial\Omega^i} \partial^k_\epsilon \Phi^i[\epsilon]\, d\sigma = \delta_{0,k}\,, \tag{8.121}$$

for all $\epsilon \in]-\epsilon_\Phi, \epsilon_\Phi[$ and all $k \in \mathbb{N}$, where we understand that $\sum_{j=0}^{k-1}$ is omitted for $k = 0$. By standard properties of real analytic maps we have that

$$(\Phi_k^o, \Phi_k^i) = (\partial_\epsilon^k \Phi^o[0], \partial_\epsilon^k \Phi^i[0]) \qquad \forall k \in \mathbb{N}\,.$$

Then, taking $\epsilon = 0$ in (8.119)–(8.121) we deduce that (Φ_0^i, Φ_0^o) is a solution of (8.113)–(8.115) and that (Φ_k^i, Φ_k^o) is a solution of (8.116)–(8.118) for all $k \in \mathbb{N} \setminus \{0\}$. The uniqueness of the solutions of (8.113)–(8.115) and of (8.116)–(8.118) can be deduced by Corollary 6.15, which implies that $\frac{1}{2}I + W^t_{\Omega^o}$ is injective on $C^{0,\alpha}(\partial\Omega^o)$, and by Theorem 6.25, which implies that $-\frac{1}{2}I + W^t_{\Omega^i}$ is injective on $C^{0,\alpha}(\partial\Omega^i)_0$. □

We now turn to the map $\epsilon \mapsto (\Psi^i[\epsilon], \Psi^o[\epsilon])$ of Proposition 8.26 and show that the coefficients of its series expansion around 0 can be computed as the solutions of a certain recursive system of boundary integral equations, which involves also the sequence $\{(\Phi_k^i, \Phi_k^o)\}_{k\in\mathbb{N}}$ of the previous Proposition 8.41.

Proposition 8.42. *There exist a positive real number $\epsilon_\Psi \in]0, \epsilon_{*2}[$ and a sequence $\{(\Psi_k^i, \Psi_k^o)\}_{k\in\mathbb{N}}$ in $C^{1,\alpha}(\partial\Omega^i)_0 \times C^{1,\alpha}(\partial\Omega^o)$ such that*

$$\Psi^i[\epsilon] = \sum_{k=0}^{\infty} \frac{\Psi_k^i}{k!}\epsilon^k \quad \text{and} \quad \Psi^o[\epsilon] = \sum_{k=0}^{\infty} \frac{\Psi_k^i}{k!}\epsilon^k \qquad \forall \epsilon \in]-\epsilon_\Psi, \epsilon_\Psi[\,, \tag{8.122}$$

where the two series converge normally in $C^{1,\alpha}(\partial\Omega^i)$ and in $C^{1,\alpha}(\partial\Omega^o)$, respectively, for all $\epsilon \in]-\epsilon_\Psi, \epsilon_\Psi[$. Moreover, the pair (Ψ_0^i, Ψ_0^o) is the unique solution in $C^{1,\alpha}(\partial\Omega^i)_0 \times C^{1,\alpha}(\partial\Omega^o)$ of the following system of integral equations,

$$\frac{1}{2}\Psi_0^i(\xi) - W_{\Omega^i}[\Psi_0^i](\xi) \tag{8.123}$$
$$= f^i(\xi) - \int_{\partial\Omega^i} f^i\Phi_0^i\, d\sigma - \int_{\partial\Omega^o} f^o\Phi_0^o\, d\sigma - \int_{\partial\Omega^o} \Psi_0^o\, \nu_{\Omega^o}\cdot\nabla S_2\, d\sigma \qquad \forall \xi \in \partial\Omega^i\,,$$

$$\frac{1}{2}\Psi_0^o(x) + W_{\Omega^o}[\Psi_0^o](x) = f^o(x) \qquad \forall x \in \partial\Omega^o\,, \tag{8.124}$$

and for each $k \in \mathbb{N}\setminus\{0\}$ the pair (Ψ_k^i, Ψ_k^o) is the unique solution in $C^{1,\alpha}(\partial\Omega^i)_0 \times C^{1,\alpha}(\partial\Omega^o)$ of the following system of integral equations which involves the densities $\{(\Phi_j^i, \Phi_j^o)\}_{j=0}^k$ and $\{(\Psi_j^i, \Psi_j^o)\}_{j=0}^{k-1}$,

$$\frac{1}{2}\Psi_k^i(\xi) - W_{\Omega^i}[\Psi_k^i](\xi) = -\int_{\partial\Omega^i} f^i\Phi_k^i\, d\sigma - \int_{\partial\Omega^o} f^o\Phi_k^o\, d\sigma \tag{8.125}$$
$$+\sum_{j=0}^{k}\binom{k}{j}(-1)^{j+1}\sum_{h=0}^{j}\binom{j}{h}\xi_1^h\xi_2^{j-h}\int_{\partial\Omega^o}\Psi_{k-j}^o\nu_{\Omega^o}\cdot\nabla\partial_1^h\partial_2^{j-h}S_2\, d\sigma$$
$$\forall \xi \in \partial\Omega^i\,,$$

$$\frac{1}{2}\Psi_k^o(x) + W_{\Omega^o}[\Psi_k^o](x) \tag{8.126}$$

$$
\begin{aligned}
&= k\sum_{j=0}^{k-1}\binom{k-1}{j}(-1)^{j+1}\sum_{h=0}^{j}\binom{j}{h} \\
&\quad \times (\nabla\partial_1^h\partial_2^{j-h}S_2)(x)\cdot\int_{\partial\Omega^i}\Psi^i_{k-1-j}(\eta)\,\nu_{\Omega^i}(\eta)\eta_1^h\eta_2^{j-h}\,d\sigma_\eta \qquad \forall x\in\partial\Omega^o .
\end{aligned}
$$

Proof. The existence of ϵ_Ψ and $\{(\Psi^o_k,\Psi^i_k)\}_{k\in\mathbb{N}}$ such that (8.122) holds true follows by the real analyticity of the map that takes ϵ to $(\Psi^o[\epsilon],\Psi^i[\epsilon])$ (cf. Proposition 8.26). Then we observe that

$$
\mathcal{M}(\epsilon)[\Psi^i[\epsilon],\Psi^o[\epsilon]] = \left(f^i - \int_{\partial\Omega^i} f^i\,\Phi^i[\epsilon]\,d\sigma - \int_{\partial\Omega^o} f^o\,\Phi^o[\epsilon]\,d\sigma\,,\,f^o\right)
$$

for all $\epsilon\in]-\epsilon_\Psi,\epsilon_\Psi[$ (cf. Proposition 8.26). Accordingly,

$$
\partial_\epsilon^k\left(\mathcal{M}(\epsilon)[\Psi^i[\epsilon],\Psi^o[\epsilon]]\right) = \partial_\epsilon^k\left(f^i - \int_{\partial\Omega^i} f^i\,\Phi^i[\epsilon]\,d\sigma - \int_{\partial\Omega^o} f^o\,\Phi^o[\epsilon]\,d\sigma\,,\,f^o\right)
$$

for all $\epsilon\in]-\epsilon_\Psi,\epsilon_\Psi[$ and all $k\in\mathbb{N}$. A straightforward calculation based on the chain rule formula (8.111) shows that

$$
\begin{aligned}
&\partial_\epsilon^k(\mathcal{M}^i(\epsilon)[\Psi^i[\epsilon],\Psi^o[\epsilon]])(\xi) = \frac{1}{2}\partial_\epsilon^k\Psi^i[\epsilon](\xi) - W_{\Omega^i}[\partial_\epsilon^k\Psi^i[\epsilon]](\xi) \qquad (8.127)\\
&\quad -\sum_{j=0}^{k}\binom{k}{j}\sum_{h=0}^{j}\binom{j}{h}\xi_1^h\xi_2^{j-h}\\
&\quad \times\int_{\partial\Omega^o}\partial_\epsilon^{k-j}\Psi^o[\epsilon](y)\,\nu_{\Omega^o}(y)\cdot(\nabla\partial_1^h\partial_2^{j-h}S_2)(\epsilon\xi-y)\,d\sigma_y\\
&\quad = \delta_{0,k}f^i(\xi) - \int_{\partial\Omega^i} f^i\partial_\epsilon^k\Phi^i[\epsilon]\,d\sigma - \int_{\partial\Omega^o} f^o\partial_\epsilon^k\Phi^o[\epsilon]\,d\sigma \qquad \forall\xi\in\partial\Omega^i ,\\
&\partial_\epsilon^k(\mathcal{M}^o(\epsilon)[\Psi^i[\epsilon],\Psi^o[\epsilon]])(x) = \frac{1}{2}\partial_\epsilon^k\Psi^o[\epsilon](x) + W_{\Omega^o}[\partial_\epsilon^k\Psi^o[\epsilon]](x) \qquad (8.128)\\
&\quad +\epsilon\sum_{j=0}^{k}\binom{k}{j}(-1)^j\sum_{h=0}^{j}\binom{j}{h}\\
&\quad \times\int_{\partial\Omega^i}\partial_\epsilon^{k-j}\Psi^i[\epsilon](\eta)\,\eta_1^h\eta_2^{j-h}\nu_{\Omega^i}(\eta)\cdot(\nabla\partial_1^h\partial_2^{j-h}S_2)(x-\epsilon\eta)\,d\sigma_\eta\\
&\quad +k\sum_{j=0}^{k-1}\binom{k-1}{j}(-1)^j\sum_{h=0}^{j}\binom{j}{h}\\
&\quad \times\int_{\partial\Omega^i}\partial_\epsilon^{k-1-j}\Psi^i[\epsilon](\eta)\,\eta_1^h\eta_2^{j-h}\nu_{\Omega^i}(\eta)\cdot(\nabla\partial_1^h\partial_2^{j-h}S_2)(x-\epsilon\eta)\,d\sigma_\eta\\
&\quad = \delta_{0,k}f^o(x) \qquad \forall x\in\partial\Omega^o ,
\end{aligned}
$$

for all $\epsilon \in]-\epsilon_\Psi, \epsilon_\Psi[$ and all $k \in \mathbb{N}$, where we understand that $\sum_{j=0}^{k-1}$ is omitted for $k = 0$. By standard properties of real analytic maps we have

$$(\Psi_k^o, \Psi_k^i) = (\partial_\epsilon^k \Psi^o[0], \partial_\epsilon^k \Psi^i[0]) \qquad \forall k \in \mathbb{N}.$$

Then, taking $\epsilon = 0$ in (8.127) and (8.128) we deduce that (Ψ_0^i, Ψ_0^o) is a solution of (8.123) and (8.124), and that (Ψ_k^i, Ψ_k^o) is a solution of (8.125) and (8.126) for all $k \in \mathbb{N} \setminus \{0\}$. The uniqueness in $C^{1,\alpha}(\partial\Omega^i)_0 \times C^{1,\alpha}(\partial\Omega^o)$ of the solutions of (8.123), (8.124) and of (8.125), (8.126) is a consequence of Corollary 6.15, which implies that $\frac{1}{2}I_{\Omega^o} + W_{\Omega^o}$ is injective on $C^{1,\alpha}(\partial\Omega^o)$, and of Theorem 6.24, that implies the injectivity of $-\frac{1}{2}I_{\Omega^i} + W_{\Omega^i}$ on $C^{1,\alpha}(\partial\Omega^i)_0$. □

8.5.2 Series Expansion of u_ϵ

We now use the sequences $\{(\Phi_k^o, \Phi_k^i)\}_{k\in\mathbb{N}}$ and $\{(\Psi_k^o, \Psi_k^i)\}_{k\in\mathbb{N}}$ of Section 8.5.1 to obtain the coefficients of the series expansions around 0 of the maps $\epsilon \mapsto u_\epsilon$ and $\epsilon \mapsto u_\epsilon(\epsilon\cdot)$. We begin by writing the coefficients of the series expansions of the maps $\mathcal{F}$ and $\mathcal{R}$ of Proposition 8.27, of the maps $\mathcal{U}_M$ and $\mathcal{V}_M$ of Theorem 8.29, and of the maps $\mathcal{U}_m$ and $\mathcal{V}_m$ of Theorem 8.30. We collect the expressions for such coefficients in the following Proposition 8.43.

Proposition 8.43. *The following statements hold.*

(i) Let $\{f_k\}_{k\in\mathbb{N}}$ be the sequence in $\mathbb{R}$ defined by

$$f_k \equiv \frac{1}{k!}\int_{\partial\Omega^i} f^i \Phi_k^i \, d\sigma + \frac{1}{k!}\int_{\partial\Omega^o} f^o \Phi_k^o \, d\sigma$$

for all $k \in \mathbb{N}$. If $\mathcal{F}$ is the function of Proposition 8.27, then there exists $\epsilon_\mathcal{F} \in]0, \epsilon_*]$ such that

$$\mathcal{F}(\epsilon) = \sum_{k=0}^{\infty} f_k \, \epsilon^k \qquad \forall \epsilon \in]-\epsilon_\mathcal{F}, \epsilon_\mathcal{F}[,$$

where the series converges absolutely for all $\epsilon \in]-\epsilon_\mathcal{F}, \epsilon_\mathcal{F}[$.

(ii) Let $\{r_k\}_{k\in\mathbb{N}}$ be the sequence in $\mathbb{R}$ defined by

$$\begin{aligned} r_k \equiv & \frac{1}{m_{n-1}(\partial\Omega^i)} \frac{1}{k!} \sum_{j=0}^{k} \binom{k}{j} (-1)^j \sum_{h=0}^{j} \binom{j}{h} \\ & \times \int_{\partial\Omega^i} \eta_1^h \eta_2^{j-h} \, d\sigma_\eta \int_{\partial\Omega^o} \Phi_{k-j}^o \partial_1^h \partial_2^{j-h} S_2 \, d\sigma \\ & + \frac{1}{m_{n-1}(\partial\Omega^i)} \frac{1}{k!} \int_{\partial\Omega^i} v_{\Omega^i}[\Phi_k^i] \, d\sigma \end{aligned}$$

for all $k \in \mathbb{N}$. *If* $\mathcal{R}$ *is the function of Proposition 8.27, then there exists* $\epsilon_{\mathcal{R}}$ *in* $]0, \epsilon_*]$ *such that*

$$\mathcal{R}(\epsilon) = \sum_{k=0}^{\infty} r_k \, \epsilon^k \qquad \forall \epsilon \in]-\epsilon_{\mathcal{R}}, \epsilon_{\mathcal{R}}[,$$

where the series converges absolutely for all $\epsilon \in]-\epsilon_{\mathcal{R}}, \epsilon_{\mathcal{R}}[$.

(iii) *Let* $\{u_{M,k}\}_{k\in\mathbb{N}}$ *be the sequence of functions from* $\overline{\Omega^o} \setminus \{0\}$ *to* $\mathbb{R}$ *defined by*

$$u_{M,0}(x) \equiv w^+_{\Omega^o}[\Psi^o_0](x) \qquad \forall x \in \overline{\Omega^o} \setminus \{0\} ,$$

and

$$\begin{aligned} u_{M,k}(x) &\equiv \frac{1}{k!} w^+_{\Omega^o}[\Psi^o_k](x) \\ &+ \frac{1}{(k-1)!} \sum_{j=0}^{k-1} \binom{k-1}{j} (-1)^j \sum_{h=0}^{j} \binom{j}{h} \\ &\times (\nabla \partial_1^h \partial_2^{j-h} S_2)(x) \cdot \int_{\partial\Omega^i} \Psi^i_{k-1-j}(\eta) \nu_{\Omega^i}(\eta) \eta_1^h \eta_2^{j-h} \, d\sigma_\eta \quad \forall x \in \overline{\Omega^o} \setminus \{0\} \end{aligned}$$

for all $k \in \mathbb{N} \setminus \{0\}$. *Let* Ω_M *be an open subset of* Ω^o *with* $0 \notin \overline{\Omega_M}$. *Then we have*

$$u_{M,k|\overline{\Omega_M}} \in C^{1,\alpha}(\overline{\Omega_M})$$

for all $k \in \mathbb{N}$ *and if* $\mathcal{U}_M$ *is the map of Theorem 8.29, then there exists* $\epsilon_{\mathcal{U}_M}$ *in* $]0, \epsilon_M]$ *such that*

$$\mathcal{U}_M[\epsilon] = \sum_{k=0}^{\infty} u_{M,k|\overline{\Omega_M}} \, \epsilon^k \qquad \forall \epsilon \in]-\epsilon_{\mathcal{U}_M}, \epsilon_{\mathcal{U}_M}[,$$

where the series converges normally in $C^{1,\alpha}(\overline{\Omega_M})$ *for all* $\epsilon \in]-\epsilon_{\mathcal{U}_M}, \epsilon_{\mathcal{U}_M}[$.

(iv) *Let* $\{v_{M,k}\}_{k\in\mathbb{N}}$ *be the sequence of functions from* $\overline{\Omega^o} \setminus \{0\}$ *to* $\mathbb{R}$ *defined by*

$$\begin{aligned} v_{M,k}(x) &\equiv \frac{1}{k!} \sum_{j=0}^{k} \binom{k}{j} (-1)^j \sum_{h=0}^{j} \binom{j}{h} (\partial_1^h \partial_2^{j-h} S_2)(x) \\ &\times \int_{\partial\Omega^i} \Phi^i_{k-j}(\eta) \eta_1^h \eta_2^{j-h} \, d\sigma_\eta + \frac{1}{k!} v^+_{\Omega^o}[\Phi^o_k](x) \qquad \forall x \in \overline{\Omega^o} \setminus \{0\} \end{aligned}$$

for all $k \in \mathbb{N} \setminus \{0\}$. *Let* Ω_M *be an open subset of* Ω^o *with* $0 \notin \overline{\Omega_M}$. *Then we have*

$$v_{M,k|\overline{\Omega_M}} \in C^{1,\alpha}(\overline{\Omega_M})$$

for all $k \in \mathbb{N}$ *and if* $\mathcal{V}_M$ *is the map of Theorem 8.29, then there exists* $\epsilon_{\mathcal{V}_M} \in]0, \epsilon_M]$ *such that*

$$\mathcal{V}_M[\epsilon] = \sum_{k=0}^{\infty} v_{M,k|\overline{\Omega_M}}\, \epsilon^k \qquad \forall \epsilon \in]-\epsilon_{\mathcal{V}_M}, \epsilon_{\mathcal{V}_M}[,$$

where the series converges normally in $C^{1,\alpha}(\overline{\Omega_M})$ *for all* $\epsilon \in]-\epsilon_{\mathcal{V}_M}, \epsilon_{\mathcal{V}_M}[$.

(v) Let $\{u_{m,k}\}_{k\in\mathbb{N}}$ *be the sequence of functions in* $C^{1,\alpha}_{\mathrm{loc}}(\mathbb{R}^2 \setminus \Omega^i)$ *defined by*

$$\begin{aligned} u_{m,k}(\xi) \equiv \frac{1}{k!} \sum_{j=0}^{k} \binom{k}{j} (-1)^j \sum_{h=0}^{j} \binom{j}{h} \xi_1^h \xi_2^{j-h} \\ \times \int_{\partial\Omega^o} \Psi^o_{k-j}\, \nu_{\Omega^o} \cdot (\nabla \partial_1^h \partial_2^{j-h} S_2)\, d\sigma - \frac{1}{k!} w^-_{\Omega^i}[\Psi^i_k](\xi) \\ \forall \xi \in \mathbb{R}^2 \setminus \Omega^i \end{aligned}$$

for all $k \in \mathbb{N}$. *Let* $\Omega_m \subseteq \mathbb{R}^2 \setminus \overline{\Omega^i}$ *be open and bounded and let* $\mathcal{U}_m$ *be the map of Theorem 8.30. Then there exists* $\epsilon_{\mathcal{U}_m} \in]0, \epsilon_m]$ *such that*

$$\mathcal{U}_m[\epsilon] = \sum_{k=0}^{\infty} u_{m,k|\overline{\Omega_m}}\, \epsilon^k \qquad \forall \epsilon \in]-\epsilon_{\mathcal{U}_m}, \epsilon_{\mathcal{U}_m}[,$$

where the series converges normally in $C^{1,\alpha}(\overline{\Omega_m})$ *for all* $\epsilon \in]-\epsilon_{\mathcal{U}_m}, \epsilon_{\mathcal{U}_m}[$.

(vi) Let $\{v_{m,k}\}_{k\in\mathbb{N}}$ *be the sequence of functions in* $C^{1,\alpha}_{\mathrm{loc}}(\mathbb{R}^2 \setminus \Omega^i)$ *defined by*

$$\begin{aligned} v_{m,k}(\xi) \equiv \frac{1}{k!} \sum_{j=0}^{k} \binom{k}{j} (-1)^j \sum_{h=0}^{j} \binom{j}{h} \xi_1^h \xi_2^{j-h} \int_{\partial\Omega^o} \Phi^o_{k-j} \partial_1^h \partial_2^{j-h} S_2\, d\sigma \\ + \frac{1}{k!} v^-_{\Omega^i}[\Phi^i_k](\xi) \qquad \forall \xi \in \mathbb{R}^2 \setminus \Omega^i \end{aligned}$$

for all $k \in \mathbb{N}$. *Let* $\Omega_m \subseteq \mathbb{R}^2 \setminus \overline{\Omega^i}$ *be open and bounded and let* $\mathcal{V}_m$ *be the map of Theorem 8.30. Then there exists* $\epsilon_{\mathcal{V}_m} \in]0, \epsilon_m]$ *such that*

$$\mathcal{V}_m[\epsilon] = \sum_{k=0}^{\infty} v_{m,k|\overline{\Omega_m}}\, \epsilon^k \qquad \forall \epsilon \in]-\epsilon_{\mathcal{V}_m}, \epsilon_{\mathcal{V}_m}[,$$

where the series converges normally in $C^{1,\alpha}(\overline{\Omega_m})$ *for all* $\epsilon \in]-\epsilon_{\mathcal{V}_m}, \epsilon_{\mathcal{V}_m}[$.

Proof. Statements (i) and (ii) can be deduced by the definition of $\mathcal{F}$ and $\mathcal{R}$ in Proposition 8.27, by the expansions of Φ^i and Φ^o in Proposition 8.41 and of Ψ^i and Ψ^o in Proposition 8.42, and by a computation based on standard properties of real analytic functions and on the chain rule formula (8.111).

Similarly, the validity of (iii)–(vi) can be deduced by the definitions (8.76) of $\mathcal{U}_M$, (8.77) of $\mathcal{V}_m$, (8.83) of $\mathcal{U}_m$, and (8.84) of $\mathcal{V}_m$, by Propositions 8.41 and 8.42, and by a straightforward computation based on (8.111). □

By the previous Proposition 8.43 and by Theorem 8.29 we can verify that for all open set $\Omega_M \subseteq \Omega^o$ with $0 \notin \overline{\Omega_M}$ and for $\epsilon > 0$ small enough we have

$$u_{\epsilon|\overline{\Omega_M}} = \sum_{k=0}^{\infty} u_{M,k|\overline{\Omega_M}}\epsilon^k + (\sum_{k=0}^{\infty} f_k\epsilon^k)\frac{\sum_{k=0}^{\infty} v_{M,k|\overline{\Omega_M}}\epsilon^k}{\sum_{k=0}^{\infty} r_k\epsilon^k + \frac{\log\epsilon}{2\pi}}. \tag{8.129}$$

Similarly, if $\Omega_m \subseteq \mathbb{R}^2 \setminus \overline{\Omega^i}$ is open and bounded, then by Proposition 8.43 and by Theorem 8.30 we have

$$u_\epsilon(\epsilon\cdot)_{|\overline{\Omega_m}} = \sum_{k=0}^{\infty} u_{m,k|\overline{\Omega_m}}\epsilon^k + (\sum_{k=0}^{\infty} f_k\epsilon^k)\frac{\sum_{k=0}^{\infty} v_{m,k|\overline{\Omega_m}}\epsilon^k + \frac{\log\epsilon}{2\pi}}{\sum_{k=0}^{\infty} r_k\epsilon^k + \frac{\log\epsilon}{2\pi}} \tag{8.130}$$

for all $\epsilon > 0$ small enough. However, because of the presence of the quotients in the right-hand sides of (8.129) and (8.130), the previous Proposition 8.43 does not provide (yet) an expression for $u_{\epsilon|\overline{\Omega_M}}$ and $u_\epsilon(\epsilon\cdot)_{|\overline{\Omega_m}}$ in terms of convergent series. Then, to deduce from (8.129) and (8.130) proper series expansions, we will exploit the following Lemma 8.44, which describes the derivatives of order m of the quotient of two real analytic functions.

Lemma 8.44. *Let $\epsilon' > 0$. Let f, g be real analytic functions from $]-\epsilon', \epsilon'[$ to $\mathbb{R}$ with $g(\epsilon) \neq 0$ for all $\epsilon \in]-\epsilon', \epsilon'[$. Let $l \in \mathbb{N} \setminus \{0\}$. Then we have*

$$\partial_\epsilon^l\left(\frac{f(\epsilon)}{g(\epsilon)}\right)$$
$$= \frac{\partial_\epsilon^l f(\epsilon)}{g(\epsilon)} + l!\sum_{k=1}^{l}\frac{\partial_\epsilon^{l-k} f(\epsilon)}{(l-k)!}\sum_{j=1}^{k}\frac{(-1)^j}{g(\epsilon)^{j+1}}\sum_{\beta\in(\mathbb{N}\setminus\{0\})^j\,,\,|\beta|=k}\frac{1}{\beta!}\prod_{h=1}^{j}\partial_\epsilon^{\beta_h} g(\epsilon)$$

for all $\epsilon \in]-\epsilon', \epsilon'[$.

Proof. If h is a real analytic map from $\mathbb{R} \setminus \{0\}$ to $\mathbb{R}$, then by the Faa di Bruno Formula we have

$$\partial_\epsilon^k (h(g(\epsilon))) = k!\sum_{j=1}^{k}\frac{1}{j!}\partial_\epsilon^j h(g(\epsilon)) \sum_{\beta\in(\mathbb{N}\setminus\{0\})^j\,,\,|\beta|=k}\frac{1}{\beta!}\prod_{h=1}^{j}\partial_\epsilon^{\beta_h} g(\epsilon)$$

for all $\epsilon \in]-\epsilon', \epsilon'[$ and for all $k \in \mathbb{N} \setminus \{0\}$. Then, taking $h(\epsilon) = 1/\epsilon$ for all $\epsilon \in \mathbb{R} \setminus \{0\}$ and exploiting equality $\partial_\epsilon^j(1/\epsilon) = (-1)^j j!(1/\epsilon^{j+1})$, we can deduce that

$$\partial_\epsilon^k\left(\frac{1}{g(\epsilon)}\right) = k!\sum_{j=1}^{k}\frac{(-1)^j}{g(\epsilon)^{j+1}}\sum_{\beta\in(\mathbb{N}\setminus\{0\})^j\,,\,|\beta|=k}\frac{1}{\beta!}\prod_{h=1}^{j}\partial_\epsilon^{\beta_h} g(\epsilon)$$

for all $\epsilon \in]-\epsilon', \epsilon'[$ and for all $k \in \mathbb{N} \setminus \{0\}$. Now the formula in the lemma follows by the Generalized Leibniz Rule. □

In the following Lemma 8.45 we compute the coefficients of the expansion around 0 of a quotient

$$\frac{\mathcal{A}(\epsilon)}{\mathcal{R}(\epsilon) + \frac{\log\epsilon}{2\pi}},$$

where $\mathcal{A}$ is a real analytic map defined in a neighborhood of 0. We will use Lemma 8.45 to handle the quotients on the right-hand side of (8.129) and (8.130).

Lemma 8.45. *Let $\epsilon' > 0$. Let $\mathcal{A}$ be a real analytic map from $]-\epsilon', \epsilon'[$ to a Banach space X such that its Taylor expansion around $\epsilon = 0$*

$$\mathcal{A}(\epsilon) = \sum_{k=0}^{\infty} a_k\, \epsilon^k$$

converges normally in X for all ϵ in $]-\epsilon', \epsilon'[$. Let $\{r_k\}_{k\in\mathbb{N}}$ be the sequence in Proposition 8.43 (ii). Let $\{\lambda_{(k,l)}\}_{(k,l)\in\mathbb{N}^2\,,\,1\le l\le k+1}$ be the family of elements of X defined by

$$\lambda_{(k,1)} \equiv a_k$$

for all $k \in \mathbb{N}$ and

$$\lambda_{(k,l)} \equiv (-1)^{l-1} \sum_{j=l-1}^{k} a_{k-j} \sum_{\beta\in(\mathbb{N}\setminus\{0\})^{l-1}\,,\,|\beta|=j} \prod_{h=1}^{l-1} r_{\beta_h}$$

for all $k, l \in \mathbb{N}$ with $2 \le l \le k+1$. Then there exists $\epsilon'_* \in]0,1[\cap]0,\epsilon_*]$ such that

$$\frac{\mathcal{A}(\epsilon)}{\mathcal{R}(\epsilon) + \frac{\log\epsilon}{2\pi}} = \sum_{k=0}^{\infty} \epsilon^k \sum_{l=1}^{k+1} \frac{\lambda_{(k,l)}}{\left(r_0 + \frac{\log\epsilon}{2\pi}\right)^l} \qquad \forall \epsilon \in]0,\epsilon'_*[\,. \tag{8.131}$$

Moreover, the series

$$\sum_{k=0}^{\infty} \sum_{l=1}^{k+1} \lambda_{(k,l)} \epsilon^k \left(\frac{t}{r_0\, t + \frac{1}{2\pi}}\right)^l \tag{8.132}$$

converges normally for all $(\epsilon, t) \in]-\epsilon'_*, \epsilon'_*[\times]1/\log\epsilon'_*, -1/\log\epsilon'_*[$.

Proof. By Proposition 8.27 the function $\mathcal{R}$ is real analytic, and therefore continuous, in a neighborhood of 0. Accordingly, there exists $\tilde{\epsilon}_* \in]0,\epsilon_*[\cap]0,1[$ such that

$$\mathcal{R}(\epsilon)\, t + \frac{1}{2\pi} \neq 0 \qquad \forall (\epsilon, t) \in]-\tilde{\epsilon}_*, \tilde{\epsilon}_*[\times]1/\log\tilde{\epsilon}_*, -1/\log\tilde{\epsilon}_*[$$

(condition $\tilde{\epsilon}_* < 1$ ensures that $-\infty < 1/\log\tilde{\epsilon}_* < 0$). Then we take $\tilde{\epsilon} \equiv \min\{\tilde{\epsilon}_*, \epsilon'\}$ and we define a function $\mathcal{B}$ from $]-\tilde{\epsilon}, \tilde{\epsilon}[\times]1/\log\tilde{\epsilon}, -1/\log\tilde{\epsilon}[$ to X by setting

$$\mathcal{B}(\epsilon, t) \equiv \frac{\mathcal{A}(\epsilon)\, t}{\mathcal{R}(\epsilon)\, t + \frac{1}{2\pi}} \qquad \forall (\epsilon, t) \in]-\tilde{\epsilon}, \tilde{\epsilon}[\times]1/\log\tilde{\epsilon}, -1/\log\tilde{\epsilon}[\,.$$

Since the map that takes a non-vanishing real analytic function to its reciprocal is real analytic and the product of real analytic functions is real analytic, we deduce that $\mathcal{B}$ is real analytic. Accordingly, for $\epsilon'_* \in]0, \tilde{\epsilon}]$ sufficiently small we have

$$\mathcal{B}(\epsilon, t) = \sum_{k,j \in \mathbb{N}} \frac{\epsilon^k t^j}{k! j!} \partial_\epsilon^k \partial_t^j \mathcal{B}(0,0) = \sum_{k=0}^{\infty} \frac{\epsilon^k}{k!} \sum_{j=0}^{\infty} \frac{t^j}{j!} \partial_\epsilon^k \partial_t^j \mathcal{B}(0,0) \tag{8.133}$$

for all $(\epsilon, t) \in]-\epsilon'_*, \epsilon'_*[\times]1/\log \epsilon'_*, -1/\log \epsilon'_*[$, where both the series converge normally in X (the second series being just one of the possible rearrangements of the first one). It follows that

$$\partial_\epsilon^k \mathcal{B}(0, t) = \sum_{j=0}^{\infty} \frac{t^j}{j!} \partial_\epsilon^k \partial_t^j \mathcal{B}(0,0) \tag{8.134}$$

for all $t \in]-\epsilon'_*, \epsilon'_*[\times]1/\log \epsilon'_*, -1/\log \epsilon'_*[$ and $k \in \mathbb{N}$. Then, by (8.133) and (8.134) we can write

$$\mathcal{B}(\epsilon, t) = \sum_{k=0}^{\infty} \frac{\epsilon^k}{k!} \partial_\epsilon^k \mathcal{B}(0, t) \tag{8.135}$$

for all $(\epsilon, t) \in]-\epsilon'_*, \epsilon'_*[\times]1/\log \epsilon'_*, -1/\log \epsilon'_*[$, where the series converges normally in X.

We now claim that the series in (8.135) coincides with that in (8.132). To prove it, we need to verify that

$$\mathcal{B}(0, t) = \frac{a_0 t}{r_0 t + \frac{1}{2\pi}} = \frac{\lambda_{(0,1)} t}{r_0 t + \frac{1}{2\pi}} \tag{8.136}$$

for all $t \in]1/\log \tilde{\epsilon}, -1/\log \tilde{\epsilon}[$ and

$$\begin{aligned} \partial_\epsilon^k \mathcal{B}(0, t) &= k! \sum_{l=1}^{k+1} \lambda_{(k,l)} \left(\frac{t}{r_0 t + \frac{1}{2\pi}} \right)^l \\ &= \frac{k! a_k t}{r_0 t + \frac{1}{2\pi}} + k! \sum_{l=1}^{k} \lambda_{(k,l+1)} \left(\frac{t}{r_0 t + \frac{1}{2\pi}} \right)^{l+1} \end{aligned} \tag{8.137}$$

for all $t \in]1/\log \tilde{\epsilon}, -1/\log \tilde{\epsilon}[$ and $k \in \mathbb{N} \setminus \{0\}$. Equality (8.136) follows by the definition of $\mathcal{B}$, a_0, r_0, and $\lambda_{(0,1)}$. By the previous Lemma 8.44 and by a straightforward computation based on the discrete Fubini Theorem, we verify that

$$
\begin{aligned}
\partial_\epsilon^k \mathcal{B}(\epsilon,t) &= \partial_\epsilon^k \frac{\mathcal{A}(\epsilon)\,t}{\mathcal{R}(\epsilon)\,t+\frac{1}{2\pi}} \\
&= \frac{\partial_\epsilon^k \mathcal{A}(\epsilon)\,t}{\mathcal{R}(\epsilon)\,t+\frac{1}{2\pi}} + k!\sum_{j=1}^{k} \frac{\partial_\epsilon^{k-j}\mathcal{A}(\epsilon)\,t}{(k-j)!} \sum_{l=1}^{j} \frac{(-1)^l}{\left(\mathcal{R}(\epsilon)\,t+\frac{1}{2\pi}\right)^{l+1}} \\
&\qquad \times \sum_{\beta\in(\mathbb{N}\setminus\{0\})^l\,,\,|\beta|=j} \frac{1}{\beta!}\prod_{h=1}^{l} \partial_\epsilon^{\beta_h}\left(\mathcal{R}(\epsilon)\,t+\frac{1}{2\pi}\right) \\
&= \frac{\partial_\epsilon^k \mathcal{A}(\epsilon)\,t}{\mathcal{R}(\epsilon)\,t+\frac{1}{2\pi}} \\
&\quad + k!\sum_{l=1}^{k} \frac{(-1)^l t^{l+1}}{\left(\mathcal{R}(\epsilon)\,t+\frac{1}{2\pi}\right)^{l+1}} \sum_{j=l}^{k} \frac{\partial_\epsilon^{k-j}\mathcal{A}(\epsilon)}{(k-j)!} \sum_{\beta\in(\mathbb{N}\setminus\{0\})^l\,,\,|\beta|=j} \frac{1}{\beta!}\prod_{h=1}^{l}\partial_\epsilon^{\beta_h}\mathcal{R}(\epsilon)
\end{aligned}
\tag{8.138}
$$

for all $(\epsilon,t)\in]-\tilde{\epsilon},\tilde{\epsilon}[\times]1/\log\tilde{\epsilon},-1/\log\tilde{\epsilon}[$ and for all $k\in\mathbb{N}\setminus\{0\}$. Then, taking $\epsilon=0$ in (8.138), we deduce the validity of (8.137).

Equalities (8.136) and (8.137) imply that the sum of the series in (8.132) coincides with $\mathcal{B}(\epsilon,t)$. Thus equality (8.131) holds true and the proof is complete. □

We are now ready to prove the main Theorems 8.46 and 8.47 of this section. In Theorem 8.46, we use the sequences $\{f_k\}_{k\in\mathbb{N}}$, $\{r_k\}_{k\in\mathbb{N}}$, $\{u_{M,k}\}_{k\in\mathbb{N}}$, and $\{v_{M,k}\}_{k\in\mathbb{N}}$ introduced in Proposition 8.43 to write the coefficients of the series expansion around zero of the map that takes ϵ to $u_{\epsilon|\overline{\Omega_M}}$.

Theorem 8.46. *Let $\{a_{M,k}\}_{k\in\mathbb{N}}$ be the sequence of functions from $\overline{\Omega^o}\setminus\{0\}$ to $\mathbb{R}$ defined by*

$$
a_{M,k} \equiv \sum_{j=0}^{k} f_{k-j}\, v_{M,j} \qquad \forall k\in\mathbb{N}\,.
$$

Let $\{\lambda_{M,(k,l)}\}_{(k,l)\in\mathbb{N}^2\,,\,l\le k+1}$ be the family of functions from $\overline{\Omega^o}\setminus\{0\}$ to $\mathbb{R}$ defined by

$$
\lambda_{M,(k,0)} \equiv u_{M,k}\,,\quad \lambda_{M,(k,1)} \equiv a_{M,k}
$$

for all $k\in\mathbb{N}$, and

$$
\lambda_{M,(k,l)} \equiv (-1)^{l-1}\sum_{j=l-1}^{k} a_{M,k-j} \sum_{\beta\in(\mathbb{N}\setminus\{0\})^{l-1}\,,\,|\beta|=j}\ \prod_{h=1}^{l-1} r_{\beta_h}
$$

for all $k,l\in\mathbb{N}$ with $2\le l\le k+1$. Let $\Omega_M\subseteq\Omega^o$ be open and such that $0\notin\overline{\Omega_M}$. Then there exists $\epsilon'_M\in]0,1[$ such that $\overline{\Omega_M}\cap\epsilon\overline{\Omega^i}=\emptyset$ for all $\epsilon\in]-\epsilon'_M,\epsilon'_M[$ and such that

$$
u_{\epsilon|\overline{\Omega_M}} = \sum_{k=0}^{\infty}\epsilon^k\sum_{l=0}^{k+1}\frac{\lambda_{M,(k,l)|\overline{\Omega_M}}}{\left(r_0+\frac{\log\epsilon}{2\pi}\right)^l} \qquad \forall\epsilon\in]0,\epsilon'_M[\,. \tag{8.139}
$$

Moreover, the series

$$\sum_{k=0}^{\infty}\sum_{l=0}^{k+1}\lambda_{M,(k,l)|\overline{\Omega_M}}\,\epsilon^k\left(\frac{t}{r_0\,t+\frac{1}{2\pi}}\right)^l$$

converges normally in $C^{1,\alpha}(\overline{\Omega_M})$ *for all* $(\epsilon,t)\in]-\epsilon'_M,\epsilon'_M[\times]1/\log\epsilon'_M,$ $-1/\log\epsilon'_M[$.

Proof. By Proposition 8.43 (i) and (iv) we have

$$\mathcal{F}(\epsilon)\mathcal{V}_M[\epsilon]=\left(\sum_{k=0}^{\infty}f_k\epsilon^k\right)\left(\sum_{k=0}^{\infty}v_{M,k}\epsilon^k\right)=\sum_{k=0}^{\infty}a_{M,k|\overline{\Omega_M}}\epsilon^k$$

for all ϵ with $|\epsilon|<\min\{\epsilon_{\mathcal{F}},\epsilon_{\mathcal{V}_M}\}$. Then we set

$$\mathcal{A}_M[\epsilon]\equiv\sum_{k=0}^{\infty}a_{M,k|\overline{\Omega_M}}\epsilon^k\qquad\forall\epsilon\in]-\min\{\epsilon_{\mathcal{F}},\epsilon_{\mathcal{V}_M}\},\min\{\epsilon_{\mathcal{F}},\epsilon_{\mathcal{V}_M}\}[$$

and we observe that the product series on the right-hand side converges normally in $C^{1,\alpha}(\overline{\Omega_M})$. Now, by the real analytic representation formula (8.73) for $u_{\epsilon|\overline{\Omega_M}}$, by the expansion for $\mathcal{U}_M[\epsilon]$ of Proposition 8.43 (iii), and by the expansion for

$$\frac{\mathcal{A}_M(\epsilon)}{\mathcal{R}(\epsilon)+\frac{\log\epsilon}{2\pi}}$$

of Lemma 8.45, we deduce the validity of the statement. □

We now turn to the rescaled function $u_\epsilon(\epsilon\cdot)$ and we show how to write the coefficients of the series expansion about 0 of $\epsilon\mapsto u_\epsilon(\epsilon\cdot)_{|\overline{\Omega_m}}$ in terms of the sequences $\{f_k\}_{k\in\mathbb{N}}$, $\{r_k\}_{k\in\mathbb{N}}$, $\{u_{m,k}\}_{k\in\mathbb{N}}$, and $\{v_{m,k}\}_{k\in\mathbb{N}}$ introduced in Proposition 8.43.

Theorem 8.47. *Let* $\{a_{m,k}\}_{k\in\mathbb{N}}$ *be the sequence of functions from* $\mathbb{R}^2\setminus\Omega^i$ *to* $\mathbb{R}$ *defined by*

$$a_{m,k}\equiv\sum_{j=0}^{k}f_{k-j}(v_{m,j}-r_j)\qquad\forall k\in\mathbb{N}.$$

Let $\{\lambda_{m,(k,l)}\}_{(k,l)\in\mathbb{N}^2\,,\,l\le k+1}$ *be the family of functions from* $\mathbb{R}^2\setminus\Omega^i$ *to* $\mathbb{R}$ *defined by*

$$\lambda_{m,(k,0)}\equiv u_{m,k}+f_k\,,\quad\lambda_{m,(k,1)}\equiv a_{m,k}\,,$$

for all $k\in\mathbb{N}$, *and*

$$\lambda_{m,(k,l)}\equiv(-1)^{l-1}\sum_{j=l-1}^{k}a_{m,k-j}\sum_{\beta\in(\mathbb{N}\setminus\{0\})^{l-1}\,,\,|\beta|=j}\ \prod_{h=1}^{l-1}r_{\beta_h}$$

for all $k, l \in \mathbb{N}$ with $2 \leq l \leq k+1$. Let $\Omega_m \subseteq \mathbb{R}^2 \setminus \overline{\Omega^i}$ be open and bounded. Then there exists $\epsilon'_m \in]0,1[$ such that $\epsilon\overline{\Omega_m} \subseteq \Omega^o$ for all $\epsilon \in]-\epsilon'_m, \epsilon'_m[$ and such that

$$u_\epsilon(\epsilon\cdot)_{|\overline{\Omega_m}} = \sum_{k=0}^{\infty} \epsilon^k \sum_{l=0}^{k+1} \frac{\lambda_{m,(k,l)|\overline{\Omega_m}}}{(r_0 + (2\pi)^{-1}\log|\epsilon|)^l} \qquad \forall \epsilon \in]0, \epsilon'_M[. \tag{8.140}$$

Moreover, the series

$$\sum_{k=0}^{\infty}\sum_{l=0}^{k+1} \lambda_{m,(k,l)|\overline{\Omega_m}}\, \epsilon^k \left(\frac{t}{r_0 t + \frac{1}{2\pi}}\right)^l$$

converges normally in $C^{1,\alpha}(\overline{\Omega_m})$ for all $(\epsilon, t) \in]-\epsilon'_m, \epsilon'_m[\times]1/\log\epsilon'_m, -1/\log\epsilon'_m[$.

Proof. We begin by observing that the real analytic representation formula (8.80) in Theorem 8.30 can be rewritten in the form

$$u_\epsilon(\epsilon\cdot)_{|\overline{\Omega_m}} = \mathcal{U}_m[\epsilon] + \mathcal{F}(\epsilon) + \frac{\mathcal{F}(\epsilon)\,(\mathcal{V}_m[\epsilon] - \mathcal{R}(\epsilon))}{\mathcal{R}(\epsilon) + \frac{\log\epsilon}{2\pi}} \qquad \forall \epsilon \in]0, \epsilon_m[. \tag{8.141}$$

Then, by Proposition 8.43 (i), (ii), and (vi) we have

$$\begin{aligned}
&\mathcal{F}(\epsilon)\,(\mathcal{V}_m[\epsilon] - \mathcal{R}(\epsilon)) \\
&\quad = \left(\sum_{k=0}^{\infty} f_k \epsilon^k\right)\left(\sum_{k=0}^{\infty} v_{M,k}\epsilon^k - \sum_{k=0}^{\infty} r_k \epsilon^k\right) = \sum_{k=0}^{\infty} a_{m,k|\overline{\Omega_M}}\epsilon^k
\end{aligned}$$

for all $\epsilon \in]-\min\{\epsilon_\mathcal{F}, \epsilon_\mathcal{R}, \epsilon_{\mathcal{V}_M}\}, \min\{\epsilon_\mathcal{F}, \epsilon_\mathcal{R}, \epsilon_{\mathcal{V}_M}\}[$. We now define

$$\mathcal{A}_m[\epsilon] \equiv \sum_{k=0}^{\infty} a_{m,k|\overline{\Omega_M}}\epsilon^k \qquad \forall \epsilon \in]-\min\{\epsilon_\mathcal{F}, \epsilon_\mathcal{R}, \epsilon_{\mathcal{V}_M}\}, \min\{\epsilon_\mathcal{F}, \epsilon_\mathcal{R}, \epsilon_{\mathcal{V}_M}\}[$$

and we observe that the product series on the right-hand side converges normally in $C^{1,\alpha}(\overline{\Omega_m})$. Then, by the representation formula (8.141) (see also Theorem 8.30), by the expansion for $\mathcal{F}(\epsilon)$ of Proposition 8.43 (i) and of $\mathcal{U}_m[\epsilon]$ of Proposition 8.43 (v), and by the expansion for

$$\frac{\mathcal{A}_m(\epsilon)}{\mathcal{R}(\epsilon) + \frac{\log\epsilon}{2\pi}}$$

of Lemma 8.45, we deduce the validity of the statement. □

We note that we could exploit Theorems 8.29 and 8.30 and Proposition 8.43 to write expansions of $\epsilon \mapsto u_{\epsilon|\overline{\Omega_M}}$ and $\epsilon \mapsto u_\epsilon(\epsilon\cdot)_{|\overline{\Omega_m}}$ as power series of ϵ and $1/\log\epsilon$ (in the fashion of (8.109)) instead of the power series of ϵ and

$$\frac{1}{r_0 + \frac{\log\epsilon}{2\pi}}$$

that we present in Theorems 8.46 and 8.47. As observed in the introduction of this section, the expansions in (8.139) and (8.140) have their own advantages. In particular, for each fixed $k' \in \mathbb{N}$, we can obtain finite sums that approximate $\epsilon \mapsto u_{\epsilon|\overline{\Omega_M}}$ and $\epsilon \mapsto u_\epsilon(\epsilon\cdot)_{|\overline{\Omega_m}}$ with remainders of order $\epsilon^{k'+1}$. Namely, for $x \in \overline{\Omega_M}$ and $\xi \in \overline{\Omega_m}$ fixed, we have

$$u_\epsilon(x) = \sum_{k=0}^{k'} \epsilon^k \sum_{l=0}^{k+1} \frac{\lambda_{M,(k,l)}(x)}{\left(r_0 + \frac{\log \epsilon}{2\pi}\right)^l} + O(\epsilon^{k'+1}) \qquad \text{as } \epsilon \to 0^+ \tag{8.142}$$

and

$$u_\epsilon(\epsilon\xi) = \sum_{k=0}^{k'} \epsilon^k \sum_{l=0}^{k+1} \frac{\lambda_{m,(k,l)}(\xi)}{\left(r_0 + \frac{\log \epsilon}{2\pi}\right)^l} + O(\epsilon^{k'+1}) \qquad \text{as } \epsilon \to 0^+ . \tag{8.143}$$

8.5.3 Principal Terms in the Series Expansion of u_ϵ

Taking $k' = 0$ in (8.142) and (8.143), we obtain

$$u_\epsilon(x) = \lambda_{M,(0,0)}(x) + \frac{\lambda_{M,(0,1)}(x)}{r_0 + \frac{\log \epsilon}{2\pi}} + O(\epsilon) \qquad \text{as } \epsilon \to 0^+$$

and

$$u_\epsilon(\epsilon\xi) = \lambda_{m,(0,0)}(\xi) + \frac{\lambda_{m,(0,1)}(\xi)}{r_0 + \frac{\log \epsilon}{2\pi}} + O(\epsilon) \qquad \text{as } \epsilon \to 0^+ .$$

By the analysis performed in the previous Subsections 8.5.1 and 8.5.2, we can reduce the computation of the coefficients r_0, $\lambda_{M,(0,0)}$, $\lambda_{M,(0,1)}$, $\lambda_{m,(0,0)}$, and $\lambda_{m,(0,1)}$ to the solution of certain systems of integral equations. In this subsection we will see that such coefficients can also be expressed in terms of solutions of certain auxiliary boundary value problems. For $\lambda_{M,(0,0)}$ and $\lambda_{m,(0,0)}$ it suffices to observe that the limiting relations (8.78) and (8.86) imply that

$$\lambda_{M,(0,0)}(x) = u^o(x) \qquad \forall x \in \overline{\Omega^o} \setminus \{0\}$$

and

$$\lambda_{m,(0,0)}(\xi) = u^i(\xi) \qquad \forall \xi \in \mathbb{R}^2 \setminus \Omega^i ,$$

with u^o and u^i solutions of (8.75) and (8.82), respectively. To obtain similar expressions also for r_0, $\lambda_{M,(0,1)}$, and $\lambda_{m,(0,1)}$ we will exploit the following Lemmas 8.48 and 8.49, where we show some properties of the functions Φ_0^i and Φ_0^o (cf. Proposition 8.41) and we introduce the auxiliary functions $H^\xi_{\Omega^i}$ and $H^{\Omega^o}_x$. In Lemma 8.48 we consider Φ_0^i.

Lemma 8.48. *Let $\xi \in \mathbb{R}^2 \setminus \partial\Omega^i$ be fixed. Let $H^{\xi}_{\Omega^i} \in C^{1,\alpha}_{\mathrm{loc}}(\mathbb{R}^2 \setminus \Omega^i)$ be the solution of problem*

$$\begin{cases} \Delta H^{\xi}_{\Omega^i} = 0 & \text{in } \mathbb{R}^2 \setminus \overline{\Omega^i}\,, \\ H^{\xi}_{\Omega^i}(\eta) = S_2(\xi - \eta) & \text{for all } \eta \in \partial\Omega^i\,, \\ H^{\xi}_{\Omega^i} \text{ is harmonic at infinity.} \end{cases}$$

Then $v_{\Omega^i}[\Phi^i_0](\xi) = \lim_{\eta\to\infty} H^{\xi}_{\Omega^i}(\eta)$. Moreover, the restriction

$$v^+_{\Omega^i}[\Phi^i_0] = v_{\Omega^i}[\Phi^i_0]_{|\overline{\Omega^i}}$$

is constant on $\overline{\Omega^i}$ and we have

$$v^+_{\Omega^i}[\Phi^i_0](\zeta) = \lim_{\eta\to\infty} H^0_{\Omega^i}(\eta) \qquad \forall \zeta \in \overline{\Omega^i}\,.$$

Proof. If $u \in C^{1,\alpha}_{\mathrm{loc}}(\mathbb{R}^2 \setminus \Omega^i)$ is harmonic in $\mathbb{R}^2 \setminus \overline{\Omega^i}$ and at infinity, then by Proposition 6.39 there exists $\psi \in C^{1,\alpha}(\partial\Omega^i)_0$ such that $u = w^-_{\Omega^i}[\psi] + \lim_{\eta\to\infty} u(\eta)$. Then, by the jump formula for $w^-_{\Omega^i}[\psi]$ of Theorem 6.6, we have

$$u_{|\partial\Omega^i} = (-\frac{1}{2}I + W_{\Omega^i})[\psi] + \lim_{\eta\to\infty} u(\eta)\,.$$

Since $-\frac{1}{2}I + W_{\Omega^i}$ and $-\frac{1}{2}I + W^t_{\Omega^i}$ are transpose to one another (cf. Theorem 6.5) we deduce that

$$\begin{aligned} \int_{\partial\Omega^i} u_{|\partial\Omega^i}\, \Phi^i_0\, d\sigma &= \int_{\partial\Omega^i} (-\frac{1}{2}\psi + W_{\Omega^i}[\psi])\, \Phi^i_0\, d\sigma + \lim_{\eta\to\infty} u(\eta) \int_{\partial\Omega^i} \Phi^i_0\, d\sigma \\ &= \int_{\partial\Omega^i} \psi\, (-\frac{1}{2}\Phi^i_0 + W^t_{\Omega^i}[\Phi^i_0])\, d\sigma + \lim_{\eta\to\infty} u(\eta) \int_{\partial\Omega^i} \Phi^i_0\, d\sigma\,. \end{aligned} \tag{8.144}$$

Then, by Equations (8.113) and (8.115) for Φ^i_0 it follows that

$$\int_{\partial\Omega^i} u_{|\partial\Omega^i}\, \Phi^i_0\, d\sigma = \lim_{\eta\to\infty} u(\eta)\,. \tag{8.145}$$

Thus, by Definition 4.18 of single layer potential and taking $u = H^{\xi}_{\Omega^i}$, we have

$$\begin{aligned} v_{\Omega^i}[\Phi^i_0](\xi) &= \int_{\partial\Omega^i} S_2(\xi - \eta)\Phi^i_0(\eta)\, d\sigma_\eta \\ &= \int_{\partial\Omega^i} H^{\xi}_{\Omega^i}(\eta)\Phi^i_0(\eta)\, d\sigma_\eta = \lim_{\eta\to\infty} H^{\xi}_{\Omega^i}(\eta)\,. \end{aligned}$$

Moreover, by Equation (8.113) and by the jump properties of the single layer potential of Theorem 6.6, we have $\nu_{\Omega^i} \cdot \nabla v^+_{\Omega^i}[\Phi^i_0]_{|\partial\Omega^i} = 0$. Then $v^+_{\Omega^i}[\Phi^i_0]$ is constant in $\overline{\Omega^i}$ and the validity of last the statement of the lemma follows. □

We also observe that the function

$$G_{\Omega^i}(\xi,\eta) \equiv S_2(\xi-\eta) - H^{\xi}_{\Omega^i}(\eta) \qquad \forall \xi,\eta \in \mathbb{R}^2 \setminus \overline{\Omega^i}\,, \xi \neq \eta\,, \tag{8.146}$$

is the Dirichlet Green's function for the exterior domain $\mathbb{R}^2 \setminus \overline{\Omega^i}$.

In the following Lemma 8.49 we consider Φ^o_0.

Lemma 8.49. *Let $x \in \Omega^o$ be fixed and let $H^{\Omega^o}_x \in C^{1,\alpha}(\overline{\Omega^o})$ be the solution of the problem*

$$\begin{cases} \Delta H^{\Omega^o}_x = 0 & \text{in } \Omega^o\,, \\ H^{\Omega^o}_x(y) = S_2(x-y) & \text{for all } y \in \partial\Omega^o\,. \end{cases}$$

Then we have $v_{\Omega^o}[\Phi^o_0](x) = -H^{\Omega^o}_x(0)$.

Proof. If $u \in C^{1,\alpha}(\overline{\Omega^o})$ and $\Delta u = 0$ in Ω^o, then Corollary 6.15 implies that there exists $\mu \in C^{1,\alpha}(\partial\Omega^o)$ such that $u_{|\partial\Omega^o} = (\frac{1}{2}I + W_{\Omega^o})[\mu]$. Moreover, by the uniqueness of the solution of the Dirichlet problem of Theorem 3.6 and by the jump formulas for the double layer potential of Theorem 6.6 we have $u = w^+_{\Omega^o}[\mu]$ in $\overline{\Omega^o}$. Since $\frac{1}{2}I + W_{\Omega^o}$ and $\frac{1}{2}I + W^t_{\Omega^o}$ are transpose to one another (cf. Theorem 6.5), Equation (8.114) for Φ^o_0 implies that

$$\begin{aligned} \int_{\partial\Omega^o} u_{|\partial\Omega^o}\Phi^o_0\, d\sigma &= \int_{\partial\Omega^o} \big(\frac{1}{2}\mu + W_{\Omega^o}[\mu]\big)\, \Phi^o_0\, d\sigma \\ &= \int_{\partial\Omega^o} \mu\, \big(\frac{1}{2}\Phi^o_0 + W^t_{\Omega^o}[\Phi^o_0]\big)\, d\sigma \\ &= -\int_{\partial\Omega^o} \mu(y)\, \nu_{\Omega^o}(y)\cdot\nabla S_2(y)\, d\sigma_y\,. \end{aligned}$$

Then

$$\int_{\partial\Omega^o} u_{|\partial\Omega^o}\Phi^o_0\, d\sigma = -w^+_{\Omega^o}[\mu](0) = -u(0) \tag{8.147}$$

by Definition 4.27 of double layer potential. Thus, by Definition 4.18 of single layer potential and taking $u = H^{\Omega^o}_x$, we have

$$v_{\Omega^o}[\Phi^o_0](x) = \int_{\partial\Omega^o} S_2(x-y)\Phi^o_0(y)\, d\sigma_y = \int_{\partial\Omega^o} H^{\Omega^o}_{x|\partial\Omega^o}\, \Phi^o_0\, d\sigma = -H^{\Omega^o}_x(0)\,.$$

□

We observe that the function

$$G^{\Omega^o}(x,y) \equiv S_2(x-y) - H^{\Omega^o}_x(y) \qquad \forall x,y \in \Omega^o\,, x \neq y\,, \tag{8.148}$$

is the Dirichlet Green's function for the domain Ω^o.

We can now describe the coefficients r_0, $\lambda_{M,(0,1)}$, and $\lambda_{m,(0,1)}$ in terms of the auxiliary functions $H^{\Omega^o}_x$, $H^{\xi}_{\Omega^i}$, of the Green's function G^{Ω^o}, and of the functions u^o and u^i (cf. (8.75) and (8.82)).

Proposition 8.50. *The following equalities hold*

$$r_0 = \lim_{\eta\to\infty} H^0_{\Omega^i}(\eta) - H^{\Omega^o}_0(0)\,, \tag{8.149}$$

$$\lambda_{M,(0,1)}(x) = \Big(\lim_{\eta\to\infty} u^i(\eta) - u^o(0)\Big) G^{\Omega^o}(x,0) \qquad \forall x \in \overline{\Omega^o} \setminus \{0\}\,, \tag{8.150}$$

and

$$\lambda_{m,(0,1)}(\xi) = \Big(\lim_{\eta\to\infty} u^i(\eta) - u^o(0)\Big) \lim_{\eta\to\infty} \Big(H^\xi_{\Omega^i}(\eta) - H^0_{\Omega^i}(\eta)\Big) \quad \forall \xi \in \mathbb{R}^2 \setminus \Omega^i\,. \tag{8.151}$$

In particular,

$$\lambda_{m,(0,1)}(\xi) = 0 \qquad \forall \xi \in \partial\Omega^i\,. \tag{8.152}$$

Proof. We first compute f_0, r_0, $v_{M,0}$, and $v_{m,0}$ and then deduce the formula for $\lambda_{M,(0,1)}$ and $\lambda_{m,(0,1)}$ by Theorems 8.46 and 8.47, respectively. By (8.82) and (8.145), we have

$$\int_{\partial\Omega^i} f^i\,\Phi^i_0\,d\sigma = \lim_{\eta\to\infty} u^i(\eta) \tag{8.153}$$

and by (8.75) and (8.147) we have

$$\int_{\partial\Omega^o} f^o\,\Phi^o_0\,d\sigma = -u^o(0)\,. \tag{8.154}$$

Thus, by the definition of f_0 in Proposition 8.43 (i) it follows that

$$f_0 = \int_{\partial\Omega^i} f^i\Phi^i_0\,d\sigma + \int_{\partial\Omega^o} f^o\Phi^o_0\,d\sigma = \lim_{\eta\to\infty} u^i(\eta) - u^o(0)\,. \tag{8.155}$$

By the definition of r_0 in Proposition 8.43 (ii) and by Definition 4.18 of single layer potential we have

$$r_0 = \int_{\partial\Omega^o} S_2\,\Phi^o_0\,d\sigma + \fint_{\partial\Omega^i} v_{\Omega^i}[\Phi^i_0]\,d\sigma = v[\Phi^o_0](0) + \fint_{\partial\Omega^i} v_{\Omega^i}[\Phi^i_0]\,d\sigma$$

and Lemmas 8.48 and 8.49 imply the validity of (8.149). By the definition of $v_{M,0}$ in Proposition 8.43 (iv), by equality (8.115), and by Lemma 8.49 it follows that

$$v_{M,0}(x) = S_2(x) + v^+_{\Omega^o}[\Phi^o_0](x) = S_2(x) - H^{\Omega^o}_x(0) = G^{\Omega^o}(x,0) \quad \forall x \in \Omega^o \setminus \{0\} \tag{8.156}$$

(see also (8.148)). Finally, by the definition of $v_{m,0}$ in Proposition 8.43 (vi) and by Definition 4.18 of single layer potential, we have

$$v_{m,0}(\xi) = \int_{\partial\Omega^o} \Phi^o_0\,S_2\,d\sigma + v^-_{\Omega^i}[\Phi^i_0](\xi) = v_{\Omega^o}[\Phi^o_0](0) + v^-_{\Omega^i}[\Phi^i_0](\xi) \quad \forall \xi \in \mathbb{R}^2 \setminus \Omega^i\,. \tag{8.157}$$

Hence, by Lemmas 8.49 and 8.48 we deduce that

$$v_{m,0}(\xi) = -H_0^{\Omega^o}(0) + \lim_{\eta\to\infty} H_{\Omega^i}^{\xi}(\eta) \qquad \forall \xi \in \mathbb{R}^2 \setminus \overline{\Omega^i} \tag{8.158}$$

and

$$v_{m,0}(\xi) = -H_0^{\Omega^o}(0) + \lim_{\eta\to\infty} H_{\Omega^i}^{0}(\eta) \qquad \forall \xi \in \partial\Omega^i. \tag{8.159}$$

Now, by Theorem 8.46 we have

$$\lambda_{M,(0,1)} = a_{M,0} = f_0\, v_{M,0}$$

and by (8.155) and (8.156) we deduce the validity of (8.150) for $x \in \Omega^o \setminus \{0\}$. Moreover, the continuity of both hand sides of (8.150) on $\overline{\Omega^o} \setminus \{0\}$ ensure the validity of the same equality on $\overline{\Omega^o} \setminus \{0\}$. Similarly, by Theorem 8.47 we have

$$\lambda_{m,(0,1)} = a_{m,0} = f_0\,(v_{m,0} - r_0)$$

and by (8.149), (8.155), and (8.158) we deduce the validity of (8.151). Then (8.152) follows by (8.151) and by equality $\lim_{\eta\to\infty} H_{\Omega^i}^{\xi}(\eta) = \lim_{\eta\to\infty} H_{\Omega^i}^{0}(\eta)$, which holds for all $\xi \in \overline{\Omega^i}$ (cf. Proposition 8.48). The proposition is now proved. □

8.5.4 Series Expansion for the Energy of u_ϵ

A result similar to that of Theorems 8.46 and 8.47 holds also for the map that takes ϵ to the energy integral $\int_{\Omega(\epsilon)} |\nabla u_\epsilon|^2 dx$. Following the footsteps of the previous Section 8.5.2, we begin by studying the series expansion of the analytic maps $\mathcal{E}_1$ and $\mathcal{E}_2$ of Theorem 8.31. We will exploit the sequence of real numbers $\{r_k\}_{k\in\mathbb{N}}$ and the sequences of functions $\{u_{M,k}\}_{k\in\mathbb{N}}$, $\{v_{M,k}\}_{k\in\mathbb{N}}$, $\{u_{m,k}\}_{k\in\mathbb{N}}$, and $\{v_{m,k}\}_{k\in\mathbb{N}}$, which have been introduced in Proposition 8.43.

Proposition 8.51. *The following statements hold.*

(i) Let $\{e_{1,k}\}_{k\in\mathbb{N}}$ be the sequence in $\mathbb{R}$ defined by

$$e_{1,k} \equiv \int_{\partial\Omega^o} f^o\, \nu_{\Omega^o} \cdot \nabla u_{M,k}\, d\sigma - \int_{\partial\Omega^i} f^i\, \nu_{\Omega^i} \cdot \nabla u_{m,k}\, d\sigma$$

for all $k \in \mathbb{N}$. Then there exists $\epsilon_{\mathcal{E}_1} \in]0, \epsilon_{\mathcal{E}}]$ such that

$$\mathcal{E}_1(\epsilon) = \sum_{k=0}^{\infty} e_{1,k}\, \epsilon^k \qquad \forall \epsilon \in]-\epsilon_{\mathcal{E}_1}, \epsilon_{\mathcal{E}_1}[$$

and the series converges absolutely for all $\epsilon \in]-\epsilon_{\mathcal{E}_1}, \epsilon_{\mathcal{E}_1}[$.

(ii) Let $\{e_{2,k}\}_{k\in\mathbb{N}}$ be the sequence in $\mathbb{R}$ defined by

$$e_{2,k} \equiv \int_{\partial\Omega^o} f^o\, \nu_{\Omega^o} \cdot \nabla v_{M,k}\, d\sigma - \int_{\partial\Omega^i} f^i\, \nu_{\Omega^i} \cdot \nabla v_{m,k}\, d\sigma$$

for all $k \in \mathbb{N}$. *Then there exists* $\epsilon_{\mathcal{E}_2} \in]0, \epsilon_{\mathcal{E}}]$ *such that*

$$\mathcal{E}_2(\epsilon) = \sum_{k=0}^{\infty} e_{2,k}\, \epsilon^k \qquad \forall \epsilon \in]-\epsilon_{\mathcal{E}_2}, \epsilon_{\mathcal{E}_2}[$$

and the series converges absolutely for all $\epsilon \in]-\epsilon_{\mathcal{E}_2}, \epsilon_{\mathcal{E}_2}[$.

Proof. It follows by the definitions of $\mathcal{E}_1$ and $\mathcal{E}_2$ in (8.90) and (8.91) and by Proposition 8.43. □

Then we have the following Theorem 8.52.

Theorem 8.52. *Let* $\{a_{E,k}\}_{k\in\mathbb{N}}$ *be the sequence in* $\mathbb{R}$ *defined by*

$$a_{E,k} \equiv \sum_{j=0}^{k} f_{k-j}\, e_{2,j} \qquad \forall k \in \mathbb{N}\,.$$

Let $\{\lambda_{E,(k,l)}\}_{(k,l)\in\mathbb{N}^2\,,\, l\leq k+1}$ *be the family of real numbers defined by*

$$\lambda_{E,(k,0)} \equiv e_{1,k}\,, \quad \lambda_{E,(k,1)} \equiv a_{E,k}$$

for all $k \in \mathbb{N}$ *and*

$$\lambda_{E,(k,l)} \equiv (-1)^{l-1} \sum_{j=l-1}^{k} a_{E,k-j} \sum_{\beta\in(\mathbb{N}\setminus\{0\})^{l-1}\,,\, |\beta|=j} \prod_{h=1}^{l-1} r_{\beta_h}$$

for all $k, l \in \mathbb{N}$ *with* $2 \leq l \leq k+1$. *Then there exists* $\epsilon'_E \in]0,1[$ *such that*

$$\int_{\Omega(\epsilon)} |\nabla u_\epsilon|^2\, d\sigma = \sum_{k=0}^{\infty} \epsilon^k \sum_{l=0}^{k+1} \frac{\lambda_{E,(k,l)}}{\left(r_0 + \frac{\log \epsilon}{2\pi}\right)^l} \qquad \forall \epsilon \in]0, \epsilon'_E[\,. \tag{8.160}$$

Moreover, the series

$$\sum_{k=0}^{\infty} \sum_{l=0}^{k+1} \lambda_{E,(k,l)|\overline{\Omega_M}}\, \epsilon^k \left(\frac{t}{r_0\, t + \frac{1}{2\pi}}\right)^l$$

converges absolutely for $(\epsilon, t) \in]-\epsilon'_E, \epsilon'_E[\times]1/\log \epsilon'_E, -1/\log \epsilon'_E[$.

Proof. By the real analytic representation formula (8.87) in Theorem 8.31, we have

$$\int_{\Omega(\epsilon)} |\nabla u_\epsilon|^2\, dx = \mathcal{E}_1(\epsilon) + \mathcal{F}(\epsilon) \frac{\mathcal{E}_2(\epsilon)}{\mathcal{R}(\epsilon) + \frac{\log \epsilon}{2\pi}} \qquad \forall \epsilon \in]0, \epsilon_E[\,.$$

By Propositions 8.43 (i) and 8.51 (ii) we have

$$\mathcal{F}(\epsilon)\mathcal{E}_2[\epsilon] = \left(\sum_{k=0}^{\infty} f_k \epsilon^k\right)\left(\sum_{k=0}^{\infty} e_{2,k}\epsilon^k\right) = \sum_{k=0}^{\infty} a_{E,k|\overline{\Omega_M}}\epsilon^k$$

for all ϵ with $|\epsilon| < \min\{\epsilon_{\mathcal{F}}, \epsilon_{\mathcal{E}_2}\}$. Then we define

$$\mathcal{A}_E(\epsilon) \equiv \sum_{k=0}^{\infty} a_{E,k|\overline{\Omega_M}}\epsilon^k \qquad \forall \epsilon \in]-\min\{\epsilon_{\mathcal{F}}, \epsilon_{\mathcal{E}_2}\}, \min\{\epsilon_{\mathcal{F}}, \epsilon_{\mathcal{E}_2}\}[$$

and we observe that the series on the right-hand side converges absolutely. Then by the expansion for $\mathcal{E}_1$ of Proposition 8.51 (i) and by the expansion for

$$\frac{\mathcal{A}_E(\epsilon)}{\mathcal{R}(\epsilon) + \frac{\log \epsilon}{2\pi}}$$

of Lemma 8.45, we deduce the validity of the statement. □

As a consequence of Theorem 8.52 we can obtain a finite sum that approximates $\int_{\Omega(\epsilon)} |\nabla u_\epsilon|^2 \, d\sigma$ with a remainder in $O(\epsilon^{k'+1})$. Namely, for all $k' \in \mathbb{N}$ we have

$$\int_{\Omega(\epsilon)} |\nabla u_\epsilon|^2 \, d\sigma = \sum_{k=0}^{k'} \epsilon^n \sum_{l=0}^{k+1} \frac{\lambda_{E,(k,l)}}{\left(r_0 + \frac{\log \epsilon}{2\pi}\right)^l} + O(\epsilon^{k'+1}) \qquad \text{as } \epsilon \to 0^+ .$$

In particular, for $k = 0$ we have

$$\int_{\Omega(\epsilon)} |\nabla u_\epsilon|^2 \, d\sigma = \lambda_{E,(0,0)} + \frac{\lambda_{E,(0,1)}}{r_0 + \frac{\log \epsilon}{2\pi}} + O(\epsilon) \qquad \text{as } \epsilon \to 0^+ .$$

By the limiting relation (8.92) we deduce that

$$\lambda_{E,(0,0)} = \int_{\Omega^o} |\nabla u^o|^2 \, dx + \int_{\mathbb{R}^2 \setminus \Omega^i} |\nabla u^i|^2 \, d\xi$$

with u^o and u^i as in (8.75) and (8.82), respectively. The term r_0 is given by (8.149). We now show that also $\lambda_{E,(0,1)}$ can be written in terms of the auxiliary functions u^o and u^i.

Proposition 8.53. *We have*

$$\lambda_{E,(0,1)} = -(\lim_{\eta\to\infty} u^i(\eta) - u^o(0))^2 .$$

Proof. By Theorem 8.52 we have

$$\lambda_{E,(0,1)} = f_0 \, e_{2,0}$$

and by (8.155) we have $f_0 = \lim_{\eta\to\infty} u^i(\eta) - u^o(0)$. Hence, it suffices to show that

$$e_{2,0} = u^o(0) - \lim_{\eta \to \infty} u^i(\eta)\,. \tag{8.161}$$

By the definition of $e_{2,0}$ in Proposition 8.51 (ii), we have

$$e_{2,0} = \int_{\partial\Omega^o} f^o\, \nu_{\Omega^o} \cdot \nabla v_{M,0}\, d\sigma - \int_{\partial\Omega^i} f^i\, \nu_{\Omega^i} \cdot \nabla v_{m,0}\, d\sigma\,.$$

Then by the formulas (8.156) and (8.157) for $v_{M,0}$ and $v_{m,0}$, we have

$$\begin{aligned} e_{2,0} = &\int_{\partial\Omega^o} f^o\, \nu_{\Omega^o} \cdot \nabla S_2\, d\sigma + \int_{\partial\Omega^o} f^o\, \nu_{\Omega^o} \cdot \nabla v^+_{\Omega^o}[\Phi^o_0]\, d\sigma \\ &- \int_{\partial\Omega^i} f^i\, \nu_{\Omega^i} \cdot \nabla v^-_{\Omega^i}[\Phi^i_0]\, d\sigma\,. \end{aligned} \tag{8.162}$$

By the jump formulas for the normal derivative of the single layer potential of Theorem 6.6 and by Equations (8.113) and (8.114) for Φ^i_0 and Φ^o_0, respectively, we have

$$\nu_{\Omega^i} \cdot \nabla v^-_{\Omega^i}[\Phi^i_0] = \Phi^i_0 + \nu_{\Omega^i} \cdot \nabla v^+_{\Omega^i}[\Phi^i_0] = \Phi^i_0$$

and

$$\nu_{\Omega^o} \cdot \nabla v^+_{\Omega^o}[\Phi^o_0] = -\Phi^o_0 + \nu_{\Omega^o} \cdot \nabla v^-_{\Omega^o}[\Phi^i_0] = -\Phi^o_0 - \nu_{\Omega^o} \cdot \nabla S_2\,.$$

Thus (8.162) implies that

$$e_{2,0} = -\int_{\partial\Omega^o} f^o\, \Phi^o_0\, d\sigma - \int_{\partial\Omega^i} f^i\, \Phi^i_0\, d\sigma$$

and the validity of (8.161) follows by (8.153) and (8.154). □

8.5.5 Series Expansions in a Circular Annulus

In this subsection we show that when Ω^i and Ω^o are balls in $\mathbb{R}^2$, the expansions of the solution $u_\epsilon(x)$, of the rescaled function $u_\epsilon(\epsilon\xi)$, and of the energy $\int_{\Omega(\epsilon)} |\nabla u_\epsilon|^2\, d\sigma$ are particularly simple. We will assume that both Ω^i and Ω^o coincide with the unit ball $\mathbb{B}_2$ of $\mathbb{R}^2$ and that $\Omega(\epsilon)$ is the annular domain $\mathbb{B}_2 \setminus \epsilon\overline{\mathbb{B}_2}$ for all $\epsilon \in]-1,1[$ (so that $\epsilon_0 = 1$). We recall that under such assumptions we have

$$\Phi^i[\epsilon](x) = \frac{1}{2\pi} \quad \text{and} \quad \Phi^o[\epsilon](x) = -\frac{1}{2\pi} \qquad \forall x \in \partial\mathbb{B}_2\,,\ \epsilon \in]-1,1[\tag{8.163}$$

(see proof of Example 8.40) which implies that

$$\Phi^i_0(x) = -\Phi^o_0(x) = \frac{1}{2\pi} \quad \forall x \in \partial\mathbb{B}_2 \quad \text{and} \quad \Phi^i_k = \Phi^o_k = 0 \quad \forall k \in \mathbb{N} \setminus \{0\}\,. \tag{8.164}$$

In the following Proposition 8.54 we consider the integral representations (8.69) and (8.79) for u_ϵ and $u_\epsilon(\epsilon\cdot)$ in the specific case of a ring domain.

Proposition 8.54. *If $\Omega^i = \Omega^o = \mathbb{B}_2$, then for all $\epsilon \in]0, \epsilon_*[$ we have*

$$u_\epsilon(x) = w^+_{\mathbb{B}_2}[\Psi^o[\epsilon]](x) + \epsilon \int_{\partial\mathbb{B}_2} \Psi^i[\epsilon](y)\, y \cdot \nabla S_2(x - \epsilon y)\, d\sigma_y$$
$$+ \left(\int_{\partial\mathbb{B}_2} f^i \, d\sigma - \int_{\partial\mathbb{B}_2} f^o \, d\sigma \right) \frac{S_2(x)}{\log \epsilon} \qquad \forall x \in \overline{\mathbb{B}_2} \setminus \epsilon\overline{\mathbb{B}_2}\,, \tag{8.165}$$

and

$$u_\epsilon(\epsilon\xi) = -w^-_{\mathbb{B}_2}[\Psi^i[\epsilon]](\xi) + w^+_{\mathbb{B}_2}[\Psi^o[\epsilon]](\epsilon\xi)$$
$$+ \left(\int_{\partial\mathbb{B}_2} f^i \, d\sigma - \int_{\partial\mathbb{B}_2} f^o \, d\sigma \right) \left(\frac{S_2(\xi)}{\log \epsilon} + \frac{1}{2\pi} \right) \quad \forall \xi \in \overline{(\epsilon^{-1}\mathbb{B}_2)} \setminus \mathbb{B}_2\,. \tag{8.166}$$

Proof. We consider the factor

$$\left(\int_{\partial\Omega^i} \Phi^i[\epsilon](\eta)\, S(x - \epsilon\eta)\, d\sigma_\eta + v^+_{\Omega^o}[\Phi^o[\epsilon]](x) \right)$$
$$\times \left(\fint_{\partial\Omega^i} v_{\Omega^i}[\Phi^i[\epsilon]](\eta) + v_{\Omega^o}[\Phi^o[\epsilon]](\epsilon\eta)\, d\sigma_\eta + \frac{\log \epsilon}{2\pi} \right)^{-1} \tag{8.167}$$

that appears in the right-hand side of the integral representation formula (8.69) for u_ϵ. By formula (8.37) for ϕ^i_ϵ and ϕ^o_ϵ in terms of τ_ϵ, by Equation (8.42) which characterizes ϕ^i_ϵ, ϕ^o_ϵ, and by equality (8.47) of $\Phi^i[\epsilon]$, $\Phi^o[\epsilon]$ and ϕ^i_ϵ, ϕ^o_ϵ for $\epsilon \in]0, \epsilon_*[$, we can see that (8.167) coincides with the quotient

$$\frac{v^+_{\Omega(\epsilon)}[\tau_\epsilon]}{\fint_{\epsilon\partial\Omega^i} V_{\Omega(\epsilon)}[\tau_\epsilon]\, d\sigma}\,,$$

where τ_ϵ is the unique solution of (8.32). By Lemma 6.29 we have

$$v^+_{\Omega(\epsilon)|\partial\Omega(\epsilon)}[\tau_\epsilon] \in \mathrm{Ker}\left(\frac{1}{2}I + W_{\Omega(\epsilon)} \right)$$

and thus Theorem 6.14 implies that $v^+_{\Omega(\epsilon)}[\tau_\epsilon]$ is constant and not zero on $\epsilon\partial\mathbb{B}_2$ and equal to zero on $\partial\mathbb{B}_2$. We deduce that the function $v_\epsilon \in C^{1,\alpha}(\overline{\Omega(\epsilon)})$ that takes x to the term in (8.167) is the unique solution of the boundary value problem

$$\begin{cases} \Delta v_\epsilon = 0 & \text{in } \Omega(\epsilon)\,, \\ v_\epsilon = 1 & \text{on } \epsilon\partial\Omega^i\,, \\ v_\epsilon = 0 & \text{on } \partial\Omega^o\,. \end{cases}$$

Then, by the assumption that $\Omega^i = \Omega^o = \mathbb{B}_2$ and by a straightforward computation we see that

$$v_\epsilon(x) = \frac{2\pi S_2(x)}{\log \epsilon} \qquad \forall x \in \overline{\mathbb{B}_2} \setminus \epsilon\mathbb{B}_2 \tag{8.168}$$

(note that the assumption on Ω^i and Ω^o is used here for the first time in the proof). Now we obtain (8.165) by (8.69), by taking $\Phi^i[\epsilon]$ and $\Phi^o[\epsilon]$ as in (8.163), and by replacing the factor (8.167) with (8.168). To prove (8.166) we observe that, by a change of variable in (8.167) and (8.168), we have

$$\begin{aligned}&\left(v_{\overline{\Omega^i}}^-[\Phi^i[\epsilon]](\xi) + v_{\Omega^o}[\Phi^o[\epsilon]](\epsilon\xi) + \frac{\log \epsilon}{2\pi}\right)\\ &\times \left(\fint_{\partial\Omega^i} v_{\Omega^i}[\Phi^i[\epsilon]](\eta) + v_{\partial\Omega^o}[\Phi^o[\epsilon]](\epsilon\eta)\, d\sigma_\eta + \frac{\log \epsilon}{2\pi}\right)^{-1} = \frac{2\pi S_2(\xi)}{\log \epsilon} + 1\,.\end{aligned} \tag{8.169}$$

Then (8.166) follows by the integral representation formula (8.79) for $u_\epsilon(\epsilon\cdot)$, by the explicit value of $\Phi^i[\epsilon]$, $\Phi^o[\epsilon]$ of (8.163), and by (8.169). □

Then we have the following Theorem 8.55, where we write the series expansions of $\epsilon \mapsto u_\epsilon(\cdot)$ and $\epsilon \mapsto u_\epsilon(\epsilon\cdot)$ by means of the sequences $\{u_{M,k}\}_{k\in\mathbb{N}}$ and $\{u_{m,k}\}_{k\in\mathbb{N}}$ introduced in Proposition 8.43.

Theorem 8.55. *If $\Omega^i = \Omega^o = \mathbb{B}_2$, then the following statements hold.*

(i) If Ω_M is an open subset of $\mathbb{B}_2$ such that $0 \notin \overline{\Omega_M}$, then there exists $\epsilon'_M \in]0,1[$ such that $\overline{\Omega_M} \cap \epsilon\overline{\mathbb{B}_2} = \emptyset$ for all $\epsilon \in]-\epsilon'_M, \epsilon'_M[$ and such that

$$u_{\epsilon|\overline{\Omega_M}} = \sum_{k=0}^{\infty} u_{M,k|\overline{\Omega_M}}\epsilon^k + \left(\int_{\partial\mathbb{B}_2} f^i\, d\sigma - \int_{\partial\mathbb{B}_2} f^o\, d\sigma\right)\frac{S_{2|\overline{\Omega_M}}}{\log \epsilon} \quad \forall \epsilon \in]0, \epsilon'_M[\,.$$

(ii) If Ω_m is an open and bounded subset of $\mathbb{R}^2 \setminus \overline{\mathbb{B}_2}$, then there exists $\epsilon'_m \in]0,1[$ such that $\epsilon\overline{\Omega_m} \subseteq \mathbb{B}_2$ for all $\epsilon \in]-\epsilon'_m, \epsilon'_m[$ and such that

$$u_\epsilon(\epsilon\cdot)_{|\overline{\Omega_m}} = \sum_{k=0}^{\infty} u_{m,k|\overline{\Omega_m}}\epsilon^k + \left(\int_{\partial\mathbb{B}_2} f^i\, d\sigma - \int_{\partial\mathbb{B}_2} f^o\, d\sigma\right)\left(\frac{S_{2|\overline{\Omega_m}}}{\log \epsilon} + \frac{1}{2\pi}\right) \quad \forall \epsilon \in]0, \epsilon'_m[\,.$$

Proof. It suffices to observe that, by the definition of $\mathcal{U}_M$ in (8.76) and $\mathcal{U}_m$ in (8.83) and by Proposition 8.54, we have

$$u_{\epsilon|\overline{\Omega_M}} = \mathcal{U}_M[\epsilon] + \left(\int_{\partial\mathbb{B}_2} f^i\, d\sigma - \int_{\partial\mathbb{B}_2} f^o\, d\sigma\right)\frac{S_{2|\overline{\Omega_M}}}{\log \epsilon} \quad \forall \epsilon \in]0, \epsilon_M[\tag{8.170}$$

and

$$u_\epsilon(\epsilon\cdot)_{|\overline{\Omega_m}} = \mathcal{U}_m[\epsilon] + \left(\int_{\partial\mathbb{B}_2} f^i\, d\sigma - \int_{\partial\mathbb{B}_2} f^o\, d\sigma\right)\left(\frac{S_{2|\overline{\Omega_m}}}{\log\epsilon} + \frac{1}{2\pi}\right) \quad \forall\epsilon \in]0,\epsilon_m[\,. \tag{8.171}$$

Then the theorem follows by Proposition 8.43 (iii), (v). □

Incidentally, we observe that Theorem 8.55 provides another proof of the statement in Example 8.40. In Theorem 8.56 below we consider the energy integral $\int_{\Omega(\epsilon)} |\nabla u_\epsilon|^2\, d\sigma$. To describe the series expansion we will use the sequence $\{e_{1,k}\}_{k\in\mathbb{N}}$ of Proposition 8.51 (i). For the proof we will exploit the following equality

$$w_{\mathbb{B}_2}[\psi](0) = \int_{\partial\mathbb{B}_2} \psi\, \nu_{\mathbb{B}_2}\cdot\nabla S_2\, d\sigma = \int_{\partial\mathbb{B}_2} \psi(x)\frac{x}{|x|}\cdot\frac{x}{2\pi|x|^2}\, d\sigma_x = \fint_{\partial\mathbb{B}_2} \psi\, d\sigma\,, \tag{8.172}$$

which holds for all $\psi \in C^{1,\alpha}(\partial\mathbb{B}_2)$.

Theorem 8.56. *If $\Omega^i = \Omega^o = \mathbb{B}_2$, then there exists $\epsilon'_E \in]0,1[$ such that*

$$\int_{\Omega(\epsilon)} |\nabla u_\epsilon|^2\, d\sigma = \sum_{k=0}^{\infty} e_{1,k}\,\epsilon^k - \frac{1}{2\pi}\left(\int_{\partial\mathbb{B}_2} f^i\, d\sigma - \int_{\partial\mathbb{B}_2} f^o\, d\sigma\right)^2 \frac{1}{\log\epsilon}$$

for all $\epsilon \in]0,\epsilon'_E[$.

Proof. We take Ω_M and Ω_m as in Theorem 8.55 and we further assume that

$$\partial\mathbb{B}_2 \subseteq \overline{\Omega_M} \qquad \text{and} \qquad \partial\mathbb{B}_2 \subseteq \overline{\Omega_m}\,.$$

Then by formula (8.89) for the energy integral, and by the above formulas (8.170) and (8.171) for $u_{\epsilon|\overline{\Omega_M}}$ and $u_\epsilon(\epsilon\cdot)_{|\overline{\Omega_m}}$, we have

$$\begin{aligned}
\int_{\Omega(\epsilon)} |\nabla u_\epsilon|^2\, dx &= \int_{\partial\mathbb{B}_2} f^o\, \nu_{\mathbb{B}_2}\cdot\nabla\mathcal{U}_M[\epsilon]\, d\sigma \\
&+ \int_{\partial\mathbb{B}_2} f^o\, \nu_{\mathbb{B}_2}\cdot\nabla\left[\left(\int_{\partial\mathbb{B}_2} f^i\, d\sigma - \int_{\partial\mathbb{B}_2} f^o\, d\sigma\right)\frac{S_2}{\log\epsilon}\right] d\sigma \\
&- \int_{\partial\mathbb{B}_2} f^i\, \nu_{\mathbb{B}_2}\cdot\nabla\mathcal{U}_m[\epsilon]\, d\sigma \\
&- \int_{\partial\mathbb{B}_2} f^i\, \nu_{\mathbb{B}_2}\cdot\nabla\left[\left(\int_{\partial\mathbb{B}_2} f^i\, d\sigma - \int_{\partial\mathbb{B}_2} f^o\, d\sigma\right)\frac{S_2}{\log\epsilon}\right] d\sigma
\end{aligned}$$

for all $\epsilon \in]0,\epsilon_E[$, with $\epsilon_E \equiv \min\{\epsilon_M,\epsilon_m\}$. Hence,

$$\begin{aligned}\int_{\Omega(\epsilon)} |\nabla u_\epsilon|^2\, dx = &\int_{\partial\mathbb{B}_2} f^o\, \nu_{\mathbb{B}_2}\cdot \nabla\mathcal{U}_M[\epsilon]\, d\sigma - \int_{\partial\mathbb{B}_2} f^i\, \nu_{\mathbb{B}_2}\cdot\nabla\mathcal{U}_m[\epsilon]\, d\sigma\\ &+\frac{1}{\log\epsilon}\left(\int_{\partial\mathbb{B}_2} f^i\, d\sigma - \int_{\partial\mathbb{B}_2} f^o\, d\sigma\right)\\ &\times\left(\int_{\partial\mathbb{B}_2} f^o\, \nu_{\mathbb{B}_2}\cdot\nabla S_2\, d\sigma - \int_{\partial\mathbb{B}_2} f^i\,\nu_{\mathbb{B}_2}\cdot\nabla S_2\, d\sigma\right)\end{aligned}$$

for all $\epsilon\in]0,\epsilon_E[$. Now the validity of the statement follows by the definition of $\mathcal{E}_1$ in (8.90), by Proposition 8.51 (i), and by the equality (8.172). □

In conclusion of this subsection, we observe that, in the specific case when $\Omega^i = \Omega^o = \mathbb{B}_2$, we can easily solve the systems of integral equation (8.123)–(8.126), which determine the sequence $\{(\Psi^i_k, \Psi^o_k)\}_{k\in\mathbb{N}}$. Indeed, we can exploit the following equality

$$\begin{aligned}W_{\mathbb{B}_2}[\psi](x) &= -\int_{\partial\mathbb{B}_2}\psi(y)\,\nu_{\mathbb{B}_2}(y)\cdot\nabla S_2(x-y)\, d\sigma_y\\ &= -\int_{\partial\mathbb{B}_2}\psi(y)\,\frac{y}{|y|}\cdot\frac{x-y}{2\pi|x-y|^2}\, d\sigma_y\\ &= \frac{1}{2\pi}\int_{\partial\mathbb{B}_2}\psi(y)\,\frac{1}{|y|}\frac{y\cdot y - x\cdot y}{x\cdot x + y\cdot y - 2x\cdot y}\, d\sigma_y = \frac{1}{2}\fint_{\partial\mathbb{B}_2}\psi\, d\sigma\end{aligned} \tag{8.173}$$

which holds for all $x\in\partial\mathbb{B}_2$ and for all $\psi\in C^{1,\alpha}(\partial\mathbb{B}_2)$ (cf. definition (4.71)). For example, for $k=0$ we deduce by (8.123), (8.124), (8.164), (8.172), and (8.173) that

$$\frac{1}{2}\Psi^i_0 - \frac{1}{2}\fint_{\partial\mathbb{B}_2}\Psi^i_0\, d\sigma = f^i - \frac{1}{2}\fint_{\partial\mathbb{B}_2} f^i\, d\sigma + \frac{1}{2}\fint_{\partial\mathbb{B}_2} f^o\, d\sigma - \fint_{\partial\mathbb{B}_2}\Psi^o_0\, d\sigma$$

and

$$\frac{1}{2}\Psi^o_0 + \frac{1}{2}\fint_{\partial\mathbb{B}_2}\Psi^o_0\, d\sigma = f^o\,.$$

Taking the integral over $\partial\mathbb{B}_2$ of the second equality we obtain

$$\fint_{\partial\mathbb{B}_2}\Psi^o_0\, d\sigma = \fint_{\partial\mathbb{B}_2} f^o\, d\sigma\,.$$

Thus, condition $\int_{\partial\mathbb{B}_2}\Psi^i_0\, d\sigma = 0$ implies that

$$\Psi^i_0 = 2f^i - 2\fint_{\partial\mathbb{B}_2} f^i\, d\sigma \quad\text{and}\quad \Psi^o_0 = 2f^o - \fint_{\partial\mathbb{B}_2} f^o\, d\sigma\,. \tag{8.174}$$

For $k=1$, (8.125) and (8.126) yield the following equations for Ψ^i_1 and Ψ^o_1,

$$\frac{1}{2}\Psi^i_1(\xi) - W_{\mathbb{B}_2}[\Psi^i_1](\xi) = -\int_{\partial\mathbb{B}_2} f^i\Phi^i_1\, d\sigma - \int_{\partial\mathbb{B}_2} f^o\Phi^o_1\, d\sigma \tag{8.175}$$

$$-\int_{\partial\mathbb{B}_2} \Psi_1^o \, \nu_{\mathbb{B}_2} \cdot \nabla S_2 \, d\sigma + \sum_{h=0}^{1} \xi_1^h \xi_2^{1-h} \int_{\partial\mathbb{B}_2} \Psi_0^o \, \nu_{\mathbb{B}_2} \cdot \nabla \partial_1^h \partial_2^{1-h} S_2 \, d\sigma \qquad \forall \xi \in \partial\Omega^i ,$$

$$\frac{1}{2}\Psi_1^o(x) + W_{\mathbb{B}_2}[\Psi_1^o](x) = -(\nabla S_2)(x) \cdot \int_{\partial\mathbb{B}_2} \Psi_0^i(\eta)\, \nu_{\partial\mathbb{B}_2}(\eta)\, d\sigma_\eta \quad \forall x \in \partial\Omega^o . \tag{8.176}$$

A straightforward computation shows that

$$\begin{aligned} &\nu_{\mathbb{B}_2}(y) \cdot \nabla \partial_1^h \partial_2^{1-h} S_2(y) = y \cdot \nabla \partial_k S_2(y) \\ &= \frac{1}{2\pi} y \cdot (\delta_{1k} - 2y_1 y_k \, , \, \delta_{2k} - 2y_2 y_k) = -\frac{1}{2\pi}\, y_k \qquad \forall y \in \partial\mathbb{B}_2 \end{aligned}$$

with $k = 2$ if $h = 0$ and $k = 1$ if $h = 1$. Then, (8.175) and (8.176) together with (8.164), (8.172), and (8.173) imply that

$$\frac{1}{2}\Psi_1^i(\xi) - \frac{1}{2} \fint_{\partial\mathbb{B}_2} \Psi_1^i \, d\sigma = -\fint_{\partial\mathbb{B}_2} \Psi_1^o \, d\sigma - \xi \cdot \fint_{\partial\mathbb{B}_2} \Psi_0^o(y)\, y\, d\sigma_y \qquad \forall \xi \in \partial\mathbb{B}_2$$

and

$$\frac{1}{2}\Psi_1^o(x) + \frac{1}{2} \fint_{\partial\mathbb{B}_2} \Psi_1^o \, d\sigma = -x \cdot \fint_{\partial\mathbb{B}_2} \Psi_0^i(\eta)\, \eta\, d\sigma \qquad x \in \partial\mathbb{B}_2 .$$

Taking the integral over $\partial\mathbb{B}_2$ of the second equality we see that

$$\fint_{\partial\mathbb{B}_2} \Psi_1^o \, d\sigma = 0 .$$

Then, condition $\int_{\partial\mathbb{B}_2} \Psi_1^i \, d\sigma = 0$ and (8.174) imply that

$$\Psi_1^i(\xi) = -4\xi \cdot \fint_{\partial\mathbb{B}_2} f^o(y)\, y\, d\sigma_y \qquad \forall \xi \in \partial\mathbb{B}_2$$

and

$$\Psi_1^o(x) = -4x \cdot \fint_{\partial\mathbb{B}_2} f^i(\eta)\, \eta\, d\sigma_\eta \qquad \forall x \in \partial\mathbb{B}_2 .$$

We can now resort to Propositions 8.43 and 8.51 to compute the first terms $u_{M,k}$, $u_{m,k}$, and $e_{1,k}$ in the series expansions of Theorems 8.55 and 8.56. For example, for $u_{M,1}$ Proposition 8.43 (iii) implies that

$$u_{M,1}(x) = w_{\mathbb{B}_2}^+[\Psi_1^o](x) + (\nabla S_2)(x) \cdot \int_{\partial\mathbb{B}_2} \Psi_0^i(\eta)\, \nu_{\partial\mathbb{B}_2}(\eta)\, d\sigma_\eta \quad \forall x \in \overline{\mathbb{B}_2} \setminus \{0\} .$$

Then, by the formula for Ψ_0^i in (8.174) and by equality (8.176) for Ψ_1^o we deduce that

$$u_{M,1}(x) = u_1^o(x) + 2\frac{x}{|x|^2} \cdot \fint_{\partial\mathbb{B}_2} f^i(\eta)\,\eta\,d\sigma_\eta \quad \forall x \in \overline{\mathbb{B}_2} \setminus \{0\}\,,$$

where u_1^o is the unique solution of the Dirichlet problem

$$\begin{cases} \Delta u_1^o = 0 & \text{in } \mathbb{B}_2\,, \\ u_1^o(x) = -2x \cdot \fint_{\partial\mathbb{B}_2} f^i(\eta)\,\eta\,d\sigma_\eta & \text{for } x \in \partial\mathbb{B}_2\,. \end{cases}$$

We could now continue the computation of the next terms of the sequence $\{(\Psi_k^i, \Psi_k^o)\}_{k\in\mathbb{N}}$ and obtain all the higher order coefficients $u_{M,k}$, $u_{m,k}$, and $e_{1,k}$. (Instead, the terms with $k = 0$ can be deduced by the limiting relations (8.78), (8.86), and (8.92). For example, by (8.78) we have $u_{M,0}(x) = u^o(x)$ for all $x \in \overline{\mathbb{B}_2} \setminus \{0\}$.)

Chapter 9
Other Problems with Linear Boundary Conditions in a Domain with a Small Hole

Abstract In Chapter 8 we introduced the Functional Analytic Approach. To do so, we considered its application to a Dirichlet problem for the Laplace equation in a bounded domain with a small hole that shrinks to an interior point. In this chapter we are going to see that the same method also works with other boundary value problems with linear boundary conditions, provided that we introduce a suitable integral representations of the solution. Instead, nonlinear problems, even with linear boundary conditions, require a specific modification of our approach. This is for example the case of the eigenvalue problems, which are not linear with respect to the pair eigenvalue-eigenfunction. To give some examples, we will consider a mixed problem for the Laplace equation, a mixed problem for the Poisson equation, and a Steklov eigenvalue problem. All of these problems will be defined in a perforated domain with a shrinking hole, so our geometric setting will still be that of Chapter 8. We will see that the analysis of the mixed problem for the Laplace equation requires no particular modification of the method introduced for the Dirichlet problem. The mixed Poisson problem can then be studied using the results obtained for the mixed Laplace problem and for the volume potentials of Chapter 7. To study the Steklov problem, on the other hand, we will have to do something new: we no longer can use Proposition A.15 on the analyticity of the inversion map and instead will resort to the Implicit Function Theorem A.19 for real analytic maps.

9.1 The Geometric Setting

The geometric setting is the same as in Chapter 8. We fix once for all a natural number

$$n \in \mathbb{N}\,,\ n \geq 2$$

and a real number

$$\alpha \in]0,1[\,.$$

M. Dalla Riva et al., *Singularly Perturbed Boundary Value Problems*,
https://doi.org/10.1007/978-3-030-76259-9_9

Then we take two sets Ω^i and Ω^o in the n-dimensional Euclidean space $\mathbb{R}^n$, the letter 'i' standing for 'inner domain' and the letter 'o' standing for 'outer domain.' We assume that Ω^i and Ω^o satisfy the following condition

$$\begin{aligned}&\Omega^i \text{ and } \Omega^o \text{ are bounded open connected subsets of } \mathbb{R}^n\\&\text{of class } C^{1,\alpha}\text{, the exterior domains } \Omega^{i-} \text{ and } \Omega^{o-} \text{ are connected,}\\&\text{and the origin } 0 \text{ of } \mathbb{R}^n \text{ belongs both to } \Omega^i \text{ and to } \Omega^o.\end{aligned}$$

Then Ω^i and Ω^o have no holes and there exists a positive real number ϵ_0 such that

$$\epsilon\overline{\Omega^i} \subseteq \Omega^o \text{ for all } \epsilon \in]-\epsilon_0, \epsilon_0[\,.$$

We denote by $\Omega(\epsilon)$ the perforated domain

$$\Omega(\epsilon) \equiv \Omega^o \setminus (\epsilon\overline{\Omega^i}) \qquad \forall \epsilon \in]-\epsilon_0, \epsilon_0[$$

and we observe that $\Omega(\epsilon)$ is an open bounded connected subset of $\mathbb{R}^n$ with boundary consisting of the disjoint union of $\partial\Omega^o$ and $\epsilon\partial\Omega^i$ (see Figure 8.1, for a proof that $\Omega(\epsilon)$ is connected see, e.g., Theorem A.10). Moreover, for $\epsilon \in]-\epsilon_0, \epsilon_0[\setminus\{0\}$ the domain $\Omega(\epsilon)$ is of class $C^{1,\alpha}$ (cf. Section 2.13). Instead, for $\epsilon = 0$ we have $\Omega(0) = \Omega^o \setminus \{0\}$ and $\Omega(0)$ is not anymore of class $C^{1,\alpha}$. More precisely, $\Omega(0)$ is not even a local strict hypograph of class C^0 (cf. Definition 2.37).

9.2 A Mixed Boundary Value Problem for the Laplace Equation

In Section 8.4 of Chapter 8 we have seen that if the dimension n is equal to two, the solution of a Dirichlet problem in $\Omega(\epsilon)$ displays a logarithmic behavior as $\epsilon \to 0$. Such behavior is strictly related to the Dirichlet condition on the boundary of the hole and might disappear with other boundary conditions. For example, it is not present in the case of the mixed boundary value problem with a Neumann condition on $\epsilon\partial\Omega^i$ that we analyze in this section. Specifically, for all $\epsilon \in]0, \epsilon_0[$ we consider a problem in the form

$$\begin{cases}\Delta u = 0 & \text{in } \Omega(\epsilon)\,,\\ \nu_{\epsilon\Omega^i}(x)\cdot\nabla u(x) = f^i(x/\epsilon) & \text{for all } x \in \epsilon\partial\Omega^i\,,\\ u = f^o & \text{on } \partial\Omega^o\end{cases} \tag{9.1}$$

for an unknown function $u \in C^{1,\alpha}(\overline{\Omega(\epsilon)})$.

Together with the dependence of the solution u upon ϵ, we study also its dependence upon the pair $(f^i, f^o) \in C^{0,\alpha}(\partial\Omega^i) \times C^{1,\alpha}(\partial\Omega^o)$ that defines the boundary conditions. To do so, we think of the triple (ϵ, f^i, f^o) as a point in the product Banach space $\mathbb{R} \times C^{0,\alpha}(\partial\Omega^i) \times C^{1,\alpha}(\partial\Omega^o)$. Then for all (ϵ, f^i, f^o) in the open subset $]0, \epsilon_0[\times C^{0,\alpha}(\partial\Omega^i) \times C^{1,\alpha}(\partial\Omega^o)$ we denote by $u[\epsilon, f^i, f^o]$ the unique solution of (9.1) (cf. Propositions 6.3 and 6.49). We will show that the maps that take (ϵ, f^i, f^o) to suitable restrictions of $u[\epsilon, f^i, f^o]$ and of $u[\epsilon, f^i, f^o](\epsilon\cdot)$ and to the energy $\int_{\Omega(\epsilon)} |\nabla u[\epsilon, f^i, f^o]|^2\, d\sigma$ have real analytic continuations on $]\epsilon', \epsilon'[\times C^{0,\alpha}(\partial\Omega^i) \times C^{1,\alpha}(\partial\Omega^o)$ for some suitable $\epsilon' > 0$. An advantage of taking into account also the dependence upon the boundary will be seen in the next Section 9.3, where we will use the results of the present section to study a Poisson problem.

We observe that the dependence upon the boundary data could be included also in the analysis of the Dirichlet problem carried out in Section 8.2. In addition, we could also consider the dependence upon the 'shape' of the domains Ω^o and Ω^i, for example, identifying the boundaries of Ω^i and Ω^o as the images of suitable diffeomorphims defined on the boundary of certain reference domains (as we have done in [162, 166]). For the sake of brevity in the exposition we have, however, chosen to treat the dependence upon the boundary data only in the specific case of a mixed problem, which is rather simple to analyze. Indeed, in this case we do not need to study separately the case of dimension $n \geq 3$ and that of dimension $n = 2$, as we did for the Dirichlet problem in Section 8.2. In particular, we can use the same integral representation of the solution regardless of the dimension n. Then, in view of Proposition 6.49, we look for solutions of the boundary value problem (9.1) in the form

$$v^-_{\epsilon\Omega^i}[\mu^i](x) + v^+_{\Omega^o}[\phi^o](x) + \rho \qquad \forall x \in \overline{\Omega(\epsilon)}\,, \tag{9.2}$$

for some $(\mu^i, \phi^o, \rho) \in C^{0,\alpha}(\epsilon\partial\Omega^i) \times C^{0,\alpha}(\partial\Omega^o)_0 \times \mathbb{R}$. By the jump formulas for the single layer potential in Theorem 6.6, the function in (9.2) is a solution of the boundary value problem (9.1) if and only if

$$\begin{aligned} &\frac{1}{2}\mu^i(x) + \nu_{\epsilon\Omega^i}(x) \cdot \int_{\epsilon\partial\Omega^i} \mu^i(y)\, \nabla S_n(x-y)\, d\sigma_y \\ &\quad + \nu_{\epsilon\Omega^i}(x) \cdot \int_{\partial\Omega^o} \phi^o(y)\, \nabla S_n(x-y)\, d\sigma_y = f^i(x/\epsilon) \qquad \forall x \in \epsilon\partial\Omega^i\,, \end{aligned} \tag{9.3}$$

$$V_{\Omega^o}[\phi^o](x) + \rho + \int_{\epsilon\partial\Omega^i} S_n(x-y)\, \mu^i(y)\, d\sigma_y = f^o(x) \qquad \forall x \in \partial\Omega^o \tag{9.4}$$

(see also definition (6.2)). By changing the variable of integration in the integrals over $\epsilon\partial\Omega^i$ we see that $(\mu^i, \phi^o, \rho) \in C^{0,\alpha}(\epsilon\partial\Omega^i) \times C^{0,\alpha}(\partial\Omega^o)_0 \times \mathbb{R}$ is a solution of (9.3) and (9.4) if and only if the triple $(\phi^i, \phi^o, \rho) \in C^{0,\alpha}(\partial\Omega^i) \times C^{0,\alpha}(\partial\Omega^o)_0 \times \mathbb{R}$ with

$$\phi^i(\xi) = \mu^i(\epsilon x) \qquad \forall x \in \partial\Omega^i$$

is a solution of

$$\begin{aligned} &\left(\frac{1}{2}I + W^t_{\Omega^i}\right)[\phi^i](\xi) \\ &\quad + \nu_{\Omega^i}(\xi) \cdot \int_{\partial\Omega^o} \phi^o(y)\, \nabla S_n(\epsilon\xi - y)\, d\sigma_y = f^i(\xi) \qquad \forall \xi \in \partial\Omega^i\,, \end{aligned} \tag{9.5}$$

$$V_{\Omega^o}[\phi^o](x) + \rho + \epsilon^{n-1} \int_{\partial\Omega^i} S_n(x - \epsilon\eta)\,\phi^i(\eta)\,d\sigma_\eta = f^o(x) \quad \forall x \in \partial\Omega^o\,. \tag{9.6}$$

Then, for all $\epsilon \in]-\epsilon_0, \epsilon_0[$ we define the operator $\mathcal{J}(\epsilon) \equiv (\mathcal{J}^i(\epsilon), \mathcal{J}^o(\epsilon))$ from $C^{0,\alpha}(\partial\Omega^i) \times C^{0,\alpha}(\partial\Omega^o)_0 \times \mathbb{R}$ to $C^{0,\alpha}(\partial\Omega^i) \times C^{1,\alpha}(\partial\Omega^o)$ that takes a triple (ϕ^i, ϕ^o, ρ) to the pair

$$(\mathcal{J}^i(\epsilon)[\phi^i, \phi^o, \rho], \mathcal{J}^o(\epsilon)[\phi^i, \phi^o, \rho])$$

defined by

$$\mathcal{J}^i(\epsilon)[\phi^i, \phi^o, \rho](\xi) \equiv \left(\frac{1}{2}I + W^t_{\Omega^i}\right)[\phi^i](\xi) \tag{9.7}$$

$$+ \nu_{\Omega^i}(\xi) \cdot \int_{\partial\Omega^o} \phi^o(y)\,\nabla S_n(\epsilon\xi - y)\,d\sigma_y \qquad \forall \xi \in \partial\Omega^i\,,$$

$$\mathcal{J}^o(\epsilon)[\phi^i, \phi^o, \rho](x) \tag{9.8}$$

$$\equiv V_{\Omega^o}[\phi^o](x) + \rho + \epsilon^{n-1} \int_{\partial\Omega^i} S_n(x - \epsilon\eta)\,\phi^i(\eta)\,d\sigma_\eta \qquad \forall x \in \partial\Omega^o\,.$$

It is then apparent that the system of Equations (9.5) and (9.6) (and thus (9.3) and (9.4)) is equivalent to the functional equation

$$\mathcal{J}(\epsilon)[\phi^i, \phi^o, \rho] = (f^i, f^o)\,. \tag{9.9}$$

Then we have the following representation formula for the solution of the boundary value problem (9.1).

Proposition 9.1. *Let* $(\epsilon, f^i, f^o) \in]0, \epsilon_0[\times C^{0,\alpha}(\partial\Omega^i) \times C^{1,\alpha}(\partial\Omega^o)$. *If the triple* $(\phi^i, \phi^o, \rho) \in C^{0,\alpha}(\partial\Omega^i) \times C^{0,\alpha}(\partial\Omega^o)_0 \times \mathbb{R}$ *is a solution of* (9.9), *then the unique solution* $u[\epsilon, f^i, f^o] \in C^{1,\alpha}(\overline{\Omega(\epsilon)})$ *of* (9.1) *is given by*

$$u[\epsilon, f^i, f^o](x) = \epsilon^{n-1} \int_{\partial\Omega^i} S_n(x - \epsilon\eta)\phi^i(\eta)\,d\sigma_\eta + \int_{\partial\Omega^o} S_n(x - y)\phi^o(y)\,d\sigma_y + \rho \tag{9.10}$$

for all $x \in \overline{\Omega(\epsilon)}$.

We now observe that the map that takes ϵ to $\mathcal{J}(\epsilon)$ is real analytic.

Lemma 9.2. *The map from* $]-\epsilon_0, \epsilon_0[$ *to*

$$\mathcal{L}(C^{0,\alpha}(\partial\Omega^i) \times C^{0,\alpha}(\partial\Omega^o)_0 \times \mathbb{R},\, C^{0,\alpha}(\partial\Omega^i) \times C^{1,\alpha}(\partial\Omega^o))$$

that takes ϵ *to* $\mathcal{J}(\epsilon)$ *is real analytic.*

Proof. It follows by Theorem 6.7 on the continuity of the linear operator $W^t_{\Omega^i}$, by Theorem 6.4 on the continuity of the linear operator V_{Ω^o}, and by the properties of integral operators with real analytic kernels of Corollary A.29. □

Our goal is now to prove that $\mathcal{J}(0)$ is an invertible operator. To do so, we first consider equation $\mathcal{J}(0)[\phi^i, \phi^o, \rho] = (f^i, f^o)$ and show that it has a unique solution.

Lemma 9.3. *For all* (f^i, f^o) *in* $C^{0,\alpha}(\partial\Omega^i) \times C^{1,\alpha}(\partial\Omega^o)$ *and all* (ϕ^i, ϕ^o, ρ) *in* $C^{0,\alpha}(\partial\Omega^i) \times C^{0,\alpha}(\partial\Omega^o)_0 \times \mathbb{R}$ *we have*

$$\mathcal{J}(0)[\phi^i, \phi^o, \rho] = (f^i, f^o) \tag{9.11}$$

if and only if

$$\begin{cases} \left(\frac{1}{2}I + W^t_{\Omega^i}\right)[\phi^i] - \nu_{\Omega^i} \cdot \int_{\partial\Omega^o} \phi^o \, \nabla S_n \, d\sigma = f^i \,, \\ V_{\Omega^o}[\phi^o] + \rho = f^o \,. \end{cases} \tag{9.12}$$

In particular, for all $(f^i, f^o) \in C^{0,\alpha}(\partial\Omega^i) \times C^{1,\alpha}(\partial\Omega^o)$ *there exists one and only one solution* $(\phi^i, \phi^o, \rho) \in C^{0,\alpha}(\partial\Omega^i) \times C^{0,\alpha}(\partial\Omega^o)_0 \times \mathbb{R}$ *of* (9.11).

Proof. By the definition of $\mathcal{J}^i$ and $\mathcal{J}^o$ in (9.7) and (9.8) we can verify that (9.11) and (9.12) are equivalent. By Theorem 6.47, the second equation in (9.12) has a unique solution $(\phi^o, \rho) \in C^{0,\alpha}(\partial\Omega^o)_0 \times \mathbb{R}$. Then, by Corollary 6.15, also the first equation in (9.12) has a unique solution $\phi^i \in C^{0,\alpha}(\partial\Omega^i)$. Now the lemma is proved. □

Then we deduce that $\mathcal{J}(0)$ is an isomorphism (i.e., a linear homeomorphism).

Lemma 9.4. *The operator* $\mathcal{J}(0)$ *is an isomorphism from* $C^{0,\alpha}(\partial\Omega^i) \times C^{0,\alpha}(\partial\Omega^o)_0 \times \mathbb{R}$ *to* $C^{0,\alpha}(\partial\Omega^i) \times C^{1,\alpha}(\partial\Omega^o)$.

Proof. Since $\mathcal{J}(0)$ is a linear bounded operator (cf. Lemma 9.2) the statement follows by Lemma 9.3 and by the Open Mapping Theorem 2.2. □

Since the set of invertible operators in open in the space of bounded linear operators and since the map that takes an invertible operator to its inverse is real analytic (cf. Proposition A.15), Lemmas 9.2 and 9.4 imply the validity of the following.

Lemma 9.5. *There exists* $\epsilon_* \in]0, \epsilon_0[$ *such that* $\mathcal{J}(\epsilon)$ *is invertible for all* $\epsilon \in]-\epsilon_*, \epsilon_*[$. *Moreover, the map from* $]-\epsilon_*, \epsilon_*[$ *to*

$$\mathcal{L}(C^{0,\alpha}(\partial\Omega^i) \times C^{1,\alpha}(\partial\Omega^o), C^{0,\alpha}(\partial\Omega^i) \times C^{0,\alpha}(\partial\Omega^o)_0 \times \mathbb{R})$$

that takes ϵ *to* $\mathcal{J}(\epsilon)^{(-1)}$ *is real analytic.*

Then by Lemma 9.5 we deduce the validity of the following.

Proposition 9.6. *There exists a real analytic map*

$$\Phi \equiv (\Phi^i, \Phi^o, \Phi^s)$$

from

$$]-\epsilon_*, \epsilon_*[\times C^{0,\alpha}(\partial\Omega^i) \times C^{1,\alpha}(\partial\Omega^o)$$

to

$$C^{0,\alpha}(\partial\Omega^i) \times C^{0,\alpha}(\partial\Omega^o)_0 \times \mathbb{R}$$

such that

$$\mathcal{J}(\epsilon)[\Phi^i[\epsilon, f^i, f^o], \Phi^o[\epsilon, f^i, f^o], \Phi^s[\epsilon, f^i, f^o]] = (f^i, f^o)$$

for all $(\epsilon, f^i, f^o) \in]-\epsilon_*, \epsilon_*[\times C^{0,\alpha}(\partial\Omega^i) \times C^{1,\alpha}(\partial\Omega^o)$.

Proof. It suffices to set

$$\Phi[\epsilon, f^i, f^o] \equiv \mathcal{J}(\epsilon)^{(-1)}[f^i, f^o]$$

for all $(\epsilon, f^i, f^o) \in]-\epsilon_*, \epsilon_*[\times C^{0,\alpha}(\partial\Omega^i) \times C^{1,\alpha}(\partial\Omega^o)$ (with ϵ_* as in Lemma 9.5). Since the evaluation map from

$$\mathcal{L}(C^{0,\alpha}(\partial\Omega^i) \times C^{1,\alpha}(\partial\Omega^o), C^{0,\alpha}(\partial\Omega^i) \times C^{0,\alpha}(\partial\Omega^o)_0 \times \mathbb{R}) \\ \times (C^{0,\alpha}(\partial\Omega^i) \times C^{1,\alpha}(\partial\Omega^o))$$

to

$$C^{0,\alpha}(\partial\Omega^i) \times C^{0,\alpha}(\partial\Omega^o)_0 \times \mathbb{R}$$

that takes a pair $(A, (g^i, g^o))$ to $A[g^i, g^o]$ is bilinear and continuous, Lemma 9.5 implies that Φ is real analytic. The last equality of the statement follows by the definition of Φ (see also Lemma 9.5). □

We note that the superscript 's' of Φ^s stands for 'scalar,' indeed $\Phi^s[\epsilon, f^i, f^o] \in \mathbb{R}$ for all $(\epsilon, f^i, f^o) \in]-\epsilon_*, \epsilon_*[\times C^{0,\alpha}(\partial\Omega^i) \times C^{1,\alpha}(\partial\Omega^o)$. We are now ready to study the maps that take (ϵ, f^i, f^o) to suitable restrictions of $u[\epsilon, f^i, f^o]$ and of $u[\epsilon, f^i, f^o](\epsilon\cdot)$ and to the energy $\int_{\Omega(\epsilon)} |\nabla u[\epsilon, f^i, f^o]|^2 \, d\sigma$. In the following Theorem 9.7 we consider $u[\epsilon, f^i, f^o]$.

Theorem 9.7. *Let Ω_M be an open subset of Ω^o such that $0 \notin \overline{\Omega_M}$. Let $\epsilon_M \in]0, \epsilon_*[$ be such that*

$$\overline{\Omega_M} \cap \epsilon\overline{\Omega^i} = \emptyset \qquad \forall \epsilon \in]-\epsilon_M, \epsilon_M[.$$

Then, there exists a real analytic map

$$\mathcal{U}_M :]-\epsilon_M, \epsilon_M[\times C^{0,\alpha}(\partial\Omega^i) \times C^{1,\alpha}(\partial\Omega^o) \to C^{1,\alpha}(\overline{\Omega_M})$$

such that

$$u[\epsilon, f^i, f^o]_{|\overline{\Omega_M}} = \mathcal{U}_M[\epsilon, f^i, f^o] \quad \forall (\epsilon, f^i, f^o) \in]0, \epsilon_M[\times C^{0,\alpha}(\partial\Omega^i) \times C^{1,\alpha}(\partial\Omega^o) . \tag{9.13}$$

Moreover, for all $(f^i, f^o) \in C^{0,\alpha}(\partial\Omega^i) \times C^{1,\alpha}(\partial\Omega^o)$ we have

$$\mathcal{U}_M[0, f^i, f^o] = u^o[f^o]_{|\overline{\Omega_M}} \tag{9.14}$$

where $u^o[f^o] \in C^{1,\alpha}(\overline{\Omega^o})$ is the unique solution of the unperturbed problem

$$\begin{cases} \Delta u^o[f^o] = 0 & \text{in } \Omega^o\,, \\ u^o[f^o] = f^o & \text{on } \partial\Omega^o\,. \end{cases} \tag{9.15}$$

Proof. We set

$$\begin{aligned} \mathcal{U}_M[\epsilon, f^i, f^o](x) \equiv \epsilon^{n-1} \int_{\partial\Omega^i} & S_n(x - \epsilon\eta)\Phi^i[\epsilon, f^i, f^o](\eta)\, d\sigma_\eta \\ & + v_{\Omega^o}[\Phi^o[\epsilon, f^i, f^o]](x) + \Phi^s[\epsilon, f^i, f^o] \qquad \forall x \in \overline{\Omega_M} \end{aligned}$$

for all $(\epsilon, f^i, f^o) \in]-\epsilon_M, \epsilon_M[\times C^{0,\alpha}(\partial\Omega^i) \times C^{1,\alpha}(\partial\Omega^o)$. Then equality (9.13) follows by Propositions 9.1 and 9.6 (see also Definition 4.18 of the single layer potential). We now prove that the map that takes (ϵ, f^i, f^o) to $\mathcal{U}_M[\epsilon, f^i, f^o] \in C^{1,\alpha}(\overline{\Omega_M})$ is real analytic. By Theorem 4.25 on the continuity of the linear operator $v^+_{\Omega^o}[\cdot]$ and Proposition 9.6, we verify that the map from $]-\epsilon_M, \epsilon_M[\times C^{0,\alpha}(\partial\Omega^i) \times C^{1,\alpha}(\partial\Omega^o)$ to $C^{1,\alpha}(\overline{\Omega_M})$ that takes (ϵ, f^i, f^o) to

$$v_{\Omega^o}[\Phi^o[\epsilon, f^i, f^o]]_{|\overline{\Omega_M}} + \Phi^s[\epsilon, f^i, f^o]$$

is real analytic. Thus, it remains to show that the map that takes (ϵ, f^i, f^o) to the function

$$\epsilon^{n-1} \int_{\partial\Omega^i} S_n(x - \epsilon\eta)\Phi^i[\epsilon, f^i, f^o](\eta)\, d\sigma_\eta$$

of $x \in \overline{\Omega_M}$ is real analytic from $]-\epsilon_M, \epsilon_M[\times C^{0,\alpha}(\partial\Omega^i) \times C^{1,\alpha}(\partial\Omega^o)$ to $C^{1,\alpha}(\overline{\Omega_M})$, a fact that follows from the properties of integral operators with real analytic kernels of Theorem A.28 (i). We now prove equality (9.14). By the definition of $\mathcal{U}_M$, we have

$$\mathcal{U}_M[0, f^i, f^o](x) = v_{\Omega^o}[\Phi^o[0, f^i, f^o]](x) + \Phi^s[0, f^i, f^o] \qquad \forall x \in \overline{\Omega_M}\,.$$

Next, we observe that

$$V_{\Omega^o}[\Phi^o[0, f^i, f^o]] + \Phi^s[0, f^i, f^o] = f^o$$

by Lemma 9.3 and we deduce that the function $v^+_{\Omega^o}[\Phi^o[0, f^i, f^o]] + \Phi^s[0, f^i, f^o]$ satisfies the unperturbed Dirichlet problem (9.15). Then the uniqueness of the solution of the Dirichlet problem for harmonic functions of Theorem 3.6 implies that $u^o[f^o] = v^+_{\Omega^o}[\Phi^o[0, f^i, f^o]] + \Phi^s[0, f^i, f^o]$. Accordingly, equality (9.14) holds true. □

We now turn to the rescaled function $u[\epsilon, f^i, f^o](\epsilon\cdot)$. We recall that $\delta_{i,j}$ denotes the Kronecker delta symbol, so that $\delta_{i,j} = 1$ for $i = j$ and $\delta_{i,j} = 0$ otherwise.

Theorem 9.8. *Let Ω_m be a bounded open subset of $\mathbb{R}^n \setminus \overline{\Omega^i}$. Let $\epsilon_m \in]0, \epsilon_*[$ be such that*

$$\epsilon\overline{\Omega_m} \subseteq \Omega^o \qquad \forall \epsilon \in]-\epsilon_m, \epsilon_m[\,.$$

Then, there exists a real analytic map

$$\mathcal{U}_m :]-\epsilon_m, \epsilon_m[\times C^{0,\alpha}(\partial\Omega^i) \times C^{1,\alpha}(\partial\Omega^o) \to C^{1,\alpha}(\overline{\Omega_m})$$

such that

$$u[\epsilon, f^i, f^o](\epsilon\xi) = \mathcal{U}_m[\epsilon, f^i, f^o](\xi) + \delta_{2,n}\frac{\epsilon\log\epsilon}{2\pi}\int_{\partial\Omega^i} f^i\, d\sigma \qquad \forall \xi \in \overline{\Omega_m} \quad (9.16)$$

for all $(\epsilon, f^i, f^o) \in]0, \epsilon_m[\times C^{0,\alpha}(\partial\Omega^i) \times C^{1,\alpha}(\partial\Omega^o)$. *Moreover,*

$$\mathcal{U}_m[0, f^i, f^o] = u^o[f^o](0) \qquad (9.17)$$

where $u^o[f^o] \in C^{1,\alpha}(\overline{\Omega^o})$ *is the unique solution of* (9.15).

Proof. Since $S_n(\epsilon\zeta) = \epsilon^{n-2}S_n(\zeta) + (\delta_{2,n}/2\pi)\log|\epsilon|$ for all $\zeta \in \mathbb{R}^n \setminus \{0\}$ and all $\epsilon \in \mathbb{R} \setminus \{0\}$, formula (9.10) for $u[\epsilon, f^i, f^o]$ and Proposition 9.6 imply that

$$\begin{aligned} &u[\epsilon, f^i, f^o](\epsilon\xi) \\ &\quad = \epsilon\, v^-_{\Omega^i}[\Phi^i[\epsilon, f^i, f^o]](\xi) + \delta_{2,n}\frac{\epsilon\log\epsilon}{2\pi}\int_{\partial\Omega^i}\Phi^i[\epsilon, f^i, f^o]\, d\sigma \\ &\qquad + \int_{\partial\Omega^o} S_n(\epsilon\xi - y)\Phi^o[\epsilon, f^i, f^o](y)\, d\sigma_y + \Phi^s[\epsilon, f^i, f^o] \end{aligned} \qquad (9.18)$$

for all $\xi \in (\epsilon^{-1}\Omega^o) \setminus \Omega^i$ and all $(\epsilon, f^i, f^o) \in]0, \epsilon_*[\times C^{0,\alpha}(\partial\Omega^i) \times C^{1,\alpha}(\partial\Omega^o)$ (see also Definition 4.18 of single layer potential). By Proposition 9.6, Φ satisfies Equation (9.5). Then Lemma 6.11 implies that

$$\begin{aligned} &\int_{\partial\Omega^i}\Phi^i[\epsilon, f^i, f^o]\, d\sigma \\ &\quad = \int_{\partial\Omega^i} f^i\, d\sigma - \int_{\partial\Omega^i}\nu_{\Omega^i}(\xi)\cdot\int_{\partial\Omega^o}\Phi^o[\epsilon, f^i, f^o](y)\,\nabla S_n(\epsilon\xi - y)\, d\sigma_y\, d\sigma_\xi\,. \end{aligned}$$

Since, for all fixed $\epsilon \in]0, \epsilon^*[$, the map that takes ξ to

$$\int_{\partial\Omega^o}\Phi^o[\epsilon, f^i, f^o](y)\,\nabla S_n(\epsilon\xi - y)\, d\sigma_y$$

is harmonic in Ω^i (and of class C^∞ on $\overline{\Omega^i}$), the first Green Identity of Theorem 4.2 implies that

$$\int_{\partial\Omega^i}\nu_{\Omega^i}(\xi)\cdot\int_{\partial\Omega^o}\Phi^o[\epsilon, f^i, f^o](y)\,\nabla S_n(\epsilon\xi - y)\, d\sigma_y\, d\sigma_\xi = 0\,.$$

Hence,

$$\int_{\partial\Omega^i}\Phi^i[\epsilon, f^i, f^o]\, d\sigma = \int_{\partial\Omega^i} f^i\, d\sigma\,. \qquad (9.19)$$

Then we set

$$\mathcal{U}_m[\epsilon, f^i, f^o](\xi) \equiv \epsilon\, v_{\Omega^i}^-[\Phi^i[\epsilon, f^i, f^o]](\xi) + \int_{\partial\Omega^o} S_n(\epsilon\xi - y)\Phi^o[\epsilon, f^i, f^o](y)\, d\sigma_y + \Phi^s[\epsilon, f^i, f^o]$$

for all $\xi \in \overline{\Omega_m}$ and for all $(\epsilon, f^i, f^o) \in]-\epsilon_m, \epsilon_m[\times C^{0,\alpha}(\partial\Omega^i) \times C^{1,\alpha}(\partial\Omega^o)$ and the validity of (9.16) follows by (9.18) and (9.19). To show that the map that takes (ϵ, f^i, f^o) to $\mathcal{U}_m[\epsilon, f^i, f^o]$ is real analytic we first observe that, by Theorem 4.25 and Proposition 9.6, the map from $]-\epsilon_m, \epsilon_m[\times C^{0,\alpha}(\partial\Omega^i) \times C^{1,\alpha}(\partial\Omega^o)$ to $C^{1,\alpha}(\overline{\Omega_m})$ that takes (ϵ, f^i, f^o) to

$$\epsilon\, v_{\Omega^i}^-[\Phi^i[\epsilon, f^i, f^o]]_{|\overline{\Omega_m}} + \Phi^s[\epsilon, f^i, f^o]$$

is real analytic. Thus, it remains to see that the map from $]-\epsilon_m, \epsilon_m[$ to $C^{1,\alpha}(\overline{\Omega_m})$ that takes ϵ to the function

$$\int_{\partial\Omega^o} S_n(\epsilon\xi - y)\Phi^o[\epsilon, f^i, f^o](y)\, d\sigma_y$$

of $\xi \in \overline{\Omega_m}$ is real analytic, a fact that follows from the properties of integral operators with real analytic kernels of Theorem A.28 (i). To prove (9.17) we observe that

$$\mathcal{U}_m[0, f^i, f^o](\xi) = v_{\Omega^o}[\Phi^o[0, f^i, f^o]](0) + \Phi^s[0, f^i, f^o] \quad \forall \xi \in \overline{\Omega_m}$$

and that

$$v_{\Omega^o}[\Phi^o[0, f^i, f^o]] + \Phi^s[0, f^i, f^o] = u^o[f^o]$$

by the second equation in (9.12) and by the uniqueness of the solution of the unperturbed Dirichlet problem (9.15) (cf. Theorem 3.6). □

Finally, we consider the energy integral $\int_{\Omega(\epsilon)} \left|\nabla u[\epsilon, f^i, f^o]\right|^2 dx$.

Theorem 9.9. *There exist $\epsilon_E \in]0, \epsilon_*[$ and a real analytic function*

$$\mathcal{E} :]-\epsilon_E, \epsilon_E[\times C^{0,\alpha}(\partial\Omega^i) \times C^{1,\alpha}(\partial\Omega^o) \to \mathbb{R}$$

such that

$$\int_{\Omega(\epsilon)} \left|\nabla u[\epsilon, f^i, f^o]\right|^2 dx = \mathcal{E}(\epsilon, f^i, f^o) \quad \forall \epsilon \in]0, \epsilon_E[\times C^{0,\alpha}(\partial\Omega^i) \times C^{1,\alpha}(\partial\Omega^o). \tag{9.20}$$

Moreover,

$$\mathcal{E}(0, f^i, f^o) = \int_{\Omega^o} |\nabla u^o[f^o]|^2\, dx\,. \tag{9.21}$$

Proof. As in the proof of Theorem 8.9 we begin by observing that by the first Green Identity of Theorem 4.2 we have

$$\int_{\Omega(\epsilon)} \left|\nabla u[\epsilon, f^i, f^o]\right|^2 dx = \int_{\partial\Omega^o} u[\epsilon, f^i, f^o]\, \nu_{\Omega^o} \cdot \nabla u[\epsilon, f^i, f^o]\, d\sigma$$
$$- \int_{\epsilon\partial\Omega^i} u[\epsilon, f^i, f^o]\, \nu_{\epsilon\Omega^i} \cdot \nabla u[\epsilon, f^i, f^o]\, d\sigma .$$

Since $u[\epsilon, f^i, f^o]$ is a solution of the mixed boundary value problem (9.1), we can write

$$\int_{\Omega(\epsilon)} \left|\nabla u[\epsilon, f^i, f^o]\right|^2 dx = \int_{\partial\Omega^o} f^o\, \nu_{\Omega^o} \cdot \nabla u[\epsilon, f^i, f^o]\, d\sigma$$
$$- \int_{\epsilon\partial\Omega^i} u[\epsilon, f^i, f^o](y)\, f^i(y/\epsilon)\, d\sigma$$

and, by a change of variable in the integral over $\epsilon\partial\Omega^i$, we deduce that

$$\int_{\Omega(\epsilon)} \left|\nabla u[\epsilon, f^i, f^o]\right|^2 dx = \int_{\partial\Omega^o} f^o\, \nu_{\Omega^o} \cdot \nabla u[\epsilon, f^i, f^o]\, d\sigma$$
$$- \epsilon^{n-1} \int_{\partial\Omega^i} u[\epsilon, f^i, f^o](\epsilon\eta)\, f^i(\eta)\, d\sigma_\eta .$$

Then we take Ω_M as in Theorem 9.7 and Ω_m as in Theorem 9.8 and we also assume that

$$\partial\Omega^o \subseteq \overline{\Omega_M} \qquad \text{and} \qquad \partial\Omega^i \subseteq \overline{\Omega_m} .$$

Using the corresponding analytic maps $\mathcal{U}_M$ and $\mathcal{U}_m$, we obtain

$$\int_{\Omega(\epsilon)} \left|\nabla u[\epsilon, f^i, f^o]\right|^2 dx = \int_{\partial\Omega^o} f^o\, \nu_{\Omega^o} \cdot \nabla \mathcal{U}_M[\epsilon, f^i, f^o]\, d\sigma$$
$$- \epsilon^{n-1} \int_{\partial\Omega^i} \mathcal{U}_m[\epsilon, f^i, f^o]\, f^i\, d\sigma$$

for all $(\epsilon, f^i, f^o) \in]0, \epsilon_E[\times C^{0,\alpha}(\partial\Omega^i) \times C^{1,\alpha}(\partial\Omega^o)$, with $\epsilon_E \equiv \min\{\epsilon_M, \epsilon_m\}$. Then (9.20) holds with $\mathcal{E}$ defined by

$$\mathcal{E}(\epsilon, f^i, f^o) \equiv \int_{\partial\Omega^o} f^o\, \nu_{\Omega^o} \cdot \nabla \mathcal{U}_M[\epsilon, f^i, f^o]\, d\sigma - \epsilon^{n-1} \int_{\partial\Omega^i} \mathcal{U}_m[\epsilon, f^i, f^o]\, f^i\, d\sigma$$

for all $(\epsilon, f^i, f^o) \in]-\epsilon_E, \epsilon_E[\times C^{0,\alpha}(\partial\Omega^i) \times C^{1,\alpha}(\partial\Omega^o)$. By (9.14) we have

$$\mathcal{E}(0, f^i, f^o) = \int_{\partial\Omega^o} f^o\, \nu_{\Omega^o} \cdot \nabla u^o\, d\sigma$$

and the validity of (9.21) follows by a computation based on the first Green Identity of Theorem 4.2 and on the fact that u^o is the solution of the unperturbed problem (9.15).

To verify that $\mathcal{E}$ is real analytic, we note that by Theorem 9.7 the map $\mathcal{U}_M$ is real analytic from $]-\epsilon_E, \epsilon_E[\times C^{0,\alpha}(\partial\Omega^i) \times C^{1,\alpha}(\partial\Omega^o)$ to $C^{1,\alpha}(\overline{\Omega_M})$. Since the map from $C^{1,\alpha}(\overline{\Omega_M})$ to $C^{0,\alpha}(\overline{\Omega_M})^n$ that takes a function to its gradient is linear and continuous, we conclude that the map $(\epsilon, f^i, f^o) \mapsto \nabla\mathcal{U}_M[\epsilon, f^i, f^o]$ is real analytic from $]-\epsilon_E, \epsilon_E[\times C^{0,\alpha}(\partial\Omega^i) \times C^{1,\alpha}(\partial\Omega^o)$ to $C^{0,\alpha}(\overline{\Omega_M})^n$. Moreover, by Theorem 9.8 and by the continuity of the embedding of $C^{1,\alpha}(\partial\Omega^i)$ into $C^{0,\alpha}(\partial\Omega^i)$ (see Theorem 2.83), the map $(\epsilon, f^i, f^o) \mapsto \mathcal{U}_m[\epsilon, f^i, f^o]$ is real analytic from $]-\epsilon_E, \epsilon_E[\times C^{0,\alpha}(\partial\Omega^i) \times C^{1,\alpha}(\partial\Omega^o)$ to $C^{0,\alpha}(\overline{\Omega_m})$. Then, by the continuity of the pointwise product in $C^{0,\alpha}(\partial\Omega^o)$, by the continuity of the linear map that takes a function of $C^{0,\alpha}(\partial\Omega^o)$ to its integral over $\partial\Omega^o$, by the continuity of the pointwise product in $C^{0,\alpha}(\partial\Omega^i)$, and by the continuity of the linear map that takes a function of $C^{0,\alpha}(\partial\Omega^i)$ to its integral over $\partial\Omega^i$, we conclude that $\mathcal{E}$ is real analytic (see also the proof of Theorem 9.8). □

9.3 A Mixed Boundary Value Problem for the Poisson Equation

In this section we show how we can use the results of the previous Section 9.2 to study the following mixed problem for the Poisson equation

$$\begin{cases} \Delta u = f & \text{in } \Omega(\epsilon)\,, \\ \nu_{\epsilon\Omega^i}(x)\cdot\nabla u(x) = f^i(x/\epsilon) & \text{for all } x \in \epsilon\partial\Omega^i\,, \\ u = f^o & \text{on } \partial\Omega^o\,. \end{cases} \tag{9.22}$$

Here $f \in C^{-1,\alpha}(\overline{\Omega^o})$, $f^i \in C^{0,\alpha}(\partial\Omega^i)$, and $f^o \in C^{1,\alpha}(\partial\Omega^o)$ are fixed functions (see Section 2.22 for the definition of $C^{-1,\alpha}(\overline{\Omega^o})$). First we show that the solution u of the problem exists, belongs to $C^{1,\alpha}(\overline{\Omega(\epsilon)})$, and is unique. To do so, we denote by $P_n[\Omega^o, f]$ the Newtonian potential

$$P_n[\Omega^o, f](x) = \int_{\Omega^o} S_n(x-y)\, f(y)\, dy \qquad \forall x \in \overline{\Omega^o}$$

(cf. definition (7.2)). Then, by Theorem 7.19, we have that $P_n[\Omega^o, f] \in C^{1,\alpha}(\overline{\Omega^o})$ and

$$\Delta(P_n[\Omega^o, f]) = f \qquad \text{in } \Omega^o \tag{9.23}$$

in the sense of distributions. Accordingly, taking

$$v \equiv u - P_n[\Omega^o, f]$$

we have that u is a solution of (9.22) if and only if v is a solution of the following mixed boundary value problem for the Laplace equation,

$$\begin{cases} \Delta v = 0 & \text{in } \Omega(\epsilon)\,, \\ \nu_{\epsilon\Omega^i}(x) \cdot \nabla v(x) = f^i(x/\epsilon) - \nu_{\epsilon\Omega^i}(x) \cdot \nabla P_n[\Omega^o, f](x) & \text{for all } x \in \epsilon\partial\Omega^i\,, \\ v(x) = f^o(x) - P_n[\Omega^o, f](x) & \text{for all } x \in \partial\Omega^o\,, \end{cases} \tag{9.24}$$

Then the existence and uniqueness of the solutions of (9.22) and (9.24) is a consequence of Propositions 6.3 and 6.49. To stress their dependence upon ϵ, we shall denote such unique solutions by u_ϵ and v_ϵ, respectively.

We now wish to analyze the dependence of u_ϵ upon ϵ. We will confine ourselves to study the 'macroscopic' behavior of u_ϵ, away from the hole $\epsilon\Omega^i$, and we plan to apply Theorem 9.7 to the specific case of problem (9.24). To do so, we have to consider the dependence of the function on the right-hand side of the second equation of (9.24) upon ϵ. We introduce the following assumption,

$$f \in C^{-1,\alpha}(\overline{\Omega^o}) \text{ is real analytic in a neighborhood of the origin } 0. \tag{9.25}$$

Then we have the following

Lemma 9.10. *If assumption* (9.25) *holds true, then there exists* $\epsilon' \in]0, \epsilon_0]$ *such that the map from* $]-\epsilon', \epsilon'[$ *to* $C^{0,\alpha}(\partial\Omega^i)$ *that takes* $\epsilon \in]-\epsilon', \epsilon'[$ *to the function*

$$f^i_\epsilon(x) \equiv f^i(x) - \nu_{\Omega^i}(x) \cdot \nabla P_n[\Omega^o, f](\epsilon x) \qquad \forall x \in \partial\Omega^i$$

is real analytic.

Proof. Since f is real analytic in a neighborhood of 0, there exist $r_1, \rho_1 > 0$ such that $\mathbb{B}_n(0, r_1) \subseteq \Omega^o$ and

$$f \in C^0_{\omega,\rho_1}(\overline{\mathbb{B}_n(0, r_1)})$$

(cf. Theorem 2.17 and definition (2.7)). Then we have

$$P_n[\Omega^o, f] = P_n[\mathbb{B}_n(0, r_1), f] + P_n[\Omega^o \setminus \mathbb{B}_n(0, r_1), f]\,.$$

Let $r_2 \in]0, r_1[$. By the regularity result for the volume potentials in Roumieu classes of Proposition 7.21 and by (7.65) (see also [162, Prop. 2.4]), there exists $\rho_2 \in]0, \rho_1]$ such that

$$P_n[\mathbb{B}_n(0, r_1), f] \in C^0_{\omega,\rho_2}(\overline{\mathbb{B}_n(0, r_2)})\,.$$

By the analyticity of S_n in $\mathbb{R}^n \setminus \{0\}$ and by the Roumieu continuity result for a linear integral operator with analytic kernel of Proposition A.30, there exist $r \in]0, r_2]$ and $\rho \in]0, \rho_2]$ such that

$$P_n[\Omega^o \setminus \mathbb{B}_n(0, r_1), f] \in C^0_{\omega,\rho}(\overline{\mathbb{B}_n(0, r)})\,.$$

It follows that $P_n[\Omega^o, f] \in C^0_{\omega,\rho}(\overline{\mathbb{B}_n(0, r)})$ and, in particular, $P_n[\Omega^o, f]$ is real analytic in $\mathbb{B}_n(0, r)$. As a consequence, also the gradient $\nabla P_n[\Omega^o, f]$ is real analytic in $\mathbb{B}_n(0, r)$. Then we take $\epsilon' \in]0, \epsilon_0]$ such that $\epsilon\overline{\Omega^i} \subseteq \mathbb{B}_n(0, r)$ for all $\epsilon \in]-\epsilon', \epsilon'[$. We observe that the map from $]-\epsilon', \epsilon'[$ to $C^{0,\alpha}(\overline{\Omega^i}, \mathbb{B}_n(0, r))$ that takes ϵ to $\epsilon I_{\overline{\Omega^i}}$ is real analytic (it is indeed the restriction of a linear continuous map). Then, by Theorem A.23 (i) we deduce that the map from $]-\epsilon',\epsilon'[$ to $C^{0,\alpha}(\overline{\Omega^i})$ that takes ϵ to the composition $\nabla P_n[\Omega^o,f](\epsilon I_{\overline{\Omega^i}})$ is real analytic. Since the restriction operator from $C^{0,\alpha}(\overline{\Omega^i})$ to $C^{0,\alpha}(\partial\Omega^i)$ is linear and continuous and since the multiplication and the sum with a fixed function of $C^{0,\alpha}(\partial\Omega^i)$ are real analytic maps from $C^{0,\alpha}(\partial\Omega^i)$ to itself, we deduce the validity of the lemma. □

We confine ourselves to describe the 'macroscopic' behavior of u_ϵ away from the boundary of the hole. From Theorem 9.7 we deduce the following.

Theorem 9.11. *If assumption* (9.25) *is satisfied and Ω_M is an open subset of Ω^o with $0 \notin \overline{\Omega_M}$, then there exist $\epsilon'_M \in]0,\epsilon'[$ such that*

$$\overline{\Omega_M} \cap \epsilon\overline{\Omega^i} = \emptyset \qquad \forall \epsilon \in]-\epsilon'_M,\epsilon'_M[$$

and a real analytic map

$$\tilde{\mathcal{U}}_M :]-\epsilon'_M,\epsilon'_M[\to C^{1,\alpha}(\overline{\Omega_M})$$

such that

$$u_{\epsilon|\overline{\Omega_M}} = \tilde{\mathcal{U}}_M[\epsilon] \quad \forall \epsilon \in]0,\epsilon'_M[. \tag{9.26}$$

Moreover,

$$\tilde{\mathcal{U}}_M[0] = u^o_{|\overline{\Omega_M}} \tag{9.27}$$

where $u^o \in C^{1,\alpha}(\overline{\Omega^o})$ is the unique solution of the unperturbed problem

$$\begin{cases} \Delta u^o = f & \text{in } \Omega^o , \\ u^o = f^o & \text{on } \partial\Omega^o . \end{cases} \tag{9.28}$$

Proof. We apply Theorem 9.7 to the specific case of problem (9.24). We deduce that

$$v_{\epsilon|\overline{\Omega_M}} = \mathcal{U}_M[\epsilon, f^i_\epsilon, f^o] \qquad \forall \epsilon \in]0,\epsilon_M[,$$

where f^i_ϵ is defined as in Lemma 9.10. Then, for $\epsilon'_M \equiv \min\{\epsilon_M,\epsilon'\}$ we have

$$u_{\epsilon|\overline{\Omega_M}} = \mathcal{U}_M[\epsilon, f^i_\epsilon, f^o] + P_n[\Omega^o,f]_{|\overline{\Omega_M}} \qquad \forall \epsilon \in]0,\epsilon'_M[$$

and (9.26) holds with

$$\tilde{\mathcal{U}}_M[\epsilon] \equiv \mathcal{U}_M[\epsilon, f^i_\epsilon, f^o] + P_n[\Omega^o,f]_{|\overline{\Omega_M}} \qquad \forall \epsilon \in]-\epsilon'_M,\epsilon'_M[.$$

Theorem 9.7 and Lemma 9.10 imply that $\tilde{\mathcal{U}}_M$ is real analytic and equality (9.27) follows by (9.14). □

9.4 A Steklov Eigenvalue Problem

In this section we consider a Steklov eigenvalue problem in the perforated domain $\Omega(\epsilon)$. Our presentation here follows that of the papers [113], [114] with Gryshchuk. The general strategy that we employ follows the guidelines of the previous sections in the present chapter and of Chapter 8. Namely, we first transform the problem into an equivalent system of integral equations and then we rewrite such a system into a functional equation of the form

$$\mathcal{Q}[\epsilon, \delta_{2,n}\epsilon \log \epsilon, \theta^i, \theta^o, c, \lambda] = 0 \tag{9.29}$$

for a certain real analytic map $\mathcal{Q}$ between suitable Banach spaces. At this point, our aim is to exploit equation (9.29) to prove that the quadruple $(\theta^i, \theta^o, c, \lambda)$ depends analytically on $(\epsilon, \delta_{2,n}\epsilon \log \epsilon)$. However, contrary to what happens for the problems considered in the previous sections and in Chapter 8, here the map $\mathcal{Q}$ is not linear with respect to $(\theta^i, \theta^o, c, \lambda)$. As a consequence, this time we cannot use the real analyticity of the inversion map of Proposition A.15 to achieve our goal. Then, to overcome this difficulty we resort to the Implicit Function Theorem A.19 and, in particular, to its last statement concerning the case of real analytic maps. As we shall see in the following chapters of the book, this version of the Functional Analytic Approach, which relies on the Implicit Function Theorem, can be conveniently used both for linear boundary value problems, of the kind considered so far, and for boundary value problems with nonlinear conditions.

We now proceed to introduce the Steklov eigenvalue problem in $\Omega(\epsilon)$. To do so, we first introduce the problem with no hole, that is, in Ω^o and then we introduce the problem in the perforated set $\Omega(\epsilon)$. So, we begin with the Steklov eigenvalue problem

$$\begin{cases} \Delta u = 0 & \text{in } \Omega^o, \\ \frac{\partial u}{\partial \nu_{\Omega^o}} = \lambda u & \text{on } \partial\Omega^o . \end{cases} \tag{9.30}$$

By definition, a Steklov eigenvalue for Δ in Ω^o is a real number λ for which problem (9.30) has a nontrivial solution in $C^{1,\alpha}(\overline{\Omega^o})$. As is well known, problem (9.30) has an increasing sequence of eigenvalues $\{\lambda_j[\Omega^o]\}_{j\in\mathbb{N}}$. Here we understand that in the sequence

$$0 = \lambda_0[\Omega^o] < \lambda_1[\Omega^o] \leq \cdots \leq \lambda_j[\Omega^o] \leq \lambda_{j+1}[\Omega^o] \leq \ldots$$

we write each eigenvalue as many times as its multiplicity (cf. e.g., Henrot [121, p. 113]). We note that problem (9.30) can be considered as a limiting Neumann problem with a density of mass concentrated near the boundary, as discussed in the paper [156] by Lamberti and Provenzano and in [83] with Provenzano.

In this section we assume that there exists $l \in \mathbb{N}$ such that the eigenvalue

$$\tilde{\lambda} \equiv \lambda_l[\Omega^o]$$

has multiplicity one. Then problem (9.30) has an eigenspace of solutions u in $C^{1,\alpha}(\overline{\Omega^o})$ of dimension one. We could also consider the case of multiplicities higher than one, but then the treatment turns out to be more elaborate. For the case of multiple eigenvalues we refer the reader to [171].

Next we consider the Steklov eigenvalue problem

$$\begin{cases} \Delta u = 0 & \text{in } \Omega(\epsilon), \\ \frac{\partial u}{\partial \nu_{\Omega(\epsilon)}} = \lambda u & \text{on } \partial\Omega(\epsilon) \end{cases} \tag{9.31}$$

in the perforated domain $\Omega(\epsilon)$ with $\epsilon \in]0, \epsilon_0[$. Again, a Steklov eigenvalue for Δ in $\Omega(\epsilon)$ is a real number λ for which problem (9.31) has a nontrivial solution, problem (9.31) has an increasing sequence of Steklov eigenvalues $\{\lambda_j[\Omega(\epsilon)]\}_{j\in\mathbb{N}}$, and we write each Steklov eigenvalue as many times as its multiplicity.

We are interested in the behavior of the eigenvalues of (9.31) which approach $\tilde{\lambda}$ as ϵ tends to 0. This type of problem has been considered for a long time.

In particular, Nazarov [237] proved that each eigenvalue of the problem in $\Omega(\epsilon)$ convergences to the corresponding eigenvalue of the unperturbed problem. Namely, we have

$$\lim_{\epsilon\to 0^+} \lambda_j[\Omega(\epsilon)] = \lambda_j[\Omega^o] \qquad \forall j \in \mathbb{N}. \tag{9.32}$$

Moreover, he showed how to use the Compound (Multiscale) Expansions Method to obtain all orders asymptotic expansions for the Steklov eigenvalues and eigenvectors and the corresponding error estimates. We also mention the explicit computations carried out by Dittmar [88] in the case of a circular annulus of inner radius ϵ and outer radius 1. By means of polar coordinates and separation of variables he could prove that there exist ϵ-dependent families of eigenvalues that tend to infinity as $\epsilon \to 0^+$ (and that coincide with $\lambda_j[\Omega(\epsilon)]$ with an index j that changes as ϵ approaches 0). Later, Nazarov [237] used again the Compound Expansions Method to obtain asymptotic expansions for the eigenvalues that tend to infinity, not only for annular domains but, more in general, for perforated domains that are not necessarily circular, as in the case of our set $\Omega(\epsilon)$. One should also say that the Steklov problem was studied also in connection with perturbations of different nature and not only for domains with shrinking holes. For example, Nazarov considered in [235] and [236] the asymptotic expansions of the Steklov eigenvalues and eigenfunctions in a domain with a peak and Nazarov and Taskinen proved in [238] that the Steklov spectrum in a domain with a peak can be either discrete or continuous, depending on the sharpness of the peak. We also mention the work of Chiado Piat, Nazarov, Piatnitski [50], Douanla [90], Mel'nik [209], Pastukhova [241], Vanninathan [271] about the case of periodic perforations. Such papers aim at understanding the limiting behavior of the eigenvalues and the existence of asymptotic expansions in terms of the size of the periodic cell.

In this section instead, we exploit the Functional Analytic Approach to show that for $\epsilon > 0$ small enough there is a simple eigenvalue $\lambda[\epsilon]$ of the Steklov problem (9.31) that approaches $\tilde{\lambda}$ as ϵ tends to 0. As usual, our aim is to write $\lambda[\epsilon]$ in terms of analytic functions and explicitly known functions of ϵ that vanish as ϵ tends to 0.

More precisely, we show that there exist $\epsilon^*, \epsilon_1^* \in]0,+\infty[$ and a real analytic map $\hat{\lambda}[\cdot,\cdot]$ from $]-\epsilon^*,\epsilon^*[\times]-\epsilon_1^*,\epsilon_1^*[$ to $]0,+\infty[$ such that

$$\lambda[\epsilon] = \hat{\lambda}[\epsilon, \delta_{2,n}\epsilon \log \epsilon] \qquad \forall \epsilon \in]0,\epsilon^*[\tag{9.33}$$

(cf. Theorem 9.22). Here as usual

$$\delta_{2,n} \equiv \begin{cases} 1 \text{ if } n = 2\,, \\ 0 \text{ if } n \geq 3\,. \end{cases}$$

As a consequence of (9.33), if $n \geq 3$ then, possibly shrinking ϵ^*, the eigenvalue $\lambda[\epsilon]$ can be continued real analytically in $\epsilon \in]-\epsilon^*,\epsilon^*[$ and the following convergent series expansion holds

$$\lambda[\epsilon] = \sum_{j=0}^{\infty} \frac{1}{j!}\partial_\epsilon^j \hat{\lambda}[0,0]\epsilon^j \tag{9.34}$$

for $\epsilon > 0$ sufficiently small.

Instead, if $n = 2$, then (9.33) implies that the following convergent series expansion holds

$$\lambda[\epsilon] = \sum_{\gamma\equiv(\gamma_1,\gamma_2)\in\mathbb{N}^2} \frac{1}{\gamma!}D^\gamma \hat{\lambda}[0,0]\epsilon^{\gamma_1}(\epsilon \log \epsilon)^{\gamma_2} \tag{9.35}$$

for $\epsilon > 0$ sufficiently small. By exploiting the continuity result (9.32) we could also prove that, possibly shrinking ϵ^*, we have $\lambda[\epsilon] = \lambda_l[\Omega(\epsilon)]$ for all $\epsilon \in]0,\epsilon^*[$. Since $\lambda_l[\Omega^o]$ has multiplicity one, one can show that problem

$$\begin{cases} \Delta u = 0 & \text{in } \Omega^o, \\ \frac{\partial u}{\partial \nu_{\Omega^o}} = \lambda_l[\Omega^o]u & \text{on } \partial\Omega^o\,, \\ \int_{\partial\Omega^o} u^2\,d\sigma = 1\,, \end{cases} \tag{9.36}$$

has exactly two solutions in $C^{1,\alpha}(\overline{\Omega^o})$ (cf. Lemma 9.14 (iii)). We select one of the two, and we denote it by $\tilde{u}$. Then we show that for ϵ sufficiently small, the Steklov problem (9.31) with $\lambda = \lambda[\epsilon]$ has exactly one solution $u[\epsilon] \in C^{1,\alpha}(\overline{\Omega(\epsilon)})$ that satisfies the condition

$$\int_{\partial\Omega^o} u\tilde{u}\,d\sigma = 1\,. \tag{9.37}$$

We analyze both the macroscopic and microscopic behavior of $u[\epsilon]$. Namely, we take a bounded open subset Ω_M of $\overline{\Omega^o}\setminus\{0\}$ such that $0 \notin \overline{\Omega_M}$ and we prove for $u[\epsilon]_{|\overline{\Omega_M}}$ a formula such as (9.33) (cf. (9.83)) and we take a bounded open subset Ω_m of $\mathbb{R}^n \setminus \Omega^i$ and we prove for $u[\epsilon](\epsilon\,\cdot)_{|\overline{\Omega_m}}$ a formula such as (9.33) (cf. (9.85)). Finally, we also study the behavior of the energy integral $\int_{\Omega(\epsilon)} |\nabla u[\epsilon]|^2\,dx$ of the eigenfunction $u[\epsilon]$.

As usual in our analysis we reduce the problem to a system of integral equations. In this sense we mention that the reduction of the Steklov problem to integral equations has also been exploited by Kuznetsov and Motygin [153] and by Shamma [260].

This section is organized as follows. In Section 9.4.1 we introduce some technical preliminaries on the Steklov problem. In Section 9.4.2 we formulate problem (9.31) in terms of integral equations. Then we analyze the dependence on ϵ of the unknown densities of the integral equations of Section 9.4.2 around the degenerate value $\epsilon = 0$. In Section 9.4.3 we show that the results of Section 9.4.2 can be used to prove the main Theorem 9.22 on the representation of $\lambda[\epsilon]$ and of $u[\epsilon]$, $u[\epsilon](\epsilon\cdot)$ in terms of real analytic functions.

We like to conclude this introduction with a few words on the case of multiplicity higher than one, a case that was studied in [171]. If the eigenvalue $\tilde{\lambda}$ has multiplicity $r > 1$, then there exists $l > 0$ such that

$$\lambda_{l-1}[\Omega^o] < \tilde{\lambda} = \lambda_l[\Omega^o] = \cdots = \lambda_{l+r-1}[\Omega^o] < \lambda_{l+r}[\Omega^o]$$

and we can resort to a scheme that was employed in the papers [155] with Lamberti and [154] by Lamberti for problems with regular perturbations (see also Buoso and Provenzano [37] for the case of the biharmonic operator with Steklov boundary conditions). In particular, when the dimension n is bigger than or equal to three, we can obtain real analytic representation formulas similar to (9.33) for the eigenvalues $\lambda_{l+j-1}[\Omega(\epsilon)]$ with $j \in \{1, \ldots, r\}$ that 'split' from $\tilde{\lambda}$. However, in the two dimensional case we do not obtain formulas like (9.33) for the eigenvalues themselves but for their elementary symmetric functions

$$\begin{aligned}
\mathrm{Sym}_1[\epsilon] &\equiv \sum_{j_1=1}^{r} \lambda_{l+j_1-1}[\Omega(\epsilon)], \\
\mathrm{Sym}_2[\epsilon] &\equiv \sum_{\substack{j_1,j_2=1,\ldots,r \\ j_1<j_2}} \lambda_{l+j_1-1}[\Omega(\epsilon)] \cdot \lambda_{l+j_2-1}[\Omega(\epsilon)], \\
&\vdots \\
\mathrm{Sym}_r[\epsilon] &\equiv \lambda_l[\Omega(\epsilon)] \cdot \ \ldots \ \cdot \lambda_{l+r-1}[\Omega(\epsilon)] .
\end{aligned}$$

Then one could ask what is the benefit of understanding the behavior of the symmetric functions, when our initial goal was studying the eigenvalues. This question links to a classical a largely investigated problem: that of understanding how the roots of a polynomial of degree r with all real roots depend upon perturbations of the coefficients, under the assumption that the perturbed polynomial still have only real roots. Indeed, the symmetric functions above are (up to the sign) the coefficients of the polynomial

$$\Pi_{j=1}^{r}(x - \lambda_{l+j-1}[\Omega(\epsilon)]) .$$

9.4.1 Some Basic Facts on Steklov Eigenvalues and Eigenfunctions

By Theorem 6.48, each harmonic function in a bounded domain can be represented as a sum of a single layer potential and a constant. Then, the classical jump formula for the normal derivative of a single layer potential of Theorem 6.6 implies the validity of the following.

Proposition 9.12. *Let Ω be a bounded open connected subset of $\mathbb{R}^n$ of class $C^{1,\alpha}$. Then the map that takes (μ, c, λ) to $(v_\Omega^+[\mu] + c, \lambda)$ is a bijection from the set*

$$F \equiv \left\{ (\mu, c, \lambda) \in C^{0,\alpha}(\partial\Omega)_0 \times \mathbb{R} \times \mathbb{R} : -\frac{1}{2}\mu + W_\Omega^t[\mu] = \lambda\big(V_\Omega[\mu] + c\big) \right\} \quad (9.38)$$

to the set of pairs $(u, \lambda) \in C^{1,\alpha}(\overline{\Omega}) \times \mathbb{R}$ that satisfy the Steklov boundary value problem

$$\Delta u = 0 \quad \text{in } \Omega, \qquad \frac{\partial u}{\partial \nu_\Omega} = \lambda u \quad \text{on } \partial\Omega\,. \quad (9.39)$$

Then we have the following classical statement for an exterior Neumann problem.

Lemma 9.13. *Let Ω be a bounded open subset of $\mathbb{R}^n$ of class $C^{1,\alpha}$ such that $\mathbb{R}^n \setminus \overline{\Omega}$ is connected. Let $g \in C^{0,\alpha}(\partial\Omega)$. Then the exterior Neumann problem*

$$\begin{cases} \Delta u = 0 & \text{in } \Omega^-\,, \\ -\frac{\partial u}{\partial \nu_\Omega}(x) = g & \text{on } \partial\Omega\,, \\ \lim_{x\to\infty} u(x) - \delta_{2,n} \int_{\partial\Omega} \frac{\partial u}{\partial \nu_\Omega}\, d\sigma \frac{1}{2\pi} \log|x| = 0\,, \end{cases} \quad (9.40)$$

has one and only one solution $u \in C^{1,\alpha}_{\mathrm{loc}}(\overline{\Omega^-})$ and $u = v_\Omega^-[\theta]$, where θ is the unique solution in $C^{0,\alpha}(\partial\Omega)$ of the integral equation

$$-\frac{1}{2}\theta - W_\Omega^t[\theta] = g \qquad \text{on } \partial\Omega\,. \quad (9.41)$$

Proof. Since Ω^- is connected, Corollary 6.15 ensures that Equation (9.41) has a unique solution θ and Lemma 6.11 ensures that $\int_{\partial\Omega} \theta\, d\sigma = -\int_{\partial\Omega} g\, d\sigma$. Then the jump formula for the normal derivative of the single layer potential of Theorem 6.6 implies that $u = v_\Omega^-[\theta]$ satisfies the first two equations of problem (9.40). Moreover, again by the jump formula for the normal derivative of a single layer potential and by Lemma 6.11, we have

$$\int_{\partial\Omega} \frac{\partial}{\partial \nu_\Omega} v_\Omega^-[\theta]\, d\sigma = \int_{\partial\Omega} \theta\, d\sigma$$

and accordingly

$$\lim_{x\to\infty} v_\Omega^-[\theta](x) - \delta_{2,n} \int_{\partial\Omega} \frac{\partial v_\Omega^-[\theta]}{\partial \nu_\Omega}\, d\sigma \frac{1}{2\pi} \log|x| \quad (9.42)$$

$$= \lim_{x\to\infty} \int_{\partial\Omega} \left[S_n(x-y) - \frac{\delta_{2,n}}{2\pi} \log|x| \right] \theta(y)\, d\sigma_y = 0$$

(see also (4.67) and (4.69)). Hence, the third equation of problem (9.40) holds. Thus we have proved that $v_\Omega^-[\theta]$ solves problem (9.40).

Conversely, we now assume that $u \in C^{1,\alpha}_{\text{loc}}(\overline{\Omega^-})$ solves problem (9.40). As we have pointed out above, the integral equation

$$-\frac{1}{2}\theta - W_\Omega^t[\theta] = -\frac{\partial u}{\partial \nu_\Omega} \qquad \text{on } \partial\Omega$$

has a unique solution $\theta \in C^{0,\alpha}(\partial\Omega)$. Moreover, the jump formula for the normal derivative of a single layer potential implies that

$$\frac{\partial v_\Omega^-[\theta]}{\partial \nu_\Omega} = \frac{\partial u}{\partial \nu_\Omega}\,. \tag{9.43}$$

Then the same computations above show that

$$\lim_{x\to\infty} v_\Omega^-[\theta](x) - \delta_{2,n} \int_{\partial\Omega} \frac{\partial v_\Omega^-[\theta]}{\partial \nu_\Omega}\, d\sigma \frac{1}{2\pi} \log|x| = 0\,.$$

Now we set

$$\begin{aligned} w(x) \equiv & \left\{ u(x) - \frac{\delta_{2,n}}{2\pi} \int_{\partial\Omega} \frac{\partial u}{\partial \nu_\Omega}\, d\sigma \log|x| \right\} \\ & - \left\{ v_\Omega^-[\theta](x) - \frac{\delta_{2,n}}{2\pi} \int_{\partial\Omega} \frac{\partial v_\Omega^-[\theta]}{\partial \nu_\Omega}\, d\sigma \log|x| \right\} \qquad \forall x \in \overline{\Omega^-}\,. \end{aligned}$$

Then the function w belongs to $C^{1,\alpha}_{\text{loc}}(\overline{\Omega^-})$ and satisfies the exterior Neumann problem

$$\begin{cases} \Delta w = 0 & \text{in } \Omega^-\,, \\ \frac{\partial w}{\partial \nu_\Omega} = 0 & \text{on } \partial\Omega\,, \\ \lim_{x\to\infty} w(x) = 0\,. \end{cases}$$

Hence, w is identically equal to zero and

$$u(x) - \frac{\delta_{2,n}}{2\pi} \int_{\partial\Omega} \frac{\partial u}{\partial \nu_\Omega}\, d\sigma \log|x| = v_\Omega^-[\theta](x) - \frac{\delta_{2,n}}{2\pi} \int_{\partial\Omega} \frac{\partial v_\Omega^-[\theta]}{\partial \nu_\Omega}\, d\sigma \log|x|$$

for all $x \in \overline{\Omega^-}$ (see the uniqueness Theorem 6.2 for the exterior Neumann problem). Then equality (9.43) implies that $u = v_\Omega^-[\theta]$. □

We also need the following technical lemma.

Lemma 9.14. *Let* $\alpha \in]0,1[$. *Let* Ω *be a bounded open subset of* $\mathbb{R}^n$ *of class* $C^{1,\alpha}$. *Then the following statements hold.*

(i) *Let* $\lambda \in]0,+\infty[$. *Let* $u \in L^2(\Omega)$ *have weak derivatives of order one in* $L^2(\Omega)$ *and be a weak solution of the Steklov eigenvalue problem (9.39). Then* $u \in C^{1,\alpha}(\overline{\Omega})$.

(ii) *If* $\lambda \in]0,+\infty[$, $u \in C^{1,\alpha}(\overline{\Omega})$ *solves (9.39), and* $\int_{\partial\Omega} u^2\, d\sigma = 0$, *then* u *vanishes identically.*

(iii) *If* $\lambda \in]0,+\infty[$ *is a Steklov eigenvalue of multiplicity one of* Δ *in* Ω, *then (9.39) has exactly two solutions* $u \in C^{1,\alpha}(\overline{\Omega})$ *such that* $\int_{\partial\Omega} u^2\, d\sigma = 1$.

Proof. Let $W^{1,2}(\Omega)$ denote the Sobolev space of functions $v \in L^2(\Omega)$ that have weak derivatives of order one in $L^2(\Omega)$. By assumption,

$$\int_\Omega \sum_{j=1}^n \frac{\partial u}{\partial x_j}\frac{\partial v}{\partial x_j}\, dx = \lambda \int_{\partial\Omega} \mathrm{Tr}\, u \mathrm{Tr}\, v\, d\sigma \qquad \forall v \in W^{1,2}(\Omega),$$

where Tr denotes the trace operator from $W^{1,2}(\Omega)$ to $L^2(\partial\Omega)$ (cf., e.g., Burenkov [38]). Then statement (i) follows by classical elliptic regularity theory for the Robin problem (cf., e.g., Agmon, Douglis and Nirenberg [2, Theorem 7.3], Gilbarg and Trudinger [107, Chap. 8]).

If u is as in (ii), then $u = 0$ on $\partial\Omega$ and thus the uniqueness Theorem 3.6 for the Dirichlet problem implies that $u = 0$. We now prove statement (iii). Let $u_1 \in C^{1,\alpha}(\overline{\Omega})$ be a nontrivial eigenfunction that corresponds to λ. By (ii), we have $\int_{\partial\Omega} u_1^2\, d\sigma \neq 0$. If $u \in C^{1,\alpha}(\overline{\Omega})$ satisfies the equality $\int_{\partial\Omega} u^2\, d\sigma = 1$ and is an eigenfunction that corresponds to λ, which has multiplicity one, then there exists $\mu \in \mathbb{R}$ such that $u = \mu u_1$ and thus $\int_{\partial\Omega} (\mu u_1)^2\, d\sigma = 1$. Hence, we have

$$\mu^2 = \left(\int_{\partial\Omega} u_1^2\, d\sigma \right)^{-1}$$

and either $u = \left(\int_{\partial\Omega} u_1^2\, d\sigma\right)^{-1/2} u_1$ or $u = -\left(\int_{\partial\Omega} u_1^2\, d\sigma\right)^{-1/2} u_1$.

On the other hand, $\left(\int_{\partial\Omega} u_1^2\, d\sigma\right)^{-1/2} u_1$ and $-\left(\int_{\partial\Omega} u_1^2\, d\sigma\right)^{-1/2} u_1$ are two distinct eigenfunctions that satisfy condition $\int_{\partial\Omega} u^2\, d\sigma = 1$. □

9.4.2 Formulation of the Steklov Problem (9.31) in Terms of Integral Equations

In order to shorten our notation, we set

$$X_{0,\alpha} \equiv C^{0,\alpha}(\partial\Omega^i) \times C^{0,\alpha}(\partial\Omega^o).$$

We now convert problem (9.31) for ϵ small enough into a system of integral equations.

Theorem 9.15. *Let $\mathcal{Q} \equiv (\mathcal{Q}_s)_{s=1,2,3}$ be the map from $]-\epsilon_0, \epsilon_0[\times\mathbb{R}\times X_{0,\alpha}\times\mathbb{R}^2$ to $X_{0,\alpha}\times\mathbb{R}$ defined by*

$$\begin{aligned}
&\mathcal{Q}_1[\epsilon,\epsilon_1,\theta^i,\theta^o,c,\lambda](\xi) \equiv -\frac{1}{2}\theta^i(\xi) - W^t_{\Omega^i}[\theta^i](\xi)\\
&\quad -\int_{\partial\Omega^o}\nu_{\Omega^i}(\xi)\cdot\nabla S_n(\epsilon\xi-y)\theta^o(y)\,d\sigma_y - \lambda\Big\{\epsilon V_{\Omega^i}[\theta^i](\xi)\\
&\quad +\epsilon_1\frac{1}{2\pi}\int_{\partial\Omega^i}\theta^i(\eta)\,d\sigma_\eta + \int_{\partial\Omega^o}S_n(\epsilon\xi-y)\theta^o(y)\,d\sigma_y + c\Big\} \qquad \forall\xi\in\partial\Omega^i\,,\\
&\mathcal{Q}_2[\epsilon,\epsilon_1,\theta^i,\theta^o,c,\lambda](x) \equiv -\frac{1}{2}\theta^o(x) + W^t_{\Omega^o}[\theta^o](x)\\
&\quad +\epsilon^{n-1}\int_{\partial\Omega^i}\nu_{\Omega^o}(x)\cdot\nabla S_n(x-\epsilon\eta)\theta^i(\eta)\,d\sigma_\eta\\
&\quad -\lambda\Big\{\epsilon^{n-1}\int_{\partial\Omega^i}S_n(x-\epsilon\eta)\theta^i(\eta)\,d\sigma_\eta + V_{\Omega^o}[\theta^o](x) + c\Big\} \qquad \forall x\in\partial\Omega^o\,,\\
&\mathcal{Q}_3[\epsilon,\epsilon_1,\theta^i,\theta^o,c,\lambda](\xi) \equiv \epsilon^{n-1}\int_{\partial\Omega^i}\theta^i\,d\sigma + \int_{\partial\Omega^o}\theta^o\,d\sigma\,,
\end{aligned} \tag{9.44}$$

for all $(\epsilon,\epsilon_1,\theta^i,\theta^o,c,\lambda)\in]-\epsilon_0,\epsilon_0[\times\mathbb{R}\times X_{0,\alpha}\times\mathbb{R}^2$. Then the following statements hold.

(i) Let $\epsilon\in]0,\epsilon_0[$. Let $u_\epsilon[\theta^i,\theta^o,c,\lambda]$ be the function of class $C^{1,\alpha}(\overline{\Omega(\epsilon)})$ from $\overline{\Omega(\epsilon)}$ to $\mathbb{R}$ defined by

$$\begin{aligned}
u_\epsilon[\theta^i,\theta^o,c,\lambda](x) &\equiv v^-_{\epsilon\Omega^i}[\theta^i(\cdot/\epsilon)](x) + v^+_{\Omega^o}[\theta^o](x) + c\\
&= \epsilon^{n-1}\int_{\partial\Omega^i}S_n(x-\epsilon\eta)\theta^i(\eta)\,d\sigma_\eta + \int_{\partial\Omega^o}S_n(x-y)\theta^o(y)\,d\sigma_y + c\,,
\end{aligned} \tag{9.45}$$

for all $x\in\overline{\Omega(\epsilon)}$ and for all $(\theta^i,\theta^o,c,\lambda)\in X_{0,\alpha}\times\mathbb{R}\times\mathbb{R}$. Then the map that takes $(\theta^i,\theta^o,c,\lambda)$ to the pair $(u_\epsilon[\theta^i,\theta^o,c,\lambda],\lambda)$ is a bijection from the set of $(\theta^i,\theta^o,c,\lambda)$ in $X_{0,\alpha}\times\mathbb{R}\times\mathbb{R}$ such that

$$\mathcal{Q}[\epsilon,\delta_{2,n}\epsilon\log\epsilon,\theta^i,\theta^o,c,\lambda] = 0\,, \tag{9.46}$$

to the set of solutions (u,λ) in $C^{1,\alpha}(\overline{\Omega(\epsilon)})\times\mathbb{R}$ of the Steklov problem (9.31).

(ii) Let $\epsilon=\epsilon_1=0$. Let $u^i[\theta^i,\theta^o,c,\lambda]\in C^{1,\alpha}_{\mathrm{loc}}(\overline{\Omega^{i-}})$ and $u^o[\theta^i,\theta^o,c,\lambda]\in C^{1,\alpha}(\overline{\Omega^o})$ be the functions defined by

$$\begin{aligned}
u^i[\theta^i,\theta^o,c,\lambda] &\equiv v^-_{\Omega^i}[\theta^i] && \text{on } \overline{\Omega^{i-}}\,,\\
u^o[\theta^i,\theta^o,c,\lambda] &\equiv v^+_{\Omega^o}[\theta^o] + c && \text{on } \overline{\Omega^o}\,,
\end{aligned} \tag{9.47}$$

for all $(\theta^i,\theta^o,c,\lambda)$ in $X_{0,\alpha}\times\mathbb{R}\times\mathbb{R}$. Then the map that takes $(\theta^i,\theta^o,c,\lambda)$ to $(u^i[\theta^i,\theta^o,c,\lambda],u^o[\theta^i,\theta^o,c,\lambda],\lambda)$, is a bijection from the set of $(\theta^i,\theta^o,c,\lambda)$ in $X_{0,\alpha}\times\mathbb{R}\times\mathbb{R}$ such that

$$\mathcal{Q}[0,0,\theta^i,\theta^o,c,\lambda] = 0\,, \tag{9.48}$$

to the set of $(u^i, u^o, \lambda) \in C^{1,\alpha}_{\mathrm{loc}}(\overline{\Omega^{i-}}) \times C^{1,\alpha}(\overline{\Omega^o}) \times \mathbb{R}$ *such that*

$$\begin{cases} \Delta u^i = 0 & \text{in } \Omega^{i-}\,, \\ -\frac{\partial u^i}{\partial \nu_{\Omega^i}} = \nu_{\Omega^i} \cdot \nabla u^o(0) + \lambda u^o(0) & \text{on } \partial\Omega^i\,, \\ \lim_{\xi\to\infty} u^i(\xi) - \frac{\delta_{2,n}}{2\pi} \int_{\partial\Omega^i} \frac{\partial u^i}{\partial \nu_{\Omega^i}}\, d\sigma \log|\xi| = 0\,, & \\ \Delta u^o = 0 & \text{in } \Omega^o\,, \\ \frac{\partial u^o}{\partial \nu_{\Omega^o}} = \lambda u^o & \text{on } \partial\Omega^o\,. \end{cases} \tag{9.49}$$

Proof. We first prove statement (i). By Proposition 9.12 the map that takes (μ, c, λ) to $(v^+_{\Omega(\epsilon)}[\mu] + c, \lambda)$ is a bijection from the set F of (9.38) with $\Omega = \Omega(\epsilon)$ to the set of (u, λ) of statement (i). By the rule of change of variables in the integrals and by the equalities

$$\nu_{\Omega(\epsilon)}(x) = -\nu_{\Omega^i}(x/\epsilon) \quad \forall x \in \epsilon\partial\Omega^i\,, \qquad \nu_{\Omega(\epsilon)}(x) = \nu_{\Omega^o}(x) \quad \forall x \in \partial\Omega^o\,,$$

for all $\epsilon \in]0, \epsilon_0[$, the map that takes $(\theta^i, \theta^o, c, \lambda)$ to (μ, c, λ), where μ is the function from $\partial\Omega(\epsilon)$ to $\mathbb{R}$ defined by

$$\mu(x) \equiv \theta^i(x/\epsilon) \qquad \forall x \in \epsilon\partial\Omega^i\,, \qquad \mu(x) \equiv \theta^o(x) \qquad \forall x \in \partial\Omega^o\,,$$

is a bijection from the set of $(\theta^i, \theta^o, c, \lambda)$, which satisfy Equation (9.46) to F with $\Omega = \Omega(\epsilon)$. Moreover,

$$u_\epsilon[\theta^i, \theta^o, c, \lambda](x) = v^+_{\Omega(\epsilon)}[\mu](x) + c \qquad \forall x \in \Omega(\epsilon)\,.$$

Then by continuity of both handsides, the same equality holds for all $x \in \overline{\Omega(\epsilon)}$.

Next we consider the proof of statement (ii). If $(\theta^i, \theta^o, c, \lambda)$ solves Equation (9.48), then the last two components of Equation (9.48) and the jump formula for the normal derivatives of a single layer potential of Theorem 6.6 imply that the function $u^o[\theta^i, \theta^o, c, \lambda]$ defined in (9.47) satisfies the last two conditions of problem (9.49). Then Lemma 9.13 with

$$g(\xi) = \nu_{\Omega^i}(\xi) \cdot \nabla u^o[\theta^i, \theta^o, c, \lambda](0) + \lambda u^o[\theta^i, \theta^o, c, \lambda](0) \qquad \forall \xi \in \partial\Omega^i$$

implies that the function $u^i[\theta^i, \theta^o, c, \lambda]$ defined in (9.47) satisfies the first three conditions of problem (9.49).

Conversely, we now assume that (u^i, u^o, λ) satisfies the boundary value problem (9.49). Then the last two equations of problem (9.49) and the representation Theorem 6.48 imply that there exists a unique $(\theta^o, c) \in C^{0,\alpha}(\partial\Omega^o)_0 \times \mathbb{R}$ such that

$$u^o = v^+_{\Omega^o}[\theta^o] + c = u^o[\theta^i, \theta^o, c, \lambda].$$

By the the jump formula for the normal derivative of a single layer potential, we have $\mathcal{Q}_j[0, 0, \theta^i, \theta^o, c, \lambda] = 0$ for $j = 2$ and $j = 3$. Then the first three conditions of problem (9.49) and Lemma 9.13 with $\Omega = \Omega^i$ and with

$$g(\xi) = \nu_{\Omega^i}(\xi) \cdot \nabla u^o(0) + \lambda u^o(0) \qquad \forall \xi \in \partial\Omega^i$$

imply that there exists a unique $\theta^i \in C^{0,\alpha}(\partial\Omega^i)$ such that $u^i = v^-_{\Omega^i}[\theta^i]$ and such that $\mathcal{Q}_1[0,0,\theta^i,\theta^o,c,\lambda] = 0$. Thus we have shown that if (u^i,u^o,λ) satisfies the boundary value problem (9.49), then there exists a unique

$$(\theta^i,\theta^o,c,\lambda) \in X_{0,\alpha} \times \mathbb{R} \times \mathbb{R}$$

such that the equalities (9.48) hold and we have $u^i = u^i[\theta^i,\theta^o,c,\lambda]$ and $u^o = u^o[\theta^i,\theta^o,c,\lambda]$. □

Next we introduce a lemma that shows that the operator L_λ whose components are the last two components of the operator in the left-hand side of Equation (9.48) is the transpose to an operator G_λ with respect to the dual system

$$(C^{0,\alpha}(\partial\Omega) \times \mathbb{R}, C^{1,\alpha}(\partial\Omega) \times \mathbb{R})$$

induced by the non-degenerate bilinear form that takes a pair $((\theta,c),(\tau,d))$ of $\left(C^{0,\alpha}(\partial\Omega) \times \mathbb{R}\right) \times \left(C^{1,\alpha}(\partial\Omega) \times \mathbb{R}\right)$ to

$$\langle(\theta,c),(\tau,d)\rangle = \int_{\partial\Omega} \theta\tau \, d\sigma + cd\,, \tag{9.50}$$

and that describes the null space of G_λ. Such a lemma reveals to be important in the analysis of Equation (9.48).

Lemma 9.16. *Let Ω be a bounded open connected subset of $\mathbb{R}^n$ of class $C^{1,\alpha}$. Let $\lambda \in \mathbb{R}$. Let $L_\lambda \equiv (L_{\lambda,1}, L_{\lambda,2})$ be the linear operator from $C^{0,\alpha}(\partial\Omega) \times \mathbb{R}$ to itself defined by*

$$\begin{aligned} L_{\lambda,1}[\theta,c](x) \equiv {} & -\frac{1}{2}\theta(x) + W^t_\Omega[\theta](x) \\ & -\lambda\left\{V_\Omega[\theta](x) + c\right\} \qquad \forall x \in \partial\Omega\,, \\ L_{\lambda,2}[\theta,c] \equiv {} & \int_{\partial\Omega} \theta \, d\sigma\,, \end{aligned} \tag{9.51}$$

for all $(\theta,c) \in C^{0,\alpha}(\partial\Omega) \times \mathbb{R}$. Then the following statements hold.

(i) L_λ coincides with the transpose with respect to the dual system

$$(C^{0,\alpha}(\partial\Omega) \times \mathbb{R}, C^{1,\alpha}(\partial\Omega) \times \mathbb{R})$$

to the operator $G_\lambda \equiv (G_{\lambda,1}, G_{\lambda,2})$ from $C^{1,\alpha}(\partial\Omega) \times \mathbb{R}$ to itself defined by

$$\begin{aligned} G_{\lambda,1}[\tau,d](y) = {} & -\frac{1}{2}\tau(y) + W_\Omega[\tau](y) \\ & -\lambda V_\Omega[\tau](y) + d \qquad \forall y \in \partial\Omega\,, \\ G_{\lambda,2}[\tau,d] = {} & -\lambda \int_{\partial\Omega} \tau \, d\sigma\,, \end{aligned} \tag{9.52}$$

for all $(\tau, d) \in C^{1,\alpha}(\partial\Omega) \times \mathbb{R}$.

(ii) *The operator* Φ *from* $C^{0,\alpha}(\partial\Omega) \times \mathbb{R}$ *to* $C^{1,\alpha}(\partial\Omega) \times \mathbb{R}$ *that maps the pair* (θ, c) *to the pair* $(V_\Omega[\theta] + c, \int_{\partial\Omega} \theta \, d\sigma)$ *restricts to a linear bijection from* $\mathrm{Ker}\, L_\lambda$ *onto* $\mathrm{Ker}\, G_\lambda$.

Proof. We first prove statement (i). By Theorems 4.25 (i) and 6.8 (ii) we know that V_Ω and W_Ω^t are linear and continous in $C^{0,\alpha}(\partial\Omega)$ and that accordingly L_λ is linear and continuous from $C^{0,\alpha}(\partial\Omega) \times \mathbb{R}$ to itself. Next we observe that L_λ is the transpose to G_λ with respect to the dual system $(C^{0,\alpha}(\partial\Omega) \times \mathbb{R}, C^{1,\alpha}(\partial\Omega) \times \mathbb{R})$ provided that

$$\int_{\partial\Omega} L_{\lambda,1}[\theta, c]\tau \, d\sigma + L_{\lambda,2}[\theta, c] d = \int_{\partial\Omega} \theta G_{\lambda,1}[\tau, d] \, d\sigma + c G_{\lambda,2}[\tau, d] \,, \quad (9.53)$$

for all (θ, c) in $C^{0,\alpha}(\partial\Omega) \times \mathbb{R}$ and (τ, d) in $C^{1,\alpha}(\partial\Omega) \times \mathbb{R}$. By Theorem 6.5 on the transpose operator W_Ω^t, by the Fubini Theorem, and by the parity of S_n, the left-hand side of (9.53) equals

$$\int_{\partial\Omega} \theta(y) \Big\{ -\frac{1}{2}\tau(y) + W_\Omega[\tau](y) \\ -\lambda \int_{\partial\Omega} S_n(y - x)\tau(x) \, d\sigma_x + d \Big\} \, d\sigma_y - \lambda \int_{\partial\Omega} \tau \, d\sigma c \,,$$

for all (θ, c) in $C^{0,\alpha}(\partial\Omega) \times \mathbb{R}$ and for all $(\tau, d) \in C^{1,\alpha}(\partial\Omega) \times \mathbb{R}$. Thus formula (9.52) implies the validity of equality (9.53). Then Theorems 4.25 (i) and 6.8 (i) ensure that V_Ω and W_Ω are linear and continuous in $C^{1,\alpha}(\partial\Omega)$ and that accordingly G_λ is linear and continuous from $C^{1,\alpha}(\partial\Omega) \times \mathbb{R}$ to itself.

Next we consider the proof of statement (ii). Theorem 4.25 (i) ensures that V_Ω is linear and continuous from $C^{0,\alpha}(\partial\Omega)$ to $C^{1,\alpha}(\partial\Omega)$. Hence the map Φ is linear and continuous from $C^{0,\alpha}(\partial\Omega) \times \mathbb{R}$ to $C^{1,\alpha}(\partial\Omega) \times \mathbb{R}$.

By Theorem 6.47, we already know that Φ restricts to an injection from $\mathrm{Ker}\, L_\lambda$ to $C^{1,\alpha}(\partial\Omega) \times \mathbb{R}$. We now prove that Φ maps $\mathrm{Ker}\, L_\lambda$ into $\mathrm{Ker}\, G_\lambda$. Let (θ, c) belong to $\mathrm{Ker}\, L_\lambda$. Then we have

$$\frac{\partial}{\partial\nu_\Omega}(v_\Omega^+[\theta]v + c) - \lambda(v_\Omega^+[\theta]_{|\partial\Omega} + c) = 0 \qquad \text{on } \partial\Omega \,. \quad (9.54)$$

Applying the single layer potential operator $v_\Omega^+[\cdot]$ on both hand sides, we obtain

$$v_\Omega^+[\frac{\partial}{\partial\nu_\Omega}(v_\Omega^+[\theta] + c)] - \lambda v_\Omega^+[v_\Omega^+[\theta]_{|\partial\Omega} + c] = 0 \qquad \text{on } \partial\Omega \,. \quad (9.55)$$

Since $v_\Omega^+[\theta] + c$ is harmonic in Ω, the third Green Identity of Theorem 6.9 implies that

$$v_\Omega^+[\theta] + c = w_\Omega^+[v_\Omega^+[\theta]_{|\partial\Omega} + c] - v_\Omega^+[\frac{\partial}{\partial\nu_\Omega}(v_\Omega^+[\theta] + c)] \qquad \text{on } \overline{\Omega} \,. \quad (9.56)$$

By the jump formulas for double layer potentials of Theorem 6.6, we have

$$w_\Omega^+[v_\Omega^+[\theta]_{|\partial\Omega} + c]_{|\partial\Omega} = (v_\Omega^+[\theta]_{|\partial\Omega} + c)_{|\partial\Omega} + w_\Omega^-[v_\Omega^+[\theta]_{|\partial\Omega} + c]_{|\partial\Omega} \tag{9.57}$$

on $\partial\Omega$. Then, combining (9.55)–(9.57) we obtain

$$w_\Omega^-[v_\Omega^+[\theta]_{|\partial\Omega} + c]_{|\partial\Omega} - \lambda v_\Omega^+[v_\Omega^+[\theta]_{|\partial\Omega} + c]_{|\partial\Omega} = 0 \qquad \text{on } \partial\Omega\,,$$

i.e., $G_{\lambda,1}[v_\Omega^+[\theta]_{|\partial\Omega} + c, 0] = 0$. Since $\int_{\partial\Omega} \theta\, d\sigma = 0$ for all $(\theta, c) \in \operatorname{Ker} L_\lambda$, we conclude that $\Phi[\theta, c] = (v_\Omega^+[\theta]_{|\partial\Omega} + c, 0)$ and that accordingly

$$G_{\lambda,1}[\Phi[\theta, c]] = G_{\lambda,1}[v_\Omega^+[\theta]_{|\partial\Omega} + c, 0] = 0 \qquad \forall(\theta, c) \in \operatorname{Ker} L_\lambda\,.$$

Then we have

$$G_{\lambda,2}[\Phi[\theta, c]] = -\lambda \int_{\partial\Omega} (v_\Omega^+[\theta]_{|\partial\Omega} + c)\, d\sigma = 0 \qquad \forall(\theta, c) \in \operatorname{Ker} L_\lambda\,.$$

Indeed, if $(\theta, c) \in \operatorname{Ker} L_\lambda$, then (9.54) holds, and by integrating both hand sides of (9.54) on $\partial\Omega$ and by exploiting the Divergence Theorem 4.1, we obtain

$$\int_{\partial\Omega} \lambda(v_\Omega^+[\theta]_{|\partial\Omega} + c)\, d\sigma = \int_{\partial\Omega} \frac{\partial}{\partial\nu_\Omega}(v_\Omega^+[\theta] + c)\, d\sigma = 0 \qquad \text{on } \partial\Omega\,.$$

Indeed, $(v_\Omega^+[\theta] + c)$ is harmonic in Ω. Hence, $\Phi[\theta, c] \in \operatorname{Ker} G_\lambda$ for all (θ, c) in $\operatorname{Ker} L_\lambda$.

Next we note that

$$L_{\lambda,2}[\theta, c] \equiv -\frac{1}{2}c - \left(-\int_{\partial\Omega} \theta\, d\sigma - \frac{1}{2}c\right) \qquad \forall(\theta, c) \in C^{0,\alpha}(\partial\Omega) \times \mathbb{R}\,.$$

Also, Theorem 6.4 and the compactness of the embedding of the space $C^{1,\alpha}(\partial\Omega)$ into $C^{0,\alpha}(\partial\Omega)$ ensure that V_Ω is compact in $C^{0,\alpha}(\partial\Omega)$ and Theorem 6.8 (ii) ensures that W_Ω^t is compact in $C^{0,\alpha}(\partial\Omega)$. Thus the operator L_λ equals the product of $-\frac{1}{2}$ and of a compact perturbation of the identity in $C^{0,\alpha}(\partial\Omega) \times \mathbb{R}$. Then the Fredholm Alternative Theorem 5.8 implies that the dimension of $\operatorname{Ker} L_\lambda$ equals the dimension of $\operatorname{Ker} G_\lambda$. Hence, Φ restricts to a linear bijection from $\operatorname{Ker} L_\lambda$ onto $\operatorname{Ker} G_\lambda$. □

Theorem 9.15 reduces the analysis of the domain dependent Steklov eigenvalue problem (9.31) to that of Equation (9.46). We now derive a corresponding equation for the Steklov eigenvalue problem (9.31) subject to the normalizing condition (9.37). We do so by means of the following immediate consequence of Theorem 9.15.

Corollary 9.17. *Let $f \in C^{1,\alpha}(\partial\Omega^o)$. Let $\mathcal{Q}_f \equiv (\mathcal{Q}_{f,s})_{s\in\{1,2,3,4\}}$ be the map from $]-\epsilon_0, \epsilon_0[\times\mathbb{R} \times X_{0,\alpha} \times \mathbb{R}^2$ to $X_{0,\alpha} \times \mathbb{R}^2$ defined by*

$$\mathcal{Q}_{f,s}[\epsilon, \epsilon_1, \theta^i, \theta^o, c, \lambda] \equiv \mathcal{Q}_s[\epsilon, \epsilon_1, \theta^i, \theta^o, c, \lambda] \qquad \forall s \in \{1, 2, 3\}\,, \tag{9.58}$$

$$\mathcal{Q}_{f,4}[\epsilon, \epsilon_1, \theta^i, \theta^o, c, \lambda] \equiv \int_{\partial\Omega^o} \left\{ \epsilon^{n-1} \int_{\partial\Omega^i} S_n(x - \epsilon\eta)\theta^i(\eta)\, d\sigma_\eta \right.$$
$$\left. + \int_{\partial\Omega^o} S_n(x-y)\theta^o(y)\, d\sigma_y + c \right\} f(x)\, d\sigma_x - 1\,,$$

for all $(\epsilon, \epsilon_1, \theta^i, \theta^o, \lambda) \in]-\epsilon_0, \epsilon_0[\times\mathbb{R} \times X_{0,\alpha} \times \mathbb{R}^2$. *Then the following statements hold.*

(i) *Let* $\epsilon \in]0, \epsilon_0[$. *Then the map that takes* $(\theta^i, \theta^o, c, \lambda)$ *to* $(u_\epsilon[\theta^i, \theta^o, c, \lambda], \lambda)$, *with* $u_\epsilon[\theta^i, \theta^o, c, \lambda]$ *as in (9.45) is a bijection from the set of* $(\theta^i, \theta^o, c, \lambda)$ *in* $X_{0,\alpha} \times \mathbb{R}^2$ *such that*

$$\mathcal{Q}_f[\epsilon, \delta_{2,n}\epsilon \log \epsilon, \theta^i, \theta^o, c, \lambda] = 0\,, \tag{9.59}$$

to the set of (u, λ) *in* $C^{1,\alpha}(\overline{\Omega(\epsilon)}) \times \mathbb{R}$ *such that*

$$\begin{cases} \Delta u = 0 & \text{in } \Omega(\epsilon), \\ \frac{\partial u}{\partial \nu_{\Omega(\epsilon)}} = \lambda u & \text{on } \partial\Omega(\epsilon)\,, \\ \int_{\partial\Omega^o} uf\, d\sigma = 1\,. \end{cases} \tag{9.60}$$

(ii) *Let* $\epsilon = \epsilon_1 = 0$. *Then the map that takes the quadruple* $(\theta^i, \theta^o, c, \lambda)$ *to the triple* $(u^i[\theta^i, \theta^o, c, \lambda], u^o[\theta^i, \theta^o, c, \lambda], \lambda)$ *as in (9.47) is a bijection from the set of* $(\theta^i, \theta^o, c, \lambda)$ *in* $X_{0,\alpha} \times \mathbb{R}^2$ *such that*

$$\mathcal{Q}_f[0, 0, \theta^i, \theta^o, c, \lambda] = 0\,, \tag{9.61}$$

to the set of $(u^i, u^o, \lambda) \in C^{1,\alpha}_{\mathrm{loc}}(\overline{\Omega^{i-}}) \times C^{1,\alpha}(\overline{\Omega^o}) \times \mathbb{R}$ *such that*

$$\begin{cases} \Delta u^i = 0 & \text{in } \Omega^{i-}\,, \\ -\frac{\partial u^i}{\partial \nu_{\Omega^i}} = \nu_{\Omega^i} \cdot \nabla u^o(0) + \lambda u^o(0) & \text{on } \partial\Omega^i\,, \\ \lim_{\xi\to\infty} u^i(\xi) - \frac{\delta_{2,n}}{2\pi} \int_{\partial\Omega^i} \frac{\partial u^i}{\partial \nu_{\Omega^i}}\, d\sigma \log|\xi| = 0\,, & \\ \Delta u^o = 0 & \text{in } \Omega^o\,, \\ \frac{\partial u^o}{\partial \nu_{\Omega^o}} = \lambda u^o & \text{on } \partial\Omega^o\,, \\ \int_{\partial\Omega^o} uf\, d\sigma - 1 = 0\,. \end{cases} \tag{9.62}$$

We now analyze some properties of the map $\mathcal{Q}_f$ of (9.58) in the specific case where f is the boundary value of a Steklov eigenfunction $\tilde{u}$ corresponding to an eigenvalue of multiplicity one.

Proposition 9.18. *Let* $\tilde{\lambda} \in [0, +\infty[$ *be a Steklov eigenvalue of* Δ *in* Ω^o *of multiplicity one (see problem (9.30)). Let* $\tilde{u} \in C^{1,\alpha}(\overline{\Omega^o})$ *be one of the two solutions of problem*

$$\begin{cases} \Delta \tilde{u} = 0 & \text{in } \Omega^o, \\ \frac{\partial \tilde{u}}{\partial \nu_{\Omega^o}} = \tilde{\lambda}\tilde{u} & \text{on } \partial\Omega^o\,, \\ \int_{\partial\Omega^o} \tilde{u}^2\, d\sigma = 1 \end{cases} \tag{9.63}$$

(cf. Lemma 9.14 (iii)). Let $\tilde{\mathcal{Q}}$ *denote the map* $\mathcal{Q}_f$ *of (9.58) with* $f \equiv \tilde{u}_{|\partial\Omega^o}$. *Then the following statements hold.*

(i) Problem (9.62) with $f = \tilde{u}_{|\partial\Omega^o}$ *and* $\lambda = \tilde{\lambda}$ *has a unique solution* $(\tilde{u}^i, \tilde{u}^o)$ *in* $C^{1,\alpha}_{\mathrm{loc}}(\overline{\Omega^{i-}}) \times C^{1,\alpha}(\overline{\Omega^o})$ *and we have* $\tilde{u}^o = \tilde{u}$.

(ii) The equation

$$\tilde{\mathcal{Q}}[0,0,\tilde{\theta}^i,\tilde{\theta}^o,\tilde{c},\tilde{\lambda}] = 0 \tag{9.64}$$

has a unique solution $(\tilde{\theta}^i, \tilde{\theta}^o, \tilde{c})$ *in* $X_{0,\alpha} \times \mathbb{R}$, *and*

$$\tilde{u}^i = u^i[\tilde{\theta}^i,\tilde{\theta}^o,\tilde{c},\tilde{\lambda}]\,, \qquad \tilde{u}^o = u^o[\tilde{\theta}^i,\tilde{\theta}^o,\tilde{c},\tilde{\lambda}]\,, \tag{9.65}$$

and $v^+_{\Omega^o}[\tilde{\theta}^o] + \tilde{c} = \tilde{u}$ *(cf. Theorem 6.48 (i)).*

(iii) $\tilde{\mathcal{Q}}$ *is real analytic.*

Proof. We first prove statement (i). By assumption, the last three equations of (9.62) have at least the solution $\tilde{u}^o = \tilde{u}$. Let $\tilde{u}^o_1$ be another solution of the last three equations of (9.62). Since $\tilde{\lambda}$ has multiplicity one and $\tilde{u}$ is a nontrivial eigenfunction, there exists $c \in \mathbb{R}$ such that $\tilde{u}^o_1 = c\tilde{u}$. Then we have $1 = \int_{\partial\Omega^o} \tilde{u}^o_1 \tilde{u}\, d\sigma = \int_{\partial\Omega^o} c\tilde{u}\tilde{u}\, d\sigma = c$ and thus $\tilde{u}^o_1 = \tilde{u}$ and the uniqueness of $\tilde{u}^o$ follows.

Since $\nu_{\Omega^i}(\cdot) \cdot \nabla\tilde{u}^o(0) + \tilde{\lambda}\tilde{u}^o(0) \in C^{0,\alpha}(\partial\Omega^i)$, Lemma 9.13 (ii) implies the existence of a unique solution $\tilde{u}^i \in C^{1,\alpha}_{\mathrm{loc}}(\overline{\Omega^{i-}})$ of the first three equations in (9.62). Hence, statement (i) is proven.

Statement (ii) is an immediate consequence of statement (i) and of Corollary 9.17 (ii) with $f \equiv \tilde{u}_{|\partial\Omega^o}$.

By the continuity of the pointwise product in Schauder spaces, by Theorem 6.4 on the continuity of the single layer and by Theorem 6.8 (ii) on the continuity of the transpose to the double layer, and by the analyticity of nonlinear integral operators with analytic kernel of Theorem A.28 (ii), we conclude that the map $\tilde{\mathcal{Q}}$ is real analytic and the proof of statement (iii) is complete. □

Then we can prove the following.

Proposition 9.19. *Let* $\tilde{\lambda} \in [0,+\infty[$ *be a Steklov eigenvalue of* Δ *in* Ω^o *of multiplicity one (see problem (9.30)). Let* $\tilde{u} \in C^{1,\alpha}(\overline{\Omega^o})$ *be one of the two solutions of problem (9.63) (cf. Lemma 9.14 (iii)). Let* $\tilde{\mathcal{Q}}$ *denote the map* $\mathcal{Q}_f$ *of (9.58) with* $f \equiv \tilde{u}_{|\partial\Omega^o}$.

Let $(\tilde{\theta}^i, \tilde{\theta}^o, \tilde{c}) \in X_{0,\alpha} \times \mathbb{R}$ *be as in Proposition 9.18 (ii). Then the Fréchet differential* $\partial_{(\theta^i,\theta^o,c,\lambda)}\tilde{\mathcal{Q}}[0,0,\tilde{\theta}^i,\tilde{\theta}^o,\tilde{c},\tilde{\lambda}]$ *of* $\tilde{\mathcal{Q}}$ *at* $(0,0,\tilde{\theta}^i,\tilde{\theta}^o,\tilde{c},\tilde{\lambda})$ *with respect to the variable* $(\theta^i,\theta^o,c,\lambda)$ *is a linear homeomorphism from* $X_{0,\alpha} \times \mathbb{R}^2$ *onto itself.*

Proof. By standard Calculus in Banach space, the differential of $\tilde{\mathcal{Q}}$ at the point $(0,0,\tilde{\theta}^i,\tilde{\theta}^o,\tilde{c},\tilde{\lambda})$ with respect to the variables $(\theta^i,\theta^o,c,\lambda)$ is delivered by the following formula.

$$\begin{aligned}
&\partial_{(\theta^i,\theta^o,c,\lambda)}\tilde{\mathcal{Q}}_1[0,0,\tilde{\theta}^i,\tilde{\theta}^o,\tilde{c},\tilde{\lambda}](\theta^i,\theta^o,c,\lambda)(\xi) \\
&\quad = -\frac{1}{2}\theta^i(\xi) - W^t_{\Omega^i}[\theta^i](\xi) \\
&\quad - \int_{\partial\Omega^o} \nu_{\Omega^i}(\xi)\nabla S_n(-y)\theta^o(y)\, d\sigma_y - \tilde{\lambda}\Big\{\int_{\partial\Omega^o} S_n(-y)\theta^o(y)\, d\sigma_y + c\Big\}
\end{aligned} \tag{9.66}$$

$$-\lambda\Big\{\int_{\partial\Omega^o} S_n(-y)\tilde{\theta}^o(y)\,d\sigma_y+\tilde{c}\Big\}\qquad \forall\xi\in\partial\Omega^i\,,$$

$$\partial_{(\theta^i,\theta^o,c,\lambda)}\tilde{\mathcal{Q}}_2[0,0,\tilde{\theta}^i,\tilde{\theta}^o,\tilde{c},\tilde{\lambda}](\theta^i,\theta^o,c,\lambda)(x)$$
$$=-\frac{1}{2}\theta^o(x)+W^t_{\Omega^o}[\theta^o](x)-\tilde{\lambda}\Big\{V_{\Omega^o}[\theta^o](x)+c\Big\}$$
$$-\lambda\Big\{V_{\Omega^o}[\tilde{\theta}^o](x)+\tilde{c}\Big\}\qquad \forall x\in\partial\Omega^o\,,$$

$$\partial_{(\theta^i,\theta^o,c,\lambda)}\tilde{\mathcal{Q}}_3[0,0,\tilde{\theta}^i,\tilde{\theta}^o,\tilde{c},\tilde{\lambda}](\theta^i,\theta^o,c,\lambda)=\int_{\partial\Omega^o}\theta^o\,d\sigma\,,$$

$$\partial_{(\theta^i,\theta^o,c,\lambda)}\tilde{\mathcal{Q}}_4[0,0,\tilde{\theta}^i,\tilde{\theta}^o,\tilde{c},\tilde{\lambda}](\theta^i,\theta^o,c,\lambda)=\int_{\partial\Omega^o}\Big\{V_{\Omega^o}[\theta^o](x)+c\Big\}\tilde{u}(x)\,d\sigma_x$$

for all $(\theta^i,\theta^o,c,\lambda)\in X_{0,\alpha}\times\mathbb{R}^2$. By the Open Mapping Theorem 2.2, it suffices to prove that $\partial_{(\theta^i,\theta^o,c,\lambda)}\tilde{\mathcal{Q}}[0,0,\tilde{\theta}^i,\tilde{\theta}^o,\tilde{c},\tilde{\lambda}]$ is a bijection. Let $(f^i,f^o,\gamma_1,\gamma_2)$ be in $X_{0,\alpha}\times\mathbb{R}^2$. We now show that there exists a unique $(\theta^i,\theta^o,c,\lambda)$ in $X_{0,\alpha}\times\mathbb{R}^2$ such that

$$\partial_{(\theta^i,\theta^o,c,\lambda)}\tilde{\mathcal{Q}}[0,0,\tilde{\theta}^i,\tilde{\theta}^o,\tilde{c},\tilde{\lambda}](\theta^i,\theta^o,c,\lambda)=(f^i,f^o,\gamma_1,\gamma_2)\,. \tag{9.67}$$

Now by the jump formula for the normal derivative of a single layer potential, we can rewrite the second and the third equation of (9.67) in the form

$$\begin{cases}\frac{\partial}{\partial\nu_{\Omega^o}}(v^+_{\Omega^o}[\theta^o]+c)-\tilde{\lambda}(v^+_{\Omega^o}[\theta^o]+c)=f^o+\lambda(v^+_{\Omega^o}[\tilde{\theta}^o]+\tilde{c})\text{ on }\partial\Omega^o\,,\\ \int_{\partial\Omega^o}\theta^o\,d\sigma=\gamma_1\,.\end{cases} \tag{9.68}$$

By exploiting the notation of (9.51) with $\Omega=\Omega^o$ and the equality $\tilde{u}=v^+_{\Omega^o}[\tilde{\theta}^o]+\tilde{c}$, we can rewrite (9.68) in the form

$$L_{\tilde{\lambda}}[\theta^o,c]=(f^o+\lambda\tilde{u}_{|\partial\Omega^o},\gamma_1)\,.$$

By the Fredholm Alternative Theorem 5.8, such an equation has a solution (θ^{o*},c^*) in $C^{0,\alpha}(\partial\Omega^o)\times\mathbb{R}$ exactly when the datum $(f^o+\lambda\tilde{u}_{|\partial\Omega^o},\gamma_1)$ is orthogonal to $\mathrm{Ker}\,G_{\tilde{\lambda}}$ with respect to the bilinear form of (9.50). Indeed, $L_{\tilde{\lambda}}$ is the transpose to $G_{\tilde{\lambda}}$ (see (9.52)). Since $\tilde{u}=v^+_{\Omega^o}[\tilde{\theta}^o]+\tilde{c}$ is a Steklov eigenfunction in Ω^o which corresponds to the eigenvalue $\tilde{\lambda}$, Proposition 9.12 with $\Omega=\Omega^o$ implies that $(\tilde{\theta}^o,\tilde{c})$ belongs to $\mathrm{Ker}\,L_{\tilde{\lambda}}$. Since $\tilde{\lambda}$ is simple, Proposition 9.12 with $\Omega=\Omega^o$ implies that $\mathrm{Ker}\,L_{\tilde{\lambda}}$ is one dimensional. Since $\mathrm{Ker}\,L_{\tilde{\lambda}}$ is one dimensional and $(\tilde{\theta}^o,\tilde{c})\neq(0,0)$, the pair $(\tilde{\theta}^o,\tilde{c})$ generates $\mathrm{Ker}\,L_{\tilde{\lambda}}$. Then Lemma 9.16 implies that $\Phi[\tilde{\theta}^o,\tilde{c}]=(v^+_{\Omega^o}[\tilde{\theta}^o]_{|\partial\Omega^o}+\tilde{c},\int_{\partial\Omega^o}\tilde{\theta}^o\,d\sigma)=(\tilde{u}_{|\partial\Omega^o},0)$ generates $\mathrm{Ker}\,G_{\tilde{\lambda}}$. Thus (9.68) has a solution $(\theta^{o*},c^*)\in C^{0,\alpha}(\partial\Omega^o)\times\mathbb{R}$ exactly when

$$\int_{\partial\Omega^o}f^o\tilde{u}\,d\sigma+\lambda\int_{\partial\Omega^o}\tilde{u}^2\,d\sigma+\gamma_1 0=0\,. \tag{9.69}$$

Condition (9.69) is equivalent to

$$\lambda = -\frac{\int_{\partial\Omega^o} f^o \tilde{u}\, d\sigma}{\int_{\partial\Omega^o} \tilde{u}^2\, d\sigma} = -\int_{\partial\Omega^o} f^o \tilde{u}\, d\sigma \tag{9.70}$$

(see (9.63)). With such a choice of λ, system (9.68) has a solution (θ^{o*}, c^*) in $C^{0,\alpha}(\partial\Omega^o) \times \mathbb{R}$. Then the affine space of solutions in $C^{0,\alpha}(\partial\Omega^o) \times \mathbb{R}$ of system (9.68) coincides with

$$\left\{(\theta^{o*}, c^*) + \varsigma(\tilde{\theta}^o, \tilde{c}) : \varsigma \in \mathbb{R}\right\}.$$

Now we show that the fourth equation of (9.67) determines uniquely ς. The fourth equation of (9.67) with

$$(\theta^o, c) = (\theta^{o*}, c^*) + \varsigma(\tilde{\theta}^o, \tilde{c}), \tag{9.71}$$

can be rewritten as

$$\int_{\partial\Omega^o} \{v^+_{\Omega^o}[\theta^{o*}] + c^*\}\, \tilde{u}\, d\sigma + \varsigma \int_{\partial\Omega^o} \left\{v^+_{\Omega^o}[\tilde{\theta}^o] + \tilde{c}\right\} \tilde{u}\, d\sigma = \gamma_2\,. \tag{9.72}$$

By (9.63), we have

$$\int_{\partial\Omega^o} \left\{v^+_{\Omega^o}[\tilde{\theta}^o] + \tilde{c}\right\} \tilde{u}\, d\sigma = \int_{\partial\Omega^o} \tilde{u}^2\, d\sigma = 1\,,$$

and accordingly we can solve uniquely (9.72) by taking

$$\varsigma \equiv \gamma_2 - \int_{\partial\Omega^o} \{v^+_{\Omega^o}[\theta^{o*}] + c^*\}\, \tilde{u}\, d\sigma\,. \tag{9.73}$$

Next we rewrite the first equation of (9.67) in the form

$$\begin{aligned}
&-\frac{1}{2}\theta^i(\xi) - W^t_{\Omega^i}[\theta^i](\xi) \qquad\qquad (9.74)\\
&= \int_{\partial\Omega^o} \nu_{\Omega^i}(\xi)\nabla S_n(-y)\theta^o(y)\, d\sigma_y + \tilde{\lambda}\left\{\int_{\partial\Omega^o} S_n(-y)\theta^o(y)\, d\sigma_y + c\right\}\\
&+\lambda\left\{\int_{\partial\Omega^o} S_n(-y)\tilde{\theta}^o(y)\, d\sigma_y + \tilde{c}\right\} + f^i(\xi) \qquad \forall \xi \in \partial\Omega^i\,,
\end{aligned}$$

with λ as in (9.70) and (θ^o, c) as in (9.71) with ς as in (9.73). Since Ω^{i-} is connected, and $\nu_{\Omega^i} \in C^{0,\alpha}(\partial\Omega^i, \mathbb{R}^n)$, $f^i \in C^{0,\alpha}(\partial\Omega^i)$, then Corollary 6.15 ensures that Equation (9.74) has a unique solution $\theta^i \in C^{0,\alpha}(\partial\Omega^i)$. Hence we have shown that if $(f^i, f^o, \gamma_1, \gamma_2) \in X_{0,\alpha} \times \mathbb{R}^2$, then there exists a uniquely determined $(\theta^i, \theta^o, c, \lambda)$ in $X_{0,\alpha} \times \mathbb{R}^2$ which solves equation (9.67). □

By Corollary 9.17, the analysis of the Steklov problem (9.31) with the normalizing condition (9.37) is reduced to that of equation

$$\tilde{\mathcal{Q}}[\epsilon, \delta_{2,n}\epsilon \log \epsilon, \theta^i, \theta^o, c, \lambda] = 0 \tag{9.75}$$

where $\tilde{\mathcal{Q}}$ denotes the operator $\mathcal{Q}_f$ of (9.58) with $f \equiv \tilde{u}_{|\partial\Omega^o}$. In order to understand the dependence of the solutions $(\theta^i, \theta^o, c, \lambda)$ of Equation (9.75) upon ϵ, we first consider the dependence of the solutions $(\theta^i, \theta^o, c, \lambda)$ of the equation

$$\tilde{\mathcal{Q}}[\epsilon, \epsilon_1, \theta^i, \theta^o, c, \lambda] = 0$$

upon the pair (ϵ, ϵ_1), and we prove the following.

Theorem 9.20. *Let $\tilde{\lambda} \in [0, +\infty[$ be a Steklov eigenvalue of Δ in Ω^o of multiplicity one (see problem (9.30)). Let $\tilde{u} \in C^{1,\alpha}(\overline{\Omega^o})$ be one of the two solutions of problem (9.63) (cf. Lemma 9.14 (iii)). Let $\tilde{\mathcal{Q}}$ denote the operator $\mathcal{Q}_f$ of (9.58) with $f \equiv \tilde{u}_{|\partial\Omega^o}$. Let $(\tilde{\theta}^i, \tilde{\theta}^o, \tilde{c}) \in X_{0,\alpha} \times \mathbb{R}$ be as in Proposition 9.18 (ii).*

Then there exist $\epsilon' \in]0, \epsilon_0[$, $\epsilon_1' \in]0, +\infty[$, $\varsigma \in]0, +\infty[$, an open neighborhood $\tilde{V}$ of $(\tilde{\theta}^i, \tilde{\theta}^o, \tilde{c})$ in the space $X_{0,\alpha} \times \mathbb{R}$, and a real analytic map $(\Theta^i, \Theta^o, C, \hat{\lambda})$ from $]-\epsilon', \epsilon'[\times]-\epsilon_1', \epsilon_1'[$ to $\tilde{V} \times]\tilde{\lambda} - \varsigma, \tilde{\lambda} + \varsigma[$ such that

$$\delta_{2,n}\epsilon \log \epsilon \in]-\epsilon_1', \epsilon_1'[\qquad \forall \epsilon \in]0, \epsilon'[, \tag{9.76}$$

$$]\tilde{\lambda} - 2\varsigma, \tilde{\lambda} + 2\varsigma[\setminus\{\tilde{\lambda}\} \text{ contains no Steklov eigenvalue of } \Delta \text{ in } \Omega^o, \tag{9.77}$$

and such that the set of zeros of $\tilde{\mathcal{Q}}$ in $]-\epsilon', \epsilon'[\times]-\epsilon_1', \epsilon_1'[\times \tilde{V} \times]\tilde{\lambda} - \varsigma, \tilde{\lambda} + \varsigma[$ coincides with the graph of $(\Theta^i, \Theta^o, C, \hat{\lambda})$. In particular,

$$\Theta^i[0,0] = \tilde{\theta}^i, \qquad \Theta^o[0,0] = \tilde{\theta}^o, \qquad C[0,0] = \tilde{c}, \qquad \hat{\lambda}[0,0] = \tilde{\lambda}. \tag{9.78}$$

Proof. We plan to apply the Implicit Function Theorem for real analytic maps in Banach spaces to equation $\tilde{\mathcal{Q}} = 0$ around the point $(0, 0, \tilde{\theta}^i, \tilde{\theta}^o, \tilde{c}, \tilde{\lambda})$ (cf. Theorem A.19).

By Proposition 9.18 (iii) and Proposition 9.19, the map $\tilde{\mathcal{Q}}$ is real analytic and the differential of $\tilde{\mathcal{Q}}$ at the point $(0, 0, \tilde{\theta}^i, \tilde{\theta}^o, \tilde{c}, \tilde{\lambda})$ with respect to the variable $(\theta^i, \theta^o, c, \lambda)$ is a linear homeomorphism from $X_{0,\alpha} \times \mathbb{R}^2$ onto itself. By our assumption on $\tilde{\theta}^i$, $\tilde{\theta}^o$, $\tilde{c}$, we have $\tilde{\mathcal{Q}}[0, 0, \tilde{\theta}^i, \tilde{\theta}^o, \tilde{c}, \tilde{\lambda}] = 0$ (cf. (9.64)).

Then the Implicit Function Theorem implies the existence of ϵ', ϵ_1', $\tilde{V}$, ς, $(\Theta^i, \Theta^o, C, \hat{\lambda})$ such that the set of zeros of $\tilde{\mathcal{Q}}$ in

$$]-\epsilon', \epsilon'[\times]-\epsilon_1', \epsilon_1'[\times \tilde{V} \times]\tilde{\lambda} - \varsigma, \tilde{\lambda} + \varsigma[$$

coincides with the graph of $(\Theta^i, \Theta^o, C, \hat{\lambda})$.

Since the Steklov eigenvalues of Δ in Ω^o are isolated, possibly shrinking ς, we can assume that (9.77) holds. Then possibly shrinking ϵ', ϵ_1', we can still assume that the set of zeros of $\tilde{\mathcal{Q}}$ in $]-\epsilon', \epsilon'[\times]-\epsilon_1', \epsilon_1'[\times \tilde{V} \times]\tilde{\lambda} - \varsigma, \tilde{\lambda} + \varsigma[$ coincides with the graph of $(\Theta^i, \Theta^o, C, \hat{\lambda})$. Finally, possibly shrinking ϵ' once more, we can also assume that (9.76) holds. □

9.4.3 Real Analytic Representations for the Simple Steklov Eigenvalues and Eigenfunctions

In this section, we exploit Theorem 9.20 in order to analyze the behavior of the simple eigenvalues and eigenfunctions of problem (9.31) for ϵ small enough. We first introduce the following.

Definition 9.21. Let the assumptions of Theorem 9.20 hold. If $\epsilon \in]0, \epsilon'[$, then we define the function $u[\epsilon]$ from $\overline{\Omega(\epsilon)}$ to $\mathbb{R}$ by the formula

$$u[\epsilon](x) \equiv \epsilon^{n-1} \int_{\partial\Omega^i} S_n(x - \epsilon\eta)\Theta^i[\epsilon, \delta_{2,n}\epsilon \log \epsilon](\eta)\, d\sigma_\eta \qquad (9.79)$$
$$+ \int_{\partial\Omega^o} S_n(x - y)\Theta^o[\epsilon, \delta_{2,n}\epsilon \log \epsilon](y)\, d\sigma_y + C[\epsilon, \delta_{2,n}\epsilon \log \epsilon]\,,$$

for all $x \in \overline{\Omega(\epsilon)}$.

Then we prove that there exists an ϵ-dependent family of simple Steklov eigenvalues in $\Omega(\epsilon)$ that satisfies the real analytic representation formula (9.33). We also see that a basis for the corresponding eigenspaces is given by the functions $u[\epsilon]$ of Definition 9.21.

Theorem 9.22. *Let $\tilde{\lambda}$ be a Steklov eigenvalue of multiplicity one for Δ in Ω^o. Let $\tilde{u} \in C^{1,\alpha}(\overline{\Omega^o})$ be one of the two solutions of problem (9.63). Then the following statements hold.*

(i) There exist $\epsilon^* \in]0, \epsilon_0[$ and $\epsilon_1^* \in]0, +\infty[$ and a real analytic function $\hat{\lambda}$ from $]-\epsilon^*, \epsilon^*[\times]-\epsilon_1^*, \epsilon_1^*[$ to $\mathbb{R}$ such that

$$\delta_{2,n}\epsilon \log \epsilon \in]-\epsilon_1^*, \epsilon_1^*[\qquad \forall \epsilon \in]0, \epsilon^*[\qquad (9.80)$$

and such that $\hat{\lambda}[\epsilon, \delta_{2,n}\epsilon \log \epsilon]$ is a simple Steklov eigenvalue of Δ in $\Omega(\epsilon)$ for all $\epsilon \in]0, \epsilon^*[$. Moreover, $\hat{\lambda}[0, 0] = \tilde{\lambda}$.

(ii) If $\epsilon \in]0, \epsilon^*[$, then the function $u[\epsilon]$ of (9.79) is the unique eigenfunction of the Steklov problem (9.31) corresponding to the eigenvalue $\hat{\lambda}[\epsilon, \delta_{2,n}\epsilon \log \epsilon]$ that satisfies the normalizing condition (9.37).

Proof. Let ϵ', ϵ_1', Θ^i, Θ^o, C, $\hat{\lambda}$, $\tilde{V}$, ς be as in Theorem 9.20. Let $\epsilon^* \equiv \epsilon'$, $\epsilon_1^* \equiv \epsilon_1'$. Then the set of zeros of $\tilde{Q}$ in $]-\epsilon^*, \epsilon^*[\times]-\epsilon_1^*, \epsilon_1^*[\times\tilde{V}\times]\tilde{\lambda} - \varsigma, \tilde{\lambda} + \varsigma[$ coincides with the graph of $(\Theta^i, \Theta^o, C, \hat{\lambda})$. If $\epsilon \in]0, \epsilon^*[$, then $\delta_{2,n}\epsilon \log \epsilon \in]-\epsilon_1^*, \epsilon_1^*[$ and

$$(\epsilon, \delta_{2,n}\epsilon \log \epsilon, \Theta^i[\epsilon, \delta_{2,n}\epsilon \log \epsilon], \Theta^o[\epsilon, \delta_{2,n}\epsilon \log \epsilon], C[\epsilon, \delta_{2,n}\epsilon \log \epsilon])$$

is a zero of $\tilde{Q}$ and solves Equation (9.59) with $f \equiv \tilde{u}_{|\partial\Omega^o}$. Then Corollary 9.17 (i) with $f \equiv \tilde{u}_{|\partial\Omega^o}$, implies that the function $u[\epsilon]$ defined in (9.79) solves the problem (9.60) with $\lambda = \hat{\lambda}[\epsilon, \delta_{2,n}\epsilon \log \epsilon]$, i.e., $u[\epsilon]$ is a Steklov eigenfunction corresponding to the eigenvalue $\hat{\lambda}[\epsilon, \delta_{2,n}\epsilon \log \epsilon]$ and satisfies the normalizing condition (9.37).

To proceed further, we find convenient to introduce some abbreviations. Namely, we set

$$\Theta^i[\epsilon] \equiv \Theta^i[\epsilon, \delta_{2,n}\epsilon \log \epsilon] , \qquad \Theta^o[\epsilon] \equiv \Theta^o[\epsilon, \delta_{2,n}\epsilon \log \epsilon] ,$$
$$C[\epsilon] \equiv C[\epsilon, \delta_{2,n}\epsilon \log \epsilon] , \qquad \lambda[\epsilon] \equiv \hat{\lambda}[\epsilon, \delta_{2,n}\epsilon \log \epsilon] ,$$

for all $\epsilon \in]0, \epsilon^*[$.

Next we prove that $\lambda[\epsilon]$ is a simple Steklov eigenvalue of Δ in $\Omega(\epsilon)$ for all $\epsilon \in]0, \epsilon^*[$. Let $\epsilon \in]0, \epsilon^*[$. We assume that $v_\epsilon \in C^{1,\alpha}(\overline{\Omega(\epsilon)})$ is an arbitrarily fixed Steklov eigenfunction of Δ in $\Omega(\epsilon)$ corresponding to the eigenvalue $\lambda[\epsilon]$. Namely, we assume that v_ϵ satisfies problem (9.31) with $\lambda = \lambda[\epsilon]$. We now prove that v_ϵ is necessarily a multiple of the eigenfunction $u[\epsilon]$ of (9.79).

By Theorem 9.15 (i), there exists a unique $(\theta^i_\epsilon, \theta^o_\epsilon, c_\epsilon) \in X_{0,\alpha} \times \mathbb{R}$ such that $(\epsilon, \delta_{2,n}\epsilon \log \epsilon, \theta^i_\epsilon, \theta^o_\epsilon, c_\epsilon, \lambda[\epsilon])$ satisfies Equation (9.46) and such that

$$v_\epsilon = u_\epsilon[\theta^i_\epsilon, \theta^o_\epsilon, c_\epsilon, \lambda[\epsilon]] .$$

Since Equation (9.46) is linear in (θ^i, θ^o, c), the triple

$$(\theta^i_{\epsilon,\rho}, \theta^o_{\epsilon,\rho}, c_{\epsilon,\rho}) \equiv (\Theta^i[\epsilon], \Theta^o[\epsilon], C[\epsilon]) + \rho\left\{(\theta^i_\epsilon, \theta^o_\epsilon, c_\epsilon) - (\Theta^i[\epsilon], \Theta^o[\epsilon], C[\epsilon])\right\} \tag{9.81}$$

satisfies Equation (9.46) for all $\rho \in [0, 1]$. Since $(\Theta^i[\epsilon], \Theta^o[\epsilon], C[\epsilon])$ satisfies condition

$$\int_{\partial\Omega^o} u_\epsilon[\Theta^i[\epsilon], \Theta^o[\epsilon], C[\epsilon], \lambda[\epsilon]]\tilde{u}\, d\sigma = 1 ,$$

there exists $\rho^*_\epsilon \in]0, 1[$ such that

$$\int_{\partial\Omega^o} u_\epsilon[\theta^i_{\epsilon,\rho}, \theta^o_{\epsilon,\rho}, c_{\epsilon,\rho}, \lambda[\epsilon]]\tilde{u}\, d\sigma > 0 ,$$

for all $\rho \in]0, \rho^*_\epsilon]$. Then we set

$$(\theta^{*i}_{\epsilon,\rho}, \theta^{*o}_{\epsilon,\rho}, c^*_{\epsilon,\rho}) \equiv \frac{(\theta^i_{\epsilon,\rho}, \theta^o_{\epsilon,\rho}, c_{\epsilon,\rho})}{\int_{\partial\Omega^o} u_\epsilon[\theta^i_{\epsilon,\rho}, \theta^o_{\epsilon,\rho}, c_{\epsilon,\rho}, \lambda[\epsilon]]\tilde{u}\, d\sigma} \tag{9.82}$$

for all $\rho \in]0, \rho^*_\epsilon]$. Clearly

$$\int_{\partial\Omega^o} u_\epsilon[\theta^{*i}_{\epsilon,\rho}, \theta^{*o}_{\epsilon,\rho}, c^*_{\epsilon,\rho}, \lambda[\epsilon]]\tilde{u}\, d\sigma = 1 \qquad \forall \rho \in]0, \rho^*_\epsilon]$$

and accordingly

$$\tilde{\mathcal{Q}}[\epsilon, \delta_{2,n}\epsilon \log \epsilon, \theta^{*i}_{\epsilon,\rho}, \theta^{*o}_{\epsilon,\rho}, c^*_{\epsilon,\rho}, \lambda[\epsilon]] = 0 \qquad \forall \rho \in]0, \rho^*_\epsilon] .$$

Since

$$\lim_{\rho \to 0}(\theta^{*i}_{\epsilon,\rho}, \theta^{*o}_{\epsilon,\rho}, c^*_{\epsilon,\rho}) = (\Theta^i[\epsilon], \Theta^o[\epsilon], C[\epsilon]) \qquad \text{in } X_{0,\alpha} \times \mathbb{R} ,$$

there exists $\rho_\epsilon \in]0, \rho_\epsilon^*]$ such that

$$(\theta^{*i}_{\epsilon,\rho}, \theta^{*o}_{\epsilon,\rho}, c^*_{\epsilon,\rho}) \in \tilde{V} \qquad \forall \rho \in]0, \rho_\epsilon] .$$

Then Theorem 9.20 implies that

$$(\theta^{*i}_{\epsilon,\rho}, \theta^{*o}_{\epsilon,\rho}, c^*_{\epsilon,\rho}) = (\Theta^i[\epsilon], \Theta^o[\epsilon], C[\epsilon]) \qquad \forall \rho \in]0, \rho_\epsilon] .$$

Then for any such ρ equalities (9.81) and (9.82) imply that $(\theta^i_\epsilon, \theta^o_\epsilon, c_\epsilon)$ is a multiple of $(\Theta^i[\epsilon], \Theta^o[\epsilon], C[\epsilon])$. Hence, v_ϵ is a constant multiple of $u[\epsilon]$ and the eigenspace corresponding to the eigenvalue $\lambda[\epsilon]$ has dimension one. □

If the index of the eigenvalue $\tilde{\lambda}$ of Theorem 9.22 is l, i.e., if $\tilde{\lambda} = \lambda_l[\Omega^o]$, then one can exploit the continuity result (9.32) of Nazarov [237], the limiting relation $\lim_{\epsilon\to 0} \hat{\lambda}[\epsilon, \delta_{2,n}\epsilon \log \epsilon] = \tilde{\lambda} = \lambda_l[\Omega^o]$, and (9.77) to show that, possibly shrinking ϵ^*, we have $\hat{\lambda}[\epsilon, \delta_{2,n}\epsilon \log \epsilon] = \lambda_l[\Omega(\epsilon)]$ for all $\epsilon \in]0, \epsilon^*[$.

We now consider the macroscopic behavior of the eigenfunction $u[\epsilon]$ away from boundary of the hole.

Theorem 9.23. *Let $\tilde{\lambda}$, $\tilde{u}$, ϵ^*, and ϵ_1^* be as in Theorem 9.22. Let Ω_M be an open subset of Ω^o such that $0 \notin \overline{\Omega_M}$. Then there exist $\epsilon'_M \in]0, \epsilon^*[$ and a real analytic map $\mathcal{U}_M$ from $]-\epsilon'_M, \epsilon'_M[\times]-\epsilon_1^*, \epsilon_1^*[$ to $C^{1,\alpha}(\overline{\Omega_M})$ such that*

$$\begin{aligned} \overline{\Omega_M} &\subseteq \overline{\Omega(\epsilon)}, \\ u[\epsilon]_{|\overline{\Omega_M}} &= \mathcal{U}_M[\epsilon, \delta_{2,n}\epsilon \log \epsilon], \end{aligned} \tag{9.83}$$

for all $\epsilon \in]0, \epsilon'_{\Omega_M}[$. Moreover, $\mathcal{U}_M[0,0] = \tilde{u}_{|\overline{\Omega_M}}$.

Proof. Let $\epsilon'_M \in]0, \epsilon^*[$ be such that $\overline{\Omega_M} \subseteq \overline{\Omega(\epsilon)}$ for all $\epsilon \in [-\epsilon'_M, \epsilon'_M]$. Then we set

$$\begin{aligned} \mathcal{U}_{\Omega(\epsilon^*_M)}[\epsilon, \epsilon_1](x) \equiv \epsilon^{n-1} &\int_{\partial\Omega^i} S_n(x - \epsilon\eta)\Theta^i[\epsilon, \epsilon_1](\eta)\, d\sigma_\eta \\ &+ \int_{\partial\Omega^o} S_n(x-y)\Theta^o[\epsilon, \epsilon_1](y)\, d\sigma_y + C[\epsilon, \epsilon_1] \qquad \forall x \in \overline{\Omega_M} \end{aligned} \tag{9.84}$$

for all $(\epsilon, \epsilon_1) \in]-\epsilon'_M, \epsilon'_M[\times]-\epsilon_1^*, \epsilon_1^*[$. Then the equality in (9.83) holds by definition of $\mathcal{U}_{\Omega(\epsilon^*_M)}$ (see also (9.79)). Also, Proposition 9.18 (ii) and equality (9.78) imply the validity of equality $\mathcal{U}_M[0,0] = \tilde{u}_{|\overline{\Omega_M}}$.

We now prove the analyticity of $\mathcal{U}_M$. By Theorem 4.25 (i), $v^+_{\Omega^o}[\cdot]$ is linear and continuous from $C^{0,\alpha}(\partial\Omega^o)$ to $C^{1,\alpha}(\overline{\Omega^o})$, and the restriction map from $C^{1,\alpha}(\overline{\Omega^o})$ to $C^{1,\alpha}(\overline{\Omega_M})$ is linear and continuous, and the function Θ^o is real analytic from the set $]-\epsilon'_M, \epsilon'_M[\times]-\epsilon_1^*, \epsilon_1^*[$ to $C^{0,\alpha}(\partial\Omega^o)$. Then we conclude that the second summand in the right-hand side of (9.84) defines a real analytic map from $]-\epsilon'_M, \epsilon'_M[\times]-\epsilon_1^*, \epsilon_1^*[$ to $C^{1,\alpha}(\overline{\Omega_M})$. We now turn to consider the first summand. By the analyticity of nonlinear integral operators with analytic kernel of Theorem A.28 (i), we deduce that the map from $]-\epsilon'_{\Omega_M}, \epsilon'_{\Omega_M}[\times L^1(\partial\Omega^i)$ to $C^{1,\alpha}(\overline{\Omega_M})$ that takes

(ϵ, f) to the function $\int_{\partial\Omega^i} S_n(x-\epsilon\eta)f(\eta)\,d\sigma_\eta$ of the variable $x \in \overline{\Omega_M}$ is real analytic. Since Θ^i is real analytic from $]-\epsilon'_M, \epsilon'_M[\times]-\epsilon_1^*, \epsilon_1^*[$ to $C^{0,\alpha}(\partial\Omega^i)$, and $C^{0,\alpha}(\partial\Omega^i)$ is continuously embedded into $L^1(\partial\Omega^i)$, we conclude that the function from $]-\epsilon'_M, \epsilon'_M[\times]-\epsilon_1^*, \epsilon_1^*[$ to $C^{1,\alpha}(\overline{\Omega_M})$ that takes (ϵ, ϵ_1) to the first summand in the right-hand side of (9.84) is real analytic. Since $C[\cdot,\cdot]$ is real analytic, the proof is complete. □

Next we turn to study the microscopic behavior of $u[\epsilon]$ close to the boundary of the hole, that is, the behavior of a suitable restriction of the rescaled function $u[\epsilon](\epsilon\cdot)$. We do so in the following theorem.

Theorem 9.24. *Let $\tilde{\lambda}$, $\tilde{u}$, ϵ^*, and ϵ_1^* be as in Theorem 9.22. Let Ω_m be a bounded open subset of $\mathbb{R}^n \setminus \overline{\Omega^i}$. Then there exist $\epsilon'_{m,1} \in]0, \epsilon^*[$ and a real analytic function $\mathcal{U}_m$ from $]-\epsilon'_{m,1}, \epsilon'_{m,1}[\times]-\epsilon_1^*, \epsilon_1^*[$ to $C^{1,\alpha}(\overline{\Omega_m})$ such that*

$$\begin{aligned} &\epsilon\overline{\Omega_m} \subseteq \Omega^o, \\ &u[\epsilon](\epsilon\xi) = \mathcal{U}_m[\epsilon, \delta_{2,n}\epsilon\log\epsilon](\xi) \qquad \forall \xi \in \overline{\Omega_m}, \end{aligned} \tag{9.85}$$

for all $\epsilon \in]0, \epsilon'_{m,1}[$. Moreover,

$$\mathcal{U}_m[0,0] = \tilde{u}(0). \tag{9.86}$$

Proof. Let $\epsilon'_{m,1} \in]0, \epsilon^*[$ be such that $\overline{\Omega_m} \subseteq \frac{1}{\epsilon}\Omega^o$ for all $\epsilon \in [-\epsilon'_{m,1}, \epsilon'_{m,1}] \setminus \{0\}$. Clearly, if $\epsilon \in]0, \epsilon'_{m,1}[$ and $\xi \in \overline{\Omega_m}$, then we have $\epsilon\xi \in \overline{\Omega(\epsilon)}$. Then the equality

$$S_n(\epsilon\xi) = \epsilon^{2-n}S_n(\xi) + \frac{\delta_{2,n}}{2\pi}\log\epsilon \qquad \forall \epsilon \in]0, +\infty[$$

and formula (9.79) imply that

$$\begin{aligned} u[\epsilon](\epsilon\xi) = {}& \epsilon\int_{\partial\Omega^i} S_n(\xi-\eta)\Theta^i[\epsilon, \delta_{2,n}\epsilon\log\epsilon](\eta)\,d\sigma_\eta \\ &+\delta_{2,n}(\epsilon\log\epsilon)\frac{1}{2\pi}\int_{\partial\Omega^i}\Theta^i[\epsilon, \delta_{2,n}\epsilon\log\epsilon](\eta)\,d\sigma_\eta \\ &+\int_{\partial\Omega^o} S_n(\epsilon\xi-y)\Theta^o[\epsilon, \delta_{2,n}\epsilon\log\epsilon](y)\,d\sigma_y + C[\epsilon, \delta_{2,n}\epsilon\log\epsilon] \end{aligned} \tag{9.87}$$

for all $\xi \in \overline{\Omega_m}$ and all $\epsilon \in]0, \epsilon'_{m,1}]$. Since Θ^i is analytic, the function $\mathcal{U}_{m,1}$ from $]-\epsilon^*, \epsilon^*[\times]-\epsilon_1^*, \epsilon_1^*[$ to $\mathbb{R}$ defined by

$$\mathcal{U}_1[\epsilon, \epsilon_1] \equiv \frac{1}{2\pi}\int_{\partial\Omega^i}\Theta^i[\epsilon, \epsilon_1]\,d\sigma \qquad \forall(\epsilon, \epsilon_1) \in]-\epsilon^*, \epsilon^*[\times]-\epsilon_1^*, \epsilon_1^*[\tag{9.88}$$

is real analytic. Then we set

$$\mathcal{U}_{m,2}[\epsilon, \epsilon_1](\xi) \equiv \epsilon\int_{\partial\Omega^i} S_n(\xi-\eta)\Theta^i[\epsilon, \epsilon_1](\eta)\,d\sigma_\eta \tag{9.89}$$

$$+\int_{\partial\Omega^o} S_n(\epsilon\xi - y)\Theta^o[\epsilon,\epsilon_1](y)\,d\sigma_y + C[\epsilon,\epsilon_1] \qquad \forall \xi \in \overline{\Omega_m}\,,$$

for all $(\epsilon,\epsilon_1) \in]-\epsilon'_{m,1},\epsilon'_{m,1}[\times]-\epsilon_1^*,\epsilon_1^*[$. The analyticity of the first summand in the right-hand side of (9.89) follows by the analyticity of Θ^i and by the linearity and continuity of $v^-_{\Omega^i}[\cdot]$ (see Theorem 4.25 (ii)).

Next we consider the second summand in the right-hand side of (9.89). By the analyticity of nonlinear integral operators with analytic kernel of Theorem A.28 (i), we deduce that the map from $]-\epsilon'_{m,1},\epsilon'_{m,1}[\times L^1(\partial\Omega^o)$ to $C^{1,\alpha}(\overline{\Omega_m})$ that takes (ϵ, f) to the function $\int_{\partial\Omega^o} S_n(\epsilon\xi-y)f(y)\,d\sigma_y$ of the variable $\xi\in\overline{\Omega_m}$ is real analytic. Since Θ^o is real analytic from $]-\epsilon'_{m,1},\epsilon'_{m,1}[\times]-\epsilon_1^*,\epsilon_1^*[$ to $C^{0,\alpha}(\partial\Omega^o)$, and $C^{0,\alpha}(\partial\Omega^o)$ is continuously embedded into $L^1(\partial\Omega^o)$, we conclude that the function from $]-\epsilon'_{m,1},\epsilon'_{m,1}[\times]-\epsilon_1^*,\epsilon_1^*[$ to $C^{1,\alpha}(\overline{\Omega_m})$, that takes (ϵ,ϵ_1) to the second summand in the right-hand side of (9.89) is real analytic. Since C is analytic, we conclude that $\mathcal{U}_{m,2}$ is real analytic. Hence the map

$$\begin{aligned}&\mathcal{U}_m[\epsilon,\epsilon_1]\\ &\quad\equiv \epsilon_1\mathcal{U}_1[\epsilon,\epsilon_1] + \mathcal{U}_{m,2}[\epsilon,\epsilon_1] \quad \forall(\epsilon,\epsilon_1)\in]-\epsilon'_{m,1},\epsilon'_{m,1}[\times]-\epsilon_1^*,\epsilon_1^*[\end{aligned}$$

is real analytic from $]-\epsilon'_{m,1},\epsilon'_{m,1}[\times]-\epsilon_1^*,\epsilon_1^*[$ to $C^{1,\alpha}(\overline{\Omega_m})$ and satisfies the equality

$$u[\epsilon](\epsilon\xi) = \mathcal{U}_m[\epsilon,\delta_{2,n}\epsilon\log\epsilon](\xi) \qquad \forall\xi\in\overline{\Omega_m}\,,$$

for all $\epsilon\in]0,\epsilon'_{m,1}[$, i.e. $\mathcal{U}_m$ satisfies equality (9.85). Finally, equality (9.78) and equality $\tilde{u} = \tilde{u}^o[\tilde{\theta}^i,\tilde{\theta}^o,\tilde{c},\tilde{\lambda}] = v^+_{\Omega^o}[\tilde{\theta}^o] + \tilde{c}$ imply that

$$\mathcal{U}_m[0,0] = \tilde{u}(0)$$

(see Proposition 9.18 (ii)). Hence, (9.86) follows and the proof is complete. □

Finally, we turn to analyze the behavior of the energy integral of the eigenfunction $u[\epsilon]$ of (9.79).

Theorem 9.25. *Let $\tilde{\lambda}$, $\tilde{u}$, ϵ^* and ϵ_1^* be as in Theorem 9.22. Then there exist ϵ_e in $]0,\epsilon^*[$ and a real analytic function $\mathcal{E}$ from $]-\epsilon_e,\epsilon_e[\times]-\epsilon_1^*,\epsilon_1^*[$ to $\mathbb{R}$ such that*

$$\int_{\Omega(\epsilon)} |\nabla u[\epsilon](x)|^2\,dx = \mathcal{E}[\epsilon,\delta_{2,n}\epsilon\log\epsilon] \qquad \forall\epsilon\in]0,\epsilon_e[\,. \tag{9.90}$$

Moreover,

$$\mathcal{E}[0,0] = \int_{\Omega^o} |\nabla\tilde{u}|^2\,dx\,. \tag{9.91}$$

Proof. By the first Green Identity of Theorem 4.2, we have

$$\int_{\Omega(\epsilon)} |\nabla u[\epsilon](x)|^2\,dx = \int_{\partial\Omega(\epsilon)} u[\epsilon]\frac{\partial}{\partial\nu_{\Omega(\epsilon)}}u[\epsilon]\,d\sigma$$

$$
\begin{aligned}
&= \hat{\lambda}[\epsilon, \delta_{2,n}\epsilon \log \epsilon] \int_{\partial\Omega(\epsilon)} u[\epsilon]^2 \, d\sigma \\
&= \hat{\lambda}[\epsilon, \delta_{2,n}\epsilon \log \epsilon] \left\{ \int_{\epsilon\partial\Omega^i} u[\epsilon]^2 \, d\sigma + \int_{\partial\Omega^o} u[\epsilon]^2 \, d\sigma \right\} \\
&= \hat{\lambda}[\epsilon, \delta_{2,n}\epsilon \log \epsilon] \left\{ \epsilon^{n-1} \int_{\partial\Omega^i} u[\epsilon]^2(\epsilon\xi) \, d\sigma_\xi + \int_{\partial\Omega^o} u[\epsilon]^2 \, d\sigma \right\}.
\end{aligned}
$$

Then we take Ω_M as in Theorem 9.23 and Ω_m as in Theorem 9.24 such that

$$
\partial\Omega^o \subseteq \overline{\Omega_M}\,, \qquad \partial\Omega^i \subseteq \overline{\Omega_m}\,.
$$

If $\mathcal{U}_M$ and $\mathcal{U}_m$ are the corresponding real analytic maps, then we have

$$
\begin{aligned}
&\int_{\Omega(\epsilon)} |\nabla u[\epsilon]|^2 \, dx \\
&\qquad = \hat{\lambda}[\epsilon, \delta_{2,n}\epsilon \log \epsilon] \left\{ \epsilon^{n-1} \int_{\partial\Omega^i} \Big(\mathcal{U}_m[\epsilon, \delta_{2,n}\epsilon \log \epsilon](\xi) \Big)^2 d\sigma_\xi \right. \\
&\qquad\qquad \left. + \int_{\partial\Omega^o} \Big(\mathcal{U}_M[\epsilon, \delta_{2,n}\epsilon \log \epsilon] \Big)^2 d\sigma \right\}
\end{aligned}
\tag{9.92}
$$

for all $\epsilon \in]0, \min\{\epsilon'_M, \epsilon'_{m,1}\}[$. Then it suffices to set

$$
\epsilon_e \equiv \min\{\epsilon'_M, \epsilon'_{m,1}\}\,,
$$

and to define $\mathcal{E}$ as the function from $]-\epsilon_e, \epsilon_e[\times]-\epsilon_1^*, \epsilon_1^*[$ to $\mathbb{R}$ given by

$$
\begin{aligned}
\mathcal{E}[\epsilon, \epsilon_1] \equiv \hat{\lambda}[\epsilon, \epsilon_1] \Bigg\{ \epsilon^{n-1} \int_{\partial\Omega^i} \Big(\mathcal{U}_m[\epsilon, \epsilon_1](\xi) \Big)^2 d\sigma_\xi \\
+ \int_{\partial\Omega^o} \Big(\mathcal{U}_M[\epsilon, \epsilon_1] \Big)^2 d\sigma \Bigg\}
\end{aligned}
$$

for all $(\epsilon, \epsilon_1) \in]-\epsilon_e, \epsilon_e[\times]-\epsilon_1^*, \epsilon_1^*[$. Then Equation (9.90) holds by the definition of $\mathcal{E}$ and by equality (9.92). Moreover, equalities $\hat{\lambda}[0,0] = \tilde{\lambda}$, $\mathcal{U}_M[0,0] = \tilde{u}_{|\overline{\Omega_M}}$ and the first Green Identity of Theorem 4.2 imply that

$$
\mathcal{E}[0,0] = \tilde{\lambda} \int_{\partial\Omega^o} \tilde{u}^2 \, d\sigma = \int_{\partial\Omega^o} \tilde{u} \frac{\partial \tilde{u}}{\partial \nu_{\Omega^o}} \, d\sigma = \int_{\Omega^o} |\nabla \tilde{u}|^2 \, dx\,.
$$

Hence, equality (9.91) holds true and the theorem is proven. □

Chapter 10
A Dirichlet Problem in a Domain with Two Small Holes

Abstract In this chapter we see an application of the Functional Analytic Approach to a domain perturbation of a somewhat different nature: instead of a domain with a single shrinking hole we consider a domain that contains two holes that collide into one another while shrinking in size. We confine ourselves to a Dirichlet problem for the Laplace equation. Different boundary value problems can also be analyzed and some examples can be found in Dalla Riva and Musolino (Comm Partial Differ Equ 41(5):812–837, 2016; Math Methods Appl Sci 41(3):986–993, 2018). Yet as occurs for the single hole problem of Chapter 8, the Dirichlet problem presents certain interesting features related to the appearance of a logarithmic behavior in the two-dimensional case. For this reason we prefer to focus on this problem and investigate it thoroughly.

Further applications of the Functional Analytic Approach in problems with different domain perturbations can be found in the papers [29] and [30] with Bonnaillie-Noël and Dambrine, which concern the case of a small hole approaching the outer boundary of a domain, and in the paper [60] with Costabel and Dauge, which deals with perturbations of a plane sector domain. The mathematics of this chapter has been in part presented also in [79].

10.1 The Geometric Setting

We begin by introducing the geometric setting of our problem. We fix a real number $\alpha \in]0,1[$ and a natural number $n \geq 2$. Then we take three sets Ω^o, Ω_1, and Ω_2 that satisfy the following condition:

$$\begin{aligned}&\Omega^o,\ \Omega_1 \text{ and } \Omega_2 \text{ are open bounded connected subsets of } \mathbb{R}^n\\ &\text{of class } C^{1,\alpha}, \text{ they contain the origin } 0 \text{ of } \mathbb{R}^n \text{ and}\\ &\text{the exterior domains } \Omega^{o-},\ \Omega_1^-, \text{ and } \Omega_2^- \text{ are connected.}\end{aligned}$$

M. Dalla Riva et al., *Singularly Perturbed Boundary Value Problems*,
https://doi.org/10.1007/978-3-030-76259-9_10

As usual the letter 'o' stands for 'outer domain' and Ω^o will play the role of the unperturbed outer domain in which we make two holes. To do so, we take two points

$$p^1, p^2 \in \mathbb{R}^n, \quad p^1 \neq p^2$$

and we choose

$$\delta_2 > 0$$

such that

$$(p^1 + \epsilon_2 \overline{\Omega_1}) \cap (p^2 + \epsilon_2 \overline{\Omega_2}) = \emptyset \qquad \forall \epsilon_2 \in [-\delta_2, \delta_2]. \tag{10.1}$$

Then we define the rescaled sets

$$\Omega_1(\epsilon_1, \epsilon_2) \equiv \epsilon_1 p^1 + \epsilon_1\epsilon_2\Omega_1, \quad \Omega_2(\epsilon_1, \epsilon_2) \equiv \epsilon_1 p^2 + \epsilon_1\epsilon_2\Omega_2, \quad \forall \epsilon_1, \epsilon_2 \in \mathbb{R},$$

which will play the role of the holes. We observe that, for $\epsilon_1, \epsilon_2 \in \mathbb{R} \setminus \{0\}$ and $i \in \{1, 2\}$, each $\Omega_i(\epsilon_1, \epsilon_2)$ is an open bounded subset of $\mathbb{R}^2$ and contains the point $\epsilon_1 p^i$. Instead, when $\epsilon_1 = 0$ or $\epsilon_2 = 0$, $\Omega_i(\epsilon_1, \epsilon_2)$ collapses to a point and we have $\Omega_i(0, \epsilon_2) = \{0\}$ and $\Omega_i(\epsilon_1, 0) = \{\epsilon_1 p^i\}$. In addition, condition (10.1) implies that

$$\overline{\Omega_1(\epsilon_1, \epsilon_2)} \cap \overline{\Omega_2(\epsilon_1, \epsilon_2)} = \emptyset \qquad \forall \epsilon_1 \in \mathbb{R} \setminus \{0\}, \ \epsilon_2 \in [-\delta_2, \delta_2].$$

Then we note that the mutual distance between $\Omega_1(\epsilon_1, \epsilon_2)$ and $\Omega_2(\epsilon_1, \epsilon_2)$ is controlled by $|\epsilon_1|$, while their size is proportional to $|\epsilon_1\epsilon_2|$. As a consequence, when both ϵ_1 and ϵ_2 approach zero, the size tends to zero at a faster rate than the mutual distance. When this happens, we say that the holes are 'moderately close' (here we follow the terminology of Bonnaillie-Noël et al. [31]). We will also consider the case when the size and the distance are comparable, i.e. when ϵ_1 tends to zero and ϵ_2 stays away from zero.

Since we want the holes to be contained in Ω^o, we have to restrict the set of the 'admissible' parameters ϵ_1 for which we define the perforated domain. Then we take

$$\delta_1 > 0$$

such that

$$\overline{\Omega_1(\epsilon_1, \epsilon_2)} \cup \overline{\Omega_2(\epsilon_1, \epsilon_2)} \subseteq \Omega^o \qquad \forall (\epsilon_1, \epsilon_2) \in [-\delta_1, \delta_1] \times [-\delta_2, \delta_2] \tag{10.2}$$

and we regard as admissible parameters the pairs (ϵ_1, ϵ_2) in the rectangular domain $[-\delta_1, \delta_1] \times [-\delta_2, \delta_2]$. For these (ϵ_1, ϵ_2) we define the perforated domain

$$\Omega(\epsilon_1, \epsilon_2) \equiv \Omega^o \setminus \left(\overline{\Omega_1(\epsilon_1, \epsilon_2)} \cup \overline{\Omega_2(\epsilon_1, \epsilon_2)}\right)$$

(see Figure 10.1). We observe that for $\epsilon_1 \in [-\delta_1, \delta_1] \setminus \{0\}$ and $\epsilon_2 \in [-\delta_2, \delta_2] \setminus \{0\}$, the domain $\Omega(\epsilon_1, \epsilon_2)$ is an open bounded connected subset of $\mathbb{R}^n$ of class $C^{1,\alpha}$ and the boundary of $\Omega(\epsilon_1, \epsilon_2)$ consists of the disjoint union of $\partial\Omega^o$, $\partial\Omega_1(\epsilon_1, \epsilon_2)$, and $\partial\Omega_2(\epsilon_1, \epsilon_2)$ (cf. Section 2.13 for the definition of class $C^{1,\alpha}$, see also Theorem A.10 for a proof that $\Omega(\epsilon_1, \epsilon_2)$ is connected). For $\epsilon_1 = 0$ the set $\Omega(0, \epsilon_2)$ equals $\Omega^o \setminus \{0\}$

Fig. 10.1 *The perforated domain* $\Omega(\epsilon_1, \epsilon_2)$ $(n = 2)$

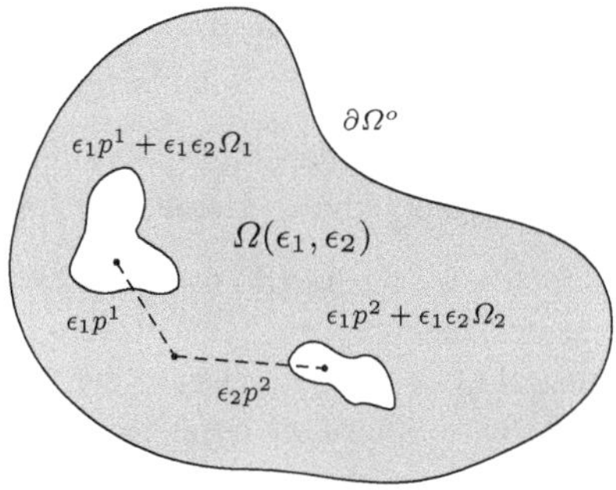

and for $\epsilon_2 = 0$ we have $\Omega(\epsilon_1, 0) = \Omega^o \setminus (\{\epsilon_1 p^1\} \cup \{\epsilon_1 p^2\})$. Thus $\Omega(0, \epsilon_2)$ and $\Omega(\epsilon_1, 0)$ are not local strict hypographs of class C^0 in the sense of Definition 2.37. We also find convenient to denote by $\Omega_\sharp(\epsilon_2)$ the set

$$\Omega_\sharp(\epsilon_2) \equiv \Omega_1(1, \epsilon_2) \cup \Omega_2(1, \epsilon_2) \qquad \forall \epsilon_2 \in [-\delta_2, \delta_2] \,. \tag{10.3}$$

We note that the set $\Omega_\sharp(\epsilon_2)$ is not necessarily a subset of Ω^o.

10.2 A Dirichlet Problem in $\Omega(\epsilon_1, \epsilon_2)$

We now introduce a Dirichlet boundary value problem in the domain $\Omega(\epsilon_1, \epsilon_2)$. To define the boundary conditions, we fix three functions

$$f^o \in C^{1,\alpha}(\partial\Omega^o)\,, \; f_1 \in C^{1,\alpha}(\partial\Omega_1)\,, \text{ and } f_2 \in C^{1,\alpha}(\partial\Omega_2)\,.$$

Then, for $\epsilon_1 \in]0, \delta_1[$ and $\epsilon_2 \in]0, \delta_2[$, we consider the Dirichlet boundary value problem

$$\begin{cases} \Delta u = 0 & \text{in } \Omega(\epsilon_1, \epsilon_2), \\ u = f^o & \text{on } \partial\Omega^o\,, \\ u(x) = f_1((x - \epsilon_1 p^1)/(\epsilon_1\epsilon_2)) & \forall x \in \partial\Omega_1(\epsilon_1, \epsilon_2)\,, \\ u(x) = f_2((x - \epsilon_1 p^2)/(\epsilon_1\epsilon_2)) & \forall x \in \partial\Omega_2(\epsilon_1, \epsilon_2) \end{cases} \tag{10.4}$$

for a function $u \in C^{1,\alpha}(\overline{\Omega(\epsilon_1, \epsilon_2)})$. The existence and uniqueness Theorem 6.27 implies that the solution of problem (10.4) exists and is unique (see also Theorem 3.6). Clearly, such solution depends on the pair (ϵ_1, ϵ_2) and for this reason we denote it by $u_{\epsilon_1,\epsilon_2}$. Then we can consider the map that takes (ϵ_1, ϵ_2) to (a suitable restriction of) $u_{\epsilon_1,\epsilon_2}$. We will analyze separately case $n \geq 3$ and $n = 2$. For $n \geq 3$ we will study both the cases of 'moderately close' holes, where the pair (ϵ_1, ϵ_2) approaches $(0, 0)$, and the case of 'close' holes, where (ϵ_1, ϵ_2) approaches a singular pair $(0, \tilde{\epsilon}_2)$ with $\tilde{\epsilon}_2 > 0$ (cf. Section 10.1 for the notion of 'close' and 'moderately close' holes). For $n = 2$ instead, we will focus on the case of moderately close holes. The reader can find the analysis of the two-dimensional problem with close holes in the paper [79]. Both for $n \geq 3$ and $n = 2$, our aim is to represent the map that takes (ϵ_1, ϵ_2) to a suitable restriction of $u_{\epsilon_1,\epsilon_2}$ in terms of two ingredients:

- Real analytic maps of (ϵ_1, ϵ_2) defined in a neighborhood of a singular pair.
- Completely known elementary functions of ϵ_1 and ϵ_2.

Similarly we wish to do for the maps that take (ϵ_1, ϵ_2) to suitable restrictions of the rescaled functions $u_{\epsilon_1,\epsilon_2}(\epsilon_1 p^1 + \epsilon_1\epsilon_2 \cdot)$ and $u_{\epsilon_1,\epsilon_2}(\epsilon_1 p^2 + \epsilon_1\epsilon_2 \cdot)$ and to the energy integral of $u_{\epsilon_1,\epsilon_2}$. We observe that here the functions $u_{\epsilon_1,\epsilon_2}(\epsilon_1 p^i + \epsilon_1\epsilon_2 \cdot)$ with $i \in \{1, 2\}$ take care of the behavior of $u_{\epsilon_1,\epsilon_2}$ close to the boundary of the holes. As we did in the previous chapters, we will use the capital subscript 'M' to indicate sets or maps that we use to describe the 'macroscopic' behavior of $u_{\epsilon_1,\epsilon_2}$ away from the holes and the small 'm' for analogous objects that concern the 'microscopic' behavior of $u_{\epsilon_1,\epsilon_2}$ near the boundary of the holes.

Both for $n \geq 3$ and $n = 2$ the analysis of this chapter follows the guidelines of the Functional Analytic Approach introduced in Chapter 8 for the Dirichlet problem in a domain with a single small hole, with some additional complications due to the additional hole and perturbation parameter. When $n \geq 3$, also the results that we obtain are similar to those of Section 8.3 for the corresponding problem with a single hole. For example, we show that a suitable restriction of $u_{\epsilon_1,\epsilon_2}$ can be continued in a neighborhood of $(0, 0)$ (or $(0, \tilde{\epsilon}_2)$ with $\tilde{\epsilon}_2 > 0$) by a real analytic function of the pair (ϵ_1, ϵ_2). More precisely, if Ω_M is an open subset of Ω^o with $0 \notin \overline{\Omega_M}$, then we prove in Theorem 10.7 that there exists a real analytic map $\mathcal{U}_M$ from a suitably small square domain $]-\delta_M, \delta_M[^2$ to $C^{1,\alpha}(\overline{\Omega_M})$ such that

$$u_{\epsilon_1,\epsilon_2|\overline{\Omega_M}} = \mathcal{U}_M[\epsilon_1, \epsilon_2] \qquad \forall (\epsilon_1, \epsilon_2) \in]0, \delta_M[^2 .$$

Then, for a possibly smaller $\delta_M > 0$, there exists a family $\{u_{(j,k)}\}_{(j,k)\in\mathbb{N}^2}$ of functions of $C^{1,\alpha}(\overline{\Omega_M})$ such that

$$u_{\epsilon_1,\epsilon_2|\overline{\Omega_M}} = \sum_{(j,k)\in\mathbb{N}^2} u_{(j,k)}\, \epsilon_1^j \epsilon_2^k \qquad \forall (\epsilon_1, \epsilon_2) \in]0, \delta_M[^2 ,$$

where the series converges normally in $C^{1,\alpha}(\overline{\Omega_M})$. We could also compute the coefficients $\{u_{(j,k)}\}_{(j,k)\in\mathbb{N}^2}$, either using the expansion methods of Asymptotic Analysis or adapting the technique based on the Functional Analytic Approach that we have introduced in Section 8.5. (We will not present such computation, which is however quite straightforward. The reader can find a similar computation in [30].) The case of dimension $n = 2$ is instead more involved: Not only we have to deal with a logarithmic behavior, which is expected for the Dirichlet problem in a planar perforated domain, but, due to the presence of two moderately close small holes, a new feature appears that was not present when dealing with a single hole. Specifically, to provide explicit expressions of the first coefficient of an expansion it is necessary to introduce some relation between the perturbation parameters ϵ_1 and ϵ_2, that is to say, between the size and the mutual distance of the holes (cf. Section 10.4).

10.3 Close and Moderately Close Holes in Dimension $n \geq 3$

As we have done for the Dirichlet problem in a domain with a single shrinking hole of Chapter 8, we split the analysis for $n \geq 3$ and $n = 2$. In particular, for $n \geq 3$ we can represent the solution in terms of a single layer potential. Indeed, Theorem 6.48 (ii) implies that there exists a density function $\theta \in C^{0,\alpha}(\partial\Omega(\epsilon_1, \epsilon_2))$ such that

$$v^+_{\Omega(\epsilon_1,\epsilon_2)}[\theta]$$

is the solutions of (10.4). Then we observe that $v^+_{\Omega(\epsilon_1,\epsilon_2)}[\theta]$ is the solution of (10.4) if and only if the following boundary conditions are satisfied:

$$\begin{aligned} v_{\Omega(\epsilon_1,\epsilon_2)}[\theta](x) &= f_1((x - \epsilon_1 p^1)/(\epsilon_1\epsilon_2)) && \text{for } x \in \partial\Omega_1(\epsilon_1, \epsilon_2)\,, \\ v_{\Omega(\epsilon_1,\epsilon_2)}[\theta](x) &= f_2((x - \epsilon_1 p^2)/(\epsilon_1\epsilon_2)) && \text{for } x \in \partial\Omega_2(\epsilon_1, \epsilon_2)\,, \\ v_{\Omega(\epsilon_1,\epsilon_2)}[\theta](x) &= f^o(x) && \text{for } x \in \partial\Omega^o\,. \end{aligned} \tag{10.5}$$

Our next step is to transform the equations on the (ϵ_1, ϵ_2)-dependent boundaries $\partial\Omega_1(\epsilon_1, \epsilon_2)$ and $\partial\Omega_2(\epsilon_1, \epsilon_2)$ into equations on the fixed boundaries $\partial\Omega_1$ and $\partial\Omega_2$. To do so, we introduce the rescaled densities

$$\begin{aligned} \theta_1(\xi) &\equiv \epsilon_1\epsilon_2\,\theta(\epsilon_1 p^1 + \epsilon_1\epsilon_2\xi) \quad \forall \xi \in \partial\Omega_1\,, \\ \theta_2(\xi) &\equiv \epsilon_1\epsilon_2\,\theta(\epsilon_1 p^2 + \epsilon_1\epsilon_2\xi) \quad \forall \xi \in \partial\Omega_2\,, \end{aligned} \tag{10.6}$$

and we set $\theta^o(x) = \theta(x)$ for all $x \in \partial\Omega^o$. We also observe that

$$S_n(\epsilon_1 p^1 + \epsilon_1\epsilon_2\xi - \epsilon_1 p^1 + \epsilon_1\epsilon_2\eta) = \epsilon_1^{2-n}\epsilon_2^{2-n} S_n(\xi - \eta)$$

for all $\epsilon_1, \epsilon_2 > 0$ and all $\xi, \eta \in \mathbb{R}^n$ with $\xi \neq \eta$ and

$$S_n(\epsilon_1 p^1 + \epsilon_1\epsilon_2\xi - \epsilon_1 p^2 + \epsilon_1\epsilon_2\eta) = \epsilon_1^{2-n} S_n(p^1 - p^2 + \epsilon_2(\xi - \eta))$$

for all $\epsilon_1, \epsilon_2 > 0$ and all $\xi, \eta \in \mathbb{R}^n$ such that $p^1 - p^2 + \epsilon_2(\xi - \eta) \neq 0$. Then, by Definition 4.18 of single layer potential and by the rule of change of variables in integrals, we obtain a new set of conditions, equivalent to (10.5) and defined on the fixed boundaries $\partial\Omega_1$, $\partial\Omega_2$, and $\partial\Omega^o$:

$$\begin{aligned} &\int_{\partial\Omega_1} S_n(\xi - \eta)\theta_1(\eta)\, d\sigma_\eta + \epsilon_2^{n-2} \int_{\partial\Omega_2} S_n\left(p^1 - p^2 + \epsilon_2(\xi - \eta)\right)\theta_2(\eta)\, d\sigma_\eta \\ &\qquad + \int_{\partial\Omega^o} S_n(\epsilon_1 p^1 + \epsilon_1\epsilon_2\xi - y)\theta^o(y)\, d\sigma_y = f_1(\xi) \qquad \forall \xi \in \partial\Omega_1\,, \end{aligned} \tag{10.7}$$

$$\int_{\partial\Omega_2} S_n(\xi - \eta)\theta_2(\eta)\, d\sigma_\eta + \epsilon_2^{n-2} \int_{\partial\Omega_1} S_n\left(p^2 - p^1 + \epsilon_2(\xi - \eta)\right)\theta_1(\eta)\, d\sigma_\eta$$

$$+ \int_{\partial\Omega^o} S_n(\epsilon_1 p^2 + \epsilon_1\epsilon_2\xi - y)\theta^o(y)\, d\sigma_y = f_2(\xi) \qquad \forall \xi \in \partial\Omega_2\,, \tag{10.8}$$

$$\begin{aligned} &\epsilon_1^{n-2}\epsilon_2^{n-2} \int_{\partial\Omega_1} S_n(x - \epsilon_1 p^1 + \epsilon_1\epsilon_2\eta)\theta_1(\eta)\, d\sigma_\eta \\ &\quad + \epsilon_1^{n-2}\epsilon_2^{n-2} \int_{\partial\Omega_2} S_n(x - \epsilon_1 p^2 + \epsilon_1\epsilon_2\eta)\theta_2(\eta)\, d\sigma_\eta \\ &\quad + \int_{\partial\Omega^o} S_n(x-y)\theta^o(y)\, d\sigma_y = f^o(x) \qquad \forall x \in \partial\Omega^o\,. \end{aligned} \tag{10.9}$$

In view of (10.7)–(10.9), we introduce for all $(\epsilon_1, \epsilon_2) \in]-\delta_1, \delta_1[\times]-\delta_2, \delta_2[$ the bounded linear operator

$$\mathcal{V}(\epsilon_1, \epsilon_2) \equiv (\mathcal{V}_1(\epsilon_1, \epsilon_2), \mathcal{V}_2(\epsilon_1, \epsilon_2), \mathcal{V}^o(\epsilon_1, \epsilon_2))$$

from $C^{0,\alpha}(\partial\Omega_1)\times C^{0,\alpha}(\partial\Omega_2) \times C^{0,\alpha}(\partial\Omega^o)$ to $C^{1,\alpha}(\partial\Omega_1)\times C^{1,\alpha}(\partial\Omega_2) \times C^{1,\alpha}(\partial\Omega^o)$ that takes a triple $(\theta_1, \theta_2, \theta^o)$ to

$$\begin{aligned} \mathcal{V}_1(\epsilon_1, \epsilon_2)[\theta_1, \theta_2, \theta^o](\xi) \equiv &\int_{\partial\Omega_1} S_n(\xi - \eta)\theta_1(\eta)\, d\sigma_\eta \\ &+ \epsilon_2^{n-2} \int_{\partial\Omega_2} S_n\left(p^1 - p^2 + \epsilon_2(\xi - \eta)\right)\theta_2(\eta)\, d\sigma_\eta \\ &+ \int_{\partial\Omega^o} S_n(\epsilon_1 p^1 + \epsilon_1\epsilon_2\xi - y)\theta^o(y)\, d\sigma_y \\ &\qquad \forall \xi \in \partial\Omega_1\,, \end{aligned} \tag{10.10}$$

$$\begin{aligned} \mathcal{V}_2(\epsilon_1, \epsilon_2)[\theta_1, \theta_2, \theta^o](\xi) \equiv &\int_{\partial\Omega_2} S_n(\xi - \eta)\theta_2(\eta)\, d\sigma_\eta \\ &+ \epsilon_2^{n-2} \int_{\partial\Omega_1} S_n\left(p^2 - p^1 + \epsilon_2(\xi - \eta)\right)\theta_1(\eta)\, d\sigma_\eta \\ &+ \int_{\partial\Omega^o} S_n(\epsilon_1 p^2 + \epsilon_1\epsilon_2\xi - y)\theta^o(y)\, d\sigma_y \\ &\qquad \forall \xi \in \partial\Omega_2\,, \end{aligned} \tag{10.11}$$

$$\begin{aligned} \mathcal{V}^o(\epsilon_1, \epsilon_2)[\theta_1, \theta_2, \theta^o](x) \equiv\ &\epsilon_1^{n-2}\epsilon_2^{n-2} \int_{\partial\Omega_1} S_n(x - \epsilon_1 p^1 + \epsilon_1\epsilon_2\eta)\theta_1(\eta)\, d\sigma_\eta \\ &+ \epsilon_1^{n-2}\epsilon_2^{n-2} \int_{\partial\Omega_2} S_n(x - \epsilon_1 p^2 + \epsilon_1\epsilon_2\eta)\theta_2(\eta)\, d\sigma_\eta \\ &+ \int_{\partial\Omega^o} S_n(x-y)\theta^o(y)\, d\sigma_y \qquad \forall x \in \partial\Omega^o\,. \end{aligned} \tag{10.12}$$

Then the three equations in (10.7)–(10.9) are equivalent to

$$\mathcal{V}(\epsilon_1, \epsilon_2)[\theta_1, \theta_2, \theta^o] = (f_1, f_2, f^o). \tag{10.13}$$

Accordingly, Equation (10.13) is equivalent to (10.5) via (10.6). Moreover, for $n \geq 3$ Theorem 6.46 implies that the single layer operator $V_{\Omega(\epsilon_1,\epsilon_2)}$ is an isomorphism (i.e., a linear homeomorphism) from $C^{0,\alpha}(\partial\Omega(\epsilon_1, \epsilon_2))$ to $C^{1,\alpha}(\partial\Omega(\epsilon_1, \epsilon_2))$. Then the system of equations in (10.5) have a unique solution $\theta \in C^{0,\alpha}(\partial\Omega(\epsilon_1, \epsilon_2))$ and we deduce the validity of the following.

Proposition 10.1. *Let* $\epsilon_1 \in]0, \delta_1[$ *and* $\epsilon_2 \in]0, \delta_2[$. *There exists one and only one triple of functions* $(\theta_1, \theta_2, \theta^o) \in C^{0,\alpha}(\partial\Omega_1) \times C^{0,\alpha}(\partial\Omega_2) \times C^{0,\alpha}(\partial\Omega^o)$ *such that* (10.13) *holds true. Moreover, the function* u *defined by*

$$\begin{aligned} u(x) \equiv \epsilon_1^{n-2}\epsilon_2^{n-2} &\int_{\partial\Omega_1} S_n(x - \epsilon_1 p^1 + \epsilon_1\epsilon_2\eta)\theta_1(\eta)\, d\sigma_\eta \\ &+ \epsilon_1^{n-2}\epsilon_2^{n-2} \int_{\partial\Omega_2} S_n(x - \epsilon_1 p^2 + \epsilon_1\epsilon_2\eta)\theta_2(\eta)\, d\sigma_\eta \\ &+ \int_{\partial\Omega^o} S_n(x - y)\theta^o(y)\, d\sigma_y \qquad \forall x \in \overline{\Omega(\epsilon_1, \epsilon_2)} \end{aligned} \tag{10.14}$$

coincides with the unique solution $u_{\epsilon_1,\epsilon_2}$ *in* $C^{1,\alpha}(\overline{\Omega(\epsilon_1, \epsilon_2)})$ *of the boundary value problem* (10.4).

10.3.1 Moderately Close Holes in Dimension $n \geq 3$

In this subsection we study the solution $u_{\epsilon_1,\epsilon_2}$ when both ϵ_1 and ϵ_2 approach 0. That is, in the case of moderately close holes (cf. Section 10.1).

By Proposition 10.1, we can consider the map from $]0, \delta_1[\times]0, \delta_2[$ to the space $C^{0,\alpha}(\partial\Omega_1) \times C^{0,\alpha}(\partial\Omega_2) \times C^{0,\alpha}(\partial\Omega^o)$ that takes (ϵ_1, ϵ_2) to the unique solution

$$(\theta_1(\epsilon_1, \epsilon_2), \theta_2(\epsilon_1, \epsilon_2), \theta^o(\epsilon_1, \epsilon_2)) \tag{10.15}$$

of (10.13). We will show that such map has a real analytic continuation in a open neighborhood of $(\epsilon_1, \epsilon_2) = (0, 0)$.

To do so, we first prove that the map that takes (ϵ_1, ϵ_2) to $\mathcal{V}(\epsilon_1, \epsilon_2)$ is real analytic and then we verify that for $(\epsilon_1, \epsilon_2) = (0, 0)$ the operator $\mathcal{V}(0, 0)$ is an isomorphism of Banach spaces. We deduce that $\mathcal{V}(\epsilon_1, \epsilon_2)$ is invertible in a neighborhood of $(0, 0)$ and that the map that takes (ϵ_1, ϵ_2) to its inverse $\mathcal{V}(\epsilon_1, \epsilon_2)^{(-1)}$ is real analytic. Then the real analytic continuation of the map

$$(\epsilon_1, \epsilon_2) \mapsto (\theta_1(\epsilon_1, \epsilon_2), \theta_2(\epsilon_1, \epsilon_2), \theta^o(\epsilon_1, \epsilon_2))$$

follows by the equality

$$(\theta_1(\epsilon_1, \epsilon_2), \theta_2(\epsilon_1, \epsilon_2), \theta^o(\epsilon_1, \epsilon_2)) = \mathcal{V}(\epsilon_1, \epsilon_2)^{(-1)}[f_1, f_2, f^o].$$

So, our first lemma states that the map $(\epsilon_1, \epsilon_2) \mapsto \mathcal{V}(\epsilon_1, \epsilon_2)$ is real analytic.

Lemma 10.2. *The map from* $]-\delta_1, \delta_1[\times]-\delta_2, \delta_2[$ *to*

$$\mathcal{L}\big(C^{0,\alpha}(\partial\Omega_1) \times C^{0,\alpha}(\partial\Omega_2) \times C^{0,\alpha}(\partial\Omega^o),\\ C^{1,\alpha}(\partial\Omega_1) \times C^{1,\alpha}(\partial\Omega_2) \times C^{1,\alpha}(\partial\Omega^o)\big)$$

that takes (ϵ_1, ϵ_2) *to* $\mathcal{V}(\epsilon_1, \epsilon_2)$ *is real analytic.*

Proof. The statement is a consequence of definition (10.10)–(10.12) of $\mathcal{V}$, of the properties of integral operators with real analytic kernel of Corollary A.29 of the Appendix, and of the mapping properties of the single layer potential of Theorem 6.4 (see also the Definition 4.18 of single layer potential). □

In the following Lemmas 10.3 we consider Equation (10.13) for $(\epsilon_1, \epsilon_2) = (0, 0)$.

Lemma 10.3. *For all* $(g_1, g_2, g^o) \in C^{1,\alpha}(\partial\Omega_1) \times C^{1,\alpha}(\partial\Omega_1) \times C^{1,\alpha}(\partial\Omega^o)$ *and all* $(\theta_1, \theta_2, \theta^o) \in C^{0,\alpha}(\partial\Omega_1) \times C^{0,\alpha}(\partial\Omega_1) \times C^{0,\alpha}(\partial\Omega^o)$ *we have*

$$\mathcal{V}(0,0)[\theta_1, \theta_2, \theta^o] = (g_1, g_2, g^o) \tag{10.16}$$

if and only if

$$\begin{cases} V_{\Omega_1}[\theta_1] + v_{\Omega^o}[\theta^o](0) = g_1\,, \\ V_{\Omega_2}[\theta_2] + v_{\Omega^o}[\theta^o](0) = g_2\,, \\ V_{\Omega^o}[\theta^o] = g^o\,. \end{cases} \tag{10.17}$$

Moreover, for all $(g_1, g_2, g^o) \in C^{1,\alpha}(\partial\Omega_1) \times C^{1,\alpha}(\partial\Omega_2) \times C^{1,\alpha}(\partial\Omega^o)$ *there exists one and only one triple* $(\theta_1, \theta_2, \theta^o) \in C^{0,\alpha}(\partial\Omega_1) \times C^{0,\alpha}(\partial\Omega_2) \times C^{0,\alpha}(\partial\Omega^o)$ *for which* (10.16) *(and, as a consequence,* (10.17)*) holds true.*

Proof. The equivalence of (10.16) and (10.17) follows by the definition of $\mathcal{V}$ in (10.10)–(10.12). To prove the last statement of the lemma we observe that Theorem 6.46 implies that the third equation in (10.17) has a unique solution $\theta^o \in C^{0,\alpha}(\partial\Omega^o)$. Then, again Theorem 6.46 implies that also the first and second equations in (10.17) have unique solutions $\theta_1 \in C^{0,\alpha}(\partial\Omega_1)$ and $\theta_2 \in C^{0,\alpha}(\partial\Omega_2)$, respectively. □

Then we verify that $\mathcal{V}(0, 0)$ is an isomorphism of Banach spaces.

Lemma 10.4. *The operator* $\mathcal{V}(0, 0)$ *is an isomorphism from* $C^{0,\alpha}(\partial\Omega_1) \times C^{0,\alpha}(\partial\Omega_2) \times C^{0,\alpha}(\partial\Omega^o)$ *to* $C^{1,\alpha}(\partial\Omega_1) \times C^{1,\alpha}(\partial\Omega_2) \times C^{1,\alpha}(\partial\Omega^o)$.

Proof. The previous Lemma 10.2 implies that $\mathcal{V}(0, 0)$ is linear and bounded. Then, by the Open Mapping Theorem 2.2, it is an a homeomorphism provided that it is bijective. Hence, we can conclude the proof by Lemma 10.3. □

Since the set of invertible operators is open in the space of bounded operators and since the map that takes an invertible operator to its inverse is real analytic, we now deduce the validity of following lemma (cf. Proposition A.15 in the Appendix).

Lemma 10.5. *There exists $0 < \delta_* < \min\{\delta_1, \delta_2\}$ such that $\mathcal{V}(\epsilon_1, \epsilon_2)$ is invertible for all $(\epsilon_1, \epsilon_2) \in]-\delta_*, \delta_*[^2$ and the map from $]-\delta_*, \delta_*[^2$ to*

$$\mathcal{L}\big(C^{1,\alpha}(\partial\Omega_1) \times C^{1,\alpha}(\partial\Omega_2) \times C^{1,\alpha}(\partial\Omega^o), \\ C^{0,\alpha}(\partial\Omega_1) \times C^{0,\alpha}(\partial\Omega_2) \times C^{0,\alpha}(\partial\Omega^o)\big)$$

that takes (ϵ_1, ϵ_2) to $\mathcal{V}(\epsilon_1, \epsilon_2)^{(-1)}$ is real analytic.

Proof. The statement follows by Lemmas 10.2 and 10.4. □

Then, by Lemma 10.5 can show that there exists a real analytic continuation of the map $(\epsilon_1, \epsilon_2) \mapsto (\theta_1(\epsilon_1, \epsilon_2), \theta_1(\epsilon_1, \epsilon_2), \theta^o(\epsilon_1, \epsilon_2))$.

Proposition 10.6. *There exists a real analytic map*

$$\Theta \equiv (\Theta_1, \Theta_2, \Theta^o) :]-\delta_*, \delta_*[^2 \to C^{0,\alpha}(\partial\Omega_1) \times C^{0,\alpha}(\partial\Omega_2) \times C^{0,\alpha}(\partial\Omega^o)$$

such that

$$\mathcal{V}(\epsilon_1, \epsilon_2)[\Theta_1[\epsilon_1, \epsilon_2], \Theta_2[\epsilon_1, \epsilon_2], \Theta^o[\epsilon_1, \epsilon_2]] = (f_1, f_2, f^o) \quad \forall (\epsilon_1, \epsilon_2) \in]-\delta_*, \delta_*[^2 .$$

In particular,

$$(\theta_1(\epsilon_1, \epsilon_2), \theta_1(\epsilon_1, \epsilon_2), \theta^o(\epsilon_1, \epsilon_2)) = \Theta[\epsilon_1, \epsilon_2] \qquad \forall (\epsilon_1, \epsilon_2) \in]0, \delta_*[^2 .$$

Proof. We set

$$\Theta[\epsilon_1, \epsilon_2] \equiv \mathcal{V}(\epsilon_1, \epsilon_2)^{(-1)}[f_1, f_2, f^o] \qquad \forall \epsilon \in]-\delta_*, \delta_*[^2$$

(with δ_* as in Lemma 10.5). Then the proposition follows by Proposition 10.1 and Lemma 10.5. Indeed, the map that takes an operator

$$A \in \mathcal{L}\big(C^{1,\alpha}(\partial\Omega_1) \times C^{1,\alpha}(\partial\Omega_2) \times C^{1,\alpha}(\partial\Omega^o), \\ C^{0,\alpha}(\partial\Omega_1) \times C^{0,\alpha}(\partial\Omega_2) \times C^{0,\alpha}(\partial\Omega^o)\big)$$

to $A[f_1, f_2, f^o] \in C^{0,\alpha}(\partial\Omega_1) \times C^{0,\alpha}(\partial\Omega_2) \times C^{0,\alpha}(\partial\Omega^o)$ is linear and continuous, and therefore real analytic. □

We can now exploit Proposition 10.6 to deduce some real analytic continuation results for the solution $u_{\epsilon_1,\epsilon_2}$ of problem (10.4). In the next Theorem 10.7 we will examine the 'macroscopic' behavior of $u_{\epsilon_1,\epsilon_2}$ away from the holes $\Omega_1(\epsilon_1, \epsilon_2)$ and $\Omega_2(\epsilon_1, \epsilon_2)$.

Theorem 10.7. *Let Ω_M be an open subset of Ω^o such that $0 \notin \overline{\Omega_M}$. Let $\delta_M \in]0, \delta_*[$ be such that*

$$\overline{\Omega_M} \cap \overline{\Omega_1(\epsilon_1, \epsilon_2)} = \emptyset \quad \text{and} \quad \overline{\Omega_M} \cap \overline{\Omega_2(\epsilon_1, \epsilon_2)} = \emptyset \quad \forall (\epsilon_1, \epsilon_2) \in]-\delta_M, \delta_M[^2.$$

Then, there exists a real analytic map

$$\mathcal{U}_M :]-\delta_M, \delta_M[^2 \to C^{1,\alpha}(\overline{\Omega_M})$$

such that

$$u_{\epsilon_1,\epsilon_2|\overline{\Omega_M}} = \mathcal{U}_M[\epsilon_1, \epsilon_2] \qquad \forall (\epsilon_1, \epsilon_2) \in]0, \delta_M[^2 . \tag{10.18}$$

Moreover,

$$\mathcal{U}_M[0,0] = u^o_{|\overline{\Omega_M}} \tag{10.19}$$

where $u^o \in C^{1,\alpha}(\overline{\Omega^o})$ *is the unique solution of the unperturbed problem*

$$\begin{cases} \Delta u^o = 0 & \text{in } \Omega^o , \\ u^o = f^o & \text{on } \partial\Omega^o . \end{cases} \tag{10.20}$$

Proof. In view of formula (10.14) for $u_{\epsilon_1,\epsilon_2}$, we set

$$\mathcal{U}_M[\epsilon_1, \epsilon_2](x) \equiv \epsilon_1^{n-2}\epsilon_2^{n-2} \int_{\partial\Omega_1} S_n(x - \epsilon_1 p^1 + \epsilon_1\epsilon_2\eta)\, \Theta_1[\epsilon_1, \epsilon_2](\eta)\, d\sigma_\eta$$
$$+ \epsilon_1^{n-2}\epsilon_2^{n-2} \int_{\partial\Omega_2} S_n(x - \epsilon_1 p^2 + \epsilon_1\epsilon_2\eta)\, \Theta_2[\epsilon_1, \epsilon_2](\eta)\, d\sigma_\eta + v_{\Omega^o}[\Theta^o[\epsilon_1, \epsilon_2]](x)$$

for all $x \in \overline{\Omega_M}$ and for all $(\epsilon_1, \epsilon_2) \in]-\delta_M, \delta_M[^2$. To show that

$$(\epsilon_1, \epsilon_2) \mapsto \mathcal{U}_M[\epsilon_1, \epsilon_2]$$

is real analytic we first observe that, by Theorem 4.25 (i) and Proposition 10.6, the map from $]-\delta_M, \delta_M[^2$ to $C^{1,\alpha}(\overline{\Omega_M})$ that takes (ϵ_1, ϵ_2) to $v_{\Omega^o}[\Theta^o[\epsilon_1, \epsilon_2]]_{|\overline{\Omega_M}}$ is real analytic. Thus, it remains to prove that the maps from $]-\delta_M, \delta_M[^2$ to $C^{1,\alpha}(\overline{\Omega_M})$ that take (ϵ_1, ϵ_2) to the functions

$$\epsilon_1^{n-2}\epsilon_2^{n-2} \int_{\partial\Omega_1} S_n(x - \epsilon_1 p^1 + \epsilon_1\epsilon_2\eta)\, \Theta_1[\epsilon_1, \epsilon_2](\eta)\, d\sigma_\eta \qquad \forall x \in \overline{\Omega_M}$$

and

$$\epsilon_1^{n-2}\epsilon_2^{n-2} \int_{\partial\Omega_2} S_n(x - \epsilon_1 p^2 + \epsilon_1\epsilon_2\eta)\, \Theta_2[\epsilon_1, \epsilon_2](\eta)\, d\sigma_\eta \qquad \forall x \in \overline{\Omega_M},$$

are also real analytic, a result that follows by the analyticity of Θ of Proposition 10.6 and by Theorem A.28 (i) on the analyticity of integral operators with a real analytic kernel.

Equality (10.18) is a consequence of Propositions 10.1 and 10.6. To prove equality (10.19), we observe that $\mathcal{U}_M[0,0](x) = v_{\Omega^o}[\Theta^o[0,0]](x)$ for all $x \in \overline{\Omega_M}$ and that $v_{\Omega^o}[\Theta^o[0,0]]$ is the only solution in $C^{1,\alpha}(\overline{\Omega^o})$ of the unperturbed problem (10.20). Indeed, Lemma 10.3 and Proposition 10.6 imply that

$$V_{\Omega^o}[\Theta^o[0,0]] = f^o .$$

□

We now analyze the behavior of $u_{\epsilon_1,\epsilon_2}$ near the boundary of the holes $\Omega_1(\epsilon_1, \epsilon_2)$ and $\Omega_2(\epsilon_1, \epsilon_2)$. We will avoid to write two very similar statements, one for $\Omega_1(\epsilon_1, \epsilon_2)$ and one for $\Omega_2(\epsilon_1, \epsilon_2)$. Instead, we will consider a hole $\Omega_i(\epsilon_1, \epsilon_2)$ with $i \in \{1, 2\}$. Then, for the rescaled function $u_{\epsilon_1,\epsilon_2}(\epsilon_1 p^i + \epsilon_1\epsilon_2\xi)$ of the variable $\xi \in [(\overline{\Omega^o} - \epsilon_1 p^i)/(\epsilon_1\epsilon_2)] \setminus \Omega_i$, we have the following.

Theorem 10.8. *Let $i \in \{1, 2\}$ be fixed. Let $\Omega_{i,m}$ be an open bounded subset of $\mathbb{R}^n \setminus \overline{\Omega_i}$. Let $\delta_m^{(i)} \in]0, \delta_*[$ be such that*

$$\epsilon_1 p^i + \epsilon_1\epsilon_2\overline{\Omega_{i,m}} \subseteq \Omega^o$$

and

$$(\epsilon_1 p^i + \epsilon_1\epsilon_2\overline{\Omega_{i,m}}) \cap \overline{\Omega_j(\epsilon_1, \epsilon_2)} = \emptyset$$

for all $(\epsilon_1, \epsilon_2) \in] - \delta_m^{(i)}, \delta_m^{(i)}[^2$, where $j = 1$ if $i = 2$ and $j = 2$ if $i = 1$. Then, there exists a real analytic map

$$\mathcal{U}_{i,m} :] - \delta_m^{(i)}, \delta_m^{(i)}[^2 \to C^{1,\alpha}(\overline{\Omega_{i,m}})$$

such that

$$u_{\epsilon_1,\epsilon_2}(\epsilon_1 p^i + \epsilon_1\epsilon_2\xi) = \mathcal{U}_{i,m}[\epsilon_1, \epsilon_2](\xi) \quad \forall \xi \in \overline{\Omega_{i,m}}\,, \ (\epsilon_1, \epsilon_2) \in]0, \delta_m^{(i)}[^2\,. \quad (10.21)$$

Moreover,

$$\mathcal{U}_{i,m}[0, 0] = v^i_{|\overline{\Omega_{i,m}}} \qquad (10.22)$$

where $v^i \in C^{1,\alpha}_{\mathrm{loc}}(\mathbb{R}^n \setminus \Omega_i)$ is the unique solution of

$$\begin{cases} \Delta v^i = 0 & \text{in } \mathbb{R}^n \setminus \overline{\Omega_i}\,, \\ v^i = f^i & \text{on } \partial\Omega_i\,, \\ \lim_{\xi\to\infty} v^i(\xi) = u^o(0) & \end{cases} \qquad (10.23)$$

and u^o is the unique solution of (10.20).

Proof. We take $j \equiv 1$ if $i = 2$ and $j \equiv 2$ if $i = 1$. Since $u_{\epsilon_1,\epsilon_2}$ is delivered by the right-hand side of (10.14), we find natural to set

$$\begin{aligned} \mathcal{U}_{i,m}[\epsilon_1, \epsilon_2](\xi) \equiv\ & v_{\Omega_i}[\Theta_i[\epsilon_1, \epsilon_2]](\xi) \\ & + \epsilon_2^{n-2} \int_{\partial\Omega_j} S_n\left(p^i - p^j + \epsilon_2(\xi - \eta)\right) \Theta_j[\epsilon_1, \epsilon_2](\eta)\, d\sigma_\eta \\ & + \int_{\partial\Omega^o} S_n\left(\epsilon_1 p^1 + \epsilon_1\epsilon_2\xi - y\right)\, \Theta^o[\epsilon_1, \epsilon_2](y)\, d\sigma_y \end{aligned}$$

for all $\xi \in \overline{\Omega_{i,m}}$ and $(\epsilon_1, \epsilon_2) \in] - \delta_m^{(i)}, \delta_m^{(i)}[^2$. To show that

$$(\epsilon_1, \epsilon_2) \mapsto \mathcal{U}_{i,m}[\epsilon_1, \epsilon_2]$$

is real analytic we first observe that, by Theorem 4.25 (ii) and Proposition 10.6, the map from $]-\delta_m^{(i)}, \delta_m^{(i)}[^2$ to $C^{1,\alpha}(\overline{\Omega_{i,m}})$ that takes (ϵ_1, ϵ_2) to $v_{\Omega_i}[\Theta_i[\epsilon_1, \epsilon_2]]$ is real analytic. Thus, it remains to prove that the map from $]-\delta_m^{(i)}, \delta_m^{(i)}[^2$ to $C^{1,\alpha}(\overline{\Omega_{i,m}})$ that takes (ϵ_1, ϵ_2) to the function

$$\epsilon_2^{n-2} \int_{\partial\Omega_j} S_n\left(p^i - p^j + \epsilon_2(\xi - \eta)\right) \Theta_j[\epsilon_1, \epsilon_2](\eta)\, d\sigma_\eta$$
$$+ \int_{\partial\Omega^o} S_n\left(\epsilon_1 p^1 + \epsilon_1\epsilon_2\xi - y\right) \Theta^o[\epsilon_1, \epsilon_2](y)\, d\sigma_y$$

of $\xi \in \overline{\Omega_{i,m}}$ is also real analytic, a fact that follows by the analyticity of Θ in Proposition 10.6 and by Theorem A.28 (i) for the integral operators with real analytic kernels. Equality (10.21) holds by the definition of $\mathcal{U}_{i,m}$. To prove equality (10.22), we observe that

$$\mathcal{U}_{i,m}[0,0](\xi) = v_{\Omega_i}[\Theta_i[0,0]](\xi) + v_{\Omega^o}[\Theta^o[0,0]](0) \qquad \forall \xi \in \overline{\Omega_{i,m}}$$

and that the function $v_{\Omega_i}[\Theta_i[0,0]](\cdot) + v_{\Omega^o}[\Theta^o[0,0]](0)$ is the only solution in $C^{1,\alpha}_{\text{loc}}(\mathbb{R}^n \setminus \Omega_i)$ of problem (10.23). Indeed, Lemma 10.3 and Proposition 10.6 imply that

$$V_{\Omega_i}[\Theta_i[0,0]] + v_{\Omega^o}[\Theta^o[0,0]](0) = f^i \qquad \text{on } \partial\Omega_i$$

and Theorem 4.23 (i) implies that $v_{\Omega_i}[\Theta_i[0,0]]$ is harmonic at infinity (see also Proposition 6.1 on the uniqueness of the solution of the exterior Dirichlet problem). □

Theorems 10.7 and 10.8 can be used to study functionals related to the solution $u_{\epsilon_1,\epsilon_2}$. For example, in the following Theorem 10.9 we consider the energy integral.

Theorem 10.9. *There exist* $\delta_E \in]0, \delta_*[$ *and a real analytic function*

$$\mathcal{E} :]-\delta_E, \delta_E[^2 \to \mathbb{R}$$

such that

$$\int_{\Omega(\epsilon)} |\nabla u_{\epsilon_1,\epsilon_2}|^2\, dx = \mathcal{E}(\epsilon_1, \epsilon_2) \qquad \forall (\epsilon_1, \epsilon_2) \in]0, \delta_E[^2 . \tag{10.24}$$

Moreover,

$$\mathcal{E}(0,0) = \int_{\Omega^o} |\nabla u^o|^2\, dx\,, \tag{10.25}$$

where u^o *is the unique solution of the unperturbed problem* (10.20).

Proof. By the first Green Identity of Theorem 4.2 and by the rule of change of variable in integrals we have

$$
\begin{aligned}
&\int_{\Omega(\epsilon_1,\epsilon_2)} |\nabla u_{\epsilon_1,\epsilon_2}|^2\, dx \\
&= \int_{\partial\Omega^o} u_{\epsilon_1,\epsilon_2}\, \nu_{\Omega^o} \cdot \nabla u_{\epsilon_1,\epsilon_2}\, d\sigma - \sum_{i=1,2} \int_{\partial\Omega_i(\epsilon_1,\epsilon_2)} u_{\epsilon_1,\epsilon_2}\, \nu_{\Omega_i(\epsilon_1,\epsilon_2)} \cdot \nabla u_{\epsilon_1,\epsilon_2}\, d\sigma \\
&= \int_{\partial\Omega^o} u_{\epsilon_1,\epsilon_2}\, \nu_{\Omega^o} \cdot \nabla u_{\epsilon_1,\epsilon_2}\, d\sigma \\
&-(\epsilon_1\epsilon_2)^{n-2} \sum_{i=1,2} \int_{\partial\Omega_i} u_{\epsilon_1,\epsilon_2}(\epsilon_1 p^i + \epsilon_1\epsilon_2\eta)\, \nu_{\Omega_i}(\eta) \cdot \nabla_\eta \left(u_{\epsilon_1,\epsilon_2}(\epsilon_1 p^i + \epsilon_1\epsilon_2\eta)\right) d\sigma_\eta
\end{aligned}
$$

for all $(\epsilon_1, \epsilon_2) \in]0, \delta_1[\times]0, \delta_2[$. Then, recalling that $u_{\epsilon_1,\epsilon_2}$ is a solution of the boundary value problem (10.4), we deduce that

$$
\begin{aligned}
\int_{\Omega(\epsilon_1,\epsilon_2)} |\nabla u_{\epsilon_1,\epsilon_2}|^2\, dx &= \int_{\partial\Omega^o} f^o\, \nu_{\Omega^o} \cdot \nabla u_{\epsilon_1,\epsilon_2}\, d\sigma \\
&- (\epsilon_1\epsilon_2)^{n-2} \sum_{i=1,2} \int_{\partial\Omega_i} f_i(\eta)\, \nu_{\Omega_i}(\eta) \cdot \nabla_\eta \left(u_{\epsilon_1,\epsilon_2}(\epsilon_1 p^i + \epsilon_1\epsilon_2\eta)\right) d\sigma_\eta
\end{aligned}
$$

for all $(\epsilon_1, \epsilon_2) \in]0, \delta_1[\times]0, \delta_2[$. We now take Ω_M as in Theorem 10.7 and $\Omega_{i,m}$ with $i \in \{1, 2\}$ as in Theorem 10.8. In addition, we assume that

$$
\partial\Omega^o \subseteq \overline{\Omega_M} \qquad \text{and} \qquad \partial\Omega_i \subseteq \overline{\Omega_{i,m}} \quad \text{for } i \in \{1, 2\}.
$$

By exploiting the corresponding maps $\mathcal{U}_M$ and $\mathcal{U}_{i,m}$, $i \in \{1, 2\}$, we can write

$$
\begin{aligned}
&\int_{\Omega(\epsilon_1,\epsilon_2)} |\nabla u_{\epsilon_1,\epsilon_2}|^2\, dx \\
&= \int_{\partial\Omega^o} f^o\, \nu_{\Omega^o} \cdot \nabla\mathcal{U}_M[\epsilon_1, \epsilon_2]\, d\sigma - (\epsilon_1\epsilon_2)^{n-2} \sum_{i=1,2} \int_{\partial\Omega_i} f_i\, \nu_{\Omega_i} \cdot \nabla\mathcal{U}_{i,m}[\epsilon_1, \epsilon_2]\, d\sigma
\end{aligned}
$$

for all $(\epsilon_1, \epsilon_2) \in]0, \delta_E[^2$, with $\delta_E \equiv \min\{\delta_M, \delta_m^{(1)}, \delta_m^{(2)}\}$. Then (10.24) holds with $\mathcal{E}$ defined by

$$
\begin{aligned}
&\mathcal{E}(\epsilon_1, \epsilon_2) \\
&\equiv \int_{\partial\Omega^o} f^o\, \nu_{\Omega^o} \cdot \nabla\mathcal{U}_M[\epsilon_1, \epsilon_2]\, d\sigma - (\epsilon_1\epsilon_2)^{n-2} \sum_{i=1,2} \int_{\partial\Omega_i} f_i\, \nu_{\Omega_i} \cdot \nabla\mathcal{U}_{i,m}[\epsilon_1, \epsilon_2]\, d\sigma
\end{aligned}
$$

for all $(\epsilon_1, \epsilon_2) \in]0, \delta_E[^2$. Moreover, equality (10.19) implies that

$$
\mathcal{E}(0, 0) = \int_{\partial\Omega^o} f^o\, \nu_{\Omega^o} \cdot \nabla u^o\, d\sigma
$$

and the validity of (10.25) follows by a computation based on the Divergence Theorem 4.1 and on the fact that u^o is the solution of the unperturbed problem (10.20).

To verify that $\mathcal{E}$ is real analytic we note that the map $\mathcal{U}_M$ is real analytic from $]-\delta_E, \delta_E[^2$ to $C^{1,\alpha}(\overline{\Omega_M})$ by Theorem 10.7. Since the map from $C^{1,\alpha}(\overline{\Omega_M})$ to $C^{0,\alpha}(\overline{\Omega_M})^n$ that takes a function to its gradient is linear and continuous, we deduce that the map $(\epsilon_1, \epsilon_2) \mapsto \nabla\mathcal{U}_M[\epsilon_1, \epsilon_2]$ is real analytic from $]-\delta_E, \delta_E[^2$ to $C^{0,\alpha}(\overline{\Omega_M})^n$. Similarly, Theorems 10.8 implies that $(\epsilon_1, \epsilon_2) \mapsto \nabla\mathcal{U}_{i,m}[\epsilon_1, \epsilon_2]$ is real analytic from $]-\delta_E, \delta_E[^2$ to $C^{0,\alpha}(\overline{\Omega_{i,m}})^n$ for all $i \in \{1,2\}$. Then, by the continuity of the linear map that takes a function of $C^{0,\alpha}(\partial\Omega^o)$ to its integral over $\partial\Omega^o$ and of the linear map that takes a function of $C^{0,\alpha}(\partial\Omega_i)$ to its integral over $\partial\Omega_i$, with $i \in \{1,2\}$, the map $\mathcal{E}$ is real analytic. □

10.3.2 Close Holes in Dimension $n \geq 3$

If ϵ_1 approaches 0 and ϵ_2 remains away from 0, then the holes shrink to one another while reducing their size at a comparable speed. In this case we say that the holes are 'close,' rather than 'moderately close.' To analyze the behavior of the solution $u_{\epsilon_1,\epsilon_2}$ under such assumption we consider the operator $\mathcal{V}(0, \tilde{\epsilon}_2)$ of (10.10)–(10.12) for a fixed $\tilde{\epsilon}_2$ in $]0, \delta_2[$. For the equation $\mathcal{V}(0, \tilde{\epsilon}_2)[\theta_1, \theta_2, \theta^o] = (g_1, g_2, g^o)$ we have the following.

Lemma 10.10. *If $\tilde{\epsilon}_2 \in]0, \delta_2[$ and (g_1, g_2, g^o) is a triple of $C^{1,\alpha}(\partial\Omega_1) \times C^{1,\alpha}(\partial\Omega_1) \times C^{1,\alpha}(\partial\Omega^o)$, then equation*

$$\mathcal{V}(0, \tilde{\epsilon}_2)[\theta_1, \theta_2, \theta^o] = (g_1, g_2, g^o). \tag{10.26}$$

is equivalent to

$$\begin{cases} V_{\Omega_\sharp(\tilde{\epsilon}_2)}[\theta_\sharp] + v_{\Omega^o}[\theta^o](0) = g_\sharp\,, \\ V_{\Omega^o}[\theta^o] = g^o\,, \end{cases} \tag{10.27}$$

with $\theta_\sharp \in C^{0,\alpha}(\partial\Omega_\sharp(\tilde{\epsilon}_2))$ and $g_\sharp \in C^{1,\alpha}(\partial\Omega_\sharp(\tilde{\epsilon}_2))$ given by

$$\theta_\sharp(x) \equiv \tilde{\epsilon}_2^{-1}\theta_i\Big(\frac{x-p^i}{\tilde{\epsilon}_2}\Big)\,,\ g_\sharp(x) \equiv g_i\Big(\frac{x-p^i}{\tilde{\epsilon}_2}\Big)\,, \quad \forall i \in \{1,2\}\,,\ x \in \partial\Omega_i(1, \tilde{\epsilon}_2)\,,$$

(see (10.3) for the definition of $\Omega_\sharp(\tilde{\epsilon}_2)$). Moreover, there exists one and only one triple $(\theta_1, \theta_2, \theta^o)$ in $C^{0,\alpha}(\partial\Omega_1) \times C^{0,\alpha}(\partial\Omega_2) \times C^{0,\alpha}(\partial\Omega^o)$ for which (10.26) (and (10.27)) holds true.

Proof. The equivalence of (10.26) and (10.27) follows by the definition of $\mathcal{V}$ in (10.10)–(10.12) and by a computation based on the rule of change of variable in integrals. To prove the last statement of the lemma we observe that, by Theorem 6.46, the third equation in (10.27) has a unique solution $\theta^o \in C^{0,\alpha}(\partial\Omega^o)$, then, again by Theorem 6.46, also the first equation in (10.27) has a unique solutions $\theta_\sharp \in C^{0,\alpha}(\partial\Omega_\sharp(\tilde{\epsilon}_2))$. □

Then we verify that the operator $\mathcal{V}(0, \tilde{\epsilon}_2)$ is an isomorphism of Banach spaces.

Lemma 10.11. *Let $\tilde{\epsilon}_2 \in]0, \delta_2[$. Then the operator $\mathcal{V}(0, \tilde{\epsilon}_2)$ is an isomorphism from $C^{0,\alpha}(\partial\Omega_1) \times C^{0,\alpha}(\partial\Omega_2) \times C^{0,\alpha}(\partial\Omega^o)$ to $C^{1,\alpha}(\partial\Omega_1) \times C^{1,\alpha}(\partial\Omega_2) \times C^{1,\alpha}(\partial\Omega^o)$.*

Proof. By Lemma 10.2 we know that $\mathcal{V}(0, \tilde{\epsilon}_2)$ is a linear bounded operator. Then, to prove that it is a homeomorphism, it suffices to show that it is a bijection and invoke the Open Mapping Theorem 2.2. Hence the statement follows by Lemma 10.10. □

At this point, we have the following.

Lemma 10.12. *Let $\tilde{\epsilon}_2 \in]0, \delta_2[$. Then there exists $0 < \delta_* < \min\{\delta_1, \delta_2 - \tilde{\epsilon}_2\}$ such that $\mathcal{V}(\epsilon_1, \epsilon_2)$ is invertible for all $(\epsilon_1, \epsilon_2) \in]-\delta_*, \delta_*[\times]\tilde{\epsilon}_2 - \delta_*, \tilde{\epsilon}_2 + \delta_*[$ and the map from $]-\delta_*, \delta_*[\times]\tilde{\epsilon}_2 - \delta_*, \tilde{\epsilon}_2 + \delta_*[$ to*

$$\mathcal{L}(C^{1,\alpha}(\partial\Omega_1) \times C^{1,\alpha}(\partial\Omega_2) \times C^{1,\alpha}(\partial\Omega^o), \\ C^{0,\alpha}(\partial\Omega_1) \times C^{0,\alpha}(\partial\Omega_2) \times C^{0,\alpha}(\partial\Omega^o))$$

that takes (ϵ_1, ϵ_2) to $\mathcal{V}(\epsilon_1, \epsilon_2)^{(-1)}$ is real analytic.

Proof. Since the set of invertible operators is open in the space of bounded operators and since the map that takes an invertible operator to its inverse is real analytic the lemma follows by Lemmas 10.2 and 10.11 (cf. Proposition A.15 in the Appendix). □

Then by Lemma 10.12 we deduce the following.

Proposition 10.13. *Let $\tilde{\epsilon}_2 \in]0, \delta_2[$. Let $0 < \delta_* < \min\{\delta_1, \delta_2 - \tilde{\epsilon}_2\}$ be as in Lemma 10.12. Then there exists a real analytic map $\Theta \equiv (\Theta_1, \Theta_2, \Theta^o)$ from the square domain*

$$]-\delta_*, \delta_*[\times]\tilde{\epsilon}_2 - \delta_*, \tilde{\epsilon}_2 + \delta_*[$$

to

$$C^{0,\alpha}(\partial\Omega_1) \times C^{0,\alpha}(\partial\Omega_2) \times C^{0,\alpha}(\partial\Omega^o)$$

such that

$$\mathcal{V}(\epsilon_1, \epsilon_2)[\Theta_1[\epsilon_1, \epsilon_2], \Theta_2[\epsilon_1, \epsilon_2], \Theta^o[\epsilon_1, \epsilon_2]] = (f_1, f_2, f^o)$$

for all $(\epsilon_1, \epsilon_2) \in]-\delta_*, \delta_*[\times]\tilde{\epsilon}_2 - \delta_*, \tilde{\epsilon}_2 + \delta_*[$. In particular,

$$(\theta_1(\epsilon_1, \epsilon_2), \theta_1(\epsilon_1, \epsilon_2), \theta^o(\epsilon_1, \epsilon_2)) = \Theta[\epsilon_1, \epsilon_2]$$

for all $\epsilon_1 \in]0, \delta_*[$ and $\epsilon_2 > 0$ in the interval $]\tilde{\epsilon}_2 - \delta_*, \tilde{\epsilon}_2 + \delta_*[$.

Proof. We take

$$\Theta[\epsilon_1, \epsilon_2] \equiv \mathcal{V}(\epsilon_1, \epsilon_2)^{(-1)}[f_1, f_2, f^o] \qquad \forall \epsilon \in]-\delta_*, \delta_*[\times]\tilde{\epsilon}_2 - \delta_*, \tilde{\epsilon}_2 + \delta_*[\,.$$

Then the proposition follows by Proposition 10.1 and Lemma 10.12 (see also the argument in the proof of Proposition 10.6). □

We observe that a priori the scalar δ_* and the map Θ do not necessarily coincide with those introduced in Proposition 10.6. We can, however, justify our notation by noting that we can choose the same δ_* in Propositions 10.6 and 10.13 and that the two corresponding maps Θ coincide in the intersection of their domains by virtue of the Identity Principle for real analytic functions of Proposition A.14.

We adopt a similar (and similarly justified) abuse of notation also in the following Theorems 10.14, 10.15, and 10.16, where we derive some consequences of Proposition 10.6 and of the integral representation (10.14). We begin with Theorem 10.14, where we study the 'macroscopic' behavior of $u_{\epsilon_1,\epsilon_2}$ away from the holes $\Omega_1(\epsilon_1,\epsilon_2)$ and $\Omega_2(\epsilon_1,\epsilon_2)$.

Theorem 10.14. *Let $\tilde{\epsilon}_2 \in]0,\delta_2[$. Let δ_* be as in Lemma 10.12. Let Ω_M be an open subset of Ω^o such that $0 \notin \overline{\Omega_M}$. Let $\delta_M \in]0,\delta_*[$ be such that*

$$\overline{\Omega_M} \cap \overline{\Omega_1(\epsilon_1,\epsilon_2)} = \emptyset \quad \text{and} \quad \overline{\Omega_M} \cap \overline{\Omega_2(\epsilon_1,\epsilon_2)} = \emptyset$$

for all $(\epsilon_1,\epsilon_2) \in]-\delta_M,\delta_M[\times]\tilde{\epsilon}_2-\delta_M,\tilde{\epsilon}_2+\delta_M[$. Then, there exists a real analytic map

$$\mathcal{U}_M :]-\delta_M,\delta_M[\times]\tilde{\epsilon}_2-\delta_M,\tilde{\epsilon}_2+\delta_M[\to C^{1,\alpha}(\overline{\Omega_M})$$

such that

$$u_{\epsilon_1,\epsilon_2|\overline{\Omega_M}} = \mathcal{U}_M[\epsilon_1,\epsilon_2] \tag{10.28}$$

for all $\epsilon_1 \in]0,\delta_M[$ and $\epsilon_2 > 0$ in the interval $]\tilde{\epsilon}_2-\delta_M,\tilde{\epsilon}_2+\delta_M[$. Moreover,

$$\mathcal{U}_M[0,\tilde{\epsilon}_2] = u^o_{|\overline{\Omega_M}} \tag{10.29}$$

where $u^o \in C^{1,\alpha}(\overline{\Omega^o})$ is the unique solution of the unperturbed problem (10.20).

Proof. The proof is very similar to that of Theorem 10.7. First we define

$$\mathcal{U}_M[\epsilon_1,\epsilon_2](x) \equiv \epsilon_1^{n-2}\epsilon_2^{n-2}\int_{\partial\Omega_1} S_n(x-\epsilon_1 p^1+\epsilon_1\epsilon_2\eta)\,\Theta_1[\epsilon_1,\epsilon_2](\eta)\,d\sigma_\eta$$
$$+\epsilon_1^{n-2}\epsilon_2^{n-2}\int_{\partial\Omega_2} S_n(x-\epsilon_1 p^2+\epsilon_1\epsilon_2\eta)\,\Theta_2[\epsilon_1,\epsilon_2](\eta)\,d\sigma_\eta + v_{\Omega^o}[\Theta^o[\epsilon_1,\epsilon_2]](x)$$

for all $x \in \overline{\Omega_M}$ and $(\epsilon_1,\epsilon_2) \in]-\delta_M,\delta_M[\times]\tilde{\epsilon}_2-\delta_M,\tilde{\epsilon}_2+\delta_M[$. Then, applying the argument in the proof of Theorem 10.7 around the point $(0,\tilde{\epsilon}_2)$ instead of $(0,0)$, we show that $(\epsilon_1,\epsilon_2) \mapsto \mathcal{U}_M[\epsilon_1,\epsilon_2]$ is real analytic. Equality (10.28) is a consequence of Propositions 10.1 and 10.13. Finally, to prove (10.29) we observe that $\mathcal{U}_M[0,\tilde{\epsilon}_2](x) = v_{\Omega^o}[\Theta^o[0,\tilde{\epsilon}_2]](x)$ for all $x \in \overline{\Omega_M}$ and that $v_{\Omega^o}[\Theta^o[0,\tilde{\epsilon}_2]]$ is the only solution in $C^{1,\alpha}(\overline{\Omega^o})$ of the unperturbed problem (10.20). Indeed, Lemma 10.10 and Proposition 10.13 imply that

$$V_{\Omega^o}[\Theta^o[0,\tilde{\epsilon}_2]] = f^o\,. \tag{10.30}$$

□

Then, for the 'microscopic' behavior of $u_{\epsilon_1,\epsilon_2}$ close to the boundary of the holes $\Omega_1(\epsilon_1,\epsilon_2)$ and $\Omega_2(\epsilon_1,\epsilon_2)$ we have the following.

Theorem 10.15. *Let $\tilde{\epsilon}_2 \in]0, \delta_2[$. Let δ_* be as in Lemma 10.12. Let $i \in \{1,2\}$. Let $\Omega_{i,m}$ be an open bounded subset of $\mathbb{R}^n \setminus \overline{\Omega_i}$. Let $\delta_m^{(i)} \in]0, \delta_*[$ be such that*

$$\epsilon_1 p^i + \epsilon_1\epsilon_2 \overline{\Omega_{i,m}} \subseteq \Omega^o$$

and

$$(\epsilon_1 p^i + \epsilon_1\epsilon_2 \overline{\Omega_{i,m}}) \cap \overline{\Omega_j(\epsilon_1,\epsilon_2)} = \emptyset$$

for all $(\epsilon_1,\epsilon_2) \in]-\delta_m^{(i)}, \delta_m^{(i)}[\times]\tilde{\epsilon}_2 - \delta_m^{(i)}, \tilde{\epsilon}_2 + \delta_m^{(i)}[$, where $j = 1$ if $i = 2$ and $j = 2$ if $i = 1$. Then, there exists a real analytic map

$$\mathcal{U}_{i,m} :]-\delta_m^{(i)}, \delta_m^{(i)}[\times]\tilde{\epsilon}_2 - \delta_m^{(i)}, \tilde{\epsilon}_2 + \delta_m^{(i)}[\to C^{1,\alpha}(\overline{\Omega_{i,m}})$$

such that

$$u_{\epsilon_1,\epsilon_2}(\epsilon_1 p^i + \epsilon_1\epsilon_2\xi) = \mathcal{U}_{i,m}[\epsilon_1,\epsilon_2](\xi) \quad \forall \xi \in \overline{\Omega_{i,m}} \tag{10.31}$$

for all $\epsilon_1 \in]0, \delta_m^{(i)}[$ and $\epsilon_2 > 0$ in the interval $]\tilde{\epsilon}_2 - \delta_m^{(i)}, \tilde{\epsilon}_2 + \delta_m^{(i)}[$. Moreover,

$$\mathcal{U}_{i,m}[0,\tilde{\epsilon}_2](\xi) = v_\sharp(p^i + \tilde{\epsilon}_2\xi) \qquad \forall \xi \in \overline{\Omega_{i,m}} \tag{10.32}$$

where $v_\sharp \in C^{1,\alpha}_{\mathrm{loc}}(\mathbb{R}^n \setminus \Omega_\sharp(\tilde{\epsilon}_2))$ is the unique solution of

$$\begin{cases} \Delta v_\sharp = 0 & \text{in } \mathbb{R}^n \setminus \overline{\Omega_\sharp(\tilde{\epsilon}_2)}\,, \\ v_\sharp = f_\sharp & \text{on } \partial\Omega_\sharp(\tilde{\epsilon}_2)\,, \\ \lim_{\xi\to\infty} v_\sharp(\xi) = u^o(0)\,, \end{cases} \tag{10.33}$$

where $u^o \in C^{1,\alpha}(\overline{\Omega^o})$ is the unique solution of the unperturbed problem (10.20), *and where $f_\sharp$ is defined by*

$$f_\sharp(x) \equiv f_k\Big(\frac{x - p^k}{\tilde{\epsilon}_2}\Big)\,, \qquad \forall k \in \{1,2\}\,,\ x \in \partial\Omega_k(1,\tilde{\epsilon}_2)\,.$$

Proof. Let $j = 1$ if $i = 2$ and $j = 2$ otherwise. We define

$$\begin{aligned} \mathcal{U}_{i,m}[\epsilon_1,\epsilon_2](\xi) &\equiv v_{\Omega_i}[\Theta_i[\epsilon_1,\epsilon_2]](\xi) \\ &\quad + \epsilon_2^{n-2}\int_{\partial\Omega_j} S_n\left(p^i - p^j + \epsilon_2(\xi - \eta)\right)\Theta_j[\epsilon_1,\epsilon_2](\eta)\,d\sigma_\eta \\ &\quad + \int_{\partial\Omega^o} S_n\left(\epsilon_1 p^i + \epsilon_1\epsilon_2\xi - y\right)\,\Theta^o[\epsilon_1,\epsilon_2](y)\,d\sigma_y \end{aligned}$$

for all $\xi \in \overline{\Omega_{i,m}}$ and $(\epsilon_1,\epsilon_2) \in]-\delta_m^{(i)}, \delta_m^{(i)}[\times]\tilde{\epsilon}_2 - \delta_m^{(i)}, \tilde{\epsilon}_2 + \delta_m^{(i)}[$. To show that

$$(\epsilon_1,\epsilon_2) \mapsto \mathcal{U}_{i,m}[\epsilon_1,\epsilon_2]$$

is real analytic, we argue as in the proof of Theorem 10.8. Equality (10.31) can be verified by a change of variable in the integral representation (10.14). Then to prove the statement it remains to verify (10.32). By a change of variables in the integrals over $\partial\Omega_i$ and $\partial\Omega_j$ we can observe that

$$\mathcal{U}_{i,m}[0,\tilde{\epsilon}_2](\xi) = v_{\Omega_\sharp(\tilde{\epsilon}_2)}[\theta_\sharp](p^i + \tilde{\epsilon}_2\xi) + v_{\Omega^o}[\Theta^o[0,\tilde{\epsilon}_2]](0) \qquad \forall \xi \in \overline{\Omega_{i,m}}$$

with $\theta_\sharp \in C^{0,\alpha}(\partial\Omega_\sharp(\tilde{\epsilon}_2))$ given by

$$\theta_\sharp(y) \equiv \tilde{\epsilon}_2^{-1}\Theta_k[0,\epsilon_2]\Big(\frac{y-p^k}{\tilde{\epsilon}_2}\Big) \qquad \forall y \in \partial\Omega_k(1,\tilde{\epsilon}_2)\,,\ k \in \{1,2\}\,.$$

Then we note that

$$v_{\Omega_\sharp(\tilde{\epsilon}_2)}[\theta_\sharp] + v_{\Omega^o}[\Theta^o[0,\tilde{\epsilon}_2]](0)$$

is the unique solution in $C^{1,\alpha}_{\mathrm{loc}}(\mathbb{R}^n\setminus\Omega_\sharp(\tilde{\epsilon}_2))$ of problem (10.33). Indeed, Lemma 10.10 and Proposition 10.13 imply that

$$V_{\Omega_\sharp(\tilde{\epsilon}_2)}[\theta_\sharp] + v_{\Omega^o}[\Theta^o[0,\tilde{\epsilon}_2]](0) = f_\sharp \qquad \text{on } \partial\Omega_\sharp(\tilde{\epsilon}_2)\,,$$

and by (10.30) we have $v^+_{\Omega^o}[\Theta^o[0,\tilde{\epsilon}_2]] = u^o$ with u^o as in (10.20). Hence, the validity of equality (10.32) follows. □

We observe that the limiting problem (10.33) is defined in the exterior of $\Omega_\sharp(\tilde{\epsilon}_2)$ and contains boundary conditions both on $\partial\Omega_1(1,\tilde{\epsilon}_2)$ and on $\partial\Omega_2(1,\tilde{\epsilon}_2)$. Thus, both the sets Ω_1 and Ω_2 and the corresponding functions f_1 and f_2 take a part in it. Instead, the corresponding problem (10.23) obtained in the case of 'moderately close' holes is defined in the exterior of Ω_i and has only a condition on the boundary $\partial\Omega_i$. In a sense, we can say that the 'microscopic' limit of the rescaled solution $u_{\epsilon_1,\epsilon_2}(\epsilon_1 p^i + \epsilon_1\epsilon_2\cdot)$ 'continues to see' both the holes if these are 'close' and 'sees' only one of the holes when these are 'moderately close' to one another.

Finally, for the energy integral of $u_{\epsilon_1,\epsilon_2}$ we have the following Theorem 10.16. The proof can be effected following the footsteps of the proof of Theorem 10.9. We just need to use Theorem 10.14 instead of Theorem 10.7 and Theorem 10.15 instead of Theorem 10.8.

Theorem 10.16. *Let $\tilde{\epsilon}_2 \in]0,\delta_2[$. Let δ_* be as in Lemma 10.12. Then there exist $\delta_E \in]0,\delta_*[$ and a real analytic function*

$$\mathcal{E} :]-\delta_E,\delta_E[\times]\tilde{\epsilon}_2-\delta_E,\tilde{\epsilon}_2+\delta_E[\to \mathbb{R}$$

such that

$$\int_{\Omega(\epsilon)} |\nabla u_{\epsilon_1,\epsilon_2}|^2\, dx = \mathcal{E}(\epsilon_1,\epsilon_2)$$

for all $\epsilon_1 \in]0,\delta_E[$ and $\epsilon_2 > 0$ in the interval $]\tilde{\epsilon}_2-\delta_E,\tilde{\epsilon}_2+\delta_E[$. Moreover,

$$\mathcal{E}(0,0) = \int_{\Omega^o} |\nabla u^o|^2\, dx\,,$$

where u^o is the unique solution of the unperturbed problem (10.20).

10.4 Moderately Close Holes in Dimension $n = 2$

We now turn to the case of dimension $n = 2$. In contrast with what we have done for $n \geq 3$, we consider only the case where the pair (ϵ_1, ϵ_2) approaches $(0, 0)$ and the holes are 'moderately close' to one another. As we have explained in Section 10.1, this means that the holes are colliding into one another while shrinking their size at a faster speed. Instead, we do not treat the case where ϵ_1 tends to zero and ϵ_2 stays away from zero. That is, the case of 'close' holes. The interested reader can find the analysis of this last case in the the paper [79].

Before entering into mathematical details, we would like to comment some of the results that we are going to prove. In particular, in Theorem 10.35 we will analyze the behavior of the solution $u_{\epsilon_1,\epsilon_2}$ away from the boundary of the holes $\partial\Omega_1(\epsilon_1, \epsilon_2)$ and $\partial\Omega_2(\epsilon_1, \epsilon_2)$. As usual, we will consider the restriction of $u_{\epsilon_1,\epsilon_2}$ to an open subset of Ω^o that does not contain the origin in its closure. If we content ourselves to pick a single point $x \in \Omega^o \setminus \{0\}$ and look at the evaluation of $u_{\epsilon_1,\epsilon_2}$ at x, then Theorem 10.35 implies that, possibly shrinking δ_1 and δ_2, we have

$$u_{\epsilon_1,\epsilon_2}(x) = u^o(x) + \epsilon_1\epsilon_2\, \mathcal{U}_x[\epsilon_1, \epsilon_2] + \mathcal{F}(\epsilon_1, \epsilon_2)^t\, \Lambda(\epsilon_1, \epsilon_2)^{-1}\, \mathcal{V}_x[\epsilon_1, \epsilon_2] \quad (10.34)$$

for all $\epsilon_1 \in]0, \delta_1[$ and all $\epsilon_2 \in]0, \delta_2[$. In the equality above u^o is the solution of the unperturbed Dirichlet problem in Ω^o with boundary datum f^o, the function $\mathcal{U}_x$ is real analytic from $] - \delta_1, \delta_1[\times] - \delta_2, \delta_2[$ to $\mathbb{R}$, the functions $\mathcal{F}$ and $\mathcal{V}_x$ are real analytic from $] - \delta_1, \delta_1[\times] - \delta_2, \delta_2[$ to $\mathbb{R}^2$, and $\Lambda(\epsilon_1, \epsilon_2)$ is a 2×2 matrix such that

$$\Lambda(\epsilon_1, \epsilon_2) - \frac{1}{2\pi}\begin{pmatrix} \log(\epsilon_1\epsilon_2) & \log \epsilon_1 \\ \log \epsilon_1 & \log(\epsilon_1\epsilon_2) \end{pmatrix} = \mathcal{R}(\epsilon_1, \epsilon_2)$$

where $\mathcal{R}$ is a real analytic matrix-valued function on $] - \delta_1, \delta_1[\times] - \delta_2, \delta_2[$. We shall see that $\Lambda(\epsilon_1, \epsilon_2)$ is invertible for every pair (ϵ_1, ϵ_2) of positive real numbers in $]0, \delta_1[\times]0, \delta_2[$. Accordingly, the inverse matrix $\Lambda(\epsilon_1, \epsilon_2)^{-1}$ in the right-hand side of (10.34) makes sense.

We also observe that the matrix $\Lambda(\epsilon_1, \epsilon_2)$ takes care of the logarithmic behavior which is expected for the Dirichlet problems in a perforated domain in dimension two. Moreover, equality (10.34) fulfills the goal stated in Section 10.2. That is, it represents the map that takes (ϵ_1, ϵ_2) to $u_{\epsilon_1,\epsilon_2}(x)$ in terms of the following two ingredients.

- Real analytic functions of the perturbation parameter, in this case the pair (ϵ_1, ϵ_2), defined in a open neighborhood of the singular point $(0, 0)$ in $\mathbb{R}^2$.
- Singular but completely known functions of the perturbation parameter, in this case $\log(\epsilon_1\epsilon_2)$ and $\log \epsilon_1$.

However, the presence of the matrix $\Lambda(\epsilon_1, \epsilon_2)^{-1}$ in (10.34) produces a specific problem that we did not meet when dealing with a single hole in Section 8.4 or with two holes in dimension $n \geq 3$ in Section 10.3. Indeed, if we want to use (10.34) to deduce an asymptotic approximation of $u_{\epsilon_1,\epsilon_2}(x)$, as we did in Section 8.5 for the one hole problem, then we have to compute the inverse matrix $\Lambda(\epsilon_1, \epsilon_2)^{-1}$. Doing so, we obtain an expression which involves the quotient

$$\frac{\log \epsilon_1}{\log(\epsilon_1 \epsilon_2)} \tag{10.35}$$

(cf. Proposition 10.37 below). Now, the problem with (10.35) is that its limit as $(\epsilon_1, \epsilon_2) \to (0, 0)$ does not exist. This is a new situation that we did not foresee when we have introduced the problem in Section 10.2. One way to overcome this difficulty is to introduce a relation between the parameters ϵ_1 and ϵ_2. For example, we can replace ϵ_1 with a positive real parameter t and set $\epsilon_2 = \gamma(t)$, with γ a function from the interval $]0, 1[$ to itself such that

$$\lim_{t \to 0^+} \gamma(t) = 0$$

and such that the limit

$$\lambda_0 \equiv \lim_{t \to 0^+} \frac{\log t}{\log(t\gamma(t))} \tag{10.36}$$

exists finite in $[0, +\infty[$. In this way, we switch from a two parameter problem to a one parameter problem and, thanks to the existence of the limit (10.36), we can compute the an asymptotic approximation of the map $t \mapsto u_{t,\gamma(t)}$ for $t > 0$ that tends to zero. In particular, if we stop our computation to the first two terms we obtain

$$\begin{aligned} u_{t,\gamma(t)}(x) &= u^o(x) \\ &+ \frac{1}{\log(t\gamma(t))} \frac{2\pi}{1 + \lambda_0} \Big(\lim_{\eta \to \infty} u_1(\eta) + \lim_{\eta \to \infty} u_2(\eta) - 2u^o(0) \Big) G^{\Omega^o}(x, 0) \\ &+ o\left(\frac{1}{\log(t\gamma(t))} \right) \end{aligned} \tag{10.37}$$

as $t > 0$ tends to zero (see Proposition 10.39). In (10.37), u_1 and u_2 denote the harmonic solution of the exterior Dirichlet problem in $\mathbb{R}^2 \setminus \overline{\Omega_1}$ and $\mathbb{R}^2 \setminus \overline{\Omega_2}$ with boundary datum f_1 and f_2, respectively, and G^{Ω^o} is the Dirichlet Green's function of Ω^o.

A particular feature of (10.37) is that the limit value λ_0 appears explicitly in the second asymptotic term of the expansion. This seems to indicate that no asymptotic approximation with at least two terms could be obtained without prescribing a limiting value for the quotient (10.35). In this sense, the necessity of introducing a specific relation between the size of the holes and their mutual distance is intrinsic in the nature of the problem and could not be avoided, no matter what method we adopt to analyze it.

It is thus no surprise that some kind of condition on the size and the distance of the holes has appeared in previous research papers dealing with the Dirichlet problem in a domain with moderately close holes, also when the adopted techniques are very different from those of the Functional Analytic Approach of the present book. For example, in [31], Bonnaillie-Noël, Dambrine, and Lacave use the Multi-Scale Approximation Method to study the Poisson problem with Dirichlet conditions in a domain with two moderately close holes. To carry on their computations they assume that the distance behaves like the size to some power $\beta \in]0,1[$, a condition that corresponds, with our notation, to the case where $\gamma(t) = t^{(1-\beta)/\beta}$ and the quotient (10.35) is constant and equal to $1-\beta$.

Somehow related to this problem is also the work of Maz'ya, Movchan, and Nieves on the Meso-Scale Approximation Method, which aims at studying boundary value problems in domains with a large (but fixed) number of small close holes, a so-called 'cloud of voids' (cf. [201]). Most of their papers deal with the three-dimensional case, but a relation between the size and the distance of the holes still plays a role in the analysis. We mention, for example, the paper [199] of Maz'ya and Movchan, on a three-dimensional Poisson problem with Dirichlet conditions, and the paper [202] of Maz'ya, Movchan, and Nieves, on a two-dimensional scattering problem for an elastic membrane.

10.4.1 Integral Representation of the Solution

We now begin our analysis of the Dirichlet problem (10.4) in dimension $n = 2$. As for the case of the one hole problem of Section 8.4, we exploit the integral representation (6.19) for the solution $u_{\epsilon_1,\epsilon_2}$. Then, by writing (6.19) for the specific case when the domain $\Omega = \Omega(\epsilon_1,\epsilon_2)$ has two holes, we obtain

$$\begin{aligned} u_{\epsilon_1,\epsilon_2} &= w^+_{\Omega(\epsilon_1,\epsilon_2)}[\mu_{\epsilon_1,\epsilon_2}] \\ &+ \sum_{i,j=1}^{2}\left(\int_{\partial\Omega(\epsilon_1,\epsilon_2)} f_{\epsilon_1,\epsilon_2}\tau^{(i)}_{\epsilon_1,\epsilon_2}\,d\sigma\right)(\Lambda^{-1}_{\Omega(\epsilon_1,\epsilon_2)^-})_{i,j}\, v^+_{\Omega(\epsilon_1,\epsilon_2)}[\tau^{(j)}_{\epsilon_1,\epsilon_2}] \end{aligned} \tag{10.38}$$

with $f_{\epsilon_1,\epsilon_2} \in C^{1,\alpha}(\partial\Omega(\epsilon_1,\epsilon_2))$ defined by

$$f_{\epsilon_1,\epsilon_2}(x) \equiv \begin{cases} f_1\left(\dfrac{x-\epsilon_1 p^1}{\epsilon_1\epsilon_2}\right) & \text{if } x \in \partial\Omega_1(\epsilon_1,\epsilon_2)\,, \\ f_2\left(\dfrac{x-\epsilon_1 p^2}{\epsilon_1\epsilon_2}\right) & \text{if } x \in \partial\Omega_2(\epsilon_1,\epsilon_2)\,, \\ f^o(x) & \text{if } x \in \partial\Omega^o, \end{cases}$$

and where $\tau^{(1)}_{\epsilon_1,\epsilon_2}, \tau^{(2)}_{\epsilon_1,\epsilon_2} \in C^{0,\alpha}(\partial\Omega(\epsilon_1,\epsilon_2))$ and $\mu_{\epsilon_1,\epsilon_2} \in C^{1,\alpha}(\partial\Omega(\epsilon_1,\epsilon_2))$ are the unique solutions of the following systems of integral equations

$$\begin{cases} \left(\frac{1}{2}I + W^t_{\Omega(\epsilon_1,\epsilon_2)}\right)[\tau^{(j)}_{\epsilon_1,\epsilon_2}] = 0\,, \\ \int_{\partial\Omega_i(\epsilon_1,\epsilon_2)} \tau^{(j)}_{\epsilon_1,\epsilon_2}\,d\sigma = \delta_{i,j} \qquad \forall i \in \{1,2\} \end{cases} \tag{10.39}$$

for $j \in \{1,2\}$, and

$$\begin{cases} \left(\frac{1}{2}I + W_{\Omega(\epsilon_1,\epsilon_2)}\right)[\mu_{\epsilon_1,\epsilon_2}] \\ \qquad = f_{\epsilon_1,\epsilon_2} - \sum_{j=1}^{2} \int_{\partial\Omega(\epsilon_1,\epsilon_2)} f_{\epsilon_1,\epsilon_2}\tau^{(j)}_{\epsilon_1,\epsilon_2}\,d\sigma\,\zeta_{\partial\Omega_j(\epsilon_1,\epsilon_2)}\,, \\ \int_{\partial\Omega_j(\epsilon_1,\epsilon_2)} \mu_{\epsilon_1,\epsilon_2}\,d\sigma = 0 \end{cases} \tag{10.40}$$

(see (6.17) for the definition of $\zeta_{\partial\Omega_j(\epsilon_1,\epsilon_2)}$). Finally, the matrix $\Lambda_{\Omega(\epsilon_1,\epsilon_2)^-}$ that appears in (10.38) is given by

$$\Lambda_{\Omega(\epsilon_1,\epsilon_2)^-} = \begin{pmatrix} \fint_{\partial\Omega_1(\epsilon_1,\epsilon_2)} V_{\Omega(\epsilon_1,\epsilon_2)}[\tau^{(1)}_{\epsilon_1,\epsilon_2}]\,d\sigma & \fint_{\partial\Omega_1(\epsilon_1,\epsilon_2)} V_{\Omega(\epsilon_1,\epsilon_2)}[\tau^{(2)}_{\epsilon_1,\epsilon_2}]\,d\sigma \\ \fint_{\partial\Omega_2(\epsilon_1,\epsilon_2)} V_{\Omega(\epsilon_1,\epsilon_2)}[\tau^{(1)}_{\epsilon_1,\epsilon_2}]\,d\sigma & \fint_{\partial\Omega_2(\epsilon_1,\epsilon_2)} V_{\Omega(\epsilon_1,\epsilon_2)}[\tau^{(2)}_{\epsilon_1,\epsilon_2}]\,d\sigma \end{pmatrix} \tag{10.41}$$

with $\tau^{(1)}_{\epsilon_1,\epsilon_2}$ and $\tau^{(2)}_{\epsilon_1,\epsilon_2}$ as in (10.39). As we have observed in Lemma 6.30, $\Lambda_{\Omega(\epsilon_1,\epsilon_2)^-}$ is invertible. Therefore, we are allowed to write $\Lambda^{-1}_{\Omega(\epsilon_1,\epsilon_2)^-}$ in (10.38).

To analyze problem (10.4) we proceed as follows. First we analyze system (10.39) and prove a prove a real analytic continuation result for $\tau^{(1)}_{\epsilon_1,\epsilon_2}$ and $\tau^{(2)}_{\epsilon_1,\epsilon_2}$, then we turn to system (10.40) and prove a similar real analytic continuation result also for $\mu_{\epsilon_1,\epsilon_2}$. At this point we will be ready to study the solution $u_{\epsilon_1,\epsilon_2}$ by means of the representation formula (10.38).

As usual our aim is to represent the maps that take (ϵ_1,ϵ_2) to suitable restrictions of $u_{\epsilon_1,\epsilon_2}$ and of the rescaled functions $u_{\epsilon_1,\epsilon_2}(\epsilon_1 p^i + \epsilon_1\epsilon_2\cdot)$ with $i \in \{1,2\}$ and to the energy $\int_{\Omega(\epsilon_1,\epsilon_2)} |\nabla u_{\epsilon_1,\epsilon_2}|^2\,dx$ in terms of real analytic maps of (ϵ_1,ϵ_2) and of elementary functions. To describe the limiting values of such analytic maps, we will introduce in Section 10.4.4 certain auxiliary functions related to the Dirichlet Green's functions of Ω^o and of the exteriors of Ω_1 and Ω_2. In Section 10.4.5, we state Theorem 10.35 on the representation of $u_{\epsilon_1,\epsilon_2}$ and the $u_{\epsilon_1,\epsilon_2}(\epsilon_1 p^i + \epsilon_1\epsilon_2\cdot)$'s in terms of real analytic maps and known functions. Then in Theorem 10.36 we consider the energy of $u_{\epsilon_1,\epsilon_2}$. In the last Section 10.4.6 of this chapter, we examine the problem of computing an asymptotic expansion of the solution. We will need to introduce a specific relation between ϵ_1 and ϵ_2 to ensure the existence of the limit of the quotient (10.35) for $(\epsilon_1,\epsilon_2) \to (0,0)$. We also verify that the second term in the asymptotic expansion of $u_{\epsilon_1,\epsilon_2}$ depends explicitly on such limit.

10.4.2 Analysis of System (10.39)

In this subsection we study system (10.39) and prove a real analytic continuation result for the densities $\tau^{(1)}_{\epsilon_1,\epsilon_2}$ and $\tau^{(2)}_{\epsilon_1,\epsilon_2}$. We exploit the same strategy that we have used in Section 8.4 to deal with the case of a single shrinking hole. Accordingly, we first split the first equation of (10.39) over $\partial\Omega_1(\epsilon_1,\epsilon_2)$, $\partial\Omega_2(\epsilon_1,\epsilon_2)$, and $\partial\Omega^o$, and then we perform a change of variable in the integrals over the (ϵ_1,ϵ_2)-dependent boundaries $\partial\Omega_1(\epsilon_1,\epsilon_2)$, $\partial\Omega_2(\epsilon_1,\epsilon_2)$. We observe that

$$\nabla S_2(\epsilon_1 p^i + \epsilon_1\epsilon_2\xi - \epsilon_1 p^i - \epsilon_1\epsilon_2\eta) = (\epsilon_1\epsilon_2)^{-1}\nabla S_2(\xi - \eta) \tag{10.42}$$

and

$$\nabla S_2(\epsilon_1 p^i + \epsilon_1\epsilon_2\xi - \epsilon_1 p^j - \epsilon_1\epsilon_2\eta) = \epsilon_1^{-1}\nabla S_2(p^i - p^j + \epsilon_2(\xi - \eta)), \tag{10.43}$$

for all $i, j \in \{1, 2\}$, $i \neq j$, and all $\xi, \eta \in \partial\Omega_i$ for which the functions in (10.42) and (10.43) are defined. Then, introducing the new densities

$$\begin{aligned} \phi^{1,(j)}_{\epsilon_1,\epsilon_2}(\xi) &\equiv \epsilon_1\epsilon_2\,\tau^{(j)}_{\epsilon_1,\epsilon_2}(\epsilon_1 p^1 + \epsilon_1\epsilon_2\xi) \quad && \forall\xi \in \partial\Omega_1\,, \\ \phi^{2,(j)}_{\epsilon_1,\epsilon_2}(\xi) &\equiv \epsilon_1\epsilon_2\,\tau^{(j)}_{\epsilon_1,\epsilon_2}(\epsilon_1 p^2 + \epsilon_1\epsilon_2\xi) \quad && \forall\xi \in \partial\Omega_2\,, \\ \phi^{o,(j)}_{\epsilon_1,\epsilon_2}(x) &\equiv \tau^{(j)}_{\epsilon_1,\epsilon_2}(x) && \forall x \in \partial\Omega^o, \end{aligned} \tag{10.44}$$

we find that system (10.39) for $j \in \{1, 2\}$ is equivalent to the system of the following four equations:

$$\begin{aligned} &\frac{1}{2}\phi^{1,(j)}_{\epsilon_1,\epsilon_2}(\xi) - W^t_{\Omega_1}[\phi^{1,(j)}_{\epsilon_1,\epsilon_2}](\xi) \\ &\quad - \epsilon_2\int_{\partial\Omega_2}\phi^{2,(j)}_{\epsilon_1,\epsilon_2}(\eta)\,\nu_{\Omega_1}(\xi)\cdot\nabla S_2(p^1 - p^2 + \epsilon_2(\xi - \eta))\,d\sigma_\eta \\ &\quad - \epsilon_1\epsilon_2\int_{\partial\Omega^o}\phi^{o,(j)}_{\epsilon_1,\epsilon_2}(y)\,\nu_{\Omega_1}(\xi)\cdot\nabla S_2(\epsilon_1 p^1 + \epsilon_1\epsilon_2\xi - y)\,d\sigma_y = 0 \qquad \forall\xi \in \partial\Omega_1\,, \end{aligned} \tag{10.45}$$

$$\begin{aligned} &\frac{1}{2}\phi^{2,(j)}_{\epsilon_1,\epsilon_2}(\xi) - W^t_{\Omega_2}[\phi^{2,(j)}_{\epsilon_1,\epsilon_2}](\xi) \\ &\quad - \epsilon_2\int_{\partial\Omega_1}\phi^{1,(j)}_{\epsilon_1,\epsilon_2}(\eta)\,\nu_{\Omega_2}(\xi)\cdot\nabla S_2(p^2 - p^1 + \epsilon_2(\xi - \eta))\,d\sigma_\eta \\ &\quad - \epsilon_1\epsilon_2\int_{\partial\Omega^o}\phi^{o,(j)}_{\epsilon_1,\epsilon_2}(y)\,\nu_{\Omega_2}(\xi)\cdot\nabla S_2(\epsilon_1 p^2 + \epsilon_1\epsilon_2\xi - y)\,d\sigma_y = 0 \qquad \forall\xi \in \partial\Omega_2\,, \end{aligned} \tag{10.46}$$

$$\frac{1}{2}\phi^{o,(j)}_{\epsilon_1,\epsilon_2}(x) + W^t_{\Omega^o}[\phi^{o,(j)}_{\epsilon_1,\epsilon_2}](x)$$

$$+\sum_{h=1}^{2}\int_{\partial\Omega_h}\phi^{(i,h)}_{\epsilon_1,\epsilon_2}(\eta)\,\nu_{\Omega^o}(x)\cdot\nabla S_2(x-\epsilon_1 p^h-\epsilon_1\epsilon_2\eta)\,d\sigma_\eta=0 \qquad \forall x\in\partial\Omega^o\,, \tag{10.47}$$

$$\int_{\partial\Omega_i}\phi^{i,(j)}_{\epsilon_1,\epsilon_2}\,d\sigma=\delta_{i,j} \qquad \forall i\in\{1,2\}\,. \tag{10.48}$$

In order to cast the system of the four equations in (10.45)–(10.48) into an operator equation in Banach space, we introduce for each (ϵ_1,ϵ_2) in $]-\delta_1,\delta_1[\times]-\delta_2,\delta_2[$ an auxiliary operator. Namely, the operator

$$\mathcal{N}(\epsilon_1,\epsilon_2)\equiv\left(\mathcal{N}^1(\epsilon_1,\epsilon_2)\,,\ \mathcal{N}^2(\epsilon_1,\epsilon_2)\,,\ \mathcal{N}^o(\epsilon_1,\epsilon_2)\,,\ \mathcal{N}^s(\epsilon_1,\epsilon_2)\right)$$

(the superscript 's' of $\mathcal{N}^s$ stands for 'scalars') from

$$C^{0,\alpha}(\partial\Omega_1)\times C^{0,\alpha}(\partial\Omega_2)\times C^{0,\alpha}(\partial\Omega^o)$$

to

$$C^{0,\alpha}(\partial\Omega_1)\times C^{0,\alpha}(\partial\Omega_2)\times C^{0,\alpha}(\partial\Omega^o)\times\mathbb{R}^2$$

defined by

$$\begin{aligned}
\mathcal{N}^1(\epsilon_1,\epsilon_2)[\phi_1,\phi_2,\phi^o](\xi)&\equiv\frac{1}{2}\phi_1(\xi)-W^t_{\Omega_1}[\phi_1](\xi) && (10.49)\\
&-\epsilon_2\int_{\partial\Omega_2}\phi_2(\eta)\,\nu_{\Omega_1}(\xi)\cdot\nabla S_2(p^1-p^2+\epsilon_2(\xi-\eta))\,d\sigma_\eta\\
&-\epsilon_1\epsilon_2\int_{\partial\Omega^o}\phi^o(y)\,\nu_{\Omega_1}(\xi)\cdot\nabla S_2(\epsilon_1p^1+\epsilon_1\epsilon_2\xi-y)\,d\sigma_y \qquad \forall\xi\in\partial\Omega_1\,,\\
\mathcal{N}^2(\epsilon_1,\epsilon_2)[\phi_1,\phi_2,\phi^o](\xi)&\equiv\frac{1}{2}\phi_2(\xi)-W^t_{\Omega_2}[\phi_2](\xi) && (10.50)\\
&-\epsilon_2\int_{\partial\Omega_1}\phi_1(\eta)\,\nu_{\Omega_2}(\xi)\cdot\nabla S_2(p^2-p^1+\epsilon_2(\xi-\eta))\,d\sigma_\eta\\
&-\epsilon_1\epsilon_2\int_{\partial\Omega^o}\phi^o(y)\,\nu_{\Omega_2}(\xi)\cdot\nabla S_2(\epsilon_1p^2+\epsilon_1\epsilon_2\xi-y)\,d\sigma_y \qquad \forall\xi\in\partial\Omega_2\,,\\
\mathcal{N}^o(\epsilon_1,\epsilon_2)[\phi_1,\phi_2,\phi^o](x)&\equiv\frac{1}{2}\phi^o(x)+W^t_{\Omega^o}[\phi^o](x) && (10.51)\\
&+\sum_{h=1}^{2}\int_{\partial\Omega_h}\phi_h(\eta)\,\nu_{\Omega^o}(x)\cdot\nabla S_2(x-\epsilon_1p^h-\epsilon_1\epsilon_2\eta)\,d\sigma_\eta \qquad \forall x\in\partial\Omega^o\,,\\
\mathcal{N}^s(\epsilon_1,\epsilon_2)[\phi_1,\phi_2,\phi^o]&\equiv\left(\int_{\partial\Omega_1}\phi_1\,d\sigma\,,\ \int_{\partial\Omega_2}\phi_2\,d\sigma\right) && (10.52)
\end{aligned}$$

for all $(\epsilon_1,\epsilon_2)\in]-\delta_1,\delta_1[\times]-\delta_2,\delta_2[$ and all $(\phi_1,\phi_2,\phi^o)\in C^{0,\alpha}(\partial\Omega_1)\times C^{0,\alpha}(\partial\Omega_2)\times C^{0,\alpha}(\partial\Omega^o)$. It is apparent that the system of Equations (10.45)–(10.48) is equivalent to

$$\mathcal{N}(\epsilon_1, \epsilon_2)[\phi^{1,(j)}_{\epsilon_1,\epsilon_2}, \phi^{2,(j)}_{\epsilon_1,\epsilon_2}, \phi^{o,(j)}_{\epsilon_1,\epsilon_2}] = (0, 0, 0, (\delta_{1,j}, \delta_{2,j})) \tag{10.53}$$

and thus system (10.39) for $\tau^{(j)}_{\epsilon_1,\epsilon_2}$ is equivalent to (10.53) with $(\phi^{1,(j)}_{\epsilon_1,\epsilon_2}, \phi^{2,(j)}_{\epsilon_1,\epsilon_2}, \phi^{o,(j)}_{\epsilon_1,\epsilon_2})$ and $\tau^{(j)}_{\epsilon_1,\epsilon_2}$ related by (10.44). Accordingly, Lemma 6.28 on the unique solvability of system (10.39) implies the validity of the following.

Proposition 10.17. *For all* $(\epsilon_1, \epsilon_2) \in]0, \delta_1[\times]0, \delta_2[$ *and* $j \in \{1, 2\}$, *there exists one and only one solution*

$$(\phi^{1,(j)}_{\epsilon_1,\epsilon_2}, \phi^{2,(j)}_{\epsilon_1,\epsilon_2}, \phi^{o,(j)}_{\epsilon_1,\epsilon_2}) \in C^{0,\alpha}(\partial\Omega_1) \times C^{0,\alpha}(\partial\Omega_2) \times C^{0,\alpha}(\partial\Omega^o)$$

of (10.53).

Proposition 10.17 implies that, for each $j \in \{1, 2\}$, there is a map from $]0, \delta_1[\times]0, \delta_2[$ to $C^{0,\alpha}(\partial\Omega_1) \times C^{0,\alpha}(\partial\Omega_2) \times C^{0,\alpha}(\partial\Omega^o)$ that takes a pair (ϵ_1, ϵ_2) to the triple

$$(\phi^{1,(j)}_{\epsilon_1,\epsilon_2}, \phi^{2,(j)}_{\epsilon_1,\epsilon_2}, \phi^{o,(j)}_{\epsilon_1,\epsilon_2}). \tag{10.54}$$

We now prove that such a map has a real analytic continuation in a neighborhood of the singular pair $(0, 0)$. To do so, we begin with the following technical lemma.

Lemma 10.18. *For all* (ϵ_1, ϵ_2) *in* $] - \delta_1, \delta_1[\times] - \delta_2, \delta_2[$ *and all* (ϕ_1, ϕ_2, ϕ^o) *in* $C^{0,\alpha}(\partial\Omega_1) \times C^{0,\alpha}(\partial\Omega_2) \times C^{0,\alpha}(\partial\Omega^o)$ *the pair of functions*

$$(\mathcal{N}^1(\epsilon_1, \epsilon_2)[\phi_1, \phi_2, \phi^o] \,,\, \mathcal{N}^2(\epsilon_1, \epsilon_2)[\phi_1, \phi_2, \phi^o]) \tag{10.55}$$

belongs to $C^{0,\alpha}(\partial\Omega_1)_0 \times C^{0,\alpha}(\partial\Omega_2)_0$.

Proof. The membership of the pair of (10.55) in $C^{0,\alpha}(\partial\Omega_1) \times C^{0,\alpha}(\partial\Omega_2)$ is a consequence of the mapping properties of the operator W^t (cf. Theorem 6.7) and of the mapping properties of the integral operators with a real analytic kernel of Theorem A.28 (ii) in the Appendix. To prove the lemma it remains to show that

$$\int_{\partial\Omega_1} \mathcal{N}^1(\epsilon_1, \epsilon_2)[\phi_1, \phi_2, \phi^o] \, d\sigma = 0 \tag{10.56}$$

and

$$\int_{\partial\Omega_2} \mathcal{N}^2(\epsilon_1, \epsilon_2)[\phi_1, \phi_2, \phi^o] \, d\sigma = 0. \tag{10.57}$$

We confine ourselves to verify (10.56). The proof of (10.57) is indeed the same. To deduce (10.56) we exploit the definition of $\mathcal{N}^1(\epsilon_1, \epsilon_2)[\phi_1, \phi_2, \phi^o]$ in (10.49). We begin by observing that, by the jump formulas for the normal derivative of the single layer potential of Theorem 6.6 and by the first Green Identity of Theorem 4.2, we have

$$\int_{\partial\Omega_1} \left(\frac{1}{2}\phi_1 - W^t_{\Omega_1}[\phi_1]\right) d\sigma = - \int_{\partial\Omega_1} \nu_{\Omega_1} \cdot \nabla v^+_{\Omega_1}[\phi_1] \, d\sigma = 0 \,.$$

If $\epsilon_2 = 0$, the equality above is sufficient to conclude the proof. If instead $\epsilon_2 \neq 0$, then we have also to deal with the second term in the right-hand side of (10.49), and if both ϵ_1 and ϵ_2 are not 0, then we have to consider also the third term.

So, we now assume that $\epsilon_2 \neq 0$ and we study the second term in the expression for $\mathcal{N}^1(\epsilon_1, \epsilon_2)[\phi_1, \phi_2, \phi^o]$ in (10.49). We observe that the classical theorem of differentiability for integrals depending on a parameter and the chain rule imply that

$$\begin{aligned}\epsilon_2 \int_{\partial\Omega_2} &\phi_2(\eta)\, \nu_{\Omega_1}(\xi) \cdot \nabla S_2(p^1 - p^2 + \epsilon_2(\xi - \eta))\, d\sigma_\eta \\ &= \nu_{\Omega_1}(\xi) \cdot \nabla_\xi \left(\int_{\partial\Omega_2} \phi_2(\eta)\, S_2(p^1 - p^2 + \epsilon_2(\xi - \eta))\, d\sigma_\eta \right).\end{aligned}$$

Since the function that takes ξ to

$$\int_{\partial\Omega_2} \phi_2(\eta)\, S_2(p^1 - p^2 + \epsilon_2(\xi - \eta))\, d\sigma_\eta$$

is harmonic in a neighborhood of $\overline{\Omega_1}$ for $\epsilon_2 \in [-\delta_2, \delta_2]$ (cf. assumption (10.1)), we deduce by the first Green Identity of Theorem 4.2 that

$$\begin{aligned}\int_{\partial\Omega_1} \epsilon_2 \int_{\partial\Omega_2} &\phi_2(\eta)\, \nu_{\Omega_1}(\xi) \cdot \nabla S_2(p^1 - p^2 + \epsilon_2(\xi - \eta))\, d\sigma_\eta\, d\sigma_\xi \\ &= \int_{\partial\Omega_1} \nu_{\Omega_1}(\xi) \cdot \nabla_\xi \left(\int_{\partial\Omega_2} \phi_2(\eta)\, S_2(p^1 - p^2 + \epsilon_2(\xi - \eta))\, d\sigma_\eta \right) d\sigma_\xi = 0.\end{aligned}$$

Finally, we assume that both ϵ_1 and ϵ_2 are different from 0 and we consider the third term in the expression for $\mathcal{N}^1(\epsilon_1, \epsilon_2)[\phi_1, \phi_2, \phi^o]$ in (10.49). First we observe that the classical theorem of differentiability for integrals depending on a parameter and the chain rule imply that

$$\begin{aligned}\epsilon_1\epsilon_2 \int_{\partial\Omega^o} &\phi^o(y)\, \nu_{\Omega_1}(\xi) \cdot \nabla S_2(\epsilon_1 p^1 + \epsilon_1\epsilon_2\xi - y)\, d\sigma_y \\ &= \nu_{\Omega_1}(\xi) \cdot \nabla_\xi \left(\int_{\partial\Omega^o} \phi^o(y)\, S_2(\epsilon_1 p^1 + \epsilon_1\epsilon_2\xi - y) \right) d\sigma_y.\end{aligned}$$

Then we note that the function that takes ξ to

$$\int_{\partial\Omega^o} \phi^o(y)\, S_2(\epsilon_1 p^1 + \epsilon_1\epsilon_2\xi - y)\, d\sigma_y$$

is harmonic in a neighborhood of $\overline{\Omega_1}$ for $(\epsilon_1, \epsilon_2) \in [-\delta_1, \delta_1] \times [-\delta_2, \delta_2]$ (cf. assumptions (10.1) and (10.2)). Then the first Green Identity of Theorem 4.2 implies that

$$\int_{\partial\Omega_1} \epsilon_1\epsilon_2 \int_{\partial\Omega^o} \phi^o(y)\, \nu_{\Omega_1}(\xi)\cdot\nabla S_2(\epsilon_1 p^1 + \epsilon_1\epsilon_2\xi - y)\, d\sigma_y\, d\sigma_\xi$$
$$= \int_{\partial\Omega_1} \nu_{\Omega_1}(\xi)\cdot\nabla_\xi\Big(\int_{\partial\Omega^o} \phi^o(y)\, S_2(\epsilon_1 p^1 + \epsilon_1\epsilon_2\xi - y)\, d\sigma_y\Big)\, d\sigma_\xi = 0.$$

The validity of the lemma is now proved. □

We now show that the map $(\epsilon_1, \epsilon_2) \mapsto \mathcal{N}(\epsilon_1, \epsilon_2)$ is real analytic.

Lemma 10.19. *The map from* $]-\delta_1, \delta_1[\times]-\delta_2, \delta_2[$ *to*

$$\mathcal{L}(C^{0,\alpha}(\partial\Omega_1) \times C^{0,\alpha}(\partial\Omega_2) \times C^{0,\alpha}(\partial\Omega^o),$$
$$C^{0,\alpha}(\partial\Omega_1)_0 \times C^{0,\alpha}(\partial\Omega_2)_0 \times C^{0,\alpha}(\partial\Omega^o) \times \mathbb{R}^2)$$

that takes (ϵ_1, ϵ_2) *to* $\mathcal{N}(\epsilon_1, \epsilon_2)$ *is real analytic.*

Proof. The analyticity of $\mathcal{N}$ follows by Corollary A.29 of the Appendix on the analyticity of integral operators with real analytic kernel, by the continuity of the linear boundary operators $W^t_{\Omega_1}$, $W^t_{\Omega_2}$, and $W^t_{\Omega^o}$ (see Theorem 6.7), and by the continuity of the linear map from $C^{0,\alpha}(\partial\Omega_i)$, $i \in \{1, 2\}$, to $\mathbb{R}$ that takes a function to its integral. □

We now consider the operator $\mathcal{N}(0, 0)$. In the following Lemma 10.20 we investigate equation $\mathcal{N}(0, 0)[\phi_1, \phi_2, \phi^o] = (g_1, g_2, g^o, (\rho_1, \rho_2))$ with $j \in \{1, 2\}$ and we show that it has a unique solution.

Lemma 10.20. *For all*

$$(g_1, g_2, g^o, (\rho_1, \rho_2)) \in C^{0,\alpha}(\partial\Omega_1)_0 \times C^{0,\alpha}(\partial\Omega_2)_0 \times C^{0,\alpha}(\partial\Omega^o) \times \mathbb{R}^2$$

and all $(\phi_1, \phi_2, \phi^o) \in C^{0,\alpha}(\partial\Omega_1) \times C^{0,\alpha}(\partial\Omega_2) \times C^{0,\alpha}(\partial\Omega^o)$ *we have*

$$\mathcal{N}(0, 0)[\phi_1, \phi_2, \phi^o] = (g_1, g_2, g^o, (\rho_1, \rho_2)) \tag{10.58}$$

if and only if

$$\begin{cases} -\frac{1}{2}\phi_1 + W^t_{\Omega_1}[\phi_1] = -g_1, \\ -\frac{1}{2}\phi_2 + W^t_{\Omega_2}[\phi_2] = -g_2, \\ \frac{1}{2}\phi^o + W^t_{\Omega^o}[\phi^o] = g^o - (\rho_1 + \rho_2)\, \nu_{\Omega^o}\cdot(\nabla S_2)_{|\partial\Omega^o}, \\ \int_{\partial\Omega_1} \phi_1\, d\sigma = \rho_1, \\ \int_{\partial\Omega_2} \phi_2\, d\sigma = \rho_2. \end{cases} \tag{10.59}$$

In particular, for all

$$(g_1, g_2, g^o, (\rho_1, \rho_2)) \in C^{0,\alpha}(\partial\Omega_1)_0 \times C^{0,\alpha}(\partial\Omega_2)_0 \times C^{0,\alpha}(\partial\Omega^o) \times \mathbb{R}^2$$

there exists one and only one $(\phi_1, \phi_2, \phi^o) \in C^{0,\alpha}(\partial\Omega_1) \times C^{0,\alpha}(\partial\Omega_2) \times C^{0,\alpha}(\partial\Omega^o)$ *solution of Equation* (10.58).

Proof. (The reader may notice that this proof follows the same steps of the proof of Lemma 8.17.) The equivalence of (10.58) and (10.59) is a consequence of the definition of $\mathcal{N}$ in (10.49)–(10.52). We now prove the last statement of the lemma. Since $\mathbb{R}^2 \setminus \overline{\Omega^o}$ is connected, Corollary 6.15 implies that the operator $\frac{1}{2}I + W^t_{\Omega^o}$ is an isomorphism from $C^{0,\alpha}(\partial\Omega^o)$ to itself. As a consequence, there exists one and only one solution ϕ^o of the third equation of (10.59) (note that $\nu^o \in C^{0,\alpha}(\partial\Omega^o)$). We now consider the system consisting of the first and fourth equation of (10.59). That is,

$$\begin{cases} -\frac{1}{2}\phi_1 + W^t_{\Omega_1}[\phi_1] = -g_1, \\ \int_{\partial\Omega_1} \phi_1 \, d\sigma = \rho_1. \end{cases} \tag{10.60}$$

By the membership of g_1 in $C^{0,\alpha}(\partial\Omega_1)_0$ we have $\int_{\partial\Omega_1} g_1 \, d\sigma = 0$. Since Ω_1 is connected, it follows that g_1 is orthogonal to the null space of $-\frac{1}{2}I + W_{\Omega_1}$, which consists of the constant functions on $\partial\Omega_1$ (cf. Lemma 6.21). Thus, by the Fredholm Alternative of Theorem 5.8 there exists $\tilde{\phi}_1 \in C^{0,\alpha}(\partial\Omega_1)$ that satisfies the first equation of (10.60). Moreover, the Fredholm Alternative Theorem 5.8 also implies that the null space of $-\frac{1}{2}I + W^t_{\Omega_1}$ is one-dimensional and is accordingly spanned by a function $\phi \in C^{0,\alpha}(\partial\Omega_1)$. By Theorem 6.25, we have $\int_{\partial\Omega_1} \phi \, d\sigma \neq 0$. Then we can verify that the function

$$\phi_1 \equiv \tilde{\phi}_1 + \frac{\rho_1 - \int_{\partial\Omega_1} \tilde{\phi}_1 \, d\sigma}{\int_{\partial\Omega_1} \phi \, d\sigma} \phi$$

is a solution of (10.60). Finally, Theorem 6.25 implies that ϕ_1 is unique. A similar argument shows that there is a unique solution ϕ_2 of the second and last equation of (10.59). □

Then, Lemma 10.20 and the Open Mapping Theorem 2.2 imply the validity of the following Lemma 10.21.

Lemma 10.21. *The operator $\mathcal{N}(0,0)$ is an isomorphism (i.e., a linear homeomorphism) from $C^{0,\alpha}(\partial\Omega_1) \times C^{0,\alpha}(\partial\Omega_2) \times C^{0,\alpha}(\partial\Omega^o)$ to $C^{0,\alpha}(\partial\Omega_1)_0 \times C^{0,\alpha}(\partial\Omega_2)_0 \times C^{0,\alpha}(\partial\Omega^o) \times \mathbb{R}^2$.*

Proof. Lemma 10.19 implies that $\mathcal{N}(0,0)$ is a linear bounded operator. Then by the Open Mapping Theorem 2.2, it is sufficient to verify that it is a bijection in order to prove that it is an isomorphism. Accordingly, the statement follows by Lemma 10.20. □

Since the set of invertible operators is open in the space of bounded operators and since the map that takes an invertible operator to its inverse is real analytic (cf. Proposition A.15 of the Appendix), we deduce the following.

Lemma 10.22. *There exists* $0 < \delta_\Phi < \min\{\delta_1, \delta_2\}$ *such that* $\mathcal{N}(\epsilon_1, \epsilon_2)$ *is invertible for all* $(\epsilon_1, \epsilon_2) \in]-\delta_\Phi, \delta_\Phi[^2$ *and the map from* $]-\delta_\Phi, \delta_\Phi[^2$ *to*

$$\mathcal{L}(C^{0,\alpha}(\partial\Omega_1)_0 \times C^{0,\alpha}(\partial\Omega_2)_0 \times C^{0,\alpha}(\partial\Omega^o) \times \mathbb{R}^2, \\ C^{0,\alpha}(\partial\Omega_1) \times C^{0,\alpha}(\partial\Omega_2) \times C^{0,\alpha}(\partial\Omega^o))$$

that takes (ϵ_1, ϵ_2) *to the inverse operator* $\mathcal{N}(\epsilon_1, \epsilon_2)^{(-1)}$ *is real analytic.*

Finally, by Lemma 10.22 we deduce that for $j \in \{1, 2\}$ the map that takes (ϵ_1, ϵ_2) to the triple in (10.54) has a real analytic continuation in a neighborhood of $(0, 0)$.

Proposition 10.23. *Let* $j \in \{1, 2\}$. *There exists a real analytic map*

$$\Phi^{(j)} \equiv (\Phi^{1,(j)}, \Phi^{2,(j)}, \Phi^{o,(j)}) :]-\delta_\Phi, \delta_\Phi[^2 \to C^{0,\alpha}(\partial\Omega_1) \times C^{0,\alpha}(\partial\Omega_2) \times C^{0,\alpha}(\partial\Omega^o)$$

such that

$$\mathcal{N}(\epsilon_1, \epsilon_2)[\Phi^{(j)}[\epsilon_1, \epsilon_2]] = (0, 0, 0, (\delta_{1,j}, \delta_{2,j})) \quad \forall(\epsilon_1, \epsilon_2) \in]-\delta_\Phi, \delta_\Phi[^2. \quad (10.61)$$

In particular,

$$(\phi^{1,(j)}_{\epsilon_1,\epsilon_2}, \phi^{2,(j)}_{\epsilon_1,\epsilon_2}, \phi^{o,(j)}_{\epsilon_1,\epsilon_2}) = \Phi^{(j)}[\epsilon_1, \epsilon_2] \quad \forall(\epsilon_1, \epsilon_2) \in]0, \delta_\Phi[^2. \quad (10.62)$$

Proof. It suffices to set

$$\Phi^{(j)}[\epsilon_1, \epsilon_2] \equiv \mathcal{N}(\epsilon_1, \epsilon_2)^{(-1)}[0, 0, 0, (\delta_{1,j}, \delta_{2,j})] \quad \forall(\epsilon_1, \epsilon_2) \in]-\delta_\Phi, \delta_\Phi[^2.$$

Then the statement is a consequence of Proposition 10.17 and of Lemma 10.22 (see also the argument in the proof of Proposition 10.6). □

In particular, Proposition 10.23 implies that

$$\int_{\partial\Omega_i} \Phi^{i,(j)}[\epsilon_1, \epsilon_2]\, d\sigma = \delta_{i,j} \qquad \forall(\epsilon_1, \epsilon_2) \in]-\delta_\Phi, \delta_\Phi[^2,\ i, j \in \{1, 2\}. \quad (10.63)$$

10.4.3 Analysis of System (10.40)

We now turn to system (10.40) and prove a real analytic continuation result for the density $\mu_{\epsilon_1,\epsilon_2}$. As for system (10.39), we begin by splitting the first equation of (10.40) over $\partial\Omega_1(\epsilon_1, \epsilon_2)$, $\partial\Omega_2(\epsilon_1, \epsilon_2)$, and $\partial\Omega^o$, and then we perform a change of variable in the integrals over the (ϵ_1, ϵ_2)-dependent boundaries $\partial\Omega_1(\epsilon_1, \epsilon_2)$ and $\partial\Omega_2(\epsilon_1, \epsilon_2)$ (see also Section 8.4.2 for the case of one hole). Taking into account equalities (10.42) and (10.43) and defining the new densities

$$
\begin{aligned}
\psi^1_{\epsilon_1,\epsilon_2}(\xi) &\equiv \mu_{\epsilon_1,\epsilon_2}(\epsilon_1 p^1 + \epsilon_1\epsilon_2\xi) && \forall \xi \in \partial\Omega_1\,,\\
\psi^2_{\epsilon_1,\epsilon_2}(\xi) &\equiv \mu_{\epsilon_1,\epsilon_2}(\epsilon_1 p^2 + \epsilon_1\epsilon_2\xi) && \forall \xi \in \partial\Omega_2\,,\\
\psi^o_{\epsilon_1,\epsilon_2}(x) &\equiv \mu_{\epsilon_1,\epsilon_2}(x) && \forall x \in \partial\Omega^o,
\end{aligned}
\tag{10.64}
$$

we find that system (10.40) is equivalent to the system of the following three equations

$$
\begin{aligned}
&\frac{1}{2}\psi^1_{\epsilon_1,\epsilon_2}(\xi) - W_{\Omega_1}[\psi^1_{\epsilon_1,\epsilon_2}](\xi)\\
&\quad + \epsilon_2 \int_{\partial\Omega_2} \psi^2_{\epsilon_1,\epsilon_2}(\eta)\, \nu_{\Omega_2}(\eta)\cdot \nabla S_2(p^1 - p^2 + \epsilon_2(\xi-\eta))\, d\sigma_\eta\\
&\quad - \int_{\partial\Omega^o} \psi^o(y)\, \nu_{\Omega^o}(y)\cdot \nabla S_2(\epsilon_1 p^1 + \epsilon_1\epsilon_2\xi - y)\, d\sigma_y\\
&\quad = f_1(\xi) - \sum_{i=1,2}\int_{\partial\Omega_i} f_i\, \phi^{i,(1)}_{\epsilon_1,\epsilon_2}\, d\sigma - \int_{\partial\Omega^o} f^o\, \phi^{o,(1)}_{\epsilon_1,\epsilon_2}\, d\sigma\\
&\qquad\qquad\qquad\qquad\qquad\qquad\qquad\qquad\qquad \forall \xi\in\partial\Omega_1,\\
&\frac{1}{2}\psi^2_{\epsilon_1,\epsilon_2}(\xi) - W_{\Omega_2}[\psi^2_{\epsilon_1,\epsilon_2}](\xi)\\
&\quad + \epsilon_2 \int_{\partial\Omega_1} \psi^1_{\epsilon_1,\epsilon_2}(\eta)\, \nu_{\Omega_1}(\eta)\cdot \nabla S_2(p^2 - p^1 + \epsilon_2(\xi-\eta))\, d\sigma_\eta\\
&\quad - \int_{\partial\Omega^o} \psi^o(y)\, \nu_{\Omega^o}(y)\cdot \nabla S_2(\epsilon_1 p^2 + \epsilon_1\epsilon_2\xi - y)\, d\sigma_y\\
&\quad = f_2(\xi) - \sum_{i=1,2}\int_{\partial\Omega_i} f_i\, \phi^{i,(2)}_{\epsilon_1,\epsilon_2}\, d\sigma - \int_{\partial\Omega^o} f^o\, \phi^{o,(2)}_{\epsilon_1,\epsilon_2}\, d\sigma\\
&\qquad\qquad\qquad\qquad\qquad\qquad\qquad\qquad\qquad \forall \xi\in\partial\Omega_2,\\
&\frac{1}{2}\psi^o_{\epsilon_1,\epsilon_2}(x) + W_{\Omega^o}[\psi^o_{\epsilon_1,\epsilon_2}](x)\\
&\quad + \epsilon_1\epsilon_2 \sum_{i=1}^{2}\int_{\partial\Omega_i} \psi^i_{\epsilon_1,\epsilon_2}(\eta)\, \nu_{\Omega_i}(\eta)\cdot \nabla S_2(x - \epsilon_1 p^i - \epsilon_1\epsilon_2\eta)\, d\sigma_\eta = f^o(x)\\
&\qquad\qquad\qquad\qquad\qquad\qquad\qquad\qquad\qquad \forall x\in\partial\Omega^o,
\end{aligned}
$$

with the additional condition that

$$
\int_{\partial\Omega_1} \psi^1_{\epsilon_1,\epsilon_2}\, d\sigma = 0 \quad\text{and}\quad \int_{\partial\Omega_2} \psi^2_{\epsilon_1,\epsilon_2}\, d\sigma = 0. \tag{10.65}
$$

Then for all $(\epsilon_1,\epsilon_2) \in]-\delta_1,\delta_1[\times]-\delta_2,\delta_2[$ we introduce the auxiliary operator

$$
\mathcal{M}(\epsilon_1,\epsilon_2) \equiv (\mathcal{M}_1(\epsilon_1,\epsilon_2),\, \mathcal{M}_2(\epsilon_1,\epsilon_2),\, \mathcal{M}^o(\epsilon_1,\epsilon_2))
$$

from $C^{1,\alpha}(\partial\Omega_1)_0 \times C^{1,\alpha}(\partial\Omega_2)_0 \times C^{1,\alpha}(\partial\Omega^o)$ to $C^{1,\alpha}(\partial\Omega_1) \times C^{1,\alpha}(\partial\Omega_2) \times C^{1,\alpha}(\partial\Omega^o)$ defined by

$$\begin{aligned}\mathcal{M}_1(\epsilon_1,\epsilon_2)[\psi_1,\psi_2,\psi^o](\xi) \equiv \frac{1}{2}\psi_1(\xi) - W_{\Omega_1}[\psi_1](\xi) \\ + \epsilon_2 \int_{\partial\Omega_2} \psi_2(\eta)\,\nu_{\Omega_2}(\eta)\cdot\nabla S_2(p^1 - p^2 + \epsilon_2(\xi-\eta))\,d\sigma_\eta \\ - \int_{\partial\Omega^o} \psi^o(y)\,\nu_{\Omega^o}(y)\cdot\nabla S_2(\epsilon_1 p^1 + \epsilon_1\epsilon_2\xi - y)\,d\sigma_y \qquad \forall\xi\in\partial\Omega_1,\end{aligned} \tag{10.66}$$

$$\begin{aligned}\mathcal{M}_2(\epsilon_1,\epsilon_2)[\psi_1,\psi_2,\psi^o](\xi) \equiv \frac{1}{2}\psi_2(\xi) - W_{\Omega_2}[\psi_2](\xi) \\ + \epsilon_2 \int_{\partial\Omega_1} \psi_1(\eta)\,\nu_{\Omega_1}(\eta)\cdot\nabla S_2(p^2 - p^1 + \epsilon_2(\xi-\eta))\,d\sigma_\eta \\ - \int_{\partial\Omega^o} \psi^o(y)\,\nu_{\Omega^o}(y)\cdot\nabla S_2(\epsilon_1 p^2 + \epsilon_1\epsilon_2\xi - y)\,d\sigma_y \qquad \forall\xi\in\partial\Omega_2,\end{aligned} \tag{10.67}$$

$$\begin{aligned}\mathcal{M}^o(\epsilon_1,\epsilon_2)[\psi_1,\psi_2,\psi^o](\xi) \equiv \frac{1}{2}\psi^o(x) + W_{\Omega^o}[\psi^o](x) \\ + \epsilon_1\epsilon_2 \sum_{i=1}^{2} \int_{\partial\Omega_i} \psi^i(\eta)\,\nu_{\Omega_i}(\eta)\cdot\nabla S_2(x - \epsilon_1 p^i - \epsilon_1\epsilon_2\eta)\,d\sigma_\eta \quad \forall x\in\partial\Omega^o,\end{aligned} \tag{10.68}$$

for all $(\psi_1,\psi_2,\psi^o) \in C^{1,\alpha}(\partial\Omega_1)_0 \times C^{1,\alpha}(\partial\Omega_2)_0 \times C^{1,\alpha}(\partial\Omega^o)$. It follows that for $(\epsilon_1,\epsilon_2) \in]0,\delta_1[\times]0,\delta_2[$ the system of equations (10.40) is equivalent to

$$\begin{aligned}\mathcal{M}(\epsilon_1,\epsilon_2)[\psi^1_{\epsilon_1,\epsilon_2},\psi^2_{\epsilon_1,\epsilon_2},\psi^o_{\epsilon_1,\epsilon_2}] \\ = \Bigg(f_1(\xi) - \sum_{i=1,2}\int_{\partial\Omega_i} f_i\,\phi^{i,(1)}_{\epsilon_1,\epsilon_2}\,d\sigma - \int_{\partial\Omega^o} f^o\,\phi^{o,(1)}_{\epsilon_1,\epsilon_2}\,d\sigma, \\ f_2(\xi) - \sum_{i=1,2}\int_{\partial\Omega_i} f_i\,\phi^{i,(2)}_{\epsilon_1,\epsilon_2}\,d\sigma - \int_{\partial\Omega^o} f^o\,\phi^{o,(2)}_{\epsilon_1,\epsilon_2}\,d\sigma,\ f^o \Bigg)\end{aligned} \tag{10.69}$$

and the membership of $\psi^1_{\epsilon_1,\epsilon_2}$ in $C^{1,\alpha}(\partial\Omega_1)_0$ and $\psi^2_{\epsilon_1,\epsilon_2}$ in $C^{1,\alpha}(\partial\Omega_2)_0$ implies condition (10.65) (see (10.44) for the definition of $\phi^{i,(j)}_{\epsilon_1,\epsilon_2}$ and $\phi^{o,(j)}_{\epsilon_1,\epsilon_2}$, $i,j\in\{1,2\}$). Then, Equation (10.69) for $(\psi^1_{\epsilon_1,\epsilon_2},\psi^2_{\epsilon_1,\epsilon_2},\psi^o_{\epsilon_1,\epsilon_2})$ is equivalent to Equation (10.40) for $\mu_{\epsilon_1,\epsilon_2}$, with $(\psi^1_{\epsilon_1,\epsilon_2},\psi^2_{\epsilon_1,\epsilon_2},\psi^o_{\epsilon_1,\epsilon_2})$ and $\mu_{\epsilon_1,\epsilon_2}$ related by (10.64). Accordingly, the existence and uniqueness result of Theorem 6.31 (i) implies that (10.69) has a unique solution. Namely, we have the following.

Proposition 10.24. *For all* $(\epsilon_1,\epsilon_2) \in]0,\delta_1[\times]0,\delta_2[$ *there exists one and only one solution* $(\psi^1_{\epsilon_1,\epsilon_2},\psi^2_{\epsilon_1,\epsilon_2},\psi^o_{\epsilon_1,\epsilon_2}) \in C^{1,\alpha}(\partial\Omega_1)_0\times C^{1,\alpha}(\partial\Omega_2)_0\times C^{1,\alpha}(\partial\Omega^o)$ *of* (10.69).

By Proposition 10.24 we can define a map from $]0,\delta_1[\times]0,\delta_2[$ to $C^{1,\alpha}(\partial\Omega_1)_0 \times C^{1,\alpha}(\partial\Omega_2)_0 \times C^{1,\alpha}(\partial\Omega^o)$ that takes a pair (ϵ_1,ϵ_2) to the unique solution

$$(\psi^1_{\epsilon_1,\epsilon_2},\psi^2_{\epsilon_1,\epsilon_2},\psi^o_{\epsilon_1,\epsilon_2}) \tag{10.70}$$

of (10.69). Our aim in this subsection is to show that such a map has a real analytic continuation in a neighborhood of $(0,0)$. We begin by proving the following.

Lemma 10.25. *The map from* $]-\delta_1,\delta_1[\times]-\delta_2,\delta_2[$ *to*

$$\mathcal{L}(C^{1,\alpha}(\partial\Omega_1)_0 \times C^{1,\alpha}(\partial\Omega_2)_0 \times C^{1,\alpha}(\partial\Omega^o),\\ C^{1,\alpha}(\partial\Omega_1) \times C^{1,\alpha}(\partial\Omega_2) \times C^{1,\alpha}(\partial\Omega^o))$$

that takes (ϵ_1,ϵ_2) *to* $\mathcal{M}(\epsilon_1,\epsilon_2)$ *is real analytic.*

Proof. The analyticity of $\mathcal{M}$ follows by definition (10.66)–(10.68), by Corollary A.29 of the Appendix on the analyticity of integral operators with real analytic kernel, and by the continuity of the linear boundary operators W_{Ω_1}, W_{Ω_2}, and W_{Ω^o} (see Theorem 4.37). □

Then we consider the operator $\mathcal{M}(0,0)$. By the definition of $\mathcal{M}$ in (10.66)–(10.68), we have the following.

Proposition 10.26. *For all* $(g_1,g_2,g^o) \in C^{1,\alpha}(\partial\Omega_1) \times C^{1,\alpha}(\partial\Omega_2) \times C^{1,\alpha}(\partial\Omega^o)$ *and all* $(\psi_1,\psi_2,\psi^o) \in C^{1,\alpha}(\partial\Omega_1)_0 \times C^{1,\alpha}(\partial\Omega_2)_0 \times C^{1,\alpha}(\partial\Omega^o)$ *we have*

$$\mathcal{M}(0,0)[\psi_1,\psi_2,\psi^o] = (g_1,g_2,g^o) \tag{10.71}$$

if and only if

$$\begin{cases} -\frac{1}{2}\psi_1 + W_{\Omega_1}[\psi_1] = w_{\Omega^o}[\psi^o](0) - g_1\,, \\ -\frac{1}{2}\psi_2 + W_{\Omega_2}[\psi_2] = w_{\Omega^o}[\psi^o](0) - g_2\,, \\ \frac{1}{2}\psi^o + W_{\Omega^o}[\psi^o] = g^o\,. \end{cases} \tag{10.72}$$

We now show that $\mathcal{M}(0,0)$ is injective but not surjective.

Lemma 10.27. *The operator* $\mathcal{M}(0,0)$ *is injective and its image* $\operatorname{Im}\mathcal{M}(0,0)$ *is the closed subspace of* $C^{1,\alpha}(\partial\Omega_1) \times C^{1,\alpha}(\partial\Omega_2) \times C^{1,\alpha}(\partial\Omega^o)$ *consisting of the triples* (g_1,g_2,g^o) *such that*

$$\int_{\partial\Omega_1} g_1 \Phi^{1,(1)}[0,0]\,d\sigma + \int_{\partial\Omega_2} g_2\,\Phi^{2,(1)}[0,0]\,d\sigma + \int_{\partial\Omega^o} g^o\,\Phi^{o,(1)}[0,0]\,d\sigma = 0 \tag{10.73}$$

and

$$\int_{\partial\Omega_1} g_1 \Phi^{1,(2)}[0,0]\,d\sigma + \int_{\partial\Omega_2} g_2\,\Phi^{2,(2)}[0,0]\,d\sigma + \int_{\partial\Omega^o} g^o\,\Phi^{o,(2)}[0,0]\,d\sigma = 0 \tag{10.74}$$

(cf. Proposition 10.23 for the definition of $\Phi^{i,(j)}[0,0]$ *and* $\Phi^{o,(j)}[0,0]$, $i,j \in \{1,2\}$*).*

Proof. To prove the lemma, we show that (10.71) has a solution (ψ_1,ψ_2,ψ^o) in $C^{1,\alpha}(\partial\Omega_1)_0 \times C^{1,\alpha}(\partial\Omega_2)_0 \times C^{1,\alpha}(\partial\Omega^o)$ if and only if the triple (g_1,g_2,g^o) in $C^{1,\alpha}(\partial\Omega_1) \times C^{1,\alpha}(\partial\Omega_2) \times C^{1,\alpha}(\partial\Omega^o)$ satisfies conditions (10.73) and (10.74). Moreover, if it exists, such a solution is unique.

We note that (10.71) is a system of three equations: $\mathcal{M}_1(0,0)[\psi_1,\psi_2,\psi^o] = g_1$, $\mathcal{M}_2(0,0)[\psi_1,\psi_2,\psi^o] = g_2$, and $\mathcal{M}^o(0,0)[\psi_1,\psi_2,\psi^o] = g^o$. We now consider the system consisting of the first two of such equations. The definitions of $\mathcal{M}_1$ and $\mathcal{M}_2$ in (10.66) and (10.67) imply that such system is equivalent to

$$\begin{cases} -\frac{1}{2}\psi_1 + W_{\Omega_1}[\psi_1] = w_{\Omega^o}[\psi^o](0) - g_1\,, \\ -\frac{1}{2}\psi_2 + W_{\Omega_2}[\psi_2] = w_{\Omega^o}[\psi^o](0) - g_2 \end{cases} \tag{10.75}$$

(see also Definition 4.27 of double layer potential). Then we denote by J the operator from $C^{1,\alpha}(\partial\Omega_1) \times C^{1,\alpha}(\partial\Omega_2)$ to $C^{1,\alpha}(\partial\Omega_1) \times C^{1,\alpha}(\partial\Omega_2)$ defined by

$$J \equiv \left(-\frac{1}{2}I + W_{\Omega_1}\,,\; -\frac{1}{2}I + W_{\Omega_2}\right)$$

and we rewrite (10.75) as

$$J[\psi_1,\psi_2] = \left(w_{\Omega^o}[\psi^o](0) - g_1\,,\; w_{\Omega^o}[\psi^o](0) - g_2\right). \tag{10.76}$$

The compactness of W_{Ω_1} and W_{Ω_2} of Theorem 6.8 and the Fredholm Alternative of Theorem 5.8 imply that the last equation has a solution in $C^{1,\alpha}(\partial\Omega_1) \times C^{1,\alpha}(\partial\Omega_2)$ if and only if

$$\left(w_{\Omega^o}[\psi^o](0) - g_1\,,\; w_{\Omega^o}[\psi^o](0) - g_2\right)$$

is orthogonal to the null space of

$$J^t \equiv \left(-\frac{1}{2}I + W^t_{\Omega_1}\,,\; -\frac{1}{2}I + W^t_{\Omega_2}\right).$$

By Lemmas 6.34 and 10.20 and by Proposition 10.23 the null space of J^t is spanned by the pairs of functions $(\Phi^{1,(1)}[0,0]\,,\, \Phi^{2,(1)}[0,0])$ and $(\Phi^{1,(2)}[0,0]\,,\, \Phi^{2,(2)}[0,0])$. It follows that (10.76) has a solution if and only if the following two conditions hold,

$$\int_{\partial\Omega_1} \left(w_{\Omega^o}[\psi^o](0) - g_1\right) \Phi^{1,(1)}[0,0]\, d\sigma + \int_{\partial\Omega_2} \left(w_{\Omega^o}[\psi^o](0) - g_2\right) \Phi^{2,(1)}[0,0]\, d\sigma = 0$$

and

$$\int_{\partial\Omega_1} \left(w_{\Omega^o}[\psi^o](0) - g_1\right) \Phi^{1,(2)}[0,0]\, d\sigma + \int_{\partial\Omega_2} \left(w_{\Omega^o}[\psi^o](0) - g_2\right) \Phi^{2,(2)}[0,0]\, d\sigma = 0.$$

Since $\int_{\partial\Omega^h} \Phi^{i,(j)}[0,0]\, d\sigma = \delta_{i,j}$ (cf. (10.63)) these conditions are equivalent to

$$\int_{\partial\Omega_1} g_1\, \Phi^{1,(1)}[0,0]\, d\sigma + \int_{\partial\Omega_2} g_2\, \Phi^{2,(1)}[0,0]\, d\sigma = w_{\Omega^o}[\psi^o](0) \tag{10.77}$$

and

$$\int_{\partial\Omega_1} g_1\, \Phi^{1,(2)}[0,0]\, d\sigma + \int_{\partial\Omega_2} g_2\, \Phi^{2,(2)}[0,0]\, d\sigma = w_{\Omega^o}[\psi^o](0)\,. \tag{10.78}$$

If conditions (10.77) and (10.78) are satisfied, then Theorem 6.24, which characterizes the null spaces of $-\frac{1}{2}I + W_{\Omega_1}$ and $-\frac{1}{2}I + W_{\Omega_2}$, implies that (10.76) (and thus (10.75)) has one and only one solution (ψ_1, ψ_2) in $C^{1,\alpha}(\partial\Omega_1)_0 \times C^{1,\alpha}(\partial\Omega_2)_0$.

We now turn to equation $\mathcal{M}^o(0,0)[\psi_1,\psi_2,\psi^o] = g^o$, which is equivalent to the third equation in (10.72). By Corollary 6.15, such equation has a unique solution $\psi^o \in C^{1,\alpha}(\partial\Omega^o)$ for all given $g^o \in C^{1,\alpha}(\partial\Omega^o)$. Moreover, by Theorem 6.5 the operator $\frac{1}{2}I + W^t_{\Omega^o}$ is transpose to $\frac{1}{2}I + W_{\Omega^o}$. As a consequence, if ψ^o is the solution of the third equation in (10.72), then we have

$$\begin{aligned}\int_{\partial\Omega^o} g^o\, \Phi^{o,(1)}[0,0]\, d\sigma &= \int_{\partial\Omega^o} \left(\frac{1}{2}\psi^o + W_{\Omega^o}[\psi^o]\right) \Phi^{o,(1)}[0,0]\, d\sigma \\ &= \int_{\partial\Omega^o} \psi^o \left(\frac{1}{2}I + W^t_{\Omega^o}\right) \left[\Phi^{o,(1)}[0,0]\right] d\sigma\,.\end{aligned} \tag{10.79}$$

Since $\Phi^{(1)}[0,0]$ is a solution of (10.61) with $(\epsilon_1,\epsilon_2) = (0,0)$, Lemma 10.20 implies that

$$\left(\frac{1}{2}I + W^t_{\Omega^o}\right)[\Phi^{o,(1)}[0,0]] = -(\delta_{1,1}+\delta_{1,2})\nu_{\Omega^o}\cdot(\nabla S_2)_{|\partial\Omega^o} = -\nu_{\Omega^o}\cdot(\nabla S_2)_{|\partial\Omega^o}$$

and thus the last integral in the right-hand side of (10.79) equals

$$-\int_{\partial\Omega^o} \psi^o\, \nu_{\Omega^o}\cdot\nabla S_2\, d\sigma\,.$$

Then, by (10.79) and by the Definition 4.27 of double layer potential we obtain that

$$\int_{\partial\Omega^o} g^o\, \Phi^{o,(1)}[0,0]\, d\sigma = -w_{\Omega^o}[\psi^o](0)\,.$$

Similarly, we can show that

$$\int_{\partial\Omega^o} g^o\, \Phi^{o,(2)}[0,0]\, d\sigma = -w_{\Omega^o}[\psi^o](0)\,.$$

Hence, conditions (10.77) and (10.78) are equivalent to (10.73) and (10.74). The lemma is now proved. □

Since $\mathcal{M}(0,0)$ is not invertible we cannot exploit the real analyticity of the map that takes an invertible operator to its inverse to prove that the map that takes (ϵ_1,ϵ_2) to the triple in (10.70) is real analytic. Then we resort to Lemma 8.25 and we obtain the following Proposition 10.28 where we prove a real analytic continuation result for the map that takes (ϵ_1, ϵ_2) to the triple $(\psi^1_{\epsilon_1,\epsilon_2}, \psi^2_{\epsilon_1,\epsilon_2}, \psi^o_{\epsilon_1,\epsilon_2})$.

Proposition 10.28. *Let δ_Φ be as in Proposition 10.23. There exist $0 < \delta_\Psi \leq \delta_\Phi$ and a real analytic map*

$$\Psi \equiv (\Psi_1, \Psi_2, \Psi^o) \, :] - \delta_\Psi, \delta_\Psi[^2 \to C^{1,\alpha}(\partial\Omega_1)_0 \times C^{1,\alpha}(\partial\Omega_2)_0 \times C^{1,\alpha}(\partial\Omega^o)$$

such that

$$\begin{aligned}
&\mathcal{M}(\epsilon_1, \epsilon_2) \left[\Psi^1[\epsilon_1, \epsilon_2], \Psi^2[\epsilon_1, \epsilon_2], \Psi^o[\epsilon_1, \epsilon_2]\right] \\
&\quad = \Bigg(f_1 - \sum_{i=1,2} \int_{\partial\Omega_i} f_i \, \Phi^{i,(1)}[\epsilon_1, \epsilon_2] \, d\sigma - \int_{\partial\Omega^o} f^o \, \Phi^{o,(1)}[\epsilon_1, \epsilon_2] \, d\sigma, \\
&\qquad f_2 - \sum_{i=1,2} \int_{\partial\Omega_i} f_i \, \Phi^{i,(2)}[\epsilon_1, \epsilon_2] \, d\sigma - \int_{\partial\Omega^o} f^o \, \Phi^{o,(2)}[\epsilon_1, \epsilon_2] \, d\sigma, \; f^o \Bigg)
\end{aligned} \tag{10.80}$$

for all $(\epsilon_1, \epsilon_2) \in] - \delta_\Psi, \delta_\Psi[^2$ (cf. Proposition 10.23 for the definition of $\Phi^{i,(j)}$ and $\Phi^{o,(j)}$, $i, j \in \{1, 2\}$). In particular,

$$(\psi^1_{\epsilon_1,\epsilon_2}, \psi^2_{\epsilon_1,\epsilon_2}, \psi^o_{\epsilon_1,\epsilon_2}) = \Psi[\epsilon] \qquad \forall (\epsilon_1, \epsilon_2) \in]0, \delta_\Psi[^2 \, . \tag{10.81}$$

Proof. To show the existence of Ψ, we plan to apply Lemma 8.25 with

$$\begin{aligned}
&E = \mathbb{R}^2 \, , \qquad U =] - \delta_\Phi, \delta_\Phi[^2 \, , \qquad \varepsilon = (\epsilon_1, \epsilon_2) \, , \\
&X = C^{1,\alpha}(\partial\Omega_1)_0 \times C^{1,\alpha}(\partial\Omega_2)_0 \times C^{1,\alpha}(\partial\Omega^o) \, , \\
&Y = C^{1,\alpha}(\partial\Omega_1) \times C^{1,\alpha}(\partial\Omega_2) \times C^{1,\alpha}(\partial\Omega^o) \, , \\
&Z = \mathbb{R}^2 \, , \qquad \mathcal{A} = \mathcal{M} \, ,
\end{aligned}$$

with the functions f (of Lemma 8.25) defined by

$$\begin{aligned}
f(\epsilon_1, \epsilon_2) = &\Bigg(f_1 - \sum_{i=1,2} \int_{\partial\Omega_i} f_i \, \Phi^{i,(1)}[\epsilon_1, \epsilon_2] \, d\sigma - \int_{\partial\Omega^o} f^o \, \Phi^{o,(1)}[\epsilon_1, \epsilon_2] \, d\sigma, \\
&\quad f_2 - \sum_{i=1,2} \int_{\partial\Omega_i} f_i \, \Phi^{i,(2)}[\epsilon_1, \epsilon_2] \, d\sigma - \int_{\partial\Omega^o} f^o \, \Phi^{o,(2)}[\epsilon_1, \epsilon_2] \, d\sigma, \; f^o \Bigg) \, ,
\end{aligned}$$

for all $(\epsilon_1, \epsilon_2) \in] - \delta_\Phi, \delta_\Phi[^2$, and with $\mathcal{B} = (\mathcal{B}_1, \mathcal{B}_2)$ given by

$$\mathcal{B}_j(\epsilon_1, \epsilon_2)[g_1, g_2, g^o] = \sum_{i=1,2} \int_{\partial\Omega_i} g_i \, \Phi^{i,(j)}[\epsilon_1, \epsilon_2] \, d\sigma + \int_{\partial\Omega^o} g^o \, \Phi^{o,(j)}[\epsilon_1, \epsilon_2] \, d\sigma \, ,$$

for all $j \in \{1,2\}$, $(\epsilon_1,\epsilon_2) \in]-\delta_\Phi,\delta_\Phi[^2$, and $(g_1,g_2,g^o) \in C^{1,\alpha}(\partial\Omega_1) \times C^{1,\alpha}(\partial\Omega_2) \times C^{1,\alpha}(\partial\Omega^o)$. Then we have to verify that the assumptions of Lemma 8.25 are satisfied with this specific choice. As a first step, we observe that our map $\mathcal{A}$ is real analytic by Lemma 10.25. Moreover, Proposition 10.23 implies that Φ is real analytic, and accordingly the maps f and $\mathcal{B}$ are real analytic. In particular, $\mathcal{B}$ is continuous. The first condition (8.59) of Lemma 8.25 concerns the equality of the null space of $\mathcal{B}(0,0)$ and the image of $\mathcal{A}(0,0)$. In our case such equality is a consequence of Lemma 10.27. The second condition (8.60) requires that

$$\mathcal{B}_j(\epsilon_1,\epsilon_2)\left[\mathcal{A}(\epsilon_1,\epsilon_2)[\psi_1,\psi_2,\psi^o] - f(\epsilon_1,\epsilon_2)\right] = 0 \tag{10.82}$$

for all $j \in \{1,2\}$, $(\epsilon_1,\epsilon_2) \in]-\delta_\Phi,\delta_\Phi[^2$ and $(\psi_1,\psi_2,\psi^o) \in C^{1,\alpha}(\partial\Omega_1)_0 \times C^{1,\alpha}(\partial\Omega_2)_0 \times C^{1,\alpha}(\partial\Omega^o)$. To prove that (10.82) holds true, we remember equality (10.63) and we deduce by a straightforward computation that

$$\mathcal{B}_j(\epsilon_1,\epsilon_2)[f(\epsilon_1,\epsilon_2)] = 0 \qquad \forall j \in \{1,2\}\,,\ (\epsilon_1,\epsilon_2) \in]-\delta_\Phi,\delta_\Phi[^2\,. \tag{10.83}$$

Moreover,

$$\begin{aligned}\mathcal{B}_j(\epsilon_1,\epsilon_2)&[\mathcal{A}(\epsilon_1,\epsilon_2)[\psi_1,\psi_2,\psi^o]] \\ &= \sum_{i=1,2}\int_{\partial\Omega_i} \mathcal{M}_i(\epsilon_1,\epsilon_2)[\psi_1,\psi_2,\psi^o]\ \Phi^{i,(j)}[\epsilon_1,\epsilon_2]\,d\sigma \\ &\quad + \int_{\partial\Omega^o} \mathcal{M}^o(\epsilon_1,\epsilon_2)[\psi_1,\psi_2,\psi^o]\ \Phi^{o,(j)}[\epsilon_1,\epsilon_2]\,d\sigma\end{aligned}$$

for all $j \in \{1,2\}$ and $(\epsilon_1,\epsilon_2) \in]-\delta_\Phi,\delta_\Phi[^2$. Then, changing the variable of integration in the integrals over $\partial\Omega_1$ and $\partial\Omega_2$ to obtain integrals over the (ϵ_1,ϵ_2)-dependent boundaries $\partial\Omega_1(\epsilon_1,\epsilon_2)$ and $\partial\Omega_2(\epsilon_1,\epsilon_2)$ and taking into account definition (4.71) of double layer potential, we can verify that

$$\begin{aligned}\mathcal{M}_i(\epsilon_1,\epsilon_2)&[\psi_1,\psi_2,\psi^o](\xi) \\ &= \frac{1}{2}\mu_{\epsilon_1,\epsilon_2}(\epsilon_1 p^i + \epsilon_1\epsilon_2\xi) + W_{\Omega(\epsilon_1,\epsilon_2)}[\mu_{\epsilon_1,\epsilon_2}](\epsilon_1 p^i + \epsilon_1\epsilon_2\xi)\,,\end{aligned}$$

for all $i \in \{1,2\}$ and $\xi \in \partial\Omega_i$, and

$$\mathcal{M}^o(\epsilon_1,\epsilon_2)[\psi_1,\psi_2,\psi^o](x) = \frac{1}{2}\mu_{\epsilon_1,\epsilon_2}(x) + W_{\Omega(\epsilon_1,\epsilon_2)}[\mu_{\epsilon_1,\epsilon_2}](x) \qquad \forall x \in \partial\Omega^o$$

with $\mu_{\epsilon_1,\epsilon_2} \in C^{1,\alpha}(\partial\Omega(\epsilon_1,\epsilon_2))$ which satisfies the conditions

$$\mu_{\epsilon_1,\epsilon_2}(\epsilon_1 p^i + \epsilon_1\epsilon_2\xi) = \psi_i(\xi) \quad \forall i \in \{1,2\}\,,\ \xi \in \partial\Omega_i$$

and

$$\mu_{\epsilon_1,\epsilon_2}(x) = \psi^o(x) \quad \forall x \in \partial\Omega^o$$

(here we have to follow backward the calculation carried out at the beginning of this section to introduce the operator $\mathcal{M}(\epsilon_1, \epsilon_2)$). By equalities (10.44) and (10.62), we have

$$\begin{aligned} \Phi^{1,(j)}_{\epsilon_1,\epsilon_2}(\xi) &= \epsilon_1\epsilon_2\, \tau^{(j)}_{\epsilon_1,\epsilon_2}(\epsilon_1 p^1 + \epsilon_1\epsilon_2\xi) && \forall \xi \in \partial\Omega_1\,, \\ \Phi^{2,(j)}_{\epsilon_1,\epsilon_2}(\xi) &= \epsilon_1\epsilon_2\, \tau^{(j)}_{\epsilon_1,\epsilon_2}(\epsilon_1 p^2 + \epsilon_1\epsilon_2\xi) && \forall \xi \in \partial\Omega_2\,, \\ \Phi^{o,(j)}_{\epsilon_1,\epsilon_2}(x) &= \tau^{(j)}_{\epsilon_1,\epsilon_2}(x) && \forall x \in \partial\Omega^o, \end{aligned}$$

for all $j \in \{1,2\}$ and $(\epsilon_1, \epsilon_2) \in]0, \delta_\Phi[^2$, where $\tau^{(j)}_{\epsilon_1,\epsilon_2} \in C^{0,\alpha}(\partial\Omega(\epsilon_1,\epsilon_2))$ is the unique solution of system (10.39). It follows that

$$\begin{aligned} &\mathcal{B}_j(\epsilon_1,\epsilon_2)[\mathcal{A}(\epsilon_1,\epsilon_2)[\psi_1,\psi_2,\psi^o]] \\ &\quad = \sum_{i=1,2} \epsilon_1\epsilon_2 \int_{\partial\Omega_i} \left(\frac{1}{2}\mu_{\epsilon_1,\epsilon_2}(\epsilon_1 p^i + \epsilon_1\epsilon_2\xi) + W_{\Omega(\epsilon_1,\epsilon_2)}[\mu_{\epsilon_1,\epsilon_2}](\epsilon_1 p^i + \epsilon_1\epsilon_2\xi)\right) \\ &\qquad\qquad \times \tau^{(j)}_{\epsilon_1,\epsilon_2}(\epsilon_1 p^i + \epsilon_1\epsilon_2\xi)\, d\sigma \\ &\quad + \int_{\partial\Omega^o} \left(\frac{1}{2}\mu_{\epsilon_1,\epsilon_2}(x) + W_{\Omega(\epsilon_1,\epsilon_2)}[\mu_{\epsilon_1,\epsilon_2}](x)\right) \tau^{(j)}_{\epsilon_1,\epsilon_2}(x)\, d\sigma \qquad \forall j \in \{1,2\}. \end{aligned}$$

Then, changing the variable of integration in the integrals over $\partial\Omega_1$ and $\partial\Omega_2$, we verify that

$$\begin{aligned} &\mathcal{B}_j(\epsilon_1,\epsilon_2)[\mathcal{A}(\epsilon_1,\epsilon_2)[\psi_1,\psi_2,\psi^o]] \\ &\qquad = \int_{\partial\Omega(\epsilon_1,\epsilon_2)} \left(\frac{1}{2}\mu_{\epsilon_1,\epsilon_2} + W_{\Omega(\epsilon_1,\epsilon_2)}[\mu_{\epsilon_1,\epsilon_2}]\right) \tau^{(j)}_{\epsilon_1,\epsilon_2}\, d\sigma \end{aligned}$$

for all $j \in \{1,2\}$ and $(\epsilon_1, \epsilon_2) \in]0, \delta_\Phi[^2$. Since the operator $\frac{1}{2}I + W^t_{\Omega(\epsilon_1,\epsilon_2)}$ is transpose to $\frac{1}{2}I + W_{\Omega(\epsilon_1,\epsilon_2)}$ (cf. Theorem 6.5), we deduce that

$$\begin{aligned} &\mathcal{B}_j(\epsilon_1,\epsilon_2)[\mathcal{A}(\epsilon_1,\epsilon_2)[\psi_1,\psi_2,\psi^o]] \\ &\qquad = \int_{\partial\Omega(\epsilon_1,\epsilon_2)} \mu_{\epsilon_1,\epsilon_2} \left(\frac{1}{2}I + W^t_{\Omega(\epsilon_1,\epsilon_2)}\right) \left[\tau^{(j)}_{\epsilon_1,\epsilon_2}\right] d\sigma \end{aligned}$$

for all $j \in \{1,2\}$ and $(\epsilon_1, \epsilon_2) \in]0, \delta_\Phi[^2$. Hence, we have

$$\mathcal{B}_j(\epsilon_1,\epsilon_2)[\mathcal{A}(\epsilon_1,\epsilon_2)[\psi_1,\psi_2,\psi^o]] = 0 \qquad \forall j \in \{1,2\}\,,\ (\epsilon_1,\epsilon_2) \in]0,\delta_\Phi[^2$$

by the first equation of (10.39). We now observe that the function from $]-\delta_\Phi, \delta_\Phi[^2$ to $\mathbb{R}$ that takes (ϵ_1, ϵ_2) to $\mathcal{B}_j(\epsilon_1,\epsilon_2)[\mathcal{A}(\epsilon_1,\epsilon_2)[\psi^i,\psi^o]]$ is real analytic. Then the Identity Principle for real analytic functions of Proposition 2.16 implies that

$$\mathcal{B}_j(\epsilon_1,\epsilon_2)[\mathcal{A}(\epsilon_1,\epsilon_2)[\psi_1,\psi_2,\psi^o]] = 0 \qquad \forall j \in \{1,2\}\,,\ (\epsilon_1,\epsilon_2) \in]-\delta_\Phi,\delta_\Phi[^2\,. \tag{10.84}$$

Now (10.82) follows by (10.83) and (10.84). Finally, to verify that all the conditions of Lemma 8.25 are verified we have to show that $\mathrm{Im}\,\mathcal{A}(0,0)$ has a closed direct complement in Y. To do so, we observe that $\mathcal{B}(0,0)$ is a bounded operator from Y to $\mathbb{R}^2$ and we recall that $\operatorname{Im}\mathcal{A}(0,0) = \operatorname{Ker}\mathcal{B}(0,0)$. It follows that $\operatorname{Im}\mathcal{A}(0,0)$ has finite positive co-dimension in Y. Accordingly, it has a closed direct complement of finite dimension in Y (see Corollary 2.7).

Now that the assumptions of Lemma 8.25 are verified, we can deduce that there exist $\delta_\Psi \in]0, \delta_\Phi]$ and a real analytic map Ψ from $]-\delta_\Psi, \delta_\Psi[^2$ to $C^{1,\alpha}(\partial\Omega_1) \times C^{1,\alpha}(\partial\Omega_2) \times C^{1,\alpha}(\partial\Omega^o)$ that satisfies (10.80). Then (10.81) follows by Proposition 10.24. □

10.4.4 The Auxiliary Functions $H^\xi_{\Omega_1}$, $H^\xi_{\Omega_2}$, and $H^{\Omega^o}_x$

In this subsection we describe the relation between the limiting densities $\Phi^{i,(j)}[0,0]$ and $\Phi^{o,(j)}[0,0]$, with $i, j \in \{1,2\}$, and certain auxiliary functions $H^\xi_{\Omega_1}$, $H^\xi_{\Omega_2}$, and $H^{\Omega^o}_x$, which will be be introduced below (see also Lemmas 8.48 and 8.49). As we shall see, such functions play a role in the description of the limit behavior of $u_{\epsilon_1,\epsilon_2}$.

For the functions $\Phi^{i,(j)}[0,0]$ we have the following.

Lemma 10.29. *Let $i, j \in \{1,2\}$ and $\xi \in \mathbb{R}^2 \setminus \partial\Omega_i$ be fixed. Let $H^\xi_{\Omega_i}$ be the only solution in $C^{1,\alpha}_{\mathrm{loc}}(\mathbb{R}^2 \setminus \Omega_i)$ of the Dirichlet boundary value problem*

$$\begin{cases} \Delta H^\xi_{\Omega_i} = 0 & \text{in } \mathbb{R}^2 \setminus \overline{\Omega_i}\,, \\ H^\xi_{\Omega_i}(\eta) = S_2(\xi - \eta) & \text{for all } \eta \in \partial\Omega_i\,, \\ H^\xi_{\Omega_i} \text{ is harmonic at infinity.} \end{cases}$$

Then we have

$$v_{\Omega_i}[\Phi^{i,(j)}[0,0]](\xi) = \delta_{i,j} \lim_{\eta\to\infty} H^\xi_{\Omega_i}(\eta)\,. \tag{10.85}$$

Moreover, the restriction $v^+_{\Omega_i}[\Phi^{i,(j)}[0,0]] = v_{\Omega_i}[\Phi^{i,(j)}[0,0]]_{|\overline{\Omega_i}}$ is constant on $\overline{\Omega_i}$ and we have

$$v^+_{\Omega_i}[\Phi^{i,(j)}[0,0]] = \delta_{i,j} \lim_{\eta\to\infty} H^0_{\Omega_i}(\eta) \qquad \text{on } \overline{\Omega_i}\,. \tag{10.86}$$

Proof. The reader may observe that the proof (and the statement) is very similar to that of Lemma 8.48. If $u \in C^{1,\alpha}_{\mathrm{loc}}(\mathbb{R}^2 \setminus \Omega_i)$ is harmonic in $\mathbb{R}^2 \setminus \overline{\Omega_i}$ and at infinity, then Proposition 6.39 implies that there exists $\psi \in C^{1,\alpha}(\partial\Omega_i)_0$ such that

$$u = w^-_{\Omega_i}[\psi] + \lim_{\eta\to\infty} u(\eta)\,.$$

Then, by the jump formula for $w^-_{\Omega_i}[\psi]$ of Theorem 6.6 we have

$$u_{|\partial\Omega_i} = \left(-\frac{1}{2}I + W_{\Omega_i}\right)[\psi] + \lim_{\eta\to\infty} u(\eta)\,.$$

Since $-\frac{1}{2}I + W_{\Omega_i}$ and $-\frac{1}{2}I + W^t_{\Omega_i}$ are transpose to one another (cf. Theorem 6.5) we deduce that

$$\begin{aligned}\int_{\partial\Omega_i} & u_{|\partial\Omega_i}\, \Phi^{i,(j)}[0,0]\, d\sigma \\ &= \int_{\partial\Omega_i} \big(-\frac{1}{2}\psi + W_{\Omega_i}[\psi]\big)\, \Phi^{i,(j)}[0,0]\, d\sigma + \lim_{\eta\to\infty} u(\eta) \int_{\partial\Omega_i} \Phi^{i,(j)}[0,0]\, d\sigma \\ &= \int_{\partial\Omega_i} \psi\, \big(-\frac{1}{2}\Phi^{i,(j)}[0,0] + W^t_{\Omega_i}[\Phi^{i,(j)}[0,0]]\big)\, d\sigma \\ &\quad + \lim_{\eta\to\infty} u(\eta) \int_{\partial\Omega_i} \Phi^{i,(j)}[0,0]\, d\sigma\,. \end{aligned} \tag{10.87}$$

By equality (10.61) the densities $\Phi^{1,(j)}[0,0]$ and $\Phi^{2,(j)}[0,0]$ solve the first, second, fourth, and fifth equation of (10.59) with $g_1 = 0$, $g_2 = 0$, $\rho_1 = \delta_{1,j}$, and $\rho_2 = \delta_{2,j}$. Accordingly we have

$$-\frac{1}{2}\Phi^{i,(j)}[0,0] + W^t_{\Omega_i}[\Phi^{i,(j)}[0,0]] = 0 \quad \text{and} \quad \int_{\partial\Omega_i} \Phi^{i,(j)}[0,0]\, d\sigma = \delta_{i,j}\,. \tag{10.88}$$

By (10.87) it follows that

$$\int_{\partial\Omega_i} u_{|\partial\Omega_i}\, \Phi^{i,(j)}[0,0]\, d\sigma = \delta_{i,j} \lim_{\eta\to\infty} u(\eta)\,. \tag{10.89}$$

Thus, by Definition 4.18 of single layer potential and by taking $u = H^{\xi}_{\Omega_i}$ in (10.89), we have

$$\begin{aligned} v_{\Omega_i}[\Phi^{i,(j)}[0,0]](\xi) &= \int_{\partial\Omega_i} S_2(\xi - \eta)\Phi^{i,(j)}[0,0](\eta)\, d\sigma_\eta \\ &= \int_{\partial\Omega_i} H^{\xi}_{\Omega_i}(\eta)\Phi^{i,(j)}[0,0](\eta)\, d\sigma_\eta = \delta_{i,j} \lim_{\eta\to\infty} H^{\xi}_{\Omega_i}(\eta)\,. \end{aligned}$$

Moreover, by the first equation of (10.88) and by the jump formula of the normal derivative of the single layer potential of Theorem 6.6 we have

$$\nu_{\Omega_i} \cdot \nabla v^+_{\Omega_i}[\Phi^{i,(j)}[0,0]]_{|\partial\Omega_i} = 0\,.$$

Then the uniqueness result for the Neumann boundary value problem of Theorem 6.2 (i) imply that $v^+_{\Omega_i}[\Phi^{i,(j)}[0,0]]$ is constant in $\overline{\Omega_i}$. Then the validity of last the statement of the lemma follows. □

As observed for (8.146), the function

$$G_{\Omega_i}(\xi,\eta) \equiv S_2(\xi - \eta) - H^{\xi}_{\Omega_i}(\eta) \qquad \forall \xi,\eta \in \mathbb{R}^2 \setminus \overline{\Omega_i}\,, \xi \neq \eta \tag{10.90}$$

is the Dirichlet Green's function for the exterior domain $\mathbb{R}^2 \setminus \overline{\Omega_i}$.

In the following Lemma 10.30 we study the limiting value of $\Phi^{o,(j)}$. Instead of confining ourselves to $\Phi^{o,(j)}[0,0]$, we consider also $\Phi^{o,(j)}[0,\epsilon_2]$ with $\epsilon_2 \in]-\delta_\Phi, \delta_\Phi[$. In this way we obtain an information that will be useful to analyze the asymptotic behavior of $u_{\epsilon_1,\epsilon_2}$ in Section 10.4.6.

Lemma 10.30. *Let $x \in \Omega^o$ be fixed. Let $H_x^{\Omega^o}$ be the unique solution in $C^{1,\alpha}(\overline{\Omega^o})$ of the Dirichlet problem*

$$\begin{cases} \Delta H_x^{\Omega^o} = 0 & \text{in } \Omega^o\,, \\ H_x^{\Omega^o}(y) = S_2(x-y) & \text{for all } y \in \partial\Omega^o\,. \end{cases}$$

Then we have

$$v_{\Omega^o}[\Phi^{o,(j)}[0,\epsilon_2]](x) = -H_x^{\Omega^o}(0)$$

for $j \in \{1,2\}$ and for all $\epsilon_2 \in]-\delta_\Phi, \delta_\Phi[$.

Proof. The proof (and the statement) is very similar to that of Lemma 8.49. If $u \in C^{1,\alpha}(\overline{\Omega^o})$ and $\Delta u = 0$ in Ω^o, then Corollary 6.15 implies that there exists $\mu \in C^{1,\alpha}(\partial\Omega^o)$ such that $u_{|\partial\Omega^o} = (\frac{1}{2}I + W_{\Omega^o})[\mu]$. Moreover, by the uniqueness of the solution of the Dirichlet problem of Theorem 3.6 and by the jump formulas for the double layer potential of Theorem 6.6 we have $u = w^+_{\Omega^o}[\mu]$ in $\overline{\Omega^o}$. Let $j \in \{1,2\}$ be fixed. Since $\frac{1}{2}I + W_{\Omega^o}$ and $\frac{1}{2}I + W^t_{\Omega^o}$ are transpose operators (cf. Theorem 6.5), we have

$$\begin{aligned} \int_{\partial\Omega^o} u_{|\partial\Omega^o}\Phi^{o,(j)}[0,\epsilon_2]\,d\sigma &= \int_{\partial\Omega^o} \big(\frac{1}{2}\mu + W_{\Omega^o}[\mu]\big)\,\Phi^{o,(j)}[0,\epsilon_2]\,d\sigma \\ &= \int_{\partial\Omega^o} \mu\,\big(\frac{1}{2}\Phi^{o,(j)}[0,\epsilon_2] + W^t_{\Omega^o}[\Phi^{o,(j)}[0,\epsilon_2]]\big)\,d\sigma \end{aligned} \tag{10.91}$$

for all $\epsilon_2 \in]-\delta_\Phi, \delta_\Phi[$. By equality (10.61) and by the definition of $\mathcal{N}^{o,(j)}$ in (10.51) we have

$$\begin{aligned} \frac{1}{2}\Phi^{o,(j)}[0,\epsilon_2] + W^t_{\Omega^o}[\Phi^{o,(j)}[0,\epsilon_2]] &= -\left(\delta_{1,j} + \delta_{2,j}\right)\nu_{\Omega^o}\cdot(\nabla S)_{|\partial\Omega^o} \\ &= -\nu_{\Omega^o}\cdot(\nabla S)_{|\partial\Omega^o} \end{aligned}$$

for all $\epsilon_2 \in]-\delta_\Phi, \delta_\Phi[$. Then (10.91), Definition 4.27 of double layer potential, and the third Green Identity of Theorem 6.9 imply that

$$\begin{aligned} &\int_{\partial\Omega^o} u_{|\partial\Omega^o}\Phi^{o,(j)}[0,\epsilon_2]\,d\sigma \\ &\quad = -\int_{\partial\Omega^o} \mu(y)\,\nu_{\Omega^o}(y)\cdot\nabla S_2(y)\,d\sigma_y = -w^+_{\Omega^o}[\mu](0) = -u(0)\,. \end{aligned} \tag{10.92}$$

Thus, by Definition 4.18 of single layer potential and by taking $u = H_x^{\Omega^o}$, we have

$$v_{\Omega^o}[\Phi^{o,(j)}[0,\epsilon_2]](x) = \int_{\partial\Omega^o} S_2(x-y)\Phi^{o,(j)}[0,\epsilon_2](y)\,d\sigma_y$$
$$= \int_{\partial\Omega^o} H^{\Omega^o}_{x|\partial\Omega^o}\,\Phi^{o,(j)}[0,\epsilon_2]\,d\sigma = -H^{\Omega^o}_x(0)\,.$$

□

As observed for (8.148), the function

$$G^{\Omega^o}(x,y) \equiv S_2(x-y) - H^{\Omega^o}_x(y) \qquad \forall x,y \in \Omega^o\,, x \neq y \tag{10.93}$$

is the Dirichlet Green's function for the domain Ω^o.

10.4.5 Representation of $u_{\epsilon_1,\epsilon_2}$ in Terms of Analytic Maps

In Theorem 10.35 below we write the maps that takes (ϵ_1,ϵ_2) to suitable restrictions of $u_{\epsilon_1,\epsilon_2}$ and of the rescaled function $u_{\epsilon_1,\epsilon_2}(\epsilon_1 p^j + \epsilon_1\epsilon_2\cdot)$ in terms of real analytic maps of ϵ_1 and ϵ_2 and of elementary functions of $\log\epsilon_1$ and $\log(\epsilon_1\epsilon_2)$. To do so, we exploit the representation formula (10.38), the real analyticity results of Propositions 10.23 and 10.28, and the auxiliary functions of Section 10.4.4.

In the following Propositions 10.31–10.34 we consider one by one the terms that appear in the representation formula (10.38) and we introduce certain auxiliary maps $U[\epsilon_1,\epsilon_2]$ and $V[\epsilon_1,\epsilon_2]$, a vector valued functions $\mathcal{F}(\epsilon_1,\epsilon_2)$, and matrix valued functions $\mathcal{R}(\epsilon_1,\epsilon_2)$ and $\Lambda(\epsilon_1,\epsilon_2)$ that we then use in the representation Theorem 10.35. In what follows, we set

$$\delta_* \equiv \delta_\Psi \tag{10.94}$$

Then, by Lemma 10.22 and Proposition 10.28 we have

$$\delta_* = \delta_\Psi \leq \delta_\Phi < \min\{\delta_1,\delta_2\}\,. \tag{10.95}$$

In Proposition 10.31 we consider the term $w^+_{\Omega(\epsilon_1,\epsilon_2)}[\mu_{\epsilon_1,\epsilon_2}]$ that appears in (10.38).

Proposition 10.31. *For each $(\epsilon_1,\epsilon_2) \in]-\delta_*,\delta_*[^2$ there exists and is unique a function $U[\epsilon_1,\epsilon_2] \in C^{1,\alpha}(\overline{\Omega(\epsilon_1,\epsilon_2)})$ such that*

$$\begin{aligned} U[\epsilon_1,\epsilon_2](x) &= w^+_{\Omega^o}[\Psi^o[\epsilon_1,\epsilon_2]](x) \\ &+ \epsilon_1\epsilon_2\sum_{k=1}^{2}\int_{\partial\Omega_k} \nu_{\Omega_k}(\eta)\cdot\nabla S_2(x-\epsilon_1 p^k - \epsilon_1\epsilon_2\eta)\Psi_k[\epsilon_1,\epsilon_2](\eta)\,d\sigma_\eta \end{aligned} \tag{10.96}$$

for all $x \in \overline{\Omega^o} \setminus \overline{\Omega_1(\epsilon_1,\epsilon_2) \cup \Omega_2(\epsilon_1,\epsilon_2)}$. Moreover, the following statements hold.

(i) For all $(\epsilon_1,\epsilon_2) \in]0,\delta_*[^2$ we have

$$U[\epsilon_1,\epsilon_2] = w^+_{\Omega(\epsilon_1,\epsilon_2)}[\mu_{\epsilon_1,\epsilon_2}]\,,$$

where $\mu_{\epsilon_1,\epsilon_2} \in C^{1,\alpha}(\partial\Omega(\epsilon_1,\epsilon_2))$ *is the solution of* (10.40).

(ii) Let Ω_M *be an open subset of* Ω^o *such that* $0 \notin \overline{\Omega_M}$. *Let* $\delta_M \in]0, \delta_*]$ *be such that* $\overline{\Omega_M} \cap \overline{\Omega_k(\epsilon_1,\epsilon_2)} = \emptyset$ *for all* $(\epsilon_1,\epsilon_2) \in]-\delta_M,\delta_M[\times]-\delta_*,\delta_*[$ *and for all* $k \in \{1,2\}$. *Then there exists a real analytic map*

$$\mathcal{U}_M :]-\delta_M,\delta_M[\times]-\delta_*,\delta_*[\to C^{1,\alpha}(\overline{\Omega_M})$$

such that

$$U[\epsilon_1,\epsilon_2]_{|\overline{\Omega_M}} = u^o_{|\overline{\Omega_M}} + \epsilon_1\epsilon_2 \mathcal{U}_M[\epsilon_1,\epsilon_2] \tag{10.97}$$

for all $(\epsilon_1,\epsilon_2) \in]-\delta_M,\delta_M[\times]-\delta_*,\delta_*[$, *where* $u^o \in C^{1,\alpha}(\overline{\Omega^o})$ *is the unique solution of*

$$\begin{cases} \Delta u^o = 0 \text{ in } \Omega^o\,, \\ u^o = f^o \quad \text{on } \partial\Omega^o\,. \end{cases} \tag{10.98}$$

(iii) Let $j \in \{1,2\}$. *Let* Ω_m *be an open bounded subset of* $\mathbb{R}^2 \setminus \overline{\Omega_j}$. *Let* $\delta_m^{(j)} \in]0,\delta_*]$ *be such that*

$$\epsilon_1 p^j + \epsilon_1\epsilon_2\overline{\Omega_m} \subseteq \Omega^o$$

and

$$(\epsilon_1 p^j + \epsilon_1\epsilon_2\overline{\Omega_m}) \cap \overline{\Omega_k(\epsilon_1,\epsilon_2)} = \emptyset$$

for all $(\epsilon_1,\epsilon_2) \in]-\delta_m^{(j)},\delta_m^{(j)}[^2$, *where* $k=1$ *if* $j=2$ *and* $k=2$ *if* $j=1$. *Then there exists a real analytic map*

$$\mathcal{U}_m^{(j)} :]-\delta_m^{(j)},\delta_m^{(j)}[^2 \to C^{1,\alpha}(\overline{\Omega_m})$$

such that

$$U[\epsilon_1,\epsilon_2](\epsilon_1 p^j + \epsilon_1\epsilon_2\xi) = \mathcal{U}_m^{(j)}[\epsilon_1,\epsilon_2](\xi) \tag{10.99}$$

for all $\xi \in \overline{\Omega_m}$ *and* $(\epsilon_1,\epsilon_2) \in (]-\delta_m^{(j)},\delta_m^{(j)}[\setminus\{0\})^2$. *Moreover,*

$$\mathcal{U}_m^{(j)}[0,0](\xi) = u^o(0) + u_j(\xi) - \lim_{\eta\to\infty} u_j(\eta) \quad \forall \xi \in \overline{\Omega_m}\,, \tag{10.100}$$

where $u_j \in C^{1,\alpha}_{\mathrm{loc}}(\mathbb{R}^2 \setminus \Omega_j)$ *is the solution of*

$$\begin{cases} \Delta u_j = 0 & \text{in } \mathbb{R}^2 \setminus \overline{\Omega_j}\,, \\ u_j = f_j & \text{on } \partial\Omega_j\,, \\ u_j \text{ is harmonic at infinity.} \end{cases} \tag{10.101}$$

Proof. We observe that we cannot use the expression on the right-hand side of (10.96) to define $U[\epsilon_1,\epsilon_2]$ on the whole of $\overline{\Omega(\epsilon_1,\epsilon_2)}$. Indeed, the integrals over $\partial\Omega_1$ and $\partial\Omega_2$ might have a singular kernel for x on $\partial\Omega_1(\epsilon_1,\epsilon_2)$ or $\partial\Omega_2(\epsilon_1,\epsilon_2)$. Then we observe that, by equality (10.81) of $(\psi^1_{\epsilon_1,\epsilon_2},\psi^2_{\epsilon_1,\epsilon_2},\psi^o_{\epsilon_1,\epsilon_2})$ and $\Psi[\epsilon_1,\epsilon_2]$, by equality (10.64) of $(\psi^1_{\epsilon_1,\epsilon_2},\psi^2_{\epsilon_1,\epsilon_2},\psi^o_{\epsilon_1,\epsilon_2})$ and $\mu_{\epsilon_1,\epsilon_2}$, and by a computation based on the rule of change of variable in integrals, the function on the right-hand side of (10.96) coincides with the double layer potential $w^+_{\Omega(\epsilon_1,\epsilon_2)}[\mu_{\epsilon_1,\epsilon_2}]$ on

$\overline{\Omega^o} \setminus \overline{\Omega_1(\epsilon_1, \epsilon_2) \cup \Omega_2(\epsilon_1, \epsilon_2)}$ (see also Definition 4.27 of double layer potential). Since $w^+_{\Omega(\epsilon_1,\epsilon_2)}[\mu_{\epsilon_1,\epsilon_2}] \in C^{1,\alpha}(\overline{\Omega(\epsilon_1, \epsilon_2)})$ (cf. Theorem 4.31), it follows that the the function on the right-hand side of (10.96) has a unique continuous extension to $\overline{\Omega(\epsilon_1, \epsilon_2)}$. Accordingly, there is a unique function $U[\epsilon_1, \epsilon_2] \in C^{1,\alpha}(\overline{\Omega(\epsilon_1, \epsilon_2)})$ for which (10.96) holds true. Moreover, we have $U[\epsilon_1, \epsilon_2] = w^+_{\Omega(\epsilon_1,\epsilon_2)}[\mu_{\epsilon_1,\epsilon_2}]$ on $\overline{\Omega(\epsilon_1, \epsilon_2)}$ and statement (i) is verified.

We now prove statement (ii). By equality (10.80) and by the definition (10.68) of $\mathcal{M}^o$, we have

$$(\frac{1}{2}I + W_{\Omega^o})[\Psi^o[\epsilon_1, \epsilon_2]] = f^o \qquad \forall(\epsilon_1, \epsilon_2) \in]-\delta_*, \delta_*[^2 \text{ such that } \epsilon_1\epsilon_2 = 0.$$

Corollary 6.15 implies that equation

$$(\frac{1}{2}I + W_{\Omega^o})[\mu^o] = f^o \tag{10.102}$$

has a unique solution $\mu^o \in C^{1,\alpha}(\partial\Omega^o)$. Accordingly, we have

$$\Psi^o[\epsilon_1, \epsilon_2] = \mu^o \text{ for all } (\epsilon_1, \epsilon_2) \in]-\delta_*, \delta_*[^2 \text{ with } \epsilon_1\epsilon_2 = 0. \tag{10.103}$$

Since the map that takes (ϵ_1, ϵ_2) to $\Psi^o[\epsilon_1, \epsilon_2]$ is real analytic from $]-\delta_*, \delta_*[$ to $C^{1,\alpha}(\partial\Omega^o)$, there exist $\delta_{**} \in]0, \delta_*]$ and a family of functions $\{\Psi^o_{i,j}\}_{i,j=0}^\infty \subseteq C^{1,\alpha}(\partial\Omega^o)$ such that

$$\Psi^o[\epsilon_1, \epsilon_2] = \sum_{i,j=0}^\infty \epsilon_1^i\epsilon_2^j\Psi^o_{i,j} \qquad \forall(\epsilon_1, \epsilon_2) \in]-\delta_{**}, \delta_{**}[^2,$$

where the series converges normally in $C^{1,\alpha}(\partial\Omega^o)$. Then (10.103) implies that $\Psi^o_{0,0} = \mu^o$ and that $\Psi^o_{i,0} = \Psi^o_{0,j} = 0$ for all $i, j \in \mathbb{N} \setminus \{0\}$. Accordingly we have

$$\Psi^o[\epsilon_1, \epsilon_2] = \mu^o + \epsilon_1\epsilon_2 \sum_{i,j=0}^\infty \epsilon_1^i\epsilon_2^j\Psi^o_{i+1,j+1} \qquad \forall(\epsilon_1, \epsilon_2) \in]-\delta_{**}, \delta_{**}[^2.$$

Now let $\tilde{\Psi}^o$ be the map from $]-\delta_*, \delta_*[^2$ to $C^{1,\alpha}(\partial\Omega^o)$ defined by

$$\tilde{\Psi}^o[\epsilon_1, \epsilon_2] = \sum_{i,j=0}^\infty \epsilon_1^i\epsilon_2^j\Psi^o_{i+1,j+1} \qquad \forall(\epsilon_1, \epsilon_2) \in]-\delta_{**}, \delta_{**}[^2$$

and

$$\tilde{\Psi}^o[\epsilon_1, \epsilon_2] = \frac{\Psi^o[\epsilon_1, \epsilon_2] - \mu^o}{\epsilon_1\epsilon_2} \qquad \forall(\epsilon_1, \epsilon_2) \in]-\delta_*, \delta_*[^2 \setminus]-\delta_{**}, \delta_{**}[^2.$$

We can verify that $\tilde{\Psi}^o$ is real analytic and we have

$$\Psi^o[\epsilon_1, \epsilon_2] = \mu^o + \epsilon_1\epsilon_2 \tilde{\Psi}^o[\epsilon_1, \epsilon_2] \quad \forall(\epsilon_1, \epsilon_2) \in]-\delta_*, \delta_*[^2 .$$

Moreover, by equality (10.102), by the jump formula for the double layer potential of Theorem 6.6 and by the uniqueness of the solution of the Dirichlet problem (cf. Theorem 3.6) we have $w^+_{\Omega^o}[\mu^o] = u^o$. It follows that

$$w^+_{\Omega^o}[\Psi^o[\epsilon_1, \epsilon_2]] = u^o + \epsilon_1\epsilon_2\, w^+_{\Omega^o}[\tilde{\Psi}^o[\epsilon_1, \epsilon_2]] \quad \forall(\epsilon_1, \epsilon_2) \in]-\delta_*, \delta_*[^2 . \quad (10.104)$$

Then we define

$$\begin{aligned} \mathcal{U}_M[\epsilon_1, \epsilon_2](x) \equiv\; & w^+_{\Omega^o}[\tilde{\Psi}^o[\epsilon_1, \epsilon_2]](x) \\ & + \sum_{k=1}^{2} \int_{\partial\Omega_k} \nu_{\Omega_k}(\eta) \cdot \nabla S_2(x - \epsilon_1 p^k - \epsilon_1\epsilon_2\eta)\Psi_k[\epsilon_1, \epsilon_2](\eta)\, d\sigma_\eta \end{aligned} \quad (10.105)$$

for all $x \in \overline{\Omega_M}$ and for all $(\epsilon_1, \epsilon_2) \in]-\delta_M, \delta_M[\times]-\delta_*, \delta_*[$. The validity of (10.97) follows by (10.96), (10.104), and (10.105). In addition, Theorem A.28 (i) on the properties of integral operators with a real analytic kernel and the mapping properties of the double layer potential of Theorem 4.31 (i) imply that the map from $]-\delta_M, \delta_M[\times]-\delta_*, \delta_*[$ to $C^{1,\alpha}(\overline{\Omega_M})$ that takes (ϵ_1, ϵ_2) to $\mathcal{U}_M[\epsilon_1, \epsilon_2]$ is real analytic.

Finally, we prove statement (iii). We take $k \equiv 1$ if $j = 2$ and $k \equiv 2$ if $j = 1$. Then we set

$$\begin{aligned} \mathcal{U}^{(j)}_m[\epsilon_1, \epsilon_2](\xi) \equiv\; & w_{\Omega^o}[\Psi^o[\epsilon_1, \epsilon_2]](\epsilon_1 p^j + \epsilon_1\epsilon_2\xi) - w^-_{\Omega_j}[\Psi_j[\epsilon_1, \epsilon_2]](\xi) \\ & + \epsilon_2 \int_{\partial\Omega_k} \nu_{\Omega_k}(\eta) \cdot \nabla S_2(p^j - p^k + \epsilon_2(\xi - \eta))\Psi_k[\epsilon_1, \epsilon_2](\eta)\, d\sigma_\eta \end{aligned}$$

for all $\xi \in \overline{\Omega_m}$ and $(\epsilon_1, \epsilon_2) \in]-\delta^{(j)}_m, \delta^{(j)}_m[^2$. Then, Theorem A.28 (i) on the properties of integral operators with a real analytic kernel and the mapping properties of the double layer potential of Theorem 4.31 (i) and (ii) imply that the map that takes (ϵ_1, ϵ_2) to $\mathcal{U}^{(j)}_m[\epsilon_1, \epsilon_2]$ is real analytic from $]-\delta^{(j)}_m, \delta^{(j)}_m[^2$ to $C^{1,\alpha}(\overline{\Omega_m})$. The validity of equality (10.99) can be deduced by a straightforward computation based on the rule of change of variables in integrals. We now verify (10.100). A straightforward computation shows that

$$\mathcal{U}^{(j)}_m[0, 0](\xi) = w_{\Omega^o}[\Psi^o[0, 0]](0) - w^-_{\Omega_j}[\Psi_j[0, 0]](\xi) \qquad \forall\xi \in \overline{\Omega_m} . \quad (10.106)$$

Equality (10.104) with $(\epsilon_1, \epsilon_2) = (0, 0)$ implies that

$$w^+_{\Omega^o}[\Psi^o[0, 0]](0) = u^o(0) . \quad (10.107)$$

Moreover, by equality (10.80) with $(\epsilon_1, \epsilon_2) = (0, 0)$ and by Proposition 10.26 we have

$$-\frac{1}{2}\Psi_j[0,0] + W_{\Omega_j}[\Psi_j[0,0]] = w_{\Omega^o}[\Psi^o[0,0]](0) \\ - f_j + \sum_{i=1}^{2} \int_{\partial\Omega_i} f_i \, \Phi^{i,(j)}[0,0] \, d\sigma + \int_{\partial\Omega^o} f^o \, \Phi^{o,(j)}[0,0] \, d\sigma \, .$$

Now, equality (10.89) and the definition (10.101) of u_i imply that

$$\int_{\partial\Omega_i} f_i \, \Phi^{i,(j)}[0,0] \, d\sigma = \int_{\partial\Omega_i} u_{i|\partial\Omega_i} \, \Phi^{i,(j)}[0,0] \, d\sigma = \delta_{i,j} \lim_{\eta\to\infty} u_i(\eta) \, ,$$

for all $i \in \{1,2\}$, and equality (10.92) and system (10.98) for u^o imply that

$$\int_{\partial\Omega^o} f^o \, \Phi^{o,(j)}[0,0] \, d\sigma = \int_{\partial\Omega^o} u^o_{|\partial\Omega^o} \, \Phi^{o,(j)}[0,0] \, d\sigma = -u^o(0) \, .$$

Hence,

$$-\frac{1}{2}\Psi_j[0,0] + W_{\Omega_j}[\Psi_j[0,0]] = u^o(0) - f_j + \sum_{i=1}^{2} \left(\delta_{i,j} \lim_{\eta\to\infty} u_i(\eta) \right) - u^o(0) \\ = -f_j + \lim_{\eta\to\infty} u_j(\eta) \, .$$

(see also (10.107)). Then, the jump formula for the double layer potential of Theorem 6.6 implies that

$$-w^-_{\Omega_j}[\Psi_j[0,0]]_{|\partial\Omega_j} = f_j - \lim_{\eta\to\infty} u_j(\eta)$$

and by the uniqueness of the solution of the exterior Dirichlet problem of Proposition 6.1 we deduce that

$$-\, w^-_{\Omega_j}[\Psi_j[0,0]] = u_j - \lim_{\eta\to\infty} u_j(\eta) \qquad \text{on } \mathbb{R}^2 \setminus \Omega_j \, . \tag{10.108}$$

Now, equality (10.100) follows by (10.106), (10.107), and (10.108). □

In the the following Proposition 10.32 we consider the factors $v^+_{\Omega(\epsilon_1,\epsilon_2)}[\tau^{(j)}_{\epsilon_1,\epsilon_2}]$ with $j \in \{1,2\}$ that appear in the representation formula (10.38).

Proposition 10.32. *For all* $(\epsilon_1,\epsilon_2) \in]-\delta_*,\delta_*[^2$ *we denote by*

$$V[\epsilon_1,\epsilon_2] \equiv (V_1[\epsilon_1,\epsilon_2], V_2[\epsilon_1,\epsilon_2])$$

the element of $C^{1,\alpha}(\overline{\Omega(\epsilon_1,\epsilon_2)})^2$ *defined by*

$$V_j[\epsilon_1,\epsilon_2](x) \\ \equiv v_{\Omega^o}[\Phi^{o,(j)}[\epsilon_1,\epsilon_2]](x) + \sum_{i=1}^{2} \int_{\partial\Omega_i} S_2(x - \epsilon_1 p^i - \epsilon_1\epsilon_2\eta) \, \Phi^{i,(j)}[\epsilon_1,\epsilon_2](\eta) \, d\sigma_\eta$$

for all $x \in \overline{\Omega(\epsilon_1, \epsilon_2)}$ *and* $j \in \{1, 2\}$. *Then the following statements hold.*

(i) *For all* $(\epsilon_1, \epsilon_2) \in]0, \delta_*[^2$ *and* $j \in \{1, 2\}$ *we have*

$$V_j[\epsilon_1, \epsilon_2] = v^+_{\Omega(\epsilon_1,\epsilon_2)}[\tau^{(j)}_{\epsilon_1,\epsilon_2}],$$

where $\tau^{(j)}_{\epsilon_1,\epsilon_2} \in C^{0,\alpha}(\partial\Omega(\epsilon_1, \epsilon_2))$ *is the solution of* (10.39).

(ii) *Let* Ω_M *be an open subset of* Ω^o *such that* $0 \notin \overline{\Omega_M}$. *Let* $\delta_M \in]0, \delta_*]$ *be such that* $\overline{\Omega_M} \cap \Omega_i(\epsilon_1, \epsilon_2) = \emptyset$ *for all* $(\epsilon_1, \epsilon_2) \in] - \delta_M, \delta_M[\times] - \delta_*, \delta_*[$ *and for all* $i \in \{1, 2\}$. *Then there exists a real analytic map*

$$\mathcal{V}_M \equiv (\mathcal{V}_{M,1}, \mathcal{V}_{M,2}) :] - \delta_M, \delta_M[\times] - \delta_*, \delta_*[\to C^{1,\alpha}(\overline{\Omega_M})^2$$

such that

$$V[\epsilon_1, \epsilon_2]_{|\overline{\Omega_M}} = \mathcal{V}_M[\epsilon_1, \epsilon_2] \qquad \forall(\epsilon_1, \epsilon_2) \in] - \delta_M, \delta_M[\times] - \delta_*, \delta_*[.$$

Moreover, for all $j \in \{1, 2\}$ *and all* $\epsilon_2 \in] - \delta_*, \delta_*[$ *we have*

$$\mathcal{V}_{M,j}[0, \epsilon_2](x) = G^{\Omega^o}(x, 0) \quad \forall x \in \overline{\Omega_M}, \tag{10.109}$$

where $G^{\Omega^o}(x, y)$ *is the Dirichlet Green function of* Ω^o *(cf.* (10.93)*).*

(iii) *Let* $h \in \{1, 2\}$. *Let* Ω_m *be an open bounded subset of* $\mathbb{R}^2 \setminus \overline{\Omega_i}$. *Let* $\delta^{(h)}_m \in]0, \delta_*]$ *be such that*

$$\epsilon_1 p^h + \epsilon_1\epsilon_2\overline{\Omega_m} \subseteq \Omega^o$$

and

$$(\epsilon_1 p^h + \epsilon_1\epsilon_2\overline{\Omega_m}) \cap \overline{\Omega_k(\epsilon_1, \epsilon_2)} = \emptyset$$

for all $(\epsilon_1, \epsilon_2) \in] - \delta^{(h)}_m, \delta^{(h)}_m[^2$, *where* $k \equiv 1$ *if* $h = 2$ *and* $k \equiv 2$ *if* $h = 1$. *Then there exists a real analytic map*

$$\mathcal{V}^{(h)}_m \equiv (\mathcal{V}^{(h)}_{m,1}, \mathcal{V}^{(h)}_{m,2}) :] - \delta^{(h)}_m, \delta^{(h)}_m[^2 \to C^{1,\alpha}(\overline{\Omega_m})^2$$

such that

$$\begin{aligned} &V_j[\epsilon_1, \epsilon_2](\epsilon_1 p^h + \epsilon_1\epsilon_2\xi) \\ &\quad = \mathcal{V}^{(h)}_{m,j}[\epsilon_1, \epsilon_2](\xi) + \delta_{h,j}\frac{1}{2\pi}\log|\epsilon_1\epsilon_2| + (1 - \delta_{h,j})\frac{1}{2\pi}\log|\epsilon_1| \quad \forall \xi \in \overline{\Omega_m} \end{aligned} \tag{10.110}$$

for all $(\epsilon_1, \epsilon_2) \in \left(] - \delta^{(h)}_m, \delta^{(h)}_m[\setminus\{0\}\right)^2$ *and* $j \in \{1, 2\}$. *Moreover,*

$$\mathcal{V}^{(h)}_{m,j}[0, 0](\xi) = -H^{\Omega^o}_0(0) + \delta_{h,j} \lim_{\eta\to\infty} H^{\xi}_{\Omega_i}(\eta) + (1-\delta_{h,j})S_2(p^h - p^k) \tag{10.111}$$

for all $\xi \in \overline{\Omega_m}$ *and* $j \in \{1, 2\}$, *where* $k \equiv 1$ *if* $h = 2$ *and* $k \equiv 2$ *if* $h = 1$.

Proof. Statement (i) can be deduced by equality (10.62) of $(\phi^{1,(j)}_{\epsilon_1,\epsilon_2}, \phi^{2,(j)}_{\epsilon_1,\epsilon_2}, \phi^{o,(j)}_{\epsilon_1,\epsilon_2})$ and $\Phi^{(j)}[\epsilon_1, \epsilon_2]$, by equality (10.44) of $\tau^{(j)}_{\epsilon_1,\epsilon_2}$ and $(\phi^{1,(j)}_{\epsilon_1,\epsilon_2}, \phi^{2,(j)}_{\epsilon_1,\epsilon_2}, \phi^{o,(j)}_{\epsilon_1,\epsilon_2})$, and by a calculation based on the rule of change of variable in integrals (see also Definition 4.18 of single layer potential).

To prove statement (ii) we set

$$\mathcal{V}_M[\epsilon_1, \epsilon_2] \equiv V[\epsilon_1, \epsilon_2]_{|\overline{\Omega_M}} \qquad \forall (\epsilon_1, \epsilon_2) \in]-\delta_M, \delta_M[\times]-\delta_2, \delta_2[\,.$$

Then, the real analyticity of $\mathcal{V}_M$ follows by Theorem A.28 (i) on the properties of the integral operators with a real analytic kernel and on the mapping properties of the single layer potential of Theorem 4.25. To prove (10.109) we observe that

$$\mathcal{V}_{M,j}[0, \epsilon_2](x) = v_{\Omega^o}[\Phi^{o,(j)}[0, \epsilon_2]](x) + S_2(x) \sum_{i=1}^{2} \int_{\partial\Omega_i} \Phi^{i,(j)}[0, \epsilon_2]\, d\sigma \quad \forall x \in \overline{\Omega_M}$$

for all $\epsilon_2 \in]-\delta_*, \delta_*[$. Then Lemma 10.30 and equality $\int_{\partial\Omega_i} \Phi^{i,(j)}[0, \epsilon_2]\, d\sigma = \delta_{i,j}$ (cf. (10.63)) imply that

$$\mathcal{V}_{M,j}[0, \epsilon_2](x) = -H^{\Omega^o}_x(0) + S_2(x) \qquad \forall x \in \overline{\Omega_M}\,,\ \epsilon_2 \in]-\delta_*, \delta_*[\,.$$

Hence, the validity of (10.109) follows by the definition of G^{Ω^o} in (10.93).

Finally, we consider statement (iii). We set $k \equiv 1$ if $h = 2$ and $k \equiv 2$ if $h = 1$. Then we observe that

$$\begin{aligned} V_j[\epsilon_1, \epsilon_2](\epsilon_1 p^h + \epsilon_1\epsilon_2\xi) &= v_{\Omega^o}[\Phi^{o,(j)}[\epsilon_1, \epsilon_2]](\epsilon_1 p^h + \epsilon_1\epsilon_2\xi) \\ &\quad + \int_{\partial\Omega_h} S_2(\epsilon_1\epsilon_2(\xi - \eta))\, \Phi^{h,(j)}[\epsilon_1, \epsilon_2](\eta)\, d\sigma_\eta \\ &\quad + \int_{\partial\Omega_k} S_2(\epsilon_1(p^h - p^k) + \epsilon_1\epsilon_2(\xi - \eta))\, \Phi^{h,(j)}[\epsilon_1, \epsilon_2](\eta)\, d\sigma_\eta \end{aligned}$$

for all $\xi \in \overline{\Omega_m}$ and $(\epsilon_1, \epsilon_2) \in]-\delta^{(h)}_m, \delta^{(h)}_m[^2$. Moreover, by Definition 3.2 of S_2 we have

$$S_2(\epsilon_1\epsilon_2(\xi - \eta)) = \frac{1}{2\pi} \log|\epsilon_1\epsilon_2| + S_2(\xi - \eta) \tag{10.112}$$

for all $\xi, \eta \in \mathbb{R}^2$, $\xi \neq \eta$, and $(\epsilon_1, \epsilon_2) \in (\mathbb{R} \setminus \{0\})^2$, and

$$S_2(\epsilon_1(p^h - p^k) + \epsilon_1\epsilon_2(\xi - \eta)) = \frac{1}{2\pi} \log|\epsilon_1| + S_2(p^h - p^k + \epsilon_2(\xi - \eta)) \tag{10.113}$$

for all $\xi, \eta \in \mathbb{R}^2$ and $(\epsilon_1, \epsilon_2) \in (\mathbb{R} \setminus \{0\})^2$ such that $p^h - p^k + \epsilon_2(\xi - \eta) \neq 0$. It follows that

$$\begin{aligned}V_j[\epsilon_1,\epsilon_2](\epsilon_1 p^h+\epsilon_1\epsilon_2\xi) &= v_{\Omega^o}[\Phi^{o,(j)}[\epsilon_1,\epsilon_2]](\epsilon_1 p^h+\epsilon_1\epsilon_2\xi)\\ &\quad + v_{\Omega_h}[\Phi^{h,(j)}[\epsilon_1,\epsilon_2]](\xi)+\frac{1}{2\pi}\log|\epsilon_1\epsilon_2|\int_{\partial\Omega_h}\Phi^{h,(j)}[\epsilon_1,\epsilon_2](\eta)\,d\sigma_\eta\\ &\quad + \int_{\partial\Omega_k} S_2(p^h-p^k+\epsilon_2(\xi-\eta))\,\Phi^{k,(j)}[\epsilon_1,\epsilon_2](\eta)\,d\sigma_\eta\\ &\quad + \frac{1}{2\pi}\log|\epsilon_1|\int_{\partial\Omega_k}\Phi^{k,(j)}[\epsilon_1,\epsilon_2](\eta)\,d\sigma_\eta\end{aligned}$$

for all $\xi\in\overline{\Omega_m}$ and $(\epsilon_1,\epsilon_2)\in]-\delta_m^{(h)},\delta_m^{(h)}[^2$ (see also Definition 4.18 of single layer potential). Then equality $\int_{\partial\Omega_h}\Phi^{h,(j)}[0,0]\,d\sigma=\delta_{h,j}$ (cf. (10.63)) implies that

$$\begin{aligned}V_j[\epsilon_1,\epsilon_2](\epsilon_1 p^h+\epsilon_1\epsilon_2\xi) &= v_{\Omega^o}[\Phi^{o,(j)}[\epsilon_1,\epsilon_2]](\epsilon_1 p^h+\epsilon_1\epsilon_2\xi)\\ &\quad + v_{\Omega_h}[\Phi^{h,(j)}[\epsilon_1,\epsilon_2]](\xi)\\ &\quad + \int_{\partial\Omega_k} S_2(p^h-p^k+\epsilon_2(\xi-\eta))\,\Phi^{k,(j)}[\epsilon_1,\epsilon_2](\eta)\,d\sigma_\eta\\ &\quad + \delta_{h,j}\frac{1}{2\pi}\log|\epsilon_1\epsilon_2|+\delta_{j,k}\frac{1}{2\pi}\log|\epsilon_1|\end{aligned}\tag{10.114}$$

for all $\xi\in\overline{\Omega_m}$ and $(\epsilon_1,\epsilon_2)\in]-\delta_m^{(h)},\delta_m^{(h)}[^2$. We now set

$$\begin{aligned}\mathcal{V}^{(h)}_{m,j}[\epsilon_1,\epsilon_2](\xi) &\equiv v_{\Omega^o}[\Phi^{o,(j)}[\epsilon_1,\epsilon_2]](\epsilon_1 p^h+\epsilon_1\epsilon_2\xi)+v_{\Omega_h}[\Phi^{h,(j)}[\epsilon_1,\epsilon_2]](\xi)\\ &\quad + \int_{\partial\Omega_k} S_2(p^h-p^k-\epsilon_2(\xi-\eta))\Phi^{k,(j)}[\epsilon_1,\epsilon_2](\eta)\,d\sigma_\eta\end{aligned}$$

for all $\xi\in\overline{\Omega_m}$ and $(\epsilon_1,\epsilon_2)\in]-\delta_m^{(h)},\delta_m^{(h)}[^2$. Then (10.110) follows by (10.114). Moreover, by Theorem A.28 (i) on the properties of the integral operators with a real analytic kernel and by the mapping properties of the single layer potential of Theorem 4.25 we can verify that the map from $]-\delta_m^{(h)},\delta_m^{(h)}[^2$ to $C^{1,\alpha}(\overline{\Omega_m})^2$ that takes (ϵ_1,ϵ_2) to $\mathcal{V}^{(h)}_m[\epsilon_1,\epsilon_2]\equiv(\mathcal{V}^{(h)}_{m,1}[\epsilon_1,\epsilon_2],\mathcal{V}^{(h)}_{m,2}[\epsilon_1,\epsilon_2])$ is real analytic. To prove (10.111) we observe that

$$\begin{aligned}\mathcal{V}^{(h)}_{m,j}[0,0](\xi) &= v_{\Omega^o}[\Phi^{o,(j)}[0,0]](0)+v_{\Omega_h}[\Phi^{h,(j)}[0,0]](\xi)\\ &\quad + S_2(p^h-p^k)\int_{\partial\Omega_k}\Phi^{k,(j)}[0,0](\eta)\,d\sigma_\eta\end{aligned}$$

for all $\xi\in\overline{\Omega_m}$. Then (10.111) follows by Lemma 10.30, by equality (10.85), and by equality $\int_{\partial\Omega_k}\Phi^{k,(j)}[0,0]\,d\sigma=\delta_{k,j}$ (cf. (10.63)). □

In the following Proposition 10.33 we consider the term $\int_{\partial\Omega(\epsilon_1,\epsilon_2)} f_{\epsilon_1,\epsilon_2}\tau^{(i)}_{\epsilon_1,\epsilon_2}\,d\sigma$ of (10.38).

Proposition 10.33. *Let $\mathcal{F}\equiv(\mathcal{F}_1,\mathcal{F}_2)$ be the function from $]-\delta_*,\delta_*[^2$ to $\mathbb{R}^2$ defined by*

$$\mathcal{F}_j(\epsilon_1, \epsilon_2) \equiv \int_{\partial\Omega^o} f^o\, \Phi^{o,(j)}[\epsilon_1, \epsilon_2]\, d\sigma + \sum_{i=1}^{2} \int_{\partial\Omega_i} f_i\, \Phi^{i,(j)}[\epsilon_1, \epsilon_2]\, d\sigma$$

for all $j \in \{1,2\}$ and $(\epsilon_1, \epsilon_2) \in]-\delta_*, \delta_*[^2$. Then we have

$$\mathcal{F}_j(\epsilon_1, \epsilon_2) = \int_{\partial\Omega(\epsilon_1,\epsilon_2)} f_{\epsilon_1,\epsilon_2} \tau^{(j)}_{\epsilon_1,\epsilon_2}\, d\sigma \tag{10.115}$$

for all $(\epsilon_1, \epsilon_2) \in]0, \delta_*[^2$ and $i \in \{1,2\}$, where $\tau^{(j)}_{\epsilon_1,\epsilon_2} \in C^{0,\alpha}(\partial\Omega(\epsilon_1, \epsilon_2))$ is the solution of (10.39)*. Moreover, $\mathcal{F}$ is real analytic and*

$$\mathcal{F}_j[0,0] = -u^o(0) + \lim_{\eta\to\infty} u_j(\eta) \qquad \forall j \in \{1,2\}, \tag{10.116}$$

where u^o is the unique solution of (10.98) *and u_j with $j \in \{1,2\}$ is the unique solution of* (10.101)*.*

Proof. Equality (10.115) can be deduced by a straightforward computation and using equality (10.62) of $(\phi^{1,(j)}_{\epsilon_1,\epsilon_2}, \phi^{2,(j)}_{\epsilon_1,\epsilon_2}, \phi^{o,(j)}_{\epsilon_1,\epsilon_2})$ and $\Phi^{(j)}[\epsilon_1, \epsilon_2]$, equality (10.44) of $\tau^{(j)}_{\epsilon_1,\epsilon_2}$ and $(\phi^{1,(j)}_{\epsilon_1,\epsilon_2}, \phi^{2,(j)}_{\epsilon_1,\epsilon_2}, \phi^{o,(j)}_{\epsilon_1,\epsilon_2})$, and the rule of change of variable in integrals. The real analyticity of $\mathcal{F}$ is a consequence of the real analyticity of $\Phi^{(j)}$ of Proposition 10.23 and of the continuity of the linear map that takes a function of $C^{0,\alpha}(\partial\Omega^o)$ to its integral over $\partial\Omega^o$ and of the linear map that takes a function of $C^{0,\alpha}(\partial\Omega_j)$ to its integral over $\partial\Omega_j$. To prove (10.116) we observe that

$$\begin{aligned}\mathcal{F}_j(0,0) &= \int_{\partial\Omega^o} f^o\, \Phi^{o,(j)}[0,0]\, d\sigma + \sum_{i=1}^{2} \int_{\partial\Omega_i} f_i\, \Phi^{i,(j)}[0,0]\, d\sigma \\ &= \int_{\partial\Omega^o} u^o_{|\partial\Omega^o}\, \Phi^{o,(j)}[0,0]\, d\sigma + \sum_{i=1}^{2} \int_{\partial\Omega_i} u_{i|\partial\Omega_i}\, \Phi^{i,(j)}[0,0]\, d\sigma\,.\end{aligned}$$

(cf. (10.98) and (10.101)). Then (10.116) follows by (10.89) and (10.92). □

Finally, we turn to the matrix $\Lambda_{\Omega(\epsilon_1,\epsilon_2)^-}$ that appears in the representation formula (10.38) and prove the following.

Proposition 10.34. *Let $\mathcal{R} \equiv (\mathcal{R}_{i,j})_{(i,j)\in\{1,2\}^2}$ be the function from $]-\delta_*, \delta_*[^2$ to the space of two by two real matrices $\mathbb{M}_2(\mathbb{R})$ defined by*

$$\begin{aligned}\mathcal{R}_{i,j}(\epsilon_1, \epsilon_2) \equiv \int_{\partial\Omega_i} & v_{\Omega^o}[\Phi^{o,(j)}[\epsilon_1, \epsilon_2]](\epsilon_1 p^i + \epsilon_1\epsilon_2\xi) + v_{\Omega_i}[\Phi^{i,(j)}[\epsilon_1, \epsilon_2]](\xi) \\ & + \left(\int_{\partial\Omega_k} S_2(p^i - p^k + \epsilon_2(\xi - \eta))\Phi^{k,(j)}[\epsilon_1, \epsilon_2](\eta)\, d\sigma_\eta\right) d\sigma_\xi\end{aligned}$$

for all $(\epsilon_1, \epsilon_2) \in]-\delta_*, \delta_*[^2$ and for all $i, j \in \{1,2\}$. Here $k \equiv 1$ if $i = 2$ and $k \equiv 2$ if $i = 1$. Let $\Lambda(\epsilon_1, \epsilon_2) \equiv \big(\Lambda_{i,j}(\epsilon_1, \epsilon_2)\big)_{(i,j)\in\{1,2\}^2}$ be the matrix of $\mathbb{M}_2(\mathbb{R})$ defined by

$$\Lambda(\epsilon_1, \epsilon_2) \equiv \frac{1}{2\pi} \begin{pmatrix} \log(\epsilon_1\epsilon_2) & \log\epsilon_1 \\ \log\epsilon_1 & \log(\epsilon_1\epsilon_2) \end{pmatrix} + \mathcal{R}(\epsilon_1, \epsilon_2) \tag{10.117}$$

for all $(\epsilon_1, \epsilon_2) \in]0, \delta_*[^2$. *Then the following statements hold.*

(i) For all $(\epsilon_1, \epsilon_2) \in]0, \delta_*[^2$ *we have*

$$\Lambda(\epsilon_1, \epsilon_2) = \Lambda_{\Omega(\epsilon_1,\epsilon_2)^-}, \tag{10.118}$$

where $\Lambda_{\Omega(\epsilon_1,\epsilon_2)^-}$ *is the matrix in* (10.41).

(ii) The matrix $\Lambda(\epsilon_1, \epsilon_2)$ *is invertible for all* $(\epsilon_1, \epsilon_2) \in]0, \delta_*[^2$.

(iii) The function from $]-\delta_*, \delta_*[^2$ *to* $\mathbb{M}_2(\mathbb{R})$ *that takes* (ϵ_1, ϵ_2) *to* $\mathcal{R}(\epsilon_1, \epsilon_2)$ *is real analytic.*

(iv) We have

$$\mathcal{R}_{i,j}[0,0] = -H_0^{\Omega^o}(0) + \delta_{i,j} \lim_{\eta\to\infty} H^0_{\Omega_j}(\eta) + (1-\delta_{i,j})\, S_2(p^i - p^j) \tag{10.119}$$

for all $i, j \in \{1, 2\}$.

Proof. To prove (10.118) we have to show that

$$\Lambda_{i,j}(\epsilon_1, \epsilon_2) = \fint_{\partial\Omega_i(\epsilon_1,\epsilon_2)} V_{\Omega(\epsilon_1,\epsilon_2)}[\tau^{(j)}_{\epsilon_1,\epsilon_2}]\, d\sigma$$

for all $i, j \in \{1, 2\}$ and all $(\epsilon_1, \epsilon_2) \in]0, \delta_*[^2$, where $\tau^{(j)}_{\epsilon_1,\epsilon_2} \in C^{0,\alpha}(\partial\Omega(\epsilon_1, \epsilon_2))$ is the solution of (10.39). This last equality can be verified exploiting Definition 4.18 of single layer potential, equalities (10.112) and (10.113) for S_2, equality (10.62) of $(\phi^{1,(j)}_{\epsilon_1,\epsilon_2}, \phi^{2,(j)}_{\epsilon_1,\epsilon_2}, \phi^{o,(j)}_{\epsilon_1,\epsilon_2})$ and $\Phi^{(j)}[\epsilon_1, \epsilon_2]$, equality (10.44) of $\tau^{(j)}_{\epsilon_1,\epsilon_2}$ and $(\phi^{1,(j)}_{\epsilon_1,\epsilon_2}, \phi^{2,(j)}_{\epsilon_1,\epsilon_2}, \phi^{o,(j)}_{\epsilon_1,\epsilon_2})$, the rule of change of variable in integrals, and equality (10.63). Moreover, Lemma 6.30 implies that $\Lambda_{\Omega(\epsilon_1,\epsilon_2)^-}$ is invertible for all (ϵ_1, ϵ_2) in $]0, \delta_*[^2$. Thus $\Lambda(\epsilon_1, \epsilon_2)$ is invertible for all $(\epsilon_1, \epsilon_2) \in]0, \delta_*[^2$ and statement (ii) is proved.

To prove statement (iii) we observe that

$$\begin{aligned} \mathcal{R}_{i,j}(\epsilon_1, \epsilon_2) = \fint_{\partial\Omega_i} \Bigg(& \int_{\partial\Omega^o} S_2(\epsilon_1 p^i + \epsilon_1\epsilon_2\xi - y)\Phi^{o,(j)}[\epsilon_1, \epsilon_2](y)\, d\sigma_y \Bigg) \\ & + V_{\Omega_i}[\Phi^{i,(j)}[\epsilon_1, \epsilon_2]](\xi) \\ & + \Bigg(\int_{\partial\Omega_k} S_2(p^i - p^k + \epsilon_2(\xi - \eta))\Phi^{k,(j)}[\epsilon_1, \epsilon_2](\eta)\, d\sigma_\eta \Bigg)\, d\sigma_\xi \end{aligned}$$

for all $(\epsilon_1, \epsilon_2) \in]-\delta_*, \delta_*[^2$ and for all $i, j \in \{1, 2\}$, with $k \equiv 1$ if $i = 2$ and $k \equiv 2$ if $i = 1$. Then, by the real analyticity of $\Phi^{(j)}$ in Proposition 10.23, by Theorem A.28 (ii) on the properties of the integral operators with real analytic kernel, by Theorem 6.4 on the continuity of the linear map that takes $\phi \in C^{0,\alpha}(\partial\Omega_i)$ to $V_{\Omega_i}[\phi] \in C^{1,\alpha}(\partial\Omega_i)$, and by the continuity of the linear map that takes a function of $C^{1,\alpha}(\partial\Omega_i)$ to its integral mean over $\partial\Omega_i$, we deduce that the function from $]-\delta_*,\delta_*[^2$ to $\mathbb{M}_2(\mathbb{R})$ that takes (ϵ_1,ϵ_2) to $\mathcal{R}(\epsilon_1,\epsilon_2)$ is real analytic.

Finally, to verify statement (iv) we observe that

$$\mathcal{R}_{i,j}(0,0) = v_{\Omega^o}[\Phi^{o,(j)}[0,0]](0) + \int_{\partial\Omega_i} v_{\Omega_i}[\Phi^{i,(j)}[0,0]](\xi)\,d\sigma_\xi$$
$$+ S_2(p^i - p^k)\int_{\partial\Omega_k} \Phi^{k,(j)}[0,0]\,d\sigma$$

for all $i,j \in \{1,2\}$, where $k \equiv 1$ if $i = 2$ and $k \equiv 2$ if $i = 1$. Then (10.119) follows by Lemma 10.30, by equality (10.86), and by equality $\int_{\partial\Omega_k} \Phi^{k,(j)}[0,0]\,d\sigma = \delta_{j,k}$ (cf. (10.63)). □

We can now prove Theorem 10.35, where we write the maps that take (ϵ_1,ϵ_2) to suitable restrictions of $u_{\epsilon_1,\epsilon_2}$ and of the rescaled functions $u_{\epsilon_1,\epsilon_2}(\epsilon_1 p^h + \epsilon_1\epsilon_2\cdot)$, with $h \in \{1,2\}$, in terms of the real analytic maps $\mathcal{U}_M$, $\mathcal{U}_m$, $\mathcal{V}_M$, $\mathcal{V}_m^{(j)}$, and $\mathcal{F}$ of Propositions 10.31, 10.32, and 10.33, of the matrix function Λ of Proposition 10.34, and of the singular functions $\log\epsilon_1$ and $\log(\epsilon_1\epsilon_2)$.

In what follows we retain the following notation: A^t denotes the transpose of a matrix A and A^{-1} denotes the inverse of an invertible matrix A. For the correct understanding of the matrix product below, we also remind that in this book all vectors and vector valued functions are understood to be columns. As a consequence, $\mathcal{F}(\epsilon_1,\epsilon_2)^t$ is a row vector for all $(\epsilon_1,\epsilon_2) \in]-\delta_*,\delta_*[^2$.

Theorem 10.35. *The following statements hold.*

(i) Let Ω_M be an open subset of Ω^o such that $0 \notin \overline{\Omega_M}$. Let $\delta_M \in]0,\delta_*]$ be such that

$$\overline{\Omega_M} \cap \overline{\Omega_k(\epsilon_1,\epsilon_2)} = \emptyset$$

for all $(\epsilon_1,\epsilon_2) \in]-\delta_M,\delta_M[\times]-\delta_*,\delta_*[$ and for all $k \in \{1,2\}$. Then we have

$$u_{\epsilon_1,\epsilon_2|\overline{\Omega_M}} = u^o + \epsilon_1\epsilon_2\,\mathcal{U}_M[\epsilon_1,\epsilon_2] + \mathcal{F}(\epsilon_1,\epsilon_2)^t\,\Lambda(\epsilon_1,\epsilon_2)^{-1}\,\mathcal{V}_M[\epsilon_1,\epsilon_2]$$

for all $(\epsilon_1,\epsilon_2) \in]0,\delta_M[\times]0,\delta_*[$, where u^o is the unique solution of (10.98).

(ii) Let $j \in \{1,2\}$. Let Ω_m be an open bounded subset of $\mathbb{R}^2 \setminus \overline{\Omega_j}$. Let $\delta_m^{(j)} \in]0,\delta_*]$ be such that

$$\epsilon_1 p^j + \epsilon_1\epsilon_2\overline{\Omega_m} \subseteq \Omega^o$$

and

$$(\epsilon_1 p^j + \epsilon_1\epsilon_2\overline{\Omega_m}) \cap \overline{\Omega_k(\epsilon_1,\epsilon_2)} = \emptyset$$

for all $(\epsilon_1,\epsilon_2) \in]-\delta_m^{(j)},\delta_m^{(j)}[^2$, where $k = 1$ if $j = 2$ and $k = 2$ if $j = 1$. Then we have

$$u_{\epsilon_1,\epsilon_2}(\epsilon_1 p^j + \epsilon_1\epsilon_2\cdot)_{|\overline{\Omega_m}}$$
$$= \mathcal{U}_m^{(j)}[\epsilon_1,\epsilon_2] + \mathcal{F}(\epsilon_1,\epsilon_2)^t\,\Lambda(\epsilon_1,\epsilon_2)^{-1}\left(\mathcal{V}_m^{(j)}[\epsilon_1,\epsilon_2] + Z^{(j)}(\epsilon_1,\epsilon_2)\right)$$

for all $(\epsilon_1, \epsilon_2) \in]0, \delta_m^{(j)}[^2$, *where* $Z^{(j)}(\epsilon_1, \epsilon_2) \equiv (Z^{(j)}(\epsilon_1, \epsilon_2)_1, Z^{(j)}(\epsilon_1, \epsilon_2)_2)$ *is the vector of* $\mathbb{R}^2$ *defined by*

$$Z^{(j)}(\epsilon_1, \epsilon_2)_i \equiv \delta_{i,j} \frac{1}{2\pi} \log(\epsilon_1 \epsilon_2) + (1 - \delta_{i,j}) \frac{1}{2\pi} \log \epsilon_1 \qquad \forall i, j \in \{1, 2\}\,.$$

Proof. Statement (i) follows by the representation formula (10.38), by Proposition 10.31 (i) and equality (10.97), by Proposition 10.32 (i) and (ii), by equality (10.115), and by Proposition 10.34 (i). Statement (ii) follows by the representation formula (10.38), by Proposition 10.31 (i) and equality (10.99), by Proposition 10.32 (i) and equality (10.110), by equality (10.115), and by Proposition 10.34 (i). □

Finally, we derive by Theorem 10.35 the following Theorem 10.36 for the energy $\int_{\Omega(\epsilon_1,\epsilon_2)} |\nabla u_{\epsilon_1,\epsilon_2}|^2 \, dx$.

Theorem 10.36. *There exists* $\delta_E \in]0, \delta_*]$ *and two real analytic functions,*

$$\mathcal{E}_1 :]-\delta_E, \delta_E[^2 \to \mathbb{R}\,,$$
$$\mathcal{E}_2 :]-\delta_E, \delta_E[^2 \to \mathbb{R}^n$$

such that

$$\int_{\Omega(\epsilon_1,\epsilon_2)} |\nabla u_{\epsilon_1,\epsilon_2}|^2 \, dx = \mathcal{E}_1(\epsilon_1, \epsilon_2) + \mathcal{F}(\epsilon_1, \epsilon_2)^t \Lambda(\epsilon_1, \epsilon_2)^{-1} \mathcal{E}_2(\epsilon_1, \epsilon_2)$$

for all $(\epsilon_1, \epsilon_2) \in]0, \delta_E[^2$. *Moreover, we have*

$$\mathcal{E}_1(0,0) = \int_{\Omega^o} |\nabla u^o|^2 \, dx + \sum_{i=1,2} \int_{\mathbb{R}^2 \setminus \Omega_i} |\nabla u_i|^2 \, d\xi\,,$$

where u^o *is the unique solution of* (10.98) *and* u_1, u_2 *are the unique solutions of* (10.101) *with* $j = 1$ *and* $j = 2$, *respectively.*

Proof. Let Ω_M and Ω_m be as in statements (i) and (ii) of Theorem 10.35 and assume in addition that

$$\partial \Omega^o \subseteq \overline{\Omega_M} \qquad \text{and} \qquad \partial \Omega_i \subseteq \overline{\Omega_{i,m}} \quad \text{for } i \in \{1, 2\}\,.$$

Let $\delta_E \equiv \min\{\delta_M, \delta_m^{(1)}, \delta_m^{(2)}\}$. A calculation based on the first Green Identity of Theorem 4.2 and on the representation formulas of Theorem 10.35 shows that

$$\begin{aligned}
&\int_{\Omega(\epsilon_1,\epsilon_2)} |\nabla u_{\epsilon_1,\epsilon_2}|^2 \, dx \\
&= \int_{\partial\Omega^o} f^o \nu_{\Omega^o} \cdot \nabla u^o \, d\sigma - \sum_{i=1,2} \int_{\partial\Omega^i} f_i \, \nu_{\Omega^i} \cdot \nabla \mathcal{U}_m^{(i)}[\epsilon_1, \epsilon_2] \, d\sigma \\
&\quad + \epsilon_1 \epsilon_2 \int_{\partial\Omega^o} f^o \, \nu_{\Omega^o} \cdot \nabla \mathcal{U}_M[\epsilon_1, \epsilon_2] \, d\sigma
\end{aligned}$$

$$
+ \mathcal{F}(\epsilon_1, \epsilon_2)^t \Lambda(\epsilon_1, \epsilon_2)^{-1}
\times \left\{ \int_{\partial\Omega^o} f^o \, \nu_{\Omega^o} \cdot \nabla \mathcal{V}_M[\epsilon_1, \epsilon_2] \, d\sigma - \sum_{i=1,2} \int_{\partial\Omega^i} f_i \, \nu_{\Omega^i} \cdot \nabla \mathcal{V}_m^{(i)}[\epsilon_1, \epsilon_2] \, d\sigma \right\}
$$

for all $(\epsilon_1, \epsilon_2) \in]-\delta_E, \delta_E[^2$ (see also the proof of Theorem 10.9). Then we define $\mathcal{E}_1(\epsilon_1, \epsilon_2)$ as the sum of the first, second, and third term on the right-hand side of the equality above and $\mathcal{E}_2(\epsilon_1, \epsilon_2)$ as the sum of the terms in braces. The real analyticity of $\mathcal{E}_1$ and $\mathcal{E}_2$ follows by the real analyticity of the maps $\mathcal{U}_M, \mathcal{U}_m, \mathcal{V}_M$, and $\mathcal{V}_m^{(j)}$ of Propositions 10.31 and 10.32, by the continuity of the linear maps from $C^{1,\alpha}(\overline{\Omega_M})$ to $C^{0,\alpha}(\overline{\Omega_M})^n$ and from $C^{1,\alpha}(\overline{\Omega_{i,m}})$ to $C^{0,\alpha}(\overline{\Omega_{i,m}})^n$ that take a function to its gradient, and by the continuity of the linear functions that take elements of $C^{0,\alpha}(\partial\Omega_M)^n$ and $C^{0,\alpha}(\partial\Omega_i)^n$ to their integral over $\partial\Omega^o$ and $\partial\Omega_i$, $i \in \{1,2\}$. The last statement of the theorem follows by equality (10.97), by the expression for $\mathcal{U}_m^{(i)}[0,0]$ of Proposition 10.31 (iii), by the first Green Identity of Theorem 4.2, and by the first Green Identity in exterior domains of Corollary 4.7. □

10.4.6 Asymptotic Behavior of $u_{\epsilon_1,\epsilon_2}$ as $(\epsilon_1, \epsilon_2) \to (0,0)$

In this section we exploit Theorem 10.35 to obtain an asymptotic approximations of the solution $u_{\epsilon_1,\epsilon_2}$ of problem (10.4) as the pair of parameters (ϵ_1, ϵ_2) approaches a singular value $(0,0)$. We will often find convenient to write the quotient $1/\log(\epsilon_1\epsilon_2)$. To do so, we have to avoid the case when $\epsilon_1\epsilon_2 = 1$ and a zero appears at the denominator. Therefore, we shrink the set $]-\delta_1, \delta_1[\times]-\delta_2, \delta_2[$ of admissible pairs (ϵ_1, ϵ_2) and we assume, without loss of generality, that we have

$$
\delta_1 \in]0, 1/\delta_2[\, .
$$

Then inequality $\delta_* < \min\{\delta_1, \delta_2\}$ (cf. (10.95)) ensures that $\epsilon_1\epsilon_2 < 1$ for all (ϵ_1, ϵ_2) in $]0, \delta_*[^2$ (we also recall that δ_M, $\delta_m^{(1)}$, and $\delta_m^{(2)}$ of Theorem 10.35 are smaller than δ_*).

In what follows, $\mathrm{cof}(A)$ denotes the cofactor matrix of a matrix $A \in \mathbb{M}_2(\mathbb{R})$, so that $A^{-1} = (\det A)^{-1}\mathrm{cof}(A)^t$ for all invertible matrices $A \in \mathbb{M}_2(\mathbb{R})$. Proposition 10.37 describes the inverse of the matrix $\Lambda(\epsilon_1, \epsilon_2)$. The proof can be obtained by a direct computation based on equality (10.117) for $\Lambda(\epsilon_1, \epsilon_2)$.

Proposition 10.37. *For all $(\epsilon_1, \epsilon_2) \in]0, \delta_*[^2$ we have*

$$
\det \Lambda(\epsilon_1, \epsilon_2) = \frac{1}{4\pi^2} \, \rho(\epsilon_1, \epsilon_2) \, \log(\epsilon_1\epsilon_2)
$$

and

$$\Lambda(\epsilon_1,\epsilon_2)^{-1} = \frac{2\pi}{\rho(\epsilon_1,\epsilon_2)}\left\{\begin{pmatrix} 1 & -\frac{\log\epsilon_1}{\log(\epsilon_1\epsilon_2)} \\ -\frac{\log\epsilon_1}{\log(\epsilon_1\epsilon_2)} & 1\end{pmatrix} + 2\pi\frac{\mathrm{cof}(\mathcal{R}(\epsilon_1,\epsilon_2))^t}{\log(\epsilon_1\epsilon_2)}\right\}, \tag{10.120}$$

where the real number $\rho(\epsilon_1,\epsilon_2)$ *is given by*

$$\begin{aligned}\rho(\epsilon_1,\epsilon_2) \equiv{}& \log\epsilon_2 + \frac{\log\epsilon_1}{\log(\epsilon_1\epsilon_2)}\log\epsilon_2 \\ &- 2\pi(\mathcal{R}_{1,2}(\epsilon_1,\epsilon_2)+\mathcal{R}_{2,1}(\epsilon_1,\epsilon_2))\frac{\log\epsilon_1}{\log(\epsilon_1\epsilon_2)} \\ &+ 2\pi(\mathcal{R}_{1,1}(\epsilon_1,\epsilon_2)+\mathcal{R}_{2,2}(\epsilon_1,\epsilon_2)) \\ &+ 4\pi^2(\mathcal{R}_{1,1}(\epsilon_1,\epsilon_2)\mathcal{R}_{2,2}(\epsilon_1,\epsilon_2) - \mathcal{R}_{1,2}(\epsilon_1,\epsilon_2)\mathcal{R}_{2,1}(\epsilon_1,\epsilon_2))\frac{1}{\log(\epsilon_1\epsilon_2)}\,.\end{aligned} \tag{10.121}$$

As observed in Proposition 10.34 (ii), $\Lambda(\epsilon_1,\epsilon_2)$ is invertible for all (ϵ_1,ϵ_2) in $]0,\delta_*[^2$. Then Proposition 10.37 implies that

$$\rho(\epsilon_1,\epsilon_2) \neq 0 \qquad \forall(\epsilon_1,\epsilon_2)\in]0,\delta_*[^2\,.$$

At first, we focus on the 'macroscopic' behavior of the solution, that is, the behavior of the map that takes (ϵ_1,ϵ_2) to a restriction $u_{\epsilon_1,\epsilon_2|\overline{\Omega_M}}$ with $\Omega_M \subseteq \Omega^o$ open and such that $0 \notin \overline{\Omega_M}$. In the following Proposition 10.38 we exploit Proposition 10.37 to refine the representation formula of Theorem 10.35 (i).

Proposition 10.38. *Let* Ω_M *be an open subset of* Ω^o *such that* $0 \notin \overline{\Omega_M}$. *Let* $\delta_M \in]0,\delta_*]$ *be such that*

$$\overline{\Omega_M} \cap \overline{\Omega_k(\epsilon_1,\epsilon_2)} = \emptyset$$

for all $(\epsilon_1,\epsilon_2) \in]-\delta_M,\delta_M[\times]-\delta_*,\delta_*[$ *and for all* $k \in \{1,2\}$. *Then there exists a real analytic map*

$$\mathcal{X}_M \equiv (\mathcal{X}_{M,1},\mathcal{X}_{M,2}) :]-\delta_M,\delta_M[\times]-\delta_*,\delta_*[\to C^{1,\alpha}(\overline{\Omega_M})^2$$

such that

$$\begin{aligned}u_{\epsilon_1,\epsilon_2}(x) ={}& u^o(x) + \epsilon_1\epsilon_2\,\mathcal{U}_M[\epsilon_1,\epsilon_2](x) \\ &+ 2\pi\frac{1}{\log(\epsilon_1\epsilon_2)}\frac{\log\epsilon_2}{\rho(\epsilon_1,\epsilon_2)}(\mathcal{F}_1(\epsilon_1,\epsilon_2)+\mathcal{F}_2(\epsilon_1,\epsilon_2))\,G^{\Omega^o}(x,0) \\ &+ 2\pi\frac{\epsilon_1}{\rho(\epsilon_1,\epsilon_2)}\mathcal{F}(\epsilon_1,\epsilon_2)^t\begin{pmatrix} 1 & -\frac{\log\epsilon_1}{\log(\epsilon_1\epsilon_2)} \\ -\frac{\log\epsilon_1}{\log(\epsilon_1\epsilon_2)} & 1\end{pmatrix}\mathcal{X}_M[\epsilon_1,\epsilon_2](x) \\ &+ 4\pi^2\frac{1}{\log(\epsilon_1\epsilon_2)}\frac{1}{\rho(\epsilon_1,\epsilon_2)}\mathcal{F}(\epsilon_1,\epsilon_2)^t\,\mathrm{cof}(\mathcal{R}(\epsilon_1,\epsilon_2))^t\,\mathcal{V}_M[\epsilon_1,\epsilon_2](x)\end{aligned} \tag{10.122}$$

for all $x \in \overline{\Omega_M}$ *and all* $(\epsilon_1,\epsilon_2) \in]0,\delta_M[\times]0,\delta_*[$.

Proof. Equality (10.109) implies that we have $\mathcal{V}_{M,j}[0, \epsilon_2] - G^{\Omega^o}(\cdot, 0)_{|\overline{\Omega^o}} = 0$ for all $\epsilon_2 \in]-\delta_*, \delta_*[$ and $j \in \{1, 2\}$. Moreover, the real analyticity of $\mathcal{V}_M$ of Proposition 10.32 (ii) implies that the map from $]-\delta_M, \delta_M[\times]-\delta_*, \delta_*[$ to $C^{1,\alpha}(\overline{\Omega_M})$ that takes (ϵ_1, ϵ_2) to $\mathcal{V}_{M,j}[\epsilon_1, \epsilon_2] - G^{\Omega^o}(\cdot, 0)_{|\overline{\Omega_M}}$ is real analytic. Then we can prove that the map from $(]-\delta_M, \delta_M[\setminus\{0\})\times]-\delta_*, \delta_*[$ to $C^{1,\alpha}(\overline{\Omega_M})$ that takes (ϵ_1, ϵ_2) to

$$\frac{\mathcal{V}_{M,j}[\epsilon_1, \epsilon_2] - G^{\Omega^o}(\cdot, 0)_{|\overline{\Omega_M}}}{\epsilon_1}$$

has a real analytic extension to $]-\delta_M, \delta_M[\times]-\delta_*, \delta_*[$. If we denote such an extension by $\mathcal{X}_{M,j}$, then the map $\mathcal{X}_M = (\mathcal{X}_{M,1}, \mathcal{X}_{M,2})$ is real analytic from $]-\delta_M, \delta_M[\times]-\delta_*, \delta_*[$ to $C^{1,\alpha}(\overline{\Omega_M})^2$ and we have

$$\mathcal{V}_{M,j}[\epsilon_1, \epsilon_2](x) = G^{\Omega^o}(x, 0) + \epsilon_1 \mathcal{X}_{M,j}[\epsilon_1, \epsilon_2](x) \quad \forall x \in \overline{\Omega_M}$$

for all $(\epsilon_1, \epsilon_2) \in]-\delta_M, \delta_M[\times]-\delta_*, \delta_*[$ and $j \in \{1, 2\}$. Then, by a straightforward computation based on Proposition 10.37 we have

$$\begin{aligned}
&\Lambda(\epsilon_1, \epsilon_2)^{-1}\mathcal{V}_M[\epsilon_1, \epsilon_2](x) \\
&\quad = 2\pi \frac{1}{\log(\epsilon_1\epsilon_2)} \frac{\log \epsilon_2}{\rho(\epsilon_1, \epsilon_2)} \begin{pmatrix} G^{\Omega^o}(x, 0) \\ G^{\Omega^o}(x, 0) \end{pmatrix} \\
&\qquad + 2\pi \frac{\epsilon_1}{\rho(\epsilon_1, \epsilon_2)} \begin{pmatrix} 1 & -\frac{\log \epsilon_1}{\log(\epsilon_1\epsilon_2)} \\ -\frac{\log \epsilon_1}{\log(\epsilon_1\epsilon_2)} & 1 \end{pmatrix} \mathcal{X}_M[\epsilon_1, \epsilon_2](x) \\
&\qquad + 4\pi^2 \frac{1}{\log(\epsilon_1\epsilon_2)} \frac{1}{\rho(\epsilon_1, \epsilon_2)} \mathrm{cof}(\mathcal{R}(\epsilon_1, \epsilon_2))^t \, \mathcal{V}_M[\epsilon_1, \epsilon_2](x) \qquad \forall x \in \overline{\Omega_M}
\end{aligned}$$

for all $(\epsilon_1, \epsilon_2) \in]0, \delta_*[^2$. To complete the proof we can now plug the expression above into the representation formula for $u_{\epsilon_1,\epsilon_2|\overline{\Omega_M}}$ of Theorem 10.35 (i). □

We plan to exploit the real analytic representation (10.122) to obtain an asymptotic expansion of $u_{\epsilon_1,\epsilon_2}$ as (ϵ_1, ϵ_2) tends to $(0, 0)$ in $]0, \delta_*[^2$. To do so, we face the problem that

$$\lim_{(\epsilon_1,\epsilon_2)\in]0,\delta_*[^2\to(0,0)} \frac{\log \epsilon_1}{\log(\epsilon_1\epsilon_2)}$$

does not exist. As announced in the introduction of this section, we can overcome this difficulty by replacing ϵ_1 with a positive parameter t and by taking $\epsilon_2 = \gamma(t)$, where γ is a function from the interval $]0, 1[$ to itself such that

$$\lim_{t\to 0^+} \gamma(t) = 0$$

and such that the limit

$$\lambda_0 \equiv \lim_{t\to 0^+} \frac{\log t}{\log(t\gamma(t))} \tag{10.123}$$

exists finite in $[0, +\infty[$ (we note that the membership of $\gamma(t)$ in $]0, 1[$ implies that $\lambda_0 \in [0, 1]$ when it exists). Then we can analyze the asymptotic behavior of the map that takes $t > 0$ small enough to $u_{t,\gamma(t)|\overline{\Omega_M}}$ as $t \to 0^+$. In the following Proposition 10.39 we compute the first and the second term of the corresponding asymptotic expansion.

Proposition 10.39. *Let Ω_M be an open subset of Ω^o such that $0 \notin \overline{\Omega_M}$. Let γ be a function from $]0, 1[$ to itself such that $\lim_{t\to 0^+} \gamma(t) = 0$ and such that the limit λ_0 in* (10.123) *exists finite in $[0, +\infty[$. Then we have*

$$\begin{aligned} u_{t,\gamma(t)|\overline{\Omega_M}} &= u^o_{|\overline{\Omega_M}} \\ &+ \frac{1}{\log(t\gamma(t))} \frac{2\pi}{1+\lambda_0} \Big(\lim_{\eta\to\infty} u_1(\eta) + \lim_{\eta\to\infty} u_2(\eta) - 2u^o(0) \Big) G^{\Omega^o}(\cdot, 0)_{|\overline{\Omega_M}} \\ &+ o\left(\frac{1}{\log(t\gamma(t))} \right) \end{aligned}$$

as $t \to 0^+$, where the approximation holds in the norm of $C^{1,\alpha}(\overline{\Omega_M})$ and where G^{Ω^o} is the Dirichlet Green function of Ω^o (cf. (10.93)*), u^o is the unique solution of* (10.98)*, and u_1, u_2 are the unique solutions of* (10.101) *with $j = 1$ and $j = 2$, respectively.*

Proof. We observe that the functions

$$t\gamma(t), \qquad \frac{t}{\log\gamma(t)}, \qquad \frac{1}{\log(t\gamma(t))} \frac{1}{\log\gamma(t)}$$

are all $o(1/\log(t\gamma(t)))$ for $t \to 0^+$. Moreover, by definition (10.121) of $\rho(\epsilon_1, \epsilon_2)$ we can verify that

$$\lim_{t\to 0^+} \frac{\log\gamma(t)}{\rho(t, \gamma(t))} = \frac{1}{1+\lambda_0}. \tag{10.124}$$

Then, taking $(\epsilon_1, \epsilon_2) = (t, \gamma(t))$ in the representation formula (10.122), we deduce that

$$\begin{aligned} u_{t,\gamma(t)|\overline{\Omega_M}}(x) &= u^o_{|\overline{\Omega_M}} \\ &+ 2\pi \frac{1}{\log(t\gamma(t))} \frac{\log\gamma(t)}{\rho(t, \gamma(t))} \left(\mathcal{F}_1(t, \gamma(t)) + \mathcal{F}_2(t, \gamma(t)) \right) G^{\Omega^o}(\cdot, 0)_{|\overline{\Omega_M}} \\ &+ o\left(\frac{1}{\log(t\gamma(t))} \right) \end{aligned}$$

for $t \to 0^+$. By (10.124) we also have

$$\frac{1}{\log(t\gamma(t))} \left(\frac{\log\gamma(t)}{\rho(t, \gamma(t))} - \frac{1}{1+\lambda_0} \right) = o\left(\frac{1}{\log(t\gamma(t))} \right) \qquad \text{as } t \to 0^+.$$

Then, we can write

$$u_{t,\gamma(t)|\overline{\Omega_M}} = u^o_{|\overline{\Omega_M}} + 2\pi \frac{1}{\log(t\gamma(t))} \frac{1}{1+\lambda_0} (\mathcal{F}_1(0,0) + \mathcal{F}_2(0,0))\, G^{\Omega^o}(\cdot,0)_{|\overline{\Omega_M}} + o\left(\frac{1}{\log(t\gamma(t))}\right)$$

as $t \to 0^+$. Since (10.116) implies that

$$\mathcal{F}_1(0,0) + \mathcal{F}_2(0,0) = \lim_{\eta\to\infty} u_1(\eta) + \lim_{\eta\to\infty} u_2(\eta) - 2u^o(0)\,,$$

we deduce the validity of statement. □

We now turn to the rescaled function $u_{\epsilon_1,\epsilon_2}(\epsilon_1 p^j + \epsilon_1\epsilon_2\cdot)$. Taking $(\epsilon_1,\epsilon_2) = (t,\gamma(t))$ with γ as above, we can compute the limit value of $u_{t,\gamma(t)}(tp^j + t\gamma(t)\cdot)$ as $t \to 0^+$. By the real analytic representation of Theorem 10.35 (ii) and by the formula for $\Lambda(\epsilon_1,\epsilon_2)^{-1}$ in Proposition 10.37 we can prove the following.

Proposition 10.40. *Let $j \in \{1,2\}$. Let Ω_m be an open bounded subset of $\mathbb{R}^2 \setminus \overline{\Omega_j}$. Let γ be a function from $]0,1[$ to itself such that $\lim_{t\to 0^+}\gamma(t) = 0$ and such that the limit λ_0 in* (10.123) *exists finite in $[0,+\infty[$. Then we have*

$$u_{t,\gamma(t)}(tp^j + t\gamma(t)\cdot)_{|\overline{\Omega_m}} = u_{j|\overline{\Omega_m}} + o(1) \qquad \textit{as } t \to 0^+\,,$$

where the approximation holds in $C^{1,\alpha}(\overline{\Omega_m})$ and where u_j is the unique solution of (10.101).

Proof. By the representation formula of Theorem 10.35 (ii) and by the formula for $\Lambda(\epsilon_1,\epsilon_2)^{-1}$ of Proposition 10.37 we can write

$$\begin{aligned}
u_{t,\gamma(t)}&(tp^j + t\gamma(t)\cdot)_{|\overline{\Omega_m}} = \mathcal{U}^{(j)}_m[t,\gamma(t)] \\
&+ \mathcal{F}(t,\gamma(t))^t \frac{2\pi}{\rho(t,\gamma(t))} \left\{ \begin{pmatrix} 1 & -\frac{\log t}{\log(t\gamma(t))} \\ -\frac{\log t}{\log(t\gamma(t))} & 1 \end{pmatrix} + 2\pi \frac{\mathrm{cof}(\mathcal{R}(t,\gamma(t)))^t}{\log(t\gamma(t))} \right\} \\
&\times \left(\mathcal{V}^{(j)}_m[t,\gamma(t)] + Z^{(j)}(t,\gamma(t))\right) \\
&= \mathcal{U}^{(j)}_m[t,\gamma(t)] \\
&+ \mathcal{F}(t,\gamma(t))^t \frac{2\pi}{\rho(t,\gamma(t))} \begin{pmatrix} 1 & -\frac{\log t}{\log(t,\gamma(t))} \\ -\frac{\log t}{\log(t,\gamma(t))} & 1 \end{pmatrix} Z^{(j)}(t,\gamma(t)) \\
&+ \mathcal{F}(t,\gamma(t))^t \frac{2\pi}{\rho(t,\gamma(t))} \begin{pmatrix} 1 & -\frac{\log t}{\log(t,\gamma(t))} \\ -\frac{\log t}{\log(t,\gamma(t))} & 1 \end{pmatrix} \mathcal{V}^{(j)}_m[t,\gamma(t)] \\
&+ \mathcal{F}(t,\gamma(t))^t \frac{4\pi^2}{\rho(t,\gamma(t))} \frac{\mathrm{cof}(\mathcal{R}(t,\gamma(t)))^t}{\log(t\gamma(t))} \left(\mathcal{V}^{(j)}_m[t,\gamma(t)] + Z^{(j)}(t,\gamma(t))\right)
\end{aligned} \tag{10.125}$$

for all $t > 0$ small enough so that $(t, \gamma(t)) \in]0, \delta_m^{(j)}[^2$. By the definition of $Z^{(j)}$ in Theorem 10.35 (ii) we have

$$\frac{2\pi}{\rho(t,\gamma(t))} \begin{pmatrix} 1 & -\frac{\log t}{\log(t\gamma(t))} \\ -\frac{\log t}{\log(t\gamma(t))} & 1 \end{pmatrix} Z^{(j)}(t,\gamma(t))$$
$$= \frac{1}{\rho(t,\gamma(t))} \left(\log(t\gamma(t)) - \frac{\log^2 t}{\log(t\gamma(t))} \right) \begin{pmatrix} \delta_{1,j} \\ \delta_{2,j} \end{pmatrix}.$$

We now compute the limit of the scalar factor on the right-hand side as $t \to 0^+$. In view of the limit relations (10.123) and (10.124), we have

$$\begin{aligned} &\lim_{t\to 0^+} \frac{1}{\rho(t,\gamma(t))} \left(\log(t\gamma(t)) - \frac{\log^2 t}{\log(t\gamma(t))} \right) \\ &= \lim_{t\to 0^+} \frac{\log\gamma(t)}{\rho(t,\gamma(t))} \left(\frac{\log(t\gamma(t))}{\log\gamma(t)} - \frac{\log t}{\log\gamma(t)} \frac{\log t}{\log(t\gamma(t))} \right) \\ &= \lim_{t\to 0^+} \frac{1}{1+\lambda_0} \left(\left(\frac{\log(t\gamma(t))}{\log\gamma(t)} - \frac{\log t}{\log\gamma(t)} \right) + \frac{\log t}{\log\gamma(t)} \left(1 - \frac{\log t}{\log(t\gamma(t))} \right) \right) \\ &= \lim_{t\to 0^+} \frac{1}{1+\lambda_0} \left(1 + \frac{\log t}{\log\gamma(t)} \frac{\log\gamma(t)}{\log(t\gamma(t))} \right) = \frac{1}{1+\lambda_0}(1+\lambda_0) = 1 . \end{aligned} \tag{10.126}$$

We also observe that the definition of ρ in (10.121) implies that

$$\lim_{t\to 0^+} \frac{1}{\rho(t,\gamma(t))} = 0 \tag{10.127}$$

and the definition of $Z^{(j)}$ in Theorem 10.35 implies that

$$\lim_{t\to 0^+} \frac{1}{\log(t\gamma(t))} Z^{(j)}(t,\gamma(t)) = \frac{1}{2\pi} \begin{pmatrix} \delta_{1,j} + (1-\delta_{1,j})\lambda_0 \\ \delta_{2,j} + (1-\delta_{2,j})\lambda_0 \end{pmatrix} . \tag{10.128}$$

Moreover, the real analyticity of $\mathcal{U}_m^{(j)}$ of Proposition 10.31 (iii), of $\mathcal{V}_m^{(j)}$ of Proposition 10.32 (iii), of $\mathcal{F}$ of Proposition 10.33, and of $\mathcal{R}$ of Proposition 10.34 (iii) imply that the maps $\mathcal{U}_m^{(j)}[t,\gamma(t)]$, $\mathcal{V}_m^{(j)}[t,\gamma(t)]$, $\mathcal{F}(t,\gamma(t))$, and $\mathcal{R}(t,\gamma(t))$ have a limiting value at $t = 0$. Then, by the limit relations (10.123), (10.126), (10.127), and (10.128) and by the representation formula (10.125) we have

$$\lim_{t\to 0^+} u_{t,\gamma(t)}(tp^j + t\gamma(t)\cdot)_{|\overline{\Omega_m}} = \mathcal{U}_m^{(j)}[0,0] + \mathcal{F}(0,0)^t \begin{pmatrix} \delta_{1,j} \\ \delta_{2,j} \end{pmatrix}$$

in $C^{1,\alpha}(\overline{\Omega_m})$. Now the statement follows by (10.100) and (10.116). □

Finally, for the energy of $u_{t,\gamma(t)}$ we have the following proposition.

Proposition 10.41. *If γ is a function from $]0,1[$ to itself such that $\lim_{t\to 0^+}\gamma(t) = 0$ and such that the limit λ_0 in (10.123) exists finite in $[0,+\infty[$, then we have*

$$\int_{\Omega(t,\gamma(t))} |\nabla u_{t,\gamma(t)}|^2\,dx = \int_{\Omega^o} |\nabla u^o|^2\,dx + \sum_{i=1,2} \int_{\mathbb{R}^2\setminus\Omega_i} |\nabla u_i|^2\,d\xi + o(1)$$

as $t \to 0^+$, where u^o is the unique solution of (10.98) and u_1, u_2 are the unique solutions of (10.101) with $j = 1$ and $j = 2$, respectively.

Proof. It follows by the representation Theorem 10.36, by noting that the matrix $\Lambda(t,\gamma(t))^{-1}$ tends to the zero matrix as $t \to 0^+$ (cf. equality (10.120)), and by the limit relations (10.123) and (10.127). □

Chapter 11
Nonlinear Boundary Value Problems in Domains with a Small Hole

Abstract This chapter is devoted to the application of the Functional Analytic Approach to nonlinear boundary value problems in domains with a small hole. We consider only the case where the nonlinearity appears in the boundary conditions. In our examples we deal with nonlinear Robin boundary conditions and nonlinear transmission conditions.

11.1 The Geometric Setting

The geometric setting is the same as in Chapter 8. We fix once for all a natural number

$$n \in \mathbb{N}, \ n \geq 2$$

and a real number

$$\alpha \in]0, 1[.$$

Then we take two sets Ω^i and Ω^o in the n-dimensional Euclidean space $\mathbb{R}^n$, the letter 'i' standing for 'inner domain' and the letter 'o' standing for 'outer domain'. We assume that Ω^i and Ω^o satisfy the following condition

$$\begin{aligned} &\Omega^i \text{ and } \Omega^o \text{ are bounded open connected subsets of } \mathbb{R}^n \\ &\text{of class } C^{1,\alpha}, \text{ the exterior domains } \Omega^{i-} \text{ and } \Omega^{o-} \text{ are connected,} \\ &\text{and the origin 0 of } \mathbb{R}^n \text{ belongs both to } \Omega^i \text{ and to } \Omega^o. \end{aligned}$$

Then Ω^i and Ω^o have no holes and there exists a positive real number ϵ_0 such that

$$\epsilon\overline{\Omega^i} \subseteq \Omega^o \qquad \forall \epsilon \in]-\epsilon_0, \epsilon_0[.$$

We denote by $\Omega(\epsilon)$ the perforated domain

$$\Omega(\epsilon) \equiv \Omega^o \setminus (\epsilon\overline{\Omega^i}) \qquad \forall \epsilon \in]-\epsilon_0, \epsilon_0[$$

M. Dalla Riva et al., *Singularly Perturbed Boundary Value Problems*,
https://doi.org/10.1007/978-3-030-76259-9_11

and we observe that $\Omega(\epsilon)$ is an open bounded connected subset of $\mathbb{R}^n$ with boundary consisting of the union of the two sets $\partial\Omega^o$ and $\epsilon\partial\Omega^i$ (cf., e.g., Theorem A.10). Moreover, for ϵ in $]-\epsilon_0, \epsilon_0[\setminus\{0\}$ the domain $\Omega(\epsilon)$ is of class $C^{1,\alpha}$ (*cf.* Section 2.13). Instead, for $\epsilon = 0$ we have $\Omega(0) = \Omega^o \setminus \{0\}$ and $\Omega(0)$ is no longer a local strict hypograph of class C^0 (*cf.* Definition 2.37).

11.2 A Nonlinear Robin Problem

In this section, we consider a nonlinear Robin problem in $\Omega(\epsilon)$. First we introduce an 'unperturbed' problem with no hole in Ω^o, and then we consider the case with a hole $\epsilon\Omega^i$.

To do so, we assign a function G^o from $\partial\Omega^o \times \mathbb{R}$ to $\mathbb{R}$ and we assume that the 'unperturbed' nonlinear Robin boundary value problem

$$\begin{cases} \Delta u = 0 & \text{in } \Omega^o, \\ \frac{\partial u}{\partial \nu_{\Omega^o}}(x) = G^o(x, u(x)) & \forall x \in \partial\Omega^o \end{cases} \tag{11.1}$$

admits at least a solution $\tilde{u}^o \in C^{1,\alpha}(\overline{\Omega^o})$. Here ν_{Ω^o} denotes the outward unit normal to $\partial\Omega^o$. Moreover, we denote by G^o_u the partial derivative of G^o with respect to the second variable in $\mathbb{R}$ and we assume that the function

$$G^o_u(x, \tilde{u}^o(x)) \qquad \forall x \in \partial\Omega^o$$

is less or equal to 0 and has a nonzero integral, an assumption which ensures that problem (11.1) does not have other solutions close to $\tilde{u}^o$.

Solutions as $\tilde{u}^o$ can be shown to exist under reasonable assumptions on G^o. For example, we might introduce a specific growth condition and exploit the topological degree, as in [164, Theorem 7.11]. Other existence results for problems such as (11.1) can be found in the works of Carleman [40], Nakamori and Suyama [231], Klingelhöfer [139–142], and Efendiev et al. [94]. We also mention the contribution of Begehr and Hsiao [18, 19] and Begehr and Hile [17] for problems in the plane.

Next we make a hole in the domain Ω^o and for each $\epsilon \in]0, \epsilon_0[$ we consider the nonlinear Robin problem

$$\begin{cases} \Delta u = 0 & \text{in } \Omega(\epsilon), \\ -\frac{\partial u}{\partial \nu_{\epsilon\Omega^i}} = 0 & \text{on } \epsilon\partial\Omega^i, \\ \frac{\partial u}{\partial \nu_{\Omega^o}}(x) = G^o(x, u(x)) & \forall x \in \partial\Omega^o, \end{cases} \tag{11.2}$$

in the perforated domain $\Omega(\epsilon)$. Here $\nu_{\epsilon\Omega^i}$ denotes the outward unit normal to $\partial(\epsilon\Omega^i)$. We will prove that possibly shrinking ϵ_0, problem (11.2) has a solution $u(\epsilon, \cdot)$ in $C^{1,\alpha}(\overline{\Omega(\epsilon)})$ for all $\epsilon \in]0, \epsilon_0[$. Moreover, the family of solutions $\{u(\epsilon, \cdot)\}_{\epsilon \in]0,\epsilon_0[}$ satisfies the limiting relation $\lim_{\epsilon\to 0^+} u(\epsilon, \cdot) = \tilde{u}^o$ in the $C^{1,\alpha}(\overline{\Omega_M})$-

norm for all bounded open subsets Ω_M of Ω^o such that $0 \notin \overline{\Omega_M}$ and it is unique in a local sense which we clarify below. Then we pose the following three questions:

(i) Let x be fixed in $\overline{\Omega^o} \setminus \{0\}$. What can be said on the map $\epsilon \mapsto u(\epsilon, x)$ when $\epsilon > 0$ is close to 0?
(ii) Let ξ be fixed in $\mathbb{R}^n \setminus \Omega^i$. What can be said on the map $\epsilon \mapsto u(\epsilon, \epsilon\xi)$ when $\epsilon > 0$ is close to 0?
(iii) What can be said on the map $\epsilon \mapsto \int_{\Omega(\epsilon)} |Du(\epsilon, y)|^2 \, dy$ when $\epsilon > 0$ is close to 0?

In a sense, question (i) concerns the 'macroscopic' behavior of $u(\epsilon, \cdot)$, whereas question (ii) concerns the 'microscopic' behavior of $u(\epsilon, \cdot)$ and question (iii) deals with the energy of $u(\epsilon, \cdot)$. We plan to answer such questions in the spirit of the present book. We also remark that the content of the present section develops from [164] and [66].

We now briefly outline our strategy. We first convert problem (11.2) into a nonlinear integral equation by exploiting classical Potential Theory. Then we observe that the corresponding integral equations can be written, after an appropriate rescaling and with an appropriate choice of the functional variables, in a form that can be analyzed by means of the Implicit Function Theorem around the degenerate case $\epsilon = 0$ and we represent the unknown densities of the integral equation in terms of real analytic functions of ϵ. Next we exploit the integral representation of the solutions of problem (11.2) in terms of the unknown densities and we deduce both the existence of $u(\epsilon, \cdot)$ and its representation in terms of ϵ.

11.2.1 Formulation of a Nonlinear Robin Problem in Terms of Integral Equations

Given a function $G \in C^0(\partial\Omega \times \mathbb{R})$, we denote by N_G the (nonlinear nonautonomous) composition operator from $C^0(\partial\Omega)$ to itself that maps $v \in C^0(\partial\Omega)$ to the function $N_G[v]$ defined by

$$N_G[v](t) \equiv G(t, v(t)) \qquad \forall t \in \partial\Omega \, .$$

We now transform our nonlinear Robin boundary value problem (11.1) into a problem for integral equations by means of the following.

Proposition 11.1. *Let Ω be an open bounded connected subset of $\mathbb{R}^n$ of class $C^{1,\alpha}$. Let $G \in C^0(\partial\Omega \times \mathbb{R})$ be such that N_G maps $C^{0,\alpha}(\partial\Omega)$ to itself. Then the map that takes a pair (μ, c) to $v^+_\Omega[\mu] + c$ is a bijection from the set of pairs (μ, c) of $C^{0,\alpha}(\partial\Omega)_0 \times \mathbb{R}$ that satisfy the equation*

$$L_{\Omega,0,+}[\mu, c] \equiv -\frac{1}{2}\mu + W^t_\Omega[\mu] = N_G[V_\Omega[\mu] + c] \, , \tag{11.3}$$

(cf. definition (6.56)) to the set of the solutions $u \in C^{1,\alpha}(\overline{\Omega})$ of the problem

$$\begin{cases} \Delta u = 0 & \text{in } \Omega\,, \\ \frac{\partial u}{\partial \nu_\Omega} = N_G[u] & \text{on } \partial\Omega\,. \end{cases} \tag{11.4}$$

Proof. If $(\mu, c) \in C^{0,\alpha}(\partial\Omega)_0 \times \mathbb{R}$ satisfies Equation (11.3), then we know that $v_\Omega^+[\mu] + c$ belongs to $C^{1,\alpha}(\overline{\Omega})$ and is harmonic in Ω (cf. Theorem 4.25). Then the jump properties for the normal derivative of a single layer potential of Theorem 6.6 imply the validity of the boundary condition in problem (11.4). Hence, $v_\Omega^+[\mu] + c$ satisfies problem (11.4).

If $u \in C^{1,\alpha}(\overline{\Omega})$ satisfies problem (11.4), then Theorem 6.48 (i) on the representation of a harmonic function as a sum of a single layer potential and of a constant ensures that there exists a unique $(\mu, c) \in C^{0,\alpha}(\partial\Omega)_0 \times \mathbb{R}$ such that $u = v_\Omega^+[\mu] + c$. The jump formulas for the normal derivative of a single layer of Theorem 6.6 and the boundary condition in (11.4) imply that the integral equation of (11.3) is satisfied. Hence, the map of the statement is a bijection. □

We also note that the assumption that N_G maps $C^{0,\alpha}(\partial\Omega)$ to itself is certainly satisfied, for example, when G is locally Lipschitz continuous (cf. Lemma A.24).

11.2.2 Formulation of Problems (11.1) and (11.2) in Terms of Integral Equations

We now provide a formulation of the Robin problem (11.1) in Ω^o and (11.2) in $\Omega(\epsilon)$ in terms of integral equations. To do so, we consider the following assumptions

$$\begin{gathered} G^o \in C^0(\partial\Omega^o \times \mathbb{R})\,, \\ N_{G^o} \text{ maps } C^{0,\alpha}(\partial\Omega^o) \text{ to itself}\,. \end{gathered} \tag{11.5}$$

In addition, we denote by G the function from $\partial\Omega(\epsilon) \times \mathbb{R}$ to $\mathbb{R}$ defined by

$$G(x,a) \equiv \begin{cases} G^o(x,a) & \text{if } (x,a) \in \partial\Omega^o \times \mathbb{R}\,, \\ 0 & \text{if } (x,a) \in \epsilon\partial\Omega^i \times \mathbb{R}\,. \end{cases} \tag{11.6}$$

We could apply directly Proposition 11.1. We note, however, that the corresponding representation formulas include an integration in $\partial\Omega(\epsilon)$ and thus on $\epsilon\partial\Omega^i$, a set that depends on ϵ. In order to get rid of such a dependence, we properly rescale the restriction of the unknown function to $\epsilon\partial\Omega^i$. Then we obtain the following Theorem 11.2. In what follows, we find convenient to use the abbreviation

$$X_{0,\alpha} \equiv C^{0,\alpha}(\partial\Omega^i) \times C^{0,\alpha}(\partial\Omega^o)\,.$$

Theorem 11.2. *Let $G^o \in C^0(\partial\Omega^o \times \mathbb{R})$ satisfy (11.5). Let $\mathcal{M} \equiv (\mathcal{M}_1, \mathcal{M}_2, \mathcal{M}_3)$ be the map from $]-\epsilon_0, \epsilon_0[\times X_{0,\alpha} \times \mathbb{R}$ to $X_{0,\alpha} \times \mathbb{R}$ defined by*

$$\mathcal{M}_1[\epsilon,\theta^i,\theta^o,c](\xi) \equiv \frac{1}{2}\theta^i(\xi) + \int_{\partial\Omega^i} \nu_{\Omega^i}(\xi)\cdot\nabla S_n(\xi-\eta)\theta^i(\eta)\,d\sigma_\eta \tag{11.7}$$

$$+\int_{\partial\Omega^o} \nu_{\Omega^i}(\xi)\cdot\nabla S_n(\epsilon\xi-y)\theta^o(y)\,d\sigma_y \qquad \forall \xi\in\partial\Omega^i\,,$$

$$\mathcal{M}_2[\epsilon,\theta^i,\theta^o,c](x) \equiv -\frac{1}{2}\theta^o(x) + \epsilon^{n-1}\int_{\partial\Omega^i} \nu_{\Omega^o}(x)\cdot\nabla S_n(x-\epsilon\eta)\theta^i(\eta)\,d\sigma_\eta$$

$$+\int_{\partial\Omega^o} \nu_{\Omega^o}(x)\cdot\nabla S_n(x-y)\theta^o(y)\,d\sigma_y$$

$$-G^o\left(\epsilon^{n-1}\int_{\partial\Omega^i} S_n(x-\epsilon\eta)\theta^i(\eta)\,d\sigma_\eta + \int_{\partial\Omega^o} S_n(x-y)\theta^o(y)\,d\sigma_y + c\right)$$

$$\forall x\in\partial\Omega^o\,,$$

$$\mathcal{M}_3[\epsilon,\theta^i,\theta^o,c] \equiv \epsilon^{n-1}\int_{\partial\Omega^i}\theta^i\,d\sigma + \int_{\partial\Omega^o}\theta^o\,d\sigma\,,$$

for all $(\epsilon,\theta^i,\theta^o,c)\in]-\epsilon_0,\epsilon_0[\times X_{0,\alpha}\times\mathbb{R}$. *Let* $u[\epsilon,\theta^i,\theta^o,c]$ *be the function from* $\overline{\Omega(\epsilon)}$ *to* $\mathbb{R}$ *defined by*

$$u[\epsilon,\theta^i,\theta^o,c] \equiv v^+_{\Omega(\epsilon)}[\mu] + c\,, \tag{11.8}$$

where

$$\mu(x)\equiv\theta^o(x) \quad \forall x\in\partial\Omega^o\,, \qquad \mu(x)\equiv\theta^i(x/\epsilon) \quad \forall x\in\epsilon\partial\Omega^i\,, \tag{11.9}$$

for all $(\epsilon,\theta^i,\theta^o,c)\in]0,\epsilon_0[\times X_{0,\alpha}\times\mathbb{R}$.

If $\epsilon\in]0,\epsilon_0[$, *then the map* $u[\epsilon,\cdot,\cdot,\cdot]$ *is a bijection from the set of solutions* $(\theta^i,\theta^o,c)\in X_{0,\alpha}\times\mathbb{R}$ *of equation*

$$\mathcal{M}[\epsilon,\theta^i,\theta^o,c]=0\,, \tag{11.10}$$

to the set of solutions $u\in C^{1,\alpha}(\overline{\Omega(\epsilon)})$ *of problem (11.2).*

Proof. A simple computation based on the rule of change of variables in integrals over $\partial\Omega^i$ shows that (θ^i,θ^o,c) solves equation (11.10) if and only the function μ of (11.9) and c solve the integral equation (11.3) with $\Omega=\Omega(\epsilon)$. Thus statement (i) follows by Proposition 11.1 (see also (11.6)). □

We note that, contrary to problem (11.2), which has been defined only for $\epsilon>0$, equation $\mathcal{M}[0,\theta^i,\theta^o,c]=0$ makes sense also for $\epsilon=0$. We now show that in the degenerate case where $\epsilon=0$, equation $\mathcal{M}[\epsilon,\theta^i,\theta^o,c]=0$ is equivalent to a 'limiting boundary value problem'.

Theorem 11.3. *Let* $G^o\in C^0(\partial\Omega^o\times\mathbb{R})$ *satisfy (11.5). A triple* (θ^i,θ^o,c) *in* $X_{0,\alpha}\times\mathbb{R}$ *satisfies equation*

$$\mathcal{M}[0,\theta^i,\theta^o,c]=0\,, \tag{11.11}$$

if and only if both the following conditions are satisfied.

(i) *The pair* $(\theta^o,c)\in C^{0,\alpha}(\partial\Omega^o)\times\mathbb{R}$ *is a solution of the system*

$$\begin{cases} -\frac{1}{2}\theta^o(x) + \int_{\partial\Omega^o} \nu_{\Omega^o}(x) \cdot \nabla S_n(x-y)\theta^o(y)\, d\sigma_y \\ \qquad -G^o\left(\int_{\partial\Omega^o} S_n(x-y)\theta^o(y)\, d\sigma_y + c\right) = 0 \qquad \forall x \in \partial\Omega^o\,, \\ \int_{\partial\Omega^o} \theta^o\, d\sigma = 0\,. \end{cases} \tag{11.12}$$

(ii) θ^i is the only solution in $C^{0,\alpha}(\partial\Omega^i)$ of the integral equation

$$\frac{1}{2}\theta^i(\xi) + \int_{\partial\Omega^i} \nu_{\Omega^i}(\xi) \cdot \nabla S_n(\xi-\eta)\theta^i(\eta)\, d\sigma_\eta \tag{11.13}$$

$$= \int_{\partial\Omega^o} \nu_{\Omega^i}(\xi) \cdot \nabla S_n(y)\theta^o(y)\, d\sigma_y \qquad \forall \xi \in \partial\Omega^i\,.$$

For all (θ^o, c) in $C^{0,\alpha}(\partial\Omega^o) \times \mathbb{R}$, let

$$(u_0^i[\theta^o, c], u_0^o[\theta^o, c]) \equiv (v^-_{\Omega^i}[\theta^i], v^+_{\Omega^o}[\theta^o] + c)\,, \tag{11.14}$$

where

$$\theta^i \equiv \left(\frac{1}{2}I + W^t_\Omega\right)^{(-1)} \left[\nu_{\Omega^i} \cdot \int_{\partial\Omega^o} \nabla S_n(y)\theta^o(y)\, d\sigma_y\right] \tag{11.15}$$

is the only solution in $C^{0,\alpha}(\partial\Omega^i)$ of Equation (11.13). Then the map $(u_0^i[\cdot,\cdot], u_0^o[\cdot,\cdot])$ is a bijection from the set of solutions (θ^o, c) in $C^{0,\alpha}(\partial\Omega^o) \times \mathbb{R}$ of system (11.12) to the set of solutions $(u^i, u^o) \in C^{1,\alpha}_{\mathrm{loc}}(\overline{\Omega^{i-}}) \times C^{1,\alpha}(\overline{\Omega^o})$ of the 'limiting boundary value problem'

$$\begin{cases} \Delta u^o = 0 & \text{in } \Omega^o\,, \\ \frac{\partial u^o}{\partial \nu_{\Omega^o}}(x) = G^o(x, u^o(x)) & \forall x \in \partial\Omega^o\,, \\ \Delta u^i = 0 & \text{in } \Omega^i\,, \\ \frac{\partial u^i}{\partial \nu_{\Omega^i}}(\xi) = -\nu_{\Omega^i}(\xi) \cdot \nabla u^o(0) & \forall \xi \in \partial\Omega^i\,, \\ \lim_{\xi\to\infty} u^i(\xi) = 0\,. \end{cases} \tag{11.16}$$

Moreover, $u_0^o[\cdot,\cdot]$ is a bijection from the set of solutions (θ^o, c) in $C^{0,\alpha}(\partial\Omega^o) \times \mathbb{R}$ of problem (11.12) to the set of solutions $u^o \in C^{1,\alpha}(\overline{\Omega^o})$ of the nonlinear Robin problem (11.1) in Ω^o.

Proof. The equivalence of equation $\mathcal{M}[0, \theta^i, \theta^o, c] = 0$ and of the system of the equations in (11.12) and (11.13) is obvious. Since Ω^{i-} is connected, Corollary 6.15 implies that $\frac{1}{2}I + W^t_{\Omega^i}$ is a linear bijection from $C^{0,\alpha}(\partial\Omega^i)$ to itself. Since ν_{Ω^i} belongs to $C^{0,\alpha}(\partial\Omega^i, \mathbb{R}^n)$, the right-hand side of equation (11.13) is of class $C^{0,\alpha}(\partial\Omega^i)$ and accordingly Equation (11.13) has a unique solution in $C^{0,\alpha}(\partial\Omega^i)$ for each fixed $\theta^o \in C^{0,\alpha}(\partial\Omega^o)$.

By Proposition 11.1 with $\Omega = \Omega^o$, it follows that $u^o[\cdot,\cdot]$ is a bijection from the set of solutions of (11.12) to the set of solutions of (11.1) and thus from the set of solutions of (11.12) to the set of solutions of the first two equations of (11.16).

If (θ^i, θ^o, c) solves equation (11.11), then $u^o[\theta^o, c]$ solves the first two equations of (11.16). Moreover, θ^i must equal the right-hand side of (11.15) and thus the jump formulas of Theorem 6.6 imply that

$$\frac{\partial}{\partial \nu_{\Omega^i}} v^-_{\Omega^i}[\theta^i] = \nu_{\Omega^i} \cdot \int_{\partial \Omega^o} \nabla S_n(y)\theta^o(y)\, d\sigma_y = -\nu_{\Omega^i} \cdot \nabla u^o[\theta^o, c](0) .$$

Then the third and fourth equation of (11.16) are satisfied. Moreover,

$$\int_{\partial\Omega^i} \theta^i \, d\sigma = \int_{\partial\Omega^i} \frac{1}{2}\theta^i + W^t_{\Omega^i}[\theta^i]\, d\sigma$$
$$= -\int_{\partial\Omega^i} \nu_{\Omega^i}(\xi) \cdot \int_{\partial\Omega^o} \nabla S_n(y)\theta^o(y)\, d\sigma_y \, d\sigma_\xi = 0 ,$$

and thus $v^-_{\Omega^i}[\theta^i]$ tends to zero at infinity (cf. Lemma 6.11 and Theorem 4.23). Hence, $(u^i_0[\theta^o, c], u^o_0[\theta^o, c])$ solves (11.16).

We now assume that a pair $(u^i, u^o) \in C^{1,\alpha}_{\mathrm{loc}}(\overline{\Omega^{i-}}) \times C^{1,\alpha}(\overline{\Omega^o})$ satisfies the limiting boundary value problem (11.16) and show that there exists a solution (θ^o, c) in $C^{0,\alpha}(\partial\Omega^o) \times \mathbb{R}$ of system (11.12) such that $(u^i, u^o) = (u^i_0[\theta^o, c], u^o_0[\theta^o, c])$. Since $u^o_0[\cdot, \cdot]$ is a bijection, then there exists a unique $(\theta^o, c) \in C^{0,\alpha}(\partial\Omega^o) \times \mathbb{R}$ such that system (11.12) is satisfied and $u^o = u^o_0[\theta^o, c]$. We now define θ^i by means of (11.15) and we show that $u^i = v^-_{\Omega^i}[\theta^i]$. Then the jump formula for the normal derivative of a single layer potential of Theorem 6.6 implies that $\frac{\partial}{\partial \nu_{\Omega^i}} v^-_{\Omega^i}[\theta^i] = \frac{1}{2}\theta^i + W^t_{\Omega^i}[\theta^i]$. Since

$$\nu_{\Omega^i} \cdot \int_{\partial\Omega^o} \nabla S_n(y)\theta^o(y)\, d\sigma_y = -\nu_{\Omega^i} \cdot \nabla u^o_0[\theta^o, c](0) = -\nu_{\Omega^i} \cdot \nabla u^o(0) ,$$

the definition of θ^i implies that the function $v^-_{\Omega^i}[\theta^i]$ satisfies the third and fourth equation of the limiting boundary value problem (11.16). As above we can prove that equality (11.15) implies that $\int_{\partial\Omega^i} \theta^i \, d\sigma = 0$. Hence, $v^-_{\Omega^i}[\theta^i]$ tends to zero at infinity (cf. Theorem 4.23). Then the difference $u^i - v^-_{\Omega^i}[\theta^i]$ is harmonic in Ω^{i-}, tends to zero at infinity, and satisfies the Neumann boundary condition $\frac{\partial}{\partial \nu_{\Omega^i}}[u^i - v^-_{\Omega^i}[\theta^i]] = 0$ on $\partial\Omega^i$. Then the uniqueness of the exterior Neumann problem for harmonic functions on the connected exterior domain Ω^{i-} implies that $u^i - v^-_{\Omega^i}[\theta^i] = 0$ (cf. Theorem 6.2). Hence, $(u^i, u^o) = (u^i_0[\theta^o, c], u^o_0[\theta^o, c])$.

The injectivity of the map $(u^i_0[\cdot, \cdot], u^o_0[\cdot, \cdot])$ is a consequence of the injectivity of $u^o_0[\cdot, \cdot]$. □

Theorems 11.2 and 11.3 reduce the analysis of the Robin problems (11.1) in Ω^o and (11.2) in $\Omega(\epsilon)$ to that of equation $\mathcal{M} = 0$. We shall show that under reasonable assumptions on the data, if problem (11.1) admits a solution $\tilde{u}^o$ satisfying certain nondegeneracy conditions, then, for ϵ sufficently small, problem (11.2) has a solution, which is unique in a local sense that we clarify in section 11.2.4. To show an existence result for problem (11.2), we apply the Implicit Function Theorem A.19 to the equation $\mathcal{M} = 0$ around a zero $(0, \tilde{\theta}^i, \tilde{\theta}^o, \tilde{c})$ of $\mathcal{M}$ such that $\tilde{u}^o_0 = u^o_0[\tilde{\theta}^o, \tilde{c}]$. Thus we prove the following.

Theorem 11.4. *Assume that*

$$G^o \text{ is real analytic from } \partial\Omega^o \times \mathbb{R} \text{ to } \mathbb{R}\,, \tag{11.17}$$

i.e., G^o has a real analytic extension on an open neighborhood of $\partial\Omega^o \times \mathbb{R}$. Assume that there exists a solution $\tilde{u}^o \in C^{1,\alpha}(\overline{\Omega^o})$ of problem (11.1) such that

$$\begin{aligned} G^o_u(x, \tilde{u}^o(x)) \leq 0 \qquad \forall x \in \partial\Omega^o\,, \\ \int_{\partial\Omega^o} G^o(y, \tilde{u}^o(y))\, d\sigma_y \neq 0\,. \end{aligned} \tag{11.18}$$

Let $(\tilde{\theta}^o, \tilde{c})$ be the unique solution of (11.12) in $C^0(\partial\Omega^o) \times \mathbb{R}$ such that

$$\tilde{u}^o = u^o_0[\tilde{\theta}^o, \tilde{c}]\,.$$

Let $\tilde{\theta}^i$ be as in (11.15) with $\theta^o \equiv \tilde{\theta}^o$. Then there exist $\epsilon' \in]0, \epsilon_0[$, an open neighborhood $\tilde{V}$ of $(\tilde{\theta}^i, \tilde{\theta}^o, \tilde{c})$ in $X_{0,\alpha} \times \mathbb{R}$, and a real analytic map (Θ^i, Θ^o, C) from $]-\epsilon', \epsilon'[$ to $\tilde{V}$ such that the set of zeros of $\mathcal{M}$ in $]-\epsilon', \epsilon'[\times \tilde{V}$ coincides with the graph of (Θ^i, Θ^o, C). In particular,

$$(\Theta^i[0], \Theta^o[0], C[0]) = (\tilde{\theta}^i, \tilde{\theta}^o, \tilde{c})\,. \tag{11.19}$$

Proof. We plan to apply the Implicit Function Theorem A.19 to Equation (11.10) around the point $(0, \tilde{\theta}^i, \tilde{\theta}^o, \tilde{c})$.

By assumption (11.17), by the analyticity of composition operators generated by analytic functions (cf. Theorem A.23 (ii)), by the linearity and continuity of the layer potential operators V_{Ω^o}, $W^t_{\Omega^i}$, $W^t_{\Omega^o}$ (cf. Theorems 6.4 and 6.8), and by the analyticity of the nonlinear integral operators with analytic kernel of Theorem A.28 (ii), the operator $\mathcal{M}$ is real analytic.

We now compute the differential $\partial_{(\theta^i,\theta^o,c)}\mathcal{M}_1[0, \tilde{\theta}^i, \tilde{\theta}^o, \tilde{c}](\theta^{\sharp i}, \theta^{\sharp o}, c^\sharp)$ of $\mathcal{M}$ at $(0, \tilde{\theta}^i, \tilde{\theta}^o, \tilde{c})$ with respect to the variable (θ^i, θ^o, c). By definition of $\mathcal{M}$, we have

$$\begin{aligned} \mathcal{M}[0, \theta^i, \theta^o, c] = \Bigg(& \frac{1}{2}\theta^i + W^t_{\Omega^i}[\theta^i] - \int_{\partial\Omega^o} \nu_{\Omega^i} \cdot \nabla S_n\, \theta^o\, d\sigma\,, \\ & -\frac{1}{2}\theta^o + W^t_{\Omega^o}[\theta^o] - G^o \circ (V_{\Omega^o}[\theta^o] + c), \\ & \int_{\partial\Omega^o} \theta^o\, d\sigma \Bigg) \qquad \forall (\theta^o, \theta^i, c) \in X_{0,\alpha} \times \mathbb{R}\,. \end{aligned}$$

The first and third components of $\mathcal{M}[0, \theta^i, \theta^o, c]$ and the term $-\frac{1}{2}I + W^t_{\Omega^o}$ that appears in the second component are linear and continuous. Accordingly, such two components and the term $-\frac{1}{2}I + W^t_{\Omega^o}$ coincide at all points with their own differential.

Instead, $G^o \circ (V_{\Omega^o}[\theta^o] + c)$ can be seen as the composition of the linear and continuous map

$$(\theta^o, c) \mapsto V_{\Omega^o}[\theta^o] + c$$

with the composition operator N_{G^o} from $C^{1,\alpha}(\partial\Omega^o)$ to itself, which is defined by

$$N_{G^o}[v] \equiv G^o \circ v \qquad \forall v \in C^{1,\alpha}(\partial\Omega^o)\,.$$

Then we observe that N_{G^o} is differentiable by Theorem A.23 (ii) and, by the chain rule, the differential of $G^o \circ (V_{\Omega^o}[\theta^o]+c)$ at $(\tilde{\theta}^o,\tilde{c})$ coincides with the composition of the differential of N_{G^o} at $V_{\Omega^o}[\tilde{\theta}^o]+\tilde{c}$, which is delivered by the formula

$$dN_{G^o}(V_{\Omega^o}[\tilde{\theta}^o]+\tilde{c})[v] = G^o_u(V_{\Omega^o}[\tilde{\theta}^o]+\tilde{c})v \quad \forall v \in C^{1,\alpha}(\partial\Omega^o)$$

with the differential of the linear and continuous operator

$$(\theta^o, c) \mapsto V_{\Omega^o}[\theta^o]+c\,,$$

which coincides with the operator itself. Namely, we have

$$d(G^o \circ (V_{\Omega^o}[\theta^o]+c))(\tilde{\theta}^o,\tilde{c})[\theta^{\sharp o},c^\sharp] = G^o_u(V_{\Omega^o}[\tilde{\theta}^o]+\tilde{c})\left(V_{\Omega^o}[\theta^{\sharp o}]+c^\sharp\right)$$

for all $(\theta^{\sharp o},c^\sharp) \in C^{0,\alpha}(\partial\Omega^o)\times\mathbb{R}$. We conclude that the differential of $\mathcal{M}$ at $(0,\tilde{\theta}^i,\tilde{\theta}^o,\tilde{c})$ with respect to the variable (θ^i,θ^o,c) is delivered by the formula

$$\begin{aligned}
&\partial_{(\theta^i,\theta^o,c)}\mathcal{M}_1[0,\tilde{\theta}^i,\tilde{\theta}^o,\tilde{c}](\theta^{\sharp i},\theta^{\sharp o},c^\sharp)(\xi) = \frac{1}{2}\theta^{\sharp i}(\xi) \\
&\qquad + \int_{\partial\Omega^i} \nu_{\Omega^i}(\xi)\cdot\nabla S_n(\xi-\eta)\theta^{\sharp i}(\eta)\,d\sigma_\eta \\
&\qquad - \int_{\partial\Omega^o} \nu_{\Omega^i}(\xi)\cdot\nabla S_n(y)\theta^{\sharp o}(y)\,d\sigma_y \qquad \forall \xi \in \partial\Omega^i, \\
&\partial_{(\theta^i,\theta^o,c)}\mathcal{M}_2[0,\tilde{\theta}^i,\tilde{\theta}^o,\tilde{c}](\theta^{\sharp i},\theta^{\sharp o},c^\sharp)(x) \\
&\qquad = -\frac{1}{2}\theta^{\sharp o}(x) + \int_{\partial\Omega^o} \nu_{\Omega^o}(x)\cdot\nabla S_n(x-y)\theta^{\sharp o}(y)\,d\sigma_y \\
&\qquad -G^o_u(x,\tilde{u}^o(x))\left\{\int_{\partial\Omega^o} S_n(x-y)\theta^{\sharp o}(y)\,d\sigma_y + c^\sharp\right\} \qquad \forall x \in \partial\Omega^o, \\
&\partial_{(\theta^i,\theta^o,c)}\mathcal{M}_3[0,\tilde{\theta}^i,\tilde{\theta}^o,\tilde{c}](\theta^{\sharp i},\theta^{\sharp o},c^\sharp) = \int_{\partial\Omega^o} \theta^{\sharp o}\,d\sigma\,,
\end{aligned} \tag{11.20}$$

for all $(\theta^{\sharp i},\theta^{\sharp o},c^\sharp) \in X_{0,\alpha}\times\mathbb{R}$. We now prove that $\partial_{(\theta^i,\theta^o,c)}\mathcal{M}[0,\tilde{\theta}^i,\tilde{\theta}^o,\tilde{c}]$ is a linear homeomorphism. By the Open Mapping Theorem 2.2, it suffices to show that $\partial_{(\theta^i,\theta^o,c)}\mathcal{M}[0,\tilde{\theta}^i,\tilde{\theta}^o,\tilde{c}]$ is a bijection from $X_{0,\alpha}\times\mathbb{R}$ onto itself. Let (f^i,f^o,a) belong to $X_{0,\alpha}\times\mathbb{R}$. We must show that there exists a unique $(\theta^{\sharp i},\theta^{\sharp o},c^\sharp) \in X_{0,\alpha}\times\mathbb{R}$ such that

$$\partial_{(\theta^i,\theta^o,c)}\mathcal{M}[0,\tilde{\theta}^i,\tilde{\theta}^o,\tilde{c}](\theta^{\sharp i},\theta^{\sharp o},c^\sharp) = (f^i,f^o,a)\,. \tag{11.21}$$

We first observe that the second and third equation of (11.21) can be rewritten in the form

$$\left(L_{\Omega^o,b^o,+}[\theta^{\sharp o},c^\sharp],\int_{\partial\Omega^o}\theta^{\sharp o}\,d\sigma\right)=(f^o,a)\,,\tag{11.22}$$

where

$$L_{\Omega^o,b^o,+}[\theta^{\sharp o},c^\sharp]=-\frac{1}{2}\theta^{\sharp o}+W^t_{\Omega^o}[\theta^{\sharp o}]+b^oV_{\Omega^o}[\theta^{\sharp o}]+b^oc^\sharp\qquad\text{on }\partial\Omega^o$$

and $b^o(x)\equiv -G^o_u(x,\tilde{u}^o(x))$ for all $x\in\partial\Omega^o$ (we have already introduced the symbol $L_{\Omega^o,b^o,+}$ in (6.56)). Since $N_{G^o_u}$ maps $C^{0,\alpha}(\partial\Omega^o)$ to itself and $\tilde{u}^o\in C^{1,\alpha}(\overline{\Omega^o})$, we have $b^o\in C^{0,\alpha}(\partial\Omega^o)$. Then Proposition 6.57 (i) ensures that Equation (11.22) has a unique solution $(\theta^{\sharp o},c^\sharp)$ in $C^{0,\alpha}(\partial\Omega^o)\times\mathbb{R}$.

We now consider the first equation of (11.21). Since the function

$$\nu_{\Omega^i}\cdot\int_{\partial\Omega^o}\nabla S_n(y)\theta^{\sharp o}(y)\,d\sigma_y$$

belongs to $C^{0,\alpha}(\partial\Omega^i)$, $f^i+\nu_{\Omega^i}\cdot\int_{\partial\Omega^o}\nabla S_n(y)\theta^{\sharp o}(y)\,d\sigma_y$ belongs to $C^{0,\alpha}(\partial\Omega^i)$. Since Ω^{i-} is connected, then $\frac{1}{2}I+W^t_{\Omega^i}$ is a linear bijection from $C^{0,\alpha}(\partial\Omega^i)$ to itself, and accordingly there exists a unique $\theta^{\sharp i}\in C^{0,\alpha}(\partial\Omega^i)$ that solves equation

$$\left(\frac{1}{2}I+W^t_{\Omega^i}\right)[\theta^{\sharp i}]=f^i+\nu_{\Omega^i}\cdot\int_{\partial\Omega^o}\nabla S_n(y)\theta^{\sharp o}(y)\,d\sigma_y\,,$$

i.e., the first equation of (11.21) (cf. Corollary 6.15). Hence, the linear operator $\partial_{(\theta^i,\theta^o,c)}\mathcal{M}[0,\tilde{\theta}^i,\tilde{\theta}^o,\tilde{c}]$ is a continuous linear bijection and the statement of the theorem follows by the Implicit Function Theorem in Banach spaces (cf. Theorem A.19). □

Remark 11.5. We can prove that solutions $\tilde{u}^o\in C^{1,\alpha}(\overline{\Omega^o})$ of problem (11.1) as in (11.17) and (11.18) exist for a large class of functions G^o. For example, Theorem 7.11 of [164] (see also the end of Appendix B and Proposition 7.13 of the same paper) shows that one can use a topological degree argument to prove that a solution exists when G^o satisfies a suitable growth condition. In particular, if there exist a function $g\in C^{0,\alpha}(\partial\Omega)$ with $g\geq 0$ and $\int_{\partial\Omega}g\,d\sigma\neq 0$, and a real number $\delta\in]0,1[$ such that

$$\sup_{(x,a)\in\partial\Omega\times\mathbb{R}}\frac{|G^o(x,a)+g(x)a|}{(1+|a|)^\delta}<+\infty\,.$$

We can now introduce an ϵ-dependent family of solutions of (11.2).

Definition 11.6. Let the assumptions of Theorem 11.4 hold. Then we set

$$u(\epsilon,x)\equiv u\left[\epsilon,\Theta^i[\epsilon],\Theta^o[\epsilon],C[\epsilon]\right](x)\qquad\forall x\in\overline{\Omega(\epsilon)}\,,$$

for all $\epsilon\in]0,\epsilon'[$ (cf. (11.8)).

We shall prove that possibly shrinking ϵ', the family of solutions $\{u(\epsilon,\cdot)\}_{\epsilon\in]0,\epsilon'[}$ is the only family of solutions of problem (11.2) that is close to $\tilde{u}^o$ in a sense which we clarify in Section 11.2.4. First we analyze its behavior as ϵ approaches 0.

11.2.3 Analytic Representation for the Family $\{u(\epsilon,\cdot)\}_{\epsilon\in]0,\epsilon'[}$

The following statement shows that $\{u(\epsilon,\cdot)\}_{\epsilon\in]0,\epsilon'[}$ can be continued real analytically for negative values of ϵ.

Theorem 11.7. *With the assumptions of Theorem 11.4, the following statements hold.*

(i) If Ω_M is a bounded open subset of $\Omega^o\setminus\{0\}$ such that $0\notin\overline{\Omega_M}$, then there exist $\epsilon_M\in]0,\epsilon'[$ and a real analytic map $\mathcal{U}_M$ from $]-\epsilon_M,\epsilon_M[$ to $C^{1,\alpha}(\overline{\Omega_M})$ such that

$$\overline{\Omega_M}\subseteq\overline{\Omega(\epsilon)}\qquad\forall\epsilon\in]-\epsilon_M,\epsilon_M[\,,\tag{11.23}$$

$$u(\epsilon,\cdot)_{|\overline{\Omega_M}}=\mathcal{U}_M[\epsilon](\cdot)\qquad\forall\epsilon\in]0,\epsilon_{\Omega_M}[\,.\tag{11.24}$$

Moreover, $\mathcal{U}_M[0]=\tilde{u}^o_{|\overline{\Omega_M}}$. In particular, $\lim_{\epsilon\to0}u(\epsilon,\cdot)_{|\overline{\Omega_M}}=\tilde{u}^o_{|\overline{\Omega_M}}$ in the $C^{1,\alpha}$-norm.

(ii) There exists a real analytic map $\mathcal{U}_{\partial\Omega^o}$ from $]-\epsilon',\epsilon'[$ to $C^{1,\alpha}(\partial\Omega^o)$ such that

$$u(\epsilon,\cdot)_{|\partial\Omega^o}=\mathcal{U}_{\partial\Omega^o}[\epsilon](\cdot)\qquad\forall\epsilon\in]0,\epsilon'[\,.\tag{11.25}$$

Moreover, $\mathcal{U}_{\partial\Omega^o}[0]=\tilde{u}^o_{|\partial\Omega^o}$.

Proof. Since 0 has a positive distance from $\overline{\Omega_M}$, there exists $\epsilon_M\in]0,\epsilon'[$ such that

$$\overline{\Omega_M}\subseteq\overline{\Omega(\epsilon)}\qquad\forall\epsilon\in[-\epsilon_M,\epsilon_M]\,.$$

If $\epsilon\in]0,\epsilon_M[$, then we have

$$\begin{aligned}u(\epsilon,x)&=u\left[\epsilon,\Theta^i[\epsilon],\Theta^o[\epsilon],C[\epsilon]\right](x)\\&=\int_{\partial\Omega(\epsilon)}S_n(x-y)\Theta_\epsilon(y)\,d\sigma_y+C[\epsilon]\qquad\forall x\in\overline{\Omega(\epsilon)}\,,\end{aligned}$$

where

$$\Theta_\epsilon(y)\equiv\Theta^o[\epsilon](y)\quad\forall y\in\partial\Omega^o\,,\qquad\Theta_\epsilon(y)\equiv\Theta^i[\epsilon](y/\epsilon)\quad\forall y\in\epsilon\partial\Omega^i\,.$$

Accordingly

$$\begin{aligned}u(\epsilon,x)=\epsilon^{n-1}&\int_{\partial\Omega^i}S_n(x-\epsilon\eta)\Theta^i[\epsilon](\eta)\,d\sigma_\eta\\&+\int_{\partial\Omega^o}S_n(x-y)\Theta^o[\epsilon](y)\,d\sigma_y+C[\epsilon]\qquad\forall x\in\overline{\Omega(\epsilon)}\,.\end{aligned}\tag{11.26}$$

Thus, we find natural to define a map $\mathcal{U}_M$ from $]-\epsilon_M,\epsilon_M[$ to $C^{1,\alpha}(\overline{\Omega_M})$ by setting

$$\mathcal{U}_M[\epsilon](x)\equiv\epsilon^{n-1}\int_{\partial\Omega^i}S_n(x-\epsilon\eta)\Theta^i[\epsilon](\eta)\,d\sigma_\eta\tag{11.27}$$

$$+\int_{\partial\Omega^o} S_n(x-y)\Theta^o[\epsilon](y)\,d\sigma_y + C[\epsilon] \qquad \forall x \in \overline{\Omega_M}\,,$$

for all $\epsilon \in]-\epsilon_M, \epsilon_M[$. It clearly suffices to show that the right-hand side of (11.27) defines a real analytic map from $]-\epsilon_M, \epsilon_M[$ to $C^{1,\alpha}(\overline{\Omega_M})$. Since $v^+_{\Omega^o}[\cdot]$ is linear and continuous from $C^{0,\alpha}(\partial\Omega^o)$ to $C^{1,\alpha}(\overline{\Omega^o})$ (cf. Theorem 4.25) and since $\overline{\Omega_M} \subseteq \overline{\Omega^o}$ and Θ^o is real analytic from $]-\epsilon_M, \epsilon_M[$ to $C^{0,\alpha}(\partial\Omega^o)$, we conclude that the second summand in the right-hand side of (11.27) defines a real analytic map from $]-\epsilon_M, \epsilon_M[$ to $C^{1,\alpha}(\overline{\Omega_M})$. We now consider the first summand. By the analyticity of the nonlinear integral operators with analytic kernel of Theorem A.28, the map from $]-\epsilon_M, \epsilon_M[\times L^1(\partial\Omega^i)$ to $C^{1,\alpha}(\overline{\Omega_M})$ that takes (ϵ, f) to $\int_{\partial\Omega^i} S_n(\cdot - \epsilon\eta)f(\eta)\,d\sigma_\eta$ is real analytic. Since Θ^i is real analytic and $C^{0,\alpha}(\partial\Omega^i) \subseteq L^1(\partial\Omega^i)$ with continuous embedding, we conclude that the function from $]-\epsilon_M, \epsilon_M[$ to $C^{1,\alpha}(\overline{\Omega_M})$ that takes ϵ to the first summand in the right-hand side of (11.27) is real analytic. Since $C[\cdot]$ is analytic, the proof of the analyticity of $\mathcal{U}_M$ is complete. By Theorems 11.3 and 11.4, we have

$$\mathcal{U}_M[0] = v^+_{\Omega^o}[\tilde{\theta}^o]_{|\overline{\Omega_M}} + \tilde{c} = \tilde{u}^o_{|\overline{\Omega_M}}\,.$$

Then the limiting relation in ϵ follows by the continuity of $\mathcal{U}_M$. We now consider statement (ii). We define $\mathcal{U}_{\partial\Omega^o}$ to be equal to the right-hand side of (11.27) for all $x \in \partial\Omega^o$ and $\epsilon \in]-\epsilon', \epsilon'[$. Then equality (11.27) for all $x \in \partial\Omega^o$ and ϵ in $]-\epsilon', \epsilon'[$ implies the validity of (11.25). The analyticity of $\mathcal{U}_{\partial\Omega^o}$ follows by the same argument used above to prove the analyticity of $\mathcal{U}_M$. Then we have

$$\mathcal{U}_{\partial\Omega^o}[0] = v^+_{\Omega^o}[\tilde{\theta}^o]_{|\partial\Omega^o} + \tilde{c} = \tilde{u}^o_{|\partial\Omega^o}\,,$$

and thus the proof is complete. □

In a sense, the previous theorem describes the 'macroscopic' behavior of the family $\{u(\epsilon,\cdot)\}_{\epsilon\in]0,\epsilon'[}$ as ϵ is close to 0 and it implies that if we fix $x \in \overline{\Omega^o} \setminus \{0\}$, then we can expand $u(\epsilon, x)$ into a convergent power expansion of ϵ for ϵ small enough.

We now wish to understand the behavior of $u(\epsilon,\cdot)$ close to the boundary $\epsilon\partial\Omega^i$ of the cavity. To do so, we apply a dilation of factor ϵ. We observe that if $\xi \in \mathbb{R}^n \setminus \Omega^i$, then

$$\epsilon\xi \subseteq \mathbb{R}^n \setminus \epsilon\Omega^i\,,$$

and accordingly

$$\epsilon\xi \in \overline{\Omega^o} \setminus \epsilon\Omega^i = \overline{\Omega(\epsilon)}\,,$$

for ϵ small enough so that $\epsilon\xi \in \Omega^o$. Then for such ϵ, it makes sense to consider the evaluation of $u(\epsilon,\cdot)$ at $\epsilon\xi$. Namely,

$$u(\epsilon, \epsilon\xi)\,.$$

We note that the function $u(\epsilon, \epsilon\xi)$ of $\xi \in \frac{1}{\epsilon}\overline{\Omega(\epsilon)}$ is a 'rescaled' version of $u(\epsilon, x)$, with $x \in \overline{\Omega(\epsilon)}$, and we can think of it as the 'microscopic' version of $u(\epsilon, x)$. Then we ask what can be said on the map $\epsilon \mapsto u(\epsilon, \epsilon\xi)$ when ϵ is close to 0. To answer in the spirit of the present book, we prove the following theorem.

Theorem 11.8. *Let the assumptions of Theorem 11.4 hold. Let Ω_m be a bounded open subset of $\mathbb{R}^n \setminus \overline{\Omega^i}$. Then there exist $\epsilon_m \in]0, \epsilon'[$, a real analytic map $\mathcal{U}^1_m$ from $]-\epsilon_m, \epsilon_m[$ to $C^{1,\alpha}(\overline{\Omega_m})$, and a real analytic map $\mathcal{U}^2$ from $]-\epsilon_m, \epsilon_m[$ to $\mathbb{R}$ such that*

$$\epsilon\overline{\Omega_m} \subseteq \Omega^o \quad \forall \epsilon \in]-\epsilon_m, \epsilon_m[\,, \tag{11.28}$$
$$u(\epsilon, \epsilon\xi) = \mathcal{U}^1_m[\epsilon](\xi) + \delta_{2,n}(\epsilon \log \epsilon)\mathcal{U}^2[\epsilon] \;\; \forall \xi \in \overline{\Omega_m} \;\; \forall \epsilon \in]0, \epsilon_m[\,.$$

Moreover, $\mathcal{U}^1_m[0] = \tilde{u}^o(0)$ and $\mathcal{U}^2[0] = 0$.

Proof. Let $\epsilon_m \in]0, \epsilon_0[$ be such that

$$\epsilon\overline{\Omega_m} \subseteq \Omega^o \qquad \forall \epsilon \in [-\epsilon_m, \epsilon_m] \setminus \{0\}\,.$$

Then we have

$$\overline{\Omega_m} \subseteq \frac{1}{\epsilon}\overline{\Omega(\epsilon)} \qquad \forall \epsilon \in [-\epsilon_m, \epsilon_m] \setminus \{0\}\,.$$

By the definition of $u(\epsilon, \cdot)$, we have

$$\begin{aligned}
u(\epsilon, \epsilon\xi) &= \int_{\partial\Omega^i} S_n(\epsilon\xi - \epsilon\eta)\Theta^i[\epsilon](\eta)\, d\sigma_\eta \epsilon^{n-1} \\
&\quad + \int_{\partial\Omega^o} S_n(\epsilon\xi - y)\Theta^o[\epsilon](y)\, d\sigma_y + C[\epsilon] \\
&= \int_{\partial\Omega^i} S_n(\xi - \eta)\Theta^i[\epsilon](\eta)\, d\sigma_\eta \epsilon + \frac{\delta_{2,n}}{2\pi}(\epsilon \log \epsilon) \int_{\partial\Omega^i} \Theta^i[\epsilon]\, d\sigma \\
&\quad + \int_{\partial\Omega^o} S_n(\epsilon\xi - y)\Theta^o[\epsilon](y)\, d\sigma_y + C[\epsilon]
\end{aligned}$$

for all $\xi \in \overline{\Omega_m}$ and for all $\epsilon \in]0, \epsilon_m[$. Then we find natural to set

$$\begin{aligned}
&\mathcal{U}^1_m[\epsilon](\xi) \\
&\quad \equiv \int_{\partial\Omega^i} S_n(\xi - \eta)\Theta^i[\epsilon](\eta)\, d\sigma_\eta \epsilon + \int_{\partial\Omega^o} S_n(\epsilon\xi - y)\Theta^o[\epsilon](y)\, d\sigma_y + C[\epsilon]
\end{aligned}$$

for all $\xi \in \overline{\Omega_m}$ and for all $\epsilon \in]-\epsilon_m, \epsilon_m[$, and

$$\mathcal{U}^2[\epsilon] \equiv \frac{1}{2\pi} \int_{\partial\Omega^i} \Theta^i[\epsilon]\, d\sigma \qquad \forall \epsilon \in]-\epsilon', \epsilon'[\,.$$

By the analyticity of Θ^i, Θ^o, C and by the linearity and continuity of $v_{\overline{\Omega^i}}^-[\cdot]_{|\overline{\Omega_m}}$ from $C^{0,\alpha}(\partial\Omega^i)$ to $C^{0,\alpha}(\overline{\Omega_m})$ (cf. Theorem 4.25) and by the analyticity of nonlinear integral operators with analytic kernel of Theorem A.28 (i), the maps $\mathcal{U}^1_m$ and $\mathcal{U}^2$ are analytic.

We now turn to prove the last part of the statement. By equality (11.19), we have $(\Theta^i[0], \Theta^o[0], C[0]) = (\tilde{\theta}^i, \tilde{\theta}^o, \tilde{c})$. Then the definition of $(\tilde{\theta}^i, \tilde{\theta}^o, \tilde{c})$, which must satisfy equality (11.15), and Lemma 6.11 imply that

$$\begin{aligned} \mathcal{U}^1_m[0] &= v^+_{\Omega^o}[\tilde{\theta}^o](0) + \tilde{c} = \tilde{u}^o(0)\,, \\ \mathcal{U}^2[0] &= \frac{1}{2\pi}\int_{\partial\Omega^i} \tilde{\theta}^i \, d\sigma = \frac{1}{2\pi}\int_{\partial\Omega^i} \frac{1}{2}\tilde{\theta}^i + W^t_{\Omega^i}[\tilde{\theta}^i] \, d\sigma \\ &= \frac{1}{2\pi}\int_{\partial\Omega^i} \nu_{\Omega^i}(\eta)\cdot\left(\int_{\partial\Omega^o} \nabla S_n(y)\tilde{\theta}^o(y)\, d\sigma_y\right) d\sigma_\eta = 0\,. \end{aligned}$$

Thus the proof is complete. □

The previous theorem implies that if $\xi \in \Omega^{i-}$ and $n \geq 3$, then we can expand $u(\epsilon, \epsilon\xi)$ into a convergent power expansion of ϵ, for $\epsilon > 0$ small enough. For $n = 2$ instead, it shows that we can expand $u(\epsilon, \epsilon\xi)$ into a convergent power expansion of ϵ and $(\epsilon \log \epsilon)$, for $\epsilon > 0$ small enough.

11.2.4 Local Uniqueness of the Family $\{u(\epsilon, \cdot)\}_{\epsilon\in]0,\epsilon_0[}$

We now turn to the question of local uniqueness of the family $\{u(\epsilon, \cdot)\}_{\epsilon\in]0,\epsilon_0[}$, and we prove the following.

Theorem 11.9. *Let the assumptions of Theorem 11.4 hold. If $\{\varepsilon_j\}_{j\in\mathbb{N}}$ is a sequence in $]0, +\infty[$ that converges to 0 and if $\{u_j\}_{j\in\mathbb{N}}$ is a sequence of functions such that*

$$\begin{gathered} \text{the function } u_j \text{ belongs to } C^{1,\alpha}(\overline{\Omega(\varepsilon_j)}) \\ \text{and solves (11.2) for } \epsilon = \varepsilon_j \text{ for all } j \in \mathbb{N}\,, \\ \lim_{j\to\infty} u_j = \tilde{u}^o \text{ in } C^{1,\alpha}(\overline{\Omega_M}) \text{ for all bounded open subsets} \\ \Omega_M \text{ of } \Omega^o \setminus \{0\} \text{ such that } 0 \notin \overline{\Omega_M}\,, \end{gathered}$$

then there exists $j_0 \in \mathbb{N}$ such that $u_j(\cdot) = u(\varepsilon_j, \cdot)$ for all $j \geq j_0$.

Proof. Possibly neglecting a finite number of indexes, we can assume that $\varepsilon_j \in]0, \epsilon'[$ for all $j \in \mathbb{N}$. Since G^o is real analytic, Lemma A.24 ensures that condition (11.5) is satisfied. Then Theorems 11.2 and 11.3 imply that there exist $(\theta^i_j, \theta^o_j, c_j)$ and $(\tilde{\theta}^i, \tilde{\theta}^o, \tilde{c}) \in X_{0,\alpha} \times \mathbb{R}$ such that

$$\begin{aligned} u_j &= u[\varepsilon_j, \theta^i_j, \theta^o_j, c_j]\,, & \mathcal{M}[\varepsilon_j, \theta^i_j, \theta^o_j, c_j] &= 0\,, \\ \tilde{u}^o &= u^o_0[\tilde{\theta}^o, \tilde{c}]\,, & \mathcal{M}[0, \tilde{\theta}^i, \tilde{\theta}^o, \tilde{c}] &= 0\,. \end{aligned}$$

We rewrite equation $\mathcal{M}[\epsilon, \theta^i, \theta^o, c] = 0$ for an arbitrary $(\epsilon, \theta^i, \theta^o, c)$ in the domain of $\mathcal{M}$ in the following form

$$\mathcal{M}_1[\epsilon, \theta^i, \theta^o, c] = 0\,, \tag{11.29}$$

$$
\begin{aligned}
&-\frac{1}{2}\theta^o(x) + \epsilon^{n-1} \int_{\partial\Omega^i} \nu_{\Omega^o}(x) \cdot \nabla S_n(x - \epsilon\eta)\theta^i(\eta)\, d\sigma_\eta \\
&\quad + \int_{\partial\Omega^o} \nu_{\Omega^o}(x) \cdot \nabla S_n(x - y)\theta^o(y)\, d\sigma_y \\
&\quad - G_u^o(x, \tilde{u}^o(x)) \Big\{ \epsilon^{n-1} \int_{\partial\Omega^i} S_n(x - \epsilon\eta)\theta^i(\eta)\, d\sigma_\eta \\
&\qquad + \int_{\partial\Omega^o} S_n(x - y)\theta^o(y)\, d\sigma_y + c \Big\} \\
&= G^o \left(x, \epsilon^{n-1} \int_{\partial\Omega^i} S_n(x - \epsilon\eta)\theta^i(\eta)\, d\sigma_\eta + \int_{\partial\Omega^o} S_n(x - y)\theta^o(y)\, d\sigma_y + c \right) \\
&\quad - G_u^o(x, \tilde{u}^o(x)) \Big\{ \epsilon^{n-1} \int_{\partial\Omega^i} S_n(x - \epsilon\eta)\theta^i(\eta)\, d\sigma_\eta \\
&\qquad + \int_{\partial\Omega^o} S_n(x - y)\theta^o(y)\, d\sigma_y + c \Big\} \qquad \forall x \in \partial\Omega^o, \\
&\mathcal{M}_3[\epsilon, \theta^i, \theta^o, c] = 0.
\end{aligned}
$$

Next we denote by $\mathcal{N}[\epsilon, \theta^i, \theta^o, c] \equiv \left(\mathcal{N}_l[\epsilon, \theta^i, \theta^o, c]\right)_{l=1,2,3}$, the function from the set $]-\epsilon_0, \epsilon_0[\times X_{0,\alpha} \times \mathbb{R}$ to $X_{0,\alpha} \times \mathbb{R}$ such that

$$
\mathcal{N}_1 \equiv \mathcal{M}_1, \qquad \mathcal{N}_3 \equiv \mathcal{M}_3,
$$

and such that $\mathcal{N}_2$ equals the left-hand side of the second equation in (11.29). Thus Equation (11.29) can be rewritten as

$$
\begin{aligned}
&\mathcal{N}_1[\epsilon, \theta^i, \theta^o, c] = 0, \qquad\qquad (11.30)\\
&\mathcal{N}_2[\epsilon, \theta^i, \theta^o, c](x) \\
&= G^o \left(x, \epsilon^{n-1} \int_{\partial\Omega^i} S_n(x - \epsilon\eta)\theta^i(\eta)\, d\sigma_\eta + \int_{\partial\Omega^o} S_n(x - y)\theta^o(y)\, d\sigma_y + c \right) \\
&\quad - G_u^o(x, \tilde{u}^o(x)) \Big\{ \epsilon^{n-1} \int_{\partial\Omega^i} S_n(x - \epsilon\eta)\theta^i(\eta)\, d\sigma_\eta \\
&\qquad + \int_{\partial\Omega^o} S_n(x - y)\theta^o(y)\, d\sigma_y + c \Big\} \qquad \forall x \in \partial\Omega^o, \\
&\mathcal{N}_3[\epsilon, \theta^i, \theta^o, c] = 0.
\end{aligned}
$$

By arguing precisely as we did to prove the analyticity of $\mathcal{M}$, we can show that $\mathcal{N}$ is analytic. We now note that $\mathcal{N}[\epsilon, \cdot, \cdot, \cdot]$ is linear and continuous in $X_{0,\alpha} \times \mathbb{R}$ for all fixed $\epsilon \in]-\epsilon_0, \epsilon_0[$. Since $\mathcal{N}$ is analytic, the map from $]-\epsilon_0, \epsilon_0[$ to $\mathcal{L}(X_{0,\alpha} \times \mathbb{R}, X_{0,\alpha} \times \mathbb{R})$ that takes ϵ to the linear operator $\mathcal{N}[\epsilon, \cdot, \cdot, \cdot]$ is analytic. We also note that

$$
\mathcal{N}[0, \cdot, \cdot, \cdot] = \partial_{(\theta^i, \theta^o, c)} \mathcal{M}[0, \tilde{\theta}^i, \tilde{\theta}^o, \tilde{c}](\cdot, \cdot, \cdot),
$$

and that the operator in the right-hand side is a linear homeomorphism (cf. the proof of Theorem 11.4 and (11.20)). Since the set of linear homeomorphisms is open in the set of linear and continuous operators, and since the map that takes a linear invertible operator to its inverse is real analytic (cf. Proposition A.15), there exists $\epsilon'' \in]0, \epsilon'[$ such that the map $\epsilon \mapsto \mathcal{N}[\epsilon, \cdot, \cdot, \cdot]^{(-1)}$ is real analytic from $]-\epsilon'', \epsilon''[$ to $\mathcal{L}(X_{0,\alpha} \times \mathbb{R}, X_{0,\alpha} \times \mathbb{R})$. Since $\mathcal{M}[\varepsilon_j, \theta^i_j, \theta^o_j, c_j] = 0$, the invertibility of $\mathcal{N}[\varepsilon_j, \cdot, \cdot, \cdot]$ and equality (11.30) guarantee that

$$(\theta^i_j, \theta^o_j, c_j) = \mathcal{N}[\varepsilon_j, \cdot, \cdot, \cdot]^{(-1)} \left[0, N_{G^o}[u_{j|\partial\Omega^o}] - N_{G^o_u}[\tilde{u}^o_{|\partial\Omega^o}]u_{j|\partial\Omega^o}, 0\right]$$

whenever $\varepsilon_j \in]0, \epsilon''[$. Since G^o is real analytic, then $N_{G^o}[\cdot]$ is also real analytic (cf. Theorem A.23 (ii)), and thus $N_{G^o}[\cdot]$ is also continuous. Accordingly,

$$\lim_{j\to\infty} N_{G^o}[u_{j|\partial\Omega^o}] - N_{G^o_u}[\tilde{u}^o_{|\partial\Omega^o}]u_{j|\partial\Omega^o} = N_{G^o}[\tilde{u}^o_{|\partial\Omega^o}] - N_{G^o_u}[\tilde{u}^o_{|\partial\Omega^o}]\tilde{u}^o_{|\partial\Omega^o}, \tag{11.31}$$

in $C^{0,\alpha}(\partial\Omega^o)$. Moreover, the analyticity of $\epsilon \mapsto \mathcal{N}[\epsilon, \cdot, \cdot, \cdot]^{(-1)}$ guarantees that

$$\lim_{j\to\infty} \mathcal{N}[\varepsilon_j, \cdot, \cdot, \cdot]^{(-1)} = \mathcal{N}[0, \cdot, \cdot, \cdot]^{(-1)}, \tag{11.32}$$

in $\mathcal{L}(X_{0,\alpha} \times \mathbb{R}, X_{0,\alpha} \times \mathbb{R})$. Since the evaluation map from

$$\mathcal{L}(X_{0,\alpha} \times \mathbb{R}, X_{0,\alpha} \times \mathbb{R}) \times X_{0,\alpha} \times \mathbb{R}$$

to $X_{0,\alpha} \times \mathbb{R}$ that takes a pair (R, v) to $R[v]$ is bilinear and continuous, the limiting relations (11.31) and (11.32) imply that

$$\begin{aligned} &\lim_{j\to\infty} (\theta^i_j, \theta^o_j, c_j) \\ &\quad = \lim_{j\to\infty} \mathcal{N}[\epsilon_j, \cdot, \cdot, \cdot]^{(-1)} \left[0, N_{G^o}[u_{j|\partial\Omega^o}] - N_{G^o_u}[\tilde{u}^o_{|\partial\Omega^o}]u_{j|\partial\Omega^o}, 0\right] \\ &\quad = \mathcal{N}[0, \cdot, \cdot, \cdot]^{(-1)} \left[0, N_{G^o}[\tilde{u}^o_{|\partial\Omega^o}] - N_{G^o_u}[\tilde{u}^o_{|\partial\Omega^o}]\tilde{u}_{|\partial\Omega^o}, 0\right] \quad \text{in } X_{0,\alpha} \times \mathbb{R}. \end{aligned} \tag{11.33}$$

Since $\mathcal{M}[0, \tilde{\theta}^i, \tilde{\theta}^o, \tilde{c}] = 0$, the right-hand side of (11.33) equals $(\tilde{\theta}^i, \tilde{\theta}^o, \tilde{c})$. Hence,

$$\lim_{j\to\infty} (\varepsilon_j, \theta^i_j, \theta^o_j, c_j) = (0, \tilde{\theta}^i, \tilde{\theta}^o, \tilde{c}),$$

in $]-\epsilon'', \epsilon''[\times X_{0,\alpha} \times \mathbb{R}$. Hence, there exists $j_0 \in \mathbb{N}$ such that every $(\varepsilon_j, \theta^i_j, \theta^o_j, c_j)$ with $j \geq j_0$ belongs to the neighborhood $]-\epsilon', \epsilon, [\times \tilde{V}$ of $(0, \tilde{\theta}^i, \tilde{\theta}^o, \tilde{c})$ introduced in Theorem 11.4. Since $(\varepsilon_j, \theta^i_j, \theta^o_j, c_j)$ are zeros of $\mathcal{M}$, Theorem 11.4 implies that

$$\theta^i_j = \Theta^i[\varepsilon_j], \qquad \theta^o_j = \Theta^o[\varepsilon_j] \qquad c_j = C[\varepsilon_j],$$

for all $j \geq j_0$. Accordingly,

$$u_j = u[\varepsilon_j, \theta_j^i, \theta_j^o, c_j] = u[\Theta^i[\varepsilon_j], \Theta^o[\varepsilon_j], C[\varepsilon_j]] = u(\varepsilon_j, \cdot)$$

for all $j \geq j_0$. □

11.2.5 Analytic Representation for the Energy Integral of the Family $\{u(\epsilon, \cdot)\}_{\epsilon \in]0,\epsilon'[}$

We now turn to the energy integral, and we prove the following.

Theorem 11.10. *Let the assumptions of Theorem 11.4 hold. Then there exists a real analytic map $\mathcal{E}$ from $]-\epsilon', \epsilon'[$ to $\mathbb{R}$ such that*

$$\int_{\Omega(\epsilon)} |\nabla_y u(\epsilon, y)|^2 \, dy = \mathcal{E}[\epsilon] \qquad \forall \epsilon \in]0, \epsilon'[\, .$$

Moreover, $\mathcal{E}[0] = \int_{\Omega^o} |\nabla \tilde{u}^o|^2 \, dy$.

Proof. By the first Green Identity of Theorem 4.2, we have

$$\begin{aligned} &\int_{\Omega(\epsilon)} |\nabla_y u(\epsilon, y)|^2 \, dy \\ &\quad = -\int_{\epsilon\partial\Omega^i} u(\epsilon, y) \frac{\partial}{\partial \nu_{\epsilon\Omega^i_y}} u(\epsilon, y) \, d\sigma_y + \int_{\partial\Omega^o} u(\epsilon, y) \frac{\partial}{\partial \nu_{\Omega^o_y}} u(\epsilon, y) \, d\sigma_y \\ &\quad = \int_{\partial\Omega^o} u(\epsilon, y) G^o(y, u(\epsilon, y)) \, d\sigma_y \, , \end{aligned}$$

for all $\epsilon \in]0, \epsilon'[$. Then it suffices to take

$$\mathcal{E}[\epsilon] \equiv \int_{\partial\Omega^o} \mathcal{U}_{\partial\Omega^o}[\epsilon](y) G^o(y, \mathcal{U}_{\partial\Omega^o}[\epsilon](y)) \, d\sigma_y$$

for all $\epsilon \in]-\epsilon', \epsilon'[$ (cf. Theorem 11.7). The analyticity of $\mathcal{U}_{\partial\Omega^o}[\cdot]$ from $]-\epsilon', \epsilon'[$ to $C^{1,\alpha}(\partial\Omega^o)$, the analyticity of N_{G^o} from $C^{0,\alpha}(\partial\Omega^o)$ to itself, which follows by the analyticity of G^o and by Theorem A.23 (ii), the continuity of the pointwise product in $C^0(\partial\Omega^o)$, and the continuity of the linear map that takes a function of $C^0(\partial\Omega^o)$ to its integral ensure that $\mathcal{E}$ is real analytic on $]-\epsilon', \epsilon'[$. Finally, equality $\mathcal{U}_{\partial\Omega^o}[0] = \tilde{u}^o_{|\partial\Omega^o}$ and the definition of $\mathcal{E}$, ensure that $\mathcal{E}[0] = \int_{\Omega^o} |\nabla \tilde{u}^o|^2 \, dy$. □

11.3 A Nonlinear Transmission Problem

In this section, we consider a nonlinear transmission problem in a domain with a small inclusion.

First we introduce an 'unperturbed' problem with no hole (and no transmission) in Ω^o, and then we consider the case with a small hole $\epsilon\Omega^i$ (cf. (8.1) and (8.2)).

To do so, we assign a Dirichlet datum

$$f^o \in C^{1,\alpha}(\partial\Omega^o) . \tag{11.34}$$

As is well known, the 'unperturbed' Dirichlet problem

$$\begin{cases} \Delta u = 0 \text{ in } \Omega^o, \\ u = f^o \quad \text{on } \partial\Omega^o , \end{cases} \tag{11.35}$$

has one and only one solution $\tilde{u}^o \in C^{1,\alpha}(\overline{\Omega^o})$ (cf. Theorem 6.27).

Next we make a hole $\epsilon\Omega^i$ in the domain Ω^o and we consider a nonlinear transmission problem in the perforated domain $\Omega(\epsilon)$ as in (8.3) and in the corresponding hole $\epsilon\Omega^i$ for each $\epsilon \in]0, \epsilon_0[$. In order to introduce the nonlinear transmission conditions on $\epsilon\partial\Omega^i$, we introduce

$$\text{an increasing real analytic diffeomorphism } F \text{ from } \mathbb{R} \text{ onto } \mathbb{R} , \tag{11.36}$$

and

$$\text{a real analytic map } G \text{ from } \mathbb{R} \text{ to } \mathbb{R} . \tag{11.37}$$

Then we consider the nonlinear transmission problem

$$\begin{cases} \Delta u^i = 0 & \text{in } \epsilon\Omega^i , \\ \Delta u^o = 0 & \text{in } \Omega(\epsilon), \\ u^o = F(u^i) & \text{on } \epsilon\partial\Omega^i , \\ \frac{\partial u^o}{\partial \nu_{\epsilon\Omega^i}}(x) = \frac{\partial u^i}{\partial \nu_{\epsilon\Omega^i}}(x) + G(u^i(x)) & \forall x \in \epsilon\partial\Omega^i , \\ u^o = f^o & \text{on } \partial\Omega^o , \end{cases} \tag{11.38}$$

where $\nu_{\epsilon\Omega^i}$ denotes the outward unit normal to $\epsilon\partial\Omega^i$. Nonlinear transmission problems of this type arise in the study of the heat conduction in composite materials with different nonconstant thermal conductivities. More specifically, when the conductivities of the two materials are (nonlinear) functions of the temperature itself, we can perform a suitable change of the unknown functions (by means of the so called 'Kirchhoff transformation') and transform the boundary value problem into a boundary value problem for the Laplace equation with a nonlinear transmission condition on $\epsilon\partial\Omega^i$ as the one in (11.38). We present an application of this strategy in the case of a periodic problem in Section 13.8.2 (see also Mityushev and Rogosin [224, Chap. 5.1]). We note that in the simpler case in which F is affine and G equals 0, (11.38) becomes a standard linear transmission problem with a temperature jump on the interface.

A priori, it is not clear why problem (11.38) should admit a classical solution. We shall prove that possibly shrinking ϵ_0, problem (11.38) has a family of solutions $(u^i(\epsilon,\cdot), u^o(\epsilon,\cdot))$ in $C^{1,\alpha}(\overline{\epsilon\Omega^i}) \times C^{1,\alpha}(\overline{\Omega(\epsilon)})$ for all $\epsilon \in]0,\epsilon_0[$, which has the property that

$$\lim_{\epsilon\to 0^+} u^i(\epsilon,\epsilon\xi) = F^{(-1)}(\tilde{u}^o(0)) \quad \text{in the } C^{1,\alpha}\text{-norm in } \xi \in \overline{\Omega^i}\,,$$

$$\lim_{\epsilon\to 0^+} u^o(\epsilon,\epsilon\xi) = \tilde{u}^o(0)$$

in the $C^{1,\alpha}$-norm for ξ in the closure of all bounded open subsets Ω_m of $\mathbb{R}^n \setminus \overline{\Omega^i}$,

$$\lim_{\epsilon\to 0^+} u^o(\epsilon,x) = \tilde{u}^o(x)$$

in the $C^{1,\alpha}$-norm for x in the closure of all bounded open subsets Ω_M of Ω^o such that $0 \notin \overline{\Omega_M}$.

This family of solutions is also unique in a local sense which we clarify in section 11.3.3. Moreover, we answer the following questions

(i) Let ξ be fixed in $\overline{\Omega^i}$. What can be said on the map $\epsilon \mapsto u^i(\epsilon,\epsilon\xi)$ when $\epsilon > 0$ is close to 0?
(ii) Let ξ be fixed in $\mathbb{R}^n \setminus \Omega^i$. What can be said on the map $\epsilon \mapsto u^o(\epsilon,\epsilon\xi)$ when $\epsilon > 0$ is close to 0?
(iii) Let x be fixed in $\overline{\Omega^o} \setminus \{0\}$. What can be said on the map $\epsilon \mapsto u^o(\epsilon,x)$ when $\epsilon > 0$ is close to 0?
(iv) What can be said on the map $\epsilon \mapsto \int_{\epsilon\Omega^i} |\nabla_y u^i(\epsilon,y)|^2\,dy$ when $\epsilon > 0$ is close to 0?
(v) What can be said on the map $\epsilon \mapsto \int_{\Omega(\epsilon)} |\nabla_y u^o(\epsilon,y)|^2\,dy$ when $\epsilon > 0$ is close to 0?

In a sense, questions (i), (ii) concern the 'microscopic' behavior of $u^i(\epsilon,\cdot)$ and $u^o(\epsilon,\cdot)$, whereas question (iii) concerns the 'macroscopic' behavior of $u^o(\epsilon,\cdot)$ and questions (iv), (v) deal with the energy of $u^i(\epsilon,\cdot)$ and $u^o(\epsilon,\cdot)$, respectively.

To analyze the problem and answer the above questions we resort to the Functional Analytic Approach of this book. Accordingly, we first convert the nonlinear transmission problem (11.38) into a system of nonlinear integral equations by exploiting classical Potential Theory. Then we observe that, by changing the variables appropriately, we can obtain a functional equation that can be analyzed by means of the Implicit Function Theorem around the degenerate case where $\epsilon = 0$. We deduce that we can represent the unknown densities of the integral equations in terms of real analytic functions of ϵ for $n \geq 3$, and of ϵ, $\log^{-1}\epsilon$, and $\epsilon\log^2\epsilon$ for $n = 2$. Next we go back to the integral representation of the solutions of problem (11.38) and we deduce both the existence of $u^i(\epsilon,\cdot)$ and of $u^o(\epsilon,\cdot)$ and the representation formulas that describe their dependence upon ϵ and the dependence upon ϵ of their energies. Doing so, we answer to questions (i)–(v).

We observe that the content of the present chapter develops from [168]. Moreover, for the analysis of a similar problem, but with nonautonomous transmission conditions, we refer to the paper [225] of Molinarolo. For an existence result in the case of a big inclusion (that is, for $\epsilon > 0$ fixed) we refer to the paper [71] with Mishuris and for a local uniqueness result for the solutions of (11.38) themselves, rather than for the family of solutions, we mention the paper [72] with Molinarolo.

11.3.1 Formulation of the Nonlinear Transmission Problem in Terms of Integral Equations

As a first step, we wish to convert the transmission problem (11.38) into a system of integral equations. By exploiting the representation Theorem 6.48 for harmonic functions, the solutions of (11.38) can be written as sums of a single layer potential and of a constant. Then by exploiting the classical jump formulas for single layer potentials (cf. Theorem 6.6), we obtain a system of integral equations on the ϵ-dependent domain $\epsilon\partial\Omega^i \cup \partial\Omega^o$. In order to get rid of ϵ in the domain, we rescale our functional variables on $\epsilon\partial\Omega^i$ and finally obtain an ϵ-dependent system of integral equations on the boundaries $\partial\Omega^i$ and $\partial\Omega^o$, which do not depend on ϵ. We present such a change of variable in the following Theorem 11.11. To do so, we find convenient to introduce the product space $Y_{0,\alpha}$ defined by

$$Y_{0,\alpha} \equiv C^{0,\alpha}(\partial\Omega^i)_0 \times \mathbb{R} \times C^{0,\alpha}(\partial\Omega^i) \times C^{0,\alpha}(\partial\Omega^o) \times \mathbb{R}\,. \tag{11.39}$$

Theorem 11.11. *Let F, G, f^o be as in (11.36), (11.37), (11.34). Let $\epsilon \in]0, \epsilon_0[$. Then the map that takes $(\psi, c^i, \theta^i, \theta^o, c^o)$ to the pair of functions*

$$\left(v^+_{\epsilon\Omega^i}[\psi(\cdot/\epsilon)] + c^i, v^+_{\Omega^o}[\theta^o] + v^-_{\epsilon\Omega^i}[\theta^i(\cdot/\epsilon)] + c^o\right)$$

is a bijection from the set of $(\psi, c^i, \theta^i, \theta^o, c^o)$ in $Y_{0,\alpha}$, that solve the following system of integral equations

$$\begin{aligned}
&\epsilon V_{\Omega^i}[\theta^i](\xi) + \frac{\delta_{2,n}}{2\pi}(\epsilon \log \epsilon)\int_{\partial\Omega^i} \theta^i \, d\sigma \qquad\qquad (11.40)\\
&\qquad + \int_{\partial\Omega^o} S_n(\epsilon\xi - y)\theta^o(y)\, d\sigma_y + c^o = F\left(\epsilon V_{\Omega^i}[\psi](\xi) + c^i\right) \qquad \forall \xi \in \partial\Omega^i\,,\\
&\frac{1}{2}\theta^i(\xi) + W^t_{\Omega^i}[\theta^i](\xi) + \int_{\partial\Omega^o} \nu_{\Omega^i}(\xi)\cdot\nabla S_n(\epsilon\xi - y)\theta^o(y)\, d\sigma_y\\
&\qquad = -\frac{1}{2}\psi(\xi) + W^t_{\Omega^i}[\psi](\xi) + G\left(\epsilon V_{\Omega^i}[\psi](\xi) + c^i\right) \qquad \forall \xi \in \partial\Omega^i\,,\\
&\epsilon^{n-1}\int_{\partial\Omega^i} S_n(x - \epsilon\eta)\theta^i(\eta)\, d\sigma_\eta + V_{\Omega^o}[\theta^o](x) + c^o = f^o(x) \qquad \forall x \in \partial\Omega^o\,,
\end{aligned}$$

$$\epsilon^{n-1}\int_{\partial\Omega^i}\theta^i\,d\sigma+\int_{\partial\Omega^o}\theta^o\,d\sigma=0\,,$$

to the set of pairs (u^i,u^o) *of* $C^{1,\alpha}(\epsilon\overline{\Omega^i})\times C^{1,\alpha}(\overline{\Omega(\epsilon)})$, *that solve the boundary value problem (11.38).*

Proof. By the representation Theorem 6.48 for harmonic functions as a sum of a single layer potential and a constant, if a pair of functions (u^i,u^o) of $C^{1,\alpha}(\epsilon\overline{\Omega^i})\times C^{1,\alpha}(\overline{\Omega(\epsilon)})$ solves the transmission problem (11.38), then there exists a unique quadruple

$$(\omega,c^i,\mu,c^o)\in C^{0,\alpha}(\epsilon\partial\Omega^i)_0\times\mathbb{R}\times C^{0,\alpha}(\partial\Omega(\epsilon))_0\times\mathbb{R}$$

such that

$$u^i=v^+_{\epsilon\Omega^i}[\omega]+c^i\,,\qquad u^o=v^+_{\Omega(\epsilon)}[\mu]+c^o\,. \tag{11.41}$$

Moreover, the jump formulas for the normal derivative of a single layer of Theorem 6.6 implies that (ω,c^i,μ,c^o) satisfies the following system of integral equations

$$\begin{aligned}
&\int_{\epsilon\partial\Omega^i}S_n(x-y)\mu(y)\,d\sigma_y+\int_{\partial\Omega^o}S_n(x-y)\mu(y)\,d\sigma_y+c^o \\
&\qquad=F\left(\int_{\epsilon\partial\Omega^i}S_n(x-y)\omega(y)\,d\sigma_y+c^i\right)\qquad\forall x\in\epsilon\partial\Omega^i\,,\\
&\frac{1}{2}\mu(x)-\int_{\epsilon\partial\Omega^i}\nu_{\Omega(\epsilon)}(x)\cdot\nabla S_n(x-y)\mu(y)\,d\sigma_y\\
&\qquad-\int_{\partial\Omega^o}\nu_{\Omega(\epsilon)}(x)\cdot\nabla S_n(x-y)\mu(y)\,d\sigma_y\\
&\qquad=-\frac{1}{2}\omega(x)+\int_{\epsilon\partial\Omega^i}\nu_{\epsilon\Omega^i}(x)\cdot\nabla S_n(x-y)\omega(y)\,d\sigma_y\\
&\qquad+G\left(\int_{\epsilon\partial\Omega^i}S_n(x-y)\omega(y)\,d\sigma_y+c^i\right)\qquad\forall x\in\epsilon\partial\Omega^i\,,\\
&\int_{\epsilon\partial\Omega^i}S_n(x-y)\mu(y)\,d\sigma_y\\
&\qquad+\int_{\partial\Omega^o}S_n(x-y)\mu(y)\,d\sigma_y+c^o=f^o(x)\qquad\forall x\in\partial\Omega^o\,.
\end{aligned}\tag{11.42}$$

Conversely, if (ω,c^i,μ,c^o) satisfies system (11.42), then the pair of functions in (11.41) solves the transmission problem (11.38) in $\Omega(\epsilon)$.

Now we observe that (ω,c^i,μ,c^o) solves (11.42) if and only if the 5-tuple $(\psi,c^i,\theta^i,\theta^o,c^o)$ of $Y_{0,\alpha}$ with

$$\begin{aligned}
&\psi(\xi)\equiv\omega(\epsilon\xi)\qquad\forall\xi\in\partial\Omega^i\,,\qquad\theta^i(\xi)\equiv\mu(\epsilon\xi)\qquad\forall\xi\in\partial\Omega^i\,,\\
&\theta^o(x)\equiv\mu(x)\qquad\forall x\in\partial\Omega^o
\end{aligned}$$

satisfies system (11.40). □

If the system of integral equations (11.40) has a family of solutions

$$\{(\psi_{[\epsilon]}, c^i_{[\epsilon]}, \theta^i_{[\epsilon]}, \theta^o_{[\epsilon]}, c^o_{[\epsilon]})\}_{\epsilon\in[0,\epsilon_0[}$$

in $Y_{0,\alpha}$ that depends continuously on the parameter ϵ, then by taking the limit as ϵ tends to 0 in (11.40), we obtain

$$\begin{cases} v^+_{\Omega^o}[\theta^o_{[0]}](0) + c^o_{[0]} = F(c^i_{[0]}) \,, \\ \frac{1}{2}\theta^i_{[0]} + W^t_{\Omega^i}[\theta^i_{[0]}] + \nu_{\Omega^i}(\xi)\cdot\int_{\partial\Omega^o} \nabla S_n(-y)\theta^o_{[0]}(y)\,d\sigma_y \\ \qquad = -\frac{1}{2}\psi_{[0]} + W^t_{\Omega^i}[\psi_{[0]}] + G(c^i_{[0]}) \,, \\ V_{\Omega_o}[\theta^o_{[0]}] + c^o_{[0]} = f^o \,, \\ \int_{\partial\Omega^o} \theta^o_{[0]}\,d\sigma = 0 \,. \end{cases} \tag{11.43}$$

We now prove that $(\psi_{[0]}, c^i_{[0]}, \theta^i_{[0]}, \theta^o_{[0]}, c^o_{[0]})$ is not the only solution in $Y_{0,\alpha}$ of system (11.43). By the representation Theorem 6.48 for harmonic functions as a sum of a single layer potential and a constant, the last two equations uniquely determine $\theta^o_{[0]}$ and $c^o_{[0]}$. Moreover, $v^+_{\Omega^o}[\theta^o_{[0]}] + c^o_{[0]}$ coincides with the unique solution $\tilde{u}^o$ of the Dirichlet problem (11.35). Then the first equation of the limit system (11.43) implies that $c^i_{[0]} = F^{(-1)}(\tilde{u}^o(0))$. However, if we choose an arbitrary $\psi_{[0]}$ in $C^{0,\alpha}(\partial\Omega^i)$, then the second equation of the limit system (11.43) has a unique solution $\theta^i_{[0]}$ in $C^{0,\alpha}(\partial\Omega^i)$ (cf. Corollary 6.15). Hence, the limit system (11.43) does not determine uniquely the limiting value of a family of solutions of (11.38), even if we assume that the limiting value of the corresponding family of solutions of (11.40) does exist. In this sense, we may regard the limit system (11.43) as singular.

To overcome this difficulty, we introduce a change of 'functional' variables in system (11.40), so that we can obtain a nonsingular limiting system for a family of solutions that does have a limit as ϵ tends to zero. To do so, we exploit the following remark.

Remark 11.12. Let F, G, f^o be as in (11.36), (11.37), (11.34). Let $(\tilde{\theta}^o, \tilde{c}^o)$ be the unique element of $C^{0,\alpha}(\partial\Omega^o)_0 \times \mathbb{R}$ such that

$$\tilde{u}^o = v^+_{\Omega^o}[\tilde{\theta}^o] + \tilde{c}^o \,, \tag{11.44}$$

where $\tilde{u}^o$ is the unique solution of the 'unperturbed' Dirichlet problem (11.35) in Ω^o. If $\epsilon \in]0, \epsilon_0[$, then the map from $Y_{0,\alpha}$ to itself that maps $(\psi, c^i_1, \theta^i, \theta^o_1, c^o_1)$ to the 5-tuple

$$\Bigg(\psi, F^{(-1)}(\tilde{u}^o(0)) + \epsilon c^i_1 + \frac{\delta_{2,n}(\epsilon\log\epsilon)}{2\pi F'(F^{(-1)}(\tilde{u}^o(0)))}\int_{\partial\Omega^i}\theta^i\,d\sigma, \tag{11.45}$$

$$\theta^i, \tilde{\theta}^o + \epsilon\theta^o_1, \tilde{c}^o + \epsilon c^o_1\Bigg)$$

is a bijection.

We exploit the 'functional' variables of Remark 11.12 to transform the system of integral equations (11.40) into a new system, which we later show to yield a non-singular limiting system when ϵ tends to zero.

Theorem 11.13. *Let F, G, f^o be as in (11.36), (11.37), and (11.34). Let*

$$\left(u^i[\epsilon,\psi,c_1^i,\theta^i,\theta_1^o,c_1^o],u^o[\epsilon,\psi,c_1^i,\theta^i,\theta_1^o,c_1^o]\right)$$

be the pair of functions of $C^{1,\alpha}(\epsilon\overline{\Omega^i})\times C^{1,\alpha}(\overline{\Omega(\epsilon)})$ defined by

$$\begin{aligned}u^i[\epsilon,\psi,c_1^i,\theta^i,\theta_1^o,c_1^o](x) &\equiv v^+_{\epsilon\Omega^i}[\psi(\cdot/\epsilon)](x)+F^{(-1)}(\tilde{u}^o(0))\\ &\quad+\epsilon c_1^i+\frac{\delta_{2,n}(\epsilon\log\epsilon)}{2\pi F'(F^{(-1)}(\tilde{u}^o(0)))}\int_{\partial\Omega^i}\theta^i\,d\sigma\\ &=\epsilon^{n-1}\int_{\partial\Omega^i}S_n(x-\epsilon\eta)\psi(\eta)\,d\sigma_\eta+F^{(-1)}(\tilde{u}^o(0))\\ &\quad+\epsilon c_1^i+\frac{\delta_{2,n}(\epsilon\log\epsilon)}{2\pi F'(F^{(-1)}(\tilde{u}^o(0)))}\int_{\partial\Omega^i}\theta^i\,d\sigma\qquad\forall x\in\epsilon\overline{\Omega^i},\\ u^o[\epsilon,\psi,c_1^i,\theta^i,\theta_1^o,c_1^o](x)&\equiv\epsilon v^+_{\Omega^o}[\theta_1^o](x)+\epsilon c_1^o+\tilde{u}^o(x)+v^-_{\epsilon\Omega^i}[\theta^i(\cdot/\epsilon)](x)\\ &=\epsilon v^+_{\Omega^o}[\theta_1^o](x)+\epsilon c_1^o+\tilde{u}^o(x)+\epsilon^{n-1}\int_{\partial\Omega^i}S_n(x-\epsilon\eta)\theta^i(\eta)\,d\sigma_\eta\quad\forall x\in\overline{\Omega(\epsilon)}\end{aligned}\tag{11.46}$$

for all $(\epsilon,\psi,c_1^i,\theta^i,\theta_1^o,c_1^o)$ in $]0,\epsilon_0[\times Y_{0,\alpha}$.

If $\epsilon\in]0,\epsilon_0[$, then the map $(u^i[\epsilon,\cdot,\cdot,\cdot,\cdot,\cdot],u^o[\epsilon,\cdot,\cdot,\cdot,\cdot,\cdot])$ is a bijection from the subset of $Y_{0,\alpha}$ consisting of the 5-tuples $(\psi,c_1^i,\theta^i,\theta_1^o,c_1^o)$ that solve the following system of integral equations,

$$\begin{aligned}&V_{\Omega^i}[\theta^i](\xi)+\int_{\partial\Omega^o}S_n(\epsilon\xi-y)\theta_1^o(y)\,d\sigma_y+c_1^o+\int_0^1\nabla\tilde{u}^o(\beta\epsilon\xi)\xi\,d\beta\\ &\quad=F'(F^{(-1)}(\tilde{u}^o(0)))\left\{V_{\Omega^i}[\psi](\xi)+c_1^i\right\}+\epsilon\int_0^1(1-\beta)F''\Big(F^{(-1)}(\tilde{u}^o(0))\\ &\qquad+\beta\epsilon\Big(V_{\Omega^i}[\psi](\xi)+c_1^i+\frac{\delta_{2,n}\log\epsilon}{2\pi F'(F^{(-1)}(\tilde{u}^o(0)))}\int_{\partial\Omega^i}\theta^i\,d\sigma\Big)\Big)\,d\beta\\ &\qquad\times\left\{V_{\Omega^i}[\psi](\xi)+c_1^i+\frac{\delta_{2,n}\log\epsilon}{2\pi F'(F^{(-1)}(\tilde{u}^o(0)))}\int_{\partial\Omega^i}\theta^i\,d\sigma\right\}^2\qquad\forall\xi\in\partial\Omega^i,\\ &\frac{1}{2}\theta^i(\xi)+W^t_{\Omega^i}[\theta^i](\xi)+\epsilon\int_{\partial\Omega^o}\nu_{\Omega^i}(\xi)\cdot\nabla S_n(\epsilon\xi-y)\theta_1^o(y)\,d\sigma_y\\ &\quad+\int_{\partial\Omega^o}\nu_{\Omega^i}(\xi)\cdot\nabla S_n(\epsilon\xi-y)\tilde{\theta}^o(y)\,d\sigma_y\\ &=-\frac{1}{2}\psi(\xi)+W^t_{\Omega^i}[\psi](\xi)\\ &\quad+G\left(\epsilon V_{\Omega^i}[\psi](\xi)+F^{(-1)}(\tilde{u}^o(0))+c_1^i\epsilon+\frac{\delta_{2,n}(\epsilon\log\epsilon)}{2\pi F'(F^{(-1)}(\tilde{u}^o(0)))}\int_{\partial\Omega^i}\theta^i\,d\sigma\right)\end{aligned}\tag{11.47}$$

$$\forall \xi \in \partial\Omega^i\,,$$

$$\epsilon^{n-2}\int_{\partial\Omega^i} S_n(x-\epsilon\eta)\theta^i(\eta)\,d\sigma_\eta + V_{\Omega^o}[\theta_1^o](x) + c_1^o = 0 \qquad \forall x \in \partial\Omega^o\,,$$

$$\epsilon^{n-2}\int_{\partial\Omega^i} \theta^i\,d\sigma + \int_{\partial\Omega^o} \theta_1^o\,d\sigma = 0\,,$$

to the set of solutions $(u^i, u^o) \in C^{1,\alpha}(\epsilon\overline{\Omega^i}) \times C^{1,\alpha}(\overline{\Omega(\epsilon)})$ *of the transmission problem (11.38).*

Proof. By Theorem 11.11 and by Remark 11.12, it suffices to show that the 5-tuple $(\psi, c_1^i, \theta^i, \theta_1^o, c_1^o)$ in $Y_{0,\alpha}$ solves the system of integral equations (11.47) if and only if the 5-tuple of (11.45) satisfies the system of integral equations (11.40).

By plugging the 5-tuple of (11.45) into system (11.40), we obtain

$$\begin{aligned}
&\epsilon V_{\Omega^i}[\theta^i](\xi) + \frac{\delta_{2,n}}{2\pi}(\epsilon\log\epsilon)\int_{\partial\Omega^i}\theta^i\,d\sigma + \int_{\partial\Omega^o} S_n(\epsilon\xi - y)\tilde{\theta}^o(y)\,d\sigma_y + \tilde{c}^o \qquad (11.48)\\
&\quad +\epsilon\int_{\partial\Omega^o} S_n(\epsilon\xi - y)\theta_1^o(y)\,d\sigma_y + \epsilon c_1^o\\
&\quad = F\left(\epsilon V_{\Omega^i}[\psi](\xi) + F^{(-1)}(\tilde{u}^o(0)) + \epsilon c_1^i + \frac{\delta_{2,n}(\epsilon\log\epsilon)}{2\pi F'(F^{(-1)}(\tilde{u}^o(0)))}\int_{\partial\Omega^i}\theta^i\,d\sigma\right)\\
&\qquad\qquad\qquad\qquad\qquad\qquad\qquad\qquad\qquad\qquad \forall \xi \in \partial\Omega^i\,,\\
&\frac{1}{2}\theta^i(\xi) + W^t_{\Omega^i}[\theta^i](\xi) + \int_{\partial\Omega^o}\nu_{\Omega^i}(\xi)\cdot\nabla S_n(\epsilon\xi - y)\tilde{\theta}^o(y)\,d\sigma_y\\
&\quad +\epsilon\int_{\partial\Omega^o}\nu_{\Omega^i}(\xi)\cdot\nabla S_n(\epsilon\xi - y)\theta_1^o(y)\,d\sigma_y\\
&\quad = -\frac{1}{2}\psi(\xi) + W^t_{\Omega^i}[\psi](\xi)\\
&\quad +G\left(\epsilon V_{\Omega^i}[\psi](\xi) + F^{(-1)}(\tilde{u}^o(0)) + \epsilon c_1^i + \frac{\delta_{2,n}(\epsilon\log\epsilon)}{2\pi F'(F^{(-1)}(\tilde{u}^o(0)))}\int_{\partial\Omega^i}\theta^i\,d\sigma\right)\\
&\qquad\qquad\qquad\qquad\qquad\qquad\qquad\qquad\qquad\qquad \forall \xi \in \partial\Omega^i\,,\\
&\epsilon^{n-1}\int_{\partial\Omega^i} S_n(x-\epsilon\eta)\theta^i(\eta)\,d\sigma_\eta + V_{\Omega^o}[\tilde{\theta}^o](x) + \tilde{c}^o + \epsilon V_{\Omega^o}[\theta_1^o](x) + \epsilon c_1^o = f^o(x)\\
&\qquad\qquad\qquad\qquad\qquad\qquad\qquad\qquad\qquad\qquad \forall x \in \partial\Omega^o\,,\\
&\epsilon^{n-1}\int_{\partial\Omega^i}\theta^i\,d\sigma + \int_{\partial\Omega^o}\tilde{\theta}^o\,d\sigma + \epsilon\int_{\partial\Omega^o}\theta_1^o\,d\sigma = 0\,.
\end{aligned}$$

By the Taylor Formula, we have

$$\begin{aligned}
F\left(F^{(-1)}(\tilde{u}^o(0)) + t\right) &= \tilde{u}^o(0) + F'(F^{(-1)}(\tilde{u}^o(0)))t\\
&\quad +t^2\int_0^1 (1-\beta)F''(F^{(-1)}(\tilde{u}^o(0)) + \beta t)\,d\beta \qquad \forall t \in \mathbb{R}\,,
\end{aligned}$$

$$\tilde{u}^o(\epsilon\xi) - \tilde{u}^o(0) = \epsilon \int_0^1 \nabla\tilde{u}^o(\beta\epsilon\xi) \cdot \xi \, d\beta \qquad \forall \xi \in \partial\Omega^i .$$

Then by taking

$$t = \epsilon V_{\Omega^i}[\psi](\xi) + \epsilon c_1^i + \frac{\delta_{2,n}(\epsilon \log \epsilon)}{2\pi F'(F^{(-1)}(\tilde{u}^o(0)))} \int_{\partial\Omega^i} \theta^i \, d\sigma ,$$

by plugging the corresponding expression in the first equation of (11.48), by replacing

$$\int_{\partial\Omega^o} S_n(\epsilon\xi - y)\tilde{\theta}^o(y) \, d\sigma_y + \tilde{c}^o = \tilde{u}^o(\epsilon\xi)$$

(cf. (11.44)) with

$$\tilde{u}^o(0) + \epsilon \int_0^1 \nabla\tilde{u}^o(\beta\epsilon\xi) \cdot \xi \, d\beta ,$$

and by dividing the corresponding expression by ϵ, we obtain the first equation of (11.47). Next, we observe that the second equation of (11.48) coincides with the second equation of (11.47). Since $V_{\Omega^o}[\tilde{\theta}^o](x) + \tilde{c}^o = \tilde{u}^o = f^o$ on $\partial\Omega^o$, the third equation of (11.48) is equivalent to the third equation of (11.47). Since $\int_{\partial\Omega^o} \tilde{\theta}^o \, d\sigma = 0$, the fourth equation of (11.48) is equivalent to the fourth equation of (11.47). □

Hence, we are now reduced to analyze system (11.47). At first glance, we can see that system (11.47) contains the term $\log \epsilon$, which is singular at 0, only in case $n = 2$. Then we understand that case $n = 2$ needs a separate treatment.

Our first step to analyze system (11.47) is to note that by letting ϵ tend to zero, we obtain a system, which we address to as the 'limiting system', and which has the following form.

$$\begin{aligned} & V_{\Omega^i}[\theta^i](\xi) + \int_{\partial\Omega^o} S_n(-y)\theta_1^o(y) \, d\sigma_y + c_1^o + \nabla\tilde{u}^o(0)\xi \\ & \qquad = F'(F^{(-1)}(\tilde{u}^o(0))) \left\{ V_{\Omega^i}[\psi](\xi) + c_1^i \right\} \qquad \forall \xi \in \partial\Omega^i , \\ & \frac{1}{2}\theta^i(\xi) + W^t_{\Omega^i}[\theta^i](\xi) + \nu_{\Omega^i}(\xi) \cdot \nabla\tilde{u}^o(0) \\ & \qquad = -\frac{1}{2}\psi(\xi) + W^t_{\Omega^i}[\psi](\xi) + G\left(F^{(-1)}(\tilde{u}^o(0))\right) \qquad \forall \xi \in \partial\Omega^i , \\ & \delta_{2,n} \int_{\partial\Omega^i} S_n(x)\theta^i(\eta) \, d\sigma_\eta + V_{\Omega^o}[\theta_1^o](x) + c_1^o = 0 \qquad \forall x \in \partial\Omega^o , \\ & \delta_{2,n} \int_{\partial\Omega^i} \theta^i \, d\sigma + \int_{\partial\Omega^o} \theta_1^o \, d\sigma = 0 . \end{aligned} \tag{11.49}$$

To analyze this limiting system, we need the following technical statement.

Lemma 11.14. *Let Ω be a bounded open connected subset of $\mathbb{R}^n$ of class $C^{1,\alpha}$. Then there exists $\tilde{\phi} \in \mathrm{Ker}\left(-\frac{1}{2}I + W^t_\Omega\right)$ such that the function $V_\Omega[\tilde{\phi}]$ is constant and $\int_{\partial\Omega} \tilde{\phi} \, d\sigma = 1$*

Proof. Since Ω is connected, the number $\varkappa^+$ of connected components of Ω equals 1 and we have $1 = \varkappa^+ = \dim \mathrm{Ker}\left(-\frac{1}{2}I + W_\Omega^t\right)$ (see Theorem 6.24 and following comment). Thus there exists $\tilde{\phi} \in \mathrm{Ker}\left(-\frac{1}{2}I + W_\Omega^t\right) \setminus \{0\}$. By the jump formula for the normal derivative of a single layer potential of Theorem 6.6, we have $\frac{\partial}{\partial \nu_\Omega} v_\Omega^+[\tilde{\phi}] = 0$ on $\partial\Omega$ and thus $v_\Omega^+[\tilde{\phi}]$ is constant in $\overline{\Omega}$. Then Lemma 6.22 and condition $\tilde{\phi} \neq 0$ imply that $\int_{\partial\Omega} \tilde{\phi}\, d\sigma \neq 0$. Possibly multiplying $\tilde{\phi}$ by a suitable constant, we can assume that $\int_{\partial\Omega} \tilde{\phi}\, d\sigma = 1$. □

We can now prove an existence and uniqueness result for the solution of a system of boundary integral equations, that we need to solve system (11.49).

Theorem 11.15. *Let Ω be a bounded open connected subset of $\mathbb{R}^n$ of class $C^{1,\alpha}$ such that the exterior of Ω is connected. Let $a \in]0, +\infty[$. Let (f, g) belong to $C^{1,\alpha}(\partial\Omega) \times C^{0,\alpha}(\partial\Omega)$. Then the system of integral equations*

$$\begin{cases} V_\Omega[\theta] - a(V_\Omega[\psi] + c) = f & \text{on } \partial\Omega\,, \\ \frac{1}{2}\theta + W_\Omega^t[\theta] + \frac{1}{2}\psi - W_\Omega^t[\psi] = g & \text{on } \partial\Omega\,, \\ \int_{\partial\Omega} \psi\, d\sigma = 0\,, \end{cases} \tag{11.50}$$

has one and only one solution $(\theta, \psi, c) \in C^{0,\alpha}(\partial\Omega) \times C^{0,\alpha}(\partial\Omega) \times \mathbb{R}$.

Proof. We first assume that problem (11.50) does have a solution (θ, ψ, c) in $C^{0,\alpha}(\partial\Omega) \times C^{0,\alpha}(\partial\Omega) \times \mathbb{R}$, and we prove that such a solution must necessarily be delivered by a certain formula. By the first equation, we must have

$$V_\Omega[\theta a^{-1} - \psi] - c = f a^{-1}\,. \tag{11.51}$$

We claim that $\theta a^{-1} - \psi$ and c are uniquely determined by $f a^{-1}$. Ultimately, this will be a consequence of Theorem 6.47, which implies that $f a^{-1}$ can be written in a unique way as the sum of a constant function and a single layer potential with density that has zero integral on $\partial\Omega$. We cannot, however, use directly Theorem 6.47, because the integral $\int_{\partial\Omega} \theta a^{-1} - \psi\, d\sigma$ might be different from zero. Then we observe that Lemma 11.14 implies that there exists $\tilde{\phi} \in C^{0,\alpha}(\partial\Omega)$ such that $V_\Omega[\tilde{\phi}]$ is constant on $\partial\Omega$ and $\int_{\partial\Omega} \tilde{\phi}\, d\sigma = 1$. Thus there exists a unique $\tilde{a} \in \mathbb{R}$ such that

$$\int_{\partial\Omega} \theta a^{-1} - \psi - \tilde{a}\tilde{\phi}\, d\sigma = 0 \tag{11.52}$$

and we have

$$V_\Omega[\theta a^{-1} - \psi - \tilde{a}\tilde{\phi}] + V_\Omega[\tilde{\phi}]\tilde{a} - c = f a^{-1}\,. \tag{11.53}$$

By (11.52), we have

$$\tilde{a} = \int_{\partial\Omega} \theta a^{-1} - \psi\, d\sigma = \int_{\partial\Omega} \theta\, d\sigma a^{-1}\,. \tag{11.54}$$

By Theorem 6.47, there exists a unique pair $(\omega, d) \in C^{0,\alpha}(\partial\Omega)_0 \times \mathbb{R}$ such that

$$V_\Omega[\omega] + d = f a^{-1}\,. \tag{11.55}$$

By (11.52) and (11.53), we must necessarily have

$$\theta = a(\psi + \omega + \tilde{a}\tilde{\phi}) , \qquad c = V_\Omega[\tilde{\phi}]\tilde{a} - d , \tag{11.56}$$

with $\tilde{a}$ as in (11.54). Then the second equation of (11.50) implies that

$$\frac{1}{2}(a+1)\psi + (a-1)W^t_\Omega[\psi] = g - a\left[\frac{1}{2}\omega + W^t_\Omega[\omega] + \left(\frac{1}{2}\tilde{\phi} + W^t_\Omega[\tilde{\phi}]\right)\tilde{a}\right] . \tag{11.57}$$

By integrating the last equation on $\partial\Omega$, by invoking Lemma 6.11, and by equalities

$$\int_{\partial\Omega} \omega \, d\sigma = 0 , \qquad \int_{\partial\Omega} \tilde{\phi} \, d\sigma = 1 , \qquad \int_{\partial\Omega} \psi \, d\sigma = 0 ,$$

we obtain

$$0 = \int_{\partial\Omega} g \, d\sigma - a\tilde{a} ,$$

and thus

$$\tilde{a} = a^{-1} \int_{\partial\Omega} g \, d\sigma . \tag{11.58}$$

By Theorem 6.51, and by inequality

$$-1 < \frac{a-1}{a+1} < 1 ,$$

which holds for all $a \in]0, +\infty[$, the operator $\frac{1}{2}I + \frac{a-1}{a+1}W^t_\Omega$ is a linear homeomorphism of $C^{0,\alpha}(\partial\Omega)$. Then equality (11.57) implies that ψ is necessarily delivered by the formula

$$\psi = \frac{1}{a+1}\left(\frac{1}{2}I + \frac{a-1}{a+1}W^t_\Omega\right)^{(-1)} \left[g - a\left(\frac{1}{2}\omega + W^t_\Omega[\omega]\right) - \left(\frac{1}{2}\tilde{\phi} + W^t_\Omega[\tilde{\phi}]\right)\int_{\partial\Omega} g \, d\sigma\right] . \tag{11.59}$$

In conclusion, if (θ, ψ, c) satisfies problem (11.50), then ψ, θ, and c are delivered by the formulas (11.56) and (11.59), with $\tilde{\phi}$ as in Lemma 11.14, ω and d as in (11.55), and $\tilde{a}$ as in (11.58). In particular, problem (11.50) has at most one solution.

Conversely, by reading backward the proof above, we can check that the triple (θ, ψ, c) delivered by (11.56), (11.58), (11.59) with $\tilde{\phi}$ as in Lemma 11.14, ω and d as in (11.55), and $\tilde{a}$ as in (11.58), actually belongs to $C^{0,\alpha}(\partial\Omega) \times C^{0,\alpha}(\partial\Omega) \times \mathbb{R}$ and solves problem (11.50). □

Then we have the following theorem, which shows the unique solvability of the limiting system (11.49) and its link with a boundary value problem which we address to as the 'limiting boundary value problem'.

Theorem 11.16. *Let F, G, f^o be as in (11.36), (11.37), (11.34). Then the following statements hold.*

(i) The limiting system (11.49) has one and only one solution in $Y_{0,\alpha}$, which we denote by $(\tilde{\psi}, \tilde{c}_1^i, \tilde{\theta}^i, \tilde{\theta}_1^o, \tilde{c}_1^o)$. Moreover,

$$\tilde{c}_1^o = 0\,, \qquad \int_{\partial\Omega^i} \tilde{\theta}^i \, d\sigma = m_{n-1}(\partial\Omega^i) G(F^{(-1)}(\tilde{u}^o(0)))\,. \tag{11.60}$$

If $n \geq 3$, then we also have $\tilde{\theta}_1^o = 0$.

(ii) The limiting boundary value problem

$$\begin{cases} \Delta u_1^{i,r} = 0 & \text{in } \Omega^i\,, \\ \Delta u_1^{o,r} = 0 & \text{in } \Omega^{i-}\,, \\ \Delta u_1^o = 0 & \text{in } \Omega^o\,, \\ u_1^{o,r}(\xi) + u_1^o(0) + \nabla\tilde{u}^o(0)\cdot\xi = F'(F^{(-1)}(\tilde{u}^o(0)))u_1^{i,r}(\xi) & \forall \xi \in \partial\Omega^i\,, \\ \frac{\partial u_1^{o,r}}{\partial\nu_{\Omega^i}}(\xi) + \nu_{\Omega^i}(\xi)\cdot\nabla\tilde{u}^o(0) = \frac{\partial u_1^{i,r}}{\partial\nu_{\Omega^i}}(\xi) + G(F^{(-1)}(\tilde{u}^o(0))) & \forall \xi \in \partial\Omega^i\,, \\ u_1^o(x) + \delta_{2,n}S_n(x)m_{n-1}(\partial\Omega^i)G(F^{(-1)}(\tilde{u}^o(0))) = 0 & \forall x \in \partial\Omega^o\,, \\ \lim_{\xi\to\infty}\left[u_1^{o,r}(\xi) - \delta_{2,n}m_{n-1}(\partial\Omega^i)G(F^{(-1)}(\tilde{u}^o(0)))S_n(\xi)\right] = 0 \end{cases} \tag{11.61}$$

has one and only one solution in $C^{1,\alpha}(\overline{\Omega^i}) \times C^{1,\alpha}_{\mathrm{loc}}(\overline{\Omega^{i-}}) \times C^{1,\alpha}(\overline{\Omega^o})$, which we denote by $(\tilde{u}_1^{i,r}, \tilde{u}_1^{o,r}, \tilde{u}_1^o)$, and which is delivered by the following formulas

$$\begin{aligned} &\tilde{u}_1^{i,r} = v^+_{\Omega^i}[\tilde{\psi}] + \tilde{c}_1^i && \text{in } \overline{\Omega^i}\,, \qquad \tilde{u}_1^{o,r} = v^-_{\Omega^i}[\tilde{\theta}^i] \qquad \text{in } \overline{\Omega^{i-}}\,, \\ &\tilde{u}_1^o = v^+_{\Omega^o}[\tilde{\theta}_1^o] + \tilde{c}_1^o && \text{in } \overline{\Omega^o}\,, \end{aligned} \tag{11.62}$$

where $(\tilde{\psi}, \tilde{c}_1^i, \tilde{\theta}^i, \tilde{\theta}_1^o, \tilde{c}_1^o)$ is the only solution of the limiting system (11.49) (see (i)).

Proof. (i) We first assume that the limiting system (11.49) has a solution

$$(\psi, c_1^i, \theta^i, \theta_1^o, c_1^o)$$

and we show that θ_1^o, c_1^o must necessarily be delivered by a certain formula. By the last two equations of (11.49), we have

$$\int_{\partial\Omega^o}\left(S_n(x-y) - \frac{\delta_{2,n}}{2\pi}\log|x|\right)\theta_1^o(y)\,d\sigma_y + c_1^o = 0 \qquad \forall x \in \partial\Omega^o\,. \tag{11.63}$$

If $n \geq 3$, then the isomorphism Theorem 6.47 implies that $\theta_1^o = 0$, $c_1^o = 0$. If instead $n = 2$, then we set

$$v(x) \equiv \int_{\partial\Omega^o}\frac{1}{2\pi}\log\frac{|x-y|}{|x|}\theta_1^o(y)\,d\sigma_y + c_1^o \qquad \forall x \in \overline{\Omega^{o-}}\,.$$

By the limiting relation (4.69), we deduce that

$$\lim_{x\to\infty} v(x) = c_1^o \in \mathbb{R}\,.$$

Since v is harmonic in Ω^{o-} and equals 0 on $\partial\Omega^o$, the uniqueness of the solution of an exterior Dirichlet problem of Proposition 6.1 implies that $v(x) = 0$ for all $x \in \overline{\Omega^{o-}}$. Hence, $c_1^o = 0$ and

$$\int_{\partial\Omega^o} S_2(x-y)\theta_1^o(y)\,d\sigma_y = S_2(x)\int_{\partial\Omega^o}\theta_1^o(y)\,d\sigma_y \qquad \forall x \in \overline{\Omega^{o-}}\,. \tag{11.64}$$

Now, the last equation of the limiting system (11.49) implies that

$$\int_{\partial\Omega^o}\theta_1^o d\sigma = -\int_{\partial\Omega^i}\theta^i\,d\sigma\,.$$

Thus, by integrating the second equality in (11.49), by Lemmas 6.11, 6.21, and by equality $\int_{\partial\Omega^i}\nu_{\Omega^i}\cdot\nabla\tilde{u}^o(0)\,d\sigma = 0$, which follows by the First Green Identity of Theorem 4.2, we obtain that

$$\int_{\partial\Omega^i}\theta^i\,d\sigma = m_{n-1}(\partial\Omega^i)G(F^{(-1)}(\tilde{u}^o(0))) \tag{11.65}$$

for all $n \geq 2$. In particular, for $n = 2$ we have

$$\int_{\partial\Omega^o}\theta_1^o\,d\sigma = -\int_{\partial\Omega^i}\theta^i\,d\sigma = -m_1(\partial\Omega^i)G(F^{(-1)}(\tilde{u}^o(0)))\,. \tag{11.66}$$

By taking the normal derivative of (11.64) and by exploiting the jump relations of Theorem 6.6, we obtain that

$$\left(\frac{1}{2}I + W_{\Omega^o}^t\right)[\theta_1^o] = -(\nu_{\Omega^o}\cdot\nabla S_2)m_1(\partial\Omega^i)G(F^{(-1)}(\tilde{u}^o(0)))\,. \tag{11.67}$$

Since Ω^{o-} is connected, Corollary 6.15 implies that $\frac{1}{2}I + W_{\Omega^o}^t$ is invertible and thus we have

$$\theta_1^o = -\left(\frac{1}{2}I + W_{\Omega^o}^t\right)^{(-1)}[\nu_{\Omega^o}\cdot\nabla S_2]m_1(\partial\Omega^i)G(F^{(-1)}(\tilde{u}^o(0)))\,. \tag{11.68}$$

Hence, we have the formulas

$$\begin{aligned}\theta_1^o &= -\delta_{2,n}\left(\frac{1}{2}I + W_{\Omega^o}^t\right)^{(-1)}[\nu_{\Omega^o}\cdot\nabla S_2]m_{n-1}(\partial\Omega^i)G(F^{(-1)}(\tilde{u}^o(0)))\\ &\qquad\qquad\qquad\qquad\qquad\qquad\qquad\qquad\qquad\qquad\text{on } \partial\Omega^o\,,\\ c_1^o &= 0\,,\end{aligned} \tag{11.69}$$

which hold both if $n \geq 3$ and $n = 2$. Next we note that Theorem 11.15 and the positivity of $F'(F^{(-1)}(\tilde{u}^o(0)))$, which follows by assumption (11.36) on F, imply that the first two equations of (11.49) with θ_1^o, c_1^o as in (11.69) have a unique solution (ψ, c_1^i, θ^i) in $C^{0,\alpha}(\partial\Omega^i)_0 \times \mathbb{R} \times C^{0,\alpha}(\partial\Omega^i)$. Hence, we have proved that the limiting system (11.49) has at most one solution.

On the other hand, if we set (θ_1^o, c_1^o) equal to the pair delivered by the right-hand sides of (11.69) and if we set (ψ, c_1^i, θ^i) equal to the only solution of the first two equations of (11.49) with (θ_1^o, c_1^o) equal to the above choice of (θ_1^o, c_1^o), we can verify that $(\psi, c_1^i, \theta^i, \theta_1^o, c_1^o)$ solves the limiting system (11.49) by reading backward the above argument.

Indeed, if $n \geq 3$, we have $\theta_1^o = 0$, $c_1^o = 0$ and the validity of (11.49) follows by our choice of (ψ, c_1^i, θ^i). If instead $n = 2$, we observe that our choice of (ψ, c_1^i, θ^i) implies the validity of the first two equations of (11.49). Then by integrating the second equation of (11.49) on $\partial\Omega^i$ we obtain (11.65). By the first equality in (11.69), we obtain (11.67). Then by integrating (11.67) on $\partial\Omega^o$ and by exploiting the equality $\int_{\partial\Omega^o} \nu_{\Omega^o} \cdot \nabla S_n \, d\sigma = 1$ and Lemma 6.11 we obtain that the left and right-hand sides of (11.66) are equal. Hence, equality (11.65) implies that the last equation of the limiting system (11.49) is satisfied. Then the formulas (11.69) imply that

$$\frac{\partial}{\partial \nu_{\Omega^o}} v(x) = 0 \qquad \text{on } \partial\Omega^o \,,$$

where v is as above. Moreover, the limiting relation (4.69) and equality $c_1^o = 0$ imply that

$$\lim_{x\to\infty} v(x) = \lim_{x\to\infty} \int_{\partial\Omega^o} \frac{1}{2\pi} \log \frac{|x-y|}{|x|} \theta_1^o(y) \, d\sigma + c_1^o = 0 \,,$$

and the unique solvability of the exterior Neumann problem for harmonic functions implies that $v = 0$ in $\overline{\Omega^{o-}}$ (cf. Theorem 6.2). Hence, the above proved equality

$$\int_{\partial\Omega^i} \theta^i \, d\sigma = - \int_{\partial\Omega^o} \theta_1^o \, d\sigma$$

(i.e., the last equation of the limiting system (11.49)), together with equality $v = 0$ in $\overline{\Omega^{o-}}$ imply the validity of the third equation of the limiting system (11.49).

We now consider statement (ii) and we show that the limiting boundary value problem (11.61) has a unique solution.

If $n \geq 3$, the third and sixth equations of (11.61) and the uniqueness for the Dirichlet problem imply that $u_1^o = 0$ (cf. Theorem 3.6). Then problem (11.61) with $u_1^o = 0$ and $n \geq 3$ is a linear transmission problem in the unknown $(u_1^{i,r}, u_1^{o,r})$, which has a unique solution in $C^{1,\alpha}(\overline{\Omega^i}) \times C^{1,\alpha}_{\text{loc}}(\overline{\Omega^{i-}})$ (see Proposition 6.54, Theorem 6.55 and the positivity of $F'(F^{(-1)}(\tilde{u}^o(0)))$, which follows by assumption (11.36) on F).

We now consider case $n = 2$ and we show that the limiting boundary value problem (11.61) has a unique solution. To do so, we introduce the auxiliary function

$$v_1^{o,r}(\xi) \equiv u_1^{o,r}(\xi) - m_1(\partial\Omega^i) G(F^{(-1)}(\tilde{u}^o(0))) S_2(\xi) \qquad \forall \xi \in \overline{\Omega^{i-}} \,. \quad (11.70)$$

Then the limiting boundary value problem is equivalent to the following auxiliary boundary value problem in the unknown $(u_1^{i,r}, v_1^{o,r}, u_1^o)$ in $C^{1,\alpha}(\overline{\Omega^i}) \times C^{1,\alpha}_{\mathrm{loc}}(\overline{\Omega^{i-}}) \times C^{1,\alpha}(\overline{\Omega^o})$,

$$\begin{cases} \Delta u_1^{i,r} = 0 & \text{in } \Omega^i\,, \\ \Delta v_1^{o,r} = 0 & \text{in } \Omega^{i-}\,, \\ \Delta u_1^o = 0 & \text{in } \Omega^o\,, \\ v_1^{o,r}(\xi) + m_1(\partial\Omega^i)G(F^{(-1)}(\tilde{u}^o(0)))S_2(\xi) + u_1^o(0) & \\ \qquad +\nabla\tilde{u}^o(0)\cdot\xi = F'(F^{(-1)}(\tilde{u}^o(0)))u_1^{i,r}(\xi) & \forall\xi\in\partial\Omega^i\,, \\ \frac{\partial v_1^{o,r}}{\partial\nu_{\Omega^i}}(\xi) + m_1(\partial\Omega^i)G(F^{(-1)}(\tilde{u}^o(0)))\nu_{\Omega^i}(\xi)\cdot\nabla S_2(\xi) & \\ \qquad +\nu_{\Omega^i}(\xi)\cdot\nabla\tilde{u}^o(0) = \frac{\partial u_1^{i,r}}{\partial\nu_{\Omega^i}}(\xi) + G(F^{(-1)}(\tilde{u}^o(0))) & \forall\xi\in\partial\Omega^i\,, \\ u_1^o(x) = -m_1(\partial\Omega^i)G(F^{(-1)}(\tilde{u}^o(0)))S_2(x) & \forall x\in\partial\Omega^o\,, \\ \lim_{\xi\to\infty} v_1^{o,r}(\xi) = 0\,. \end{cases} \tag{11.71}$$

By the existence and uniqueness Theorem 6.27 for the interior Dirichlet problem, the third and sixth equations have a unique solution $u_1^o \in C^{1,\alpha}(\overline{\Omega^o})$. Next we note that the equality $\int_{\partial\Omega^i} \nu_{\Omega^i}(\xi)\cdot\nabla S_2(\xi)\,d\sigma_\xi = 1$ implies that

$$\begin{aligned} \int_{\partial\Omega^i} \Big[& G(F^{(-1)}(\tilde{u}^o(0))) \\ & -m_1(\partial\Omega^i)G(F^{(-1)}(\tilde{u}^o(0)))\nu_{\Omega^i}(\xi)\cdot\nabla S_2(\xi) - \nu_{\Omega^i}(\xi)\cdot\nabla\tilde{u}_1^o(0)\Big]\,d\sigma_\xi \\ & = \int_{\partial\Omega^i} G(F^{(-1)}(\tilde{u}^o(0)))\,d\sigma_\xi \\ & \quad -m_1(\partial\Omega^i)G(F^{(-1)}(\tilde{u}^o(0)))\int_{\partial\Omega^i}\nu_{\Omega^i}(\xi)\cdot\nabla S_2(\xi)\,d\sigma_\xi \\ & \quad -\int_{\partial\Omega^i}\nu_{\Omega^i}(\xi)\cdot\nabla\tilde{u}_1^o(0)\,d\sigma_\xi = 0\,. \end{aligned} \tag{11.72}$$

Then Remark 6.53, Proposition 6.54, Theorem 6.55, and the positivity of the constant $F'(F^{(-1)}(\tilde{u}^o(0)))$, which follows by assumption (11.36) on F, ensure the existence of a unique solution

$$(u_1^{i,r}, v_1^{o,r}) \in C^{1,\alpha}(\overline{\Omega^i}) \times C^{1,\alpha}_{\mathrm{loc}}(\overline{\Omega^{i-}})$$

of the linear transmission problem of the first, second, fourth, fifth, and seventh equation of the auxiliary problem (11.71). Hence, the limiting boundary value problem has a unique solution.

Finally, if $(\tilde{\psi}, \tilde{c}_1^i, \tilde{\theta}^i, \tilde{\theta}_1^o, \tilde{c}_1^o)$ is the only solution of the limiting system (11.49), then the jump properties of the single layer potential of Theorem 6.6 imply that the triple $(\tilde{u}_1^{i,r}, \tilde{u}_1^{o,r}, \tilde{u}_1^o)$ as in (11.62) solves the first five equations of the limiting boundary value problem (11.61). Then (11.65) and the fourth equation of the limiting system (11.49) imply the validity of the sixth equation of the limiting boundary value problem (11.61). Finally, (11.65) implies that

$$\lim_{\xi\to\infty} v^-_{\Omega^i}[\tilde{\theta}^i](\xi) - \delta_{2,n} m_{n-1}(\partial\Omega^i) G(F^{(-1)}(\tilde{u}^o(0))) S_n(\xi)$$
$$= \lim_{\xi\to\infty} \int_{\partial\Omega^i} S_n(\xi-\eta)\tilde{\theta}^i(\eta)\, d\sigma_\eta - \delta_{2,n} \int_{\partial\Omega^i} \tilde{\theta}^i(\eta)\, d\sigma_\eta S_n(\xi) = 0$$

(cf. Theorem 4.23 and (4.69)). □

We now introduce an abstract formulation of system (11.47) and of the limiting system (11.49). We note that in case $n = 2$ system (11.47) contains a logarithmic term, which is not analytic around $\epsilon = 0$. Here the idea is to group all terms containing $\log \epsilon$ in terms of the form $\delta_{2,n} \log^{-1} \epsilon$, $\delta_{2,n}\epsilon \log^2 \epsilon$, which are not real analytic around $\epsilon = 0$, but which at least have the virtue to tend to zero as ϵ tends to zero. Thus we now introduce the new independent variables

$$\epsilon_1 = \delta_{2,n} \log^{-1} \epsilon\,, \qquad \epsilon_2 = \delta_{2,n}\epsilon \log^2 \epsilon$$

and observe that if we replace $\delta_{2,n} \log^{-1} \epsilon$, $\delta_{2,n}\epsilon \log^2 \epsilon$ in (11.47) with ϵ_1, ϵ_2 we obtain an equation which is not singular in ϵ, ϵ_1, ϵ_2 around the degenerate point $\epsilon = \epsilon_1 = \epsilon_2 = 0$ and that is accordingly easier to analyze. We do so by means of the following Theorem 11.74. Here and in what follows we find convenient to set

$$Z_{0,\alpha} \equiv C^{1,\alpha}(\partial\Omega^i) \times C^{0,\alpha}(\partial\Omega^i) \times C^{1,\alpha}(\partial\Omega^o) \times \mathbb{R}\,. \tag{11.73}$$

Theorem 11.17. *Let F, G, f^o satisfy the conditions in (11.36), (11.37), and (11.34). Let $\mathcal{M} \equiv (\mathcal{M}_l)_{l=1,2,3,4}$ be the map from $]-\epsilon_0, \epsilon_0[\times \mathbb{R}^2 \times Y_{0,\alpha}$ to $Z_{0,\alpha}$ defined by*

$$\begin{aligned}
&\mathcal{M}_1[\epsilon,\epsilon_1,\epsilon_2,\psi,c_1^i,\theta^i,\theta_1^o,c_1^o](\xi) \qquad\qquad (11.74)\\
&\equiv V_{\Omega^i}[\theta^i](\xi) + \int_{\partial\Omega^o} S_n(\epsilon\xi - y)\theta_1^o(y)\, d\sigma_y + c_1^o + \int_0^1 \nabla\tilde{u}^o(\beta\epsilon\xi)\xi\, d\beta \\
&-F'(F^{(-1)}(\tilde{u}^o(0)))\left\{V_{\Omega^i}[\psi](\xi) + c_1^i\right\} - \int_0^1 (1-\beta)F''\Big(F^{(-1)}(\tilde{u}^o(0)) \\
&+\beta\epsilon\left(V_{\Omega^i}[\psi](\xi) + c_1^i\right) + \frac{\beta\epsilon_1\epsilon_2}{2\pi F'(F^{(-1)}(\tilde{u}^o(0)))}\int_{\partial\Omega^i}\theta^i\, d\sigma\Big)\, d\beta \\
&\times\Bigg\{\epsilon\left(V_{\Omega^i}[\psi](\xi) + c_1^i\right)^2 + \left(V_{\Omega^i}[\psi](\xi) + c_1^i\right)\frac{2\epsilon_1\epsilon_2}{2\pi F'(F^{(-1)}(\tilde{u}^o(0)))}\int_{\partial\Omega^i}\theta^i\, d\sigma \\
&+\frac{\epsilon_2\left(\int_{\partial\Omega^i}\theta^i\, d\sigma\right)^2}{\left(2\pi F'(F^{(-1)}(\tilde{u}^o(0)))\right)^2}\Bigg\} \qquad \forall\xi\in\partial\Omega^i\,,
\end{aligned}$$

$$\begin{aligned}
&\mathcal{M}_2[\epsilon,\epsilon_1,\epsilon_2,\psi,c_1^i,\theta^i,\theta_1^o,c_1^o](\xi) \\
&\equiv \frac{1}{2}\theta^i(\xi) + W^t_{\Omega^i}[\theta^i](\xi) + \epsilon\int_{\partial\Omega^o} \nu_{\Omega^i}(\xi)\cdot\nabla S_n(\epsilon\xi - y)\theta_1^o(y)\, d\sigma_y \\
&+\int_{\partial\Omega^o} \nu_{\Omega^i}(\xi)\cdot\nabla S_n(\epsilon\xi - y)\tilde{\theta}^o(y)\, d\sigma_y
\end{aligned}$$

$$+\frac{1}{2}\psi(\xi) - W^t_{\Omega^i}[\psi](\xi)$$
$$-G\left(\epsilon V_{\Omega^i}[\psi](\xi) + F^{(-1)}(\tilde{u}^o(0)) + c^i_1\epsilon + \frac{\epsilon_1\epsilon_2}{2\pi F'(F^{(-1)}(\tilde{u}^o(0)))}\int_{\partial\Omega^i}\theta^i\, d\sigma\right)$$
$$\forall \xi \in \partial\Omega^i\,,$$

$$\mathcal{M}_3[\epsilon, \epsilon_1, \epsilon_2, \psi, c^i_1, \theta^i, \theta^o_1, c^o_1](x)$$
$$\equiv \epsilon^{n-2}\int_{\partial\Omega^i} S_n(x-\epsilon\eta)\theta^i(\eta)\, d\sigma_\eta + V_{\Omega^o}[\theta^o_1](x) + c^o_1 \qquad \forall x \in \partial\Omega^o\,,$$

$$\mathcal{M}_4[\epsilon, \epsilon_1, \epsilon_2, \psi, c^i_1, \theta^i, \theta^o_1, c^o_1] \equiv \epsilon^{n-2}\int_{\partial\Omega^i}\theta^i\, d\sigma + \int_{\partial\Omega^o}\theta^o_1\, d\sigma$$

for all $(\epsilon, \epsilon_1, \epsilon_2, \psi, c^i_1, \theta^i, \theta^o_1, c^o_1) \in]-\epsilon_0, \epsilon_0[\times\mathbb{R}^2\times Y_{0,\alpha}$ *(see (11.39) for the definition of* $Y_{0,\alpha}$*). Then the following statements hold.*

(i) If $\epsilon = \epsilon_1 = \epsilon_2 = 0$*, then equation*

$$\mathcal{M}[0,0,0,\psi, c^i_1, \theta^i, \theta^o_1, c^o_1] = 0$$

is equivalent to the limiting system (11.49) and has one and only one solution $(\tilde{\psi}, \tilde{c}^i_1, \tilde{\theta}^i, \tilde{\theta}^o_1, \tilde{c}^o_1)$ *in* $Y_{0,\alpha}$ *(see Theorem 11.16 (i)).*

(ii) If $\epsilon \in]0, \inf\{\epsilon_0, 1\}[$, $\epsilon_1 = \delta_{2,n}\log^{-1}\epsilon$, $\epsilon_2 = \delta_{2,n}\epsilon\log^2\epsilon$*, then equation*

$$\mathcal{M}[\epsilon, \epsilon_1, \epsilon_2, \psi, c^i_1, \theta^i, \theta^o_1, c^o_1] = 0$$

is equivalent to system (11.47).

(iii) The map $\mathcal{M}$ *is real analytic and the differential*

$$\partial_{(\psi, c^i_1, \theta^i, \theta^o_1, c^o_1)}\mathcal{M}[0,0,0,\tilde{\psi}, \tilde{c}^i_1, \tilde{\theta}^i, \tilde{\theta}^o_1, \tilde{c}^o_1]$$

of $\mathcal{M}$ *at* $(0,0,0,\tilde{\psi}, \tilde{c}^i_1, \tilde{\theta}^i, \tilde{\theta}^o_1, \tilde{c}^o_1)$ *with respect to the variable* $(\psi, c^i_1, \theta^i, \theta^o_1, c^o_1)$ *is a linear homeomorphism from* $Y_{0,\alpha}$ *onto* $Z_{0,\alpha}$*.*

(iv) There exist $\epsilon' \in]0, \inf\{\epsilon_0, 1\}[$*, an open neighborhood* $\tilde{U}$ *of* $(0,0)$ *in* $\mathbb{R}^2$*, an open neighborhood* $\tilde{V}$ *of* $(\tilde{\psi}, \tilde{c}^i_1, \tilde{\theta}^i, \tilde{\theta}^o_1, \tilde{c}^o_1)$ *in* $Y_{0,\alpha}$*, and a real analytic map* $(\Psi, C^i_1, \Theta^i, \Theta^o_1, C^o_1)$ *from* $]-\epsilon', \epsilon'[\times\tilde{U}$ *to* $\tilde{V}$ *such that*

$$(\delta_{2,n}\log^{-1}\epsilon, \delta_{2,n}\epsilon\log^2\epsilon) \in \tilde{U} \qquad \forall \epsilon \in]0, \epsilon'[$$

and such that the set of zeros of $\mathcal{M}$ *in* $]-\epsilon', \epsilon'[\times\tilde{U}\times\tilde{V}$ *coincides with the graph of the map* $(\Psi, C^i_1, \Theta^i, \Theta^o_1, C^o_1)$*. In particular*

$$(\Psi[0,0,0], C^i_1[0,0,0], \Theta^i[0,0,0], \Theta^o_1[0,0,0], C^o_1[0,0,0]) = (\tilde{\psi}, \tilde{c}^i_1, \tilde{\theta}^i, \tilde{\theta}^o_1, \tilde{c}^o_1)\,. \tag{11.75}$$

Proof. If $(\epsilon, \epsilon_1, \epsilon_2, \psi, c^i_1, \theta^i, \theta^o_1, c^o_1)$ belongs to the domain of $\mathcal{M}$, then classical regularity properties of single layer potentials and of the derivatives of single layer potentials ensure that $\mathcal{M}[\epsilon, \epsilon_1, \epsilon_2, \psi, c^i_1, \theta^i, \theta^o_1, c^o_1]$ belongs to $Z_{0,\alpha}$ (cf. Theorems 4.25, 6.6).

Statements (i) and (ii) hold true by the definition of $\mathcal{M}$. We now prove (iii). Since F and G are analytic, then the composition operators N_F, N_G are both analytic (cf. Theorem A.23). Then, by the linearity and continuity of the operators V_Ω, W^t_Ω (cf. Theorems 6.4, 6.8 (ii)) and by the analyticity of the nonlinear integral operators with analytic kernel of Theorem A.28, the operator $\mathcal{M}$ is analytic. In particular, we note that the analyticity of the sixth term of the first component of $\mathcal{M}$ follows by applying Theorem A.28 (iii) to the analytic function

$$(1-\beta)F''\Big(F^{(-1)}(\tilde{u}^o(0))+\beta a\Big)$$

of the variable $(a,\beta)\in\mathbb{R}^2$. By standard differentiation rules, the differential of $\mathcal{M}$ at $(0,0,0,\tilde{\psi},\tilde{c}^i_1,\tilde{\theta}^i,\tilde{\theta}^o_1,\tilde{c}^o_1)$ with respect to the variable $(\psi,c^i_1,\theta^i,\theta^o_1,c^o_1)$ is delivered by the formula

$$\begin{aligned}
&\partial_{(\psi,c^i_1,\theta^i,\theta^o_1,c^o_1)}\mathcal{M}_1[0,0,0,\tilde{\psi},\tilde{c}^i_1,\tilde{\theta}^i,\tilde{\theta}^o_1,\tilde{c}^o_1](\psi^\sharp,c^{\sharp i}_1,\theta^{\sharp i},\theta^{\sharp o}_1,c^{\sharp o}_1)(\xi) \qquad (11.76)\\
&\quad = V_{\Omega^i}[\theta^{\sharp i}](\xi)+v^+_{\Omega^o}[\theta^{\sharp o}_1](0)+c^{\sharp o}_1\\
&\qquad -F'(F^{(-1)}(\tilde{u}^o(0)))\left\{V_{\Omega^i}[\psi^\sharp](\xi)+c^{\sharp i}_1\right\} \qquad \forall\xi\in\partial\Omega^i\,,\\
&\partial_{(\psi,c^i_1,\theta^i,\theta^o_1,c^o_1)}\mathcal{M}_2[0,0,0,\tilde{\psi},\tilde{c}^i_1,\tilde{\theta}^i,\tilde{\theta}^o_1,\tilde{c}^o_1](\psi^\sharp,c^{\sharp i}_1,\theta^{\sharp i},\theta^{\sharp o}_1,c^{\sharp o}_1)(\xi)\\
&\quad = \frac{1}{2}\theta^{\sharp i}(\xi)+W^t_{\Omega^i}[\theta^{\sharp i}](\xi)+\frac{1}{2}\psi^\sharp(\xi)-W^t_{\Omega^i}[\psi^\sharp](\xi) \qquad \forall\xi\in\partial\Omega^i\,,\\
&\partial_{(\psi,c^i_1,\theta^i,\theta^o_1,c^o_1)}\mathcal{M}_3[0,0,0,\tilde{\psi},\tilde{c}^i_1,\tilde{\theta}^i,\tilde{\theta}^o_1,\tilde{c}^o_1](\psi^\sharp,c^{\sharp i}_1,\theta^{\sharp i},\theta^{\sharp o}_1,c^{\sharp o}_1)(x)\\
&\quad = \delta_{2,n}S_n(x)\int_{\partial\Omega^i}\theta^{\sharp i}\,d\sigma+V_{\Omega^o}[\theta^{\sharp o}_1](x)+c^{\sharp o}_1 \qquad \forall x\in\partial\Omega^o\,,\\
&\partial_{(\psi,c^i_1,\theta^i,\theta^o_1,c^o_1)}\mathcal{M}_4[0,0,0,\tilde{\psi},\tilde{c}^i_1,\tilde{\theta}^i,\tilde{\theta}^o_1,\tilde{c}^o_1](\psi^\sharp,c^{\sharp i}_1,\theta^{\sharp i},\theta^{\sharp o}_1,c^{\sharp o}_1)\\
&\quad = \delta_{2,n}\int_{\partial\Omega^i}\theta^{\sharp i}\,d\sigma+\int_{\partial\Omega^o}\theta^{\sharp o}_1\,d\sigma
\end{aligned}$$

for all $(\psi^\sharp,c^{\sharp i}_1,\theta^{\sharp i},\theta^{\sharp o}_1,c^{\sharp o}_1)\in Y_{0,\alpha}$.

We now prove that the linear and continuous operator

$$\partial_{(\psi,c^i_1,\theta^i,\theta^o_1,c^o_1)}\mathcal{M}[0,0,0,\tilde{\psi},\tilde{c}^i_1,\tilde{\theta}^i,\tilde{\theta}^o_1,\tilde{c}^o_1]$$

is a homeomorphism. By the Open Mapping Theorem 2.2, it suffices to show that $\partial_{(\psi,c^i_1,\theta^i,\theta^o_1,c^o_1)}\mathcal{M}[0,0,0,\tilde{\psi},\tilde{c}^i_1,\tilde{\theta}^i,\tilde{\theta}^o_1,\tilde{c}^o_1]$ is a bijection from $Y_{0,\alpha}$ onto $Z_{0,\alpha}$. Let $(f^{\sharp i},g^\sharp,f^{\sharp o},a)\in Z_{0,\alpha}$. We must show that there exists a unique

$$(\psi^\sharp,c^{\sharp i}_1,\theta^{\sharp i},\theta^{\sharp o}_1,c^{\sharp o}_1)\in Y_{0,\alpha}$$

such that

$$\partial_{(\psi,c^i_1,\theta^i,\theta^o_1,c^o_1)}\mathcal{M}[0,0,0,\tilde{\psi},\tilde{c}^i_1,\tilde{\theta}^i,\tilde{\theta}^o_1,\tilde{c}^o_1](\psi^\sharp,c^{\sharp i}_1,\theta^{\sharp i},\theta^{\sharp o}_1,c^{\sharp o}_1)=(f^{\sharp i},g^\sharp,f^{\sharp o},a)\,. \qquad (11.77)$$

We first assume that (11.77) has a solution $(\psi^\sharp, c_1^{\sharp i}, \theta^{\sharp i}, \theta_1^{\sharp o}, c_1^{\sharp o})$ and we show that $\theta_1^{\sharp o}$, $c_1^{\sharp o}$ are uniquely determined. By the last two equations of (11.77), we have

$$\int_{\partial\Omega^o} \left(S_n(x-y) - \frac{\delta_{2,n}}{2\pi} \log|x| \right) \theta_1^{\sharp o}(y)\, d\sigma_y + c_1^{\sharp o} = f^{\sharp o}(x) - a\frac{\delta_{2,n}}{2\pi} \log|x| \,, \tag{11.78}$$

for all $x \in \partial\Omega^o$. Now let $n \geq 3$. By Lemma 11.14 there exists

$$\tilde{\phi} \in \mathrm{Ker}\left(-\frac{1}{2}I + W^t_{\Omega^o} \right)$$

such that $V_{\Omega^o}[\tilde{\phi}]$ is constant and $\int_{\partial\Omega^o} \tilde{\phi}\, d\sigma = 1$. Then Theorem 6.47 implies that there exists a unique $(\theta_1^\dagger, c_1^\dagger)$ in $C^{0,\alpha}(\partial\Omega^o)_0 \times \mathbb{R}$ such that

$$V_{\Omega^o}[\theta_1^\dagger] + c_1^\dagger = -aV_{\Omega^o}[\tilde{\phi}] + f^{\sharp o} \qquad \text{on } \partial\Omega^o\,. \tag{11.79}$$

Thus if we set

$$\theta_1^{\sharp o} \equiv \theta_1^\dagger + a\tilde{\phi}\,, \qquad c_1^{\sharp o} \equiv c_1^\dagger\,, \tag{11.80}$$

$(\theta_1^{\sharp o}, c_1^{\sharp o})$ is the only solution of Equation (11.78) in $C^{0,\alpha}(\partial\Omega^o) \times \mathbb{R}$ such that $\int_{\partial\Omega^o} \theta_1^{\sharp o}\, d\sigma = a$ and thus the only solution of the last two equations of (11.77). We now consider case $n = 2$ and we set

$$v(x) \equiv \int_{\partial\Omega^o} \frac{1}{2\pi} \log\frac{|x-y|}{|x|} \theta_1^{\sharp o}(y)\, d\sigma_y + c_1^{\sharp o} \qquad \forall x \in \overline{\Omega^{o-}}\,.$$

By (4.69), v satisfies the limiting relation

$$\lim_{x\to\infty} v(x) = c_1^{\sharp o} \in \mathbb{R}\,.$$

Since v is harmonic in Ω^{o-}, v equals the unique solution in $C^{1,\alpha}_{\mathrm{loc}}(\overline{\Omega^{o-}}) \cap C^2(\Omega^{o-})$ of the exterior Dirichlet problem

$$\begin{cases} \Delta v = 0 & \text{in } \Omega^{o-}\,, \\ v = f^{\sharp o} - aS_2 & \text{on } \partial\Omega^o\,, \\ v \text{ is harmonic at } \infty\,, \end{cases} \tag{11.81}$$

solution which we now denote by $b[f^{\sharp o}, a]$ (cf. Theorem 6.37). Hence,

$$\int_{\partial\Omega^o} \frac{1}{2\pi} \log|x-y| \theta_1^{\sharp o}(y)\, d\sigma_y - \frac{1}{2\pi} \int_{\partial\Omega^o} \theta_1^{\sharp o}(y)\, d\sigma_y \log|x| + c_1^{\sharp o} = b[f^{\sharp o}, a](x)\,, \tag{11.82}$$

for all $x \in \overline{\Omega^{o-}}$ and thus

$$\begin{aligned} c_1^{\sharp o} &= -\lim_{x\to\infty} \int_{\partial\Omega^o} \frac{1}{2\pi} \log\frac{|x-y|}{|x|} \theta_1^{\sharp o}(y)\, d\sigma_y + \lim_{x\to\infty} b[f^{\sharp o}, a](x) \\ &= \lim_{x\to\infty} b[f^{\sharp o}, a](x)\,. \end{aligned} \tag{11.83}$$

By the last equation of (11.77), we have

$$\int_{\partial\Omega^o} \theta_1^{\sharp o}(y)\, d\sigma_y = -\int_{\partial\Omega^i} \theta^{\sharp i}(y)\, d\sigma_y + a\,.$$

By integrating the second equation of (11.77) on $\partial\Omega^i$ and by Lemmas 6.11 and 6.21, we obtain

$$\int_{\partial\Omega^i} \theta^{\sharp i}(y)\, d\sigma_y = \int_{\partial\Omega^i} g^\sharp\, d\sigma\,. \tag{11.84}$$

Hence,

$$\int_{\partial\Omega^o} \theta_1^{\sharp o}(y)\, d\sigma_y = -\int_{\partial\Omega^i} \theta^{\sharp i}(y)\, d\sigma_y + a = -\int_{\partial\Omega^i} g^\sharp\, d\sigma + a\,. \tag{11.85}$$

Now by taking the normal derivative of (11.82) and by exploiting the jump relations of Theorem 6.6, we obtain

$$\frac{1}{2}\theta_1^{\sharp o} + W_{\Omega^o}^t[\theta_1^{\sharp o}] = -\nu_{\Omega^o}\cdot\nabla S_2 \int_{\partial\Omega^i} g^\sharp\, d\sigma + \nu_{\Omega^o}\cdot\nabla S_2 a + \frac{\partial}{\partial\nu_{\Omega^o}} b[f^{\sharp o}, a]\,. \tag{11.86}$$

Since Ω^{o-} is connected, Corollary 6.15 implies that the operator $\frac{1}{2}I + W_{\Omega^o}^t$ is invertible and thus we obtain

$$\theta_1^{\sharp o} = \left(\frac{1}{2}I + W_{\Omega^o}^t\right)^{(-1)} \left[-\nu_{\Omega^o}\cdot\nabla S_2 \left(\int_{\partial\Omega^i} g^\sharp\, d\sigma - a\right) + \frac{\partial}{\partial\nu_{\Omega^o}} b[f^{\sharp o}, a]\right]. \tag{11.87}$$

In conclusion, if (11.77) has a solution, then $\theta_1^{\sharp o}$ and $c_1^{\sharp o}$ must be as in (11.79), (11.80) for $n \geq 3$ and must equal the right-hand side of (11.83), (11.87), respectively for $n = 2$.

Next we note that Theorem 11.15 and the positivity of $F'(F^{(-1)}(\tilde{u}^o(0)))$, which follows by assumption (11.36) on F, imply that the first two equations of (11.77) with $\theta_1^{\sharp o}$ and $c_1^{\sharp o}$ as in (11.79), (11.80) in case $n \geq 3$ and as in the right-hand sides of (11.83), (11.87) for $n = 2$ have a unique solution

$$(\psi^\sharp, c_1^{\sharp i}, \theta^{\sharp i}) \in C^{0,\alpha}(\partial\Omega^i)_0 \times \mathbb{R} \times C^{0,\alpha}(\partial\Omega^i)\,.$$

Hence, we have proved that problem (11.77) has at most one solution.

On the other hand, if we take $\theta_1^{\sharp o}$, $c_1^{\sharp o}$ as in (11.79) and (11.80) in case $n \geq 3$ and as in right-hand sides of (11.83) and (11.87) for $n = 2$, and if we set $(\psi^\sharp, c_1^{\sharp i}, \theta^{\sharp i})$ equal to the only solution in $C^{0,\alpha}(\partial\Omega^i)_0 \times \mathbb{R} \times C^{0,\alpha}(\partial\Omega^i)$ of the first two equations of (11.77) in which $\theta_1^{\sharp o}$, $c_1^{\sharp o}$ equal the above defined $\theta_1^{\sharp o}$, $c_1^{\sharp o}$, we can verify that $(\theta_1^{\sharp o}, c_1^{\sharp o}, \psi^\sharp, c_1^{\sharp i}, \theta^{\sharp i})$ solves (11.77) by reading backward the above argument.

Indeed, if $n \geq 3$, the last two equations of (11.77) hold by our choice of $\theta_1^{\sharp o}$, $c_1^{\sharp o}$ and the first two equations of (11.77) hold by our choice of $(\psi^\sharp, c_1^{\sharp i}, \theta^{\sharp i})$.

If instead $n = 2$, then the first two equations of (11.77) hold for our choice of $(\psi^\sharp, c_1^{\sharp i}, \theta^{\sharp i})$. Then, by integrating the second equation of (11.77) on $\partial\Omega^i$, we obtain (11.84). On the other hand, our definition (11.87) of $\theta_1^{\sharp o}$ implies the validity of (11.86). Then, by integrating both hand sides of equality (11.86) on $\partial\Omega^o$, by exploiting equality $\int_{\partial\Omega^o} \nu_{\Omega^o} \cdot \nabla S_n \, d\sigma = 1$, by Lemma 6.11, and by the equality

$$\int_{\partial\Omega^o} \frac{\partial}{\partial\nu_{\Omega^o}} b[f^{\sharp o}, a] \, d\sigma = 0 \,,$$

which follows by the first Green Identity for an exterior domain of Corollary 4.7, we obtain that the left-hand side of (11.85) equals the right-hand side of (11.85). Then (11.86) implies that

$$\frac{\partial}{\partial\nu_{\Omega^o}} v = \frac{\partial}{\partial\nu_{\Omega^o}} b[f^{\sharp o}, a] \,,$$

with $v(x) \equiv \int_{\partial\Omega^o} \frac{1}{2\pi} \log \frac{|x-y|}{|x|} \theta_1^{\sharp o}(y) \, d\sigma_y + c_1^{\sharp o}$ and with $c_1^{\sharp o}$ as in (11.83). By (4.69), we have

$$\lim_{x\to\infty} v(x) - b[f^{\sharp o}, a](x) = 0 \,.$$

Since $v - b[f^{\sharp o}, a]$ is harmonic in Ω^{o-}, the uniqueness Theorem 6.2 (ii) for the exterior Neumann problem implies that $v = b[f^{\sharp o}, a]$ in $\overline{\Omega^{o-}}$ and the validity of the third equation of (11.77) follows (see also (11.84) and (11.85)). Then by (11.84) and by the equality of the left and right-hand side of (11.85), we deduce the validity of the fourth equation of (11.77).

Hence, we have proved that $\partial_{(\psi, c_1^i, \theta^i, \theta_1^o, c_1^o)} \mathcal{M}[0, 0, 0, \tilde{\psi}, \tilde{c}_1^i, \tilde{\theta}^i, \tilde{\theta}_1^o, \tilde{c}_1^o]$ is a bijection from $Y_{0,\alpha}$ onto $Z_{0,\alpha}$ and statement (iii) holds true.

Then the existence of ϵ', $\tilde{U}$, $\tilde{V}$, $(\Psi, C_1^i, \Theta^i, \Theta_1^o, C_1^o)$ as in statement (iv) is a consequence of statements (i), (iii) and of the Implicit Function Theorem in Banach spaces (cf. Theorem A.19). Possibly shrinking ϵ', we can assume that $(\delta_{2n} \log^{-1} \epsilon, \delta_{2n} \epsilon \log^2 \epsilon)$ belongs to $\tilde{U}$ for ϵ in $]0, \epsilon'[$. □

In order to simplify our notation, we set

$$\Xi_n[\epsilon] \equiv (\delta_{2,n} \log^{-1} \epsilon, \delta_{2,n} \epsilon \log^2 \epsilon) \qquad \forall \epsilon \in]0, \epsilon'[\,. \tag{11.88}$$

We are now in the position to introduce the following.

Definition 11.18. Let the assumptions of Theorem 11.17 hold. Let $\epsilon \in]0, \epsilon'[$. Let $u^i[\epsilon, \cdot, \cdot, \cdot, \cdot, \cdot]$, $u^o[\epsilon, \cdot, \cdot, \cdot, \cdot, \cdot]$ be as in Theorem 11.13. Then we set

$$\begin{aligned} &u^i(\epsilon, x) \\ &\quad \equiv u^i[\epsilon, \Psi[\epsilon, \Xi_n[\epsilon]], C_1^i[\epsilon, \Xi_n[\epsilon]], \Theta^i[\epsilon, \Xi_n[\epsilon]], \Theta_1^o[\epsilon, \Xi_n[\epsilon]], C_1^o[\epsilon, \Xi_n[\epsilon]]](x) \\ &\qquad\qquad \forall x \in \epsilon\overline{\Omega^i} \,, \end{aligned}$$

$$\begin{aligned} &u^o(\epsilon, x) \\ &\quad \equiv u^o[\epsilon, \Psi[\epsilon, \Xi_n[\epsilon]], C_1^i[\epsilon, \Xi_n[\epsilon]], \Theta^i[\epsilon, \Xi_n[\epsilon]], \Theta_1^o[\epsilon, \Xi_n[\epsilon]], C_1^o[\epsilon, \Xi_n[\epsilon]]](x) \\ &\qquad\qquad \forall x \in \overline{\Omega(\epsilon)} \,. \end{aligned}$$

We will prove that, for ϵ sufficiently small and positive, $(u^i(\epsilon,\cdot), u^o(\epsilon,\cdot))$ is the only solution of the transmission problem (11.38) which is close to $(F^{(-1)}(\tilde{u}^o(0)), \tilde{u}^o)$ in a sense which we clarify in Section 11.3.3.

11.3.2 Analytic Representation for the Family of Solutions $\{(u^i(\epsilon,\cdot), u^o(\epsilon,\cdot))\}_{\epsilon\in]0,\epsilon'[}$

We first analyze the macroscopic behavior of our family of solutions. We do so by means of the following.

Theorem 11.19. *Let the assumptions of Theorem 11.17 hold. Let Ω_M be a bounded open subset of $\Omega^o \setminus \{0\}$ such that $0 \notin \overline{\Omega_M}$. Then there exist $\epsilon_M \in]0,\epsilon'[$ and an analytic map $\mathcal{U}_M$ from $]-\epsilon_M,\epsilon_M[\times\tilde{U}$ to $C^{1,\alpha}(\overline{\Omega_M})$ such that*

$$\overline{\Omega_M} \subseteq \overline{\Omega(\epsilon)} \qquad \forall \epsilon \in]-\epsilon_M,\epsilon_M[\,, \tag{11.89}$$

$$u^o(\epsilon,\cdot)_{|\overline{\Omega_M}} = \tilde{u}^o(\cdot)_{|\overline{\Omega_M}} + \epsilon\mathcal{U}_M[\epsilon,\Xi_n[\epsilon]] \qquad \forall \epsilon \in]0,\epsilon_M[\tag{11.90}$$

(see (11.88) for the definition of Ξ_n). Moroever,

$$\mathcal{U}_M[0,0,0](x) = \tilde{u}^o_1(x) + \delta_{2,n}S_n(x)m_{n-1}(\partial\Omega^i)G(F^{(-1)}(\tilde{u}^o(0))) \qquad \forall x \in \overline{\Omega_M}\,, \tag{11.91}$$

where $(\tilde{u}^{i,r}_1, \tilde{u}^{o,r}_1, \tilde{u}^o_1)$ is the only solution of the limiting boundary value problem (11.61) (cf. Theorem 11.16).

Proof. By the Definition 11.18 of $u^o(\epsilon,\cdot)$, we have

$$\begin{aligned} u^o(\epsilon,x) &= u^o[\epsilon,\Psi[\epsilon,\Xi_n[\epsilon]], C^i_1[\epsilon,\Xi_n[\epsilon]], \Theta^i[\epsilon,\Xi_n[\epsilon]], \Theta^o_1[\epsilon,\Xi_n[\epsilon]], C^o_1[\epsilon,\Xi_n[\epsilon]]](x) \\ &= \tilde{u}^o(x) + \epsilon v^+_{\Omega^o}[\Theta^o_1[\epsilon,\Xi_n[\epsilon]]](x) + \epsilon C^o_1[\epsilon,\Xi_n[\epsilon]] \\ &\quad + \epsilon^{n-1}\int_{\partial\Omega^i} S_n(x-\epsilon\eta)\Theta^i[\epsilon,\Xi_n[\epsilon]](\eta)\,d\sigma_\eta \qquad \forall x \in \overline{\Omega(\epsilon)}\,, \end{aligned} \tag{11.92}$$

for all $\epsilon \in]0,\epsilon'[$ (see also the second definition of (11.46)). Since 0 has a positive distance from $\overline{\Omega_M}$, there exists $\epsilon_M \in]0,\epsilon'[$ such that

$$\overline{\Omega_M} \subseteq \overline{\Omega(\epsilon)} \qquad \forall \epsilon \in [-\epsilon_M,\epsilon_M]\,.$$

By equality (11.92), we find natural to define a map $\mathcal{U}_M$ from $]-\epsilon_M,\epsilon_M[\times\tilde{U}$ to $C^{1,\alpha}(\overline{\Omega_M})$ by setting

$$\begin{aligned} \mathcal{U}_M[\epsilon,\epsilon_1,\epsilon_2](x) &\equiv v^+_{\Omega^o}[\Theta^o_1[\epsilon,\epsilon_1,\epsilon_2]](x) + C^o_1[\epsilon,\epsilon_1,\epsilon_2] \\ &\quad + \epsilon^{n-2}\int_{\partial\Omega^i} S_n(x-\epsilon\eta)\Theta^i[\epsilon,\epsilon_1,\epsilon_2](\eta)\,d\sigma_\eta \qquad \forall x \in \overline{\Omega_M}\,, \end{aligned} \tag{11.93}$$

for all $(\epsilon, \epsilon_1, \epsilon_2) \in]-\epsilon_M, \epsilon_M[\times\tilde{U}$ (see Theorem 11.17 (iv)). It clearly suffices to show that the right-hand side of (11.93) defines a real analytic map from $]-\epsilon_M, \epsilon_M[\times\tilde{U}$ to $C^{1,\alpha}(\overline{\Omega_M})$. Since the operator $v^+_{\Omega^o}[\cdot]$ is linear and continuous from $C^{0,\alpha}(\partial\Omega^o)$ to $C^{1,\alpha}(\overline{\Omega^o})$ and $\overline{\Omega_M} \subseteq \overline{\Omega^o}$ and Θ^o_1 is real analytic from the set $]-\epsilon_M, \epsilon_M[\times\tilde{U}$ to $C^{0,\alpha}(\partial\Omega^o)$, we conclude that the first summand in the right-hand side of (11.93) defines a real analytic map from $]-\epsilon_M, \epsilon_M[\times\tilde{U}$ to $C^{1,\alpha}(\overline{\Omega_M})$. We now consider the third summand. By the analyticity of nonlinear integral operators with analytic kernel of Theorem A.28, the map from $]-\epsilon_M, \epsilon_M[\times L^1(\partial\Omega^i)$ to $C^{1,\alpha}(\overline{\Omega_M})$ that takes (ϵ, f) to $\int_{\partial\Omega^i} S_n(\cdot - \epsilon\eta) f(\eta)\, d\sigma_\eta$ is real analytic. Since Θ^i is real analytic and $C^{0,\alpha}(\partial\Omega^i)$ is continuously embedded into $L^1(\partial\Omega^i)$, we conclude that the function from $]-\epsilon_M, \epsilon_M[\times\tilde{U}$ to $C^{1,\alpha}(\overline{\Omega_M})$ that takes $(\epsilon, \epsilon_1, \epsilon_2)$ to the third summand in the right-hand side of (11.93) is analytic. Since C^o_1 is analytic, the proof of the analyticity of $\mathcal{U}_M$ is complete.

By Theorem 11.17 (iv), we have

$$\mathcal{U}_M[0,0,0](x) = \delta_{2,n} S_n(x) \int_{\partial\Omega^i} \tilde{\theta}^i\, d\sigma + v^+_{\Omega^o}[\tilde{\theta}^o_1] + \tilde{c}^o_1 \qquad \forall x \in \overline{\Omega_M}\,.$$

Hence, equalities (11.60) and (11.62) imply the validity of (11.91). □

The previous theorem implies that if we fix $x \in \overline{\Omega^o} \setminus \{0\}$, then in case $n \geq 3$ we can expand $u^o(\epsilon, x)$ into a convergent power series of ϵ, while in case $n = 2$ we can expand $u^o(\epsilon, x)$ into a convergent triple power series of ϵ, $\log^{-1}\epsilon$, $\epsilon\log^2\epsilon$.

We are also interested in the 'macroscopic' behavior of $\{u^i(\epsilon,\cdot))\}_{\epsilon\in]0,\epsilon'[}$ as ϵ is close to 0. Since the only point which belongs to the domain of all the functions $u^i(\epsilon,\cdot)$ as $\epsilon \in]0, \epsilon'[$ is $x = 0$, here we mean that we are interested in the behavior of $\{u^i(\epsilon, 0))\}_{\epsilon\in]0,\epsilon'[}$ as ϵ is close to 0.

Such a behavior is a specific case of the microscopic behavior of the family $\{u^i(\epsilon,\cdot))\}_{\epsilon\in]0,\epsilon'[}$, i.e., of the behavior of $\{u^i(\epsilon, \epsilon\xi))\}_{\epsilon\in]0,\epsilon'[}$ in case $\xi = 0$, a case that we now analyze by considering the microscopic behavior of our family of solutions, i.e., the behavior of the rescaled family

$$\{(u^i(\epsilon, \epsilon\cdot), u^o(\epsilon, \epsilon\cdot))\}_{\epsilon\in]0,\epsilon'[}\,.$$

Theorem 11.20. *With the assumptions of Theorem 11.17, the following statements hold.*

(i) There exist a real analytic map $\mathcal{U}^i_1$ from $]-\epsilon', \epsilon'[\times\tilde{U}$ to $C^{1,\alpha}(\overline{\Omega^i})$ and a real analytic map $\mathcal{U}^i_2$ from $]-\epsilon', \epsilon'[\times\tilde{U}$ to $\mathbb{R}$ such that

$$\begin{aligned} u^i(\epsilon, \epsilon\xi) = F^{(-1)}(\tilde{u}^o(0)) + \epsilon\,\mathcal{U}^i_1[\epsilon, \Xi_n[\epsilon]](\xi) \\ +(\delta_{2,n}\epsilon\log\epsilon)\,\mathcal{U}^i_2[\epsilon, \Xi_n[\epsilon]] \qquad \forall \xi \in \overline{\Omega^i} \end{aligned} \tag{11.94}$$

(cf. (11.88) for the definition of Ξ_n). Moreover,

$$\mathcal{U}_1^i[0,0,0] = \tilde{u}_1^{i,r}, \qquad \mathcal{U}_2^i[0,0,0] = \frac{m_{n-1}(\partial\Omega^i)G(F^{(-1)}(\tilde{u}^o(0)))}{2\pi F'(F^{(-1)}(\tilde{u}^o(0)))}, \quad (11.95)$$

where $\tilde{u}_1^{i,r}$ has been defined in Theorem 11.16 (ii) and $\tilde{u}^o$ has been defined in Remark 11.12.

(ii) Let Ω_m be a bounded open subset of $\mathbb{R}^n \setminus \overline{\Omega^i}$. Then there exist $\epsilon_m \in]0,\epsilon'[$, a real analytic map $\mathcal{U}_{m,1}^o$ from $]-\epsilon_m,\epsilon_m[\times\tilde{U}$ to $C^{1,\alpha}(\overline{\Omega_m})$, and a real analytic map $\mathcal{U}_2^o$ from $]-\epsilon',\epsilon'[\times\tilde{U}$ to $\mathbb{R}$ such that

$$\epsilon\overline{\Omega_m} \subseteq \Omega^o \qquad \forall\epsilon \in]-\epsilon_m,\epsilon_m[, \quad (11.96)$$

$$u^o(\epsilon,\epsilon\xi) = \tilde{u}^o(0) + \epsilon\mathcal{U}_{m,1}^o[\epsilon,\Xi_n[\epsilon]](\xi) \quad (11.97)$$
$$+(\delta_{2,n}\epsilon\log\epsilon)\mathcal{U}_2^o[\epsilon,\Xi_n[\epsilon]] \qquad \forall\xi \in \overline{\Omega_m},\ \epsilon \in]0,\epsilon_m[.$$

Moreover,

$$\mathcal{U}_{m,1}^o[0,0,0](\xi) = \nabla\tilde{u}_1^o(0)\cdot\xi + \tilde{u}_1^o(0) + \tilde{u}_1^{o,r}(\xi) \qquad \forall\xi \in \overline{\Omega_m}, \quad (11.98)$$
$$\mathcal{U}_2^o[0,0,0] = \frac{m_{n-1}(\partial\Omega^i)G(F^{(-1)}(\tilde{u}^o(0)))}{2\pi},$$

where $\tilde{u}_1^o$ and $\tilde{u}_1^{o,r}$ have been defined in Theorem 11.16 (ii) and $\tilde{u}^o$ has been defined in Remark 11.12.

Proof. (i) By the Definition 11.18 of $u^i(\epsilon,\cdot)$ and by the equalities (8.30) and

$$\int_{\partial\Omega^i} \Psi[\epsilon,\Xi_n[\epsilon]](\eta)\,d\sigma_\eta = 0 \qquad \forall\epsilon \in]0,\epsilon'[,$$

we have

$$\begin{aligned} u^i(\epsilon,\epsilon\xi) &= F^{(-1)}(\tilde{u}^o(0)) + \epsilon^{n-1}\int_{\partial\Omega^i} S_n(\epsilon\xi-\epsilon\eta)\Psi[\epsilon,\Xi_n[\epsilon]](\eta)\,d\sigma_\eta \quad (11.99)\\ &\quad +\epsilon C_1^i[\epsilon,\Xi_n[\epsilon]] + \frac{\delta_{2,n}\epsilon\log\epsilon}{2\pi F'(F^{(-1)}(\tilde{u}^o(0)))}\int_{\partial\Omega^i}\Theta^i[\epsilon,\Xi_n[\epsilon]]\,d\sigma_\eta\\ &= F^{(-1)}(\tilde{u}^o(0)) + \epsilon v_{\Omega^i}^+[\Psi[\epsilon,\Xi_n[\epsilon]]](\xi) + \epsilon C_1^i[\epsilon,\Xi_n[\epsilon]]\\ &\quad +\frac{\delta_{2,n}\epsilon\log\epsilon}{2\pi F'(F^{(-1)}(\tilde{u}^o(0)))}\int_{\partial\Omega^i}\Theta^i[\epsilon,\Xi_n[\epsilon]]\,d\sigma_\eta \qquad \forall\xi \in \overline{\Omega^i}, \end{aligned}$$

for all $\epsilon \in]0,\epsilon'[$ (see also (11.46)). Thus, we find natural to set

$$\mathcal{U}_1^i[\epsilon,\epsilon_1,\epsilon_2](\xi) \equiv v_{\Omega^i}^+[\Psi[\epsilon,\epsilon_1,\epsilon_2]](\xi) + C_1^i[\epsilon,\epsilon_1,\epsilon_2] \qquad \forall\xi \in \overline{\Omega^i},$$
$$\mathcal{U}_2^i[\epsilon,\epsilon_1,\epsilon_2] \equiv \frac{1}{2\pi F'(F^{(-1)}(\tilde{u}^o(0)))}\int_{\partial\Omega^i}\Theta^i[\epsilon,,\epsilon_1,\epsilon_2]\,d\sigma$$

for all $(\epsilon,\epsilon_1,\epsilon_2) \in]-\epsilon',\epsilon'[\times\tilde{U}$. Since $v_{\Omega^i}^+[\cdot]$ is a linear and continuous operator from $C^{0,\alpha}(\partial\Omega^i)$ to $C^{1,\alpha}(\overline{\Omega^i})$ (cf. Theorem 4.25), the analyticity of Ψ, C_1^i implies that $\mathcal{U}_1^i$ is real analytic (cf. Theorem 11.17). By the analyticity of Θ^i, the map $\mathcal{U}_2^i$ is analytic (cf. Theorem 11.17). Then formula (11.94) holds by equality (11.99) and by our definition of $\mathcal{U}_1^i, \mathcal{U}_2^i$. Moreover

$$\mathcal{U}_1^i[0,0,0] = v_{\Omega^i}^+[\tilde{\psi}] + \tilde{c}_1^i = \tilde{u}_1^{i,r},$$
$$\mathcal{U}_2^i[0,0,0] = \frac{1}{2\pi F'(F^{(-1)}(\tilde{u}^o(0)))} \int_{\partial\Omega^i} \tilde{\theta}^i \, d\sigma = \frac{m_{n-1}(\partial\Omega^i) G(F^{(-1)}(\tilde{u}^o(0)))}{2\pi F'(F^{(-1)}(\tilde{u}^o(0)))}$$

(see Theorem 11.16, (11.60), (11.62), and (11.75)).

We now consider statement (ii). Let ϵ_m be such that

$$\epsilon\overline{\Omega_m} \subseteq \Omega^o \qquad \forall \epsilon \in [-\epsilon_m, \epsilon_m] .$$

Then we have

$$\overline{\Omega_m} \subseteq \frac{1}{\epsilon}\overline{\Omega(\epsilon)} \qquad \forall \epsilon \in [-\epsilon_m, \epsilon_m] \setminus \{0\} .$$

By the Definition 11.18 of $u^o(\epsilon, \cdot)$, we have

$$\begin{aligned} u^o(\epsilon, \epsilon\xi) & \qquad (11.100) \\ &= u^o[\epsilon, \Psi[\epsilon, \Xi_n[\epsilon]], C_1^i[\epsilon, \Xi_n[\epsilon]], \Theta^i[\epsilon, \Xi_n[\epsilon]], \Theta_1^o[\epsilon, \Xi_n[\epsilon]], C_1^o[\epsilon, \Xi_n[\epsilon]]](\epsilon\xi) \\ &= \tilde{u}^o(\epsilon\xi) + \epsilon \int_{\partial\Omega^o} S_n(\epsilon\xi - y)\Theta_1^o[\epsilon, \Xi_n[\epsilon]](y)\, d\sigma_y + \epsilon C_1^o[\epsilon, \Xi_n[\epsilon]] \\ &\quad + \epsilon^{n-1} \int_{\partial\Omega^i} S_n(\epsilon\xi - \epsilon\eta)\Theta^i[\epsilon, \Xi_n[\epsilon]](\eta)\, d\sigma_\eta \\ &= \tilde{u}^o(0) + \epsilon \int_0^1 \nabla\tilde{u}^o(\beta\epsilon\xi)\cdot\xi \, d\beta + \epsilon \int_{\partial\Omega^o} S_n(\epsilon\xi - y)\Theta_1^o[\epsilon, \Xi_n[\epsilon]](y)\, d\sigma_y \\ &\quad + \epsilon C_1^o[\epsilon, \Xi_n[\epsilon]] + \epsilon \int_{\partial\Omega^i} S_n(\xi - \eta)\Theta^i[\epsilon, \Xi_n[\epsilon]](\eta)\, d\sigma_\eta \\ &\quad + \frac{\delta_{2,n}\epsilon\log\epsilon}{2\pi} \int_{\partial\Omega^i} \Theta^i[\epsilon, \Xi_n[\epsilon]]\, d\sigma \qquad \forall \xi \in \overline{\Omega_m}, \end{aligned}$$

for all $\epsilon \in]0, \epsilon_m[$ (see also the second definition of (11.46)). Thus we find natural to set

$$\begin{aligned} \mathcal{U}_{m,1}^o[\epsilon, \epsilon_1, \epsilon_2](\xi) \equiv & \int_0^1 \nabla\tilde{u}^o(\beta\epsilon\xi)\cdot\xi\, d\beta \qquad (11.101) \\ & + \int_{\partial\Omega^o} S_n(\epsilon\xi - y)\Theta_1^o[\epsilon, \epsilon_1, \epsilon_2](y)\, d\sigma_y + C_1^o[\epsilon, \epsilon_1, \epsilon_2] \\ & + v_{\Omega^i}^-[\Theta^i[\epsilon, \epsilon_1, \epsilon_2]](\xi) \qquad \forall \xi \in \overline{\Omega_m}, \end{aligned}$$

for all $(\epsilon, \epsilon_1, \epsilon_2) \in]-\epsilon_m, \epsilon_m[\times \tilde{U}$ and we prove that $\mathcal{U}_{m,1}^o$ is analytic.

Since $\tilde{u}^o$ is analytic in Ω^o and S_n is analytic in $\mathbb{R}^n \setminus \{0\}$, Theorem A.28 (i), (iv) on the analyticity of nonlinear integral operators with analytic kernel implies that the map from $]-\epsilon_m, \epsilon_m[$ to $C^{1,\alpha}(\overline{\Omega_m})$ that takes ϵ to $\int_0^1 \nabla\tilde{u}^o(\beta\epsilon\xi)\cdot\xi\, d\beta$ is analytic and that the map from $]-\epsilon_m, \epsilon_m[\times L^1(\partial\Omega^o)$ to $C^{1,\alpha}(\overline{\Omega_m})$ that takes (ϵ, h) to $\int_{\partial\Omega^o} S_n(\epsilon\cdot -y))h(y)\,d\sigma_y$ is analytic.

By Theorem 11.17 (iv), Θ_1^o is analytic from $]-\epsilon_m, \epsilon_m[\times\tilde{U}$ to $L^1(\partial\Omega^o)$, and thus the sum of the first two summands in the right-hand side of (11.101) is analytic. Theorem 11.17 (iv) implies that Θ^i is analytic from $]-\epsilon_m, \epsilon_m[\times\tilde{U}$ to $C^{0,\alpha}(\partial\Omega^i)$ and Theorem 4.25 (ii) implies that $v^-_{\Omega^i}[\cdot]$ is linear and continuous from $C^{0,\alpha}(\partial\Omega^i)$ to $C^{1,\alpha}(\overline{\Omega_m})$. Hence, the fourth summand in the right-hand side of (11.101) is analytic from $]-\epsilon_m, \epsilon_m[\times\tilde{U}$ to $C^{1,\alpha}(\overline{\Omega_m})$. Since C_1^o is analytic, we conclude that $\mathcal{U}^o_{m,1}$ is analytic. By Theorems 11.16, 11.17 (iv), we have

$$
\begin{aligned}
\mathcal{U}^o_{m,1}[0,0,0](\xi) &= \nabla\tilde{u}^o(0)\cdot\xi + v^+_{\Omega^o}[\tilde{\theta}^o_1](0) + \tilde{c}^o_1 + v^-_{\Omega^i}[\tilde{\theta}^i](\xi) \\
&= \nabla\tilde{u}^o(0)\cdot\xi + \tilde{u}^o_1(0) + \tilde{u}^{o,r}_1(\xi) \qquad \forall\xi\in\overline{\Omega_m}\,.
\end{aligned}
$$

Then if we set

$$
\mathcal{U}^o_2[\epsilon,\epsilon_1,\epsilon_2] \equiv \frac{1}{2\pi F'(F^{(-1)}(\tilde{u}^o(0)))}\int_{\partial\Omega^i}\Theta^i[\epsilon,\epsilon_1,\epsilon_2]\,d\sigma \qquad \forall(\epsilon,\epsilon_1,\epsilon_2)\in]-\epsilon',\epsilon'[\times\tilde{U}\,,
$$

equality (11.97) holds by the definition of $\mathcal{U}^o_{m,1}$, $\mathcal{U}^o_2$ and by equality (11.100). Then the analyticity of Θ^i, equality $\Theta^i[0,0,0] = \tilde{\theta}^i$, and equalities (11.60) imply the analyticity of $\mathcal{U}^o_2$ and the validity of the second equality in (11.98). The proof of statement (ii) is now complete. □

11.3.3 A Property of Local Uniqueness for the Family of Solutions $\{(u^i(\epsilon,\cdot), u^o(\epsilon,\cdot))\}_{\epsilon\in]0,\epsilon'[}$

We now turn to the question of local uniqueness for the family of solutions

$$
\{(u^i(\epsilon,\cdot), u^o(\epsilon,\cdot))\}_{\epsilon\in]0,\epsilon'[}
$$

and prove the following.

Theorem 11.21. *Let the assumptions of Theorem 11.17 hold. Let $\{\varepsilon_j\}_{j\in\mathbb{N}}$ be a sequence that converges to 0 in $]0,\epsilon_0[$. If $\{(u^i_j, u^o_j)\}_{j\in\mathbb{N}}$ is a sequence of pairs of functions such that*

for each $j\in\mathbb{N}$ the pair (u^i_j, u^o_j) belongs to $C^{1,\alpha}(\varepsilon_j\overline{\Omega^i})\times C^{1,\alpha}(\overline{\Omega(\varepsilon_j)})$ and solves the transmission problem (11.38) for $\epsilon=\varepsilon_j$, and

$$
\lim_{j\to\infty}\frac{u^i(\varepsilon_j\xi) - F^{(-1)}(\tilde{u}^o(0))}{\varepsilon_j^{1/2}} = 0 \qquad \text{in } C^{1,\alpha}(\partial\Omega^i)\,, \tag{11.102}
$$

then there exists $j_0\in\mathbb{N}$ such that

$$(u_j^i, u_j^o) = (u^i(\varepsilon_j, \cdot), u^o(\varepsilon_j, \cdot))$$

for all $j \geq j_0$.

Proof. Possibly neglecting a finite number of indexes, we can assume that the pair $(\varepsilon_j, \Xi_n[\varepsilon_j])$ belongs to $]0, \epsilon'[\times \tilde{U}$ for all $j \in \mathbb{N}$. By Theorems 11.13 and 11.17 (ii), there exists a unique $(\psi_j, c_{1j}^i, \theta_j^i, \theta_{1j}^o, c_{1j}^o)$ in $Y_{0,\alpha}$ such that

$$u_j^i = u^i[\varepsilon_j, \psi_j, c_{1j}^i, \theta_j^i, \theta_{1j}^o, c_{1j}^o], u_j^o = u^o[\varepsilon_j, \psi_j, c_{1j}^i, \theta_j^i, \theta_{1j}^o, c_{1j}^o], \tag{11.103}$$

$$\mathcal{M}[\varepsilon_j, \Xi_n[\varepsilon_j], \psi_j, c_{1j}^i, \theta_j^i, \theta_{1j}^o, c_{1j}^o] = 0, \tag{11.104}$$

for each $j \in \mathbb{N}$. We now try to show that

$$\lim_{j\to\infty} (\psi_j, c_{1j}^i, \theta_j^i, \theta_{1j}^o, c_{1j}^o) = (\tilde{\psi}, \tilde{c}_1^i, \tilde{\theta}^i, \tilde{\theta}_1^o, \tilde{c}_1^o) \qquad \text{in } Y_{0,\alpha} \tag{11.105}$$

(cf. Theorem 11.16 (i)). To do so, we now rewrite equation

$$\mathcal{M}[\epsilon, \epsilon_1, \epsilon_2, \psi, c_1^i, \theta^i, \theta_1^o, c_1^o] = 0$$

for an arbitrary $(\epsilon, \epsilon_1, \epsilon_2, \psi, c_1^i, \theta^i, \theta_1^o, c_1^o)$ in the domain of $\mathcal{M}$ in the following form

$$\begin{aligned}
&V_{\Omega^i}[\theta^i](\xi) + \int_{\partial\Omega^o} S_n(\epsilon\xi - y)\theta_1^o(y)\, d\sigma_y + c_1^o \\
&\quad -F'(F^{(-1)}(\tilde{u}^o(0)))\left\{V_{\Omega^i}[\psi](\xi) + c_1^i\right\} \\
&\quad = -\int_0^1 \nabla\tilde{u}^o(\beta\epsilon\xi)\xi\, d\beta + \int_0^1 (1-\beta)F''\Big(F^{(-1)}(\tilde{u}^o(0)) \\
&\qquad +\beta\epsilon\left(V_{\Omega^i}[\psi](\xi) + c_1^i\right) + \frac{\beta\epsilon_1\epsilon_2}{2\pi F'(F^{(-1)}(\tilde{u}^o(0)))}\int_{\partial\Omega^i}\theta^i\, d\sigma\Big)\, d\beta \\
&\quad \times\Bigg\{\epsilon\left(V_{\Omega^i}[\psi](\xi) + c_1^i\right)^2 + \left(V_{\Omega^i}[\psi](\xi) + c_1^i\right)\frac{2\epsilon_1\epsilon_2}{2\pi F'(F^{(-1)}(\tilde{u}^o(0)))}\int_{\partial\Omega^i}\theta^i\, d\sigma \\
&\qquad + \frac{\epsilon_2\left(\int_{\partial\Omega^i}\theta^i\, d\sigma\right)^2}{\left(2\pi F'(F^{(-1)}(\tilde{u}^o(0)))\right)^2}\Bigg\} \qquad \forall\xi \in \partial\Omega^i, \\
&\frac{1}{2}\theta^i(\xi) + W_{\Omega^i}^t[\theta^i](\xi) + \epsilon\int_{\partial\Omega^o} \nu_{\Omega^i}(\xi)\cdot\nabla S_n(\epsilon\xi - y)\theta_1^o(y)\, d\sigma_y \\
&\quad +\frac{1}{2}\psi(\xi) - W_{\Omega^i}^t[\psi](\xi) \\
&\quad = G\left(\epsilon V_{\Omega^i}[\psi](\xi) + F^{(-1)}(\tilde{u}^o(0)) + c_1^i\epsilon + \frac{\epsilon_1\epsilon_2}{2\pi F'(F^{(-1)}(\tilde{u}^o(0)))}\int_{\partial\Omega^i}\theta^i\, d\sigma\right) \\
&\quad -\int_{\partial\Omega^o} \nu_{\Omega^i}(\xi)\cdot\nabla S_n(\epsilon\xi - y)\tilde{\theta}^o(y)\, d\sigma_y \qquad \forall\xi \in \partial\Omega^i, \\
&\epsilon^{n-2}\int_{\partial\Omega^i} S_n(x - \epsilon\eta)\theta^i(\eta)\, d\sigma_\eta + V_{\Omega^o}[\theta_1^o](x) + c_1^o = 0 \qquad \forall x \in \partial\Omega^o,
\end{aligned} \tag{11.106}$$

$$\epsilon^{n-2}\int_{\partial\Omega^i}\theta^i\,d\sigma+\int_{\partial\Omega^o}\theta^o_1\,d\sigma=0\,.$$

Next we denote by $\mathcal{N}\equiv(\mathcal{N}_l)_{l=1,2,3,4}$ and $\mathcal{S}\equiv(\mathcal{S}_l)_{l=1,2,3,4}$ the functions from $]-\epsilon',\epsilon'[\times\tilde{U}\times Y_{0,\alpha}$ to $Z_{0,\alpha}$ defined by the left-hand side and by the right-hand side of (11.106), respectively. Then Equation (11.106) can be rewritten as

$$\mathcal{N}[\epsilon,\epsilon_1,\epsilon_2,\psi,c^i_1,\theta^i,\theta^o_1,c^o_1]=\mathcal{S}[\epsilon,\epsilon_1,\epsilon_2,\psi,c^i_1,\theta^i,\theta^o_1,c^o_1]\,. \tag{11.107}$$

By arguing precisely as we did to prove the analyticity of $\mathcal{M}$, we can show that $\mathcal{N}$ and $\mathcal{S}$ are analytic. We now note that $\mathcal{N}[\epsilon,\epsilon_1,\epsilon_2,\cdot,\cdot,\cdot,\cdot,\cdot]$ is linear and continuous from $Y_{0,\alpha}$ to $Z_{0,\alpha}$ for all $(\epsilon,\epsilon_1,\epsilon_2)$ in $]-\epsilon',\epsilon'[\times\tilde{U}$. Since $\mathcal{N}$ is analytic, the map from $]-\epsilon',\epsilon'[\times\tilde{U}$ to $\mathcal{L}(Y_{0,\alpha},Z_{0,\alpha})$ that takes $(\epsilon,\epsilon_1,\epsilon_2)$ to $\mathcal{N}[\epsilon,\epsilon_1,\epsilon_2,\cdot,\cdot,\cdot,\cdot,\cdot]$ is also analytic. Indeed, such a map coincides with the partial differential of $\mathcal{N}$ with respect to its last five variables. We also note that

$$\mathcal{N}[0,0,0,\cdot,\cdot,\cdot,\cdot,\cdot]=\partial_{(\psi,c^i_1,\theta^i,\theta^o_1,c^o_1)}\mathcal{M}[0,0,0,\tilde{\psi},\tilde{c}^i_1,\tilde{\theta}^i,\tilde{\theta}^o_1,\tilde{c}^o_1]\,, \tag{11.108}$$

and that we already know that the operator in the right-hand side is a linear homeomorphism (cf. Theorem 11.17 (iii)). Since the set of linear homeomorphisms is open in the set of linear and continuous operators, and since the map that takes a linear invertible operator to its inverse is real analytic (cf. Proposition A.15), there exists an open neighborhood $\tilde{W}$ of $(0,0,0)$ in $]-\epsilon',\epsilon'[\times\tilde{U}$ such that the map from $\tilde{W}$ to $\mathcal{L}(Z_{0,\alpha},Y_{0,\alpha})$ that takes $(\epsilon,\epsilon_1,\epsilon_2)$ to $\mathcal{N}[\epsilon,\epsilon_1,\epsilon_2,\cdot,\cdot,\cdot,\cdot,\cdot]^{(-1)}$ is analytic. Clearly, there exists $j_0\in\mathbb{N}$ such that

$$(\varepsilon_j,\Xi_n[\varepsilon_j])\in\tilde{W}\qquad\forall j_0\le j\in\mathbb{N}\,.$$

Then by Equation (11.104), by the invertibility of $\mathcal{N}[\varepsilon_j,\Xi_n[\varepsilon_j],\cdot,\cdot,\cdot,\cdot,\cdot]$ for $j\ge j_0$, and by the equivalence of equation $\mathcal{M}=0$ and of equality (11.107), we have

$$\begin{aligned}&(\psi_j,c^i_{1j},\theta^i_j,\theta^o_{1j},c^o_{1j})\\&\quad=\mathcal{N}[\varepsilon_j,\Xi_n[\varepsilon_j],\cdot,\cdot,\cdot,\cdot,\cdot]^{(-1)}\circ\mathcal{S}[\varepsilon_j,\Xi_n[\varepsilon_j],\psi_j,c^i_{1j},\theta^i_j,\theta^o_{1j},c^o_{1j}]\end{aligned} \tag{11.109}$$

for $j\ge j_0$. By (11.46) and (11.103), and equality $\int_{\partial\Omega^i}\psi_j\,d\sigma=0$, we have

$$\begin{aligned}\varepsilon_j\left(V_{\Omega^i}[\psi_j](\xi)+c^i_{1j}\right)+\frac{(\delta_{2,n}\log^{-1}\varepsilon_j)(\delta_{2,n}\varepsilon_j\log^2\varepsilon_j)}{2\pi F'(F^{(-1)}(\tilde{u}^o(0)))}\int_{\partial\Omega^i}\theta^i_j\,d\sigma&\\=u^i_j(\varepsilon_j\xi)-F^{(-1)}(\tilde{u}^o(0))&\end{aligned}$$

for $j\ge j_0$. Then by the definition of $\mathcal{S}$, and by the equality

$$\int_{\partial\Omega^o}\nu_{\Omega^i}(\xi)\cdot\nabla S_n(\varepsilon_j\xi-y)\tilde{\theta}^o(y)\,d\sigma_y=\nu_{\Omega^i}(\xi)\cdot\nabla\tilde{u}^o(\varepsilon_j\xi)\qquad\forall\xi\in\partial\Omega^i$$

(cf. Remark 11.12), we have

$$
\begin{aligned}
&\mathcal{S}[\varepsilon_j, \Xi_n[\varepsilon_j], \psi_j, c^i_{1j}, \theta^i_j, \theta^o_{1j}, c^o_{1j}](\xi) \\
&= \Bigg(-\int_0^1 \nabla\tilde{u}^o(\beta\varepsilon_j\xi)\xi\, d\beta + \int_0^1 (1-\beta)F''\Big(F^{(-1)}(\tilde{u}^o(0)) \\
&\quad +\beta\left(u^i_j(\varepsilon_j\xi) - F^{(-1)}(\tilde{u}^o(0))\right)\Big)\, d\beta\varepsilon_j^{-1}\{u^i_j(\varepsilon_j\xi) - F^{(-1)}(\tilde{u}^o(0))\}^2, \\
&\quad G(u^i_j(\varepsilon_j\xi)) - \nu_{\Omega^i}(\xi)\cdot\nabla\tilde{u}^o(\varepsilon_j\xi), 0, 0\Bigg) \qquad \forall\xi\in\partial\Omega^i\,,\ j_0\le j\in\mathbb{N}\,.
\end{aligned}
$$

Since $\nabla\tilde{u}^o$ is analytic in Ω^o, the analyticity of nonlinear integral operators with analytic kernel of Theorem A.28 (iii) and the continuity of the pointwise product in Schauder spaces imply that

$$
\lim_{j\to\infty}\int_0^1 \nabla\tilde{u}^o(\beta\varepsilon_j\xi)\xi\, d\beta = \int_0^1 \nabla\tilde{u}^o(0)\xi\, d\beta \qquad \text{in } C^{1,\alpha}(\partial\Omega^i)\,. \tag{11.110}
$$

By assumption (11.102), we have

$$
\lim_{j\to\infty} u^i_j(\varepsilon_j\cdot) - F^{(-1)}(\tilde{u}^o(0)) = 0 \qquad \text{in } C^{1,\alpha}(\partial\Omega^i)\,. \tag{11.111}
$$

Since $(1-\beta)F''\Big(F^{(-1)}(\tilde{u}^o(0)) + \beta\left(a - F^{(-1)}(\tilde{u}^o(0))\right)\Big)$ is analytic in the variable $(a,\beta)\in\mathbb{R}^2$, the analyticity of nonlinear integral operators with analytic kernel of Theorem A.28 (iii) and the limiting relation (11.111) imply that

$$
\begin{aligned}
&\lim_{j\to\infty}\int_0^1 (1-\beta)F''\Big(F^{(-1)}(\tilde{u}^o(0)) + \beta\left(u^i_j(\varepsilon_j\xi) - F^{(-1)}(\tilde{u}^o(0))\right)\Big)\, d\beta \\
&\qquad = \int_0^1 (1-\beta)F''\Big(F^{(-1)}(\tilde{u}^o(0))\Big)\, d\beta = \frac{1}{2}F''\Big(F^{(-1)}(\tilde{u}^o(0))\Big)
\end{aligned} \tag{11.112}
$$

in $C^{1,\alpha}(\partial\Omega^i)$. Since G is analytic, Theorem A.23 (ii) implies that the composition operator N_G is analytic and accordingly continuous in $C^{1,\alpha}(\partial\Omega^i)$. Then assumption (11.102) implies that

$$
\lim_{j\to\infty} G(u^i_j(\varepsilon_j\xi)) = G(F^{(-1)}(\tilde{u}^o(0))) \qquad \text{in } C^{1,\alpha}(\partial\Omega^i)\,, \tag{11.113}
$$

and accordingly in $C^{0,\alpha}(\partial\Omega^i)$. Since $\nabla\tilde{u}^o$ is analytic in a neighborhood of 0, Theorem A.23 (ii) implies that the composition operator $N_{\frac{\partial\tilde{u}^o}{\partial x_j}}$ is analytic in $C^{0,\alpha}(\partial\Omega^i)$ for all $\in\{1,\ldots,n\}$. Then the continuity of the pointwise product in Schauder spaces implies that

$$
\lim_{j\to\infty} \nu_{\Omega^i}\cdot\nabla\tilde{u}^o(\varepsilon_j\cdot) = \nu_{\Omega^i}\cdot\nabla\tilde{u}^o(0) \qquad \text{in } C^{0,\alpha}(\partial\Omega^i)\,. \tag{11.114}
$$

Then assumption (11.102) and the above limiting relations (11.110)–(11.114) imply that

$$\lim_{j\to\infty} \mathcal{S}[\varepsilon_j, \Xi_n[\varepsilon_j], \psi_j, c^i_{1j}, \theta^i_j, \theta^o_{1j}, c^o_{1j}] \\ = (-\nabla\tilde{u}^o(0)\cdot\xi, G(F^{(-1)}(\tilde{u}^o(0))) - \nu_{\Omega^i}\cdot\nabla\tilde{u}^o(0), 0, 0) \tag{11.115}$$

in $Z_{0,\alpha}$. Then (11.109) and (11.115), the analyticity of the map that takes $(\epsilon, \epsilon_1, \epsilon_2)$ to $\mathcal{N}[\epsilon, \epsilon_1, \epsilon_2, \cdot, \cdot, \cdot, \cdot, \cdot]^{(-1)}$, and the continuity of the bilinear map from

$$\mathcal{L}(Z_{0,\alpha}, Y_{0,\alpha}) \times Z_{0,\alpha}$$

to $Y_{0,\alpha}$ that takes a pair (A_1, A_2) to $A_1[A_2]$ imply that

$$\lim_{j\to\infty} (\psi_j, c^i_{1j}, \theta^i_j, \theta^o_{1j}, c^o_{1j}) \\ = \mathcal{N}[0,0,0,\cdot,\cdot,\cdot,\cdot,\cdot]^{(-1)}(-\nabla\tilde{u}^o(0)\cdot\xi, G(F^{(-1)}(\tilde{u}^o(0))) - \nu_{\Omega^i}\cdot\nabla\tilde{u}^o(0), 0, 0) \tag{11.116}$$

in $Y_{0,\alpha}$. By Theorem 11.16, we have $\mathcal{M}[0,0,0,\tilde{\psi}, \tilde{c}^i_1, \tilde{\theta}^i, \tilde{\theta}^o_1, \tilde{c}^o_1] = 0$. Then the equivalence of equation $\mathcal{M} = 0$ and of equality (11.107) and equality

$$\int_{\partial\Omega^o} \nu_{\Omega^i}(\xi)\cdot\nabla S_n(0-y)\tilde{\theta}^o(y)\,d\sigma_y = \nu_{\Omega^i}(\xi)\cdot\nabla\tilde{u}^o(0) \qquad \forall \xi \in \partial\Omega^i$$

(cf. Remark 11.12) imply that

$$\mathcal{N}[0,0,0,\tilde{\psi}, \tilde{c}^i_1, \tilde{\theta}^i, \tilde{\theta}^o_1, \tilde{c}^o_1] = \mathcal{S}[0,0,0,\tilde{\psi}, \tilde{c}^i_1, \tilde{\theta}^i, \tilde{\theta}^o_1, \tilde{c}^o_1] \\ = (-\nabla\tilde{u}^o(0)\cdot\xi, G(F^{(-1)}(\tilde{u}^o(0))) - \nu_{\Omega^i}\cdot\nabla\tilde{u}^o(0), 0, 0)\,,$$

and thus the right-hand side of (11.116) coincides with $(\tilde{\psi}, \tilde{c}^i_1, \tilde{\theta}^i, \tilde{\theta}^o_1, \tilde{c}^o_1)$ and the limiting relation (11.105) holds true.

Then there exists $j_0 \in \mathbb{N}$ such that $(\psi_j, c^i_{1j}, \theta^i_j, \theta^o_{1j}, c^o_{1j})$ belongs to the neighborhood $\tilde{V}$ of $(\tilde{\psi}, \tilde{c}^i_1, \tilde{\theta}^i, \tilde{\theta}^o_1, \tilde{c}^o_1)$ in $Y_{0,\alpha}$ of Theorem 11.17 (iv). Then Theorem 11.17 (iv) and equation (11.104) imply that

$$(\psi_j, c^i_{1j}, \theta^i_j, \theta^o_{1j}, c^o_{1j}) \\ = (\Psi[\varepsilon_j, \Xi_n[\varepsilon_j]], C^i_1, \Theta^i[\varepsilon_j, \Xi_n[\varepsilon_j]], \Theta^o_1[\varepsilon_j, \Xi_n[\varepsilon_j]], C^o_1[\varepsilon_j, \Xi_n[\varepsilon_j]])\,,$$

for all $j_0 \le j \in \mathbb{N}$, and accordingly the Definition 11.18 of our family of solutions implies that $(u^i_j, u^o_j) = (u^i(\varepsilon_j, \cdot), u^o(\varepsilon_j, \cdot))$ for all $j \ge j_0$. □

11.3.4 Analytic Representation for the Energy Integrals of the Family of Solutions $\{(u^i(\epsilon,\cdot), u^o(\epsilon,\cdot))\}_{\epsilon\in]0,\epsilon'[}$

We now consider the energy integral and prove the following.

Theorem 11.22. *Let the assumptions of Theorem 11.17 hold. Then the following statements hold.*

(i) There exists an analytic map $\mathcal{E}^i$ from $]-\epsilon', \epsilon'[\times\tilde{U}$ to $\mathbb{R}$ such that

$$\int_{\epsilon\Omega^i} |\nabla_y u^i(\epsilon, y)|^2 \, dy = \epsilon^n \mathcal{E}^i[\epsilon, \Xi_n[\epsilon]] \qquad \forall \epsilon \in]0, \epsilon'[\,.$$

Moreover,

$$\mathcal{E}^i[0,0,0] = \int_{\Omega^i} |\nabla \tilde{u}_1^{i,r}(\eta)|^2 \, d\eta$$

(for the definition of $\tilde{u}_1^{i,r}$, see Theorem 11.16 (ii)).

(ii) There exist $\epsilon'' \in]0, \epsilon'[$ and an analytic map $\mathcal{E}^o$ from $]-\epsilon'', \epsilon''[\times\tilde{U}$ to $\mathbb{R}$ such that

$$\int_{\Omega(\epsilon)} |\nabla_y u^o(\epsilon, y)|^2 \, dy = \mathcal{E}^o[\epsilon, \Xi_n[\epsilon]] \qquad \forall \epsilon \in]0, \epsilon''[\,.$$

Moreover,

$$\mathcal{E}^o[0,0,0] = \int_{\Omega^o} |\nabla \tilde{u}^o(y)|^2 \, dy \,,$$

(for the definition of $\tilde{u}^o$, see (11.35)).

Proof. By the first Green Identity of Theorem 4.2 and by the representation formula (11.99) for $u^i(\epsilon, \epsilon\cdot)$, we have

$$\begin{aligned}
\int_{\epsilon\Omega^i} |\nabla_y u^i(\epsilon, y)|^2 \, dy &= \int_{\epsilon\partial\Omega^i} u^i(\epsilon, y) \frac{\partial}{\partial\nu_{\epsilon\Omega^i}} u^i(\epsilon, y) \, d\sigma_y \\
&= \int_{\partial\Omega^i} u^i(\epsilon, \epsilon\eta) \nu_{\Omega^i}(\eta) \cdot \nabla_\eta (u^i(\epsilon, \epsilon\eta)) \epsilon^{-1} \, d\sigma_\eta \epsilon^{n-1} \\
&= \epsilon^{n-2} \int_{\partial\Omega^i} \Big\{ F^{(-1)}(\tilde{u}^o(0)) + \epsilon v^+_{\Omega^i}[\Psi[\epsilon, \Xi_n[\epsilon]]] \\
&\quad + \epsilon C_1^i[\epsilon, \Xi_n[\epsilon]] + \frac{\delta_{2,n}\epsilon \log \epsilon}{2\pi F'(F^{(-1)}(\tilde{u}^o(0)))} \int_{\partial\Omega^i} \Theta^i[\epsilon, \Xi_n[\epsilon]] \, d\sigma \Big\} \\
&\quad \times \epsilon \nu_{\Omega^i} \cdot \nabla v^+_{\Omega^i}[\Psi[\epsilon, \Xi_n[\epsilon]]] \, d\sigma \qquad \forall \epsilon \in]0, \epsilon'[\,.
\end{aligned}$$

Next we note that $v^+_{\Omega^i}[\Psi[\epsilon, \Xi_n[\epsilon]]]$ is harmonic in Ω^i and that accordingly the Divergence Theorem 4.1 implies that

$$\int_{\partial\Omega^i} \nu_{\Omega^i} \cdot \nabla v^+_{\Omega^i}[\Psi[\epsilon, \Xi_n[\epsilon]]] \, d\sigma = 0\,.$$

Thus we have

$$\int_{\epsilon\Omega^i} |\nabla_y u^i(\epsilon,y)|^2\,dy = \epsilon^n \int_{\partial\Omega^i} v^+_{\Omega^i}[\Psi[\epsilon,\Xi_n[\epsilon]]] \frac{\partial}{\partial\nu_{\Omega^i}} v^+_{\Omega^i}[\Psi[\epsilon,\Xi_n[\epsilon]]]\,d\sigma$$
$$= \epsilon^n \int_{\Omega^i} |\nabla v^+_{\Omega^i}[\Psi[\epsilon,\Xi_n[\epsilon]]](\eta)|^2\,d\eta \qquad \forall \epsilon \in]0,\epsilon'[\,.$$

Thus it suffices to take

$$\mathcal{E}^i[\epsilon,\epsilon_1,\epsilon_2] \equiv \int_{\Omega^i} |\nabla v^+_{\Omega^i}[\Psi[\epsilon,\epsilon_1,\epsilon_2]](\eta)|^2\,d\eta \qquad \forall (\epsilon,\epsilon_1,\epsilon_2) \in]-\epsilon',\epsilon'[\times\tilde{U}\,.$$

Since $v^+_{\Omega^i}[\cdot]$ is linear and continuous from $C^{0,\alpha}(\partial\Omega^i)$ to $C^{1,\alpha}(\overline{\Omega^i})$ (cf. Theorem 4.25), the pointwise product is continuous in $C^{0,\alpha}(\overline{\Omega^i})$, the integration in Ω^i is linear and continuous on $C^{0,\alpha}(\overline{\Omega^i})$, and Ψ is analytic from $]-\epsilon',\epsilon'[\times\tilde{U}$ to $C^{0,\alpha}(\partial\Omega^i)$, we conclude that $\mathcal{E}^i$ is analytic. Moreover, we have

$$\mathcal{E}^i[0,0,0] = \int_{\Omega^i} |\nabla v^+_{\Omega^i}[\tilde{\psi}](\eta)|^2\,d\eta = \int_{\Omega^i} |\nabla \tilde{u}^{i,r}_1(\eta)|^2\,d\eta$$

(see (11.62) and Theorem 11.17 (iv)).

(ii) By the first Green Identity of Theorem 4.2, we have

$$\int_{\Omega(\epsilon)} |\nabla_y u^o(\epsilon,y)|^2\,dy = -\int_{\epsilon\partial\Omega^i} u^o(\epsilon,y)\frac{\partial}{\partial\nu_{\epsilon\Omega^i}} u^o(\epsilon,y)\,d\sigma_y$$
$$+\int_{\partial\Omega^o} u^o(\epsilon,y)\frac{\partial}{\partial\nu_{\Omega^o}} u^o(\epsilon,y)\,d\sigma_y$$
$$= -\int_{\partial\Omega^i} u^o(\epsilon,\epsilon\eta)\nu_{\Omega^i}(\eta)\cdot\nabla_\eta(u^o(\epsilon,\epsilon\eta))\,d\sigma_\eta\epsilon^{n-2}$$
$$+\int_{\partial\Omega^o} u^o(\epsilon,y)\frac{\partial}{\partial\nu_{\Omega^o}} u^o(\epsilon,y)\,d\sigma_y \qquad \forall\epsilon\in]0,\epsilon'[\,.$$

We now take Ω_M as in Theorem 11.19, Ω_m as in Theorem 11.20 (ii), and we also assume that $\partial\Omega^o \subseteq \overline{\Omega_M}$ and $\partial\Omega^i \subseteq \overline{\Omega_m}$. Then by exploiting the corresponding maps $\mathcal{U}_M, \mathcal{U}^o_{m,1}$, and $\mathcal{U}^o_2$, we have

$$\int_{\Omega(\epsilon)} |\nabla_y u^o(\epsilon,y)|^2\,dy$$
$$= -\epsilon^{n-1}\int_{\partial\Omega^i}\Big\{\tilde{u}^o(0)+\epsilon\mathcal{U}^o_{m,1}[\epsilon,\Xi_n[\epsilon]]$$
$$+(\delta_{2,n}\epsilon\log\epsilon)\mathcal{U}^o_2[\epsilon,\Xi_n[\epsilon]]\Big\}\nu_{\Omega^i}\cdot\nabla\mathcal{U}^o_{m,1}[\epsilon,\Xi_n[\epsilon]]\,d\sigma$$
$$+\int_{\partial\Omega^o}\Big\{\tilde{u}^o_{|\overline{\Omega_M}}+\epsilon\mathcal{U}_M[\epsilon,\Xi_n[\epsilon]]\Big\}\nu_{\Omega^o}\cdot\Big\{\nabla\tilde{u}^o_{|\overline{\Omega_M}}+\epsilon\nabla\mathcal{U}_M[\epsilon,\Xi_n[\epsilon]]\Big\}\,d\sigma\,,$$

for all $\epsilon\in]0,\epsilon''[$ with $\epsilon''\equiv\min\{\epsilon_M,\epsilon_m\}$. Thus we find natural to set

$$\mathcal{E}^o[\epsilon,\epsilon_1,\epsilon_2] \equiv -\epsilon^{n-1}\int_{\epsilon\partial\Omega^i}\Big\{\tilde{u}^o(0)+\epsilon\mathcal{U}^o_{m,1}[\epsilon,\epsilon_1,\epsilon_2]$$
$$+\epsilon_1\epsilon_2\mathcal{U}^o_2[\epsilon,\epsilon_1,\epsilon_2]\Big\}\nu_{\Omega^i}\cdot\nabla\mathcal{U}^o_{m,1}[\epsilon,\epsilon_1,\epsilon_2]\,d\sigma$$
$$+\int_{\partial\Omega^o}\Big\{\tilde{u}^o_{|\overline{\Omega_M}}+\epsilon\mathcal{U}_M[\epsilon,\epsilon_1,\epsilon_2]\Big\}\nu_{\Omega^o}\cdot\Big\{\nabla\tilde{u}^o_{|\overline{\Omega_M}}+\epsilon\nabla\mathcal{U}_M[\epsilon,\epsilon_1,\epsilon_2]\Big\}\,d\sigma\,,$$

for all $(\epsilon,\epsilon_1,\epsilon_2)\in]-\epsilon'',\epsilon''[\times\tilde{U}$. By Theorems 11.19 and 11.20 (ii), the maps $\mathcal{U}_M$ and $\mathcal{U}^o_{m,1}$, $\mathcal{U}^o_2$ are analytic. Hence, the continuity of the (bilinear) pointwise product in Schauder spaces, and the continuity of the linear map that takes a function to its gradient in Schauder spaces, the continuity of the linear map that takes a function of $C^{0,\alpha}(\partial\Omega^i)$ and of $C^{0,\alpha}(\partial\Omega^o)$ to its integral on $\partial\Omega^i$ and on $\partial\Omega^o$, respectively, imply that $\mathcal{E}^o$ is analytic. Moreover,

$$\mathcal{E}^o[0,0,0]=\int_{\partial\Omega^o}\tilde{u}^o\frac{\partial}{\partial\nu_{\Omega^o}}\tilde{u}^o\,d\sigma=\int_{\Omega^o}|\nabla\tilde{u}^o|^2\,dy\,.$$

□

Chapter 12
Boundary Value Problems in Periodic Domains, a Potential Theoretic Approach

Abstract In this chapter we consider boundary value problems for the Laplace equation in periodic domains obtained by removing a periodic set of holes from $\mathbb{R}^n$. To do so, we present a periodic version of Potential Theory based on layer potentials that are defined by replacing the fundamental solution of the Laplace equation with a periodic analog.

12.1 A Periodic Analog of the Fundamental Solution

Boundary value problems in periodic domains have a large variety of applications (cf., e.g., Ammari and Kang [8, Chaps. 2, 8], Milton [211, Chap. 1], Drygaś et al. [91], Gluzman et al. [110], and Mityushev et al. [221]). In particular, they are relevant in the computation of effective properties of composite materials, properties that in turn can be justified by an argument based on Homogenization Theory (cf., e.g., Allaire [4, Chap. 1], Bakhvalov and Panasenko [16, Chap. 1], Bensoussan et al. [20, Chap. 1], Jikov et al. [128, Chap. 1]). As is well known, potential theoretic methods have proven to be a powerful tool to analyze boundary value problems for elliptic differential operators and in particular to analyze periodic versions of such problems (cf., e.g., Ammari and Kang [8, Chaps. 2, 8], Berdichevskij [23], Mityushev and Nawalaniec [220], Shcherbina [261, 262]).

In particular, to extend potential theoretic methods to the case of periodic problems, we first need to construct a periodic analog of the fundamental solution. However, we shall see that such a construction is essentially the same for any strongly elliptic differential operator with complex coefficients. Therefore, we are going to present it in such a generality, before confining ourselves to the case of the Laplace equation. Our presentation here follows that of [174]. We note that even if both in Chapters 12 and 13 we will be dealing with boundary value problems for real-valued functions, in this section we will need to consider functions and distributions with values in $\mathbb{C}$. This is mainly because we will use tools of Fourier analysis for which the complex setting is a natural choice. With some abuse of notation, only in

M. Dalla Riva et al., *Singularly Perturbed Boundary Value Problems*,
https://doi.org/10.1007/978-3-030-76259-9_12

this section, we will retain the same symbols that we have used for some spaces of real-valued functions also for the case of complex-valued functions.

Unless stated otherwise, we fix a natural number

$$n \in \mathbb{N}\,,\; n \geq 2\,.$$

We take

$$(q_{11}, \ldots, q_{nn}) \in]0, +\infty[^n \tag{12.1}$$

and we introduce a periodicity cell

$$Q \equiv \prod_{j=1}^{n}]0, q_{jj}[\,.$$

Then we denote by q the diagonal matrix

$$q \equiv \begin{pmatrix} q_{11} & 0 & \ldots & 0 \\ 0 & q_{22} & \ldots & 0 \\ \ldots & \ldots & \ldots & \ldots \\ 0 & 0 & \ldots & q_{nn} \end{pmatrix}. \tag{12.2}$$

Clearly, $m_n(Q) = \prod_{j=1}^{n} q_{jj}$ and

$$q\mathbb{Z}^n \equiv \{qz : z \in \mathbb{Z}^n\} \tag{12.3}$$

is the set of vertices of a periodic subdivision of $\mathbb{R}^n$ corresponding to the fundamental cell Q. Next, we introduce a class of elliptic differential operators. Let N_2 denote the number of multi-indexes $\alpha \in \mathbb{N}^n$ with $|\alpha| \leq 2$. For each $\mathbf{a} \equiv (a_\alpha)_{|\alpha|\leq 2} \in \mathbb{C}^{N_2}$, we set

$$\mathbf{a}^{(2)} \equiv (a_{lj}^{(2)})_{l,j=1,\ldots,n} \qquad \mathbf{a}^{(1)} \equiv (a_j)_{j=1,\ldots,n}$$

with $a_{lj}^{(2)} \equiv 2^{-1} a_{e_l+e_j}$ for $j \neq l$, $a_{jj}^{(2)} \equiv a_{e_j+e_j}$, and $a_j \equiv a_{e_j}$, where

$$\{e_j : j = 1, \ldots, n\}$$

is the canonical basis of $\mathbb{R}^n$. We note that the matrix $\mathbf{a}^{(2)}$ is symmetric. If $\mathbf{a} \in \mathbb{C}^{N_2}$, then we set

$$P[\mathbf{a}, x] \equiv \sum_{|\alpha|\leq 2} a_\alpha x^\alpha \qquad \forall x \in \mathbb{R}^n\,.$$

We also set

$$\mathcal{E} \equiv \left\{ \mathbf{a} \equiv (a_\alpha)_{|\alpha|\leq 2} \in \mathbb{C}^{N_2} : \inf_{x\in\mathbb{R}^n, |x|=1} \mathrm{Re}\left\{ \sum_{|\alpha|=2} a_\alpha x^\alpha \right\} > 0 \right\}.$$

As usual, if $\alpha = (\alpha_1, \ldots, \alpha_n) \in \mathbb{N}^n$ and $x = (x_1, \ldots, x_n) \in \mathbb{R}^n$, then $x^\alpha \equiv x_1^{\alpha_1} \ldots x_n^{\alpha_n}$. Clearly, $\mathcal{E}$ coincides with the set of coefficients $\mathbf{a} \equiv (a_\alpha)_{|\alpha|\leq 2}$ such that the differential operator

$$P[\mathbf{a}, D] \equiv \sum_{|\alpha|\leq 2} a_\alpha D^\alpha$$

is strongly elliptic and has complex coefficients. If $\mathbf{a} \in \mathcal{E}$, a q-periodic distribution G is a q-periodic fundamental solution of $P[\mathbf{a}, D]$ provided that

$$P[\mathbf{a}, D]G = \sum_{z\in\mathbb{Z}^n} \delta_{qz}\,, \tag{12.4}$$

where δ_{qz} denotes the Dirac measure with mass at qz, for all $z \in \mathbb{Z}^n$ (see (12.7) below for the definition of q-periodic distribution).

Unfortunately however, not all operators $P[\mathbf{a}, D]$ admit q-periodic fundamental solutions, not even in the case where $P[\mathbf{a}, D]$ is the Laplace operator.

Instead, if we denote by $E_{2\pi i q^{-1}z}$ the function defined by

$$E_{2\pi i q^{-1}z}(x) \equiv e^{2\pi i(q^{-1}z)\cdot x} \qquad \forall x \in \mathbb{R}^n\,,$$

for all $z \in \mathbb{Z}^n$ and if $\mathbf{a} \in \mathcal{E}$, then we can show that the set

$$\mathbb{Z}(\mathbf{a}) \equiv \{z \in \mathbb{Z}^n : P[\mathbf{a}, 2\pi i q^{-1}z] = 0\}$$

is finite and that the q-periodic distribution

$$G_{\mathbf{a}} \equiv \sum_{z\in\mathbb{Z}^n\setminus\mathbb{Z}(\mathbf{a})} \frac{1}{m_n(Q)} \frac{1}{P[\mathbf{a}, 2\pi i q^{-1}z]} E_{2\pi i q^{-1}z} \tag{12.5}$$

satisfies the equality

$$P[\mathbf{a}, D](G_{\mathbf{a}}) = \sum_{z\in\mathbb{Z}^n} \delta_{qz} - \sum_{z\in\mathbb{Z}(\mathbf{a})} \frac{1}{m_n(Q)} E_{2\pi i q^{-1}z} \tag{12.6}$$

(cf. Theorem 12.2 below). Equality (12.6) can be considered as an effective substitute of equality (12.4), and the distribution $G_{\mathbf{a}}$ can be used to introduce layer potentials, which can be employed to analyze boundary value problems on q-periodic domains. Therefore, we find convenient to introduce the following definition.

Definition 12.1. Let $\mathbf{a} \in \mathcal{E}$. We say that a distribution S in $\mathbb{R}^n$ is an analog of a q-periodic fundamental solution of $P[\mathbf{a}, D]$ if S is q-periodic and $P[\mathbf{a}, D](S)$ equals the right-hand side of (12.6).

Before proving the existence of a periodic analog of a fundamental solution of a strongly elliptic operator with complex coefficients, we introduce some notation. If $y \in \mathbb{R}^n$ and f is a function defined in $\mathbb{R}^n$, we set $\tau_y f(x) \equiv f(x-y)$ for all $x \in \mathbb{R}^n$. If u is a distribution in $\mathbb{R}^n$, then we set

$$\langle \tau_y u, f\rangle = \langle u, \tau_{-y} f\rangle \qquad \forall f \in \mathcal{D}(\mathbb{R}^n)\,.$$

We say that a distribution u in $\mathbb{R}^n$ is q-periodic if

$$\tau_{q_{jj}e_j} u = u \qquad \forall j \in \{1, \ldots, n\}. \tag{12.7}$$

We denote by $\mathcal{S}(\mathbb{R}^n)$ the Schwartz space of rapidly decreasing complex-valued functions (cf. Section 2.3). Then we denote by

$$\hat{\varphi}(x) \equiv \int_{\mathbb{R}^n} \varphi(y) e^{-2\pi i x \cdot y}\, dy \qquad \forall x \in \mathbb{R}^n$$

the Fourier transform of $\varphi \in L^1(\mathbb{R}^n)$. We now verify that the generalized series $G_{\mathbf{a}}$ in (12.5) defines an analog of a q-periodic fundamental solution of $P[\mathbf{a}, D]$.

Theorem 12.2. *Let* $\mathbf{a} \in \mathcal{E}$. *Then the generalized series* $G_{\mathbf{a}}$ *in* (12.5) *defines a q-periodic tempered distribution in* $\mathbb{R}^n$ *which is an analog of a q-periodic fundamental solution of* $P[\mathbf{a}, D]$.

Proof. By assumption, there exists $y > 0$ such that

$$\mathrm{Re}\left\{ \sum_{|\alpha|=2} a_\alpha x^\alpha \right\} > y|x|^2 \qquad \forall x \in \mathbb{R}^n \setminus \{0\}.$$

Hence,

$$\begin{aligned} |P[\mathbf{a}, ix]| &= |x|^2 \left| \frac{1}{|x|^2} x^t \mathbf{a}^{(2)} x - \frac{i}{|x|^2} x^t \mathbf{a}^{(1)} - \frac{a_0}{|x|^2} \right| \\ &\geq |x|^2 \left(y - \frac{1}{|x|} \left(1 + \frac{1}{|x|} \right) \max\{|\mathbf{a}^{(1)}|, |a_0|\} \right) \\ &\geq |x|^2 y 2^{-1} \end{aligned}$$

for all $|x| \geq \max\{1, 4y^{-1} \max\{|\mathbf{a}^{(1)}|, |a_0|\}\}$. Thus, there exists $c_{\mathbf{a}} > 0$ such that

$$\frac{1}{|P[\mathbf{a}, 2\pi i q^{-1} z]|} \leq c_{\mathbf{a}} \qquad \forall z \in \mathbb{Z}^n \setminus \mathbb{Z}(\mathbf{a}).$$

Now let $\varphi \in \mathcal{S}(\mathbb{R}^n)$. Then we have

$$\begin{aligned} &\sum_{z \in \mathbb{Z}^n \setminus \mathbb{Z}(\mathbf{a})} \frac{1}{m_n(Q)} \frac{1}{|P[\mathbf{a}, 2\pi i q^{-1} z]|} | < E_{2\pi i q^{-1} z}, \varphi > | \\ &\quad = \sum_{z \in \mathbb{Z}^n \setminus \mathbb{Z}(\mathbf{a})} \frac{1}{m_n(Q)} \frac{1}{|P[\mathbf{a}, 2\pi i q^{-1} z]|} |\hat{\varphi}(-q^{-1} z)| \\ &\quad \leq c_{\mathbf{a}} \sum_{z \in \mathbb{Z}^n} \frac{1}{m_n(Q)} |\hat{\varphi}(q^{-1} z)| < +\infty. \end{aligned}$$

Indeed, by Poisson's summation formula, we have

$$\sum_{z\in\mathbb{Z}^n} \varphi(qz) = \sum_{z\in\mathbb{Z}^n} \frac{1}{m_n(Q)}\hat{\varphi}(q^{-1}z)\,,$$

and the generalized series in the right-hand side converges absolutely (cf., e.g., Folland [103, p. 254]).

Hence, the evaluation of the partial sums of the series in the right-hand side of (12.5) at an element φ of $\mathcal{S}(\mathbb{R}^n)$ converges to the evaluation of an element of the algebraic dual of $\mathcal{S}(\mathbb{R}^n)$ at φ. Since $\mathcal{S}(\mathbb{R}^n)$ is a Fréchet space, then the Banach-Steinhaus Theorem (cf., e.g., Trèves [266, pp. 347, 348]) ensures that such an element of the algebraic dual is actually an element of the topological dual of $\mathcal{S}(\mathbb{R}^n)$ and accordingly that $G_{\mathbf{a}}$ is a tempered distribution in $\mathbb{R}^n$. Since $G_{\mathbf{a}}$ is the sum of a generalized series of q-periodic distributions, we conclude that $G_{\mathbf{a}}$ is also q-periodic.

Finally, by the continuity of $P[\mathbf{a}, D]$ in $\mathcal{S}'(\mathbb{R}^n)$ and by the Poisson's summation formula, we have

$$\begin{aligned}
P[\mathbf{a}, D](G_{\mathbf{a}}) &= \sum_{z\in\mathbb{Z}^n\setminus\mathbb{Z}(\mathbf{a})} \frac{1}{m_n(Q)}\frac{1}{P[\mathbf{a}, 2\pi i q^{-1}z]} P[\mathbf{a}, D]E_{2\pi i q^{-1}z} \\
&= \frac{1}{m_n(Q)} \sum_{z\in\mathbb{Z}^n\setminus\mathbb{Z}(\mathbf{a})} E_{2\pi i q^{-1}z} \\
&= \frac{1}{m_n(Q)} \sum_{z\in\mathbb{Z}^n} E_{2\pi i q^{-1}z} - \frac{1}{m_n(Q)} \sum_{z\in\mathbb{Z}(\mathbf{a})} E_{2\pi i q^{-1}z} \\
&= \sum_{z\in\mathbb{Z}^n} \delta_{qz} - \frac{1}{m_n(Q)} \sum_{z\in\mathbb{Z}(\mathbf{a})} E_{2\pi i q^{-1}z}\,.
\end{aligned}$$

□

By the classical elliptic regularity theory, each analog of a q-periodic fundamental solution is analytic in $\mathbb{R}^n \setminus q\mathbb{Z}^n$. Indeed, the following holds.

Theorem 12.3. *Let* $\mathbf{a} \in \mathcal{E}$. *Let* $S_{\mathbf{a}}$ *be a (classical) fundamental solution of* $P[\mathbf{a}, D]$. *Let* S *be an analog of a q-periodic fundamental solution of* $P[\mathbf{a}, D]$. *Then the following statements hold.*

(i) S *is real analytic in* $\mathbb{R}^n \setminus q\mathbb{Z}^n$.
(ii) The difference $S - S_{\mathbf{a}}$ *admits a real analytic extension to* $(\mathbb{R}^n \setminus q\mathbb{Z}^n) \cup \{0\}$. *Moreover,*

$$P[\mathbf{a}, D](S - S_{\mathbf{a}}) = \sum_{z\in\mathbb{Z}^n\setminus\{0\}} \delta_{qz} - \frac{1}{m_n(Q)} \sum_{z\in\mathbb{Z}(\mathbf{a})} E_{2\pi i q^{-1}z} \qquad \textit{in } \mathcal{D}'(\mathbb{R}^n)\,.$$

(iii) $S \in L^1_{\mathrm{loc}}(\mathbb{R}^n)$ *(and has complex values).*

Proof. Since $\sum_{z\in\mathbb{Z}(\mathbf{a})} \frac{1}{m_n(Q)} E_{2\pi i q^{-1}z}$ is an analytic function in $\mathbb{R}^n$ and the distribution $\sum_{z\in\mathbb{Z}^n\setminus\{0\}} \delta_{qz}$ vanishes in $(\mathbb{R}^n \setminus q\mathbb{Z}^n) \cup \{0\}$ and since $P[\mathbf{a}, D]$ is strongly elliptic, the classical interior regularity theory implies the validity of statements (i), (ii) (cf., e.g., Friedman [104, Theorem 1.2, p. 205]). Statement (iii) follows by the real analyticity of the analytic extension of $S - S_{\mathbf{a}}$ in $(\mathbb{R}^n \setminus q\mathbb{Z}^n) \cup \{0\}$, by the q-periodicity of S, and by the local integrability of $S_{\mathbf{a}}$ in $\mathbb{R}^n$ (cf., e.g., John [129, Chap. III]). □

We note that an analyticity result in Sobolev-Bessel potential spaces for the periodic analog of the fundamental solution of a general elliptic partial differential operator upon the parameter q (which determines the periodicity cell) has been obtained in [180]. Moreover, in [180] it has been shown that the periodic analogs of the fundamental solution of the Helmholtz and of the Laplace operators are jointly analytic in the spatial variable and in the parameters which determine the size of the periodicity cell. Furthermore, concerning a periodic analog of a fundamental solution of a differential operator, we also mention the paper [192] by Luzzini, where the author has constructed a periodic fundamental solution for the heat equation and investigated the properties of the corresponding layer potentials (see also [193] for the application to transmission problems for the heat equation in periodic domains).

We now confine our attention to the case of the Laplace operator and, in the following Theorem 12.4, we use Theorems 12.2 and 12.3 to introduce a corresponding q-periodic analog of the fundamental solution. To emphasize its dependence on n and q we denote it by $S_{q,n}$, whereas we keep the notation S_n for the classical (non-periodic) fundamental solution introduced in (3.2). For similar definitions we mention, for example, Hasimoto [119], Shcherbina [261], Poulton et al. [245], Ammari et al. [12], and Ammari and Kang [8, p. 53]. We also mention the papers of Lin and Wang [189], Mityushev and Adler [218], and Mamode [197], where the authors use elliptic functions to obtain a constructive formula in the two-dimensional case.

Theorem 12.4. *The generalized series*

$$S_{q,n} \equiv \sum_{z\in\mathbb{Z}^n\setminus\{0\}} \frac{1}{-4\pi^2|q^{-1}z|^2 m_n(Q)} E_{2\pi i q^{-1}z} \tag{12.8}$$

defines a q-periodic tempered distribution in $\mathbb{R}^n$ *and*

$$\Delta S_{q,n} = \sum_{z\in\mathbb{Z}^n} \delta_{qz} - \frac{1}{m_n(Q)} \,. \tag{12.9}$$

Moreover, the following statements hold.

(i) $S_{q,n}$ *is real analytic in* $\mathbb{R}^n \setminus q\mathbb{Z}^n$.
(ii) $S_{q,n} - S_n$ *admits a real analytic extension* $R_{q,n}$ *in* $(\mathbb{R}^n \setminus q\mathbb{Z}^n) \cup \{0\}$, *which we address to as the regular part of* $S_{q,n}$, *and we have*

$$\Delta R_{q,n} = \sum_{z\in\mathbb{Z}^n\setminus\{0\}} \delta_{qz} - \frac{1}{m_n(Q)} \,, \tag{12.10}$$

in the sense of distributions.
(iii) $S_{q,n}$ *belongs to* $L^1_{\mathrm{loc}}(\mathbb{R}^n)$ *and has real values.*
(iv) $S_{q,n}(x) = S_{q,n}(-x)$ *for all* $x \in \mathbb{R}^n \setminus q\mathbb{Z}^n$.
(v) $R_{q,n}(x) = R_{q,n}(-x)$ *for all* $x \in (\mathbb{R}^n \setminus q\mathbb{Z}^n) \cup \{0\}$ *and*

$$D^\beta R_{q,n}(0) = 0 \qquad \text{for all } \beta \in \mathbb{N}^n \text{ such that } |\beta| \text{ is odd}.$$

Proof. By Theorems 12.2 and 12.3, we deduce the validity of statements (i), (ii). We now consider statement (iii). By Theorem 12.3 (iii), $S_{q,n}$ is a locally integrable complex-valued function. In order to prove that $S_{q,n}$ is real valued, it suffices to show that

$$\overline{\langle S_{q,n}, \overline{\phi}\rangle} = \langle S_{q,n}, \phi\rangle \qquad \forall \phi \in \mathcal{S}(\mathbb{R}^n), \tag{12.11}$$

i.e.,

$$\overline{\int_{\mathbb{R}^n} S_{q,n}(x)\overline{\phi(x)}\, dx} = \int_{\mathbb{R}^n} S_{q,n}(x)\phi(x)\, dx \qquad \forall \phi \in \mathcal{S}(\mathbb{R}^n).$$

By the definition of $S_{q,n}$, by the equality

$$\overline{\int_{\mathbb{R}^n} E_{2\pi i q^{-1} z}(x)\overline{\phi(x)}\, dx} = \int_{\mathbb{R}^n} E_{2\pi i q^{-1}(-z)}(x)\phi(x)\, dx \qquad \forall \phi \in \mathcal{S}(\mathbb{R}^n), z \in \mathbb{Z}^n \setminus \{0\},$$

and by the obvious identity

$$\frac{1}{-4\pi^2|-q^{-1}z|^2 m_n(Q)} = \frac{1}{-4\pi^2|q^{-1}z|^2 m_n(Q)} \qquad \forall z \in \mathbb{Z}^n \setminus \{0\},$$

we can conclude that equality (12.11) holds and accordingly that $S_{q,n}$ is actually a real-valued function. We now turn to the proof of (iv). By a straightforward computation based on the definition (12.8) of $S_{q,n}$, we have

$$\int_{\mathbb{R}^n} S_{q,n}(x)\phi(-x)\, dx = \int_{\mathbb{R}^n} S_{q,n}(x)\phi(x)\, dx \qquad \forall \phi \in \mathcal{S}(\mathbb{R}^n),$$

and thus $S_{q,n}(x) = S_{q,n}(-x)$ for all $x \in \mathbb{R}^n \setminus q\mathbb{Z}^n$. Then the validity of (v) is a consequence of (iv) and of the parity of S_n. Hence, the proof is complete. □

12.2 Periodic Layer Potentials for the Laplace Equation

In this section, we present the integral operators that we use in order to transform periodic boundary value problems into an equivalent system of integral equations. Integral equation techniques have long been exploited for solving periodic boundary value problems and we mention, as an example, the fundamental book by Grigolyuk and Fil'shtinskij [111], where the authors have proposed a method of integral equations for planar doubly periodic problems in the frame of elasticity (see also Fil'shtinskij [100] and the more recent work Filshtinsky and Mityushev [101]). Other contributions where the authors have developed and exploited integral equations to solve periodic boundary value problems are, for example, the monographs by Ammari and Kang [8] and by Ammari et al. [10].

The method we use here is based on a periodic version of classical potential theory and thus we now introduce the periodic single and double layer potentials for the Laplace equation. For their construction, it suffices to replace the classical fundamental solution S_n with the periodic analog $S_{q,n}$. Then all the main properties of the periodic layer potentials can be deduced by the classical ones. Indeed, the layer potentials associated to the regular part $R_{q,n}$ of $S_{q,n}$ display a high degree of regularity because $R_{q,n}$ is analytic in a neighborhood of 0.

We note that in the present section, as well as in the rest of present chapter and in Chapter 13, we will consider only real-valued functions. In particular, since Theorem 12.4 ensures that $S_{q,n}$ has real values, we can construct real-valued layer potentials by using the periodic analog of a fundamental solution $S_{q,n}$.

12.2.1 Geometric Setting

We introduce here the geometric setting which will be used until the end of the present chapter. We remind that, unless stated otherwise,

$$n \in \mathbb{N}, \, n \geq 2,$$

is a fixed natural number. We take $(q_{11}, \ldots, q_{nn})$, Q, q, and $q\mathbb{Z}^n$ as in (12.1)–(12.3) and we fix

$$\alpha \in]0,1[.$$

We denote by Ω_Q an open bounded subset of $\mathbb{R}^n$ of class $C^{1,\alpha}$ such that $\overline{\Omega_Q} \subseteq Q$, and by Ω_Q^- the exterior of Ω_Q, namely

$$\Omega_Q^- \equiv \mathbb{R}^n \setminus \overline{\Omega_Q}.$$

For the sake of simplicity, we also assume that

$$\text{both } \Omega_Q \text{ and } \Omega_Q^- \text{ are connected.}$$

Then we introduce the periodic domains where we are going to consider our boundary value problems. We set

$$\mathbb{S}[\Omega_Q] \equiv \bigcup_{z \in \mathbb{Z}^n} (qz + \Omega_Q) = q\mathbb{Z}^n + \Omega_Q,$$
$$\mathbb{S}[\Omega_Q]^- \equiv \mathbb{R}^n \setminus \overline{\mathbb{S}[\Omega_Q]}$$

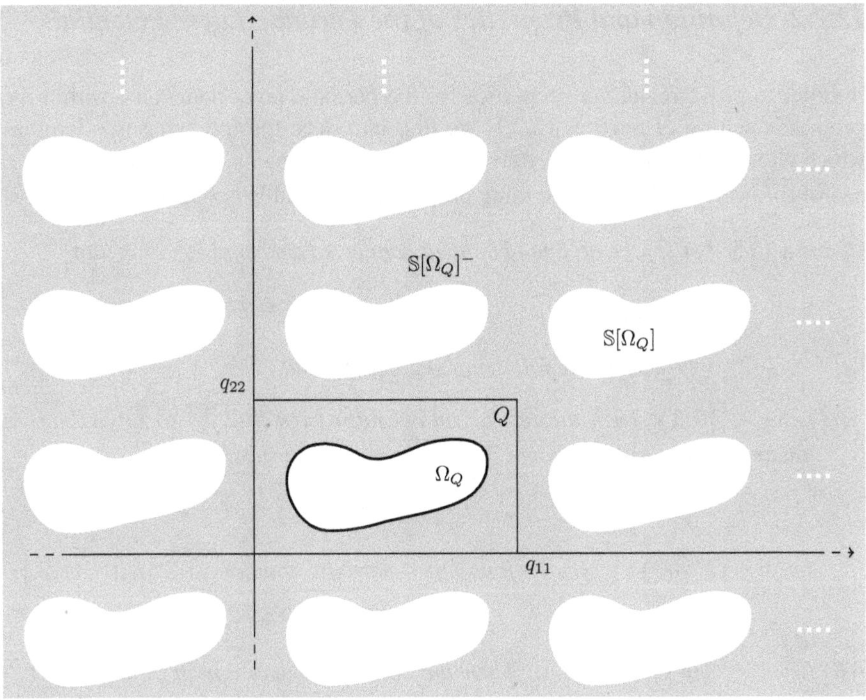

Fig. 12.1 In white (a part of) the set $\mathbb{S}[\Omega_Q]$; in grey (a part of) the set $\mathbb{S}[\Omega_Q]^-$ $(n = 2)$

(see Figure 12.1). Since $\mathbb{R}^n \setminus \overline{\Omega_Q}$ is connected, then the set $\mathbb{S}[\Omega_Q]^-$ is connected as well (cf., e.g., Theorem A.10).

If $k \in \mathbb{N}$, $\beta \in]0,1]$, then we introduce the spaces of real-valued periodic functions by setting:

$$
\begin{aligned}
C_q^k(\overline{\mathbb{S}[\Omega_Q]}) &\equiv \left\{u \in C^k(\overline{\mathbb{S}[\Omega_Q]}) : u \text{ is } q\text{-periodic}\right\} , \\
C_q^{k,\beta}(\overline{\mathbb{S}[\Omega_Q]}) &\equiv \left\{u \in C^{k,\beta}(\overline{\mathbb{S}[\Omega_Q]}) : u \text{ is } q\text{-periodic}\right\} , \\
C_q^k(\overline{\mathbb{S}[\Omega_Q]^-}) &\equiv \left\{u \in C^k(\overline{\mathbb{S}[\Omega_Q]^-}) : u \text{ is } q\text{-periodic}\right\} , \\
C_q^{k,\beta}(\overline{\mathbb{S}[\Omega_Q]^-}) &\equiv \left\{u \in C^{k,\beta}(\overline{\mathbb{S}[\Omega_Q]^-}) : u \text{ is } q\text{-periodic}\right\} .
\end{aligned}
$$

We regard $C_q^k(\overline{\mathbb{S}[\Omega_Q]})$, $C_q^{k,\beta}(\overline{\mathbb{S}[\Omega_Q]})$, $C_q^k(\overline{\mathbb{S}[\Omega_Q]^-})$, and $C_q^k(\overline{\mathbb{S}[\Omega_Q]^-})$ as Banach subspaces of $C_b^k(\overline{\mathbb{S}[\Omega_Q]})$, of $C_b^{k,\beta}(\overline{\mathbb{S}[\Omega_Q]})$, of $C_b^k(\overline{\mathbb{S}[\Omega_Q]^-})$, and of $C_b^{k,\beta}(\overline{\mathbb{S}[\Omega_Q]^-})$, respectively (cf. Sections 2.6 and 2.11). Similarly, we set

$$
\begin{aligned}
C_q^k(\mathbb{R}^n) &\equiv \{u \in C^k(\mathbb{R}^n) : u \text{ is } q\text{-periodic}\} , \\
C_q^{k,\beta}(\mathbb{R}^n) &\equiv \{u \in C^{k,\beta}(\mathbb{R}^n) : u \text{ is } q\text{-periodic}\} ,
\end{aligned}
$$

which we regard as Banach subspaces of $C_b^k(\mathbb{R}^n)$ and of $C_b^{k,\beta}(\mathbb{R}^n)$, respectively.

12.2.2 Definition and Properties of the Periodic Layer Potentials

In order to prove regularity properties for the periodic layer potentials, within the geometric setting of Section 12.2.1, we first introduce the following two lemmas which allow us to deduce properties of periodic functions on the whole of $\overline{\mathbb{S}[\Omega_Q]^-}$ and of $\overline{\mathbb{S}[\Omega_Q]}$ from the corresponding properties on suitable bounded sets.

Lemma 12.5. *Let $\Omega_\#$ be an open bounded connected subset of $\mathbb{R}^n$ such that*

$$\overline{Q} \subseteq \Omega_\#, \qquad \overline{\Omega_\#} \cap (qz + \overline{\Omega_Q}) = \emptyset \quad \forall z \in \mathbb{Z}^n \setminus \{0\}.$$

Let $\Omega^* \equiv \Omega_\# \setminus \overline{\Omega_Q}$. Then the following statements hold.

(i) Let $k \in \{0, 1\}$. Then the restriction operator from $\overline{\mathbb{S}[\Omega_Q]^-}$ to $\overline{\Omega^*}$ induces a linear homeomorphism from $C^k_q(\overline{\mathbb{S}[\Omega_Q]^-})$ onto the subspace

$$C^k_q(\overline{\Omega^*}) \equiv \left\{ v \in C^k(\overline{\Omega^*}) : v = u_{|\overline{\Omega^*}} \text{ for some } q\text{-periodic function } u \colon \overline{\mathbb{S}[\Omega_Q]^-} \to \mathbb{R} \right\}$$

of $C^k(\overline{\Omega^*})$.

(ii) Let $k \in \{0, 1\}$, $\beta \in]0, 1]$. Then the restriction operator of $\overline{\mathbb{S}[\Omega_Q]^-}$ to $\overline{\Omega^*}$ induces a linear homeomorphism from $C^{k,\beta}_q(\overline{\mathbb{S}[\Omega_Q]^-})$ onto the subspace

$$C^{k,\beta}_q(\overline{\Omega^*}) \equiv \left\{ v \in C^{k,\beta}(\overline{\Omega^*}) : v = u_{|\overline{\Omega^*}} \text{ for some } q\text{-periodic function } u \colon \overline{\mathbb{S}[\Omega_Q]^-} \to \mathbb{R} \right\}$$

of $C^{k,\beta}(\overline{\Omega^*})$.

Proof. Since the uniform limit of a periodic function is periodic, we deduce that $C^k_q(\overline{\Omega^*})$ is a closed subspace of the Banach space $C^k_b(\overline{\Omega^*})$. As a consequence $C^k_q(\overline{\Omega^*})$ is a Banach space too. Since $\overline{Q} \setminus \Omega_Q \subseteq \overline{\Omega^*}$, we deduce that the restriction operator is injective. On the other hand, if $v \in C^k_q(\overline{\Omega^*})$ then there exists a unique q-periodic function u from $\overline{\mathbb{S}[\Omega_Q]^-}$ to $\mathbb{R}$ such that $v = u$ on $\overline{\Omega^*}$ (and accordingly $u \in C^k_q(\overline{\mathbb{S}[\Omega_Q]^-})$). Hence, the restriction operator is a continuous linear bijection and the Open Mapping Theorem 2.2 implies the validity of statement (i).

We now consider statement (ii). By definition of norm in $C^{k,\beta}_q(\overline{\mathbb{S}[\Omega_Q]^-})$ and by statement (i), it clearly suffices to consider case $k = 0$. Indeed, by statement (i), a function in $C^{1,\beta}_q(\overline{\Omega^*})$ has a unique periodic extension to $\overline{\mathbb{S}[\Omega_Q]^-}$ and such an extension belongs to $C^1_q(\overline{\mathbb{S}[\Omega_Q]^-})$.

Obviously, if $u \in C_q^{0,\beta}(\overline{\mathbb{S}[\Omega_Q]^-})$, then its restriction $u_{|\overline{\Omega^*}}$ belongs to $C_q^{0,\beta}(\overline{\Omega^*})$. Conversely, if $v \in C_q^{0,\beta}(\overline{\Omega^*})$, then there exists a q-periodic function u from the set $\overline{\mathbb{S}[\Omega_Q]^-}$ to $\mathbb{R}$ such that $v = u_{|\overline{\Omega^*}}$. Clearly,

$$|u(x) - u(y)| \leq |v : \overline{\Omega^*}|_\beta |x-y|^\beta \qquad \forall x, y \in \overline{\Omega^*},$$

where $|v : \overline{\Omega^*}|_\beta$ denotes the β-Hölder constant of v on $\overline{\Omega^*}$. Next we set

$$d \equiv \inf\{|x-y| : (x,y) \in \overline{Q} \times (\mathbb{R}^n \setminus \Omega_\#)\}.$$

Clearly, $d > 0$, and we have

$$|u(x) - u(y)| \leq \max\left\{2d^{-\beta} \sup_{\overline{\Omega^*}} |v|, |v : \overline{\Omega^*}|_\beta\right\} |x-y|^\beta$$

for all $x, y \in \overline{\mathbb{S}[\Omega_Q]^-}$. Indeed, if there exists $z_0 \in \mathbb{Z}^n$ such that

$$x, y \in \overline{\Omega^*} + qz_0,$$

then, by periodicity, we have

$$|u(x) - u(y)| \leq |v : \overline{\Omega^*}|_\beta |x-y|^\beta.$$

If not, let $z_1 \in \mathbb{Z}^n$ be such that

$$x \in (\overline{Q} + qz_1), \qquad y \notin (\overline{\Omega^*} + qz_1).$$

Then

$$d \leq |x-y|, \qquad |u(x) - u(y)| \leq 2 \sup_{\overline{\Omega^*}} |v|.$$

As a consequence,

$$|u(x) - u(y)| \leq 2d^{-\beta} \sup_{\overline{\Omega^*}} |v| d^\beta \leq 2d^{-\beta} \sup_{\overline{\Omega^*}} |v||x-y|^\beta.$$

Accordingly, $u \in C_q^{0,\beta}(\overline{\mathbb{S}[\Omega_Q]^-})$.

Hence, the restriction operator is a bijection from $C_q^{0,\beta}(\overline{\mathbb{S}[\Omega_Q]^-})$ onto $C_q^{0,\beta}(\overline{\Omega^*})$. Since the restriction operator is obviously linear and continuous, then the Open Mapping Theorem 2.2 implies that the restriction operator is a homeomorphism. □

Then we have a corresponding lemma for q-periodic functions on $\overline{\mathbb{S}[\Omega_Q]}$.

Lemma 12.6. *The following statements hold.*

(i) Let $k \in \{0, 1\}$. Then the restriction operator from $C_q^k(\overline{\mathbb{S}[\Omega_Q]})$ to $C^k(\overline{\Omega_Q})$ is a linear homeomorphism.

(ii) Let $k \in \{0,1\}$, $\beta \in]0,1]$. Then the restriction operator from $C_q^{k,\beta}(\overline{\mathbb{S}[\Omega_Q]})$ to $C^{k,\beta}(\overline{\Omega_Q})$ is a linear homeomorphism.

Proof. Since $(qz_1 + \overline{\Omega_Q}) \cap (qz_2 + \overline{\Omega_Q}) = \emptyset$ whenever z_1, $z_2 \in \mathbb{Z}^n$, $z_1 \neq z_2$, the restriction operator from $C_q^k(\overline{\mathbb{S}[\Omega_Q]})$ to $C^k(\overline{\Omega_Q})$ is a bijection. Then statement (i) is an immediate consequence of the Open Mapping Theorem 2.2. We now turn to statement (ii). It clearly suffices to consider case $k = 0$. Obviously, if $u \in C_q^{0,\beta}(\overline{\mathbb{S}[\Omega_Q]})$, then its restriction $u_{|\overline{\Omega_Q}}$ belongs to $C^{0,\beta}(\overline{\Omega_Q})$. Conversely, if $v \in C^{0,\beta}(\overline{\Omega_Q})$, then there exists a q-periodic function u from $\overline{\mathbb{S}[\Omega_Q]}$ to $\mathbb{R}$ such that $v = u_{|\overline{\Omega_Q}}$ and

$$|u(x) - u(y)| \leq |v : \overline{\Omega_Q}|_\beta |x - y|^\beta \qquad \forall x, y \in \overline{\Omega_Q}.$$

Then we set

$$d \equiv \inf\{|x - y| : (x, y) \in \overline{\Omega_Q} \times (\mathbb{R}^n \setminus Q)\}.$$

Clearly, $d > 0$. If $x, y \in \overline{\mathbb{S}[\Omega_Q]}$, then there exist two points z_x, $z_y \in \mathbb{Z}^n$ such that $x \in qz_x + \overline{\Omega_Q}$, $y \in qz_y + \overline{\Omega_Q}$. By considering separately the case in which $z_x \neq z_y$ and the case in which $z_x = z_y$ and by exploiting the q-periodicity of u, we can conclude that

$$|u(x) - u(y)| \leq \max\left\{2d^{-\beta} \sup_{\overline{\Omega_Q}} |v|, |v : \overline{\Omega_Q}|_\beta\right\} |x - y|^\beta$$

for all $x, y \in \overline{\mathbb{S}[\Omega_Q]}$. Accordingly, $u \in C_q^{0,\beta}(\overline{\mathbb{S}[\Omega_Q]})$. Hence, the restriction operator is a bijection from $C_q^{0,\beta}(\overline{\mathbb{S}[\Omega_Q]})$ onto $C^{0,\beta}(\overline{\Omega_Q})$. Since the restriction operator is linear and continuous, the Open Mapping Theorem 2.2 implies that it is a homeomorphism. □

In order to construct periodic layer potentials, it suffices to replace the classical fundamental solution S_n with its periodic analog $S_{q,n}$. As a first step, we now introduce the periodic single layer potential.

Definition 12.7. If $\phi \in C^{0,\alpha}(\partial\Omega_Q)$, then $v_{q,\Omega_Q}[\phi]$ denotes the q-periodic single layer potential with density ϕ defined by

$$v_{q,\Omega_Q}[\phi](x) \equiv \int_{\partial\Omega_Q} \phi(y) S_{q,n}(x - y)\, d\sigma_y \qquad \forall x \in \mathbb{R}^n.$$

We are now in the position to prove the following theorem, where we collect some properties of the periodic single layer potential $v_{q,\Omega_Q}[\phi]$.

Theorem 12.8. *The following statements hold.*

(i) If $\phi \in C^{0,\alpha}(\partial\Omega_Q)$, then the function $v_{q,\Omega_Q}[\phi]$ is continuous. Moreover, $v_{q,\Omega_Q}[\phi]$ is q-periodic and

$$\Delta v_{q,\Omega_Q}[\phi](x) = -\frac{1}{m_n(Q)} \int_{\partial\Omega_Q} \phi \, d\sigma \tag{12.12}$$

for all $x \in \mathbb{R}^n \setminus \partial\mathbb{S}[\Omega_Q]$.

(ii) If $\phi \in C^{0,\alpha}(\partial\Omega_Q)$*, then the function* $v^+_{q,\Omega_Q}[\phi] \equiv v_{q,\Omega_Q}[\phi]_{|\overline{\mathbb{S}[\Omega_Q]}}$ *belongs to* $C^{1,\alpha}_q(\overline{\mathbb{S}[\Omega_Q]})$ *and the linear operator* $v^+_{q,\Omega_Q}[\cdot]$ *is bounded from* $C^{0,\alpha}(\partial\Omega_Q)$ *to* $C^{1,\alpha}_q(\overline{\mathbb{S}[\Omega_Q]})$.

(iii) If $\phi \in C^{0,\alpha}(\partial\Omega_Q)$*, then the function* $v^-_{q,\Omega_Q}[\phi] \equiv v_{q,\Omega_Q}[\phi]_{|\overline{\mathbb{S}[\Omega_Q]^-}}$ *belongs to* $C^{1,\alpha}_q(\overline{\mathbb{S}[\Omega_Q]^-})$ *and the linear operator* $v^-_{q,\Omega_Q}[\cdot]$ *is bounded from* $C^{0,\alpha}(\partial\Omega_Q)$ *to* $C^{1,\alpha}_q(\overline{\mathbb{S}[\Omega_Q]^-})$.

Proof. We first introduce some preliminary remarks. Theorem 12.4 ensures that $R_{q,n}$ is real analytic in $(\mathbb{R}^n \setminus q\mathbb{Z}^n) \cup \{0\}$. By the classical Lemma 2.70, there exists a bounded open connected subset $\Omega_\#$ of class C^∞ of $\mathbb{R}^n$ such that

$$\overline{Q} \subseteq \Omega_\#, \qquad \overline{\Omega_\#} \cap (qz + \overline{\Omega_Q}) = \emptyset \quad \forall z \in \mathbb{Z}^n \setminus \{0\}. \tag{12.13}$$

If $x \in \overline{\Omega_\#}$ and $y \in \partial\Omega_Q$, then the difference $x - y$ cannot belong to $q\mathbb{Z}^n \setminus \{0\}$, otherwise $x \in \partial\Omega_Q + (q\mathbb{Z}^n \setminus \{0\})$ and thus there would exist $z \in \mathbb{Z}^n \setminus \{0\}$ with

$$\overline{\Omega_\#} \cap (\partial\Omega_Q + qz) \neq \emptyset,$$

a contradiction. Hence,

$$\overline{\Omega_\#} - \partial\Omega_Q \subseteq (\mathbb{R}^n \setminus q\mathbb{Z}^n) \cup \{0\}.$$

Since $R_{q,n}$ is real analytic in $(\mathbb{R}^n \setminus q\mathbb{Z}^n) \cup \{0\}$, the analyticity of the integral operators with analytic kernel of Theorem A.28 (i) (with $\epsilon_1 = \epsilon_2 = 1$) or the continuity result for linear integral operators with analytic kernel of Proposition A.30 implies that the function $v_{R_{q,n},\Omega_Q}[\phi]$ from $\overline{\Omega_\#}$ to $\mathbb{R}$ defined by

$$v_{R_{q,n},\Omega_Q}[\phi](x) \equiv \int_{\partial\Omega_Q} \phi(y) R_{q,n}(x-y)\, d\sigma_y \qquad \forall x \in \overline{\Omega_\#}, \tag{12.14}$$

belongs to $C^\infty(\overline{\Omega_\#})$. Next we note that

$$v_{q,\Omega_Q}[\phi](x) = v_{\Omega_Q}[\phi](x) + v_{R_{q,n},\Omega_Q}[\phi](x) \qquad \forall x \in \overline{\Omega_\#}. \tag{12.15}$$

We now turn to statements (i)–(iii). We first consider statement (i). Since $S_{q,n}$ is q-periodic, $v_{q,\Omega_Q}[\phi]$ is easily seen to be q-periodic. By equality (12.9), we have

$$\Delta S_{q,n} = -\frac{1}{m_n(Q)} \qquad \text{in } \mathbb{R}^n \setminus q\mathbb{Z}^n.$$

We note that if $x \in \mathbb{R}^n \setminus \partial\mathbb{S}[\Omega_Q]$, then $x \notin \partial\mathbb{S}[\Omega_Q] = q\mathbb{Z}^n + \partial\Omega_Q$ and thus

$$x - y \notin q\mathbb{Z}^n \qquad \forall y \in \partial\Omega_Q .$$

Since $S_{q,n}$ is analytic in $\mathbb{R}^n \setminus q\mathbb{Z}^n$, classical differentiation theorems of integrals depending on a parameter imply that

$$\begin{aligned}\Delta v_{q,\Omega_Q}[\phi](x) &= \int_{\partial\Omega_Q} \phi(y)(\Delta S_{q,n})(x-y)\, d\sigma_y \\ &= -\frac{1}{m_n(Q)} \int_{\partial\Omega_Q} \phi(y)\, d\sigma_y ,\end{aligned}$$

and thus equality (12.12) follows.

As is well known, $v_{\Omega_Q}[\phi]$ is continuous in $\mathbb{R}^n$ (cf., e.g., Theorem 4.22 (ii)). Since the function in (12.14) is of class $C^\infty(\overline{\Omega_\#})$, equality (12.15) implies that the periodic function $v_{q,\Omega_Q}[\phi]$ is continuous on $\overline{\Omega_\#}$ and thus, by periodicity, in the whole of $\mathbb{R}^n$.

We now consider (ii), (iii). By Theorem 4.25, we have $v_{\Omega_Q}[\phi]_{|\overline{\Omega_Q}} \in C^{1,\alpha}(\overline{\Omega_Q})$ and $v_{\Omega_Q}[\phi]_{|\overline{\mathbb{B}_n(0,R)}\setminus\Omega_Q} \in C^{1,\alpha}(\overline{\mathbb{B}_n(0,R)}\setminus\Omega_Q)$ for all $R \in]0,+\infty[$ such that $\overline{\Omega_Q} \subseteq \mathbb{B}_n(0,R)$. Since the function defined in (12.14) belongs to the space $C^\infty(\overline{\Omega_\#}) \subseteq C^{1,\alpha}(\overline{\Omega_\#})$, equality (12.15) implies that $v^+_{q,\Omega_Q}[\phi]_{|\overline{\Omega_Q}} \in C^{1,\alpha}(\overline{\Omega_Q})$ and also that $v^-_{q,\Omega_Q}[\phi]_{|\overline{\Omega_\#}\setminus\Omega_Q} \in C^{1,\alpha}(\overline{\Omega_\#}\setminus\Omega_Q)$. Then Lemmas 12.5, 12.6 imply that $v^+_{q,\Omega_Q}[\phi] \in C^{1,\alpha}_q(\overline{\mathbb{S}[\Omega_Q]})$ and that $v^-_{q,\Omega_Q}[\phi] \in C^{1,\alpha}_q(\overline{\mathbb{S}[\Omega_Q]^-})$.

In order to prove the boundedness of the operators in (ii), (iii), we exploit again equality (12.15) and the boundedness of $v_{R_{q,n},\Omega_Q}[\cdot]$. The boundedness Theorem 4.25 for the single layer potential implies that the operators $v^+_{\Omega_Q}[\cdot]_{|\overline{\Omega_Q}}$ and $v^-_{\Omega_Q}[\cdot]_{|\overline{\Omega_\#}\setminus\Omega_Q}$ are bounded from $C^{0,\alpha}(\partial\Omega_Q)$ to $C^{1,\alpha}(\overline{\Omega_Q})$ and to $C^{1,\alpha}(\overline{\Omega_\#}\setminus\Omega_Q)$, respectively. Then the analyticity of the integral operators with analytic kernel of Theorem A.28 (i) (with $\epsilon_1 = \epsilon_2 = 1$) or the continuity result for linear integral operators with analytic kernel of Proposition A.30 implies that the operators from $C^{0,\alpha}(\partial\Omega_Q)$ to $C^{1,\alpha}(\overline{\Omega_Q})$ and to $C^{1,\alpha}(\overline{\Omega_\#}\setminus\Omega_Q)$ that take ϕ to the restriction of the function in (12.14) to $\overline{\Omega_Q}$ and to $\overline{\Omega_\#}\setminus\Omega_Q$ are bounded, respectively. Equality (12.15) implies that the operators $v^+_{q,\Omega_Q}[\cdot]_{|\overline{\Omega_Q}}$ and $v^-_{q,\Omega_Q}[\cdot]_{|\overline{\Omega_\#}\setminus\Omega_Q}$ are bounded from $C^{0,\alpha}(\partial\Omega_Q)$ to $C^{1,\alpha}(\overline{\Omega_Q})$ and to $C^{1,\alpha}(\overline{\Omega_\#}\setminus\Omega_Q)$, respectively. Then Lemmas 12.5, 12.6 imply that the operators in (ii), (iii) are bounded. □

We now introduce the periodic double layer potential.

Definition 12.9. If $\psi \in C^{1,\alpha}(\partial\Omega_Q)$, then $w_{q,\Omega_Q}[\psi]$ denotes the q-periodic double layer potential with density ψ defined by

$$w_{q,\Omega_Q}[\psi](x) \equiv -\int_{\partial\Omega_Q} \psi(y)\, \nu_{\Omega_Q}(y) \cdot \nabla S_{q,n}(x-y)\, d\sigma_y \qquad \forall x \in \mathbb{R}^n .$$

As we have done for the periodic single layer potential, we collect some properties of the periodic double layer potential.

Theorem 12.10. *The following statements hold.*

(i) If $\psi \in C^{1,\alpha}(\partial\Omega_Q)$, then the function $w_{q,\Omega_Q}[\psi]$ is q-periodic and

$$\Delta w_{q,\Omega_Q}[\psi](x) = 0 \tag{12.16}$$

for all $x \in \mathbb{R}^n \setminus \partial\mathbb{S}[\Omega_Q]$.

(ii) If $\psi \in C^{1,\alpha}(\partial\Omega_Q)$, then the function $w_{q,\Omega_Q}[\psi]_{|\mathbb{S}[\Omega_Q]}$ extends to a function $w^+_{q,\Omega_Q}[\psi] \in C^{1,\alpha}_q(\overline{\mathbb{S}[\Omega_Q]})$. Moreover, the linear operator $w^+_{q,\Omega_Q}[\cdot]$ is bounded from $C^{1,\alpha}(\partial\Omega_Q)$ to $C^{1,\alpha}_q(\overline{\mathbb{S}[\Omega_Q]})$.

(iii) If $\psi \in C^{1,\alpha}(\partial\Omega_Q)$, then the function $w_{q,\Omega_Q}[\psi]_{|\mathbb{S}[\Omega_Q]^-}$ extends to a function $w^-_{q,\Omega_Q}[\psi] \in C^{1,\alpha}_q(\overline{\mathbb{S}[\Omega_Q]^-})$. Moreover, the linear operator $w^-_{q,\Omega_Q}[\cdot]$ is bounded from $C^{1,\alpha}(\partial\Omega_Q)$ to $C^{1,\alpha}_q(\overline{\mathbb{S}[\Omega_Q]^-})$.

(iv) If $\psi \in C^{1,\alpha}(\partial\Omega_Q)$, then we have

$$w_{q,\Omega_Q}[\psi](x) = -\sum_{j=1}^{n} \frac{\partial}{\partial x_j} \int_{\partial\Omega_Q} S_{q,n}(x-y)(\nu_{\Omega_Q}(y))_j \psi(y)\, d\sigma_y \tag{12.17}$$

for all $x \in \mathbb{R}^n \setminus \partial\mathbb{S}[\Omega_Q]$.

Proof. We first consider (i). Since $S_{q,n}$ is q-periodic, the function $w_{q,\Omega_Q}[\psi]$ is easily seen to be q-periodic as well. First we note that

$$x - y \notin q\mathbb{Z}^n \qquad \forall (x,y) \in (\mathbb{R}^n \setminus \partial\mathbb{S}[\Omega_Q]) \times \partial\Omega_Q\,.$$

Indeed, if $(x,y) \in (\mathbb{R}^n \setminus \partial\mathbb{S}[\Omega_Q]) \times \partial\Omega_Q$ and $x - y \in q\mathbb{Z}^n$, then we would have $x \in \partial\Omega_Q + q\mathbb{Z}^n = \partial\mathbb{S}[\Omega_Q]$, contrary to our assumption on x. By the equality (12.9), we have

$$\begin{aligned} \Delta(S_{q,n}(x-y)) &= -\frac{1}{|Q|}\,, \\ \Delta(\nabla S_{q,n}(x-y)) &= 0\,, \end{aligned}$$

for all $(x,y) \in (\mathbb{R}^n \setminus \partial\mathbb{S}[\Omega_Q]) \times \partial\Omega_Q$. Then by standard differentiation theorems for integrals depending on a parameter, we have

$$\begin{aligned} &\Delta w_{q,\Omega_Q}[\phi](x) \\ &\quad = -\int_{\partial\Omega_Q} \psi(y)\, \nu_{\Omega_Q}(y) \cdot (\Delta\nabla S_{q,n})(x-y)\, d\sigma_y = 0 \qquad \forall x \in \partial\mathbb{R}^n \setminus \mathbb{S}[\Omega_Q]\,, \end{aligned}$$

and equality (12.16) holds true.

We now consider statement (ii). As in the proof of Theorem 12.8, we take a bounded open connected subset $\Omega_\#$ of class C^∞ of $\mathbb{R}^n$ that satisfies the conditions in (12.13). Since $R_{q,n}$ is real analytic in $(\mathbb{R}^n \setminus q\mathbb{Z}^n) \cup \{0\}$, the analyticity of the integral operators with analytic kernel of Theorem A.28 (i) (with $\epsilon_1 = \epsilon_2 = 1$) or the continuity result for linear integral operators with analytic kernel of Proposition A.30 implies that the map defined by

$$w_{R_{q,n},\Omega_Q}[\psi](x) \equiv -\int_{\partial\Omega_Q} \psi(y)\, \nu_{\Omega_Q}(y) \cdot \nabla R_{q,n}(x-y)\, d\sigma_y \qquad \forall x \in \overline{\Omega_\#}\,,$$

is of class $C^\infty(\overline{\Omega_\#})$. Next we note that

$$w_{q,\Omega_Q}[\psi] = w_{\Omega_Q}[\psi] + w_{R_{q,n},\Omega_Q}[\psi] \qquad \text{in } \overline{\Omega_\#}\,. \tag{12.18}$$

By Theorem 4.31, $w_{\Omega_Q}[\psi]_{|\Omega_Q}$ can be uniquely extended to an element $w^+_{\Omega_Q}[\psi]$ of $C^{m,\alpha}(\overline{\Omega_Q})$ and $w_{\Omega_Q}[\psi]_{|\mathbb{R}^n\setminus\Omega_Q}$ can be uniquely extended to an element $w^-_{\Omega_Q}[\psi]$ of $C^{1,\alpha}_{\mathrm{loc}}(\overline{\Omega_Q^-})$. Since $w_{R_{q,n},\Omega_Q}[\psi] \in C^\infty(\overline{\Omega_\#})$, equality (12.18) implies that the function $w_{q,\Omega_Q}[\psi]_{|\Omega_Q}$ can be extended uniquely to an element of $C^{1,\alpha}(\overline{\Omega_Q})$. Then the periodicity of $w_{q,\Omega_Q}[\psi]$ and Lemma 12.6 imply that $w_{q,\Omega_Q}[\psi]_{|\mathbb{S}[\Omega_Q]}$ admits a unique extension $w^+_{q,\Omega_Q}[\psi]$ in $C^{1,\alpha}_q(\overline{\mathbb{S}[\Omega_Q]})$. Since $w_{R_{q,n},\Omega_Q}[\psi] \in C^\infty(\overline{\Omega_\#})$, equality (12.18) implies that $w_{q,\Omega_Q}[\psi]_{|\overline{\Omega_\#}\setminus\overline{\Omega_Q}}$ can be extended to a unique element of $C^{1,\alpha}(\overline{\Omega_\# \setminus \Omega_Q})$. Then the periodicity of $w_{q,\Omega_Q}[\psi]$ and Lemma 12.5 imply that $w_{q,\Omega_Q}[\psi]$ admits a unique extension $w^-_{q,\Omega_Q}[\psi]$ in $C^{1,\alpha}_q(\overline{\mathbb{S}[\Omega_Q]^-})$. In order to prove the boundedness of the operators in (ii), (iii), we exploit again equality (12.18). The boundedness Theorem 4.31 for the double layer potential implies that the operators $w^+_{\Omega_Q}[\cdot]_{|\overline{\Omega_Q}}$ and $w^-_{\Omega_Q}[\cdot]_{|\overline{\Omega_\#}\setminus\Omega_Q}$ are bounded from $C^{1,\alpha}(\partial\Omega_Q)$ to $C^{1,\alpha}(\overline{\Omega_Q})$ and to $C^{1,\alpha}(\overline{\Omega_\#} \setminus \Omega_Q)$, respectively. Then the analyticity of the integral operators with analytic kernel of Theorem A.28 (i) (with $\epsilon_1 = \epsilon_2 = 1$) or the continuity result for linear integral operators with analytic kernel of Proposition A.30 implies that the operators from $C^{1,\alpha}(\partial\Omega_Q)$ to $C^{1,\alpha}(\overline{\Omega_Q})$ and to $C^{1,\alpha}(\overline{\Omega_\#} \setminus \Omega_Q)$ that take ψ to the restriction of $w_{R_{q,n},\Omega_Q}[\psi]$ to $\overline{\Omega_Q}$ and to $\overline{\Omega_\#} \setminus \Omega_Q$ are bounded, respectively. Equality (12.18) implies that the operators $w^+_{q,\Omega_Q}[\cdot]_{|\overline{\Omega_Q}}$ and $w^-_{q,\Omega_Q}[\cdot]_{|\overline{\Omega_\#}\setminus\Omega_Q}$ are bounded from $C^{1,\alpha}(\partial\Omega_Q)$ to $C^{1,\alpha}(\overline{\Omega_Q})$ and to $C^{1,\alpha}(\overline{\Omega_\#} \setminus \Omega_Q)$, respectively.

Then Lemmas 12.5, 12.6 imply that the operators in (ii), (iii) are bounded.

Statement (iv) is an immediate consequence of the definition of $w_{q,\Omega_Q}[\psi]$, of classical differentiation theorems for integrals depending on a parameter, of the analyticity of $S_{q,n}(x-y)$ for $x \in \overline{\Omega_\#} \setminus \partial\Omega_Q$, $y \in \partial\Omega_Q$, of the q-periodicity of $w_{q,\Omega_Q}[\psi]$, and of the right-hand side of (12.17). □

In order to describe the boundary behavior of the periodic layer potentials, we denote by W_{q,Ω_Q} and W^t_{q,Ω_Q} the integral operators defined by

$$W_{q,\Omega_Q}[\psi](x) \equiv -\int_{\partial\Omega_Q} \psi(y)\, \nu_{\Omega_Q}(y) \cdot \nabla S_{q,n}(x-y)\, d\sigma_y \qquad \forall x \in \partial\Omega_Q\,,$$

for all $\psi \in C^{1,\alpha}(\partial\Omega_Q)$, and

$$W^t_{q,\Omega_Q}[\phi](x) \equiv \int_{\partial\Omega_Q} \phi(y)\, \nu_{\Omega_Q}(x) \cdot \nabla S_{q,n}(x-y)\, d\sigma_y \qquad \forall x \in \partial\Omega_Q\,,$$

for all $\phi \in C^{0,\alpha}(\partial\Omega_Q)$. We note that by following the lines of the proof of Theorem 6.5 on the classical layer potentials, we can verify that the operator W^t_{q,Ω_Q} is transpose to W_{q,Ω_Q} with respect to the dual system $\langle C^{0,\alpha}(\partial\Omega_Q), C^{1,\alpha}(\partial\Omega_Q)\rangle$ (cf. Theorem 12.14 below). By exploiting the operators W_{q,Ω_Q} and W^t_{q,Ω_Q} we can write the jump formulas for the periodic single and double layer potentials.

Theorem 12.11. *Let $\phi \in C^{0,\alpha}(\partial\Omega_Q)$ and $\psi \in C^{1,\alpha}(\partial\Omega_Q)$. Then the following jump formulas hold on $\partial\Omega_Q$.*

$$\begin{aligned}
\nu_{\Omega_Q} \cdot \nabla v^+_{q,\Omega_Q}[\phi] &= -\frac{1}{2}\phi + W^t_{q,\Omega_Q}[\phi]\,,\\
\nu_{\Omega_Q} \cdot \nabla v^-_{q,\Omega_Q}[\phi] &= \frac{1}{2}\phi + W^t_{q,\Omega_Q}[\phi]\,,\\
w^+_{q,\Omega_Q}[\psi]_{|\partial\Omega_Q} &= \frac{1}{2}\psi + W_{q,\Omega_Q}[\psi]\,,\\
w^-_{q,\Omega_Q}[\psi]_{|\partial\Omega_Q} &= -\frac{1}{2}\psi + W_{q,\Omega_Q}[\psi]\,,\\
\nu_{\Omega_Q} \cdot \nabla w^+_{q,\Omega_Q}[\psi]_{|\partial\Omega_Q} &= \nu_{\Omega_Q} \cdot \nabla w^-_{q,\Omega_Q}[\psi]_{|\partial\Omega_Q}\,.
\end{aligned}$$

Proof. The proofs of the five equalities are all similar. Therefore, for the sake of brevity, we prove only the third one and we leave the others to the reader. So let $\Omega_\#$ and $w_{R_{q,n},\Omega_Q}[\psi]$ be as in the proof of Theorem 12.10. Since $w_{R_{q,n},\Omega_Q}[\psi]$ belongs to $C^\infty(\overline{\Omega_\#})$, the equality

$$w^+_{q,\Omega_Q}[\psi]_{|\partial\Omega_Q} = \frac{1}{2}\psi + W_{q,\Omega_Q}[\psi]$$

on $\partial\Omega_Q$ follows immediately by the corresponding classical jump formula for $w_{\Omega_Q}[\psi]$ and by equality

$$w_{q,\Omega_Q}[\psi] = w_{\Omega_Q}[\psi] + w_{R_{q,n},\Omega_Q}[\psi] \qquad \text{in } \overline{\Omega_\#}$$

(cf. Theorem 4.30). □

Proposition 12.12. *The following equalities hold.*

$$w_{q,\Omega_Q}[1](x) = \frac{1}{2} - \frac{m_n(\Omega_Q)}{m_n(Q)} \qquad \forall x \in \partial\mathbb{S}[\Omega_Q]\,, \tag{12.19}$$

$$w_{q,\Omega_Q}[1](x) = 1 - \frac{m_n(\Omega_Q)}{m_n(Q)} \qquad \forall x \in \mathbb{S}[\Omega_Q]\,, \tag{12.20}$$

$$w_{q,\Omega_Q}[1](x) = -\frac{m_n(\Omega_Q)}{m_n(Q)} \qquad \forall x \in \mathbb{S}[\Omega_Q]^-\,. \tag{12.21}$$

Proof. We first prove equality (12.21). By the q-periodicity of $w_{q,\Omega_Q}[1]$, we can assume $x \in \overline{Q} \setminus \overline{\Omega_Q}$. Since $x \in \overline{Q}$, $\overline{\Omega_Q} + qz \subseteq Q + qz$ for all $z \in \mathbb{Z}^n$, and $(Q + qz) \cap \overline{Q} = \emptyset$ for all $z \in \mathbb{Z}^n \setminus \{0\}$, we have

$$(x - q(\mathbb{Z}^n \setminus \{0\})) \cap \overline{\Omega_Q} = \emptyset .$$

Since $x \notin \overline{\Omega_Q}$, we have

$$(x - q0) \notin \overline{\Omega_Q} .$$

Hence, $(x - q\mathbb{Z}^n) \cap \overline{\Omega_Q} = \emptyset$ and $S_{q,n}(x - \cdot) \in C^\infty(\overline{\Omega_Q})$. By equality (12.9), we have $\Delta S_{q,n} = -\frac{1}{m_n(Q)}$ in $\mathbb{R}^n \setminus q\mathbb{Z}^n$. Then the first Green Identity of Theorem 4.2 implies that

$$-\int_{\partial\Omega_Q} \nu_{\Omega_Q}(y) \cdot \nabla S_{q,n}(x - y)\, d\sigma_y = \int_{\Omega_Q} \Delta_y(S_{q,n}(x - y))\, dy = -\frac{m_n(\Omega_Q)}{m_n(Q)} ,$$

and accordingly (12.21) holds true. By the jump formula for $w^-_{q,\Omega_Q}[1]$ of Theorem 12.11, we have

$$-\frac{m_n(\Omega_Q)}{m_n(Q)} = w^-_{q,\Omega_Q}[1] = -\frac{1}{2} + w_{q,\Omega_Q}[1] \qquad \text{on } \partial\mathbb{S}[\Omega_Q] ,$$

and thus equality (12.19) follows. By the jump formula for $w^+_{q,\Omega_Q}[1]$ of Theorem 12.11 and by equality (12.19), we have

$$w^+_{q,\Omega_Q}[1] = \frac{1}{2} + w_{q,\Omega_Q}[1] = \frac{1}{2} + \left(\frac{1}{2} - \frac{m_n(\Omega_Q)}{m_n(Q)}\right) = 1 - \frac{m_n(\Omega_Q)}{m_n(Q)} \qquad \text{on } \partial\mathbb{S}[\Omega_Q] .$$

By Theorem 12.10, $w^+_{q,\Omega_Q}[1]$ is harmonic in Ω_Q. Since $w^+_{q,\Omega_Q}[1]$ equals the constant $1 - \frac{m_n(\Omega_Q)}{m_n(Q)}$ on $\partial\Omega_Q$, the uniqueness for the Dirichlet problem implies that $w^+_{q,\Omega_Q}[1] = 1 - \frac{m_n(\Omega_Q)}{m_n(Q)}$ on $\overline{\Omega_Q}$ (cf. Theorem 3.6). Since $w^+_{q,\Omega_Q}[1]$ is q-periodic, equality (12.20) follows. □

By the identity $S_{q,n} = S_n + R_{q,n}$ and the corresponding results for the classical layer potentials, we deduce the validity of the following Theorems 12.13 and 12.14 (cf. Theorems 6.5 and 6.8).

Theorem 12.13. *The following statements hold.*

(i) The operator W_{q,Ω_Q} is compact from $C^{1,\alpha}(\partial\Omega_Q)$ to itself.
(ii) The operator W^t_{q,Ω_Q} is compact from $C^{0,\alpha}(\partial\Omega_Q)$ to itself.

Proof. We prove only statement (i). The proof of (ii) is very similar and it is left to the reader. As in the proof of Theorem 12.8, we take a bounded open connected subset $\Omega_\#$ of class C^∞ of $\mathbb{R}^n$ such that

$$\overline{Q} \subseteq \Omega_\#, \qquad \overline{\Omega_\#} \cap (qz + \overline{\Omega_Q}) = \emptyset \quad \forall z \in \mathbb{Z}^n \setminus \{0\} .$$

We first note that if we set

$$w_{R_{q,n},\Omega_Q}[\psi](x) \equiv -\int_{\partial\Omega_Q} \psi(y)\,\nu_{\Omega_Q}(y)\cdot\nabla R_{q,n}(x-y)\,d\sigma_y \qquad \forall x \in \overline{\Omega_\#}\,,$$

for all $\psi \in C^{1,\alpha}(\partial\Omega_Q)$, then we have

$$W_{q,\Omega_Q}[\psi](x) = W_{\Omega_Q}[\psi](x) + w_{R_{q,n},\Omega_Q}[\psi](x) \qquad \forall x \in \partial\Omega_Q\,, \tag{12.22}$$

for all $\psi \in C^{1,\alpha}(\partial\Omega_Q)$. Since $R_{q,n}$ is real analytic in $(\mathbb{R}^n \setminus q\mathbb{Z}^n) \cup \{0\}$, the analyticity of the integral operators with analytic kernel of Theorem A.28 (i) (with $\epsilon_1 = \epsilon_2 = 1$) or the continuity result for linear integral operators with analytic kernel of Proposition A.30 implies that $w_{R_{q,n},\Omega_Q}[\cdot]$ defines a bounded linear operator from $C^{1,\alpha}(\partial\Omega_Q)$ to $C^2(\overline{\Omega_\#})$. Then by the compactness of the embedding of $C^2(\overline{\Omega_\#})$ into $C^{1,\alpha}(\overline{\Omega_\#})$ (cf. Corollary 2.57 (iii)) and the continuity of the restriction operator from $C^{1,\alpha}(\overline{\Omega_\#})$ to $C^{1,\alpha}(\partial\Omega_Q)$ we deduce that the map

$$\psi \mapsto w_{R_{q,n},\Omega_Q}[\psi]_{|\partial\Omega_Q}$$

is compact from $C^{1,\alpha}(\partial\Omega_Q)$ to itself. Then, by the compactness of W_{Ω_Q} from $C^{1,\alpha}(\partial\Omega_Q)$ to itself (cf. Theorem 6.8 (i)) and equality (12.22), we deduce the validity of (i). □

Theorem 12.14. *The operator W^t_{q,Ω_Q} is transpose to W_{q,Ω_Q} with respect to the dual system $\langle C^{0,\alpha}(\partial\Omega_Q), C^{1,\alpha}(\partial\Omega_Q)\rangle$.*

Proof. We proceed as in the proof of Theorem 6.5 concerning the corresponding statement for classical layer potentials. By Theorem 12.13, we already know that the linear operator W_{q,Ω_Q} is bounded from $C^{1,\alpha}(\partial\Omega_Q)$ to itself and that the linear operator W^t_{q,Ω_Q} is bounded from $C^{0,\alpha}(\partial\Omega_Q)$ to itself. We first note that by Lemma 2.76 (ii), by the identity

$$S_{q,n} = S_n + R_{q,n}$$

and by the analyticity of $R_{q,n}$ in $(\mathbb{R}^n \setminus q\mathbb{Z}^n) \cup \{0\}$, there exists $c_{q,\Omega_Q,\alpha} > 0$ such that

$$\begin{aligned} &|\phi(x)\psi(y)\nu_{\Omega_Q}(x)\cdot\nabla S_{q,n}(x-y)| \\ &\qquad \le \frac{c_{q,\Omega_Q,\alpha}\sup_{\partial\Omega_Q}|\phi|\,\sup_{\partial\Omega_Q}|\psi|}{|x-y|^{(n-1)-\alpha}} \quad \forall (x,y) \in (\partial\Omega_Q)^2,\ x \ne y \end{aligned}$$

(see also the proof of Theorem 6.5). Then the inequality $(n-1)-\alpha < (n-1)$, the Fubini-Tonelli Theorem, and Theorem 12.4 (iv) on the parity of $S_{q,n}$ imply that

$$\langle \phi, W_{q,\Omega_Q}[\psi] \rangle = -\int_{\partial\Omega_Q} \phi(x) \int_{\partial\Omega_Q} \nu_{\Omega_Q}(y) \cdot \nabla S_{q,n}(x-y)\psi(y)\, d\sigma_y\, d\sigma_x$$
$$= \int_{\partial\Omega_Q} \psi(y) \int_{\partial\Omega_Q} \nu_{\Omega}(y) \cdot \nabla S_{q,n}(y-x)\phi(x)\, d\sigma_x\, d\sigma_y = \langle W^t_{q,\Omega_Q}[\phi], \psi \rangle\,,$$

for all $(\phi, \psi) \in C^{0,\alpha}(\partial\Omega_Q) \times C^{1,\alpha}(\partial\Omega_Q)$. Hence, the operator W^t_{q,Ω_Q} is transpose to W_{q,Ω_Q} as in the statement. □

Then we deduce the validity of the following.

Proposition 12.15. *Let $\phi \in C^{0,\alpha}(\partial\Omega_Q)$. Then the following equalities hold:*

$$\int_{\partial\Omega_Q} W^t_{q,\Omega_Q}[\phi]\, d\sigma = \left(\frac{1}{2} - \frac{m_n(\Omega_Q)}{m_n(Q)}\right) \int_{\partial\Omega_Q} \phi\, d\sigma\,, \tag{12.23}$$

$$\int_{\partial\Omega_Q} \left(-\frac{1}{2}I + W^t_{q,\Omega_Q}\right)[\phi]\, d\sigma = -\frac{m_n(\Omega_Q)}{m_n(Q)} \int_{\partial\Omega_Q} \phi\, d\sigma\,, \tag{12.24}$$

$$\int_{\partial\Omega_Q} \left(\frac{1}{2}I + W^t_{q,\Omega_Q}\right)[\phi]\, d\sigma = \left(1 - \frac{m_n(\Omega_Q)}{m_n(Q)}\right) \int_{\partial\Omega_Q} \phi\, d\sigma. \tag{12.25}$$

In particular, each of the two conditions

$$\int_{\partial\Omega_Q} \left(\frac{1}{2}I + W^t_{q,\Omega_Q}\right)[\phi]\, d\sigma = 0\,,$$

and

$$\int_{\partial\Omega_Q} \left(-\frac{1}{2}I + W^t_{q,\Omega_Q}\right)[\phi]\, d\sigma = 0\,,$$

is equivalent to the equality

$$\int_{\partial\Omega_Q} \phi\, d\sigma = 0\,.$$

Proof. We have

$$\int_{\partial\Omega_Q} W^t_{q,\Omega_Q}[\phi]\, d\sigma = \int_{\partial\Omega_Q} \phi\, W_{q,\Omega_Q}[1]\, d\sigma$$
$$= \int_{\partial\Omega_Q} \phi \left(\frac{1}{2} - \frac{m_n(\Omega_Q)}{m_n(Q)}\right) d\sigma\,,$$

and thus the validity of (12.23) follows. Equalities (12.24) and (12.25) are then immediate consequence of (12.23). Finally, the last part of the statement follows by the following inequalities

$$-\frac{m_n(\Omega_Q)}{m_n(Q)} \neq 0\,, \qquad \left(1 - \frac{m_n(\Omega_Q)}{m_n(Q)}\right) \neq 0\,.$$

□

12.3 Uniqueness Results for Periodic Boundary Value Problems

In this section, we start our analysis of periodic boundary value problems for real-valued periodic functions in the domains $\mathbb{S}[\Omega_Q]^-$ and $\mathbb{S}[\Omega_Q]$. We retain the geometric setting of Section 12.2.1. In particular, we recall that we are assuming that Ω_Q and $\mathbb{R}^n \setminus \overline{\Omega_Q}$ are connected. As a consequence, in particular, the set $\mathbb{S}[\Omega_Q]^-$ is also connected.

Theorem 12.16. *The following statements hold.*

(i) Let $u \in C^{1,\alpha}_q(\overline{\mathbb{S}[\Omega_Q]^-})$ satisfy the following q-periodic Dirichlet boundary value problem

$$\begin{cases} \Delta u = 0 & \text{in } \mathbb{S}[\Omega_Q]^- , \\ u(x+qz) = u(x) & \forall x \in \overline{\mathbb{S}[\Omega_Q]^-} , z \in \mathbb{Z}^n , \\ u = 0 & \text{on } \partial\Omega_Q . \end{cases} \tag{12.26}$$

Then

$$u = 0 \qquad \text{in } \overline{\mathbb{S}[\Omega_Q]^-} .$$

(ii) Let $u \in C^{1,\alpha}_q(\overline{\mathbb{S}[\Omega_Q]^-})$ satisfy the following q-periodic Neumann boundary value problem

$$\begin{cases} \Delta u = 0 & \text{in } \mathbb{S}[\Omega_Q]^- , \\ u(x+qz) = u(x) & \forall x \in \overline{\mathbb{S}[\Omega_Q]^-} , z \in \mathbb{Z}^n , \\ \nu_{\Omega_Q} \cdot \nabla u = 0 & \text{on } \partial\Omega_Q . \end{cases} \tag{12.27}$$

Then there exists a constant $c \in \mathbb{R}$ such that

$$u = c \qquad \text{in } \overline{\mathbb{S}[\Omega_Q]^-} .$$

Proof. Since a continuous q-periodic function in a q-periodic domain has both a maximum and a minimum, one could prove statement (i) by applying the Maximum Principle and statement (ii) by applying the Hopf-Oleĭnik Lemma 3.15.

Here, instead, we provide an energy argument based on the first Green Identity of Theorem 4.2 which has the virtue of implying the validity of both statements.

So let $u \in C^{1,\alpha}_q(\overline{\mathbb{S}[\Omega_Q]^-})$ be harmonic in $\mathbb{S}[\Omega_Q]^-$. Then the first Green identity implies that

$$\begin{aligned} \int_{Q\setminus\overline{\Omega_Q}} |\nabla u|^2 \, dx &= \int_{\partial(Q\setminus\overline{\Omega_Q})} u \, \nu_{Q\setminus\overline{\Omega_Q}} \cdot \nabla u \, d\sigma \\ &= \int_{\partial Q} u \, \nu_Q \cdot \nabla u \, d\sigma - \int_{\partial\Omega_Q} u \, \nu_{\Omega_Q} \cdot \nabla u \, d\sigma . \end{aligned}$$

We now consider $\int_{\partial Q} u \, \nu_Q \cdot \nabla u \, d\sigma$ and we note that by the periodicity of u (and of ∇u) the integrals of $u \, \nu_Q \cdot \nabla u$ on opposite sides of the unit cell Q cancel each other. Therefore, we have that

$$\int_{\partial Q} u\, \nu_Q \cdot \nabla u \, d\sigma = 0 \,,$$

and accordingly

$$\int_{Q \setminus \overline{\Omega_Q}} |\nabla u|^2 \, dx = - \int_{\partial \Omega_Q} u\, \nu_{\Omega_Q} \cdot \nabla u \, d\sigma \,.$$

Hence, both in the case where u solves problem (12.26) and in the case where u solves problem (12.27), we conclude that $\int_{Q \setminus \overline{\Omega_Q}} |\nabla u|^2 \, dx = 0$ and that accordingly u is constant. In case of the Dirichlet problem (12.26) such constant must necessarily be equal to zero. □

Then we have a corresponding results for periodic boundary value problems in the periodic domain $\mathbb{S}[\Omega_Q]$, which is immediately deduced by the uniqueness results for the same problems in the bounded connected domain Ω_Q and by the periodicity of the solutions.

Theorem 12.17. *The following statements hold.*

(i) Let $u \in C^{1,\alpha}_q(\overline{\mathbb{S}[\Omega_Q]})$ satisfy the following q-periodic Dirichlet boundary value problem

$$\begin{cases} \Delta u = 0 & \text{in } \mathbb{S}[\Omega_Q] \,, \\ u(x+qz) = u(x) & \forall x \in \overline{\mathbb{S}[\Omega_Q]} \,, z \in \mathbb{Z}^n \,, \\ u = 0 & \text{on } \partial\Omega_Q \,. \end{cases} \tag{12.28}$$

Then

$$u = 0 \qquad \text{in } \overline{\mathbb{S}[\Omega_Q]} \,.$$

(ii) Let $u \in C^{1,\alpha}_q(\overline{\mathbb{S}[\Omega_Q]})$ satisfy the following q-periodic Neumann boundary value problem

$$\begin{cases} \Delta u = 0 & \text{in } \mathbb{S}[\Omega_Q] \,, \\ u(x+qz) = u(x) & \forall x \in \overline{\mathbb{S}[\Omega_Q]} \,, z \in \mathbb{Z}^n \,, \\ \nu_{\Omega_Q} \cdot \nabla u = 0 & \text{on } \partial\Omega_Q \,. \end{cases} \tag{12.29}$$

Then there exists a constant $c \in \mathbb{R}$ such that

$$u = c \qquad \text{in } \overline{\mathbb{S}[\Omega_Q]} \,.$$

In Section 12.5, we shall prove existence results for periodic boundary value problems. Before doing that, we show that the integral on $\partial\Omega_Q$ of the normal derivative of a periodic harmonic function in $\mathbb{S}[\Omega_Q]$ or in $\mathbb{S}[\Omega_Q]^-$ is zero.

Proposition 12.18. *The following statements hold.*

(i) Let $u \in C^{1,\alpha}_q(\overline{\mathbb{S}[\Omega_Q]})$ *be harmonic, then*

$$\int_{\partial\Omega_Q} \nu_{\Omega_Q} \cdot \nabla u \, d\sigma = 0 \, .$$

(ii) Let $u \in C^{1,\alpha}_q(\overline{\mathbb{S}[\Omega_Q]^-})$ *be harmonic, then*

$$\int_{\partial\Omega_Q} \nu_{\Omega_Q} \cdot \nabla u \, d\sigma = 0 \, .$$

Proof. We first consider (i). Since u is harmonic in Ω_Q, the first Green Identity of Theorem 4.2 implies that

$$0 = \int_{\Omega_Q} \Delta u \, dx = \int_{\partial\Omega_Q} \nu_{\Omega_Q} \cdot \nabla u \, d\sigma \, .$$

Thus (i) follows. We now consider statement (ii). Since u is harmonic in $Q \setminus \overline{\Omega_Q}$, the first Green Identity of Theorem 4.2 implies that

$$\begin{aligned} 0 = \int_{Q\setminus\overline{\Omega_Q}} \Delta u \, dx &= \int_{\partial(Q\setminus\overline{\Omega_Q})} \nu_{Q\setminus\overline{\Omega_Q}} \cdot \nabla u \, d\sigma \\ &= \int_{\partial Q} \nu_Q \cdot \nabla u \, d\sigma - \int_{\partial\Omega_Q} \nu_{\Omega_Q} \cdot \nabla u \, d\sigma \, . \end{aligned}$$

Then the periodicity of u implies that ∇u is q-periodic as well and accordingly

$$\int_{\partial Q} \nu_Q \cdot \nabla u \, d\sigma = 0 \, .$$

Thus the validity of statement (ii) follows. □

12.4 Mapping Properties of $\pm\frac{1}{2}I + W_{q,\Omega_Q}$ and $\pm\frac{1}{2}I + W^t_{q,\Omega_Q}$

We begin with the following technical statement.

Theorem 12.19. *The following statements hold.*

(i) The operator $-\frac{1}{2}I + W_{q,\Omega_Q}$ *is a linear homeomorphism from* $C^{1,\alpha}(\partial\Omega_Q)$ *onto itself.*
(ii) The operator $-\frac{1}{2}I + W^t_{q,\Omega_Q}$ *is a linear homeomorphism from* $C^{0,\alpha}(\partial\Omega_Q)$ *onto itself.*

Proof. By the Fredholm Alternative Theorem 5.8 and by the Open Mapping Theorem 2.2, both in case (i) and (ii), it suffices to show that the operator $-\frac{1}{2}I + W_{q,\Omega_Q}$ is injective. So let $\psi \in C^{1,\alpha}(\partial\Omega_Q)$ be such that $-\frac{1}{2}\psi + W_{q,\Omega_Q}[\psi] = 0$. Then by Theorems 12.10, 12.11, $w^-_{q,\Omega_Q}[\psi]$ is a solution of the homogeneous periodic Dirichlet problem (12.26) in $\mathbb{S}[\Omega_Q]^-$. Then the uniqueness Theorem 12.16 implies that $w^-_{q,\Omega_Q}[\psi] = 0$ in $\overline{\mathbb{S}[\Omega_Q]^-}$. Hence, the jump formulas of Theorem 12.11 imply that

$$\nu_{\Omega_Q} \cdot \nabla w^+_{q,\Omega_Q}[\psi]_{|\partial\Omega_Q} = \nu_{\Omega_Q} \cdot \nabla w^-_{q,\Omega_Q}[\psi]_{|\partial\Omega_Q} = 0 .$$

Then, $w^+_{q,\Omega_Q}[\psi]$ is a solution of the homogeneous periodic Neumann problem (12.29) in $\mathbb{S}[\Omega_Q]$ and thus Theorem 12.17 ensures the existence of a constant $c \in \mathbb{R}$ such that $w^+_{q,\Omega_Q}[\psi] = c$ in $\overline{\mathbb{S}[\Omega_Q]}$. Then the jump properties of the trace of the periodic double layer potential of Theorem 12.11 imply that

$$\psi = w^+_{q,\Omega_Q}[\psi]_{|\partial\Omega_Q} - w^-_{q,\Omega_Q}[\psi]_{|\partial\Omega_Q} = c - 0 = c .$$

On the other hand,

$$0 = w^-_{q,\Omega_Q}[\psi] = w^-_{q,\Omega_Q}[c] = -\frac{m_n(\Omega_Q)}{m_n(Q)} c \qquad \text{in } \overline{\mathbb{S}[\Omega_Q]^-} ,$$

and thus $c = 0$. Hence, $\psi = 0$ and the injectivity of $-\frac{1}{2}I + W_{q,\Omega_Q}$ follows. □

Theorem 12.20. *The following statements hold.*

(i) The operator $\frac{1}{2}I + W_{q,\Omega_Q}$ is a linear homeomorphism from $C^{1,\alpha}(\partial\Omega_Q)$ onto itself.

(ii) The operator $\frac{1}{2}I + W^t_{q,\Omega_Q}$ is a linear homeomorphism from $C^{0,\alpha}(\partial\Omega_Q)$ onto itself.

Proof. We proceed as in the proof of Theorem 12.19. By the Fredholm Alternative Theorem 5.8 and by the Open Mapping Theorem 2.2, both in case (i) and (ii), it suffices to show that the operator $\frac{1}{2}I + W_{q,\Omega_Q}$ is injective. So let $\psi \in C^{1,\alpha}(\partial\Omega_Q)$ be such that $\frac{1}{2}\psi + W_{q,\Omega_Q}[\psi] = 0$. Then by Theorems 12.10, 12.11, $w^+_{q,\Omega_Q}[\psi]$ is a solution of the homogeneous periodic Dirichlet problem (12.28) in $\mathbb{S}[\Omega_Q]$.

Then the uniqueness Theorem 12.17 implies that $w^+_{q,\Omega_Q}[\psi] = 0$ in $\overline{\mathbb{S}[\Omega_Q]}$. Hence, the jump formulas of Theorem 12.11 imply that

$$\nu_{\Omega_Q} \cdot \nabla w^-_{q,\Omega_Q}[\psi]_{|\partial\Omega_Q} = \nu_{\Omega_Q} \cdot \nabla w^+_{q,\Omega_Q}[\psi]_{|\partial\Omega_Q} = 0 .$$

Then, $w^-_{q,\Omega_Q}[\psi]$ is a solution of a homogeneous periodic Neumann problem in $\mathbb{S}[\Omega_Q]^-$ and thus Theorem 12.16 ensures the existence of a constant $c \in \mathbb{R}$ such that $w^-_{q,\Omega_Q}[\psi] = c$ in $\overline{\mathbb{S}[\Omega_Q]^-}$. Then the jump properties of the trace of the periodic double layer potential of Theorem 12.11 imply that

$$\psi = w^+_{q,\Omega_Q}[\psi]_{|\partial\Omega_Q} - w^-_{q,\Omega_Q}[\psi]_{|\partial\Omega_Q} = 0 - c = -c .$$

On the other hand,

$$0 = w^+_{q,\Omega_Q}[\psi] = -w^+_{q,\Omega_Q}[c] = -\left(1 - \frac{m_n(\Omega_Q)}{m_n(Q)}\right)c \qquad \text{in } \overline{\mathbb{S}[\Omega_Q]},$$

and thus $c = 0$. Hence, $\psi = 0$ and the injectivity of $\frac{1}{2}I + W_{q,\Omega_Q}$ follows. □

Taking into account the value of $w_{q,\Omega_Q}[1]$, which has been computed in Proposition 12.12, and by the isomorphism Theorems 12.19, 12.20 for $\pm\frac{1}{2}I + W_{q,\Omega_Q}$, we deduce the validity of the following proposition.

Proposition 12.21. *The following statements hold.*

(i) The map from $C^{1,\alpha}(\partial\Omega_Q)_0 \times \mathbb{R}$ to $C^{1,\alpha}(\partial\Omega_Q)$ that takes a pair (ψ, c) to $\frac{1}{2}\psi + W_{q,\Omega_Q}[\psi] + c$ is a linear homeomorphism.
(ii) The map from $C^{1,\alpha}(\partial\Omega_Q)_0 \times \mathbb{R}$ to $C^{1,\alpha}(\partial\Omega_Q)$ that takes a pair (ψ, c) to $-\frac{1}{2}\psi + W_{q,\Omega_Q}[\psi] + c$ is a linear homeomorphism.

Proof. By Proposition 12.12 both $\frac{1}{2}I+W_{q,\Omega_Q}$ and $-\frac{1}{2}I+W_{q,\Omega_Q}$ map constant functions to constant functions. Since $C^{1,\alpha}(\partial\Omega_Q)$ is the direct sum of $C^{1,\alpha}(\partial\Omega_Q)_0$ and of the constant functions on $\partial\Omega_Q$, the isomorphism Theorems 12.19 (i) and 12.20 (i), imply the validity of statements (i) and (ii). □

Taking into account the value of the integral of W^t_{q,Ω_Q} on $\partial\Omega_Q$, which has been computed in Proposition 12.15, and by the isomorphism Theorems 12.19, 12.20 for $\pm\frac{1}{2}I + W^t_{q,\Omega_Q}$, we deduce the validity of the following proposition.

Proposition 12.22. *The following statements hold.*

(i) The restriction of the map $-\frac{1}{2}I + W^t_{q,\Omega_Q}$ to $C^{0,\alpha}(\partial\Omega_Q)_0$ is a linear homeomorphism from $C^{0,\alpha}(\partial\Omega_Q)_0$ onto itself.
(ii) The restriction of the map $\frac{1}{2}I + W^t_{q,\Omega_Q}$ to $C^{0,\alpha}(\partial\Omega_Q)_0$ is a linear homeomorphism from $C^{0,\alpha}(\partial\Omega_Q)_0$ onto itself.

Proof. By Proposition 12.15, $\left(-\frac{1}{2}\phi + W^t_{q,\Omega_Q}[\phi]\right)$ belongs to $C^{0,\alpha}(\partial\Omega_Q)_0$ if and only if $\phi \in C^{0,\alpha}(\partial\Omega_Q)_0$. Then Theorem 12.19 (ii) implies the validity of (i). The proof of (ii) is similar and we leave it to the reader. □

12.5 Existence Results for Periodic Boundary Value Problems

In this section, we continue our analysis of boundary value problems for real-valued periodic functions in the domains $\mathbb{S}[\Omega_Q]^-$ and $\mathbb{S}[\Omega_Q]$ (cf. the geometric assumption of Section 12.2.1).

Theorem 12.23. *The following statements hold.*

(i) Let $g \in C^{1,\alpha}(\partial\Omega_Q)$. Then there exists a unique function $u \in C^{1,\alpha}_q(\overline{\mathbb{S}[\Omega_Q]^-})$ that satisfies the periodic Dirichlet boundary value problem

$$\begin{cases} \Delta u = 0 & \text{in } \mathbb{S}[\Omega_Q]^-\,, \\ u(x+qz) = u(x) & \forall x \in \overline{\mathbb{S}[\Omega_Q]^-}\,,\ z \in \mathbb{Z}^n\,, \\ u = g & \text{on } \partial\Omega_Q\,. \end{cases} \tag{12.30}$$

Moreover,

$$u = w^-_{q,\Omega_Q}[\psi] \qquad \text{in } \overline{\mathbb{S}[\Omega_Q]^-}\,, \tag{12.31}$$

where ψ is the unique function in $C^{1,\alpha}(\partial\Omega_Q)$ such that

$$g = \Big(-\frac{1}{2}I + W_{q,\Omega_Q}\Big)[\psi]\,. \tag{12.32}$$

(ii) Let $g \in C^{0,\alpha}(\partial\Omega_Q)_0$. Then the set of the functions $u \in C^{1,\alpha}_q(\overline{\mathbb{S}[\Omega_Q]^-})$ that satisfy the periodic Neumann boundary value problem

$$\begin{cases} \Delta u = 0 & \text{in } \mathbb{S}[\Omega_Q]^-\,, \\ u(x+qz) = u(x) & \forall x \in \overline{\mathbb{S}[\Omega_Q]^-}\,,\ z \in \mathbb{Z}^n\,, \\ \nu_{\Omega_Q} \cdot \nabla u = g & \text{on } \partial\Omega_Q\,, \end{cases} \tag{12.33}$$

is delivered by

$$\Big\{ v^-_{q,\Omega_Q}[\phi] + c \colon c \in \mathbb{R} \Big\}\,, \tag{12.34}$$

where ϕ is the unique function in $C^{0,\alpha}(\partial\Omega_Q)_0$ such that

$$g = \Big(\frac{1}{2}I + W^t_{q,\Omega_Q}\Big)[\phi]\,. \tag{12.35}$$

Proof. We first consider statement (i) and we note that, by the uniqueness Theorem 12.16 (i), problem (12.30) has at most one solution. By the isomorphism Theorem 12.19, there exists a unique function $\psi \in C^{1,\alpha}(\partial\Omega_Q)$ such that (12.32) holds. Then by the properties of the periodic double layer potential (cf. Theorem 12.10), the function in the right-hand side of (12.31) belongs to $C^{1,\alpha}_q(\overline{\mathbb{S}[\Omega_Q]^-})$, is harmonic in $\mathbb{S}[\Omega_Q]^-$, and satisfies the Dirichlet boundary condition in (12.30).

We now turn to the proof of (ii) and we note that, by the uniqueness Theorem 12.16 (ii), it suffices to show the existence of one solution in $C^{1,\alpha}_q(\overline{\mathbb{S}[\Omega_Q]^-})$ of problem (12.33), which is unique up to additive constants. By the membership of g in $C^{0,\alpha}(\partial\Omega_Q)_0$, Proposition 12.22 implies that there exists a unique function $\phi \in C^{0,\alpha}(\partial\Omega_Q)_0$ such that equation (12.35) holds. Then by the properties of the periodic single layer potential (cf. Theorem 12.8), the function $v^-_{q,\Omega_Q}[\phi]$ belongs to $C^{1,\alpha}_q(\overline{\mathbb{S}[\Omega_Q]^-})$ and satisfies the Neumann boundary condition in (12.33). Moreover, we have

$$\Delta v^-_{q,\Omega_Q}[\phi](x) = -\frac{1}{m_n(Q)} \int_{\partial\Omega_Q} \phi \, d\sigma = 0$$

for all $x \in \mathbb{S}[\Omega_Q]^-$ (cf. (12.12)). Indeed, $\int_{\partial\Omega_Q} \phi \, d\sigma = 0$. Thus $v^-_{q,\Omega_Q}[\phi]$ solves problem (12.33). Hence, by Theorem 12.16, the set of functions in (12.34) coincides with the set of the solutions of problem (12.33). □

By replacing Theorem 12.19 (i) with the representation Proposition 12.21 (ii) of a function in $C^{1,\alpha}(\partial\Omega_Q)$ as a sum of the trace of a periodic double layer and a constant in the proof of Theorem 12.23 (i), we can deduce an alternative representation formula for the solution of the Dirichlet problem in $\mathbb{S}[\Omega_Q]^-$.

Proposition 12.24. *Let $g \in C^{1,\alpha}(\partial\Omega_Q)$. Then there exists a unique function u in $C^{1,\alpha}_q(\overline{\mathbb{S}[\Omega_Q]^-})$ that satisfies the periodic Dirichlet boundary value problem*

$$\begin{cases} \Delta u = 0 & \text{in } \mathbb{S}[\Omega_Q]^- , \\ u(x+qz) = u(x) & \forall x \in \overline{\mathbb{S}[\Omega_Q]^-} , z \in \mathbb{Z}^n , \\ u = g & \text{on } \partial\Omega_Q . \end{cases}$$

Moreover,

$$u = w^-_{q,\Omega_Q}[\psi] + c \qquad \text{in } \overline{\mathbb{S}[\Omega_Q]^-} ,$$

where (ψ, c) is the unique pair in $C^{1,\alpha}(\partial\Omega_Q)_0 \times \mathbb{R}$ such that

$$g = \Big(-\frac{1}{2} I + W_{q,\Omega_Q} \Big)[\psi] + c .$$

We now turn to periodic boundary value problems in $\mathbb{S}[\Omega_Q]$.

Theorem 12.25. *The following statements hold.*

(i) Let $g \in C^{1,\alpha}(\partial\Omega_Q)$. Then there exists a unique function $u \in C^{1,\alpha}_q(\overline{\mathbb{S}[\Omega_Q]})$ that satisfies the periodic Dirichlet boundary value problem

$$\begin{cases} \Delta u = 0 & \text{in } \mathbb{S}[\Omega_Q] , \\ u(x+qz) = u(x) & \forall x \in \overline{\mathbb{S}[\Omega_Q]} , z \in \mathbb{Z}^n , \\ u = g & \text{on } \partial\Omega_Q . \end{cases} \tag{12.36}$$

Moreover,

$$u = w^+_{q,\Omega_Q}[\psi] \qquad \text{in } \overline{\mathbb{S}[\Omega_Q]} , \tag{12.37}$$

where ψ is the unique function in $C^{1,\alpha}(\partial\Omega_Q)$ such that

$$g = \Big(\frac{1}{2} I + W_{q,\Omega_Q} \Big)[\psi] . \tag{12.38}$$

(ii) Let $g \in C^{0,\alpha}(\partial\Omega_Q)_0$. Then the set of the functions $u \in C^{1,\alpha}_q(\overline{\mathbb{S}[\Omega_Q]})$ that satisfy the periodic Neumann boundary value problem

$$\begin{cases} \Delta u = 0 & \text{in } \mathbb{S}[\Omega_Q]\,, \\ u(x+qz) = u(x) & \forall x \in \overline{\mathbb{S}[\Omega_Q]}\,, z \in \mathbb{Z}^n\,, \\ \nu_{\Omega_Q} \cdot \nabla u = g & \text{on } \partial\Omega_Q\,, \end{cases} \tag{12.39}$$

is delivered by

$$\left\{ v^+_{q,\Omega_Q}[\phi] + c \colon c \in \mathbb{R} \right\}, \tag{12.40}$$

where ϕ is the unique function in $C^{0,\alpha}(\partial\Omega_Q)_0$ such that

$$g = \Big(-\frac{1}{2} I + W^t_{q,\Omega_Q} \Big)[\phi]\,. \tag{12.41}$$

Proof. We first consider statement (i) and we note that, by Theorem 12.17 (i), problem (12.36) has at most one solution. By Theorem 12.20 (i) there exists a unique function $\psi \in C^{1,\alpha}(\partial\Omega_Q)$ such that (12.38) holds. Then by the properties of the periodic double layer potential (cf. Theorem 12.10), the function in the right-hand side of (12.37) belongs to $C^{1,\alpha}_q(\overline{\mathbb{S}[\Omega_Q]})$, is harmonic in $\mathbb{S}[\Omega_Q]$ and satisfies the Dirichlet boundary condition in (12.36).

We now turn to the proof of (ii) and we note that, by Theorem 12.17 (ii), it suffices to show the existence of one solution in $C^{1,\alpha}_q(\overline{\mathbb{S}[\Omega_Q]})$ of problem (12.39), which is unique up to additive constants. By the membership of g in $C^{0,\alpha}(\partial\Omega_Q)_0$, Proposition 12.22 (i) implies that there exists a unique function $\phi \in C^{0,\alpha}(\partial\Omega_Q)_0$ such that equation (12.41) holds. Then by the properties of the periodic single layer potential (cf. Theorem 12.8), the function $v^+_{q,\Omega_Q}[\phi]$ belongs to $C^{1,\alpha}_q(\overline{\mathbb{S}[\Omega_Q]})$ and satisfies the Neumann boundary condition in (12.39). Moreover, we have

$$\Delta v^+_{q,\Omega_Q}[\phi](x) = -\frac{1}{m_n(Q)} \int_{\partial\Omega_Q} \phi \, d\sigma = 0$$

for all $x \in \mathbb{S}[\Omega_Q]$ (cf. (12.12)). Indeed, $\int_{\partial\Omega_Q} \phi \, d\sigma = 0$. Thus $v^+_{q,\Omega_Q}[\phi]$ solves problem (12.39). Hence, by Theorem 12.17, the set in (12.40) coincides with the set of the solutions of problem (12.39). □

Again, by replacing Theorem 12.20 (i) with the representation Proposition 12.21 (i) of a function in $C^{1,\alpha}(\partial\Omega_Q)$ as a sum of the trace of a periodic double layer and a constant in the proof of Theorem 12.25 (i), we deduce an alternative representation formula for the solution of the Dirichlet problem in $\mathbb{S}[\Omega_Q]$.

Proposition 12.26. *Let $g \in C^{1,\alpha}(\partial\Omega_Q)$. Then there exists a unique function u in $C^{1,\alpha}_q(\overline{\mathbb{S}[\Omega_Q]})$ that satisfies the periodic Dirichlet boundary value problem*

$$\begin{cases} \Delta u = 0 & \text{in } \mathbb{S}[\Omega_Q]\,, \\ u(x+qz) = u(x) & \forall x \in \overline{\mathbb{S}[\Omega_Q]}\,, z \in \mathbb{Z}^n\,, \\ u = g & \text{on } \partial\Omega_Q\,. \end{cases}$$

Moreover,

$$u = w^{+}_{q,\Omega_Q}[\psi] + c \qquad \textit{in } \overline{\mathbb{S}[\Omega_Q]}\,,$$

where (ψ, c) *is the unique pair in* $C^{1,\alpha}(\partial\Omega_Q)_0 \times \mathbb{R}$ *such that*

$$g = \left(\frac{1}{2}I + W_{q,\Omega_Q}\right)[\psi] + c\,.$$

Chapter 13
Singular Perturbation Problems in Periodic Domains

Abstract In this chapter, we study the asymptotic behavior of the solutions of singularly perturbed boundary value problems in periodic domains. We introduce the domain $\mathbb{S}[\Omega_{p,\epsilon}]^-$ obtained by removing from the Euclidean space a periodic set of inclusions, each of them of a size controlled by a positive parameter ϵ. Then, for each positive and sufficiently small ϵ, we consider a boundary value problem in $\mathbb{S}[\Omega_{p,\epsilon}]^-$ or a transmission problem in the pair of domains consisting of $\mathbb{S}[\Omega_{p,\epsilon}]^-$ and its complementary set $\mathbb{S}[\Omega_{p,\epsilon}]$, and we analyze the behavior of the solution as the parameter ϵ tends to $\epsilon = 0$.

13.1 Introduction

In this chapter, we will study the asymptotic behavior of the solutions of singular perturbation problems in periodic domains. More precisely, we will introduce the domain $\mathbb{S}[\Omega_{p,\epsilon}]^-$ obtained by removing from the Euclidean space a periodic set of inclusions, each of them of size controlled by a positive parameter ϵ. Then, for each ϵ positive and small enough, we will consider a boundary value problem in $\mathbb{S}[\Omega_{p,\epsilon}]^-$ (or a transmission problem in the pair of domains consisting of $\mathbb{S}[\Omega_{p,\epsilon}]^-$ and its complementary set $\mathbb{S}[\Omega_{p,\epsilon}]$) and we will analyze the behavior of the solution (and of related functionals) as the parameter ϵ approaches the degenerate value $\epsilon = 0$, in correspondence of which the holes or the inclusions collapse to points.

As we shall see, boundary value problems in periodic domains arise in connection with several applications, concerning, e.g., the permeability of periodic arrays of cylinders or the effective conductivity of periodic composite materials. For this reason, in this chapter we will consider several problems which are relevant also in applications. We recall that in the whole of the present chapter, we will consider boundary value problems for functions with values in $\mathbb{R}$.

M. Dalla Riva et al., *Singularly Perturbed Boundary Value Problems*,
https://doi.org/10.1007/978-3-030-76259-9_13

13.2 The Geometric Setting

To define our singularly perturbed boundary value problems, we first need to introduce our geometric setting.

Unless stated otherwise, we fix a natural number

$$n \in \mathbb{N}\,,\; n \geq 2\,,$$

we take

$$(q_{11}, \ldots, q_{nn}) \in]0, +\infty[^n\,,$$

and we retain the notation of Chapter 12 for the periodicity cell Q, the matrix q, and the periodic domains $\mathbb{S}[\Omega_Q]$ and $\mathbb{S}[\Omega_Q]^-$, with Ω_Q a subset of Q (see (12.1)–(12.3) in Sections 12.1 and 12.2.1). We fix once for all

$$\alpha \in]0, 1[\,.$$

Then we take

$$p \in Q$$

and a set Ω in the n-dimensional Euclidean space $\mathbb{R}^n$. We assume that Ω satisfies the following condition

$$\begin{aligned}&\Omega \text{ is a bounded open connected subset of } \mathbb{R}^n \text{ of class } C^{1,\alpha}\\ &\text{such that the exterior domain } \Omega^- \text{ is connected,}\\ &\text{and the origin 0 of } \mathbb{R}^n \text{ belongs to } \Omega.\end{aligned} \tag{13.1}$$

Here we recall that

$$\Omega^- \equiv \mathbb{R}^n \setminus \overline{\Omega}\,.$$

Condition (13.1) implies that Ω has no holes and that there exists a positive real number ϵ_0 such that

$$p + \epsilon\overline{\Omega} \subseteq Q \text{ for all } \epsilon \in]-\epsilon_0, \epsilon_0[\,.$$

For the sake of brevity, we find convenient to set

$$\Omega_{p,\epsilon} \equiv p + \epsilon\Omega \qquad \forall \epsilon \in \mathbb{R}\,.$$

In this chapter we want to consider boundary value problems in the singularly perturbed periodic domains $\mathbb{S}[\Omega_{p,\epsilon}]$ and $\mathbb{S}[\Omega_{p,\epsilon}]^-$, as $\epsilon \to 0$. We note that in correspondence of the degenerate value $\epsilon = 0$, we have

$$\mathbb{S}[\Omega_{p,\epsilon}] = p + q\mathbb{Z}^n\,, \qquad \mathbb{S}[\Omega_{p,\epsilon}]^- = \mathbb{R}^n \setminus (p + q\mathbb{Z}^n)\,.$$

If $\epsilon \in]-\epsilon_0, \epsilon_0[\setminus\{0\}$ a topological argument shows that $\mathbb{S}[\Omega_{p,\epsilon}]^-$ is a periodic open connected subset of $\mathbb{R}^n$ (cf., e.g., Theorem A.10). Moreover, the boundary

$\partial\mathbb{S}[\Omega_{p,\epsilon}]^-$ of $\mathbb{S}[\Omega_{p,\epsilon}]^-$ (or equivalently the boundary $\partial\mathbb{S}[\Omega_{p,\epsilon}]$ of $\mathbb{S}[\Omega_{p,\epsilon}]$) consists of the union

$$\cup_{z\in\mathbb{Z}^n}(qz + \partial\Omega_{p,\epsilon}),$$

for all $\epsilon \in]-\epsilon_0, \epsilon_0[$.

13.3 Perturbed Problems in Periodic Domains

In this chapter, we will consider, for example, a certain boundary value problem in the domain $\mathbb{S}[\Omega_{p,\epsilon}]^-$ and, for ϵ small enough, we will denote the solution by u_ϵ. Then we will be interested in studying the behavior of a functional related to u_ϵ, say $F(\epsilon)$, as the parameter ϵ tends to 0. The functional $F(\epsilon)$ may represent quantities such as the longitudinal flow around an array of cylinders or the effective conductivity of a periodic dilute composite.

As it happens for the problems with a single hole of our previous chapters, also in the case of periodic domains, a common approach to study the asymptotic behavior of functionals related to the solutions of singularly perturbed problems with small holes and inclusions is that of Asymptotic Analysis, which enables to write out asymptotic expansions for the functional $F(\epsilon)$. In this sense, in Ammari et al. [12] the authors compute an asymptotic expansion of the effective electrical conductivity of a periodic dilute composite with ideal contact condition (see also the monograph Ammari and Kang [8]). We also mention Ammari et al. [9] where the authors consider anisotropic heat conductors, Ammari et al. [11] where effective elastic properties have been investigated, and Ammari et al. [7] for the analysis of effective viscosity properties. For the application of Asymptotic Analysis to dilute and densely packed composites, we refer to Movchan et al. [226]. We note that the above mentioned techniques allow to prove asymptotic expansions of the type

$$F(\epsilon) = \sum_{l=0}^{r} a_l \epsilon^l + R(\epsilon) \qquad \text{as } \epsilon \to 0\,,$$

for some $r \in \mathbb{N}$, $a_0, \ldots, a_r \in \mathbb{R}$, and some (asymptotically small) remainder function $R(\cdot)$. However, such techniques have not been conceived to show that the associated power series $\sum_{l=0}^{\infty} a_l \epsilon^l$ is convergent and equal to $F(\epsilon)$, although they do provide an algorithm to compute one or more of the coefficients a_l.

On the other hand, another technique, the so-called Functional Equation Method, enables to study boundary value problems in periodic domains and to express quantities as the longitudinal flow or the effective conductivity in terms of convergent power series of the diameter of the inclusion (cf., e.g., Castro and Pesetskaya [44, 46], Castro et al. [47], Drygaś and Mityushev [92], Kapanadze et al. [134, 135], Mityushev [216, 217, 223], Mityushev and Adler [218, 219], Mityushev and Rylko [222]), also in the case of random composites (cf., e.g., Berlyand and Mityushev [25, 26]). Such method is based on a complex variable technique and applies to specific two dimensional geometries, for example to circular inclusions. For the Functional Equation Method in connection with composite media, we also refer to the monographs Gluzman et al. [110] and Drygaś et al. [91].

Finally, we also mention the monographs of Kolpakov and Kolpakov [144] and of Berlyand et al. [24] for the application of the Network Approximation Method for composite materials. For the analysis of doubly periodic media with such a method we refer to Nawalaniec et al. [233].

In this chapter, we apply the Functional Analytic Approach of this book to investigate the behavior of singular perturbation problems in periodically perforated domains. Our main aim will be to describe the solution or related functionals in terms of real analytic maps of the singular perturbation parameter. Representations of this type enable to write the solution or related functionals in terms of convergent power series and to justify corresponding asymptotic expansions.

We observe that singularly perturbed Dirichlet problems for the Laplace and the Poisson equation have been considered in [228, 229], whereas in [179] we have analyzed a Neumann problem for the Poisson equation. Nonlinear Robin boundary conditions have been investigated in [176] and [68] and quasilinear heat transmission problems have been analyzed in [177] and [70]. Finally, we mention that nonlinear traction problems in periodically perforated domains for the Lamé equations have been studied [76, 77].

13.4 Preliminaries and Notation

In this section, we introduce some notation and preliminaries that we need in what follows.

As we have seen in the non-periodic case, when dealing with boundary value problems for the Poisson equation, it is convenient to introduce a suitable space of real analytic functions (the Roumieu class) and we can resort to volume potentials. Here we present the periodic counterpart of those spaces and we introduce some properties of the periodic volume potentials.

If $\rho \in]0, +\infty[$, then we set

$$C^0_{q,\omega,\rho}(\mathbb{R}^n) \equiv \left\{ u \in C^\infty_q(\mathbb{R}^n) : \sup_{\beta \in \mathbb{N}^n} \frac{\rho^{|\beta|}}{|\beta|!} \|D^\beta u\|_{C^0(\overline{Q})} < +\infty \right\}, \tag{13.2}$$

where $C^\infty_q(\mathbb{R}^n)$ denotes the set of q-periodic functions of $C^\infty(\mathbb{R}^n)$, and we define the norm

$$\|u\|_{C^0_{q,\omega,\rho}(\mathbb{R}^n)} \equiv \sup_{\beta \in \mathbb{N}^n} \frac{\rho^{|\beta|}}{|\beta|!} \|D^\beta u\|_{C^0(\overline{Q})} \qquad \forall u \in C^0_{q,\omega,\rho}(\mathbb{R}^n) .$$

The Roumieu class $\left(C^0_{q,\omega,\rho}(\mathbb{R}^n), \|\cdot\|_{C^0_{q,\omega,\rho}(\mathbb{R}^n)}\right)$ is known to be a Banach space.

In order to deal with non-homogenous problems, we need to recall the definition of the Newtonian potential. If $\tilde{\Omega}$ is a bounded open subset of $\mathbb{R}^n$ and $f \in L^\infty(\tilde{\Omega})$, then

$$P_n[\tilde{\Omega}, f](x) \equiv \mathcal{P}[S_n, f](x) = \int_{\tilde{\Omega}} S_n(x-y) f(y)\, dy \qquad \forall x \in \mathbb{R}^n$$

(cf. (7.2)). We now find convenient to introduce the periodic Newtonian potential, which plays the role of a periodic counterpart of $P_n[\tilde{\Omega}, f]$. Thus, we further assume that $\tilde{\Omega}_Q$ is a bounded open subset of $\mathbb{R}^n$ such that $\tilde{\Omega}_Q \subseteq Q + \tilde{x}$ for some $\tilde{x} \in \mathbb{R}^n$ and we set

$$P_{q,n}[\tilde{\Omega}_Q, f](x) \equiv \int_{\tilde{\Omega}_Q} S_{q,n}(x-y) f(y)\, dy \qquad \forall x \in \mathbb{R}^n . \tag{13.3}$$

We now turn to some properties of periodic volume potentials. Before doing so, we introduce the following elementary lemma on the integral of periodic functions. For a proof, we refer, e.g., to Cioranescu and Donato [56, Lemma 2.3, p. 27].

Lemma 13.1. *Let $u \in L^1_{\mathrm{loc}}(\mathbb{R}^n)$ be such that*

$$u(x+qz) = u(x) \qquad \textit{for a.e. } x \in \mathbb{R}^n, \quad \forall z \in \mathbb{Z}^n .$$

Let $y \in \mathbb{R}^n$. Then

$$\int_{y+Q} u(x)\, dx = \int_Q u(x)\, dx .$$

Then we have the following two statements concerning periodic volume potentials (cf. [179, Prop. A.1, A.2]).

Proposition 13.2. *Let $f \in C^{0,\alpha}_q(\mathbb{R}^n)$. Then $P_{q,n}[Q, f] \in C^{2,\alpha}_q(\mathbb{R}^n)$ and*

$$\Delta P_{q,n}[Q, f](x) = f(x) - \frac{1}{m_n(Q)} \int_Q f(y)\, dy \qquad \forall x \in \mathbb{R}^n . \tag{13.4}$$

Proof. Since $S_{q,n}$ is q-periodic, then $P_{q,n}[Q, f]$ is q-periodic as well and thus it suffices to show that, for each $x_0 \in \mathbb{R}^n$, there exists an open neighborhood U_{x_0} of x_0 such that $P_{q,n}[Q, f]_{|U_{x_0}} \in C^{2,\alpha}(U_{x_0})$ and (13.4) holds for $x \in U_{x_0}$. So let $x_0 \in \mathbb{R}^n$ and set

$$U_{x_0} \equiv \mathbb{B}_n\left(x_0, \min\{q_{11}, \dots, q_{nn}\}/4\right) ,$$
$$U'_{x_0} \equiv \mathbb{B}_n\left(x_0, \min\{q_{11}, \dots, q_{nn}\}/3\right) ,$$

and

$$\tilde{Q}_{x_0} \equiv \prod_{j=1}^{n}]-\frac{q_{jj}}{2}, \frac{q_{jj}}{2}[+ x_0 .$$

Clearly,

$$U_{x_0} \subseteq \overline{U_{x_0}} \subseteq U'_{x_0} \subseteq \overline{U'_{x_0}} \subseteq \tilde{Q}_{x_0} .$$

By the periodicity of $S_{q,n}$ and f and by Lemma 13.1, we have

$$P_{q,n}[Q,f] = P_{q,n}[\tilde{Q}_{x_0}, f] \,.$$

Moreover,

$$P_{q,n}[\tilde{Q}_{x_0}, f](x) = \int_{\tilde{Q}_{x_0} \setminus U'_{x_0}} S_n(x-y) f(y)\, dy + \int_{U'_{x_0}} S_n(x-y) f(y)\, dy$$
$$+ \int_{\tilde{Q}_{x_0}} R_{q,n}(x-y) f(y)\, dy \qquad \forall x \in U_{x_0} \,.$$

By the continuity result of a linear integral operator with analytic kernel in Roumieu spaces of Proposition A.30 and by the analyticity of S_n in $\mathbb{R}^n \setminus \{0\}$ and of $R_{q,n}$ in $(\mathbb{R}^n \setminus q\mathbb{Z}^n) \cup \{0\}$, we have that the functions

$$U_{x_0} \ni x \mapsto \int_{\tilde{Q}_{x_0} \setminus U'_{x_0}} S_n(x-y) f(y)\, dy$$

and

$$U_{x_0} \ni x \mapsto \int_{\tilde{Q}_{x_0}} R_{q,n}(x-y) f(y)\, dy \,,$$

belong to $C^\infty(U_{x_0})$. On the other hand, by Theorem 7.18 on the classical volume potential the function

$$U_{x_0} \ni x \mapsto \int_{U'_{x_0}} S_n(x-y) f(y)\, dy \,,$$

belongs to $C^{2,\alpha}(U_{x_0})$. As a consequence, $P_{q,n}[Q,f]_{|U_{x_0}} = P_{q,n}[\tilde{Q}_{x_0}, f]_{|U_{x_0}}$ belongs to $C^{2,\alpha}(U_{x_0})$. To conclude we note that the classical theorem of differentiation under the integral sign, the equalities

$$\Delta S_n = 0 \quad \text{in } \mathbb{R}^n \setminus \{0\} \,, \quad \Delta R_{q,n} = -\frac{1}{m_n(Q)} \quad \text{in } (\mathbb{R}^n \setminus q\mathbb{Z}^n) \cup \{0\}$$

(cf. equality (12.10)), and Theorem 7.18 imply that

$$\begin{aligned} \Delta_x P_{q,n}[\tilde{Q}_{x_0}, f](x) =& \Delta_x \int_{\tilde{Q}_{x_0} \setminus U'_{x_0}} S_n(x-y) f(y)\, dy \\ &+ \Delta_x \int_{U'_{x_0}} S_n(x-y) f(y)\, dy \\ &+ \Delta_x \int_{\tilde{Q}_{x_0}} R_{q,n}(x-y) f(y)\, dy \\ =& f(x) - \frac{1}{m_n(Q)} \int_{\tilde{Q}_{x_0}} f(y)\, dy \end{aligned}$$

$$= f(x) - \frac{1}{m_n(Q)} \int_Q f(y)\, dy \qquad \forall x \in U_{x_0} .$$

□

Proposition 13.3. *Let $\rho > 0$. Then there exists $\rho' \in]0, \rho]$ such that the linear map from $C^0_{q,\omega,\rho}(\mathbb{R}^n)$ to $C^0_{q,\omega,\rho'}(\mathbb{R}^n)$ that takes f to $P_{q,n}[Q, f]$ is continuous.*

Proof. It suffices to show that for each $x_0 \in \mathbb{R}^n$, there exist r_{x_0} and ρ_{x_0} in $]0, +\infty[$ such that $P_{q,n}[Q, \cdot]_{|\overline{\mathbb{B}_n(x_0, r_{x_0})}}$ is a continuous operator from $C^0_{q,\omega,\rho}(\mathbb{R}^n)$ to $C^0_{q,\omega,\rho_{x_0}}(\overline{\mathbb{B}_n(x_0, r_{x_0})})$. Indeed, if this is true, the compactness of $\overline{Q}$ implies that there exist $x_1, \ldots, x_k$ in $\mathbb{R}^n$ such that

$$\overline{Q} \subseteq \cup_{j=1}^k \mathbb{B}_n(x_j, r_{x_j}) .$$

Then if we set

$$\rho' \equiv \min_{j=1,\ldots,k} \rho_{x_j} ,$$

we have

$$\frac{\rho'^{|\beta|}}{|\beta|!} \|D^\beta u\|_{C^0(\overline{Q})} \leq \max_{j=1,\ldots,k} \frac{\rho_{x_j}^{|\beta|}}{|\beta|!} \|D^\beta u\|_{C^0(\overline{\mathbb{B}_n(x_j, r_{x_j})})} .$$

If for each $j = 1, \ldots, k$ the linear operator $P_{q,n}[Q, \cdot]_{|\overline{\mathbb{B}_n(x_j, r_{x_j})}}$ is continuous from $C^0_{q,\omega,\rho}(\mathbb{R}^n)$ to $C^0_{q,\omega,\rho_{x_j}}(\overline{\mathbb{B}_n(x_j, r_{x_j})})$, then for each $j = 1, \ldots, k$ there exists $C_j > 0$ such that

$$\sup_{\beta \in \mathbb{N}^n} \frac{\rho_{x_j}^{|\beta|}}{|\beta|!} \|D^\beta u\|_{C^0(\overline{\mathbb{B}_n(x_j, r_{x_j})})} \leq C_j \|u\|_{C^0_{q,\omega,\rho}(\mathbb{R}^n)} ,$$

for all $u \in C^0_{q,\omega,\rho}(\mathbb{R}^n)$. As a consequence, if we set

$$C \equiv \max_{j=1,\ldots,k} C_j ,$$

we deduce that

$$\sup_{\beta \in \mathbb{N}^n} \max_{j=1,\ldots,k} \frac{\rho_{x_j}^{|\beta|}}{|\beta|!} \|D^\beta u\|_{C^0(\overline{\mathbb{B}_n(x_j, r_{x_j})})} \leq C \|u\|_{C^0_{q,\omega,\rho}(\mathbb{R}^n)} ,$$

for all $u \in C^0_{q,\omega,\rho}(\mathbb{R}^n)$, and accordingly

$$\sup_{\beta \in \mathbb{N}^n} \frac{\rho'^{|\beta|}}{|\beta|!} \|D^\beta u\|_{C^0(\overline{Q})} \leq C \|u\|_{C^0_{q,\omega,\rho}(\mathbb{R}^n)} ,$$

for all $u \in C^0_{q,\omega,\rho}(\mathbb{R}^n)$.

So we now fix $x_0 \in \mathbb{R}^n$ and set

$$\tilde{Q}_{x_0} \equiv \prod_{j=1}^{n}]-\frac{q_{jj}}{2}, \frac{q_{jj}}{2}[+x_0\,,$$
$$r_{x_0} \equiv \min\{q_{11}, \ldots, q_{nn}\}/4\,,$$
$$\tilde{r}_{x_0} \equiv \min\{q_{11}, \ldots, q_{nn}\}/3\,.$$

By the periodicity of $S_{q,n}$ and f and by Lemma 13.1, we have $P_{q,n}[Q, f] = P_{q,n}[\tilde{Q}_{x_0}, f]$. As a consequence,

$$\begin{aligned} P_{q,n}[Q, f](x) =& P_n[\mathbb{B}_n(x_0, \tilde{r}_{x_0}), f](x) + \int_{\tilde{Q}_{x_0}\setminus\overline{\mathbb{B}_n(x_0,\tilde{r}_{x_0})}} S_n(x-y)f(y)\,dy \\ &+ \int_{\tilde{Q}_{x_0}} R_{q,n}(x-y)f(y)\,dy \qquad \forall x \in \overline{\mathbb{B}_n(x_0, r_{x_0})}\,, \end{aligned} \tag{13.5}$$

for all $f \in C^0_{q,\omega,\rho}(\mathbb{R}^n)$. By the regularity result of volume potentials in Roumieu classes of Proposition 7.21, we deduce that there exists $\tilde{\rho}_{x_0} \in]0, \rho]$ such that the linear operator $P_n[\mathbb{B}_n(x_0, \tilde{r}_{x_0}), \cdot]_{|\overline{\mathbb{B}_n(x_0,r_{x_0})}}$ is continuous from $C^0_{q,\omega,\rho}(\mathbb{R}^n)$ to $C^0_{\omega,\tilde{\rho}_{x_0}}(\overline{\mathbb{B}_n(x_0, r_{x_0})})$. By the analyticity of the functions S_n and $R_{q,n}$ in $\mathbb{R}^n \setminus \{0\}$ and $(\mathbb{R}^n \setminus q\mathbb{Z}^n) \cup \{0\}$, respectively, and by the continuity result of a linear integral operator with real analytic kernel in Roumieu spaces of Proposition A.30, there exists $\rho_{x_0} \in]0, \tilde{\rho}_{x_0}]$ such that the linear operators from $C^0_{q,\omega,\rho}(\mathbb{R}^n)$ to $C^0_{q,\omega,\rho_{x_0}}(\overline{\mathbb{B}_n(x_0, r_{x_0})})$ which map f to $\int_{\tilde{Q}_{x_0}\setminus\overline{\mathbb{B}_n(x_0,\tilde{r}_{x_0})}} S_n(\cdot - y)f(y)\,dy$ and to $\int_{\tilde{Q}_{x_0}} R_{q,n}(\cdot - y)f(y)\,dy$ are continuous. Hence, equality (13.5) implies that $P_{q,n}[Q, \cdot]_{|\overline{\mathbb{B}_n(x_0,r_{x_0})}}$ is continuous from $C^0_{q,\omega,\rho}(\mathbb{R}^n)$ to $C^0_{q,\omega,\rho_{x_0}}(\overline{\mathbb{B}_n(x_0, r_{x_0})})$. □

In order to study perturbation problems by means of layer and volume potentials, we need to understand how these operators behave when we rescale the support of integration. To do that, we need to look at the behavior of the kernels, which is related to the fundamental solution. Therefore, we note that the following elementary equality holds,

$$S_{q,n}(\epsilon x) = \epsilon^{2-n} S_n(x) + \frac{1}{2\pi}(\delta_{2,n} \log \epsilon) + R_{q,n}(\epsilon x)\,, \tag{13.6}$$

for all $x \in \mathbb{R}^n \setminus \epsilon^{-1}q\mathbb{Z}^n$ and $\epsilon \in]0, +\infty[$.

Next we introduce the following technical lemma (cf. [179, Lemma A.7]).

Lemma 13.4. *Let $\Omega^\sharp$ be a bounded open subset of $\mathbb{R}^n$ of class C^1. Then the following statements hold.*

(i) Let $\epsilon_\sharp \in]0, +\infty[$ be such that

$$\epsilon(\overline{\Omega^\sharp} - \overline{\Omega}) \subseteq (\mathbb{R}^n \setminus q\mathbb{Z}^n) \cup \{0\} \qquad \forall \epsilon \in]-\epsilon_\sharp, \epsilon_\sharp[\,. \tag{13.7}$$

Then the maps from $]-\epsilon_\sharp,\epsilon_\sharp[\times L^1(\partial\Omega)$ *to* $C^{1,\alpha}(\overline{\Omega^\sharp})$ *that take* (ϵ,θ) *to the functions*

$$\int_{\partial\Omega} R_{q,n}(\epsilon(\xi-\eta))\theta(\eta)\,d\sigma_\eta \qquad \forall\xi\in\overline{\Omega^\sharp}\,, \tag{13.8}$$

$$\int_{\partial\Omega} \partial_{x_j} R_{q,n}(\epsilon(\xi-\eta))\theta(\eta)\,d\sigma_\eta \quad \forall\xi\in\overline{\Omega^\sharp}\,, j\in\{1,\ldots,n\}\,, \tag{13.9}$$

and the map from $]-\epsilon_\sharp,\epsilon_\sharp[$ *to* $C^{1,\alpha}(\overline{\Omega^\sharp})$ *that takes* ϵ *to the function*

$$\int_\Omega R_{q,n}(\epsilon(\xi-\eta))\,d\eta \qquad \forall\xi\in\overline{\Omega^\sharp}\,, \tag{13.10}$$

are real analytic.

(ii) Let $\rho\in]0,+\infty[$. *Let* $p\in Q$. *Then the maps from* $\mathbb{R}\times C^0_{q,\omega,\rho}(\mathbb{R}^n)$ *to* $C^{1,\alpha}(\overline{\Omega^\sharp})$ *that take* (ϵ,f) *to the functions*

$$P_{q,n}[Q,f](p+\epsilon\xi) \qquad \forall\xi\in\overline{\Omega^\sharp}\,, \tag{13.11}$$

$$\partial_{x_j} P_{q,n}[Q,f](p+\epsilon\xi) \quad \forall\xi\in\overline{\Omega^\sharp}\,, j\in\{1,\ldots,n\}\,, \tag{13.12}$$

$$\int_0^1 \partial_{x_j} P_{q,n}[Q,f](p+\epsilon\xi\tau)\,d\tau \quad \forall\xi\in\overline{\Omega^\sharp}\,, j\in\{1,\ldots,n\}\,, \tag{13.13}$$

are real analytic.

Proof. We first consider (i). By the real analyticity of $R_{q,n}$ in $(\mathbb{R}^n\setminus q\mathbb{Z}^n)\cup\{0\}$, by (13.7), and by the analyticity of nonlinear integral operators with analytic kernel of Theorem A.28 (i), the map delivered by (13.8) is real analytic from $]-\epsilon_\sharp,\epsilon_\sharp[\times L^1(\partial\Omega)$ to $C^{1,\alpha}(\overline{\Omega^\sharp})$. The proof of the analyticity of the maps in (13.9) and (13.10) follows the lines of the proof of the analyticity of the map delivered by (13.8).

Next we consider statement (ii) and we prove the analyticity of the map delivered by (13.11). By Proposition 13.3 there exists $\rho'\in]0,\rho]$ such that the linear map from $C^0_{q,\omega,\rho}(\mathbb{R}^n)$ to $C^0_{q,\omega,\rho'}(\mathbb{R}^n)$ that takes f to $P_{q,n}[Q,f]$ is continuous. Since the map from $\mathbb{R}$ to $C^{1,\alpha}(\overline{\Omega^\sharp})$ that takes ϵ to $p+\epsilon I_{\overline{\Omega^\sharp}}$ is analytic, Proposition A.25 on the analyticity of the composition operator implies that the map from $\mathbb{R}\times C^0_{q,\omega,\rho}(\mathbb{R}^n)$ to $C^{1,\alpha}(\overline{\Omega^\sharp})$ delivered by (13.11) is real analytic.

We now consider the analyticity of the map delivered by (13.12). We can verify that there exists $0<\tilde\rho\le\rho'$ such that ∂_{x_j} is continuous from $C^0_{q,\omega,\rho'}(\mathbb{R}^n)$ to $C^0_{q,\omega,\tilde\rho}(\mathbb{R}^n)$. Then, by the same argument used to prove the analyticity of the map in (13.11), we deduce the analyticity of the map delivered by (13.12).

We now consider the analyticity of the map delivered by (13.13). Since the restriction map is linear and continuous, there is no loss of generality in assuming that $\Omega^\sharp$ is an open ball with center at the origin. By Proposition 13.3 and by the continuity of ∂_{x_j} there exists $\rho'\in]0,\rho]$ such that the linear map from $C^0_{q,\omega,\rho}(\mathbb{R}^n)$ to $C^0_{q,\omega,\rho'}(\mathbb{R}^n)$ that takes f to $\partial_{x_j}P_{q,n}[Q,f]$ is continuous. Since the map from $\mathbb{R}$ to $C^{1,\alpha}(\overline{\Omega^\sharp})$ that takes ϵ to $p+\epsilon I_{\overline{\Omega^\sharp}}$ is analytic, Proposition A.25 on the analyticity of the composition operator implies that the map from $\mathbb{R} \times C^0_{q,\omega,\rho}(\mathbb{R}^n)$ to $C^{1,\alpha}(\overline{\Omega^\sharp})$ that takes (ϵ, f) to $\partial_{x_j} P_{q,n}[Q,f](p+\epsilon\cdot)$ is real analytic. By the classical differentiability theorem for integrals depending on a parameter, there exists $c \in]0,+\infty[$ such that

$$\left\| \int_0^1 h(\beta I_{\overline{\Omega^\sharp}})\, d\beta \right\|_{C^{1,\alpha}(\overline{\Omega^\sharp})} \leq c\|h\|_{C^{1,\alpha}(\overline{\Omega^\sharp})} \qquad \forall h \in C^{1,\alpha}(\overline{\Omega^\sharp}) .$$

Hence, the map from $C^{1,\alpha}(\overline{\Omega^\sharp})$ to itself that maps h to $\int_0^1 h(\beta I_{\overline{\Omega^\sharp}})\, d\beta$ is linear and continuous and thus the composite map in (13.13) is real analytic. □

In this chapter we will consider nonhomogeneous equations and we will address the problem of transforming them into homogeneous ones. The standard way to do it would be to exploit the q-periodic Newtonian potential $P_{q,n}[Q,f]$ with a convenient density f. Unfortunately however, in general $\int_Q f\, dx \neq 0$ and accordingly the q-periodic Newtonian potential $P_{q,n}[Q,f]$ satisfies the equation

$$\Delta\left(P_{q,n}[Q,f]\right) = f - \fint_Q f\, dx ,$$

and not the Poisson equation with datum f. Then we adopt a different strategy and we need the following technical statement.

Lemma 13.5. *Let $\epsilon \in]0,\epsilon_0[$. Then the following statements hold.*

(i) The function a_ϵ from $\overline{\mathbb{S}[\Omega_{p,\epsilon}]}$ to $\mathbb{R}$ defined by

$$a_\epsilon(x) \equiv \int_{\Omega_{p,\epsilon}} S_n(x - qz - y)\, dy \qquad \forall x \in qz + \Omega_{p,\epsilon}\, , z \in \mathbb{Z}^n ,$$

belongs to $C^{2,\alpha}_q(\overline{\mathbb{S}[\Omega_{p,\epsilon}]})$ and satisfies the equalities

$$\Delta a_\epsilon(x) = 1 \qquad \forall x \in \mathbb{S}[\Omega_{p,\epsilon}] , \tag{13.14}$$

and

$$\frac{\partial}{\partial x_j} a_\epsilon(x) = -\int_{\partial\Omega_{p,\epsilon}} S_n(x - qz - y)(\nu_{\Omega_{p,\epsilon}})_j(y)\, d\sigma_y \quad \forall x \in qz + \Omega_{p,\epsilon}\, , \; z \in \mathbb{Z}^n . \tag{13.15}$$

(ii) The function b_ϵ from $\mathbb{R}^n$ to $\mathbb{R}$ defined by

$$b_\epsilon(x) \equiv \int_{\Omega_{p,\epsilon}} S_{q,n}(x-y)\, dy \qquad \forall x \in \mathbb{R}^n ,$$

belongs to $C^1_q(\mathbb{R}^n)$ and

$$\frac{\partial}{\partial x_j} b_\epsilon(x) = -\int_{\partial \Omega_{p,\epsilon}} S_{q,n}(x-y)(\nu_{\Omega_{p,\epsilon}})_j(y)\, d\sigma_y \qquad \forall x \in \mathbb{R}^n\,. \tag{13.16}$$

Moreover,

$$b_{\epsilon|\overline{\mathbb{S}[\Omega_{p,\epsilon}]}} \in C_q^{2,\alpha}(\overline{\mathbb{S}[\Omega_{p,\epsilon}]})\,, \tag{13.17}$$

$$\Delta b_\epsilon(x) = 1 - \frac{m_n(\Omega_{p,\epsilon})}{m_n(Q)} \qquad \forall x \in \mathbb{S}[\Omega_{p,\epsilon}]\,, \tag{13.18}$$

and

$$b_{\epsilon|\overline{\mathbb{S}[\Omega_{p,\epsilon}]^-}} \in C_q^{2,\alpha}(\overline{\mathbb{S}[\Omega_{p,\epsilon}]^-})\,, \tag{13.19}$$

$$\Delta b_\epsilon(x) = -\frac{m_n(\Omega_{p,\epsilon})}{m_n(Q)} \qquad \forall x \in \mathbb{S}[\Omega_{p,\epsilon}]^-\,. \tag{13.20}$$

(iii) Let $f \in C_q^{0,\alpha}(\mathbb{R}^n)$. Then the function $F_\epsilon^i[f]$ from $\overline{\mathbb{S}[\Omega_{p,\epsilon}]}$ to $\mathbb{R}$ defined by

$$F_\epsilon^i[f](x) \equiv \int_Q S_{q,n}(x-y)f(y)\, dy + \frac{1}{m_n(Q)} a_\epsilon(x) \int_Q f(y)\, dy$$

for all $x \in \overline{\mathbb{S}[\Omega_{p,\epsilon}]}$ belongs to $C_q^{2,\alpha}(\overline{\mathbb{S}[\Omega_{p,\epsilon}]})$ and satisfies the equality

$$\Delta(F_\epsilon^i[f])(x) = f(x) \qquad \forall x \in \mathbb{S}[\Omega_{p,\epsilon}]\,.$$

(iv) Let $f \in C_q^{0,\alpha}(\mathbb{R}^n)$. Then the function $F_\epsilon^o[f]$ from $\overline{\mathbb{S}[\Omega_{p,\epsilon}]^-}$ to $\mathbb{R}$ defined by

$$F_\epsilon^o[f](x) \equiv \int_Q S_{q,n}(x-y)f(y)\, dy - \frac{1}{m_n(\Omega_{p,\epsilon})} b_\epsilon(x) \int_Q f(y)\, dy$$

for all $x \in \overline{\mathbb{S}[\Omega_{p,\epsilon}]^-}$ belongs to $C_q^{2,\alpha}(\overline{\mathbb{S}[\Omega_{p,\epsilon}]^-})$ and satisfies the equality

$$\Delta(F_\epsilon^o[f])(x) = f(x) \qquad \forall x \in \mathbb{S}[\Omega_{p,\epsilon}]^-\,.$$

Proof. We first consider statement (i). The function a_ϵ is the q-periodic extension of the Newtonian potential with moment 1, which is well known to be of class $C^{2,\alpha}(\overline{\Omega_{p,\epsilon}})$ and to satisfy equation (13.14) in $\Omega_{p,\epsilon}$ and equation (13.15) on $\Omega_{p,\epsilon}$ for $z = 0$ (cf. Lemma 7.7 and Theorem 7.18). Hence, $a_\epsilon \in C_q^{2,\alpha}(\overline{\mathbb{S}[\Omega_{p,\epsilon}]})$ and equations (13.14) and (13.15) hold.

Next we consider statement (ii). We have

$$b_\epsilon(x) = P_n[\Omega_{p,\epsilon}, 1](x) + \int_{\Omega_{p,\epsilon}} R_{q,n}(x-y)\, dy \qquad \forall x \in \mathbb{R}^n\,. \tag{13.21}$$

Since $R_{q,n}$ is real analytic in $(\mathbb{R}^n \setminus q\mathbb{Z}^n) \cup \{0\}$, the classical differentiation theorem for integrals depending on a parameter imply that the map $\int_{\Omega_{p,\epsilon}} R_{q,n}(\cdot - y)\, dy$ is of class C^1 in an open neighborhood of $\overline{Q}$. Then the regularity result of Proposition 7.5 (ii) for the volume potential imply that $P_n[\Omega_{p,\epsilon}, 1]$ belongs to $C^1(\mathbb{R}^n)$. Then b_ϵ is of class C^1 in an open neighborhood of $\overline{Q}$ and the periodicity of b_ϵ implies that $b_\epsilon \in C^1_q(\mathbb{R}^n)$. Moreover, by exploiting the equality in (13.21) in a neighborhood of $\overline{Q}$, by the classical Theorem 7.18 for the Newtonian potential, by classical theorems of differentiation under the integral sign, by formula (12.10), and by the periodicity of b_ϵ, we deduce the validity of equalities (13.18) and (13.20).

By exploiting the equality in (13.21) for $x \in \overline{Q} \setminus \partial\Omega_{p,\epsilon}$, we have that

$$\frac{\partial}{\partial x_j} b_\epsilon(x) = \frac{\partial}{\partial x_j} P_n[\Omega_{p,\epsilon}, 1](x) + \frac{\partial}{\partial x_j} \int_{\Omega_{p,\epsilon}} R_{q,n}(x-y)\, dy \quad \forall x \in \overline{Q} \setminus \partial\Omega_{p,\epsilon} .$$

By the Divergence Theorem 4.1, we verify that

$$\begin{aligned} \frac{\partial}{\partial x_j} \int_{\Omega_{p,\epsilon}} R_{q,n}(x-y)\, dy &= \int_{\Omega_{p,\epsilon}} \frac{\partial}{\partial x_j} R_{q,n}(x-y)\, dy \\ &= -\int_{\Omega_{p,\epsilon}} \frac{\partial}{\partial y_j} R_{q,n}(x-y)\, dy \\ &= -\int_{\partial\Omega_{p,\epsilon}} R_{q,n}(x-y)(\nu_{\Omega_{p,\epsilon}})_j(y)\, d\sigma_y \qquad \forall x \in \overline{Q} \setminus \partial\Omega_{p,\epsilon} . \end{aligned}$$

Then by the formula of Lemma 7.7 for the first order derivatives of the Newtonian potential and by the continuity and the periodicity of $\frac{\partial}{\partial x_j} b_\epsilon$, we deduce the validity of equality (13.16).

Since $(\nu_{\Omega_{p,\epsilon}})_j \in C^{0,\alpha}(\partial\Omega_{p,\epsilon})$ for all $j \in \{1, \ldots, n\}$, equality (13.16) and classical properties of the q-periodic single layer potential imply that $\frac{\partial}{\partial x_j} b_{\epsilon|\overline{\mathbb{S}[\Omega_{p,\epsilon}]}}$ belongs to $C^{1,\alpha}_q(\overline{\mathbb{S}[\Omega_{p,\epsilon}]})$ and that $\frac{\partial}{\partial x_j} b_{\epsilon|\overline{\mathbb{S}[\Omega_{p,\epsilon}]^-}} \in C^{1,\alpha}_q(\overline{\mathbb{S}[\Omega_{p,\epsilon}]^-})$ (cf., e.g., Theorem 12.8). Hence, the membership relations (13.17) and (13.19) follow.

Statements (iii) and (iv) are immediate consequences of (i), (ii), and of Proposition 13.2 on the q-periodic Newtonian potential. □

Then we have the following elementary technical lemma about the periodic domain $\mathbb{S}[\Omega_{p,\epsilon}]^-$.

Lemma 13.6. *Let* $\epsilon_1 \in]0, \epsilon_0[$.

(i) Let $\tilde{\Omega}$ *be an open subset of* $\mathbb{R}^n$ *with a nonzero distance from* $p + q\mathbb{Z}^n$. *Then there exist* $\epsilon^*_{\tilde{\Omega}} \in]0, \epsilon_1[$ *such that*

$$\tilde{\Omega} \subseteq \mathbb{S}[\Omega_{p,\epsilon}]^- \qquad \forall \epsilon \in [-\epsilon^*_{\tilde{\Omega}}, \epsilon^*_{\tilde{\Omega}}] ,$$

and $\epsilon_{\tilde{\Omega}} \in]0, \epsilon^*_{\tilde{\Omega}}[$ *such that*

$$\overline{\mathbb{S}[\Omega_{p,\epsilon^*_{\tilde{\Omega}}}]^-} \subseteq \mathbb{S}[\Omega_{p,\epsilon}]^- \qquad \forall \epsilon \in [-\epsilon_{\tilde{\Omega}}, \epsilon_{\tilde{\Omega}}] .$$

(ii) Let $\Omega^\sharp$ be a bounded open subset of $\mathbb{R}^n$ such that $\Omega^\sharp \subseteq \mathbb{R}^n \setminus \overline{\Omega}$. Then there exists $\epsilon_{\Omega^\sharp,r} \in]0, \epsilon_1[$ such that

$$p+\epsilon\overline{\Omega^\sharp} \subseteq Q\,, \qquad p+\epsilon\Omega^\sharp \subseteq \mathbb{S}[\Omega_{p,\epsilon}]^- \qquad \forall \epsilon \in [-\epsilon_{\Omega^\sharp,r}, \epsilon_{\Omega^\sharp,r}]\setminus\{0\}\,. \quad (13.22)$$

Proof. We first consider statement (i). Since $\overline{\tilde{\Omega}}$ has a positive distance $d_{\tilde{\Omega}}$ from $p + q\mathbb{Z}^n$, it suffices to take $\epsilon^*_{\tilde{\Omega}} \in]0, \epsilon_1[$ such that

$$\epsilon\overline{\Omega} \subseteq \mathbb{B}_n(0, 2^{-1}d_{\tilde{\Omega}}) \qquad \forall \epsilon \in [-\epsilon^*_{\tilde{\Omega}}, \epsilon^*_{\tilde{\Omega}}]\,.$$

Since Ω is an open neighborhood of 0, there exists $\epsilon_{\tilde{\Omega}} \in]0, \epsilon^*_{\tilde{\Omega}}[$ such that

$$\epsilon\overline{\Omega} \subseteq \epsilon^*_{\tilde{\Omega}}\Omega \qquad \forall \epsilon \in [-\epsilon_{\tilde{\Omega}}, \epsilon_{\tilde{\Omega}}]\,,$$

or equivalently such that

$$p + qz + \epsilon\overline{\Omega} \subseteq p + qz + \epsilon^*_{\tilde{\Omega}}\Omega \qquad \forall \epsilon \in [-\epsilon_{\tilde{\Omega}}, \epsilon_{\tilde{\Omega}}]\,,$$

for all $z \in \mathbb{Z}^n$. Then statement (i) follows.

We now consider statement (ii). By taking $\epsilon_{\Omega^\sharp,r}$ small enough, we can clearly assume that

$$p + \epsilon\overline{\Omega^\sharp} \subseteq Q \qquad \forall \epsilon \in [-\epsilon_{\Omega^\sharp,r}, \epsilon_{\Omega^\sharp,r}]\,.$$

Then if $\epsilon \in [-\epsilon_{\Omega^\sharp,r}, \epsilon_{\Omega^\sharp,r}] \setminus \{0\}$, we have

$$p + \epsilon\Omega^\sharp = p + \epsilon\big(\Omega^\sharp \cap (\mathbb{R}^n \setminus \overline{\Omega})\big) = (p + \epsilon\Omega^\sharp) \setminus (p + \epsilon\overline{\Omega})$$

and

$$(p + \epsilon\Omega^\sharp) \setminus (p + \epsilon\overline{\Omega}) \subseteq Q \setminus (p + \epsilon\overline{\Omega}) \subseteq \mathbb{S}[\Omega_{p,\epsilon}]^-\,.$$

As a consequence,

$$p + \epsilon\Omega^\sharp \subseteq \mathbb{S}[\Omega_{p,\epsilon}]^-$$

for all $\epsilon \in [-\epsilon_{\Omega^\sharp,r}, \epsilon_{\Omega^\sharp,r}]\setminus\{0\}$ and thus (13.22) holds. □

Finally, we state the following known technical lemma, whose validity can be immediately deduced by Theorems 12.23 (ii) and 12.25 (ii).

Lemma 13.7. *Let $\epsilon \in]0, \epsilon_0[$. Then the following statements hold.*

(i) The map that takes a pair (μ, c) to $v^-_{q,\Omega_{p,\epsilon}}[\mu] + c$ is a linear homeomorphism from $C^{0,\alpha}(\partial\Omega_{p,\epsilon})_0 \times \mathbb{R}$ to the subspace of those functions of $C^{1,\alpha}_q(\overline{\mathbb{S}[\Omega_{p,\epsilon}]^-})$ that are harmonic in $\mathbb{S}[\Omega_{p,\epsilon}]^-$.

(ii) The map that takes a pair (μ, c) to $v^+_{q,\Omega_{p,\epsilon}}[\mu] + c$ is a linear homeomorphism from $C^{0,\alpha}(\partial\Omega_{p,\epsilon})_0 \times \mathbb{R}$ to the subspace of those functions of $C^{1,\alpha}_q(\overline{\mathbb{S}[\Omega_{p,\epsilon}]})$ that are harmonic in $\mathbb{S}[\Omega_{p,\epsilon}]$.

Proof. We prove statement (i). We first note that if $u \in C^{1,\alpha}_q(\overline{\mathbb{S}[\Omega_{p,\epsilon}]^-})$ is harmonic in $\mathbb{S}[\Omega_{p,\epsilon}]^-$, then Proposition 12.18 (ii) implies

$$\int_{\partial\Omega_{p,\epsilon}} \nu_{\Omega_{p,\epsilon}} \cdot \nabla u \, d\sigma = 0 \, .$$

As a consequence, since $\nu_{\Omega_{p,\epsilon}} \cdot \nabla u \in C^{0,\alpha}(\partial\Omega_{p,\epsilon})$, we have

$$\nu_{\Omega_{p,\epsilon}} \cdot \nabla u \in C^{0,\alpha}(\partial\Omega_{p,\epsilon})_0 \, .$$

Accordingly, Theorem 12.23 (ii) ensures that there exists a pair (μ, c) in the space $C^{0,\alpha}(\partial\Omega_{p,\epsilon})_0 \times \mathbb{R}$ such that

$$u = v^-_{q,\Omega_{p,\epsilon}}[\mu] + c \qquad \text{in } \overline{\mathbb{S}[\Omega_{p,\epsilon}]^-} \, .$$

In particular, μ is the unique function in $C^{0,\alpha}(\partial\Omega_{p,\epsilon})_0$ such that

$$\nu_{\Omega_{p,\epsilon}} \cdot \nabla u = \Big(\frac{1}{2} I + W^t_{q,\Omega_{p,\epsilon}}\Big)[\mu] \, . \tag{13.23}$$

Thus, equation (13.23) uniquely determines μ. Then, since u and $v^-_{q,\Omega_{p,\epsilon}}[\mu]$ solve the same Neumann problem in $\overline{\mathbb{S}[\Omega_{p,\epsilon}]^-}$, they differ by a constant, and hence c is uniquely determined. The proof of statement (ii) is very similar and is accordingly omitted. □

13.5 Asymptotic Behavior of the Longitudinal Flow of a Periodic Array of Thin Cylinders

In this section, we study the longitudinal flow of a Newtonian fluid at low Reynolds numbers around a periodic array of cylinders (cf. Mityushev and Adler [218, 219], Sangani and Yao [252, 253], Schmid [259]). We assume that the diameter of the cylinders is proportional to a nonzero parameter ϵ (close to 0) and that the driving pressure gradient is parallel to the cylinders. Then the velocity field has only one non-zero component which by virtue of the Stokes equations satisfies a Poisson equation (cf. problem (13.24) below). Knowing the longitudinal component of the velocity field, one can define the average longitudinal flow $\Sigma_{II}[\epsilon]$ for each ϵ close to 0. Here, we are interested in studying the behavior of $\Sigma_{II}[\epsilon]$ as ϵ approaches the degenerate value $\epsilon = 0$, i.e., when the diameter of the cylinders collapses. (Our presentation here follows that of the paper [230] with Mityushev.)

Several authors have studied the longitudinal flow of arrays of cylinders by exploiting different techniques.

Happel [118] provided an analysis for the dilute case on the basis that two concentric cylinders can serve as the model for fluid moving through an assemblage of cylinders.

In the seminal paper [119], Hasimoto investigated the viscous flow past a cubic array of spheres, in particular, for the case when the radius of the spheres is small compared to the mutual distance. He also applied his results to the two-dimensional flow past a square array of circular cylinders. His techniques are based on the construction of a spatially periodic fundamental solution for the Stokes system and can be applied to specific shapes (circular/spherical obstacles and square/cubic arrays).

Schmid [259] investigated the longitudinal laminar flow in an infinite square array of circular cylinders. Sangani and Yao [252] extended a method described in [253] to find the permeability of random arrays of infinitely long cylinders. In [253], the transverse and longitudinal permeabilities averaged over several configurations of random arrays of circular cylinders have been expressed as a function of the area fraction of the cylinders.

Mityushev and Adler [218, 219] considered the longitudinal permeability of periodic rectangular arrays of circular cylinders. The longitudinal permeability is defined as the integral of the opposite of the flow velocity, which, in its turn, is the solution of a Dirichlet problem for the Poisson equation. By methods of complex variable, the authors transformed such boundary value problem into a functional equation. Then they expressed the solution to such a functional equation in terms of a power series of the radius of the cylinders, which can then be used to derive a formula for the longitudinal flow as a logarithmic term and a power series in the radius of the cylinder. The authors have first considered the case of a single circular cylinder in the unit cell in [218] and then they have turned to study the case of an arbitrary (finite) distribution of circular cylinders inside the periodicity cell in [219].

Here instead, we investigate the asymptotic behavior of the longitudinal flow of a rectangular array of thin cylinders, corresponding to the so called dilute case. Our method allows to treat general shapes of the cross section of the cylinders and we do not need to confine ourselves to circular cylinders. Our aim is to extend the results of Mityushev and Adler [218, 219] to very general shapes and to prove rigorously that the longitudinal flow can be represented as the sum of a logarithmic term and a power series of the square of the 'size parameter' of the cylinders. This is deduced by an analyticity result for a singularly perturbed boundary value problem.

Instead of the Functional Equation Method of Mityushev and Adler [218, 219], here we exploit our Functional Analytic Approach to investigate to behavior of the longitudinal flow in the dilute case.

We now introduce the mathematical problem. All through this section, we fix

$$n = 2\,,$$

and

$$l \in]0, +\infty[\,, \qquad Q \equiv]0, l[\times]0, 1/l[\,, \qquad q \equiv \begin{pmatrix} l & 0 \\ 0 & 1/l \end{pmatrix}.$$

In other words, in this section, we confine to the case of dimension two and we take

$$q_{11} = l\,, \qquad q_{22} = 1/l\,,$$

(cf. (12.2)). If $\epsilon \in]-\epsilon_0, \epsilon_0[$, the set $\overline{\mathbb{S}[\Omega_{p,\epsilon}]} \times \mathbb{R}$ represents an infinite array of parallel cylinders. Instead, the set $\mathbb{S}[\Omega_{p,\epsilon}]^- \times \mathbb{R}$ is the region where a Newtonian fluid of viscosity μ is flowing at low Reynolds number. Then we assume that the pressure gradient is constant and parallel to the axis of the cylinders. As a consequence, by a standard argument based on the particular geometry of the problem (cf., e.g., Adler [1, Chap. 4], Sangani and Yao [253], and Mityushev and Adler [218, 219]), we can reduce the Stokes' system to a Poisson equation for the non-zero component of the velocity field. Since we are working with dimensionless quantities, we may assume that the viscosity of the fluid and the pressure gradient are both set equal to 1. (For a more complete discussion on spatially periodic structures, we refer to Adler [1, Chap. 4].) Accordingly, if $\epsilon \in]-\epsilon_0, \epsilon_0[\setminus\{0\}$, we consider the following Dirichlet problem for the Poisson equation

$$\begin{cases} \Delta u = 1 & \text{in } \mathbb{S}[\Omega_{p,\epsilon}]^- , \\ u(x+qz) = u(x) & \forall x \in \overline{\mathbb{S}[\Omega_{p,\epsilon}]^-} , z \in \mathbb{Z}^2 , \\ u(x) = 0 & \forall x \in \partial\Omega_{p,\epsilon} . \end{cases} \tag{13.24}$$

As we shall see, if $\epsilon \in]-\epsilon_0, \epsilon_0[\setminus\{0\}$, then the solution of problem (13.24) in $C^{1,\alpha}_q(\overline{\mathbb{S}[\Omega_{p,\epsilon}]^-})$ exists and is unique (cf. Theorem 12.23 and formula (13.28) below). Therefore, we shall denote it by u_ϵ. From the physical point of view, the function u_ϵ represents the non-zero component of the velocity field (cf. Mityushev and Adler [218, Sect. 2]). By means of the function u_ϵ, we can introduce the average longitudinal velocity $\Sigma_{II}[\epsilon]$ which we define as the integral of the flow velocity over the unit cell (cf. Adler [1], Mityushev and Adler [218, Sect. 3]), i.e.,

$$\Sigma_{II}[\epsilon] \equiv \int_{Q\setminus\overline{\Omega_{p,\epsilon}}} u_\epsilon(x)\,dx \qquad \forall \epsilon \in]-\epsilon_0, \epsilon_0[\setminus\{0\} .$$

We pose the following question:

$$\text{What can be said on the map } \epsilon \mapsto \Sigma_{II}[\epsilon] \text{ when } \epsilon \text{ is close to } 0? \tag{13.25}$$

Here, we answer the question in (13.25) by showing that

$$\Sigma_{II}[\epsilon] = \frac{\log|\epsilon|}{2\pi} + a_0 + a_1\epsilon^2 + \epsilon^4 \mathcal{J}[\epsilon] \tag{13.26}$$

for $\epsilon \neq 0$ close to 0, where a_0, a_1 are suitable real numbers that we compute in terms of the solution of explicit boundary value problems and integral equations and $\mathcal{J}$ is a real analytic function defined in a neighborhood of 0. Moreover, we prove that $\mathcal{J}$ is an even function. We observe that equality (13.26) and the parity of $\mathcal{J}$ (cf. Proposition 13.17) imply that

$$\Sigma_{II}[\epsilon] = \frac{\log|\epsilon|}{2\pi} + a_0 + a_1\epsilon^2 + \epsilon^4 \sum_{j=0}^{+\infty} b_j \epsilon^{2j} ,$$

for a suitable sequence of coefficients $\{b_j\}_{j=0}^{+\infty}$, where the series converges absolutely on a whole neighborhood of the degenerate value $\epsilon = 0$.

We observe that a real analyticity result for the dependence of the longitudinal flow upon regular perturbation of the shape of the cross section of the cylinders and the periodicity cell has been obtained by Luzzini, Pukhtaievych, and the third-named author in [195].

13.5.1 Asymptotic Behavior of $\Sigma_{II}[\epsilon]$

First of all, we transform the boundary value problem (13.24) for the Poisson equation into a problem for the Laplace equation. To do so, we need to introduce a periodic function B_ϵ such that $\Delta B_\epsilon = 1$. This is done in the following lemma, which follows immediately by Theorem 12.4.

Lemma 13.8. *Let $\epsilon \in]-\epsilon_0, \epsilon_0[\setminus\{0\}$. Let B_ϵ be the function from $\mathbb{R}^2 \setminus (p + q\mathbb{Z}^2)$ to $\mathbb{R}$ defined by*

$$B_\epsilon(x) \equiv -S_{q,2}(x-p) + \frac{\log|\epsilon|}{2\pi} \qquad \forall x \in \mathbb{R}^2 \setminus (p + q\mathbb{Z}^2)\,.$$

Then

$$\begin{aligned} &B_{\epsilon|\overline{\mathbb{S}[\Omega_{p,\epsilon}]^-}} \in C^{1,\alpha}_q(\overline{\mathbb{S}[\Omega_{p,\epsilon}]^-})\,, \qquad \Delta B_\epsilon = 1 \qquad \text{in } \mathbb{S}[\Omega_{p,\epsilon}]^-\,,\\ &B_\epsilon(p+\epsilon\xi) = -S_2(\xi) - R_{q,2}(\epsilon\xi) \qquad \forall \xi \in \partial\Omega\,. \end{aligned}$$

By means of the function B_ϵ introduced in Lemma 13.8, we are in the position to convert the homogeneous Dirichlet problem (13.24) for the Poisson equation into a non-homogeneous Dirichlet problem for the Laplace equation. If $\epsilon \in]-\epsilon_0, \epsilon_0[\setminus\{0\}$, we denote by $u_{\#,\epsilon}$ the unique solution in $C^{1,\alpha}_q(\overline{\mathbb{S}[\Omega_{p,\epsilon}]^-})$ of the following auxiliary boundary value problem

$$\begin{cases} \Delta u = 0 & \text{in } \mathbb{S}[\Omega_{p,\epsilon}]^-\,,\\ u(x+qz) = u(x) \ \forall x \in \overline{\mathbb{S}[\Omega_{p,\epsilon}]^-}\,, z \in \mathbb{Z}^2\,,\\ u(x) = -B_\epsilon(x) \quad \forall x \in \partial\Omega_{p,\epsilon}\,. \end{cases} \tag{13.27}$$

Clearly, the function

$$u_{\#,\epsilon} + B_\epsilon \qquad \text{in } \overline{\mathbb{S}[\Omega_{p,\epsilon}]^-}\,,$$

solves problem (13.24). Since Theorem 12.16 implies that problem (13.24) may have at most one solution, we have that the function u_ϵ defined by

$$u_\epsilon \equiv u_{\#,\epsilon} + B_\epsilon \qquad \text{in } \overline{\mathbb{S}[\Omega_{p,\epsilon}]^-}\,, \tag{13.28}$$

is the unique solution of (13.24). Moreover,

$$\Sigma_{II}[\epsilon] = \int_{Q\setminus\overline{\Omega_{p,\epsilon}}} u_{\#,\epsilon}\,dx + \int_{Q\setminus\overline{\Omega_{p,\epsilon}}} B_\epsilon\,dx\,. \tag{13.29}$$

Now, for each $\epsilon \in]-\epsilon_0, \epsilon_0[$, we define the function $\Gamma[\epsilon] \in C^{1,\alpha}(\partial\Omega)$ by setting

$$\Gamma[\epsilon](\xi) \equiv S_2(\xi) + R_{q,2}(\epsilon\xi) \qquad \forall \xi \in \partial\Omega\,. \tag{13.30}$$

As a consequence, if $\epsilon \in]-\epsilon_0, \epsilon_0[\setminus\{0\}$, the Dirichlet condition in problem (13.27) can be rewritten as

$$u(x) = \Gamma[\epsilon]\big((x-p)/\epsilon\big) \qquad \forall x \in \partial\Omega_{p,\epsilon}\,.$$

Now we would like to study the behavior of the integral of $u_{\#,\epsilon}$ in the perforated cell $Q \setminus \overline{\Omega_{p,\epsilon}}$. As a first step, we verify the analytic dependence of the Dirichlet datum of the problem defining $u_{\#,\epsilon}$ (i.e., of $\Gamma[\epsilon]$) upon ϵ. We do so in the following lemma.

Lemma 13.9. *The map Γ from $]-\epsilon_0, \epsilon_0[$ to $C^{1,\alpha}(\partial\Omega)$, which takes ϵ to the function of the variable $\xi \in \partial\Omega$ defined by* (13.30)*, is real analytic.*

Proof. We first note that if $\epsilon \in]-\epsilon_0, \epsilon_0[$, then $\epsilon\xi \in (\mathbb{R}^2 \setminus q\mathbb{Z}^2) \cup \{0\}$ for all $\xi \in \overline{\Omega}$. Clearly, the map from $]-\epsilon_0, \epsilon_0[$ to $C^{1,\alpha}(\overline{\Omega}, (\mathbb{R}^2 \setminus q\mathbb{Z}^2) \cup \{0\})$ that takes ϵ to the function $\epsilon\xi$ of the variable $\xi \in \overline{\Omega}$ is real analytic. Then, by the analyticity of $R_{q,2}$ in $(\mathbb{R}^2 \setminus q\mathbb{Z}^2) \cup \{0\}$ and by analyticity results for the composition operator (cf. Theorem A.23 or Remark 2.18 and Proposition A.25), we deduce that the map from $]-\epsilon_0, \epsilon_0[$ to $C^{1,\alpha}(\overline{\Omega})$ that takes ϵ to the function $R_{q,2}(\epsilon\xi)$ of the variable $\xi \in \overline{\Omega}$ is real analytic. Finally, by the continuity of the trace operator from $C^{1,\alpha}(\overline{\Omega})$ to $C^{1,\alpha}(\partial\Omega)$ we immediately deduce the validity of the lemma. □

In the following proposition, we provide a formulation in terms of integral equations of the auxiliary problem (13.27).

Proposition 13.10. *Let $\mathcal{M}$ be the map from $]-\epsilon_0, \epsilon_0[\times C^{1,\alpha}(\partial\Omega)_0 \times \mathbb{R}$ to $C^{1,\alpha}(\partial\Omega)$ defined by*

$$\begin{aligned}\mathcal{M}[\epsilon,\theta,c](\xi) \equiv &-\frac{1}{2}\theta(\xi) - \int_{\partial\Omega} \nu_\Omega(\eta)\cdot\nabla S_2(\xi-\eta)\theta(\eta)\,d\sigma_\eta \\ &-\epsilon\int_{\partial\Omega} \nu_\Omega(\eta)\cdot\nabla R_{q,2}(\epsilon(\xi-\eta))\theta(\eta)\,d\sigma_\eta + c - \Gamma[\epsilon](\xi) \\ &\qquad\qquad \forall \xi \in \partial\Omega\,,\end{aligned}$$

for all $(\epsilon,\theta,c) \in]-\epsilon_0, \epsilon_0[\times C^{1,\alpha}(\partial\Omega)_0 \times \mathbb{R}$. Then the following statements hold.

(i) If $\epsilon \in]-\epsilon_0, \epsilon_0[$, then there exists a unique pair (θ, c) in $C^{1,\alpha}(\partial\Omega)_0 \times \mathbb{R}$ such that $\mathcal{M}[\epsilon,\theta,c] = 0$, and we denote such a pair by $(\theta_\epsilon, c_\epsilon)$. In particular, θ_0 is the unique function in $C^{1,\alpha}(\partial\Omega)_0$ such that

$$-\frac{1}{2}\theta_0(\xi) - \int_{\partial\Omega} \nu_\Omega(\eta)\cdot\nabla S_2(\xi-\eta)\theta_0(\eta)\,d\sigma_\eta = S_2(\xi) - \lim_{\eta\to\infty} H^o_\Omega(\eta) \qquad \forall \xi \in \partial\Omega\,, \tag{13.31}$$

and

$$c_0 = \lim_{\eta\to\infty} H^o_\Omega(\eta) + R_{q,2}(0)\,, \tag{13.32}$$

where H^o_Ω is the unique function in $C^{1,\alpha}_{\mathrm{loc}}(\mathbb{R}^2 \setminus \Omega)$ such that

$$\begin{cases} \Delta H^o_\Omega(\eta) = 0 & \forall \eta \in \mathbb{R}^2 \setminus \overline{\Omega}\,, \\ H^o_\Omega(\eta) = S_2(\eta) & \forall \eta \in \partial\Omega\,, \\ H^o_\Omega \text{ is harmonic at infinity} \end{cases} \tag{13.33}$$

(see also Lemma 8.48). Moreover,

$$w^-_\Omega[\theta_0] = H^o_\Omega - \lim_{\eta\to\infty} H^o_\Omega(\eta) \qquad \text{in } \mathbb{R}^2 \setminus \Omega\,. \tag{13.34}$$

(ii) There exist $\epsilon_1 \in]0,\epsilon_0[$ and a real analytic map (Θ, C) from $]-\epsilon_1,\epsilon_1[$ to $C^{1,\alpha}(\partial\Omega)_0 \times \mathbb{R}$ such that $(\Theta[\epsilon], C[\epsilon]) = (\theta_\epsilon, c_\epsilon)$ for all $\epsilon \in]-\epsilon_1,\epsilon_1[$. Moreover,

$$(\Theta[\epsilon], C[\epsilon]) = (\Theta[-\epsilon], C[-\epsilon]) \qquad \forall \epsilon \in]-\epsilon_1,\epsilon_1[\,. \tag{13.35}$$

(iii) If $\epsilon \in]-\epsilon_1,\epsilon_1[\setminus\{0\}$, then

$$u_{\#,\epsilon}(x) = w^-_{q,\Omega_{p,\epsilon}}[\Theta[\epsilon]((\cdot - p)/\epsilon)](x) + C[\epsilon] \qquad \forall x \in \overline{\mathbb{S}[\Omega_{p,\epsilon}]^-}\,.$$

As a consequence,

$$u_{\#,\epsilon}(x) = -\epsilon \int_{\partial\Omega} \nu_\Omega(\eta)\cdot\nabla S_{q,2}(x-p-\epsilon\eta)\Theta[\epsilon](\eta)\,d\sigma_\eta + C[\epsilon] \quad \forall x \in \mathbb{S}[\Omega_{p,\epsilon}]^-\,.$$

Proof. We first consider statement (i). If $\epsilon \in]-\epsilon_0,\epsilon_0[\setminus\{0\}$, the existence and uniqueness of a pair $(\theta_\epsilon, c_\epsilon)$ in $C^{1,\alpha}(\partial\Omega)_0 \times \mathbb{R}$ such that $\mathcal{M}[\epsilon,\theta_\epsilon,c_\epsilon] = 0$ follows by the theorem of change of variables in integrals and by Proposition 12.21 (ii). Indeed, we first note that if $\epsilon \in]-\epsilon_0,\epsilon_0[\setminus\{0\}$, equality

$$\mathcal{M}[\epsilon,\theta,c] = 0$$

implies that

$$\begin{aligned} -\frac{1}{2}\theta(\xi) - \int_{\partial\Omega} \nu_\Omega(\eta)\cdot\nabla S_2(\xi-\eta)\theta(\eta)\,d\sigma_\eta & \\ -\epsilon\int_{\partial\Omega} \nu_\Omega(\eta)\cdot\nabla R_{q,2}(\epsilon(\xi-\eta))\theta(\eta)\,d\sigma_\eta + c = \Gamma[\epsilon](\xi) & \\ \forall \xi \in \partial\Omega\,. & \end{aligned} \tag{13.36}$$

By the theorem of change of variable in integrals if $\epsilon \in]-\epsilon_0,\epsilon_0[\setminus\{0\}$ we have

$$\int_{\partial\Omega_{p,\epsilon}} \nu_{\Omega_{p,\epsilon}}(y)\cdot\nabla S_{q,2}(p+\epsilon\xi-y)\Theta[\epsilon]((y-p)/\epsilon)\,d\sigma_y$$

$$= |\epsilon| \int_{\partial\Omega} \mathrm{sgn}(\epsilon)\nu_\Omega(\eta) \cdot \nabla S_{q,2}(p + \epsilon\xi - p - \epsilon\eta)\Theta[\epsilon](\eta)\, d\sigma_\eta$$
$$= \epsilon \int_{\partial\Omega} \nu_\Omega(\eta) \cdot \nabla S_{q,2}(\epsilon(\xi - \eta))\Theta[\epsilon](\eta)\, d\sigma_\eta$$

for all $\xi \in \partial\Omega$. Moreover, if $\xi, \eta \in \partial\Omega$ and $\epsilon \in]-\epsilon_0, \epsilon_0[\setminus\{0\}$, then

$$\begin{aligned}\nabla S_{q,2}(\epsilon(\xi - \eta)) &= \nabla S_2(\epsilon(\xi - \eta)) + \nabla R_{q,2}(\epsilon(\xi - \eta)) \\ &= \frac{1}{s_2} \frac{\epsilon(\xi - \eta)}{|\epsilon(\xi - \eta)|^2} + \nabla R_{q,2}(\epsilon(\xi - \eta)) \\ &= \frac{1}{s_2} \frac{\epsilon(\xi - \eta)}{\epsilon^2 |\xi - \eta|^2} + \nabla R_{q,2}(\epsilon(\xi - \eta)) \\ &= \frac{1}{\epsilon} \nabla S_2(\xi - \eta) + \nabla R_{q,2}(\epsilon(\xi - \eta)) .\end{aligned}$$

As a consequence, equality (13.36) can be rewritten as

$$\begin{aligned}-\frac{1}{2}\theta((x - p)/\epsilon) - \int_{\partial\Omega_{p,\epsilon}} \nu_{\Omega_{p,\epsilon}}(y) \cdot \nabla S_{q,2}(x - y)\theta((y - p)/\epsilon)\, d\sigma_y + c \\ = \Gamma[\epsilon]((x - p)/\epsilon) \qquad \forall x \in \partial\Omega_{p,\epsilon} .\end{aligned}$$

Then Proposition 12.21 (ii) implies the existence and uniqueness of a pair $(\theta_\epsilon, c_\epsilon)$ in $C^{1,\alpha}(\partial\Omega)_0 \times \mathbb{R}$ such that $\mathcal{M}[\epsilon, \theta_\epsilon, c_\epsilon] = 0$.

If instead $\epsilon = 0$, the existence and uniqueness of a pair (θ_0, c_0) in $C^{1,\alpha}(\partial\Omega)_0 \times \mathbb{R}$ such that $\mathcal{M}[0, \theta_0, c_0] = 0$ follows by Lemma 6.38. Next we denote by τ_0 the unique solution in $C^{0,\alpha}(\partial\Omega)$ of the following problem

$$\begin{cases} -\frac{1}{2}\tau(\xi) + \int_{\partial\Omega} \nu_\Omega(\xi) \cdot \nabla S_2(\xi - \eta)\tau(\eta)\, d\sigma_\eta = 0 & \forall \xi \in \partial\Omega , \\ \int_{\partial\Omega} \tau\, d\sigma = 1 , \end{cases} \tag{13.37}$$

and we observe that τ_0 is a generator of $\mathrm{Ker}\big(-\frac{1}{2}I + W_\Omega^t\big)$ (cf. Lemma 6.34). Accordingly, by the Fredholm Alternative Theorem 5.8,

$$\int_{\partial\Omega} \Big(-\frac{1}{2}I + W_\Omega\Big)[\theta_0]\tau_0\, d\sigma = 0 .$$

Then, by multiplying equality $\mathcal{M}[0, \theta_0, c_0] = 0$ by τ_0 and integrating on $\partial\Omega$, we deduce that

$$c_0 = \int_{\partial\Omega} \Gamma[0]\tau_0\, d\sigma .$$

Next we note that

$$\int_{\partial\Omega} \Gamma[0]\tau_0\, d\sigma = \int_{\partial\Omega} S_2(\eta)\tau_0(\eta)\, d\sigma_\eta + R_{q,2}(0) .$$

Then by Lemma 8.48, one shows that

$$\int_{\partial\Omega} S_2\tau_0\, d\sigma = \int_{\partial\Omega} S_2(0-\eta)\tau_0(\eta)\, d\sigma_\eta = \lim_{\eta\to\infty} H^o_\Omega(\eta)\,, \tag{13.38}$$

and, as a consequence, the validity of (13.32) follows. Then by equality (13.32) we can verify that θ_0 is the unique solution of (13.31). By the jump formula of Theorem 6.6 and by equation (13.31), the double layer potential $w^-_\Omega[\theta_0]$ and $H^o_\Omega - \lim_{\eta\to\infty} H^o_\Omega(\eta)$ coincide on $\partial\Omega$. Since both the double layer potential $w^-_\Omega[\theta_0]$ and $H^o_\Omega - \lim_{\eta\to\infty} H^o_\Omega(\eta)$ are harmonic in $\mathbb{R}^2\setminus\overline{\Omega}$ and vanish at infinity, then the validity of (13.34) follows.

We now turn to prove statement (ii). By Theorem 4.31 (i), the operator $w^-_\Omega[\cdot]_{|\partial\Omega}$ is linear and continuous from $C^{1,\alpha}(\partial\Omega)$ to itself. Then Lemma 13.9 implies that the map from $]-\epsilon_0,\epsilon_0[\times C^{1,\alpha}(\partial\Omega)_0\times\mathbb{R}$ to $C^{1,\alpha}(\partial\Omega)$, that takes (ϵ,θ,c) to the function

$$\begin{aligned} w^-_\Omega[\theta]_{|\partial\Omega}(\xi)&+c-\Gamma[\epsilon](\xi)\\ &= -\frac{1}{2}\theta(\xi) - \int_{\partial\Omega} \nu_\Omega(\eta)\cdot\nabla S_2(\xi-\eta)\theta(\eta)\, d\sigma_\eta + c - \Gamma[\epsilon](\xi) \end{aligned}$$

of the variable $\xi\in\partial\Omega$, is real analytic. Then we note that

$$\epsilon\partial\Omega - \epsilon\partial\Omega \subseteq (\mathbb{R}^n\setminus q\mathbb{Z}^n)\cup\{0\} \qquad \forall\epsilon\in]-\epsilon_0,\epsilon_0[\,.$$

Accordingly, Lemma 13.4 (i) and the continuity of the embedding of $C^{1,\alpha}(\partial\Omega)$ in $L^1(\partial\Omega)$ imply that the map from $]-\epsilon_0,\epsilon_0[\times C^{1,\alpha}(\partial\Omega)_0$ to $C^{1,\alpha}(\partial\Omega)$, that takes (ϵ,θ) to the function

$$-\epsilon\int_{\partial\Omega} \nu_\Omega(\eta)\cdot\nabla R_{q,2}(\epsilon(\xi-\eta))\theta(\eta)\, d\sigma_\eta$$

of the variable $\xi\in\partial\Omega$, is real analytic. As a consequence, $\mathcal{M}$ is a real analytic map from $]-\epsilon_0,\epsilon_0[\times C^{1,\alpha}(\partial\Omega)_0\times\mathbb{R}$ to $C^{1,\alpha}(\partial\Omega)$. By standard Calculus in Banach space, the differential $\partial_{(\theta,c)}\mathcal{M}[0,\theta_0,c_0]$ of $\mathcal{M}$ at $(0,\theta_0,c_0)$ with respect to (θ,c) is delivered by the following formula:

$$\begin{aligned} \partial_{(\theta,c)}&\mathcal{M}[0,\theta_0,c_0](\theta^\#,c^\#)(\xi)\\ &= -\frac{1}{2}\theta^\#(\xi) - \int_{\partial\Omega} \nu_\Omega(\eta)\cdot\nabla S_2(\xi-\eta)\theta^\#(\eta)\, d\sigma_\eta + c^\# \qquad \forall\xi\in\partial\Omega\,, \end{aligned}$$

for all $(\theta^\#,c^\#)\in C^{1,\alpha}(\partial\Omega)_0\times\mathbb{R}$. We now want to show that $\partial_{(\theta,c)}\mathcal{M}[0,\theta_0,c_0]$ is a linear homeomorphism from $C^{1,\alpha}(\partial\Omega)_0\times\mathbb{R}$ onto $C^{1,\alpha}(\partial\Omega)$. We note that

$$\partial_{(\theta,c)}\mathcal{M}[0,\theta_0,c_0](\theta^\#,c^\#) = \Big(-\frac{1}{2}I + W_\Omega\Big)[\theta^\#] + c^\#\,,$$

for all $(\theta^\#,c^\#)\in C^{1,\alpha}(\partial\Omega)_0\times\mathbb{R}$. As a consequence, by Lemma 6.38, we deduce that $\partial_{(\theta,c)}\mathcal{M}[0,\theta_0,c_0]$ is a linear homeomorphism from $C^{1,\alpha}(\partial\Omega)_0\times\mathbb{R}$ onto

$C^{1,\alpha}(\partial\Omega)$. Then by the Implicit Function Theorem for real analytic maps in Banach spaces (cf. Theorem A.19), we immediately deduce the existence of $\epsilon_1 \in]0, \epsilon_0[$ and of a real analytic map (Θ, C) from $]-\epsilon_1, \epsilon_1[$ to $C^{1,\alpha}(\partial\Omega)_0 \times \mathbb{R}$ such that

$$(\Theta[\epsilon], C[\epsilon]) = (\theta_\epsilon, c_\epsilon)$$

for all $\epsilon \in]-\epsilon_1, \epsilon_1[$. Since

$$\mathcal{M}[\epsilon, \cdot, \cdot] = \mathcal{M}[-\epsilon, \cdot, \cdot] \qquad \forall \epsilon \in]-\epsilon_0, \epsilon_0[\,,$$

we deduce the validity of (13.35).

To prove statement (iii), we first note that if $\epsilon \in]-\epsilon_0, \epsilon_0[\setminus\{0\}$, equality

$$\mathcal{M}[\epsilon, \Theta[\epsilon], C[\epsilon]] = 0\,,$$

implies that

$$\begin{aligned} -\frac{1}{2}\Theta[\epsilon](\xi) - \int_{\partial\Omega} \nu_\Omega(\eta) \cdot \nabla S_2(\xi - \eta)\Theta[\epsilon](\eta)\, d\sigma_\eta \\ - \epsilon \int_{\partial\Omega} \nu_\Omega(\eta) \cdot \nabla R_{q,2}(\epsilon(\xi - \eta))\Theta[\epsilon](\eta)\, d\sigma_\eta + C[\epsilon] = \Gamma[\epsilon](\xi) \\ \forall \xi \in \partial\Omega\,. \end{aligned} \tag{13.39}$$

By the theorem of change of variable in integrals if $\epsilon \in]-\epsilon_0, \epsilon_0[\setminus\{0\}$ and by the same argument of the proof of statement (i), equality (13.39) can be rewritten as

$$\begin{aligned} -\frac{1}{2}\Theta[\epsilon]((x-p)/\epsilon) - \int_{\partial\Omega_{p,\epsilon}} \nu_{\Omega_{p,\epsilon}}(y) \cdot \nabla S_{q,2}(x-y)\Theta[\epsilon]((y-p)/\epsilon)\, d\sigma_y + C[\epsilon] \\ = \Gamma[\epsilon]((x-p)/\epsilon) \qquad \forall x \in \partial\Omega_{p,\epsilon}\,. \end{aligned}$$

Then, by Theorem 12.24 the function

$$w^-_{q,\Omega_{p,\epsilon}}[\Theta[\epsilon]((\cdot - p)/\epsilon)](x) + C[\epsilon]$$

solves problem (13.27) and accordingly

$$u_{\#,\epsilon}(x) = w^-_{q,\Omega_{p,\epsilon}}[\Theta[\epsilon]((\cdot - p)/\epsilon)](x) + C[\epsilon]\,.$$

To complete the proof of (iii) we just note that the theorem of change of variable in integrals implies that

$$\begin{aligned} w^-_{q,\Omega_{p,\epsilon}}[\Theta[\epsilon]((\cdot - p)/\epsilon)](x) \\ = -|\epsilon| \int_{\partial\Omega} \mathrm{sgn}(\epsilon)\nu_\Omega(\eta) \cdot \nabla S_{q,2}(x - p - \epsilon\eta)\Theta[\epsilon](\eta)\, d\sigma_\eta \end{aligned}$$

$$= -\epsilon \int_{\partial\Omega} \nu_\Omega(\eta) \cdot \nabla S_{q,2}(x - p - \epsilon\eta)\Theta[\epsilon](\eta)\, d\sigma_\eta$$

for all $x \in \mathbb{S}[\Omega_{p,\epsilon}]^-$.

□

By Proposition 13.10, we can provide an asymptotic expansion of $C[\epsilon]$ for ϵ close to 0.

Lemma 13.11. *There exist $\epsilon_2 \in]0, \epsilon_1[$ and a real analytic function $\tilde{C}$ from $]-\epsilon_2, \epsilon_2[$ to $\mathbb{R}$ such that*

$$\begin{aligned} C[\epsilon] &= \lim_{\eta\to\infty} H^o_\Omega(\eta) + R_{q,2}(0) \\ &+ \epsilon^2\Big(\int_{\partial\Omega}\int_{\partial\Omega} (\xi-\eta)^t \nabla^2 R_{q,2}(0)\nu_\Omega(\eta)\theta_0(\eta)\, d\sigma_\eta \tau_0(\xi)\, d\sigma_\xi \\ &+ \sum_{\beta\in\mathbb{N}^2, |\beta|=2} \frac{D^\beta R_{q,2}(0)}{\beta!} \int_{\partial\Omega} \xi^\beta \tau_0(\xi)\, d\sigma_\xi\Big) + \epsilon^4 \tilde{C}[\epsilon] \qquad \forall \epsilon \in]-\epsilon_2, \epsilon_2[\,, \end{aligned}$$

where τ_0 is the unique solution in $C^{0,\alpha}(\partial\Omega)$ of (13.37).

Here $\nabla^2 R_{q,2}$ denotes the Hessian matrix of second order partial derivatives of $R_{q,2}$. We also recall that the exponent ^{t} denotes the transpose, and $\xi^\beta \equiv \xi_1^{\beta_1}\xi_2^{\beta_2}$ for all $\beta = (\beta_1, \beta_2) \in \mathbb{N}^2$ and $\xi = (\xi_1, \xi_2) \in \mathbb{R}^2$.

Proof. By Proposition 13.10, we deduce the existence of $\epsilon_2 \in]0, \epsilon_1[$, $c_* \in \mathbb{R}$, and of a real analytic even function $\tilde{C}$ from $]-\epsilon_2, \epsilon_2[$ to $\mathbb{R}$ such that

$$C[\epsilon] = c_0 + \epsilon^2 c_* + \epsilon^4 \tilde{C}[\epsilon] \qquad \forall \epsilon \in]-\epsilon_2, \epsilon_2[\,.$$

As a consequence,

$$\begin{aligned} &-\frac{1}{2}\Theta[\epsilon](\xi) - \int_{\partial\Omega} \nu_\Omega(\eta) \cdot \nabla S_2(\xi-\eta)\Theta[\epsilon](\eta)\, d\sigma_\eta \\ &-\epsilon \int_{\partial\Omega} \nu_\Omega(\eta) \cdot \nabla R_{q,2}(\epsilon(\xi-\eta))\Theta[\epsilon](\eta)\, d\sigma_\eta + \lim_{\eta\to\infty} H^o_\Omega(\eta) + R_{q,2}(0) + \epsilon^2 c_* \\ &+ \epsilon^4\tilde{C}[\epsilon] = S_2(\xi) + R_{q,2}(\epsilon\xi) \qquad \forall \xi \in \partial\Omega\,, \epsilon \in]-\epsilon_2, \epsilon_2[\,. \end{aligned}$$

Since $R_{q,2}$ is an even function which is analytic in $(\mathbb{R}^2 \setminus q\mathbb{Z}^2) \cup \{0\}$, there exists $R > 0$ such that $\overline{\mathbb{B}_2(0, R)} \subseteq (\mathbb{R}^2 \setminus q\mathbb{Z}^2) \cup \{0\}$ and

$$R_{q,2}(x) = R_{q,2}(0) + \sum_{\beta\in\mathbb{N}^2, |\beta|=2} \frac{D^\beta R_{q,2}(0)}{\beta!} x^\beta + \sum_{k=2}^{+\infty} \sum_{\beta\in\mathbb{N}^2, |\beta|=2k} \frac{D^\beta R_{q,2}(0)}{\beta!} x^\beta\,,$$

for all $x \in \overline{\mathbb{B}_2(0, R)}$. Now let $\tilde{\epsilon} \in]0, \epsilon_2[$ be such that

$$\overline{\tilde{\epsilon}\mathbb{B}_2(0, \mathrm{diam}\,(\overline{\Omega}))} \subseteq \mathbb{B}_2(0, R)\,.$$

Then for all $(\epsilon,\xi)\in]-\tilde{\epsilon},\tilde{\epsilon}[\times\mathbb{B}_2(0,\mathrm{diam}\,(\overline{\Omega}))$

$$R_{q,2}(\epsilon\xi) = R_{q,2}(0)+\epsilon^2\sum_{\beta\in\mathbb{N}^2,\,|\beta|=2}\frac{D^\beta R_{q,2}(0)}{\beta!}\xi^\beta + \epsilon^4\sum_{k=2}^{+\infty}\sum_{\beta\in\mathbb{N}^2,\,|\beta|=2k}\frac{D^\beta R_{q,2}(0)}{\beta!}\epsilon^{4-|\beta|}\xi^\beta .$$

By Definition 2.15 of real analytic functions, possibly taking a smaller $\tilde{\epsilon}>0$, the expression

$$\sum_{k=2}^{+\infty}\sum_{\beta\in\mathbb{N}^2,\,|\beta|=2k}\frac{D^\beta R_{q,2}(0)}{\beta!}\epsilon^{4-|\beta|}\xi^\beta$$

in the variable (ϵ,ξ) defines a real analytic function $\tilde{R}_1$ from

$$]-\tilde{\epsilon},\tilde{\epsilon}[\times\mathbb{B}_2(0,\mathrm{diam}\,(\overline{\Omega}))$$

to $\mathbb{R}$ such that

$$R_{q,2}(\epsilon\xi) = R_{q,2}(0)+\epsilon^2\sum_{\beta\in\mathbb{N}^2,\,|\beta|=2}\frac{D^\beta R_{q,2}(0)}{\beta!}\xi^\beta+\epsilon^4\tilde{R}_1(\epsilon,\xi)$$

for all $(\epsilon,\xi)\in]-\tilde{\epsilon},\tilde{\epsilon}[\times\mathbb{B}_2(0,\mathrm{diam}\,(\overline{\Omega}))$. Similarly, possibly taking a smaller $\tilde{\epsilon}>0$, there exists an analytic function $\tilde{R}_2$ from $]-\tilde{\epsilon},\tilde{\epsilon}[\times\mathbb{B}_2(0,\mathrm{diam}\,(\overline{\Omega}))$ to $\mathbb{R}^2$ such that

$$\nu_\Omega(\eta)\cdot\nabla R_{q,2}(\epsilon(\xi-\eta)) = \epsilon(\xi-\eta)^t\nabla^2 R_{q,2}(0)\nu_\Omega(\eta)+\epsilon^2\nu_\Omega(\eta)\cdot\tilde{R}_2(\epsilon,\xi-\eta) \qquad \forall\epsilon\in]-\tilde{\epsilon},\tilde{\epsilon}[,\ \xi,\eta\in\partial\Omega .$$

As a consequence,

$$\begin{aligned}
-\frac{1}{2}\Theta[\epsilon](\xi) &- \int_{\partial\Omega}\nu_\Omega(\eta)\cdot\nabla S_2(\xi-\eta)\Theta[\epsilon](\eta)\,d\sigma_\eta \\
&-\epsilon^2\int_{\partial\Omega}(\xi-\eta)^t\nabla^2 R_{q,2}(0)\nu_\Omega(\eta)\Theta[\epsilon](\eta)\,d\sigma_\eta \\
&-\epsilon^3\int_{\partial\Omega}\nu_\Omega(\eta)\cdot\tilde{R}_2(\epsilon,\xi-\eta)\Theta[\epsilon](\eta)\,d\sigma_\eta+\lim_{\eta\to\infty}H^o_\Omega(\eta) \\
&+R_{q,2}(0)+\epsilon^2 c_*+\epsilon^4\tilde{C}[\epsilon] \\
&= S_2(\xi)+R_{q,2}(0)+\epsilon^2\sum_{\beta\in\mathbb{N}^2,\,|\beta|=2}\frac{D^\beta R_{q,2}(0)}{\beta!}\xi^\beta+\epsilon^4\tilde{R}_1(\epsilon,\xi) \\
&\qquad \forall\xi\in\partial\Omega\,,\epsilon\in]-\tilde{\epsilon},\tilde{\epsilon}[\,.
\end{aligned} \tag{13.40}$$

Now let τ_0 be the unique solution in $C^{0,\alpha}(\partial\Omega)$ of problem (13.37). Then

$$\int_{\partial\Omega}\Big(-\frac{1}{2}\Theta[\epsilon](\xi)-\int_{\partial\Omega}\nu_\Omega(\eta)\cdot\nabla S_2(\xi-\eta)\Theta[\epsilon](\eta)\,d\sigma_\eta\Big)\tau_0(\xi)\,d\sigma_\xi$$
$$=\int_{\partial\Omega}\left(-\frac{1}{2}I_\Omega+W_\Omega\right)[\Theta[\epsilon]]\tau_0\,d\sigma$$
$$=\int_{\partial\Omega}\Theta[\epsilon]\left(\frac{1}{2}I_\Omega+W^t_\Omega\right)[\tau_0]\,d\sigma=0\,,$$

for all $\epsilon\in]-\tilde{\epsilon},\tilde{\epsilon}[$ (cf. Theorem 6.5). By equality (13.38), we have

$$\lim_{\eta\to\infty}H^o_\Omega(\eta)+R_{q,2}(0)=\int_{\partial\Omega}S_2\tau_0\,d\sigma+R_{q,2}(0)\,.$$

Hence, by multiplying (13.40) by $\epsilon^{-2}\tau_0(\xi)$ and integrating on $\partial\Omega$ with respect to the variable ξ, we obtain

$$-\int_{\partial\Omega}\int_{\partial\Omega}(\xi-\eta)^t\nabla^2R_{q,2}(0)\nu_\Omega(\eta)\Theta[\epsilon](\eta)\,d\sigma_\eta\tau_0(\xi)\,d\sigma_\xi$$
$$-\epsilon\int_{\partial\Omega}\int_{\partial\Omega}\nu_\Omega(\eta)\cdot\tilde{R}_2(\epsilon,\xi-\eta)\Theta[\epsilon](\eta)\,d\sigma_\eta\tau_0(\xi)\,d\sigma_\xi+c_*+\epsilon^2\tilde{C}[\epsilon]$$
$$=\sum_{\beta\in\mathbb{N}^2,\,|\beta|=2}\frac{D^\beta R_{q,2}(0)}{\beta!}\int_{\partial\Omega}\xi^\beta\tau_0(\xi)\,d\sigma_\xi$$
$$+\epsilon^2\int_{\partial\Omega}\tilde{R}_1(\epsilon,\xi)\tau_0(\xi)\,d\sigma_\xi\qquad\forall\xi\in\partial\Omega\,,\epsilon\in]-\tilde{\epsilon},\tilde{\epsilon}[\,.$$

Thus, by letting ϵ tend to 0, we deduce that

$$c_*=\int_{\partial\Omega}\int_{\partial\Omega}(\xi-\eta)^t\nabla^2R_{q,2}(0)\nu_\Omega(\eta)\theta_0(\eta)\,d\sigma_\eta\tau_0(\xi)\,d\sigma_\xi$$
$$+\sum_{\beta\in\mathbb{N}^2,\,|\beta|=2}\frac{D^\beta R_{q,2}(0)}{\beta!}\int_{\partial\Omega}\xi^\beta\tau_0(\xi)\,d\sigma_\xi\,.$$

Hence, the validity of the statement follows. □

Remark 13.12. Let $H_{2,\beta}$ be the unique solution in $C^{1,\alpha}_{\mathrm{loc}}(\mathbb{R}^2\setminus\Omega)$ of the exterior Dirichlet problem

$$\begin{cases}\Delta H_{2,\beta}(\xi)=0 & \forall\xi\in\mathbb{R}^2\setminus\overline{\Omega}\,,\\ H_{2,\beta}(\xi)=\xi^\beta & \forall\xi\in\partial\Omega\,,\\ H_{2,\beta}\text{ is harmonic at infinity}\,,\end{cases}$$

for all $\beta\in\mathbb{N}^2$ such that $|\beta|=2$ (cf. Theorem 6.33). Let τ_0 be the unique solution in $C^{0,\alpha}(\partial\Omega)$ of (13.37). By arguing as in the proof of Lemma 8.48, we can verify that

$$\int_{\partial\Omega} \xi^\beta \tau_0(\xi)\, d\sigma_\xi = \lim_{\xi\to\infty} H_{2,\beta}(\xi)\,,$$

for all $\beta \in \mathbb{N}^2$ such that $|\beta| = 2$ (cf. (8.145)).

In the following proposition, we study the behavior of $\int_{Q\setminus\overline{\Omega_{p,\epsilon}}} u_{\#,\epsilon}\, dx$.

Proposition 13.13. *There exist $\epsilon_3 \in]0, \epsilon_2[$ and a real analytic function $\mathcal{J}_{\#,1}$ from $]-\epsilon_3, \epsilon_3[$ to $\mathbb{R}$, such that*

$$\begin{aligned}
&\int_{Q\setminus\overline{\Omega_{p,\epsilon}}} u_{\#,\epsilon}(x)\, dx = \lim_{\xi\to\infty} H^o_\Omega(\xi) + R_{q,2}(0)\\
&+\epsilon^2\Bigg(- m_2(\Omega)\Big(\lim_{\xi\to\infty} H^o_\Omega(\xi) + R_{q,2}(0)\Big)\\
&+\int_{\partial\Omega}\int_{\partial\Omega} (\xi-\eta)^t \nabla^2 R_{q,2}(0)\nu_\Omega(\eta)\theta_0(\eta)\, d\sigma_\eta \tau_0(\xi)\, d\sigma_\xi\\
&+\sum_{\beta\in\mathbb{N}^2, |\beta|=2} \frac{D^\beta R_{q,2}(0)}{\beta!}\int_{\partial\Omega} \xi^\beta \tau_0(\xi)\, d\sigma_\xi\\
&+\sum_{j=1}^{2}\int_{\partial\Omega}\Big(\int_{\partial\Omega} S_2(\xi-\eta)\theta_0(\eta)(\nu_\Omega(\eta))_j\, d\sigma_\eta\Big)(\nu_\Omega(\xi))_j\, d\sigma_\xi\Bigg)\\
&+\epsilon^4 \mathcal{J}_{\#,1}[\epsilon] \qquad \forall \epsilon \in]-\epsilon_3, \epsilon_3[\setminus\{0\}\,,
\end{aligned} \tag{13.41}$$

where τ_0 is the unique solution in $C^{0,\alpha}(\partial\Omega)$ of (13.37) *and H^o_Ω is the unique solution of (13.33).*

We recall that in equality (13.41) the symbol $(\nu_\Omega(\cdot))_j$ denotes the j-th component of $\nu_\Omega(\cdot)$.

Proof. Let $\epsilon \in]-\epsilon_2, \epsilon_2[\setminus\{0\}$. By the formula for $u_{\#,\epsilon}$ of Proposition 13.10 (iii) and by the formula for $C[\epsilon]$ of Lemma 13.11, we have

$$\begin{aligned}
&\int_{Q\setminus\overline{\Omega_{p,\epsilon}}} u_{\#,\epsilon}(x)\, dx = \int_{Q\setminus\overline{\Omega_{p,\epsilon}}} w^-_{q,\Omega_{p,\epsilon}}\big[\Theta[\epsilon]((\cdot-p)/\epsilon)\big](x)\, dx\\
&+\Big(\lim_{\xi\to\infty} H^o_\Omega(\xi) + R_{q,2}(0)\Big)\,(1-\epsilon^2 m_2(\Omega))\\
&+\epsilon^2\Big(\int_{\partial\Omega}\int_{\partial\Omega}(\xi-\eta)^t\nabla^2 R_{q,2}(0)\nu_\Omega(\eta)\theta_0(\eta)\, d\sigma_\eta\tau_0(\xi)\, d\sigma_\xi\\
&+\sum_{\beta\in\mathbb{N}^2, |\beta|=2}\frac{D^\beta R_{q,2}(0)}{\beta!}\int_{\partial\Omega}\xi^\beta\tau_0(\xi)\, d\sigma_\xi\Big)(1-\epsilon^2 m_2(\Omega))\\
&+\epsilon^4\tilde{C}[\epsilon](1-\epsilon^2 m_2(\Omega))\,.
\end{aligned}$$

By equality (12.17), we have

$$
\begin{aligned}
&w^{-}_{q,\Omega_{p,\epsilon}}\big[\Theta[\epsilon]((\cdot-p)/\epsilon)\big](x)\\
&\quad=-\sum_{j=1}^{2}\frac{\partial}{\partial x_j}v^{-}_{q,\Omega_{p,\epsilon}}\big[\Theta[\epsilon]((\cdot-p)/\epsilon)(\nu_{\Omega_{p,\epsilon}}(\cdot))_j\big](x)\quad \forall x\in\overline{Q}\setminus\overline{\Omega}_{p,\epsilon}\,.
\end{aligned}
$$

Now let $j\in\{1,2\}$. By the Divergence Theorem 4.1 and the periodicity of the periodic single layer potential and by the rule of change of variables of (2.60), (2.61), we have

$$
\begin{aligned}
&\int_{Q\setminus\overline{\Omega_{p,\epsilon}}}\frac{\partial}{\partial x_j}v^{-}_{q,\Omega_{p,\epsilon}}\big[\Theta[\epsilon]((\cdot-p)/\epsilon)(\nu_{\Omega_{p,\epsilon}}(\cdot))_j\big](x)\,dx\\
&\quad=\int_{\partial Q}v^{-}_{q,\Omega_{p,\epsilon}}\big[\Theta[\epsilon]((\cdot-p)/\epsilon)(\nu_{\Omega_{p,\epsilon}}(\cdot))_j\big](x)(\nu_Q(x))_j\,d\sigma_x\\
&\qquad-\int_{\partial\Omega_{p,\epsilon}}v^{-}_{q,\Omega_{p,\epsilon}}\big[\Theta[\epsilon]((\cdot-p)/\epsilon)(\nu_{\Omega_{p,\epsilon}}(\cdot))_j\big](x)(\nu_{\Omega_{p,\epsilon}}(x))_j\,d\sigma_x\\
&\quad=-\epsilon\int_{\partial\Omega}v^{-}_{q,\Omega_{p,\epsilon}}\big[\Theta[\epsilon]((\cdot-p)/\epsilon)(\nu_{\Omega_{p,\epsilon}}(\cdot))_j\big](p+\epsilon\xi)(\nu_{\Omega}(\xi))_j\,d\sigma_\xi\,.
\end{aligned}
$$

Then, again by the rule of change of variables of (2.60), (2.61), we have

$$
\begin{aligned}
&v^{-}_{q,\Omega_{p,\epsilon}}\big[\Theta[\epsilon]((\cdot-p)/\epsilon)(\nu_{\Omega_{p,\epsilon}}(\cdot))_j\big](p+\epsilon\xi)\\
&\quad=\epsilon\int_{\partial\Omega}S_2(\epsilon(\xi-\eta))\Theta[\epsilon](\eta)(\nu_\Omega(\eta))_j\,d\sigma_\eta\\
&\qquad+\epsilon\int_{\partial\Omega}R_{q,2}(\epsilon(\xi-\eta))\Theta[\epsilon](\eta)(\nu_\Omega(\eta))_j\,d\sigma_\eta\qquad\forall\xi\in\partial\Omega\,.
\end{aligned}
$$

We now observe that if $\epsilon\neq 0$ and $\rho\in\mathbb{R}^2\setminus\{0\}$, then we have

$$
S_2(\epsilon\rho)=S_2(\rho)+\frac{1}{2\pi}\log|\epsilon|\,. \tag{13.42}
$$

Moreover, by the Divergence Theorem 4.1 we have

$$
\begin{aligned}
&\int_{\partial\Omega}\Big(\int_{\partial\Omega}\Theta[\epsilon](\eta)(\nu_\Omega(\eta))_j\,d\sigma_\eta\Big)(\nu_\Omega(\xi))_j\,d\sigma_\xi\\
&\quad=\Big(\int_{\partial\Omega}\Theta[\epsilon](\eta)(\nu_\Omega(\eta))_j\,d\sigma_\eta\Big)\Big(\int_{\partial\Omega}(\nu_\Omega(\xi))_j\,d\sigma_\xi\Big)=0\,.
\end{aligned} \tag{13.43}
$$

Hence, by equalities (13.42) and (13.43), if $\epsilon\in]-\epsilon_2,\epsilon_2[\setminus\{0\}$, we have

$$\int_{Q\setminus\overline{\Omega_{p,\epsilon}}} w^{-}_{q,\Omega_{p,\epsilon}}\big[\Theta[\epsilon]((\cdot-p)/\epsilon)\big](x)\,dx$$
$$=\sum_{j=1}^{2}\epsilon^2\Big[\int_{\partial\Omega}\Big(\int_{\partial\Omega}S_2(\xi-\eta)\Theta[\epsilon](\eta)(\nu_\Omega(\eta))_j\,d\sigma_\eta\Big)(\nu_\Omega(\xi))_j\,d\sigma_\xi$$
$$+\int_{\partial\Omega}\Big(\int_{\partial\Omega}R_{q,2}(\epsilon(\xi-\eta))\Theta[\epsilon](\eta)(\nu_\Omega(\eta))_j\,d\sigma_\eta\Big)(\nu_\Omega(\xi))_j\,d\sigma_\xi\Big].$$

Thus we set

$$\tilde{\mathcal{J}}_{\#,1}[\epsilon]\equiv\sum_{j=1}^{2}\Big[\int_{\partial\Omega}\Big(\int_{\partial\Omega}S_2(\xi-\eta)\Theta[\epsilon](\eta)(\nu_\Omega(\eta))_j\,d\sigma_\eta\Big)(\nu_\Omega(\xi))_j\,d\sigma_\xi$$
$$+\int_{\partial\Omega}\Big(\int_{\partial\Omega}R_{q,2}(\epsilon(\xi-\eta))\Theta[\epsilon](\eta)(\nu_\Omega(\eta))_j\,d\sigma_\eta\Big)(\nu_\Omega(\xi))_j\,d\sigma_\xi\Big],$$

for all $\epsilon\in]-\epsilon_2,\epsilon_2[$. Then the analyticity of Θ, the continuity of the linear map from $C^{0,\alpha}(\partial\Omega)$ to $C^{1,\alpha}(\partial\Omega)$ that takes f to the function $\int_{\partial\Omega}S_2(\xi-\eta)f(\eta)\,d\sigma_\eta$ of the variable $\xi\in\partial\Omega$ (cf. Theorem 4.25), the continuity of the pointwise product in Schauder spaces, the analyticity of nonlinear integral operators with analytic kernel of Theorem A.28 (i), and standard Calculus in Banach spaces imply that $\tilde{\mathcal{J}}_{\#,1}$ is a real analytic function from $]-\epsilon_2,\epsilon_2[$ to $\mathbb{R}$. Moreover, equality $\int_{\partial\Omega}(\nu_\Omega(\xi))_j\,d\sigma_\xi=0$ implies that

$$\tilde{\mathcal{J}}_{\#,1}[0]=\sum_{j=1}^{2}\int_{\partial\Omega}\Big(\int_{\partial\Omega}S_2(\xi-\eta)\theta_0(\eta)(\nu_\Omega(\eta))_j\,d\sigma_\eta\Big)(\nu_\Omega(\xi))_j\,d\sigma_\xi\,.$$

Now, by Theorem 12.4 (v) and by (13.35), $(\Theta[\epsilon],C[\epsilon])$ is even in ϵ and $R_{q,2}(x)$ is even in x. Then $\tilde{\mathcal{J}}_{\#,1}[\epsilon]=\tilde{\mathcal{J}}_{\#,1}[-\epsilon]$ for all $\epsilon\in]-\epsilon_2,\epsilon_2[$. As a consequence, $\tilde{\mathcal{J}}_{\#,1}$ can be expanded into a series of even powers of ϵ (cf. Definition 2.15 of real analytic functions) and there exist $\epsilon_3\in]0,\epsilon_2[$ and a real analytic function $\tilde{\mathcal{J}}_{\#}$ from $]-\epsilon_3,\epsilon_3[$ to $\mathbb{R}$ such that

$$\int_{Q\setminus\overline{\Omega_{p,\epsilon}}} w^{-}_{q,\Omega_{p,\epsilon}}\big[\Theta[\epsilon]((\cdot-p)/\epsilon)\big](x)\,dx$$
$$=\epsilon^2\sum_{j=1}^{2}\int_{\partial\Omega}\Big(\int_{\partial\Omega}S_2(\xi-\eta)\theta_0(\eta)(\nu_\Omega(\eta))_j\,d\sigma_\eta\Big)(\nu_\Omega(\xi))_j\,d\sigma_\xi$$
$$+\epsilon^4\tilde{\mathcal{J}}_{\#}[\epsilon]\qquad\forall\epsilon\in]-\epsilon_3,\epsilon_3[\setminus\{0\}\,.$$

Hence, if we set

$$\mathcal{J}_{\#,1}[\epsilon] = \tilde{\mathcal{J}}_{\#}[\epsilon] + \tilde{C}[\epsilon](1-\epsilon^2 m_2(\Omega))$$
$$- m_2(\Omega)\bigg(\int_{\partial\Omega}\int_{\partial\Omega}(\xi-\eta)^t\nabla^2 R_{q,2}(0)\nu_\Omega(\eta)\theta_0(\eta)\,d\sigma_\eta\tau_0(\xi)\,d\sigma_\xi$$
$$+ \sum_{\beta\in\mathbb{N}^2,\,|\beta|=2}\frac{D^\beta R_{q,2}(0)}{\beta!}\int_{\partial\Omega}\xi^\beta\tau_0(\xi)\,d\sigma_\xi\bigg),$$

for all $\epsilon\in]-\epsilon_3,\epsilon_3[$, we immediately deduce that $\mathcal{J}_{\#,1}$ is a real analytic function from $]-\epsilon_3,\epsilon_3[$ to $\mathbb{R}$ such that equality (13.41) holds, and thus the proof is complete. □

Then we turn to the behavior of $\int_{Q\setminus\overline{\Omega_{p,\epsilon}}} B_\epsilon(x)\,dx$ and we prove the following.

Proposition 13.14. *There exist $\epsilon_4\in]0,\epsilon_0[$ and a real analytic function $\mathcal{J}_{\#,2}$ from $]-\epsilon_4,\epsilon_4[$ to $\mathbb{R}$, such that*

$$\int_{Q\setminus\overline{\Omega_{p,\epsilon}}} B_\epsilon(x)\,dx = -\int_Q S_{q,2}(x)\,dx + \frac{\log|\epsilon|}{2\pi}$$
$$+\epsilon^2\Big(R_{q,2}(0)m_2(\Omega)+\int_\Omega S_2(\xi)\,d\xi\Big)+\epsilon^4\mathcal{J}_{\#,2}[\epsilon], \tag{13.44}$$

for all $\epsilon\in]-\epsilon_4,\epsilon_4[\setminus\{0\}$.

Proof. We first note that

$$\int_{Q\setminus\overline{\Omega_{p,\epsilon}}} B_\epsilon(x)\,dx = \int_Q B_\epsilon(x)\,dx - \int_{\Omega_{p,\epsilon}} B_\epsilon(x)\,dx$$
$$= (1-\epsilon^2 m_2(\Omega))\frac{\log|\epsilon|}{2\pi} - \int_Q S_{q,2}(x-p)\,dx + \epsilon^2\int_\Omega S_{q,2}(\epsilon\xi)\,d\xi,$$

for all $\epsilon\in]-\epsilon_0,\epsilon_0[\setminus\{0\}$. Then we note that, by the periodicity of $S_{q,2}$ and by Lemma 13.1, we have $\int_Q S_{q,2}(x-p)\,dx = \int_Q S_{q,2}(x)\,dx$. Moreover,

$$\int_\Omega S_{q,2}(\epsilon\xi)\,d\xi = \int_\Omega S_2(\xi)\,d\xi + \frac{m_2(\Omega)\log|\epsilon|}{2\pi} + \int_\Omega R_{q,2}(\epsilon\xi)\,d\xi,$$

for all $\epsilon\in]-\epsilon_0,\epsilon_0[\setminus\{0\}$. By the analyticity of $R_{q,2}$ in $(\mathbb{R}^2\setminus q\mathbb{Z}^2)\cup\{0\}$ and by analyticity results for the composition operator (cf. Theorem A.23 or Remark 2.18 and Proposition A.25), the map from $]-\epsilon_0,\epsilon_0[$ to $C^0(\overline{\Omega})$ that takes ϵ to the function $R_{q,2}(\epsilon\xi)$ of the variable $\xi\in\overline{\Omega}$ is real analytic. Since $C^0(\overline{\Omega})$ is continuously embedded into $L^1(\Omega)$ and the map from $L^1(\Omega)$ to $\mathbb{R}$ that takes an integrable function to its integral on Ω is linear and continuous, then the map $\tilde{\mathcal{J}}$ from $]-\epsilon_0,\epsilon_0[$ to $\mathbb{R}$ that takes ϵ to

$$\tilde{\mathcal{J}}[\epsilon]\equiv\int_\Omega R_{q,2}(\epsilon\xi)\,d\xi,$$

is real analytic. Now by Theorem 12.4 (v) the map $R_{q,2}$ is even and thus $\tilde{\mathcal{J}}[\epsilon]=\tilde{\mathcal{J}}[-\epsilon]$ for all $\epsilon\in]-\epsilon_0,\epsilon_0[$. Moreover, the definition of $\tilde{\mathcal{J}}$ implies that

$$\tilde{\mathcal{J}}[0] = R_{q,2}(0)m_2(\Omega)\,.$$

As a consequence, there exist $\epsilon_4 \in]0, \epsilon_0[$ and a real analytic function $\mathcal{J}_{\#,2}$ from $]-\epsilon_4, \epsilon_4[$ to $\mathbb{R}$ such that

$$\tilde{\mathcal{J}}[\epsilon] = R_{q,2}(0)m_2(\Omega) + \epsilon^2 \mathcal{J}_{\#,2}[\epsilon] \qquad \forall \epsilon \in]-\epsilon_4, \epsilon_4[\,.$$

Hence, equality (13.44) holds and the proof is complete. □

By equality (13.29) and Propositions 13.13 and 13.14, we can immediately deduce the validity of our main result on the behavior of $\Sigma_{II}[\epsilon]$.

Theorem 13.15. *Let H^o_Ω and ϵ_3 be as in (13.33) and in Proposition 13.13, respectively. Let ϵ_4 be as in Theorem 13.14. Let $\epsilon_\# \equiv \min\{\epsilon_3, \epsilon_4\}$. Let τ_0 be the unique solution in $C^{0,\alpha}(\partial\Omega)$ of (13.37). Then there exists a real analytic function $\mathcal{J}$ from $]-\epsilon_\#, \epsilon_\#[$ to $\mathbb{R}$ such that*

$$\begin{aligned}
\Sigma_{II}[\epsilon] = &\frac{\log|\epsilon|}{2\pi} + \lim_{\xi\to\infty} H^o_\Omega(\xi) + R_{q,2}(0) - \int_Q S_{q,2}(x)\,dx \\
&- \epsilon^2 \Bigg(m_2(\Omega) \lim_{\xi\to\infty} H^o_\Omega(\xi) - \int_\Omega S_2(\xi)\,d\xi \\
&- \int_{\partial\Omega}\int_{\partial\Omega} (\xi-\eta)^t \nabla^2 R_{q,2}(0)\nu_\Omega(\eta)\theta_0(\eta)\,d\sigma_\eta \tau_0(\xi)\,d\sigma_\xi \\
&- \sum_{\beta\in\mathbb{N}^2, |\beta|=2} \frac{D^\beta R_{q,2}(0)}{\beta!} \int_{\partial\Omega} \xi^\beta \tau_0(\xi)\,d\sigma_\xi \\
&- \sum_{j=1}^{2} \int_{\partial\Omega}\Big(\int_{\partial\Omega} S_2(\xi-\eta)\theta_0(\eta)(\nu_\Omega(\eta))_j\,d\sigma_\eta\Big)(\nu_\Omega(\xi))_j\,d\sigma_\xi \Bigg) \\
&+ \epsilon^4 \mathcal{J}[\epsilon] \qquad \forall \epsilon \in]-\epsilon_\#, \epsilon_\#[\setminus\{0\}\,.
\end{aligned}$$

Now we want to investigate the parity properties of the function $\mathcal{J}$ of Theorem 13.15. To do so, we need the following lemma on Σ_{II}.

Lemma 13.16. *We have*

$$\Sigma_{II}[\epsilon] = \Sigma_{II}[-\epsilon] \qquad \forall \epsilon \in]-\epsilon_0, \epsilon_0[\setminus\{0\}\,.$$

Proof. We first show that if $\epsilon \in]-\epsilon_0, \epsilon_0[\setminus\{0\}$, then $u_\epsilon(x) = u_{-\epsilon}(-x+2p)$ for all $x \in \overline{\mathbb{S}[\Omega_{p,\epsilon}]^-}$. To do so, we prove that the function $u_{-\epsilon}(-\cdot+2p)$ solves problem (13.24). Indeed, if $\epsilon \in]-\epsilon_0, \epsilon_0[\setminus\{0\}$, then

$$\overline{\mathbb{S}[\Omega_{p,\epsilon}]^-} = -(\overline{\mathbb{S}[\Omega_{p,-\epsilon}]^-}) + 2p\,.$$

and thus $u_{-\epsilon}(-\cdot+2p) \in C^{1,\alpha}_q(\overline{\mathbb{S}[\Omega_{p,\epsilon}]^-})$. Moreover,

$$\Delta\big(u_{-\epsilon}(-x+2p)\big) = \big(\Delta u_{-\epsilon}\big)(-x+2p) = 1 \qquad \forall x \in \mathbb{S}[\Omega_{p,\epsilon}]^-\,.$$

Also,

$$u_{-\epsilon}(-(p+\epsilon\xi)+2p) = u_{-\epsilon}(-\epsilon\xi+p) = 0 \qquad \forall \xi \in \partial\Omega\,.$$

As a consequence, $u_{-\epsilon}(-\cdot+2p)$ solves problem (13.24) and accordingly

$$u_{\epsilon}(x) = u_{-\epsilon}(-x+2p)$$

for all $x \in \overline{\mathbb{S}[\Omega_{p,\epsilon}]^{-}}$. Then, by the theorem of change of variable in integrals, we have

$$\begin{aligned}\Sigma_{II}[\epsilon] = \int_{Q\setminus\overline{\Omega_{p,\epsilon}}} u_{\epsilon}(x)\,dx &= \int_{Q\setminus\overline{\Omega_{p,\epsilon}}} u_{-\epsilon}(-x+2p)\,dx \\ &= \int_{-(Q\setminus\overline{\Omega_{p,\epsilon}})+2p} u_{-\epsilon}(x)\,dx \\ &= \int_{(-Q+2p)\setminus\overline{\Omega_{p,-\epsilon}}} u_{-\epsilon}(x)\,dx\,.\end{aligned}$$

Now we introduce the function $\widetilde{u_{-\epsilon}} \in L^1_{\mathrm{loc}}(\mathbb{R}^2)$ defined by

$$\widetilde{u_{-\epsilon}}(x) \equiv \begin{cases} u_{-\epsilon}(x) & \text{if } x \in \overline{\mathbb{S}[\Omega_{p,-\epsilon}]^{-}}\,, \\ 0 & \text{if } x \in \mathbb{R}^2 \setminus \overline{\mathbb{S}[\Omega_{p,-\epsilon}]^{-}}\,. \end{cases}$$

Clearly,

$$\widetilde{u_{-\epsilon}}(x+qz) = \widetilde{u_{-\epsilon}}(x) \qquad \forall x \in \mathbb{R}^2\,, z \in \mathbb{Z}^2\,.$$

By a straightforward computation, we can verify that

$$\overline{\Omega_{p,-\epsilon}} \subseteq (-Q+2p)\,.$$

Accordingly,

$$\int_{(-Q+2p)\setminus\overline{\Omega_{p,-\epsilon}}} u_{-\epsilon}(x)\,dx = \int_{(-Q+2p)} \widetilde{u_{-\epsilon}}(x)\,dx\,.$$

Then we note that

$$-Q+2p = Q - qe_1 - qe_2 + 2p\,.$$

Therefore, by Lemma 13.1 we have

$$\int_{(-Q+2p)} \widetilde{u_{-\epsilon}}(x)\,dx = \int_{Q} \widetilde{u_{-\epsilon}}(x)\,dx = \int_{Q\setminus\overline{\Omega_{p,-\epsilon}}} u_{-\epsilon}(x)\,dx = \Sigma_{II}[-\epsilon]\,.$$

As a consequence, $\Sigma_{II}[\epsilon] = \Sigma_{II}[-\epsilon]$. □

By Theorem 13.15 and Lemma 13.16, we deduce the validity of the following.

Proposition 13.17. *Let ϵ_5 and $\mathcal{J}$ be as in Theorem 13.15. Then*

$$\mathcal{J}[\epsilon] = \mathcal{J}[-\epsilon] \qquad \forall \epsilon \in]-\epsilon_5, \epsilon_5[\,.$$

We conclude that for $|\epsilon|$ small enough $\mathcal{J}[\epsilon]$ can be expanded into a series of even powers of ϵ.

13.6 A Singularly Perturbed Non-ideal Transmission Problem and Application to the Effective Conductivity of a Periodic Composite

In this section, we consider the heat conduction in a periodic two-phase dilute composite with thermal resistance at the two-phase interface and we follow the presentation of [75]. The composite consists of a matrix and of a periodic set of inclusions. The matrix and the inclusions are filled with two (possibly different) homogeneous and isotropic heat conductor materials. We assume that each inclusion occupies a bounded domain of $\mathbb{R}^n$ of diameter proportional to a parameter $\epsilon > 0$. The normal component of the heat flux is assumed to be continuous at the two-phase interface, while we impose that the temperature field displays a jump proportional to the normal heat flux by means of a parameter $\rho(\epsilon) > 0$. In physics, the appearance of such a discontinuity in the temperature field is a well known phenomenon and has been largely investigated since 1941, when Kapitza carried out the first systematic study of the thermal interface behavior in liquid helium (see, e.g., Swartz and Pohl [263], Lipton [190] and references therein). The aim of this section is to study the behavior of the effective conductivity of the composite when the parameter ϵ tends to 0 and the size of the inclusions collapses. The expression defining the effective conductivity of a composite with imperfect contact conditions has been introduced by Benveniste and Miloh in [22] by generalizing the dual theory of the effective behavior of composites with perfect contact (see also Benveniste [21] and for a review Drygaś and Mityushev [92]). By the argument of Benveniste and Miloh, in order to evaluate the effective conductivity, one has to study the thermal distribution of the composite when so called 'homogeneous conditions' are prescribed. To do so, we now introduce a particular transmission problem with non-ideal contact conditions where we impose that the temperature field displays a fixed jump along a certain direction and is periodic in all the other directions. For the sake of completeness, we also investigate non-homogeneous boundary conditions at the two-phase interface.

We take two positive constants λ^+, λ^-, a function $f \in C^{0,\alpha}(\partial\Omega)_0$, a function $g \in C^{0,\alpha}(\partial\Omega)$, and a function ρ from $]0, \epsilon_0[$ to $]0, +\infty[$, and for each $k \in \{1, \dots, n\}$ we consider the following transmission problem for the unknown pair of functions $(u_k^+, u_k^-) \in C^{1,\alpha}_{\mathrm{loc}}(\overline{\mathbb{S}[\Omega_{p,\epsilon}]}) \times C^{1,\alpha}_{\mathrm{loc}}(\overline{\mathbb{S}[\Omega_{p,\epsilon}]^-})$

$$\begin{cases} \Delta u_k^+ = 0 & \text{in } \mathbb{S}[\Omega_{p,\epsilon}]\,, \\ \Delta u_k^- = 0 & \text{in } \mathbb{S}[\Omega_{p,\epsilon}]^-\,, \\ u_k^+(x+qz) = u_k^+(x) + q_{kk}z_k & \forall x \in \overline{\mathbb{S}[\Omega_{p,\epsilon}]}\,, z \in \mathbb{Z}^n\,, \\ u_k^-(x+qz) = u_k^-(x) + q_{kk}z_k & \forall x \in \overline{\mathbb{S}[\Omega_{p,\epsilon}]^-}\,, z \in \mathbb{Z}^n\,, \\ \lambda^- \nu_{\Omega_{p,\epsilon}}(x) \cdot \nabla u_k^-(x) - \lambda^+ \nu_{\Omega_{p,\epsilon}}(x) \cdot \nabla u_k^+(x) & \\ \qquad = f((x-p)/\epsilon) & \forall x \in \partial\Omega_{p,\epsilon}\,, \\ \lambda^+ \nu_{\Omega_{p,\epsilon}}(x) \cdot \nabla u_k^+(x) + \frac{1}{\rho(\epsilon)}(u_k^+(x) - u_k^-(x)) & \\ \qquad = g((x-p)/\epsilon) & \forall x \in \partial\Omega_{p,\epsilon}\,, \\ \int_{\partial\Omega_{p,\epsilon}} u_k^+(x)\, d\sigma_x = 0\,, & \end{cases} \tag{13.45}$$

for all $\epsilon \in]0, \epsilon_0[$. In contrast with the previous section, we recall that here

$$n \in \mathbb{N} \setminus \{0, 1\}\,.$$

In other words, we do not confine ourselves to the case of dimension $n = 2$. We observe that the functions u_k^+ and u_k^- represent the temperature field in the inclusions occupying the periodic set $\mathbb{S}[\Omega_{p,\epsilon}]$ and in the matrix occupying $\mathbb{S}[\Omega_{p,\epsilon}]^-$, respectively. The parameters λ^+ and λ^- play the role of thermal conductivity of the materials which fill the inclusions and the matrix, respectively. The quantity $\rho(\epsilon)$ plays the role of the interfacial thermal resistivity. The fifth condition in (13.45) describes the jump of the normal heat flux across the two-phase interface and the sixth condition describes the jump of the temperature field. In particular, if f and g are identically 0, then the normal heat flux is continuous and the temperature field has a jump proportional to the normal heat flux by means of the parameter $\rho(\epsilon)$. The third and fourth conditions in (13.45) imply that the temperature distributions u_k^+ and u_k^- have a jump equal to q_{kk} in the direction e_k and are periodic in all the other directions. Indeed, for u_k^+ the condition

$$u_k^+(x+qz) = u_k^+(x) + q_{kk}z_k \qquad \forall x \in \overline{\mathbb{S}[\Omega_{p,\epsilon}]}\,, z \in \mathbb{Z}^n\,,$$

implies that

$$u_k^+(x+qe_k) = u_k^+(x) + q_{kk} \qquad \forall x \in \overline{\mathbb{S}[\Omega_{p,\epsilon}]}\,,$$

and for $l \in \{1, \dots, n\} \setminus \{k\}$ that

$$u_k^+(x+qe_l) = u_k^+(x) \qquad \forall x \in \overline{\mathbb{S}[\Omega_{p,\epsilon}]}$$

(and similarly for u^-). Finally, the seventh condition in (13.45) is an auxiliary condition which we introduce to guarantee the uniqueness of the solution (u_k^+, u_k^-). Such a condition does not interfere in the definition of the effective conductivity, which is invariant for constant modifications of the temperature field.

We observe that the boundary value problem in (13.45) generalizes transmission problems that have been largely investigated in connection with the theory of heat conduction in two-phase periodic composites with imperfect contact conditions (cf., e.g., Castro and Pesetskaya [44], Castro et al. [47], Drygaś and Mityushev [92], Lipton [190], Mityushev [217]). Other transmission problems with imperfect contact conditions have been analyzed for example in Castro et al. [42, 43] and in Castro and Pesetskaya [45]. We also note that Ammari et al. [12] have exploited integral equation techniques to compute an asymptotic expansion of the effective electrical conductivity of a periodic dilute composite with ideal contact condition (see also Ammari et al. [9] for anisotropic heat conductors, Ammari et al. [11] for effective elastic properties, and Ammari et al. [7] for the analysis of effective viscosity properties). Moreover, a scheme for the effective properties of unidirectional fibre-reinforced media can be found in Joyce et al. [131]. Finally, we mention that in the paper [249] Pukhatievych has applied the functional analytic approach to a different singularly perturbed transmission problem in a periodic domain. Finally, we mention that a real analyticity result for the effective conductivity of an ideal two-phase periodic composite upon regular perturbation of the domain and the contrast parameter has been obtained by Luzzini and the third named author in [194].

Due to the presence of the factor $1/\rho(\epsilon)$, the boundary condition may display a singularity as ϵ tends to 0. In this section, we consider the case in which

$$\text{the limit } \lim_{\epsilon\to 0^+} \frac{\epsilon}{\rho(\epsilon)} \text{ exists finite in } \mathbb{R}. \tag{13.46}$$

Assumption (13.46) will allow us to analyze problem (13.45) around the degenerate value $\epsilon = 0$. We also note that we make no regularity assumption on the function ρ. If assumption (13.46) holds, then we set

$$r_* \equiv \lim_{\epsilon\to 0^+} \frac{\epsilon}{\rho(\epsilon)} .$$

If $\epsilon \in]0, \epsilon_0[$, then the solution in $C^{1,\alpha}_{\mathrm{loc}}(\overline{\mathbb{S}[\Omega_{p,\epsilon}]}) \times C^{1,\alpha}_{\mathrm{loc}}(\overline{\mathbb{S}[\Omega_{p,\epsilon}]^-})$ of problem (13.45) exists and is unique and we denote it by $(u^+_{k,\epsilon}, u^-_{k,\epsilon})$. Then we introduce the effective conductivity matrix $\lambda^{\mathrm{eff}}[\epsilon]$ with (j,k)-entry $\lambda^{\mathrm{eff}}_{jk}[\epsilon]$ defined by means of the following.

Definition 13.18. We set

$$\begin{aligned}\lambda^{\mathrm{eff}}_{jk}[\epsilon] \equiv & \frac{1}{m_n(Q)}\left(\lambda^+ \int_{\Omega_{p,\epsilon}} \frac{\partial u^+_{k,\epsilon}(x)}{\partial x_j}\, dx + \lambda^- \int_{Q\setminus\overline{\Omega_{p,\epsilon}}} \frac{\partial u^-_{k,\epsilon}(x)}{\partial x_j}\, dx\right) \\ & + \frac{1}{m_n(Q)} \int_{\partial\Omega_{p,\epsilon}} f((x-p)/\epsilon) x_j \, d\sigma_x \qquad \forall \epsilon \in]0, \epsilon_0[.\end{aligned}$$

We observe that Definition 13.18 extends that of Benveniste and Miloh to the case of non-homogeneous boundary conditions and coincides with the classical definition when f and g are identically 0 (cf. Benveniste [21] and Benveniste and Miloh [22]).

Next, if $(j,k) \in \{1, \ldots, n\}^2$, we pose the following question.

$$\text{What can be said on the map } \epsilon \mapsto \lambda^{\mathrm{eff}}_{jk}[\epsilon] \text{ when } \epsilon \text{ is close to 0 and positive?} \tag{13.47}$$

Here we answer the question in (13.47) by showing that

$$\lambda^{\text{eff}}_{jk}[\epsilon] = \lambda^- \delta_{j,k} + \epsilon^n \mathcal{L}_{jk}[\epsilon, \epsilon/\rho(\epsilon)]$$

for $\epsilon > 0$ small, where $\mathcal{L}_{jk}$ is a real analytic map defined in a neighborhood of the pair $(0, r_*)$.

So for example, if we know that for $\epsilon > 0$ the function $\epsilon/\rho(\epsilon)$ equals a real analytic function defined in a whole neighborhood of $\epsilon = 0$, then the above formula implies that $\lambda^{\text{eff}}_{jk}[\epsilon]$ can be expanded into a convergent power series for ϵ small. This is the case if for example $\rho(\epsilon) = \epsilon$ or ρ is a constant function. These two specific cases will be thoroughly investigated in Section 13.7, where we will show how to compute the coefficients of the power series expansion of $\lambda^{\text{eff}}_{jk}[\epsilon]$. We also observe that Pukhtaievych has studied in [250] the effective conductivity of a periodic dilute composite with perfect contact and has obtained explicit expressions for the coefficients of its series expansion.

13.6.1 Transmission Problems with Non-ideal Contact Conditions

In this subsection we present some preliminary results concerning transmission problems with non-ideal contact conditions. We will consider a bounded open connected subset Ω_Q of $\mathbb{R}^n$ of class $C^{1,\alpha}$ such that $\mathbb{R}^n \setminus \overline{\Omega_Q}$ is connected and $\overline{\Omega_Q} \subseteq Q$ (see also the geometric assumptions of Section 12.2.1).

We first have the following uniqueness result for a periodic transmission problem, whose proof is based on a standard energy argument for periodic harmonic functions.

Proposition 13.19. *Let $\gamma^\# \in]0, +\infty[$. Let*

$$(v^+, v^-) \in C^{1,\alpha}_q(\overline{\mathbb{S}[\Omega_Q]}) \times C^{1,\alpha}_q(\overline{\mathbb{S}[\Omega_Q]^-})$$

be such that

$$\begin{cases} \Delta v^+ = 0 & \text{in } \mathbb{S}[\Omega_Q]\,, \\ \Delta v^- = 0 & \text{in } \mathbb{S}[\Omega_Q]^-\,, \\ v^+(x+qz) = v^+(x) & \forall x \in \overline{\mathbb{S}[\Omega_Q]}\,, z \in \mathbb{Z}^n\,, \\ v^-(x+qz) = v^-(x) & \forall x \in \overline{\mathbb{S}[\Omega_Q]^-}\,, z \in \mathbb{Z}^n\,, \\ \lambda^- \nu_{\Omega_Q}(x) \cdot \nabla v^-(x) - \lambda^+ \nu_{\Omega_Q}(x) \cdot \nabla v^+(x) = 0 \ \forall x \in \partial\Omega_Q\,, \\ \lambda^+ \nu_{\Omega_Q}(x) \cdot \nabla v^+(x) + \gamma^\# (v^+(x) - v^-(x)) = 0 \ \forall x \in \partial\Omega_Q\,, \\ \int_{\partial\Omega_Q} v^+(x)\, d\sigma_x = 0\,. \end{cases} \tag{13.48}$$

Then $v^+ = 0$ on $\overline{\mathbb{S}[\Omega_Q]}$ and $v^- = 0$ on $\overline{\mathbb{S}[\Omega_Q]^-}$.

Proof. We note that the periodicity of v^- implies that

$$\int_{\partial Q} v^-(x) \frac{\partial v^-}{\partial \nu_Q}(x)\, d\sigma_x = 0\,.$$

Then by the first Green identity of Theorem 4.2 we have

$$\begin{aligned} 0 \le \int_{Q\setminus\overline{\Omega_Q}} |\nabla v^-(x)|^2\, dx &= \int_{\partial Q} v^-(x) \frac{\partial v^-}{\partial \nu_Q}(x)\, d\sigma_x - \int_{\partial \Omega_Q} v^-(x) \frac{\partial v^-}{\partial \nu_{\Omega_Q}}(x)\, d\sigma_x \\ &= -\int_{\partial \Omega_Q} v^-(x) \frac{\partial v^-}{\partial \nu_{\Omega_Q}}(x)\, d\sigma_x\,. \end{aligned}$$

Accordingly, by the fifth and sixth condition in (13.48) we have

$$\begin{aligned} 0 \le &- \int_{\partial \Omega_Q} v^-(x) \frac{\partial v^-}{\partial \nu_{\Omega_Q}}(x)\, d\sigma_x \\ &= -\left[\frac{\lambda^+}{\lambda^-} \int_{\partial \Omega_Q} v^+(x) \frac{\partial v^+}{\partial \nu_{\Omega_Q}}(x)\, d\sigma_x + \frac{1}{\gamma^\# \lambda^-} \int_{\partial \Omega_Q} \left|\lambda^+ \frac{\partial v^+}{\partial \nu_{\Omega_Q}}(x)\right|^2 d\sigma\right] \\ &= -\left[\frac{\lambda^+}{\lambda^-} \int_{\Omega_Q} |\nabla v^+(x)|^2\, dx + \frac{1}{\gamma^\# \lambda^-} \int_{\partial \Omega_Q} \left|\lambda^+ \frac{\partial v^+}{\partial \nu_{\Omega_Q}}(x)\right|^2 d\sigma\right] \le 0\,. \end{aligned}$$

As a consequence, $\nabla v^- = 0$ and $\nabla v^+ = 0$. Thus there exists $c^- \in \mathbb{R}$ such that $v^-(x) = c^-$ for all $x \in \overline{\mathbb{S}[\Omega_Q]^-}$. Similarly, there exists $c^+ \in \mathbb{R}$ such that $v^+(x) = c^+$ for all $x \in \overline{\mathbb{S}[\Omega_Q]}$. Then by the sixth condition in (13.48) we deduce that $c^- = c^+$, and by the seventh condition in (13.48) we conclude that $c^+ = 0$ and that accordingly also $c^- = 0$. Thus the proof is complete. □

Then we have the following technical lemma.

Lemma 13.20. *The following statements hold.*

(i) If $\mu \in C^{0,\alpha}(\partial\Omega_Q)_0$, then $W^t_{q,\Omega_Q}[\mu] \in C^{0,\alpha}(\partial\Omega_Q)_0$.
(ii) The map from $C^{0,\alpha}(\partial\Omega_Q)_0$ to $C^{0,\alpha}(\partial\Omega_Q)_0$ that takes μ to $W^t_{q,\Omega_Q}[\mu]$ is compact.

Proof. Statement (i) follows by Theorem 12.13 (ii) and Proposition 12.15. Statement (ii) is a straightforward consequence of Theorem 12.13 (ii) and of statement (i). □

We now study an integral operator that we need to solve a periodic transmission problem related to (13.45) by means of periodic single layer potentials.

Proposition 13.21. *Let $\gamma^\# \in]0, +\infty[$. Let $J_{\gamma^\#} \equiv (J_{\gamma^\#,1}, J_{\gamma^\#,2})$ be the operator from $(C^{0,\alpha}(\partial\Omega_Q)_0)^2$ to $(C^{0,\alpha}(\partial\Omega_Q)_0)^2$ defined by*

$$\begin{aligned} J_{\gamma^\#,1}[\mu^i, \mu^o] \equiv& \lambda^-\Big(\frac{1}{2}\mu^o + W^t_{q,\Omega_Q}[\mu^o]\Big) - \lambda^+\Big(-\frac{1}{2}\mu^i + W^t_{q,\Omega_Q}[\mu^i]\Big)\,, \\ J_{\gamma^\#,2}[\mu^i, \mu^o] \equiv& \lambda^+\Big(-\frac{1}{2}\mu^i + W^t_{q,\Omega_Q}[\mu^i]\Big) \end{aligned}$$

$$+\gamma^{\#}\Big(v^{+}_{q,\Omega_Q}[\mu^i]_{|\partial\Omega_Q}-\int\!\!\!\!-_{\partial\Omega_Q} v^{+}_{q,\Omega_Q}[\mu^i]\,d\sigma$$

$$-v^{-}_{q,\Omega_Q}[\mu^o]_{|\partial\Omega_Q}+\int\!\!\!\!-_{\partial\Omega_Q} v^{-}_{q,\Omega_Q}[\mu^o]\,d\sigma\Big)\,,$$

for all $(\mu^i,\mu^o)\in(C^{0,\alpha}(\partial\Omega_Q)_0)^2$. *Then* $J_{\gamma\#}$ *is a linear homeomorphism.*

Proof. Let $\hat{J}_{\gamma\#}\equiv(\hat{J}_{\gamma\#,1},\hat{J}_{\gamma\#,2})$ be the linear operator from $(C^{0,\alpha}(\partial\Omega_Q)_0)^2$ to $(C^{0,\alpha}(\partial\Omega_Q)_0)^2$ defined by

$$\hat{J}_{\gamma\#,1}[\mu^i,\mu^o]\equiv(\lambda^-/2)\mu^o+(\lambda^+/2)\mu^i\,,\quad \hat{J}_{\gamma\#,2}[\mu^i,\mu^o]\equiv-(\lambda^+/2)\mu^i$$

for all $(\mu^i,\mu^o)\in(C^{0,\alpha}(\partial\Omega_Q)_0)^2$. Clearly, $\hat{J}_{\gamma\#}$ is a linear homeomorphism from $(C^{0,\alpha}(\partial\Omega_Q)_0)^2$ to $(C^{0,\alpha}(\partial\Omega_Q)_0)^2$. Next, let $\tilde{J}_{\gamma\#}\equiv(\tilde{J}_{\gamma\#,1},\tilde{J}_{\gamma\#,2})$ be the operator from $(C^{0,\alpha}(\partial\Omega_Q)_0)^2$ to $(C^{0,\alpha}(\partial\Omega_Q)_0)^2$ defined by

$$\begin{aligned}\tilde{J}_{\gamma\#,1}[\mu^i,\mu^o]\equiv&\lambda^-W^t_{q,\Omega_Q}[\mu^o]-\lambda^+W^t_{q,\Omega_Q}[\mu^i]\,,\\ \tilde{J}_{\gamma\#,2}[\mu^i,\mu^o]\equiv&\lambda^+W^t_{q,\Omega_Q}[\mu^i]\\ &+\gamma^{\#}\Big(v^{+}_{q,\Omega_Q}[\mu^i]_{|\partial\Omega_Q}-\int\!\!\!\!-_{\partial\Omega_Q} v^{+}_{q,\Omega_Q}[\mu^i]\,d\sigma\\ &-v^{-}_{q,\Omega_Q}[\mu^o]_{|\partial\Omega_Q}+\int\!\!\!\!-_{\partial\Omega_Q} v^{-}_{q,\Omega_Q}[\mu^o]\,d\sigma\Big)\end{aligned}$$

for all $(\mu^i,\mu^o)\in(C^{0,\alpha}(\partial\Omega_Q)_0)^2$. Then, by Lemma 13.20, by the boundedness of the operator from $C^{0,\alpha}(\partial\Omega_Q)_0$ to $C^{1,\alpha}(\partial\Omega_Q)_0$ that takes μ to

$$v_{q,\Omega_Q}[\mu]_{|\partial\Omega_Q}-\int\!\!\!\!-_{\partial\Omega_Q} v_{q,\Omega_Q}[\mu]\,d\sigma$$

(cf. Theorem 12.8), and by the compactness of the embedding of $C^{1,\alpha}(\partial\Omega_Q)_0$ into $C^{0,\alpha}(\partial\Omega_Q)_0$ (cf. Proposition 2.25 and Theorem 2.83 (ii)), the operator $\tilde{J}_{\gamma\#}$ is compact. Now, since $J_{\gamma\#}=\hat{J}_{\gamma\#}+\tilde{J}_{\gamma\#}$ and since compact perturbations of isomorphisms are Fredholm operators of index 0 (cf. Theorem 5.3), we deduce that $J_{\gamma\#}$ is a Fredholm operator of index 0. Thus to show that $J_{\gamma\#}$ is a linear homeomorphism it suffices to show that it is injective. So, let $(\mu^i,\mu^o)\in(C^{0,\alpha}(\partial\Omega_Q)_0)^2$ be such that $J_{\gamma\#}[\mu^i,\mu^o]=(0,0)$. Then by the jump formulas for the normal derivative of the periodic single layer potential we verify that the pair $(v^+,v^-)\in C^{1,\alpha}_q(\overline{\mathbb{S}[\Omega_Q]})\times C^{1,\alpha}_q(\overline{\mathbb{S}[\Omega_Q]^-})$ defined by

$$v^+\equiv v^{+}_{q,\Omega_Q}[\mu^i]-\int\!\!\!\!-_{\partial\Omega_Q} v^{+}_{q,\Omega_Q}[\mu^i]\,d\sigma\,,$$

$$v^-\equiv v^{-}_{q,\Omega_Q}[\mu^o]-\int\!\!\!\!-_{\partial\Omega_Q} v^{-}_{q,\Omega_Q}[\mu^o]\,d\sigma\,,$$

is a solution of the boundary value problem in (13.48) (see the jump formula of Theorem 12.11). Accordingly, Proposition 13.19 implies that $v^- = 0$ and $v^+ = 0$. In particular,

$$v^+_{q,\Omega_Q}[\mu^i] - \fint_{\partial\Omega_Q} v^+_{q,\Omega_Q}[\mu^i]\, d\sigma = 0 \qquad \text{in } \overline{\mathbb{S}[\Omega_Q]}\,, \tag{13.49}$$
$$v^-_{q,\Omega_Q}[\mu^o] - \fint_{\partial\Omega_Q} v^-_{q,\Omega_Q}[\mu^o]\, d\sigma = 0 \qquad \text{in } \overline{\mathbb{S}[\Omega_Q]^-}\,.$$

Then, by the jump formulas for the normal derivative of the periodic single layer potential we have

$$\nu_{\Omega_Q}(x) \cdot \nabla v^-(x) = \frac{1}{2}\mu^o(x) + W^t_{q,\Omega_Q}[\mu^o](x) = 0 \qquad \forall x \in \partial\Omega_Q$$

(cf. Theorem 12.11). Since $\frac{1}{2}I + W^t_{q,\Omega_Q}$ is an isomorphism, we deduce that $\mu^o = 0$ (cf. Theorem 12.20 (ii)). Moreover, by (13.49) and by the continuity of the single layer potential we have

$$v^-_{q,\Omega_Q}[\mu^i]_{|\partial\Omega_Q} = v^+_{q,\Omega_Q}[\mu^i]_{|\partial\Omega_Q} = \fint_{\partial\Omega_Q} v^+_{q,\Omega_Q}[\mu^i]\, d\sigma = \fint_{\partial\Omega_Q} v^-_{q,\Omega_Q}[\mu^i]\, d\sigma\,.$$

Since $\int_{\partial\Omega_Q} \mu^i\, d\sigma = 0$, the q-periodic function $v^-_{q,\Omega_Q}[\mu^i]$ is harmonic in $\mathbb{S}[\Omega_Q]^-$. Then the uniqueness Theorem 12.16 for the exterior periodic Dirichlet problem implies that

$$v^-_{q,\Omega_Q}[\mu^i] - \fint_{\partial\Omega_Q} v^-_{q,\Omega_Q}[\mu^i]\, d\sigma = 0 \qquad \text{in } \overline{\mathbb{S}[\Omega_Q]^-}\,,$$

and by arguing as above we conclude that $\mu^i = 0$. Accordingly, $J_{\gamma\#}$ is injective, and, as a consequence, it is a linear homeomorphism. □

By the jump formulas for the normal derivative of the periodic single layer potential of Theorem 12.11, we can now deduce the validity of the following theorem.

Theorem 13.22. *Let* $\gamma^\# \in]0, +\infty[$. *Let* $(\Phi, \Gamma, c) \in C^{0,\alpha}(\partial\Omega_Q)_0 \times C^{0,\alpha}(\partial\Omega_Q) \times \mathbb{R}$. *Let* $J_{\gamma\#}$ *be as in Proposition 13.21. Then a pair* $(\mu^i, \mu^o) \in (C^{0,\alpha}(\partial\Omega_Q)_0)^2$ *satisfies the equality*

$$J_{\gamma\#}[\mu^i, \mu^o] = \left(\Phi, \Gamma - \fint_{\partial\Omega_Q} \Gamma\, d\sigma\right)$$

if and only if the pair $(v^+, v^-) \in C^{1,\alpha}_q(\overline{\mathbb{S}[\Omega_Q]}) \times C^{1,\alpha}_q(\overline{\mathbb{S}[\Omega_Q]^-})$ *defined by*

$$v^+ \equiv v^+_{q,\Omega_Q}[\mu^i] - \fint_{\partial\Omega_Q} v^+_{q,\Omega_Q}[\mu^i]\, d\sigma + \frac{1}{m_{n-1}(\partial\Omega_Q)} c\,,$$
$$v^- \equiv v^-_{q,\Omega_Q}[\mu^o] - \fint_{\partial\Omega_Q} v^-_{q,\Omega_Q}[\mu^o]\, d\sigma$$

$$+\frac{1}{m_{n-1}(\partial\Omega_Q)}c - \frac{1}{\gamma^{\#}}\int_{\partial\Omega_Q}\Gamma\, d\sigma\,,$$

is a solution of

$$\begin{cases} \Delta v^{+} = 0 & \text{in } \mathbb{S}[\Omega_Q]\,,\\ \Delta v^{-} = 0 & \text{in } \mathbb{S}[\Omega_Q]^{-}\,,\\ v^{+}(x+qz) = v^{+}(x) & \forall x \in \overline{\mathbb{S}[\Omega_Q]}\,, z\in\mathbb{Z}^n\,,\\ v^{-}(x+qz) = v^{-}(x) & \forall x \in \overline{\mathbb{S}[\Omega_Q]^{-}}\,, z\in\mathbb{Z}^n\,,\\ \lambda^{-}\nu_{\Omega_Q}(x)\cdot\nabla v^{-}(x) - \lambda^{+}\nu_{\Omega_Q}(x)\cdot\nabla v^{+}(x) = \Phi(x) & \forall x\in\partial\Omega_Q\,,\\ \lambda^{+}\nu_{\Omega_Q}(x)\cdot\nabla v^{+}(x) + \gamma^{\#}(v^{+}(x)-v^{-}(x)) = \Gamma(x) & \forall x\in\partial\Omega_Q\,,\\ \int_{\partial\Omega_Q} v^{+}(x)\,d\sigma_x = c\,. \end{cases} \tag{13.50}$$

Propositions 13.19, 13.21 and Theorem 13.22 imply that for each triple (Φ,Γ,c) in $C^{0,\alpha}(\partial\Omega_Q)_0 \times C^{0,\alpha}(\partial\Omega_Q)\times\mathbb{R}$, the solution in $C^{1,\alpha}_q(\overline{\mathbb{S}[\Omega_Q]})\times C^{1,\alpha}_q(\overline{\mathbb{S}[\Omega_Q]^{-}})$ of problem (13.50) exists and is unique.

We now turn to non-periodic problems and we prove some results which we use in what follows to analyze problem (13.45) around the degenerate case $\epsilon = 0$. We first have the following uniqueness result, whose validity can be deduced by a standard energy argument (we recall the assumptions in (13.1) where in particular we require Ω and $\mathbb{R}^n\setminus\overline{\Omega}$ to be connected).

Proposition 13.23. *Let* $\tilde\gamma\in[0,+\infty[$. *Let* $(v^{+},v^{-})\in C^{1,\alpha}(\overline{\Omega})\times C^{1,\alpha}_{\mathrm{loc}}(\mathbb{R}^n\setminus\Omega)$ *be such that*

$$\begin{cases} \Delta v^{+} = 0 & \text{in } \Omega\,,\\ \Delta v^{-} = 0 & \text{in } \mathbb{R}^n\setminus\overline{\Omega}\,,\\ \lambda^{-}\nu_{\Omega}(\xi)\cdot\nabla v^{-}(\xi) - \lambda^{+}\nu_{\Omega}(\xi)\cdot\nabla v^{+}(\xi) = 0 & \forall\xi\in\partial\Omega\,,\\ \lambda^{+}\nu_{\Omega}(\xi)\cdot\nabla v^{+}(\xi) + \tilde\gamma\Big(v^{+}(\xi)-v^{-}(\xi)\Big) = 0 & \forall\xi\in\partial\Omega\,,\\ \int_{\partial\Omega} v^{+}(\xi)\,d\sigma_\xi = 0\,,\\ \int_{\partial\Omega} v^{-}(\xi)\,d\sigma_\xi = 0\,,\\ \lim_{\xi\to\infty} v^{-}(\xi)\in\mathbb{R}\,. \end{cases} \tag{13.51}$$

Then $v^{+} = 0$ *on* $\overline{\Omega}$ *and* $v^{-} = 0$ *on* $\mathbb{R}^n\setminus\Omega$.

Proof. We first observe that if $u^{-}\in C^{1,\alpha}_{\mathrm{loc}}(\mathbb{R}^n\setminus\Omega)$ is harmonic in $\Omega^{-}\equiv\mathbb{R}^n\setminus\overline{\Omega}$ and harmonic at infinity, then the first Green Identity in exterior domains of Corollary 4.7 implies that

$$0\le\int_{\Omega^{-}}|\nabla u^{-}(x)|^2\,dx = -\int_{\partial\Omega}u^{-}(x)\frac{\partial u^{-}}{\partial\nu_\Omega}(x)\,d\sigma_x\,. \tag{13.52}$$

We now set

$$v^{-}_{\infty}\equiv\lim_{\xi\to\infty}v^{-}(\xi)\,.$$

We note that the existence in $\mathbb{R}$ of the limit above and Theorem 3.22 imply that if $n=2$, then v^{-} is harmonic at infinity, whereas, if $n\ge 3$, then $v^{-}-v^{-}_{\infty}$ is harmonic at infinity. Therefore, if $n=2$ equality (13.52) with u^- replaced with v^- implies that

$$0 \leq \int_{\Omega^-} |\nabla v^-(x)|^2 \, dx = -\int_{\partial\Omega} v^-(x) \frac{\partial v^-}{\partial \nu_\Omega}(x) \, d\sigma_x \,. \tag{13.53}$$

Instead, if $n=3$ equality (13.52) with u^- replaced with $v^- - v^-_\infty$ implies that

$$0 \leq \int_{\Omega^-} |\nabla(v^-(x) - v^-_\infty)|^2 \, dx = -\int_{\partial\Omega} (v^-(x) - v^-_\infty) \frac{\partial(v^-(x) - v^-_\infty)}{\partial \nu_\Omega} \, d\sigma_x \,. \tag{13.54}$$

On the other hand,

$$\int_{\Omega^-} |\nabla(v^-(x) - v^-_\infty)|^2 \, dx = \int_{\Omega^-} |\nabla v^-(x)|^2 \, dx$$

and

$$\int_{\partial\Omega} (v^-(x) - v^-_\infty) \frac{\partial(v^-(x) - v^-_\infty)}{\partial \nu_\Omega} \, d\sigma_x = \int_{\partial\Omega} (v^-(x) - v^-_\infty) \frac{\partial v^-}{\partial \nu_\Omega}(x) \, d\sigma_x \,. \tag{13.55}$$

Since v^+ is harmonic in Ω, we have

$$\int_{\partial\Omega} \frac{\partial v^+}{\partial \nu_\Omega} d\sigma = 0 \,,$$

an equality that, together with the third condition in (13.51), implies that

$$\int_{\partial\Omega} \frac{\partial v^-}{\partial \nu_\Omega} d\sigma = 0 \,. \tag{13.56}$$

Then equations (13.54)–(13.55) and (13.56) imply that also when $n \geq 3$ we have

$$0 \leq \int_{\Omega^-} |\nabla v^-(x)|^2 \, dx = -\int_{\partial\Omega} v^-(x) \frac{\partial v^-}{\partial \nu_\Omega}(x) \, d\sigma_x \,. \tag{13.57}$$

We now consider separately case $\tilde{\gamma} \in]0, +\infty[$ and case $\tilde{\gamma} = 0$.
We first assume that $\tilde{\gamma} \in]0, +\infty[$. Using (13.53) if $n=2$ and (13.57) if $n \geq 3$ and combining the third and fourth condition in (13.51), we have

$$\begin{aligned}
0 &\leq -\int_{\partial\Omega} v^-(x) \frac{\partial v^-}{\partial \nu_\Omega}(x) \, d\sigma_x \\
&= -\left[\frac{\lambda^+}{\lambda^-} \int_{\partial\Omega} v^+(x) \frac{\partial v^+}{\partial \nu_\Omega}(x) \, d\sigma_x + \frac{1}{\tilde{\gamma}\lambda^-} \int_{\partial\Omega} \left|\lambda^+ \frac{\partial v^+}{\partial \nu_\Omega}(x)\right|^2 d\sigma\right] \\
&= -\left[\frac{\lambda^+}{\lambda^-} \int_{\Omega} |\nabla v^+(x)|^2 \, dx + \frac{1}{\tilde{\gamma}\lambda^-} \int_{\partial\Omega} \left|\lambda^+ \frac{\partial v^+}{\partial \nu_\Omega}(x)\right|^2 d\sigma\right] \leq 0 \,.
\end{aligned}$$

As a consequence, we deduce that $\nabla v^- = 0$ and $\nabla v^+ = 0$. Thus there exist c^-, $c^+ \in \mathbb{R}$ such that $v^-(x) = c^-$ for all $x \in \overline{\Omega^-}$ and $v^+(x) = c^+$ for all $x \in \overline{\Omega}$. Then by the fifth and sixth conditions in (13.51) we conclude that $c^- = c^+ = 0$.

We now consider the case where $\tilde{\gamma} = 0$. By the fourth condition, v^+ solves the homogeneous Neumann problem in Ω and therefore is constant. Then, by the fifth condition, we deduce that $v^+ = 0$ in $\overline{\Omega}$. Accordingly, by the third condition in (13.51) and by (13.53) if $n = 2$ and (13.57) if $n \geq 3$, we have

$$\int_{\Omega^-} |\nabla v^-(x)|^2 \, dx = 0 .$$

As a consequence v^- is constant in Ω^- and by the sixth condition in (13.51) it is identically equal to 0. Thus the proof is complete. □

We now study an integral operator that we need to solve (non-periodic) transmission problems in terms of single layer potentials.

Proposition 13.24. *Let $\tilde{\gamma} \in [0, +\infty[$. Let $K_{\tilde{\gamma}} \equiv (K_{\tilde{\gamma},1}, K_{\tilde{\gamma},2})$ be the operator from $(C^{0,\alpha}(\partial\Omega)_0)^2$ to $(C^{0,\alpha}(\partial\Omega)_0)^2$ defined by*

$$\begin{aligned} K_{\tilde{\gamma},1}[\theta^i, \theta^o] \equiv& \lambda^- \Big(\frac{1}{2}\theta^o + W^t_\Omega[\theta^o]\Big) - \lambda^+ \Big(-\frac{1}{2}\theta^i + W^t_\Omega[\theta^i]\Big) , \\ K_{\tilde{\gamma},2}[\theta^i, \theta^o] \equiv& \lambda^+ \Big(-\frac{1}{2}\theta^i + W^t_\Omega[\theta^i]\Big) + \tilde{\gamma}\Big(v^+_\Omega[\theta^i]_{|\partial\Omega} - \fint_{\partial\Omega} v^+_\Omega[\theta^i] \, d\sigma \\ & - v^-_\Omega[\theta^o]_{|\partial\Omega} + \fint_{\partial\Omega} v^-_\Omega[\theta^o] \, d\sigma \Big) , \end{aligned}$$

for all $(\theta^i, \theta^o) \in (C^{0,\alpha}(\partial\Omega)_0)^2$. Then $K_{\tilde{\gamma}}$ is a linear homeomorphism.

Proof. By the same argument used in the proof of Proposition 13.21 for $J_{\gamma\#}$, by replacing Lemma 13.20 with Theorem 6.8, and by Lemma 6.11, we can prove that $K_{\tilde{\gamma}}$ is a Fredholm operator of index 0. Thus, to show that $K_{\tilde{\gamma}}$ is a linear homeomorphism it suffices to show that it is injective. So, let $(\theta^i, \theta^o) \in (C^{0,\alpha}(\partial\Omega)_0)^2$ be such that $K_{\tilde{\gamma}}[\theta^i, \theta^o] = (0, 0)$. Then by the jump formulas for the normal derivative of the single layer potential of Theorem 6.6, the pair $(v^+, v^-) \in C^{1,\alpha}(\overline{\Omega}) \times C^{1,\alpha}_{\mathrm{loc}}(\mathbb{R}^n \setminus \Omega)$ defined by

$$\begin{aligned} v^+ &\equiv v^+_\Omega[\theta^i] - \fint_{\partial\Omega} v^+_\Omega[\theta^i] \, d\sigma , \\ v^- &\equiv v^-_\Omega[\theta^o] - \fint_{\partial\Omega} v^-_\Omega[\theta^o] \, d\sigma , \end{aligned}$$

is a solution of the boundary value problem in (13.51). Accordingly, Proposition 13.23 implies that $v^- = 0$ and $v^+ = 0$. Since the integrals of θ^i and of θ^o on $\partial\Omega$ equal zero, then the representation Theorem 6.47 of a function on $\partial\Omega$ in terms of a single layer and a constant implies that $\theta^i = 0$ and $\theta^o = 0$. Accordingly, $K_{\tilde{\gamma}}$ is injective, and, as a consequence, it is a linear homeomorphism. □

By Propositions 13.23, 13.24, and by the jump formulas for the normal derivative of the classical single layer potential (cf. Theorem 6.6), we immediately deduce the validity of the following result concerning the solvability of a (non-periodic) transmission problem.

Theorem 13.25. *Let $\tilde{\gamma} \in [0, +\infty[$. Let $K_{\tilde{\gamma}}$ be as in Proposition 13.24. Let (Φ, Γ) be in $(C^{0,\alpha}(\partial\Omega)_0)^2$. Let $(\theta^i, \theta^o) \in (C^{0,\alpha}(\partial\Omega)_0)^2$ be such that*

$$K_{\tilde{\gamma}}[\theta^i, \theta^o] = (\Phi, \Gamma) .$$

Let $(v^+, v^-) \in C^{1,\alpha}(\overline{\Omega}) \times C^{1,\alpha}_{\mathrm{loc}}(\mathbb{R}^n \setminus \Omega)$ be defined by

$$v^+ \equiv v^+_\Omega[\theta^i] - \fint_{\partial\Omega} v^+_\Omega[\theta^i]\, d\sigma ,$$
$$v^- \equiv v^-_\Omega[\theta^o] - \fint_{\partial\Omega} v^-_\Omega[\theta^o]\, d\sigma .$$

Then (v^+, v^-) is the unique solution in $C^{1,\alpha}(\overline{\Omega}) \times C^{1,\alpha}_{\mathrm{loc}}(\mathbb{R}^n \setminus \Omega)$ of

$$\begin{cases} \Delta v^+ = 0 & \text{in } \Omega , \\ \Delta v^- = 0 & \text{in } \mathbb{R}^n \setminus \overline{\Omega} , \\ \lambda^- \nu_\Omega(\xi) \cdot \nabla v^-(\xi) - \lambda^+ \nu_\Omega(\xi) \cdot \nabla v^+(\xi) = \Phi(\xi) & \forall \xi \in \partial\Omega , \\ \lambda^+ \nu_\Omega(\xi) \cdot \nabla v^+(\xi) + \tilde{\gamma}\Big(v^+(\xi) - v^-(\xi)\Big) = \Gamma(\xi) & \forall \xi \in \partial\Omega , \\ \int_{\partial\Omega} v^+(\xi)\, d\sigma_\xi = 0 , \\ \int_{\partial\Omega} v^-(\xi)\, d\sigma_\xi = 0 , \\ \lim_{\xi \to \infty} v^-(\xi) \in \mathbb{R} . \end{cases}$$

13.6.2 Formulation of the Singularly Perturbed Transmission Problem in Terms of Integral Equations

In Proposition 13.26 below, we formulate problem (13.45) in terms of integral equations on $\partial\Omega$. To do so, we exploit Theorem 13.22 and the rule of change of variables in integrals. Indeed, if $\epsilon \in]0, \epsilon_0[$ we can convert problem (13.45) into a periodic transmission problem by a simple computation (see problem (13.62) below). Then, by Theorem 13.22, we can reformulate such a problem in terms of a system of integral equations defined on the ϵ-dependent domain $\partial\Omega_{p,\epsilon}$. Finally, by exploiting an appropriate change of variable, we can get rid of such a dependence and obtain an equivalent system of integral equations defined on the fixed domain $\partial\Omega$, as Proposition 13.26 shows.

We now find convenient to introduce the following notation. We denote

$$\mathcal{M} \equiv (\mathcal{M}_1, \mathcal{M}_2)$$

the map from $] -\epsilon_0, \epsilon_0[\times \mathbb{R} \times (C^{0,\alpha}(\partial\Omega)_0)^2$ to $(C^{0,\alpha}(\partial\Omega)_0)^2$ defined by

$$
\begin{aligned}
&\mathcal{M}_1[\epsilon,\epsilon',\theta^i,\theta^o](\xi)\\
&\equiv \lambda^-\Big(\frac{1}{2}\theta^o(\xi)+W^t_\Omega[\theta^o](\xi)+\epsilon^{n-1}\int_{\partial\Omega}\nu_\Omega(\xi)\cdot\nabla R_{q,n}(\epsilon(\xi-\eta))\theta^o(\eta)\,d\sigma_\eta\Big)\\
&-\lambda^+\Big(-\frac{1}{2}\theta^i(\xi)+W^t_\Omega[\theta^i](\xi)+\epsilon^{n-1}\int_{\partial\Omega}\nu_\Omega(\xi)\cdot\nabla R_{q,n}(\epsilon(\xi-\eta))\theta^i(\eta)\,d\sigma_\eta\Big)\\
&-f(\xi)+(\lambda^--\lambda^+)(\nu_\Omega(\xi))_k \qquad \forall\xi\in\partial\Omega\,, \qquad (13.58)
\end{aligned}
$$

$$
\begin{aligned}
&\mathcal{M}_2[\epsilon,\epsilon',\theta^i,\theta^o](\xi)\\
&\equiv \lambda^+\Big(-\frac{1}{2}\theta^i(\xi)+W^t_\Omega[\theta^i](\xi)+\epsilon^{n-1}\int_{\partial\Omega}\nu_\Omega(\xi)\cdot\nabla R_{q,n}(\epsilon(\xi-\eta))\theta^i(\eta)\,d\sigma_\eta\Big)\\
&+\epsilon'\Bigg(v^+_\Omega[\theta^i](\xi)+\epsilon^{n-2}\int_{\partial\Omega}R_{q,n}(\epsilon(\xi-\eta))\theta^i(\eta)\,d\sigma_\eta\\
&\quad-\fint_{\partial\Omega}\Big(v^+_\Omega[\theta^i](\zeta)+\epsilon^{n-2}\int_{\partial\Omega}R_{q,n}(\epsilon(\zeta-\eta))\theta^i(\eta)d\sigma_\eta\Big)d\sigma_\zeta\\
&\quad-v^-_\Omega[\theta^o](\xi)-\epsilon^{n-2}\int_{\partial\Omega}R_{q,n}(\epsilon(\xi-\eta))\theta^o(\eta)\,d\sigma_\eta\\
&\quad+\fint_{\partial\Omega}\Big(v^-_\Omega[\theta^o](\zeta)+\epsilon^{n-2}\int_{\partial\Omega}R_{q,n}(\epsilon(\zeta-\eta))\theta^o(\eta)d\sigma_\eta\Big)d\sigma_\zeta\Bigg)\\
&-g(\xi)+\fint_{\partial\Omega}g\,d\sigma+\lambda^+(\nu_\Omega(\xi))_k \qquad \forall\xi\in\partial\Omega\,, \qquad (13.59)
\end{aligned}
$$

for all $(\epsilon,\epsilon',\theta^i,\theta^o)\in]-\epsilon_0,\epsilon_0[\times\mathbb{R}\times(C^{0,\alpha}(\partial\Omega)_0)^2$. We note that, in order to ensure that $\mathcal{M}[\epsilon,\epsilon',\theta^i,\theta^o]\in(C^{0,\alpha}(\partial\Omega)_0)^2$, we need to verify that

$$
\int_{\partial\Omega}\mathcal{M}_1[\epsilon,\epsilon',\theta^i,\theta^o]\,d\sigma=0\,, \qquad \int_{\partial\Omega}\mathcal{M}_2[\epsilon,\epsilon',\theta^i,\theta^o]\,d\sigma=0\,,
$$

for all $(\epsilon,\epsilon',\theta^i,\theta^o)\in]-\epsilon_0,\epsilon_0[\times\mathbb{R}\times(C^{0,\alpha}(\partial\Omega)_0)^2$. To do so, we first observe that if $(\epsilon,\epsilon',\theta^i,\theta^o)\in]-\epsilon_0,\epsilon_0[\times\mathbb{R}\times(C^{0,\alpha}(\partial\Omega)_0)^2$, then Proposition 12.15, the membership of f to $C^{0,\alpha}(\partial\Omega)_0$, the rule of change of variable in integrals, and equality

$$
\int_{\partial\Omega}(\nu_\Omega(\xi))_k\,d\sigma_\xi=0 \qquad (13.60)
$$

imply that

$$
\int_{\partial\Omega}\mathcal{M}_1[\epsilon,\epsilon',\theta^i,\theta^o]\,d\sigma=0\,.
$$

Similarly, Proposition 12.15, equality (13.60), the rule of change of variable in integrals, the equalities

$$\int_{\partial\Omega}\Bigg(v^+_{\Omega}[\theta^i](\xi)+\epsilon^{n-2}\int_{\partial\Omega}R_{q,n}(\epsilon(\xi-\eta))\theta^i(\eta)\,d\sigma_\eta$$
$$-\fint_{\partial\Omega}\Big(v^+_{\Omega}[\theta^i](\zeta)+\epsilon^{n-2}\int_{\partial\Omega}R_{q,n}(\epsilon(\zeta-\eta))\theta^i(\eta)d\sigma_\eta\Big)d\sigma_\zeta\Bigg)\,d\sigma_\xi=0\,,$$

$$\int_{\partial\Omega}\Bigg(-v^-_{\Omega}[\theta^o](\xi)-\epsilon^{n-2}\int_{\partial\Omega}R_{q,n}(\epsilon(\xi-\eta))\theta^o(\eta)\,d\sigma_\eta$$
$$+\fint_{\partial\Omega}\Big(v^-_{\Omega}[\theta^o](\zeta)+\epsilon^{n-2}\int_{\partial\Omega}R_{q,n}(\epsilon(\zeta-\eta))\theta^o(\eta)d\sigma_\eta\Big)d\sigma_\zeta\Bigg)\,d\sigma_\xi=0\,,$$

and

$$\int_{\partial\Omega}\Big(g(\xi)-\fint_{\partial\Omega}g\,d\sigma\Big)\,d\sigma_\xi=0\,,$$

imply that

$$\int_{\partial\Omega}\mathcal{M}_2[\epsilon,\epsilon',\theta^i,\theta^o]\,d\sigma=0\,.$$

Then we have the following proposition.

Proposition 13.26. *Let $\epsilon\in]0,\epsilon_0[$. Let $k\in\{1,\dots,n\}$. Then there exists a unique solution $(u^+_{k,\epsilon},u^-_{k,\epsilon})$ in $C^{1,\alpha}_{\mathrm{loc}}(\overline{\mathbb{S}[\Omega_{p,\epsilon}]})\times C^{1,\alpha}_{\mathrm{loc}}(\overline{\mathbb{S}[\Omega_{p,\epsilon}]^-})$ of problem* (13.45). *Such a solution is delivered by*

$$\begin{aligned}
u^+_{k,\epsilon}(x)\equiv{}& v^+_{q,\Omega_{p,\epsilon}}[\theta^i_{k,\epsilon}((\cdot-p)/\epsilon)](x)\\
&-\frac{\epsilon^{1-n}}{m_{n-1}(\partial\Omega)}\int_{\partial\Omega_{p,\epsilon}}v^+_{q,\Omega_{p,\epsilon}}[\theta^i_{k,\epsilon}((\cdot-p)/\epsilon)]\,d\sigma\\
&+x_k-\frac{\epsilon^{1-n}}{m_{n-1}(\partial\Omega)}\int_{\partial\Omega_{p,\epsilon}}y_k\,d\sigma_y\qquad\forall x\in\overline{\mathbb{S}[\Omega_{p,\epsilon}]}\,,\\
u^-_{k,\epsilon}(x)\equiv{}& v^-_{q,\Omega_{p,\epsilon}}[\theta^o_{k,\epsilon}((\cdot-p)/\epsilon)](x)\\
&-\frac{\epsilon^{1-n}}{m_{n-1}(\partial\Omega)}\int_{\partial\Omega_{p,\epsilon}}v^-_{q,\Omega_{p,\epsilon}}[\theta^o_{k,\epsilon}((\cdot-p)/\epsilon)]\,d\sigma\\
&-\rho(\epsilon)\frac{\epsilon^{1-n}}{m_{n-1}(\partial\Omega)}\int_{\partial\Omega_{p,\epsilon}}g((y-p)/\epsilon)\,d\sigma_y\\
&+x_k-\frac{\epsilon^{1-n}}{m_{n-1}(\partial\Omega)}\int_{\partial\Omega_{p,\epsilon}}y_k\,d\sigma_y\quad\forall x\in\overline{\mathbb{S}[\Omega_{p,\epsilon}]^-}\,,
\end{aligned}$$

where $(\theta^i_{k,\epsilon},\theta^o_{k,\epsilon})$ denotes the unique solution (θ^i,θ^o) in $(C^{0,\alpha}(\partial\Omega)_0)^2$ of

$$\mathcal{M}\Big[\epsilon,\frac{\epsilon}{\rho(\epsilon)},\theta^i,\theta^o\Big]=0\,.\tag{13.61}$$

Proof. We first note that a pair (u_k^+, u_k^-) in the product space $C^{1,\alpha}_{\mathrm{loc}}(\overline{\mathbb{S}[\Omega_{p,\epsilon}]}) \times C^{1,\alpha}_{\mathrm{loc}}(\overline{\mathbb{S}[\Omega_{p,\epsilon}]^-})$ solves problem (13.45) if and only if

$$
\begin{aligned}
u_k^+(x) &= v_k^+(x) + x_k && \forall x \in \overline{\mathbb{S}[\Omega_{p,\epsilon}]}\,,\\
u_k^-(x) &= v_k^-(x) + x_k && \forall x \in \overline{\mathbb{S}[\Omega_{p,\epsilon}]^-}\,,
\end{aligned}
$$

where the pair (v_k^+, v_k^-) is in $C^{1,\alpha}_q(\overline{\mathbb{S}[\Omega_{p,\epsilon}]}) \times C^{1,\alpha}_q(\overline{\mathbb{S}[\Omega_{p,\epsilon}]^-})$ and solves the following periodic problem

$$
\begin{cases}
\Delta v^+ = 0 & \text{in } \mathbb{S}[\Omega_{p,\epsilon}]\,,\\
\Delta v^- = 0 & \text{in } \mathbb{S}[\Omega_{p,\epsilon}]^-\,,\\
v^+(x+qz) = v^+(x) & \forall x \in \overline{\mathbb{S}[\Omega_{p,\epsilon}]}\,, z \in \mathbb{Z}^n\,,\\
v^-(x+qz) = v^-(x) & \forall x \in \overline{\mathbb{S}[\Omega_{p,\epsilon}]^-}\,, z \in \mathbb{Z}^n\,,\\
\lambda^- \nu_{\Omega_{p,\epsilon}}(x) \cdot \nabla v^-(x) - \lambda^+ \nu_{\Omega_{p,\epsilon}}(x) \cdot \nabla v^+(x) & \\
\quad = f((x-p)/\epsilon) + (\lambda^+ - \lambda^-)(\nu_{\Omega_{p,\epsilon}}(x))_k & \forall x \in \partial\Omega_{p,\epsilon}\,,\\
\lambda^+ \nu_{\Omega_{p,\epsilon}}(x) \cdot \nabla v^+(x) + \frac{1}{\rho(\epsilon)}(v^+(x) - v^-(x)) & \\
\quad = g((x-p)/\epsilon) - \lambda^+ (\nu_{\Omega_{p,\epsilon}}(x))_k & \forall x \in \partial\Omega_{p,\epsilon}\,,\\
\int_{\partial\Omega_{p,\epsilon}} v^+(x)\, d\sigma_x = -\int_{\partial\Omega_{p,\epsilon}} y_k\, d\sigma_y\,. &
\end{cases}
\tag{13.62}
$$

On the other hand, Propositions 13.19, 13.21 and Theorem 13.22 imply that the solution in $C^{1,\alpha}_q(\overline{\mathbb{S}[\Omega_Q]}) \times C^{1,\alpha}_q(\overline{\mathbb{S}[\Omega_Q]^-})$ of problem (13.62) exists and is unique. As a consequence, also problem (13.45) has a unique solution in the product space $C^{1,\alpha}_{\mathrm{loc}}(\overline{\mathbb{S}[\Omega_{p,\epsilon}]}) \times C^{1,\alpha}_{\mathrm{loc}}(\overline{\mathbb{S}[\Omega_{p,\epsilon}]^-})$ and we denote such a solution by $(u^+_{k,\epsilon}, u^-_{k,\epsilon})$. Moreover, the pair $(u^+_{k,\epsilon}, u^-_{k,\epsilon})$ is delivered by

$$
\begin{aligned}
u^+_{k,\epsilon}(x) &\equiv v^+_{k,\epsilon}(x) + x_k && \forall x \in \overline{\mathbb{S}[\Omega_{p,\epsilon}]}\,,\\
u^-_{k,\epsilon}(x) &\equiv v^-_{k,\epsilon}(x) + x_k && \forall x \in \overline{\mathbb{S}[\Omega_{p,\epsilon}]^-}\,,
\end{aligned}
$$

where $(v^+_{k,\epsilon}, v^-_{k,\epsilon})$ is the unique solution in $C^{1,\alpha}_q(\overline{\mathbb{S}[\Omega_{p,\epsilon}]}) \times C^{1,\alpha}_q(\overline{\mathbb{S}[\Omega_{p,\epsilon}]^-})$ of the periodic problem in (13.62). Then by Proposition 13.21, by Theorem 13.22, and by a simple computation based on the rule of change of variables in integrals, we deduce the validity of the proposition. □

By Proposition 13.26, we are reduced to analyze the system of integral equations (13.61). We note that equation (13.61) does not make sense for $\epsilon = 0$, but we can consider the equation

$$
\mathcal{M}[\epsilon, \epsilon', \theta^i, \theta^o] = 0\,, \tag{13.63}
$$

around $(\epsilon, \epsilon') = (0, r_*)$, and equation (13.63) makes perfectly sense if

$$
\epsilon = 0\,, \quad \epsilon' = r_*\,. \tag{13.64}
$$

By Proposition 13.26, we already know that if $\epsilon \in]0, \epsilon_0[$ and $\epsilon' = \epsilon/\rho(\epsilon)$, then equation (13.63) is equivalent to problem (13.45). We now consider equation (13.63) under condition (13.64) by means of the following.

Theorem 13.27. *Let $k \in \{1,\dots,n\}$. Then there exists a unique pair (θ^i,θ^o) in $(C^{0,\alpha}(\partial\Omega)_0)^2$ such that*

$$\mathcal{M}[0,r_*,\theta^i,\theta^o]=0\,, \tag{13.65}$$

and we denote such a pair by $(\tilde\theta^i_k,\tilde\theta^o_k)$. Moreover, the pair of functions $(\tilde u^+_k,\tilde u^-_k)$ in $C^{1,\alpha}(\overline{\Omega})\times C^{1,\alpha}_{\mathrm{loc}}(\mathbb{R}^n\setminus\Omega)$, defined by

$$\tilde u^+_k \equiv v^+_\Omega[\tilde\theta^i_k]-\fint_{\partial\Omega} v^+_\Omega[\tilde\theta^i_k]\,d\sigma\,, \tag{13.66}$$

$$\tilde u^-_k \equiv v^-_\Omega[\tilde\theta^o_k]-\fint_{\partial\Omega} v^-_\Omega[\tilde\theta^o_k]\,d\sigma\,, \tag{13.67}$$

is the unique solution of the following 'limiting boundary value problem'

$$\begin{cases} \Delta u^+=0 & \text{in } \Omega\,,\\ \Delta u^-=0 & \text{in } \mathbb{R}^n\setminus\overline{\Omega}\,,\\ \lambda^-\nu_\Omega(\xi)\cdot\nabla u^-(\xi)-\lambda^+\nu_\Omega(\xi)\cdot\nabla u^+(\xi) & \\ \qquad =f(\xi)+(\lambda^+-\lambda^-)(\nu_\Omega(\xi))_k & \forall\xi\in\partial\Omega\,,\\ \lambda^+\nu_\Omega(\xi)\cdot\nabla u^+(\xi)+r_*\big(u^+(\xi)-u^-(\xi)\big) & \\ \qquad =g(\xi)-\fint_{\partial\Omega} g\,d\sigma-\lambda^+(\nu_\Omega(\xi))_k & \forall\xi\in\partial\Omega\,,\\ \int_{\partial\Omega} u^+(\xi)\,d\sigma_\xi=0\,, & \\ \int_{\partial\Omega} u^-(\xi)\,d\sigma_\xi=0\,, & \\ \lim_{\xi\to\infty} u^-(\xi)\in\mathbb{R}\,. & \end{cases} \tag{13.68}$$

Proof. We first note that equation (13.65) can be rewritten as

$$K_{r_*}[\theta^i,\theta^o]=\Big(f+(\lambda^+-\lambda^-)(\nu_\Omega)_k\,,\;g-\fint_{\partial\Omega} g\,d\sigma-\lambda^+(\nu_\Omega)_k\Big)$$

(see Proposition 13.24). Then, by Proposition 13.24, there exists a unique pair $(\theta^i,\theta^o)\in(C^{0,\alpha}(\partial\Omega)_0)^2$ such that (13.65) holds. Then by Theorem 13.25 and by Theorem 4.23 on the behavior at infinity of the single layer potential, the pair of functions delivered by (13.66)–(13.67) is the unique solution of problem (13.68). □

Let $\tilde\theta^o_k$, $\tilde u^-_k$ be as in Theorem 13.27. Then by the limiting behavior of the single layer potential at infinity of Theorem 4.23 and by equality $\int_{\partial\Omega}\tilde\theta^o_k\,d\sigma=0$, we have

$$\tilde l^-_k \equiv \lim_{\xi\to\infty}\tilde u^-_k(\xi) = -\fint_{\partial\Omega} v^-_\Omega[\tilde\theta^o_k]\,d\sigma\,. \tag{13.69}$$

We now analyze equation (13.63) for (ϵ,ϵ') in a neighborhood of $(0,r_*)$ by means of the following.

Theorem 13.28. *Let $k\in\{1,\dots,n\}$. Let $\mathcal{M}$ be as in (13.58)–(13.59). Let $(\theta^i_{k,\epsilon},\theta^o_{k,\epsilon})$ be as in Proposition 13.26. Let $(\tilde\theta^i_k,\tilde\theta^o_k)$ be as in Theorem 13.27. Then there exist $\epsilon_1\in]0,\epsilon_0]$, an open neighborhood U_{r_*} of r_* in $\mathbb{R}$, an open neighborhood V of*

$(\tilde{\theta}^i_k, \tilde{\theta}^o_k)$ in $(C^{0,\alpha}(\partial\Omega)_0)^2$, and a real analytic map (Θ^i_k, Θ^o_k) from $]-\epsilon_1, \epsilon_1[\times U_{r_*}$ to V such that

$$\epsilon/\rho(\epsilon) \in U_{r_*} \qquad \forall \epsilon \in]0, \epsilon_1[\,,$$

and such that the set of zeros of $\mathcal{M}$ in $]-\epsilon_1, \epsilon_1[\times U_{r_*} \times V$ coincides with the graph of (Θ^i_k, Θ^o_k). In particular,

$$\left(\Theta^i_k\left[\epsilon, \frac{\epsilon}{\rho(\epsilon)}\right], \Theta^o_k\left[\epsilon, \frac{\epsilon}{\rho(\epsilon)}\right]\right) = \left(\theta^i_{k,\epsilon}, \theta^o_{k,\epsilon}\right) \; \forall \epsilon \in]0, \epsilon_1[\,,$$
$$(\Theta^i_k[0, r_*], \Theta^o_k[0, r_*]) = (\tilde{\theta}^i_k, \tilde{\theta}^o_k)\,.$$

Proof. We plan to apply the Implicit Function Theorem A.19 to equation (13.63) around the point $(0, r_*, \tilde{\theta}^i_k, \tilde{\theta}^o_k)$. By the properties of integral operators with real analytic kernels of Theorem A.28, and by classical mapping properties of layer potentials (cf. Lemma 13.4 (i) and Theorems 4.25 and 6.8), we conclude that $\mathcal{M}$ is real analytic. By definition of $(\tilde{\theta}^i_k, \tilde{\theta}^o_k)$, we have $\mathcal{M}[0, r_*, \tilde{\theta}^i_k, \tilde{\theta}^o_k] = 0$. By standard Calculus in Banach spaces, the differential of $\mathcal{M}$ at the point $(0, r_*, \tilde{\theta}^i_k, \tilde{\theta}^o_k)$ with respect to the variables (θ^i, θ^o) is delivered by the equality

$$\partial_{(\theta^i,\theta^o)}\mathcal{M}[0, r_*, \tilde{\theta}^i_k, \tilde{\theta}^o_k] = K_{r_*}$$

(see Proposition 13.24). Then by Proposition 13.24, $\partial_{(\theta^i,\theta^o)}\mathcal{M}[0, r_*, \tilde{\theta}^i_k, \tilde{\theta}^o_k]$ is a linear homeomorphism from $(C^{0,\alpha}(\partial\Omega)_0)^2$ onto $(C^{0,\alpha}(\partial\Omega)_0)^2$. Hence the existence of ϵ_1, U_{r_*}, V, Θ^i_k, Θ^o_k as in the statement follows by the Implicit Function Theorem for real analytic maps in Banach spaces (cf. Theorem A.19). □

13.6.3 A Functional Analytic Representation Theorem for the Solutions of the Singularly Perturbed Transmission Problem

In the following Theorem 13.29 we investigate the microscopic behavior of $u^+_{k,\epsilon}$ for ϵ small and positive.

Theorem 13.29. *Let the assumptions of Theorem 13.28 hold. Then there exists a real analytic map $\mathcal{U}^+_k$ from $]-\epsilon_1, \epsilon_1[\times U_{r_*}$ to $C^{1,\alpha}(\overline{\Omega})$ such that*

$$u^+_{k,\epsilon}(p + \epsilon\xi) = \epsilon\mathcal{U}^+_k\left[\epsilon, \frac{\epsilon}{\rho(\epsilon)}\right](\xi) \qquad \forall \xi \in \overline{\Omega}\,,$$

for all $\epsilon \in]0, \epsilon_1[$. Moreover,

$$\mathcal{U}^+_k[0, r_*](\xi) = \tilde{u}^+_k(\xi) + \xi_k - \int_{\partial\Omega} \eta_k \, d\sigma_\eta \qquad \forall \xi \in \overline{\Omega}\,, \tag{13.70}$$

where $\tilde{u}^+_k$ is defined as in Theorem 13.27.

Proof. By the formula for $u^+_{k,\epsilon}$ of Proposition 13.26, by the formula for $\theta^i_{k,\epsilon}$ of Theorem 13.28, and by the rule of change of variables in integrals, we have

$$\begin{aligned} u^+_{k,\epsilon}(p+\epsilon\xi) =& \epsilon v^+_\Omega\Big[\Theta^i_k\Big[\epsilon, \frac{\epsilon}{\rho(\epsilon)}\Big]\Big](\xi) \\ &+ \epsilon^{n-1}\int_{\partial\Omega} R_{q,n}(\epsilon(\xi-\eta))\Theta^i_k\Big[\epsilon, \frac{\epsilon}{\rho(\epsilon)}\Big](\eta)\, d\sigma_\eta \\ &- \epsilon \fint_{\partial\Omega}\Big(v^+_\Omega\Big[\Theta^i_k\Big[\epsilon, \frac{\epsilon}{\rho(\epsilon)}\Big]\Big](\zeta) \\ &+ \epsilon^{n-2}\int_{\partial\Omega} R_{q,n}(\epsilon(\zeta-\eta))\Theta^i_k\Big[\epsilon, \frac{\epsilon}{\rho(\epsilon)}\Big](\eta) d\sigma_\eta\Big) d\sigma_\zeta \\ &+ \epsilon\xi_k - \epsilon\fint_{\partial\Omega}\eta_k\, d\sigma_\eta \qquad \forall \xi\in\overline{\Omega}, \end{aligned}$$

for all $\epsilon\in]0,\epsilon_1[$. Therefore it is natural to set

$$\begin{aligned} &\mathcal{U}^+_k[\epsilon,\epsilon'](\xi) \equiv v^+_\Omega\Big[\Theta^i_k[\epsilon,\epsilon']\Big](\xi) + \epsilon^{n-2}\int_{\partial\Omega} R_{q,n}(\epsilon(\xi-\eta))\Theta^i_k[\epsilon,\epsilon'](\eta)\, d\sigma_\eta \\ &- \fint_{\partial\Omega}\Big(v^+_\Omega\Big[\Theta^i_k[\epsilon,\epsilon']\Big](\zeta) + \epsilon^{n-2}\int_{\partial\Omega} R_{q,n}(\epsilon(\zeta-\eta))\Theta^i_k[\epsilon,\epsilon'](\eta) d\sigma_\eta\Big) d\sigma_\zeta \\ &+ \xi_k - \fint_{\partial\Omega}\eta_k\, d\sigma_\eta \qquad\qquad \forall\xi\in\overline{\Omega}, \end{aligned} \tag{13.71}$$

for all $(\epsilon,\epsilon')\in]-\epsilon_1,\epsilon_1[\times U_{r_*}$. By the properties of integral operators with real analytic kernel of Theorem A.28 (i), (ii), by the classical mapping properties of the layer potentials (cf. Lemma 13.4 (i) and Theorem 4.25 (i)), and by Theorem 13.28, we conclude that $\mathcal{U}^+_k$ is real analytic. Moreover, Theorem 13.28 implies that $\Theta^i_k[0,r_*] = \tilde{\theta}^i_k$ and thus the validity of equality (13.70) follows (see also Theorem 13.27). □

In the following Theorem 13.30 we investigate the behavior of $u^-_{k,\epsilon}$ for ϵ small and positive.

Theorem 13.30. *Let the assumptions of Theorem 13.28 hold. Let $(u^+_{k,\epsilon}, u^-_{k,\epsilon})$ be the unique solution of problem (13.45) for all $\epsilon\in]0,\epsilon_0[$. Let $\tilde{l}^-_k$ be as in (13.69). Then there exists a real analytic map $\mathcal{C}^-_k$ from $]-\epsilon_1,\epsilon_1[\times U_{r_*}$ to $\mathbb{R}$ such that*

$$\mathcal{C}^-_k[0,r_*] = -\fint_{\partial\Omega} g\, d\sigma + r_*\tilde{l}^-_k - r_*\fint_{\partial\Omega}\eta_k\, d\sigma_\eta\,, \tag{13.72}$$

and such that the following statements hold.

(i) Let $\tilde{\Omega}$ be an open bounded subset of $\mathbb{R}^n$ such that $\overline{\tilde{\Omega}}\cap(p+q\mathbb{Z}^n)=\emptyset$. Let $j\in\mathbb{N}$. Then there exist $\epsilon_{\tilde{\Omega}}\in]0,\epsilon_1[$ and a real analytic map $\mathcal{U}^-_{k,\tilde{\Omega}}$ from $]-\epsilon_{\tilde{\Omega}},\epsilon_{\tilde{\Omega}}[\times U_{r_*}$ to $C^j(\overline{\tilde{\Omega}})$ such that $\overline{\tilde{\Omega}}\subseteq \mathbb{S}[\Omega_{p,\epsilon}]^-$ for all $\epsilon\in]-\epsilon_{\tilde{\Omega}},\epsilon_{\tilde{\Omega}}[$, and such that

$$u^-_{k,\epsilon}(x) = x_k - p_k + \rho(\epsilon)\mathcal{C}^-_k\left[\epsilon, \frac{\epsilon}{\rho(\epsilon)}\right] + \epsilon^n \mathcal{U}^-_{k,\tilde{\Omega}}\left[\epsilon, \frac{\epsilon}{\rho(\epsilon)}\right](x) \quad \forall x \in \overline{\tilde{\Omega}} \quad (13.73)$$

for all $\epsilon \in]0, \epsilon_{\tilde{\Omega}}[$. *Moreover,*

$$\mathcal{U}^-_{k,\tilde{\Omega}}[0, r_*](x) = \nabla S_{q,n}(x-p) \cdot \left(\int_{\partial\Omega} \nu_\Omega(\eta)\tilde{u}^-_k(\eta)\, d\sigma_\eta - \int_{\partial\Omega} \eta \nu_\Omega(\eta) \cdot \nabla \tilde{u}^-_k(\eta)\, d\sigma_\eta \right) \quad \forall x \in \overline{\tilde{\Omega}}, \quad (13.74)$$

where $\tilde{u}^-_k$ *is defined as in Theorem 13.27.*

(ii) Let $\tilde{\Omega}$ *be a bounded open subset of* $\mathbb{R}^n \setminus \overline{\Omega}$. *Then there exist* $\epsilon^\#_{\tilde{\Omega}} \in]0, \epsilon_1[$ *and a real analytic map* $\mathcal{V}^-_{k,\tilde{\Omega}}$ *from* $]-\epsilon^\#_{\tilde{\Omega}}, \epsilon^\#_{\tilde{\Omega}}[\times U_{r_*}$ *to* $C^{1,\alpha}(\overline{\tilde{\Omega}})$ *such that* $p + \epsilon\overline{\tilde{\Omega}} \subseteq \overline{\mathbb{S}[\Omega_{p,\epsilon}]^-}$ *for all* $\epsilon \in]-\epsilon^\#_{\tilde{\Omega}}, \epsilon^\#_{\tilde{\Omega}}[$, *and*

$$u^-_{k,\epsilon}(p+\epsilon\xi) = \rho(\epsilon)\mathcal{C}^-_k\left[\epsilon, \frac{\epsilon}{\rho(\epsilon)}\right] + \epsilon \mathcal{V}^-_{k,\tilde{\Omega}}\left[\epsilon, \frac{\epsilon}{\rho(\epsilon)}\right](\xi) \qquad \forall \xi \in \overline{\tilde{\Omega}}$$

for all $\epsilon \in]0, \epsilon^\#_{\tilde{\Omega}}[$. *Moreover,*

$$\mathcal{V}^-_{k,\tilde{\Omega}}[0, r_*](\xi) = \tilde{u}^-_k(\xi) - \tilde{l}^-_k + \xi_k \qquad \forall \xi \in \overline{\tilde{\Omega}}, \quad (13.75)$$

where $\tilde{u}^-_k$ *is defined as in Theorem 13.27.*

(iii) There exists a real analytic map $\mathcal{V}^-_{k,\partial\Omega}$ *from* $]-\epsilon_1, \epsilon_1[\times U_{r_*}$ *to* $C^{1,\alpha}(\partial\Omega)$ *such that*

$$u^-_{k,\epsilon}(p+\epsilon\xi) = \rho(\epsilon)\mathcal{C}^-_k\left[\epsilon, \frac{\epsilon}{\rho(\epsilon)}\right] + \epsilon \mathcal{V}^-_{k,\partial\Omega}\left[\epsilon, \frac{\epsilon}{\rho(\epsilon)}\right](\xi) \qquad \forall \xi \in \partial\Omega, \quad (13.76)$$

for all $\epsilon \in]0, \epsilon_1[$. *Moreover,*

$$\mathcal{V}^-_{k,\partial\Omega}[0, r_*](\xi) = \tilde{u}^-_k(\xi) - \tilde{l}^-_k + \xi_k \qquad \forall \xi \in \partial\Omega, \quad (13.77)$$

where $\tilde{u}^-_k$ *is defined as in Theorem 13.27.*

Proof. We set

$$\mathcal{C}^-_k[\epsilon, \epsilon'] \equiv -\!\!\!\!\!\!\int_{\partial\Omega} g\, d\sigma - \epsilon' \,-\!\!\!\!\!\!\int_{\partial\Omega} \Big(v^-_\Omega\Big[\Theta^o_k[\epsilon, \epsilon']\Big](\xi) + \epsilon^{n-2} \int_{\partial\Omega} R_{q,n}(\epsilon(\xi-\eta))\Theta^o_k[\epsilon, \epsilon'](\eta) d\sigma_\eta \Big) d\sigma_\xi - \epsilon' \,-\!\!\!\!\!\!\int_{\partial\Omega} \eta_k\, d\sigma_\eta \quad (13.78)$$

for all $(\epsilon, \epsilon') \in]-\epsilon_1, \epsilon_1[\times U_{r_*}$. By the properties of integral operators with real analytic kernel of Theorem A.28 (ii), by the classical mapping properties of the layer potentials (cf. Lemma 13.4 (i) and Theorem 4.25 (ii)), and by Theorem 13.28, we deduce that $\mathcal{C}_k^-$ is real analytic. Then, by (13.69) and by equality $\Theta_k^o[0, r_*] = \tilde{\theta}_k^o$ we can verify the validity of (13.72).

We now consider the proof of statement (i). Without loss of generality, we can assume that $\tilde{\Omega}$ is of class C^∞. Indeed, if this is not the case, we can consider an open bounded connected set $\tilde{\Omega}'$ of class C^∞ such that

$$\overline{\tilde{\Omega}} \subseteq \tilde{\Omega}'$$

and that

$$\overline{\tilde{\Omega}'} \cap (p + q\mathbb{Z}^n) = \emptyset$$

(cf. Lemma 2.70). Then if we prove the validity of statement (i) with $\tilde{\Omega}$ replaced with $\tilde{\Omega}'$, the continuity of the restriction operator from $C^j(\overline{\tilde{\Omega}'})$ to $C^j(\overline{\tilde{\Omega}})$ implies that statement (i) holds also for $\tilde{\Omega}$. Therefore, we can assume that $\tilde{\Omega}$ is of class C^∞ and, by taking $\epsilon_{\tilde{\Omega}}$ small enough, we can also assume that $\overline{\tilde{\Omega}} \subseteq \mathbb{S}[\Omega_{p,\epsilon}]^-$ for all $\epsilon \in [-\epsilon_{\tilde{\Omega}}, \epsilon_{\tilde{\Omega}}]$. By the formula for $u_{k,\epsilon}^-$ of Proposition 13.26 and the formula for $\theta_{k,\epsilon}^o$ of Theorem 13.28, we have

$$\begin{aligned} u_{k,\epsilon}^-(x) =& \epsilon^{n-1} \int_{\partial\Omega} S_{q,n}(x - p - \epsilon\eta) \Theta_k^o\Big[\epsilon, \frac{\epsilon}{\rho(\epsilon)}\Big](\eta)\, d\sigma_\eta \\ & - \epsilon \fint_{\partial\Omega} \Big(v_{\Omega}^-\Big[\Theta_k^o\Big[\epsilon, \frac{\epsilon}{\rho(\epsilon)}\Big]\Big](\xi) \\ & + \epsilon^{n-2} \int_{\partial\Omega} R_{q,n}(\epsilon(\xi - \eta)) \Theta_k^o\Big[\epsilon, \frac{\epsilon}{\rho(\epsilon)}\Big](\eta) d\sigma_\eta \Big) d\sigma_\xi \\ & - \rho(\epsilon) \fint_{\partial\Omega} g\, d\sigma + x_k - p_k - \epsilon \fint_{\partial\Omega} \eta_k\, d\sigma_\eta \qquad \forall x \in \overline{\tilde{\Omega}}, \end{aligned}$$

for all $\epsilon \in]0, \epsilon_{\tilde{\Omega}}[$. Then we note that if $\epsilon \in [-\epsilon_{\tilde{\Omega}}, \epsilon_{\tilde{\Omega}}]$, then $\partial\mathbb{S}[\Omega_{p,\epsilon}] \cap \overline{\tilde{\Omega}} = \emptyset$. Also, we observe that if $x \in \overline{\tilde{\Omega}}$, then $x - p - \epsilon\beta\eta$ does not belong to $q\mathbb{Z}^n$ for any $\eta \in \partial\Omega$, $\epsilon \in] - \epsilon_{\tilde{\Omega}}, \epsilon_{\tilde{\Omega}}[$, and $\beta \in [0, 1]$. Accordingly, we can invoke the Taylor formula with integral remainder and write

$$S_{q,n}(x - p - \epsilon\eta) - S_{q,n}(x - p) = -\epsilon \int_0^1 \eta \cdot \nabla S_{q,n}(x - p - \beta\epsilon\eta)\, d\beta\,,$$

for all $(x, \eta) \in \overline{\tilde{\Omega}} \times \partial\Omega$ and $\epsilon \in]0, \epsilon_{\tilde{\Omega}}[$. Since $\int_{\partial\Omega} \Theta_k^o[\epsilon, \epsilon/\rho(\epsilon)]\, d\sigma = 0$ for all $\epsilon \in]0, \epsilon_{\tilde{\Omega}}[$, we conclude that

$$
\begin{aligned}
u^-_{k,\epsilon}(x) = & -\epsilon^n \int_{\partial\Omega} \left(\int_0^1 \eta \cdot \nabla S_{q,n}(x-p-\beta\epsilon\eta)\, d\beta \right) \Theta^o_k\Big[\epsilon, \frac{\epsilon}{\rho(\epsilon)}\Big](\eta)\, d\sigma_\eta \\
& - \epsilon \fint_{\partial\Omega} \Bigg(v^-_{\Omega}\Big[\Theta^o_k\Big[\epsilon, \frac{\epsilon}{\rho(\epsilon)}\Big]\Big](\xi) \\
& + \epsilon^{n-2} \int_{\partial\Omega} R_{q,n}(\epsilon(\xi-\eta))\Theta^o_k\Big[\epsilon, \frac{\epsilon}{\rho(\epsilon)}\Big](\eta) d\sigma_\eta \Bigg) d\sigma_\xi \\
& - \rho(\epsilon) \fint_{\partial\Omega} g\, d\sigma + x_k - p_k - \epsilon \fint_{\partial\Omega} \eta_k\, d\sigma_\eta
\end{aligned} \tag{13.79}
$$

for all $x \in \overline{\tilde{\Omega}}$ and for all $\epsilon \in]0, \epsilon_{\tilde{\Omega}}[$. Thus it is natural to set

$$
\begin{aligned}
& \mathcal{U}^-_{k,\tilde{\Omega}}[\epsilon, \epsilon'](x) \\
& \equiv - \int_{\partial\Omega} \left(\int_0^1 \eta \cdot \nabla S_{q,n}(x-p-\beta\epsilon\eta)\, d\beta \right) \Theta^o_k[\epsilon, \epsilon'](\eta)\, d\sigma_\eta \quad \forall x \in \overline{\tilde{\Omega}}
\end{aligned} \tag{13.80}
$$

for all $(\epsilon, \epsilon') \in]-\epsilon_{\tilde{\Omega}}, \epsilon_{\tilde{\Omega}}[\times U_{r_*}$. Then the validity of (13.73) follows by definitions (13.78), (13.80) and equality (13.79).

We first note that the function $\Phi(x, \eta, \epsilon, \beta) \equiv \eta \cdot \nabla S_{q,n}(x - p - \beta\epsilon\eta)$ is analytic when $(x, \eta, \epsilon, \beta)$ belongs to an open neighborhood of $\overline{\tilde{\Omega}} \times \overline{\Omega} \times [-\epsilon_{\tilde{\Omega}}, \epsilon_{\tilde{\Omega}}] \times [0, 1]$. Then we take a bounded open set W in $\mathbb{R}^n \times \mathbb{R}^n \times \mathbb{R}$, regular in the sense of Whitney, such that

$$
\overline{\tilde{\Omega} \times \Omega \times] - \epsilon_{\tilde{\Omega}}, \epsilon_{\tilde{\Omega}}[} \subseteq W\,,
$$

and such that the function $\Phi(x, \eta, \epsilon, \beta) \equiv \eta \cdot \nabla S_{q,n}(x - p - \beta\epsilon\eta)$ is analytic when $(x, \eta, \epsilon, \beta)$ belongs to an open neighborhood of $\overline{W} \times [0, 1]$ (cf. Proposition 2.47 and Lemma 2.70). Remark 2.18 implies that there exists $\rho \in]0, +\infty[$ such that

$$
\Phi \in C^0_{\omega,\rho}(\overline{W \times]0, 1[})\,.
$$

Then by integrating Φ in the variable $\beta \in]0, 1[$, Proposition 2.19 implies that $\int_0^1 \Phi(\cdot, \cdot, \cdot, \beta)\, d\beta \in C^0_{\omega,\rho}(\overline{W})$. Then, by Proposition 2.32, the function

$$
\int_0^1 \Phi(\cdot, \cdot, \cdot, \beta)\, d\beta
$$

is analytic in an open neighborhood of $\overline{W}$. As a consequence, Theorem A.27 (i) and the continuity of the embedding of $C^{j,1}(\overline{\tilde{\Omega}})$ into $C^j(\overline{\tilde{\Omega}})$ imply that the map from $]-\epsilon_{\tilde{\Omega}}, \epsilon_{\tilde{\Omega}}[\times L^1(\partial\Omega)$ to $C^j(\overline{\tilde{\Omega}})$ that takes (ϵ, θ) to

$$
\int_{\partial\Omega} \int_0^1 \Phi(\cdot, \eta, \epsilon, \beta)\, d\beta\theta(\eta)\, d\sigma_\eta
$$

is analytic. Then the analyticity of Θ_k^o of Theorem 13.28 imply that $\mathcal{U}^-_{k,\tilde{\Omega}}$ is real analytic from $]-\epsilon_{\tilde{\Omega}}, \epsilon_{\tilde{\Omega}}[\times U_{r_*}$ to $C^j(\overline{\tilde{\Omega}})$.

Next we prove formula (13.74). Theorem 13.28 implies that $\Theta_k^o[0, r_*] = \tilde{\theta}_k^o$. Now let $l \in \{1, \dots, n\}$. By the jump formulas for the normal derivative of the classical single layer potential of Theorem 6.6, we have

$$\int_{\partial\Omega} \eta_l \tilde{\theta}_k^o(\eta)\, d\sigma_\eta = \int_{\partial\Omega} \eta_l \nu_\Omega(\eta)\cdot\nabla v^-_{\Omega}[\tilde{\theta}_k^o](\eta)\, d\sigma_\eta - \int_{\partial\Omega} \eta_l \nu_\Omega(\eta)\cdot\nabla v^+_{\Omega}[\tilde{\theta}_k^o](\eta)\, d\sigma_\eta\,.$$

Since both η_l and $v^+_{\Omega}[\tilde{\theta}_k^o]$ are harmonic in Ω, the second Green Identity of Theorem 4.3 implies that

$$\int_{\partial\Omega} \eta_l \nu_\Omega(\eta)\cdot\nabla v^+_{\Omega}[\tilde{\theta}_k^o](\eta)\, d\sigma_\eta = \int_{\partial\Omega} (\nu_\Omega(\eta))_l v^+_{\Omega}[\tilde{\theta}_k^o](\eta)\, d\sigma_\eta\,.$$

Moreover,

$$\nu_\Omega \cdot \nabla \tilde{u}_k^- = \nu_\Omega \cdot \nabla v^-_{\Omega}[\tilde{\theta}_k^o] \qquad \text{on } \partial\Omega\,.$$

As a consequence, the equality $v^+_{\Omega}[\tilde{\theta}_k^o]_{|\partial\Omega} = v^-_{\Omega}[\tilde{\theta}_k^o]_{|\partial\Omega}$ implies that

$$\begin{aligned}\int_{\partial\Omega} \eta_l \tilde{\theta}_k^o(\eta)\, d\sigma_\eta &= \int_{\partial\Omega} \eta_l \nu_\Omega(\eta)\cdot\nabla \tilde{u}_k^-(\eta)\, d\sigma_\eta - \int_{\partial\Omega} (\nu_\Omega(\eta))_l v^+_{\Omega}[\tilde{\theta}_k^o](\eta)\, d\sigma_\eta \\ &= \int_{\partial\Omega} \eta_l \nu_\Omega(\eta)\cdot\nabla \tilde{u}_k^-(\eta)\, d\sigma_\eta - \int_{\partial\Omega} (\nu_\Omega(\eta))_l v^-_{\Omega}[\tilde{\theta}_k^o](\eta)\, d\sigma_\eta\,.\end{aligned}$$

On the other hand, since

$$\tilde{u}_k^-(\eta) = v^-_{\Omega}[\tilde{\theta}_k^o](\eta) - \fint_{\partial\Omega} v^-_{\Omega}[\tilde{\theta}_k^o]\, d\sigma \qquad \forall \eta \in \partial\Omega\,,$$

and

$$\int_{\partial\Omega} (\nu_\Omega(\eta))_l\, d\sigma_\eta = 0\,,$$

we have

$$\int_{\partial\Omega} (\nu_\Omega(\eta))_l v^-_{\Omega}[\tilde{\theta}_k^o](\eta)\, d\sigma_\eta = \int_{\partial\Omega} (\nu_\Omega(\eta))_l \tilde{u}_k^-(\eta)\, d\sigma_\eta\,.$$

Accordingly, (13.74) holds.

We now consider statement (ii). By assumption, there exists $R > 0$ such that $(\overline{\tilde{\Omega}} \cup \overline{\Omega}) \subseteq \mathbb{B}_n(0, R)$. Then we set $\Omega^* \equiv \mathbb{B}_n(0, R) \setminus \overline{\Omega}$. By Lemma 13.6 (ii), there exists $\epsilon^\#_{\Omega^*} \in]0, \epsilon_1[$ such that $p + \epsilon\overline{\Omega^*} \subseteq Q$, and $p + \epsilon\Omega^* \subseteq \mathbb{S}[\Omega_{p,\epsilon}]^-$, for all $\epsilon \in [-\epsilon^\#_{\Omega^*}, \epsilon^\#_{\Omega^*}] \setminus \{0\}$. Then we set $\epsilon^\#_{\tilde{\Omega}} \equiv \epsilon^\#_{\Omega^*}$. It clearly suffices to show that $\mathcal{V}^-_{k,\Omega^*}$ exists and then to set $\mathcal{V}^-_{k,\tilde{\Omega}}$ equal to the composition of the restriction of $C^{1,\alpha}(\overline{\Omega^*})$ to $C^{1,\alpha}(\overline{\tilde{\Omega}})$ with $\mathcal{V}^-_{k,\Omega^*}$. The advantage of Ω^* with respect to $\tilde{\Omega}$ is that Ω^* is of class C^1 and that accordingly $C^2(\overline{\Omega^*})$ is continuously embedded into $C^{1,\alpha}(\overline{\Omega^*})$, a fact which we exploit below (cf. Proposition 2.54 (ii)).

By the formula for u_k^- of Proposition 13.26, by the formula for $\theta^o_{k,\epsilon}$ of Theorem 13.28, by (13.78), by equality $\int_{\partial\Omega} \Theta^o_k[\epsilon, \epsilon/\rho(\epsilon)](\eta)\, d\sigma_\eta = 0$, and by a simple computation based on the rule of change of variables in integrals, we have

$$
\begin{aligned}
& u_k^-[\epsilon](p+\epsilon\xi) \\
&= \epsilon^{n-1} \int_{\partial\Omega} S_{q,n}(\epsilon(\xi-\eta))\Theta^o_k\Big[\epsilon, \frac{\epsilon}{\rho(\epsilon)}\Big](\eta)\, d\sigma_\eta + \rho(\epsilon)\mathcal{C}_k^-\Big[\epsilon, \frac{\epsilon}{\rho(\epsilon)}\Big] + \epsilon\xi_k \\
&= \epsilon\Big(\int_{\partial\Omega} S_n(\xi-\eta)\Theta^o_k\Big[\epsilon, \frac{\epsilon}{\rho(\epsilon)}\Big](\eta)\, d\sigma_\eta \\
&\quad + \epsilon^{n-2} \int_{\partial\Omega} R_{q,n}(\epsilon(\xi-\eta))\Theta^o_k\Big[\epsilon, \frac{\epsilon}{\rho(\epsilon)}\Big](\eta)\, d\sigma_\eta \Big) \\
&\quad + \rho(\epsilon)\mathcal{C}_k^-\Big[\epsilon, \frac{\epsilon}{\rho(\epsilon)}\Big] + \epsilon\xi_k \qquad \forall \xi \in \overline{\Omega^*},
\end{aligned}
$$

for all $\epsilon \in]0, \epsilon^\#_{\tilde{\Omega}}[$. Thus it is natural to set

$$
\begin{aligned}
\mathcal{V}^-_{k,\Omega^*}[\epsilon,\epsilon'](\xi) \equiv & \int_{\partial\Omega} S_n(\xi-\eta)\Theta^o_k[\epsilon,\epsilon'](\eta)\, d\sigma_\eta \\
& + \epsilon^{n-2} \int_{\partial\Omega} R_{q,n}(\epsilon(\xi-\eta))\Theta^o_k[\epsilon,\epsilon'](\eta)\, d\sigma_\eta + \xi_k \qquad \forall \xi \in \overline{\Omega^*},
\end{aligned}
\tag{13.81}
$$

for all $(\epsilon,\epsilon') \in]-\epsilon^\#_{\tilde{\Omega}}, \epsilon^\#_{\tilde{\Omega}}[\times U_{r_*}$. Since $v_{\tilde{\Omega}}[\cdot]_{|\overline{\Omega^*}}$ is linear and continuous from $C^{0,\alpha}(\partial\Omega)$ to $C^{1,\alpha}(\overline{\Omega^*})$ and Θ^o_k is real analytic, the map from $]-\epsilon^\#_{\tilde{\Omega}}, \epsilon^\#_{\tilde{\Omega}}[\times U_{r_*}$ to $C^{1,\alpha}(\overline{\Omega^*})$ that takes (ϵ,ϵ') to the function $\int_{\partial\Omega} S_n(\xi-\eta)\Theta^o_k[\epsilon,\epsilon'](\eta)\, d\sigma_\eta$ of the variable $\xi \in \overline{\Omega^*}$ is real analytic (cf. Theorem 4.25 (ii)). Clearly, we have $\epsilon\overline{\Omega^*} - \epsilon\partial\Omega \subseteq (\mathbb{R}^n \setminus q\mathbb{Z}^n) \cup \{0\}$ for all $\epsilon \in]-\epsilon^\#_{\tilde{\Omega}}, \epsilon^\#_{\tilde{\Omega}}[$. As a consequence, the properties of integral operators with real analytic kernel of Theorem A.28 (i) imply that the map from $]-\epsilon^\#_{\tilde{\Omega}}, \epsilon^\#_{\tilde{\Omega}}[\times L^1(\partial\Omega)$ to $C^2(\overline{\Omega^*})$ which takes (ϵ, f) to the function $\int_{\partial\Omega} R_{q,n}(\epsilon(\xi-\eta))f(\eta)\, d\sigma_\eta$ of the variable $\xi \in \overline{\Omega^*}$ is real analytic (cf. Lemma 13.4). Then by the analyticity of Θ^o_k and by the continuity of the embeddings of $C^{0,\alpha}(\partial\Omega)_0$ into $L^1(\partial\Omega)$ and of $C^2(\overline{\Omega^*})$ into $C^{1,\alpha}(\overline{\Omega^*})$, we conclude that the map from $]-\epsilon^\#_{\tilde{\Omega}}, \epsilon^\#_{\tilde{\Omega}}[\times U_{r_*}$ to $C^{1,\alpha}(\overline{\Omega^*})$ that takes (ϵ,ϵ') to the second term in the right-hand side of (13.81) is real analytic. Then, by standard Calculus in Banach space, we deduce that $\mathcal{V}^-_{k,\Omega^*}$ is real analytic. By Theorem 13.27, by equality $\Theta^o_k[0, r_*] = \tilde{\theta}^o_k$, and by (13.69), the validity of (13.75) follows.

Finally, in order to prove statement (iii), we set

$$
\begin{aligned}
\mathcal{V}^-_{k,\partial\Omega}[\epsilon,\epsilon'](\xi) \equiv & \int_{\partial\Omega} S_n(\xi-\eta)\Theta^o_k[\epsilon,\epsilon'](\eta)\, d\sigma_\eta \\
& + \epsilon^{n-2} \int_{\partial\Omega} R_{q,n}(\epsilon(\xi-\eta))\Theta^o_k[\epsilon,\epsilon'](\eta)\, d\sigma_\eta + \xi_k \qquad \forall \xi \in \partial\Omega,
\end{aligned}
\tag{13.82}
$$

for all $(\epsilon, \epsilon') \in]-\epsilon_1, \epsilon_1[\times U_{r_*}$. By arguing as in the proof of statement (ii), we can show that $\mathcal{V}^-_{k,\partial\Omega}$ is real analytic from $]-\epsilon_1, \epsilon_1[\times U_{r_*}$ to $C^{1,\alpha}(\partial\Omega)$ and that (13.76) and (13.77) hold. Thus the proof is complete. □

13.6.4 A Functional Analytic Representation Theorem for the Effective Conductivity

In following theorem we answer to the question in (13.47) on the behavior of $\lambda^{\mathrm{eff}}_{jk}[\epsilon]$ when ϵ is close to 0 and positive.

Theorem 13.31. *Let the assumptions of Theorem 13.28 hold. Let $j \in \{1, \dots, n\}$. Then there exist $\epsilon_2 \in]0, \epsilon_1[$ and a real analytic function $\mathcal{L}_{jk}$ from $]-\epsilon_2, \epsilon_2[\times U_{r_*}$ to $\mathbb{R}$ such that*

$$\lambda^{\mathrm{eff}}_{jk}[\epsilon] = \lambda^- \delta_{j,k} + \epsilon^n \mathcal{L}_{jk}\left[\epsilon, \frac{\epsilon}{\rho(\epsilon)}\right], \tag{13.83}$$

for all $\epsilon \in]0, \epsilon_2[$. Moreover,

$$\begin{aligned}
&\mathcal{L}_{jk}[0, r_*] \\
&\quad = \frac{1}{m_n(Q)}\left(\lambda^+ \int_{\partial\Omega} \tilde{u}^+_k(\xi)(\nu_\Omega(\xi))_j \, d\sigma_\xi - \lambda^- \int_{\partial\Omega} \tilde{u}^-_k(\xi)(\nu_\Omega(\xi))_j \, d\sigma_\xi\right) \\
&\quad\quad + \frac{m_n(\Omega)}{m_n(Q)}(\lambda^+ - \lambda^-)\delta_{j,k} + \frac{1}{m_n(Q)} \int_{\partial\Omega} f(\xi)\xi_j \, d\sigma_\xi \, ,
\end{aligned} \tag{13.84}$$

where $m_n(\Omega)$ denotes the n-dimensional measure of Ω and where $\tilde{u}^+_k$, $\tilde{u}^-_k$ are defined as in Theorem 13.27.

Proof. Let $\mathcal{U}^+_k$ be as in Theorem 13.29. We first note that if $\epsilon \in]0, \epsilon_1[$, then, by a computation based on the Divergence Theorem 4.1, we have

$$\begin{aligned}
\int_{\Omega_{p,\epsilon}} \frac{\partial u^+_{k,\epsilon}(x)}{\partial x_j} \, dx &= \int_{\partial\Omega_{p,\epsilon}} u^+_{k,\epsilon}(x)(\nu_{\Omega_{p,\epsilon}}(x))_j \, d\sigma_x \\
&= \epsilon^n \int_{\partial\Omega} \mathcal{U}^+_k\left[\epsilon, \frac{\epsilon}{\rho(\epsilon)}\right](\xi)(\nu_\Omega(\xi))_j \, d\sigma_\xi \, .
\end{aligned}$$

Then we set

$$\mathcal{L}^+_{jk}[\epsilon, \epsilon'] \equiv \frac{1}{m_n(Q)} \int_{\partial\Omega} \mathcal{U}^+_k[\epsilon, \epsilon'](\xi)(\nu_\Omega(\xi))_j \, d\sigma_\xi \, , \tag{13.85}$$

for all $(\epsilon, \epsilon') \in]-\epsilon_1, \epsilon_1[\times U_{r_*}$. By Theorem 13.29, $\mathcal{L}^+_{jk}$ is a real analytic function from $]-\epsilon_1, \epsilon_1[\times U_{r_*}$ to $\mathbb{R}$. Moreover, by equalities (13.60), (13.70), and by the

Divergence Theorem we have

$$\mathcal{L}^+_{jk}[0,r_*] = \frac{1}{m_n(Q)}\int_{\partial\Omega} \tilde{u}^+_k(\xi)(\nu_\Omega(\xi))_j\,d\sigma_\xi + \frac{1}{m_n(Q)}\int_{\partial\Omega} \xi_k(\nu_\Omega(\xi))_j\,d\sigma_\xi$$
$$= \frac{1}{m_n(Q)}\int_{\partial\Omega} \tilde{u}^+_k(\xi)(\nu_\Omega(\xi))_j\,d\sigma_\xi + \frac{m_n(\Omega)}{m_n(Q)}\delta_{j,k}\,.$$

Now let $R>0$ be such that $\overline{\Omega}\subseteq\mathbb{B}_n(0,R)$ and set $\tilde{\Omega}\equiv\mathbb{B}_n(0,R)\setminus\overline{\Omega}$. Let $\mathcal{V}^-_{k,\tilde{\Omega}}$ and $\epsilon^\#_{\tilde{\Omega}}$ be as in Theorem 13.30 (ii). Then a computation based on the Divergence Theorem 4.1, on the periodicity of the function that takes x to $u^-_{k,\epsilon}(x)-x_k$, and on equality (13.60) shows that

$$\int_{Q\setminus\overline{\Omega_{p,\epsilon}}} \frac{\partial u^-_{k,\epsilon}(x)}{\partial x_j}\,dx$$
$$= \int_{Q\setminus\overline{\Omega_{p,\epsilon}}} \frac{\partial\big(u^-_{k,\epsilon}(x)-x_k\big)}{\partial x_j}\,dx + \delta_{j,k}\big(m_n(Q)-\epsilon^n m_n(\Omega)\big)$$
$$= \int_{\partial(Q\setminus\overline{\Omega_{p,\epsilon}})} \big(u^-_{k,\epsilon}(x)-x_k\big)(\nu_{Q\setminus\overline{\Omega_{p,\epsilon}}}(x))_j\,d\sigma_x + \delta_{j,k}\big(m_n(Q)-\epsilon^n m_n(\Omega)\big)$$
$$= -\int_{\partial\Omega_{p,\epsilon}} \big(u^-_{k,\epsilon}(x)-x_k\big)(\nu_{\Omega_{p,\epsilon}}(x))_j\,d\sigma_x + \delta_{j,k}\big(m_n(Q)-\epsilon^n m_n(\Omega)\big)$$
$$= -\epsilon^n\int_{\partial\Omega} \mathcal{V}^-_{k,\tilde{\Omega}}\Big[\epsilon,\frac{\epsilon}{\rho(\epsilon)}\Big](\xi)(\nu_\Omega(\xi))_j\,d\sigma_\xi + \delta_{j,k}m_n(Q) \qquad \forall\epsilon\in]0,\epsilon^\#_{\tilde{\Omega}}[\,.$$

Then we set $\epsilon_2\equiv\epsilon^\#_{\tilde{\Omega}}$ and

$$\mathcal{L}^-_{jk}[\epsilon,\epsilon'] \equiv -\frac{1}{m_n(Q)}\int_{\partial\Omega} \mathcal{V}^-_{k,\tilde{\Omega}}[\epsilon,\epsilon'](\xi)(\nu_\Omega(\xi))_j\,d\sigma_\xi\,, \tag{13.86}$$

for all $(\epsilon,\epsilon')\in]-\epsilon_2,\epsilon_2[\times U_{r_*}$. By Theorem 13.30 (ii), $\mathcal{L}^-_{jk}$ is a real analytic function from $]-\epsilon_2,\epsilon_2[\times U_{r_*}$ to $\mathbb{R}$. Moreover, by equalities (13.60), (13.75), we have

$$\mathcal{L}^-_{jk}[0,r_*] = -\frac{1}{m_n(Q)}\int_{\partial\Omega} \tilde{u}^-_k(\xi)(\nu_\Omega(\xi))_j\,d\sigma_\xi - \frac{m_n(\Omega)}{m_n(Q)}\delta_{j,k}\,. \tag{13.87}$$

Therefore, if we set

$$\mathcal{L}_{jk}[\epsilon,\epsilon'] \equiv \lambda^+\mathcal{L}^+_{jk}[\epsilon,\epsilon'] + \lambda^-\mathcal{L}^-_{jk}[\epsilon,\epsilon'] + \frac{1}{m_n(Q)}\int_{\partial\Omega} f(\xi)\xi_j\,d\sigma_\xi\,, \tag{13.88}$$

for all $(\epsilon,\epsilon')\in]-\epsilon_2,\epsilon_2[\times U_{r_*}$, we deduce that $\mathcal{L}_{jk}$ is a real analytic function from $]-\epsilon_2,\epsilon_2[\times U_{r_*}$ to $\mathbb{R}$ such that equality (13.83) holds for all $\epsilon\in]0,\epsilon_2[$. Finally, by equalities (13.85), (13.87), we deduce the validity of (13.84). □

13.7 Series Expansion for the Effective Conductivity

In this section we continue the analysis of Section 13.6 on the asymptotic behavior of the effective conductivity of a two-phase composite consisting of a matrix and of a periodic set of inclusions, each of them of size $\epsilon > 0$. Our presentation here follows that of the paper with Pukhtaievych [81]. In Section 13.6, we have proved that under suitable assumptions the effective conductivity $\lambda^{\mathrm{eff}}[\epsilon]$ of the composite can be represented in terms of a real analytic function for ϵ close to 0 (cf. Definition 13.18 and Theorem 13.31). As a consequence, $\lambda^{\mathrm{eff}}[\epsilon]$ can be expanded into a convergent power series of ϵ for ϵ small. The aim of this section is to compute explicitly *all* the coefficients of such a power series, providing therefore a constructive method to approximate the effective conductivity at *any* desired degree of precision and which is valid for *all* the shapes of the inclusions.

For the sake of simplicity, in this section we consider only the two-dimensional case, i.e., we take

$$n = 2\,.$$

Next, we take two positive constants λ^+, λ^-, a function $f \in C^{0,\alpha}(\partial\Omega)_0$, a function $g \in C^{0,\alpha}(\partial\Omega)$, and a function ρ from $]0,\epsilon_0[$ to $]0,+\infty[$. For each $k \in \{1,2\}$ we consider problem (13.45), which in the two-dimensional case reads as

$$\begin{cases}
\Delta u_k^+ = 0 & \text{in } \mathbb{S}[\Omega_{p,\epsilon}]\,,\\
\Delta u_k^- = 0 & \text{in } \mathbb{S}[\Omega_{p,\epsilon}]^-\,,\\
u_k^+(x+qz) = u_k^+(x) + q_{kk}z_k & \forall x \in \overline{\mathbb{S}[\Omega_{p,\epsilon}]}\,, z \in \mathbb{Z}^2\,,\\
u_k^-(x+qz) = u_k^-(x) + q_{kk}z_k & \forall x \in \overline{\mathbb{S}[\Omega_{p,\epsilon}]^-}\,, z \in \mathbb{Z}^2\,,\\
\lambda^-\nu_{\Omega_{p,\epsilon}}(x)\cdot\nabla u_k^-(x) - \lambda^+\nu_{\Omega_{p,\epsilon}}(x)\cdot\nabla u_k^+(x) & \\
\quad = f((x-p)/\epsilon) & \forall x \in \partial\Omega_{p,\epsilon}\,,\\
\lambda^+\nu_{\Omega_{p,\epsilon}}(x)\cdot\nabla u_k^+(x) + \frac{1}{\rho(\epsilon)}\left(u_k^+(x) - u_k^-(x)\right) & \\
\quad = g((x-p)/\epsilon) & \forall x \in \partial\Omega_{p,\epsilon}\,,\\
\int_{\partial\Omega_{p,\epsilon}} u_k^+(x)\,d\sigma_x = 0\,. &
\end{cases} \tag{13.89}$$

Then, for each $\epsilon \in]0,\epsilon_0[$, problem (13.89) has a unique solution $(u_{k,\epsilon}^+, u_{k,\epsilon}^-)$ in $C^{1,\alpha}_{\mathrm{loc}}(\overline{\mathbb{S}[\Omega_{p,\epsilon}]}) \times C^{1,\alpha}_{\mathrm{loc}}(\overline{\mathbb{S}[\Omega_{p,\epsilon}]^-})$ (cf. Proposition 13.26 and Theorem 13.28).

As we have already mentioned, Theorem 13.31 shows that under the assumption (13.46) the effective conductivity can be continued real analytically in the parameter ϵ around the degenerate value $\epsilon = 0$ if

$$\epsilon \mapsto \frac{\epsilon}{\rho(\epsilon)}$$

is the restriction of an analytic function on a neighborhood of 0. This is the case, for example, if $\rho(\epsilon) \equiv 1/r_\#$ or $\rho(\epsilon) \equiv \epsilon/r_\#$, where $r_\#$ is a positive real number. In particular, if $\rho(\epsilon)$ is as above, then $\lambda^{\mathrm{eff}}_{jk}[\epsilon]$ can be expanded into a (convergent) power series for ϵ small. The aim of this section is to present a fully constructive method to compute *all* the coefficients of such power series.

In our analysis we investigate two specific cases: $\rho(\epsilon) \equiv 1/r_\#$ and $\rho(\epsilon) \equiv \epsilon/r_\#$. We observe that the first case corresponds to the situation where the thermal boundary resistance $\rho(\epsilon)$ is independent of ϵ, whereas in the second case the resistance is proportional to the size of the contact interface $\partial\Omega_{p,\epsilon}$. This latter case has been considered also in the works of Castro et al. [47] and of Drygaś and Mityushev [92]. In the first case, our main result is represented by Theorem 13.40, where we prove that

$$\lambda^{\mathrm{eff}}_{jk}[\epsilon] = \lambda^{-}\delta_{j,k} + \epsilon^2 \frac{1}{m_2(Q)} \sum_{l=0}^{+\infty} \frac{c_{(j,k),l}}{l!}\epsilon^l ,$$

for explicitly defined coefficients $c_{(j,k),l}$. In the second case, Theorem 13.44 shows that $\lambda^{\mathrm{eff}}_{jk}[\epsilon]$ can be expressed by means of a convergent power series of ϵ^2, namely

$$\lambda^{\mathrm{eff}}_{jk}[\epsilon] = \lambda^{-}\delta_{j,k} + \epsilon^2 \frac{1}{m_2(Q)} \sum_{l=0}^{+\infty} \frac{d_{(j,k),2l}}{(2l)!}\epsilon^{2l} .$$

for explicitly defined coefficients $d_{(j,k),2l}$. We mention that, in both situations, it is possible to simplify the expressions of the coefficients in the specific case where Ω is the unit ball (cf. [81]).

13.7.1 Preliminaries

We will exploit the fact that the regular part $R_{q,2}$ of the periodic analog $S_{q,2}$ of the fundamental solution is an even function and that

$$(D^\gamma R_{q,2})(0) = 0 \quad \forall \gamma \in \mathbb{N}^2 \text{ with } |\gamma| \text{ odd} \tag{13.90}$$

(cf. Theorem 12.4 (v)).

We shall also need the following classical result of Potential Theory, that follows from a combination of Theorem 6.8, Lemma 6.11, Corollary 6.15, and Theorem 6.24.

Lemma 13.32. *The operators* $\frac{1}{2}I + W^t_\Omega$ *and* $-\frac{1}{2}I + W^t_\Omega$ *are linear homeomorphisms from* $C^{0,\alpha}(\partial\Omega)_0$ *to itself.*

Proof. By Theorem 6.8 (ii) and Lemma 6.11, the operators $\frac{1}{2}I + W^t_\Omega$ and $-\frac{1}{2}I + W^t_\Omega$ are Fredholm operators of index 0 from $C^{0,\alpha}(\partial\Omega)_0$ to itself. Then, by Corollary 6.15, we deduce that $\frac{1}{2}I + W^t_\Omega$ is an isomorphism from $C^{0,\alpha}(\partial\Omega)_0$ to itself. Instead, by Theorem 6.24, we deduce that $-\frac{1}{2}I + W^t_\Omega$ is injective in $C^{0,\alpha}(\partial\Omega)_0$, and therefore, by Fredholm Alternative Theorem 5.8 an isomorphism from $C^{0,\alpha}(\partial\Omega)_0$ to itself. □

As shown in Section 13.6, we can convert problem (13.89) into an equivalent system of integral equations, by means of the periodic single layer potential.

Now, let $k \in \{1,2\}$ and let $\rho(\cdot)$ be either $\epsilon \mapsto 1/r_\#$ or $\epsilon \mapsto \epsilon/r_\#$. To provide an integral equation formulation of problem (13.89), we define the map $\mathcal{N}_\rho \equiv (\mathcal{N}_{\rho,1}, \mathcal{N}_{\rho,2})$ from $]-\epsilon_0, \epsilon_0[\times (C^{0,\alpha}(\partial\Omega)_0)^2$ to $(C^{0,\alpha}(\partial\Omega)_0)^2$ by setting

$$\mathcal{N}_\rho[\epsilon, \theta^i, \theta^o] \equiv \mathcal{M}[\epsilon, \epsilon/\rho(\epsilon), \theta^i, \theta^o]$$

for all $(\epsilon, \theta^i, \theta^o) \in]-\epsilon_0, \epsilon_0[\times (C^{0,\alpha}(\partial\Omega)_0)^2$, where $\mathcal{M}$ is as in Section 13.6.2.

By means of the operator $\mathcal{N}_\rho$, we can convert problem (13.89) into a system of integral equations. This is done in the following reformulation of Proposition 13.26, where we represent the solution of (13.89) using the periodic single layer potential.

Proposition 13.33. *Let either $\rho(\epsilon) \equiv 1/r_\#$ for all $\epsilon \in]-\epsilon_0, \epsilon_0[$ or $\rho(\epsilon) \equiv \epsilon/r_\#$ for all $\epsilon \in]-\epsilon_0, \epsilon_0[$. Let $\epsilon \in]0, \epsilon_0[$. Let $k \in \{1,2\}$. Then the unique solution $(u^+_{k,\epsilon}, u^-_{k,\epsilon})$ in $C^{1,\alpha}_{\mathrm{loc}}(\overline{\mathbb{S}[\Omega_{p,\epsilon}]}) \times C^{1,\alpha}_{\mathrm{loc}}(\overline{\mathbb{S}[\Omega_{p,\epsilon}]^-})$ of problem (13.89) is delivered by*

$$\begin{aligned}
u^+_{k,\epsilon}(x) \equiv\, & v^+_{q,\Omega_{p,\epsilon}}[\theta^i_{k,\epsilon}((\cdot - p)/\epsilon)](x) \\
& - \fint_{\partial\Omega_{p,\epsilon}} v^+_{q,\Omega_{p,\epsilon}}[\theta^i_{k,\epsilon}((\cdot - p)/\epsilon)]\, d\sigma \\
& + x_k - \fint_{\partial\Omega_{p,\epsilon}} y_k\, d\sigma_y \qquad \forall x \in \overline{\mathbb{S}[\Omega_{p,\epsilon}]}, \\
u^-_{k,\epsilon}(x) \equiv\, & v^-_{q,\Omega_{p,\epsilon}}[\theta^o_{k,\epsilon}((\cdot - p)/\epsilon)](x) \\
& - \fint_{\partial\Omega_{p,\epsilon}} v^-_{q,\Omega_{p,\epsilon}}[\theta^o_{k,\epsilon}((\cdot - p)/\epsilon)]\, d\sigma \\
& - \rho(\epsilon) \fint_{\partial\Omega_{p,\epsilon}} g((y-p)/\epsilon)\, d\sigma_y + x_k - \fint_{\partial\Omega_{p,\epsilon}} y_k\, d\sigma_y \qquad \forall x \in \overline{\mathbb{S}[\Omega_{p,\epsilon}]^-},
\end{aligned}$$

where $(\theta^i_{k,\epsilon}, \theta^o_{k,\epsilon})$ denotes the unique solution (θ^i, θ^o) in $(C^{0,\alpha}(\partial\Omega)_0)^2$ of

$$\mathcal{N}_\rho\Big[\epsilon, \theta^i, \theta^o\Big] = 0\,.$$

In order to investigate the asymptotic behavior of the (j,k)-entry $\lambda^{\mathrm{eff}}_{jk}[\epsilon]$ of the effective conductivity matrix as $\epsilon \to 0^+$, we need to study the functions $u^+_{k,\epsilon}$ and $u^-_{k,\epsilon}$ for ϵ close to the degenerate value 0. On the other hand, Proposition 13.33 tells us how to represent $u^+_{k,\epsilon}$ and $u^-_{k,\epsilon}$ in terms of the densities $\theta^i_{k,\epsilon}$ and $\theta^o_{k,\epsilon}$. Therefore, the analysis of $\lambda^{\mathrm{eff}}_{jk}[\epsilon]$ for ϵ close to 0 can be deduced by the analytic representation behavior of Theorem 13.28 for $\theta^i_{k,\epsilon}$ and $\theta^o_{k,\epsilon}$. As a first step, we present in the following theorem a regularity result for $\theta^i_{k,\epsilon}$ and $\theta^o_{k,\epsilon}$ for ϵ small and positive (cf. Theorem 13.28). We note that Proposition 13.34 here below is an immediate consequence of Theorem 13.28 if we set

$$\mathcal{N}_\rho\Big[\epsilon, \cdot, \cdot\Big] = \mathcal{M}\Big[\epsilon, \epsilon/\rho(\epsilon), \cdot, \cdot\Big]$$

and

$$\left(\Theta^i_{\rho,k}[\epsilon], \Theta^o_{\rho,k}[\epsilon]\right) = \left(\Theta^i_k[\epsilon, \epsilon/\rho(\epsilon)], \Theta^o_k[\epsilon, \epsilon/\rho(\epsilon)]\right)$$

for all $\epsilon \in]-\epsilon_0, \epsilon_0[$, where $\rho(\epsilon) \equiv 1/r_\#$ for all $\epsilon \in]-\epsilon_0, \epsilon_0[$ or $\rho(\epsilon) \equiv \epsilon/r_\#$ for all $\epsilon \in]-\epsilon_0, \epsilon_0[$ (with $\mathcal{M}$ and $(\Theta^i_k[\cdot,\cdot], \Theta^o_k[\cdot,\cdot])$ as in Theorem 13.28).

Proposition 13.34. *Let either $\rho(\epsilon) \equiv 1/r_\#$ for all $\epsilon \in]-\epsilon_0, \epsilon_0[$ or $\rho(\epsilon) \equiv \epsilon/r_\#$ for all $\epsilon \in]-\epsilon_0, \epsilon_0[$. Let $k \in \{1, 2\}$. The following statements hold.*

(i) $\mathcal{N}_\rho$ is a real analytic map from $]-\epsilon_0, \epsilon_0[\times(C^{0,\alpha}(\partial\Omega)_0)^2$ to $(C^{0,\alpha}(\partial\Omega)_0)^2$.

(ii) There exists a unique pair $(\tilde{\theta}^i_k, \tilde{\theta}^o_k) \in (C^{0,\alpha}(\partial\Omega)_0)^2$ such that

$$\mathcal{N}_\rho\left[0, \tilde{\theta}^i_k, \tilde{\theta}^o_k\right] = 0\,.$$

(iii) There exist $\epsilon_1 \in]0, \epsilon_0[$ and a real analytic map $(\Theta^i_{\rho,k}[\cdot], \Theta^o_{\rho,k}[\cdot])$ from $]-\epsilon_1, \epsilon_1[$ to $(C^{0,\alpha}(\partial\Omega)_0)^2$ such that

$$\mathcal{N}_\rho\left[\epsilon, \Theta^i_{\rho,k}[\epsilon], \Theta^o_{\rho,k}[\epsilon]\right] = 0 \qquad \forall \epsilon \in]-\epsilon_1, \epsilon_1[\,. \tag{13.91}$$

In particular,

$$(\Theta^i_{\rho,k}[\epsilon], \Theta^o_{\rho,k}[\epsilon]) = (\theta^i_{k,\epsilon}, \theta^o_{k,\epsilon}) \quad \forall \epsilon \in]0, \epsilon_1[$$

and

$$(\Theta^i_{\rho,k}[0], \Theta^o_{\rho,k}[0]) = (\tilde{\theta}^i_k, \tilde{\theta}^o_k)\,,$$

where the pair $(\theta^i_{k,\epsilon}, \theta^o_{k,\epsilon})$ has been defined in Proposition 13.33 and the pair $(\tilde{\theta}^i_k, \tilde{\theta}^o_k)$ is as in statement (ii).

Now we observe that the real analyticity result of Proposition 13.34 (iii) implies that there exists $\epsilon_2 \in]0, \epsilon_1[$ small enough that we can expand $\Theta^i_{\rho,k}[\epsilon]$ and $\Theta^o_{\rho,k}[\epsilon]$ into power series of ϵ, i.e.,

$$\Theta^i_{\rho,k}[\epsilon] = \sum_{m=0}^{+\infty} \frac{\theta^i_{k,m}}{m!}\epsilon^m\,, \qquad \Theta^o_{\rho,k}[\epsilon] = \sum_{m=0}^{+\infty} \frac{\theta^o_{k,m}}{m!}\epsilon^m\,, \tag{13.92}$$

for some $\{\theta^i_{k,m}\}_{m\in\mathbb{N}}$, $\{\theta^o_{k,m}\}_{m\in\mathbb{N}}$ and for all $\epsilon \in]-\epsilon_2, \epsilon_2[$. Moreover,

$$\theta^i_{k,m} = \left(\partial^m_\epsilon \Theta^i_{\rho,k}[\epsilon]\right)_{|\epsilon=0}\,, \qquad \theta^o_{k,m} = \left(\partial^m_\epsilon \Theta^o_{\rho,k}[\epsilon]\right)_{|\epsilon=0}\,,$$

for all $m \in \mathbb{N}$. As a consequence,

$$\theta^i_{k,\epsilon} = \sum_{m=0}^{+\infty} \frac{\theta^i_{k,m}}{m!}\epsilon^m\,, \qquad \theta^o_{k,\epsilon} = \sum_{m=0}^{+\infty} \frac{\theta^o_{k,m}}{m!}\epsilon^m\,,$$

for all $\epsilon \in]0, \epsilon_2[$. Therefore, in order to obtain a power series expansion for $\lambda^{\text{eff}}_{jk}[\epsilon]$ for ϵ close to 0, we want to exploit the expansion of $(\Theta^i_{\rho,k}[\epsilon], \Theta^o_{\rho,k}[\epsilon])$ (or equivalently of $(\theta^i_{k,\epsilon}, \theta^o_{k,\epsilon})$). Since the coefficients of the expansions in (13.92) are given by the derivatives with respect to ϵ of $\Theta^i_{\rho,k}[\epsilon]$ and $\Theta^o_{\rho,k}[\epsilon]$, we would like to obtain some equations in order to compute $\left(\partial^m_\epsilon \Theta^i_{\rho,k}[\epsilon]\right)_{|\epsilon=0}$ and $\left(\partial^m_\epsilon \Theta^o_{\rho,k}[\epsilon]\right)_{|\epsilon=0}$. The plan is to obtain such equations by differentiating with respect to ϵ equality (13.91), which then leads to

$$\partial^m_\epsilon \left(\mathcal{N}_\rho \left[\epsilon, \Theta^i_{\rho,k}[\epsilon], \Theta^o_{\rho,k}[\epsilon]\right] \right) = 0 \qquad \forall \epsilon \in]-\epsilon_1, \epsilon_1[\,, \qquad \forall m \in \mathbb{N}\,. \quad (13.93)$$

Then by taking $\epsilon = 0$ in (13.93), we will obtain integral equations which uniquely determine $\left(\partial^m_\epsilon \Theta^i_{\rho,k}[\epsilon]\right)_{|\epsilon=0}$ and $\left(\partial^m_\epsilon \Theta^o_{\rho,k}[\epsilon]\right)_{|\epsilon=0}$.

In order to compute the derivative in (13.93), we recall the chain rule formula

$$\partial^m_\epsilon (F(\epsilon\xi)) = \sum_{l=0}^{m} \binom{m}{l} \xi_1^l \xi_2^{m-l} (\partial_1^l \partial_2^{m-l} F)(\epsilon\xi) \qquad (13.94)$$

which holds for all $m \in \mathbb{N}$, $\epsilon \in \mathbb{R}$, $\xi \equiv (\xi_1, \xi_2) \in \mathbb{R}^2$, and for all functions F that are analytic in a neighborhood of $\epsilon\xi$ (cf. (8.111)). Here, $(\partial_1 F)(\eta)$ and $(\partial_2 F)(\eta)$ denote the partial derivative with respect to the first and to the second variable evaluated at $\eta \equiv (\eta_1, \eta_2) \in \mathbb{R}^2$.

To compute the derivatives in (13.93), we find convenient introduce the maps $\mathcal{R}$ and $\mathcal{R}_\nu$ from $]-\epsilon_0, \epsilon_0[\times C^{0,\alpha}(\partial\Omega)_0$ to $C^{1,\alpha}(\partial\Omega)$ and to $C^{0,\alpha}(\partial\Omega)$ defined by

$$\mathcal{R}[\epsilon, \theta](\xi) \equiv \int_{\partial\Omega} R_{q,2}(\epsilon(\xi - \eta))\theta(\eta)\,d\sigma_\eta \qquad \forall \xi \in \partial\Omega\,,$$

and by

$$\mathcal{R}_\nu[\epsilon, \theta](\xi) \equiv \int_{\partial\Omega} \nu_\Omega(\xi) \cdot \nabla R_{q,2}(\epsilon(\xi - \eta))\theta(\eta)\,d\sigma_\eta \qquad \forall \xi \in \partial\Omega\,,$$

for all $(\epsilon, \theta) \in]-\epsilon_0, \epsilon_0[\times C^{0,\alpha}(\partial\Omega)_0$, respectively. We begin with a preliminary lemma.

Lemma 13.35. *Let $\epsilon'_0 \in]0, \epsilon_0[$. Let $\Theta[\cdot]$ be a real analytic map from $]-\epsilon'_0, \epsilon'_0[$ to $C^{0,\alpha}(\partial\Omega)_0$. Possibly shrinking ϵ'_0, we can assume that there exists a sequence $\{\theta_m\}_{m\in\mathbb{N}}$ in $C^{0,\alpha}(\partial\Omega)_0$ such that*

$$\Theta[\epsilon] = \sum_{m=0}^{+\infty} \frac{\theta_m}{m!}\epsilon^m \qquad \forall \epsilon \in]-\epsilon'_0, \epsilon'_0[\,,$$

where the series converges in $C^{0,\alpha}(\partial\Omega)_0$ for $\epsilon \in]-\epsilon'_0, \epsilon'_0[$. Then the following statements hold.

(i) The map from $]-\epsilon'_0, \epsilon'_0[$ to $C^{1,\alpha}(\partial\Omega)$ that takes ϵ to $\mathcal{R}[\epsilon, \Theta[\epsilon]]$ is real analytic and we have

$$\mathcal{R}[0, \Theta[0]] = 0\,, \qquad \left(\partial_\epsilon \left(\mathcal{R}[\epsilon, \Theta[\epsilon]]\right)\right)_{|\epsilon=0} = 0\,,$$

and

$$\left(\partial_\epsilon^m \left(\mathcal{R}[\epsilon,\Theta[\epsilon]]\right)(\xi)\right)_{|\epsilon=0}$$
$$= \sum_{l=2}^{m} \binom{m}{l} \sum_{h=0}^{l} \binom{l}{h} (\partial_1^h \partial_2^{l-h} R_{q,2})(0) \int_{\partial\Omega} (\xi_1-\eta_1)^h (\xi_2-\eta_2)^{l-h} \theta_{m-l}(\eta)\, d\sigma_\eta$$
$$\forall \xi \in \partial\Omega\,, \quad m \in \mathbb{N}\setminus\{0,1\}\,. \quad (13.95)$$

(ii) *The map from* $]-\epsilon_0',\epsilon_0'[$ *to* $C^{0,\alpha}(\partial\Omega)$ *that takes* ϵ *to* $\mathcal{R}_\nu[\epsilon,\Theta[\epsilon]]$ *is real analytic and we have*

$$\mathcal{R}_\nu[0,\Theta[0]] = 0\,,$$

and

$$\int_{\partial\Omega} \partial_\epsilon^m \left(\epsilon \mathcal{R}_\nu[\epsilon,\Theta[\epsilon]]\right) d\sigma = 0 \quad \forall \epsilon \in]-\epsilon_0',\epsilon_0'[\,, m \in \mathbb{N}\,. \quad (13.96)$$

Moreover,

$$\left(\partial_\epsilon \left(\epsilon \mathcal{R}_\nu[\epsilon,\Theta[\epsilon]]\right)\right)_{|\epsilon=0} = 0\,,$$

and

$$\left(\partial_\epsilon^m \left(\epsilon \mathcal{R}_\nu[\epsilon,\Theta[\epsilon]]\right)(\xi)\right)_{|\epsilon=0}$$
$$= m \sum_{l=1}^{m-1} \binom{m-1}{l} \sum_{h=0}^{l} \binom{l}{h} (\partial_1^h \partial_2^{l-h} \nabla R_{q,2})(0) \cdot \nu_\Omega(\xi)$$
$$\times \int_{\partial\Omega} (\xi_1-\eta_1)^h (\xi_2-\eta_2)^{l-h} \theta_{m-1-l}(\eta) d\sigma_\eta \quad \forall \xi \in \partial\Omega\,, m \in \mathbb{N}\setminus\{0,1\}\,.$$
$$(13.97)$$

Proof. We first consider statement (i). The analyticity of the map

$$\epsilon \mapsto \mathcal{R}[\epsilon,\Theta[\epsilon]]$$

from $]-\epsilon_0',\epsilon_0'[$ to $C^{1,\alpha}(\partial\Omega)$ follows by the analyticity of $\Theta[\cdot]$ and by the properties of integral operators with real analytic kernel of Theorem A.28 (cf. Lemma 13.4 (i)). Clearly,

$$\mathcal{R}[0,\Theta[0]](\xi) = R_{q,2}(0) \int_{\partial\Omega} \Theta[0](\eta) d\sigma_\eta = 0 \qquad \forall \xi \in \partial\Omega\,.$$

The validity of equality (13.95) follows by standard Calculus in Banach spaces and by the chain rule formula (13.94). Moreover, by formula (13.94) and equality (13.90), one verifies that $\left(\partial_\epsilon \left(\mathcal{R}[\epsilon,\Theta[\epsilon]]\right)\right)_{|\epsilon=0} = 0$.

We now turn to prove (ii). Again, by the analyticity of $\Theta[\cdot]$ and by the properties of integral operators with real analytic kernel of Theorem A.28 (cf. Lemma 13.4

(i)), we deduce the analyticity of the map

$$\epsilon \mapsto \mathcal{R}_\nu[\epsilon, \Theta[\epsilon]]$$

from $]-\epsilon_0', \epsilon_0'[$ to $C^{0,\alpha}(\partial\Omega)$. Moreover, by standard Calculus in Banach spaces and by the chain rule formula (13.94), we deduce the validity of (13.97). Clearly,

$$\mathcal{R}_\nu[0, \Theta[0]](\xi) = \nu_\Omega(\xi) \cdot \nabla R_{q,2}(0) \int_{\partial\Omega} \Theta[0](\eta) d\sigma_\eta = 0 \qquad \forall \xi \in \partial\Omega,$$

and accordingly, $(\partial_\epsilon (\epsilon \mathcal{R}_\nu[\epsilon, \Theta[\epsilon]]) (\xi))_{|\epsilon=0} = 0$ for all $\xi \in \partial\Omega$. To prove (13.96), we note that the map H from $]-\epsilon_0', \epsilon_0'[$ to $C^{1,\alpha}(\overline{\Omega})$ that takes ϵ to the function

$$H[\epsilon](\xi) \equiv \int_{\partial\Omega} R_{q,2}(\epsilon(\xi - \eta))\Theta[\epsilon](\eta) d\sigma_\eta \qquad \forall \xi \in \overline{\Omega},$$

is real analytic. Moreover,

$$\begin{aligned} \Delta_\xi \int_{\partial\Omega} R_{q,2}(\epsilon(\xi - \eta))\Theta[\epsilon](\eta) d\sigma_\eta &= \epsilon^2 \int_{\partial\Omega} (\Delta R_{q,2})(\epsilon(\xi - \eta))\Theta[\epsilon](\eta) d\sigma_\eta \\ &= -\frac{\epsilon^2}{m_2(Q)} \int_{\partial\Omega} \Theta[\epsilon](\eta) d\sigma_\eta = 0 \qquad \forall \xi \in \overline{\Omega}, \end{aligned}$$

and thus $H[\epsilon]$ is harmonic in Ω for all $\epsilon \in]-\epsilon_0', \epsilon_0'[$ (cf. formula (12.10) for $\Delta R_{q,2}$). Therefore,

$$\int_{\partial\Omega} \nu_\Omega \cdot \nabla H[\epsilon] \, d\sigma = 0 \qquad \forall \epsilon \in]-\epsilon_0', \epsilon_0'[.$$

On the other hand, a straightforward computation shows that

$$\nu_\Omega \cdot \nabla H[\epsilon] = \epsilon \mathcal{R}_\nu[\epsilon, \Theta[\epsilon]] \qquad \forall \epsilon \in]-\epsilon_0', \epsilon_0'[.$$

As a consequence,

$$\int_{\partial\Omega} \epsilon \mathcal{R}_\nu[\epsilon, \Theta[\epsilon]] \, d\sigma = 0 \qquad \forall \epsilon \in]-\epsilon_0', \epsilon_0'[. \tag{13.98}$$

By differentiating equality (13.98) with respect to ϵ, we deduce that

$$0 = \partial_\epsilon^m \left(\int_{\partial\Omega} \epsilon \mathcal{R}_\nu[\epsilon, \Theta[\epsilon]] \, d\sigma \right) = \int_{\partial\Omega} \partial_\epsilon^m (\epsilon \mathcal{R}_\nu[\epsilon, \Theta[\epsilon]]) \, d\sigma \qquad \forall \epsilon \in]-\epsilon_0', \epsilon_0'[.$$

Thus the proof is complete. □

In order to shorten the presentation of the formulas of Lemma 13.35, we find convenient to introduce the following notation:

$$\mathcal{R}^m[\theta_0, \ldots, \theta_{m-2}](\xi)$$

$$\equiv \sum_{l=2}^{m} \binom{m}{l} \sum_{h=0}^{l} \binom{l}{h} (\partial_1^h \partial_2^{l-h} R_{q,2})(0) \int_{\partial\Omega} (\xi_1 - \eta_1)^h (\xi_2 - \eta_2)^{l-h} \theta_{m-l}(\eta) d\sigma_\eta$$

$$\forall \xi \in \partial\Omega\,, (\theta_0, \ldots, \theta_{m-2}) \in (C^{0,\alpha}(\partial\Omega)_0)^{m-1}\,, \quad (13.99)$$

for all $m \in \mathbb{N} \setminus \{0,1\}$, and

$$\mathcal{R}_\nu^m[\theta_0, \ldots, \theta_{m-2}](\xi)$$
$$\equiv m \sum_{l=1}^{m-1} \binom{m-1}{l} \sum_{h=0}^{l} \binom{l}{h} (\partial_1^h \partial_2^{l-h} \nabla R_{q,2})(0) \cdot \nu_\Omega(\xi)$$
$$\times \int_{\partial\Omega} (\xi_1 - \eta_1)^h (\xi_2 - \eta_2)^{l-h} \theta_{m-1-l}(\eta) d\sigma_\eta$$
$$\forall \xi \in \partial\Omega\,, (\theta_0, \ldots, \theta_{m-2}) \in (C^{0,\alpha}(\partial\Omega)_0)^{m-1}\,, \quad (13.100)$$

for all $m \in \mathbb{N} \setminus \{0,1\}$. We observe that by (13.90), $\mathcal{R}^m[\theta_0, \ldots, \theta_{m-2}]$ and $\mathcal{R}_\nu^m[\theta_0, \ldots, \theta_{m-2}]$ depend only on the θ_l's with l odd if m is odd, and only on the θ_l's with l even if m is even. Then, by a linearity argument we can deduce the validity of the following technical result on $\mathcal{R}^{2m+1}$ and $\mathcal{R}_\nu^{2m+1}$ for $m \in \mathbb{N} \setminus \{0\}$.

Lemma 13.36. *The following statements hold.*

(i) Let $m \in \mathbb{N} \setminus \{0\}$. If $\theta_0, \ldots, \theta_{2m-1} \in C^{0,\alpha}(\partial\Omega)_0$ are such that $\theta_{2l-1} = 0$ for all $l \in \{1, \ldots, m\}$, then $\mathcal{R}^{2m+1}[\theta_0, \ldots, \theta_{2m-1}](\xi) = 0$ for all $\xi \in \partial\Omega$.
(ii) Let $m \in \mathbb{N} \setminus \{0\}$. If $\theta_0, \ldots, \theta_{2m-1} \in C^{0,\alpha}(\partial\Omega)_0$ are such that $\theta_{2l-1} = 0$ for all $l \in \{1, \ldots, m\}$, then $\mathcal{R}_\nu^{2m+1}[\theta_0, \ldots, \theta_{2m-1}](\xi) = 0$ for all $\xi \in \partial\Omega$.

13.7.2 Power Series Expansion for $\rho(\epsilon) \equiv 1/r_\#$

Throughout this subsection, we consider the case where

$$\rho(\epsilon) \equiv 1/r_\# \qquad \forall \epsilon \in]-\epsilon_0, \epsilon_0[\,. \quad (13.101)$$

In order to compute the asymptotic expansion of the effective conductivity under assumption (13.101), we start with the following proposition where we identify the coefficients of the power series expansions of $\Theta^i_{\rho,k}[\epsilon]$ and of $\Theta^o_{\rho,k}[\epsilon]$ in terms of solutions of systems of integral equations.

Proposition 13.37. *Let $k \in \{1,2\}$. Let ϵ_1, $\Theta^i_{\rho,k}[\cdot]$, and $\Theta^o_{\rho,k}[\cdot]$ be as in Proposition 13.34. Then there exist $\epsilon_2 \in]0, \epsilon_1[$ and a sequence $\{(\theta^i_{k,m}, \theta^o_{k,m})\}_{m\in\mathbb{N}}$ in $(C^{0,\alpha}(\partial\Omega)_0)^2$ such that*

$$\Theta^i_{\rho,k}[\epsilon] = \sum_{m=0}^{+\infty} \frac{\theta^i_{k,m}}{m!} \epsilon^m \quad \textit{and} \quad \Theta^o_{\rho,k}[\epsilon] = \sum_{m=0}^{+\infty} \frac{\theta^o_{k,m}}{m!} \epsilon^m \qquad \forall \epsilon \in]-\epsilon_2, \epsilon_2[\,,$$
$$(13.102)$$

where the two series converge in $C^{0,\alpha}(\partial\Omega)_0$ for $\epsilon \in]-\epsilon_2, \epsilon_2[$. Moreover, the following statements hold.

(i) *The pair of functions $(\theta^i_{k,0}, \theta^o_{k,0})$ is the unique solution in $(C^{0,\alpha}(\partial\Omega)_0)^2$ of the following system of integral equations*

$$-\frac{1}{2}\theta^i_{k,0}(\xi) + W^t_\Omega[\theta^i_{k,0}](\xi) = \frac{1}{\lambda^+}\left(g(\xi) - \int_{\partial\Omega} g\,d\sigma\right) - (\nu_\Omega(\xi))_k \;\forall \xi \in \partial\Omega\,, \tag{13.103}$$

$$\frac{1}{2}\theta^o_{k,0}(\xi) + W^t_\Omega[\theta^o_{k,0}](\xi) = \frac{1}{\lambda^-}\left(g(\xi) - \int_{\partial\Omega} g\,d\sigma + f(\xi)\right) - (\nu_\Omega(\xi))_k \qquad \forall \xi \in \partial\Omega\,. \tag{13.104}$$

(ii) *The pair of functions $(\theta^i_{k,1}, \theta^o_{k,1})$ is the unique solution in $(C^{0,\alpha}(\partial\Omega)_0)^2$ of the following system of integral equations*

$$-\frac{1}{2}\theta^i_{k,1}(\xi) + W^t_\Omega[\theta^i_{k,1}](\xi) = -\frac{r_\#}{\lambda^+}\Bigg(v^+_\Omega[\theta^i_{k,0}](\xi) - \int_{\partial\Omega} v^+_\Omega[\theta^i_{k,0}]d\sigma - v^-_\Omega[\theta^o_{k,0}](\xi) + \int_{\partial\Omega} v^-_\Omega[\theta^o_{k,0}]d\sigma\Bigg) \qquad \forall \xi \in \partial\Omega\,, \tag{13.105}$$

$$\frac{1}{2}\theta^o_{k,1}(\xi) + W^t_\Omega[\theta^o_{k,1}](\xi) = -\frac{r_\#}{\lambda^-}\Bigg(v^+_\Omega[\theta^i_{k,0}](\xi) - \int_{\partial\Omega} v^+_\Omega[\theta^i_{k,0}]d\sigma - v^-_\Omega[\theta^o_{k,0}](\xi) + \int_{\partial\Omega} v^-_\Omega[\theta^o_{k,0}]d\sigma\Bigg) \qquad \forall \xi \in \partial\Omega\,. \tag{13.106}$$

(iii) *The pair of functions $(\theta^i_{k,2}, \theta^o_{k,2})$ is the unique solution in $(C^{0,\alpha}(\partial\Omega)_0)^2$ of the following system of integral equations*

$$-\frac{1}{2}\theta^i_{k,2}(\xi) + W^t_\Omega[\theta^i_{k,2}](\xi) = -\mathcal{R}^2_\nu\left[\theta^i_{k,0}\right](\xi) - \frac{2r_\#}{\lambda^+}\Bigg(v^+_\Omega[\theta^i_{k,1}](\xi) - \int_{\partial\Omega} v^+_\Omega[\theta^i_{k,1}]d\sigma - v^-_\Omega[\theta^o_{k,1}](\xi) + \int_{\partial\Omega} v^-_\Omega[\theta^o_{k,1}]d\sigma\Bigg) \qquad \forall \xi \in \partial\Omega\,, \tag{13.107}$$

$$\frac{1}{2}\theta^o_{k,2}(\xi) + W^t_\Omega[\theta^o_{k,2}](\xi) = -\mathcal{R}^2_\nu[\theta^o_{k,0}](\xi)$$

$$
\begin{aligned}
&- \frac{2r_\#}{\lambda^-}\Bigg(v^+_\Omega[\theta^i_{k,1}](\xi) - \fint_{\partial\Omega} v^+_\Omega[\theta^i_{k,1}]d\sigma \\
&- v^-_\Omega[\theta^o_{k,1}](\xi) + \fint_{\partial\Omega} v^-_\Omega[\theta^o_{k,1}]d\sigma \Bigg) \qquad \forall \xi \in \partial\Omega \qquad (13.108)
\end{aligned}
$$

(cf. equation (13.100) *for the definition of* $\mathcal{R}^2_\nu$*).*

(iv) For all $m \in \mathbb{N} \setminus \{0, 1, 2\}$ *the pair of functions* $(\theta^i_{k,m}, \theta^o_{k,m})$ *is the unique solution in* $(C^{0,\alpha}(\partial\Omega)_0)^2$ *of the following system of integral equations which involves* $\{(\theta^i_{k,h}, \theta^o_{k,h})\}_{h=0}^{m-1}$,

$$
\begin{aligned}
&-\frac{1}{2}\theta^i_{k,m}(\xi) + W^t_\Omega[\theta^i_{k,m}](\xi) = -\mathcal{R}^m_\nu\left[\theta^i_{k,0}, \ldots, \theta^i_{k,m-2}\right](\xi) \\
&- \frac{m r_\#}{\lambda^+}\Bigg(v^+_\Omega[\theta^i_{k,m-1}](\xi) - \fint_{\partial\Omega} v^+_\Omega[\theta^i_{k,m-1}]d\sigma \\
&- v^-_\Omega[\theta^o_{k,m-1}](\xi) + \fint_{\partial\Omega} v^-_\Omega[\theta^o_{k,m-1}]d\sigma \Bigg) \\
&- \frac{m r_\#}{\lambda^+}\Bigg(\mathcal{R}^{m-1}[\theta^i_{k,0}, \ldots, \theta^i_{k,m-3}](\xi) - \fint_{\partial\Omega} \mathcal{R}^{m-1}[\theta^i_{k,0}, \ldots, \theta^i_{k,m-3}]d\sigma \\
&- \mathcal{R}^{m-1}[\theta^o_{k,0}, \ldots, \theta^o_{k,m-3}](\xi) + \fint_{\partial\Omega} \mathcal{R}^{m-1}[\theta^o_{k,0}, \ldots, \theta^o_{k,m-3}]d\sigma \Bigg) \\
&\qquad \forall \xi \in \partial\Omega\,,
\end{aligned} \qquad (13.109)
$$

$$
\begin{aligned}
&\frac{1}{2}\theta^o_{k,m}(\xi) + W^t_\Omega[\theta^o_{k,m}](\xi) = -\mathcal{R}^m_\nu\left[\theta^o_{k,0}, \ldots, \theta^o_{k,m-2}\right](\xi) \\
&- \frac{m r_\#}{\lambda^-}\Bigg(v^+_\Omega[\theta^i_{k,m-1}](\xi) - \fint_{\partial\Omega} v^+_\Omega[\theta^i_{k,m-1}]d\sigma \\
&- v^-_\Omega[\theta^o_{k,m-1}](\xi) + \fint_{\partial\Omega} v^-_\Omega[\theta^o_{k,m-1}]d\sigma \Bigg) \\
&- \frac{m r_\#}{\lambda^-}\Bigg(\mathcal{R}^{m-1}[\theta^i_{k,0}, \ldots, \theta^i_{k,m-3}](\xi) - \fint_{\partial\Omega} \mathcal{R}^{m-1}[\theta^i_{k,0}, \ldots, \theta^i_{k,m-3}]d\sigma \\
&- \mathcal{R}^{m-1}[\theta^o_{k,0}, \ldots, \theta^o_{k,m-3}](\xi) + \fint_{\partial\Omega} \mathcal{R}^{m-1}[\theta^o_{k,0}, \ldots, \theta^o_{k,m-3}]d\sigma \Bigg) \\
&\qquad \forall \xi \in \partial\Omega
\end{aligned} \qquad (13.110)
$$

(cf. equation (13.99) *for the definition of* $\mathcal{R}^{m-1}$ *and equation* (13.100) *for the definition of* $\mathcal{R}^m_\nu$*).*

Proof. We first note that Proposition 13.34 (iii) implies the existence of ϵ_2 and of a family $\{(\theta^i_{k,m}, \theta^o_{k,m})\}_{m\in\mathbb{N}}$ such that (13.102) holds. By standard properties of real analytic maps, we have

$$(\theta^i_{k,m}, \theta^o_{k,m}) = \left(\left(\partial^m_\epsilon \Theta^i_{\rho,k}[\epsilon]\right)_{|\epsilon=0}, \left(\partial^m_\epsilon \Theta^o_{\rho,k}[\epsilon]\right)_{|\epsilon=0}\right) \quad \forall m \in \mathbb{N}.$$

By equality $\rho(\epsilon) \equiv 1/r_\#$ and by taking $\epsilon = 0$, equation (13.91) can be written as the system of integral equations (13.103)–(13.104). The existence and the uniqueness of the solution for this system are then ensured by Proposition 13.34 (ii). Then we observe that $\mathcal{N}_\rho\left[\epsilon, \Theta^i_{\rho,k}[\epsilon], \Theta^o_{\rho,k}[\epsilon]\right] = 0$ for all $\epsilon \in]-\epsilon_2, \epsilon_2[$ (cf. Proposition 13.34 (iii)). Accordingly, the map that takes ϵ to $\mathcal{N}_\rho\left[\epsilon, \Theta^i_{\rho,k}[\epsilon], \Theta^o_{\rho,k}[\epsilon]\right]$ has derivatives which are equal to zero, i.e., $\partial^m_\epsilon(\mathcal{N}_\rho\left[\epsilon, \Theta^i_{\rho,k}[\epsilon], \Theta^o_{\rho,k}[\epsilon]\right]) = 0$ for all $\epsilon \in]-\epsilon_2, \epsilon_2[$ and all $m \in \mathbb{N}\setminus\{0\}$. Keeping equality $\rho(\epsilon) \equiv 1/r_\#$ in mind, a straightforward computation shows that

$$\begin{aligned}
&\partial^m_\epsilon \left(\mathcal{N}_{\rho,1}\left[\epsilon, \Theta^i_{\rho,k}[\epsilon], \Theta^o_{\rho,k}[\epsilon]\right]\right)(\xi) \\
&\quad = \lambda^-\Big(\frac{1}{2}\partial^m_\epsilon \Theta^o_{\rho,k}[\epsilon](\xi) + W^t_\Omega\left[\partial^m_\epsilon \Theta^o_{\rho,k}[\epsilon]\right](\xi) + \partial^m_\epsilon\left(\epsilon\mathcal{R}_\nu\left[\epsilon, \Theta^o_{\rho,k}[\epsilon]\right]\right)(\xi)\Big) \\
&\quad - \lambda^+\Big(-\frac{1}{2}\partial^m_\epsilon \Theta^i_{\rho,k}[\epsilon](\xi) + W^t_\Omega\left[\partial^m_\epsilon \Theta^i_{\rho,k}[\epsilon]\right](\xi) \\
&\quad + \partial^m_\epsilon\left(\epsilon\mathcal{R}_\nu\left[\epsilon, \Theta^i_{\rho,k}[\epsilon]\right]\right)(\xi)\Big) = 0 \qquad \forall \xi \in \partial\Omega,
\end{aligned} \tag{13.111}$$

$$\begin{aligned}
&\partial^m_\epsilon \left(\mathcal{N}_{\rho,2}\left[\epsilon, \Theta^i_{\rho,k}[\epsilon], \Theta^o_{\rho,k}[\epsilon]\right]\right)(\xi) \\
&\quad = \lambda^+\Big(-\frac{1}{2}\partial^m_\epsilon \Theta^i_{\rho,k}[\epsilon](\xi) + W^t_\Omega\left[\partial^m_\epsilon \Theta^i_{\rho,k}[\epsilon]\right](\xi) \\
&\quad + \partial^m_\epsilon\left(\epsilon\mathcal{R}_\nu\left[\epsilon, \Theta^i_{\rho,k}[\epsilon]\right]\right)(\xi)\Big) \\
&\quad + \epsilon r_\#\Bigg(v^+_\Omega\left[\partial^m_\epsilon \Theta^i_{\rho,k}[\epsilon]\right](\xi) + \partial^m_\epsilon\left(\mathcal{R}\left[\epsilon, \Theta^i_{\rho,k}[\epsilon]\right]\right)(\xi) \\
&\quad - \fint_{\partial\Omega}\Big(v^+_\Omega\left[\partial^m_\epsilon \Theta^i_{\rho,k}[\epsilon]\right] + \partial^m_\epsilon\left(\mathcal{R}\left[\epsilon, \Theta^i_{\rho,k}[\epsilon]\right]\right)\Big)d\sigma \\
&\quad - v^-_\Omega\left[\partial^m_\epsilon \Theta^o_{\rho,k}[\epsilon]\right](\xi) - \partial^m_\epsilon\left(\mathcal{R}\left[\epsilon, \Theta^o_{\rho,k}[\epsilon]\right]\right)(\xi) \\
&\quad + \fint_{\partial\Omega}\Big(v^-_\Omega\left[\partial^m_\epsilon \Theta^o_{\rho,k}[\epsilon]\right] + \partial^m_\epsilon\left(\mathcal{R}\left[\epsilon, \Theta^o_{\rho,k}[\epsilon]\right]\right)\Big)d\sigma\Bigg)
\end{aligned}$$

$$
\begin{aligned}
&+ m r_{\#} \Bigg(v_{\Omega}^{+} \left[\partial_{\epsilon}^{m-1} \Theta_{\rho,k}^{i}[\epsilon] \right] (\xi) + \partial_{\epsilon}^{m-1} \left(\mathcal{R} \left[\epsilon, \Theta_{\rho,k}^{i}[\epsilon] \right] \right) (\xi) \\
&- \int_{\partial\Omega} \left(v_{\Omega}^{+} \left[\partial_{\epsilon}^{m-1} \Theta_{\rho,k}^{i}[\epsilon] \right] + \partial_{\epsilon}^{m-1} \left(\mathcal{R} \left[\epsilon, \Theta_{\rho,k}^{i}[\epsilon] \right] \right) \right) d\sigma \\
&- v_{\Omega}^{-} \left[\partial_{\epsilon}^{m-1} \Theta_{\rho,k}^{o}[\epsilon] \right] (\xi) - \partial_{\epsilon}^{m-1} \left(\mathcal{R} \left[\epsilon, \Theta_{\rho,k}^{o}[\epsilon] \right] \right) (\xi) \\
&+ \int_{\partial\Omega} \left(v_{\Omega}^{-} \left[\partial_{\epsilon}^{m-1} \Theta_{\rho,k}^{o}[\epsilon] \right] + \partial_{\epsilon}^{m-1} \left(\mathcal{R} \left[\epsilon, \Theta_{\rho,k}^{o}[\epsilon] \right] \right) \right) d\sigma \Bigg) = 0 \quad \forall \xi \in \partial\Omega ,
\end{aligned}
\tag{13.112}
$$

for all $\epsilon \in]-\epsilon_2, \epsilon_2[$ and all $m \in \mathbb{N} \setminus \{0\}$. Then, by Lemma 13.35, we verify that system (13.111)–(13.112) with $\epsilon = 0$ can be rewritten as (13.105)–(13.106) if $m = 1$, as (13.107)–(13.108) if $m = 2$, and as (13.109)–(13.110) for all $m \in \mathbb{N} \setminus \{0, 1, 2\}$.

We can verify that the integral on $\partial\Omega$ of the functions in right-hand sides of (13.105) and (13.106) vanishes. Then, by Lemma 13.32 we prove that the solution $(\theta_{k,1}^{i}, \theta_{k,1}^{o}) \in (C^{0,\alpha}(\partial\Omega)_0)^2$ of (13.105)–(13.106) exists and is unique (we have already observed that the existence is granted by Proposition 13.34 (iii)). By Lemma 13.35 (ii) we also have that

$$
\int_{\partial\Omega} \mathcal{R}_{\nu}^{m} \left[\theta_{k,0}^{i}, \dots, \theta_{k,m-2}^{i} \right] d\sigma = 0
$$

and

$$
\int_{\partial\Omega} \mathcal{R}_{\nu}^{m} \left[\theta_{k,0}^{o}, \dots, \theta_{k,m-2}^{o} \right] d\sigma = 0
$$

for all $m \in \mathbb{N} \setminus \{0, 1\}$. Then a straightforward computation shows that the right-hand sides of (13.107)–(13.110) belong to $C^{0,\alpha}(\partial\Omega)_0$ for all $m \in \mathbb{N} \setminus \{0, 1\}$. Hence, Lemma 13.32 ensures that the solution $(\theta_{k,m}^{i}, \theta_{k,m}^{o}) \in (C^{0,\alpha}(\partial\Omega)_0)^2$ of system (13.107)–(13.108) for $m = 2$ and of system (13.109)–(13.110) for all $m \in \mathbb{N} \setminus \{0, 1, 2\}$ exists and is unique (the existence also follows by Proposition 13.34 (iii)). The proof is now complete. □

To compute the series expansion for the effective conductivity under assumption (13.101), we need the following two lemmas, where we compute the power series expansions of two auxiliary maps. In particular, the map $\tilde{\mathcal{U}}_k^{+}[\epsilon]$ of Lemma 13.38 below equals the restriction of $\mathcal{U}_k^{+}[\epsilon, \epsilon/\rho(\epsilon)]$ on $\partial\Omega$ under the assumption that $\rho(\epsilon) \equiv 1/r_{\#}$, where $\mathcal{U}_k^{+}$ is as in Theorem 13.29. Instead, the map $\tilde{\mathcal{V}}_k^{-}[\epsilon]$ of Lemma 13.39 below equals the restriction of $\mathcal{V}_{k,\partial\Omega}^{-}[\epsilon, \epsilon/\rho(\epsilon)]$ on $\partial\Omega$ under the assumption that $\rho(\epsilon) \equiv 1/r_{\#}$, where $\mathcal{V}_{k,\partial\Omega}^{-}$ is as in Theorem 13.30 (iii).

Lemma 13.38. *Let $k \in \{1, 2\}$. Let $\Theta_{\rho,k}^{i}[\cdot]$ and $\Theta_{\rho,k}^{o}[\cdot]$ be as in Proposition 13.34. Let ϵ_2 and $\{(\theta_{k,m}^{i}, \theta_{k,m}^{o})\}_{m\in\mathbb{N}}$ be as in Proposition 13.37. Let $\tilde{\mathcal{U}}_k^{+}$ be the map from $]-\epsilon_2, \epsilon_2[$ to $C^{1,\alpha}(\partial\Omega)$ defined by*

$$\tilde{\mathcal{U}}_k^+[\epsilon](\xi) \equiv v_\Omega^+\left[\Theta^i_{\rho,k}[\epsilon]\right](\xi) + \mathcal{R}\left[\epsilon, \Theta^i_{\rho,k}[\epsilon]\right](\xi) \\ - \fint_{\partial\Omega}\left(v_\Omega^+\left[\Theta^i_{\rho,k}[\epsilon]\right] + \mathcal{R}\left[\epsilon, \Theta^i_{\rho,k}[\epsilon]\right]\right)d\sigma + \xi_k - \fint_{\partial\Omega}\eta_k\, d\sigma_\eta \qquad \forall \xi \in \partial\Omega\,.$$

Then $\tilde{\mathcal{U}}_k^+$ is real analytic and there exists $\epsilon_3 \in]0, \epsilon_2[$ such that

$$\int_{\partial\Omega}\tilde{\mathcal{U}}_k^+[\epsilon](\xi)(\nu_\Omega(\xi))_j\, d\sigma_\xi = m_2(\Omega)\delta_{j,k} + \int_{\partial\Omega}v_\Omega^+[\theta^i_{k,0}](\xi)(\nu_\Omega(\xi))_j\, d\sigma_\xi \\ + \epsilon\int_{\partial\Omega}v_\Omega^+[\theta^i_{k,1}](\xi)(\nu_\Omega(\xi))_j\, d\sigma_\xi \\ + \sum_{m=2}^{+\infty}\frac{1}{m!}\left(\int_{\partial\Omega}\left(v_\Omega^+[\theta^i_{k,m}](\xi) + \mathcal{R}^m\left[\theta^i_{k,0}, \dots, \theta^i_{k,m-2}\right](\xi)\right)(\nu_\Omega(\xi))_j\, d\sigma_\xi\right)\epsilon^m \\ \forall \epsilon \in]-\epsilon_3, \epsilon_3[\,, \tag{13.113}$$

for all $j \in \{1, 2\}$, where the series converges absolutely for ϵ in $]-\epsilon_3, \epsilon_3[$.

Proof. Since we are assuming that $\rho(\epsilon) \equiv 1/r_\#$ for all $\epsilon \in]-\epsilon_0, \epsilon_0[$, we first note that

$$\tilde{\mathcal{U}}_k^+[\epsilon] = \mathcal{U}_k^+[\epsilon, r_\#\epsilon]_{|\partial\Omega} \tag{13.114}$$

for all $\epsilon \in]-\epsilon_2, \epsilon_2[$, where $\mathcal{U}_k^+$ is as in Theorem 13.29 (cf. (13.71)). Accordingly, Theorem 13.29 implies that $\tilde{\mathcal{U}}_k^+$ is a real analytic map from $]-\epsilon_2, \epsilon_2[$ to $C^{1,\alpha}(\partial\Omega)$. Therefore, there exist $\epsilon_3 \in]0, \epsilon_2[$ and a sequence $\{a_{k,m}\}_{m\in\mathbb{N}}$ in $C^{1,\alpha}(\partial\Omega)$ such that

$$\tilde{\mathcal{U}}_k^+[\epsilon](\xi) = \sum_{m=0}^{+\infty}\frac{a_{k,m}(\xi)}{m!}\epsilon^m \qquad \forall \epsilon \in]-\epsilon_3, \epsilon_3[\,, \quad \xi \in \partial\Omega\,,$$

where the series converges in $C^{1,\alpha}(\partial\Omega)$ for $\epsilon \in]-\epsilon_3, \epsilon_3[$. By taking $\epsilon = 0$ and by Lemma 13.35 (i), we can verify that

$$a_{k,0}(\xi) = v_\Omega^+[\theta^i_{k,0}](\xi) - \fint_{\partial\Omega}v_\Omega^+[\theta^i_{k,0}]d\sigma + \xi_k - \fint_{\partial\Omega}\eta_k\, d\sigma_\eta \qquad \forall \xi \in \partial\Omega\,.$$

In order to compute the other coefficients, we take the derivative of order m in $\mathbb{N} \setminus \{0\}$ of the map $\epsilon \mapsto \tilde{\mathcal{U}}_k^+[\epsilon]$ and we obtain

$$\partial_\epsilon^m\left(\tilde{\mathcal{U}}_k^+[\epsilon]\right)(\xi) = v_\Omega^+\left[\partial_\epsilon^m\Theta^i_{\rho,k}[\epsilon]\right](\xi) + \partial_\epsilon^m\left(\mathcal{R}[\epsilon, \Theta^i_{\rho,k}[\epsilon]]\right)(\xi) \\ - \fint_{\partial\Omega}\left(v_\Omega^+\left[\partial_\epsilon^m\Theta^i_{\rho,k}[\epsilon]\right] + \partial_\epsilon^m\left(\mathcal{R}[\epsilon, \Theta^i_{\rho,k}[\epsilon]]\right)\right)d\sigma \qquad \forall \xi \in \partial\Omega\,.$$

Again, by taking $\epsilon = 0$ and by Lemma 13.35 (i), we find that

$$a_{k,1}(\xi) = v_\Omega^+[\theta^i_{k,1}](\xi) - \fint_{\partial\Omega}v_\Omega^+\left[\theta^i_{k,1}\right]d\sigma \qquad \forall \xi \in \partial\Omega,$$

$$a_{k,m}(\xi) = v_{\Omega}^{+}[\theta_{k,m}^{i}](\xi) + \mathcal{R}^{m}\left[\theta_{k,0}^{i},\dots,\theta_{k,m-2}^{i}\right](\xi) - \fint_{\partial\Omega}\Big(v_{\Omega}^{+}[\theta_{k,m}^{i}] + \mathcal{R}^{m}\left[\theta_{k,0}^{i},\dots,\theta_{k,m-2}^{i}\right]\Big)d\sigma \qquad \forall \xi \in \partial\Omega\,, m \in \mathbb{N}\setminus\{0,1\}\,.$$

As a consequence, possibly shrinking ϵ_3, we have

$$\int_{\partial\Omega} \tilde{\mathcal{U}}_k^{+}[\epsilon](\xi)(\nu_{\Omega}(\xi))_j\, d\sigma_{\xi} = \int_{\partial\Omega}\sum_{m=0}^{+\infty}\frac{a_{k,m}(\xi)}{m!}\epsilon^{m}(\nu_{\Omega}(\xi))_j\, d\sigma_{\xi}$$
$$= \sum_{m=0}^{+\infty}\frac{1}{m!}\left(\int_{\partial\Omega} a_{k,m}(\xi)(\nu_{\Omega}(\xi))_j\, d\sigma_{\xi}\right)\epsilon^{m}\,,$$

where the series converges absolutely for $\epsilon \in]-\epsilon_3,\epsilon_3[$. Then we consider separately the cases $m=0$, $m=1$, and $m \in \mathbb{N}\setminus\{0,1\}$, and we have

$$\begin{aligned}
&\int_{\partial\Omega} a_{k,0}(\xi)(\nu_{\Omega}(\xi))_j\, d\sigma_{\xi}\\
&= \int_{\partial\Omega}\Big(v_{\Omega}^{+}[\theta_{k,0}^{i}](\xi) - \fint_{\partial\Omega} v_{\Omega}^{+}[\theta_{k,0}^{i}]d\sigma + \xi_k - \fint_{\partial\Omega}\eta_k\, d\sigma_{\eta}\Big)(\nu_{\Omega}(\xi))_j\, d\sigma_{\xi}\\
&= \int_{\partial\Omega} v_{\Omega}^{+}[\theta_{k,0}^{i}](\xi)(\nu_{\Omega}(\xi))_j\, d\sigma_{\xi} - \fint_{\partial\Omega} v_{\Omega}^{+}[\theta_{k,0}^{i}]d\sigma\int_{\partial\Omega}(\nu_{\Omega}(\xi))_j\, d\sigma_{\xi}\\
&\quad + \int_{\partial\Omega}\xi_k(\nu_{\Omega}(\xi))_j\, d\sigma_{\xi} - \fint_{\partial\Omega}\eta_k\, d\sigma_{\eta}\int_{\partial\Omega}(\nu_{\Omega}(\xi))_j\, d\sigma_{\xi}\\
&= \int_{\partial\Omega} v_{\Omega}^{+}[\theta_{k,0}^{i}](\xi)(\nu_{\Omega}(\xi))_j\, d\sigma_{\xi} + \int_{\partial\Omega}\xi_k(\nu_{\Omega}(\xi))_j\, d\sigma_{\xi}\,,
\end{aligned}$$

$$\begin{aligned}
\int_{\partial\Omega} a_{k,1}(\xi)(\nu_{\Omega}(\xi))_j\, d\sigma_{\xi} &= \int_{\partial\Omega}\Big(v_{\Omega}^{+}[\theta_{k,1}^{i}](\xi) - \fint_{\partial\Omega} v_{\Omega}^{+}[\theta_{k,1}^{i}]d\sigma\Big)(\nu_{\Omega}(\xi))_j\, d\sigma_{\xi}\\
&= \int_{\partial\Omega} v_{\Omega}^{+}[\theta_{k,1}^{i}](\xi)(\nu_{\Omega}(\xi))_j\, d\sigma_{\xi}\,,
\end{aligned}$$

and

$$\begin{aligned}
\int_{\partial\Omega} a_{k,m}(\xi)(\nu_{\Omega}(\xi))_j\, d\sigma_{\xi} &= \int_{\partial\Omega}\Bigg(v_{\Omega}^{+}[\theta_{k,m}^{i}](\xi) + \mathcal{R}^{m}\left[\theta_{k,0}^{i},\dots,\theta_{k,m-2}^{i}\right](\xi)\\
&\qquad - \fint_{\partial\Omega}\Big(v_{\Omega}^{+}[\theta_{k,m}^{i}] + \mathcal{R}^{m}\left[\theta_{k,0}^{i},\dots,\theta_{k,m-2}^{i}\right]\Big)d\sigma\Bigg)(\nu_{\Omega}(\xi))_j\, d\sigma_{\xi}\\
&= \int_{\partial\Omega}\Big(v_{\Omega}^{+}[\theta_{k,m}^{i}](\xi) + \mathcal{R}^{m}\left[\theta_{k,0}^{i},\dots,\theta_{k,m-2}^{i}\right](\xi)\Big)(\nu_{\Omega}(\xi))_j\, d\sigma_{\xi}\,.
\end{aligned}$$

Moreover, by the Divergence Theorem 4.1, we have

$$\int_{\partial\Omega}\xi_k(\nu_{\Omega}(\xi))_j\, d\sigma_{\xi} = \int_{\Omega}\frac{\partial\xi_k}{\partial\xi_j}\, d\xi = m_2(\Omega)\delta_{j,k}\,.$$

Accordingly, the validity of (13.113) follows. □

Lemma 13.39. *Let $k \in \{1,2\}$. Let $\Theta^i_{\rho,k}[\cdot]$ and $\Theta^o_{\rho,k}[\cdot]$ be as in Proposition 13.34. Let ϵ_2 and $\{(\theta^i_{k,m}, \theta^o_{k,m})\}_{m\in\mathbb{N}}$ be as in Proposition 13.37. Let $\tilde{\mathcal{V}}^-_k$ be the map from $]-\epsilon_2, \epsilon_2[$ to $C^{1,\alpha}(\partial\Omega)$ defined by*

$$\tilde{\mathcal{V}}^-_k[\epsilon](\xi) \equiv v^-_{\Omega}\left[\Theta^o_{\rho,k}[\epsilon]\right](\xi) + \mathcal{R}\left[\epsilon, \Theta^o_{\rho,k}[\epsilon]\right](\xi) + \xi_k \qquad \forall \xi \in \partial\Omega .$$

Then $\tilde{\mathcal{V}}^-_k$ is real analytic and there exists $\epsilon_4 \in]0, \epsilon_2[$ such that

$$\begin{aligned}
&\int_{\partial\Omega} \tilde{\mathcal{V}}^-_k[\epsilon](\xi)(\nu_\Omega(\xi))_j \, d\sigma_\xi \\
&= m_2(\Omega)\delta_{j,k} + \int_{\partial\Omega} v^-_{\Omega}[\theta^o_{k,0}](\xi)(\nu_\Omega(\xi))_j \, d\sigma_\xi + \epsilon \int_{\partial\Omega} v^-_{\Omega}[\theta^o_{k,1}](\xi)(\nu_\Omega(\xi))_j \, d\sigma_\xi \\
&+ \sum_{m=2}^{+\infty} \frac{1}{m!}\left(\int_{\partial\Omega}\left(v^-_{\Omega}[\theta^o_{k,m}](\xi) + \mathcal{R}^m\left[\theta^o_{k,0}, \ldots, \theta^o_{k,m-2}\right](\xi)\right)(\nu_\Omega(\xi))_j \, d\sigma_\xi\right)\epsilon^m ,
\end{aligned} \tag{13.115}$$

for all $j \in \{1,2\}$, where the series converges absolutely for ϵ in $]-\epsilon_4, \epsilon_4[$.

Proof. Since we are assuming that $\rho(\epsilon) \equiv 1/r_\#$ for all $\epsilon \in]-\epsilon_0, \epsilon_0[$, we first note that

$$\tilde{\mathcal{V}}^-_k[\epsilon] = \mathcal{V}^-_{k,\partial\Omega}[\epsilon, r_\#\epsilon] \tag{13.116}$$

for all $\epsilon \in]-\epsilon_2, \epsilon_2[$, where $\mathcal{V}^-_{k,\partial\Omega}$ is as in Theorem 13.30 (iii) (cf. (13.82)). Accordingly, Theorem 13.30 (iii) implies that $\tilde{\mathcal{V}}^-_k$ is a real analytic map from $]-\epsilon_2, \epsilon_2[$ to $C^{1,\alpha}(\partial\Omega)$. Therefore, there exist $\epsilon_4 \in]0, \epsilon_2[$ and a sequence $\{b_{k,m}\}_{m\in\mathbb{N}}$ in $C^{1,\alpha}(\partial\Omega)$ such that

$$\tilde{\mathcal{V}}^-_k[\epsilon](\xi) = \sum_{m=0}^{+\infty} \frac{b_{k,m}(\xi)}{m!}\epsilon^m \qquad \forall \epsilon \in]-\epsilon_4, \epsilon_4[, \xi \in \partial\Omega,$$

where the series converges in $C^{1,\alpha}(\partial\Omega)$ for $\epsilon \in]-\epsilon_4, \epsilon_4[$. By taking $\epsilon = 0$ and by Lemma 13.35 (i), we can verify that

$$b_{k,0}(\xi) = v^-_{\Omega}[\theta^o_{k,0}](\xi) + \xi_k \qquad \forall \xi \in \partial\Omega$$

(we recall that $\int_{\partial\Omega} \Theta^o_{\rho,k}[\epsilon] \, d\sigma = 0$ for all $\epsilon \in]-\epsilon_1, \epsilon_1[$). In order to compute the other coefficients, we take the derivative of order $m \in \mathbb{N}\setminus\{0\}$ of the map $\epsilon \mapsto \tilde{\mathcal{V}}^-_k[\epsilon]$ and we obtain

$$\partial^m_\epsilon\left(\tilde{\mathcal{V}}^-_k[\epsilon]\right)(\xi) = v^-_{\Omega}[\partial^m_\epsilon \Theta^o_{\rho,k}[\epsilon]](\xi) + \partial^m_\epsilon\left(\mathcal{R}[\epsilon, \Theta^o_{\rho,k}[\epsilon]]\right)(\xi) \qquad \forall \xi \in \partial\Omega .$$

Again, by taking $\epsilon = 0$ and by Lemma 13.35 (i), we find

$$b_{k,1}(\xi) = v_{\Omega}^{-}[\theta_{k,1}^{o}](\xi) \qquad \forall \xi \in \partial\Omega,$$
$$b_{k,m}(\xi) = v_{\Omega}^{-}[\theta_{k,m}^{o}](\xi) + \mathcal{R}^{m}\left[\theta_{k,0}^{o}, \dots, \theta_{k,m-2}^{o}\right](\xi) \qquad \forall \xi \in \partial\Omega$$

and for all $m \in \mathbb{N} \setminus \{0, 1\}$.

As a consequence, possibly shrinking ϵ_4, we have

$$\int_{\partial\Omega} \tilde{\mathcal{V}}_{k}^{-}[\epsilon](\xi)(\nu_{\Omega}(\xi))_j \, d\sigma_{\xi} = \int_{\partial\Omega} \sum_{m=0}^{+\infty} \frac{b_{k,m}(\xi)}{m!} \epsilon^{m} (\nu_{\Omega}(\xi))_j \, d\sigma_{\xi}$$
$$= \sum_{m=0}^{+\infty} \frac{1}{m!} \left(\int_{\partial\Omega} b_{k,m}(\xi)(\nu_{\Omega}(\xi))_j \, d\sigma_{\xi} \right) \epsilon^{m}$$

where the series converges absolutely for $\epsilon \in]-\epsilon_4, \epsilon_4[$. Then we consider cases $k = 0$, $k = 1$, and $m \in \mathbb{N} \setminus \{0, 1\}$ separately and we have

$$\int_{\partial\Omega} b_{k,0}(\xi)(\nu_{\Omega}(\xi))_j \, d\sigma_{\xi} = \int_{\partial\Omega} (v_{\Omega}^{-}[\theta_{k,0}^{o}](\xi) + \xi_k)(\nu_{\Omega}(\xi))_j \, d\sigma_{\xi}$$
$$= \int_{\partial\Omega} v_{\Omega}^{-}[\theta_{k,0}^{o}](\xi)(\nu_{\Omega}(\xi))_j \, d\sigma_{\xi} + m_2(\Omega)\delta_{j,k} \, ,$$
$$\int_{\partial\Omega} b_{k,1}(\xi)(\nu_{\Omega}(\xi))_j \, d\sigma_{\xi} = \int_{\partial\Omega} v_{\Omega}^{-}[\theta_{k,1}^{o}](\xi)(\nu_{\Omega}(\xi))_j \, d\sigma_{\xi}$$
$$\int_{\partial\Omega} b_{k,m}(\xi)(\nu_{\Omega}(\xi))_j \, d\sigma_{\xi}$$
$$= \int_{\partial\Omega} \left(v_{\Omega}^{-}[\theta_{k,m}^{o}](\xi) + \mathcal{R}^{m}\left[\theta_{k,0}^{o}, \dots, \theta_{k,m-2}^{o}\right](\xi)\right) (\nu_{\Omega}(\xi))_j \, d\sigma_{\xi} \, .$$

Accordingly, the validity of (13.115) follows. □

We are now ready to prove the main result of this section, where we expand $\lambda_{jk}^{\mathrm{eff}}[\epsilon]$ as a power series and we provide an explicit and constructive expression for the coefficients of the series.

Theorem 13.40. *Let $j, k \in \{1, 2\}$. Let ϵ_2 and $\{(\theta_{k,m}^{i}, \theta_{k,m}^{o})\}_{m\in\mathbb{N}}$ be as in Proposition 13.37. Then there exists $\epsilon_5 \in]0, \epsilon_2[$ such that*

$$\lambda_{jk}^{\mathrm{eff}}[\epsilon] = \lambda^{-}\delta_{j,k} + \epsilon^{2} \frac{1}{m_2(Q)} \sum_{m=0}^{+\infty} \frac{c_{(j,k),m}}{m!} \epsilon^{m} \tag{13.117}$$

for all $\epsilon \in]0, \epsilon_5[$, where

$$c_{(j,k),0} = \lambda^{+} \int_{\partial\Omega} v_{\Omega}^{+}[\theta_{k,0}^{i}](\xi)(\nu_{\Omega}(\xi))_j \, d\sigma_{\xi} + (\lambda^{+} - \lambda^{-}) m_2(\Omega)\delta_{j,k}$$
$$- \lambda^{-} \int_{\partial\Omega} v_{\Omega}^{-}[\theta_{k,0}^{o}](\xi)(\nu_{\Omega}(\xi))_j \, d\sigma_{\xi} + \int_{\partial\Omega} f(\xi)\xi_j \, d\sigma_{\xi},$$

$$
\begin{aligned}
c_{(j,k),1} =&\lambda^+ \int_{\partial\Omega} v^+_\Omega[\theta^i_{k,1}](\xi)(\nu_\Omega(\xi))_j \, d\sigma_\xi \\
&- \lambda^- \int_{\partial\Omega} v^-_\Omega[\theta^o_{k,1}](\xi)(\nu_\Omega(\xi))_j \, d\sigma_\xi \,, \\
c_{(j,k),m} =&\lambda^+ \int_{\partial\Omega} \Big(v^+_\Omega[\theta^i_{k,m}](\xi) + \mathcal{R}^m \left[\theta^i_{k,0}, \dots, \theta^i_{k,m-2}\right](\xi)\Big)(\nu_\Omega(\xi))_j \, d\sigma_\xi \\
&- \lambda^- \int_{\partial\Omega} \left(v^-_\Omega[\theta^o_{k,m}](\xi) + \mathcal{R}^m \left[\theta^o_{k,0}, \dots, \theta^o_{k,m-2}\right](\xi)\right)(\nu_\Omega(\xi))_j \, d\sigma_\xi \,,
\end{aligned}
$$

for all $m \in \mathbb{N} \setminus \{0, 1\}$. *The series on the right-hand side of* (13.117) *converges absolutely for all* $\epsilon \in]-\epsilon_5, \epsilon_5[$.

Proof. By the identities (13.85), (13.86), and (13.88) in Theorem 13.31 and by the equalities (13.114) and (13.116) concerning $\tilde{\mathcal{U}}^+_k$ and $\tilde{\mathcal{V}}^-_k$, respectively, if we set

$$
\begin{aligned}
\mathcal{R}_{jk}[\epsilon] \equiv& \frac{\lambda^+}{m_2(Q)} \int_{\partial\Omega} \tilde{\mathcal{U}}^+_k[\epsilon](\xi)(\nu_\Omega(\xi))_j \, d\sigma_\xi \\
&- \frac{\lambda^-}{m_2(Q)} \int_{\partial\Omega} \tilde{\mathcal{V}}^-_k[\epsilon](\xi)(\nu_\Omega(\xi))_j \, d\sigma_\xi + \frac{1}{m_2(Q)} \int_{\partial\Omega} f(\xi)\xi_j \, d\sigma_\xi \,,
\end{aligned}
$$

for all $\epsilon \in]-\epsilon_2, \epsilon_2[$, then by Theorem 13.31 we have

$$
\lambda^{\text{eff}}_{jk}[\epsilon] = \lambda^- \delta_{j,k} + \epsilon^2 \mathcal{R}_{jk}[\epsilon] \qquad \forall \epsilon \in]0, \epsilon_2[\,.
$$

Then the definition of $\{c_{(j,k),m}\}_{m\in\mathbb{N}}$ and Lemmas 13.38, 13.39 imply the validity of the statement. □

13.7.3 Power Series Expansions for $\rho(\epsilon) \equiv \epsilon/r_\#$

Throughout this subsection we consider the case where

$$
\rho(\epsilon) \equiv \epsilon/r_\# \qquad \forall \epsilon \in]-\epsilon_0, \epsilon_0[\,. \tag{13.118}
$$

As done in Section 13.7.2, in order to compute the asymptotic expansion of the effective conductivity under assumption (13.118), we start with the following proposition, where we identify the coefficients of the power series expansions of $\Theta^i_{\rho,k}[\epsilon]$ and of $\Theta^o_{\rho,k}[\epsilon]$ in terms of the solutions of certain systems of integral equations.

Proposition 13.41. *Let* ϵ_1, $\Theta^i_{\rho,k}[\cdot]$, *and* $\Theta^o_{\rho,k}[\cdot]$ *be as in Proposition 13.34. Then there exist* $\epsilon_2 \in]0, \epsilon_1[$ *and a sequence* $\{(\theta^i_{k,m}, \theta^o_{k,m})\}_{m\in\mathbb{N}}$ *in* $(C^{0,\alpha}(\partial\Omega)_0)^2$ *such that*

$$\Theta^i_{\rho,k}[\epsilon] = \sum_{m=0}^{+\infty} \frac{\theta^i_{k,m}}{m!}\epsilon^m \quad and \quad \Theta^o_{\rho,k}[\epsilon] = \sum_{m=0}^{+\infty} \frac{\theta^o_{k,m}}{m!}\epsilon^m \quad \forall \epsilon \in]-\epsilon_2, \epsilon_2[, \tag{13.119}$$

where the two series converge in $C^{0,\alpha}(\partial\Omega)_0$ for $\epsilon \in]-\epsilon_2, \epsilon_2[$. Moreover, the following statements hold.

(i) The pair of functions $(\theta^i_{k,0}, \theta^o_{k,0})$ is the unique solution in $(C^{0,\alpha}(\partial\Omega)_0)^2$ of the following system of integral equations

$$\begin{aligned}\lambda^-\Big(\frac{1}{2}\theta^o_{k,0}(\xi) + W^t_\Omega[\theta^o_{k,0}](\xi)\Big) - \lambda^+\Big(-\frac{1}{2}\theta^i_{k,0}(\xi) + W^t_\Omega[\theta^i_{k,0}](\xi)\Big)&\\ - f(\xi) + (\lambda^- - \lambda^+)(\nu_\Omega(\xi))_k = 0 \qquad \forall \xi \in \partial\Omega\,,&\end{aligned} \tag{13.120}$$

$$\begin{aligned}\lambda^+\Big(-\frac{1}{2}\theta^i_{k,0}(\xi) + W^t_\Omega[\theta^i_{k,0}](\xi)\Big) + r_\#\Big(v^+_\Omega[\theta^i_{k,0}](\xi) - \int_{\partial\Omega} v^+_\Omega[\theta^i_{k,0}]d\sigma&\\ - v^-_\Omega[\theta^o_{k,0}](\xi) + \int_{\partial\Omega} v^-_\Omega[\theta^o_{k,0}]d\sigma\Big)&\\ - g(\xi) + \int_{\partial\Omega} g\, d\sigma + \lambda^+(\nu_\Omega(\xi))_k = 0 \qquad \forall \xi \in \partial\Omega\,.&\end{aligned} \tag{13.121}$$

(ii) For all $m \in \mathbb{N}$, we have $(\theta^i_{k,2m+1}, \theta^o_{k,2m+1}) = (0,0)$.

(iii) For all $m \in \mathbb{N}\setminus\{0\}$, the pair of functions $(\theta^i_{k,2m}, \theta^o_{k,2m})$ is the unique solution in $(C^{0,\alpha}(\partial\Omega)_0)^2$ of the following system of integral equations which involves $\{(\theta^i_{k,2h}, \theta^o_{k,2h})\}_{h=0}^{m-1}$,

$$\begin{aligned}&\lambda^-\Big(\frac{1}{2}\theta^o_{k,2m}(\xi) + W^t_\Omega[\theta^o_{k,2m}](\xi)\Big) - \lambda^+\Big(-\frac{1}{2}\theta^i_{k,2m}(\xi) + W^t_\Omega[\theta^i_{k,2m}](\xi)\Big)\\ &\quad = \lambda^+\mathcal{R}^{2m}_\nu[\theta^i_{k,0},\dots,\theta^i_{k,2m-2}](\xi) - \lambda^-\mathcal{R}^{2m}_\nu[\theta^o_{k,0},\dots,\theta^o_{k,2m-2}](\xi)\\ &\qquad\qquad \forall \xi \in \partial\Omega\,,\end{aligned}$$

$$\begin{aligned}&\lambda^+\Big(-\frac{1}{2}\theta^i_{k,2m}(\xi) + W^t_\Omega[\theta^i_{k,2m}](\xi)\Big) + r_\#\Bigg(v^+_\Omega[\theta^i_{k,2m}](\xi)\\ &\quad - \int_{\partial\Omega} v^+_\Omega[\theta^i_{k,2m}]d\sigma - v^-_\Omega[\theta^o_{k,2m}](\xi) + \int_{\partial\Omega} v^-_\Omega[\theta^o_{k,2m}]d\sigma\Bigg)\\ &\quad = -\lambda^+\mathcal{R}^{2m}_\nu[\theta^i_{k,0},\dots,\theta^i_{k,2m-2}](\xi)\\ &\quad - r_\#\Bigg(\mathcal{R}^{2m}[\theta^i_{k,0},\dots,\theta^i_{k,2m-2}](\xi) - \int_{\partial\Omega} \mathcal{R}^{2m}[\theta^i_{k,0},\dots,\theta^i_{k,2m-2}]d\sigma\end{aligned}$$

$$- \mathcal{R}^{2m}[\theta^o_{k,0}, \ldots, \theta^o_{k,2m-2}](\xi) + \int_{\partial\Omega} \mathcal{R}^{2m}[\theta^o_{k,0}, \ldots, \theta^o_{k,2m-2}] d\sigma \Big) \qquad \forall \xi \in \partial\Omega .$$

Proof. The existence of ϵ_2 and $\{(\theta^i_{k,m}, \theta^o_{k,m})\}_{m\in\mathbb{N}}$ for which (13.119) holds true is granted by Proposition 13.34 (iii). By equality $\rho(\epsilon) \equiv \epsilon/r_\#$ and by taking $\epsilon = 0$, equation (13.91) can be written as the system of integral equations (13.120)–(13.121). The uniqueness of the solution for this system is then ensured by Proposition 13.34 (ii). Next, we observe that $\mathcal{N}_\rho\left[\epsilon, \Theta^i_{\rho,k}[\epsilon], \Theta^o_{\rho,k}[\epsilon]\right] = 0$ for all $\epsilon \in]-\epsilon_2, \epsilon_2[$ (cf. Proposition 13.34 (iii)). Accordingly, the map that takes ϵ to $\mathcal{N}_\rho\left[\epsilon, \Theta^i_{\rho,k}[\epsilon], \Theta^o_{\rho,k}[\epsilon]\right]$ has all the derivatives equal to zero, i.e.,

$$\partial^m_\epsilon (\mathcal{N}_\rho\left[\epsilon, \Theta^i_{\rho,k}[\epsilon], \Theta^o_{\rho,k}[\epsilon]\right]) = 0$$

for all $\epsilon \in]-\epsilon_2, \epsilon_2[$ and all $m \in \mathbb{N} \setminus \{0\}$. Keeping equality $\rho(\epsilon) \equiv \epsilon/r_\#$ in mind, a straightforward computation shows that

$$\begin{aligned}
&\partial^m_\epsilon \left(\mathcal{N}_{\rho,1}\left[\epsilon, \Theta^i_{\rho,k}[\epsilon], \Theta^o_{\rho,k}[\epsilon]\right]\right)(\xi) \\
&= \lambda^-\Big(\frac{1}{2}\partial^m_\epsilon \Theta^o_{\rho,k}[\epsilon](\xi) + W^t_\Omega\left[\partial^m_\epsilon \Theta^o_{\rho,k}[\epsilon]\right](\xi) + \partial^m_\epsilon\left(\epsilon \mathcal{R}_\nu\left[\epsilon, \Theta^o_{\rho,k}[\epsilon]\right]\right)(\xi)\Big) \\
&- \lambda^+\Big(-\frac{1}{2}\partial^m_\epsilon \Theta^i_{\rho,k}[\epsilon](\xi) + W^t_\Omega\left[\partial^m_\epsilon \Theta^i_{\rho,k}[\epsilon]\right](\xi) + \partial^m_\epsilon\left(\epsilon \mathcal{R}_\nu\left[\epsilon, \Theta^i_{\rho,k}[\epsilon]\right]\right)(\xi)\Big) = 0 \\
&\qquad \forall \xi \in \partial\Omega ,
\end{aligned} \tag{13.122}$$

$$\begin{aligned}
&\partial^m_\epsilon \left(\mathcal{N}_{\rho,2}\left[\epsilon, \Theta^i_{\rho,k}[\epsilon], \Theta^o_{\rho,k}[\epsilon]\right]\right)(\xi) \\
&= \lambda^+\Big(-\frac{1}{2}\partial^m_\epsilon \Theta^i_{\rho,k}[\epsilon](\xi) + W^t_\Omega\left[\partial^m_\epsilon \Theta^i_{\rho,k}[\epsilon]\right](\xi) + \partial^m_\epsilon\left(\epsilon \mathcal{R}_\nu\left[\epsilon, \Theta^i_{\rho,k}[\epsilon]\right]\right)(\xi)\Big) \\
&+ r_\# \Bigg(v^+_\Omega\left[\partial^m_\epsilon \Theta^i_{\rho,k}[\epsilon]\right](\xi) + \partial^m_\epsilon\left(\mathcal{R}\left[\epsilon, \Theta^i_{\rho,k}[\epsilon]\right]\right)(\xi) \\
&- \int_{\partial\Omega}\Big(v^+_\Omega\left[\partial^m_\epsilon \Theta^i_{\rho,k}[\epsilon]\right] + \partial^m_\epsilon\left(\mathcal{R}[\epsilon, \Theta^i_{\rho,k}[\epsilon]]\right)\Big) d\sigma - v^-_\Omega\left[\partial^m_\epsilon \Theta^o_{\rho,k}[\epsilon]\right](\xi) \\
&- \partial^m_\epsilon\left(\mathcal{R}\left[\epsilon, \Theta^o_{\rho,k}[\epsilon]\right]\right)(\xi) + \int_{\partial\Omega}\Big(v^-_\Omega\left[\partial^m_\epsilon \Theta^o_{\rho,k}[\epsilon]\right] + \partial^m_\epsilon\left(\mathcal{R}\left[\epsilon, \Theta^o_{\rho,k}[\epsilon]\right]\right)\Big) d\sigma \Bigg) = 0 \\
&\qquad \forall \xi \in \partial\Omega ,
\end{aligned} \tag{13.123}$$

for all $\epsilon \in]-\epsilon_2, \epsilon_2[$ and all $m \in \mathbb{N} \setminus \{0\}$. By taking $\epsilon = 0$ in (13.122)–(13.123) and noting that $(\theta^i_{k,m}, \theta^o_{k,m}) = (\partial^m_\epsilon \Theta^i_{\rho,k}[0], \partial^m_\epsilon \Theta^o_{\rho,k}[0])$ for all $m \in \mathbb{N}$, we deduce that $(\theta^i_{k,1}, \theta^o_{k,1})$ is a solution of the following system

$$\lambda^-\Big(\frac{1}{2}\theta^o_{k,1}(\xi)+W^t_\Omega[\theta^o_{k,1}](\xi)\Big)-\lambda^+\Big(-\frac{1}{2}\theta^i_{k,1}(\xi)+W^t_\Omega[\theta^i_{k,1}](\xi)\Big)=0 \quad \forall \xi\in\partial\Omega\,, \tag{13.124}$$

$$\lambda^+\Big(-\frac{1}{2}\theta^i_{k,1}(\xi)+W^t_\Omega[\theta^i_{k,1}](\xi)\Big)+r_\#\Bigg(v^+_\Omega[\theta^i_{k,1}](\xi)-\int_{\partial\Omega}v^+_\Omega[\theta^i_{k,1}]d\sigma \\ -v^-_\Omega[\theta^o_{k,1}](\xi)+\int_{\partial\Omega}v^-_\Omega[\theta^o_{k,1}]d\sigma\Bigg)=0 \qquad \forall\xi\in\partial\Omega \tag{13.125}$$

(see also Lemma 13.35), and that, for $m\in\mathbb{N}\setminus\{0,1\}$, the pair $(\theta^i_{k,m},\theta^o_{k,m})$ is a solution of the following system

$$\lambda^-\Big(\frac{1}{2}\theta^o_{k,m}(\xi)+W^t_\Omega[\theta^o_{k,m}](\xi)\Big)-\lambda^+\Big(-\frac{1}{2}\theta^i_{k,m}(\xi)+W^t_\Omega[\theta^i_{k,m}](\xi)\Big) \\ =\lambda^+\mathcal{R}^m_\nu\left[\theta^i_{k,0},\ldots,\theta^i_{k,m-2}\right](\xi)-\lambda^-\mathcal{R}^m_\nu\left[\theta^o_{k,0},\ldots,\theta^o_{k,m-2}\right](\xi) \quad \forall\xi\in\partial\Omega\,, \tag{13.126}$$

$$\lambda^+\Big(-\frac{1}{2}\theta^i_{k,m}(\xi)+W^t_\Omega[\theta^i_{k,m}](\xi)\Big)+r_\#\Bigg(v^+_\Omega[\theta^i_{k,m}](\xi)-\int_{\partial\Omega}v^+_\Omega[\theta^i_{k,m}]d\sigma \\ -v^-_\Omega[\theta^o_{k,m}](\xi)+\int_{\partial\Omega}v^-_\Omega[\theta^o_{k,m}]d\sigma\Bigg)=-\lambda^+\mathcal{R}^m_\nu\left[\theta^i_{k,0},\ldots,\theta^i_{k,m-2}\right](\xi) \\ -r_\#\Bigg(\mathcal{R}^m\left[\theta^i_{k,0},\ldots,\theta^i_{k,m-2}\right](\xi)-\int_{\partial\Omega}\mathcal{R}^m\left[\theta^i_{k,0},\ldots,\theta^i_{k,m-2}\right]d\sigma \\ -\mathcal{R}^m\left[\theta^o_{k,0},\ldots,\theta^o_{k,m-2}\right](\xi)+\int_{\partial\Omega}\mathcal{R}^m\left[\theta^o_{k,0},\ldots,\theta^o_{k,m-2}\right]d\sigma\Bigg) \quad \forall\xi\in\partial\Omega\,. \tag{13.127}$$

Since the right-hand sides of equalities (13.124)–(13.125) and (13.126)–(13.127) belong to the space $C^{0,\alpha}(\partial\Omega)_0$, the uniqueness of the solution to (13.124)–(13.125) and of the solution to (13.126)–(13.127) follows by Proposition 13.21 (we have already observed that the existence is granted by Proposition 13.34 (iii)). Moreover, $(\theta^i_{k,1},\theta^o_{k,1})=(0,0)$. Also, by Lemma 13.36 and by the uniqueness of the solution to system (13.126)–(13.127), we can verify that $(\theta^i_{2m+1},\theta^o_{2m+1})=(0,0)$ for all $m\in\mathbb{N}\setminus\{0\}$. The validity of the proposition is now proved. □

By exploiting Proposition 13.41, we can prove the following Lemmas 13.42 and 13.43. The proofs can be effected by a straightforward modification of the proofs of the analogous Lemmas 13.38 and 13.39 and are accordingly omitted.

Lemma 13.42. *Let $k\in\{1,2\}$. Let $\Theta^i_{\rho,k}[\cdot]$ and $\Theta^o_{\rho,k}[\cdot]$ be as in Proposition 13.34. Let ϵ_2 and $\{(\theta^i_{k,m},\theta^o_{k,m})\}_{m\in\mathbb{N}}$ be as in Proposition 13.41. Let $\mathcal{U}^{\#,+}_k$ be the map from $]-\epsilon_2,\epsilon_2[$ to $C^{1,\alpha}(\partial\Omega)$ defined by*

$$\mathcal{U}_k^{\#,+}[\epsilon](\xi) \equiv v_\Omega^+\left[\Theta_{\rho,k}^i[\epsilon]\right](\xi) + \mathcal{R}\left[\epsilon, \Theta_{\rho,k}^i[\epsilon]\right](\xi) \\ - \int_{\partial\Omega} \left(v_\Omega^+\left[\Theta_{\rho,k}^i[\epsilon]\right] + \mathcal{R}\left[\epsilon, \Theta_{\rho,k}^i[\epsilon]\right]\right) d\sigma + \xi_k - \int_{\partial\Omega} \eta_k \, d\sigma_\eta \qquad \forall \xi \in \partial\Omega .$$

Then $\mathcal{U}_k^{\#,+}$ is real analytic and there exists $\epsilon_3 \in]0, \epsilon_2[$ such that

$$\int_{\partial\Omega} \mathcal{U}_k^{\#,+}[\epsilon](\xi)(\nu_\Omega(\xi))_j \, d\sigma_\xi = \int_{\partial\Omega} v_\Omega^+[\theta_{k,0}^i](\xi)(\nu_\Omega(\xi))_j \, d\sigma_\xi + m_2(\Omega)\delta_{j,k} \\ + \sum_{m=1}^{+\infty} \frac{1}{(2m)!} \Bigg(\int_{\partial\Omega} \Big(v_\Omega^+[\theta_{k,2m}^i](\xi) \\ + \mathcal{R}^{2m}[\theta_{k,0}^i, \dots, \theta_{k,2m-2}^i](\xi) \Big) (\nu_\Omega(\xi))_j \, d\sigma_\xi \Bigg) \epsilon^{2m} \qquad \forall \epsilon \in]-\epsilon_3, \epsilon_3[,$$

for all $j \in \{1, 2\}$, where the series converges absolutely for ϵ in $]-\epsilon_3, \epsilon_3[$.

Lemma 13.43. *Let $k \in \{1, 2\}$. Let $\Theta_{\rho,k}^i[\cdot]$ and $\Theta_{\rho,k}^o[\cdot]$ be as in Proposition 13.34. Let ϵ_2 and $\{(\theta_{k,m}^i, \theta_{k,m}^o)\}_{m\in\mathbb{N}}$ be as in Proposition 13.41. Let $\mathcal{V}_k^{\#,-}$ be the map from $]-\epsilon_2, \epsilon_2[$ to $C^{1,\alpha}(\partial\Omega)$ defined by*

$$\mathcal{V}_k^{\#,-}[\epsilon](\xi) \equiv v_\Omega^-\left[\Theta_{\rho,k}^o[\epsilon]\right](\xi) + \mathcal{R}\left[\epsilon, \Theta_{\rho,k}^o[\epsilon]\right](\xi) + \xi_k \qquad \forall \xi \in \partial\Omega .$$

Then $\mathcal{V}_k^{\#,-}$ is real analytic and there exists $\epsilon_4 \in]0, \epsilon_2[$ such that

$$\int_{\partial\Omega} \mathcal{V}_k^{\#,-}[\epsilon](\xi)(\nu_\Omega(\xi))_j \, d\sigma_\xi = \int_{\partial\Omega} v_\Omega^-[\theta_{k,0}^o](\xi)(\nu_\Omega(\xi))_j \, d\sigma_\xi + m_2(\Omega)\delta_{j,k} \\ + \sum_{m=1}^{+\infty} \frac{1}{(2m)!} \Bigg(\int_{\partial\Omega} \Big(v_\Omega^+[\theta_{k,2m}^o](\xi) \\ + \mathcal{R}^{2m}[\theta_{k,0}^o, \dots, \theta_{k,2m-2}^o](\xi) \Big) (\nu_\Omega(\xi))_j \, d\sigma_\xi \Bigg) \epsilon^{2m} \qquad \forall \epsilon \in]-\epsilon_4, \epsilon_4[,$$

for all $j \in \{1, 2\}$, where the series converges absolutely for ϵ in $]-\epsilon_4, \epsilon_4[$.

By Lemmas 13.42 and 13.43 and by arguing as in the proof of Theorem 13.40, one deduces the validity of the following result concerning the expansion of $\lambda_{jk}^{\text{eff}}[\epsilon]$.

Theorem 13.44. *Let $j, k \in \{1, 2\}$. Let ϵ_2 and $\{(\theta_{k,m}^i, \theta_{k,m}^o)\}_{m\in\mathbb{N}}$ be as in Proposition 13.41. Then there exists $\epsilon_5 \in]0, \epsilon_2[$ such that*

$$\lambda_{jk}^{\text{eff}}[\epsilon] = \lambda^- \delta_{j,k} + \epsilon^2 \frac{1}{m_2(Q)} \sum_{m=0}^{+\infty} \frac{d_{(j,k),2m}}{(2m)!} \epsilon^{2m} \tag{13.128}$$

for all $\epsilon \in]0, \epsilon_5[$, where

$$
\begin{aligned}
d_{(j,k),0} =& \lambda^{+} \int_{\partial\Omega} v^{+}_{\Omega}[\theta^{i}_{k,0}](\xi)(\nu_{\Omega}(\xi))_j \, d\sigma_\xi + (\lambda^{+} - \lambda^{-}) m_2(\Omega)\delta_{j,k} \\
& -\lambda^{-} \int_{\partial\Omega} v^{-}_{\Omega}[\theta^{o}_{k,0}](\xi)(\nu_{\Omega}(\xi))_j \, d\sigma_\xi + \int_{\partial\Omega} f(\xi)\xi_j \, d\sigma_\xi , \\
d_{(j,k),2m} =& \lambda^{+} \int_{\partial\Omega} \Big(v^{+}_{\Omega}[\theta^{i}_{k,2m}](\xi) + \mathcal{R}^{2m}[\theta^{i}_{k,0}, \ldots, \theta^{i}_{k,2m-2}](\xi) \Big) (\nu_{\Omega}(\xi))_j \, d\sigma_\xi \\
& -\lambda^{-} \int_{\partial\Omega} \left(v^{-}_{\Omega}[\theta^{o}_{k,2m}](\xi) + \mathcal{R}^{2m}[\theta^{o}_{k,0}, \ldots, \theta^{o}_{k,2m-2}](\xi) \right) (\nu_{\Omega}(\xi))_j \, d\sigma_\xi
\end{aligned}
$$

for all $m \in \mathbb{N} \setminus \{0\}$. *The series in the right-hand side of* (13.128) *converges absolutely for* $\epsilon \in]-\epsilon_5, \epsilon_5[$.

13.8 A Quasilinear Heat Transmission Problem

In this section, we study the asymptotic behavior of the solution of a quasilinear heat transmission problem in a periodic two-phase dilute composite. Our presentation here follows that of [177].

13.8.1 Introduction

In this subsection we first introduce the data of our problem.

We assume that

$$
\begin{aligned}
& K_i,\ K_o \ \text{ are real analytic diffeomorphisms from } \mathbb{R} \text{ onto itself}, \\
& \kappa_i(\tau) \equiv K_i'(\tau) > 0 , \qquad \kappa_o(\tau) \equiv K_o'(\tau) > 0 \qquad \forall \tau \in \mathbb{R} .
\end{aligned}
$$

The functions κ_i and κ_o represent the heat conductivities of the materials occupying the sets $\mathbb{S}[\Omega_{p,\epsilon}]$ and $\mathbb{S}[\Omega_{p,\epsilon}]^-$, respectively. Next we fix

$$
\rho \in]0, +\infty[
$$

and we recall that we denote by $C^0_{q,\omega,\rho}(\mathbb{R}^n)$ the corresponding Roumieu Banach space of q-periodic analytic functions in $\mathbb{R}^n$ (cf. (13.2)). Then we assume that

$$
\{f_\epsilon\}_{\epsilon \in]-\epsilon_0, \epsilon_0[} \text{ is a real analytic family in } C^0_{q,\omega,\rho}(\mathbb{R}^n) , \tag{13.129}
$$

i.e., that the map from $]-\epsilon_0, \epsilon_0[$ to $C^0_{q,\omega,\rho}(\mathbb{R}^n)$ that takes ϵ to f_ϵ is real analytic, and that

$$
\{g_\epsilon\}_{\epsilon \in]-\epsilon_0, \epsilon_0[} \text{ is a real analytic family in } C^{0,\alpha}(\partial\Omega) . \tag{13.130}
$$

For a fixed value $\epsilon \in]0, \epsilon_0[$ the function f_ϵ represents the opposite of the external heat source per volume unit applied to the composite material. The function g_ϵ represents the external heat supply per surface unit applied at the interface between the materials of $\mathbb{S}[\Omega_{p,\epsilon}]$ and $\mathbb{S}[\Omega_{p,\epsilon}]^-$. Then we introduce a function $k(\cdot)$ such that

$$k(\cdot) \text{ is a real analytic function from }]-\epsilon_0, \epsilon_0[\text{ to } \mathbb{R}\,. \qquad (13.131)$$

The role of $k(\epsilon)$ is that of an ϵ-dependent normalizing condition for the temperature in $\mathbb{S}[\Omega_{p,\epsilon}]$.

Then for each $\epsilon \in]0, \epsilon_0[$, we consider the following quasilinear transmission problem.

$$\begin{cases} \operatorname{div}\left(\kappa_i(T^i(x))\nabla T^i(x)\right) = f_\epsilon(x) & \forall x \in \mathbb{S}[\Omega_{p,\epsilon}]\,, \\ \operatorname{div}\left(\kappa_o(T^o(x))\nabla T^o(x)\right) = f_\epsilon(x) & \forall x \in \mathbb{S}[\Omega_{p,\epsilon}]^-\,, \\ T^i(x+qz) = T^i(x) & \forall x \in \overline{\mathbb{S}[\Omega_{p,\epsilon}]}\,, z \in \mathbb{Z}^n\,, \\ T^o(x+qz) = T^o(x) & \forall x \in \overline{\mathbb{S}[\Omega_{p,\epsilon}]^-}\,, z \in \mathbb{Z}^n\,, \\ T^o(x) = T^i(x) & \forall x \in \partial\Omega_{p,\epsilon}\,, \\ \kappa_o(T^o(x))\nu_{\Omega_{p,\epsilon}}(x)\cdot\nabla T^o(x) & \\ \quad = \kappa_i(T^i(x))\nu_{\Omega_{p,\epsilon}}(x)\cdot\nabla T^i(x) & \\ \qquad + g_\epsilon\left(\dfrac{x-p}{\epsilon}\right) & \forall x \in \partial\Omega_{p,\epsilon}\,, \\ \fint_{\partial\Omega_{p,\epsilon}} K_i(T^i(x))\,d\sigma_x = k(\epsilon)\,, & \end{cases} \qquad (13.132)$$

where $\nu_{\Omega_{p,\epsilon}}$ denotes the outward unit normal to $\Omega_{p,\epsilon}$, and where the functions T^i and T^o from $\overline{\mathbb{S}[\Omega_{p,\epsilon}]}$ to $\mathbb{R}$ and from $\overline{\mathbb{S}[\Omega_{p,\epsilon}]^-}$ to $\mathbb{R}$ represent the temperature distribution of the material in $\mathbb{S}[\Omega_{p,\epsilon}]$ and in $\mathbb{S}[\Omega_{p,\epsilon}]^-$, respectively. The pair of functions (T^i, T^o) is the unknown of the problem, and we look for solutions in the Schauder space of q-periodic functions $C^{1,\alpha}_q(\overline{\mathbb{S}[\Omega_{p,\epsilon}]}) \times C^{1,\alpha}_q(\overline{\mathbb{S}[\Omega_{p,\epsilon}]^-})$ (cf. Section 13.8.3).

As we shall see, by applying the Divergence Theorem 4.1, one can easily show that problem (13.132) can have a classical solution (T^i, T^o) only if

$$\int_Q f_\epsilon\,dx + \epsilon^{n-1}\int_{\partial\Omega} g_\epsilon\,d\sigma = 0 \qquad (13.133)$$

(see Lemma 13.45).

One may wonder about the role of the normalizing condition involving the constant $k(\epsilon)$ in the last equation of problem (13.132). In the linear case, in which κ_i and κ_o are constant, the solution of problem (13.132) without the last equation is well known to be determined up to an additive constant, which can be uniquely determined by the last equation. If κ_i and κ_o are not constant, problem (13.132) without the last equation is not linear and solutions are not determined up to an additive constant. Nevertheless one may expect that there is no uniqueness for problem (13.132) without the last equation. Our analysis below shows that for each choice of $k(\cdot)$, one has at least a solution $(T^i(\epsilon,\cdot), T^o(\epsilon,\cdot))$ of problem (13.132), at least for all $\epsilon \in]0, \epsilon'[$ with $\epsilon' > 0$ small, and that accordingly our expectations are true and that the solutions of the problem (13.132) without the last equation are not unique.

We also observe that one could prove a local uniqueness result for the family of solutions $\{(T^i(\epsilon,\cdot),T^o(\epsilon,\cdot))\}_{\epsilon\in]0,\epsilon'[}$ by a similar strategy of that of Section 11.3.3 for the solutions of a nonlinear transmission problem in a bounded domain with a small hole (cf. [70]).

We note that a priori, it is not clear why problem (13.132) should admit a classical solution. We prove that possibly shrinking ϵ_0, problem (13.132) has a solution $(T^i(\epsilon,\cdot),T^o(\epsilon,\cdot))$ in $C^{1,\alpha}_q(\overline{\mathbb{S}[\Omega_{p,\epsilon}]})\times C^{1,\alpha}_q(\overline{\mathbb{S}[\Omega_{p,\epsilon}]^-})$ that satisfies the limiting conditions

$$
\begin{aligned}
&\lim_{\epsilon\to 0} T^i(\epsilon,p+\epsilon\xi) \text{ exists uniformly when } \xi \text{ ranges in } \overline{\Omega}\,,\\
&\lim_{\epsilon\to 0} T^o(\epsilon,p+\epsilon\xi) \text{ exists uniformly when } \xi \text{ ranges in the compact subsets}\\
&\text{of } \mathbb{R}^n\setminus\Omega,\\
&\lim_{\epsilon\to 0} T^o(\epsilon,x) \text{ exists uniformly when } x \text{ ranges in the compact subsets of}\\
&\mathbb{R}^n\setminus(p+q\mathbb{Z}^n)\,.
\end{aligned}
$$

Next we pose the following three questions.

(i) Let ξ be a fixed point in $\overline{\Omega}$. What can be said on the function $\epsilon\mapsto T^i(\epsilon,p+\epsilon\xi)$ when ϵ is close to 0 and positive?
(ii) Let ξ be a fixed point in $\mathbb{R}^n\setminus\Omega$. What can be said on the function $\epsilon\mapsto T^o(\epsilon,p+\epsilon\xi)$ when ϵ is close to 0 and positive?
(iii) Let x be a fixed point in $\mathbb{R}^n\setminus(p+q\mathbb{Z}^n)$. What can be said on the function $\epsilon\mapsto T^o(\epsilon,x)$ when ϵ is close to 0 and positive?

In a sense, question (iii) concerns the 'macroscopic' behavior of $T^o(\epsilon,\cdot)$, whereas questions (i) and (ii) concern the 'microscopic' behavior of $T^o(\epsilon,\cdot)$ and $T^i(\epsilon,\cdot)$, respectively.

Here we wish to represent the functions in (i)–(iii) in terms of real analytic maps and in terms of known functions of ϵ (such as $\epsilon\log\epsilon$).

We note that nonlinear transmission problems in periodic domains have been analyzed with the Functional Equation Method, for example, in Kapanadze et al. [134] and in Kapanadze et al. [136], in connection with the effective properties of nonlinear doubly periodic 2D composites. An homogenization result for a class of nonlinear stationary diffusion problems whose diffusion coefficients depend on the unknown function can be found in Artola and Duvaut [15]. Moreover, a general scheme for the homogenization of quasilinear composites is provided by Gałka et al. [106].

We now briefly outline our strategy. To apply the Functional Analytic Approach of this book we first use the Kirchoff transformation to convert problem (13.132) into a periodic problem for the Laplace operator with nonlinear boundary conditions. Then we can transform our problem into a system of nonlinear integral equations by exploiting layer potential representations of the solutions. Then we observe that, by changing the unknown functions appropriately, we can obtain a system that can be analyzed around the degenerate case $\epsilon=0$ by means of the Implicit Function

Theorem A.19, and we represent the unknowns of the system of integral equations in terms of analytic functions of ϵ and $\epsilon \log \epsilon$. Next we exploit the integral representations of the solutions in terms of the unknowns of the system of integral equations, and we deduce both the existence of the solution $(T^i(\epsilon,\cdot), T^o(\epsilon,\cdot))$ and its representation in terms of real analytic maps of ϵ and $\epsilon \log \epsilon$.

13.8.2 An Equivalent Formulation of Problem (13.132)

We exploit the so called Kirchoff transformation to convert our quasilinear boundary value problem (13.132) into a problem for a linear equation with a nonlinear boundary condition (cf., e.g., Mityushev and Rogosin [224, Chap. 5]). We observe that if $\epsilon \in]0, \epsilon_0[$, a pair $(T^i, T^o) \in C^{1,\alpha}_q(\overline{\mathbb{S}[\Omega_{p,\epsilon}]}) \times C^{1,\alpha}_q(\overline{\mathbb{S}[\Omega_{p,\epsilon}]^-})$ satisfies problem (13.132) if and only if the pair $(v^i, v^o) \in C^{1,\alpha}_q(\overline{\mathbb{S}[\Omega_{p,\epsilon}]}) \times C^{1,\alpha}_q(\overline{\mathbb{S}[\Omega_{p,\epsilon}]^-})$ defined by

$$v^i = K_i \circ T^i\,, \qquad v^o = K_o \circ T^o\,,$$

satisfies the following nonlinear transmission problem.

$$\begin{cases} \Delta v^i(x) = f_\epsilon(x) & \forall x \in \mathbb{S}[\Omega_{p,\epsilon}]\,,\\ \Delta v^o(x) = f_\epsilon(x) & \forall x \in \mathbb{S}[\Omega_{p,\epsilon}]^-\,,\\ v^i(x+qz) = v^i(x) & \forall x \in \overline{\mathbb{S}[\Omega_{p,\epsilon}]}\,, z \in \mathbb{Z}^n\,,\\ v^o(x+qz) = v^o(x) & \forall x \in \overline{\mathbb{S}[\Omega_{p,\epsilon}]^-}\,, z \in \mathbb{Z}^n\,,\\ K_o^{(-1)} \circ v^o(x) = K_i^{(-1)} \circ v^i(x) & \forall x \in \partial\Omega_{p,\epsilon}\,,\\ \nu_{\Omega_{p,\epsilon}}(x) \cdot \nabla v^o(x) & \\ \quad = \nu_{\Omega_{p,\epsilon}}(x) \cdot \nabla v^i(x) + g_\epsilon\left(\frac{x-p}{\epsilon}\right) & \forall x \in \partial\Omega_{p,\epsilon}\,,\\ \fint_{\partial\Omega_{p,\epsilon}} v^i(x)\, d\sigma_x = k(\epsilon)\,. \end{cases} \tag{13.134}$$

Next we exploit a classical argument to show that the data f_ϵ and g_ϵ are subject to the compatibility condition (13.133). Indeed, by applying the Divergence Theorem 4.1 to v^o in $Q \setminus \overline{\Omega}_{p,\epsilon}$ and by exploiting the equality $\int_{\partial Q} \frac{\partial v^o}{\partial \nu_Q}\, d\sigma = 0$, which follows by the q-periodicity of v^o, we can prove the following.

Lemma 13.45. *Let $\epsilon \in]0, \epsilon_0[$. If problem (13.134) admits at least a solution (v^i, v^o) in $C^{1,\alpha}_q(\overline{\mathbb{S}[\Omega_{p,\epsilon}]}) \times C^{1,\alpha}_q(\overline{\mathbb{S}[\Omega_{p,\epsilon}]^-})$, then condition (13.133) holds.*

We are now ready to convert our problem (13.134) into a homogeneous problem, and we do so by means of the following theorem, which is an immediate consequence of the previous lemma.

Theorem 13.46. *Let $\epsilon \in]0, \epsilon_0[$. Then the pair of functions*

$$(v^i, v^o) \in C^{1,\alpha}_q(\overline{\mathbb{S}[\Omega_{p,\epsilon}]}) \times C^{1,\alpha}_q(\overline{\mathbb{S}[\Omega_{p,\epsilon}]^-})$$

satisfies problem (13.134) if and only if the pair of functions (u^i, u^o) *in the space* $C^{1,\alpha}_q(\overline{\mathbb{S}[\Omega_{p,\epsilon}]}) \times C^{1,\alpha}_q(\overline{\mathbb{S}[\Omega_{p,\epsilon}]^-})$ *defined by*

$$u^i(x) \equiv v^i(x) - F^i_\epsilon[f_\epsilon](x) \qquad \forall x \in \overline{\mathbb{S}[\Omega_{p,\epsilon}]}\,,$$
$$u^o(x) \equiv v^o(x) - F^o_\epsilon[f_\epsilon](x) \qquad \forall x \in \overline{\mathbb{S}[\Omega_{p,\epsilon}]^-}\,,$$

satisfies the following nonlinear transmission problem

$$\begin{cases} \Delta u^i(x) = 0 & \forall x \in \mathbb{S}[\Omega_{p,\epsilon}]\,, \\ \Delta u^o(x) = 0 & \forall x \in \mathbb{S}[\Omega_{p,\epsilon}]^-\,, \\ u^i(x+qz) = u^i(x) & \forall x \in \overline{\mathbb{S}[\Omega_{p,\epsilon}]}\,, z \in \mathbb{Z}^n\,, \\ u^o(x+qz) = u^o(x) & \forall x \in \overline{\mathbb{S}[\Omega_{p,\epsilon}]^-}\,, z \in \mathbb{Z}^n\,, \\ u^o(x) + P_{q,n}[Q, f_\epsilon](x) - \frac{1}{m_n(\Omega_{p,\epsilon})} P_{q,n}[\Omega_{p,\epsilon}, 1](x) \int_Q f_\epsilon(y)\,dy & \\ = K_o \circ K_i^{(-1)} \left(u^i(x) + P_{q,n}[Q, f_\epsilon](x) + P_n[\Omega_{p,\epsilon}, 1](x) \fint_Q f_\epsilon(y)\,dy \right) & \\ \qquad \forall x \in \partial\Omega_{p,\epsilon}\,, & \\ \nu_{\Omega_{p,\epsilon}}(x) \cdot \nabla u^o(x) - \frac{1}{m_n(\Omega_{p,\epsilon})} \nu_{\Omega_{p,\epsilon}}(x) \cdot \nabla P_{q,n}[\Omega_{p,\epsilon}, 1](x) \int_Q f_\epsilon(y)\,dy & \\ = \nu_{\Omega_{p,\epsilon}}(x) \cdot \nabla u^i(x) & \\ + \nu_{\Omega_{p,\epsilon}}(x) \cdot \nabla P_n[\Omega_{p,\epsilon}, 1](x) \fint_Q f_\epsilon(y)\,dy + g_\epsilon\left(\frac{x-p}{\epsilon}\right) & \forall x \in \partial\Omega_{p,\epsilon}\,, \\ \fint_{\partial\Omega_{p,\epsilon}} u^i(x)\,d\sigma_x = k(\epsilon) - \fint_{\partial\Omega_{p,\epsilon}} P_{q,n}[Q, f_\epsilon](x)\,d\sigma_x & \\ \qquad - \fint_{\partial\Omega_{p,\epsilon}} P_n[\Omega_{p,\epsilon}, 1](x)\,d\sigma_x \fint_Q f_\epsilon(y)\,dy\,. & \end{cases} \tag{13.135}$$

(For the definition of $F^i_\epsilon[f_\epsilon]$ and $F^o_\epsilon[f_\epsilon]$, we refer to Lemma 13.5 (iii), (iv). For the definition of $P_{q,n}[\cdot,\cdot]$, we refer to (13.3).)

13.8.3 Formulation of Problem (13.135) in Terms of Integral Equations

In order to convert our problem (13.135) into a system of integral equations, we exploit Lemma 13.7 and we have the following.

Theorem 13.47. *Let* $\epsilon \in]0, \epsilon_0[$. *Let* $K \equiv K_o \circ K_i^{(-1)}$. *Let*

$$m^i[\epsilon, \psi](\xi) \equiv v_\Omega[\psi](\xi) + \epsilon^{n-2} \int_{\partial\Omega} R_{q,n}(\epsilon(\xi - \eta))\psi(\eta)\,d\sigma_\eta$$
$$- \fint_{\partial\Omega} \left(v_\Omega[\psi](\tau) + \epsilon^{n-2} \int_{\partial\Omega} R_{q,n}(\epsilon(\tau - \eta))\psi(\eta)\,d\sigma_\eta \right) d\sigma_\tau \qquad \forall \xi \in \partial\Omega\,,$$

$$m_1^i[\epsilon](\xi) \equiv \int_0^1 \xi \cdot \nabla P_{q,n}[Q, f_\epsilon](p + \epsilon\xi\tau)\, d\tau + \epsilon P_n[\Omega, 1](\xi) \fint_Q f_\epsilon\, dx \ \forall \xi \in \partial\Omega\,,$$

$$m^o[\epsilon, \theta, c](\xi) \equiv v_\Omega[\theta](\xi) + \epsilon^{n-2} \int_{\partial\Omega} R_{q,n}(\epsilon(\xi - \eta))\theta(\eta)\, d\sigma_\eta + c \quad \forall \xi \in \partial\Omega\,,$$

$$G[\epsilon, \epsilon_1] \equiv K\Bigg(k(\epsilon) - \fint_{\partial\Omega} P_{q,n}[Q, f_\epsilon](p + \epsilon\xi)\, d\sigma_\xi$$
$$- \fint_Q f_\epsilon\, dy \fint_{\partial\Omega} \epsilon^2 P_n[\Omega, 1](\xi)\, d\sigma_\xi - \fint_Q f_\epsilon\, dy \epsilon \frac{m_n(\Omega)}{2\pi}\epsilon_1 + P_{q,n}[Q, f_\epsilon](p)\Bigg)\,,$$

for all $(\epsilon, \epsilon_1, \psi, \theta, c) \in]-\epsilon_0, \epsilon_0[\times \mathbb{R} \times C^{0,\alpha}(\partial\Omega)^2 \times \mathbb{R}$.
Then the map $(u^i[\epsilon, \cdot, \cdot, \cdot], u^o[\epsilon, \cdot, \cdot, \cdot])$ *that takes the triple* (ψ, θ, c) *to the pair* $(u^i[\epsilon, \psi, \theta, c], u^o[\epsilon, \psi, \theta, c])$ *defined by*

$$\begin{aligned} & u^i[\epsilon, \psi, \theta, c](x) \\ & \equiv v^+_{q,\Omega_{p,\epsilon}}[\psi(\tfrac{\cdot - p}{\epsilon})](x) - \fint_{\partial\Omega_{p,\epsilon}} v^+_{q,\Omega_{p,\epsilon}}[\psi(\tfrac{\cdot - p}{\epsilon})]\, d\sigma \\ & + k(\epsilon) - \fint_{\partial\Omega} P_{q,n}[Q, f_\epsilon](p + \epsilon\xi)\, d\sigma_\xi + \frac{\epsilon^{n-1}}{m_n(Q)} \int_{\partial\Omega} g_\epsilon\, d\sigma \\ & \times \fint_{\partial\Omega} P_n[\Omega, 1](\xi)\epsilon^2 + \frac{m_n(\Omega)}{2\pi}\epsilon(\delta_{2,n}\epsilon \log \epsilon)\, d\sigma_\xi \quad \forall x \in \overline{\mathbb{S}[\Omega_{p,\epsilon}]}\,, \end{aligned} \tag{13.136}$$

and

$$\begin{aligned} u^o[\epsilon,\psi, \theta, c](x) & \equiv v^-_{q,\Omega_{p,\epsilon}}[\theta(\tfrac{\cdot - p}{\epsilon})](x) + \epsilon c - P_{q,n}[Q, f_\epsilon](p) \\ & + G[\epsilon, \delta_{2,n}\epsilon \log \epsilon] + \int_Q f_\epsilon\, dy \frac{1}{2\pi}\delta_{2,n} \log \epsilon \quad \forall x \in \overline{\mathbb{S}[\Omega_{p,\epsilon}]^-}\,, \end{aligned} \tag{13.137}$$

is a bijection from the set of the triples $(\psi, \theta, c) \in (C^{0,\alpha}(\partial\Omega)_0)^2 \times \mathbb{R}$ *that solve the integral equations*

$$
\begin{aligned}
& m^o[\epsilon,\theta,c](\xi) + \int_0^1 \nabla(P_{q,n}[Q,f_\epsilon])(p+\tau\epsilon\xi)\cdot\xi\,d\tau \\
&\quad + \frac{1}{m_n(\Omega)}\int_{\partial\Omega} g_\epsilon\,d\sigma\left\{P_n[\Omega,1](\xi) + \epsilon^{n-2}\int_\Omega R_{q,n}(\epsilon(\xi-\eta))\,d\eta\right\} \\
&\quad = K'(K^{(-1)}(G[\epsilon,\delta_{2,n}\epsilon\log\epsilon])) \\
&\quad \times\left[m^i[\epsilon,\psi](\xi) + m^i_1[\epsilon](\xi) + \frac{m_n(\Omega)}{2\pi m_n(Q)}(\delta_{2,n}\epsilon\log\epsilon)\int_Q f_\epsilon\,dx\right] \\
&\quad + \epsilon\left[m^i[\epsilon,\psi](\xi) + m^i_1[\epsilon](\xi) + \frac{m_n(\Omega)}{2\pi m_n(Q)}(\delta_{2,n}\epsilon\log\epsilon)\int_Q f_\epsilon\,dx\right]^2 \\
&\quad \times\int_0^1 (1-\beta)K''\Bigg(K^{(-1)}(G[\epsilon,\delta_{2,n}\epsilon\log\epsilon]) \\
&\quad + \beta\epsilon\left[m^i[\epsilon,\psi](\xi) + m^i_1[\epsilon](\xi) + \frac{m_n(\Omega)}{2\pi m_n(Q)}(\delta_{2,n}\epsilon\log\epsilon)\int_Q f_\epsilon\,dx\right]\Bigg)\,d\beta \\
&\qquad\qquad\qquad\qquad\qquad\qquad\qquad\qquad \forall\xi\in\partial\Omega\,,
\end{aligned}
\tag{13.138}
$$

$$
\begin{aligned}
& \frac{1}{2}\theta(\xi) + W^t_\Omega[\theta](\xi) + \epsilon^{n-1}\int_{\partial\Omega}\nu_\Omega(\xi)\cdot\nabla R_{q,n}(\epsilon(\xi-\eta))\theta(\eta)\,d\sigma_\eta \\
&\quad - \frac{1}{m_n(\Omega)}\int_{\partial\Omega} g_\epsilon\,d\sigma\bigg\{\int_{\partial\Omega} S_n(\xi-\eta)\nu_\Omega(\xi)\cdot\nu_\Omega(\eta)\,d\sigma_\eta \\
&\quad + \epsilon^{n-2}\int_{\partial\Omega} R_{q,n}(\epsilon(\xi-\eta))\nu_\Omega(\xi)\cdot\nu_\Omega(\eta)\,d\sigma_\eta\bigg\} \\
&\quad = -\frac{1}{2}\psi(\xi) + W^t_\Omega[\psi](\xi) + \epsilon^{n-1}\int_{\partial\Omega}\nu_\Omega(\xi)\cdot\nabla R_{q,n}(\epsilon(\xi-\eta))\psi(\eta)\,d\sigma_\eta \\
&\quad + \frac{\epsilon^n}{m_n(Q)}\int_{\partial\Omega} g_\epsilon\,d\sigma\int_{\partial\Omega} S_n(\xi-\eta)\nu_\Omega(\xi)\cdot\nu_\Omega(\eta)\,d\sigma_\eta + g_\epsilon(\xi)\quad \forall\xi\in\partial\Omega\,,
\end{aligned}
\tag{13.139}
$$

to the set of pairs (u^i,u^o) *of* $C^{1,\alpha}_q(\overline{\mathbb{S}[\Omega_{p,\epsilon}]})\times C^{1,\alpha}_q(\overline{\mathbb{S}[\Omega_{p,\epsilon}]^-})$ *which solve problem (13.135).*

Proof. Assume that the pair (u^i,u^o) of $C^{1,\alpha}_q(\overline{\mathbb{S}[\Omega_{p,\epsilon}]})\times C^{1,\alpha}_q(\overline{\mathbb{S}[\Omega_{p,\epsilon}]^-})$ solves problem (13.135). Then by Lemma 13.7 there exist unique pairs (ψ,c_i), (θ,c_o) in $C^{0,\alpha}(\partial\Omega)_0\times\mathbb{R}$ such that

$$
\begin{aligned}
u^i(x) &\equiv v^+_{q,\Omega_{p,\epsilon}}[\psi(\frac{\cdot-p}{\epsilon})](x) + c_i \qquad \forall x\in\overline{\mathbb{S}[\Omega_{p,\epsilon}]}\,,\\
u^o(x) &\equiv v^-_{q,\Omega_{p,\epsilon}}[\theta(\frac{\cdot-p}{\epsilon})](x) + c_o \qquad \forall x\in\overline{\mathbb{S}[\Omega_{p,\epsilon}]^-}\,.
\end{aligned}
$$

Then there exists a unique $c\in\mathbb{R}$ such that equation (13.137) holds, i.e., such that

$$
v^-_{q,\Omega_{p,\epsilon}}[\theta(\frac{\cdot-p}{\epsilon})] + c_o = v^-_{q,\Omega_{p,\epsilon}}[\theta(\frac{\cdot-p}{\epsilon})] + \epsilon c - P_{q,n}[Q,f_\epsilon](p)
$$

$$+G[\epsilon,\delta_{2,n}\epsilon\log\epsilon]+\int_Q f_\epsilon\,dy\frac{1}{2\pi}\delta_{2,n}\log\epsilon\,,$$

and we have

$$c=\epsilon^{-1}\Big(c_o+P_{q,n}[Q,f_\epsilon](p)-G[\epsilon,\delta_{2,n}\epsilon\log\epsilon]-\int_Q f_\epsilon\,dy\frac{1}{2\pi}\delta_{2,n}\log\epsilon\Big)\,.$$

Since u^i satisfies the last equation of (13.135), we must necessarily have

$$\begin{aligned}&\fint_{\partial\Omega_{p,\epsilon}} v^+_{q,\Omega_{p,\epsilon}}[\psi(\frac{\cdot-p}{\epsilon})]\,d\sigma+c_i=\fint_{\partial\Omega_{p,\epsilon}} u^i\,d\sigma\\ &=k(\epsilon)-\fint_{\partial\Omega}P_{q,n}[Q,f_\epsilon](p+\epsilon\xi)d\sigma_\xi-\fint_Q f_\epsilon\,dy\epsilon^n\fint_{\partial\Omega}\int_\Omega S_n(\epsilon(\xi-\eta))\,d\eta d\sigma_\xi\,,\end{aligned}$$

and accordingly

$$\begin{aligned}c_i=&-\fint_{\partial\Omega_{p,\epsilon}} v^+_{q,\Omega_{p,\epsilon}}[\psi(\frac{\cdot-p}{\epsilon})]\,d\sigma+k(\epsilon)\\ &-\fint_{\partial\Omega}P_{q,n}[Q,f_\epsilon](p+\epsilon\xi)\,d\sigma_\xi-\fint_Q f_\epsilon\,dy\epsilon^n\fint_{\partial\Omega}\int_\Omega S_n(\epsilon(\xi-\eta))\,d\eta d\sigma_\xi\,.\end{aligned}$$

Since $\int_Q f_\epsilon\,dx=-\epsilon^{n-1}\int_{\partial\Omega}g_\epsilon\,d\sigma$, equality (13.136) holds.

Next we verify that (ψ,θ,c) solves equations (13.138) and (13.139). To do so, we note that condition $\int_{\partial\Omega}\psi\,d\sigma=0$ and equality (13.6) for $S_{q,n}(\epsilon\cdot)$ imply that

$$\begin{aligned}u^i(p+\epsilon\xi)=&\,\epsilon\Big[v_\Omega[\psi](\xi)+\epsilon^{n-2}\int_{\partial\Omega}R_{q,n}(\epsilon(\xi-\eta))\psi(\eta)\,d\sigma_\eta\\ &-\fint_{\partial\Omega}\Big(v_\Omega[\psi](\tau)+\epsilon^{n-2}\int_{\partial\Omega}R_{q,n}(\epsilon(\tau-\eta))\psi(\eta)\,d\sigma_\eta\Big)\,d\sigma_\tau\Big]\\ &+k(\epsilon)-\fint_{\partial\Omega}P_{q,n}[Q,f_\epsilon](p+\epsilon\tau)\,d\sigma_\tau\\ &-\fint_Q f_\epsilon\,dy\fint_{\partial\Omega}\Big(\epsilon^2P_n[\Omega,1]+\frac{m_n(\Omega)}{2\pi}\epsilon(\delta_{2,n}\epsilon\log\epsilon)\Big)\,d\sigma\\ =&\,\epsilon m^i[\epsilon,\psi](\xi)+K^{(-1)}(G[\epsilon,\delta_{2,n}\epsilon\log\epsilon])-P_{q,n}[Q,f_\epsilon](p)\,,\end{aligned}\tag{13.140}$$

for all $\xi\in\partial\Omega$. Similarly, condition $\int_{\partial\Omega}\theta\,d\sigma=0$ implies that

$$
\begin{aligned}
u^o(p+\epsilon\xi) = & \epsilon\left[v_\Omega[\theta](\xi) + \epsilon^{n-2}\int_{\partial\Omega} R_{q,n}(\epsilon(\xi-\eta))\theta(\eta)\,d\sigma_\eta + c\right] \\
& - P_{q,n}[Q, f_\epsilon](p) + G[\epsilon, \delta_{2,n}\epsilon\log\epsilon] + \int_Q f_\epsilon\,dy\frac{1}{2\pi}\delta_{2,n}\log\epsilon \\
= & \epsilon m^o[\epsilon,\theta,c](\xi) - P_{q,n}[Q, f_\epsilon](p) + G[\epsilon, \delta_{2,n}\epsilon\log\epsilon] + \int_Q f_\epsilon\,dy\frac{1}{2\pi}\delta_{2,n}\log\epsilon,
\end{aligned}
\tag{13.141}
$$

for all $\xi \in \partial\Omega$. Next we want to rewrite the fifth equation in (13.135) in a different way. To do so, we note that

$$
\begin{aligned}
& P_{q,n}[Q, f_\epsilon](p+\epsilon\xi) - \frac{1}{m_n(\Omega_{p,\epsilon})}P_{q,n}[\Omega_{p,\epsilon}, 1](p+\epsilon\xi)\int_Q f_\epsilon(y)\,dy \\
& \quad = \epsilon\int_0^1 \xi\cdot\nabla P_{q,n}[Q, f_\epsilon](p+\xi\tau\epsilon)\,d\tau + P_{q,n}[Q, f_\epsilon](p) \\
& \quad\quad - \frac{1}{m_n(\Omega)}\int_\Omega S_{q,n}(\epsilon(\xi-\eta))\,d\eta\int_Q f_\epsilon(y)\,dy \\
& \quad = \epsilon\int_0^1 \xi\cdot\nabla P_{q,n}[Q, f_\epsilon](p+\xi\tau\epsilon)\,d\tau + P_{q,n}[Q, f_\epsilon](p) \\
& \quad\quad - \frac{1}{m_n(\Omega)}\int_Q f_\epsilon(y)\,dy\Big\{\epsilon^{2-n}P_n[\Omega, 1](\xi) \\
& \quad\quad + \frac{m_n(\Omega)}{2\pi}\delta_{2,n}\log\epsilon + \int_\Omega R_{q,n}(\epsilon(\xi-\eta))\,d\eta\Big\},
\end{aligned}
$$

for all $\xi \in \partial\Omega$, and that

$$
\begin{aligned}
& P_{q,n}[Q, f_\epsilon](p+\epsilon\xi) + P_n[\Omega_{p,\epsilon}, 1](p+\epsilon\xi)\fint_Q f_\epsilon(y)\,dy \\
& \quad = \epsilon\int_0^1 \xi\cdot\nabla P_{q,n}[Q, f_\epsilon](p+\xi\tau\epsilon)d\tau + P_{q,n}[Q, f_\epsilon](p) \\
& \quad\quad + \int_\Omega S_n(\epsilon(\xi-\eta))d\eta\epsilon^n\fint_Q f_\epsilon dy,
\end{aligned}
$$

for all $\xi \in \partial\Omega$. Then we rewrite the fifth equation of problem (13.135) in the following form

$$
\begin{aligned}
& \epsilon m^o[\epsilon,\theta,c](\xi) - P_{q,n}[Q, f_\epsilon](p) + G[\epsilon, \delta_{2,n}\epsilon\log\epsilon] \\
& \quad + \int_Q f_\epsilon\,dy\frac{1}{2\pi}\delta_{2,n}\log\epsilon + \epsilon\int_0^1 \xi\cdot\nabla P_{q,n}[Q, f_\epsilon](p+\xi\tau\epsilon)\,d\tau + P_{q,n}[Q, f_\epsilon](p) \\
& \quad - \frac{1}{m_n(\Omega)}\int_Q f_\epsilon\,dy\Big\{\epsilon^{2-n}P_n[\Omega, 1](\xi) + \frac{m_n(\Omega)}{2\pi}\delta_{2,n}\log\epsilon + \int_\Omega R_{q,n}(\epsilon(\xi-\eta))d\eta\Big\} \\
& = K_o\circ K_i^{(-1)}\Big(\epsilon m^i[\epsilon,\psi](\xi) + K_i\circ K_o^{(-1)}(G[\epsilon, \delta_{2,n}\epsilon\log\epsilon])
\end{aligned}
$$

$$
\begin{aligned}
&- P_{q,n}[Q, f_\epsilon](p) + \epsilon \int_0^1 \xi \cdot \nabla P_{q,n}[Q, f_\epsilon](p + \xi\tau\epsilon)\, d\tau + P_{q,n}[Q, f_\epsilon](p) \\
&+ \int_\Omega S_n(\xi - \eta)\, d\eta \epsilon^2 \fint_Q f_\epsilon\, dy + \frac{m_n(\Omega)}{2\pi} \epsilon(\delta_{2,n}\epsilon \log \epsilon) \fint_Q f_\epsilon\, dy \Big) , \qquad (13.142)
\end{aligned}
$$

for all $\xi \in \partial\Omega$. We note that the right-hand side equals

$$
\begin{aligned}
K\Big(&\epsilon m^i[\epsilon, \psi](\xi) + K^{(-1)}(G[\epsilon, \delta_{2,n}\epsilon \log \epsilon]) + \epsilon m_1^i[\epsilon](\xi) \\
&+ \frac{m_n(\Omega)}{2\pi} \epsilon(\delta_{2,n}\epsilon \log \epsilon) \fint_Q f_\epsilon\, dy \Big) \\
&= K\Big(K^{(-1)}(G[\epsilon, \delta_{2,n}\epsilon \log \epsilon]) \\
&+ \epsilon \Big(m^i[\epsilon, \psi](\xi) + m_1^i[\epsilon](\xi) + \frac{m_n(\Omega)}{2\pi} (\delta_{2,n}\epsilon \log \epsilon) \fint_Q f_\epsilon\, dy \Big)\Big) , \qquad (13.143)
\end{aligned}
$$

for all $\xi \in \partial\Omega$. Next we show that equation (13.142) can be rewritten as equation (13.138). By the Taylor formula, we have

$$
\begin{aligned}
K(K^{(-1)}&(G[\epsilon, \delta_{2,n}\epsilon \log \epsilon]) + x) = G[\epsilon, \delta_{2,n}\epsilon \log \epsilon] \\
&+ K'(K^{(-1)}(G[\epsilon, \delta_{2,n}\epsilon \log \epsilon]))x \\
&+ x^2 \int_0^1 (1 - \beta) K''(K^{(-1)}(G[\epsilon, \delta_{2,n}\epsilon \log \epsilon]) + \beta x)\, d\beta ,
\end{aligned}
$$

for all $x \in \mathbb{R}$. Then by assumption (13.133) and by dividing both hand sides of (13.142)–(13.143) by ϵ, we obtain

$$
\begin{aligned}
&m^o[\epsilon, \theta, c](\xi) + \int_0^1 \xi \cdot \nabla P_{q,n}[Q, f_\epsilon](p + \xi\tau\epsilon)\, d\tau \\
&+ \frac{1}{m_n(\Omega)} \int_{\partial\Omega} g_\epsilon\, d\sigma \epsilon^{n-2} \Big\{ \epsilon^{2-n} P_n[\Omega, 1](\xi) + \int_\Omega R_{q,n}(\epsilon(\xi - \eta))\, d\eta \Big\} \\
&= K'(K^{(-1)}(G[\epsilon, \delta_{2,n}\epsilon \log \epsilon])) \\
&\times \Big[m^i[\epsilon, \psi](\xi) + m_1^i[\epsilon](\xi) + \frac{m_n(\Omega)}{2\pi m_n(Q)} (\delta_{2,n}\epsilon \log \epsilon) \int_Q f_\epsilon\, dx \Big] \\
&+ \epsilon \Big[m^i[\epsilon, \psi](\xi) + m_1^i[\epsilon](\xi) + \frac{m_n(\Omega)}{2\pi m_n(Q)} (\delta_{2,n}\epsilon \log \epsilon) \int_Q f_\epsilon\, dx \Big]^2 \\
&\times \int_0^1 (1 - \beta) K'' \Big(K^{(-1)}(G[\epsilon, \delta_{2,n}\epsilon \log \epsilon]) \\
&+ \beta\epsilon \Big[m^i[\epsilon, \psi](\xi) + m_1^i[\epsilon](\xi) + \frac{m_n(\Omega)}{2\pi m_n(Q)} (\delta_{2,n}\epsilon \log \epsilon) \int_Q f_\epsilon\, dx \Big] \Big)\, d\beta
\end{aligned}
$$

for all $\xi \in \partial\Omega$. Hence, equation (13.138) follows. Next we prove that (θ, ψ, c) satisfies equation (13.139). By setting $x = p + \epsilon\xi$ in the sixth equation of problem (13.135), by the equalities (13.133), and by the equalities

$$\nu_{\Omega_{p,\epsilon}}(x) \cdot \nabla P_n[\Omega_{p,\epsilon}, 1](x) = -\int_{\partial\Omega_{p,\epsilon}} S_n(x-y)\nu_{\Omega_{p,\epsilon}}(x) \cdot \nu_{\Omega_{p,\epsilon}}(y)\, d\sigma_y\,,$$

$$\nu_{\Omega_{p,\epsilon}}(x) \cdot \nabla P_{q,n}[\Omega_{p,\epsilon}, 1](x) = -\int_{\partial\Omega_{p,\epsilon}} S_{q,n}(x-y)\nu_{\Omega_{p,\epsilon}}(x) \cdot \nu_{\Omega_{p,\epsilon}}(y)\, d\sigma_y\,,$$

for all $x \in \partial\Omega_{p,\epsilon}$ (see Lemma 7.7 and (13.16)), we obtain

$$\begin{aligned}
&\frac{1}{2}\theta(\xi) + W^t_\Omega[\theta](\xi) + \epsilon^{n-1}\int_{\partial\Omega} \nu_\Omega(\xi)\cdot\nabla R_{q,n}(\epsilon(\xi-\eta))\theta(\eta)\, d\sigma_\eta \\
&\quad - \frac{\epsilon^{-1}}{m_n(\Omega)}\int_{\partial\Omega} g_\epsilon\, d\sigma \int_{\partial\Omega} S_{q,n}(\epsilon(\xi-\eta))\nu_\Omega(\xi)\cdot\nu_\Omega(\eta)\, d\sigma_\eta \epsilon^{n-1} \\
&= -\frac{1}{2}\psi(\xi) + W^t_\Omega[\psi](\xi) + \epsilon^{n-1}\int_{\partial\Omega} \nu_\Omega(\xi)\cdot\nabla R_{q,n}(\epsilon(\xi-\eta))\psi(\eta)\, d\sigma_\eta \\
&\quad + \frac{\epsilon^{n-1}}{m_n(Q)}\int_{\partial\Omega} g_\epsilon\, d\sigma \int_{\partial\Omega} S_n(\epsilon(\xi-\eta))\nu_\Omega(\xi)\cdot\nu_\Omega(\eta)\, d\sigma_\eta \epsilon^{n-1} + g_\epsilon(\xi) \quad \forall \xi \in \partial\Omega\,.
\end{aligned} \tag{13.144}$$

By formula (13.6), we have

$$\begin{aligned}
&\int_{\partial\Omega} S_{q,n}(\epsilon(\xi-\eta))\nu_\Omega(\xi)\cdot\nu_\Omega(\eta)\, d\sigma_\eta \epsilon^{n-1} = \epsilon\int_{\partial\Omega} S_n(\xi-\eta)\nu_\Omega(\xi)\cdot\nu_\Omega(\eta)\, d\sigma_\eta \\
&+ \frac{1}{2\pi}(\delta_{2,n}\epsilon\log\epsilon)\int_{\partial\Omega} \nu_\Omega(\xi)\cdot\nu_\Omega(\eta)\, d\sigma_\eta \\
&+ \epsilon^{n-1}\int_{\partial\Omega} R_{q,n}(\epsilon(\xi-\eta))\nu_\Omega(\xi)\cdot\nu_\Omega(\eta)\, d\sigma_\eta, \\
&\int_{\partial\Omega} S_n(\epsilon(\xi-\eta))\nu_\Omega(\xi)\cdot\nu_\Omega(\eta)\, d\sigma_\eta \epsilon^{n-1} = \epsilon\int_{\partial\Omega} S_n(\xi-\eta)\nu_\Omega(\xi)\cdot\nu_\Omega(\eta)\, d\sigma_\eta \\
&\qquad\qquad + \frac{1}{2\pi}(\delta_{2,n}\epsilon\log\epsilon)\int_{\partial\Omega} \nu_\Omega(\xi)\cdot\nu_\Omega(\eta)\, d\sigma_\eta\,,
\end{aligned}$$

for all $\xi \in \partial\Omega$. By the Divergence Theorem 4.1, we have

$$\int_{\partial\Omega} \nu_\Omega(\xi)\cdot\nu_\Omega(\eta)\, d\sigma_\eta = 0 \qquad \forall \xi \in \partial\Omega\,.$$

Hence, equation (13.144) implies the validity of equation (13.139). Conversely, by reading backward the above argument, one can easily show that if (ψ, θ, c) belongs to $(C^{0,\alpha}(\partial\Omega)_0)^2 \times \mathbb{R}$ and solves equations (13.138)–(13.139), then the pair of functions defined by (13.136)–(13.137) satisfies problem (13.135). □

Hence, we are reduced to analyze system (13.138)–(13.139). As a first step, we note that if $(\psi, \theta, c) \in (C^{0,\alpha}(\partial\Omega)_0)^2 \times \mathbb{R}$ and if we let ϵ tend to 0, we obtain a system which we address to as the 'limiting system', and which has the following form

$$\begin{aligned} & v_\Omega[\theta](\xi) + c + \nabla(P_{q,n}[Q, f_0])(p) \cdot \xi \\ & \quad + \frac{1}{m_n(\Omega)} \int_{\partial\Omega} g_0 \, d\sigma \{P_n[\Omega, 1](\xi) + \delta_{2,n} m_n(\Omega) R_{q,n}(0)\} \\ & = K'(k(0)) \left[v_\Omega[\psi](\xi) - \fint_{\partial\Omega} v_\Omega[\psi] \, d\sigma + \xi \cdot \nabla(P_{q,n}[Q, f_0])(p) \right] \quad \forall \xi \in \partial\Omega , \end{aligned} \tag{13.145}$$

$$\begin{aligned} & \frac{1}{2}\theta(\xi) + W^t_\Omega[\theta](\xi) \\ & \quad - \frac{1}{m_n(\Omega)} \int_{\partial\Omega} g_0 \, d\sigma \int_{\partial\Omega} S_n(\xi - \eta) \nu_\Omega(\xi) \cdot \nu_\Omega(\eta) \, d\sigma_\eta \\ & = -\frac{1}{2}\psi(\xi) + W^t_\Omega[\psi](\xi) + g_0(\xi) \qquad \forall \xi \in \partial\Omega . \end{aligned} \tag{13.146}$$

We first introduce the following variant of a well known results of classical Potential Theory, which follows immediately by Lemma 6.11 and Theorem 6.51.

Lemma 13.48. *Let* $a \in]-1, 1[$. *Then the following statements hold.*

(i) Let $\theta \in C^{0,\alpha}(\partial\Omega)$. *Then*

$$\int_{\partial\Omega} \left(\frac{1}{2}\theta(\xi) - a W^t_\Omega[\theta](\xi) \right) d\sigma_\xi = \frac{1}{2}(1 - a) \int_{\partial\Omega} \theta \, d\sigma .$$

(ii) The map

$$\frac{1}{2} I - a W^t_\Omega[\cdot]$$

is a linear homeomorphism from $C^{0,\alpha}(\partial\Omega)$ *onto* $C^{0,\alpha}(\partial\Omega)$.

Next we introduce the following variant of Theorem 11.15.

Proposition 13.49. *Let* $a_1, a_2 \in]0, +\infty[$. *If* $(f, g) \in C^{1,\alpha}(\partial\Omega) \times C^{0,\alpha}(\partial\Omega)_0$, *then the system*

$$\begin{cases} a_1 \left(v_\Omega[\psi](\xi) - \fint_{\partial\Omega} v_\Omega[\psi] \, d\sigma \right) \\ \qquad\qquad - \left(v_\Omega[\theta](\xi) + c \right) = f(\xi) \quad \forall \xi \in \partial\Omega , \\ \frac{1}{2}\theta(\xi) + W^t_\Omega[\theta](\xi) + \frac{1}{2} a_2 \psi(\xi) - a_2 W^t_\Omega[\psi](\xi) = g(\xi) \ \forall \xi \in \partial\Omega , \end{cases} \tag{13.147}$$

has one and only one solution $(\psi, \theta, c) \in (C^{0,\alpha}(\partial\Omega)_0)^2 \times \mathbb{R}$.

Proof. To prove the statement, we shall assume that system (13.147) has a solution and prove that such a solution must necessarily be delivered by a certain formula.

Thus uniqueness will follow. Then we shall exploit such a formula to show existence. By Theorem 6.47, there exists a unique pair (η,τ) in $C^{0,\alpha}(\partial\Omega)_0 \times \mathbb{R}$, such that

$$f = V_\Omega[\eta] + \tau \qquad \text{on } \partial\Omega .$$

As a consequence, the first equation of (13.147) implies that

$$a_1\psi(\xi) - \theta(\xi) = \eta(\xi) \qquad \forall \xi \in \partial\Omega . \tag{13.148}$$

Moreover, by integrating both hand sides of the first equation of (13.147), we obtain

$$c = -\frac{1}{\int_{\partial\Omega} d\sigma}\Big(\int_{\partial\Omega} v_\Omega[\theta]\, d\sigma + \int_{\partial\Omega} f\, d\sigma\Big) . \tag{13.149}$$

By (13.148), we can rewrite the second equation of (13.147) in the following form

$$\psi(\xi) - 2\frac{a_2 - a_1}{a_2 + a_1} W^t_\Omega[\psi](\xi) = \frac{2}{a_2 + a_1}\Big(g(\xi) + \frac{1}{2}\eta(\xi) + W^t_\Omega[\eta](\xi)\Big) \;\; \forall \xi \in \partial\Omega . \tag{13.150}$$

Since $\int_{\partial\Omega} g\, d\sigma = 0$ and $\int_{\partial\Omega} \eta\, d\sigma = 0$, by integrating (13.150) on $\partial\Omega$ and by applying the Fubini Theorem, we obtain

$$\int_{\partial\Omega}\Big(g(\xi) + \frac{1}{2}\eta(\xi) + W^t_\Omega[\eta](\xi)\Big)\, d\sigma_\xi = 0 .$$

By (13.150) and by Theorem 6.51 (ii) with $\lambda = -2\frac{a_2 - a_1}{a_2 + a_1}$, we have

$$\psi = \frac{2}{a_2 + a_1}\Big(I - 2\frac{a_2 - a_1}{a_2 + a_1} W^t_\Omega[\cdot]\Big)^{(-1)}\Big(g + \frac{1}{2}\eta + W^t_\Omega[\eta]\Big) . \tag{13.151}$$

Then by Lemma 6.11, we must have $\int_{\partial\Omega} \psi\, d\sigma = 0$. Accordingly, ψ in $C^{0,\alpha}(\partial\Omega)_0$ is uniquely determined. Hence, equalities (13.148), (13.149) uniquely determine θ in $C^{0,\alpha}(\partial\Omega)_0$ and c in $\mathbb{R}$. Conversely, by reading backward the proof above, we can easily check that the triple (ψ, θ, c) delivered by formulas (13.148), (13.149), (13.151) belongs to $(C^{0,\alpha}(\partial\Omega)_0)^2 \times \mathbb{R}$ and solves system (13.147). □

Then we have the following theorem, which shows the unique solvability of the system of equations (13.145), (13.146) and its link with a boundary value problem which we shall address to as the 'limiting boundary value problem'.

Theorem 13.50. *Let $K \equiv K_o \circ K_i^{(-1)}$. Then the following statements hold.*

(i) The limiting system (13.145)–(13.146) has one and only one solution $(\tilde\psi, \tilde\theta, \tilde c)$ in the space $(C^{0,\alpha}(\partial\Omega)_0)^2 \times \mathbb{R}$. Moreover,

$$\tilde{c} = -\fint_{\partial\Omega} v_\Omega[\tilde{\theta}]\,d\sigma + \fint_{\partial\Omega} \xi \cdot \nabla(P_{q,n}[Q, f_o])(p)\,d\sigma_\xi (K'(k(0)) - 1) \\ - \frac{1}{m_n(\Omega)} \int_{\partial\Omega} g_o\,d\sigma \Big\{ \fint_{\partial\Omega} P_n[\Omega, 1]\,d\sigma + \delta_{2,n} m_n(\Omega) R_{q,n}(0) \Big\}. \tag{13.152}$$

(ii) The 'limiting boundary value problem'

$$\begin{cases} \Delta u^i = 0 & \text{in } \Omega\,, \\ \Delta u^o = 0 & \text{in } \mathbb{R}^n \setminus \overline{\Omega}\,, \\ u^o(\xi) - \fint_{\partial\Omega} u^o\,d\sigma + \xi \cdot \nabla(P_{q,n}[Q, f_o])(p) & \\ \quad + \fint_{\partial\Omega} \eta \cdot \nabla(P_{q,n}[Q, f_o])(p)\,d\sigma_\eta (K'(k(0)) - 1) & \\ \quad + \frac{1}{m_n(\Omega)} \int_{\partial\Omega} g_o\,d\sigma \left\{ P_n[\Omega, 1](\xi) - \fint_{\partial\Omega} P_n[\Omega, 1]\,d\sigma \right\} & \\ \quad = K'(k(0))[u^i(\xi) + \xi \cdot \nabla(P_{q,n}[Q, f_o])(p)] & \forall \xi \in \partial\Omega\,, \\ \nu_\Omega(\xi) \cdot \nabla u^o(\xi) = \nu_\Omega(\xi) \cdot \nabla u^i(\xi) + g_o(\xi) & \\ \quad + \frac{1}{m_n(\Omega)} \int_{\partial\Omega} g_o\,d\sigma \int_{\partial\Omega} S_n(\xi - \eta) \nu_\Omega(\xi) \cdot \nu_\Omega(\eta)\,d\sigma_\eta & \forall \xi \in \partial\Omega\,, \\ \fint_{\partial\Omega} u^i\,d\sigma = 0\,, & \\ \lim_{\xi \to \infty} u^o(\xi) = 0\,, & \end{cases} \tag{13.153}$$

has one and only one solution $(\tilde{u}^i, \tilde{u}^o) \in C^{1,\alpha}(\overline{\Omega}) \times C^{1,\alpha}_{\mathrm{loc}}(\mathbb{R}^n \setminus \Omega)$*, and the following formulas hold,*

$$\tilde{u}^i(\xi) = v_\Omega[\tilde{\psi}](\xi) - \fint_{\partial\Omega} v_\Omega[\tilde{\psi}]\,d\sigma \qquad \forall \xi \in \overline{\Omega}\,, \tag{13.154}$$

$$\tilde{u}^o(\xi) = v_\Omega[\tilde{\theta}](\xi) \qquad \forall \xi \in \mathbb{R}^n \setminus \Omega\,, \tag{13.155}$$

where $\tilde{\psi}$ *and* $\tilde{\theta}$ *are as in statement* (i).

Proof. We first consider statement (i). Lemma 7.7 on the derivative of volume potentials, implies that

$$\int_{\partial\Omega} S_n(\xi - \eta) \nu_\Omega(\xi) \cdot \nu_\Omega(\eta)\,d\sigma_\eta = -\nu_\Omega(\xi) \cdot \nabla P_n[\Omega, 1](\xi) \qquad \forall \xi \in \partial\Omega\,,$$

and accordingly

$$\int_{\partial\Omega} \left\{ \int_{\partial\Omega} S_n(\xi - \eta) \nu_\Omega(\xi) \cdot \nu_\Omega(\eta)\,d\sigma_\eta \right\} d\sigma_\xi \\ = -\int_{\partial\Omega} \nu_\Omega \cdot \nabla\left(P_n[\Omega, 1]\right) d\sigma = -\int_\Omega \Delta(P_n[\Omega, 1])\,dx = -m_n(\Omega)\,.$$

Hence,

$$\int_{\partial\Omega}\left\{g_0(\xi)+\frac{1}{m_n(\Omega)}\int_{\partial\Omega}g_0\,d\sigma\int_{\partial\Omega}S_n(\xi-\eta)\nu_\Omega(\xi)\cdot\nu_\Omega(\eta)\,d\sigma_\eta\right\}d\sigma_\xi=0\,.$$

Since $K'>0$, Proposition 13.49 implies that the limiting system (13.145)–(13.146) has one and only one solution $(\tilde{\psi},\tilde{\theta},\tilde{c})\in(C^{0,\alpha}(\partial\Omega)_0)^2\times\mathbb{R}$. Then by integrating both hand sides of (13.145) on $\partial\Omega$, we obtain equality (13.152).

Next we consider statement (ii). By (i) and by classical Potential Theory, the functions $\tilde{u}^i$, $\tilde{u}^o$ delivered by the right-hand side of (13.154), (13.155) belong to $C^{1,\alpha}(\overline{\Omega})$ and to $C^{1,\alpha}_{\text{loc}}(\mathbb{R}^n\setminus\Omega)$, respectively, and solve the limiting problem (13.153) (cf., e.g., Theorems 4.23–4.25). We now consider the uniqueness for the linear problem (13.153), and we proceed by a classical argument. Let the pair $(w^i,w^o)\in\left(C^1(\overline{\Omega})\cap C^2(\Omega)\right)\times\left(C^1(\mathbb{R}^n\setminus\Omega)\cap C^2(\mathbb{R}^n\setminus\overline{\Omega})\right)$ solve the homogeneous problem

$$\begin{cases}\Delta w^i=0 & \text{in }\Omega\,,\\ \Delta w^o=0 & \text{in }\mathbb{R}^n\setminus\Omega\,,\\ w^o(\xi)-\fint_{\partial\Omega}w^o\,d\sigma=K'(k(0))w^i(\xi) & \forall\xi\in\partial\Omega\,,\\ \nu_\Omega(\xi)\cdot\nabla w^o(\xi)=\nu_\Omega(\xi)\cdot\nabla w^i(\xi) & \forall\xi\in\partial\Omega\,,\\ \fint_{\partial\Omega}w^i\,d\sigma=0\,,\\ \lim_{\xi\to\infty}w^o(\xi)=0\,.\end{cases}\tag{13.156}$$

We first note that the last condition in (13.156) and Theorem 3.22 imply that w^o is harmonic at infinity. By the first and third equation of system (13.156) and by the first Green Identity of Theorem 4.2, we have

$$\int_{\partial\Omega}\nu_\Omega\cdot\nabla w^o\,d\sigma=\int_{\partial\Omega}\nu_\Omega\cdot\nabla w^i\,d\sigma=\int_\Omega\Delta w^i\,dx=0\,.$$

Then by the first, third and fourth equations of system (13.156), we have

$$\begin{aligned}0\le\int_\Omega|\nabla w^i|^2\,dx&=\int_{\partial\Omega}w^i\nu_\Omega\cdot\nabla w^i\,d\sigma\\&=\int_{\partial\Omega}\left\{K'(k(0))^{-1}[w^o-\fint_{\partial\Omega}w^o\,d\sigma]\right\}\nu_\Omega\cdot\nabla w^o\,d\sigma\\&=K'(k(0))^{-1}\left\{\int_{\partial\Omega}w^o\nu_\Omega\cdot\nabla w^o\,d\sigma-\fint_{\partial\Omega}w^o\,d\sigma\int_{\partial\Omega}\nu_\Omega\cdot\nabla w^o\,d\sigma\right\}\\&=K'(k(0))^{-1}\int_{\partial\Omega}w^o\nu_\Omega\cdot\nabla w^o\,d\sigma\,.\end{aligned}\tag{13.157}$$

Since w^o is harmonic in $\mathbb{R}^n\setminus\overline{\Omega}$ and tends to zero at infinity, the first Green Identity in exterior domains of Corollary 4.7 imply that

$$\int_{\partial\Omega}w^o\nu_\Omega\cdot\nabla w^o\,d\sigma=-\int_{\mathbb{R}^n\setminus\overline{\Omega}}|\nabla w^o|^2\,dx\le0\,.$$

Then inequality (13.157) implies that

$$0 \leq \int_{\Omega} |\nabla w^i|^2 \, dx = -K'(k(0))^{-1} \int_{\mathbb{R}^n \setminus \overline{\Omega}} |\nabla w^o|^2 \, dx \leq 0 \, .$$

Since w^o has limiting value 0 at infinity and $\mathbb{R}^n \setminus \overline{\Omega}$ is connected, we have

$$w^o = 0 \qquad \text{in } \mathbb{R}^n \setminus \overline{\Omega} \, .$$

Since $\fint_{\partial\Omega} w^i \, d\sigma = 0$ and Ω is connected, we also have $w^i = 0$ in $\overline{\Omega}$. □

We are now ready to analyze equations (13.138)–(13.139) around the degenerate case in which $\epsilon = 0$. To do so, we introduce the following.

Theorem 13.51. *Let* $K \equiv K_o \circ K_i^{(-1)}$. *Let* $\mathcal{M} \equiv (\mathcal{M}_j)_{j=1,2}$ *be the map from* $]-\epsilon_0, \epsilon_0[\times \mathbb{R} \times (C^{0,\alpha}(\partial\Omega)_0)^2 \times \mathbb{R}$ *to* $C^{1,\alpha}(\partial\Omega) \times C^{0,\alpha}(\partial\Omega)$ *defined by*

$$\begin{aligned}
&\mathcal{M}_1[\epsilon, \epsilon_1, \psi, \theta, c](\xi) \\
&\equiv m^o[\epsilon, \theta, c](\xi) + \int_0^1 \nabla(P_{q,n}[Q, f_\epsilon])(p + \tau\epsilon\xi) \cdot \xi \, d\tau \\
&+ \frac{1}{m_n(\Omega)} \int_{\partial\Omega} g_\epsilon \, d\sigma \left\{ P_n[\Omega, 1](\xi) + \epsilon^{n-2} \int_{\Omega} R_{q,n}(\epsilon(\xi - \eta)) \, d\eta \right\} \\
&-K'(K^{(-1)}(G[\epsilon, \epsilon_1])) \\
&\times \left[m^i[\epsilon, \psi](\xi) + m_1^i[\epsilon](\xi) + \frac{m_n(\Omega)}{2\pi m_n(Q)} \epsilon_1 \int_Q f_\epsilon \, dx \right] \\
&-\epsilon \left[m^i[\epsilon, \psi](\xi) + m_1^i[\epsilon](\xi) + \frac{m_n(\Omega)}{2\pi m_n(Q)} \epsilon_1 \int_Q f_\epsilon \, dx \right]^2 \\
&\times \int_0^1 (1-\beta) K'' \Big(K^{(-1)}(G[\epsilon, \epsilon_1]) \\
&+\beta\epsilon \left[m^i[\epsilon, \psi](\xi) + m_1^i[\epsilon](\xi) + \frac{m_n(\Omega)}{2\pi m_n(Q)} \epsilon_1 \int_Q f_\epsilon \, dx \right] \Big) \, d\beta \qquad \forall \xi \in \partial\Omega \, , \\
&\mathcal{M}_2[\epsilon, \epsilon_1, \psi, \theta, c](\xi) \\
&\equiv \frac{1}{2}\theta(\xi) + W_\Omega^t[\theta](\xi) + \epsilon^{n-1} \int_{\partial\Omega} \nu_\Omega(\xi) \cdot \nabla R_{q,n}(\epsilon(\xi - \eta)) \theta(\eta) \, d\sigma_\eta \\
&- \frac{1}{m_n(\Omega)} \int_{\partial\Omega} g_\epsilon \, d\sigma \left\{ \int_{\partial\Omega} S_n(\xi - \eta) \nu_\Omega(\xi) \cdot \nu_\Omega(\eta) \, d\sigma_\eta \right. \\
&\left. + \epsilon^{n-2} \int_{\partial\Omega} R_{q,n}(\epsilon(\xi - \eta)) \nu_\Omega(\xi) \cdot \nu_\Omega(\eta) \, d\sigma_\eta \right\} \\
&+ \frac{1}{2}\psi(\xi) - W_\Omega^t[\psi](\xi) - \epsilon^{n-1} \int_{\partial\Omega} \nu_\Omega(\xi) \cdot \nabla R_{q,n}(\epsilon(\xi - \eta)) \psi(\eta) \, d\sigma_\eta \\
&- \frac{\epsilon^n}{m_n(Q)} \int_{\partial\Omega} g_\epsilon \, d\sigma \int_{\partial\Omega} S_n(\xi - \eta) \nu_\Omega(\xi) \cdot \nu_\Omega(\eta) \, d\sigma_\eta - g_\epsilon(\xi) \qquad \forall \xi \in \partial\Omega \, ,
\end{aligned}$$

for all $(\epsilon, \epsilon_1, \psi, \theta, c) \in]-\epsilon_0, \epsilon_0[\times\mathbb{R}\times(C^{0,\alpha}(\partial\Omega)_0)^2\times\mathbb{R}$. *Then the following statements hold.*

(i) Equation $\mathcal{M}[0,0,\psi,\theta,c] = 0$ *is equivalent to the limiting system (13.145), (13.146) and has one and only one solution* $(\tilde{\psi},\tilde{\theta},\tilde{c})$ *(cf. Theorem 13.50 (i)).*

(ii) If $\epsilon \in]0,\epsilon_0[$*, then the equation* $\mathcal{M}[\epsilon, \delta_{2,n}\epsilon\log\epsilon, \psi,\theta,c] = 0$ *is equivalent to the system (13.138)–(13.139) in the unknown* (ψ,θ,c)*.*

(iii) The map $\mathcal{M}$ *is real analytic. Moreover, the image of* $\mathcal{M}$ *is contained in the space* $C^{1,\alpha}(\partial\Omega)\times C^{0,\alpha}(\partial\Omega)_0$*.*

(iv) The differential $\partial_{(\psi,\theta,c)}\mathcal{M}[0,0,\tilde{\psi},\tilde{\theta},\tilde{c}]$ *of* $\mathcal{M}$ *at* $(0,0,\tilde{\psi},\tilde{\theta},\tilde{c})$ *with respect to the variable* (ψ,θ,c) *is a homeomorphism from* $(C^{0,\alpha}(\partial\Omega)_0)^2\times\mathbb{R}$ *onto* $C^{1,\alpha}(\partial\Omega)\times C^{0,\alpha}(\partial\Omega)_0$*.*

(v) There exist $(\epsilon', \epsilon^\sharp) \in]0,\epsilon_0[\times]0,+\infty[$*, an open neighborhood* U *of* $(\tilde{\psi},\tilde{\theta},\tilde{c})$ *in the space* $(C^{0,\alpha}(\partial\Omega)_0)^2\times\mathbb{R}$*, and a real analytic map* $(\Psi[\cdot,\cdot],\Theta[\cdot,\cdot],C[\cdot,\cdot])$ *from* $]-\epsilon',\epsilon'[\times]-\epsilon^\sharp,\epsilon^\sharp[$ *to* U *such that* $\delta_{2,n}\epsilon\log\epsilon \in]-\epsilon^\sharp,\epsilon^\sharp[$ *for all* $\epsilon\in]0,\epsilon'[$ *and such that the set of zeros of the map* $\mathcal{M}$ *in* $]-\epsilon',\epsilon'[\times]-\epsilon^\sharp,\epsilon^\sharp[\times U$ *coincides with the graph of* $(\Psi[\cdot,\cdot],\Theta[\cdot,\cdot],C[\cdot,\cdot])$*. In particular,*

$$(\Psi[0,0],\Theta[0,0],C[0,0]) = (\tilde{\psi},\tilde{\theta},\tilde{c})\,.$$

Proof. Statements (i) and (ii) are an immediate consequence of the definition of $\mathcal{M}$ and of Theorems 13.47 and 13.50.

Next we turn to statement (iii) and we prove that $\mathcal{M}$ is real analytic. We first observe that the maps from $]-\epsilon_0,\epsilon_0[\times L^1(\partial\Omega)$ to $C^{1,\alpha}(\partial\Omega)$ that take (ϵ,μ) to the functions

$$\int_{\partial\Omega} R_{q,n}(\epsilon(\xi-\eta))\mu(\eta)\,d\sigma_\eta \qquad \forall\xi\in\partial\Omega\,,$$
$$\int_{\partial\Omega} \partial_{x_j}R_{q,n}(\epsilon(\xi-\eta))\mu(\eta)\,d\sigma_\eta \;\; \forall\xi\in\partial\Omega\,,\; j\in\{1,\dots,n\}\,,$$

respectively, and the map from $]-\epsilon_0,\epsilon_0[$ to $C^{1,\alpha}(\partial\Omega)$ that takes ϵ to the function

$$\int_{\Omega} R_{q,n}(\epsilon(\xi-\eta))\,ds \qquad \forall\xi\in\partial\Omega\,,$$

are real analytic (cf. Lemma 13.4 (i)). Moreover the maps from $\mathbb{R}\times C^0_{q,\omega,\rho}(\mathbb{R}^n)$ to $C^{1,\alpha}(\partial\Omega)$ that take (ϵ,f) to the functions

$$P_{q,n}[Q,f](p+\epsilon\xi) \qquad \forall\xi\in\partial\Omega\,,$$
$$\int_0^1 \partial_{x_j}P_{q,n}[Q,f](p+\epsilon\xi\tau)\,d\tau \;\; \forall j\in\{1,\dots,n\},\; \xi\in\partial\Omega\,,$$

are real analytic, respectively (cf. Lemma 13.4 (ii)).

Then the analyticity of $\mathcal{M}$ follows by the linearity and continuity of $v_\Omega[\cdot]_{|\partial\Omega}$ from $C^{0,\alpha}(\partial\Omega)$ to $C^{1,\alpha}(\partial\Omega)$, by the linearity and continuity of the operator $W^t_\Omega[\cdot]$ from $C^{0,\alpha}(\partial\Omega)$ to itself, by assumptions (13.129)–(13.131), by the analyticity of

K, and by the analyticity of the map from $\mathbb{R} \times C^{1,\alpha}(\partial\Omega)$ to $C^{1,\alpha}(\partial\Omega)$ that takes (η_1, φ) to the function defined by

$$\int_0^1 (1-\beta)K''(K^{(-1)}(\eta_1) + \beta\varphi(\xi))\, d\beta \qquad \forall \xi \in \partial\Omega\,.$$

To prove the analyticity of this last map, we observe that the map from $\mathbb{R}\times\mathbb{R}\times[0,1]$ to $\mathbb{R}$ that takes (η_1, η_2, β) to $(1-\beta)K''(K^{(-1)}(\eta_1)+\beta\eta_2)$ is real analytic, and then we exploit Theorem A.28 (iii) on the analyticity of nonlinear integral operators with analytic kernel.

Next we turn to prove that

$$\int_{\partial\Omega} \mathcal{M}_2[\epsilon, \epsilon_1, \psi, \theta, c]\, d\sigma = 0 \quad \forall(\epsilon, \epsilon_1, \psi, \theta, c) \in]-\epsilon_0, \epsilon_0[\times\mathbb{R}\times(C^{0,\alpha}(\partial\Omega)_0)^2\times\mathbb{R}\,. \tag{13.158}$$

We first prove that (13.158) holds with the supplementary assumption that $\epsilon > 0$. Let $(\epsilon, \epsilon_1, \psi, \theta, c)$ belong to $]0, \epsilon_0[\times\mathbb{R} \times (C^{0,\alpha}(\partial\Omega)_0)^2 \times \mathbb{R}$. By the definition of $\mathcal{M}_2$ and by adding and subtracting $\epsilon^{1-n}\int_{\partial\Omega_{p,\epsilon}} \nu_{\Omega_{p,\epsilon}} \cdot \nabla P_{q,n}[Q, f_\epsilon]\, d\sigma$, we obtain

$$\begin{aligned}
&\int_{\partial\Omega} \mathcal{M}_2[\epsilon, \epsilon_1, \psi, \theta, c]\, d\sigma \\
&= \epsilon^{1-n}\int_{\partial\Omega_{p,\epsilon}} \Bigg\{\Bigg[\nu_{\Omega_{p,\epsilon}}(x)\cdot\nabla v^-_{q,\Omega_{p,\epsilon}}[\theta(\epsilon^{-1}(\cdot - p))](x) \\
&\quad -\frac{1}{m_n(\Omega_{p,\epsilon})}\int_Q f_\epsilon\, dy \nu_{\Omega_{p,\epsilon}}(x)\cdot\nabla P_{q,n}[\Omega_{p,\epsilon}, 1](x)\Bigg] \\
&\quad -\Bigg[\nu_{\Omega_{p,\epsilon}}(x)\cdot\nabla v^+_{q,\Omega_{p,\epsilon}}[\psi(\epsilon^{-1}(\cdot - p))](x) \\
&\quad +\fint_Q f_\epsilon\, dy\, \nu_{\Omega_{p,\epsilon}}(x)\cdot\nabla P_n[\Omega_{p,\epsilon}, 1](x)\Bigg] - g_\epsilon(\epsilon^{-1}(x-p))\Bigg\}\, d\sigma_x \\
&= \epsilon^{1-n}\int_{\partial\Omega_{p,\epsilon}} \Bigg\{\Bigg[\nu_{\Omega_{p,\epsilon}}(x)\cdot\nabla v^-_{q,\Omega_{p,\epsilon}}[\theta(\epsilon^{-1}(\cdot - p))](x) \\
&\quad +\nu_{\Omega_{p,\epsilon}}(x)\cdot\nabla\Bigg(P_{q,n}[Q, f_\epsilon] - \frac{1}{m_n(\Omega_{p,\epsilon})}\int_Q f_\epsilon\, dy P_{q,n}[\Omega_{p,\epsilon}, 1]\Bigg)(x)\Bigg] \\
&\quad -\Bigg[\nu_{\Omega_{p,\epsilon}}(x)\cdot\nabla v^+_{q,\Omega_{p,\epsilon}}[\psi(\epsilon^{-1}(\cdot - p))](x) \\
&\quad +\nu_{\Omega_{p,\epsilon}}(x)\cdot\nabla\Bigg(P_{q,n}[Q, f_\epsilon] + \fint_Q f_\epsilon\, dy\, P_n[\Omega_{p,\epsilon}, 1]\Bigg)(x)\Bigg] \\
&\quad -g_\epsilon(\epsilon^{-1}(x-p))\Bigg\}\, d\sigma_x\,.
\end{aligned}$$

Then by setting

$$v^i \equiv v^+_{q,\Omega_{p,\epsilon}}[\psi(\epsilon^{-1}(\cdot - p))] + F^i_\epsilon[f_\epsilon]\,,$$
$$v^o \equiv v^-_{q,\Omega_{p,\epsilon}}[\theta(\epsilon^{-1}(\cdot - p))] + F^o_\epsilon[f_\epsilon]\,,$$

we conclude that

$$\begin{aligned}
&\epsilon^{n-1}\int_{\partial\Omega} \mathcal{M}_2[\epsilon,\epsilon_1,\psi,\theta,c]\,d\sigma \\
&= \int_{\partial\Omega_{p,\epsilon}} \nu_{\Omega_{p,\epsilon}}\cdot\nabla v^o\,d\sigma - \int_{\partial\Omega_{p,\epsilon}} \nu_{\Omega_{p,\epsilon}}\cdot\nabla v^i\,d\sigma - \int_{\partial\Omega_{p,\epsilon}} g_\epsilon(\epsilon^{-1}(x-p))\,d\sigma_x\,.
\end{aligned}$$

By Theorem 12.8 and by Lemma 13.5, the pair (v^i, v^o) belongs to the space $C^{1,\alpha}_q(\overline{\mathbb{S}[\Omega_{p,\epsilon}]}) \times C^{1,\alpha}_q(\overline{\mathbb{S}[\Omega_{p,\epsilon}]^-})$ and

$$\Delta v^i = f_\epsilon \quad \text{in } \mathbb{S}[\Omega_{p,\epsilon}]\,, \qquad \Delta v^o = f_\epsilon \quad \text{in } \mathbb{S}[\Omega_{p,\epsilon}]^-\,.$$

Then by applying the first Green Identity of Theorem 4.2 to v^o in $Q\setminus\overline{\Omega}_{p,\epsilon}$ and to v^i in $\Omega_{p,\epsilon}$, and by assumption (13.133), we deduce that

$$\begin{aligned}
&\epsilon^{n-1}\int_{\partial\Omega} \mathcal{M}_2[\epsilon,\epsilon_1,\psi,\theta,c]\,d\sigma \\
&= \int_{\partial\Omega_{p,\epsilon}} \nu_{\Omega_{p,\epsilon}}\cdot\nabla v^o\,d\sigma - \int_{\partial\Omega_{p,\epsilon}} \nu_{\Omega_{p,\epsilon}}\cdot\nabla v^i\,d\sigma - \int_{\partial\Omega_{p,\epsilon}} g_\epsilon(\epsilon^{-1}(x-p))\,d\sigma_x \\
&= -\Big\{\int_{\partial Q} \nu_Q\cdot\nabla v^o\,d\sigma - \int_{\partial\Omega_{p,\epsilon}} \nu_{\Omega_{p,\epsilon}}\cdot\nabla v^o\,d\sigma\Big\} \\
&\quad - \int_{\Omega_{p,\epsilon}} \Delta v^i\,dy - \int_{\partial\Omega_{p,\epsilon}} g_\epsilon(\epsilon^{-1}(x-p))\,d\sigma_x \\
&= -\int_{Q\setminus\overline{\Omega}_{p,\epsilon}} \Delta v^o\,dy - \int_{\Omega_{p,\epsilon}} \Delta v^i\,dy - \int_{\partial\Omega_{p,\epsilon}} g_\epsilon(\epsilon^{-1}(x-p))\,d\sigma_x \\
&= -\int_Q f_\epsilon\,dy - \epsilon^{n-1}\int_{\partial\Omega} g_\epsilon\,d\sigma = 0\,.
\end{aligned}$$

Next we note that the map $\int_{\partial\Omega}\mathcal{M}_2[\cdot,\epsilon_1,\psi,\theta,c]\,d\sigma$ is real analytic in $]-\epsilon_0,\epsilon_0[$ and vanishes in $]0,\epsilon_0[$. Hence, the Identity Principle of Proposition 2.16 implies that $\epsilon^{n-1}\int_{\partial\Omega}\mathcal{M}_2[\epsilon,\epsilon_1,\psi,\theta,c]\,d\sigma$ vanishes in $]-\epsilon_0,\epsilon_0[$ and we conclude that (13.158) holds. The proof of statement (iii) is complete.

We now turn to the proof of statement (iv). By standard Calculus in Banach space, the differential of $\mathcal{M}$ at $(0,0,\tilde\psi,\tilde\theta,\tilde c)$ with respect to the variable (ψ,θ,c) is delivered by the following formula,

$$\begin{aligned}
&\partial_{(\psi,\theta,c)}\mathcal{M}_1[0,0,\tilde\psi,\tilde\theta,\tilde c](\psi^\#,\theta^\#,c^\#)(\xi) \\
&= v_\Omega[\theta^\#] + c^\# - K'(k(0))\Big[v_\Omega[\psi^\#](\xi) - \fint_{\partial\Omega} v_\Omega[\psi^\#]\,d\sigma\Big] \qquad \forall \xi\in\partial\Omega\,, \\
&\partial_{(\psi,\theta,c)}\mathcal{M}_2[0,0,\tilde\psi,\tilde\theta,\tilde c](\psi^\#,\theta^\#,c^\#)(\xi)
\end{aligned}$$

$$= \frac{1}{2}\theta^{\#}(\xi) + W^t_{\Omega}[\theta^{\#}](\xi) + \frac{1}{2}\psi^{\#}(\xi) - W^t_{\Omega}[\psi^{\#}](\xi) \qquad \forall \xi \in \partial\Omega \,,$$

for all $(\psi^{\#}, \theta^{\#}, c^{\#}) \in (C^{0,\alpha}(\partial\Omega)_0)^2 \times \mathbb{R}$. By Proposition 13.49, the linear and continuous operator $\partial_{(\psi,\theta,c)}\mathcal{M}[0,0,\tilde{\psi},\tilde{\theta},\tilde{c}]$ is a bijection onto $C^{1,\alpha}(\partial\Omega) \times C^{0,\alpha}(\partial\Omega)_0$. Then, by the Open Mapping Theorem 2.2, $\partial_{(\psi,\theta,c)}\mathcal{M}[0,0,\tilde{\psi},\tilde{\theta},\tilde{c}]$ is a homeomorphism and the proof of statement (iv) is complete. Statement (v) is an immediate consequence of statements (iii) and (iv) and of the Implicit Function Theorem in Banach spaces (cf. Theorem A.19). □

We are now in the position to introduce the following.

Definition 13.52. Let the assumptions of Theorem 13.51 hold. Let both $\epsilon' \in]0, \epsilon_0[$ and $(\Psi[\cdot,\cdot], \Theta[\cdot,\cdot], C[\cdot,\cdot])$ be as in Theorem 13.51 (v). Let $\epsilon \in]0, \epsilon'[$. Let the map $(u^i[\epsilon,\cdot,\cdot,\cdot], u^o[\epsilon,\cdot,\cdot,\cdot])$ be as in Theorem 13.47. Then we set

$$u^i(\epsilon, x) \equiv u^i[\epsilon, \Psi[\epsilon, \delta_{2,n}\epsilon\log\epsilon], \Theta[\epsilon, \delta_{2,n}\epsilon\log\epsilon], C[\epsilon, \delta_{2,n}\epsilon\log\epsilon]](x) \qquad \forall x \in \overline{\mathbb{S}[\Omega_{p,\epsilon}]} \,,$$

$$u^o(\epsilon, x) \equiv u^o[\epsilon, \Psi[\epsilon, \delta_{2,n}\epsilon\log\epsilon], \Theta[\epsilon, \delta_{2,n}\epsilon\log\epsilon], C[\epsilon, \delta_{2,n}\epsilon\log\epsilon]](x) \qquad \forall x \in \overline{\mathbb{S}[\Omega_{p,\epsilon}]^-} \,.$$

By definition, the pair $(u^i(\epsilon,\cdot), u^o(\epsilon,\cdot))$ is a solution of problem (13.135) for all $\epsilon \in]0, \epsilon'[$. Then Theorem 13.46 implies that we can define a corresponding solution of our original problem (13.132). We do so in the following.

Definition 13.53. Let the assumptions of Theorem 13.51 hold. Let $\epsilon' \in]0, \epsilon_0[$ be as in Theorem 13.51 (v). Let $\epsilon \in]0, \epsilon'[$. Let $(u^i(\epsilon,\cdot), u^o(\epsilon,\cdot))$ be as in Definition 13.52. Then we set

$$T^i(\epsilon, x) \equiv K_i^{(-1)}(F^i_\epsilon[f_\epsilon](x) + u^i(\epsilon, x)) \qquad \forall x \in \overline{\mathbb{S}[\Omega_{p,\epsilon}]} \,,$$

$$T^o(\epsilon, x) \equiv K_o^{(-1)}(F^o_\epsilon[f_\epsilon](x) + u^o(\epsilon, x)) \qquad \forall x \in \overline{\mathbb{S}[\Omega_{p,\epsilon}]^-} \,.$$

By Theorem 13.46 and by Definition 13.52, the pair $(T^i(\epsilon,\cdot), T^o(\epsilon,\cdot))$ is a solution of problem (13.132) for all $\epsilon \in]0, \epsilon'[$.

13.8.4 A Representation Theorem for the Family of Solutions of Problem (13.132)

In this section, we exploit Theorem 13.51 in order to analyze the behavior of the family $\{(T^i(\epsilon,\cdot), T^o(\epsilon,\cdot))\}_{\epsilon\in]0,\epsilon'[}$ for ϵ small enough.

Theorem 13.54. *Let the assumptions of Theorem 13.51 hold. Let ϵ', $\epsilon^\sharp$ be as in Theorem 13.51 (v). Then the following statements hold.*

(i) There exists a real analytic map $\mathcal{T}^{i,r}$ *from* $]-\epsilon',\epsilon'[\times]-\epsilon^\sharp,\epsilon^\sharp[$ *to* $C^{1,\alpha}(\overline{\Omega})$ *such that*

$$T^i(\epsilon,p+\epsilon\xi)=\mathcal{T}^{i,r}[\epsilon,\delta_{2,n}\epsilon\log\epsilon](\xi)\qquad\forall\xi\in\overline{\Omega},$$

for all $\epsilon\in]0,\epsilon'[$*. Moreover,*

$$\mathcal{T}^{i,r}[0,0](\xi)=K_i^{(-1)}(k(0))\qquad\forall\xi\in\overline{\Omega}.\tag{13.159}$$

(ii) Let $\tilde{\Omega}$ *be a bounded open subset of* $\mathbb{R}^n\setminus\overline{\Omega}$*. Then there exist* $\epsilon_{\tilde{\Omega},r}\in]0,\epsilon'[$ *and a real analytic map* $\mathcal{T}^{o,r}_{\tilde{\Omega}}$ *from* $]-\epsilon_{\tilde{\Omega},r},\epsilon_{\tilde{\Omega},r}[\times]-\epsilon^\sharp,\epsilon^\sharp[$ *to* $C^{1,\alpha}(\overline{\tilde{\Omega}})$ *such that*

$$p+\epsilon\overline{\tilde{\Omega}}\subseteq\overline{\mathbb{S}[\Omega_{p,\epsilon}]^-}\qquad\forall\epsilon\in]-\epsilon_{\tilde{\Omega},r},\epsilon_{\tilde{\Omega},r}[,$$

$$T^o(\epsilon,p+\epsilon\xi)=\mathcal{T}^{o,r}_{\tilde{\Omega}}[\epsilon,\delta_{2,n}\epsilon\log\epsilon](\xi)\qquad\forall\xi\in\overline{\tilde{\Omega}},$$

for all $\epsilon\in]0,\epsilon_{\tilde{\Omega},r}[$*. Moreover,*

$$\mathcal{T}^{o,r}_{\tilde{\Omega}}[0,0](\xi)=K_i^{(-1)}(k(0))\qquad\forall\xi\in\overline{\tilde{\Omega}}.\tag{13.160}$$

(iii) Let $\tilde{\Omega}$ *be an open subset of* $\mathbb{R}^n$ *with nonzero distance from* $p+q\mathbb{Z}^n$*. Then there exist* $\epsilon^*_{\tilde{\Omega}}\in]0,\epsilon'[$ *such that*

$$\tilde{\Omega}\subseteq\mathbb{S}[\Omega_{p,\epsilon}]^-\qquad\forall\epsilon\in[-\epsilon^*_{\tilde{\Omega}},\epsilon^*_{\tilde{\Omega}}],$$

$\epsilon_{\tilde{\Omega}}\in]0,\epsilon^*_{\tilde{\Omega}}[$ *such that* $\overline{\mathbb{S}[\Omega_{p,\epsilon^*_{\tilde{\Omega}}}]^-}\subseteq\mathbb{S}[\Omega_{p,\epsilon}]^-$ *for all* $\epsilon\in[-\epsilon_{\tilde{\Omega}},\epsilon_{\tilde{\Omega}}]$*, and a real analytic map* $\mathcal{T}^o_{\mathbb{S}[\Omega_{p,\epsilon^*_{\tilde{\Omega}}}]^-}$ *from* $]-\epsilon_{\tilde{\Omega}},\epsilon_{\tilde{\Omega}}[\times]-\epsilon^\sharp,\epsilon^\sharp[$ *to* $C^{1,\alpha}_q(\overline{\mathbb{S}[\Omega_{p,\epsilon^*_{\tilde{\Omega}}}]^-})$ *such that*

$$T^o(\epsilon,x)=\mathcal{T}^o_{\mathbb{S}[\Omega_{p,\epsilon^*_{\tilde{\Omega}}}]^-}[\epsilon,\delta_{2,n}\epsilon\log\epsilon](x)\qquad\forall x\in\overline{\mathbb{S}[\Omega_{p,\epsilon^*_{\tilde{\Omega}}}]^-}$$

for all $\epsilon\in]0,\epsilon_{\tilde{\Omega}}[$*. Moreover,*

$$\begin{aligned}&\mathcal{T}^o_{\mathbb{S}[\Omega_{p,\epsilon^*_{\tilde{\Omega}}}]^-}[0,0](x)\\&\quad=K_o^{(-1)}\Big(P_{q,n}[Q,f_0](x)-P_{q,n}[Q,f_0](p)+K_o(K_i^{(-1)}(k(0)))\Big),\end{aligned}\tag{13.161}$$

for all $x\in\overline{\mathbb{S}[\Omega_{p,\epsilon^*_{\tilde{\Omega}}}]^-}$.

Proof. We first prove statement (i). By the definition of $T^i(\epsilon,\cdot)$ and by the definition of $u^i(\epsilon,\cdot)$ (see also Lemma 13.5 (iii) and (13.140)), we have

$$T^i(\epsilon,p+\epsilon\xi)=K_i^{(-1)}\Big(P_{q,n}[Q,f_\epsilon](p+\epsilon\xi)$$

$$
\begin{aligned}
&+P_n[\Omega_{p,\epsilon},1](p+\epsilon\xi)\int_Q f_\epsilon(y)\,dy + u^i(\epsilon,p+\epsilon\xi)\Big)\\
&= K_i^{(-1)}\Big(P_{q,n}[Q,f_\epsilon](p+\epsilon\xi)\\
&+\Big[\epsilon^2 P_n[\Omega,1](\xi)+\frac{m_n(\Omega)}{2\pi}\epsilon(\delta_{2,n}\epsilon\log\epsilon)\Big]\int_Q f_\epsilon(y)\,dy\\
&+\epsilon\Big[v_\Omega[\Psi[\epsilon,\delta_{2,n}\epsilon\log\epsilon]](\xi)\\
&+\epsilon^{n-2}\int_{\partial\Omega} R_{q,n}(\epsilon(\xi-\eta))\Psi[\epsilon,\delta_{2,n}\epsilon\log\epsilon](\eta)\,d\sigma_\eta\\
&-\int_{\partial\Omega}\Big(v_\Omega[\Psi[\epsilon,\delta_{2,n}\epsilon\log\epsilon]](\tau)\\
&+\epsilon^{n-2}\int_{\partial\Omega} R_{q,n}(\epsilon(\tau-\eta))\Psi[\epsilon,\delta_{2,n}\epsilon\log\epsilon](\eta)\,d\sigma_\eta\Big)\,d\sigma_\tau\Big]\\
&+k(\epsilon)-\int_{\partial\Omega} P_{q,n}[Q,f_\epsilon](p+\epsilon\tau)\,d\sigma_\tau\\
&-\int_Q f_\epsilon\,dy\int_{\partial\Omega}\Big(\epsilon^2 P_n[\Omega,1]+\frac{m_n(\Omega)}{2\pi}\epsilon(\delta_{2,n}\epsilon\log\epsilon)\Big)\,d\sigma\Big),
\end{aligned}
$$

for all $\xi \in \overline{\Omega}$ and for all $\epsilon \in]0,\epsilon'[$. Thus it is natural to set

$$
\begin{aligned}
\mathcal{T}^{i,r}[\epsilon,\epsilon_1](\xi) \equiv{}& K_i^{(-1)}\Big(P_{q,n}[Q,f_\epsilon](p+\epsilon\xi)\\
&+\Big[\epsilon^2 P_n[\Omega,1](\xi)+\frac{m_n(\Omega)}{2\pi}\epsilon\epsilon_1\Big]\int_Q f_\epsilon(y)\,dy\\
&+\epsilon\Big[v_\Omega[\Psi[\epsilon,\epsilon_1]](\xi)+\epsilon^{n-2}\int_{\partial\Omega} R_{q,n}(\epsilon(\xi-\eta))\Psi[\epsilon,\epsilon_1](\eta)\,d\sigma_\eta\\
&-\int_{\partial\Omega}\Big(v_\Omega[\Psi[\epsilon,\epsilon_1]](\tau)+\epsilon^{n-2}\int_{\partial\Omega} R_{q,n}(\epsilon(\tau-\eta))\Psi[\epsilon,\epsilon_1](\eta)\,d\sigma_\eta\Big)\,d\sigma_\tau\Big]\\
&+k(\epsilon)-\int_{\partial\Omega} P_{q,n}[Q,f_\epsilon](p+\epsilon\tau)\,d\sigma_\tau\\
&-\int_Q f_\epsilon\,dy\int_{\partial\Omega}\Big(\epsilon^2 P_n[\Omega,1]+\frac{m_n(\Omega)}{2\pi}\epsilon\epsilon_1\Big)\,d\sigma\Big) \qquad \forall \xi\in\overline{\Omega},
\end{aligned}
\tag{13.162}
$$

for all $(\epsilon,\epsilon_1) \in]-\epsilon',\epsilon'[\times]-\epsilon^\sharp,\epsilon^\sharp[$. We now turn to prove that $\mathcal{T}^{i,r}$ is real analytic. We observe that

the map from $]-\epsilon_0,\epsilon_0[$ to $C^{1,\alpha}(\overline{\Omega})$ that takes ϵ to the function $P_{q,n}[Q,f_\epsilon](p+\epsilon\xi)$ of the variable $\xi\in\overline{\Omega}$ is analytic (13.163)

and that

$$\text{the map from }]-\epsilon_0,\epsilon_0[\times C^{0,\alpha}(\partial\Omega) \text{ to } C^{1,\alpha}(\overline{\Omega}) \qquad (13.164)$$
$$\text{that takes } (\epsilon,\psi) \text{ to the function } \int_{\partial\Omega} R_{q,n}(\epsilon(\xi-\eta))\psi(\eta)\,d\sigma_\eta$$
$$\text{of the variable } \xi\in\overline{\Omega} \text{ is analytic}$$

(cf. Lemma 13.4 (i), (ii)). Since $v_\Omega[\cdot]_{|\overline{\Omega}}$ is linear and continuous by Theorem 4.25 and Ψ is analytic, assumption (13.129) and statements (13.163), (13.164) imply that the argument of $K_i^{(-1)}$ in the right-hand side of (13.162) defines a real analytic map from $]-\epsilon',\epsilon'[\times]-\epsilon^\sharp,\epsilon^\sharp[$ to $C^{1,\alpha}(\overline{\Omega})$. By the analyticity of the map from $C^{1,\alpha}(\overline{\Omega})$ to itself that takes h to the composite function $K_i^{(-1)}\circ h$, we conclude that $\mathcal{T}^{i,r}$ is real analytic (cf. Theorem A.23). Then by setting $\epsilon=0=\epsilon_1$ in (13.162), we obtain (13.159).

Next we consider statement (ii). By assumption, there exists $R>0$ such that $\overline{\tilde{\Omega}}\subseteq\mathbb{B}_n(0,R)$. Then we set $\Omega^*\equiv\mathbb{B}_n(0,R)\setminus\overline{\Omega}$. Let $\epsilon_{\Omega^*,r}$ be as in Lemma 13.6 (ii) with $\epsilon_1=\epsilon'$, $\Omega^\sharp=\Omega^*$. Then we take $\epsilon_{\tilde{\Omega},r}\equiv\epsilon_{\Omega^*,r}$. It clearly suffices to show that $\mathcal{T}^{o,r}_{\Omega^*}$ exists and then to set $\mathcal{T}^{o,r}_{\tilde{\Omega}}$ equal to the composition of the restriction from $C^{1,\alpha}(\overline{\Omega^*})$ to $C^{1,\alpha}(\overline{\tilde{\Omega}})$ with $\mathcal{T}^{o,r}_{\Omega^*}$.

By Definition 13.53 of $T^o(\epsilon,\cdot)$ and by the definition of $u^o(\epsilon,\cdot)$ (see also (13.133), Lemma 13.5 (iv), (13.141)), we have

$$\begin{aligned}
T^o(\epsilon,p+\epsilon\xi) &= K_o^{(-1)}\Bigg(P_{q,n}[Q,f_\epsilon](p+\epsilon\xi) \\
&\quad -\frac{1}{m_n(\Omega_{p,\epsilon})}\int_{\Omega_{p,\epsilon}} S_{q,n}(p+\epsilon\xi-y)\,dy\int_Q f_\epsilon(y)\,dy+u^o(\epsilon,p+\epsilon\xi)\Bigg) \\
&= K_o^{(-1)}\Bigg(P_{q,n}[Q,f_\epsilon](p+\epsilon\xi)-\frac{1}{m_n(\Omega_{p,\epsilon})}\int_{\Omega_{p,\epsilon}} S_{q,n}(p+\epsilon\xi-y)\,dy\int_Q f_\epsilon(y)\,dy \\
&\quad +\epsilon m^o[\epsilon,\Theta[\epsilon,\delta_{2,n}\epsilon\log\epsilon],C[\epsilon,\delta_{2,n}\epsilon\log\epsilon]](\xi)-P_{q,n}[Q,f_\epsilon](p) \\
&\quad +G[\epsilon,\delta_{2,n}\epsilon\log\epsilon]+\int_Q f_\epsilon\,dy\frac{1}{2\pi}\delta_{2,n}\log\epsilon\Bigg) \\
&= K_o^{(-1)}\Bigg(P_{q,n}[Q,f_\epsilon](p+\epsilon\xi)-P_{q,n}[Q,f_\epsilon](p) \\
&\quad +\frac{1}{m_n(\Omega)}\int_{\partial\Omega} g_\epsilon\,d\sigma\,\epsilon^{n-1}\Bigg\{\epsilon^{2-n}P_n[\Omega,1](\xi)+\frac{m_n(\Omega)}{2\pi}\delta_{2,n}\log\epsilon \\
&\quad +\int_\Omega R_{q,n}(\epsilon(\xi-\eta))\,d\eta\Bigg\}+\epsilon\Bigg[v_\Omega[\Theta[\epsilon,\delta_{2,n}\epsilon\log\epsilon]](\xi) \\
&\quad +\epsilon^{n-2}\int_{\partial\Omega} R_{q,n}(\epsilon(\xi-\eta))\Theta[\epsilon,\delta_{2,n}\epsilon\log\epsilon](\eta)\,d\eta+C[\epsilon,\delta_{2,n}\epsilon\log\epsilon]\Bigg] \\
&\quad +K\Bigg(k(\epsilon)-\fint_{\partial\Omega} P_{q,n}[Q,f_\epsilon](p+\epsilon\tau)\,d\sigma_\tau
\end{aligned}$$

$$-\int_Q f_\epsilon\, dy \int_{\partial\Omega} \epsilon^2 P_n[\Omega,1]\, d\sigma - \int_Q f_\epsilon\, dy \epsilon \frac{m_n(\Omega)}{2\pi}\delta_{2,n}\epsilon\log\epsilon$$
$$+P_{q,n}[Q,f_\epsilon](p)\Big) - \int_{\partial\Omega} g_\epsilon\, d\sigma \epsilon^{n-2}\frac{1}{2\pi}(\delta_{2,n}\epsilon\log\epsilon)\Big),$$

for all $\xi \in \overline{\Omega^*}$. Thus it is natural to set

$$\mathcal{T}^{o,r}_{\mathbb{S}[\Omega_{p,\epsilon^*_{\tilde\Omega}}]^-}[\epsilon,\epsilon_1](\xi) \equiv K_o^{(-1)}\Big(P_{q,n}[Q,f_\epsilon](p+\epsilon\xi) - P_{q,n}[Q,f_\epsilon](p) \tag{13.165}$$
$$+\frac{1}{m_n(\Omega)}\int_{\partial\Omega} g_\epsilon\, d\sigma\Big\{\epsilon P_n[\Omega,1](\xi) + \frac{m_n(\Omega)}{2\pi}\epsilon_1 + \epsilon^{n-1}\int_\Omega R_{q,n}(\epsilon(\xi-\eta))\, d\eta\Big\}$$
$$+\epsilon\Big[v_\Omega[\Theta[\epsilon,\epsilon_1]](\xi) + \epsilon^{n-2}\int_{\partial\Omega} R_{q,n}(\epsilon(\xi-\eta))\Theta[\epsilon,\epsilon_1](\eta)\, d\eta + C[\epsilon,\epsilon_1]\Big]$$
$$+K\Big(k(\epsilon) - \int_{\partial\Omega} P_{q,n}[Q,f_\epsilon](p+\epsilon\tau)\, d\sigma_\tau$$
$$-\int_Q f_\epsilon\, dy \int_{\partial\Omega} \epsilon^2 P_n[\Omega,1]\, d\sigma - \int_Q f_\epsilon\, dy \epsilon\frac{m_n(\Omega)}{2\pi}\epsilon_1$$
$$+P_{q,n}[Q,f_\epsilon](p)\Big) - \int_{\partial\Omega} g_\epsilon\, d\sigma\epsilon^{n-2}\frac{1}{2\pi}\epsilon_1\Big),$$

for all $\xi \in \overline{\Omega^*}$, for all $(\epsilon,\epsilon_1) \in]-\epsilon_{\tilde\Omega,r},\epsilon_{\tilde\Omega,r}[\times]-\epsilon^\sharp,\epsilon^\sharp[$. By the analyticity of the map in (13.163) and by (13.129), we deduce that the argument of K in (13.165) defines an analytic map from $]-\epsilon_{\tilde\Omega,r},\epsilon_{\tilde\Omega,r}[\times]-\epsilon^\sharp,\epsilon^\sharp[$ to $C^{1,\alpha}(\overline{\Omega^*})$. Since K is real analytic, Theorem A.23 implies that the map from $C^{1,\alpha}(\overline{\Omega^*})$ to itself that takes h to $K\circ h$ is real analytic. Then the second last summand in the term in the argument of $K_o^{(-1)}$ in (13.165) defines an analytic map from $]-\epsilon_{\tilde\Omega,r},\epsilon_{\tilde\Omega,r}[\times]-\epsilon^\sharp,\epsilon^\sharp[$ to $C^{1,\alpha}(\overline{\Omega^*})$.

Since $R_{q,n}$ is analytic in $(\mathbb{R}^n\setminus q\mathbb{Z}^n)\cup\{0\}$, then Lemma 13.4 (i) implies that the map from $]-\epsilon_0,\epsilon_0[$ to $C^{m+1}(\overline{\Omega^*})$ that takes ϵ to the function $\int_\Omega R_{q,n}(\epsilon(\xi-\eta))\, d\sigma_\eta$ of the variable $\xi\in\overline{\Omega^*}$ is analytic.

By assumption (13.130), and by the linearity and continuity of $v_{\overline{\Omega}}[\cdot]_{|\overline{\Omega^*}}$ from the space $C^{0,\alpha}(\partial\Omega)$ to $C^{1,\alpha}(\overline{\Omega^*})$ (cf. Theorem 4.25), and by the analyticity of Θ, we conclude that the map from $]-\epsilon_{\tilde\Omega,r},\epsilon_{\tilde\Omega,r}[\times]-\epsilon^\sharp,\epsilon^\sharp[$ to $C^{1,\alpha}(\overline{\Omega^*})$ that takes (ϵ,ϵ_1) to $v_{\overline{\Omega}}[\Theta[\epsilon,\epsilon_1]]_{|\overline{\Omega^*}}$ is analytic. We observe that the map from $]-\epsilon_0,\epsilon_0[$ to the space $C^{1,\alpha}(\overline{\Omega^*})$ that takes ϵ to the function $P_{q,n}[Q,f_\epsilon](p+\epsilon\xi)$ of the variable $\xi\in\overline{\Omega^*}$ is real analytic, and that the map from $]-\epsilon_0,\epsilon_0[\times C^{0,\alpha}(\partial\Omega)$ to $C^{1,\alpha}(\overline{\Omega^*})$ that takes (ϵ,θ) to the function $\int_{\partial\Omega} R_{q,n}(\epsilon(\xi-\eta))\theta(\eta)\, d\sigma_\eta$ of the variable ξ in $\overline{\Omega^*}$ is real analytic (cf. Lemma 13.4 (i)). Then we conclude that the function from $]-\epsilon_{\tilde\Omega,r},\epsilon_{\tilde\Omega,r}[\times]-\epsilon^\sharp,\epsilon^\sharp[$ to $C^{1,\alpha}(\overline{\Omega^*})$ that takes (ϵ,ϵ_1) to the argument of $K_o^{(-1)}$ in the right-hand side of (13.165) is analytic. Since $K_o^{(-1)}$ is analytic, the map from $C^{1,\alpha}(\overline{\Omega^*})$ to itself that takes h to the composite function $K_o^{(-1)}\circ h$ is analytic (cf. Theorem A.23). Then we conclude that $\mathcal{T}^{o,r}_{\Omega^*}$ is analytic from the set $]-\epsilon_{\tilde{\Omega},r},\epsilon_{\tilde{\Omega},r}[\times]-\epsilon^\sharp,\epsilon^\sharp[$ to $C^{1,\alpha}(\overline{\Omega^*})$. Finally, by setting $\epsilon=0=\epsilon_1$ in (13.165), we obtain (13.160).

Next we consider statement (iii). Let $\epsilon^*_{\tilde{\Omega}}$, $\epsilon_{\tilde{\Omega}}$ be as in Lemma 13.6 (i). By (13.133), Lemma 13.5 (iv), (13.137) for $u^o[\epsilon,\cdot,\cdot]$, Definition 13.53 of $T^o(\epsilon,\cdot)$, and equality (13.6) for $S_{q,n}(\epsilon\cdot)$, we have

$$
\begin{aligned}
T^o(\epsilon,x) &= K_o^{(-1)}\Bigg(P_{q,n}[Q,f_\epsilon](x)\\
&\quad -\frac{1}{m_n(\Omega_{p,\epsilon})}P_{q,n}[\Omega_{p,\epsilon},1](x)\int_Q f_\epsilon(y)\,dy + u^o(\epsilon,x)\Bigg)\\
&= K_o^{(-1)}\Bigg(P_{q,n}[Q,f_\epsilon](x) - \frac{1}{m_n(\Omega_{p,\epsilon})}\int_{\Omega_{p,\epsilon}} S_{q,n}(x-y)\,dy\int_Q f_\epsilon(y)\,dy\\
&\quad +\epsilon^{n-1}\int_{\partial\Omega} S_{q,n}(x-p-\epsilon\eta)\Theta[\epsilon,\delta_{2,n}\epsilon\log\epsilon](\eta)\,d\sigma_\eta + \epsilon C[\epsilon,\delta_{2,n}\epsilon\log\epsilon]\\
&\quad -P_{q,n}[Q,f_\epsilon](p) + G[\epsilon,\delta_{2,n}\epsilon\log\epsilon] + \int_Q f_\epsilon\,dy\frac{\delta_{2,n}}{2\pi}\log\epsilon\Bigg)\\
&= K_o^{(-1)}\Bigg(P_{q,n}[Q,f_\epsilon](x) - P_{q,n}[Q,f_\epsilon](p)\\
&\qquad +\int_{\partial\Omega} g_\epsilon\,d\sigma\epsilon^{n-1}\frac{1}{m_n(\Omega)}\int_\Omega S_{q,n}(x-p-\epsilon\eta)\,d\eta\\
&\quad +\epsilon^{n-1}\int_{\partial\Omega} S_{q,n}(x-p-\epsilon\eta)\Theta[\epsilon,\delta_{2,n}\epsilon\log\epsilon](\eta)\,d\sigma_\eta\\
&\quad +\epsilon C[\epsilon,\delta_{2,n}\epsilon\log\epsilon] + G[\epsilon,\delta_{2,n}\epsilon\log\epsilon] - \int_{\partial\Omega} g_\epsilon\,d\sigma\epsilon^{n-1}\frac{\delta_{2,n}}{2\pi}\log\epsilon\Bigg)
\end{aligned}
$$

for all $x\in\overline{\mathbb{S}[\Omega_{p,\epsilon^*_{\tilde{\Omega}}}]^-}$ and all $\epsilon\in]0,\epsilon_{\tilde{\Omega}}[$. Thus it is natural to set

$$
\begin{aligned}
\mathcal{T}^o_{\mathbb{S}[\Omega_{p,\epsilon^*_{\tilde{\Omega}}}]^-}[\epsilon,\epsilon_1](x) &\equiv K_o^{(-1)}\Bigg(P_{q,n}[Q,f_\epsilon](x) - P_{q,n}[Q,f_\epsilon](p) \qquad (13.166)\\
&\quad +\int_{\partial\Omega} g_\epsilon\,d\sigma\epsilon^{n-1}\frac{1}{m_n(\Omega)}\int_\Omega S_{q,n}(x-p-\epsilon\eta)\,d\eta\\
&\quad +\epsilon^{n-1}\int_{\partial\Omega} S_{q,n}(x-p-\epsilon\eta)\Theta[\epsilon,\epsilon_1](\eta)\,d\sigma_\eta\\
&\quad +\epsilon C[\epsilon,\epsilon_1] + G[\epsilon,\epsilon_1] - \int_{\partial\Omega} g_\epsilon\,d\sigma\frac{\epsilon_1}{2\pi}\Bigg)
\end{aligned}
$$

for all $x\in\overline{\mathbb{S}[\Omega_{p,\epsilon^*_{\tilde{\Omega}}}]^-}$ and all $(\epsilon,\epsilon_1)\in]-\epsilon_{\tilde{\Omega}},\epsilon_{\tilde{\Omega}}[\times]-\epsilon^\sharp,\epsilon^\sharp[$. By assumption (13.129), the map from $]-\epsilon_0,\epsilon_0[$ to $C^{1,\alpha}_q(\overline{\mathbb{S}[\Omega_{p,\epsilon^*_{\tilde{\Omega}}}]^-})$ that takes ϵ to $P_{q,n}[Q,f_\epsilon]_{|\overline{\mathbb{S}[\Omega_{p,\epsilon^*_{\tilde{\Omega}}}]^-}}$ is analytic (cf. Lemma 13.4 (ii)). Similarly, the map from $]-\epsilon_0,\epsilon_0[$ to $\mathbb{R}$ that takes ϵ to $P_{q,n}[Q,f_\epsilon](p)$ is analytic. Then the result on integral operators with analytic kernels of Theorem A.28 (i) implies that the map from $]-\epsilon_{\tilde{\Omega}}, \epsilon_{\tilde{\Omega}}[\times L^1(\partial\Omega)$ to $C^{1,\alpha}_q(\overline{\mathbb{S}[\Omega_{p,\epsilon^*_{\tilde{\Omega}}}]^-})$ that takes (ϵ, h) to $\int_{\partial\Omega} S_{q,n}(\cdot - p - \epsilon\eta)h(\eta)\, d\sigma_\eta$ is real analytic. Then by the analyticity of Θ and by the continuity of the embedding from $C^{0,\alpha}(\partial\Omega)$ to $L^1(\partial\Omega)$ and by Lemma 12.5 (ii), we deduce that the map from $]-\epsilon_{\tilde{\Omega}}, \epsilon_{\tilde{\Omega}}[\times]-\epsilon^\sharp, \epsilon^\sharp[$ to $C^{1,\alpha}_q(\overline{\mathbb{S}[\Omega_{p,\epsilon^*_{\tilde{\Omega}}}]^-})$ that takes (ϵ, ϵ_1) to the function $\int_{\partial\Omega} S_{q,n}(\cdot - p - \epsilon\eta)\Theta[\epsilon, \epsilon_1](\eta)\, d\sigma_\eta$ is real analytic. Then the map from $]-\epsilon_{\tilde{\Omega}}, \epsilon_{\tilde{\Omega}}[\times]-\epsilon^\sharp, \epsilon^\sharp[$ to $C^{1,\alpha}_q(\overline{\mathbb{S}[\Omega_{p,\epsilon^*_{\tilde{\Omega}}}]^-})$ that takes (ϵ, ϵ_1) to the argument of $K_o^{(-1)}$ in the right-hand side of (13.166) is real analytic. Since the map $K_o^{(-1)}$ is real analytic, Theorem A.23 implies that the composition operator from $C^{1,\alpha}_q(\overline{\mathbb{S}[\Omega_{p,\epsilon^*_{\tilde{\Omega}}}]^-})$ that takes h to $K_o^{(-1)} \circ h$ is real analytic. Hence, $\mathcal{T}^o_{\mathbb{S}[\Omega_{p,\epsilon^*_{\tilde{\Omega}}}]^-}$ is real analytic. Then by setting $\epsilon = 0 = \epsilon_1$ in (13.166), we obtain (13.161). □

To conclude this section, we note that one could prove a local uniqueness result for the family of solutions $\{(T^i(\epsilon,\cdot), T^o(\epsilon,\cdot))\}_{\epsilon\in]0,\epsilon'[}$ of Definition 13.53 by a similar strategy of that of Section 11.3.3 for the solutions of a nonlinear transmission problem in a bounded domain with a small hole. More details can be found in [70].

Appendix A

Here we present some technical concepts and results that we utilize in the book. Specifically, the Homomorphism Theorem, the notion of inductive topology, the Lebesgue number of an open cover, the Vitali Convergence Theorem, the Rademacher Theorem, elements of calculus in Banach spaces and the Implicit Function Theorem, some results on nonlinear composition operators and on linear and nonlinear integral operators, Sard's Theorem, the Theorem of Invariance of Domain, mollifiers, and the partition of unity.

A.1 The Homomorphism Theorem

Theorem A.1 (of the Homomorphism). *Let X, Y, Z be vector spaces on $\mathbb{K}$. Let L_1 be a linear map from Y to Z. Let L_2 be a linear surjective map from Y onto X. Then the following statements are equivalent.*

(i) There exists a linear map Λ from X to Z such that

$$L_1 = \Lambda \circ L_2 .$$

(ii) $\mathrm{Ker}\, L_2$ *is a subspace of* $\mathrm{Ker}\, L_1$.

Here $\mathbb{K}$ denotes either the field $\mathbb{R}$ or $\mathbb{C}$. For a proof, we refer for example to Kostrikin and Manin [145, Prop. 6.8, p. 46].

A.2 The Inductive Topology

In addition to the standard notions of topology, we need the inductive topology, in particular to deal with quotient spaces.

M. Dalla Riva et al., *Singularly Perturbed Boundary Value Problems*,
https://doi.org/10.1007/978-3-030-76259-9

Theorem A.2. *Let $\{(X_i, \mathcal{T}_i)\}_{i\in I}$ be a family of topological spaces. Let Y be a set. Let f_i be a function from X_i to Y for each $i \in I$. Let*

$$\mathcal{T} \equiv \{A \in \mathcal{P}(Y) : f_i^{\leftarrow}(A) \in \mathcal{T}_i, \ \forall i \in I\} .$$

Then $\mathcal{T}$ is a topology on Y and f_i is continuous from $(X_i, \mathcal{T}_i)$ to $(Y, \mathcal{T})$ for all $i \in I$.

Moreover, if $\mathcal{T}'$ is a topology on Y such that f_i is continuous from $(X_i, \mathcal{T}_i)$ to $(Y, \mathcal{T}')$ for all $i \in I$, then $\mathcal{T}' \subseteq \mathcal{T}$.

Proof. By the known set theoretic properties of the preimage of sets and by the assumption that $\mathcal{T}_i$ is a topology for each $i \in I$, we can easily verify that $\mathcal{T}$ is a topology on Y. By the definition of $\mathcal{T}$, f_i is continuous from $(X_i, \mathcal{T}_i)$ to $(Y, \mathcal{T})$ for all $i \in I$.

We now consider the last part of the statement. Since f_i is continuous from $(X_i, \mathcal{T}_i)$ to $(Y, \mathcal{T}')$, then $f_i^{\leftarrow}(A)$ belongs to $\mathcal{T}_i$ for all $i \in I$ and for each element A of $\mathcal{T}'$. Hence, $\mathcal{T}' \subseteq \mathcal{T}$. □

The topology $\mathcal{T}$ of the previous statement is said to be the inductive topology on Y with respect to the family $\{f_i\}_{i\in I}$, and is the strongest topology on Y for which f_i is continuous for all $i \in I$.

Proposition A.3. *Let $\{(X_i, \mathcal{T}_i)\}_{i\in I}$ be a family of topological spaces. Let Y be a set. Let f_i be a function from X_i to Y for each $i \in I$. Let $\mathcal{T}$ be the inductive topology on Y with respect to the family $\{f_i\}_{i\in I}$. Let $(Z, \mathcal{T}_Z)$ be a topological space. Let g be a function from $(Y, \mathcal{T})$ to $(Z, \mathcal{T}_Z)$.*

Then g is continuous if and only if $g \circ f_i$ is continuous from $(X_i, \mathcal{T}_i)$ to $(Z, \mathcal{T}_Z)$ for each $i \in I$.

Proof. The necessity follows by the continuity of compositions of continuous functions. We now prove the sufficiency. Let $A \in \mathcal{T}_Z$. We must prove that $g^{\leftarrow}(A) \in \mathcal{T}$. By the definition of $\mathcal{T}$, it suffices to show that $f_i^{\leftarrow}(g^{\leftarrow}(A)) \in \mathcal{T}_i$ for all $i \in I$. Since $g \circ f_i$ is continuous, we have $f_i^{\leftarrow}(g^{\leftarrow}(A)) = (g \circ f_i)^{\leftarrow}(A) \in \mathcal{T}_i$ for all $i \in I$. □

Definition A.4. Let $(X, \mathcal{T})$ be a topological space. Let $\sim$ be an equivalence relation on X. We define as topological quotient space of X over $\sim$, the quotient space $X/\sim$ with the inductive topology with respect to the canonical projection from X to $X/\sim$. Such an inductive topology is said to be the quotient topology on $X/\sim$

Then we have the following immediate consequence of Proposition A.3.

Proposition A.5. *Let X and Z be topological spaces. Let $\sim$ be an equivalence relation on X and let π be the canonical projection of X onto $X/\sim$. Then, a function f from the topological quotient space $X/\sim$ to Z is continuous if and only if $f \circ \pi$ is continuous from X to Z.*

A.3 Lebesgue Number of an Open Cover

We also make a large use of the notion of the Lebesgue number of an open cover.

Theorem A.6. *Let (X,d) be a compact metric space. Let $\{U_\lambda\}_{\lambda\in\Lambda}$ be an open cover of X, i.e., a family of open subsets $\{U_\lambda\}_{\lambda\in\Lambda}$ of X such that $X = \bigcup_{\lambda\in\Lambda} U_\lambda$. Then there exists $\delta > 0$ such that every subset A of X with diameter*

$$\mathrm{diam}\,(A) \equiv \sup\{d(x,y) : x,y \in A\}$$

smaller than or equal to δ is contained in U_λ for at least a $\lambda \in \Lambda$.

For a proof, we refer for example to Dugundji [93, Theorem 4.5, Chap. XI].

A number δ as in the previous statement is said to be a Lebesgue number for the cover $\{U_\lambda\}_{\lambda\in\Lambda}$ of X.

A.4 Perforated Connected Domains Are Connected

In this section we assume that

$$n \in \mathbb{N},\ n \geq 2$$

and we show that perforated domains such as $\Omega(\epsilon)$ in Chapters 8, 9, 11 and $\Omega(\epsilon_1,\epsilon_2)$ in Chapter 10 are connected, a fact that can be considered as folklore. In what follows we use the Jordan Separation Theorem in the form presented (and proved) in Deimling's book [87, Theorem 5.2]:

Theorem A.7 (Jordan Separation). *Let Γ_1 and Γ_2 be compact subsets of $\mathbb{R}^n$ that are homeomorphic to each other, then $\mathbb{R}^n \setminus \Gamma_1$ and $\mathbb{R}^n \setminus \Gamma_2$ have the same number of connected components.*

We also need two elementary lemmas.

Lemma A.8. *Let Ω' and Ω'' be nonempty open subsets of $\mathbb{R}^n$ with $\Omega' \subseteq \Omega''$. If Ω' is connected and Ω'' contains no point of the boundary of Ω', then Ω' is a connected component of Ω''.*

Proof. If not there would be a connect component $\tilde{\Omega}$ of Ω'' that properly contains Ω'. Then it would be $\overline{\Omega'} \cap \tilde{\Omega} = \Omega'$ and Ω' would be a nonempty closed and open proper subset of $\tilde{\Omega}$, contradicting the fact that $\tilde{\Omega}$ is connected. □

Lemma A.9. *Let $k \in \mathbb{N} \setminus \{0,1\}$. Let $\Gamma_1, \ldots, \Gamma_k$ be compact subsets of $\mathbb{R}^n$ with $\mathbb{R}^n \setminus \Gamma_j$ connected for all $j \in \{1,\ldots,k\}$. Assume that there are elements*

$$a_0 \equiv -\infty < a_1 < \cdots < a_{k-1} < a_k \equiv +\infty$$

of $\mathbb{R}\cup\{-\infty,+\infty\}$ *such that*

$$\Gamma_j \subseteq J_{a_{j-1},a_j} \equiv \{x\in\mathbb{R}^n \,:\, a_{j-1} < x\cdot e_n < a_j\}$$

for all $j\in\{1,\ldots,k\}$. *Then*

$$\mathbb{R}^n \setminus \cup_{j=1}^k \Gamma_j$$

is (path) connected.

Proof. The proof can be effected in two steps.
Step 1. First we note that for all $j\in\{1,\ldots,k\}$ the perforated stripe

$$J_{a_{j-1},a_j}\setminus\Gamma_j$$

is path connected. This can be verified exhibiting a homeomorphism from the open set $J_{a_{j-1},a_j}\setminus\Gamma_j$ onto $\mathbb{R}^n\setminus\Gamma_j$ (we leave the details to the reader).
Step 2. Then we turn to $\mathbb{R}^n\setminus\cup_{j=1}^k\Gamma_j$. To show that it is path connected we take two points x and y in $\mathbb{R}^n\setminus\cup_{j=1}^k\Gamma_j$. If x and y are contained in a same perforated stripe $J_{a_{h-1},a_h}\setminus\Gamma_h$, then, by Step 1, there exists a path in $\mathbb{R}^n\setminus\cup_{j=1}^k\Gamma_j$ that joins x with y. If not, then we take a real number $\delta>\max\{z\cdot e_1 \,:\, z\in\cup_{j=1}^k\Gamma_j\}$ and we observe that $\xi\equiv(\delta,0,\ldots,0,x\cdot e_n)$ belongs to the same perforated stripe of x and $\eta\equiv(\delta,0,\ldots,0,y\cdot e_n)$ belongs to the same perforated stripe of y. So there exist paths $\gamma_{x,\xi}$ and $\gamma_{\eta,y}$ in $\mathbb{R}^n\setminus\cup_{j=1}^k\Gamma_j$ that join x with ξ and η with y, respectively. Moreover, the path $\gamma_{\xi,\eta}(t)\equiv t\xi+(1-t)\eta$ with $t\in[0,1]$ is contained in $\mathbb{R}^n\setminus\cup_{j=1}^k\Gamma_j$ and joins ξ with η. Then the composition (in the sense of paths) of $\gamma_{x,\xi}$, $\gamma_{\xi,\eta}$, and $\gamma_{\eta,y}$ is contained in $\mathbb{R}^n\setminus\cup_{j=1}^k\Gamma_j$ and joins x with y. □

We can now prove the main theorem of this section.

Theorem A.10. *Let* $k\in\mathbb{N}\setminus\{0\}$. *Let* Ω_0, Ω_1, ..., Ω_k *be open bounded connected subsets of* $\mathbb{R}^n$ *with connected exteriors and such that*

$$\overline{\Omega_i}\cap\overline{\Omega_j}=\emptyset \quad \forall i,j\in\{1,\ldots,k\}\,,\ i\neq j\,, \quad \textit{and} \quad \overline{\Omega_i}\subseteq\Omega_0 \qquad \forall i\in\{1,\ldots,k\}\,.$$

Then $\Omega_0\setminus\cup_{j=1}^k\overline{\Omega_j}$ *is connected.*

Proof. We split the proof in two steps. In the first one we show that

$$\mathbb{R}^n\setminus\cup_{j=0}^k\partial\Omega_j \text{ has } k+2 \text{ connected components} \tag{A.1}$$

and in the second one we complete the proof of the theorem.

Step 1. We can verify that there exist vectors ξ_0, ξ_1, ..., $\xi_k\in\mathbb{R}^n$ and elements $a_0\equiv-\infty<a_1<\cdots<a_k<a_{k+1}\equiv+\infty$ of $\mathbb{R}\cup\{-\infty,+\infty\}$ such that

$$\overline{\Omega_{j,\xi_j}}\subseteq J_{a_j,a_{j+1}} \qquad \forall j\in\{0,1,\ldots,k\}$$

where $\Omega_{j,\xi_j}\equiv\Omega_j+\xi_j$. Since translations are homeomorphisms, $\cup_{j=0}^k\partial\Omega_j$ is homeomorphic to $\cup_{j=0}^k\partial\Omega_{j,\xi_j}$ and thus the Jordan Separation Theorem A.7 implies that

$\mathbb{R}^n \setminus \cup_{j=0}^k \partial\Omega_j$ and $\mathbb{R}^n \setminus \cup_{j=0}^k \partial\Omega_{j,\xi_j}$ have the same number of connected components. Accordingly, we can prove (A.1) by showing that

$$\mathbb{R}^n \setminus \cup_{j=0}^k \partial\Omega_{j,\xi_j} \text{ has } k+2 \text{ connected components.} \tag{A.2}$$

We first verify that $\mathbb{R}^n \setminus \cup_{j=0}^k \partial\Omega_{j,\xi_j}$ has at least $k+2$ components. Since $\cup_{j=0}^k \partial\Omega_{j,\xi_j}$ is bounded, $\mathbb{R}^n \setminus \cup_{j=0}^k \partial\Omega_{j,\xi_j}$ has one unbounded connected component, which we denote by Ω_∞. Moreover, every set Ω_{i,ξ_i} with $i \in \{0, 1, \dots, k\}$ is open, bounded, connected, contained in $\mathbb{R}^n \setminus \cup_{j=0}^k \partial\Omega_{j,\xi_j}$, and no point of its boundary belongs to $\mathbb{R}^n \setminus \cup_{j=0}^k \partial\Omega_{j,\xi_j}$. By Lemma A.8 it follows that Ω_{i,ξ_i} is a connected component of $\mathbb{R}^n \setminus \cup_{j=0}^k \partial\Omega_{j,\xi_j}$. In particular, Ω_{i,ξ_i} cannot intersect Ω_∞. otherwise $\Omega_{i,\xi_i} = \Omega_\infty$, that is contrary to the assumption that Ω_{i,ξ_i} is bounded. Since Ω_∞, $\Omega_{0,\xi_0}, \dots, \Omega_{k,\xi_k}$ are disjoint it follows that $\mathbb{R}^n \setminus \cup_{j=0}^k \partial\Omega_{j,\xi_j}$ has at least $k+2$ connected components.

Next we prove (A.2) by showing that $\mathbb{R}^n \setminus \cup_{j=0}^k \partial\Omega_{j,\xi_j}$ has no connected component other than Ω_∞ and the Ω_{j,ξ_j}'s. Indeed, if Ω' is a connected component of $\mathbb{R}^n \setminus \cup_{j=0}^k \partial\Omega_{j,\xi_j}$, than either Ω' is bounded or Ω' is unbounded. If Ω' is unbounded then it coincides with Ω_∞, because there is only one unbounded component. If Ω' is bounded, then it is not Ω_∞ and thus $\Omega' \cap \Omega_\infty = \emptyset$. Moreover, Ω_∞ contains $\mathbb{R}^n \setminus \cup_{j=0}^k \overline{\Omega_{j,\xi_j}}$, because $\mathbb{R}^n \setminus \cup_{j=0}^k \overline{\Omega_{j,\xi_j}}$ is unbounded and connected by Lemma A.9 (here we invoke the assumption on the exterior of the Ω_j's). Thus $\Omega' \cap \left(\mathbb{R}^n \setminus \cup_{j=0}^k \overline{\Omega_{j,\xi_j}}\right) = \emptyset$ or, equivalently, $\Omega' \subseteq \cup_{j=0}^k \overline{\Omega_{j,\xi_j}}$. Since Ω' is connected and the compact sets $\overline{\Omega_{0,\xi_0}}, \dots, \overline{\Omega_{k,\xi_k}}$ are pairwise disjoint, it follows that there exists exactly one $i \in \{0, \dots, k\}$ such that $\Omega' \subseteq \overline{\Omega_{i,\xi_i}}$. Then $\Omega' \subseteq \Omega_{i,\xi_i}$ because Ω' is open, and $\Omega' = \Omega_{i,\xi_i}$ because both Ω' and Ω_{i,ξ_i} are (maximal) connected components.

Step 2. The set Ω_0^- is open, connected, contained in $\mathbb{R}^n \setminus \cup_{j=0}^k \partial\Omega_j$, and no point of its boundary $\partial\Omega_0$ is in $\mathbb{R}^n \setminus \cup_{j=0}^k \partial\Omega_j$. By Lemma A.8 it follows that Ω_0^- is a connected component of $\mathbb{R}^n \setminus \cup_{j=0}^k \partial\Omega_j$. Similarly we can prove that every Ω_i with $i \in \{1, \dots, k\}$ is a connected component of $\mathbb{R}^n \setminus \cup_{j=0}^k \partial\Omega_j$. Then $k+1$ of the $k+2$ connected components of $\mathbb{R}^n \setminus \cup_{j=0}^k \partial\Omega_j$ are given by $\Omega_0^-, \Omega_1, \dots, \Omega_k$. Since $\mathbb{R}^n \setminus \cup_{j=0}^k \partial\Omega_j$ is the disjoint union of $\Omega_0^-, \Omega_1, \dots, \Omega_k$, and $\Omega_0 \setminus \cup_{j=0}^k \overline{\Omega_j}$, it follows that $\Omega_0 \setminus \cup_{j=0}^k \overline{\Omega_j}$ is connected. □

A.5 Measure Theory

Together with the classical Monotone Convergence Theorem and the Lebesgue Dominated Convergence, we also exploit the following Vitali Convergence Theorem (cf., e.g., Folland [102, Ex. 15, p. 187], Fichera [97]).

Theorem A.11 (of Convergence of Vitali). *Let $(X, \mathcal{M}, \mu)$ be a measured space with finite measure. Let $\{f_j\}_{j\in\mathbb{N}}$ be a sequence of functions in $L^1_\mu(X)$ which con-*

verges pointwise to a measurable function f from X to $\mathbb{R}$. If for each $\epsilon > 0$ there exists $\delta > 0$ such that

$$\int_E |f_j|\, d\mu \leq \epsilon \qquad \forall E \in \mathcal{M}, \mu(E) \leq \delta, \forall j \in \mathbb{N},$$

then $f \in L^1_\mu(X)$ and $\lim_{j\to\infty} f_j = f$ in $L^1_\mu(X)$.

Moreover, we need the Rademacher Theorem.

Theorem A.12 (of Rademacher). *Let Ω be an open subset of $\mathbb{R}^n$. If f is a Lipschitz continuous function from Ω to $\mathbb{R}$, then there exists a subset N of measure zero of Ω such that f is differentiable at all points of $\Omega \setminus N$.*

For a proof, we refer for example to Federer [96, 3.1.6, p. 216].

A.6 Calculus in Banach Spaces and the Implicit Function Theorem

In this section, we provide a brief account of the basic elements of differential calculus in normed spaces that are necessary to read the book. A more complete account can be found in Bourbaki [34], Cartan [41], Deimling [87], Krasnosel'skiĭ et al. [148], Nirenberg [239], Prodi and Ambrosetti [248], Ambrosetti and Prodi [6].

Let X, Y be normed spaces on the field $\mathbb{K}$. As usual, $\mathbb{K}$ denotes either $\mathbb{R}$ or $\mathbb{C}$. Let Ω be a subset of X. We say that a map f from Ω to Y, is differentiable at a point p of Ω provided that there exist $T \in \mathcal{L}(X,Y)$ and a map σ_p from Ω to Y, with σ_p continuous at p and $\sigma_p(p) = 0$, such that

$$f(x) = f(p) + T[x-p] + \sigma_p(x)\|x-p\|_X \qquad \forall x \in \Omega .$$

In general, T is not unique. However, T is unique if for example p is in the interior of Ω, or if $X = \Omega = \{0\}$ and $p = 0$. When it is unique, we denote T with the symbol $df(p)$ and we say that $df(p)$ is the differential of f at p. Moreover, we have

$$df(p)[u] = \partial_u f(p) \equiv \lim_{t\to 0} \frac{f(p+tu) - f(p)}{t} \qquad \forall u \in X .$$

We say that f is differentiable in Ω provided that f is differentiable at all points of Ω. If $f \in \mathcal{L}(X,Y)$, then f is differentiable at all points of X and $df(p) = f$ for all points p of X.

If $f(\Omega) \subseteq E \subseteq Y$ and if g is a map from the open subset E of Y to another normed space Z and is differentiable at $q \equiv f(p)$, with $dg(q)$ as differential, then the composite function $g \circ f$ is also differentiable at p and the following chain rule formula holds

$$d(g \circ f)(p) = dg(q) \circ df(p) .$$

We denote by $\mathcal{L}^{(2)}(X\times Y, Z)$ the space of bilinear and continuous maps from $X\times Y$ to Z and we set

$$\|\beta\|_{\mathcal{L}^{(2)}(X\times Y,Z)} \equiv \sup_{(x,y)\in B_X(0,1)\times B_Y(0,1)} \|\beta[x,y]\|_Z \qquad \forall \beta\in\mathcal{L}^{(2)}(X\times Y,Z).$$

If Z is complete, then $\left(\mathcal{L}^{(2)}(X\times Y,Z), \|\cdot\|_{\mathcal{L}^{(2)}(X\times Y,Z)}\right)$ is well known to be a Banach space. If $\beta\in\mathcal{L}^{(2)}(X\times Y,Z)$, then β is differentiable at all points (x_0,y_0) of $X\times Y$ and

$$d\beta(x_0,y_0)[(\xi,\eta)] = \beta[x_0,\eta]+\beta[\xi,y_0] \quad \forall(\xi,\eta)\in X\times Y.$$

If Ω is an open subset of X and if f is differentiable from Ω to Y, then we say that f is twice differentiable at a point $p\in\Omega$ provided that the map df from Ω to $\mathcal{L}(X,Y)$, which takes x to $df(x)$ is differentiable at p. If so, then the differential $d(df)(p)$ of df at p is an element of $\mathcal{L}(X,\mathcal{L}(X,Y))$. Since $\mathcal{L}(X,\mathcal{L}(X,Y))$ can be canonically identified with $\mathcal{L}^{(2)}(X\times X,Y)$, which we abbreviate as $\mathcal{L}^{(2)}(X;Y)$, it is customary to define as second differential of f at p the element

$$d^2 f(p)[\cdot,\cdot]$$

of $\mathcal{L}^{(2)}(X;Y)$ that is canonically associated to $d(df)(p)\in\mathcal{L}(X,\mathcal{L}(X,Y))$. It is known that $d^2f(p)[\cdot,\cdot]$ is symmetric, i.e., its value does not change if we permute its two arguments.

The subspace of the symmetric elements of $\mathcal{L}^{(2)}(X;Y)$ is denoted $\mathcal{L}_s^{(2)}(X;Y)$.

If $k\in\mathbb{N}\setminus\{0,1\}$, $\mathcal{L}^{(k)}(X;Y)$ denotes the space of k-linear and continuous maps from X^k to Y, and we set

$$\|\gamma\|_{\mathcal{L}^{(k)}(X;Y)} \equiv \sup_{(x_1,\dots,x_k)\in B_X(0,1)^k} \|\gamma[x_1,\dots,x_k]\|_Y \qquad \forall\gamma\in\mathcal{L}^{(k)}(X;Y). \tag{A.3}$$

If Y is complete, then $\left(\mathcal{L}^{(k)}(X;Y), \|\cdot\|_{\mathcal{L}^{(k)}(X;Y)}\right)$ is well known to be a Banach space. We sometimes use the standard abbreviation

$$\gamma[v]^k \equiv \gamma[\overbrace{v,\dots,v}^{k \text{ times}}] \qquad \forall v\in X.$$

Then we define the k-th differential of a function f by induction on $k\geq 1$. Let f be a function from an open subset Ω of X to Y and $p\in\Omega$. If $k=1$ we say that f is 1-time differentiable at p provided that f is differentiable at p and we set $d^1f(p)\equiv df(p)$. If $k>1$, then we say that f is k-times differentiable at p provided that f is $k-1$ times differentiable in an open neighborhood W of p and that the map $d^{k-1}f$ from W to $\mathcal{L}^{(k-1)}(X;Y)$, which takes x to $d^{k-1}f(x)$, is differentiable at p. Since $\mathcal{L}(X,\mathcal{L}^{(k-1)}(X;Y))$ can be canonically identified with $\mathcal{L}^{(k)}(X;Y)$, it is customary to define as k-th order differential of f at p, the element

$$d^k f(p)[\overbrace{\cdot,\ldots,\cdot}^{k \text{ times}}]$$

of $\mathcal{L}^{(k)}(X;Y)$ that is canonically associated to the differential

$$d(d^{k-1}f)(p) \in \mathcal{L}(X, \mathcal{L}^{(k-1)}(X;Y)),$$

of the map $d^{k-1}f$ at p. It is known that $d^k f(p)[\cdot,\ldots,\cdot]$ is symmetric, i.e., its value does not change if we permute its k arguments. The subspace of the symmetric elements of $\mathcal{L}^{(k)}(X;Y)$ is denoted $\mathcal{L}_s^{(k)}(X;Y)$. It is also customary to say that all functions f from Ω to Y are 0-times differentiable at all points of Ω, and to set $d^0 f \equiv f$.

If $k \in \mathbb{N}$, we say that f is k times continuously differentiable in an open set Ω, or of class C^k in Ω provided that f is k times differentiable in Ω, i.e., at all points of Ω, and that all differentials of f in Ω up to order k are continuous in Ω. The linear space of functions of class C^k from Ω to Y is denoted $C^k(\Omega, Y)$. Then we set

$$C^\infty(\Omega, Y) \equiv \bigcap_{k\in\mathbb{N}} C^k(\Omega, Y).$$

If $k \in \mathbb{N} \cup \{\infty\}$, the composition of maps of class C^k is of class C^k. All k-linear and continuous maps are of class C^∞.

Let $f \in C^\infty(\Omega, Y)$. We say that f is analytic at a point $p \in \Omega$ provided that there exists $r \in]0, +\infty[$ such that

$$\sum_{j=0}^{\infty} \frac{\|d^j f(p)\|_{\mathcal{L}^{(j)}(X;Y)}}{j!} r^j < +\infty, \tag{A.4}$$

and

$$f(x) = \sum_{j=0}^{\infty} \frac{1}{j!} d^j f(p)[x-p]^j \qquad \forall x \in B_X(p,r) \cap \Omega. \tag{A.5}$$

In case $\mathbb{K} = \mathbb{R}$, then we may say 'real analytic' instead of analytic in order to emphasize that the field is $\mathbb{R}$. We observe that condition (A.4) implies that

$$\|d^j f(p)\|_{\mathcal{L}^{(j)}(X;Y)} \leq C j! (1/r)^j \qquad \forall j \in \mathbb{N} \tag{A.6}$$

for some constant $C > 0$. On the other hand, if Y is complete and (A.6) holds with r replaced with some $r_1 > r$, then we can prove the convergence of the series in (A.4). In particular, we have the following equivalent conditions.

Lemma A.13. *Let X, Y be Banach spaces on $\mathbb{K}$. Let Ω be an open subset of X. Let f be a function of class C^∞ from Ω to Y. The following statements are equivalent.*

(i) f is analytic at $p \in \Omega$.
(ii) There exist real numbers $C, r > 0$ such that (A.5) *and* (A.6) *hold true.*

(iii) There exist multilinear symmetric operators $a_j \in \mathcal{L}^{(j)}(X;Y)$ and real numbers $C, M > 0$ such that

$$\|a_j\|_{\mathcal{L}^{(j)}(X;Y)} \leq CM^j \qquad \forall j \in \mathbb{N}$$

and

$$f(x) = \sum_{j=0}^{\infty} a_j[x-p]^j \qquad \forall x \in B_X(p, 1/M) \cap \Omega\,.$$

We say that f is analytic in Ω, provided that f is analytic at all points of Ω. We denote by $C^\omega(\Omega, Y)$ the set of analytic functions from Ω to Y.

By the classical theorems on the sum and product with scalars of limits, we can prove that the set $C^\omega(\Omega, Y)$ is a vector space. The series in the right-hand side of equality (A.5) is known as the Taylor series of f with initial point p.

All k-linear and continuous operators between Banach spaces are analytic and the composition of analytic functions is analytic.

Just as for real-valued analytic functions defined on an open subset of the Euclidean space, we have the following Identity Principle.

Proposition A.14. *Let X, Y be Banach spaces. Let Ω be a connected open subset of X. Let $f \in C^\omega(\Omega, Y)$. If the set of zeros $Z(f)$ of f has at least an interior point, then f is equal to zero at all points of Ω.*

A map f from an open subset Ω of X to an open subset E of Y is said to be a diffeomorphism of class C^k or C^ω, provided that f is a bijection from Ω onto E and that both f and the inverse function $f^{(-1)}$ of f are of class C^k or C^ω, respectively.

Then we have the following proposition. For a proof, we refer to Hille and Phillips [124, Theorems 4.3.2 and 4.3.4], Prodi and Ambrosetti [248, Theorem 13.1, Remark 13.2].

Proposition A.15. *Let X, Y be Banach spaces on $\mathbb{K}$. Then the set*

$$\mathcal{I}(X,Y) \equiv \{T \in \mathcal{L}(X,Y) : T \text{ is a bijection}\}$$

is open in $\mathcal{L}(X,Y)$ and $T^{(-1)} \in \mathcal{L}(Y,X)$ for all $T \in \mathcal{I}(X,Y)$. Moreover, the map J from $\mathcal{I}(X,Y)$ to $\mathcal{I}(Y,X)$ that takes T to the inverse operator $T^{(-1)}$ is analytic. Finally, if $T_0 \in \mathcal{I}(X,Y)$, then

$$dJ(T_0)[S] = -T_0^{(-1)} \circ S \circ T_0^{(-1)} \qquad \forall S \in \mathcal{L}(X,Y)\,.$$

We now turn to the Implicit Function Theorem in Banach spaces. First, we need some preliminaries. If X, Y, Z are normed spaces and if f is a function from an open subset Ω of $X \times Y$ to Z, and $(x^o, y^o) \in \Omega$, then we say that f is partially differentiable with respect to the variable x at (x_0, y_0) provided that the function $f(\cdot, y_0)$ from $\{x \in X : (x, y_0) \in \Omega\}$ to Z is differentiable at the point x_0. If this is the case, then $d_X f(x_0, y_0)$ denotes the differential of $f(\cdot, y_0)$ at the point $x = x_0$.

Similarly, we say that f is partially differentiable with respect to the variable y at (x_0, y_0) provided that the function $f(x_0, \cdot)$ from $\{y \in Y : (x_0, y) \in \Omega\}$ to

Z is differentiable at the point y_0. If this is the case, then $d_Y f(x_0, y_0)$ denotes the differential of $f(x_0, \cdot)$ at the point $y = y_0$.

We are now ready to introduce the following three statements related to the Implicit Function Theorem of Dini. For a proof we refer to Prodi and Ambrosetti [248, Theorem 11.6], Nirenberg [239, Theorem 2.7.2], Deimling [87, Theorem 15.3], Antman [13, Appendix]. The first statement concerns the existence and continuity of an implicitly defined function.

Theorem A.16. *Let X, Y, Z be normed spaces. Let Y and Z be complete. Let Ω be an open subset of $X \times Y$. Let $(x^o, y^o) \in \Omega$. Let f be a continuous function from Ω to Z such that*

(i) $f(x^o, y^o) = 0$.
(ii) $d_Y f(x, y)$ *exists for all* $(x, y) \in \Omega$ *and the function* $d_Y f$ *from* Ω *to* $\mathcal{L}(Y, Z)$ *is continuous at the point* (x^o, y^o).
(iii) $d_Y f(x^o, y^o)$ *is a linear homeomorphism from* Y *onto* Z.

Then there exist r, $\delta \in]0, +\infty[$ and a continuous function φ from $B_X(x^o, r)$ to $B_Y(y^o, \delta)$ such that

$$
\begin{aligned}
&\overline{B_X(x^o, r)} \times \overline{B_Y(y^o, \delta)} \subseteq \Omega\,, \\
&\{(x, y) \in B_X(x^o, r) \times B_Y(y^o, \delta) : f(x, y) = 0\} \\
&\qquad\qquad = \{(x, \varphi(x)) : x \in B_X(x^o, r)\} = \text{graph}\,(\varphi)\,,
\end{aligned}
$$

and

$$
d_Y f(x, \varphi(x)) \text{ is a linear homeomorphism from } Y \text{ onto } Z \qquad \text{for all } x \in B_X(x^o, r). \tag{A.7}
$$

In particular, $\varphi(x^o) = y^o$.

Remark A.17. The proof of the above Dini Theorem is based on the application of the Banach-Caccioppoli fixed point Theorem to the function

$$
g(x, y) \equiv y - Q^{(-1)} \circ f(x, y)
$$

of the variable $y \in \overline{B_Y(y^o, \delta)}$ for all fixed $x \in \overline{B_X(x^o, r)}$, where $Q \equiv d_Y f(x^o, y^o)$. One can check that it suffices to choose r, $\delta \in]0, +\infty[$ such that

$$
\begin{aligned}
&\overline{B_X(x^o, r)} \times \overline{B_Y(y^o, \delta)} \subseteq \Omega\,, \\
&a \equiv \sup_{(\xi,\eta) \in \overline{B_X(x^o, r)} \times \overline{B_Y(y^o, \delta)}} \|I_Y - Q^{(-1)} \circ d_Y f(\xi, \eta)\|_{\mathcal{L}(Y)} < 1\,, \\
&\sup_{\xi \in \overline{B_X(x^o, r)}} \|f(\xi, y^o)\|_Z < \frac{(1-a)\delta}{\|Q^{(-1)}\|_{\mathcal{L}(Z,Y)}}\,,
\end{aligned}
\tag{A.8}
$$

in order that φ as in the statement of the Dini Theorem exists (cf., e.g., Nirenberg [239, proof of Theorem 2.7.2]). We also note that if r, $\delta \in]0, +\infty[$ satisfy conditions (A.8), then the assumption that Ω is open and the continuity of $\|f(\cdot, \cdot)\|_Z$

and $\|I_Y - Q^{(-1)} \circ d_Y f(\cdot,\cdot)\|_{\mathcal{L}(Y)}$ at (x^o, y^o) ensure that there exists $r' > r$ such that r', δ still satisfy the same conditions in (A.8) and that accordingly φ has a continuous extension to $B_X(x^o, r')$ which maps $B_X(x^o, r')$ to $B_Y(y^o, \delta)$. In particular, φ has a uniquely determined continuous extension on $\overline{B_X(x^o, r)}$ and we have

$$\varphi(\overline{B_X(x^o, r)}) \subseteq B_Y(y^o, \delta),$$

where we still denote by φ the extension of φ.

Under the assumptions of the first part of the previous theorem, the implicitly defined function φ is continuous. Next we see that a function φ whose graph is contained in the set of zeros of a differentiable function f that satisfies certain conditions is necessarily differentiable and that φ carries the same regularity of f. More precisely, we have of the following Dini regularization statement. For a proof, we refer again to the above mentioned textbooks.

Proposition A.18. *Let X, Y, Z be normed spaces. Let Y and Z be complete. Let U be an open subset of X. Let V be an open subset of Y. Let f be a function from $U \times V$ to Z. Let φ be a function from U to V such that*

$$f(x, \varphi(x)) = 0 \qquad \forall x \in U. \tag{A.9}$$

Then the following statements hold.

(i) If $x^o \in U$, φ is continuous at x^o, f is differentiable at $(x^o, \varphi(x^o))$, and $d_Y f(x^o, \varphi(x^o))$ is a linear homeomorphism from Y onto Z, then φ is differentiable at x^o and

$$d\varphi(x^o) = -\left(d_Y f(x^o, y^o)\right)^{(-1)} \circ d_X f(x^o, y^o).$$

(ii) Let $k \in \{\infty\} \cup \mathbb{N} \setminus \{0\}$. If φ is of class C^0, f is of class C^k and $d_Y f(x, \varphi(x))$ is a linear homeomorphism from Y onto Z for all $x \in U$, then φ is of class C^k and

$$d\varphi(x) = -\left(d_Y f(x, \varphi(x))\right)^{(-1)} \circ d_X f(x, \varphi(x)) \qquad \forall x \in U. \tag{A.10}$$

By combining Theorem A.16 of Dini with the regularization Proposition A.18 of Dini, we deduce that if we choose f of class C^k for some k, then φ is also of class C^k. One could also prove that if f is analytic, then φ is also analytic. Indeed, we have the following.

Theorem A.19 (Implicit Function Theorem, of Dini). *Let X, Y, Z be normed spaces. Let Y and Z be complete. Let Ω be an open subset of $X \times Y$. Let (x^o, y^o) be in Ω. Let f be a function from Ω to Z such that*

(i) $f(x^o, y^o) = 0$.
(ii) f is differentiable in Ω and $d_Y f$ is continuous from Ω to $\mathcal{L}(Y, Z)$ at the point (x^o, y^o).
(iii) $d_Y f(x^o, y^o)$ is a linear homeomorphism from Y onto Z.

Then there exist $r,\ \delta \in]0,+\infty[$ *and a differentiable function* φ *from* $B_X(x^o,r)$ *to* $B_Y(y^o,\delta)$ *such that*

$$\overline{B_X(x^o,r)} \times \overline{B_Y(y^o,\delta)} \subseteq \Omega\,,$$

and

$$\begin{aligned}\{(x,y) \in B_X(x^o,r) \times B_Y(y^o,\delta) : f(x,y) = 0\} \\ = \{(x,\varphi(x)) : x \in B_X(x^o,r)\} = \mathrm{graph}\,(\varphi)\,.\end{aligned} \tag{A.11}$$

Moreover,

$$d\varphi(x) = -\left(d_Y f(x,\varphi(x))\right)^{(-1)} \circ d_X f(x,\varphi(x)) \qquad \forall x \in B_X(x^o,r)\,. \tag{A.12}$$

If f *is of class* C^k *for some* $k \in \{\infty\} \cup \mathbb{N} \setminus \{0\}$*, then* φ *is of class* C^k*. If we further assume that* f *is analytic, then* φ *is analytic.*

We now need to introduce a technical corollary of the Implicit Function Theorem, which can be considered as 'folklore' and which provides a uniform bound for the $C^{m,\alpha}$ norm of the implicitly defined functions corresponding to a suitable family of equations. To do so, we introduce the following preliminary definition of a class of functions that determines the family of equations.

Definition A.20. Let $n \in \mathbb{N}\setminus\{0,1\}$. Let $m \in \mathbb{N}\setminus\{0\}$, $\alpha \in [0,1]$. Let ι, $\sigma \in]0,+\infty[$, $\iota < \sigma$. Let Ω be a Whitney regular bounded open subset of $\mathbb{R}^{n-1} \times \mathbb{R}$. Let (x^o,y^o) belong to Ω.

(i) If $(m,\alpha) \neq (1,0)$, $\mathcal{F}_{m,\alpha}$ denotes the set of functions $f \in C^{m,\alpha}(\overline{\Omega})$ such that

$$f(x^o,y^o) = 0\,, \qquad \iota < \inf_{\Omega}\left|\frac{\partial f}{\partial y}\right|\,, \qquad \|f\|_{C^{m,\alpha}(\overline{\Omega})} \leq \sigma\,. \tag{A.13}$$

(ii) Let $\omega_{1,0}$ be an increasing function from $[0,+\infty[$ to itself such that $\omega_{1,0}(t) > 0$ for $t \in]0,+\infty[$ and

$$\lim_{t\to 0^+} \omega_{1,0}(t) = 0\,.$$

Then $\mathcal{F}_{1,0}$ denotes of functions $f \in C^1(\overline{\Omega})$ such that conditions (A.13) hold for $m = 1$ and $\alpha = 0$ and such that

$$\left|\nabla_{(x,y)} f(x',y') - \nabla_{(x,y)} f(x'',y'')\right| \leq \omega_{1,0}(|(x',y') - (x'',y'')|) \tag{A.14}$$

for all $(x',y'), (x'',y'') \in \Omega$.

We note that the class $\mathcal{F}_{1,0}$ depends also on the choice of $\omega_{1,0}$. However, just for the sake of brevity, we have chosen not to display the symbol $\omega_{1,0}$ in the notation for $\mathcal{F}_{1,0}$. Then we have the following.

Corollary A.21. *Let* $n \in \mathbb{N}\setminus\{0,1\}$*. Let* $m \in \mathbb{N}\setminus\{0\}$*,* $\alpha \in [0,1]$*. Let* ι*,* $\sigma \in]0,+\infty[$*,* $\iota < \sigma$*. Let* Ω *be a Whitney regular bounded open subset of* $\mathbb{R}^{n-1} \times \mathbb{R}$*. Let* (x^o,y^o) *belong to* Ω*.*

Let $\omega_{1,0}$ be an increasing function from $[0,+\infty[$ to itself such that $\omega_{1,0}(t) > 0$ for $t \in]0,+\infty[$ and

$$\lim_{t\to 0^+} \omega_{1,0}(t) = 0 .$$

Let $\omega_{1,\alpha}$ be the function from $[0,+\infty[$ to itself defined by

$$\omega_{1,\alpha}(t) \equiv \sigma t^{\alpha} \qquad \forall t \in [0,+\infty[$$

for all $\alpha \in]0,1]$.

Let $\omega_{k,\alpha}$ be the function from $[0,+\infty[$ to itself defined by

$$\omega_{k,\alpha}(t) \equiv 2\sigma c[\Omega] t \qquad \forall t \in [0,+\infty[,$$

for all $\alpha \in [0,1]$ and $k \in \mathbb{N} \setminus \{0,1\}$. Here $c[\Omega]$ is the Whitney constant of Ω (cf. (2.13)). Then the following statements hold.

(i) The set S of pairs $(r,\delta) \in]0,+\infty[^2$ that satisfy the following conditions

$$\overline{\mathbb{B}_{n-1}(x^o,r)} \times [y^o-\delta, y^o+\delta] \subseteq \Omega , \tag{A.15}$$
$$\omega_{m,\alpha}(r+\delta) < \iota ,$$
$$r\sigma + \omega_{m,\alpha}(r+\delta)\delta < \iota\delta ,$$

is not empty.

(ii) If $(f,r,\delta) \in \mathcal{F}_{m,\alpha} \times S$, then there exists one and only one

$$\varphi_f \in C^{m,\alpha}(\overline{\mathbb{B}_{n-1}(x^o,r)})$$

such that

$$\varphi_f\left(\overline{\mathbb{B}_{n-1}(x^o,r)}\right) \subseteq]y^o-\delta, y^o+\delta[, \tag{A.16}$$

$$\{(x,y) \in \mathbb{B}_{n-1}(x^o,r) \times]y^o-\delta, y^o+\delta[: \ f(x,y)=0\} \tag{A.17}$$
$$= \{(x,\varphi_f(x)) : x \in \mathbb{B}_{n-1}(x^o,r)\} ,$$

$$\frac{\partial \varphi_f}{\partial x_l}(x) = -\left(\frac{\partial f}{\partial y}(x,\varphi_f(x))\right)^{-1} \frac{\partial f}{\partial x_l}(x,\varphi_f(x)) \tag{A.18}$$
$$\forall x \in \mathbb{B}_{n-1}(x^o,r) ,\ l \in \{1,\ldots,n-1\} .$$

Moreover, if $(r,\delta) \in S$, then there exists $M_{(r,\delta)} \in]0,+\infty[$ such that

$$\|\varphi_f\|_{C^{m,\alpha}(\overline{\mathbb{B}_{n-1}(x^o,r)})} \le M_{(r,\delta)} \qquad \forall f \in \mathcal{F}_{m,\alpha} . \tag{A.19}$$

Proof. (i) Since Ω is open, there exists $(r,\delta) \in]0,+\infty[^2$ such that

$$\overline{\mathbb{B}_{n-1}(x^o,r)} \times [y^o-\delta, y^o+\delta] \subseteq \Omega .$$

Since $\lim_{t\to 0} \omega_{m,\alpha}(t) = 0$, possibly shrinking both r and δ, we can assume that $\omega_{m,\alpha}(r+\delta) < \iota$. Then we also have $0 < \iota - \omega_{m,\alpha}(r+\delta)$. Since the function

$\omega_{m,\alpha}(\rho+\delta)$ of the variable ρ decreases as ρ decreases to 0, possibly shrinking r we can assume that

$$\sigma r < (\iota - \omega_{m,\alpha}(r+\delta))\delta\,,$$

and thus we have proved that the set S is not empty.

(ii) We first prove that

$$\left|\nabla_{(x,y)} f(x',y') - \nabla_{(x,y)} f(x'',y'')\right| \qquad \text{(A.20)}$$
$$\leq \omega_{m,\alpha}(|(x',y')-(x'',y'')|) \quad \forall (x',y'),(x'',y'') \in \Omega\,,$$

for all $f \in \mathcal{F}_{m,\alpha}$. If $m=1$ and $\alpha=0$, inequality (A.20) holds by definition of $\mathcal{F}_{m,\alpha}$ (cf. (A.14)). If $m=1$ and $\alpha \in]0,1]$, then we have

$$\begin{aligned}
&\left|\nabla_{(x,y)} f(x',y') - \nabla_{(x,y)} f(x'',y'')\right| \\
&\quad \leq \left(\sum_{l=1}^{n}\left|\frac{\partial f}{\partial x_l}(x',y') - \frac{\partial f}{\partial x_l}(x'',y'')\right|^2\right)^{1/2} \\
&\quad \leq \sum_{l=1}^{n}\left|\frac{\partial f}{\partial x_l}(x',y') - \frac{\partial f}{\partial x_l}(x'',y'')\right| \\
&\quad \leq \|f\|_{C^{1,\alpha}(\overline{\Omega})}|(x',y')-(x'',y'')|^\alpha \leq \sigma|(x',y')-(x'',y'')|^\alpha \\
&\quad = \omega_{1,\alpha}(|(x',y')-(x'',y'')|) \quad \forall (x',y'),(x'',y'') \in \Omega\,,
\end{aligned}$$

and thus inequality (A.20) holds true for $m=1$, $\alpha \in]0,1]$. If $m>1$, then

$$\begin{aligned}
&\left|\nabla_{(x,y)} f(x',y') - \nabla_{(x,y)} f(x'',y'')\right| \\
&\quad \leq \left(\sum_{l=1}^{n}\left|\frac{\partial f}{\partial x_l}(x',y') - \frac{\partial f}{\partial x_l}(x'',y'')\right|^2\right)^{1/2} \\
&\quad \leq \left(\sum_{l=1}^{n}\sup_{\Omega}\left|\nabla_{(x,y)}\frac{\partial f}{\partial x_l}\right|^2\right)^{1/2} c[\Omega]\,|(x',y')-(x'',y'')| \\
&\quad \leq \left(\sum_{l=1}^{n}\sum_{j=1}^{n}\sup_{\Omega}\left|\frac{\partial^2 f}{\partial x_j \partial x_l}\right|^2\right)^{1/2} c[\Omega]\,|(x',y')-(x'',y'')| \\
&\quad \leq \sum_{l=1}^{n}\sum_{j=1}^{n}\sup_{\Omega}\left|\frac{\partial^2 f}{\partial x_j \partial x_l}\right| c[\Omega]\,|(x',y')-(x'',y'')| \\
&\quad \leq 2\|f\|_{C^{m,\alpha}(\overline{\Omega})}c[\Omega]\,|(x',y')-(x'',y'')| \leq 2\sigma c[\Omega]\,|(x',y')-(x'',y'')| \\
&\quad = \omega_{m,\alpha}(|(x',y')-(x'',y'')|) \quad \forall (x',y'),(x'',y'') \in \Omega\,,
\end{aligned}$$

and thus inequality (A.20) holds true also for $m>1$ and $\alpha \in [0,1]$. We now exploit the Implicit Function Theorem A.19 with $X \equiv \mathbb{R}^{n-1}$ and $Y \equiv Z \equiv \mathbb{R}$. Let $(f,r,\delta) \in \mathcal{F}_{m,\alpha} \times S$. Then we observe that

$$\begin{aligned}
a \equiv & \sup_{(\xi,\eta)\in\overline{\mathbb{B}_{n-1}(x^o,r)}\times\overline{\mathbb{B}_1(y^o,\delta)}} \|I_Y - (d_Y f(x^o,y^o))^{(-1)} \circ d_Y f(\xi,\eta)\|_{\mathcal{L}(Y)} \\
& = \sup_{(\xi,\eta)\in\overline{\mathbb{B}_{n-1}(x^o,r)}\times\overline{\mathbb{B}_1(y^o,\delta)}} \left| 1 - \left(\frac{\partial f}{\partial y}(x^o,y^o)\right)^{-1} \frac{\partial f}{\partial y}(\xi,\eta) \right| \\
& \leq \sup_{(\xi,\eta)\in\overline{\mathbb{B}_{n-1}(x^o,r)}\times\overline{\mathbb{B}_1(y^o,\delta)}} \iota^{-1} \left| \frac{\partial f}{\partial y}(x^o,y^o) - \frac{\partial f}{\partial y}(\xi,\eta) \right| \\
& \leq \iota^{-1}\omega_{m,\alpha}(r+\delta) < 1 .
\end{aligned} \tag{A.21}$$

Indeed,

$$\omega_{m,\alpha}(r+\delta) < \iota .$$

In particular, (A.21) implies that

$$-\iota^{-1}\omega_{m,\alpha}(r+\delta) \leq -a .$$

Next we note that

$$\begin{aligned}
\| (d_Y f(x^o,y^o))^{(-1)} \|_{\mathcal{L}(Z,Y)} &= \left| \left(\frac{\partial f}{\partial y}(x^o,y^o)\right)^{-1} \right| < \iota^{-1} , \\
\sup_{x\in\mathbb{B}_{n-1}(x^o,r)} |f(x,y^o)| &= \sup_{x\in\mathbb{B}_{n-1}(x^o,r)} |f(x,y^o) - f(x^o,y^o)| \\
&\leq \sup_{x\in\mathbb{B}_{n-1}(x^o,r)} |\nabla_{(x,y)} f(x,y^o)|\, |x - x^o| \leq \sigma r .
\end{aligned}$$

Then we have

$$\begin{aligned}
\sup_{x\in\overline{\mathbb{B}_{n-1}(x^o,r)}} \|f(x,y^o)\|_Z &= \sup_{x\in\mathbb{B}_{n-1}(x^o,r)} |f(x,y^o)| \\
&\leq \sigma r < (\iota - \omega_{m,\alpha}(r+\delta))\delta = \frac{1-\iota^{-1}\omega_{m,\alpha}(r+\delta)}{\iota^{-1}}\delta \\
&\leq \frac{(1-a)\delta}{\| (d_Y f(x^o,y^o))^{(-1)} \|_{\mathcal{L}(Z,Y)}} ,
\end{aligned}$$

and by Theorem A.16 of Dini and Remark A.17 we deduce the existence of an implicitly defined continuous function φ_f from $\mathbb{B}_{n-1}(x^o,r)$ to $]y^o-\delta, y^o+\delta[$ which has a continuous extension to $\overline{\mathbb{B}_{n-1}(x^o,r)}$ and which satisfies the inequality

$$|\varphi_f(x) - y^o| < \delta \qquad \forall x \in \overline{\mathbb{B}_{n-1}(x^o,r)} . \tag{A.22}$$

Then Proposition A.18 implies that φ_f is of class $C^m(\mathbb{B}_{n-1}(x^o,r))$. Next we fix $(r,\delta) \in S$, and we show the existence of $M_{(r,\delta)} \in]0,+\infty[$ such that (A.19) holds true. Our plan is to estimate the $C^{m-1,\alpha}$-norm of $\frac{\partial \varphi_f}{\partial x_l}$ by exploiting the known formula

$$\frac{\partial \varphi_f}{\partial x_l}(x) = -\left(\frac{\partial f}{\partial y}(x, \varphi_f(x))\right)^{-1} \frac{\partial f}{\partial x_l}(x, \varphi_f(x)) \quad \forall x \in \mathbb{B}_{n-1}(x^o, r) \tag{A.23}$$

for all $l \in \{1, \ldots, n-1\}$ (cf. (A.18)). As a first step, we want to show the existence of a uniform upper bound for the function

$$\left\| -\left(\frac{\partial f}{\partial y}\right)^{-1} \frac{\partial f}{\partial x_l} \right\|_{C^{m-1,\alpha}(\overline{\mathbb{B}_{n-1}(x^o,r) \times \mathbb{B}_1(y^o,\delta)})}$$

with $f \in \mathcal{F}_{m,\alpha}$ and $l \in \{1, \ldots, n-1\}$. We begin observing that

$$\left\| -\left(\frac{\partial f}{\partial y}\right)^{-1} \frac{\partial f}{\partial x_l} \right\|_{C^{m-1,\alpha}(\overline{\mathbb{B}_{n-1}(x^o,r) \times \mathbb{B}_1(y^o,\delta)})} \leq \left\| \left(\frac{\partial f}{\partial y}\right)^{-1} \frac{\partial f}{\partial x_l} \right\|_{C^{m-1,\alpha}(\overline{\Omega})}$$

for all $f \in \mathcal{F}_{m,\alpha}$ and for all $l \in \{1, \ldots, n-1\}$. Since the pointwise product in $C^{m-1,\alpha}(\overline{\Omega})$ is bilinear and continuous (cf. Proposition 2.54 (iii)), it suffices to note that

$$\left\| \frac{\partial f}{\partial x_l} \right\|_{C^{m-1,\alpha}(\overline{\Omega})} \leq \sigma \qquad \forall l \in \{1, \ldots, n-1\}, \tag{A.24}$$

(cf. (2.32), (A.13)), and to show the existence of an upper bound for

$$\left\| \left(\frac{\partial f}{\partial y}\right)^{-1} \right\|_{C^{m-1,\alpha}(\overline{\Omega})},$$

for all $f \in \mathcal{F}_{m,\alpha}$. Since the function $1/x$ belongs to $C^\infty([-2\sigma, -\iota/2] \cup [\iota/2, 2\sigma])$, which is continuously embedded into $C^{m-1,1}([-2\sigma, -\iota/2] \cup [\iota/2, 2\sigma])$, and the composition maps bounded sets of

$$C^{m-1,1}([-2\sigma, -\iota/2] \cup [\iota/2, 2\sigma]) \times C^{m-1,\alpha}(\overline{\Omega}, [-\sigma, -\iota] \cup [\iota, \sigma]),$$

to bounded subsets of $C^{m-1,\alpha}(\overline{\Omega})$ (cf. Theorem 2.59), there exists $c' > 0$ such that

$$\left\| \left(\frac{\partial f}{\partial y}\right)^{-1} \right\|_{C^{m-1,\alpha}(\overline{\Omega})} \leq c', \tag{A.25}$$

for all $f \in \mathcal{F}_{m,\alpha}$. Then the continuity of the pointwise product and inequality (A.24) imply that there exists $c'' > 0$ such that

$$\left\| \left(\frac{\partial f}{\partial y}\right)^{-1} \frac{\partial f}{\partial x_l} \right\|_{C^{m-1,\alpha}(\overline{\Omega})} \leq c'' \qquad \forall l \in \{1, \ldots, n-1\} \tag{A.26}$$

for all $f \in \mathcal{F}_{m,\alpha}$.

Next we prove that if $j \in \{1, \ldots, m\}$, then there exists $M_{(r,\delta),j} > 0$ such that

$$\|\varphi_f\|_{C^{j,\alpha}(\overline{\mathbb{B}_{n-1}(x^o,r)})} \leq M_{(r,\delta),j}\,,$$

for all $f \in \mathcal{F}_{m,\alpha}$. We argue by finite induction on $j \in \{1,\dots,m\}$. Let $j = 1$. By formula (A.23) for the partial derivatives of φ_f, and by the second and third inequality of (A.13), it follows that the first order partial derivatives of φ_f can be extended with continuity to $\overline{\mathbb{B}_{n-1}(x^o,r)}$ and that

$$\sup_{\mathbb{B}_{n-1}(x^o,r)} |\nabla\varphi_f| \leq \iota^{-1}\sigma\,. \tag{A.27}$$

By inequalities (A.22) and (A.27), we have

$$\|\varphi_f\|_{C^{0,1}(\overline{\mathbb{B}_{n-1}(x^o,r)})} \leq |y^o| + \delta + \sup_{\mathbb{B}_{n-1}(x^o,r)} |\nabla\varphi_f| \leq |y^o| + \delta + \iota^{-1}\sigma\,. \tag{A.28}$$

Next we find convenient to set

$$\Phi_f(\eta) \equiv (\eta, \varphi_f(\eta)) \qquad \forall \eta \in \overline{\mathbb{B}_{n-1}(x^o,r)}\,.$$

Then we have

$$\begin{aligned}\|\Phi_f\|_{C^{0,1}(\overline{\mathbb{B}_{n-1}(x^o,r)},\mathbb{R}^n)} &\leq \sum_{l=1}^{n-1} \|\eta_l\|_{C^{0,1}(\overline{\mathbb{B}_{n-1}(x^o,r)})} + \|\varphi_f\|_{C^{0,1}(\overline{\mathbb{B}_{n-1}(x^o,r)})} \\ &\leq (n-1)(r + |x^o| + 1) + |y^o| + \delta + \iota^{-1}\sigma\end{aligned}$$

for all $f \in \mathcal{F}_{m,\alpha}$. Since the composition maps bounded subsets of

$$C^{0,\alpha}(\overline{\Omega}) \times C^{0,1}(\overline{\mathbb{B}_{n-1}(x^o,r)}, \overline{\Omega})$$

to bounded subsets of $C^{0,\alpha}(\overline{\mathbb{B}_{n-1}(x^o,r)})$ (cf. Theorem 2.59), formula (A.18) for the partial derivatives of φ_f and inequality (A.26) together with the continuity of the embedding of $C^{m-1,\alpha}(\overline{\Omega})$ into $C^{0,\alpha}(\overline{\Omega})$ imply the existence of $A_1 > 0$ such that

$$\left\|\frac{\partial\varphi_f}{\partial x_l}\right\|_{C^{0,\alpha}(\overline{\Omega})} \leq A_1 \qquad \forall l \in \{1,\dots,n-1\}\,,$$

for all $f \in \mathcal{F}_{m,\alpha}$. Then by the definition of $\|\cdot\|_{C^{1,\alpha}(\overline{\Omega})}$ norm, we have

$$\begin{aligned}\|\varphi_f\|_{C^{1,\alpha}(\overline{\Omega})} &= \sup_{\mathbb{B}_{n-1}(x^o,r)} |\varphi_f| + \sum_{l=1}^{n-1} \left\|\frac{\partial\varphi_f}{\partial x_l}\right\|_{C^{0,\alpha}(\overline{\Omega})} \\ &\leq |y^o| + \delta + (n-1)A_1 \equiv M_{(r,\delta),1}\end{aligned}$$

for all $f \in \mathcal{F}_{m,\alpha}$. We now assume that $M_{(r,\delta),j}$ does exist for $1 \leq j < m$, and we prove the existence of $M_{(r,\delta),j+1}$. By the inductive assumption, we have

$$\|\Phi_f\|_{C^{j,\alpha}(\overline{\mathbb{B}_{n-1}(x^o,r)},\mathbb{R}^n)} \leq \sum_{l=1}^{n-1} \|\eta_l\|_{C^{j,\alpha}(\overline{\mathbb{B}_{n-1}(x^o,r)})} + \|\varphi_f\|_{C^{j,\alpha}(\overline{\mathbb{B}_{n-1}(x^o,r)})} \tag{A.29}$$
$$\leq \sum_{l=1}^{n-1} \|\eta_l\|_{C^{j,\alpha}(\overline{\mathbb{B}_{n-1}(x^o,r)})} + M_{(r,\delta),j}$$

for all $f \in \mathcal{F}_{m,\alpha}$. Since the composition maps bounded subsets of

$$\left\{(\Psi,\Phi) \in C^{j,\alpha}(\overline{\Omega}) \times C^{j,\alpha}(\overline{\mathbb{B}_{n-1}(x^o,r)},\overline{\Omega}) : \Phi(\mathbb{B}_{n-1}(x^o,r)) \subseteq \Omega\right\}$$

to bounded subsets of $C^{j,\alpha}(\overline{\mathbb{B}_{n-1}(x^o,r)})$ (cf. Theorem 2.59), and $C^{m-1,\alpha}(\overline{\Omega})$ is continuously embedded into $C^{j,\alpha}(\overline{\Omega})$, formula (A.18) for the partial derivatives of φ_f and inequality (A.26) imply that there exists $A_{j+1} > 0$ such that

$$\left\|\frac{\partial \varphi_f}{\partial x_l}\right\|_{C^{j,\alpha}(\overline{\mathbb{B}_{n-1}(x^o,r)})} \leq A_{j+1}. \qquad \forall l \in \{1,\dots,n-1\}$$

for all $f \in \mathcal{F}_{m,\alpha}$. Hence, we have

$$\|\varphi_f\|_{C^{j+1,\alpha}(\overline{\mathbb{B}_{n-1}(x^o,r)})} \leq \sup_{\mathbb{B}_{n-1}(x^o,r)} |\varphi_f| + \sum_{l=1}^{n-1} \left\|\frac{\partial \varphi_f}{\partial x_l}\right\|_{C^{j,\alpha}(\overline{\mathbb{B}_{n-1}(x^o,r)})}$$
$$\leq |y^o| + \delta + (n-1)A_{j+1} \equiv M_{(r,\delta),j+1}\,,$$

for all $f \in \mathcal{F}_{m,\alpha}$ (cf. (2.32)). Hence, the constant $M_{(r,\delta),j}$ does exist for all j in $\{1,\dots,m\}$ and thus in particular for $j = m$. Accordingly, we can take $M_{(r,\delta)} \equiv M_{(r,\delta),m}$ and the proof is complete. □

Then we can readily deduce the following simplified form of Corollary A.21.

Corollary A.22. *Let $n \in \mathbb{N}\setminus\{0,1\}$. Let $m \in \mathbb{N}\setminus\{0\}$, $\alpha \in [0,1]$. Let ι, $\sigma \in]0,+\infty[$, $\iota < \sigma$. Let Ω be a Whitney regular bounded open subset of $\mathbb{R}^{n-1} \times \mathbb{R}$. Let (x^o,y^o) belong to Ω.*

In case $(m,\alpha) = (1,0)$ we also introduce an increasing function $\omega_{1,0}$ from $[0,+\infty[$ to itself such that $\omega_{1,0}(t) > 0$ for $t \in]0,+\infty[$ and

$$\lim_{t\to 0^+} \omega_{1,0}(t) = 0\,.$$

Let $\mathcal{F}_{m,\alpha}$ be defined as in Definition A.20. Let $\delta^* \in]0,+\infty[$. Then there exist $r_1 \in]0,+\infty[$, $\delta \in]0,\delta^*[$, $M \in]0,+\infty[$ such that

$$\overline{\mathbb{B}_{n-1}(x^o,r)} \times [y^o-\delta, y^o+\delta] \subseteq \Omega \qquad \forall r \in]0,r_1]$$

and such that for each $(f,r) \in \mathcal{F}_{m,\alpha}\times]0,r_1]$ there exists one and only one function $\varphi_f \in C^{m,\alpha}(\overline{\mathbb{B}_{n-1}(x^o,r)})$ such that (A.16)–(A.18) hold true and

$$\|\varphi_f\|_{C^{m,\alpha}(\overline{\mathbb{B}_{n-1}(x^o,r)})} \leq M \qquad \forall (f,r) \in \mathcal{F}_{m,\alpha}\times]0,r_1]\,. \tag{A.30}$$

Proof. By Corollary A.21 (i), the set S is not empty. Possibly shrinking first δ and then r, we can see that there exist $r_1 \in]0, +\infty[$ and $\delta < \delta^*$ such that $(r, \delta) \in S$ for all $r \in]0, r_1]$. Then Corollary A.21 (ii) implies the existence and uniqueness of φ_f and by applying Corollary A.21 (ii) with the pair (r_1, δ), we deduce the existence of M such that $\|\varphi_f\|_{C^{m,\alpha}(\overline{\mathbb{B}_{n-1}(x^o,r_1)})} \leq M$ for all $f \in \mathcal{F}_{m,\alpha}$. Since $\|\varphi_f\|_{C^{m,\alpha}(\overline{\mathbb{B}_{n-1}(x^o,r)})} \leq \|\varphi_f\|_{C^{m,\alpha}(\overline{\mathbb{B}_{n-1}(x^o,r_1)})}$, inequality (A.30) holds true. □

A.7 Composition Operators

Let Ω be a bounded open subset of $\mathbb{R}^n$. Let Ω_1 be an open subset of $\mathbb{R}$. Let $G \in C^0(\overline{\Omega} \times \Omega_1)$. We denote by N_G the (nonautonomous) composition operator from $C^0(\overline{\Omega}, \Omega_1)$ to $C^0(\overline{\Omega})$ which maps $v \in C^0(\overline{\Omega}, \Omega_1)$ to the function $N_G[v]$ defined by

$$N_G[v](x) \equiv G(x, v(x)) \qquad \forall x \in \overline{\Omega}\,.$$

Here the letter 'N' stands for Nemytskii operator. Similarly, if G belongs to $C^0(\partial\Omega \times \Omega_1)$, we denote by N_G the (nonautonomous) composition operator from $C^0(\partial\Omega, \Omega_1)$ to $C^0(\partial\Omega)$ which maps $v \in C^0(\partial\Omega, \Omega_1)$ to the function $N_G[v]$ defined by

$$N_G[v](x) \equiv G(x, v(x)) \qquad \forall x \in \partial\Omega\,.$$

If G is sufficiently regular, then N_G is known to map a Schauder space $C^{m,\alpha}$ to itself. Here we mention the work of Berkolajko, Bondarenko, Rutitskij, Zabrejko, Drábek, and of Sobolevskij mainly for case $m = 0$, as reported in the extensive monograph of Appell and Zabrejko [14]. We need the following result on (nonautonomous) composition operators.

Theorem A.23. *Let $n \in \mathbb{N} \setminus \{0, 1\}$. Let $\alpha \in]0, 1]$. Let Ω be a bounded open connected subset of $\mathbb{R}^n$ of class $C^{1,\alpha}$. Let Ω_1 be an open subset of $\mathbb{R}$. Let $m \in \{0, 1\}$.*

(i) If G is an analytic function from $\overline{\Omega} \times \Omega_1$ to $\mathbb{R}$, then the composition operator N_G is analytic from $C^{m,\alpha}(\overline{\Omega}, \Omega_1)$ to $C^{m,\alpha}(\overline{\Omega})$ and the Fréchet differential of N_G at a point $\tilde{u} \in C^{m,\alpha}(\overline{\Omega}, \Omega_1)$ is delivered by the formula

$$dN_G(\tilde{u})[v](x) = G_u(x, \tilde{u}(x))v(x) \qquad \forall x \in \overline{\Omega}\,,$$

for all $v \in C^{m,\alpha}(\overline{\Omega})$, where G_u denotes the partial derivative of G with respect to its last variable.

(ii) If G is an analytic function from $\partial\Omega \times \Omega_1$ to $\mathbb{R}$, then the composition operator N_G is analytic from $C^{m,\alpha}(\partial\Omega, \Omega_1)$ to $C^{m,\alpha}(\partial\Omega)$, and the Fréchet differential of N_G at a point $\tilde{u} \in C^{m,\alpha}(\partial\Omega, \Omega_1)$ is delivered by the formula

$$dN_G(\tilde{u})[v](x) = G_u(x, \tilde{u}(x))v(x) \qquad \forall x \in \partial\Omega\,,$$

for all $v \in C^{m,\alpha}(\partial\Omega)$, where G_u denotes the partial derivative of G with respect to its last variable.

In the above statement, the assumption that G is an analytic function from $\overline{\Omega} \times \Omega_1$ to $\mathbb{R}$ means that there exist an open neighborhood W of $\overline{\Omega} \times \Omega_1$ in $\mathbb{R}^n \times \mathbb{R}$ and an analytic function $\tilde{G}$ from W to $\mathbb{R}$ such that the restriction $\tilde{G}_{|\overline{\Omega} \times \Omega_1}$ of $\tilde{G}$ to $\overline{\Omega} \times \Omega_1$ coincides with G.

Similarly, we understand that G being analytic on $\partial\Omega \times \Omega_1$ means that G has a real analytic extension on an open neighborhood of $\partial\Omega \times \Omega_1$ in $\mathbb{R}^n \times \mathbb{R}$. In the case where G does not depend on its first variable x—the so-called autonomous case—the proof of statement (i) can be found in Böhme and Tomi [28, p. 10] and Henry [122, pp. 29, 54]. For the proof of statement (i) with G that depends also on x—the so-called nonautonomous case—we refer to Valent [269, Theorem 5.2]. Statement (ii) can be proved by the same argument of the proof of statement (i).

For a treatment of the continuity and differentiability of autonomous composition operators in Banach algebras with applications to function spaces, we refer to [158, 159].

We also need the following elementary lemma.

Lemma A.24. *Let $n \in \mathbb{N} \setminus \{0\}$. Let $\alpha \in]0, 1]$. Let $\mathbb{D}$ be a compact subset of $\mathbb{R}^n$. Let G be a locally Lipschitz continuous function from $\mathbb{D} \times \mathbb{R}$ to $\mathbb{R}$. Then the composition operator N_G from $C^{0,\alpha}(\mathbb{D})$ to itself defined by*

$$N_G[v](x) \equiv G(x, v(x)) \qquad \forall x \in \mathbb{D}$$

for all $v \in C^{0,\alpha}(\mathbb{D})$ maps bounded subsets of $C^{0,\alpha}(\mathbb{D})$ to bounded subsets of $C^{0,\alpha}(\mathbb{D})$.

Proof. Let A be a bounded subset of $C^{0,\alpha}(\mathbb{D})$. Let $a \in]0, +\infty[$ be such that

$$|v| + 1 \leq a \qquad \forall v \in A\,.$$

Since G is Lipschitz continuous on the compact set $\mathbb{D} \times [-a, a]$, the statement follows by the elementary inequality

$$\begin{aligned}|G(x, v(x)) - G(y, v(y))| &\leq \mathrm{Lip}\,(G_{|\mathbb{D}\times[-a,a]})(|x - y| + |v(x) - v(y)|)\\ &\leq \mathrm{Lip}\,(G_{|\mathbb{D}\times[-a,a]})[(\mathrm{diam}\,(\mathbb{D}))^{1-\alpha} + |v : \mathbb{D}|_\alpha]|x - y|^\alpha \qquad \forall x, y \in \mathbb{D}\,,\end{aligned}$$

for all $v \in A$. □

In order to consider the dependence upon both the functions of a composition, we need an extension of the composition operator of Theorem A.23. Then we introduce the following (slightly restated) result of Preciso [247, Prop. 1.1, p. 101] (see also Preciso [246, Proposition 4.2.16]), which also involves the notion of Roumieu class (cf. (2.7)).

Proposition A.25. *Let $n_1, n_2 \in \mathbb{N} \setminus \{0\}$, $\rho \in]0, +\infty[$. Let Ω_1 be a bounded open subset of $\mathbb{R}^{n_1}$. Let Ω_2 be a bounded open connected subset of $\mathbb{R}^{n_2}$ of class C^1. Then the composition operator T from $C^0_{\omega,\rho}(\overline{\Omega_1}) \times C^{1,\alpha}(\overline{\Omega_2}, \Omega_1)$ to $C^{1,\alpha}(\overline{\Omega_2})$ defined by*

$$T[u, v] \equiv u \circ v \quad \forall (u, v) \in C^0_{\omega,\rho}(\overline{\Omega_1}) \times C^{1,\alpha}(\overline{\Omega_2}, \Omega_1)\,,$$

is real analytic.

For other results on the differentiability of the composition operator T in Schauder spaces, we refer to Valent [269], and to [157–159].

A.8 Integral Operators with Real Analytic Kernel

In the book we need several times to prove real analyticity results for integral operators with real analytic kernels. The first time it happens is in the proof of Proposition 8.2, where we introduce the family of operators $\tilde{\mathcal{V}}^i(\epsilon)$ which take a density $\theta^o \in C^{0,\alpha}(\partial\Omega^o)$ to the function

$$\tilde{\mathcal{V}}^i(\epsilon)[\theta^o](\xi) \equiv \int_{\partial\Omega^o} S_n(\epsilon\xi - y)\theta^o(y)\, d\sigma_y \qquad \forall \xi \in \partial\Omega^i \tag{A.31}$$

and we want to show that the map that takes $\epsilon \in]-\epsilon_0, \epsilon_0[$ to

$$\tilde{\mathcal{V}}^i(\epsilon) \in \mathcal{L}(C^{0,\alpha}(\partial\Omega^o)\,,\; C^{1,\alpha}(\partial\Omega^i))$$

is real analytic. As we shall see, a result of this kind can be deduced from a proposition proven in [175] in a more general setting (cf. Theorem A.27 below, see also Corollary A.29). For the sake of completeness and to illustrate the kind of arguments that are involved, we provide a direct proof of the following Lemma A.26.

Lemma A.26. *Let $n \in \mathbb{N} \setminus \{0, 1\}$. Let $\alpha \in]0, 1[$. Let Ω^i and Ω^o be bounded open connected subsets of $\mathbb{R}^n$ of class $C^{1,\alpha}$, with connected exteriors Ω^{i-} and Ω^{o-}, and such that the origin 0 of $\mathbb{R}^n$ belongs both to Ω^i and to Ω^o. Assume that $\epsilon_0 \in]0, +\infty[$ and that $\epsilon\overline{\Omega^i} \subseteq \Omega^o$ for all $\epsilon \in]-\epsilon_0, \epsilon_0[$.*

Then the map from $]-\epsilon_0, \epsilon_0[$ to $\mathcal{L}(C^{0,\alpha}(\partial\Omega^o)\,,\; C^{1,\alpha}(\partial\Omega^i))$ that takes ϵ to the linear operator $\tilde{\mathcal{V}}^i(\epsilon)$ defined in (A.31) *is real analytic.*

Proof. We fix $\epsilon_1 \in]0, \epsilon_0[$. By the continuity of the multiplication by a scalar in $\mathbb{R}^n$ and by the compactness of the set $[-\epsilon_1, \epsilon_1] \times \overline{\Omega^i}$, the set

$$\left\{\epsilon x : (\epsilon, x) \in [-\epsilon_1, \epsilon_1] \times \overline{\Omega^i}\right\} = \cup_{\epsilon\in[-\epsilon_1,\epsilon_1]}\epsilon\overline{\Omega^i}$$

is a compact subset of the open set Ω^o. Accordingly, we have

$$\left(\cup_{\epsilon\in[-\epsilon_1,\epsilon_1]}\epsilon\overline{\Omega^i}\right) \cap \partial\Omega^o = \emptyset\,.$$

Then

$$\mathrm{dist}\left(\cup_{\epsilon\in[-\epsilon_1,\epsilon_1]}\epsilon\overline{\Omega^i}\,,\; \partial\Omega^o\right) > \delta.$$

for some positive δ. Hence,

$$\text{dist}\left(\epsilon\overline{\Omega^i}\,,\,\partial\Omega^o\right) > \delta$$

for all $\epsilon \in [-\epsilon_1, \epsilon_1]$. In particular,

$$\text{dist}\left(\epsilon\partial\Omega^i\,,\,\partial\Omega^o\right) > \delta$$

for all $\epsilon \in [-\epsilon_1, \epsilon_1]$. Then, if we set

$$(\partial\Omega^i)_{\delta/(4\epsilon_1)} \equiv \{x \in \mathbb{R}^n\,:\,\text{dist}(x, \partial\Omega^i) < \delta/(4\epsilon_1)\}$$

and

$$(\partial\Omega^o)_{\delta/4} \equiv \{x \in \mathbb{R}^n\,:\,\text{dist}(x, \partial\Omega^o) < \delta/4\},$$

the triangular inequality implies that

$$\text{dist}\left(\epsilon(\partial\Omega^i)_{\delta/(4\epsilon_1)}\,,\,(\partial\Omega^o)_{\delta/4}\right) > \delta/2$$

for all $\epsilon \in]-\epsilon_1, \epsilon_1[$. In particular,

$$\epsilon\xi - y \neq 0 \qquad \forall(\epsilon, \xi, y) \in]-\epsilon_1, \epsilon_1[\times(\partial\Omega^i)_{\delta/(4\epsilon_1)} \times (\partial\Omega^o)_{\delta/4}\,.$$

Since the fundamental solution S_n is real analytic on $\mathbb{R}^n \setminus \{0\}$, and since the map from $]-\epsilon_1, \epsilon_1[\times(\partial\Omega^i)_{\delta/(4\epsilon_1)} \times (\partial\Omega^o)_{\delta/4}$ to $\mathbb{R}^n \setminus \{0\}$ that takes (ϵ, ξ, y) to $\epsilon\xi - y$ is real analytic, it follows that the map from $]-\epsilon_1, \epsilon_1[\times(\partial\Omega^i)_{\delta/(4\epsilon_1)} \times (\partial\Omega^o)_{\delta/4}$ to $\mathbb{R}$ that takes (ϵ, ξ, y) to $S_n(\epsilon\xi - y)$ is real analytic. Hence, for each $(\epsilon_\#, \xi_\#, y_\#)$ in $]-\epsilon_1, \epsilon_1[\times(\partial\Omega^i)_{\delta/(4\epsilon_1)} \times (\partial\Omega^o)_{\delta/4}$ there exist real numbers $C_\# > 0$, $M_\# > 0$ and $a^\#_{j,\beta,\gamma}$ for all $j \in \mathbb{N}$ and $\beta, \gamma \in \mathbb{N}^n$, such that

$$\left|a^\#_{j,\beta,\gamma}\right| \leq C_\# M_\#^{j+|\beta|+|\gamma|} \tag{A.32}$$

and such that

$$S_n(\epsilon\xi - y) = \sum_{l=0}^{\infty} \sum_{(j,\beta,\gamma)\in\mathbb{N}\times\mathbb{N}^n\times\mathbb{N}^n, |(j,\beta,\gamma)|=l} a^\#_{j,\beta,\gamma}(\epsilon - \epsilon_\#)^j(\xi - \xi_\#)^\beta(y - y_\#)^\gamma$$

for all (ϵ, ξ, y) in the open neighborhood

$$]\epsilon_\# - 1/M_\#, \epsilon_\# + 1/M_\#[\times\mathbb{B}_n(\xi_\#, 1/M_\#) \times \mathbb{B}_n(y_\#, 1/M_\#)$$

of $(\epsilon_\#, \xi_\#, y_\#)$ (cf. Theorem 2.17). Then, by setting

$$a^\#_j(\xi, y) \equiv \sum_{(\beta,\gamma)\in\mathbb{N}^n\times\mathbb{N}^n} a^\#_{j,\beta,\gamma}(\xi - \xi_\#)^\beta(y - y_\#)^\gamma \tag{A.33}$$

for all $(\xi, y) \in \mathbb{B}_n(\xi_\#, 1/M_\#) \times \mathbb{B}_n(y_\#, 1/M_\#)$ and all $j \in \mathbb{N}$, we see that

$$S_n(\epsilon\xi - y) = \sum_{j=0}^{\infty} a_j^{\#}(\xi, y)(\epsilon - \epsilon_{\#})^j$$

for $|\epsilon - \epsilon_{\#}| < 1/M_{\#}$. A computation based on (A.32) shows that

$$\sum_{(\beta,\gamma)\in\mathbb{N}^n\times\mathbb{N}^n} \left|a^{\#}_{j,\beta,\gamma}\right| |\xi - \xi_{\#}|^{|\beta|} |y - y_{\#}|^{|\gamma|}$$
$$\leq C_{\#} M_{\#}^j \sum_{(\beta,\gamma)\in\mathbb{N}^n\times\mathbb{N}^n} (M_{\#} |\xi - \xi_{\#}|)^{|\beta|} (M_{\#} |y - y_{\#}|)^{|\gamma|} .$$

Since $M_{\#} |\xi - \xi_{\#}|$ and $M_{\#} |y - y_{\#}|$ are smaller than one when (ξ, y) belongs to $\mathbb{B}_n(\xi_{\#}, 1/M_{\#}) \times \mathbb{B}_n(y_{\#}, 1/M_{\#})$, we deduce that the series on the right-hand side converges on $\mathbb{B}_n(\xi_{\#}, 1/M_{\#}) \times \mathbb{B}_n(y_{\#}, 1/M_{\#})$. Accordingly, the functions $a_j^{\#}$ are real analytic on $\mathbb{B}_n(\xi_{\#}, 1/M_{\#}) \times \mathbb{B}_n(y_{\#}, 1/M_{\#})$ (cf. (A.33)). Moreover, we have

$$\partial_\xi^{\beta'} \partial_y^{\gamma'} a_j^{\#}(\xi, y)$$
$$= \sum_{(\beta,\gamma)\in\mathbb{N}^n\times\mathbb{N}^n,\, \beta\geq\beta',\gamma\geq\gamma'} a^{\#}_{j,\beta,\gamma} \frac{\beta!}{(\beta - \beta')!}(\xi - \xi_{\#})^{\beta-\beta'} \frac{\gamma!}{(\gamma - \gamma')!}(y - y_{\#})^{\gamma-\gamma'}$$

for all $\beta', \gamma' \in \mathbb{N}^n$ and all $(\xi, y) \in \mathbb{B}_n(\xi_{\#}, 1/M_{\#}) \times \mathbb{B}_n(y_{\#}, 1/M_{\#})$. By inequality (A.32) we deduce that

$$\left|\partial_\xi^{\beta'} \partial_y^{\gamma'} a_j^{\#}(\xi, y)\right|$$
$$\leq C_{\#} M_{\#}^{j+|\beta'|+|\gamma'|} \sum_{(\beta,\gamma)\in\mathbb{N}^n\times\mathbb{N}^n,\, \beta\geq\beta',\gamma\geq\gamma'} \frac{\beta!}{(\beta - \beta')!}\eta_{\#}^{\beta-\beta'} \frac{\gamma!}{(\gamma - \gamma')!} z_{\#}^{\gamma-\gamma'}$$
$$= C_{\#} M_{\#}^{j+|\beta'|+|\gamma'|} \sum_{\beta\in\mathbb{N}^n,\, \beta\geq\beta'} \frac{\beta!}{(\beta - \beta')!}\eta_{\#}^{\beta-\beta'} \sum_{\gamma\in\mathbb{N}^n,\, \gamma\geq\gamma'} \frac{\gamma!}{(\gamma - \gamma')!} z_{\#}^{\gamma-\gamma'} \tag{A.34}$$

for all $(\xi, y) \in \mathbb{B}_n(\xi_{\#}, 1/M_{\#}) \times \mathbb{B}_n(y_{\#}, 1/M_{\#})$, where $\eta_{\#}$ and $z_{\#}$ are the vectors of $\mathbb{R}^n$ defined by

$$\eta_{\#} \equiv (M_{\#}|\xi - \xi_{\#}|, \ldots, M_{\#}|\xi - \xi_{\#}|)$$

and

$$z_{\#} \equiv (M_{\#}|y - y_{\#}|, \ldots, M_{\#}|y - y_{\#}|) .$$

Since $|\eta_{\#,l}| < 1$ for all $l \in \{1, \ldots, n\}$, a computation based on the equality

$$\sum_{\beta\in\mathbb{N}^n} \eta_{\#}^{\beta} = \prod_{l=1}^{n} \frac{1}{(1 - \eta_{\#,l})}$$

shows that

$$\sum_{\beta\in\mathbb{N}^n,\,\beta\geq\beta'} \frac{\beta!}{(\beta-\beta')!}\eta_\#^{\beta-\beta'} = \left(\partial_\eta^{\beta'}\prod_{l=1}^n \frac{1}{(1-\eta_l)}\right)_{|\eta=\eta_\#}$$
$$= \prod_{l=1}^n \frac{\beta_l'!}{(1-\eta_{\#,l})^{\beta_l'+1}} = \frac{\beta'!}{(1-M_\#|\xi-\xi_\#|)^{|\beta'|+n}}$$

(cf. John [130, Chap. 3]). Similarly, we verify that

$$\sum_{\gamma\in\mathbb{N}^n,\,\gamma\geq\gamma'} \frac{\gamma!}{(\gamma-\gamma')!}z_\#^{\gamma-\gamma'} = \frac{\gamma'!}{(1-M_\#|y-y_\#|)^{|\gamma'|+n}}.$$

Then, by inequality (A.34) we deduce that

$$\left|\partial_\xi^{\beta'}\partial_y^{\gamma'}a_j^\#(\xi,y)\right|$$
$$\leq C_\# M_\#^{j+|\beta'|+|\gamma'|}\frac{\beta'!}{(1-M_\#|\xi-\xi_\#|)^{|\beta'|+n}}\frac{\gamma'!}{(1-M_\#|y-y_\#|)^{|\gamma'|+n}}$$

for all $(\xi,y)\in\mathbb{B}_n(\xi_\#,1/M_\#)\times\mathbb{B}_n(y_\#,1/M_\#)$. In particular, if m' belongs to $\{0,\ldots,m\}$ and $|\beta'|+|\gamma'|=m'$, $(\xi,y)\in\overline{\mathbb{B}_n(\xi_\#,1/(2M_\#))\times\mathbb{B}_n(y_\#,1/(2M_\#))}$, we have

$$\left|\partial_\xi^{\beta'}\partial_y^{\gamma'}a_j^\#(\xi,y)\right|\leq C_\# m'!2^{2n+m'}M_\#^{m'}M_\#^j.$$

We conclude that, for each $m\in\mathbb{N}$, there exists $C_{\#,m}>0$ such that

$$\|a_j^\#\|_{C^m\left(\overline{\mathbb{B}_n(\xi_\#,1/(2M_\#))\times\mathbb{B}_n(y_\#,1/(2M_\#))}\right)} < C_{\#,m}M_\#^j \qquad \forall j\in\mathbb{N}.$$

Then we consider a finite cover of the compact subset $\overline{(\partial\Omega^i)_{\delta/(5\epsilon_1)}\times(\partial\Omega^o)_{\delta/5}}$ of $(\partial\Omega^i)_{\delta/(4\epsilon_1)}\times(\partial\Omega^o)_{\delta/4}$ consisting of open sets

$$\mathbb{B}_n(\xi_\#^{(k)},1/(2M_\#^{(k)}))\times\mathbb{B}_n(y_\#^{(k)},1/(2M_\#^{(k)})),\ \text{with } k\in\{1,\ldots,\varkappa\},$$

constructed as $\mathbb{B}_n(\xi_\#,1/(2M_\#))\times\mathbb{B}_n(y_\#,1/(2M_\#))$ above. For each index k in $\{1,\ldots,\varkappa\}$ we have functions $a_j^{\#,(k)}$ with $j\in\mathbb{N}$ which are analytic from the set

$$\mathbb{B}_n(\xi_\#^{(k)},1/M_\#^{(k)})\times\mathbb{B}_n(y_\#^{(k)},1/M_\#^{(k)})$$

to $\mathbb{R}$ and such that

$$\|a_j^{\#,(k)}\|_{C^m\left(\overline{\mathbb{B}_n(\xi_\#^{(k)},1/(2M_\#^{(k)}))}\times\overline{\mathbb{B}_n(y_\#^{(k)},1/(2M_\#^{(k)}))}\right)} < C_{\#,m,k}\,(M_\#^{(k)})^j \quad \forall j\in\mathbb{N}$$

for some $C_{\#,m,k}>0$. Moreover,

$$S_n(\epsilon\xi-y)$$

$$
= \sum_{j=0}^{\infty} a_j^{\#,(k)}(\xi, y)(\epsilon - \epsilon_{\#})^j \quad \forall (\xi, y) \in \mathbb{B}_n(\xi_{\#}^{(k)}, 1/M_{\#}^{(k)}) \times \mathbb{B}_n(y_{\#}^{(k)}, 1/M_{\#}^{(k)})
$$

and thus a standard inductive argument shows that each two functions $a_j^{\#,(h)}$ and $a_j^{\#,(k)}$ with $h, k \in \{1, \ldots, \varkappa\}$ must coincide in the intersection of their domains. Namely, we have

$$
a_j^{\#,(k)}(\xi, y) = a_j^{\#,(h)}(\xi, y)
$$

for all (ξ, y) in

$$
\left(\mathbb{B}_n(\xi_{\#}^{(k)}, 1/M_{\#}^{(k)}) \times \mathbb{B}_n(y_{\#}^{(k)}, 1/M_{\#}^{(k)})\right) \cap \left(\mathbb{B}_n(\xi_{\#}^{(h)}, 1/M_{\#}^{(h)}) \times \mathbb{B}_n(y_{\#}^{(h)}, 1/M_{\#}^{(h)})\right) \qquad \forall j \in \mathbb{N}
$$

(the equality being vacuously true when the intersection is empty). Accordingly, for each $j \in \mathbb{N}$ we can glue together the functions $a_j^{\#,(k)}$ with $k \in \{1, \ldots, \varkappa\}$ and define a function $A_j^{\#}$ from $\overline{(\partial\Omega^i)_{\delta/(5\epsilon_1)} \times (\partial\Omega^o)_{\delta/5}}$ to $\mathbb{R}$ such that

$$
A_j^{\#}(\xi, y) = a_j^{\#,(k)}(\xi, y)
$$

for all (ξ, y) in

$$
\left(\mathbb{B}_n(\xi_{\#}^{(k)}, 1/(2M_{\#}^{(k)})) \times \mathbb{B}_n(y_{\#}^{(k)}, 1/(2M_{\#}^{(k)}))\right) \cap \left(\overline{(\partial\Omega^i)_{\delta/(5\epsilon_1)} \times (\partial\Omega^o)_{\delta/5}}\right)
$$

and for all $k \in \{1, \ldots, \varkappa\}$. Then we can verify that there exist $C_m > 0$ and $M > 0$ such that

$$
\|A_j^{\#}\|_{C^m\left(\overline{(\partial\Omega^i)_{\delta/(5\epsilon_1)} \times (\partial\Omega^o)_{\delta/5}}\right)} < C_m M^j \qquad \forall j \in \mathbb{N}\,.
$$

Moreover, we have

$$
S_n(\epsilon\xi - y) = \sum_{j=0}^{\infty} A_j^{\#}(\xi, y)(\epsilon - \epsilon_{\#})^j
$$

for all $\epsilon \in]\epsilon_{\#} - 1/M, \epsilon_{\#} + 1/M[$ and all $(\xi, y) \in \overline{(\partial\Omega^i)_{\delta/(5\epsilon_1)} \times (\partial\Omega^o)_{\delta/5}}$. It follows that the map from $]-\epsilon_1, \epsilon_1[$ to $C^m\left(\overline{\partial\Omega^i_{\delta/(5\epsilon_1)} \times \partial\Omega^o_{\delta/5}}\right)$ defined by

$$
\epsilon \mapsto \left((\xi, y) \mapsto S_n(\epsilon\xi - y)\right) \tag{A.35}
$$

is real analytic. Then, by exploiting the classical theorem on differentiation under integral sign we verify that the map from the space $C^m\left(\overline{(\partial\Omega^i)_{\delta/(5\epsilon_1)} \times (\partial\Omega^o)_{\delta/5}}\right)$ to the space $\mathcal{L}\left(C^{0,\alpha}(\partial\Omega^o), C^m\left(\overline{(\partial\Omega^i)_{\delta/(5\epsilon_1)}}\right)\right)$ defined by

$$h \mapsto \left(\theta \mapsto \int_{\partial\Omega^o} h(\cdot, y)\theta(y)\, d\sigma_y\right) \tag{A.36}$$

is linear and bounded (it would be possible to consider a larger space for θ, e.g., $L^\infty(\partial\Omega^o)$, but we do not need it for our proof). Hence, by taking $m \geq 2$, by considering the composition of the maps in (A.35) and (A.36), and by the boundedness of the restriction operator from $C^m(\overline{(\partial\Omega^i)_{\delta/(5\epsilon_1)}})$ to $C^{1,\alpha}(\partial\Omega^i)$, we deduce that the operator $\tilde{\mathcal{V}}^i$ defined by (A.31) is real analytic from $]-\epsilon_1, \epsilon_1[$ to $\mathcal{L}(C^{0,\alpha}(\partial\Omega^o), C^{1,\alpha}(\partial\Omega^i))$ for all $\epsilon_1 \in]0, \epsilon_0[$. Thus, it is real analytic from the interval $]-\epsilon_0, \epsilon_0[$ to $\mathcal{L}(C^{0,\alpha}(\partial\Omega^o), C^{1,\alpha}(\partial\Omega^i))$ and the validity of the lemma is proved. □

We now introduce a more general result of [175, Prop. 4.1], which we do not prove here.

Theorem A.27. *Let $h_1, h_2 \in \mathbb{N} \setminus \{0\}$. Let $m \in \mathbb{N}$, $\alpha \in]0,1]$. Let Y be a topological space. Let $\mathcal{M}$ be a σ-algebra of parts of Y containing the Borel sets of Y. Let μ be measure on $\mathcal{M}$. Let Z be a Banach space. Let W be a nonempty open subset of $\mathbb{R}^{h_1} \times \mathbb{R}^{h_2} \times Z$. Let G be a real analytic map from W to $\mathbb{R}$. Then the following statements hold.*

(i) Let $n \in \mathbb{N} \setminus \{0\}$. Let Ω_1 be a bounded open connected subset of $\mathbb{R}^n$. Let Ω_1 be regular in the sense of Whitney. Let

$$\tilde{F} \equiv \Big\{(\Psi, \phi, z) \in C^{m,\alpha}(\overline{\Omega_1}, \mathbb{R}^{h_1}) \times C^0_b(Y, \mathbb{R}^{h_2}) \times Z : \\ \Psi(\overline{\Omega_1}) \times \overline{\phi(Y)} \times \{z\} \subseteq W\Big\}.$$

Then the map $\tilde{\mathcal{H}}_G$ from $\tilde{F} \times L^1(Y)$ to $C^{m,\alpha}(\overline{\Omega_1})$ defined by

$$\tilde{\mathcal{H}}_G[\Psi, \phi, z, f](x) \equiv \int_Y G(\Psi(x), \phi(y), z) f(y)\, d\mu_y \qquad \forall x \in \overline{\Omega_1}$$

for all $(\Psi, \phi, z, f) \in \tilde{F} \times L^1(Y)$, is real analytic.

(ii) Let $n \in \mathbb{N} \setminus \{0,1\}$. Let Ω_1 be a bounded open connected subset of $\mathbb{R}^n$ of class $C^{\max\{1,m\},\alpha}$. Let

$$F^\sharp \equiv \Big\{(\psi, \phi, z) \in C^{m,\alpha}(\partial\Omega_1, \mathbb{R}^{h_1}) \times C^0_b(Y, \mathbb{R}^{h_2}) \times Z : \\ \psi(\partial\Omega_1) \times \overline{\phi(Y)} \times \{z\} \subseteq W\Big\}.$$

Then the map $\mathcal{H}^\sharp_G$ from $F^\sharp \times L^1(Y)$ to $C^{m,\alpha}(\partial\Omega_1)$ defined by

$$\mathcal{H}^\sharp_G[\psi, \phi, z, f](x) \equiv \int_Y G(\psi(x), \phi(y), z) f(y)\, d\mu_y \qquad \forall x \in \partial\Omega_1\,,$$

for all $(\psi, \phi, z, f) \in F^{\sharp} \times L^1(Y)$, *is real analytic.*

We now show that Theorem A.27 implies the validity of the following statement, which we actually exploit a number of times in this book.

Theorem A.28. *Let* $n \in \mathbb{N} \setminus \{0,1\}$. *Let* $\alpha \in]0,1]$. *Let* Ω *be a bounded open subset of* $\mathbb{R}^n$ *of class* $C^{1,\alpha}$.

(i) Let Ω_1 *be a bounded open subset of* $\mathbb{R}^n$. *Let* W *be an open subset of* $\mathbb{R}^n$. *Let* J_1, J_2 *be open intervals of* $\mathbb{R}$ *such that*

$$\left\{\epsilon_1 x - \epsilon_2 y : \epsilon_1 \in J_1, \epsilon_2 \in J_2, x \in \overline{\Omega_1}, y \in \partial\Omega\right\} \subseteq W .$$

If G *is an analytic function from* W *to* $\mathbb{R}$ *and* $m \in \mathbb{N}$, *then the map from* $J_1 \times J_2 \times L^1(\partial\Omega)$ *to* $C^{m,\alpha}(\overline{\Omega_1})$ *that takes* $(\epsilon_1, \epsilon_2, \theta)$ *to the function*

$$\int_{\partial\Omega} G(\epsilon_1 x - \epsilon_2 y)\theta(y)\, d\sigma_y \qquad \forall x \in \overline{\Omega_1}$$

is real analytic.

(ii) Let Ω_1 *be a bounded open subset of* $\mathbb{R}^n$ *of class* $C^{1,\alpha}$. *Let* W *be an open subset of* $\mathbb{R}^n$. *Let* J_1, J_2 *be open intervals of* $\mathbb{R}$ *such that*

$$\{\epsilon_1 x - \epsilon_2 y : \epsilon_1 \in J_1, \epsilon_2 \in J_2, x \in \partial\Omega_1, y \in \partial\Omega\} \subseteq W .$$

If G *is an analytic function from* W *to* $\mathbb{R}$, *then the map from* $J_1 \times J_2 \times L^1(\partial\Omega)$ *to* $C^{1,\alpha}(\partial\Omega_1)$ *that takes* $(\epsilon_1, \epsilon_2, \theta)$ *to the function*

$$\int_{\partial\Omega} G(\epsilon_1 x - \epsilon_2 y)\theta(y)\, d\sigma_y \qquad \forall x \in \partial\Omega_1$$

is real analytic.

(iii) Let Ω_1 *be a bounded open subset of* $\mathbb{R}^n$ *of class* $C^{1,\alpha}$. *Let* $h_1 \in \mathbb{N} \setminus \{0\}$. *Let* $m \in \{0,1\}$. *Let* W *be an open subset of* $\mathbb{R}^{h_1} \times \mathbb{R} \times \mathbb{R}$. *Let* G *be an analytic function from* W *to* $\mathbb{R}$. *Then the map from the set*

$$\mathcal{A} \equiv \left\{(\psi, \epsilon) \in C^{m,\alpha}(\partial\Omega_1, \mathbb{R}^{h_1}) \times \mathbb{R} : \psi(\partial\Omega_1) \times [0,1] \times \{\epsilon\} \subseteq W\right\}$$

to $C^{m,\alpha}(\partial\Omega_1)$ *that takes* (ψ, ϵ) *to the function*

$$\int_0^1 G(\psi(x), \beta, \epsilon)\, d\beta \qquad \forall x \in \partial\Omega_1$$

is analytic.

(iv) Let Ω_1 *be a bounded open subset of* $\mathbb{R}^n$. *Let* $m \in \mathbb{N}$. *Let* W *be an open subset of* $\mathbb{R}^n \times \mathbb{R} \times \mathbb{R}$. *Let* G *be an analytic function from* W *to* $\mathbb{R}$. *Then the map from the set*

$$J \equiv \left\{\epsilon \in \mathbb{R} : \overline{\Omega_1} \times [0,1] \times \{\epsilon\} \subseteq W\right\}$$

to $C^{m,\alpha}(\overline{\Omega_1})$ *that takes* ϵ *to the function*

$$\int_0^1 G(x,\beta,\epsilon)\,d\beta \qquad \forall x \in \overline{\Omega_1}$$

is analytic.

Proof. In order to prove statement (i) it suffices to show that if $(\tilde{\epsilon}_1, \tilde{\epsilon}_2) \in J_1 \times J_2$, then there exists $\eta \in]0,+\infty[$ such that

$$\mathbb{B}_1(\tilde{\epsilon}_1,\eta) \times \mathbb{B}_1(\tilde{\epsilon}_2,\eta) \subseteq J_1 \times J_2\,,$$

and such that the map of statement (i) is analytic on $\mathbb{B}_1(\tilde{\epsilon}_1,\eta) \times \mathbb{B}_1(\tilde{\epsilon}_2,\eta) \times L^1(\partial\Omega)$. To shorten our notation, we set

$$\tilde{J}_1 \equiv \mathbb{B}_1(\tilde{\epsilon}_1,\eta) \qquad \tilde{J}_2 \equiv \mathbb{B}_1(\tilde{\epsilon}_2,\eta)\,.$$

Since $\{\tilde{\epsilon}_1\} \times \{\tilde{\epsilon}_2\} \times \overline{\Omega_1} \times \partial\Omega$ is compact and contained in the open subset

$$\{(\epsilon_1,\epsilon_2,x,y) \in \mathbb{R}^{2+2n} : \epsilon_1 x - \epsilon_2 y \subseteq W\}$$

of $\mathbb{R}^{2+2n}$, then there exists $\eta \in]0,+\infty[$ such that

$$\{\epsilon_1 x - \epsilon_2 y : \epsilon_1 \in \tilde{J}_1, \epsilon_2 \in \tilde{J}_2, x \in \overline{\Omega_1} + \mathbb{B}_n(0,2\eta), y \in (\partial\Omega) + \mathbb{B}_n(0,2\eta)\} \subseteq W\,.$$

By Lemma 2.70, there exists a bounded open subset $\tilde{\Omega}_1$ of $\mathbb{R}^n$ of class C^∞ such that

$$\overline{\Omega_1} \subseteq \tilde{\Omega}_1 \subseteq \overline{\tilde{\Omega}_1} \subseteq \overline{\Omega_1} + \mathbb{B}_n(0,\eta)\,,$$

and thus we have

$$\left\{\epsilon_1 x - \epsilon_2 y : \epsilon_1 \in \tilde{J}_1, \epsilon_2 \in \tilde{J}_2, x \in \overline{\tilde{\Omega}_1}, y \in \partial\Omega\right\} \subseteq W\,.$$

Let $\tilde{G}$ be the map from $\tilde{W} \equiv \{(\xi,\eta) \in \mathbb{R}^n \times \mathbb{R}^n : \xi - \eta \in W\}$ defined by

$$\tilde{G}(\xi,\eta) \equiv G(\xi-\eta) \qquad \forall (\xi,\eta) \in \tilde{W}\,.$$

Since the finitely many connected components of $\tilde{\Omega}_1$ have pairwise disjoint closures (cf. Lemma 2.38) and are regular in the sense of Whitney and Ω is of class $C^{1,\alpha}$, Theorem A.27 (i) implies that the map from

$$\tilde{F} \equiv \left\{(\Psi,\phi) \in C^{m,\alpha}(\overline{\tilde{\Omega}_1}) \times C^0(\partial\Omega) : \Psi\left(\overline{\tilde{\Omega}_1}\right) - \phi(\partial\Omega) \subseteq W\right\}\,,$$

to $C^{m,\alpha}(\overline{\tilde{\Omega}_1})$ defined by

$$\tilde{\mathcal{H}}_{\tilde{G}}[\Psi,\phi,\theta](x) \equiv \int_{\partial\Omega} G(\Psi(x)-\phi(y))\theta(y)\,d\sigma_y \qquad \forall x \in \overline{\tilde{\Omega}_1}$$

for all $(\Psi,\phi,\theta) \in \tilde{F} \times L^1(\partial\Omega)$ is analytic. Since the map from $\tilde{J}_1 \times \tilde{J}_2 \times L^1(\partial\Omega)$ to $\tilde{F} \times L^1(\partial\Omega)$ that takes $(\epsilon_1,\epsilon_2,\theta)$ to $(\epsilon_1 I_{\overline{\tilde{\Omega}_1}}, \epsilon_2 I_{\partial\Omega}, \theta)$ is analytic and the restriction is linear and continuous in Schauder spaces, then the composite function that takes $(\epsilon_1, \epsilon_2, \theta)$ to the restriction $\tilde{\mathcal{H}}_{\tilde{G}}[\epsilon_1 I_{\overline{\tilde{\Omega}_1}}, \epsilon_2 I_{\partial\Omega}, \theta]_{|\overline{\Omega_1}}$ is analytic, i.e., the map of statement (i) is analytic on $\tilde{J}_1 \times \tilde{J}_2 \times L^1(\partial\Omega)$.

The proof of statement (ii) follows the lines of the proof of the analyticity of the map of statement (i) on $\tilde{J}_1 \times \tilde{J}_2 \times L^1(\partial\Omega)$ by exploiting Theorem A.27 (ii) instead of Theorem A.27 (i). In this case we do not need to introduce the set $\tilde{\Omega}_1$ because we know that the finitely many connected components of Ω_1 satisfy the assumptions of Theorem A.27 (ii) (cf. Lemma 2.38).

(iii) Since Ω_1 has finitely many connected components of class $C^{1,\alpha}$ (cf. Lemma 2.38), Theorem A.27 (ii) implies that the map from the set

$$F^\sharp \equiv \Big\{ (\psi, \phi, \epsilon) \in C^{m,\alpha}(\partial\Omega_1, \mathbb{R}^{h_1}) \times C^0([0,1]) \times \mathbb{R} : \\ \psi(\partial\Omega_1) \times \phi([0,1]) \times \{\epsilon\} \subseteq W \Big\},$$

to $C^{m,\alpha}(\partial\Omega_1)$ defined by

$$\mathcal{H}^\sharp_G[\psi, \phi, \epsilon, \theta](x) \equiv \int_{[0,1]} G(\psi(x), \phi(\beta), \epsilon)\theta(\beta)\, d\beta \qquad \forall x \in \partial\Omega_1$$

for all $(\psi, \phi, \epsilon, \theta) \in F^\sharp \times L^1([0,1])$ is analytic. Since the map from the set $\mathcal{A}$ to $F^\sharp \times L^1([0,1])$ that takes (ψ, ϵ) to the quadruple $(\psi, I_{[0,1]}, \epsilon, 1)$ is analytic, then the composite function that takes (ψ, ϵ) to $\mathcal{H}^\sharp_G[\psi, I_{[0,1]}, \epsilon, 1] = \int_0^1 G(\psi(\cdot), \beta, \epsilon)\, d\beta$ is analytic, i.e., the map of statement (iii) is analytic.

(iv) It suffices to show that if $\tilde{\epsilon}_1 \in J$, then there exists $\eta \in]0, +\infty[$ such that

$$\mathbb{B}_1(\tilde{\epsilon}_1, \eta) \subseteq J$$

and such that the map of statement (iv) is analytic on $\mathbb{B}_1(\tilde{\epsilon}_1, \eta)$. To shorten our notation, we set

$$\tilde{J}_1 \equiv \mathbb{B}_1(\tilde{\epsilon}_1, \eta).$$

Since the compact set $\overline{\Omega_1} \times [0,1] \times \{\tilde{\epsilon}_1\}$ is contained in the open set W, then there exists $\eta \in]0, +\infty[$ such that

$$(\overline{\Omega_1} + \mathbb{B}_n(0, 2\eta)) \times ([0,1] + \mathbb{B}_1(0, 2\eta)) \times \mathbb{B}_1(\tilde{\epsilon}_1, 2\eta) \subseteq W.$$

By Lemma 2.70, there exists an open subset $\tilde{\Omega}_1$ of $\mathbb{R}^n$ of class C^∞ such that

$$\overline{\Omega_1} \subseteq \tilde{\Omega}_1 \subseteq \overline{\tilde{\Omega}_1} \subseteq \overline{\Omega_1} + \mathbb{B}_n(0, \eta),$$

and thus we have

$$(\overline{\tilde{\Omega}_1} + \mathbb{B}_n(0, \eta)) \times [0,1] \times \mathbb{B}_1(\tilde{\epsilon}_1, \eta) \subseteq W.$$

Since the finitely many connected components of $\tilde{\Omega}_1$ have pairwise disjoint closures (cf. Lemma 2.38) and are regular in the sense of Whitney and Ω is of class $C^{1,\alpha}$, Theorem A.27 (i) implies that the map from

$$\tilde{F}_1 \equiv \left\{ (\Psi, \phi, \epsilon) \in C^{m,\alpha}(\overline{\tilde{\Omega}_1}) \times C^0([0,1]) \times \mathbb{R} : \Psi\left(\overline{\tilde{\Omega}_1}\right) \times \phi([0,1]) \times \{\epsilon\} \subseteq W \right\}$$

to $C^{m,\alpha}(\overline{\tilde{\Omega}_1})$ defined by

$$\tilde{\mathcal{H}}_{\tilde{G},1}[\Psi, \phi, \epsilon, \theta](x) \equiv \int_{[0,1]} G(\Psi(x), \phi(\beta), \epsilon)\theta(\beta)\, d\beta \qquad \forall x \in \overline{\tilde{\Omega}_1}$$

for all $(\Psi, \phi, \epsilon, \theta) \in \tilde{F}_1 \times L^1(\partial\Omega)$ is analytic. Since the map from $\tilde{J}_1$ to $\tilde{F}_1 \times L^1(\partial\Omega)$ that takes ϵ to $(I_{\overline{\tilde{\Omega}_1}}, I_{[0,1]}, \epsilon, 1)$ is analytic and the restriction is linear and continuous in Schauder spaces, then the composite function that takes ϵ to the restriction $\tilde{\mathcal{H}}_{\tilde{G},1}[I_{\overline{\tilde{\Omega}_1}}, I_{[0,1]}, \epsilon, 1]_{|\overline{\Omega_1}}$ is analytic, i.e., the map of statement (iv) is analytic on $\tilde{J}_1$. □

In addition, we have the following corollary of Theorem A.28 (ii).

Corollary A.29. *Let $n \in \mathbb{N} \setminus \{0, 1\}$. Let $\alpha \in]0, 1]$. Let Ω, Ω_1 be bounded open subsets of $\mathbb{R}^n$ of class $C^{1,\alpha}$. Let W be an open subset of $\mathbb{R}^n$. Let J_1, J_2 be open intervals of $\mathbb{R}$ such that*

$$\{\epsilon_1 x - \epsilon_2 y : \epsilon_1 \in J_1, \epsilon_2 \in J_2, x \in \partial\Omega_1, y \in \partial\Omega\} \subseteq W .$$

Let G be an analytic function from W to $\mathbb{R}$. For all $(\epsilon_1, \epsilon_2) \in J_1 \times J_2$, let $H(\epsilon_1, \epsilon_2)$ be the linear operator that takes $\theta \in L^1(\partial\Omega)$ to the function

$$\int_{\partial\Omega} G(\epsilon_1 x - \epsilon_2 y)\theta(y)\, d\sigma_y \qquad \forall x \in \partial\Omega_1 .$$

Then the map from $J_1 \times J_2$ to $\mathcal{L}(L^1(\partial\Omega), C^{1,\alpha}(\partial\Omega_1))$ that takes (ϵ_1, ϵ_2) to $H(\epsilon_1, \epsilon_2)$ is real analytic.

Proof. By Theorem A.28 (ii), the map $\mathcal{H}$ from $J_1 \times J_2 \times L^1(\partial\Omega)$ to $C^{1,\alpha}(\partial\Omega_1)$ that takes $(\epsilon_1, \epsilon_2, \theta)$ to $\mathcal{H}(\epsilon_1, \epsilon_2, \theta) \equiv H(\epsilon_1, \epsilon_2)[\theta]$ is real analytic. Since $\mathcal{H}$ is linear and continuous with respect to the variable θ, we have

$$H(\epsilon_1, \epsilon_2) = d_\theta \mathcal{H}(\epsilon_1, \epsilon_2, \theta) \qquad \forall (\epsilon_1, \epsilon_2, \theta) \in J_1 \times J_2 \times L^1(\partial\Omega) .$$

Since the right-hand side equals a partial Frèchet differential of an analytic map, the right-hand side is analytic. Hence H is analytic on $J_1 \times J_2 \times L^1(\partial\Omega)$. Since $H(\cdot, \cdot)$ is independent of its last variable, we conclude that H is analytic on $J_1 \times J_2$. □

For the case of a linear integral operator in Roumieu spaces, we prove the following.

Proposition A.30. *Let $n \in \mathbb{N} \setminus \{0\}$. Let Ω, Ω_1 be bounded open subsets of $\mathbb{R}^n$. Let W be an open subset of $\mathbb{R}^n$ which contains the set*

$$\overline{\Omega_1} - \overline{\Omega} = \left\{x - y : x \in \overline{\Omega_1},\, y \in \overline{\Omega}\right\} .$$

Let G be an analytic map from W to $\mathbb{R}$. Then there exists $\rho_0 \in]0, +\infty[$ such that the linear operator from $L^1(\Omega)$ to $C^0_{\omega,\rho_0}(\overline{\Omega_1})$ that takes f to the function

$$\int_\Omega G(x-y) f(y)\, dy \qquad \forall x \in \overline{\Omega_1}$$

is continuous.

Proof. Since $\overline{\Omega_1} - \overline{\Omega}$ is a compact subset of W and G is analytic in W, the analiticity criterion of Theorem 2.17 implies that there exist ρ_0, $M \in]0, +\infty[$ such that

$$\sup_{\xi \in \overline{\Omega_1} - \overline{\Omega}} |D^\beta G(\xi)| \le M \frac{|\beta|!}{\rho_0^{|\beta|}} \qquad \forall \beta \in \mathbb{N}^n .$$

Then we have

$$\begin{aligned}
&\sup_{\beta \in \mathbb{N}^n} \frac{\rho_0^{|\beta|}}{|\beta|!} \sup_{x \in \overline{\Omega_1}} \left| D_x^\beta \int_\Omega G(x-y) f(y)\, dy \right| \\
&\quad \le \sup_{\beta \in \mathbb{N}^n} \frac{\rho_0^{|\beta|}}{|\beta|!} \sup_{x \in \overline{\Omega_1}} \left| \int_\Omega D^\beta G(x-y) f(y)\, dy \right| \\
&\quad = \sup_{\beta \in \mathbb{N}^n} \frac{\rho_0^{|\beta|}}{|\beta|!} \sup_{\xi \in \overline{\Omega_1} - \overline{\Omega}} |D^\beta G(\xi)| \|f\|_{L^1(\Omega)} \le M \|f\|_{L^1(\Omega)} \qquad \forall f \in L^1(\Omega) ,
\end{aligned}$$

and thus the statement follows by the definition of norm in Roumieu spaces. □

A.9 Sard's Theorem

Theorem A.31 (of Sard). *Let $n \in \mathbb{N}$. Let Ω be an open subset of $\mathbb{R}^n$. If f is a continuously differentiable function from Ω to $\mathbb{R}$, then the Lebesgue measure of the set*

$$f\left(\{x \in \Omega : \nabla f(x) = 0\}\right)$$

is equal to zero.

For a proof, we refer for example to Bröcker and Jänich [36, (6.1), p. 56].

A.10 Theorem of Invariance of Domain

Theorem A.32 (of Invariance of Domain). *Let* $n \in \mathbb{N}$. *Let* Ω *be an open subset of* $\mathbb{R}^n$. *If a function* f *from* Ω *to* $\mathbb{R}^n$ *is continuous and injective, then* f *maps open subsets of* Ω *to open subsets of* $\mathbb{R}^n$.

For a proof, we refer for example to Deimling [87, Theorem 4.3].

A.11 Mollifiers

Let $n \in \mathbb{N} \setminus \{0\}$. Let $\eta \in \mathcal{D}(\mathbb{R}^n)$ be such that

$$\operatorname{supp} \eta \subseteq \overline{\mathbb{B}_n(0,1)}\,, \qquad \eta \geq 0\,, \qquad \int_{\mathbb{R}^n} \eta\, dx = 1\,.$$

In order to prove the existence of a function η as above, we consider the function h from $\mathbb{R}^n$ to $[0,+\infty[$ defined by

$$h(x) \equiv \begin{cases} e^{-\frac{1}{1-|x|^2}} & \text{if } |x| < 1\,, \\ 0 & \text{if } |x| \geq 1\,. \end{cases}$$

It is well-known that $h \in C_c^\infty(\mathbb{R}^n)$. Then we set

$$\eta(x) \equiv \frac{h(x)}{\int_{\mathbb{R}^n} h(y)\, dy} \qquad \forall x \in \mathbb{R}^n\,,$$

and we also have $\int_{\mathbb{R}^n} \eta\, dx = 1$. Next we set

$$\eta_\epsilon(x) \equiv \epsilon^{-n}\eta(x/\epsilon) \qquad \forall x \in \mathbb{R}^n$$

for all $\epsilon \in]0,+\infty[$. Then the family $\{\eta_\epsilon\}_{\epsilon\in]0,+\infty[}$ is said to be a family of standard mollifiers. Clearly,

$$\operatorname{supp} \eta_\epsilon \subseteq \overline{\mathbb{B}_n(0,\epsilon)}\,, \qquad \eta_\epsilon \geq 0\,, \qquad \int_{\mathbb{R}^n} \eta_\epsilon\, dx = 1 \qquad \forall \epsilon \in]0,+\infty[$$

and we have the following known result.

Proposition A.33. *Let* $n \in \mathbb{N} \setminus \{0\}$. *Let* $\{\eta_\epsilon\}_{\epsilon\in]0,+\infty[}$ *be a family of standard mollifiers. Let* f *be an essentially bounded and measurable function from* $\mathbb{R}^n$ *to* $\mathbb{R}$. *Let* U *be an open subset of* $\mathbb{R}^n$. *If* f *is continuous on* U, *then*

$$\lim_{\epsilon\to 0} \eta_\epsilon * f = f$$

uniformly on the compact subsets of U.

For a proof, we refer for example to Folland [103, Theorem 8.14].

A.12 The Partition of Unity

Lemma A.34 (of the Partition of Unity). *Let $n \in \mathbb{N} \setminus \{0\}$. Let $\{V_\lambda\}_{\lambda\in\Lambda}$ be a family of open subsets of $\mathbb{R}^n$ such that $V \equiv \bigcup_{\lambda\in\Lambda} V_\lambda$ is not empty.*

Then there exists a sequence $\{\theta_j\}_{j\in\mathbb{N}}$ of functions of $C_c^\infty(V)$ such that

(i) $\theta_j \geq 0$ for all $j \in \mathbb{N}$.
(ii) If $j \in \mathbb{N}$, then there exists $\lambda(j) \in \Lambda$ such that $\operatorname{supp}\theta_j \subseteq V_{\lambda(j)}$.
(iii) The family of sets $\{\operatorname{supp}\theta_j\}_{j\in\mathbb{N}}$ is locally finite. Namely, if $x \in V$, then there exists a neighborhood W_x of x in V such that

$$W_x \cap \operatorname{supp}\theta_j \neq \emptyset$$

holds true for at most a finite number of $j \in \mathbb{N}$. (Equivalently, if K is a compact subset of V, we have $K \cap \operatorname{supp}\theta_j \neq \emptyset$ for at most a finite number of $j \in \mathbb{N}$.)
(iv) $\sum_{j=0}^{+\infty} \theta_j(x) = 1$ for all $x \in V$.

For a proof, we refer for example to Mitrea [214, Theorem 13.34].

The sequence $\{\theta_j\}_{j\in\mathbb{N}}$ is said to be a partition of unity subordinate to the open cover $\{V_\lambda\}_{\lambda\in\Lambda}$ of V. We also note that $V \subseteq \cup_{j\in\mathbb{N}} \operatorname{supp}\theta_j \subseteq \cup_{j\in\mathbb{N}} V_{\lambda(j)} \subseteq V$. Then we have the following immediate consequence of the lemma of the partition of unity in the case of compact subsets of $\mathbb{R}^n$.

Lemma A.35 (of the Partition of Unity on a Compact Set). *Let $n \in \mathbb{N} \setminus \{0\}$. Let K be a compact subset of $\mathbb{R}^n$. Let $\{V_l\}_{l=1}^k$ be a finite family of open subsets of $\mathbb{R}^n$ such that $K \subseteq \bigcup_{l=1}^k V_l$. Then there exist $\varphi_1 \in \mathcal{D}(V_1), \ldots, \varphi_k \in \mathcal{D}(V_k)$ with $\varphi_l \geq 0$ for all $l \in \{1, \ldots, k\}$ such that*

$$\sum_{l=1}^{k} \varphi_l(x) = 1 \qquad \forall x \in K\,.$$

Proof. Let $\{\theta_j\}_{j\in\mathbb{N}}$ be a partition of unity subordinate to the cover $\{V_l\}_{l=1}^k$ of $\bigcup_{l=1}^k V_l$. Let $m \in \mathbb{N}$ be such that $K \cap \operatorname{supp}\theta_j = \emptyset$ for all $j > m$. Then we have $\sum_{j=0}^m \theta_j(x) = \sum_{j=0}^\infty \theta_j(x) = 1$ for all $x \in K$. Next we define inductively the sets $\{\Lambda_l\}_{l=1}^k$ as follows

$$\Lambda_1 \equiv \{j \in \{0, \ldots, m\} : \operatorname{supp}\theta_j \subseteq V_1\}\,,$$

$$\Lambda_{l+1} \equiv \{j \in \{0, \ldots, m\} : \operatorname{supp}\theta_j \subseteq V_{l+1}\} \setminus \bigcup_{s=1}^{l} \Lambda_s \quad \text{if } l < k\,,$$

and we set $\varphi_l \equiv \sum_{j\in\Lambda_l} \theta_j$ for all $l \in \{1, \ldots, k\}$. □

Then we can deduce the validity of the following classical lemma on the existence of a cut off function.

Lemma A.36 (of the Cut Off Function). *Let Ω be an open subset of $\mathbb{R}^n$. Let K be a compact subset of Ω. Then there exists $\varphi \in \mathcal{D}(\Omega)$ such that $\varphi = 1$ on K.*

Proof. It suffices to set $V_1 \equiv \Omega$ and to apply the previous lemma. □

References

1. P. M. Adler. *Porous media: geometry and transports*. Butterworth-Heinemann, 1992.
2. S. Agmon, A. Douglis, and L. Nirenberg. Estimates near the boundary for solutions of elliptic partial differential equations satisfying general boundary conditions. I. *Comm. Pure Appl. Math.*, 12:623–727, 1959.
3. L. V. Ahlfors. Remarks on the Neumann-Poincaré integral equation. *Pacific J. Math.*, 2:271–280, 1952.
4. G. Allaire. *Shape optimization by the homogenization method*, volume 146 of *Applied Mathematical Sciences*. Springer-Verlag, New York, 2002.
5. R. Alvarado, D. Brigham, V. Maz'ya, M. Mitrea, and E. Ziadé. On the regularity of domains satisfying a uniform hour-glass condition and a sharp version of the Hopf-Oleinik boundary point principle. *J. Math. Sci. (N.Y.)*, 176(3):281–360, 2011. Problems in mathematical analysis. No. 57.
6. A. Ambrosetti and G. Prodi. *A primer of nonlinear analysis*, volume 34 of *Cambridge Studies in Advanced Mathematics*. Cambridge University Press, Cambridge, 1995. Corrected reprint of the 1993 original.
7. H. Ammari, P. Garapon, H. Kang, and H. Lee. Effective viscosity properties of dilute suspensions of arbitrarily shaped particles. *Asymptot. Anal.*, 80(3-4):189–211, 2012.
8. H. Ammari and H. Kang. *Polarization and moment tensors*, volume 162 of *Applied Mathematical Sciences*. Springer, New York, 2007. With applications to inverse problems and effective medium theory.
9. H. Ammari, H. Kang, and K. Kim. Polarization tensors and effective properties of anisotropic composite materials. *J. Differential Equations*, 215(2):401–428, 2005.
10. H. Ammari, H. Kang, and H. Lee. *Layer potential techniques in spectral analysis*, volume 153 of *Mathematical Surveys and Monographs*. American Mathematical Society, Providence, RI, 2009.
11. H. Ammari, H. Kang, and M. Lim. Effective parameters of elastic composites. *Indiana Univ. Math. J.*, 55(3):903–922, 2006.

M. Dalla Riva et al., *Singularly Perturbed Boundary Value Problems*,
https://doi.org/10.1007/978-3-030-76259-9

12. H. Ammari, H. Kang, and K. Touibi. Boundary layer techniques for deriving the effective properties of composite materials. *Asymptot. Anal.*, 41(2):119–140, 2005.
13. S. S. Antman. *Nonlinear problems of elasticity*, volume 107 of *Applied Mathematical Sciences*. Springer, New York, second edition, 2005.
14. J. Appell and P. P. Zabrejko. *Nonlinear superposition operators*, volume 95 of *Cambridge Tracts in Mathematics*. Cambridge University Press, Cambridge, 1990.
15. M. Artola and G. Duvaut. Un résultat d'homogénéisation pour une classe de problèmes de diffusion non linéaires stationnaires. *Ann. Fac. Sci. Toulouse Math. (5)*, 4(1):1–28, 1982.
16. N. Bakhvalov and G. Panasenko. *Homogenisation: averaging processes in periodic media*, volume 36 of *Mathematics and its Applications (Soviet Series)*. Kluwer Academic Publishers Group, Dordrecht, 1989. Mathematical problems in the mechanics of composite materials, Translated from the Russian by D. Leĭtes.
17. H. Begehr and G. N. Hile. Nonlinear Riemann boundary value problems for a nonlinear elliptic system in the plane. *Math. Z.*, 179(2):241–261, 1982.
18. H. Begehr and G. C. Hsiao. Nonlinear boundary value problems for a class of elliptic systems. Wiss. Beitr., Martin-Luther-Univ. Halle-Wittenberg 1980, 41(M18), 90–102 (1980), 1980.
19. H. Begehr and G. C. Hsiao. Nonlinear boundary value problems of Riemann-Hilbert type. In *Plane Ellipticity and Related Problems*, volume 11 of *Contemporary Mathematics*, pages 139–153. American Mathematical Society, 1982.
20. A. Bensoussan, J.-L. Lions, and G. Papanicolaou. *Asymptotic analysis for periodic structures*, volume 5 of *Studies in Mathematics and its Applications*. North-Holland Publishing Co., Amsterdam-New York, 1978.
21. Y. Benveniste. Effective thermal conductivity of composites with a thermal contact resistance between the constituents: Nondilute case. *J. Appl. Phys.*, 61(8):2840–2844, 1987.
22. Y. Benveniste and T. Miloh. The effective conductivity of composites with imperfect thermal contact at constituent interfaces. *Internat. J. Engrg. Sci.*, 24(9):1537–1552, 1986.
23. V. L. Berdichevskij. Variational principles in continuum mechanics. (Variatsionnye printsipy mekhaniki sploshnoj sredy). Moskva: "Nauka" Glavnaya Redaktsiya Fiziko-Matematicheskoj Literatury. 448 p. R. 4.10 (1983), 1983.
24. L. Berlyand, A. G. Kolpakov, and A. Novikov. *Introduction to the network approximation method for materials modeling*, volume 148. Cambridge: Cambridge University Press, 2013.
25. L. Berlyand and V. Mityushev. Generalized Clausius-Mossotti formula for random composite with circular fibers. *J. Statist. Phys.*, 102(1-2):115–145, 2001.
26. L. Berlyand and V. Mityushev. Increase and decrease of the effective conductivity of two phase composites due to polydispersity. *J. Stat. Phys.*, 118(3-4):481–509, 2005.

27. J.-E. Björk. On extensions of Lipschitz functions. *Ark. Mat.*, 7:513–515, 1969.
28. R. Böhme and F. Tomi. Zur Struktur der Lösungsmenge des Plateauproblems. *Math. Z.*, 133:1–29, 1973.
29. V. Bonnaillie-Noël, M. Dalla Riva, M. Dambrine, and P. Musolino. A Dirichlet problem for the Laplace operator in a domain with a small hole close to the boundary. *J. Math. Pures Appl. (9)*, 116:211–267, 2018.
30. V. Bonnaillie-Noël, M. Dalla Riva, M. Dambrine, and P. Musolino. Global representation and multiscale expansion for the Dirichlet problem in a domain with a small hole close to the boundary. *Comm. Partial Differential Equations*, 46(2):282–309, 2021.
31. V. Bonnaillie-Noël, M. Dambrine, and C. Lacave. Interactions between moderately close inclusions for the two-dimensional Dirichlet-Laplacian. *Appl. Math. Res. Express. AMRX*, (1):1–23, 2016.
32. V. Bonnaillie-Noël, M. Dambrine, S. Tordeux, and G. Vial. Interactions between moderately close inclusions for the Laplace equation. *Math. Models Methods Appl. Sci.*, 19(10):1853–1882, 2009.
33. M. Borsuk and V. Kondratiev. *Elliptic boundary value problems of second order in piecewise smooth domains*, volume 69 of *North-Holland Mathematical Library*. Elsevier Science B.V., Amsterdam, 2006.
34. N. Bourbaki. *Éléments de mathématique. Fasc. XXXIII. Variétés différentielles et analytiques. Fascicule de résultats (Paragraphes 1 à 7).* Actualités Scientifiques et Industrielles, No. 1333. Hermann, Paris, 1967.
35. H. Brezis. *Functional analysis, Sobolev spaces and partial differential equations*. Universitext. Springer, New York, 2011.
36. T. Bröcker and K. Jänich. *Introduction to differential topology*. Cambridge University Press, Cambridge-New York, 1982. Translated from the German by C. B. Thomas and M. J. Thomas.
37. D. Buoso and L. Provenzano. A few shape optimization results for a biharmonic Steklov problem. *J. Differential Equations*, 259(5):1778–1818, 2015.
38. V. I. Burenkov. *Sobolev spaces on domains*, volume 137 of *Teubner-Texte zur Mathematik [Teubner Texts in Mathematics]*. B. G. Teubner Verlagsgesellschaft mbH, Stuttgart, 1998.
39. T. Carleman. *Über das Neumann-Poincaresche problem für ein Gebiet mit Ecken*. Almquist & Wiksells, Uppsala, 1916.
40. T. Carleman. Über eine nichtlineare Randwertaufgabe bei der Gleichung $\Delta u = 0$. *Math. Z.*, 9(1-2):35–43, 1921.
41. H. Cartan. Cours de calcul différentiel. Nouvelle ed., refondue et corrigee. Collection Methodes. Mathematique. Paris: Hermann. 356 p. (1977)., 1977.
42. L. P. Castro, D. Kapanadze, and E. Pesetskaya. Effective conductivity of a composite material with stiff imperfect contact conditions. *Math. Methods Appl. Sci.*, 38(18):4638–4649, 2015.
43. L. P. Castro, D. Kapanadze, and E. Pesetskaya. A heat conduction problem of 2D unbounded composites with imperfect contact conditions. *ZAMM Z. Angew. Math. Mech.*, 95(9):952–965, 2015.

44. L. P. Castro and E. Pesetskaya. A transmission problem with imperfect contact for an unbounded multiply connected domain. *Math. Methods Appl. Sci.*, 33(4):517–526, 2010.
45. L. P. Castro and E. Pesetskaya. Properties of a composite material with mixed imperfect contact conditions. *Math. Model. Anal.*, 21(3):283–303, 2016.
46. L. P. Castro and E. Pesetskaya. A composite material with inextensible-membrane-type interface. *Math. Mech. Solids*, 24(2):499–510, 2019.
47. L. P. Castro, E. Pesetskaya, and S. V. Rogosin. Effective conductivity of a composite material with non-ideal contact conditions. *Complex Var. Elliptic Equ.*, 54(12):1085–1100, 2009.
48. G. Cherepanov. *Mechanics of large deconstructions*. Nauka, Moscow, 1974.
49. G. Cherepanov. *Mechanics of the fracture of composite materials* (Mekhanika razrusheniya kompozitsionnykh materialov). Moskva: "Nauka", Glavnaya Redaktsiya Fiziko-Matematicheskoj Literatury. 296 p. R. 3.50 (1983)., 1983.
50. V. Chiado Piat, S. S. Nazarov, and A. L. Piatnitski. Steklov problems in perforated domains with a coefficient of indefinite sign. *Netw. Heterog. Media*, 7(1):151–178, 2012.
51. A. Cialdea. On the oblique derivation problem for the Laplace equation, and related topics. *Rend. Accad. Naz. Sci. XL Mem. Mat. (5)*, 12(1):181–200, 1988.
52. A. Cialdea. A general theory of hypersurface potentials. *Ann. Mat. Pura Appl. (4)*, 168:37–61, 1995.
53. A. Cialdea. *Appunti di teoria del potenziale*. Corso di Analisi Superiore, A.A. 2001–02. Dipartimento di Matematica, Università degli Studi della Basilicata, 2000.
54. A. Cialdea, E. Dolce, V. Leonessa, and A. Malaspina. On the potential theory in Cosserat elasticity. *Bull. TICMI*, 18(2):67–81, 2014.
55. A. Cialdea, V. Leonessa, and A. Malaspina. On the Dirichlet problem for the Stokes system in multiply connected domains. *Abstr. Appl. Anal.*, pages Art. ID 765020, 12, 2013.
56. D. Cioranescu and P. Donato. *An introduction to homogenization*, volume 17 of *Oxford Lecture Series in Mathematics and its Applications*. The Clarendon Press, Oxford University Press, New York, 1999.
57. J. D. Cole. *Perturbation methods in applied mathematics*. Blaisdell Publishing Co. Ginn and Co., Waltham, Mass.-Toronto, Ont.-London, 1968.
58. C. Constanda. *Direct and indirect boundary integral equation methods*, volume 107 of *Chapman & Hall/CRC Monographs and Surveys in Pure and Applied Mathematics*. Chapman & Hall/CRC, Boca Raton, FL, 2000.
59. C. Constanda, D. Doty, and W. Hamill. *Boundary integral equation methods and numerical solutions*, volume 35 of *Developments in Mathematics*. Springer, Cham, 2016. Thin plates on an elastic foundation.
60. M. Costabel, M. Dalla Riva, M. Dauge, and P. Musolino. Converging expansions for Lipschitz self-similar perforations of a plane sector. *Integral Equations Operator Theory*, 88(3):401–449, 2017.

61. M. Dalla Riva. A family of fundamental solutions of elliptic partial differential operators with real constant coefficients. *Integral Equations Operator Theory*, 76(1):1–23, 2013.
62. M. Dalla Riva. Stokes flow in a singularly perturbed exterior domain. *Complex Var. Elliptic Equ.*, 58(2):231–257, 2013.
63. M. Dalla Riva and M. Lanza de Cristoforis. Hypersingularly perturbed loads for a nonlinear traction boundary value problem. A functional analytic approach. *Eurasian Math. J.*, 1(2):31–58, 2010.
64. M. Dalla Riva and M. Lanza de Cristoforis. Microscopically weakly singularly perturbed loads for a nonlinear traction boundary value problem: a functional analytic approach. *Complex Var. Elliptic Equ.*, 55(8-10):771–794, 2010.
65. M. Dalla Riva and M. Lanza de Cristoforis. A perturbation result for the layer potentials of general second order differential operators with constant coefficients. *J. Appl. Funct. Anal.*, 5(1):10–30, 2010.
66. M. Dalla Riva and M. Lanza de Cristoforis. A singularly perturbed nonlinear traction boundary value problem for linearized elastostatics. A functional analytic approach. *Analysis (Munich)*, 30(1):67–92, 2010.
67. M. Dalla Riva and M. Lanza de Cristoforis. Weakly singular and microscopically hypersingular load perturbation for a nonlinear traction boundary value problem: a functional analytic approach. *Complex Anal. Oper. Theory*, 5(3):811–833, 2011.
68. M. Dalla Riva, M. Lanza de Cristoforis, and P. Musolino. On a singularly perturbed periodic nonlinear Robin problem. In *Analytic methods of analysis and differential equations: AMADE 2012. Selected papers of the 7th international workshop, Minsk, Belarus, September 10–15, 2012*, pages 73–91. Cambridge: Cambridge Scientific Publishers, 2014.
69. M. Dalla Riva, M. Lanza de Cristoforis, and P. Musolino. Analytic dependence of volume potentials corresponding to parametric families of fundamental solutions. *Integral Equations Operator Theory*, 82(3):371–393, 2015.
70. M. Dalla Riva, M. Lanza de Cristoforis, and P. Musolino. A local uniqueness result for a quasi-linear heat transmission problem in a periodic two-phase dilute composite. In *Recent trends in operator theory and partial differential equations*, volume 258 of *Oper. Theory Adv. Appl.*, pages 193–227. Birkhäuser/Springer, Cham, 2017.
71. M. Dalla Riva and G. Mishuris. Existence results for a nonlinear transmission problem. *J. Math. Anal. Appl.*, 430(2):718–741, 2015.
72. M. Dalla Riva, R. Molinarolo, and P. Musolino. Local uniqueness of the solutions for a singularly perturbed nonlinear nonautonomous transmission problem. *Nonlinear Anal.*, 191:111645, 2020.
73. M. Dalla Riva, J. Morais, and P. Musolino. A family of fundamental solutions of elliptic partial differential operators with quaternion constant coefficients. *Math. Methods Appl. Sci.*, 36(12):1569–1582, 2013.
74. M. Dalla Riva and P. Musolino. Real analytic families of harmonic functions in a domain with a small hole. *J. Differential Equations*, 252(12):6337–6355, 2012.

75. M. Dalla Riva and P. Musolino. A singularly perturbed nonideal transmission problem and application to the effective conductivity of a periodic composite. *SIAM J. Appl. Math.*, 73(1):24–46, 2013.
76. M. Dalla Riva and P. Musolino. Local uniqueness for singularly perturbed periodic nonlinear traction problems. *Electron. J. Differential Equations*, 242:1–14, 2014.
77. M. Dalla Riva and P. Musolino. A singularly perturbed nonlinear traction problem in a periodically perforated domain: a functional analytic approach. *Math. Methods Appl. Sci.*, 37(1):106–122, 2014.
78. M. Dalla Riva and P. Musolino. A mixed problem for the Laplace operator in a domain with moderately close holes. *Comm. Partial Differential Equations*, 41(5):812–837, 2016.
79. M. Dalla Riva and P. Musolino. The Dirichlet problem in a planar domain with two moderately close holes. *J. Differential Equations*, 263(5):2567–2605, 2017.
80. M. Dalla Riva and P. Musolino. Moderately close Neumann inclusions for the Poisson equation. *Math. Methods Appl. Sci.*, 41(3):986–993, 2018.
81. M. Dalla Riva, P. Musolino, and R. Pukhtaievych. Series expansion for the effective conductivity of a periodic dilute composite with thermal resistance at the two-phase interface. *Asymptot. Anal.*, 111(3-4):217–250, 2019.
82. M. Dalla Riva, P. Musolino, and S. V. Rogosin. Series expansions for the solution of the Dirichlet problem in a planar domain with a small hole. *Asymptot. Anal.*, 92(3-4):339–361, 2015.
83. M. Dalla Riva and L. Provenzano. On vibrating thin membranes with mass concentrated near the boundary: an asymptotic analysis. *SIAM J. Math. Anal.*, 50(3):2928–2967, 2018.
84. M. Dauge. *Elliptic boundary value problems on corner domains*, volume 1341 of *Lecture Notes in Mathematics*. Springer-Verlag, Berlin, 1988. Smoothness and asymptotics of solutions.
85. M. Dauge, S. Tordeux, and G. Vial. Selfsimilar perturbation near a corner: matching versus multiscale expansions for a model problem. In *Around the research of Vladimir Maz'ya. II*, volume 12 of *Int. Math. Ser. (N. Y.)*, pages 95–134. Springer, New York, 2010.
86. R. Dautray and J.-L. Lions. *Mathematical analysis and numerical methods for science and technology. Vol. 1.* Springer-Verlag, Berlin, 1990. Physical origins and classical methods, With the collaboration of Philippe Bénilan, Michel Cessenat, André Gervat, Alain Kavenoky and Hélène Lanchon, Translated from the French by Ian N. Sneddon, With a preface by Jean Teillac.
87. K. Deimling. *Nonlinear functional analysis*. Springer-Verlag, Berlin, 1985.
88. B. Dittmar. Zu einem Stekloffschen Eigenwertproblem in Ringgebieten. *Mitt. Math. Sem. Giessen*, (228):1–7, 1996.
89. F. Dondi and M. Lanza de Cristoforis. Regularizing properties of the double layer potential of second order elliptic differential operators. *Mem. Differ. Equ. Math. Phys.*, 71:69–110, 2017.
90. H. Douanla. Two-scale convergence of Stekloff eigenvalue problems in perforated domains. *Bound. Value Probl.*, pages 1–15, Art. ID 853717, 2010.

91. P. Drygaś, S. Gluzman, V. Mityushev, and W. Nawalaniec. *Applied Analysis of Composite Media. Analytical and Computational Results for Materials Scientists and Engineers*. Woodhead Publishing Series in Composites Science and Engineering. Woodhead Publishing, 2020.
92. P. Drygaś and V. Mityushev. Effective conductivity of unidirectional cylinders with interfacial resistance. *Quart. J. Mech. Appl. Math.*, 62(3):235–262, 2009.
93. J. Dugundji. *Topology*. Allyn and Bacon, Inc., Boston, Mass.-London-Sydney, 1978. Reprinting of the 1966 original, Allyn and Bacon Series in Advanced Mathematics.
94. M. A. Efendiev, H. Schmitz, and W. L. Wendland. On some nonlinear potential problems. *Electron. J. Differential Equations*, 18:1–18, 1999.
95. L. C. Evans. *Partial differential equations*, volume 19 of *Graduate Studies in Mathematics*. American Mathematical Society, Providence, RI, second edition, 2010.
96. H. Federer. *Geometric measure theory*. Die Grundlehren der mathematischen Wissenschaften, Band 153. Springer-Verlag New York Inc., New York, 1969.
97. G. Fichera. Intorno al passaggio al limite sotto il segno d'integrale. *Portugaliae Math.*, 4:1–20, 1943.
98. G. Fichera. Alcuni recenti sviluppi della teoria dei problemi al contorno per le equazioni alle derivate parziali lineari. In *Convegno Internazionale sulle Equazioni Lineari alle Derivate Parziali, Trieste, 1954*, pages 174–227. Edizioni Cremonese, Roma, 1955.
99. G. Fichera. Boundary Value Problems of Elasticity with Unilateral Constraints. In *Mechanics of solids, vol. II Encyclopedia of Physics*, pages 391–424. Springer-Verlag Berlin, *etc.*, 1984.
100. L. A. Fil'shtinskii. Stresses and displacements in an elastic sheet weakened by a doubly-periodic set of equal circular holes. *J. Appl. Math. Mech.*, 28:530–543, 1964.
101. L. Filshtinsky and V. Mityushev. Mathematical models of elastic and piezoelectric fields in two-dimensional composites. In *Mathematics without boundaries*, pages 217–262. Springer, New York, 2014.
102. G. B. Folland. *Introduction to partial differential equations*. Princeton University Press, Princeton, NJ, second edition, 1995.
103. G. B. Folland. *Real analysis*. Pure and Applied Mathematics (New York). John Wiley & Sons, Inc., New York, second edition, 1999. Modern techniques and their applications, A Wiley-Interscience Publication.
104. A. Friedman. *Partial differential equations*. Holt, Rinehart and Winston, Inc., New York-Montreal, Que.-London, 1969.
105. F. D. Gakhov. *Boundary value problems*. Translation edited by I. N. Sneddon. Pergamon Press, Oxford-New York-Paris; Addison-Wesley Publishing Co., Inc., Reading, Mass.-London, 1966.
106. A. Gałka, J. J. Telega, and S. Tokarzewski. Heat equation with temperature-dependent conductivity coefficients and macroscopic properties of microheterogeneous media. *Math. Comput. Modelling*, 33(8-9):927–942, 2001.

107. D. Gilbarg and N. S. Trudinger. *Elliptic partial differential equations of second order*, volume 224 of *Grundlehren der Mathematischen Wissenschaften [Fundamental Principles of Mathematical Sciences]*. Springer-Verlag, Berlin, second edition, 1983.
108. G. Giraud. Sur certains problèmes non linéaires de Neumann et sur certains problèmes non linéaires mixtes. *Ann. Sci. École Norm. Sup. (3)*, 49:1–104, 1932.
109. G. Giraud. Sur certains problèmes non linéaires de Neumann et sur certains problèmes non linéaires mixtes (suite). *Ann. Sci. École Norm. Sup. (3)*, 49:245–309, 1932.
110. S. Gluzman, V. Mityushev, and W. Nawalaniec. *Computational analysis of structured media*. Mathematical Analysis and Its Applications. Academic Press, London, 2018.
111. E. I. Grigolyuk and L. A. Fil'shtinskij. *Periodicheskie kusochno- odnorodnye uprugie struktury*. Moskva: Nauka, 1992.
112. P. Grisvard. *Elliptic problems in nonsmooth domains*, volume 24 of *Monographs and Studies in Mathematics*. Pitman (Advanced Publishing Program), Boston, MA, 1985.
113. S. Gryshchuk and M. Lanza de Cristoforis. Asymptotic behaviour of the energy integral of simple eigenfunctions for the Steklov problem in a domain with a small hole. A functional analytic approach. *Zb. Pr. Inst. Mat. NAN Ukr.*, 9(2):113–121, 2012.
114. S. Gryshchuk and M. Lanza de Cristoforis. Simple eigenvalues for the Steklov problem in a domain with a small hole. A functional analytic approach. *Math. Methods Appl. Sci.*, 37(12):1755–1771, 2014.
115. N. M. Günter. *Potential theory and its applications to basic problems of mathematical physics*. Translated from the Russian by John R. Schulenberger. Frederick Ungar Publishing Co., New York, 1967.
116. J. Gwinner and E. P. Stephan. *Advanced boundary element methods*, volume 52 of *Springer Series in Computational Mathematics*. Springer, Cham, 2018. Treatment of boundary value, transmission and contact problems.
117. W. Hackbusch. *Integral equations*, volume 120 of *International Series of Numerical Mathematics*. Birkhäuser Verlag, Basel, 1995. Theory and numerical treatment, Translated and revised by the author from the 1989 German original.
118. J. Happel. Viscous flow relative to arrays of cylinders. *AIChE J.*, 5:174–177, 1959.
119. H. Hasimoto. On the periodic fundamental solutions of the Stokes' equations and their application to viscous flow past a cubic array of spheres. *J. Fluid Mech.*, 5:317–328, 1959.
120. U. Heinemann. Die regularisierende Wirkung der Randintegraloperatoren der klassischen Potentialtheorie in den Räumen hölderstetiger Funktionen. Diplomarbeit, Universität Bayreuth, 1992.
121. A. Henrot. *Extremum problems for eigenvalues of elliptic operators*. Frontiers in Mathematics. Birkhäuser Verlag, Basel, 2006.

122. D. Henry. *Topics in Nonlinear Analysis*, volume 192 of *Trabalho de Matematica*. Universidade de Brasilia, 1982.
123. M. R. Hestenes. Extension of the range of a differentiable function. *Duke Math. J.*, 8:183–192, 1941.
124. E. Hille and R. S. Phillips. *Functional analysis and semi-groups*. American Mathematical Society Colloquium Publications, vol. 31. American Mathematical Society, Providence, R. I., 1957. rev. ed.
125. S. Hofmann, M. Mitrea, and M. Taylor. Singular integrals and elliptic boundary problems on regular Semmes-Kenig-Toro domains. *Int. Math. Res. Not. IMRN*, (14):2567–2865, 2010.
126. G. C. Hsiao and W. L. Wendland. *Boundary integral equations*, volume 164 of *Applied Mathematical Sciences*. Springer-Verlag, Berlin, 2008.
127. A. M. Il'in. *Matching of asymptotic expansions of solutions of boundary value problems*, volume 102 of *Translations of Mathematical Monographs*. American Mathematical Society, Providence, RI, 1992. Translated from the Russian by V. Minachin [V. V. Minakhin].
128. V. V. Jikov, S. M. Kozlov, and O. A. Oleĭnik. *Homogenization of differential operators and integral functionals*. Springer-Verlag, Berlin, 1994. Translated from the Russian by G. A. Yosifian [G. A. Iosif'yan].
129. F. John. *Plane waves and spherical means applied to partial differential equations*. Interscience Publishers, New York-London, 1955.
130. F. John. *Partial differential equations*, volume 1 of *Applied Mathematical Sciences*. Springer-Verlag, New York, fourth edition, 1982.
131. D. Joyce, W. J. Parnell, R. C. Assier, and I. D. Abrahams. An integral equation method for the homogenization of unidirectional fibre-reinforced media; antiplane elasticity and other potential problems. *Proc. A.*, 473(2201):20170080, 25, 2017.
132. L. I. Kamynin and B. N. Himčenko. The maximum principle for a second order elliptic-parabolic equation. *Sibirsk. Mat. Ž.*, 13:773–789, 1972.
133. L. I. Kamynin and B. N. Himčenko. Theorems of Giraud type for second order equations with a weakly degenerate non-negative characteristic part. *Sibirsk. Mat. Ž.*, 18(1):103–121, 238, 1977.
134. D. Kapanadze, G. Mishuris, and E. Pesetskaya. Exact solution of a nonlinear heat conduction problem in a doubly periodic 2D composite material. *Arch. Mech. (Arch. Mech. Stos.)*, 67(2):157–178, 2015.
135. D. Kapanadze, G. Mishuris, and E. Pesetskaya. Improved algorithm for analytical solution of the heat conduction problem in doubly periodic 2D composite materials. *Complex Var. Elliptic Equ.*, 60(1):1–23, 2015.
136. D. Kapanadze, W. Miszuris, and E. Pesetskaya. Relationship between the effective thermal properties of linear and nonlinear doubly periodic composites. *ZAMM Z. Angew. Math. Mech.*, 96(7):780–790, 2016.
137. O. D. Kellogg. *Foundations of potential theory*. Reprint from the first edition of 1929. Die Grundlehren der Mathematischen Wissenschaften, Band 31. Springer-Verlag, Berlin-New York, 1967.

138. D. Khavinson, M. Putinar, and H. S. Shapiro. Poincaré's variational problem in potential theory. *Arch. Ration. Mech. Anal.*, 185(1):143–184, 2007.
139. K. Klingelhöfer. Über nichtlineare Randwertaufgaben der Potentialtheorie. *Mitt. Math. Sem. Giessen Heft*, 76:70, 1967.
140. K. Klingelhöfer. Nonlinear harmonic boundary value problems. I. *Arch. Rational Mech. Anal.*, 31:364–371, 1968/1969.
141. K. Klingelhöfer. Modified Hammerstein integral equations and nonlinear harmonic boundary value problems. *J. Math. Anal. Appl.*, 28:77–87, 1969.
142. K. Klingelhöfer. Nonlinear harmonic boundary value problems. II. Modified Hammerstein integral equations. *J. Math. Anal. Appl.*, 25:592–606, 1969.
143. M. Kohr and I. Pop. *Viscous incompressible flow for low Reynolds numbers*, volume 16 of *Advances in Boundary Elements*. WIT Press, Southampton, 2004.
144. A. A. Kolpakov and A. G. Kolpakov. *Capacity and transport in contrast composite structures. Asymptotic and applications.* Boca Raton, FL: CRC Press, 2010.
145. A. I. Kostrikin and Y. I. Manin. *Linear algebra and geometry*, volume 1 of *Algebra, Logic and Applications*. Gordon and Breach Science Publishers, New York, 1989. Translated from the second Russian edition by M. E. Alferieff.
146. V. Kozlov, V. Maz'ya, and A. Movchan. *Asymptotic analysis of fields in multi-structures*. Oxford Mathematical Monographs. The Clarendon Press, Oxford University Press, New York, 1999. Oxford Science Publications.
147. S. G. Krantz and H. R. Parks. *A primer of real analytic functions*. Birkhäuser Advanced Texts: Basler Lehrbücher. [Birkhäuser Advanced Texts: Basel Textbooks]. Birkhäuser Boston, Inc., Boston, MA, second edition, 2002.
148. M. A. Krasnosel'skiĭ, G. M. Vaĭnikko, P. P. Zabreĭko, Y. B. Rutitskii, and V. Y. Stetsenko. *Approximate solution of operator equations*. Wolters-Noordhoff Publishing, Groningen, 1972. Translated from the Russian by D. Louvish.
149. R. Kress. *Linear integral equations*, volume 82 of *Applied Mathematical Sciences*. Springer, New York, third edition, 2014.
150. A. Kufner, O. John, and S. Fučík. *Function spaces*. Noordhoff International Publishing, Leyden; Academia, Prague, 1977. Monographs and Textbooks on Mechanics of Solids and Fluids; Mechanics: Analysis.
151. V. D. Kupradze, T. G. Gegelia, M. O. Basheleĭshvili, and T. V. Burchuladze. *Three-dimensional problems of the mathematical theory of elasticity and thermoelasticity*, volume 25 of *North-Holland Series in Applied Mathematics and Mechanics*. North-Holland Publishing Co., Amsterdam-New York, Russian edition, 1979. Edited by V. D. Kupradze.
152. K. Kuratowski. *Topology. Vol. I.* New edition, revised and augmented. Translated from the French by J. Jaworowski. Academic Press, New York-London; Państwowe Wydawnictwo Naukowe, Warsaw, 1966.
153. N. G. Kuznetsov and O. V. Motygin. The Steklov problem in symmetric domains with infinitely extended boundary. In *Proceedings of the St. Petersburg Mathematical Society. Vol. XIV*, volume 228 of *Amer. Math. Soc. Transl. Ser. 2*, pages 45–78. Amer. Math. Soc., Providence, RI, 2009.

154. P. D. Lamberti. Steklov-type eigenvalues associated with best Sobolev trace constants: domain perturbation and overdetermined systems. *Complex Var. Elliptic Equ.*, 59(3):309–323, 2014.
155. P. D. Lamberti and M. Lanza de Cristoforis. A real analyticity result for symmetric functions of the eigenvalues of a domain dependent Dirichlet problem for the Laplace operator. *J. Nonlinear Convex Anal.*, 5(1):19–42, 2004.
156. P. D. Lamberti and L. Provenzano. Viewing the Steklov eigenvalues of the Laplace operator as critical Neumann eigenvalues. In *Current trends in analysis and its applications*, Trends Math., pages 171–178. Birkhäuser/Springer, Cham, 2015.
157. M. Lanza de Cristoforis. Higher order differentiability properties of the composition and of the inversion operator. *Indag. Math. (N.S.)*, 5(4):457–482, 1994.
158. M. Lanza de Cristoforis. Differentiability properties of a composition operator. *Rend. Circ. Mat. Palermo (2) Suppl.*, (56):157–165, 1998. International Workshop on Operator Theory (Cefalù, 1997).
159. M. Lanza de Cristoforis. Differentiability properties of an abstract autonomous composition operator. *J. London Math. Soc. (2)*, 61(3):923–936, 2000.
160. M. Lanza de Cristoforis. Asymptotic behaviour of the conformal representation of a Jordan domain with a small hole in Schauder spaces. *Comput. Methods Funct. Theory*, 2(1, [On table of contents: 2003]):1–27, 2002.
161. M. Lanza de Cristoforis. Asymptotic behaviour of the conformal representation of a Jordan domain with a small hole, and relative capacity. In *Complex analysis and dynamical systems*, volume 364 of *Contemp. Math.*, pages 155–167. Amer. Math. Soc., Providence, RI, 2004.
162. M. Lanza de Cristoforis. A domain perturbation problem for the Poisson equation. *Complex Var. Theory Appl.*, 50(7-11):851–867, 2005.
163. M. Lanza de Cristoforis. A singular perturbation Dirichlet boundary value problem for harmonic functions on a domain with a small hole. In *Finite or infinite dimensional complex analysis and applications*, pages 205–212. Kyushu University Press, Fukuoka, 2005.
164. M. Lanza de Cristoforis. Asymptotic behavior of the solutions of a nonlinear Robin problem for the Laplace operator in a domain with a small hole: a functional analytic approach. *Complex Var. Elliptic Equ.*, 52(10-11):945–977, 2007.
165. M. Lanza de Cristoforis. Perturbation problems in potential theory, a functional analytic approach. *J. Appl. Funct. Anal.*, 2(3):197–222, 2007.
166. M. Lanza de Cristoforis. Asymptotic behavior of the solutions of the Dirichlet problem for the Laplace operator in a domain with a small hole. A functional analytic approach. *Analysis (Munich)*, 28(1):63–93, 2008.
167. M. Lanza de Cristoforis. A singular domain perturbation problem for the Poisson equation. In *More Progresses in Analysis: Proceedings of the 5th International Isaac Congress (Catania, Italy, 25–30 July 2005), Eds. H. Begehr and F. Nicolosi*, pages 955–965. World Sci. Publ., Hackensack, NJ, 2008.

168. M. Lanza de Cristoforis. Asymptotic behaviour of the solutions of a non-linear transmission problem for the Laplace operator in a domain with a small hole. A functional analytic approach. *Complex Var. Elliptic Equ.*, 55(1-3):269–303, 2010.
169. M. Lanza de Cristoforis. Simple Neumann eigenvalues for the Laplace operator in a domain with a small hole. A functional analytic approach. *Rev. Mat. Complut.*, 25(2):369–412, 2012.
170. M. Lanza de Cristoforis. An inequality for Hölder continuous functions generalizing a result of Carlo Miranda. In *Computational and Analytic Methods in Science and Engineering, C. Constanda (ed.)*, pages 185–206. Birkhäuser/Springer, New York, 2020.
171. M. Lanza de Cristoforis. Multiple eigenvalues for the Steklov problem in a domain with a small hole. A functional analytic approach. *Asymptot. Anal.*, 121(3–4):335–365, 2021.
172. M. Lanza de Cristoforis and P. Luzzini. Time dependent boundary norms for kernels and regularizing properties of the double layer heat potential. *Eurasian Math. J.*, 8(1):76–118, 2017.
173. M. Lanza de Cristoforis and P. Luzzini. Tangential derivatives and higher-order regularizing properties of the double layer heat potential. *Analysis (Berlin)*, 38(4):167–193, 2018.
174. M. Lanza de Cristoforis and P. Musolino. A perturbation result for periodic layer potentials of general second order differential operators with constant coefficients. *Far East J. Math. Sci. (FJMS)*, 52(1):75–120, 2011.
175. M. Lanza de Cristoforis and P. Musolino. A real analyticity result for a non-linear integral operator. *J. Integral Equations Appl.*, 25(1):21–46, 2013.
176. M. Lanza de Cristoforis and P. Musolino. A singularly perturbed nonlinear Robin problem in a periodically perforated domain: a functional analytic approach. *Complex Var. Elliptic Equ.*, 58(4):511–536, 2013.
177. M. Lanza de Cristoforis and P. Musolino. A quasi-linear heat transmission problem in a periodic two-phase dilute composite. A functional analytic approach. *Commun. Pure Appl. Anal.*, 13(6):2509–2542, 2014.
178. M. Lanza de Cristoforis and P. Musolino. A functional analytic approach to homogenization problems. In *Integral methods in science and engineering*, pages 353–359. Birkhäuser/Springer, Cham, 2015.
179. M. Lanza de Cristoforis and P. Musolino. A singularly perturbed Neumann problem for the Poisson equation in a periodically perforated domain. A functional analytic approach. *ZAMM Z. Angew. Math. Mech.*, 96(2):253–272, 2016.
180. M. Lanza de Cristoforis and P. Musolino. Analytic dependence of a periodic analog of a fundamental solution upon the periodicity parameters. *Ann. Mat. Pura Appl. (4)*, 197(4):1089–1116, 2018.
181. M. Lanza de Cristoforis and P. Musolino. Two-parameter anisotropic homogenization for a Dirichlet problem for the Poisson equation in an unbounded periodically perforated domain. A functional analytic approach. *Math. Nachr.*, 291(8-9):1310–1341, 2018.

182. M. Lanza de Cristoforis and P. Musolino. Two-parameter homogenization for a nonlinear periodic Robin problem for the Poisson equation: a functional analytic approach. *Rev. Mat. Complut.*, 31(1):63–110, 2018.
183. M. Lanza de Cristoforis and P. Musolino. Asymptotic behaviour of the energy integral of a two-parameter homogenization problem with nonlinear periodic Robin boundary conditions. *Proc. Edinb. Math. Soc. (2)*, 62(4):985–1016, 2019.
184. M. Lanza de Cristoforis and L. Preciso. On the analyticity of the Cauchy integral in Schauder spaces. *J. Integral Equations Appl.*, 11(3):363–391, 1999.
185. M. Lanza de Cristoforis and L. Rossi. Real analytic dependence of simple and double layer potentials upon perturbation of the support and of the density. *J. Integral Equations Appl.*, 16(2):137–174, 2004.
186. M. Lanza de Cristoforis and L. Rossi. Real analytic dependence of simple and double layer potentials for the Helmholtz equation upon perturbation of the support and of the density. In *Analytic methods of analysis and differential equations: AMADE 2006*, pages 193–220. Camb. Sci. Publ., Cambridge, 2008.
187. N. N. Lebedev. *Special functions and their applications.* Dover Publications, Inc., New York, 1972. Revised edition, translated from the Russian and edited by Richard A. Silverman, Unabridged and corrected republication.
188. G. M. Lieberman. Regularized distance and its applications. *Pacific J. Math.*, 117(2):329–352, 1985.
189. C.-S. Lin and C.-L. Wang. Elliptic functions, Green functions and the mean field equations on tori. *Ann. of Math. (2)*, 172(2):911–954, 2010.
190. R. Lipton. Heat conduction in fine scale mixtures with interfacial contact resistance. *SIAM J. Appl. Math.*, 58(1):55–72, 1998.
191. A. Lunardi and V. Vespri. Hölder regularity in variational parabolic nonhomogeneous equations. *J. Differential Equations*, 94(1):1–40, 1991.
192. P. Luzzini. Regularizing properties of space-periodic layer heat potentials and applications to boundary value problems in periodic domains. *Math. Methods Appl. Sci.*, 43(8):5273–5294, 2020.
193. P. Luzzini and P. Musolino. Periodic transmission problems for the heat equation. In *Integral methods in science and engineering*, pages 211–223. Birkhäuser/Springer, Cham, 2019.
194. P. Luzzini and P. Musolino. Perturbation analysis of the effective conductivity of a periodic composite. *Netw. Heterog. Media*, 15(4):581–603, 2020.
195. P. Luzzini, P. Musolino, and R. Pukhtaievych. Shape analysis of the longitudinal flow along a periodic array of cylinders. *J. Math. Anal. Appl.*, 477(2):1369–1395, 2019.
196. A. J. Majda and A. L. Bertozzi. *Vorticity and incompressible flow*, volume 27 of *Cambridge Texts in Applied Mathematics*. Cambridge University Press, Cambridge, 2002.
197. M. Mamode. Fundamental solution of the Laplacian on flat tori and boundary value problems for the planar Poisson equation in rectangles. *Bound. Value Probl.*, pages 1–9, 2014:221, 2014.

198. J. Mateu, J. Orobitg, and J. Verdera. Extra cancellation of even Calderón-Zygmund operators and quasiconformal mappings. *J. Math. Pures Appl. (9)*, 91(4):402–431, 2009.
199. V. Maz'ya and A. Movchan. Asymptotic treatment of perforated domains without homogenization. *Math. Nachr.*, 283(1):104–125, 2010.
200. V. Maz'ya, A. Movchan, and M. Nieves. Green's kernels for transmission problems in bodies with small inclusions. In *Operator theory and its applications*, volume 231 of *Amer. Math. Soc. Transl. Ser. 2*, pages 127–160. Amer. Math. Soc., Providence, RI, 2010.
201. V. Maz'ya, A. Movchan, and M. Nieves. *Green's kernels and meso-scale approximations in perforated domains*, volume 2077 of *Lecture Notes in Mathematics*. Springer, Heidelberg, 2013.
202. V. Maz'ya, A. Movchan, and M. Nieves. On meso-scale approximations for vibrations of membranes with lower-dimensional clusters of inertial inclusions. *Algebra i Analiz*, 32(3):219–237, 2020.
203. V. Maz'ya, S. Nazarov, and B. Plamenevskij. *Asymptotic theory of elliptic boundary value problems in singularly perturbed domains. Vol. I*, volume 111 of *Operator Theory: Advances and Applications*. Birkhäuser Verlag, Basel, 2000. Translated from the German by Georg Heinig and Christian Posthoff.
204. V. Maz'ya, S. Nazarov, and B. Plamenevskij. *Asymptotic theory of elliptic boundary value problems in singularly perturbed domains. Vol. II*, volume 112 of *Operator Theory: Advances and Applications*. Birkhäuser Verlag, Basel, 2000. Translated from the German by Plamenevskij.
205. V. G. Maz'ya and A. A. Soloviev. *Boundary integral equations on contours with peaks*, volume 196 of *Operator Theory: Advances and Applications*. Birkhäuser Verlag, Basel, 2010. Translated into English and edited by Tatyana Shaposhnikova.
206. W. McLean. *Strongly elliptic systems and boundary integral equations*. Cambridge University Press, Cambridge, 2000.
207. E. J. McShane. Extension of range of functions. *Bull. Amer. Math. Soc.*, 40(12):837–842, 1934.
208. D. Medková. *The Laplace equation*. Springer, Cham, 2018. Boundary value problems on bounded and unbounded Lipschitz domains.
209. T. A. Mel'nik. Asymptotic expansions of eigenvalues and eigenfunctions for elliptic boundary-value problems with rapidly oscillating coefficients in a perforated cube. *J. Math. Sci.*, 75(3):1646–1671, 1995.
210. S. G. Mikhlin. *Multidimensional singular integrals and integral equations*. Translated from the Russian by W. J. A. Whyte. Translation edited by I. N. Sneddon. Pergamon Press, Oxford-New York-Paris, 1965.
211. G. W. Milton. *The theory of composites*, volume 6 of *Cambridge Monographs on Applied and Computational Mathematics*. Cambridge University Press, Cambridge, 2002.
212. C. Miranda. Sulle proprietà di regolarità di certe trasformazioni integrali. *Atti Accad. Naz. Lincei Mem. Cl. Sci. Fis. Mat. Natur. Sez. I (8)*, 7:303–336, 1965.

213. C. Miranda. *Partial differential equations of elliptic type*. Ergebnisse der Mathematik und ihrer Grenzgebiete, Band 2. Springer-Verlag, New York-Berlin, 1970. Second revised edition. Translated from the Italian by Zane C. Motteler.
214. D. Mitrea. *Distributions, partial differential equations, and harmonic analysis*. Universitext. Springer, New York, 2013.
215. I. Mitrea and M. Mitrea. *Multi-layer potentials and boundary problems for higher-order elliptic systems in Lipschitz domains*, volume 2063 of *Lecture Notes in Mathematics*. Springer, Heidelberg, 2013.
216. V. Mityushev. Functional equations in a class of analytic functions and composite materials. *Demonstratio Math.*, 30(1):63–70, 1997.
217. V. Mityushev. Transport properties of doubly periodic arrays of circular cylinders and optimal design problems. *Appl. Math. Optim.*, 44(1):17–31, 2001.
218. V. Mityushev and P. M. Adler. Longitudinal permeability of spatially periodic rectangular arrays of circular cylinders. I. A single cylinder in the unit cell. *ZAMM Z. Angew. Math. Mech.*, 82(5):335–345, 2002.
219. V. Mityushev and P. M. Adler. Longitudinal permeability of spatially periodic rectangular arrays of circular cylinders. II. An arbitrary distribution of cylinders inside the unit cell. *Z. Angew. Math. Phys.*, 53(3):486–517, 2002.
220. V. Mityushev and W. Nawalaniec. Effective conductivity of a random suspension of highly conducting spherical particles. *Appl. Math. Model.*, 72:230–246, 2019.
221. V. Mityushev, E. Pesetskaya, and S. Rogosin. Analytical Methods for Heat Conduction in Composites and Porous Media. In A. Öchsner, G. E. Murch, and M. J. S. de Lemos, editors, *Cellular and Porous Materials: Thermal Properties Simulation and Prediction*. Wiley-VCH Verlag GmbH & Co. KGaA, Weinheim, Germany, 2008.
222. V. Mityushev and N. Rylko. Maxwell's approach to effective conductivity and its limitations. *Quart. J. Mech. Appl. Math.*, 66(2):241–251, 2013.
223. V. V. Mityushev. Transport properties of double-periodic arrays of circular cylinders. *Z. Angew. Math. Mech.*, 77(2):115–120, 1997.
224. V. V. Mityushev and S. V. Rogosin. *Constructive methods for linear and nonlinear boundary value problems for analytic functions*, volume 108 of *Chapman & Hall/CRC Monographs and Surveys in Pure and Applied Mathematics*. Chapman & Hall/CRC, Boca Raton, FL, 2000. Theory and applications.
225. R. Molinarolo. Existence of solutions for a singularly perturbed nonlinear non-autonomous transmission problem. *Electron. J. Differential Equations*, 53:1–29, 2019.
226. A. B. Movchan, N. V. Movchan, and C. G. Poulton. *Asymptotic models of fields in dilute and densely packed composites*. Imperial College Press, London, 2002.
227. N. I. Muskhelishvili. *Singular integral equations*. Dover Publications, Inc., New York, 1992. Boundary problems of function theory and their application to mathematical physics, Translated from the second (1946) Russian edition

and with a preface by J. R. M. Radok, Corrected reprint of the 1953 English translation.
228. P. Musolino. A singularly perturbed Dirichlet problem for the Laplace operator in a periodically perforated domain. A functional analytic approach. *Math. Methods Appl. Sci.*, 35(3):334–349, 2012.
229. P. Musolino. A singularly perturbed Dirichlet problem for the Poisson equation in a periodically perforated domain. A functional analytic approach. In *Advances in harmonic analysis and operator theory*, volume 229 of *Oper. Theory Adv. Appl.*, pages 269–289. Birkhäuser/Springer Basel AG, Basel, 2013.
230. P. Musolino and V. Mityushev. Asymptotic behavior of the longitudinal permeability of a periodic array of thin cylinders. *Electron. J. Differential Equations*, 290:1–20, 2015.
231. K. Nakamori and Y. Suyama. On a nonlinear boundary problem for the equations $\Delta u = 0$ and $\Delta u = f(x, y)$. *Mem. Fac. Sci. Kyūsyū Univ. A.*, 5:99–106, 1950.
232. J. Naumann and C. G. Simader. *Measure and integration on Lipschitz-manifolds*. EPub Bayreuth, Bayreuth, 2007.
233. W. Nawalaniec, K. Necka, and V. Mityushev. Effective conductivity of densely packed disks and energy of graphs. *Mathematics*, 8 (12), 2161, 2020.
234. A. H. Nayfeh. *Perturbation methods*. John Wiley & Sons, New York-London-Sydney, 1973. Pure and Applied Mathematics.
235. S. A. Nazarov. Asymptotic behavior of the Steklov spectral problem in a domain with a blunted peak. *Mat. Zametki*, 86(4):571–587, 2009.
236. S. A. Nazarov. On the spectrum of the Steklov problem in peak-shaped domains. In *Proceedings of the St. Petersburg Mathematical Society. Vol. XIV*, volume 228 of *Amer. Math. Soc. Transl. Ser. 2*, pages 79–131. Amer. Math. Soc., Providence, RI, 2009.
237. S. A. Nazarov. Asymptotic expansions of eigenvalues of the Steklov problem in singularly perturbed domains. *Algebra i Analiz*, 26(2):119–184, 2014.
238. S. A. Nazarov and J. Taskinen. On the spectrum of the Steklov problem in a domain with a peak. *Vestnik St. Petersburg Univ. Math.*, 41(1):45–52, 2008.
239. L. Nirenberg. *Topics in nonlinear functional analysis*, volume 6 of *Courant Lecture Notes in Mathematics*. New York University, Courant Institute of Mathematical Sciences, New York; American Mathematical Society, Providence, RI, 2001. Chapter 6 by E. Zehnder, Notes by R. A. Artino, Revised reprint of the 1974 original.
240. A. A. Novotny and J. Sokołowski. *Topological derivatives in shape optimization*. Interaction of Mechanics and Mathematics. Springer, Heidelberg, 2013.
241. S. E. Pastukhova. On the error of averaging for the Steklov problem in a punctured domain. *Differentsial'nye Uravneniya*, 31(6):1042–1054, 1102, 1995.
242. B. A. Plamenevskij. Elliptic boundary value problems in domains with piecewise smooth boundary. In *Partial differential equations, IX*, volume 79 of *Encyclopaedia Math. Sci.*, pages 217–281. Springer, Berlin, 1997. Translated from the Russian by the author.

243. J. Plemelj. *Potentialtheoretische Untersuchungen*. Preisschriften der Fiirstlich Jablonowskischen Gesellschaft zu Leipzig. Teubner-Verlag, Leipzig, 1911.
244. H. Poincaré. La méthode de Neumann et le problème de Dirichlet. *Acta Math.*, 20(1):59–142, 1897.
245. C. G. Poulton, L. C. Botten, R. C. McPhedran, and A. B. Movchan. Source-neutral Green's functions for periodic problems in electrostatics, and their equivalents in electromagnetism. *R. Soc. Lond. Proc. Ser. A Math. Phys. Eng. Sci.*, 455(1983):1107–1123, 1999.
246. L. Preciso. Perturbation Analysis of the Conformal Sewing Problem and Related Problems. Doctoral Dissertation, Università degli Studi di Padova (http://paduaresearch.cab.unipd.it/12923/), 1998.
247. L. Preciso. Regularity of the composition and of the inversion operator and perturbation analysis of the conformal sewing problem in Roumieu type spaces. *Tr. Inst. Mat., Minsk*, 5:99–104, 2000.
248. G. Prodi and A. Ambrosetti. Analisi non lineare. I quaderno. Pisa: Scuola Normale Superiore Pisa, Classe di Scienze. 174 p. L. 3.600 (1973), 1973.
249. R. Pukhtaievych. Asymptotic behavior of the solution of singularly perturbed transmission problems in a periodic domain. *Math. Methods Appl. Sci.*, 41(9):3392–3413, 2018.
250. R. Pukhtaievych. Effective conductivity of a periodic dilute composite with perfect contact and its series expansion. *Z. Angew. Math. Phys.*, 69(3):Art. 83, 22, 2018.
251. S. Rogosin and T. Vaitekhovich. On a biharmonic problem in a circular ring. *Analysis (Munich)*, 30(1):93–106, 2010.
252. S. A. Sangani and C. Yao. Transport processes in random arrays of cylinders. I. Thermal conduction. *Phys. Fluids*, 31:2426–2434, 09 1988.
253. S. A. Sangani and C. Yao. Transport properties in random arrays of cylinders. II. Viscous flow. *Phys. Fluids*, 31:2435–2444, 09 1988.
254. S. A. Sauter and C. Schwab. *Boundary element methods*, volume 39 of *Springer Series in Computational Mathematics*. Springer-Verlag, Berlin, 2011. Translated and expanded from the 2004 German original.
255. H. H. Schaefer. *Topological vector spaces*. Springer-Verlag, New York-Berlin, 1971. Third printing corrected, Graduate Texts in Mathematics, Vol. 3.
256. J. Schauder. Potentialtheoretische Untersuchungen. *Math. Z.*, 33(1):602–640, 1931.
257. J. Schauder. Bemerkung zu meiner Arbeit "Potentialtheoretische Untersuchungen I (Anhang)". *Math. Z.*, 35(1):536–538, 1932.
258. M. Schiffer. The Fredholm eigen values of plane domains. *Pacific J. Math.*, 7:1187–1225, 1957.
259. J. Schmid. Longitudinal laminar flow in an array of circular cylinders. *Int. J. Heat Mass Transfer*, 9(9):925 – 937, 1966.
260. S. E. Shamma. Asymptotic behavior of Stekloff eigenvalues and eigenfunctions. *SIAM J. Appl. Math.*, 20:482–490, 1971.

261. V. A. Shcherbina. Boundary value problems with a triply periodic solution for the Laplace equation in $\mathbf{R}^3$. *Teor. Funktsiĭ Funktsional. Anal. i Prilozhen.*, (45):132–139, iii, 1986.
262. V. A. Shcherbina. Fundamental solutions of three-dimensional cellular problems for some equations of mathematical physics. *Zh. Vychisl. Mat. i Mat. Fiz.*, 29(2):298–304, 319, 1989.
263. E. Swartz and R. Pohl. Thermal boundary resistance. *Rev. Mod. Phys.*, 61(3):605–668, 1989.
264. L. Tartar. *An introduction to Sobolev spaces and interpolation spaces*, volume 3 of *Lecture Notes of the Unione Matematica Italiana*. Springer, Berlin; UMI, Bologna, 2007.
265. A. E. Taylor and D. C. Lay. *Introduction to functional analysis*. John Wiley & Sons, New York-Chichester-Brisbane, second edition, 1980.
266. F. Trèves. *Topological vector spaces, distributions and kernels*. Academic Press, New York-London, 1967.
267. H. Triebel. *Interpolation theory, function spaces, differential operators*, volume 18 of *North-Holland Mathematical Library*. North-Holland Publishing Co., Amsterdam-New York, 1978.
268. G. M. Troianiello. *Elliptic differential equations and obstacle problems*. The University Series in Mathematics. Plenum Press, New York, 1987.
269. T. Valent. *Boundary value problems of finite elasticity*, volume 31 of *Springer Tracts in Natural Philosophy*. Springer-Verlag, New York, 1988. Local theorems on existence, uniqueness, and analytic dependence on data.
270. M. Van Dyke. *Perturbation methods in fluid mechanics*. Applied Mathematics and Mechanics, Vol. 8. Academic Press, New York-London, 1964.
271. M. Vanninathan. Homogenization of eigenvalue problems in perforated domains. *Proc. Indian Acad. Sci. Math. Sci.*, 90(3):239–271, 1981.
272. W. von Wahl. Abschätzungen für das Neumann-Problem und die Helmholtz-Zerlegung von L^p. *Nachr. Akad. Wiss. Göttingen Math.-Phys. Kl. II*, (2):29, 1990.
273. W. Wendland. Die Fredholmsche Alternative für Operatoren, die bezüglich eines bilinearen Funktionals adjungiert sind. *Math. Z.*, 101:61–64, 1967.
274. W. Wendland. Bemerkungen über die Fredholmschen Sätze. Methoden Verfahren Math. Phys. 3, B.I.-Hochschulskripten 722/722a, 141–176 (1970)., 1970.
275. H. Weyl. The method of orthogonal projection in potential theory. *Duke Math. J.*, 7:411–444, 1940.
276. J. T. Wloka, B. Rowley, and B. Lawruk. *Boundary value problems for elliptic systems*. Cambridge University Press, Cambridge, 1995.
277. W. P. Ziemer. *Weakly differentiable functions*, volume 120 of *Graduate Texts in Mathematics*. Springer-Verlag, New York, 1989. Sobolev spaces and functions of bounded variation.

Index

Symbols

M. Dalla Riva et al., *Singularly Perturbed Boundary Value Problems*,
https://doi.org/10.1007/978-3-030-76259-9

W

Kolibri 360 GmbH
Ferdinand-Jühlke-Straße 7
99095 Erfurt, Deutschland
produktsicherheit@kolibri360.de